Recent Advances in Material, Manufacturing, and Machine Learning

Proceedings of 1st International Conference (RAMMML-22), Volume 2

About the conference

The international conference on recent advances in material, manufacturing, and machine learning processes-2022 (RAMMML-22), is organized by Yeshwantrao Chavan College of Engineering, Nagpur, Maharashtra, India during April 26–27, 2022. The conference received more than 640 papers. More than 177 papers were accepted and orally presented in the conference.

The role of manufacturing in the country's economy and societal development has long been established through their wealth generating capabilities. To enhance and widen our knowledge of materials and to increase innovation and responsiveness to ever-increasing international needs, more in-depth studies of functionally graded materials/tailor-made materials, recent advancements in manufacturing processes and new design philosophies are needed at present. The objective of this conference is to bring together experts from academic institutions, industries and research organizations and professional engineers for sharing of knowledge, expertise and experience in the emerging trends related to design, advanced materials processing and characterization, advanced manufacturing processes.

The conference is structured with plenary lectures followed by parallel sessions. The plenary lectures introduces the theme of the conference delivered by eminent personalities of international repute. Each parallel session starts with an invited talk on specific topic followed by contributed papers. Papers are invited from the prospective authors from industries, academic institutions and R&D organizations and from professional engineers. This conference brings academicians, industrial experts, researchers, and scholars together from areas of Mechanical Design Engineering, Materials Engineering and Manufacturing Processes. The topics of interest includes Design, Materials and Manufacturing engineering and other related areas such as Mechatronics, Prosthetic design and Bio inspired design and Smart materials. This conference is to provide a platform for learning, exchange of ideas and networking with fellow colleagues and participants across the globe in the field of Mechanical Design, Materials and Manufacture.

This conference RAMMML-2022 have paved the way to understand the latest technological and innovative advancements especially in the fields of manufacturing, design and materials engineering. The conference has developed the solutions to physical problems, questions how things work, make things work better, and create ideas for doing things in new and different ways in the manufacturing, design and materials engineering. The conference focuses on the frontier themes of recent advances in manufacturing, design and materials engineering, as applied to multiple disciplines of engineering. Researchers, Academicians, Industrialist and Students has benefitted with the latest trends and developments in design, manufacturing and materials engineering applied to various disciplines of engineering. The objective of the conference is to have the orientation of research and practice of professionals towards attaining global supremacy in manufacturing, design and materials engineering. Also this conference aims in understanding the recent trends in manufacturing, design and materials engineering including optimization and innovation. This conference RAMMML-2022 also aims in improving Research culture in the minds of faculty in exploring the knowledge base, establishing better insights and maintaining dynamism in the teaching - learning process.

Recent Advances in Material, Manufacturing, and Machine Learning

Subtitle- Proceedings of 1st International Conference (RAMMML-22), Volume 2

Edited by

Dr. Rajiv Gupta
North Carolina State University, Raleigh, NC, United States
ORCID 0000-0003-2684-1994

Dr. Devendra Deshmukh
Indian Institute of Technology, Indore
ORCID 0000-0003-4636-3301

Dr. Awanikumar P. Patil
Visvesvaraya National Institute of Technology, Nagpur
ORCID 0000-0002-4511-7558

Dr. Naveen Kumar Shrivastava
Birla Institute of Technology and Science, Pilani, Dubai (UAE)
ORCID 0000-0002-5773-473X

Dr. Jayant Giri
Associate Professor, Department of Mechanical Engineering, Yeshwantrao
Chavan College of Engineering, Nagpur
ORCID 0000-0003-4438-2613

Dr. R.B. Chadge
Assistant Professor, Department of Mechanical Engineering, Yeshwantrao
Chavan College of Engineering, Nagpur
ORCID 0000-0001-9072-0607

CRC Press
Taylor & Francis Group
Boca Raton London New York

CRC Press is an imprint of the
Taylor & Francis Group, an **informa** business

First edition published 2023
by CRC Press
4 Park Square, Milton Park, Abingdon, Oxon, OX14 4RN

and by CRC Press
6000 Broken Sound Parkway NW, Suite 300, Boca Raton, FL 33487-2742

British Library Cataloguing-in-Publication Data
A catalogue record for this book is available from the British Library

Library of Congress Cataloging-in-Publication Data

ISBN: 9781032441313 (pbk)
ISBN: 9781003370628 (ebk)
DOI: 10.1201/9781003370628

Typeset in Sabon
by HBK Digital

Table of Contents

List of Tables and Figures

Tables

Figures

Foreword

India as a growing economy has made its significant impact on global affairs in the last few decades. The country has survived many economic, social and political challenges in the past and is determined to become a stronger economy in coming future. The path of this growth is full of challenges of Material research, manufacturing technologies, design and development while ensuring inclusive growth for every citizen of the country. The role of manufacturing in the country's economy and societal development has long been established through their wealth generating capabilities. To enhance and widen our knowledge of materials and to increase innovation and responsiveness to ever-increasing international needs, more in-depth studies of functionally graded materials/ tailor- made materials, recent advancements in manufacturing processes and new design philosophies are needed at present. The objective of this conference is to bring together experts from academic institutions, industries and research organizations and professional engineers for sharing of knowledge, expertise and experience in the emerging trends related to design, advanced materials processing and characterization, advanced manufacturing processes. Since its inception, Yeshwantrao Chavan College of Engineering, Nagpur has been working as an institution committed to contribute in the field of Materials, design, automation and sustainable development through its various academic and non-academic activities. Different departments of YCCE have been conducting various academic conferences, seminars and workshops to discuss on contemporary issues being faced by Design, development and material research sectors and bring together academicians, researchers, technologists, industry experts and policy-makers together on common platforms.

Yeshwantrao Chavan College of Engineering, Nagpur is the pioneer in organizing academic conferences to share and disseminate academic research in the field of engineering & technology with the industry and policy makers since its establishment (1984).The department of mechanical engineering, YCCE is organizing its 1st international conference on recent advances in material, manufacturing, and machine learning processes-2022 (RAMMML-22), April 26–27, 2022 with an objective & scope to deliberate, discuss and document the latest technological and innovative advancements especially in the fields of manufacturing, design and materials engineering.

It was good news to hear from the Conference Organizing Committee that they have received more than 640 papers on different topics related to different subthemes of the conference. These papers cover topics related to manufacturing, design, materials engineering, machine learning, simulation, civil engineering and many more interdisciplinary topics. We believe that this conference will become a common platform to disseminate new researches done by various researchers from different universities/ institutes before industry professionals and policy makers in the government. We hope that these new researches will suggest new directions to innovations in Material research, design, development, manufacturing industries and government policies pertaining to these sectors.

We wish all the best to our conference participants who are the real knowledge champions of their universities/institutes/organizations. We strongly believe that all of us at YCCE will make this conference a good experience for every participant of the conference and this conference will achieve its objectives effectively.

We welcome you all to RAMMML-2022.

Patrons

Chief Patron	Hon'ble Shri. Dattaji Meghe	Chairman, Nagar Yuwak Shikshan Sanstha, Founder Chancellor of Datta Meghe Institute Of Medical sciences.
Chief Patron	Hon'ble Shri. Sagarji Meghe	Secretary, Nagar Yuwak Shikshan Sanstha
Chief Patron	Hon'ble Shri. Sameerji Meghe	Treasurer, Nagar Yuwak Shikshan Sanstha
Chief Patron	Hon'ble Mrs. Vrinda Meghe	Chief Advisor, Nagar Yuwak Shikshan Sanstha
Chief Patron	Hon'ble Dr. Hemant Thakare	COO: Ceinsys Tech. LTD, President, IEI, India
Patron	Dr. U.P. Waghe	Principal, Yeshwantrao Chavan College of Engineering, Nagpur
Patron	Dr. Manali Kshirsagar	Principal, Rajiv Gandhi College of Engineering, Nagpur

Preface

The main aim of the 1st international conference on recent advances in material, manufacturing, and machine learning processes-2022 (RAMMML-22) is to bring together all interested academic researchers, scientists, engineers, and technocrats and provide a platform for continuous improvement of manufacturing, machine learning, design and materials engineering research. RAMMML 2022 received an overwhelming response with more than 640 full paper submissions. After due and careful scrutiny, about 177 of them have been selected for presentation. The papers submitted have been reviewed by experts from renowned institutions, and subsequently, the authors have revised the papers, duly incorporating the suggestions of the reviewers. This has led to significant improvement in the quality of the contributions, Taylor & Francis publications, CRC Press have agreed to publish the selected proceedings of the conference in their book series of Advances in Mechanical Engineering and Interdisciplinary Sciences. This enables fast dissemination of the papers worldwide and increases the scope of visibility for the research contributions of the authors.

This book comprises four parts, viz. Materials, Manufacturing, Machine learning and interdisciplinary sciences. Each part consists of relevant full papers in the form of chapters. The Materials part consists of chapters on research related to Advanced Materials, Ceramics, Shape Memory Alloys and Nano materials, Materials for Aerospace applications, Polymers and Polymer Composites, Glasses and Amorphous Systems, Material characterization and testing, MEMS/NEMS, Bio Materials, Optical/Electronic Materials, Magnetic Materials, 3D Materials, Cryogenic Materials, Materials applications, performance and life cycle etc. The Manufacturing part consists of chapters on Micro/Nano Machining, Metal Forming, Green Manufacturing, Non-Conventional Machining Processes, Additive Manufacturing, Subtractive Manufacturing, Industry 4.0, Sustainable Manufacturing Technologies, Casting Technology, Joining Technology, Plastic processing technology, CAD/CAM/CAE/CIM/HVAC, Product Design and Development, Multi Objective Optimization, Modelling, Analysis and Simulation, Process Monitoring and Control, Vibration Noise Analysis and Control, Thermal Optimization, Energy Analysis etc. The Machine learning part consists of chapters on Machine learning, knowledge discovery, and data mining, Artificial intelligence in biomedical engineering and informatics, Artificial neural networks and algorithms, Knowledge acquisition, representation and reasoning methodologies, Genetic algorithms, Probability-based systems and fuzzy systems, Healthcare process management, Imaging, signal processing and text analysis, Bioinformatics and neurosciences. And the Interdisciplinary part consists of chapters on Condition Monitoring, NDT, Soft Computing, VLSI, Embedded System, Computer Vision, Environment Sustainability, Water Management, Advanced Mechatronics System and Control, Structural and Geo-technical Engineering areas. This book provides a snapshot of the current research in the field of Materials, Manufacturing, Machine learning and interdisciplinary sciences and hence will serve as valuable reference material for the research community.

Details of programme committee

International Advisory Committee

S.No	Name	Details
1.	Dr. Rajiv Gupta	North Carolina State University, Raleigh, NC, United States
2.	Mr. James Barret	HE Lecturer, Turno College, Plymouth University, England
3.	Abhiram Dapke	Deuce Drone, Albama, USA
4.	Nakul Vadalkar	T&I, Celanese, Germany
5.	Mr. Laxmikant Kolekar	Assistant General Manager-Operations Alam Steel, Al Salmiyah Hawalli, Kuwait
6.	Dr. Naveen Kumar Shrivastava	Birla Institute of Technology and Science, Pilani, Dubai (UAE)
7.	Mr. Aniket Mandlekar	Field Engineer, Intertape polymer group, Canada

National Advisory Committee

S.No	Name	Details
1.	Dr. Prashant P. Datey	Indian Institute of Technology, Bombay
2.	Dr. Milind Atre	Indian Institute of Technology, Bombay
3.	Dr. Rakesh G. Mote	Indian Institute of Technology, Bombay
4.	Dr. Harekrishna Yadav	Indian Institute of Technology, Indore
5.	Dr. I. A. Palani	Indian Institute of Technology, Indore
6.	Dr. Anand Parey	Indian Institute of Technology, Indore
7.	Dr. Jitendra Sangwai	Indian Institute of Technology, Madras
8.	Dr. Devendra Deshmukh	Indian Institute of Technology, Indore
9.	Dr.Rajesh Ranganathan	Coimbatore Institute of Technology, Coimbatore, Tamilnadu
10.	Dr. Anupam Agnihotri	Director, Jawaharlal Nehru Aluminium Research Development And Design Centre, Bombay
11.	Dr. P. M. Padole	Director, Visvesvaraya National Institute of Technology, Nagpur
12.	Dr. Awanikumar P. Patil	Visvesvaraya National Institute of Technology, Nagpur
13.	Dr. Vilas R. Kalamkar	Visvesvaraya National Institute of Technology, Nagpur
14.	Prof. G. S. Dangayach	Malaviya National Institute of Technology, Jaipur
15.	Dr. Y. M. Puri	Visvesvaraya National Institute of Technology, Nagpur
16.	Dr Rakesh Shrivastava	Former Professor of Mechanical Engineering, Consultant and Trainer, Ex Chairman IEI, Nagpur Chapter
17.	Dr. D. M. Kulkarni	Birla Institute of Technology and Science, Goa
18.	Dr. B. Rajiv	College of Engineering, Pune
19.	Dr. T. N. Desai	Sardar Vallabhbhai National Institute of Technology, Surat
20.	Mr. Saquib Anwar	Manager BEL, Bombay
21.	Dr. Prakash Pantawane	College of Engineering, Pune
22.	Dr. Nalinaksh S. Vyas	Indian Institute of Technology, Kanpur

Conference Chair & Organizing Secretary

S.No	Commitee	Name	Details
1.	Conference Chair	Dr. J.P. Giri	Head of The Department, Department of Mechanical Engineering, Yeshwantrao Chavan College of Engineering, Nagpur
2.	Conference Chair	Dr. R.B. Chadge	Department of Mechanical Engineering, Yeshwantrao Chavan College of Engineering, Nagpur
3.	Organizing Secretary	Dr. S.R. Jachak	Department of Mechanical Engineering, Yeshwantrao Chavan College of Engineering, Nagpur
4.	Organizing Secretary	Dr. S.P. Ambade	Department of Mechanical Engineering, Yeshwantrao Chavan College of Engineering, Nagpur

Contact Persons

Prof. A.P. Edlabadkar   Prof. Neeraj Sunheriya

Organizing Committee Members

Prof. D. I. Sangotra	Prof. G. H. Waghmare	Prof. P. S. Barve
Prof. N. J. Giradkar	Prof. R. G. Bodkhe	Prof. N. D. Gedam
Prof. V. M. Korde	Prof. D. Y. Shahare	Prof. C. A. Mahatme
Prof. A. S. Bonde	Prof. A. P. Edlabadkar	Prof. P. V. Lande
Dr. S. T. Bagde	Dr. S. S. Khedkar	Prof. G. M. Dhote
Dr. S. S. Chaudhari	Dr. P. D. Kamble	Dr. V. R. Khawale
Dr. S. V. Prayagi	Prof. R. V. Adakane	Prof. P. A. Hatwalne
Dr. A. P. Kedar	Prof. D. N. Kashyap	Prof. Dipak M. Hajare
Prof. V. G. Thakre	Prof. A. R. Narkhede	Prof. Praful Shirpurkar
Prof. M. S. Tufail	Prof. S. P. Kamble	Prof. Ritu Shrivastava
Prof. P. N. Shende	Prof. Y. Y. Nandurkar	Prof. Sujata A Kimmatkar
Prof. A. B. Amale	Prof. M. M. Dakhore	Prof. Albela H. Pundkar

85 Ensemble method for multi-label classification on intrusion detection system

Mayur V. Tayde[a], Rahul B. Adhao[b], and Vinod K. Pachghare[c]

Department of Computer Engineering and IT, College of Engineering Pune, Pune, India

Abstract

The advancement of computer science and telecommunication engineering has provided humanity with amazing potential to succeed at all levels of their lives. The number of internet users is increasing day by day. Hence, network security has become an important part of human life. Intrusion detection system (IDS) plays a vital role in network security. It detects malicious activity from the network and sends an alert message to an administrator. It is important to select the relevant features to enhance the accuracy and mitigate the computational cost. The proposed model uses the NSL-KDD and CICIDS-2017 dataset, and arithmetic operations are performed between Information Gain and Pearson Correlation. Then, the top 25 and top 25 features are selected from CICIDS-2017 and NSL-KDD dataset respectively based on their arithmetic operation result. These reduced features are fed to Synthetic Minority Oversampling Technique (SMOTE) to handle the unbalanced dataset after oversampling, the model is run on a random forest classifier, which provides an accuracy of 99.7978% and 99.8967% for CICIDS2017 and NSL-KDD respectively.

Keywords: Feature selection, information gain, intrusion detection system, machine learning, random forest.

Introduction

Multiple security layers are used to protect the network traffic, such as Firewall, Content filtering, antimalware, etc. But If an attack originates from the organisation itself, this security mechanism won't handle such an attack. Various devices are available sitting inside the network to deal with this type of attack (Sparks et al., 2009). Their sole purpose is to detect and alert the attack IDS play an important role in detecting insider threats. IDS is if there is an attack on the network, the IDS will be sitting there and analysing traffic that passes through the network. If it sees any abnormal or malicious activity, it alerts the administrator. IDS runs through a network normalisation process where it learns about the normal functioning of the network. Any internet request is going to our local area network will have to pass from IDS, and any requests which will go out from our local area network have to be passed from IDS (Pachghare et al., 2012). IDS can be deployed on our network and the web application to detect intruder activity.

With the advent of computer networking worldwide, many new applications are being created daily. One emerging field settling in today's computing industry is cloud computing. Security attacks are targeted at organisational computing environments. Many such organisations are facing severe security problems. Hackers exploit system flaws protocol vulnerabilities after gaining access to the target system. Hackers take the first path to compromise the system most of the time through intrusion in the company network. Intrusion can be done in though manipulating data flowing through packets. With unparalleled focus along with expert talent gathered, advanced methods are employed not to get flagged by security mechanisms placed in the forefront of company IT architecture while intruding the system. We need to utilise the upcoming technology and concepts to our advantage. Machine learning is being implemented daily to adapt and recognise different security attacks. Machine learning techniques trained on a vast amount of data can detect malicious activities easily. Though challenges persist, one can decrease the number of features needed for the model to classify data as benign correctly. In doing so, one can flag any suspicious activity early, thwarting future debacles.

IDS is broadly classified into two types based on their architecture Network-based IDS (NIDS) and Host-based IDS (HIDS). HIDS monitors the system event and audits the event logs. It takes the snap of existing conditions of the system and matches it with the last snap. If any of these files is found modified or deleted, it sends an alert message to the administrator. NIDS analyses the traffic of the entire subnet every packet is monitored, and if found any malicious activity, it sends an alert to the administrator (Sawaisarje et al., 2018). The proposed model works on NIDS. According to the status 85% of attacks are made by the user

[a]taydemv20.comp@coep.ac.in; [b]rba.comp@coep.ac.in; [c]vkp.comp@coep.ac.in

who belongs to the same network, and the firewall will not detect this attack. This attacker might be authorised users of the respective firm; hence IDS is required to prevent these attacks. For complete supervision of the network, IDS plays an important role (Pachghare, 2019).

The proposed model provides a statistical feature selection-based IDS. Here the arithmetic operation is performed in between Pearson correlation and information gain. Then top features are selected, which are fed to SMOTE to handle the unbalanced dataset (Derhab et al., 2020). Finally, this reduced set of features with the balanced dataset is given to the random forest classifier, providing better accuracy than full features of NSL-KDD and CICIDS-2017. The proposed experiment's contribution provides a model that gives better accuracy for a reduced set of features than the total independent features of the dataset. The result of the model is also compared with other state of the artwork and found better in terms of accuracy. The paper is arranged to discuss a review of related work briefly. Section three starts with a discussion of the proposed model's preliminary concept, followed by a discussion of the model's architecture. The method for experiments conducted and assessing the results is provided.

Literature Survey

As the number of attacks rises regularly, it impacts research into applying deep and machine learning to detect attacks. This section looked at some recent publications that precisely used IDS to detect attacks

Adhao and Pachghare (2020)proposed a model, in this author used the CICIDS2017 dataset to detect an intrusion with an aim to select the best independent features from a set of 78 input features. The proposed model has combined and analysed different outputs from Genetic algorithm (GA) and principal component analysis (PCA), respectively, in terms of selecting the best independent features and then passed through decision tree (DT) classifier to achieve optimal results. As a result, the classifier had optimised prediction results to 99.53% accuracy with the best 40 independent features with PCA-GA-DT feature filtering technique from 99.48% accuracy with all 78 input features. The proposed model also shows results saying that removing noise from the dataset helps in increasing overall accuracy.

Siddiqi and Pak (2020)proposed a feature selection technique after performing different transformation techniques, including Yeo-Johnson power transformation, standard scalar, robust scalar, and L2 normalising transformation techniques with min-max scaling. As a result, the author has marked the Yeo-Johnson transformation technique, which outperformed other transformation techniques when used with deep neural network (DNN), giving the following optimal results on two different datasets. On the CICIDS2017 dataset, the proposed model selected the best 35 independent features out of 78 input features with 99.95% overall accuracy with considering 11 classes of attacks. Similarly, on ISCX-IDS 2012 dataset, the author has demonstrated the selection of the best 37 input features from the set of 82 original independent features with 98% overall accuracy for predicting five different classes of attacks. The author has used the SMOTE to regulate the biasness available in the dataset.

Haq et al. (2015), used three wrapper search methods Genetic Search, Best first, and Rank search NSL-KDD dataset for feature selection and used three basic classifiers Naive Bayes (NB), Bayesian network (BN) and J48. This study proposes an ensemble model with a hybrid feature selection method based on the present research framework. A reliable model is developed by selecting 12 relevant features out of 41 features to differentiate anomaly and normal with the help of a hybrid feature selection method. This model shows that the ensemble approach results better than BN, NB, and J48 classifier. The FPR of the proposed model is 0.021, with an accuracy 98.0%; this result further improved by Aghdam and Kabiri (2016). In this model, they developed a computationally efficient and effective id by performing optimal feature selection with the help of ant colony optimisation. Here they used KDD-CUP99 and NSL-KDD datasets to perform tests and comparison, providing higher accuracy in detecting intrusion attempts and low false alarm with a minimal set of features. The proposed model uses the ACO algorithm and a nearest neighbour classifier to identify a new attack. The proposed model gives an accuracy 98.9% for KDD Cup 99. We can perform feature selection based on packet payload for further improvement to improve the detection rate.

Sudar et al. (2021) designed an IDS used in software-defined networking by using hybrid ML techniques. For this experiment, the author used the NSL-KDD dataset for the classifier author developed a hybrid model using K-means and C4.5 algorithm, which provide accuracy 97.66%.This accuracy is further improved by Uikey and Gyanchandani (2019), IDS provides network security by preventing unauthorised access from the internet. In this paper, authors use NSL-KDD and KDD cup 99 datasets and use various classifiers such as KNN, Artificial Fish Swarm and Artificial Bee Colony (ABC-AFS), SVM, Naïve Bayes, Random Forest to calculate the accuracy of each label (Normal, U2R, DoS, R2L, Probe) present in the NSL-KDD dataset ABC-AFS classifier gives better accuracy 99.00% as compared to other classifiers.

The literature review clearly shows that the used datasets are imbalanced because of the large difference between attack instances. Some attacks have a more instances in the dataset, and some attacks have very few instances to handle this imbalanced dataset concept of SMOTE used in the proposed model.

Dataset

In this experiment, NSL_KDD and CICIDS-2017 datasets are used. NSL-KDD is an advance version of the KDD 99 dataset. According to Tavallaee et al. (2009) three main goals are to design the NSL-KDD dataset. The first goal was to remove redundant features from the KDD9 dataset because this irrelevant feature unnecessarily increases the model's computation cost. The second is to select the most reliable instances from the KDD99 dataset to get good results in terms of accuracy, and the third is to remove unbalancing hurdles from the dataset. This dataset contains 42 features and five labels, benign label, and four attack labels, Probe, DoS, R2L, U2R, are present in the dependent feature.

CICIDS_2017 dataset designed by The Canadian Institute For Cyber Security (CICIDS_2017) dataset contains 80 features, out of which two features are redundant; hence, these two features are removed from the dataset (Sharafaldin et al., 2018). The final dataset used in the proposed model contains 78 features. Total 15 labels are present in the dependent feature out of which one is benign, and the remaining fourteen labels are Bot,DSoS, DoSGoldenEye, DoS Hulk, DoSSlowris, DoSSlowhttptest, Heartbleed, FTP_Patator, Infiltration, SSH-Patator, Web Attack Brute Force, Web Attak-SQL PortScan, Injection,Web Attack-XSS attack label. Total instance available for each attack label and Benign label for NSLKDD and CICIDS-2017 are shown in Tables 85.1 and 85.2 respectively.

Feature Selection and Classification Technique

In ML, various feature selection techniques are available, specifically used to reduce the computational cost by choosing the most relevant features based on their functionality. In dataset pre-processing, FS is an important aspect of improving accuracy by mitigating noise from the model. In the proposed two features selection techniques are used:

1. **Information gain:** Information gain is calculated with the help of Shannon entropy. Entropy gives the average amount of information provided by the source. Entropy is the amount of impurity or measurement

Table 85.1 Total labels and there instances of NSL-KDD dataset

Labels	Count
Total Instances	125974
Benign	67344
DoS	45928
Probe	11657
R2L	996
U2R	53

Table 85.2 Total labels and there instances of CICIDS_2017 dataset

Labels	Count
Total Instances	565677
Benign	454513
DoS_Hulk	45882
Port_Scan	31703
DDoS	25617
DoS_Golden_Eye	2055
FTP-Patator	1529
SSH-Patator	1196
DoS_Slowris	1140
DoS_Slowhttptest	1134
BoT	398
Web_Attack_Brute_Force	312
Web_Attack_XSS	141
Infiltration	37
Web_Attack_SQL_Injection	22
Heartbleed	12

of information in the set of data on which entropy is applied. Hence, gaining the information from the most important features from the independent features entropy plays a major role. The formula for entropy and Information gain is shown in equations (1) and (2) (Adhao and Pachghare, 2021).

$$H(S) = \sum_{k=1}^{q} Pk = \log = 1\frac{1}{Pk} \tag{1}$$

Where H(s) = Entropy and Pk = Probability of each independent feature.

$$IG(S, A) = H(S) - \left[\frac{|Sv|}{|S|} \times H(S) \right] \tag{2}$$

Where S = Total Subset, Sv = Subset After Splitting and H(S) = Entropy.

2. **Pearson's correlation:** This feature selection technique is used to show the relationship between dependent features and independent features of the dataset, the values of correlation between features lie between -1 to 1 (Brownlee, 2019).
   a.　+1 represent a Positive correlation.
   b.　0 represent No correlation.
   c.　−1 represent a Negative correlation.
3. **SMOTE:** As we can see in Table 85.2 first four attack instances are more than the remaining attack instances; hence our dataset becomes biased toward this majority attack. To handle this biasness concept of SMOTE is used, which divides the attack instances into majority and minority classes and balances the dataset (Zhang et al., 2020). SMOTE selects the minority class instance, let us say A, and finds its k nearest neighbours, and then it generates a synthetic instance by selecting one of the nearest neighbours. Let us say B and form a line segment in the feature space by connecting A and B. This convex combination of instances A and B creates new instances. In this way, SMOTE generates new instances and handles the biasness of the dataset.

Architecture of Proposed Model

This work presents a methodology for the classification network traffic using CICIDS2017 and NSLKDD Dataset. The finest features are chosen from the dataset with the help of the IG-PC approach. This feature selection approach reduced the dataset size and selected the top 25 features from both datasets. After this, the dataset is spilt using 7030% spilt as training and test sets. Then the training data has been passed through SMOTE technique. This approach divides the dataset into majority and minority groups to solve the issue raised by the imbalance dataset. The training dataset will be divided into two categories, normal traffic + majority attack traffic and normal traffic + minority attack traffic. Then, the resulting dataset acquired from the above technique will be fed to a RF classifier. After training the classifier model, the testing data has been tested to detect normal, majority attack, and minority attack classes. The architecture of the proposed model is shown in Figure 85.1.

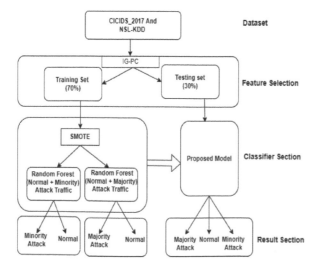

Figure 85.1 Arichitecture of proposed model

Experimental Steup

In the proposed model, two datasets, CICIDS-2017 and NSL-KDD, are used to experiment. This experiment has been done on the following platform: Windows 10, 64 bit OS, 32GB ram, and i5 processor Core(TM). 70% train and 30% test split mechanism is used to calculate accuracy, precision, F1 score, Test time, and Build time by employing a random forest classifier. The formulae for Accuracy, Precision, Recall, F1-score are taken from (Pachghare, 2019).

Result Analysis

The result on the total dataset and fewer features after applying SMOTE are shown in Table 85.3 for both CICIDS_2017 and NSL-KDD, respectively.

The comparison between proposed methodology with the existing state of art method is shown in Tables 85.4 and 85.5 with pictorial representation in Figure 85.2 and 85.3 for CICIDS_2017 and NSL_KDD dataset, respectively.

Table 85.3 Accuracy comparison with total dataset for NSL_KDD and CICIDS 2017 dataset

Number of Features	CICIDS2017		NSL-KDD	
	Total 78 Features	*Top 25 Features*	*Total 41 Features*	*Top 25 Features*
Accuracy (%)	99.7914	99.7978	99.8862	99.8967
Precision (%)	99.8802	99.9375	99.8012	99.9949
F1 score (%)	99.8883	99.8894	99.9014	99.9025
Recall (%)	99.8802	99.8107	99.8057	99.8101
FAR	0.42	0.254	0.002	0.0005

Table 85.4 Comparison between proposed work and existing state of art technique for CICIDS_2017 dataset

Author Name	Selected features (Out of 78)	Classifier Name	Accuracy (%)
Adhao and Pachghare (2020)	40	Decision Tree	99.53
Siddiqi and Pak (2020)	35	Deep Neural Network	99.73
Vijayanand and Devaraj (2020)	35	Support Vector Machine	95.91
Prasad et al. (2020)	40	Bayesian Rough Set	98.08
Proposed Model	25	Random Forest	99.7978

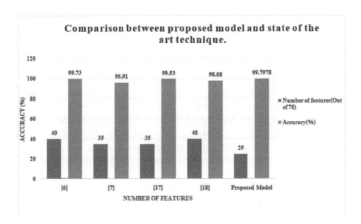

Figure 85.2 Comparison between proposed methodology with existing state of art technique for CICIDS_2017 dataset

Table 85.5 Comparison between proposed work and existing state of art technique for NSL_KDD dataset

Author Name	Selected features (Out of 41)	Classifier	Accuracy (%)
Shrivas and Dewangan (2014)	29	ANN, Bayes NET	97.78
Ingre and Yadav (2015)	29	Artificial Neural Network	81.2
Heba et al. (2010)	23	Support Vector Machine	99.5
Proposed Model	25	Random Forest	99.8967

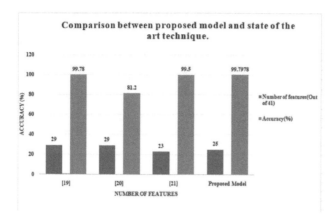

Figure 85.3 Comparison between proposed methodology with existing state of art technique for NSL_KDD dataset

Conclusion and Future Scope

The proposed work presented a feature selection-based IDS model using the Synthetic Minority Oversampling Technique.CICIDS-2017 and NSL-KDD datasets are used to experiment and calculate Information Gain and Pearson's Correlation for the independent features of the dataset. Based on the feature selection technique result, the arithmetic operation is performed between IG and PC. Then, the top 25 features from CICIDS-2017 and 25 features from the NSL-KDD dataset are selected from the top-ranked feature for both datasets, respectively. Finally, the reduced set of features is fed to a random forest classifier. This model result is compared with total dataset features for both dataset and other states of the art technique, and it shows that the proposed model shows good accuracy. In the future, proposed system performance can be compared with other datasets.

References

Adhao, R. and Pachghare, V. (2020). Feature selection using principal component analysis and genetic algorithm. J. Discret. Math. Sci. Cryptogr. 23(2):595–602.

Adhao, R. B. and Pachghare, V. K. (2021). Network traffic classification using feature selections and two-tier stacked classifier. Int. J. Next-Gen. Comput. 12(5).

Aghdam, M. H. and Kabiri, P. (2016). Feature selection for intrusion detection system using ant colony optimisation. Int. J. Netw. Secur. 18(3):420–432.

Brownlee, J. (2019). How to choose a feature selection method for machine learning. Mach. Learn. Mastery 10.

Derhab, A., Aldweesh, A., Emam, A. Z. and Khan, F. A. (2020). Intrusion detection system for internet of things based on temporal convolution neural network and efficient feature engineering. Wirel. Commun. Mob. Comput.

Haq, N. F., Onik, A. R., Hridoy, M. A. K., Rafni, M., Shah, F. M. and Farid, D. M. (2015). Application of machine learning approaches in intrusion detection system: a survey. IJARAI-Int. J. Adv. Res. Artif. Intell. 4(3):9–18.

Heba, F. E., Darwish, A., Hassanien, A. E. and Abraham, A. (2010). Principle components analysis and support vector machine based intrusion detection system. In 2010 10th international conference on intelligent systems design and applications, (pp. 363–367). IEEE.

Ingre, B. and Yadav, A. (2015). Performance analysis of NSL-KDD dataset using ANN. In 2015 international conference on signal processing and communication engineering systems, (pp. 92–96). IEEE.

Pachghare, V. K. (2019). Cryptography and information security. PHI Learning Pvt. Ltd.

Pachghare, V. K., Khatavkar, V. K. and Kulkarni, P. A. (2012). Pattern based network security using semi-supervised learning. Int. J. Info. Netw. Secur. 1(3):228.

Prasad, M., Tripathi, S. and Dahal, K. (2020). An efficient feature selection based bayesian and rough set approach for intrusion detection. Appl. Soft Comput. 87:105980.

Sawaisarje, S. K., Pachghare, V. K. and Kshirsagar, D. D. (2018). Malware detection based on string length histogram using machine learning. In 2018 3rd IEEE international conference on recent trends in electronics, information & communication technology (RTEICT), (pp. 1836–1841). IEEE.

Sharafaldin, I., Lashkari, A. H. and Ghorbani, A. A. (2018). Toward generating a new intrusion detection dataset and intrusion traffic characterisation. ICISSp 1:108–116.

Shrivas, A. K. and Dewangan, A. K. (2014). An ensemble model for classification of attacks with feature selection based on KDD99 and NSL-KDD data set. Int. J. Comput. Appl. 99(15):8–13.

Siddiqi, M. A. and Pak, W. (2020). Optimizing filter-based feature selection method flow for intrusion detection system. Electronics 9(12):2114.

Sparks, S., Embleton, S. and Zou, C. C. (2009). A chipset level network backdoor: bypassing host-based firewall & ids. In Proceedings of the 4th international symposium on information, computer, and communications security, (pp. 125–134).

Sudar, K. M., Beulah, M., Deepalakshmi, P., Nagaraj, P. and Chinnasamy, P. (2021). Detection of distributed denial of service attacks in SDN using machine learning techniques. In 2021 International conference on computer communication and informatics (ICCCI), (pp. 1–5). IEEE.

Tavallaee, M., Bagheri, E., Lu, W. and Ghorbani, A. A. (2009). A detailed analysis of the KDD CUP 99 data set. In 2009 IEEE symposium on computational intelligence for security and defense applications, (pp. 1–6). IEEE.

Uikey, R. and Gyanchandani, M. (2019). Survey on classification techniques applied to intrusion detection system and its comparative analysis. In 2019 International conference on communication and electronics systems (ICCES), (pp. 1451–1456). IEEE.

Vijayanand, R. and Devaraj, D. (2020). A novel feature selection method using whale optimisation algorithm and genetic operators for intrusion detection system in wireless mesh network. IEEE Access 8:56847–56854.

Zhang, H., Huang, L., Wu, C. Q. and Li, Z. (2020). An effective convolutional neural network based on SMOTE and Gaussian mixture model for intrusion detection in imbalanced dataset. Comput. Netw. 177:107315.

86 A review on microstructure study and properties of al7075 particle reinforced composites

Hemanth Raju T.[1,a], Vaishyag Syam Sundar[1,b], Karthik C.[1,c], Jagdish Kumar M.[d], Abhinaya V.[1,e], and Udayashankar S.[2,f]

[1]Mechanical Engineering, New Horizon College of Enginering Bengaluru, India

[2]Mechanical Engineering, Visvesvaraya Technogical University, Belgaum, India

Abstract

Due to their superior mechanical properties, low density, and high strength to weight ratio, aluminium alloys of the 7075 series are often employed in transportation applications such as aerospace, aviation, marine, and automotive. The goal of this work is to discuss the mechanical characteristics of Al7075 alloy metal matrix composites with suitable reinforcements. Also the goal of this work is to conduct a literature study on the production of aluminium MMC materials using a combination of alloys and reinforcements. Reinforcements may be in the form of particle SiC, Al_2O_3, Gr, TiO_2, or Bagase ash. Al7075 alloy can be widely used as a matrix material and it can be combined with different reinforcement particles to form a composite material. Al7075 alloy based particulate reinforced composites are successfully developed using stir casting method. The stir casting procedure incorporates these particle reinforcements. The results indicate that the mechanical characteristics and wear resistance are significantly enhanced when compared to typical base alloys. Additionally, the reinforcing particles are distributed uniformly throughout the aluminium matrix.

Keywords: Al7075, composites, density, reinforcements.

Introduction

Aluminium MMC materials are made up of two or more components, one being the matrix and another of which is filler material. Aluminium MMCs can be laminated, fibre reinforced, or particle reinforced (Singla and Mediratta, 2013). Typically, these materials are manufactured by powder metallurgy, liquid cast metal technology, or through the use of a unique production technique (Kishore, 2019). Processes for treating discontinuous particulate metal matrix materials include powder metallurgy and liquid cast metal technique. Powder metallurgy has significant drawbacks, such as operational cost and component size (Arya et al., 2018). As a result, only the casting technique should be recognised as the most efficient and cost-effective way of processing aluminium composite materials (Kumar and Rao, 2012). The synthesis of ceramic particle reinforced aluminium alloy composites resulted in the birth of a new class of tailorable engineered materials with increased performance characteristics. In addition to the type and size of reinforcement used, as well as the bonding procedure employed, the structure and properties of these MMCs are defined (Subramanian and Aravinth, 2020). Many researchers contributed to the manufacturing of these composites by developing procedures like powder metallurgy, stir casting, plasma spraying, and squeeze casting (Bhowmik et al., 2020). Stir casting method is a extensively used and cost-effective technique of creating and modifying MMC materials (Sambathkumar et al., 2017). The characteristics of these materials are dependent on a variety of production settings and the matrix and reinforcements used (Nataraja et al., 2020). The following procedures are the most significant, and the liquid metallurgy technology has been extensively researched in recent years (Pal et al., 2017). As a result, the present article presents the findings of numerous investigators under the headings mechanical and tribological behaviour (Anvesh et al., 2019). Among the factors that determine composite attributes are the volume fraction, microstructure and homogeneity of the system, which are all affected by the quantities and quality of the matrix and reinforcements used in the composite (Abdulajees et al., 2020). Additionally, an overview of the stir casting technique, process parameters, and production of AMC material using aluminium as the matrix material and different proportions of ceramic materials as reinforcement (Dasa et al., 2016).

The Aluminium 7075 alloy (Al7075)-zircon particulate reinforced metal matrix composites are extensively used in automobile applications. In automobile components they are used in parts such as connecting

---

[a]hemanth.bhadravathi@gmail.com; [b]vaishag2@gmail.com; [c]karthik1505raj@gmail.com; [d]mjkjagadishkumarofficial@gmail; [e]abhinaya07v@gmail.com com; [f]udaya_creative@yahoo.com

rods, pistons, piston rings, cylinder liners, bearings and brake drums. They are also used in aircraft applications. In aircraft they are used in fuselage, wing, and support structures.

Literature Review

Al7075 aluminium alloy-based composites literature review has been done over the last decades and are as discussed as below.

Singla and Mediratta (2013) worked on the mechanical behaviour of Al7075-Fly Ash Composites and discovered that they may be successfully made using a Stir Casting process with homogeneous ash particle dispersion throughout the sample. The Izod and Charpy tests were used to evaluate the composites' toughness. As the quantity of ash is increased, the hardness value steadily increases up to a certain point, i.e. Sample 2, but then decreases. The composites' hardness and tensile strength also exhibited similar results to those for toughness. As the quantity of ash increased up to Sample 2, it increased and then decreased. The density of the composites dropped as the ash concentration increased. Thus, these lightweight composites can be employed in applications in which the weight of an object is critical, such as the aeronautics and space sectors. According to the foregoing data, Sample 2 exhibits high toughness, hardness, and tensile strength, as well as a low density in comparison to alloys without reinforcing.

Kishore (2019) investigated Al7075 reinforced with WC MMCs and demonstrated the successful fabrication of Al7075-WC MMCs. Microhardness of the composites increased from 83HV to 121HV as the weight fraction of WC particles increased. Aluminium MMCs with WC reinforcement have increased their tensile strength from 217MPa to 298MPa. With the insertion of WC reinforcement, the impact strength of Aluminium MMCs was enhanced from 6J to 2J.

Arya et al. (2018) worked on the Production of Al7075-SiC-ZrO2 Hybrid Composites and the Impact of Reinforced Particles on the Mechanical characteristics. They concluded that the reinforced particles experimentally demonstrate an increase in the composite's mechanical characteristics when contrasted to the monolithic material. Stir casting was used to manufacture an aluminium alloy hybrid composite. Stir casting is the most cost-effective and traditional approach of casting composites. Increases in the weight % of reinforcements result in a rise in experimental density, as well as in the proportion of void contents. It is also discovered experimentally that increasing the wt. % of reinforcements enhances the impact strength of the composite by up to 10% (i.e, 6% SiC + 4% ZrO2), depending on the reinforcements used.

Kumar and Rao (2012) carried out the work on mechanical properties of Al7075-SiC composites and demonstrated that liquid metallurgical techniques (stir casting approach) were successfully used to manufacture Al7075-SiC composites comprising up to 6% filler. The developed composites' densities are discovered to be greater than theirs underlying Al7075 matrix. The density improved by 1.24% when SiC concentration increased from 0 to 6 wt. % in the Al7075-SiC composite material. Microstructural examinations revealed that the particulates in the Al7075 matrix system were dispersed uniformly. According to the findings of the testing, Al7075-6wt.% SiC exhibits exceptional mechanical and wear qualities.

Subramanian and Aravinth (2020) worked on the mechanical behaviour of Al7075-SiC-TiC composites. The authors discovered that the composites were successfully produced using stir casting methods, and that scanning electron microscopy (SEM) revealed a reasonably standardised distribution of reinforcement particles within the Al7075 metal matrix. It has been demonstrated that when the percentage of reinforcing materials in the matrix weight increases, the hardness increases. Due to the existence of TiC and SiC particulates, the ultimate tensile strength, impact strength and stiffness of the composites have all improved. When the weight % of reinforcing particles in the Al7075 alloy was adjusted, the ultimate tensile strength of the alloy increased. As a result of technological progress in functional features, meanwhile, the launch of these reinforcing goods was also made possible. Tensile strength, impact strength, and hardness all improve with increasing reinforcement percentage up to 15% TiC and SiC reinforcement, but they begin to drop dramatically at % TiC and SiC reinforcement, and they continue to reduce significantly beyond that. Following the experiment, it was determined that carbon fibre composites reinforced with 15% titanium carbide and 15% silicon carbide exhibited improved mechanical characteristics.

Bhowmik et al. (2020) studied the Mechanical behaviour of SiC/TiB2 dispersed Aluminium MMCs and discovered the occurrence of ceramic reinforcement particles throughout the aluminium matrix by X-Ray diffraction. As a result of TiB2 having a higher density than SiC, the final weight density of the composite material will be higher than it would be otherwise. During the manufacturing process, the SiC and TiB2 ceramic particles were uniformly dispersed throughout the molten aluminium matrix 7075, as evidenced by the micrographs. It was discovered that the addition of SiC and TiB2 significantly reduced the rate of wear. The COF and wear rate of Al7075-SiC- TiB2 composites were found to be lesser than those of Al7075 matrix alloy, indicating that they are more durable. Additionally, due to the hardness of TiB2 reinforced composite, it was shown to have a lower wear rate than SiC reinforced composite.

Sambathkumar et al. (2017) carried out the work on the mechanical characteristics of Al7075 (Hybrid) MMCs utilizing a two step stir casting method and demonstrated that the created composite displayed greater hardness value and ultimate tensile strength values than the basic alloy. The composite material's optical photomicrographs demonstrated a homogeneous dispersion throughout the matrix material. According to Archimedes' principle, composites have higher measured densities than their underlying matrix. The microhardness of the developed composites was improved by increasing the reinforcing content from 0 to 15%. Tensile testing revealed that the composite consisting of 10% SiC and TiC had a maximum tensile strength of 240 MPa, an increase of approximately 60 MPa (33%), above the basic alloy. The corrosion resistance of the developed composites was enhanced by increasing the wt. % of reinforcement particles.

Nataraja et al. (2020) worked the mechanical behaviour of Al7075 hybrid composites and found that while tests to evaluate ultimate tensile strength revealed no clear trends, there was an increase in UTS owing to the existence of carbon nanotubes and zirconium oxide in contrasted to base metal. The hardness rose as the percentage of CNT and ZrO_2 reinforcement increased up to 1%. The MMCs of Al7075 reinforced with carbon nanotubes and zirconium dioxide particles exhibit increased tensile strength as contrasted to the Al7075 alloy alone.

Pal et al. (2017) compared the mechanical characteristics of Al7075-Al_2O_3 nano and micro composites at varying percentages via the stir casting technique and discovered that the composites have an increased hardness and tensile strength than the base matrix, while the percentage elongation reduces as the reinforcement weight percentage increases. Nano composites outperform conventional composites in terms of hardness, tensile strength, and specific wear rate. Additionally, it is shown that micro and nano-composites outperform Al7075 in terms of wear resistance.

Anvesh et al. (2019) worked on the mechanical behaviour of Al7075 MMCs reinforced with ceramic particles and concluded that stir casting is preferable to other production procedures due to the possibility of high particle loading and complicated shaped components. Tensile testing reveals that Al7075-5% B4C has a higher ultimate tensile strength of 220 MPa than Al7075-5% SiC, which has strength of 200 MPa. Compressive strength of Al7075-5% B4C is 495.5 MPa, while Al7075-5% SiC is 336 MPa. Ceramic reinforced MMCs have hardness higher than that of the underlying Al7075 alloy. Al7075-5% B4C composite materials exhibit a higher hardness of 115 BHN, whereas Al7075-5% SiC composite materials exhibit a hardness of 110 BHN. Ceramic reinforced MMCs have higher impact strength than the underlying Al7075 alloy. Boron carbide (B4C) has higher impact strength of 8 Nm than silicon carbide (SiC), which has strength of 6 Nm. The density of Al7075-5% B4C composite material reduced little as compared to the base material, whereas the density of Al7075-5% SiC composite material decreased dramatically. Because Al7075 reinforced with SiC is easier to manufacture and less expensive than Al7075 reinforced with B4C, and because Al7075 reinforced with B4C has certain casting complications, silicon carbide can be used in areas with less loading and boron carbide can be used in areas with more loading, silicon carbide can be used in areas with more loading.

Abdulajees et al. (2020) investigated the Evaluation of Aluminium MMCs Fabricated by Stir Casting process and discovered that the maximum tensile strength of the hybrid composite is 87.02 N/mm². Tensile strength grew progressively as reinforcement weight percentage increased, as shown in the following outcomes. Tensile strength grew progressively as reinforcement weight percentage increased, as shown in the following findings.

Dasa et al. (2016)., optimised the synthesis technique for Al7075/SiC MMCs and found that optical images of Al7075-SiC composites revealed consistent dispersion of SiC particles in the Al7075 matrix. The T6 condition of heat treatment enhanced all of the mechanical characteristics examined. Linear regression models were built to predict mechanical characteristics for a range of manufacturing process parameters. The model's efficiency was determined using coefficients of determination (higher values). The normal distribution of the residuals and the model significance were determined using normal probability plots of the residuals. During multiple performance optimisation, the MMC with a mean particle size of 6.18 m and a SiC reinforcement content of 25% reached the best mean grey relational grade. It is the suggested combination of manufacturing process parameter levels for Al7075/SiC MMCs under discussion for the multiple response criterions. Validation experiments obtained approximately 42% enhancement in grey relationship grade. The ANOVA outcomes for grey relational grade reveal that the mean particle size of SiC has a greater effect on the process than its wt. % in MMCs.

Chandrasheker and Raju (2019) carried out the evaluation of mechanical behaviour of Al7075/B4C/Gr Hybrid MMCs and found that the inclusion of B4C and Graphite particles boosted the hardness of the composites. The inclusion of B4C and graphite particles increased the hardness of specimen-3, i.e. at an Al7075 + B4C 14% + Gr 3% composition. In comparison to the other two samples, the ultimate load is greatest for specimen-3, which is composed of Al7075 + B4C 14% + Gr 3%. The tensile strength is greatest for specimen-3, which is composed of Al 7075 + B4C 14% + Gr 3%. The microstructures of all samples revealed an uniform combination of reinforcing particles and molten metal. The microstructure of the polished surface resulted in the clustering of reinforcement particles in specimen-1, i.e. at an Al7075 + B4C 6% + Gr 2% composition. The etching process in specimen-2 led to a significant identification of non-metallic particles distributed between metallic particles. In specimen 3, the solidification process results in the production of dendritic structure. The inclusion of Al7075 reinforcements causes a decrease in the material's elongation.

Anbazhagan et al. (2019) carried out the development and characteristics of Al7075-SiC composites produced using stir casting method. It is reported that the composites' mechanical characteristics are found to be superior to those of their parent matrix. Microstructural examinations revealed that the particles in the Al7075 matrix were distributed uniformly. Micro-hardness of the composites revealed an increase in hardness, indicating a reduced reinforcing weight percentage. Also the tensile strength of Al7050-SiC particulate composites were found to be higher when compared to certain various 7000 series of Aluminium alloy MMCs, with 145.83 (Mpa) stronger tensile strength than that of the Al7075 matrix. Altogether, the research indicates that Al7075-SiC possesses exceptional mechanical and tribological capabilities while producing composites, the process variables that influencing the stir casting method were studied.

Sudhindra and Murali (2020) studied the mechanical behaviour of Al7075-TiC-Si metal matrix hybrid composites. In their research, they discovered that the dispersion of Titanium carbide and Silica powder particulates in an Al7075 matrix boosts the hardness value and tensile behaviour of Al7075 alloy, respectively. The weight proportion of reinforcement utilised in the composites had the greatest influence on the hardness of the stir-cast composites. As a consequence, the Al7075 matrix comprising % TiC and % Si particles achieved the maximum microhardness, followed by the Al7075 matrix having 5% Si particles. In TiC and Si reinforced hybrid composites, the tensile strength increases in direct proportion to the amount of TiC and Si employed in the composites. The UTS of the composite was improved by adding 5% TiC and 5% Si to the Al7075 base material, which increased the overall strength of the composite. All in all, the outcomes of the study reveal that the alloy Al7075TiC-Si possesses exceptional mechanical properties.

Imran et al. (2016) using stir casting method, studied the mechanical behaviour Aluminium 7075 composites graphite-bagasse-ash composites, and they reported on the effective fabrication of Al7075-bagasse ash-Gr hybrid MMC samples. When the amount of reinforcement particulates in the base alloy is increased, the hardness value of composites increases by a significant amount. It was discovered that BHN increased in proportion to the increase in reinforcement, according to the 95% confidence interval. The ductility of composites decreases when the amount of reinforcing material in the Al7075 matrix is increased. The tensile strength and yield strength of the resulting Al7075-bagasse ash-graphite hybrid MMC increase as the amount of reinforcement particles increases.

Patil and Haneef (2019) worked on the Al7075-Graphene Nanoplatelets-Beryl composites fabricated by using stir casting process. The researchers worked on the microstructure study, tensile strength and hardness value behaviour of Al7075-Graphene Nanoplatelets-Beryl composites manufactured by the stir casting technique. As a result of the SEM investigation, it was discovered that the GNPS and Beryl particles were evenly dispersed throughout the Al7075 alloy. In the presence of 6% Be and 2% GNPs particulate, the UTS of Al7075-Beryl-GNPs composites is 231.587 MPa, representing a 77.09% increase over the Al7075 matrix. Using 6% Beryl and 2% GNPs particulate, the hardness value of Al7075-Beryl-GNPs composites is demonstrated to be extremely high at 126.7 BHN, representing a 49.41% increase over the hardness of the Al7075 alloy.

From the above literature survey, it is concluded that Aluminium 7075 alloy can be widely used as a matrix material and it can be combined with different reinforcement particles to form a composite material. Aluminium 7075 alloy based particulate reinforced composites are successfully developed using stir casting method. The reinforcement particles are uniformly distributed in the Aluminium 7075 alloy matrix. The mechanical and wear properties of Aluminium 7075 alloy is increased with the addition of reinforcement particles. Also it is noted that the development and microstructure study, mechanical and wear characterisis Aluminium 7075 alloy -zircon composites has not been attended so far. Hence Aluminium 7075 matrix can be combined with zircon particles to form a composite material by using stir casting method and the developed composites are tested for microstructure study, mechanical and wear characteristics.

Materials and Fabrication Process

Matrix material: Aluminium 7075 alloy

Al7075 is a zinc-based aluminium alloy. This is a very strong aluminium alloy with comparable strength to various steels, outstanding fatigue strength, and moderate machinability; nonetheless, it exhibits less corrosion resistance than several other Al alloys (Chandrasheker and Raju, 2019). Due of its high cost, it is only useful in situations where less expensive alloys are ineffective.7075 aluminium alloy contains around

5.6 to 6.1% zn, 2.1 to 2.5% mg, 1.2 to 1.6% cu, and insignificant amounts of Si, Ti, Mn, Fe, Cr, and other metals (Anbazhagan et al., 2019). Aluminium Alloy 7075 is the strongest of the aluminium alloys usually used in screw machine construction (Sudhindra and Murali, 2020). The higher corrosion resistance of alloy 7075 makes it a suitable option for alloys such as 2024, 2014, and 2017 in a wide range of high-stress applications (Imran et al., 2016). Because of its high strength, alloy 7075 is frequently employed in the aerospace and defence industries.

Table 86.1 presents the chemical composition of Al7050 alloy.

Reinforcement material: zirconium silicate

Mineral zircon is made up of zirconium silicate, which occurs naturally in the earth's atmosphere. Various procedures are used to concentrate the ore once it has been mined from its natural source (Patil and Haneef 2019; Raju et al., 2021). Electrostatic and electromagnetic technologies are used to remove it from the sand during the separation. A zirconium salt can be combined with sodium silicate in an aqueous solution to produce the compound, as can the fusing of SiO_2 and ZrO_2 in an arc furnace (Raju et al., 2021).

In the manufacture of refractory materials, zirconium silicate is used to provide resistance to corrosion by alkali materials in situations where corrosion resistance is needed (Chandra et al., 2021). A variety of ceramics, enamels, and ceramic glazes are made with it as well as some enamels. It is also used as milling and grinding beads, which is another application for zirconium silicate (Adhikary et al., 2018; Nagendra et al., 2021; Nagabhushana et al., 2020). The Table 86.2 presents the physical properties of Zirconium silicate.

Stir casting process

Stir casting is now the most widely used and commercially successful method in liquid-state processing because it is the most cost-effective method when contrasted to other fabrication techniques (Puneeth and Prasad, 2019). It also results in a rather homogeneous distribution of reinforcements in the matrix, improved wettability, and reduced porosity, among other benefits. Generally speaking, stir-casting is concerned with the mixing of a dispersed phase in a matrix phase, which is made possible by the use of a stirring mechanism (Bopanna and Prasad, 2020; Bellie et al., 2021). Electrification of the stir-casting furnace is frequently accomplished by the use of electrical energy, and electrical resistance heating is the most frequently employed technique of heat generating (Jayanth et al., 2021; Subbian et al., 2021; Rajesh et al., 2020). The procedure consists in heating the matrix, which has been placed in the crucible, until it reaches the melting point. Crucible is constructed to be chemically inert to the matrix and reinforcements (Jose et al., 2017; Madhusudan et al., 2020; Padmini et al., 2019). Preheating of the reinforcements is frequently used to enhance the mixing of the components in the final assembly. In order to limit the likelihood of casting flaws, the mixing process should be carried out in a molten state, with the inert state maintained throughout the stirring and pouring of the charge(Nagendra et al., 2018). Particulate reinforcements are often delivered by an injection gun in order to limit the likelihood of gas trapping during the manufacturing process (Srinath and Prasad, 2018). The stirrer's propeller blades are joined to a shaft that is connected to the output of the electrical motor, which causes the stirrer to rotate (Adhikar et al., 2015). Using a lead screw arrangement powered by another electrical motor, it is possible to control the vertical motion of the stirrer with great precision (Sathish et al., 2021). A typical method of controlling the spinning speed of the stirrer is to employ stepper motors (Srinath and Prasad, 2019). The wettability between the matrix and the reinforcement should be enough in order to achieve a homogeneous mixture by this procedure (Raju et al., 2021), as previously stated.

Table 86.1 Chemical composition of Al7050 alloy

Component	Al	Mg	Si	Fe	Cu	Zn	Ti	Mn	Zr	Cr
Amount (wt.%)	Balance	2.29	0.021	0.051	2.11	5.9	0.026	0.008	0.13	0.009

Table 86.2 Physical properties of zirconium silicate (ZRSIO$_4$)

Molecular formula	$ZrSiO_4$
Molecular mass	183.31 g mol^{-1}
Appearance	Colourless crystals
Density	4.56 g cm^{-3}
Melting point	2550°C

The casted aluminium alloy particulate reinforced composite is shown in the below Figure 86.1.

Comparative Analysis

Table 86.3 shows the mechanical properties such As BHN, Tensile strength and Compression strength of Al7075 alloy combined with different reinforcement particles.

Table 86.4 shows the Wear Rate of Al7075 alloy combined with different reinforcement particles.

Table 86.5 shows the applications of Al7075 alloy combined with different reinforcement particles.

Figure 86.1 Casted aluminium alloy particulate reinforced composite

Table 86.3 Comparative analysis of mechanical characterisation of most recent research work on Al77075 based particulate reinforced composites

Author/Year	Matrix and reinforcement	BHN	Tensile strength MPa	Compression strength MPa
Singla (2013)	Al 7075+ Fly Ash	77	140	525
Kishore (2019)	Al 7075+WC	110	235	643
Arya (2018)	Al 7075 + SiC+ZiC	130	265	710
Kumar (2012)	Al 7075+ SiC	100	245	580
Subramanian (2020)	Al 7075 + SiC+TiC	83	270	736

Table 86.4 Comparative analysis of wear rate of most recent research work on Al77075 based particulate reinforced composites

Author/Year	Matrix and Reinforcement	Wear Rate ($mm^{3///}/m$)		
		Load		
		10N	20N	30N
Bhowmik1 (2020)	Al 7075+ Fly Ash	0.0028	0.0034	0.0037
Siddesh Matti/ 2020	Al 7075+Mica+Fly Ash+Red Mud	0.0013	0.0017	0.0021
Anbazhagan1 (2018)	Al 7075 + SiC	0.0031	0.0035	0.0039
Imran (2016)	Al 7075+ Graphite	0.0024	0.0027	0.0032

Table 86.5 Comparative analysis of applications of most recent research work on Al77075 based particulate reinforced composites

Author/Year	Matrix and reinforcement	Applications
Singla (2013)	Al 7075+ Fly ash	In automobile components like cylinder liners, bearings
Kishore (2019)	Al 7075+WC	In automobile components like piston, pistons rings
Arya (2018)	Al 7075 + SiC+ZiC	In automobile components like connecting rod
Kumar (2012)	Al 7075+ SiC	In aircraft they are used in fuselage, wing, and support structures

Conclusions

Many researchers have been working on Aluminium 7075 alloy- MMCs in recent years, and this review summarizes their points of view, theoretical and experimental data, and conclusions obtained during that time.

- A reduction in the density of Al7075 composites was discovered by introducing reinforcement into the Al7075 matrix.
- It has been discovered that the hardness value of Al7075 alloy based composites increases with the amount of reinforcing material present in the matrix material.
- A higher elastic modulus and tensile strength were discovered in the Al7075 matrix composites when contrasted to the basic alloys.
- When the load applied and speed increases, it was revealed that the wear rate of composites increases. It was also discovered that the highest wear rate may be achieved with the smallest particle size.

Acknowledgements

The authors wish to express their gratitude to Management and Principal of NHCE, Bengaluru-560103, for their assistance in conducting this research.

Declaration of Competing Interest

The authors claim to be unaware of any financial or personal ties that may have affected their study.

References

Abdulajees, R., Sivagami, S. M., and Vijay, R. T. (2020). Characterization of aluminium metal matrix composites fabricated by stir casting. Int. Res. J. Eng. Technol. 7(9).

Adhikary, P., Bandyopadhyay, S., Mazumdar, A., (2018) C.F.D analysis of air-cooled HVAC chiller compressors. ARPN J. Eng. Appl. Sci. 13.

Adhikary, P., Roy, P. K., and Mazumdar, A. (2015) Optimal Renewable Energy Project Selection: A Multi-Criteria Optimization Technique Approach. Glob. J. Pure Appl. Math. 11.

Anbazhagan, R., Rekka, G., Kalidoss, R., Kumar, N. M., M. and Sanjeev, M. (2019). Mechanical and micro structural analysis of Al7050-SiC composite prepared by stir casting method. TAGA J. 14:12821290.

Anvesh, D., Adam, G. A. Satyanarayana, K., and Veni, M. N. V. K, (2019). Investigation on mechanical properties of Al 7075 MMC's reinforced with ceramic particles. J. Compos. Theory. 12(8):94101.

Arya, P., Gangwar, S., and Sharma, S. (2018). Fabrication of aluminium alloy (Al7075) hybrid composite and effect of reinforced particles on physical and mechanical properties. Int. J. Adv. Res. Sci. Eng. Technol. 7:702709.

Bellie, V., Gokulraju, R., Rajasekar, C., Vinoth, S., Mohankumar, V., and Gunapriya, B. (2021). Laser induced Breakdown Spectroscopy for new product development in mining industry. ARPN J. Eng. Appl. Sci. 16.

Bhowmik1, A., Dey, D., and Biswas, A. (2020). Characteristics study of physical, mechanical and tribological behaviour of SiC/TiB2 dispersed aluminium matrix composite. Nature, 114.

Bopanna, K. D. and Prasad, G. M. S. (2020). Thermal characterization of aluminium-based composite structures using laser flash analysis. J. Ins. Eng. (India): C. 101.

Chandrasheker, J. and Raju, Nvs. (2019). Fabrication and investigation of mechanical properties of Al 7075/B4C/Gr hybrid metal matrix composites. Pramana Res. J. 9(6)13481357.

Chandra, R. P., Balachandra, R., Halemani, S., Chandrasekhar, K. M., Ravitej, Y. P., Hemanth Raju, T., and Udayshankar, S. (2021). Investigation and analysis for mechanical properties of banana and e glass fiber reinforced hybrid epoxy composites. 47.

Dasa, D. K., Mishra, P. C., Chaubey, A. K., and Singha, S., (2016). Fabrication process optimization for improved mechanical properties of Al7075/SiC metal matrix composites. Manag. Sci. Lett. 6:297308.

Imran, M., Khan, A. R. A, Megeri, S., and Sadik, S. (2016). Study of hardness and tensile strength of Aluminium-7075 percentage varying reinforced with graphite and bagasse-ash composites. Resource-Eff. Technol. 2:8188.

Jayanth, B. V., Depoures, M. V., Kaliyaperumal, G., Dillikannan, D., Jawahar, D., Palani, K., and Shivappa, G. P. M. (2021). Laser induced Breakdown Spectroscopy for new product development in mining industry. Energy Sources A: Recovery Util. Environ. Eff.

Jose, S., Athijayamani, A., Ramanathan, A., and Sidhardhan, K. S. (2017). Effects of aspect ratio and loading on the mechanical properties of prosopis juliflora fibre-reinforced phenol formaldehyde composites. Fibres Text East. Eur. 25.

Kishore, T. V. (2019). Al 7075 reinforced with WC Metal Matrix Composites. J. Sci. Technol. 4(2):4550.

Kumar, G. B. V. and Rao, C. S. P., and Selvaraj, N. (2012). Mechanical and dry sliding wear behavior of Al7075 alloy-reinforced with SiC particles. J. Compos.Mater. 10:19.

Madhusudan, M., Kumar, S., Kurse, S., Shanmuganatan S. P., John, J., and Haseebuddin, M. R. (2020). Behavioral studies of process parameters and transient numerical analysis on friction stir welded dissimilar alloys, Mater. Today: Proce.

Nagabhushana, N., Rajanna, S., and Ramesh, M. R. (2020). Influence of temperature on friction and wear behavior of aps sprayed NiCrBSi/Flyash and NiCrBSi/Flyash//TiO$_2$ coatings. J. Green Eng. 10.

Nagendra, J., Srinath, M. K., Sujeeth, S., Naresh, K. S., and Prasad, G. M. S. (2021). Optimization of process parameters and evaluation of surface roughness for 3D printed nylon-aramid composite Mater. Today: Proce. 44.

Nagendra, J., Prasad, M.S.G., Shashank, S., Md Ali, S. (2018) Comparison of tribological behavior of nylon aramid polymer composite fabricated by fused deposition modeling and injection molding process, materials and manufacturing. Int. J. Mech. Eng. Technol. 9.

Nataraja, M. M., Vinod, K. L, and Balaji, J. (2020). Investigation on mechanical properties of aluminium 7075 hybrid composites. Int. J. Res. Trends Innov. 5(9):9397.

Padmini, B. V., Niranjan, H. B., Kumar, R., Padmavathi, G., Nagabhushana, N., and Mohan, N. (2019). Influence of substrate roughness on the wear behaviour of kinetic spray coating. Mater. Today: Proce. 27.

Pal, B., Sharma, C., Dhraik, A., and Ucharia, V. (2017). Study of Comparison of Mechanical Properties of Al7075 with Al7075 reinforced with nano and micro particles of Al$_2$O$_3$ at varying percentage via stir casting route. Int. J. Sci. Dev. Res. 2(8).

Patil, S. and Haneef, M., (2019). Microstructure, tensile properties and hardness behavior of Al7075 matrix composites reinforced with graphene nanoplatelets and beryl fabricated by stir casting method. Int. J. Eng. Adv. Technol. 9(1).

Puneeth, H. V. and Prasad, G. M. S. (2019). Biological factors influencing the degradation of water-soluble metal working fluid. Sustainable Water Res. Manag. 5.

Rajesh, A., Gopal, K., Victor, M. D. P., Kumar, R. B., Sathiyagnanam, A. P., and Damodharan, D. (2020) Effect of anisole addition to waste cooking oil methyl ester on combustion, emission and performance characteristics of a DI diesel engine without any modifications. Fuel. 278.

Raju, T. H., Chandan, B. B., Belagavi, V., Udayashankar, S., Jegadesha, T., Gajakosh, A. M. (2021). Influence of zircon particles on the characterization of Al7050-Zircon composites. Mater. Today: Proce. 47.

Raju, T. H., Chandan, B. B., Vinaya, B., Udayashankar, S., Jagadeesha, T., and Gajakosh, A. K. (2021). Influence of zircon particles on the characterization of Al7050-zircon composites. Mater. Today: Proce. 47:22412246.

Raju, T. H., Kumar, R. S., Udayashankar, S., and Gajakosh A. (2021). Influence of dual reinforcement on mechanical characteristics of hot rolled AA7075/Si3N4/graphite. MMCs. J. Inst. Eng.: D. 102.

Sambathkumar, M., Navaneetha, P., K., Ponappa, K., and Sasikumar,K. S. K. (2017). Mechanical and corrosion behavior of Al7075 (Hybrid) metal matrix composites by two step stir casting process. Lat. Am. J. Solids Struct. 14:243255.

Sathish, T., Arul, S. J., GopalKaliyaperumal, G., Velmurugan, G., and Nanthakumar, P. (2021). Comparison of yield strength, ultimate tensile strength and shear strength on the annealed and heat-treated composites of stainless steel with fly ash and ZnO. Mater. Today: Proce. 46:31654348.

Singla, D. and Mediratta, S. R (2013). Evaluation of mechanical properties of Al7075-Fly ash composite material. Int. J. Innov. Res. Technol 2(4):951959.

Srinath, M. K. and Prasad, G. M. S. (2018) Numerical analysis of heat treatment of TiCN coated AA7075 aluminium alloy, AIP Conference Proceedings. 1943.

Srinath, M. K. and Prasad, G. M. S. (2019). Surface Morphology and Hardness Analysis of TiCN Coated AA7075 Aluminium Alloy. J. Ins. Eng. (India): C., 100.

Subbian, V., Kumar, S., Chaithanya, K., Arul, S. J., Kaliyaperumal, G., and Adam, K. M. (2021). Optimization of solar tunnel dryer for mango slice using response surface methodology. Mater. Today Proce. 46.

Subramanian, M. and Aravinth, B. (2020). Effect on the mechanical properties of Al7075 reinforced with SiC and TiC particles. Int. J. Eng. Adv. Res. Tech. 10(1):287290.

Sudhindra, S. K. and Murali, M. M. (2020). Experimental investigation of mechanical properties of ceramic reinforced Al-7075 metal matrix hybrid composites. Mater. Sci. Forum. 979:3439.

87 Study and analysis of various characteristics of innovative solar crop dryer

V. R. Khawale[1], S. V. Prayagi[1], and B. N. Kale[2]

[1]Mechanical Department, YCCE, Nagpur, India

[2]Dr Babasaheb Ambedkar College of Engineering and Research, Nagpur, India

Abstract

In India, farmers produce seasonal product. It is very essential to dry the products so that it will be available in the market throughout the year. Drying increases the life and cost of product. The aim of this work is to suggest the innovative solar dryer which will helps to harness the maximum solar emission. The various performance characteristics of the solar dryer were evaluated such as thermal efficiency of the collector or the drying rate of agricultural products. This model was made adjustable such that it can work as a single pass, double pass, and triple pass with reversed absorber and reflector. The type of solar dryer used was a solar dryer with reversed absorber and reflector. The average collector efficiency (η_c) values are 26.11%, 29.56% and 34% for the SDR1, SDR2 and SDR3 respectively. The indicated efficiency of heat collection of SDR3 is remarkable than SDR1 and SDR2. The drying efficiency (ηd) values of SDR1, SDR2 and SDR3 were 41%, 44% and 49%, respectively, and the picking efficiency (ηp) values were 51%, 55% and 64%, respectively. These value shows the SDR3 is more capable than SDR1 and SDR2. Compared with Newton and Henderson and Pabis models, the experimental data is more suitable for Page's model. Page's model gives the best results with the maximum value of R^2 and the minimum value of MBE and RMSE. Finally, it has been found that SDR3 is most effective solar dryer than SDR1 and SDR2.

Keywords: solar air heater, solar dryer, reversed absorber, Reflector, Red Chilies.

Introduction

Drying is the process of removing moisture. Drying has been very important process to preserve an agriculture product for long time. Because of that agriculture product may get available in all seasons. Open sun drying method is always used by farmer to dry all agriculture products. Due to limitations, like no uniform drying of agricultural product, long time required for drying and uncleanness during drying process this method is not suitable for mass production. Conventional dryers can be used to dry a agriculture product. In conventional dryer can be maintained hygienic condition but it is a costly due to use of fuel as an energy source. Solar dryer is emerging technique which has been overcome these limitations. Solar dryer technology is various advantages like simple in design, low-cost energy source and maintained hygienic condition. The critical study of solar drying processes has having great practical and economic importance. In designing the process, study of fundamentals and mechanisms, moisture level in product and temperature required to dry the product are critical factors (McMinn, 2006). Many researchers have been presents theoretical models of drying process in which heat and mass transfer studied. Thin layer models and simulations are helps for designing new dryer with improving usefulness in existing applications (Kardum et al., 2001). In thin layer model the moisture in agriculture product can be measured at any time and correlation can be developed with drying period. Most of the Researcher has developed a specific type of solar dryer and conducted experimentation for specific type of agriculture product. Analysis has been done without putting any evident that this dryer can be used for other agriculture product. Many researchers have proposed a different type of solar dryer which are classified into direct, indirect and hybrid solar dryer (Garg and Sharma, 1990; Hallak et al., 1996). SCD is an example of domestic and DSD where solar radiation direct falls on the food product (Lawand, 1996). In a direct solar dryer, the food product placed in thin layer on perforated trays and opened to the direct solar emission. In ISD, the air gets heated initially in the collector and then it has passes through the product Indirect solar dyer has functions more effectively and can be control over the drying process. Shell dryer is an example of ISD present in the literature (Fournier and Guinebault, 1995). Indirect solar dryer with natural air circulation has been specially designed for particular type of agriculture product and may be used with additional source of energy to improve the performance. Singh et al. (2006) has been developed an efficient solar dryer specially used to dry Fruits, Spices and vegetables and drying capacity of about 1kg per day. The solar dryer is featured by direct, indirect and hybrid with forced or natural circulation of air. For collecting maximum energy, the solar air collector always keeps inclined due south so that sunrays must strike perpendicular throughout the day on collector. Manual tracking may be alternative but it is not a practical. Solar air heater integrated with tracking system can be used to collect maximum heat energy but it is very expensive. The solar air heater is designed to absorb the maximum solar radiation in sunshine day. Using of single glazing, double glazing and multi pass solar air collector can also improve the efficiency. Many researchers proposed new type solar dryer model and present a nature of drying behaviour of agriculture product (Simal et al., 2005; Midilli and Kucuk, 2003; El-Beltagy et al., 2007; Kavak Akpinar et al., 2004). Thin layer drying model can be used easily as well it gives accurate results. However, the rate of drying depends on the temperature of air and drying rate can be kept constant if temperature of air is maintained. But to maintain the temperature constant throughout the day is not possible. Glazing the dryer and solar air heater can be a practical solution Mwithiga and Kigo has been noted the effectiveness of tracking the sun on increase the drying rate but it is costly affaire. Bentayeb et al. (2007) revealed a wooden SD and noted the effect of thickness of timber and partition of glass on drying rate Fatouh et al. (2006) observed the effect of drying air temperature and air velocity on the drying characteristics of agricultural product. The drying rate can be achieved if considered parameters are within limits. Singh et al. (2005) noted that the leafy vegetables are dried in short period than other vegetables. But if it can be spread in thin layer on perforated trays in slice shape, the rate of drying can be increased. Initially drying start from surface and then inside the materials. Drying rate depends on exposed area, temperature of air and flow rate condition. A rate of drying to agricultural products is being reduced due to the decrease in water content on surface of matter. During this process the rate of drying controlled by dispersal of moisture from surface to inner layer of the product Agriculture product with thin layered on trays, uniform air temperature at collector exit and low velocity of air enables to maintained a constant drying rate due to dispersal of water particle from internal layer to surface.

The intention of this paper writing is to recommend a SCD with reversed absorber and reflector to collect the maximum possible solar radiation. A novel SD has been projected by taking a reference of reported solar dryer in literature (Jain and Jain, 2004; Jain, 2007), and introduces some new features which help to maximize the solar energy absorption. The principal modification was that, it includes in indirect type of solar dryer using forced convection heat transfer with single pass, double pass and multi pass arrangement Firstly it was very important to reduce the variation in air temperature throughout the day Secondly the drying period was long because drying started from 8 am to 5 pm, tried to used whole sunshine day. The dryer was constructed, performed the experiment and compared with the other solar dryer performance. The drying characteristic was investigated by performing an experiment by using chillies.

Theoretical Analysis:

The performance of the solar dryer may be assessing from thermal efficiency of collector or drying rate of the agricultural product (Joshi et al., 2004).

Drying rate

Drying rate depends on various factors whose measuring and calculation are very monotonous and time consuming. Pathak et al. (1991) in his study noted that, temperature of air, initial moisture percentage in product, velocity and relative humidity of air are various factors over which drying rate vary. For various products, drying rates has been successfully analysed by using Page's equation (Yunfei et al., 1987). The equation is as follows;

$$Mr = e^{-zt^n} \tag{1}$$

Where,
Mr- Moisture ratio, t- drying time in hours, z and n are drying constants.
Moisture ratio given by equation:

$$\text{Moisture Ratio} = M_r = \frac{M_t - M_e}{M_o - M_e} = e^{-zt} \tag{2}$$

Where
M_t- Instantaneous moisture content (% db),
M_e- Equilibrium moisture content (% db),
M_o-Initial moisture content (% db).

Following equation may be used to calculate moisture content on dry basis M_t(% db) (Martinez, 2001):

$$M_t = \frac{(W_t - W_d)}{W_d} \tag{3}$$

And the equation used to calculate moisture content on wet basis (% wb):

$$M_t = \frac{(W_t - W_d)}{W_t} \tag{4}$$

Where
W_t = Weight of chillies at a time (t) (kg)
M_d = Mass of dry chillies (kg)
From exponential and Newtonian model (Sun and Wood, 1994), page introduced an equation as

$$\text{Drying rate} \left(\frac{dm}{dt}\right) = -z(M_t - M_e) \tag{5}$$

z is drying rate constant, per hour.
By distinguishing, the drying model can be explained as follows

$$M_z = e^{-zt^n} \tag{6}$$

In deriving this equation, the resistance of water movement and gradient in the material is ignored. If the drying rate decreases at a constant temperature, pressure and humidity of the air, this equation is valid (Nellist, 1976), which is characteristic of products with low moisture content (e.g. grains), negative sign of is important which shows the characteristics of drying process (Fatouh et al., 2006).

$$\text{Moisture Ratio } M_r = \frac{Mt - Me}{Mo - Me} = e^{-zt} \tag{7}$$

Collector efficiency

By using formula

$$\eta_c = \frac{\dot{m}a C_{pa}(T_o - T_i)}{\left[(I \times Ac1) + (Ieff \times Ac2)\right]} \tag{8}$$

Where
$\dot{m}a$ - Mass of air flowing in collector and dryer per unit time
C_{pa} - Specific heat of air at constant pressure
T_o - Air temperature at collector exit
T_i - Inlet temp. of air at collector entrance
I - solar emission on absorber plate 1 (hourly)
A_{C1} - Area of Absorber Plate-I
I_{eff} - Actual solar emission on reversed absorber plate-II (hourly) ($I_{eff} = \rho \times I \times A_R$).
ρ - Reflectivity of reflector
A_R - Area of reflector
A_{C2} - Area of absorber plate-II.
Following equation used to calculated dryer thermal efficiency

$$\eta_d = \frac{M_w H_{fg}}{\dot{m}a \, C_{pa}(T_d - T_i)} \tag{9}$$

Where
$\dot{m}a$ - Mass of air flowing per unit in the dryer
C_{pa} - Specific heat of air at constant pressure
T_d - Temperature of air in dryer
T_i - Inlet air temperature at collector equal to ambient temp.
H_{fg} - Latent heat of water (evaporation)
M_w - Mass of evaporated water.

Following equation used to calculated the dryer pick-up efficiency

$$\eta_p = \frac{M_w}{\dot{m}a\,\Delta_t\left(W_{ce} - W_{ci}\right)} \tag{10}$$

M_w- Mass of evaporated water in time

$\dot{m}a$ - air mass flow rate

W_{ce}- Humidity of Air at Dryer Exit (Absolute)

W_{pa}- Humidity of Air at Dryer Inlet (Absolute)

Use the following model to find the best model described the characteristics of drying curves for drying of chilies in solar dryer (Cakmak and Yıldız, 2011; Midilli and Kucuk, 2003; Ong, 1995; Doymaz, 2005).

Table 87.1 Drying models

SN	Model	Equation
1	NEWTON	$M_r = e^{-zt}$
2	PAGE	$M_r = e^{-zt^n}$
3	HENDERSON & PABIS	$M_r = a \times e^{-zt^n}$

Following equations were used to evaluate MBE and RMSE.

$$MBE = \frac{1}{N}\sum_{i=1}^{N}\left(M_{rpre},i - M_{rexp},i\right)^2 \tag{11}$$

$$RMSE = \sqrt{\left[\frac{1}{N}\sum_{i=1}^{N}\left(M_{rpre},i - M_{rexp},i\right)^2\right]} \tag{12}$$

The best model is selected from the mean deviation error and the root mean square error of the determination coefficient R^2. The higher R^2 value and the lower MBE and RMSE values are used to select the best model (Günhan et al., 2005).

Experimental Apparatus

The size of solar air collector was $2 \times 1 \times 0.10$ meter. Two absorber plate of aluminium material with black painted were used. First absorber plate was inclined at 30° to earth in south direction. Second absorber plate was horizontal and parallel with earth which was used as a bottom of collector and surrounded by reflector .4 mm thin clear glass has been used as a glassing over top of solar collector. The solar collector integrated with drying box which was made up in size $1 \times 1 \times 1$ meter. A ll side of box are insulated because of wood material. Two perforated trays of 5 kg capacity were used to store the chillies. Air fan were provided at the top side of dryer to control the air flow. This model made adjustable such that it can be work as a single pass, double pass and triple pass with reflector. Reflector was of aluminium material. The solar heater is just south, tilted 30° to the earth.

Experimental Procedure

During sunshine hours, from 8 am to 6 pm experimentation was carried out on designed solar dryer by using red chillies and evaluation was done of the performance of solar dryer. The solar dryer was used i.e. solar dryer with reversed absorber and reflector. 10 kg red chillies were washed by potable water and then it stores in thin layered on two perforated trays. Fan was started and set such that it will maintain a 1 m/s velocity of air during drying process. Experimental readings were taken at the interval of 1 hour and noted a solar intensity, ambient air temperature, collector outlet air temperature, weight of chillies, air temperature at dryer outlet. Measure the relative humidity of ambient air and dry air with a wet bulb hygrometer.

Result and Discussion

Three different types of solar dryers are used to dry the red pepper. The test was conducted on a sunny day. In typical experimental run the deviation in ambient temperature, relative humidity and air temperature at collector outlet of different type of solar dryer are shown in Figure 87.5. Graphical presentation shows

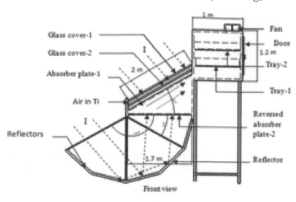

Figure 87.1 Solar dryer with reversed absorber and reflector

Figure 87.2 Photo of solar dryer with reversed absorber and reflector

Figure 87.3 Drying chamber

Figure 87.4 Perforated tray with chillies

that the temperature of air rise in SDR3 is more than others causing a drying period of agriculture product reduced.

The average temperature of drying air was recorded about 57°C and 50°C for SDR3 and SDR2 respectively at inlet of the drier. During peak daylight hours maximum temperature of drying air was recorded about 71°C at the drier inlet. From the air temperature curves it can be observed that the invented dryer may be used to dry a variety farming product.

During peak sunshine hours maximum intensity of solar energy was recorded about 905 W/m² at location Wanadongri, Nagpur (21.0987 ° N and 78.9841 ° E). The average dry and wet bulb temperature were recorded 32°C and 25.8°C correspondingly. The recorded relative humidity of air was about 32% at inlet and 70% at the exit of solar dryer chamber. Figures 87.6–87.8 shows a variation in collector, drying and pick-up efficiency with a solar flux for all the dryers.

From the Figure 87.6–87.8, it was observed that at an average value of I_s = 675 w/m², T_a = 34°C and ma = 1.614 kg/minute, collector efficiency (η_c) of SDR3 were 36%. It was noted that at an average value I_s = 427 w/m², T_a = 28°C and ma = 1.614 kg/minute, collector efficiency (η_c) values of SDR2 were 29.7%. And the collector efficiency values of SDR1 were 26%.

It was observed that dryer thermal efficiency (η_d) and Pick-up efficiency (η_p) at an average value I_s = 675w/m², T_a = 34°C and ma = 1.614 kg/min, of SDR1 were 49%; 64% respectively. At an average value I_s = 427w/m², T_a = 28°C and ma = 1.614 kg/min the dryer thermal efficiency (η_d) and pick-up efficiency (η_p) of SDR2 were 45.5% and 54% respectively. Dryer thermal efficiency (η_d) and pick-up efficiency (η_p) of SDR1 55% and 45% respectively. From the above experiment it was analysed that solar air heater with reversed absorber and reflector was more efficient. This observation was presented in earlier studies that the heat gain by reversed absorber plate is noteworthy (Goyal and Tiwari, 1997; Forsona et al., 2007).

The change of MC in red chilli with time for SDR1, SDR2 and SDR3 is graphically represented in Figure 87.9. The MC in fresh red chilli was nearly similar during all test whereas the initial value was 4.1 kg/kg (db) for SDR1. From the Figure 87.10, it is manifest that drying rate more in SDR3. These results ensure that SDR3 used forced convection attained higher air temperature which outcomes, increase moisture

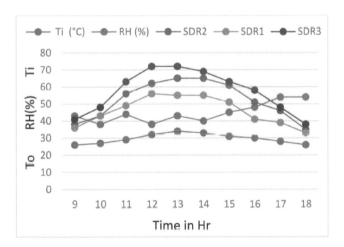

Figure 87.5 Deviation of relative humidity, ambient temperature and final temperature with day time (hour)

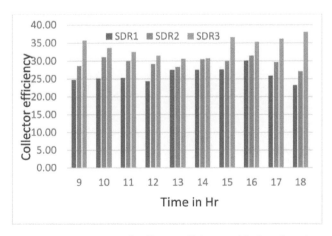

Figure 87.6 Variation of collector efficiency with time (hour)

removal rate from red chilli. From Figure 87.9 it is noted that, drying rate is continuously goes on decreasing in all the cases and drying period is less in SD.

Figure 87.6 shows that in SDR3, MC were for shorter period in the red chillies above the equilibrium MC. Figure 87.10 shows the contour of drying rate vs drying time. It was observed that the drying rate in SDR3 is greater due to higher drying air temperature and increases vapour pressure in the red chillies which reduces the effect of relative humidity during sunny day (Garg et al., 1985).

From Figure 87.9 and 87.10 it is also observed that the MC in red chilli above the equilibrium MC in SDR3 is reduced remarkably than SDR1 and SDR2. Drying rate is found higher in SDR3 than SDR1 and SDR2 because of high drying temperature and vapour pressure in the chillies which increased the rate of evaporation.

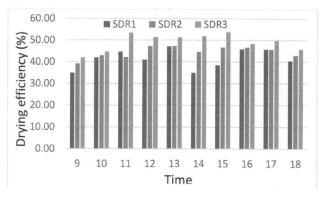

Figure 87.7 Variation of drying thermal efficiency with time (hour)

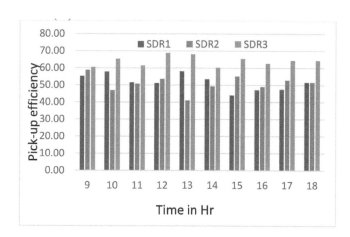

Figure 87.8 Variation of pickup thermal efficiency with time (hour)

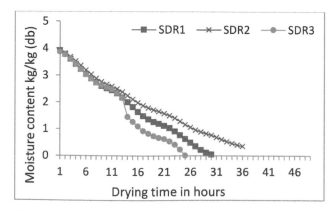

Figure 87.9 Profile of the moisture content vs drying time

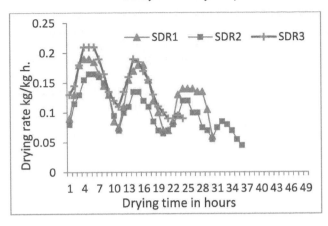

Figure 87.10 Profile of drying rate vs drying time

Table 87.1 Constants value fitting of DM

SN	Solar dryer type	Thin layer drying model	Constant	Model constants	MBE	RMSE	R^2
		NEWTON	k	0.070	0.00382	0.062	0.974
		PAGE	k	0.029	0.00026	0.016	0.997
1	SDR1		n	1.252			
		HENDERSON & PABIS	a	1.492	0.00332	0.058	0.974
			k	0.075			
		NEWTON	k	0.08	0.00303	0.055	0.97
		PAGE	k	0.038	0.00114	0.0034	0.98
2	SDR2		n	1.161			
		HENDERSON & PABIS	a	1.230	0.00301	0.055	0.974
			k	0.080			
		NEWTON	k	0.1	0.0058	0.243	0.96
		PAGE	k	0.0398	0.004	0.021	0.995
3	SDR3		n	1.267			
		HENDERSON & PABIS	a	1.276	0.00379	0.062	0.969
			k	0.102			

Experimental data of drying chillies was used to describe the DM. Newton DM, Page DM and Henderson and Pabis DM was fitted with the experimental data of drying chillies. In DM, the data of MC and drying time has been used. In all models, Excel software used to find the changes in moisture content with the time as well as constants by graphical method. The best DM was decided from higher value of and lowest value of MBA and RMSE (Devahastin and Pitaksuriyarat, 2006). The best drying model were selected out of three drying models by drying a red chillies where constant mass flow rate was maintained. Table 87.1 shows the experimental constant values which fitting in the DM. From table 87.1 it has been noted that Pages DM is most fitted DM to describe the drying curves of red chillies. The experimental data was fit to the Page's DM than Newton's and Henderson and Pabis DM. The Page's DM gives best result with higher value of and lower values of MBE and RMSE.

Conclusions

After experimentation and comparison, it has been observed that the significant improvement found in solar dryer with reversed absorber and reflector in thermal, drying and pick-up efficiency. Further it has pointed out that, at constant air flow rate, the solar collector inlet and exit air temperature difference increases with the rise in solar irradiation.

The average collector efficiency (η_c) values are 26.11%, 29.56% and 34% for the SDR1, SDR2 and SDR3 respectively. It displays that the collector efficiency (η_c) of SDR3 is more than SDR1 and SDR2. The drying efficiency (η_d) values are 41%, 44% and 49% and Pick-up efficiency (η_p) value are 51%, 55% and 64% for the SDR1, SDR2 and SDR3 respectively. These value shows the SDR3 is more capable than SDR1 and SDR2.

Experimental data of drying chillies was used to describe the drying model. In all models, Excel software used to find the changes in MC with the time as well as constants by graphical method. The best DM was decided from the highest value of R² and lowest value of MBA and RMSE. Table 87.1 shows the experimental constant values which fitting in the DM. It has been noted that Pages DM is most fitted DM. The experimental data was appropriate to the Page's DM than Newton's and Henderson & Pabis DM. The Page's DM gives best result with the highest value of and lowest values of MBE and RMSE.

Finally, it has been found that SDR3 is 14–17% more effective solar dryer than SDR1, and SDR2.

References

Bentayeb, F., Bekkioui, N., and Zeghmati, B. (2007). Modeling and simulation of a wood solar dryer in a Moroccan climate. Renew Energy. doi:10.1016/ j.renene .03.030.

Cakmak, G. and Yıldız, C. (2011). The drying kinetics of seeded grape in solar dryer with PCM-based solar integrated collector. Food Bioproducts. Proces. 89:103–108.

Devahastin. S. and Pitaksuriyarat. S. (2006). Use of latent heat storage to conserve energy during drying and its effect on drying kinetics of a food product. Appl. Therm. Eng. 26:1705–13.

Doymaz, I. (2005). Sun drying of figs: an experimental study. J. Food. Eng. 71(4):403–407.

El-Beltagy, A., Gamea,G. R, and Amer Essa, A. H. (2007). Solar drying characteristics of strawberry. J. Food Eng. 78(2):456–64.

Fatouh, M., Metwally, M. N., Helali, A. B., and Shedid, M. H. (2006). Herbs drying using a heat pump dryer. Energy Convers. Manag. 47:2629–43.

Forsona, F. K., Nazhab, M. A. A., Akuffoa, F. O., and Rajakarunab, H. (2007). Design of mixed-mode natural convection solar crop dryers: Application of principles and rules of thumb. Renewable Energy. 32:2306–2319.

Fournier, M. and Guinebault A. (1995). The 'shell' dryer-modeling and experimentation. Renew Energy. 6:459–463.

Garg, H. P. and Sharma, S. (1990). Mathematical modeling and experimental evaluation of a natural convection solar dryer. In Proceedings of 1st word renewable energy congress, Reading, UK. 2:904–908.

Garg, H. P, Sharma, V. K, Mahajan, R. B, and Bhargave, A. K. (1985). Experimental study of an inexpensive solar collector cum storage system for agricultural uses. Solar Energy. 35(4):321–31.

Goyal, R. K. and Tiwari, G. N. (1997). Parametric study of a reverse flat plate absorber cabinet dryer: A new concept. Solar Energy. 60(1):41–48.

Günhan, T., Demir, V., Hancioglu, E., and Hepbasli, A. (2005). Mathematical modeling of drying of bay leaves. Energy Convers. Manag. 46(1112):1667–1679.

Hallak, H., Hilal, J., Hilal, F., and Rahhal, R. (1996). The staircase solar dryer: design and characteristics. Renew Energy. 7:177–183.

Joshi, C. B., Gewali, M. B., and Bhadari, R. C. (2004). Performance of solar drying systems: a case study of Nepal. J. Inst. Eng. 85:53–57.

Jain, D. and Jain, R. K. (2004). Performance evaluation of an inclined multi-pass solar air heater with in-built thermal storage on deep-bed drying application. J. Food Eng. 65:497–509.

Jain, D. (2007). Modeling the performance of the reversed absorber with packed bed thermal storage natural convection solar crop dryer. J. Food. Eng. 78: 637–647.

Kardum, J. P., Sander, A., and Skansi, D. (2001). Comparison of convective, vacuum and microwave drying of chlorpropamide. Drying Technol. 9(1):167–83.

Kavak Akpinar, E., Sarsilmaz. C., and Yildiz, C. (2004). Mathematical modeling of a thin layer drying of apricots in a solar energized rotary dryer. International J. Ener. Res. 28(8):739–52.

Lawand, T. A. (1996). A solar cabinet dryer. Solar Energy. 10:158–64.

Martinez, C. (2001). ASHRAE handbook, 2001: fundamentals. Atlanta, GA.: ASHRAE, 2021.

McMinn, W. A. M. (2006). Thin-layer modelling of the convective, microwave, microwave convective and microwave-vacuum drying of lactose powder. J. Food Eng. 72:113–23.

Midilli, A. and Kucuk, H. (2003). Mathematical modeling of thin layer drying of pistachio by using solar energy. Energy Convers. Manag. 44(7):1111–1122.

Mwithiga, G. and Kigo, S. N. (2006). Performance of a solar dryer with limited sun tracking capability. J. Food Eng. 74:247–52.

Nellist, M. E. (1976). Exposed layer drying of ryegrass seeds. J. Agric. Eng. Res. 21:49–66.

Ong, K. S. (1995). Thermal performance of solar air heaters: mathematical model and solution procedure. Solar Energy. 55(2):93–109.

Pathak, P. K., Agrawal, Y. C., and Singh, B. P. N. (1991). Thin layer drying model for rapeseed. Transact. Am. Soci. Agr. Eng. 34(6):2505–2508.

Simal, S., Femenia, A., Garau, M. C., and Rossello, C. (2005). Use of exponential, pages and diffusional models to simulate the drying kinetics of kiwi fruit. J. Food Eng. 66(3):323–328.

Singh, H., Singh, A. K., Chaurasia, P. B. L., and Singh, A. (2005). Solar energy utilization: a key to employment generation in the Indian Thar Desert. Int. J. Sustain. Energy. 24(3):129–42.

Singh, P. P., Singh, S., and Dhaliwal, S. S. (2006). Multi-shelf domestic solar dryer. Energy Conver. Manag. 47:1799–815.

Sun, D. and Wood, J. L. (1994). Low-temperature moisture transfer characteristics of barley: thin-layer models and equilibrium isotherms. J. Agric. Eng. Res. 59:273–83.

Yunfei, L. R, Morey, V., and Afinrud, M. (1987). Thin-layer drying rates of oilseed sunflower. Transact. Am. Soci. Agr. Eng. 30(4):1172–1175.

88 Case study on quality circle implementation at a large scale manufacturing unit

Shantanu Kulkarni[1,a], A. K. Jha[1,c], Arun Kedar[2,b], and Rajkumar Chadge[2,d]

[1]Mechanical Department, Shri. Ramdeobaba College of Engineering & Management, Nagpur, India

[2]Department of Mechanical Engineering, Yeshwantrao Chavan College of Engineering, Nagpur, India

Abstract

Quality circle (QC) is tool for ensuring workers' participation in the problem solving and organisation's development. In many organisations however the quality circle programme is implemented in a routine manner without creating the awareness, enthusiasm and excitement about it. As a result the QC programmes either fail or do not yield the expected outcomes. The paper discusses the case study of implementing the QC programme at a large-scale manufacturing unit in Maharashtra, keeping in view the Critical Success Factors of the Quality Circles. With the involvement of top management and the expertise of the external consultant the programme was customised considering the organisational culture and implemented successfully to give excellent results. The implementation of the QCs greatly increased employees' morale and participation in the organisations improvement programmes eg Kaizens, 5S etc. It can therefore be concluded that if the QC programme is implemented properly with the full and visible involvement of the top management very good results are obtained. The study can be useful to the organizations planning to implement QC activities in the way appropriate to their existing organizational culture.

Keywords: Critical success factors, quality circles, quality circle implementation.

Introduction

Quality circle (QC) is a group of 7–12 employees working in same section or work area of an organisation, formed voluntarily and trained suitably to identify the problems or possible improvements in the work carried out by them, to come out with innovative solutions by doing a systematic analysis and to propose the solutions to the management. If accepted the QC members themselves implement the solutions. Once a problem is solved the circle members pick up the next problem or area of improvement and the process continues. The benefits of the QC implementation are two folds. The time and energies of the workers are used for a positive purpose of solving the problems through which many a basic level problem do get actually solved resulting into the increase in quality, productivity and safety, reduction in wastages, costs and accidents. On the other hand the workers get motivated as they get an opportunity to play a more meaningful and creative role in the organization and also get recognition and rewards.

Literature Review

A lot of literature related to the success and failure of the QC programmes across the world has been published. It has been observed that quality circles can fail as naturally as they can, when not implemented with strong commitment, proper planning, adequate support and sustained enthusiasm. The literature regarding the success or failure of the QCs gives the insights into the causes of failure of QCs and the Critical Success Factors of the QCs

A. Causes of failure of QC

Sillince et al. (1996) have used the classification for separating the problems related to the implementation of quality circles as Major problems with starting QCs and the Major problems while running the QCs. Goh (2000) has identified key implementation barriers and critical success factors for the quality circle through study of Singapore Housing Developing Board. He identified reluctance to change on the part of employees, initial lack of identification with the QC movement, and organisational size as the major barriers, while active CEO support, continuous publicity for the movement to create awareness and acceptance, conducive organisational culture, and appropriate reward and incentive schemes as the critical success

[a]kulkarnisr@rknec.edu; [b]jhaak1@rknec.edu; [c]arun_kedar@yahoo.com; [d]chadgerb@rediffmail.com

factors. Majumdar et al. (2011) have pointed out that there are broadly three areas in which problems can come causing the failure of the QCs 1) Organisational issues 2) Implementation issues 3) Operations issues. The various causes of failure of the QCs as pointed out by some of the researchers are listed below.

Sr. No	Causes of Failure	Reference
1	• Rejection of the concept by top management • Uncertainties caused by redundancies and company restructuring • Labour turnover (Transfers, Promotions etc.) • Lack of cooperation from middle and first line management • Failure by circle leaders to find enough time to organise meetings • Circles running out of projects to tackle • Delay in responding to circle recommendations • Circle members lacking time to carry out activities • Lack of recognition • Groups spread over too wide a work area • In adequate training • Lack of extrinsic rewards/motivation • Lack of cooperation from functional specialists • Opposition from the trade unions • Change in Management	Hayward et al. (1985)
2	• Lack of interest from top management • Lack of interest of facilitator • Poor attendance of members • Trade unions restrict their members to attend the meetings • Lack of proper quality Information system • Middle level managers believe that QCs dilute their authority and importance. • Managers believe that QCs ought to solve all problems • Management's frequent inability to accept the recommendation of QCs.	Aravindan et al. (1996)
3	• Reluctance to change on the part of employees • Initial lack of identification with the QC movement • Organisational size	Goh, (2000)

B.   Critical success factors of QCs:

Kulkarni et al. (2018) have discussed the step by step procedure for the effective implementation of the QCs and usefulness of the QC implementation for improving the productivity in the industry by taking an example of a medium scale manufacturing industry. Dasgupta (2011) carried out a study at the five industrial units in West Bengal for the QCs and has developed an organization oriented QCs effectiveness model considering 23 critical success factors (CSFs). He has concluded that when the QCs are implemented in the right industrial culture they can give success beyond expectations but when badly managed they result in failures, disappointment and cynicism and distrust about the concept of the QCs. The various critical success factors of the QCs as pointed out by some of the researchers are listed below.

Sr. No	Reference	Critical success factors
1	Sillince et al. (1996)	Focus on educating workers Focus on formulating goals Focus on participation Talks must lead to action Rewards integrated into organisational reward system Top management support Clear goals Empowerment Good organisational Communication system Members commitment training Organisational stability briefing non QC members about work of QCs

Sr. No	Reference	Critical success factors
2	Goh (2000)	Active CEO support
		Continuous publicity for the movement to create awareness and acceptance
		Conducive organizational culture
		Appropriate reward and incentive schemes
3	Salaheldin (2009)	Commitment and support from top management
		Commitment and support from middle management and first line supervisors
		Training of circle members
		Involvement and support of employees
		Training of circle leaders
4	Dasgupta (2011)	Top and middle management support
		Regularity of QC meetings and QC activities
		Voluntary and active participation of employees
		Commitment and support of facilitators
		Problem/project selection
		Suitable and adequate training
		Involvement of departmental heads
		Effective implementation of QC programme
		Effective communication system
		Positive attitude and commitment of employees

Researhc gaps:

Though the reasons for the failure of the QCs have been well researched and also the success factors have been reasonably identified the actual implementation of the QCs by taking care f those success factors in a large scale manufacturing industry demonstrating the claimed benefits has not been reported.

This study has therefore reported the experiences of actual implementation of QCs in a large scale manufacturing plant in the way it's recommended to be implemented and provides the evidences of the benefits obtained.

Case study

A case study on the actual implementation of the quality circles right from the beginning stage to the full establishment of the QCs has been conducted in a large scale manufacturing plant in Maharashtra. The plant was having lot of labour problems and QCs and hence the management decided to implement the QCs for improving the attitude and morale of the labours and increasing their participation in the organizational improvement activities. They appointed a renowned QC consultant for the complete implementation of the QC activities. The researcher hd the privilege to be part of the team of the QC consultant and personally witnessed the implementation and the results obtained. The study was carried during the period of FY 2015–2016 to FY 2017–2018. During the period the researcher visited the plant multiple times, interacted with the workers, supervisors, managers, collected the data and observed the improvements.

QC implementation

Good research papers have described the step by step procedure to implement the QC activities. Jayakumar and Krishnaraj, (2015) have discussed the step by step procedure from the formation and launch of the QCs to the operation and implementation of the QCs.

Bajpai and Malik (2015) have identified that the more commitment from the both the side as well as by increasing the awareness level of the worker the QCs can be implemented successfully.

Shrivastava and Jain (2020) have discussed steps to improve the organisation productivity improvement with the help of Quality Circle and have listed the possible benefits of the same.

Subbulakshmi (2018) has stated that human resources can be effectively utilised through participative management and this could be achieved by implementing quality circle in organization and carried out a study to know the influence of quality circle activities on productivity and innovation in manufacturing companies.

Keeping in mind the various causes of failure and critical success factors of the quality circles and experience of various organisations in the implementation of the Quality Circles, the programme was carefully implemented in a large scale manufacturing plant in Maharashtra. As per the policy of the company the study is presented in a generic form in this paper.

Circle formation:

The Quality Circle programme was implemented in the plant with a strong commitment by the top management, which is one of the most important factors critical to the success of the QCs. Quality Circles were formed with small groups of five to six employees on voluntary basis, working in the same work area or doing similar type of work.

Structure:

The initial teams had one leader selected by the members during training itself. The teams were supported, guided by guides nominated by the department head. The manager of that department was the facilitator, who looked after the functioning of the team and its progress. The head of department (HOD) had the responsibility to ensures smooth running of teams and helps if team faces any difficulty. Coordinators were appointed to organise trainings and monitor every activity of team and publish all reports related to Quality Circles. They organised reviews of QCs progress with plant leadership management team, organised internal Plant competitions, ensured rewards & recognition of teams, helped teams to participate in external competition. The structure is shown in Figure 88.2.

Circle meetings

The circle members met regularly every week at decided day and time for half or an hour to identify, analyse and resolve work related problems in the areas like quality, cost, delivery, safety, MURI, productivity, 5S, morale, environment.

Concept:

Concept behind the launch of the programme was that every person, no matter where in the organisation, desires to do quality work would like to be respected, encouraged, recognised and would like to contribute to the growth, welfare of his family, society and organisation. It was recognised that the workman on the job has the most knowledge on that particular job and team work can be more effective and can contribute more than individual effort.

Objectives:

The objective for the organization was to ensure that each member initially practices the learning of Quality Circle at home with family members then in society through which he gains confidence, happiness and satisfaction of doing some satisfactory/creative accomplishment, past that he can utilise the same learning at his work place. Apart from above basic objective some other are mentioned below:-

- Self and mutual development.
- Change in attitude.

Figure 88.1 QC organisational structure

- To enrich quality in life of employees.
- To tap the hidden potential of the employees.
- Improvement in communication.
- Improve participation.
- Group working.
- Problem solving at grass root level.
- Job satisfaction.

Aim

The company aimed to achieve high % of People Involvement in improvement activities in the plant.

Earlier the People Involvement key performance indicator (KPI) was measured through people's involvement in Kaizen, CLIT and QIP activities. With the introduction of the QCs in the organisation, it was decided to measure the score considering the people involvement percentage in the various improvement and development activities such as quality circle, Kaizen and QIP.

This activity was also essential for building/strengthening trust among workmen and management.

Training

Management decided to launch quality circle activity from May 2015. Activity plan was drafted and a renowned trainer and consultant with rich experience in quality circle field was approached for training to company employees. The training for plant leadership management team, facilitator and guides was scheduled.

Training calendar for workmen was drafted. A Batch of ten teams, each team consisting of five to six members were formed in every department and were imparted training as per the schedule fixed. Seven days training was scheduled for every batch. The nominated Guides from staff/management were imparted training along with team members. The concept of seven-days training was established keeping in mind the need of trust building, past history of the plant and mindset/attitude of workers of plant having average age of 43 and generally above 20 plus years of experience. The seven-days training is spread in three different phase. In first phase, two-days training, in second phase, three-days training and third phase, two-days training was scheduled.

In first phase training, the present market scenario, organizations need of an hour, survival of fittest was elaborated. Importance of quality circle in personal/family life and society was explained with lot of examples and homework. The existence of quality circle in ancient India, presently in India and abroad countries success stories was shared and why those organisation/country opted for quality circle is shared. So a structured approach to amend the attitude, approach of workmen to get involved in such activities is done.

In second phase the actual detailed working was taught along with many personal and work place related examples. Twelve steps problem solving methodology was taught in detail. A matrix of tools versus steps was taught. In third phase the actual problem solving is practiced. By the end of third phase it was seen that participants were fully motivated and the quality circle methodology was grasped by them thoroughly.

Shop floor meetings

Every team was asked to declare during third phase training the name of team, nominate the leader of team &particular day of department meeting. On this fixed day and decided time the meetings of team members started taking place. Guides of the teams guided the teams to carry out meeting and in problem solving steps if required.

Review mechanism

The role of Guide, facilitator was well defined during training. QBM members played role of coordinator who kept daily track of department meetings. The details of daily meeting were published daily through various media (created WhatsApp group). Weekly/monthly meeting compliance was published through mail and facilitator/heads of departments were asked to intervene in case of low meeting compliance. Initially a target of 80% was kept as minimum requirement of meeting compliance. The review of project progress was done and regular review of facilitator/HOD with plant leadership management teams was scheduled. Every month, facilitator's report card of each team was submitted by the facilitators.

Presentation in front top management

The teams whose projects got completed were given opportunity to present their projects in front Plant Head every week and month. Many teams got opportunity to present in front Senior VP/ED/MD. The team members were then felicitated by the managers with HERO/SUPER HERO badges.

Internal quality circle competition

Sharing of details was done with plant management team for getting approval, after approval an activity plan was prepared and accordingly within the plant competition was organised between teams who had completed their projects.

In FY17 at the end of the financial year the first competition was organised. An event with title 'Quality Circle Premier League 1' was organised. Total 57 teams participated out of 72 teams. Competition took place in a span of 5 to 6 days having gap in between, every day 10 to 12 teams presented their project and were felicitated with GOLD/SILVER and BRONZE trophies and a certificate to every team member. One internal and one external Judge were invited from different plants and organization to assess the team's project presentation.

In FY18 the company organised 'Quality Circle Premier League 2'. Total 100 teams participated out of 102 teams. A well in advance promotion campaign was carried out before the competition to build the enthusiasm among teams and rest of employees.

Competition details were displayed all over plant for awareness among all employees. Dates of competition were published well in advance. After the competition the winning teams' photos were displayed all over plant.

External quality circle competitio

The company always encouraged their quality circle teams to participate in various external competition events such as CCQC and NCQC. Following table (Table 88.1) shows the participation details. The received awards were displayed in plant and these teams were then felicitated in front of all employees in the plant.

Teams demonstration during various plant visitors' visit/audits:

During the top management shop floor visits and various audits, the Quality circle members were given opportunities to demonstrate their project work in front of them. A great positive difference was observed in terms of the enthusiasm and interest in showcasing their work in front of the top management.

Improvement in various plant key performance indicator:

The sustained and focused approach by the plant management resulted in formation of 107 teams in a span of less than three year time, involving 520 workmen which was 37.6% of total employees. Every team is active and together all the teams have submitted 398 projects from FY16 to FY18. The target of second projects/team/year has been achieved from last two years. The same achievement is shown in graphical form below.

Results

The implementation of the quality circle programme was very successful and following benefits were obtained by the organization. The benefits were not restricted to quality circle activities only but due to

Table 88.1 Details of awards won

Year	Event	Place	Teams number	Award
FY16	CCQC	Nagpur	2	2 Gold
FY17	CCQC	Nagpur/Mumbai	9	6 gold, 2 silver and 1 bronze
	NCQC	Raipur	3	3 Gold
FY18	CCQC	Nagpur/Mumbai	9	8 gold and 1 silver
	NCQC	Mysore	3	3 Gold

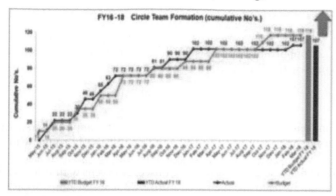

Figure 88.2 From FY16 to FY18, 107 teams got formed, trained and rolled out in shop floor

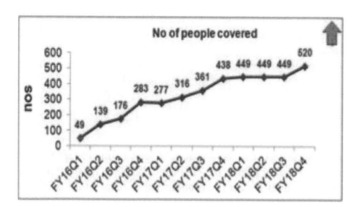

Figure 88.3 Total 520 workmen are involved in circle activity out of 1521, 37.6% of total employee's involvement

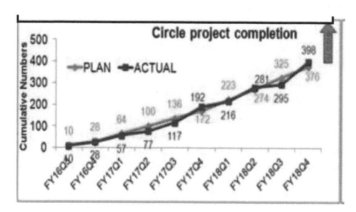

Figure 88.4 Cumulative 398 projects completed from FY16 to FY18

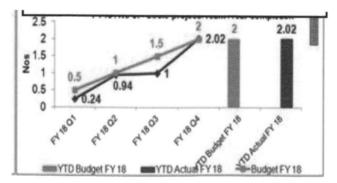

Figure 88.5 Two projects/team/year, target has been achieved in FY17 land FY18. Above graph is of FY18

increased motivation and morale of the workers across the entire organisation and their participation in performance improvement activities, even the 5S and Kaizen got a major boost.

- The people involvement in executing Kaizen enhanced (Figure 88.6)
- The plant 5S score improved exceptionally well, as shown in (Figure 88.7).
- % People involvement in improvement activities drastically improved by 86.5% (Figure 88.8)

Intangible benefits to the employees were observed like,

- Employee's got motivational boost and their morale improved.
- Good change in attitude was witnessed.

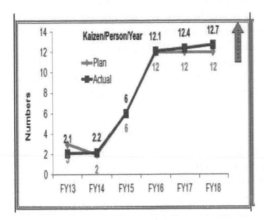

Figure 88.6 Kaizen /person/year improved 53% since FY15

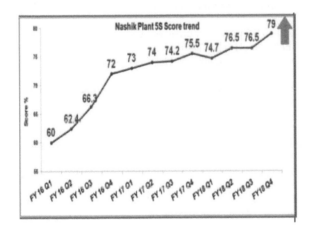

Figure 88.7 Plant 5S score improved 24% since FY15

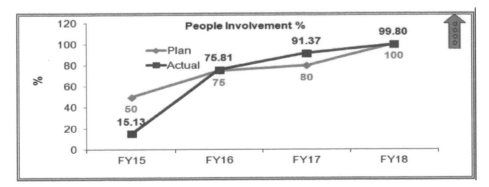

Figure 88.8 People involvement % improved to 99.80, 86.5% improvement since FY15

- Workmen eagerly participated to show case their improvement work in front all delegates visiting the premise.
- Workmen shared their personal self-development examples which they experienced with their family, with officers in various forums.
- Some of the workmen started writing poems/slogans on quality circle activities running in plant and shared during quality circle competitions.

Conclusion and Future Scope

Quality circle programme was launched at the manufacturing plant taking good care to keep it interesting, enjoyable, participative and innovative. The basic philosophy and method of implementation of the quality circles was fine-tuned as per the organisational culture so as to perfectly blend with the working of the organization. The programme proved to be a great success due to visible commitment and involvement of the top management, continuous training of the members and leaders, recognition and rewards to the members and positive attitude of the workers. There were tangible benefits like increased people involvement in Kaizen, improved 5S score and various performance indicators of the plant. The graphs show the improvement in various indices at the start of QC implementation vs the performance in next three years.

The success of the programme has created interest in the QC activities among the non-members also. Looking at it the company has aimed to involve 100% of employees in quality circle activity by year 2021.

References

Aravindan, P., Devadasan, S. R., Narasimha Reddy, E., and Selladurai, V. (1996). An expert system for implementing successful quality circle programmes in manufacturing firms. Int. J. Qual. Reliab. Manag. 13(7):57–68.

Bajpai, M. C. and Malik, M. P. (2015). Worker participation in management with special reference to the quality circle as an instrument. Int. J. Interdiscip. Res. 2(10):12–17.

Dasgupta, R. (2011). QCs' effectiveness in selected industrial enterprises in West Bengal. Ind. J. Ind. Relat. 46(4).

Goh, M. (2000). Quality circles: journey of an Asian public enterprise. Int. J. Qual. Reliab. Manag. 17(7):784–799.

Hayward, S. G., Dale, B. C., and Frazer, V. C. M. (1985). Quality circle failure and how to avoid it. Eur. Manag. J. 3(2):103–111.

Jayakumar, A. and Krishnaraj, C. (2015). Quality Circle–Formation and Implementation. *Int.*J. Emerg. Res. Eng. Sci. Technol. 2(2).

Kulkarni, S., Welekar, S., and Kedar, A. (2017). Quality circle to improve productivity: A case study in a medium scale aluminium coating industry. Int. J. Mech. Eng. Technol. 800–816.

Kulkarni, S. and Kedar, A. (2018). Productivity improvement through quality circle: a case study at calderys nagpur. Int. J. Mech. Produc. Eng. Res. Develop. 957–968.

Majumdar, P., Jyoti, and Manohar, B. M. (2011). How to make quality circle a success in manufacturing industries. Asian J Sci Res. 12(3:244–253.

Salaheldin, S. I. (2009). Problems, success factors and benefits of QCs implementation: a case of QASCO. TTQM J. 21(1):87–100.

Sillince, J. A. A. Sykes, G. M. H., Singh, D. P. (1996). Implementation, problems, success and longevity of quality circle programmes: A study of 95 UK organizations. Int. J. Oper. Prod. Manag. 16(4):88–111.

Shrivastava, N. and Jain, A. K. (2020). Quality Control Circle on Indian Manufacturing Industries to Improve Productivity. Int. J. *Appl. Sci.* Eng. Technol. 5(6):236–241.

Subbulakshmi, S. (2018). The Study about Influences of Quality Circle Activities on Productivity and Innovation in Manufacturing Companies. Int. J. Scient. Res. Comput. Sci. Appl. Manag. Studies. 7(5).

89 Analysis and design of sewer network for gadchiroli district using sewergems

Prajakta Wanjari[1,a], Khalid Ansari[1,b], H.R. Nikhade[1,c], and Sujin George[2,d]

[1]Department of Civil Engineering, Yashwantrao Chavan College of Engineering, Nagpur, India

[2]Department of Civil Engineering, St. Vincent Pallotti College of Engineering and Technology, Nagpur, India

Abstract

In the paper, SewerGEMS V8i, a sophisticated modelling and development tool software for sewage system networks. It allows the designer to accomplish tasks of designing at record speed, with efficiency, and at low cost. The paper will discuss the use of software, including SewerGEMS V8i and Autodesk Civil 3D, to evaluate sewerage systems for Gadchiroli district, Maharashtra. A sewer system was evaluated based on the overall system flow. The savings made in the design of the network will affect the overall cost of the network. The savings of network is done by reduction of cost to 2 to 90 lakhs depending upon the network length; this is done by providing an efficient number of manholes to serve the entire network and by a selection of diameters of conduit, the cost can be reduced significantly. Adjustments towards the design variables would require a full re-analysis of all assessments that can be done using software that will help the designer understand the network by stimulations on the software and cost analysis for District Sewer Network. Software allows a variety of mathematical models that are very effective in terms of scheduling, optimisation, modification, portrayal, and assessment. The district is prone to unsanitary sewage conditions, which lead to water-borne diseases. The lack of adequate sanitation facilities has resulted in widespread contamination of nullahs and surface water in and around town, therefore maintaining sanitary diseases like E. coli, shigellosis, typhoid fever, salmonella, and cholera by the design of a network, which will be adequate for district. The system allows calculations for various models and provides longitudinal sections, reports, and cross-sections to understand the designing of hydraulic networks for sanitary and storm sewers. The design of sewers generally includes manholes, pumps, conduits, and other sewer appurtenances, whereas, in the study, the economic design with the help of gravity and therefore pumps are not required.

Keywords: CAD, Gadchiroli District, SewerGEMS V8i, Sewer network.

Introduction

The technicians can use SewerGEMS to assess, design, and operate properly sanitized or mixed conveyance sewer systems by utilising built-in hydrodynamic and hydro geologic tools in an easy-to-use interface. The interconnection of sewer GEMS is higher. The goal of drain system plan is to develop the best layout as well as size of the system. A conduit is required to carry this same flow and possibly and provide a smooth surface. Gradient keeps the hydrostatic integrity requirements inside the parameters of the hydraulic design requirements constraints (Abbas et al., 2019). When especially in comparison to other public infrastructure such as power lines and water supply. The drainage system and water infrastructure are now more robust. The cost of laying sewerage system is considerably high compared to the water supply system (Punam Harising Rajpurohit et al., 2016). According to the WHO, water-borne and water-related diseases account for 80% of all human diseases. There is no public transportation in country places or some village. Wastewater that flows into low-lying areas, causing ponds, which cause mosquito breeding and a foul odour, It is primarily caused by water pollution or contamination. By accordance with CPHEEO (Central Public Health and Environmental Engineering Organization), approximately 80% of the water provided is highly probable to be wasted transformed into wastewater water (Pawar et al., 2021).

Ease of Use General Information abou Gadchiroli District

Selecting a Gadchiroli District

Gadchiroli district situated between 8° 43 to 21° 50' North latitude and 79° 45' to 80° 53' East longitude, existing 104.5 km of road line.

---

[a]prajaktawanjari117@gmail.com; [b]khalidshamim86@rediffmail.com; [c]nikhade65@gmail.com; [d]sgeorge@stvincentngp.edu.in

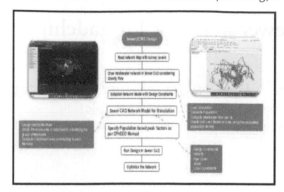

Figure 89.1 workflow

Figure 89.2 District's with existing 104.5 km of road

Existing water supply and sewerage arrangement

There is no under-ground drainage system in Gadchiroli town. The sewage and wastewater is disposed of using different on-site sanitation methods. Septic tanks are the most prevalent on-site wastewater treatment and disposal method used in town. In most recently developed areas and areas of already developed cities, wastewater from water sources is linked to septic tanks in houses with sufficient space within their premises.

Necessity of the project

Gadchiroli city has no good sewerage collection system, treatment and disposal. The existing individual septic tanks and absorption pits are a source of odour, and septic seepage finds its way into the river, elevating the nutrient content to the detriment of river life & ground water, increase infiltration to empty water lines and enter poorly sealed septic tanks and absorption pits causing overflow. The town's general drainage is through open drains, and these drains frequently find their way to nearby rivers Wainganga, growing the pollution load on water sources and, as a result, deteriorating water quality and harming the underwater ecology of the these waterways. The lack of adequate sanitation facilities has resulted in widespread contamination of nallahs and surface water in and around town, posing serious health risks from water-borne diseases. The current method of disposing of industrial wastewaters is via drainage channels or septic tanks. These drains discharge into lower areas and ponds, causing severe pollution of the environment and water bodies. Only a few homeowners have septic tanks installed in their homes. Septic tank effluent is discharged into the open. Carrying sullage through open drainage system is entirely unsatisfactory and unhygienic. Generally, disposal through septic tanks are recommended only for isolated dwellings away from habitat. With the increasing population and rise in population density of residential area, the problem may become serious for disposal of effluent from septic tanks, because of poor soaking capacity of the soil. Beside due to increasing pressure on land for residential purposes, sufficient land may not be available for providing on site disposal system in the already developed residential areas.

Materials and Method Data Collection

- Collection of the population of last five decades of Gadchiroli town.
- Collection of the existing water supply detail.
- Topographical map of Gadchiroli town.
- Road map of Gadchiroli town.
- Data of previous existing Sewer network and invert level of existing manhole (Rajpurohit, 2016).

Design criteria and sewage demand

The drainage system in Gadchiroli city is designed independently for the projected population of 2047. The per head of population waste water flow is assumed to be 80% of the rate water system of 135 lpcd for the residents and 45 Ipcd again for floating population, with an infiltration rate of 5000 Lit per km/day of sewer duration. The infiltration quantity is approximately 5% of the sewerage load, which would be the minimum infiltration rate specified in the Central Public Health and Environmental Engineering Organization (CHPEEO) manual for software architecture. The entire project area is divided into two sewerage zones based on the topographical and gradient of the ground. Nos. of ward comprising the zone 1 are Ward No 1, 2, 3, 4, 5, 6, 8, 9, 19 and partially 7, 10 and 11.It is observed from the nature profile of the ground. There is a fall of 18m, as such the waste water flows from these areas can be easily conveyed through main sewer pipeline and finally conveys to Lifting Station (LS-I) located at Kharpundi road and pumped to falls under sewerage zone no.2 and then finally to terminal pumping station located in STP plot area. In this zone there will be following main sewer lines coming from Norther part of project area having G.L 218.52 at the starting point and end of the zone with G.L as 202.70 at the end giving a fall of 16.00.

Population Forecast

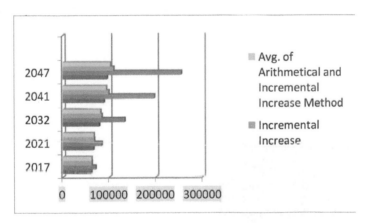

Figure 89.3 Chart for population forecast (on Y-axis = Years and X-axis = Population

Sewage network design

Laying out a network

SewerGEMS V8i is a powerful tool for designing a sanitary sewer network. It is easy to prepare a schematic or scaled model and let SewerGEMS V8i take care of the link-node connectivity. Assigning labels to pipe work but also nodes when building the infrastructure for this lesson because the software does so automatically (Rajpurohit, 2016).The above Figure 89.4 shows this same zone 1 details of the Gachirlroi neighborhood based on elevation differences, with colour-coding from pink, blue, and green to help understand the existing and planned conduits. The network connection is planned based on gravity for the economy. Figure 89.4 is prepared in Autocad, and the entire network is savedin .dxf structure so that it could be imported into SewerGEMS V8i. By importing the file into the software, you can easily plan and lay conduit manholes by assigning the conduit system and manhole position. (Rai at al., 2020).

Hydraulic Demand

The forecasted population is followed by calculations of demands for installing networks. As per CPHEEO Manual suggests 80% of water supply is expected to turn into wastewater (Rai at el.,2020).The normal rate of supply of water is about 135 lpcd as some percentage of water supplied is turning into sewage water therefore about 108 lpcd is considered as sewerage system. The entire network is designed for the sanitary sewers and therefore the residential population and water supplied data is governing factor. In addition, the CPHEEO guidelines suggests that the infiltration 10% of average dry weather flow (Rai et al., 2022).

Peak factor

The CPHEEO Manual, 2013 suggests that peaking factor for the design for sewerage system and this system is for domestic sewage. The following are different values of peak factor as per the population.

Coefficient of rughness

The coefficient of roughness is influenced by the material of pipe and for HDPE is 0.01 to 0.02 which based on sewer pipe.

Design capacity of sewer

The sewers are designed in a way that they are able to carry sewage in peak flow therefore; the sewers are designed by considering the 80 percentage full capacity at time of peak flow.

Self-cleaning velocity

The self-cleaning velocity is parameter, which will ensure that there is no deposition of suspended solids inside the sewer. Therefore, this is calculated by using following formulation (Rai at el., 2020)

$$V = \frac{1}{n} \times R_6^1 \left\{ K(S-1) \times d \right\}$$

Where,
S Specific gravity of particle
d particle size in mm
K dimensionless constant
R Hydraulic mean radius in m
n Manning's coefficient considering.

Figure 89.4 Layout of network for zone-1 and zone-2

Table 89.1 Recommended peak factors for estimation of domestic sewage

SL	Contributing population	Peak factor
1	Upto 20,000	3
2	20,000 to 50,000	2.5
3	50,00 to 7,50,000	2.25
4	Above 7,50,000	2

Source: CPHEEO Manual

Design Formula Manning's formula

Manning's formula would be adopted as per CPHEEO Manual (Naveen Kumar Rai at el., 2020)

$$Q_f = f_f \times A$$
$$V_f = \frac{1}{N} \times R_3^2 \times S_2^1$$

Where,
Q_f = Flow rate (in m3/sec)
A = Cross sectional area of pipe (sq. m.)
V_f = Velocity (in m/s)
N = Manning's roughness coefficient R = Hydraulic radius (m).
S = Slope of energy gradient.

Design and Analysis of Sewer System

The survey data and contour mas lines, manholes, conduit, outfall, etc. were drawn using the tools available in SewerGEMS software. The network is designed for sanitary loading therefore the loading is taken as 240L/day per single household this is eventually based on the population. The Load Builder is used to connect the property connected to the nearest manhole or conduit. The TRex tool is used to give the elements i.e., each element is given with the required attribution using a .shp file for contour map which is used by Arc-GIS Software. Ground Elevations and Invert Elevations is the start node of invert elevation of the outfall was taken as 0 by default and the top node of invert elevation was taken as 216.29 meters (Chaudhary et al., 2020). To check the existence of any errors validation is done to validate the network in terms of any errors by using validate the tool.

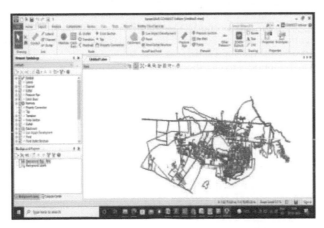

Figure 89.5 Sewer line network mapped using ModelBuilder and LoadBuilder

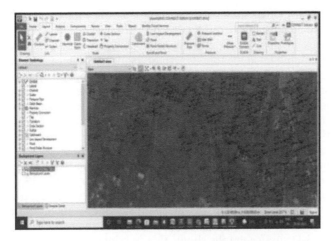

Figure 89.6 The Bing Map key is used to locate features

Table 89.2 Design parameters for design

Design parameters	Year 2047	Year 2032	Unit
Zone -1 (Lift station)			
Average Flow	8.105	5.301	MLD
Peak factor	2.25	2.25	---
Peak flow (including infiltration)	18.025	12.999	MLD
Dia. of Sewer end pipe (RCC pipe)	600	600	mm
Dia. of pumping main (DI pipe)	350	350	mm
Length of pumping main from Lifting station to MH3350	1800	1800	m
Highest RL	209	209	m
Zone-2 (STP zone)			
Average Flow	9.169	6.051	MLD
Peak factor	2.25	2.25	---
Peak flow (including infiltration)	20.199	13.186	MLD
Dia. of Sewer end pipe (RCC pipe)	800	800	mm

Table 89.3 Observations made by tender documentations

Hydraulic Design as per Tender	Reasons to Revise Hydraulic Design of Collection System
Temporary bench mark (TBM)	The TBM considered while designing the scheme was not traceable. Also the record about the TBM marked was not available for the execution of sewerage scheme. Hence, it was not possible to execute the work without the TBM.
The network layout was designed (2015) as per the road network demarcation available in the development plan.	It was found that the development plan consist some demarcated roads, almost 6Km length, which are proposed but yet not constructed. Hence there was no alignment available to lay most of the trunk mains

Results and Observation

The SewerGEMS is used to design the network which is divided into two zones, the design constraints are set up by using Table 89.2 and the loadings in the sanitary sewers load. The computation was successful done without errors moreover; the errors are given by validation of network in software.

The network is designed by adding the manual lengths and selection of GVF-Convex solver is selected for zone 1 and zone 2. The zone 1 is designed for 4144 conduits and zone is designed with 4193 conduits. The following are results from the SewerGEMS after computation. The results are approved in the flex table and manhole data of software. The software often shows red and yellow notifications according to urgency the issues are resolved with the help of guidance provided by the software. The results given above are from computations and these computations are done on the existing network which is designed for the sanitary purposes of city. The redesign has accommodated the total length of the network in each diameter of conduits. This is done by adapting the conditions on site. The manhole number which is proposed is reduced to make the economy of the network this is achieved by reducing the existing manholes which are not required, also the new manholes locations are planned in advanced as per the design constraints. The manholes are provided at every 500 m^{-1} km and therefore sometimes a junction makes the process way easier as it connects two different locations together making the number of manholes to be reduced. The Table 89.10 shows the difference in the desgined and propsed conduits length. The parameter will help save the cost for project by 15.3%. The total lenghth propsed was 1,02,680 and the designed length by considerations for diameter and manholes required at the specified location gave length of 86,947 which saved about 15,733 m of conduit length. Figure 89.7 is a 3-D graphic surface which gives colour variation graphs with different shades of colour represents as low and high values in the graphs, the graph gives length saving, % of length designed with respect to daimeter whereas in Figure 89.8 the proposed length and designed length the shades of red in the 3-D surafce graphs shows that maximum length of 7000 m and under is present in design.Figure 89.9 gives equivalent values for the daimters and saving difference of length in m. The graph gives the actual interpretation of values of saved length in m on y -axis with respect to daimeter equilvalent on the x-axis.

Figure 89.10 is network desgined as econimcal aspects considerations the green colour coding in the Figure 89.10 is 150 mm daimeter conduit and the maximum plan is covered by green coding shows that the new desgin is economical, the purple colour coding in the network is 200 mm daimeter conduit which is about 4,999 m.The pipe with colour coding as purple is 800 mm which used least for 72 m for maintaining the economy of network, HDPE is high-density polyethylene.

Table 89.4 Zone-1 conduits results

Start node	Stop node	L (m)	Material	Dia. (mm)	Elevation ground (Start) (m)	Invert (Start) (m)
1	2	30.00	HDPE	150	217.53	216.47
2	3	30.00	HDPE	150	217.48	216.29
3	4	25.80	HDPE	150	217.40	216.11
4	5	34.70	HDPE	150	217.26	215.96
5	6	18.90	HDPE	150	217.04	215.76
6	7	30.00	HDPE	150	216.98	215.65
7	8	21.10	HDPE	150	216.87	215.47
8	9	30.00	HDPE	150	216.69	215.35
9	10	20.00	HDPE	150	216.62	215.17
10	11	30.00	HDPE	150	216.50	215.05

Table 89.5 Zone-1 conduits results

Start Node	Slope (1/S)	Flow (MLD)	V (m/s)	d/D ratio (%)
1	170	0.01	0.3	6.5
2	170	0.02	0.3	9.0
3	170	0.03	0.4	10.9
4	170	0.04	0.4	12.5
5	170	0.05	0.4	13.9
6	170	0.07	0.5	15.2
7	170	0.08	0.5	16.4
8	170	0.09	0.5	17.5
9	170	0.10	0.5	18.6
10	170	0.11	0.5	19.6

Table 89.6 Zone-1 manhole results

Label	Depth (Structure) (m)	Elevation (Ground) (m)	Elevation (Invert) (m)	Elevation (Invert in 1) (m)
1	1.06	217.53	216.47	(N/A)
2	1.19	217.48	216.29	216.29
3	1.29	217.40	216.11	216.11
4	1.30	217.26	215.96	215.96
5	1.28	217.04	215.76	215.76
6	1.33	216.98	215.65	215.65
7	1.40	216.87	215.47	215.47
8	1.34	216.69	215.35	215.35
9	1.45	216.62	215.17	215.17
10	1.45	216.50	215.05	215.05
11	1.55	216.43	214.88	214.88

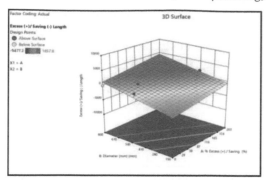

Figure 89.7 3-D graph of lenghth saving , % of length designed

Table 89.7 Zone-2 conduits results

Start node	Stop node	L (m)	Material	Dia. (mm)	Elevation Ground (Start) (m)	Elevation Ground (Stop) (m)
1211	1212	31.30	HDPE	150	215.85	216.02
1212	1213	29.00	HDPE	150	216.02	216.07
1213	1214	29.80	HDPE	150	216.07	215.95
1214	1215	31.50	HDPE	150	215.95	215.92
1215	1216	29.10	HDPE	150	215.92	215.69
1216	1217	22.50	HDPE	150	215.69	215.71
1217	1218	21.70	HDPE	150	215.71	215.85
1218	1219	26.80	HDPE	150	215.85	215.85
1219	1220	19.90	HDPE	150	215.85	215.81
1220	1221	18.30	HDPE	150	215.81	215.59

Table 89.8 Zone-2 conduits results

Start Node	Slope (1/S)	Flow (MLD)	V (m/s)	d/D ratio (%)
1211	220	0.01	0.2	7.0
1212	220	0.02	0.3	9.0
1213	220	0.03	0.3	10.6
1214	220	0.03	0.3	11.9
1215	220	0.04	0.4	13.2
1216	220	0.05	0.4	14.3
1217	220	0.06	0.4	15.3
1218	220	0.07	0.4	16.3
1219	220	0.07	0.4	17.2
1220	220	0.08	0.4	18.1

Table 89.7 Zone-2 manhole results

Label	Depth (Structure) (m)	Elevation (Ground) (m)	Elevation (Invert) (m)	Elevation (Invert in 1) (m)
1211	1.06	215.85	214.79	(N/A)
1212	1.38	216.02	214.64	214.64
1213	1.56	216.07	214.51	214.51
1214	1.57	215.95	214.38	214.38
1215	1.69	215.92	214.23	214.23
1216	1.59	215.69	214.10	214.10
1217	1.71	215.71	214.00	214.00
1218	1.95	215.85	213.90	213.90
1219	2.07	215.85	213.78	213.78
1220	2.12	215.81	213.69	213.69

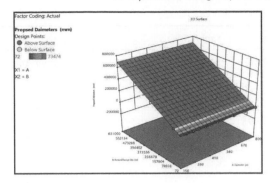

Figure 89.8 The equivalent graph

Table 89.8 Length of conduit w.r.t velocity

Diameter	Length of Conduit w.r.t Velocity						Total Length
	<=0.2	<=0.3	<=0.4	<=0.5	<=0.6	>0.6	
150	413	10328	16440	13346	11655	11814	63997
200	0	21	25	39	514	4400	4999
250	0	0	219	27	529	3765	4540
300	0	0	0	0	0	2286	2286
350	0	0	0	58	242	2052	2352
400	0	0	0	0	652	2121	2773
450	0	0	0	0	0	871	871
500	0	0	0	0	0	1688	1688
600	0	0	0	0	0	1667	1667
700	0	0	0	0	0	1702	1702
800	0	0	0	0	0	72	72
Total	413	10349	16684	13471	13592	32438	86946
%	0.48%	11.90%	19.19%	15.49%	15.63%	37.31%	100.0%
Cumulative %	0.48%	12.38%	31.57%	47.06%	62.69%	100.0%	

Table 89.9 The proposed manholes quantity and revised design manhole quantity

MH Depth		Total No. of MH (m)		Excess (+) / Saving (−) MH (No.)	% Excess (+) / Saving (−)
		Proposed Qty.	Revised Design Qty.		
>=0	<=1.5	1652	1377	−275	−16.65%
>1.5	<=2.0	783	702	−81	−10.34%
>2.0	<=2.5	285	286	1	0.35%
>2.5	<=3.0	276	265	−11	−3.99%
>3.0	<=3.5	188	186	−2	−1.06%
>3.5	<=4.0	167	209	42	25.15%
>4.0	<=4.5	69	153	84	121.74%
>4.5	<=5.0	79	100	21	26.58%
>5.0	<=5.5	19	68	49	257.89%
>5.5	<=6.0	28	40	12	42.86%
>6.0	<=6.5	10	10	0	0.00%
>6.5	<=7.0	32	9	−23	−71.88%
>7.0	<=7.5	10	9	−1	−10.00%
>7.5	<=8.0	0	1	1	100.00%

Table 89.10 Overall saving in lakhs

Sr. No.	Item	Unit	Quantity As per		Saving Qty.	Amount (Lakh)	
			Proposed	Design		Excess Amt.	Saving Amt.
1	Providing HDPE pipe	Rmt	94,059	73,536	20,523	-	149.93
2	L/L HDPE pipe	Rmt	94,059	73,536	20,523	-	9.11
3	Providing RCC pipe	Rmt	3,676	6,998	-	59.21	-
4	Providing RCC pipe	Rmt	4,945	4,127	818	-	11.85
5	L/L RCC pipe	Rmt	8,621	11,125	136	7.56	0.18
6	BB Manhole	No.	3,162	3,090	282	131.95	60.47
7	RCC Manhole	No.	436	325	111	-	56.85
	Total					198.72	288.39
	Overall Saving						89.67

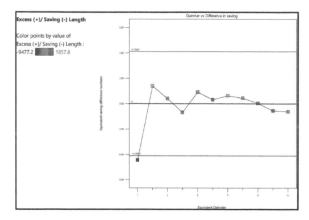

Figure 89.9

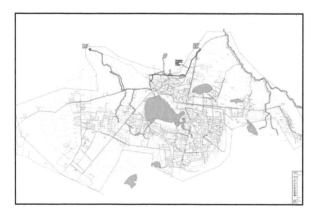

Figure 89.10 Final network design

Conclusions

The usage of software for the design made the entire process efficient concerning time and reliability. The Mode builder and load Builder serves an important role as it helps to directly import the data required on the workspace for designing purpose. The errors that can further be spotted on site are rectified and corrected using Software. The economical factor for any network design can be maintained by using various parameters, which are under the control of the user due to incorporated tools in the software. The Software helps to redesign of existing network system, there are repair works which are often carried, the software helps to assign the different data under colour coding scheme which will allow a user to differentiate in the existing and proposed network and elements. The proposed design will reduce the sanitary conditions of the district, as the network designing will allow the optimization of the network as per land conditions and reduce the diseases like typhoid fever and cholera in District. The economy is achieved by decreasing the length of conduits by 15% and the number of manholes is reduced to 5.09% as proposed requirements. The overall saving by providing optimization on the no. of manholes and diameter of conduits is 89.67 lakhs.

References

Abbas, A., Salloom, G., Ruddock, F., Alkhaddar, R., Hammoudi, S., Andoh, R., and Carnacina, L. (2019). Modelling data of an urban drainage design using a Geographic Information System (GIS) database. J. Hydrol. 574.

CPHEEO (2013) Manual on Sewer and Sewerage System. New Delhi: India: Ministry of Urban Development, Government of India.

Katti, M. (2022). Design of Sanitary Sewer Network using Sewer GEMS V8i Software. Int. J. Sci. Technol. Eng. 2(1):254–258.

Kumari, T., Bamiri, S., Robert. B.R.G., and Chaudhary, U. (2020). Sanitary Gravity Sewer Design using Sewer GEMS Software Connect Edition for Utsav Vihar, Karala. Int. J. Recent Technol. Eng. 9(2).

Pawar, A., Patil, Y., Mane, S., Patil, S., mane, S., and Ingale, Y. (2021). Design of Sewer System for Village using SewerGEMS. Int. Res. J. Eng. Technol. 8(7).Rai, N. K. (2020). Sewerage System Assessment Using Sewer Gems V8i and Autocad Civil 3d. Int. J. Eng. Sci. Invent. 9(5):24–29..

Rajpurohit, P. H. S. (2016). Design of Sanitary Sewer Network for Gandhinagar City using SewerGEMS V8i Software. Int. J. Adv. Res. Eng. Sci. Technol. 3(2).

90 Implementation of environmental management system during the construction phase of underground metro station and tunnelling works with control measures

Shruti Sontakke[1,a], Sanket Kalamkar[1,b], Nikhil Borkar[2,c], and Neelam Chorey[3,d]

[1]Civil Engineering Department, Yeshwantrao Chavan College of Engineering, Nagpur, India
[2]JKumar InfraProjects Ltd, UGC- 06 Underground Mumbai Metro Line 3, Mumbai, India
[3]Assistant Professor, PRMIT&R, Badnera, Amravati, India

Abstract

Mumbai is one of India's most populous cities, with a population of approximately 23 million people. Public transportation is overcrowded because it is the fastest-growing metropolitan region. The underground transportation system is a unique way to alleviate traffic congestion and land scarcity. During the construction phase of an underground metro station and tunnelling work, environmental issues are considered into account such as air pollution, noise pollution, water pollution, and waste generation. The purpose of this paper is to investigate the potential environmental impacts on the environment caused by construction work, as well as the implementation of an environmental management system with control measures. The study shows the noise level is above the permissible limits during night time and the C & D waste found at great extent. If all safety precautions are taken on-site and proper maintenance is performed, then metro construction sites are almost able to reduce pollution.

Keywords: Control measures, environmental impacts, environmental management system, underground metro station and tunnelling.

Introduction

The 33.5-kilometer-long Aqua Line 3 as well as the Colaba-Bandra-Sepz Line, which is operated by MMRC is Mumbai's first underground Metro line. The original estimate project costs of Rs. 23,136 crore in 2016 which has escalated to Rs. 33,406 crores in 2021. MMRC is a joint venture between the Government of India and the Government of Maharashtra. This project is supported by JICA (Japan International Cooperation Agency) as a soft loan, with the remaining funds provided by the governments of India and Maharashtra as equity and sub-debt. This underground metro line – 03 has total of 27 stations from Cuffe Parade to SEEPZ which is shown in figure 90.1.

The rising population of Mumbai city is finding it difficult to commute due to overburdened transit systems. To address this issue, an underground metro project is being developed to reduce traffic congestion and pollution. Although the project is built on a green concept, considerable environmental challenges develop throughout the construction phase. The problems can fall into several areas, including harmful air pollution, noise pollution, waste pollution, and water contamination. The generation of pollution is due to the on-going construction activities such as mining, excavation, drilling, blasting, core cutting, shotcrete, movement of machineries, spillage of hazardous materials etc. The generation of pollution influences environment, humans and other lives. To avoid environmental Impact environmental management system plays an important role. Environmental management system at construction site includes the systematic planning, policies, commitments, monitoring, auditing, control mitigation measures, compliances, and spreading the environmental awareness among the people etc.

Methodology

Study area:

Mumbai metro line (3) having eight civil packages including depot, under the package six there are three underground stations, one is casting yard and one is RMC plant. The names of three stations are CSIA International station, CSIA Domestic Station and Sahar Road Station. The casting yard is situated at Mahul

[a]sontakke.shruti2011@gmail.com; [b]sanketgk1@gmail.com; [c]nikhil.borkar@jkumarcrtg.com; [d]chorey.neelam@gmail.com

Figure 90.1 Colaba to SEEPZ underground Metro Line-03 project

Gaon where the tunnel segment is being cast and the RMC plant is situated at JVLR where concrete is being produced for the construction works.

Identification of environmental aspect and impact:

Several construction activities are currently underway at the underground metro station, which may have an impact on the environment and pose risks. It is essential to identify the sources of environmental factors associated with each project activity before implementing an environmental management system. The environmental aspect and impact can be assessed or measured in terms of likelihood and severity.

The considering factors are:

*   Air:
    The emission of pollutants are due construction, machinery and movements of vehicles, and so on. The pollutants released in the air are: Particulate matter (PM2.5 and PM10), SO_x, $NO2_x$, CO, lead, O_3, ammonia, benzene, and benzo-a-pyrene.
*   Noise:
    The underground metro construction's tunnel and station activities including usage of machineries for work, transportation, generators etc., which creates high amount of noise generation.
*   Water:
    Generally, the water pollution on site is caused by muck, the spillage of oil and diesel, paint, solvents, cleaners and other harmful chemicals, construction debris and dirt as well.
*   Waste:
    During the construction of tunnel and station, there is a large amount of waste generation a good way to immediately effect on environment and human. The different types of wastes generated are municipal, hazardous andnon-hazardous, bio-medical and C&D wastes.

The ongoing metro construction could have a significant adverse influence on both the environment and human health. It raises the risk of a number of disorders.

Various impacts on environment and humans are shown in following Table 90.1.

Environmental management system

The aim of the Environmental management system is to avoid, minimize, or mitigate negative effects on the environment and surrounding areas. Environmental impact assessment (EIA) is one of the major part of environmental management system. The study's goal is to assess the effects of the metro underground construction and tunnelling works. There are positive and negative impacts are assessed in order to determine the end result. The EIA concept starts with the baseline conditions and result of construction. It includes impacts of project location, design, construction & operation.

During construction phase, environmental management of different pollutants with control measures are as follows:

Air management:

Air pollution caused by construction activities may include the generation of dust as a result of various construction activities, air emission from construction machinery/vehicles, and so on. Particulate Matter (PM_{10}), Particulate Matter ($PM_{2.5}$), Sulphur Dioxide (SO_2), Nitrogen Oxides (NO_x), Carbon Monoxide (CO), Benzene (C_6H_6), Ammonia (NH_3), Ozone (O_3), Benzo-a-Pyrene (BaP), Lead (Pb), Arsenic (As), Nickel (Ni) (CPCB guidelines, 2019). The (AAQM) Ambient Air Quality Monitoring standards are shown in Table 90.2. To avoid or reduce the impact of air pollutants on the construction site, some mitigation measures are implemented for the protection of the environment and the health of the workers. Sprinkling is done on the construction site because drilling and blasting cause the release of dust and toxic gases. To control the quality of the surrounding air, a High Volume Sampler (Fine dust sampler, model- NPM-FDS 2.5u and respirable dust sampler, model NPM-HVS/R) is installed at a height of approximately 3.0 m above ground level and continuously measures the concentration for 24 hours. The air quality is monitored on a regular basis three times a week to evaluate the impact of construction activities on the surrounding environment.

Control measure

The underground construction work releases toxic gases for less impact curtains are use and water sprinkling carried out from time to time. During transport of construction material, storage of construction

Table 90.1 Impacts of affecting factors on environment and human

SN	Affecting factors	Impact on environment	Impact on humans
1.	Ambient air (Parameters: PM_{10}, $PM_{2.5}$, SO_x, NO_x, CO, etc.)	Adversely affect climate, ecosystems and materials etc.	Premature mortality, acute and chronic bronchitis, asthma attacks, respiratory symptoms, lung diseases, chronic heart issues etc.
2.	Ambient noise	Can disrupt breeding cycles and rearing, and may even hasten the extinction of some species.	Hearing disorders both chronic and acute, blood pressure problems, heart disease, disturbance in sleep, tension/trauma, and other factors can all contribute to hearing loss.
3.	Contaminated Water	Aquatic life may be directly affected, suffocates plants and trees, etc.	Waterborne diseases.
4.	Waste	Odour, methane gas contributes to greenhouse effect, etc.	Headache, eye irritation, nausea, illness, etc.

Table 90.2 Ambient air quality monitoring standards for particular parameters

Parameters	PM 10	PM 2.5	NO_2	SO_2	CO
AQQM Standards	100	60	80	80	2
Units	$\mu g/m^3$	$\mu g/m^3$	$\mu g/m^3$	$\mu g/m^3$	mg/m^3

Figure 90.2 High volume sampler machine

material and at RMC plant and casting yard air control measures taken. Using of mask is mandatory during construction work.

Noise management

During the construction of the underground metro station there is lots of diverse activities are occurring including drilling, blasting, demolition, transportation, etc., which creates high amount of noise generation. For controlling noise pollution some measures are taken minimization of noise generation through site planning, minimization of noise through activity scheduling, regular noise monitoring and minimization of impact of noise. The noise level should be according to the permissible limit as per CPCB are shown in table 90.3.

Control measure

Factors considering while measuring noise pollution control will includes site planning, site operation, engineering controls, administrative controls, personal protective equipment, noise monitoring with sound lLevel meter (Model Lutron SL- 4023SD) and impact minimsation. Noise levels are reduce through regular maintenance of equipment and vehicles. Earmuffs or earplugs are mandatory for workers during construction.

Water management

Water is a necessity on the construction site for both construction work and drinking. Bore wells or construction water tanker suppliers will provide water for construction purposes. Tankers (RO) and BMC water connections will provide drinking water. At each location, a daily water consumption checklist is kept. Muck, diesel, paint, solvents, oils, cleaners, and other harmful chemicals and construction debris and dirt are all sources of water pollution on construction sites. Water monitoring is performed once a month

Table 90.3 Ambient noise quality standard table as per area

Area code	Category of area/zone	Limits in dB(A) Leq*	
		Day time	Night time
(A)	Industrial area	75	70
(B)	Commercial area	65	55
(C)	Residential area	55	45
(D)	Silence zone	50	40

Figure 90.3 Sound level meter

Figure 90.4 Collection of water for sample

to avoid water contamination. The majority of construction water is reused and used for other purposes such as washing, dusting, and watering plants and trees.

Control measures

Factors to consider when measuring water pollution control include the construction of a proper drainage system to drain all surface water from the construction site, as well as a sedimentation tank for water storage. There are slit and grit removal facilities available. During the dry season, the water collected in the sedimentation tank is used. There is a separate storage area for diesel, and hazardous materials are monitored to prevent spills and leaks. Separate slurries are collected from Bentonite/polymer slurries or other grouts in order to reuse them to the greatest extent possible. The most crucial regular water quality monitoring is also carried out.

Waste management

During the construction of Underground station and tunnels, there is a large amount of waste generated like construction material, metals waste, plastics, rubber and glass, ordinary combustibles, waste oil, food waste, empty cement bags, fuel contaminated soil, used cartridges, lead acid batteries, disposal, hazardous, bio-medical waste, C&D waste etc., which is a good way to directly or indirectly effect on environment and human. As per CPCB there is separate dustbins with colour coding are followed.

Control measures

The important factors include engineering controls such as proper planning, selection of right materials, proper usage and identification of designated storages for each type of waste etc. The Operation Control Procedures (OCP)/standard operation procedure (SOP) will be implemented for different activity. On the other hand administrative control such as inventory control, selection of right materials etc. is essential. Adopting the three R's is the best waste management solution. Figure 90.6 depicts a process chart of the solid waste disposal system.

Observation and Results

Air monitoring results:

Because there is a lot of fugitive dust and pollution generated during the underground construction phase, ambient air monitoring is required at all locations. Air Quality Monitoring: Periodic inspections are carried

Figure 90.5 Collection of sedimentation tank water

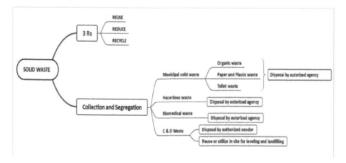

Figure 90.6 Process chart of the solid waste disposal

out by external agencies (NETEL and NABL approved) using high volume sampler (HVS) instruments. There are various parameters of ambient air but PM_{10}, $PM_{2.5}$, SO_2, NO_X, CO, are considered for the study. Tables 90.4, 90.5, 90.6 and 90.7 displays the results of air monitoring with graphs 90.1, 90.2, 90.3 and 90.4.

This all Parameters compared with Ambient air quality monitoring Standard's.

- CSIA International Station

Table 90.4 Air monitoring results of CSIA international station

Date	Time	PM 10 $\mu g/m^3$	PM 2.5 $\mu g/m3$	NO2 $\mu g/m3$	SO2 $\mu g/m3$	CO $mg/m3$
100 $\mu g/m^3$		*AAQM standard*				
		60 $\mu g/m3$	80 $\mu g/m3$	80 $\mu g/m3$	2 $mg/m3$	
30.11.21	24 H	63.1	26.3	35.6	15.2	0.40
7.12.21	24 H	62.2	27.1	19.2	15.2	0.40
14.12.21	24 H	69.2	37.7	45.3	15.2	0.50
20.12.21	24 H	58.1	31.8	29.5	12.0	0.50

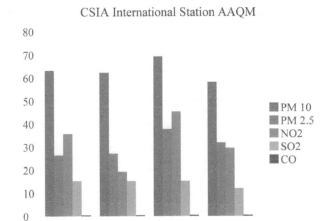

Graph 90.1 Graphical representation of CSIA International Station AAQM

- CSIA domestic station

Table 90.5 Air monitoring results of CSIA domestic station

Date	Time	PM 10 $\mu g/m^3$	PM 2.5 $\mu g/m^3$	NO2 $\mu g/m^3$	SO2 $\mu g/m^3$	CO mg/m^3
100 $\mu g/m^3$		*AAQM standard*				
		60 $\mu g/m^3$	80 $\mu g/m^3$	80 $\mu g/m^3$	2 mg/m^3	
30.11.21	24 H	63.8	30.5	17.4	9.5	0.40
7.12.21	24 H	55.5	23.4	29.2	13.0	0.40
14.12.21	24 H	56.6	31.1	29.1	12.7	0.40
20.12.21	24 H	70.0	37.5	16.4	8,3	0.43

CSIA Domestic Station AAQM

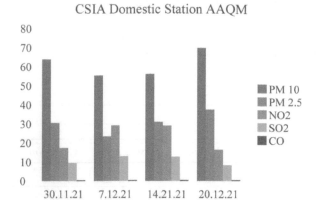

Graph 90.2 Graphical representation of CSIA Domestic Station AAQM

• Sahar Station

Table 90.6 Air monitoring results of Sahar station

Date	Time	PM 10 µg/m³	PM 2.5 µg/m³	NO2 µg/m³	SO2 µg/m³	CO mg/m³
100 µg/m3		AAQM standard				
		60 µg/m³	80 µg/m³	80 µg/m³	2 mg/m³	
30.11.21	24 H	56.1	26.1	21.6	11.2	0.60
7.12.21	24 H	61.2	24.0	20.2	11.1	0.40
14.12.21	24 H	68.7	28.2	42.6	20.3	0.50
21.12.21	24 H	59.5	28.8	24.2	11.1	0.30

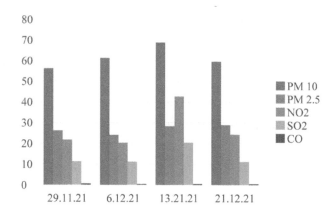

Graph 90.3 Graphical representation of Sahar Station AAQM

• JVLR

Table 90.7 Air monitoring results of JVLR station

Date	Time	PM 10 µg/m³	PM 2.5 µg/m³	NO2 µg/m³	SO2 µg/m³	CO mg/m³
100 µg/m³		AAQM standard				
		60 µg/m³	80 µg/m³	80 µg/m³	2 mg/m³	
30.11.21	24 H	70.1	35.3	34.6	16.6	0.32
6.12.21	24 H	57.0	26.0	13.2	11.1	0.40
13.12.21	24 H	72.4	31.5	45.3	18.5	0.30
16.12.21	24 H	77.6	39.8	20.0	5.6	0.60

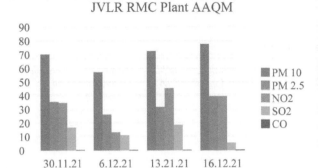

Graph 90.4 Graphical representation of CSIA Domestic Station AAQM

Discussion

Comparing with the National Ambient Air Quality (NAAQ) of Central Pollution Control Board (CPCB), all the parameters of different location of construction site's measured value is under the permissible limit.

Noise monitoring results:

Noise is generated by machinery used in the construction of an underground station, tunnelling activity, and other construction activities. To monitor or limit noise disturbance, monitoring is performed twice a week at each location. Noise is monitored during the day from 6:00 a.m. to 10:00 p.m., and at night from 10:00 p.m. to 6:00 a.m. Lmin, Lmax, Leq Day, Leq, and LDN are the ambient noise parameters. Every month, the average noise level at all construction sites is measured in order to better understand the construction site's maximum average noise level. The noise monitoring results are shown in table 90.8, 90.9, 90.10 and 90.11 with graphs 90.5, 90.6, 90.7 and 90.8.

- CSIA international station

Table 90.8 Noise monitoring result at international station

Date	Day Time Leq dB(A)	Night Time Leq dB(A)	Day Time Permissible Limit Leq 65 dB (A)	Night Time Permissible Limit Leq 55 dB (A)
30.11.2021	66.6	65.2	65	55
07.12.2021	68.1	65.4	65	55
14.12.2021	66.8	66.0	65	55
21.12.2021	66.9	66.9	65	55

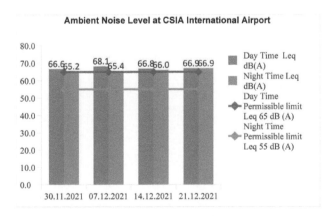

Graph 90.5 Graphical representation of ambient noise level at CSIA International station

• Domestic station

Table 90.9 Noise monitoring result at domestic station

Date	Day Time Leq dB(A)	Night Time Leq dB(A)	Day Time Permissible Limit Leq 65 dB (A)	Night Time Permissible Limit Leq 55 dB (A)
30.11.2021	65.8	62.1	65	55
07.12.2021	66.9	64.9	65	55
14.12.2021	65.3	62.1	65	55
21.12.2021	63	62	65	55

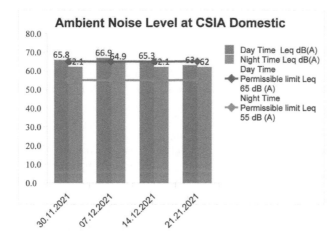

Graph 90.6 Graphical representation of ambient noise level at domestic station

• Sahar station

Table 90.10 Noise monitoring result at Sahar station

Date	Day Time Leq dB(A)	Night Time Leq dB(A)	Day Time Permissible Limit Leq 65 dB (A)	Night Time Permissible Limit Leq 55 dB (A)
29.11.2021	66.0	64.0	65	55
06.12.2021	66.8	65.5	65	55
13.12.2021	67.1	65.6	65	55
20.12.2021	66.6	64.7	65	55

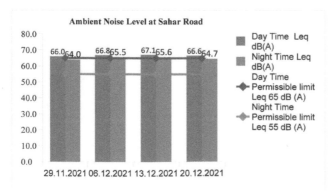

Graph 90.7 Graphical representation of ambient noise level at Sahar Station

- JVLR RMC plant

Table 90.11 Noise monitoring result at JVLR RMC plant

Date	Day Time Leq dB(A)	Night Time Leq dB(A)	Day Time Permissble Limit Leq 65 dB (A)	Night Time Permissible Limit Leq 55 dB (A)
30.11.2021	68.0	63.2	65	55
06.12.2021	66.8	61.2	65	55
13.12.2021	64.8	60.8	65	55
20.12.2021	64.4	61.5	65	55

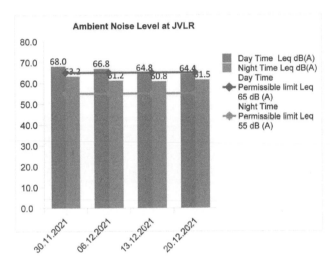

Graph 90.8 Graphical representation of ambient noise level at JVLR

Comparing with the ambient noise quality standard, all the location's measured value is mostly over the permissible limit of CPCB. During night hours it is maximum over the limit. Hence, take proper mitigation measure.

Water quality monitoring results

Water quality analysis has been carried out in all locations of UGC-06 on Monthly basis. A base reading taken in the site so that we can compare whether there are any remarkable changes in the ground water quality due to our construction activity. There are various water quality parameters, for study only some the parameters are taken into consideration.

The external association NETEL India is India's leading organisation which provides filed environment services. It analyse various sample monitoring results with the standard limits.

The result of drinking water analysis by external association NETEL India are shown in below Table 90.12.

The result of sedimentation tank water analysis by external association NETEL India are shown in below Table 90.13.

Comparing with the legal permissible limits the drinking water sample's turbidity at Sahar station is above limit.

In sedimentation tank water analysis at Domestic Station total dissolved solids are found much greater than the given standard limit (500 <3190). Hence, proper mitigation need to take.

Waste management results:

Underground metro construction work consists of various construction stages which generates large amount of wastes. For storage of refuse wastes separate dustbins are used at different sites. There is a separate storage for chemical and hazardous waste which is not mixed with other wastes. Housekeeping carried out on regular basis. The plastic and recycle waste is recycled every month with the help of tie up

Table 90.12 Result of drinking water analysis

Parameters	Units	Results				Limits
		International station	Domestic station	Sahar station	JVLR RMC Plant	
Color	Hazen	<5	<5	<5	<5	5
Odor	-	Agreeable	Agreeable	Agreeable	Agreeable	Agreeable
Taste	-	Agreeable	Agreeable	Agreeable	Agreeable	Agreeable
pH (@25)	-	7.55	7.60	7.57	7.98	6.5–8.5
Turbidity	NTU	1.1	1.0	1.9	1.0	1
Total Alkalinity	mg/l	39.3	41.4	39.3	47.6	200
Total Hardness	mg/l	43.1	43.1	43.1	49	200
Total dissolved solids	mg/l	78	84	83	84	500
Calcium	mg/l	10.2	11	10.2	12.5	75
Magnesium	mg/l	<5	<5	<5	<5	30
Fluoride	mg/l	<0.2	<0.2	<0.2	<0.2	1
Mineral oil	mg/l	<0.5	<0.5	<0.5	<0.5	0.5
Iron	mg/l	<0.1	<0.1	<0.1	<0.1	0.3
Lead	mg/l	<0.01	<0.01	<0.01	0.1	0.01
Mercury	mg/l	<0.001	<0.001	<0.001	<0.001	0.001
Arsenic	mg/l	<0.001	<0.001	<0.001	<4	0.01
Total coliform (MPN)	mpn/100 ml	0	0	0	<0.001	Absent
E coli	-	Absent	Absent	Absent	0	Absent

Table 90.13 Result of sedimentation tank water analysis

Parameter	Unit	Result		Limits
		Domestic Station	Sahar	
Colour	Hazen	<5	<5	5
Odour	-	Agreeable	Agreeable	Agreeable
Taste	-	Agreeable	Agreeable	Agreeable
pH (@25)	-	8.34	8.17	6.5–8.5
Turbidity	NTU	1.5	1.2	1
Total Alkalinity	mg/l	44	37.4	200
Total Hardness	mg/l	34.7	34.7	200
Total dissolved solids	mg/l	3190	395	500
Calcium	mg/l	7.9	7.9	75
Magnesium	mg/l	<5	<5	30
Fluoride	mg/l	0.8	1.0	1
Mineral oil	mg/l	<0.5	<0.5	0.5
Iron	mg/l	0.1	0.1	0.3
Lead	mg/l	<0.01	<0.01	0.01
Mercury	mg/l	<0.001	<0.001	0.001
Arsenic	mg/l	<0.001	<0.001	0.01
Total coliform (MPN)	mpn/100 ml	0	0	Absent
E coli	-	Absent	Absent	Absent

agency. The amount which comes from the recycled waste paper will go for the NGO for educating the orphanage child.

During the site visit a lot of C and D and scrap waste at great extent found in the surrounding.

Result

It is found that almost waste is reused, recycled or sold to the tie up agencies. The C & D is going to be used for levelling up the land and the scrap waste found the surrounding is going to be reused as developmental activities, stored safely in scrap yard or it is sold as scrap.

Conclusion

The Mumbai metro aqua line three going to be the good source of transportation without traffic congestion. It is green solution as well with gives environmental benefits. During the construction phase it is essential need for the monitoring and maintenance of site especially the environmental aspects air, noise, water and waste. The environmental management system plays important role in the site management and near surrounding area protects workers, environment and livelihoods.

References

Aadal, H., Rad, K. G., Fard, A. B., Sabet, P G. P., and Harirchian, E. (2013). Implementing 3R concept in construction waste management at construction site. J. Appl. Biol. Sci. 3(10):160–166.

Attanayake, P. M. and Waterman, M. K. (2006). Identifying environmental impacts of underground construction. Hydrogeol. J. 14(7):1160–1170.

Chinmay, H., Waghmare, S. M., Natthu, S., and Bhagat, R.(2021). Environmental Impact & Mitigation Measures during the Construction of Underground Station and Tunneling at Mumbai Metro Line-03. Int. J. Res. 9(6).

Jain, G., Gupta, V., and Pandey, M. (2016). Case study of construction pollution impact on environment. Int. J. Emerging. Technol. Eng. Res. 4(6).

Maskar, H., Randive, N., Singh, S., and Khaire, S. (2018). Study of baseline data for environmental impact assessment of Mumbai Metro line IV. Int. J. Scient. Res. Dev. 6(2).

Mumbai Metro Line 3 (Colamba-Bandra-SEEPZ) (2011, November). Mumbai Metro Corporation Limited/RITES (GOI). Revision Report, (2018).

National Ambient Air Quality Standard, (2019)

91 Web service recommendation based on user's demographic attributes

Mamta Bhamare[a], Pradnya V. Kulkarni[b], Sarika Bobde[c], Rashmi Rane[d], and Ruhi Patankar[e]

School of Computer Engineering andTechnology- MIT-WPU, Pune, India

Abstract

Web services are a simple but effective method of connecting to network resources. To offer end-users data from their virtualized resources, all modern organisations provide web service interfaces. With the increased usage of multi-channel communication modes (mobile phones, tablets, and laptops), web services are excellent for guaranteeing seamless data flow across all platforms. While Web services address many difficulties, there are a few issues that consumers or requesters have when they locate a web service that satisfies their requirements. With the increased use of online services, there is a greater demand for an effective and almost flawless web service recommendation system. Cold start, sparsity, and overspecialisation are some of the problems with recommender systems. To address this issue, in this paper, collaboration filtering and demographic filtering based on models are proposed as a new approach. Using this technique, recommendations can be generated based on the demographic characteristics of users, such as age, gender, occupation, etc. We propose using web services based on the results of both tests. This paper attempts to propose a solution to the cold start problem by using a hybrid filtering method.

Keywords: Collaborative filtering, demography based filtering, hybrid filtering, sentiment analysis, web service recommendation.

Introduction

The internet's use of Web services technology has increased tremendously. It has increased developers' productivity and throughput when creating applications. The internet and current online services have grown in popularity over the previous few decades, and an abundance of information is now available to everyone (Beel et al., 2013). We frequently confuse a website and a web service because both provide some kind of service to the end-user over the network, but there is a distinction between the two: A website is for human consumption, but web service is for code or application-level consumption. There are several interpretations of web services, but we can say that web services are meant to compose multiple software components and offer machine-machine interactions through a network, and the term implies that it is a form of practice that is implemented on the internet. It enables two separate programs operating on different servers to communicate with one another across the network, and these applications can be implemented in a variety of languages.

Web service recommendation is the practice of identifying and proposing relevant Web services to end-users on a proactive basis. Following are the methods which are used for web service.

A. Content-based filtering

This type of algorithm suggests goods to customers based on characteristics taken from the same client's past data. Content-based filtering can be understood as-

f: (user - profile; content - profile) ->T

Where T is a metric used to measure how much the item is liked by a user.

B. Collaborative filtering

Through collaborative filtering, the active user's services are linked with similar profiles of past visitors. There are two types of collaborative filtering algorithms (Kulkarni et al., 2020). Memory-based collaborative filtering: These methods are further subdivided into user-based and item-based techniques. User-based

[a]mamta.bhamare@mitwpu.edu.in; [b]pradnyav.kulkarni@mitwpu.edu.in; [c]sarika.bobde@mitwpu.edu.in; [d]rashmi.rane@mitwpu.edu.in; [e]ruhi.patankar@mitwpu.edu.in

methodologies utilise the evaluations of other comparative clients to anticipate the ratings of clients, and item-based methodologies utilize the likeness of the item to foresee the evaluations of the clients.

Model-based collaborative filtering: These techniques use statistical and machine learning approaches to train a model from rating data.

C. Demographic filtering

This sort of framework categorises clients or items in view of their own traits and makes suggestions in view of demographic categorisations. As such we can say that in demographic information-based filters clients are classified by their features, and a recommendation on suggestion is given to the class of demographic data. It is easy and efficient to make a recommendation based on demographics.

Table 91.1 list the advantages and disadvantages of content-based, collaborative filtering, and demographic-based methods.

Table 91.1 Summary of different filtering methods

	Content-based	*Collaborative filtering*	*Demographic-based*
Based on	Description of item and profile of user's interest	The similarity of user rating	Demographic attributes similarity
Advantages	Newly-deployed web services can be recommended (Wang and Tang, 2015). User independence.	Can perform in domains where there is not much content associated with items or with user profiles (Kang et al., 2012).	The history of users' ratings, the textual description, or the knowledge of items is not required. The cold start problem can be resolved (Erkin et al., 2012)
Disadvantages	Can only recommend items that score highly against a user's profile (Kang et al., 2012). Depends on the availability of descriptive data.	Cold start problem (Yao Et al., 2015). Scalability Dai et al., 2009, Gray sheep (Yao Et al., 2015).	Privacy problems in demographic information gathering (Isinkaye et al., 2015).

Literature survey

Many authors proposed different approaches to implementing recommendation systems. Liu proposed a system that recommends services using similarity of rating preference and personal attribute-based clustering. Mining is performed on service usage patterns using the generalized sequential patterns (GSP) algorithm (Liu et al., 2015). Recommended systems may use data mining techniques such as grouping and classification (Ben, 2005). The principle of the voting theory is used with clustering algorithms to solve the scalability problem of Collaborative Filtering (Das et al., 2014). Authors also proposed a data mining technique like the Apriori algorithm to recommend the web service based on users' history (Sultan et al., 2013). Different recommender systems with privacy issues are studied and possible solutions are given (Wang and Tang, 2015; Isinkaye et al., 2015). The authors give the details of content-based methods and explain the problems associated with collaborative filtering and their possible solutions (Lops et al., 2011; Su and Khoshgoftaar, 2009).

Demographic attributes have a significant impact on many types of recommendations (Isinkaye et al., 2015; Yao et al., 2015; Beel et al., 2013). Demographic attributes, Weather, and Online Reviews are useful for restaurant recommendations (Bakhshi et al., 2014). Effective results of demographic attributes on recommendation systems are given (Safoury and Salah, 2013. Also, to improve the performance of music recommendations, demographic attributes are used (Yapriady and Uitdenbogerd, 2005). This paper said that collaborative filtering has a number of drawbacks and constraints, including the necessity for a large number of rating values, and their co-occurrence across several objects and users is scarce (Pérez-Almaguer, 2001). To address these issues in this paper the content-based group recommendation systems (CB-GRS) model is built. They also analysed the hybrid CB-GRS which combines all three models. Here we have used demography-based filtering and collaborative filtering to address the issues mentioned above.

The paper (Kulkarni et al., 2020) is reviewed different recommendation system techniques for the application of eLearning. The authors here proposed a hybrid approach (collaborative filtering and content-based filtering) for a better recommendation. Also, deep learning techniques are used to train the system. Explicit feedback is used for the suggestion, but the explicit feedback is always not true feedback so to overcome this problem in this paper we have used the proposed approach. Authors incorporate both functional and non-functional characteristics of Web services (Kang et al., 2012). Through active user feedback and active monitoring, multiple Quality of Service (QoS) values of web services are considered and gathered (Li et al.,

2010). A new study (Sulikowski and Zdziebko, 2020) uses a deep neural network method to develop a user interface for an e-commerce website depending on user behavior. Their research indicated the impact of a website layout on the recommendation of things based on the user's behavior. Customer feedback is widely used in recommendation systems; the new survey is devoted to the analysis of the sentiment of text reviews. They showed that there are three types of recommendation systems based on the word, object, and opinion-based text-based reviews. Even though Recommendation systems have long been used in e-commerce, problems persist in this area. For example, there is the problem of scalability and data latency. A real-time response should be provided by recommendation systems on a website with a significant number of users (Srifi et al., 2020). Data is also missing when using data sets for evaluation as not all customers will evaluate all products. Furthermore, various stakeholders, such as farmers (Madhusree et al., 2018), can profit from recommendations in their decision-making process. By predicting the user's interests and suggesting the services to them based on their needs, web service recommendation serves as a means of saving the user's time. A novel way to measure the quality of service is proposed in which the author combines content-based and collaborative filtering techniques (Barod et al., 2016). Most of the systems used collaborative filtering for recommendation. Existing systems required ratings for similarity measures. Demographic attributes are an important feature in the recommendation system.

Methodology

In the proposed system we assumed web services recommendations based on user demographics and sentiment analysis as

RS be the Service Recommendation System

$$RS = \{U, S, F, O\} \tag{1}$$

where user attribute serves as inputs to the system and recommendation of services is an output O. Processing function consists of f = {f1, f2, f3} and each one is responsible for dedicated functionality. U = Set of users of recommendation system, {U1, U2, U3...Un}

U = Set of users of recommendation systems, {U1, U2.... Un}

$$U = \left\{ (U_c, U_a) \mid U_c \in (ID, Pw) U_a \in \left(U_{prof}, U_{age}, U_{gender} \right) \right\} \tag{2}$$

Where Uc contains Authentication credential which includes ID = User Name and PW = Password
Ua contains user attributes which include Uprof, Uage, Usal

If a target user Ui asks for recommendations of web services, the system suggests an appropriate service from the set of services S.
A set of Web services S = S1, S2... Sn and A set of users U = Ul, U2...Um
F=clustering {I, J, O}
Clustering is performed on users' demographic attributes using the Improved K-Means clustering algorithm.
  The input I: Set of tuples n where each having attributes {age, gender, profession}
  Output O: k number of Clusters where n tuples distributed properly.
  J: Objective function.
The objective is to minimize total intra-cluster variance, or, the squared error function:

$$I = \sum_{i=1}^{n} \left\| x_i^j - c_j \right\|^2 \tag{3}$$

Where $\|xi-cj\|^2$ is a chosen distance measure between a data point x_i^j and the cluster center
F2= Classification {I, C, O} where
I= {Iage, Ugender, Uprof, Dhistory}
O= {Service category}
Dhistory is history data of a user.

$$C : \mathbf{p}(\mathbf{C} / \mathbf{x}) = \frac{P(x / \mathbf{C}) P(\mathbf{C})}{P(x)} \tag{4}$$

Where P(c/x) is the posterior probability of class (target) given predictor (attribute)
P(c) is the prior probability of class.
P(x|c) is the likelihood which is the probability of predictor given class.
  P(x) is the prior probability of predictor.
  F2=Sentiment Analysis {I, P, O}

Input I: (Output of f3 i.e. Sc: Service category, Cc: Comments of category).

$S_c = \{s_1 \ldots \ldots s_n\}$

$C_c = \{C_1 \ldots \ldots C_n\}$

Output O: $\{S_{pos}, S_{neg}, S_{neu}\}$

Where S_{pos}=Services with a positive score

$\qquad S_{neg}$=Services with a negative score

$\qquad S_{neu}$=Services with a neutral score

P = (processing: Tokenize, filtering, Score count)

Final output O of the system: Recommended service with a maximum positive score.

In the suggested technique, we employ the notion of categorising users based on demographic factors and sorting online services based on user feedback or comments. The algorithm guesses what the user would want and suggests online services based on that prediction. In our system, data mining along with sentiment analysis is used for recommendation. The proposed architecture of the system is shown in Figure 91.1. In our system, three basic modules are present as:

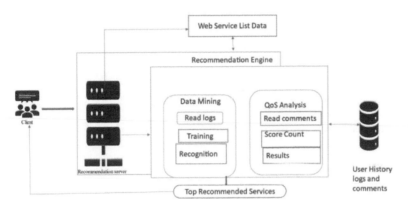

Figure 91.1 Proposed system architecture

D.   Module I

In this module, the demographic attributes of users are collected by registering the user's details. Registered users can give feedback on services. This feedback is stored along with the demographic attributes of users in the database. These demographic attributes are used to measure the similarity of users that are calculated in the next module.

E.   Module II

In this module, demographic attributes of all users are taken as historical data, and normalisation is performed on that data. After that, an improved K means clustering algorithm is used to reduce the candidate set and then classification is performed using the Naïve Bayes need (Watson, 2001). classification algorithm. As K-Means has some limitations, for example, we need to specify the number of clusters as input, but Enhanced K-Means (Wang and Su, 2011) does not require this. Here, to identify which user is belonging to which category, classification is used. According to this training model, web services are recommended to new users or clients. This approach can be understood by following Figure 91.2.

Figure 91.2 Data mining approach

F. Module III

In this module, the output of classification is taken as input and all comments on classified categories are processed. Comments are first tokeniSed and then filtering is performed to remove the irrelevant words associated with the data. In our system clustering is used to reduce the candidate set in collaborative filtering. For similarity measures, a wide dictionary of positive, negative words are referred. After that comments are categorised as positive, negative, and neutral.

When a user wants the recommendation list, web services having a maximum positive score are recommended according to their profile.

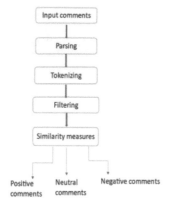

Figure 91.3 Sentiment analysis approach

Data analysis and interpretation

A. Datasets

Users' demographic information, as well as their usage history, is saved in a database. The dataset Movie lens 100k is utilised in this case (kaggle). This data collection comprises simple demographic information such as a user's age, gender, and employment connected with movie ids that have previously been watched by comparable users, and movies are also classified into distinct zones.

B. The k-fold cross validation procedure

To create the model, the target consumers in each of the test datasets were divided into two parts: one for training model preparation and the other for testing. The forecasting model was developed using the trained client's history data during training, and the forecasting performance was evaluated using the target customers in the forecasting target period.

C. Evaluation measures

Here two measures are used for evaluation. The precision is the proportion of recommendations that are good recommendations. Recall: A measure of completeness, determines the fraction of relevant items retrieved out of all relevant items. It can be given as below

$$Precision = \frac{True\ Positive}{True\ positive\ +\ False\ Positive} \tag{5}$$

$$Recall = \frac{True\ Positive}{True\ Positive\ +\ False\ Negative} \tag{6}$$

$$f - measure = 2\frac{P.R}{P+R} \tag{7}$$

D. Experimental results

The result of existing approach and our approach are as follows (Safoury and Salah, 2013).

Table 91.2 Results of precision (existing and proposed system)

Attribute	Value	Precision (existing system) %	Precision (our system) %
Gender	Female	16	18
	Male	24	26
Occupation	Student	8	9
	Engineer	11	9
	Admin	22	25
	Educator	29	32
Age	Teenager (10–19)	8	17
	Adult (20–45)	22	19
	Mid age (45–60)	25	24
	Old (>60)	17	21

The result of precision and recall of different existing approach and proposed approach are as follows, UBCF refers to user based collaborative filtering and PICF refers to popular item-based CF, by Mustafa and Frommholz (2015). Table 91.2 shows the result of precision and recall of proposed PB for movie lens (100k) dataset. Table 91.2 shows the precision, recall and f-measure on different sets of inputs. Table 91.3 shows the precision and recall of proposed system.

Table 91.3 Precision and recall of proposed system

Number of inputs	Precision (P)	Recall (R)	F-measure (F)
220	0.83	0.76	0.79
300	0.80	0.70	0.74
435	0.73	0.63	0.66

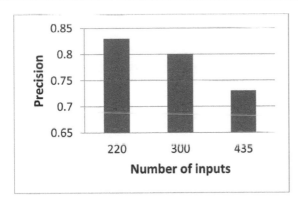

Figure 91.4 Number of input vs precision graph

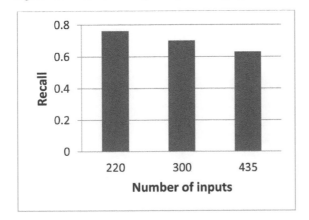

Figure 91.5 Number of input vs recall graph

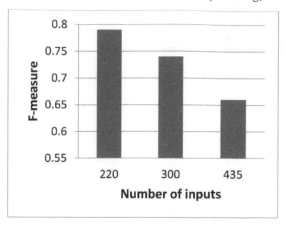

Figure 91.6 Number of input vs F-measure graph

Table 91.4 Precision and recall for UBCF approach

Precision	0.69	0.68	0.67	0.67	0.66	0.65	0.65	0.64	0.64	0.63
Recall	0.01	0.02	0.03	0.04	0.05	0.06	0.07	0.08	0.08	0.09

Table 91.5 Precision and recall for popular items-based CF approach

Precision	0.42	0.42	0.42	0.42	0.42	0.42	0.41	0.41	0.41	0.40
Recall	0.04	0.08	0.12	0.16	0.20	0.24	0.28	0.31	0.34	0.37

Table 91.6 Precision and recall for proposed approach

Precision	0.66	0.66	0.63	0.63	0.63	0.60	0.60	0.60	0.60	0.60
Recall	0.05	0.10	0.15	0.20	0.25	0.30	0.35	0.40	0.45	0.50

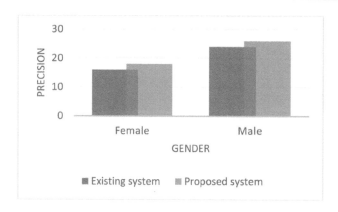

Figure 91.7 Precision of existing and proposed system on gender

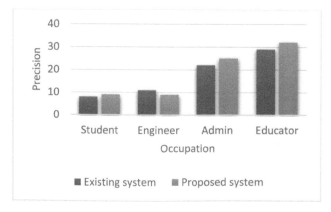

Figure 91.8 Precision of existing and proposed system on occupation

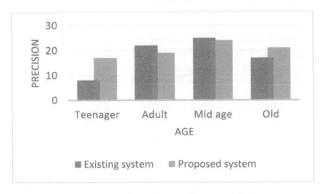

Figure 91.9 Precision of existing and proposed systems on age

E. Comparision with existing system

- Similarity measure: Most of the existing system counts the similarity by using cosine similarity or Pearson correlation using ratings however here clustering is used that can reduce the candidate set in collaborative filtering.
- QoS: Existing system considered many attributes for QoS value but it is very difficult to obtain all attributes values. Here proposed system is developed from an end-user perspective that uses comments of another user to calculate the QoS value.

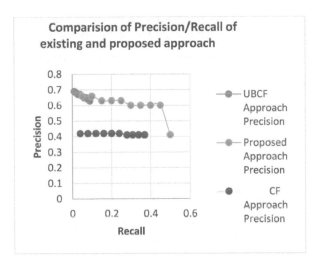

Figure 91.10 Precision recall of existing and proposed system

- Improved k-means: In the proposed system, the limitation of the number of clusters as input is removed.
- Hybrid approach: A new approach is proposed using model-based Collaborative filtering and demographic filtering. It removes the cold-start problem of CF.

Conclusion

Users are having difficulty discovering relevant services that meet their needs as the number of web services grows. We created a recommender system to make the job easier. In this system, we try to provide users with suggestions based on the user's opinion records and demographic data, enabling the selection of the appropriate service. The existing approaches have cold-start and scalability problems and require many QoS for a recommendation. We have overcome all these problems in our system and in our system clustering algorithm for similarity computation and preference computation is used by which the accuracy of assessment of user interest and preference is improved. In the future, the proposed approach can also be used in different domains. Also, the results may be compared with those of traditional methods with a larger group of users in the future. It can be further extended using deep learning approaches to get better results.

References

Aznag, M., Quafafou, M., Durand, N., and Zarir, J. (2013). Web Services Discovery and Recommendation Based on Information Extraction and Symbolic Reputation. Int. J. Web Service Comput. 4(1): 1–18.

Bakhshi, S., Kanuparthy, P., and Gilbert, E. 2014. Demographics, weather and online reviews: a study of restaurant recommendations. Proceedings of the 23rd international conference on World wide web, (pp. 443–454). https://doi.org/10.1145/2566486.2568021

Barod, P., Bhamare, M. S., and Patankar, R. 2016. A Novel Approach for Web Service Recommendation. In Proceedings of the Second International Conference on Information and Communication Technology for Competitive Strategies (ICTCS '16). Association for Computing Machinery. New York (N Y), 129:1–4. doi:https://doi.org/10.1145/2905055.2905345

Beel, J., Langer, S., Genzmehr, M., Gipp, B., Breitinger, C., and *Nürnberger, A.* (2013). Research paper recommender system evaluation: A quantitative literature survey. In Proceedings of the International Workshop on Reproducibility and Replication in Recommender Systems Evaluation, Hong Kong, China, (pp. 15–22).

Beel J., Langer S., Nürnberger A., and Genzmehr M. (2013). The impact of demographics (age and gender) and other user-characteristics on evaluating recommender systems. In Research and Advanced Technology for Digital Libraries. Ed. Aalberg T., Papatheodorou C., Dobreva M., Tsakonas G., Farrugia C.J., (pp. pp 396–400). Lecture Notes in Computer Science. 8092. Berlin, Heidelberg: Springer. https://doi.org/10.1007/978-3-642-40501-3_45

Ben, S. J. 2005. Application of Data Mining to Recommender Systems. Encyclopedia of Data Warehousing and Mining. ed. J. Wang, (pp. 44–48). http://doi:10.4018/978-1-59140-557-3.ch009

Bielza, C. and P. Larrañaga. (2014). Discrete Bayesian Network Classifiers: A Survey. 1.

Cremonesi, P., Turrin, R., Airoldi, F.: Hybrid algorithms for recommending new items. In: Proceedings of the 2nd International Workshop on Information Heterogeneity and Fusion in Recommender Systems—HetRec '11 (2011)

Dai, Y., Ye, H., and Gong, S. 2009. Personalized Recommendation Algorithm Using User Demography Information. Knowledge Discovery and Data Mining, WKDD 2009, (pp. 100103). Second International Workshop on. doi: 10.1109/WKDD.2009.156.

Das, J., Mukherjee, P., Majumder, J., and Gupta, P. 2014. Clustering-based recommender system using principles of voting theory. 2014 International Conference on Contemporary Computing and Informatics (IC3I), (pp. 230235). doi: 10.1109/IC3I.2014.7019655.

Erkin, Z., Beye, M., Veugen, T., and Lagendijk, R. L. (2012). Privacy-preserving content-based recommender system. In: Proceedings of the on Multimedia and security (MM&Sec 2012), (pp. 77–84). New York: ACM.

Harper, F. M. and Konstan, J. A. (2015). The movielens datasets: History and context. ACM Trans. Interact. Intell. Syst. 5(4):1–19.

Isinkaye, F.O., Folajimi, Y. O., and Ojokoh, B. A.(2015). Recommendation systems: Principles, methods and evaluation. Egypt. Inform. J.. 16(3):261273. https://doi.org/10.1016/j.eij.2015.06.005.

kaggle. Movie lens-100K dataset. https://www.kaggle.com/rajmehra03/movielens100k.

Kang, G., Liu, J., Tang, M., Liu, X., Cao, B., and Xu, Y. (2012). AWSR: Active Web Service Recommendation Based on Usage History. IEEE 19th International Conference on Web Services, (pp. 186193). doi: 10.1109/ICWS.2012.86.

Kulkarni, P. V., Rai, S., and Kale, R. (2020). Recommender system in eLearning: a survey. In Proceeding of International Conference on Computational Science and Applications (pp. 119–126). Singapore: Springer.

Li, S., Chen, H., and Chen, X. 2010. A Mechanism for Web Service Selection and Recommendation Based on Multi-QoS Constraints. 2010 6th World Congress on Services, (pp. 221228). doi: 10.1109/SERVICES.2010.31.Liu, L., Lecue, F., and Mehandjiev, N. (2013). Semantic content-based recommendation of software services using context. ACM Trans. Web. 7(3):17.

Liu, R., Xu, X., and Wang, Z. 2015. Service Recommendation Using Customer Similarity and Service Usage Pattern. Int. Conf. Web Se. IEEE. (pp. 408—415). doi: 10.1109/ICWS.2015.61.

Lops P., de Gemmis M., and Semeraro G. (2011). Content-based Recommender Systems: State of the Art and Trends. In Recommender Systems Handbook. ed. F., Ricci, L., Rokach, B., Shapira, and P. Kantor. Boston, MA: Springer. https://doi.org/10.1007/978-0-387-85820-3_3

Madhusree, K., Rath, B. K., and Mohanty, S. N. (2018). Crop Recommender System for the Farmers using Mamdani Fuzzy Inference Model. Int. J. Eng. Technol. 7:277–280.

Mustafa G. and Frommholz, I. 2015. Performance comparison of top N recommendation algorithms. 2015 Fourth International Conference on Future Generation Communication Technology (FGCT), (pp. 16). doi: 10.1109/FGCT.2015.7300256.

Pérez-Almaguer, Y., Yera, R., Alzahrani, A. A., Martínez, L. (2001). Content-based group recommender systems: a general taxonomy and further improvements. Expert Syst. Appl. 115444. https://doi.org/10.1016/j.eswa.2021.115444.

Safoury, L. and Salah, A. (2013). Exploiting User Demographic Attributes for Solving Cold-Start Problem in Recommender System. Lecture Notes on Software Engineering. 1(3).

Srifi, M., Oussous, A., Ait Lahcen, A., and Mouline, S. (2020). Recommender Systems Based on Collaborative Filtering Using Review Texts—A Survey. Information. 11:317.

Sulikowski, P. and Zdziebko, T.(2020). Deep Learning-Enhanced Framework for Performance Evaluation of a Recommending Interface with Varied Recommendation Position and Intensity Based on Eye-Tracking Equipment Data Processing. Electronics. 9:266

Sultan, T. I., Khedr, A. E., and Alsheref, F. K. (2013). Adaptive Model for Web Service Recommendation. Int. J. Web Servic. Comput. 4(4):2133.

Su, X. and Khoshgoftaar, T. M. (2009). A Survey of Collaborative Filtering Techniques. Adv. Artificial Intell. https://doi.org/10.1155/2009/421425

Wang, J. and Su, X. 2011. An improved K-Means clustering algorithm. 2011 IEEE 3rd International Conference on Communication Software and Networks, 2011, pp. 44–46, doi: 10.1109/ICCSN.2011.6014384.

Wang, J. and Tang, Q. (2015). Recommender Systems and their Security Concerns. ACR Cryptol. ePrint Arch. 1108.

Watson, T. J. (2001). An empirical study of the naive Bayes classifier. CiteSeerX.

Yao, L., Sheng, Q. Z., Ngu, A. H. H., Yu, J., and Segev, A. (2015). Unified Collaborative and Content-Based Web Service Recommendation. IEEE Transactions on Services Computing. 8(3):453–466. doi: 10.1109/TSC.2014.2355842.

Yapriady B. and Uitdenbogerd A. L. 2005. Combining Demographic Data with Collaborative Filtering for Automatic Music Recommendation. In Knowledge-Based Intelligent Information and Engineering Systems. KES 2005. Lecture Notes in Computer Science. ed. R., Khosla, R. J., Howlett, and L. C., Jain. (3684). Berlin, Heidelberg: Springer. https://doi.org/10.1007/11554028_29

92 Optimisation of variable thickness circular cross-section thin-walled beams for crashworthiness under bending collapse

Sanjay Patil[a] and Dilip Pangavhane[b]

Department of Automobile Engg, Govt. College of Engg and Reserch Avasari, Pune, India

Abstract

Thin-walled beams (TWB) are frequently employed in vehicles as energy absorbers due to their light weight and significant energy absorption capacity. In the case of a side collision, this beam deforms and absorbs the most impact energy. Because of this, researchers are intrigued by the collapse behaviour of TWB under static and dynamic loads. Circular and square TWB are often utilised in automobile side doors. In this study, a variable thickness (VT) circular cross section TWB is used instead of the traditional uniform thickness (UT) circular cross section TWB to increase crashworthiness performance. Finite element numerical simulations are used to evaluate crashworthiness indicators like energy absorption (EA), specific energy absorption (SEA), and crash force efficiency (CFE). These simulations are performed on ABAQUS explicit dynamic software. To validate the finite element simulation model, an experimental test was performed. Multi-objective optimisation is performed in order to study the most efficient VT TWB with the highest SEA and CFE. A surrogate model applying response surface techniques is used to create the regression function of these indicators. In addition, the effect of variable wall thickness on SEA and CFE is investigated. The non-dominated sorting genetic algorithm (NSGA-II) is then used to find the Pareto optimum solutions with the highest SEA and CFE. The results show that the VT TWB performs significantly better than the UT TWB and saves approximately 19% of the material.

Keywords: Crash force efficiency, crashworthiness, multi objective optimisation, response surface method, specific energy absorption, surrogate model, thin wall tube.

Introduction

Thin-walled beams (TWB) are the most common passive energy absorption components used in aerospace, automotive, special purpose vehicles, and other areas where safety is critical. It is because of its simple design and low cost. At the time of impact, the TWB deforms and absorbs a significant amount of energy (Luzon-Narroa et al., 2014). In side doors and the frontal crash box of automobiles TWB is used to protect the car from side and front impacts, respectively. The TWB in the side door is anticipated to bend and absorb impact energy during a collision, so the impact force should not move into the passenger compartment. As a result, the design of TWB is critical and draws researchers' attention (Long et al., 2019; Zhang et al., 2015). There have been a number of numerical and experimental researches carried out to determine and improve the effectiveness of TWB under bending collapse. TWB come in a broad variety of cross sections, from round to rectangular to square to tapered to foam filled to elliptical. Research shows that circular cross section TWB is most commonly used compared to other cross section TWB due to manufacturing easiness and effectiveness (Shaharuzaman et al., 2018).

Numerous strategies have been investigated to enhance the crashworthiness of TWB under bending collapse. The crashworthiness of TWB under the bending load is influenced by TWB parameters like thickness, loading angle, type cross section, material, loading direction etc. Among all these parameters, thickness shows the greatest influence (Kotelko et al., 2002; Lee et al., 2017a; Poonaya et al., 2009). Thus, researchers are interested in the thickness parameter of TWB. The research community considers the use of uniform thickness of TWB and employing variation in thickness of TWB to be one of the research areas of interest. On the other hand, manufacturing variable thickness TWB by modifying the parameters of radial axial rolling or using taper mandrel in tube drawing is now becoming more economical and easier (Arthington et al., 2020). Recent research on variable thickness square TWB has revealed that variable thickness TWB perform better than uniform thickness TWB (Zhang et al., 2016). At the same weight, the crashworthiness of variable thickness TWB is superior then uniform thickness TWB (Zhang et al., 2016; Prabhaharan et al., 2022).

In order to understand the bending collapse performance of TWB, various analytical, experimental, and numerical studies have been performed. Watson and Cronin (2011), Kecman (1983), and Kotelko et al. (2002) were developed an analytical model of bending collapse. Analytical models and experimental investigations were used to develop hypotheses such as static and moving plastic lines, localised hinge lines, yield line movement, and so on. TWB with variable cross sections were investigated to find a more advantageous cross section for improving crashworthiness by Liu (2010); Tang et al. (2016); and Shojaeefard (2014). Performing the studies on circular, square, rectangular, hexagonal, hat section, elliptical, polygonal, etc. observed that the bending collapse behaviour of all these cross- section TWBs is similar, the crash forces reach their initial peak value, then they start to drop and then fluctuate as plastic hinges are being formed. Among these cross-sections, circular tubes have gotten a lot of interest since they are easy to construct and have a steady progressive collapse mode for energy absorption (Poonaya et al., 2009; Tang et al., 2016). On the other hand, various hybrid structures are used by researchers to enhance the crashworthiness of TWB, Ghadianlou and Abdullah (2013) used different rib arrangements in TWB. To optimise the shape of the reinforcement rib, Zhang et al. (2019) used a multi-objective optimisation approach. Lee et al. (2017b) used form-filled material to strengthen the TWB. Li et al. (2015) employ a functionally graded thickness tube (FGT). The FGT tube absorbs more energy than the uniform thickness tube, according to the experimental and numerical data. A bending collapse behaviour of variable thickness square TWB was explored by Zhang el al. (2016). The crash force and energy absorption function were evaluated using surrogate modelling. Quasi-static simulations were performed to analyse square tubes having different flanges and webs. The simulation results were validated with an experimental three point bending test. Using multi-objective optimisation findings, it is possible to improve the crashworthiness of TWB by lowering the thickness of webs and increasing the thickness of flanges. Huang et al. (2019) used multi-cell aluminium/CFRP hybrid tubes to enhance crashworthiness. Under axial loading, variable thickness circular TWB was explored by Gao et al. (2017) and it was concluded that compared to the uniform thickness tube, the variable thickness tube performed better for crashworthiness.

Considering recent investigations in detail, it has been observed that, variable thickness cross-section TWB under bending collapse is less explored. The bending collapse behaviour of variable thickness TWB is investigated in this work. The following is the structure of this research: First, material properties are evaluated by a tensile test as per ASTM E8M-04. The uniformed thickness TWB is used for validating and comparing crashworthiness indicators in a subsequent study. On ABAQUS quasi-statics software, finite element numerical simulations are conducted, and crashworthiness indicators such as SEA and CFE are evaluated. Three-point experimental tests were performed to validate the finite element simulation model, and crashworthiness indications from experimental and numerical simulations have been compared. The same numerical model is used in the second part of this research. In the second phase, the crashworthiness of variable thickness tubes is examined. A surrogate model is created by analysing sample points of VT. Then, variations in SEA and CFE along the wall thicknesses are investigated. This study is advanced by addressing a multi objective optimisation problem with the non-dominated sorting genetic algorithm (NSGA-II) to maximise the SEA and CFE.

Experimental and numerical testing of uniform thickness tube

The VT TWB is studied using a numerical simulation model. Therefore, before beginning the investigation, the numerical simulation setup must be validated. In this section, uniform thickness TWB was used for experimental testing and validate the simulation model.

A. Material TESTINg

The uniform thickness TWB is taken for experimental study. The aluminium alloy AA1100-O material is widely used for TWB, so same material is taken in this work. To evaluate material properties for further study, a tensile test is performed on a universal testing machine (UTM), and a stress-strain relationship is obtained. The dimensions of the specimens are taken from ASTM E8M-04 and are depicted in Figure 92.1. The tensile test is performed at room temperature. Figure 92.2 shows the engineering stress-strain curve generated from the test and material properties are tabulated in Table 92.2.

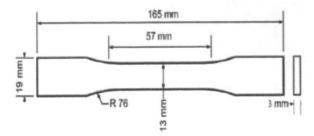

Figure 92.1 Tensile test specimen dimensions (ASTM E8M-04)

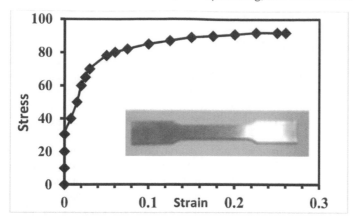

Figure 92.2 Stress-strain curve of AA1100-O

Table 92.1 Material properties

Sr. no.	Material Properties	Values
1	Young's modulus E	67.5 GPa
2	Initial yield stress σ_y	30 MPa
3	Ultimate tensile stress σ_u	90 MPa
4	Poisson's ratio ν	0.33
5	Density	1.1 kg/m3

B. Experimental setup

The numerical simulation setup is validated with experimental testing before beginning the experiments on variable thickness TWB. As uniform thickness circular cross-section TWB is easily available so, it is employed for experimental testing and validation of simulation setup. A three-point bending test using quasi-statics loading was carried out on the UTM platform. Figure 92.3 (a) and (b) illustrate the experimental setup and schematics, respectively. The outer diameter and thickness of UT TWB are 40 and 3 mm, respectively. The length L of the TWB is 400 mm, and it is supported by a cylindrical support with a diameter of 50 mm and a span D of 300 mm. A cylindrical punch with a diameter of 50 mm is moved vertically downward at a speed of 0.5 mm/second in the middle of the beam.

Figure 92.3 (a) Experimental setup of three point bending test

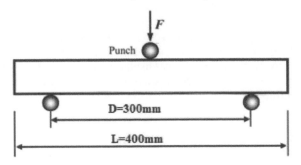

Figure 92.3 (b) Schematics of experimental setup

Punch displacement was taken 100mm to obtained information about specific energy absorption and crash force efficiency.

C. Numerical simulation setup

To simulate the three point bending test, quasi-static numerical simulation is performed in the ABAQUS quasi-static explicit dynamic module. The beam is considered a solid body, whereas the punch and supports are treated as rigid bodies. A single direction translation displacement is permitted for the punch, however all degrees of freedom for the supports are considered fixed. A coefficient of friction of 0.2 is considered at all surface-to-surface contacts between the punch, beam, and support. In ABAQUS, the 'All with self' surface friction pair is used to avoid interference of surfaces during beam bending. A four-node shell continuum (S4R) elements with five integration points were employed for all simulations. The punch has a 100 mm displacement. The finite element model is shown in Figure 92.4. Due to the fact that mesh size has an effect on the accuracy of simulation results, mesh convergence analysis is conducted to understand the optimal mesh size. The mesh size of the circular TWB was adjusted from 6 to 2 mm, and the CFE and SEA parameters were examined. Figure 92.5 depicts the findings of the mesh conversion study. The difference in SEA and CFE with mesh sizes of 3 and 2 is around 2%, so the mesh size of the beam has been set at 2 mm, and the mesh size of the punch and support has been set at 5 mm.

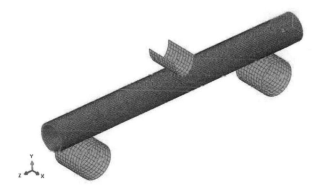

Figure 92.4 Numerical simulation model

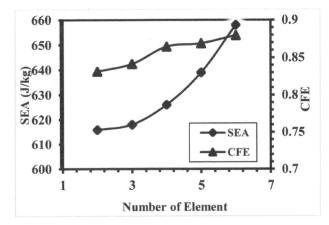

Figure 92.5 Mesh conversion analysis

D. Comparison of experimental and simulations results

This section compares the deformation patterns and force–displacement responses of UT TWB in simulation and experimental testing. Figure 92.6 (a) and (b) illustrate the curricular beam's deformation pattern during experimental and simulation testing, respectively. While, the force-displacement curves for experimental and simulation testing are depicted in Figure 92.7. In both figures exhibit a significant degree of correlation between experimental and numerical results.

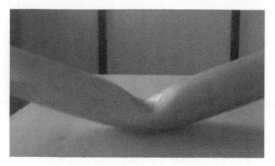

Figure 92.6 (a) Experimental test specimen

Figure 92.6 (b) Numerical test specimen

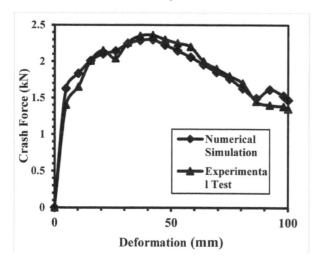

Figure 92.7 Force-displacement diagram obtained from experimental and simulation test

Generally, multiple indications have been utilised to determine a crashworthiness. Specific energy absorption (SEA) and crash force efficiency (CFE) are critical antecedents for evaluating crashworthiness performance while considering the multi-objective optimisation problem. They are calculated with the use of equations 1, 2, 3, and 4.

$$EA = \int_0^d F(x).dx \tag{1}$$

$$SEA = \frac{EA}{Mass\ of\ beam} \tag{2}$$

$$F_{avg} = \frac{EA}{d} \tag{3}$$

$$CFE = \frac{F_{avg}}{F_{max}} \tag{4}$$

Where d indicates beam deformation, F(x) denotes crash force, and Fmax represents the maximum crash force experienced during the bending test. In Figure 92.8, the SEA and CFE indicators of uniformed thickness TWB measured from experimental and simulated tests are compared.

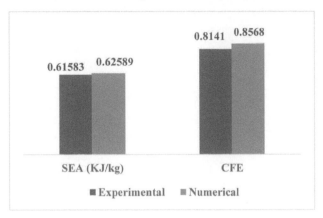

Figure 92.8 Crashworthiness indicators numerical vs experimental test

As seen in Figures 92.6 and 92.7, the deformation pattern and force-displacement curve of the experimental and simulation setups are strikingly comparable. Whereas the crashworthiness indication, which is based on experimental and simulation data, has a less than 5% errors, as seen in Figure 92.8. As a result, this simulation system is being considered for further research.

Multi objective optimisation analysis of vt twb

A uniform thickness circular TWB was experimentally and numerically explored in the previous section. An energy absorbent structure was designed to be able to carry in more energy with the least amount of material or weight. The utilisation of circular TWB with an efficient thickness distribution along their circumference is the most effective technique for reducing material usage. This type of TWB is known as 'varying thickness TWB.' In a subsequent investigation, the multi objective optimisation approach was used to determine the effective thickness distribution variable thickness TWB. The SEA and CFE key indicators are considered for the optimisation objective function.

A. Optimisation Definition

To create a variable thickness (VT) circular beam, the outside diameter of a circular TWB is retained, and the interior form is made elliptical. Based on geometric constraints, wall thickness ranges from 3mm to 0.5mm. In Figure 92.9(a), t_1 is 3 mm along the horizontal axis of the beam, and t_2 is 0.5mm along the vertical axis of the beam. Similarly, the thicknesses t_1 and t_2 in Figure 92.9(b) are 0.5 mm and 3 mm, respectively.

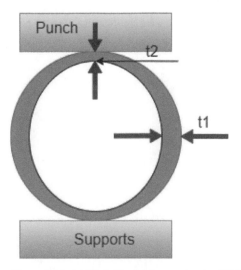

Figure 92.9 (a) Variable thickness circular TWB 1

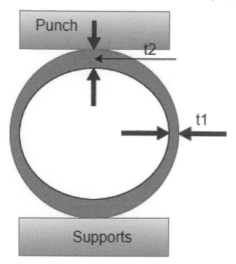

Figure 92.9 (b) Variable thickness circular TWB 2

Accordingly, in an optimisation problem, there are two thickness variables, t_1 and t_2, as illustrated in Figure 92.9. The multi objective optimisation problem for the VT circular TWB in this study may be formulated as follows, taking into consideration the two different design criteria:

$$\text{Maximum SEA} = \frac{EA}{Mass\ of\ beam} \tag{5}$$

$$\text{Maximum CFE} = \frac{F_{avg}}{F_{max}} \tag{6}$$

$$\text{Subjected to } 0.5mm < t_1, t_2 < 3mm \tag{7}$$

B. Surrogate model

The surrogate approach is used in this work to address the crashworthiness optimisation problem because of its properties such as effective higher non-linear problem, high efficiency, and accuracy. The response surface method (RSM) (Alaswad et al., 2011) is one of the most successful and efficient ways for solving such problems in surrogate modelling. RSM was widely employed in a wide range of engineering applications, including crashworthiness. According to the literature, central composite design-based DOE was frequently employed for two variable problems. As a result, the Response Surface Methodology (RSM) with the central composite design is used in the current research optimisation problem.

A RSM optimisation process outlined as follows: First, the RSM sampling is done with a central composite experimental design because it is powerful when it is used with a full quadratic polynomial model. The central composite design with two design variables necessitates a total of nineexperimental points, as illustrated in Figure 92.10. The sampling points are listed in Table 92.2.

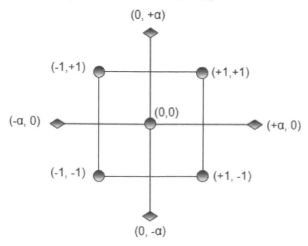

Figure 92.9 Central composite experimental design for sample points

Table 92.2 Sample points of center composite method

Exp. No.	Sample point	t_1	t_2
1	(0,0)	1.75	1.75
2	(0,-α)	1.75	0.5
3	(+1 -1)	2.634	0.866
4	(+α,0)	3	1.75
5	(+1,+1)	2.634	2.634
6	(0,+ α)	1.75	3
7	(-1,+1)	0.866	2.634
8	(-α,0)	0.5	1.75
9	(-1,-1)	0.866	0.866

Furthermore, the numerical analyses are carried out at the selected points and the SEA and CFE values of all nine experiments are evaluated. The geometry and numerical parameters are the same as described in section I (C). A full set of quadratic polynomial functions is chosen as the basic functions for the construction of an optimisation function. The variables of these basic functions are 1, t_1, t_2, t_{12}, t_1 t_2, t_{22}. By least-squares data fitting, the approximate response surface function of SEA and CFE is expressed as in Equations 8 and 9.

$$SEA = 2.9159 + 340.5521\, t_1 + 28.7345\, t_2 - 47.0323\, t_1^2 - 4.1516\, t_1 t_2 - 4.0848\, t_2^2 \tag{8}$$

$$CFE = 0.6104 + 0.1911\, t_1 - 0.0287\, t_2 - 0.0411\, t_1^2 + 0.0218\, t_1 t^2 - 0.0122\, t^2 \tag{9}$$

Table 92.3 the shows the SEA and CFE values estimated from numerical simulation and surrogate model (Equations 8 and 9).

Table 92.3 Numerical and surrogate model results

Exp.no	SEA (J/kg)		CFE	
	Numerical Simulation	*Least sq. techq*	*Numerical Simulation*	*Least sq. techq*
1	476.08	479.90	0.797	0.798
2	452.14	464.55	0.837	0.820
3	584.37	585.97	0.837	0.844
4	636.58	617.26	0.840	0.840
5	563.56	592.16	0.818	0.8196
6	509.76	482.49	0.748	0.737
7	286.84	300.43	0.608	0.634
8	192.45	195.57	0.663	0.627
9	298.03	281.26	0.704	0.727

C. Accuracy of surrogate mode

It is necessary to verify the accuracy of the surrogate model before employing it in further optimisation. The accuracy of the surrogate model is evaluated by computing fitting errors such as R Square (R2), relative average absolute error (RAAE), and root mean square error (RMSE) (Shan and Wang, 2010). The following formulae are used to estimate these fitting errors.

$$R^2 = 1 - \frac{\sum_{i=1}^{m}(y_i - \hat{y}_i)^2}{\sum_{i=1}^{m}(y_i - \bar{y}_i)^2} \tag{10}$$

$$RAAE = 1 - \frac{\sum_{i=1}^{m}|y_i - \hat{y}_i|}{\sum_{i=1}^{m}|y_i - \bar{y}_i|} \tag{11}$$

$$RMAE = \max \frac{\{|y_i - \hat{y}_i|, \dots, |y_m - \hat{y}_i|\}}{\sum_{i=1}^{m}(|y_i - \bar{y}_i|/m)} \tag{12}$$

where n is the number of test samples, $\bar{y}$ is the average value of the responses, $\hat{y}i$ is the value of responses obtained from surrogate model, yi is the actual measure value of the responses obtained from numerical simulation, and (Shan and Wang, 2010). The fitting error evaluated from above equations are tabled in Table 92.4.

Table 92.4 Fitting errors

Response	Errors		
	R2	RAAE	RMSE
SEA	0.9875575	0.09973006	15.683
CFE	0.9685022	0.16475667	0.01479984

To get a better level of accuracy, the R^2 score should be close to 1, however the RAAE score should not be as close to 1. RMSE is widely used to evaluate the performance of the surrogate model. It has the same units as the response variable; hence, a lower value of RMSE implies higher accuracy (Shan and Wang, 2010). Based on the results of Table 92.4, it can be concluded that the evaluated surrogate model is accurate and can be used for additional optimisation studies.

Result and discussion

Finding a VT TWB that has the highest SEA and CFE indicators is the goal of this work. In this such TWB is investigated as well as insight on the influence of variable thickness on SFE and CFE.

A.　Effect of VT on SEA

The influence of VT on SEA is investigated using the surrogate model that was developed in the preceding sections. The SEA Equation 8 is set objective function and the optimisation problem is solve for maximisation. The t_1 and t_2 are the independent parameter those value with values ranging from 3 mm to 0.5 mm. MATLAB's constrain nonlinear maximisation algorithm is employed to solve this optimisation problem. The relationship between the independent variables (t_1 and t_2) and the dependent variable (SEA) can be understood by the surface plot shown in Figure 92.11.

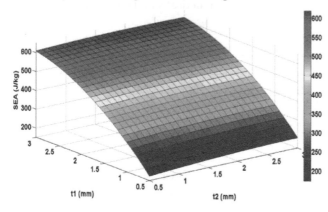

Figure 92.11 SEA surface contour

The SEA value is found to be greatest at t_1 of 3 mm and t_2 of 2 mm. The results are checked on the numerical simulation model as per the process described in section II. The uniform thickness and variable thickness, as well as surrogate and numerical simulation SEA values, are all compared in Table 92.5.

Table 92.5 SEA comparison

Type of TWB	t_1 (mm)	t_2 (mm)	SEA (J/kg)	
			Numerical simulation	Surrogate model
UT	3	3	615.85	613.357
VT	3	2	623.56	617.502

The difference between simulation and optimisation solutions is around 1%. According to the optimisation results, adding material along the horizontal axis of TWB while removing it along the vertical axis results in the highest SEA. When comparing the SEA values of uniform thickness TWB and variable thickness TWB, the SEA of variable thickness TWB is 1.23 % higher, and the material savings in variable thickness TWB is 15.13%.

B. Effect of VT on CFE

Likewise, as indicated in the prior section, the influence of VT on CFE is being studied. Equation 9 of CFE sets the objective function of the maximisation problem. Other optimisation processes and parameters are considered in the same way as in the SEA problem. The relationship between the independent variables (t_1 and t_2) and the dependent variable (CFE) is represented in the surface plot shown in Figure 92.12.

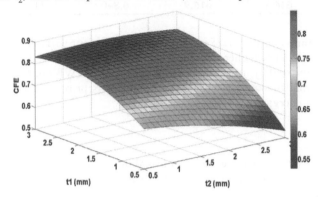

Figure 92.12 CFE contour

At t_1 2.638 mm and t_2 1.181 mm, the CFE value is at its highest. The results are checked against the numerical simulation model as per the process described in section 2.

It is found that at t_1 2.638 mm and t_2 1.181 mm the CFE value is maximum. The result are check on numerical simulation model as per the process describe in the section 2. The CFE values of uniform and variable thickness TWB, as well as CFE value surrogate and numerical simulation, are all compared in Table 92.6.

Table 92.6 CFE comparison

Type of TWB	t_1(mm)	t_2(mm)	CFE	
			Numerical Simulation	Surrogate model
UT	3	3	0.8315	0.8141
VT	3	1.181	0.863	0.8455

The table shows that increasing material on the horizontal axis of TWB while decreasing it on the vertical axis results in the maximum CFE. The CEF of variable thickness is 3.71% higher than uniform thickness TWB with 34.06% material saving.

C. Multiobjective optimisation for maximum SEA and CFE

To find an efficient variable thickness TWB with the highest SEA and CFE indicators. The Non dominated Sorting Genetic Algorithm (NSGA-II) Shan and Wang (2010) is used to search for the Pareto fronts of multiobjective optimasation problem. The SEA and CFE surrogate models are optimised by NSGA-II. Figure 92.13 shows the Pareto fronts. There are 100 iterations in the NSGA-II.

The solution of the multiobjective optimisation problem shows that maximum SEA and CFE are found at thicknesses of t_1 is 3mm and t_2 is 1.75mm Similar to the preceding section, numerical simulations are used to verify the results of optimisation. The uniform thickness and variable thickness, as well as surrogate and numerical simulation SEA and CFE indicator values, are all compared in Table 92.7.

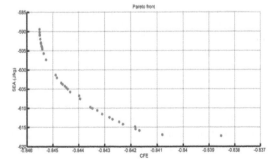

Figure 92.13 Pareto fronts of multiobjective problem

Table 92.7 Multiobjective optimisation results comparison

Type of TWB	t_1 (mm)	t_2 (mm)	SEA (J/kg)		CFE	
			Num. meth	Opti. meth	Num. meth	Opti. meth
UT	3	3	615.8	613.3	0.831	0.814
VT	3	1.75	636.5	616.9	0.840	0.840

From the table, it can be concluded that t_1 at 3mm and t_2 at 1.75mm are the most efficient variable thickness TWBs. which gives 3.36% higher SEA and 1.11% higher CFE and a uniform thickness TWB with a 19.14% material saving. Crashworthiness is improved by increasing the thickness of the TWB's horizontal axis with. TWB's deformation pattern shows that thickness along the horizontal axis provides the most resistance to deformation and absorbs the most energy.

Conclusion

In this work, variable thickness circular TWB is introduced over the traditional uniform thickness circular TWB, and the crashworthiness performance of VT circular TWB is numerically investigated. The following are summaries of the key works and conclusions:

- An experimental three-point bending test was performed on uniform-thickness TWB to validate the numerical simulation setup. The experiment and the numerical results are close to one another.
- A surrogate model for the SEA and CFE crashworthiness indicators has been developed using a response surface technique.
- From the optimised design, it is observed that there is a 1.23% increment in SEA and a 15.13% material saving by using variable thickness TWB with t_1 of 3mm and t_2 of 2mm.
- Similarly, a variable thickness TWB with t_1 2.68 mm and t_2 1.181 mm CFE can be increased by 3.71% and material can be reduced by 34.06%.
- Finally, in order to improve the crashworthiness indicators SEA and CFE, a multiobjective optimisation has been performed. The results show that variable thickness TWB having t_1 of 3mm and t_2 1.75mm have 3.36% higher SEA and 1.11% higher CFE than uniform thickness TWB.
- According to the optimisation solutions, adding more material on the horizontal axis made the TWB more crash-resistant, which is what happened.
- Hence, variable thickness circular TWB is superior in crashworthiness performance than uniform thickness circular TWB.

References

Alaswad, A., Benyounis, K., and Olabi, A. (2011). Employment of finite element analysis and response surface methodology to investigate the geometrical factors in T-type bi-layered tube hydroforming. Adv. Eng. Softw. 42(11):917–926.

Aljibori, H. S. S., Al-Qrimli, H. F., Ramli, R., Mahdi, E., Tarlochan, F., and Chong, W. P. (2010). A comparative analysis of experimental and numerical investigations of composite, tubes under axial and lateral loading. Aust. J. Basic Appl. Sci. 4(8):3077–3085.

Arthington, M. R., Havinga, J., and Duncan, S. R. (2020). Control of ring rolling with variable thickness and curvature. Int. J. Mater. Forming 13:161–175.

Gao, D., Zhang, N., and Feng, J. (2017). Multi-objective optimisation of crashworthiness for mini-bus body structures. Adv. Mech. Eng. https://doi.dox.org/10.1177/1687814017711854.

Ghadianlou, A. and Abdullah, S. B. (2013). Crashworthiness design of vehicle side door beams under low-speed pole side impacts. Thin-Walled Struct. 67:25–33.

Huang, Z., Zhang, X., and Yang, C. (2019). Experimental and numerical studies on the bending collapse of multi-cell Aluminum/CFRP hybrid tubes. Composites Part B. doi: https://doi.org/10.1016/j.compositesb.2019.107527.

Kausalyah, V., Shasthri, S., Abdullah, K. A., Idres, M. M., Shah, Q. H., and Wong, S. V. (2015). Vehicle profile optimisation using central composite design for pedestrian injury mitigation. Appl. Math. Inf. Sci. 9(1):197–204.

Kecman, D. (1983). Bending collapse of rectangular and square section tubes. Int. J. Mech. Sci. 25(9):623–636.

Kotelko, M., Lim, T. H., and Rhodes, J. (2002). Post-failure behaviour of box section beams under pure bending (an experimental study). Thin-Walled Struct. 38(2).

Lee, H., Huh, M., Kang, S., and Yun, S.-Il. (2017a). Design optimisation of thin-walled circular tubular structures with graded thickness under later impact loading. Int. J. Automot. Technol. 18(3):439–449.

Lee, H., Huh, M., Kang, S., and Yun, S. (2017b). Compressive behavior of automotive side impact beam with continuous glass fiber reinforced thermoplastics incorporating long fiber thermoplastics ribs. Fibers Polym. 18(8):1609–1613.

Li, G., Xu, F., Sun, G., and Li, Q. (2015). Crashworthiness study on functionally graded thin walled structures, Int. J. Crashworthiness. 280300. https://doi.org/10.1080/13588265.2015.1010396

Liu, Y. C. (2010). Thin-walled curved hexagonal beams in crashes FEA and design. Int. J. Crashworthiness 15(2):151e9.

Long, C. R., Yuena, S. C. K., and Nuricka, G. N. (2019). Analysis of a car door subjected to side pole impact. Lat. Am. J. Solids Struct. 16(8):e226.

Luzon-Narroa, J., Arregui-Dalmasesb, C., Hernando, L. M., Core, E., Narbona, A., and Selgasg, C. (2014). Innovative passive and active countermeasures for near side crash safety. Int. J. Crashworthiness 19(3):209–221.

Poonaya, S., Teeboonma, U., and Thinvongpituk, C. (2009). Plastic collapse analysis of thinwalled circular tubes subjected to bending. Thin-Walled Struct. 47(6):637–645.

Prabhaharan S. A., Balaji, G., and Annamalai, K. (2022). Numerical simulation of crashworthiness parameters for design optimisation of an automotive crash-box. Int. J. Simul. Multidisci. Des. Optim. 13:3.

Qin, R., Zhou, J., and Chen, B. (2019). Crashworthiness design and multiobjective optimisation for hexagon honeycomb structure with functionally graded thickness. Adv. Mater. Sci. Eng. https://doi.org/10.1155/2019/8938696

Shaharuzaman, M. A., Sapuan, S. M., Mansor, M. R., and Zuhri, M. Y. M. (2018). Passenger car's side door impact beam: A review. J. Eng. Technol. 9(1).

Shan, S. and Wang, G. G. (2010). Survey of modeling and optimisation strategies to solve high-dimensional design problems with computationally-expensive black-box functions. Struct. Multidisc. Optim. 41:219–241.

Shojaeefard, M. H. (2014). Experimental and numerical crashworthiness investigation of combined circular and square sections. J. Mech. Sci. Technol. 28(3):999–1006.

Tang, T., Zhang, W., Yin, H., and Wang, H. (2016). Crushing analysis of thin-walled beams with various section geometries under lateral impact. Thin-Walled Struct. 102:43–57.

Tarlochan, F., Samer, F., Hamouda, A. M. S., Ramesh, S., and Khalid, K. (2013). Design of thin wall structures for energy absorption applications: Enhancement of crashworthiness due to axial and oblique impact forces. Thin-Walled Struct. 71:7–17.

Watson, B. and Cronin, D. S. (2011). Side impact occupant response with varying positions. Int. J. Crashworthiness 16(5):569–582.

Yang, J.-H. (2011). Optimisation of the aluminum door impact beam considering the side door strength and the side impact capability. J. Korea Acad.-Ind. Cooperation Soc. 12(5):2025–2030.

Zhang, G., Yu, F., OuYang, Z., and Chen, H. (2015). Study on vehicle collision predicting using vehicle acceleration and angular velocity of brake pedal. SAE Technical Paper 2015-01-1405, 2015.

Zhang, X., Zhang, H., and Wang, Z. (2016). Bending collapse of square tubes with variable thickness. Int. J. Mech. Sci. 106:107–116.

Zhang, Z., Liu, S., and Tang, Z. (2019). Design optimisation of cross-sectional configuration of rib-reinforced thin-walled beam. Thin-Walled Struct. 47:868–878.

93 A comprehensive study on powder catchment efficiency in direct metal deposition

Pratheesh Kumar S.[a], Naveen Anthuvan R.[c], Dinesh R.[b], and Mukesh K.[d]

Department of Production Engineering, PSG College of Technology Coimbatore, India

Abstract

The term 'additive manufacturing' (AM) refers to a wide range of processes. One such process is direct metal deposition (DMD), which produces parts through a layer-by-layer fusion method. This article discusses the efficiency of powder catchment on parts manufactured using various powder catchment processes (PCP). The variation in powder catchment efficiency (PCE) and other operation parameters has been investigated for a variety of materials. This study aids in the determination of appropriate process parameters for the material chosen to achieve the highest possible PCE. The obtained results indicate that magnetic-assisted direct metal deposition outperforms PCP. The laser power, scanning speed, and powder flow rate (PFR) were determined to be the process parameters that influenced the outcome. Materials used in a variety of industries could benefit from this study's findings, which could identify or predict the best catchment process and parameters.

Keywords: Additive manufacturing, direct metal deposition, powder catchment efficiency, powder catchment processes, surface roughness.

Introduction

Additive manufacturing (AM) is a term for a three-dimensional printing process that is used to create complex-shaped products (Kumar et al., 2021b). It is possible to manufacture components with intricate internal channels and honeycomb structures. AM has a subtype called direct metal deposition (DMD). The material feeding system is available in two configurations: powder and wire. powder catchment efficiency (PCE) continues to be a limiting factor in DMD process efficiency when it comes to powder catchment processes (PCP). Prior research indicated that with proper control of PCE, it would be easier to process expensive materials and complex shape products (Gibson et al., 2009). This paper aims to conduct a comparative analysis of PCE and the quality of products manufactured using various types of PCP.

A. Direct metal deposition

DMD is a method of metal fusion that entails the formation of a melt pool on top of an existing surface and the injection of metal powder. Metal addition occurs layer by layer, resulting in the complex-shaped product depicted in Figure 93.1 (Gibson et al., 2009). It is divided into five stages, namely process planning, input computer-aided design (CAD), path planning, path generation, and implementation .

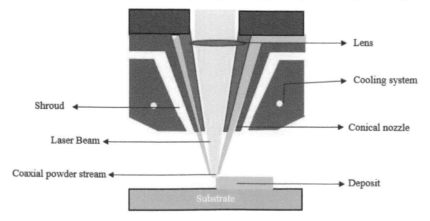

Figure 93.1 Direct metal deposition process

[a]spratheeshkumarth@gmail.com; [b]naveenanthuvan@gmail.com; [c]dineshrjrj@gmail.com; [d]mukeshkumaravelan@gmail.com

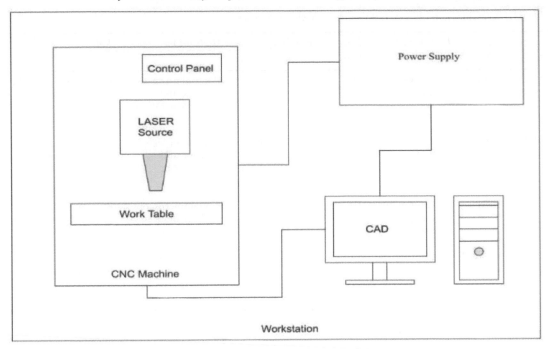

Figure 93.2 Process equipment of DMD

Process equipment

DMD is composed of many interconnected subsystems. Throughout the process, these subsystems maintain complete control of the laser's power, speed, spot position, PFR, and molten pool size (Ghosal et al., 2018). The subsystems are depicted in Figure 93.2 as a control workstation, a laser source, and multi-axis computer numerical control (CNC) systems.

A. Workstation

The term 'workstation' refers to an area where the process is effectively carried out. Workstations should be designed with ergonomics, safety, and operational efficiency in mind. It refers to the personal computer (PC). To reach the highest level of human-machine integration, quantitative data on these human traits must be linked with data on DMD machine features. The theory and data from the DMD process can be utilised to create models and simulations that provide precise information about the workplace and people (Chua et al., 2017). This machine must be closely monitored at all times. It should be used in a clean, noise-free environment. Regular checks must be included in the maintenance schedule and must occur at regular intervals, as the majority of operations are performed with little human-machine interaction (Del Rio Vilas et al., 2013).

B. Laser source

A laser is created a light source that has been enhanced using a technique known as stimulated emission. The atoms within the lasing medium were excited by a beam of light generated by a light source. The lasing media can be solid, liquid, or gaseous. Depending on the application, the type of wave used could be continuous or pulsed (Kumar et al., 2021c). The laser would have been used for cutting as well as joining operations. To melt the metal substrate in DMD, a laser is required. The laser produces a highly concentrated and collimated beam of energy, which is why it is used. The laser beam is focused on an optimal spot size after passing through the lens. They can be applied in a controlled manner when directional mirrors are used. In DMD, two types of laser processing are used: curing and heating. Lasers are critical in the production of distinctive parts. Table 93.1 (Ghosal et al. 2018; Chua et al., 2017; Del Rio Vilas et al., 2013; Kumar et al., 2021c; Rawlins et al., 2016; Pinkerton, 2010; Lee et al., 2017) lists the type, wavelength, power, efficiency, and uses of lasers.

Table 93.1 Classification of laser and their specific characteristics

Type of Laser	Wavelength (nm)	Operation Source	Power (kW)	Efficiency (%)	Application	Ref.
CO_2	1064	Continuous and pulsed wave	20	5 to 20	Material processing, medicine and isotope separation	Ghosal et al. (2018), Rawlins et al. (2016), Pinkerton (2010)
Nd-YAG	1064	Continuous and pulsed wave	16	1 to 3 (Lamp diode) and 10 to 20 (Diode pump)	Material processing, medicine and measurement	Rawlins et al. (2016), Pinkerton (2010)
YB fibre	1078	Continuous and pulsed wave	10	10 to 30	Photon microscopy and heat-induced sample degradation	Rawlins et al. (2016), Lee et al. (2017)
Excimer	193, 248 and 308	Continuous and pulsed wave	1 to 100000	1 to 4	Micro-machining, medicine and laser chemistry	Ghosal et al. (2018), Rawlins et al. (2016), Lee et al. (2017)

Powder catchment efficiency

PCE is calculated as the tracks consolidated mass dividedbythetotalmassdeliveredasshowninEquation1.

According to a recent study, between 40% and 65% of PCE is considered to be significantly good (Kumar et al., 2021d). The particle-substrate bonding conditions dictate the catchment process (Partes, 2009). The various phase interactions include solid- solid surface interactions, solid particle-liquid surface interactions, liquid particle-solid surface interactions, and liquid particle-liquid surface interactions (Steen, 1986; Liu and Kovacevic, 2014). A ricochet occurs as a result of the solid-solid surface interaction. The catchment occurs at the solid particle-liquid and liquid particle-liquid interfaces. Similarly, a liquid particle-solid surface also results in the catchment, but the structure is rapid quench (Eisenbarth et al., 2019). Table 93.2 illustrates the relationship between PCE and various parameters.

PCE = *Consolidated mass track / Total mass delivered .... (1)*

Table 93.2 Relationship between PCE and various process parameters

Relationship of PCE	Description	Ref.
	The greater the laser power, the greater the PCE. This shows that laser power has a positive impact on PCE.	Liu and Kovacevic (2014)
	Increasing the stand-off distance until it reached the centre value improved the PCE. Then, the PCE descended as the stand-off distance was continuously increased.	Liu and Kovacevic (2014)
	The graph exhibits that carrier gas flow rate (CGFR) has a negative impact. As CGFR decreases, PCE also decreases.	Liu and Kovacevic (2014)

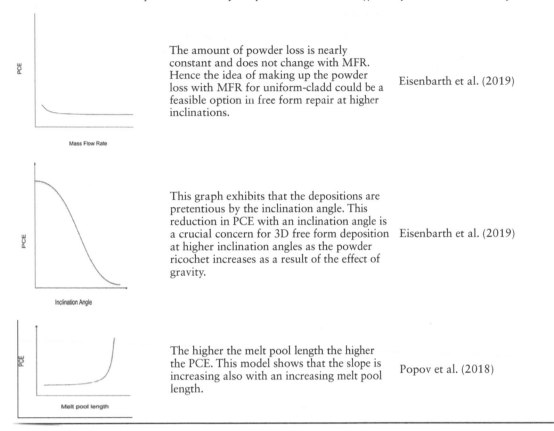

The amount of powder loss is nearly constant and does not change with MFR. Hence the idea of making up the powder loss with MFR for uniform-cladd could be a feasible option in free form repair at higher inclinations.

Eisenbarth et al. (2019)

This graph exhibits that the depositions are pretentious by the inclination angle. This reduction in PCE with an inclination angle is a crucial concern for 3D free form deposition at higher inclination angles as the powder ricochet increases as a result of the effect of gravity.

Eisenbarth et al. (2019)

The higher the melt pool length the higher the PCE. This model shows that the slope is increasing also with an increasing melt pool length.

Popov et al. (2018)

Primary contributing process parameters

Several coating parameters must be changed when using the freshly developed laser powder deposition process. The first collection of parameters has to do with the laser beam and includes things like beam length and separation, laser beam power, laser beam distribution power, and beam area size. The interaction of the laser beam with the powder is the subject of the second set of parameters. The powder coating manager's dividing boundaries, laser beam geometry, radiation distribution, and powder particle method are all included, as are assisting gases, powdered particles, and gaseous gases, as well as powder flow rate, environment, composition, powder particle size, environment, speed, and flow rate. The final set of parameters depicts the interaction of the laser beam, powder and gas jets, melting pool, and substrate at the substrate's surface (Kumar et al., 2021a). The various process parameters are shown in Figure 93.3.

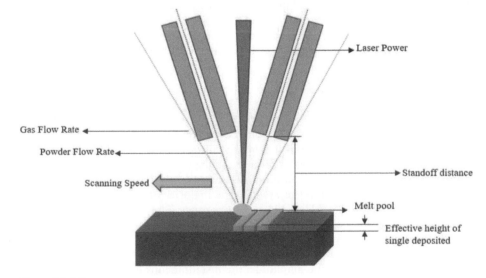

Figure 93.3 Process parameters in DMD

A. Radiation of Wavelength and Polarisation

Radiation is the emission or transfer of energy in the form of waves or particles between two unconnected bodies separated by an aerated or vacuum space. This electromagnetic phenomenon is caused by a temperature difference (Gong et al., 2020). Radiant heat is the technical term for this type of energy transfer. Even though all bodies emit radiant heat at the speed of light, it's simple to dismiss it at low temperatures. However, as the temperature rises, it becomes increasingly relevant. The nanometer (nm) is the wavelength unit.

B. Powder flow rate

The PFR indicates the amount of material leaving the delivery nozzle. It is intimately tied to the nozzle's geometry. The amount of powder transported to the laser spot region is controlled by the PFR parameters (Mahamood, 2017). Powder capture efficiency is linked to the flow rate of the powder; the higher the flow rate, the better the powder capture efficiency. The larger powder had a lower velocity due to inertia, resulting in reduced catchment efficiency. PFR is measured in grams per minute (g/min).

C. Wettability

When using coaxial nozzles, the particles are not supposed to pass through the high-intensity portions of the laser beam before reaching the melt pool. The melt and particle wetting mechanisms can then occur (Popov et al., 2018). The liquid and solid components of components play a significant role in the wetting process. Wetting is a process that involves liquid (l), solid (s), and gaseous (g) components in the surrounding environment. The wetting angle (θ) can be expressed in terms of surface energy (s_g), surface tension (l_g), and interfacial energy (l_s).

D. Temperature gradients in the liquid surface tension

The buoyant forces created by density differences in the liquid metal pool, as well as the Marangoni forces caused by temperature gradients across a melt pool were identified. The flow created by the latter, also known as thermo- capillary convection or surface-tension driven convection, has been examined and predicted in the welding process, laser surface melting, and surface alloying. In most circumstances, Marangoni forces determine melt pool movement, and the strength of the flow may be approximated using the Marangoni number, which is proportional to the reciprocal of the molten material's viscosity and thermal diffusivity. Kelvin (K) is the temperature unit.

E. Scanning speed

The time it takes for a laser beam to interact with the substrate and deposited materials is measured in scanning speed. The combination of translation speed (V) and powder feed rate determines the amount of powder deposited per unit length (F). The relationship (FV1) demonstrates a significant coefficient of determination (Dadbakhsh et al., 2010). When the translation speed of the laser is increased, the time required for laser-material interaction decreases. As a result, the proportion of powder reaching its melting point lowers, resulting in a shorter clad. The scanning speed is measured in millimetres per minute (mm/min).

Types of powder catchment processes

A. Ultrasonic assisted DMD

There are three pieces to the ultrasonic vibration fixture. One screw secures the substrate to the shaft; four pins function as feet to keep the plate on top from striking the ultrasonic cleaner's bottom as one plate lowers into the water to boost ultrasonic vibration amplitude (Zhang et al., 2019). Figure 93.4 depicts the experimental setup for vibration laser powder deposition. A laser and powder nozzle, as well as a CNC cutting tool and vibration table, are included in this system. The majority of ultrasonic equipment has a fixed vibration frequency. Between 11 and 50 kHz is the frequency range. The amount of energy consumed by different equipment varies depending on their use. A water-based cleaning machine uses between 100 and 200 W and runs at a frequency of roughly 40 kHz (Gorunov, 2020). To test the notion, the lab's cleaning equipment was used. The cleaning machine operates at a frequency of 42 kHz and has a maximum output power of 100W. Despite the little amplitude, the instantaneous energy is enormous (a few micromicrometres), and systallisation is one way to generate that energy. The sample produced with ultrasonic vibration had a finer microstructure than the sample produced with no vibration (Ning and Cong, 2020). The grain morphology of ultrasonic-deposited samples differs from that of non-vibrated samples.

Ultrasonic-deposited samples had a higher micro-hardness than non- vibrated samples. The characteristics of parts produced by ultrasonic assisted DMD of various materials are shown in Table 93.3 (Chen et al., 2012; Zhang et al., 2019; Ning and Cong, 2016; Alavi and Harimkar, 2015; Xia et al., 2018).

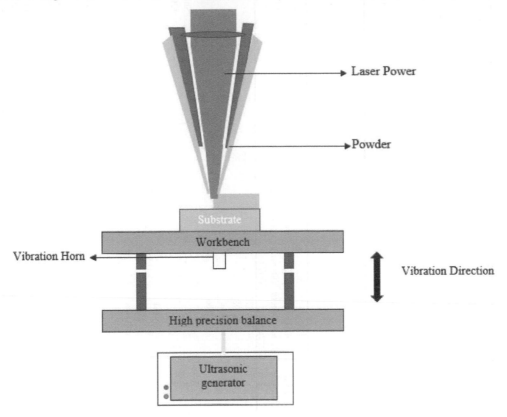

Figure 93.4 Ultrasonic assisted DMD

B. Magnetically aDMD

Increased efficiency could be achieved by managing powder catchment directly at the melt pools (Kiefer, 2004). This would be advantageous in terms of reducing flow irregularities between the nozzle tip and the melt pool, which can occur due to the laser's influence on the powder flow path in flight (Smith et al., 2021). Previous studies of magnetically assisted DMD and laser-DMD have been conducted, but Rather than regulating the powder trajectory via ferromagnetism, the work has focused on the impact of fields on the characteristics of paramagnetic materials., except for a feasibility study conducted by the authors. When large-area magnetic fields are applied, rebounding powder is attracted and magnetically preloaded to the latter stages of the deposition process before the nozzle touches it, according to this research (Smith et al., 2022). Magnetic fields induce a magnetic field around ferromagnetic materials (such as ferritic steel and nickel) when they interact with them. The magnetic forces generated by the two forces will act on these materials, and an unconstrained object will move toward the magnetic field source.

Magnetic fields are made using 12 readily available V solenoids. The frustum cone was put beneath the substrate, with the tip touching the substrate's bottom and the base touching the solenoid core, as illustrated in Figure 93.5. The diameter of the magnetic field travelling substrate was reduced from 24 mm to roughly 4 mm utilizing. The magnetic field intensity was significantly reduced when the cone tip's diameter was reduced. The magnetic field strength was controlled using pulse width modulation, and the magnetic field was automated using an Arduino Mega. The track dimensions were dynamically changed during operation using a programmable solenoid, demonstrating that magnetic assistance is available adaptive control technology. PCE can be increased by up to 25% with magnetic assistance aligned coaxially. The magnetic field strength at the substrate surface has a positive linear relationship with PCE. By attaching a solenoid to a precise place along the tool path, the cross-sectional area of tracks can be increased by up to 83% at defined spots as indicated in Table 93.4 (Smith et al., 2021; Smith et al., 2022; Chen et al., 2018; Lu et al., 2019). Even when complex tool paths are utilised, the magnetically improved catchment efficiency is maintained.

Table 93.3 PCE of various materials in ultrasonic assisted DMD

Material	No. of run	Type of laser	Scan speed (mm/min)	PFR (g/min)	Layer thickness (μm)	Laser power (W)	Ultrasonic frequency (kHz)	M (g/min)	Rand (in/min)	Grain number	Micro hardness (μm)	UTS (MPa)	YS (MPa)	Young's modulus (10^{-1} GPa)	Resolidified film thickness (μm)	Recast layer thickness (μm)	Ref.
316L	1 to 6	Coherent diode	-	-	-	750 - 1000	0 - 42	0.6 - 1	10	35 - 54	215.8 -228.9	-	-	-	-	-	Chen et al. (2012)
Al 4047	1 to 8	YLS	360	1.2	300	1100 - 1500	20	-	-	-	-	-	-	-	-	-	Zhang et al. (2019)
FeCr	-	-	500	4	-	350	-	-	-	-	-	2300	1000	300	-	-	Ning and Cong (2016)
Titanium alloy	1 to 3	CO_2	-	-	-	900	20	-	-	-	-	-	-	-	13 - 20	-	Alavi and Harimkar (2015)
GH 4037	-	Nd-YAG	-	-	-	0 - 610	-	-	-	-	-	-	-	-	-	0.8 - 5.2	Xia et al. (2018)

a. PFR – Powder flow rate; M – Powder feed rate; UTS – Ultimate tensile strength; YS – Yield strength

Table 93.4 PCE of various materials in magnetic assisted DMD

Material	Scanning speed (mm/min)	Type of laser	WaveLength (nm)	Laser spot size (mm)	Stand-off distance (mm)	Gas flow rate (l/min)	Powder flow rate (g/min)	Laser power (W)	Inclination angle (deg)	Track mass per unit length (g/min)	Magnetic field intensity (mT)	PCE (%)	Roughness S_a (μm)	MTW (mm)	Hardness(N/mm²)	Ref.
AISI 4340on mild steel substrate	500	Ytterbium doped fibre	1070	1	10	10	9.2 - 22.5	600	0	0.003 -0.008	0.5	38 - 42	10 - 27	-	-	Smith et al. (2021)
	200 -650	-	-	-	-	-	1.9 - 16.6	600 -1000	90	0.004 -0.016	0.5	22 - 38	7 - 13	-	-	
									Coaxially aligned to laser		0 - 32.2			1.45 - 3.1		
AISI 4340on SS 304substrate	500	Ytterbium doped fibre	1070	1	10	10	9.2 - 22.5	600	0	0.004 -0.012	0.5	22 - 23	12 - 14	-	-	
									90	0.006 -0.012	0.5	28 - 38	6 - 12	-	-	
AISI 4340 on nickel substrate	500	-	-	-	-	-	15.7	900	Coaxially aligned to laser	-	0 - 95	42 - 63	-	2.45 - 2.9	-	Smith et al. (2022)
Almg3	500	Disc laser	1030	200	18	300	4	16000	-	45	0 - 2	-	3.4 - 9.8	-	-	Chen et al. (2018)
316 SS	900	Quasi-continuous wave	-	-	-	10	4.3	-	-	-	62.5	-	-	-	183.3	Lu et al. (2019)

b.Sa–Arithmetic mean height; MTW–Mean track width; MCSA–Mean cross-sectional area

Table 93.5 PCE of various materials in electrostatic assisted DMD

Material	Infill speed (mm/s)	Infill overlap (mm)	Operating voltage (V)	Density (kg/m³)	Powder radii (mm)	DLT (mm)	Operating distance (mm)	Time (sec)	Layer thickness (mm)	Tensile modulus (GPa)	Equilibrium toughness (MJ/m³)	Flexural strength (MPa)	Flexural modulus (GPa)	Developing efficiency (%)	Transfer efficiency (%)	Ref.
Nylon 645 on PA6substrate	50	0.2	-	-	-	-	-	-	-	1.56 -2.94	8.999 - 25.04	33.93 -40.31	0.77 -1.17	-	-	Arigbabowo and Tate (2021)
18Ni300	-	-	12000	8.1	10	23.62	18	0 - 7	0 - 100	-	-	-	-	25	-	Tsao and Chang (2018)
PA66	-	-	1000 -2000	-	-	-	-	-	-	-	-	-	-	10 - 45	3 - 40	Barletta et al. (2006)

c. DLT – Dielectric layer thickness

Table 93.6 PCE of various materials in vacuum assisted DMD

Material	Scanning speed (mm/min)	Type of laser	Wavelength (nm)	Laser spot size (mm)	Hatching space (mm)	PFR (g/min)	Laser power (W)	Layer thickness (mm)	Chamber pressure (Pa)	VED (%)	Roughness S_a (μm)	Hardness (HV)	UTS (MPa)	YS (MPa)	Ref.
TiN on SS 316Lsubstrate	10	-	-	100	25	-	150 -200	50	5	1.9 -16	2 - 60	220 - 270	-	-	Yunin et al. (2021)
AlSi10Mg	583	Ytterbium fiber	-	-	30	-	450	-	200	-	-	-	476.8	287.2	Dai et al. (2018)
18Ni300 on 304Lsubstrate	300 - 900	Ytterbium doped fiber	1070	12	-	10 - 30	1600 -2000	-	-	-	-	-	780	-	Yao et al. (2018)

d.PFR–Powder flow rate; VED –Volumetric energy density; HV–Vickers hardness; UTS –Ultimate tensile strength; YS –Yield strength.

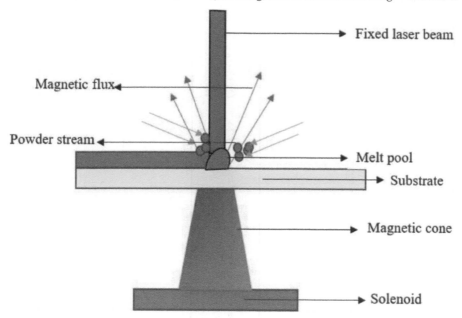

Figure 93.5 Magnetic assisted DMD

C. Electrostatic assisted DMD

The finishing industries employ coatings that are primarily designed to prevent metal corrosion, particularly in extreme heat and/or moisture environments (Arigbabowo and Tate, 2021). Electrostatic coating provides immeasurable cost savings due to increased productivity and shorter processing times, as well as being environmentally and regulatory-friendly. In the last 40 to 50 years, additional coating issues have been addressed, such as the ability to manufacture semi-conductive and corrosion-resistant coatings for electronics, as well as the ability to increase temperature resistance, increase efficiency, and improve the aesthetic appearance of the coating layer shown in Figure 93.6. A strong electric field polarises the neutral powder particles with electrodes parallel to the flow direction, resulting in a charge imbalance (Tsao and Chang, 2018). Powder particles stick to the glass walls as a result of the charge imbalance, especially around the flow channel's electrode. The exit hole is further narrowed as a result of the powder particles' adherence, resulting in a lower mass flow rate as indicated in Table 93.5 (Arigbabowo and Tate, 2021; Tsao and Chang, 2018; Barletta et al., 2006). This polarity shift pushes and pulls the particles in the right direction producing the part geometry layer by layer.

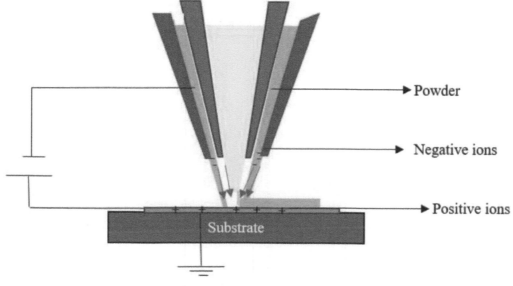

Figure 93.6 Electrostatic assisted DMD

D. Vacuum assisted DMD

The substrate's surface is smoothed by the vacuum- assisted powder deposition sheet. Figure 93.7 demonstrates how a vacuum deposition method fills the substrate with species from a source. The surface bond may move or reflect for a short distance (Yunin et al., 2021). A jet of monomer vapour coalesces on the base to form a replete liquid film that encloses the base material and characteristics, rather than growing atom by atom upward from the substrate. The VPD layer can be coupled with physical or chemical vapour deposition layers to build ultra-smooth thin-film structures (Dai et al., 2018). VPD technology allows for the ultrafast deposition of polymer films in the same vacuum as traditional PVD thin films (sputtered or evaporated). There are two types of VPD processes: evaporative and non-evaporative. The functional monomer typically is a sensitive organic fluid, that is initially desecrated. In the evaporative process, the monomer is fed into a heated tube via an ultrasonic atomiser, where it's flash evaporates and escapes as a monomer gas through a nozzle. The degassed fluid monomers are projected onto the substrate by a slotted die aperture in the non-evaporative method. The system is then connected in exactly the evaporation procedure operated. Salts, graphite, and certain other non-volatile elements can be deposited with the monomer in a homogenous mixture. The VPD process eliminates all unnecessary laminations and handling as compared to tool procedures as shown in Table 93.6 (Yunin et al., 2021; Dai et al., 2018; Yao et al., 2018). Improved layer adhesion and defect-free interfaces are the results. These four different types of PCPs are also used in combinations (Tang et al., 2021; Costa and de Almeida Neto, 2020; Fardi et al., 2020; Liu and Mandler, 2020; Han and Dong, 2017; Sato et al., 2016; Srisawadi et al., 2020).

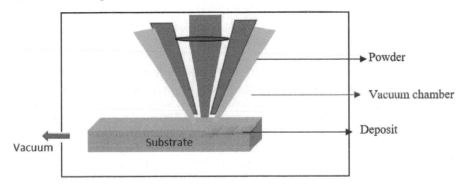

Figure 93.7 Vacuum-assisted DMD

Ultrasonic assisted DMD

The influence of various processing parameters on the surface hardness as well as the yield strength of various materials is investigated in Figure 93.8. For 316L, we can deduce that a 1000 W ultrasonic vibration produces a microhardness of 215.8 μm (Chen et al., 2012). For Al 4047 powder (Zhang et al., 2019), if the ultrasonic vibration power is 1100 W, the resulting recast layer thickness is 3 mm. The ultrasonic-assisted DMD layer contains around 45 to 55 per cent primary Al phases with diameters ranging from 10 to 90 μm. Similarly, with FeCr, a 350W ultrasonic vibration produces a tensile strength of 2300 MPa (Ning and Cong, 2016). In comparison, if the ultrasonic vibration frequency is set to 20 kHz for Ti alloy, the resulting yield strength is in the 1000 MPa range (Alavi and Harimkar, 2015). If the ultrasonic vibration power is set to 610 W in the example of GH037 Xia et al. (2018), the resulting recast layer thickness is 0.8 mm.

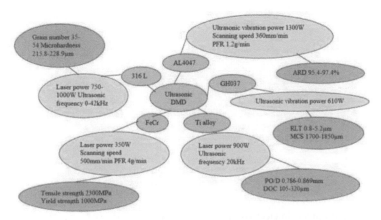

Figure 93.8 Impact of process parameters in ultrasonic-assisted DMD

Magnetically assisted DMD

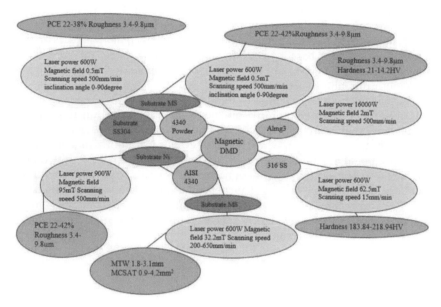

Figure 93.9 Impact of process parameters in magnetic-assisted DMD

If the laser power is 600 W and the magnetic field is 0.5 mT, PCE is 48 per cent for 4340 powders on a mild steel substrate. If the laser power is 800 W and the magnetic field is 0.5 mT, the PCE is 56 per cent for the substrate of SS304 (Smith et al., 2021). The impacts of different process variables on the surface quality of different materials are explored in Figure 93.9. If the scanning speed is 500 mm/min and the magnetic field is 95 mT, the resulting PCE is 63 per cent for AISI 4340 Smith et al. (2022) with nickel substrate. When a magnetic field of 2 mT is applied to Almg3 Chen et al. (2018), the resulting roughness is 3.4 μm and the hardness is 114.2 N/mm². Similarly, for 316 SS material Lu et al. (2019), a magnetic field of 62.5 mT results in a hardness of 183.2 N/mm².

Electrostatic assisted DMD

The influence of different process variables on the development efficiency and tensile modulus of various materials is investigated in Figure 93.10. For PA66 Barletta et al. (2006), we may deduce that at 1.2 V operating voltage, the resulting development efficiency is 53 %. Similarly, when Nylon 65 Arigbabowo and Tate (2021) is in filled at a rate of 50 mm/s, the resulting flexural strength is 40.31 MPa and the tensile modulus is 1.56 MPa. Impact of process parameters in electrostatic assisted DMD, in the case of maraging steel, Tsao and Chang (2018), a 12 V working voltage results in a development efficiency of 45 per cent and transfer efficiency of 40%.

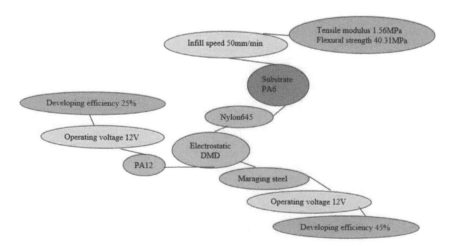

Figure 93.10 Impact of process parameters in electrostatic-assisted DMD

Vaccum assisted DMD

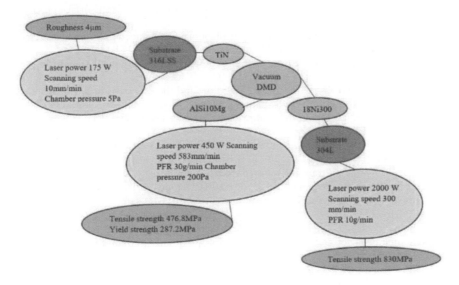

Figure 93.11 Impact of process parameters in vacuum-assisted DMD

As seen in Figure 93.11, the effect of various process parameters on the roughness achieved for various materials is explored. If the laser power is 2000 W and the chamber pressure is 5 Pa, the resultant roughness is 4 μm and the hardness is 260 HV for TiN Yunin et al. (2021) on 316L SS substrate. Similarly, for AlSi10Mg material Dai et al. (2018), a chamber pressure of 200 Pa results in tensile and yield strengths of 476.8 MPa and 287.2 MPa, respectively. When TiAl is used with the substrate of SUS 304 and the chamber pressure is set to 5 Pa, the resulting roughness is 1 μm. The tensile strength of 18Ni300 is 830 MPa if the laser power is 2000 W, the scanning speed is 300 mm/min, and the PFR is 10 g/min Yao et al. (2018).

Summary

DMD is a type of fast touring process that uses molten metal powder to manufacture parts and moulds that are then cemented in place with a laser. Components, moulds, and moulds made of genuine end materials such as tool steel and aluminium can be fabricated or restored using the DMD. To develop a new part or partial reconstruction, always start with a CAD drawing. Extend the weld pool with a little spray of powdered tool steel. This method rapidly cools and solidifies the weld pool, resulting in the best quality and strongest metal products without material loss associated with traditional machining. Furthermore, at substrate inclination slopes of up to 45 degrees, magnetic help was observed to improve particle collection efficiency. On mild steel, proof of compaction of embedded tracks was also visible in a magnetic field. As a result, magnetic aid has a huge potential as a production enhancer in DED, and it may be retrofitted to existing setups. During the LENS process, the cross-sectional form and internal microstructure of the components were modified using ultrasonic vibration.

Conclusion

The PCE differs depending on the type of PCP used. PCE is used to determine the equality of DMD parts. While earlier research has looked at these PCPs in general, this work compares and contrasts the features of the parts generated by various PCPs. We can raise the PCE of ferromagnetic powder by using magnetic aid; significant efficiency gains are conceivable. The study's principal findings are summarised here.

- The type of substrate material has a significant effect on the magnet's efficacy in capturing powder.
- Ferromagnetic mild steel facilitated the magnetic manipulation of powder and resulted in a greater PCE.
- Additionally, the larger the mass flow rate, the more constant the PCE. PCE is minimised at high values of influencing parameters such as stand-off distance, inclination angle, and carrier gas flow rate.
- A high PCE value results in a clean and transparent surface finish and texture, as well as improved mechanical property results.
- The PCE increases as rose both the laser power and the melt pool length.

References

Alavi, S. H. and Harimkar, S. P. (2015). Melt expulsion during ultrasonic vibration-assisted laser surface processing of austenitic stainless steel. Ultrasonics 59:21–30.

Arigbabowo, O. K. and Tate, J. S. (2021). Additive manufacturing of polyamide nanocomposites for electrostatic charge dissipation applications. Mater. Sci. Eng. B: Solid State Mater. Adv. Technol. 271:115–251.

Barletta, M., Gisario, A. and Tagliaferri, V. (2006). Electrostatic spray deposition (ESD) of polymeric powders on thermoplastic (PA66) substrate. Surf. Coat. Technol. 201(1–2):296–308.

Chen, J., Wei, Y., Zhan, X., Li, Y., Ou, W., and Zhang, T. (2018). Melt flow and thermal transfer during magnetically supported laser beam welding of thick aluminium alloy plates. J. Mater. Process. Technol. 254:325–337.

Chen, X., Sparks, T., Ruan, J., and Liou, F. (2012). Study of ultrasonic vibration laser metal deposition process. Int. Symp. Flex. Autom.

Chua, Z. Y., Ahn, I. H., and Moon, S. K. (2017). Process monitoring and inspection systems in metal additive manufacturing: Statusandapplications. Int. J. Precis. Eng. Manuf.-Green Technol. 4(2):235–245.

Costa, J. M. and de Almeida Neto, A. F. (2020). Ultrasound-assisted electrode position and synthesis of alloys and composite materials: A review. Ultrason. Sonochem. 68:105193. doi: 10.1016/j.ultsonch.2020.105193.

Dadbakhsh, S., Hao, L., and Kong, C. Y. (2010). Surfacefinish improvement of LMD samples using laser polishing. Virtual Phys. Prototyp. 5(4):215–221.

Dai, D., Gu, D., Zhang, H., Zhang, Z., Du, Y., Zhao, T., Hong, C, Gasser, and A., Poprawe. (2018). Heat-induced molten pool boundary softening behavior and its effect on tensile properties of laser additive manufactured aluminum alloy. Vacuum 154:341–350.

Del Rio Vilas, D., Longo, F. and Monteil, N. R. (2013). A general framework for the manufacturing workstation design optimisation: a combined ergonomic and operational approach. Simu. Trans. Soci. Modl. Simu. Inter. 89(3):306–329.

Eisenbarth, D., Borges Esteves, P. M., Wirth, F., and Wegener, K. (2019). Spatial powder flow measurement and efficiency prediction for laser direct metal deposition. Surf. Coat. Technol. 362:397–408. doi: 10.1016/j.surfcoat.2019.02.009.

Fardi, S. R., Khorsand, H., Askarnia, R., Pardehkhorram, R., and Adabifiroozjaei, E. (2020). Improvement of biomedical functionality of titanium by ultrasound-assisted electrophoretic deposition of hydroxyapatite-graphene oxide nanocomposites. Ceram. Int. 46(11):18297–18307. doi: 10.1016/j.ceramint.2020.05.049.

Ghosal, P., Majumder, M. C., and Chattopadhyay, A. (2018). Study on direct laser metal deposition. Mater. Today 5(5):12509–12518.

Gibson, I., Stucker, B. and Rosen, D. W. (2009). Additive manufacturing technologies: Rapid prototyping to direct digital manufacturing. NY: Springer.

Gong, X., You, W., Li, X., and Wang, L. (2020). Modeling the influence of injection parameters on powder efficiency in laser cladding. Weld. World 64(8):1437–1448. doi:10.1007/s40194-020-00955-7. (Journal article)

Gorunov, A. I. (2020). Additive manufacturing of Ti6Al4V parts using ultrasonic assisted direct energy deposition. J. Manuf. Process. 59:545–556.

Han, Y. and Dong, J. (2017). High-resolution electrohydrodynamic (EHD) direct printing of molten metal. Procedia Manuf. 10:845–850. doi: 10.1016/j.promfg.2017.07.070.

Kiefer, S. L. (2004). Powder coating material developments promise new opportunities for finishers. Met. Finish. 102(1):35–37.

Kumar, S. P., Elangovan, S., Mohanraj, R., and Ramakrishna, J. R. (2021a). A review on properties of Inconel 625 and Inconel 718 fabricated using direct energy deposition. Mater. Today: Proc. 46(17):7892–7906.

Kumar, S. P., Elangovan, S., Mohanraj, R., and Ramakrishna, J. R. (2021b). Review on the evolution and technology of state-of-the-art metal additive manufacturing processes. Mater. Today: Proc. 46(11):5187–5710.

Kumar, S. P., Elangovan, S., Mohanraj, R., and Sathya Narayanan, V. (2021c). Significance of continuous wave and pulsed wave laser in direct metal deposition. Mater. Today: Proc. 46(17):8086–8096.

Kumar, S. P., Elangovan, S., Mohanraj, R., and Srihari, B. (2021d). Critical review of off-axial nozzle and coaxial nozzle for powder metal deposition. Mater. Today : Proc. 46(13):8066–8079.

Lee, H., Lim, C. H. J., Low, M. J., Tham, N., Murukeshan, V. M., and Kim, Y. J. (2017). Lasers in additive manufacturing: A review. Int. J. Precis. Eng. Manuf. Green Technol. 4(3):307–322.

Liu, L. and Mandler, D. (2020). Using nanomaterials as building blocks for electrochemical deposition: A mini review. Electrochem. Commun. 120(106830):106830. doi: 10.1016/j.elecom.2020.106830.

Liu, S. and Kovacevic, R. (2014). Statistical analysis of processing parameters in high-power direct diode laser cladding. Int. J. Adv. Manuf. Technol. 74(5–8):867–878.

Lu, Y., Sun, G. F., Wang, Z., and Zhang, Y. (2019). Effects of electromagnetic field on the laser direct metal deposition of austenitic stainless steel. Opt. Laser Technol. 119:105586.

Mahamood, R. M. (2017). Laser metal deposition process of metals, alloys, and composite materials. California: Springer International Publishing.

Ning, F. and Cong, W. (2016). Microstructures and mechanical properties of Fe-Cr stainless steel parts fabricated by ultrasonic vibration-assisted laser engineered net shaping process. Mater. Lett. 179:61–64.

Ning, F. and Cong, W. (2020). Ultrasonic vibration-assisted (UV-A) manufacturing processes: State of the art and future perspectives. J. Manuf. Process. 51:174–190.

Partes, K. (2009). Analytical model of the catchment efficiency in high speed laser cladding. Surf. Coat. Technol. 204(3):366–371.

Pinkerton, A. J. (2010). Laser direct metal deposition: Theory and applications in manufacturing and maintenance. Adv. Laser Mater. Process. 461–49.

Popov Jr, V. V., Katz-Demyanetz, A., Garkun, A., and Bamberger, M. (2018). The effect of powder recycling on the mechanical properties and microstructure of electron beam melted Ti-6Al-4 V specimens. Addit. Manuf. 22:834–843. doi: 10.1016/j.addma.2018.06.003.

Rawlins, J., Din, J. N., Talwar, S., and O'Kane, P. (2016). Coronary intervention with the excimer laser: Review of the technology and outcome data. Interv. Cardiol. 11(1):27–32.

Sato, Y., Tsukamoto, M., Masuno, S., and Yamashita, Y. (2016). Investigation of the microstructure and surface morphology of a Ti6Al4V plate fabricated by vacuum selective laser melting. Appl. Phys. A: Mater. Sci. Process. 122(4):1–5. doi: 10.1007/s00339-016-9996-8.

Smith, P. H., Murray, J. W., Jackson-Crisp, A., Segal, J., and Clare, A. T. (2022). Magnetic manipulation in directed energy deposition using a programmable solenoid. J. Mater. Process. Technol. 299:117342.

Smith, P. H., Murray, J. W., Jones, D. O., Segal, J., and Clare, T. (2021). Magnetically assisted directed energy deposition. J. Mater. Process. Technol. 288:116892.

Srisawadi, S., Tanprayoon, D., Sato, Y., Tsukamoto, M., and Suga, T. (2020). Fabrication of 316L stainless steel with TiN addition by vacuum laser powder bed fusion. Opt. Laser Technol. 126:106116. doi: 10.1016/j.optlastec.2020.106116.

Steen, W. M. (1986). Laser surface cladding. Laser Surf. Treat. Met. 369–387.

Tang, Y., Wan, N., Shen, M., Jiao, H., Liu, D., Tang, X., and Zhao, L. (2021). A comparison of the microstructures and hardness values of non-equiatomic (FeNiCo)-(AlCrSiTi) high entropy alloys having thermal histories related to laser direct metal deposition or vacuum remelting. J. Jpn. Res. Inst. Adv. Copper-Base Mater. Technol. 15:696–707. doi: 10.1016/j.jmrt.2021.08.046.

Tsao, T. Y. and Chang, J. Y. (2018). Application of electrostatic adhesion method in metal-powder-based additive manufacturing layer-forming process. In International mechanical engineering congress and exposition. https://doi.org/10.1115/IMECE2018-88741

Xia, K., Ren, N., Wang, H., and Shi, C. (2018). Analysis for effects of ultrasonic power on ultrasonic vibration- assisted single-pulse laser drilling. Opt. Lasers Eng. 110:279–287.

Yao, Y., Huang, Y., Chen, B., and Tan, C. (2018). Influence of processing parameters and heat treatment on the mechanical properties of 18Ni300 manufactured by laser based directed energy deposition. Opt. Laser Technol. 105:171–179.

Yunin, P. A., Sachkov, Y. I., Travkin, V. V., Skorokhodov, E. V., and Pakhomov, G. L. (2021). Nanostructuring of Mn(II) Pc thin films by vacuum deposition in a weak magnetic field. Vacuum 194:110584.

Zhang, Y., Anfu, G., Chen, G., and Kang, J. W. (2019). Ultrasonic-assisted laser metal deposition of the Al 4047Alloy. Metals 9(10):1111. doi: 10.3390/met9101111.

94 Adverse effects of the electrolytes in electrochemical discharge machining process: A review

Prithviraj Dhanajirao Shinde Patil[a], Maneetkumar R. Dhanvijay[b], and Sudhir Madhav Patil[c]

Department of Manufacturing and Engineering and Industrial, Management, College of Engineering Pune, Pune, India

Abstract

Electrochemical discharged machining (ECDM) combines electrical discharge machining (EDM) and electrochemical machining (ECM) processes. The application of the ECDM process is conducting and non-conducting material. In recent years, several attempts have been made to use the ECDM technique for micro-machining of glass, composites, ceramics, steel, super-alloys, and other materials. The machining is affected by the parameters that involve voltage, current, and electrolyte concentration; during the machining operation, it is observed that hasardous gases are released into the environment. These are hasardous gases, they cause severe health issues to the machine operator. In this paper, a collective study on different hasardous gases is present in Electrochemical discharged machining. The primary study focuses on different gas collection techniques to understand the behavior of gases emitted from the machining process.

Keywords: ECD, EDM, electrolyte, mixed electrolyte, gases collection method, human effect, environmental effect.

Introduction

In recent events manufacturing process are attracted the non-conventional method of machining that involved Electrochemical discharged machining (ECDM), which is adopted for machining glasses, plastics, and composite materials that has nonconductive material properties.

The ECDM machining process is a combination of electrochemical machining (ECM) and electrical discharge machining (EDM) machining processes (Dhanvijay and Ahuja, 2014). Due to this time and energy of the machining process are reduced. Different electrolyte types are used during machining, such as alkaline, acidic, neutral salt, and mixed electrolyte. In beginning, this process is carried out on glass material (Sabahi and Razfar, 2017). For the machining process in ECDM, various tools such as copper, tungsten, high-speed steel, stainless steel, and so on are used, which produces successful outcomes (Jain and Adhikary, 2008).

ECDM has a wide application in non-conducting, conducting, and semi-conducting materials. In micro machining, the ECDM process is used in different industries involving mechanical, medical, defense, aircraft, space, nuclear, electrical and electronics, automobile, robotics, machine manufacturing, this method is employed in a wide range of sectors, including micropump, a micro tool, micro die, inkjet nozzles, biochip, microsensor, microreactor, microchip, and microelectromechanical system (MEMS) (Sabahi and Razfar, 2017).

Industrial revolutions have played a significant role in human development, which enabled people to enjoy a better life at their own pace and in comfort. Technological and industrialisation are two key tools that support the industrial revolution (Gupta, 2020). However, this development has also resulted in bringing to certain harmful outcomes, causing environmental degradation. The many industrial operation chemicals are used that involve electro chemical discharge machining, the use of chemicals in such process is a matter of great concern. It will directly or indirectly affect the operator's health when working closely with the substances or at a later stage (El-Hofy and Youssef, 2009). It is seen that the people are working closely with the ECDM setup experience coughing and sneezing, but very little literature is available on these aspects, and hence this paper is an attempt to make aware the researchers in the ECDM field of the harmful effects of the ECDM process. This will help the researchers be more alert and aware while working on the setup, not discouraging them from this novel machining process.

[a]prithvirajshindepatil@gmail.com; [b]mrd.mfg@coep.ac.in; [c]smp.mfg@coep.ac.in

ECDM working

ECDM is the hybrid micro-manufacturing process used for micro-holes and microchannels. There are two-electrode one is termed a cathode tool electrode, and the another is an auxiliary anode electrode (Figure 94.1). The auxiliary anode electrode has more surface area than the cathode electrode and both are connected to a dc power supply (West and Jadhav, 2007). For a smooth machining operation, the gap between the tool electrode (cathode) and counter electrode (anode) is kept at about 25–60 mm, while the tool electrode is immersed inside the electrolyte at around 1–2 mm (Wüthrich, 2009). Electrolysis occurs at the end of the cathode and anode when DC voltage is applied (Harugade and Waigaonkar, 2018). In the presence of either an acidic or alkaline electrolyte, the following reaction produces oxygen gas at the surface of the auxiliary electrode:

$$+ 4H^+ + 4e\text{- (In acidic electrolyte)}$$
$$+ 4e^- \text{ (In alkaline electrolyte)}$$

Similarly, hydrogen gas is generated near the cathode (tool) surface, as shown by the reaction:

$$\text{(In acidic electrolyte)}$$
$$\text{(In alkaline Electrolyte)}$$

Then bubbles formation takes place at the cathode and anode electrode. This increases the bubbles due to which collision of bubbles takes place, which causes the crash of bubbles, and forms a gas film at the end of the tool. This gas film does not allow electricity from the cathode tool electrode to the electrolyte. As a result, a high electric field is produced and generates a spark between the tool and the workpiece (Arya and Dvivedi, 2019). During spark, high-speed caters are formed, and material is removed. For spark generation, a 2.0 to 2.5 mm gap is required between the cathode tool electrode and workpiece (Bhattacharya et al., 1999).

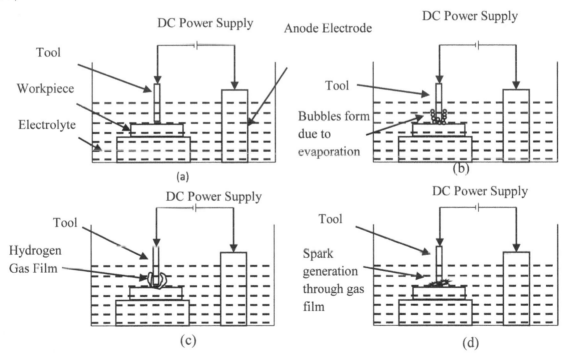

Figure 94.1 Working principle of ECDM process (a) Basic ECDM setup (b) Generation of hydrogen bubbles around tool vicinity (c) Gas film formation around tool (d) Spark initiation

A.   The effects of process variables on machining performance

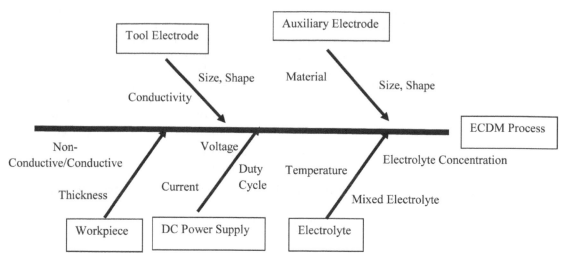

Figure 94.2 Fishbone diagram of different parameters of ECDM

1. Power supply:

The machining operation relies heavily on voltage, pulse-on time, pulse-off time, current, or duty cycle are the variable electrical characteristics related to power supply, affecting the machine's efficiency (Singh and Dvivedi, 2018). When machining with ECDM, a pulsed DC power source is widely utilised since it allows for the release of concentrated spark energy and higher spark stability than a DC power supply (Figure 94.2). The sparking in ECDM is improved by increasing the gap between tool and workpiece, voltage, and current (Kumar et al., 2020). The duty cycle defined as a percentage ratio of pulse-on time to total pulse processing times. MRR, HAZ, depth of penetration, and surface roughness improve when the duty cycle is enhanced due to an increase in current and voltage time (Singh and Singh, 2020a).

2. Electrolyte:

An electrolyte is an essential component affecting the ECDM machining efficiency. It performs various activities, including electrochemical etching, the generation of gas bubbles, and the evacuation of waste from the sparks zone. The right electrolyte should be used for stable machining processes with the ECDM method. The electrolyte is selected depending on the machining conditions, such as the material to be machined and the required output parameters involving MRR and surface finish. Electrolytes are classified into three types

- Acidic
- Neutral
- Alkaline.

The temperature of the electrolyte has a significant impact on the performance of the ECDM. As the temperature increases, the conductivity of the electrolyte improves, and the rate of the electrolysis process increases. It results in the production of increased hydrogen gas and the percentage of electrolyte has a considerable impact on machining quality. Which improves the intensity of the spark in relation to the concentration (Paul and Hiremath, 2016; Saravanan et al., 2021).

3. Tool electrode:

In ECDM, the tool electrode's negative potential (electrode) is generally used for machining and is referred to as ECDM with direct polarity. However, a few studies have been conducted by changing the tool polarity, such as attaching the tool to a positive terminal (electrode), defined as ECDM with opposite polarity. The tool tip shape plays an important role in improving the quality of gas film generation and the area to be machined during the ECDM process. The density of the hydrogen gas layer influences the rate and strength of the spark. A thin layer of hydrogen gas is suitable for improving the geometrical accuracy of the finalised product (Saini et al., 2020; Joshi et al., 2019).

B.   **The common ways of gas entering to human body**

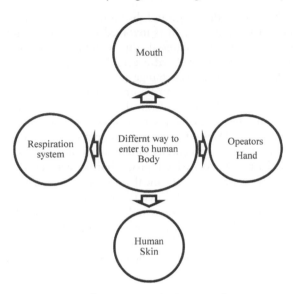

Figure 94.3 Different ways of infection to human

Figure 94.3 shows a different way to humathe n body due to the chemical present in the ECDM process. In the ECDM process, gases are formed due to different electrolytes during the ECDM machining, those gases can also infect human body. These dangerous gases can enter the humans body in multiple ways, due to dipping ungloves hand into the electrolyte and handling the part, tools without gloves, which are directly in contact with the skin. Electrolyte Concentration can splash into the worker's skin if the machine is not adequately guarded or has no guard at all. If there is a wound or cut on the skin, these electrolytes can also enter the bloodstream. If workers eat without adequately sanitizing their hands, these gases can enter the body through the mouth. Most gases enter the human body by inhaling gas vapor or mist. The severity of the inhalation is determined by the distance between the operator and the machine, operational parameters, machine enclosure, ventilation, and other factors (Rajurkar et al., 2017; Haider, 2014).

Adverse effects on humans

The operator of the ECDM machine and the surrounding people who are directly affected by the by-products of the ECDM process.

A.   Handling and storage of electrolytes

Keep this electrolyte in a tight closed container and protect from physical damage. This electrolyte can only store in a cool, dry, ventilated area and away from heat, moisture, and incompatibilities. Whenever these electrolytes contact humans, there is a short-term effect that may occur on the human body, and it immediately burns the skin and eye with possible eye damage (Viswanth et al., 2018).

B.   Inhaling of the Gases

The hydrogen gas produced during the electrolysis process has the potential to be explosive. An engineering control system such as local exhaust ventilation must be provided to remove hydrogen gas from the working regions and mist from the worker's breathing zone. Exposure to chromium compounds via inhalation, ingestion, eye or skin contact can harm the skin, liver, kidneys which causes contact dermatitis in people. The nitrates and nitrites in the electrolyte solution can cause various health problems, including interactions with hemoglobin in the blood, impaired thyroid gland function, vitamin A deficiency, and even cancer.

C.   Size of setup

There are different methods of machining which has different size of setup could be one of the contributing factors affecting operator health. Strong electrolyte, more gas film, and increasing applied voltage is required for those different methods leading to more gases formation which automatically effects operator health.

Adverse effect on the environment

ECDM has significant environmental consequences because of the usage and disposal of hasardous chemicals like cleaning solutions, etchants, and strippers. Furthermore, the costs involved with their treatment and disposal are highly significant. The substances also harm many materials due to acidic corrosion and the etchants employed in ECDM affect the acidity and alkalinity levels, affecting flora and fauna in soil and water (Bhattacharyya and Doloi, 2020). If regular water's pH level changes due to pollution, aquatic life faces a survival difficulty. In chemical machining, corrosive gases can be produced from aerosols of solid (nitrogen and sulfuric oxides) or liquid corrosive chemicals. Wastes from the ECDM process, which involve electrolyte solutions, have an impact on soil quality/fertility, which is a serious matter that must not be overlooked (Chen et al., 2015; Vipulanandan et al., 2006). Due to the creation of acid when these gases interact with water, they may be a source of pollution in the environment, including water and soil. ECDM is associated with various health problems, including irritation, corrosive injuries, burns, fast, severe, and sometimes irreversible eye damage, and throat and lung cancer. The chemical properties, which involve concentration, and the duration of time in contact with acids and alkalis determine health concerns (Liu et al., 2019). The use of more ecologically friendly chemicals is a modern industry trend, furthermore, research has been conducted on the regeneration of waste etchants and the recovery of etched metal. It was discovered that a suitable regeneration/recovery system for various etchants, such as FeCl3, CuCl2, and alkaline etchants, could be produced. To avoid contact with the hasardous chemicals, it is recommended that you use personal protective equipment.

Different gases collection methods

A. Fume mass concentration with different electrolyte

FMC is one of the fume collection methods and is used for the collection of fume particles during machining, for that required a filter, filter holder, tubing, and sampling pump to collect gases. For NaOH electrolyte applied voltage is varies from 40–70 (Singh and Singh, 2020b). In this method, two essential parameters are important electrolyte concentration and the second is applied voltage. In the case of NaOH, NaCl, and HCl applied voltage is constant at 50 V, and electrolyte concentration is varied from 20 % to 50 % Singh et al. (2020) for KOH electrolyte applied voltage varies from 40–70; electrolyte concentration varies 20 to 50% (Singh and Singh, 2020c).

As shown in Table 94.1, we observed that when electrolyte concentration and applied voltage gradually increase, there are increases in the fume mass concentration, these fume particles are harmful to the operators during ECDM machining

Table 94.1 Mass concentration of gases at various electrolyte concentrations (Singh and Singh, 2020b; Singh et al., 2020; Singh and Singh, 2020c)

Sr. No	Electrolyte Concentration %	Fume mass concentration (mg/m3)				
		NaOH	NaOH	NaCl	HCl	KOH
1	20	120	30.35	15.10	40.56	138.2
2	30	150	70.40	40.56	90.98	280.5
3	40	327	150.12	70.43	181.23	467
4	50	390	220.76	98.34	290.0	520

B. Morphological analysis -SEM

The Morphological fume particles are trapped in the glass filter during machining. In morphological analysis, different shapes of particles are available where the size of these particles is measured in image J software in SEM picture (Singh and Singh, 2020b). In NaOH, NaCl, and HCl, different shapes of particles are available in gases like Spherical, Irregular, Agglomeration, rectangular. The effect of breathing these particles hugely in lungs and dissolution in blood the severe effect on operators health (Singh et al., 2020). In the morphological analysis, electrolyte concentration and the applied voltage is influence parameters present in the different electrolytes (Singh and Singh, 2020c).

From Table 94.2 We observed that different particle shapes are made through different electrolytes.

Table 94.2 Different particles at different electrolyte (Singh and Singh, 2020b; Singh et al., 2020; Singh and Singh, 2020).

Sr.No	Electrolyte	Shape of particles	Size of particles	Average size
1	NaOH	Spherical, rectangular, irregular	9–10 µm.	9.46 µm.
2	NaOH	Spherical Irregular, Agglomeration	70–100 nm	81.96 nm.
3	NaCl	Spherical, irregular, rectangular, and agglomeration shapes	35–55 nm	43.71 nm.
4	HCl	agglomerations with tiny spherical particles	90–105 nm	91.26 nm.
5	KOH	Spherical, irregular, rectangular, hexagonal, flower shell structure and agglomeration shapes	-	7–8 µm.

C. Chemical composition – EDS

EDS method is used for the chemical composition of fume particles, which shows the different percentages of an element in the electrolyte. The particle of gases is entirely dependent on an electrolyte concentration (Singh and Singh, 2020b).

During high temp of machining, there is vaporisation, melting of the workpiece material is form due to this different residual is available which are hasardous to operator and the proportion of residual varies with various electrolyte Singh et al. (2020), Singh and Singh (2020c) and it is shown in the Table 94.3.

Table 94.3 Different chemical composition at different electrolyte (Singh and Singh, 2020b; Singh et al., 2020; Singh and Singh, 22020c)

Sr No	Electrolyte	Element Present	Weight %	Element Present	Weight %
1	NaOH	Na, Mg, Al,	-	Si, K, Ca, Cr	
2	NaOH	Carbon	41.38%	Cl	1.23%
3	NaCl	Carbon	14.37%	Cl	44.96%
4	HCl	Carbon	7.23%	Cl	4.80%
5	KOH	K	193.5%	Cr	92.31%
		Fe	114.56 %		

D Breathing Air Analysis-FTIR

Fourier transform infrared (FTIR) analyses gas and compound air (Singh and Singh, 2020b). With different electrolyte concentrations and applied voltage, these gases are produced during machining (Singh et al., 2020).

Table 94.4 Different gases at different electrolyte (Singh and Singh, 2020b; Singh et al., 2020; Singh and Singh, 2020c)

Electrolyte	NaOH	NaOH	NaCl	HCl	KOH
Compound gases available	H_2O	H_2O	Amines and O–H group	Amines and O–H group	Amines
	CH_4 and ethers	CH_4 and ethers	CH_4	CH_4	CH_4
	SO_2	SO_2	Aldehydes, NO_2	Aldehydes, NO_2	NO_2
	Methanol, halogenated	Halogenated hydrocarbon	Nitro compounds	Nitro compounds	Nitro compounds
	NO_2		Anhydrides	SO2, H_2O	
	Amines		SO2, H_2O	CO2	SO_2
	N-H stretching		Aromatic compounds	Thio compound (S–H stretch)	
			Di sulphides (S–S stretch)	Disulphides (S–S stretch)	

E.   Different gases effect on operator health

Sulfur dioxide (SO_2), methane (CH_4), nitrogen dioxide (NO_2), and amines are common gases that are emitted due to the electrolytes such as NaOH, NaCl, HCl, KOH. As these gases are reported toxic gases with harmful effects on the environment. The effect of these gases on the operator is listed below in Table 94.5.

Table 94.5 Effect on operators at different electrolyte (Sivapirakasam et al., 2011; Tantet et al., 1995b; Bierwirth, 2021; Wang et al., 2020; Tantet et al., 1995a)

Sr. No	Gases Present	Effect on operators
1	Sulfur dioxide (SO_2)	Sulfur dioxide SO_2 has an effect on the respiratory system, particularly lung function, and can cause eye irritation.
2	Methane (CH_4)	If methane inhaling a long time can cause memory loss nausea, vomiting, and headache.
3	Amines	The amines cause respiratory disorders and headaches.
4	Halogenated hydrocarbon	Halogenated hydrocarbon causes lung damage.
5	Nitrogen dioxide (NO_2)	When inhaled, NO_2 can cause eye, nose, and throat discomfort and lung irritation, and decreased lung function.
6	Carbon dioxide (CO_2)	Breathing of CO_2 gases may affect operators like headaches, dizziness, tiredness, etc.
7	Chromium (Cr)	Operators breathing of chromium (Cr) elements promotes carcinogen reactions in the human body.

Personal protection

A.   Ventilation system

A ventilation system with local and/or general exhaust is recommended to keep operators' emissions within the airborne exposure limits. Local exhaust ventilation is frequently chosen because it has the potential to manage pollutant emissions at their source, preventing them from migrating into the main work area.

B.   Skin protection

To avoid skin contact, wear resistant safety equipment such as boots, gloves, lab coats, aprons, or coveralls.

C.   Eye protection

Wear chemical safety goggles and/or a full-face shield.

Potential for future ECDM research and activity

New technologies in the machining process improve the application of sustainable-ECDM in various research and development fields. Today, the most promising technology for meeting the needs of green manufacturing is sustainable-ECDM. However, several issues must be addressed before sustainable-ECDM can be considered an efficient kind of machining. The following points describe the newly recognised research difficulties, possibilities, and suggestions related to sustainable-ECDM.

- In the research, the electrolyte used in the ECDM process produced various gases during machining, harmful to human health and the environment. So, there will be scope for eco-friendly electrolytes to be used in future ECDM processes (Goud et al., 2016).
- There is a possibility that by generating less harmful gases during the ECDM process abrasive electrolytes can be used in the future ECDM process.

Conclusion

This article provides an in-depth discussion of electrolytes based on recent research activity. It focuses on the impact on humans, the effect on the environment, other fume collecting methods, and different gases that are detrimental to the operator's health. This article mentioned that different types of gases enter the human body differently. Two effects occur after entering the human body: the first is an instantaneous effect, and the second is a major long-term effect in the human body. Skin, eye, lung problems, and other issues arise due to the instant effect. Long-term exposure to significant diseases such as cancer, asthma attacks, and memory loss impacts the operator's health. Few studies provide through information on the fume's particles present with different electrolytes during ECDM. These particles have the following shapes: spherical, irregular, rectangular, and shell. Provide the average particle size, elements present in the electrolyte, and various gases present in the multiple electrolytes. Finally, we determined that various particles, gases, and elements are present during machining. These elements are highly damaging to the human body. In this connection, are proposed some recommendations for personal protection operators during ECDM machining.

References

Arya, R. K. and Dvivedi, A. (2019). Investigations on quantification and replenishment of vaporised electrolyte during deep micro- holes drilling using pressurised flow-ECDM process. J. Mater. Process. Technol. 266:217–229. DOI: 10.1016/j.jmatprotec.2018.10.035.

Bhattacharya, B., Doloi, B. N., and Sorkhel, S. K. (1999). Experimental investigation into ECDM of non conductive ceramic materials. J. Mater. Process. Technol. 95:145–154.

Bhattacharyya, B. and Doloi, B. (2020). Modern machining technology. Advanced, hybrid, micro machining and super finishing technology. (pp. 365–460). Cambridge: Academic Press. https://doi.org/10.1016/B978-0-12-812894-7.00005-0.

Bierwirth, P. (2021). Carbon dioxide toxicity and climate change: A major unapprehended risk for human health. Canberra: Australian National University. doi:10.13140/RG.2.2.16787.48168.

Chen, R., Shi, X., Bai, R., Rang, W., Huo, L., Zhao, L., Long, D., Pui, D. Y. H., and Chen, C. (2015). Airborne nanoparticle pollution in a wire electrical discharge machining workshop and potential health risks. Aerosol Air Qual. Res. 15(1):284–294.

Dhanvijay, M. R. and Ahuja, B. B. (2014). Micromachining of ceramics by Electrochemical Discharge process considering stagnant and electrolyte flow method. Proc. Technol. 14:165–172.

El-Hofy, H. and Youssef, H. (2009). Environmental hasards of nontraditional machining production engineering department faculty of engineering, Alexandria University. EE'09: Proceedings of the 4th IASME/WSEAS international conference on Energy & environment, (pp. 140–145).

Goud, M., Sharma, A. K., and Jawalkar, C. (2016). A review on material removal mechanism in electrochemical discharge machining (ECDM) and possibilities to enhance the material removal rate. Precis. Eng. 45:1–17.

Gupta, K. (2020). A Review On Green Machining Techniques. Proce. Manufact. 51(4):1730–1736.

Haider, J. and Hashmi, M. S. J. (2014). Health and Environmental Impacts in Metal Machining Processes. Mater. Sci. 10.1016/B978-0-08-096532-1.00804-9

Harugade, M. L. and Waigaonkar, S. D. (2018). Effect of different electrolytes on material removal rate, diameter of hole, and spark in electrochemical discharge machining. In Advances in manufacturing, lecture notes in mechanical engineering, eds. A. Hamrol, O. Ciszak, S. Legutko, M. Jurczyk, (1st ed.) (pp. 427–437), Cham: Springer.

Jain, V. K. and Adhikary, S. (2008), On the mechanism of material removal in electrochemical spark machining of quartz under different polarity conditions. J. Mater. Process. Technol. 200(1–3):460–470.

Joshi, T., Jawalkar, C., and Charak, A. (2019). A critical review on different types of Tools used in ECDM Process. Int. J. Tec. Innov. Mod. Eng. Sci. 5:18–25.

Kumar, N., Mandal, N., and Das, A. K. (2020). Micro-machining through electrochemical discharge processes: A review. Mater. Manuf. Process. 35(4):363–404. doi:10.1080/10426914.2020.1711922.

Liu, X., Yang, T., Li, H., and Wu, L. (2019). Effects of interactions between soil particles and electrolytes on saturated hydraulic conductivity. Eur. J. Soil Sci. 71(2). doi: 10.1111/cjss.12855.

Paul, L. and Hiremath, S. S. (2016). Improvement in machining rate with mixed electrolyte in ECDM process. Procedia Technol. 25:1250–1256.

Rajurkar, K. P., Hadidi, H., Pariti, J., and Reddy, G.C. (2017). Review of sustainability issue in non-traditional machining. Proc. Manufact.7:714–720

Sabahi, N. and Razfar, M. R. (2017). Investigating the effect of mixed alkaline electrolyte (NaOH + KOH) on the improvement of machining efficiency in 2D electrochemical discharge machining (ECDM). Int. J. Adv. Manufact. Technol.1–4.

Sabahil, N., Hajian, M., and Razfar, M. R. (2018). Experimental study on the heat-affected zone of glass substrate machined by electrochemical discharge machining (ECDM) process. Int. J. Adv. Manufact. Technol. 97:1557–1564.

Saini, G., Manna, A., and Sethi, A. S. (2020). Investigations on performance of ECDM process using different tool electrode while machining e-glass fibre reinforced polymer composite. Mater. Today: Proc. 28:1622–1628. https://doi.org/10.1016/j.matpr.2020.04.853.

Saravanan, K. G., Prabu, R., Venkataramanan, A. R., and Beyessa, E. T. (2021). Impact of different electrolytes on the machining rate in ECM process Hindawi. Adv. Mater. Sci. Eng. https://doi.org/10.1155/2021/1432300.

Singh, M and Singh, S. (2018). Electrochemical discharge machining: A review on preceding and perspective research. Proc Inst Mech Eng B J Eng Manuf. 233(4):095440541879886

Singh, M. and Singh, S. (2020b) Sustainable electrochemical discharge machining process: Characterisation of emission products and occupational risks to operator. Machining Sci. Technol. 24(1).

Singh, M. and Singh, S. (2020c). Electrochemical discharge machining: gases generations, properties, and biological effects. Int. J. Adv. Manufact. Technol. 106(5–8).

Singh, M., Singh, S., and Kumar, S. (2020). Environmental aspects of various electrolytes used in electrochemical discharge machining process. J. Braz. Soc. Mech. Sci. Eng. 42(8)

Singh, T. and Dvivedi, A. (2018). On pressurised feeding approach for effective control on working gap in ECDM. Mater. Manuf. Process. 33(4):462–473. DOI: 10.1080/10426914.2017.1339319.

Sivapirakasam, S. P., Mathew, J., and Suria Narayanan, M. (2011). Effect of process parameters on the breathing zone concentration of gaseous hydrocarbons—a study of an electrical discharge machining process. Human Ecol. Risk Assess.. 17(6):1247–1262

Viswanth, V. S., Ramanujam, R., and Rajyalakshmi, G. (2018). A review of research scope on sustainable and Eco-friendly electrical discharge machining (E-EDM). Mater. Today: Proceed. 5(5):12525–12533.

Tantet, J., Eic, M., and Desai, R. (1995a). Breakthrough study of the adsorption and separation of sulfur dioxide from wet gas using hydrophobic zeolites. Gas Separ. Purif. 9:213–220.

Tantet, J., Eric, M., and Desai, R. (1995b). Breakthrough study of the adsorption and separation of sulfur dioxide from wet gas using hydrophobic zeolites. Fuel Energy Abstr. 36:457.

Vipulanandan, C., Kim, J. W., Oh, M. H., and Park, J. (2006). Effects of surfactants and electrolyte solutions on the properties of soil. Environ. Geol. 49:977–989.

Wang, B., Zhu, Y., Qin, Q., Liu, H., and Zhu, J. (2020). Development on hydrophobic modification of aluminosilicate and titano silicate zeolite molecular sieves, Appl. Catal. Gen. 117952.

West, J. and Jadhav, A. (2007). ECDM Methods for fluidic interfacing through thin glass substrates and the formation of spherical microcavities. J. Micro mech. Micro Eng. 17(2):403. doi: 10.1088/0960-1317/17/2/028.

Wüthrich, R. (2009). Micromachining using electrochemical discharge phenomenon. UK: William Andrew Book Company.

95 Comparative analysis of vibration and acoustic signals for fault diagnosis of plastic spur gears

Behara Santosh Sagar[a], Swapnil Gundewar[b], Prasad V. Kane[d], and Shubham Tiwari[c]

Mechanical Engineering, Visvesvaraya National Institute of Technology, Nagpur, India

Abstract

Fast Fourier transform (FFT) is the conventional tool used for fault diagnosis of elements of machines. For the steel gearbox, FFT is widely applied to identify faults due to a rise in the amplitude of gear mesh frequency peaks of FFT signature and its sidebands. In this work, the ability of FFT of vibration signals and acoustics signals to identify the fault is studied in case of plastic gears. With the wide usage of materials such as Nylon 66 and acetylene for fabrication of plastic gear for low torque transmission applications, it is needed to monitor the defects and identify faults at an early stage to reduce downtime. In this work, the faults seeded in plastic spur gears are simulated in an experimental setup. The FFT is applied to the acquired vibration and acoustics signals and the ability to identify faults using both are compared. It is observed that the FFT of the acoustic signal has an edge over the vibration signal to identify the fault.

Keywords: Defect Identification, FFT, nylon, plastic spur gear, vibration and acoustic signal.

Introduction

The gearbox remains an important device used to transmit power according to the requirement. The application of gearbox is in machines as well as in an automobile. It can be used in gear setups in small electrical appliances also, and are expected to work for long durations in the production system. Any failure in the gearbox may introduce unwanted downtime, increase expenses and increase human casualties (Saravanan et al., 2009). That is why it is important to detect the fault and diagnose the faults in the initial stage. From the production point of view, gears play an important role as there is no occurrence of slip, can be used to reduce speed and also used to change the direction of rotation. These qualities of gear make them an important mechanical component.

The essential properties of materials required for the manufacturing of gears are service conditions and strength. Service conditions are like noise, wear etc. Metallic and non-metallic materials both are used for the gears manufacturing. The metallic gears are widely made of materials like steel, steel with different alloys, cast iron, and bronze. Non-metallic materials like rawhide, synthetic resins like nylon, compressed paper, and wood are used for gears, specifically for reducing the weight and noise of the equipment (Mao et al., 2009).

There are wide ranges of applications of metallic and non-metallic (plastic) gears. Metallic gears are used for high power transmission. These gears are used in automobiles, machines and other high-end power transmission works.

Whereas plastic gears are used in light duty works and machines like lathes, grinding machines, and milling machines. Other application of plastic gears include wiper systems, variable induction system gears, automotive motor fan, seating and tracking headlight to actuator used in braking, lift gates, electronic throttle bodies, CD ROM, turbo controls, printers and washing machines.

Although there is a huge difference between the torque transmission of both metallic and non-metallic gears, because of the lightweight, less cost plastic gears are used more. And continues work is going on in the field to achieve same level of torque as that in metallic gear.

There are so many traditional methods present which are used for condition monitoring of metallic gears. Like vibration technique, noise, lubrication, and temperature monitoring. These techniques are used for gearbox machine fault identification. Gearbox conditions can be examined by measuring vibration, acoustic, thermal, electrical, and oil-based signals (Qu et al., 2014). In this work, efforts are made to apply

[a]beharasantoshsagar@gmail.com; [b]swapnilgundewar32@gmail.com; [c]prasadkane20@gmail.com; [d]tiwarishubham2050@gmail.com

vibration and acoustic technique (Kane and Andhare, 2020) for fault identification of non-metallic gears for which less literature was found.

Noise and vibration measurement and signal analysis are important tools when experimentally investigating gear condition because gears create noise at specific frequencies, related to number of teeth and the rotational speed of the gear.

At present, there is a trend for reducing weight and increase efficiency. It is known that gears are so important for day to day work. Although they are not so visible, all the automobiles including transportation vehicles and plenty of other household equipment's use gears. Various developments along the ages have come to a point where nothing new can be created. In search of betterment, research is still carried out for determining advancement in pre-existing facts. Gears are having application in various fields. For example - wristwatch, automobile, power-drive equipment, heavy machines and work part transfer machines. Various industries are carrying out various research works for substitutions of plastic gears. Using plastic gears in place of metal gears reduces weight and also reduces power consumption and increases efficiency. And because of the above mentioned reasons, plastic gears are being preferred to use more nowadays.

With the increased usage of the plastic and polymer gears, researchers are studying the compatibility of various fault diagnosis techniques for the above mentioned category of gears. In this context, Singh et al. (2018) presented a comprehensive review of the research on the low and medium loaded polymer spur gears. In the paper, various design features and modification techniques are also proposed for the improvement of durability and performance of the same. Kumar et al. (2021) had studied and compared several statistical features like crest factor, impulse indicator and kurtosis for fault diagnosis of polymer gears.

The most common methods used in the industry for data analysis are fast Fourier transform (FFT) signal analysis techniques. It was assumed that the spectral content is more important than the vibration amplitude for determining the exact location of the fault in the machine. Therefore, instead of directly analysing the vibration signal in the time domain, the FFT of the vibration and acoustic signal is used. However, the FFT technique serves good only for stationary signals.

To extract and analyse the non-stationary features for fault diagnosis, there are several other Time-Frequency analysis techniques which maps the signal to the two dimensional function of time and frequency. To name a few are Wavelet Transforms, Wigner–Ville distribution (WVD), Choi–Willams distribution (CWD) cone-shaped distribution (CSD), etc. Of all the time-frequency techniques, wavelet transform is the most adoptable linear transform which uses a series of oscillating functions with different frequencies as window functions. The fundamental advantages and disadvantages of various time-frequency analysis methods were presented in (Feng et al., 2013).

Numerous experimentations and research studies were conducted to diagnose the faults of gears using the above mentioned techniques. And most of them dealt with the conventional metallic gears. However, the vibration and acoustic characteristics of the non-metallic gears are relatively new. With the evolution of various polymer gears and corresponding tooth profiles, an extensive investigation and standardisation of the vibration and acoustic signals is very much needed.

In the current work, frequency-domain signals were obtained through FFT techniques. They use the difference of power spectral density of the signal due to the fault of gear to identify the damage of elements.

FFT is a category of special algorithms that use discrete Fourier transform with significant savings during compilation. FFT is not a different transform from DFT, but is a tool for calculating DFT with a significant reduction in the number of required calculations.

For a continuous function of time variable f (t), the Fourier Transform F (f) is defined as:

$$F(f) = \int_{-\infty}^{\infty} f(t) e^{-2j\pi ft}$$

Where F (f) – Signal in the frequency domain
F (t) – Signal in the time domain
f – Frequency of signal
t – Time

Experimental setup

The experimental setup consists of plastic spur gear, motor, DC motor controller, shaft, flexible coupling, bearing, flange and base. The RPM of motor is set to 720 rpm. The flexible coupling is used to connect motor shaft with gearbox. One gear is rigidly fixed with shaft and supported by bearing. While the other gear is free to adjust for various working conditions purposes.

The vibration signals were acquired by the vibration sensor ABRO 102-1A and the data acquisition was done by DAQ card of National Instruments (NI9178). The accelerometer was held on the bearing housing of the gearbox by a magnetic bar. Whereas, the acoustics signals were captured with the help of an array microphone of type 40 PH of make of AERO brand and model AB 102-1A. with a sensitivity of 52.14 mV/Pa. The microphone was kept at a distance of 1 m from the experimental setup to avoid any unwanted impulsive disturbances. The DAQ card (NI9234) is used for data acquisition. A sampling rate of 40 KHz was set fulfilling the Nyquist criteria. LabVIEW software was used to study and obtain various features from vibration and acoustic signals.

Different fault conditions like gear with broken tooth and gear with crack are seeded in nylon gear for the study purpose. As shown in Figure 95.1.a., a small crack was introduced in the root of the gear tooth by using a HSGM heat cutter. And in another gear Figure 95.1.b., the tooth length of one tooth is reduced from 4 mm to 2 mm by filing the tooth manually. This gear is treated as broken tooth gear. The healthy gear and gears with faults are shown below:

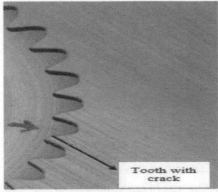

a. Gear with crack

b. Gear with broken tooth

c. Healthy gear

Figure 95.1 Experimental setup

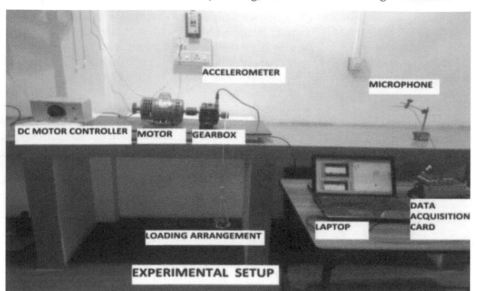

Figure 95.2 TK

Results

Two types of signals are obtained for gear with healthy condition and faulty condition i.e. vibration and acoustic signals with the help of sensors. The signals obtained for nylon gear with healthy and faulty conditions are shown in below plots. Figures 95.3 and 95.4 shows the time domain plots for vibration signals corresponding to healthy and faulty gears respectively. X and Y axes of the plot indicate time in second and vibration acceleration level in g (m/s²), these signals are processed under FFT method, these time domain signals are convert into frequency domain signals for detecting the faults as it is difficult to detect clear indications of any defect in the gear with only the time domain signals, especially if the defect is at the initial stage.

Figures 95.5 and 95.6 are the frequency domain plots of healthy and faulty gears respectively. From the figures, it is clear that variations in the frequency spectrums because of the presence of fault are remarkable. As already mentioned, all the plots and signals till now are for vibration signals of healthy and faulty gear and next are the signals of acoustics for healthy and faulty gears.

The Figures 95.7 and 95.8 are the representations of the acoustic signals in time domain and Figures 95.9 and 10 are the frequency spectrums of the acoustic time domain signals.

A comparison of acoustic signals with vibration signals in frequency domain plot is show in below Table 95.1. This table shows the spectral amplitude of various signals in frequency domain for healthy and faulty gears.

From the experiment and analysis of frequency domain plots it is observed that the presence of tooth fault in the gearbox gives rise to peaks and generates sidebands. In the frequency domain it is clearly noticeable that the sidebands are present at high amplitude frequency in the faulty gear. The amplitudes of harmonics of gearbox with fault are more than the amplitude of harmonics of healthy gearbox. Thus by using the FFT technique the fault location can be identified but it needs the human interpretation.

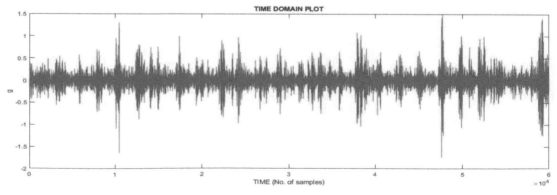

Figure 95.3 Time domain plot of vibration signals for healthy gear

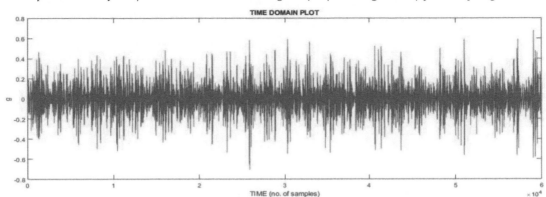

Figure 95.4 Time domain plot of vibration signals for faulty gear

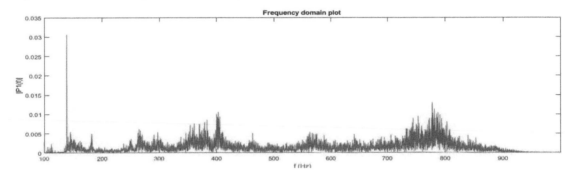

Figure 95.5 Vibration frequency spectrum of healthy gear

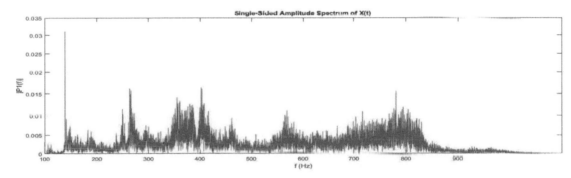

Figure 95.6 Vibration frequency spectrum of faulty gear

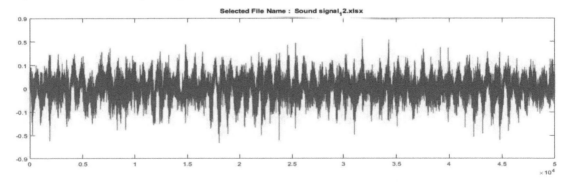

Figure 95.7 Time domain acoustic signal of healthy gear

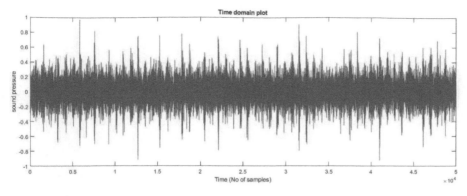

Figure 95.8 Time domain acoustic signal of faulty gear

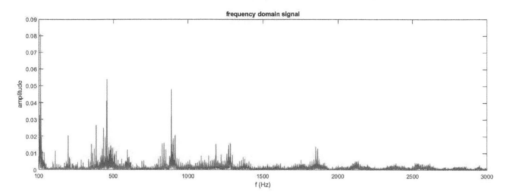

Figure 95.9 Frequency spectrum of acoustic signal of healthy gear

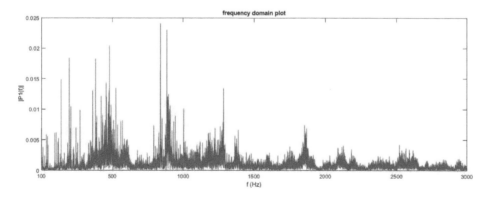

Figure 95.10 Frequency spectrum of acoustic signal of faulty gear

Table 95.1 Spectral amplitude of vibration and acoustic signal

Signal type	Gear condition	1st Peak of amplitude at around 400 Hz frequency	2nd Peak of amplitude at around 800 Hz frequency
Vibration	healthy	0.01190 m/s^2	0.01310 m/s^2
	faulty	0.01614 m/s^2	0.01689 m/s^2
Acoustic	healthy	0.01721 Pa	0.01483 Pa
	faulty	0.02065 Pa	0.02342 Pa

Conclusion

From the analysis it's clear that the vibration and acoustic signals have shown remarkable results. But if we see the frequency spectrums, it was found that the acoustic signals have shown more accurate and easily noticeable changes in the plot as far as fault is concern. Fault detection and clarity of signals is considerably good for non-metallic gear fault detection, there is a huge scope for developing it into a more accurate and automatic diagnosis system. This can be done by extracting the more weighted statistical features from the obtained data and applying the Machine Learning algorithms (Kane and Andhare, 2016). To achieve this, the current technique is being integrated with Artificial Neural Network.

References

Feng, Z., Liang, M. and Chu, F. (2013). Recent advances in time–frequency analysis methods for machinery fault diagnosis: A review with application examples. Mech. Syst. Signal Process. 38(1):165–205.

Kane, P. and Andhare, A. (2016). Application of psychoacoustics for gear fault diagnosis using artificial neural network. J. Low Freq. Noise Vib. Act. Control 35(3):207–220.

Kane, P. and Andhare, A. (2020). Critical evaluation and comparison of psychoacoustics, acoustics and vibration features for gear fault correlation and classification. Measurement 154:107495.

Kumar, A., Parey, A. and Kankar, P. K. (2021). Vibration based fault detection of polymer gear. Mater. Today: Proc. 44:2116–2120.

Mao, K., Li, W., Hooke, C. J., and Walton, D. (2009). Friction and wear behaviour of acetal and nylon gears. Wear 267(14):639–645.

Qu, Y., He, D., Yoon, J., Hecke, B. V., Bechhoefer, E., and Zhu, J. (2014). Gearbox tooth cut fault diagnostics using acoustic emission and vibration sensors—A comparative study. Sensors 14(1):1372–1393.

Saravanan, N., Cholairajan, S. and Ramachandran, K. (2009). Vibration-based fault diagnosis of spur bevel gear box using fuzzy technique. Expert Syst. Appl. 36(2):3119–3135.

Singh, A. K., Siddhartha, and Singh, P. K. (2018). Polymer spur gears behaviors under different loading conditions: A review. Proc. Inst. Mech. Eng. Part J: J. Eng. Tribol. 232(2):210–228.

96 Optimisation of turning process parameters using Deform-3D for cutting forces

Mahesh R. Jadhav[1,a], Krishnakumar D. Joshi[1,b], Pramod V. Mulik[1,c], Prashant J. Patil[1,d], Pradyumna S. Mane[1,e], and Prashant S. Jadhav[2,f]

[1]Department of Mechanical Engineering, TKIET, Warananagar, Kolhapur, India

[2]Department of Mechanical Engineering, Rajarambapu Institute of Technology, Rajaramnagar, Sangli, India

Abstract

Now a days simulation process is becoming very much important in production/manufacturing process. The most important advantages of using simulation is that it helps in reducing the expensive shop floor work, redesigning of tooling, improving die design and shortening the lead time. In future, composite of aluminium material will be leading in almost every manufacturing and production field, especially in aviation and automobile industries. Thus it is necessary to understand the various cutting conditions affecting. In current investigation, Deform 3D software and Taguchi method have been employed to conduct experimentation and analysis on turning process of Al/SiCp applied on the most machining parameters. Tool temperature and forces which come across during machining are to found by using simulation. Minitab software analyses the S/N ratio analysis and ANOVA is carried out in Minitab, where cutting force and temperature of tool are found to be controlled by depth of cut and feed rate respectively. Optimum cutting conditions for cutting forces and tool temperature have been obtained from S/N ratio Analysis. Optimum cutting conditions for cutting forces and tool temperature are , Cutting speed of 200 m/min, feed rate of 0.07 mm/rev and 0.75 mm depth of cut and Cutting speed of 150 m/min, feed rate of 0.1 mm/rev and 0.5 mm depth of cut for turning of AlSiCp/20p.

Keywords: Al/SiCp MMC, ANOVA, cutting forcecomponent, DEFORM 3D, S/N Ratio.

Introduction

In recent years, for modelling and simulation of machining process, FEA has become the principal tool. It is playing a very much important role in accurately predicting the physical parameters such as temperature in heat affected zone, stress distribution, machining forces and conceivably its breaking can be resolved quicker than utilizing exorbitant and time consuming experiments. Modelling 3D cutting process by utilizing finite element analysis is a zone of continuous exploration movement because of huge reduction in cost and gives bits of knowledge into the procedure which are not easily estimated in test (Dabade and Jadhav, 2016; Jadhav and Dabade, 2016b). This paper focus on modelling the turning procedure for Al/SiCp metal matrix composite as a workpiece material and TiN coated carbide as insert.

In this study, effect of machining parameters on temperature of tool and cutting forces have been analysed using an FEA software Deform-3D. Deform-3D is robust simulation package used for the examination of machining analysis. It consist of three steps, in which the pre-processor step is used for creating, assembling or modifying the information for the simulation and creating necessary database file and the Simulation engine step is used for playing out the mathematical estimations necessary to dissect the procedure and composing the outcomes to the database record, while a post-processor step is required for analysis the data record from the simulation step and showing the outcomes graphic and numeric form. M. R. Jadhav and U. A. Dabade performed experimentation using L16 orthogonal array that showed higher values of radial and feed force with experimental results than in Simulation with a very good agreement between them (Jadhav and Dabade, 2016a). Jadhav et al. (2020) performed surface roughness and flank wear analysis in machining of MMCs with coated carbide insert. The authors conclude that cutting speed is significant factor on surface roughness and tool wear. Form ANOVA they find out optimum machining conditions for both response variables (Jadhav et al., 2020; Garder et al., 2005). They carried out comparative study of FEM software packages like DEFORM, Advantage and Abaqus. After this they came to

[a]mrjadhav@tkietwarana.ac.in; [b]kdjmech@tkietwarana.ac.in; [c]pvmmmech@tkietwarana.ac.in; [d]pjpmech@tkietwarana.ac.in; [e]pradyumnamane123@gmail.com; [f]prashant.jadhav@ritindia.edu

know about the pros and cons about this software's and from this analysis they conclude that DEFORM is very much useful in simulating metal cutting processes (Gardner et al., 2005; Yanda et al. 2010). Performed finite element analysis of ductile cast iron with DNMA432 cutting tool. From the results they observed that rake angle is a substantial parameter in machining and that cutting forces are inversely proportional to it (Yanda et al. 2010; Constantin et al., 2010). They used DEFORM 3D software for a milling process and experimental validation for the simulated cutting forces. The tool type used by the researcher is sandvik R365-Q80-Q27-515M of material AISI 1045 steel (Constantin et al., 2010; Swamy et al., 2012). They used DEFORM 3D software for turning AISI 1045 steel with PCBN tool. They observed that at primary deformation zone very deep deformation is found and effective stress is inversely proportional to cutting speed (Swamy et al., 2012; Senthil Kumar and Tamizharasan, 2012). They used DEFORM 2D software to simulate orthogonal machining process for analysis. They concluded that with decrease in included angle between the cutting edge, the wear depth and cutting forces are getting reduced (Senthil Kumar and Tamizharassan, 2012; Al-Zkeri et al., 2012). Carried out simulation of the turning process using DEFORM 3D software with AISI 1045 steel material and carbide insert. From simulation results they concluded that at primary shear zone of the work piece more stress and strain found as compared to other zones this is due to maximum deformation in the primary shear zone (Al-Zkeri, 2012). Constantin et al. (2012) makes comparison between different finite element packages and tries to find out the pros and cons of these packages through results. They used Ti 6AL4V material for the machining process. They found almost similar results for the results obtained from the simulation (Constantin et al., 2012). Zamani et al. (2013) used DEFORM 3D software for simulation of new concept of hot machining with laser on Ti6A14V alloy using modified Johnson-crook material model (Zamani et al., 2013; Attanasio et al., 2008). They used AISI 1045 steel as workpiece material to carry out the turning operation. They also carried out the simulation by using DEFORM 3D to find out the tool wear during operation. They concluded that a very good agreement was observed between the results of simulation and experimental work (Attanasio et al., 2008; Santhosh Kumar et al., 2017). Performed experimental and simulation of machining of AISI STEEL on horizontal CNC Lathe by using DNMA432 uncoated insert. In simulation they analyse correlation of tool wear to the machining parameters. Author concludes that optimisation of tool geometry and structure for tool designer is possible by using this study (Santhosh Kumar et al., 2017). Pervaiz et al. (2014) conducted a study to analyse energy consumption during machining process by using FEA and experimental work. The comparison shows good agreement between these two methods with maximum variation in 8%. Pervaiz et al., (2014) and Vijayaraghavan et al., (2016) conducted a combined finite element based data analytics model for investigating cutting force and their relationship with process parameters. The study shows that optimising power consumption can be made by using this investigation without performing trial experiments (Vijayaraghavan et al., 2016). Mohruni et al. (2019) investigated the machinability of Ti6Al4V alloy by using FEA experimental. Author concluded that FEA results not getting good agreement with experimental results due to some susceptible reasons (Mohruni et al., 2019). Wanga et al. (2014) applied Deform 3D on milling of steel material to predict the milling force. The study reveals that prediction of milling force with experimentally gives good agreement. They also suggest that reference of this information is useful for determination optimisation of tool and process parameters (Wanga et al., 2014). Bhoyar and Kamble (2013) performed FEA of turning process on En31 material. They measure tool temperature distribution on different positions on insert. Also they compare results with experimental work and comparison shows good agreement between them (Bhoyar and Kamble, 2013).

The literature review shows that various investigations have been carried out of FEA of machining process on conventional materials. A very few researches have done the work associated with FEA modelling on composite materials for prediction of cutting forces and temperature.

Machining of MMCs is a tough task as compared to machining of the conventional metals in industry. It is because MMCs have hard ceramic reinforcements with almost same hardness property as that of the cutting tools. During cutting operation, these hard ceramic reinforcement particles are rubbed with edges of the cutting tool and damage the tool surface, result to extreme wear of tool. Causing various types of damages, such as facture of particles and deboning at the contact of particle and matrix. Thus, machining of MMCs is treated to be a challenging process. The main difficulties are seen in obtaining an optimal combination of process parameters to achieve minimum cutting forces, tool temperature, better tool life and surface finish. Therefore in the present study Al/SiCp/20p/220 MMCs composite material selected.

Simulation modelling

FEM is known for a comprehensive method for breaking down chip formation processes and anticipating the parameters like temperatures, forces, stresses, and so forth. In the present experiment, the FEM is utilised to simulate the turning of Al/SiCp/20p/220 MMCs using carbide insert with tungsten coating.

Simulation process involves preprocessing which carries out tool mesh generation, work piece setup and database generation as shown in Figure 96.1 whose essential parameters have been enlisted in Table 96.1

Table 96.1 Parameters used in Deform 3D

Sr. No.	Parameters	Values
1	An initial temperature	20°C
2	shear friction factor	0.6
3	Heat transfer coefficient	45 N/sec/mm/°C
4	Coolant	No
5	The size proportion for tool insert	4
6	The size proportion for the workpiece	7
7	Element selected for FEA	Tetrahedral
8	Size of mesh	20000
9	Workpiece	Plastic material
10	Tool insert	Rigid material
11	Cutting length for workpiece	20 mm.
12	The simulation steps	5000

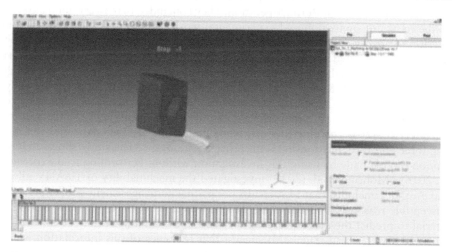

Figure 96.1 Database generation and simulation

Table 96.2 Simulation results

Expt. No.	Cutting speed (m/min)	Feed (mm/rev)	Depth of cut (mm)	Feed force (N)	Cutting force (N)	Thrust force (N)	Temp. of tool (°C)
1	150	0.05	0.5	43.31	435.27	125.23	49
2	150	0.07	0.75	26.78	330.81	97.79	45
3	150	0.1	1	105.97	843.67	221.90	43
4	200	0.05	0.75	42.75	490.72	136.38	61
5	200	0.07	1	30.39	374.88	106.14	60
6	200	0.1	0.5	26.78	330.81	97.79	48
7	250	0.05	1	105.97	843.67	231.90	92
8	250	0.07	0.5	42.75	490.72	136.38	52
9	250	0.1	0.75	30.39	374.88	106.14	61

Table 96.3 ANOVA (*P* and % contribution) for cutting forces, feed force, thurst force and tool temperature

Source	Cutting Force		Feed Force		Thrust Force		Temp. of tool	
	P value	*% contribution*	*P value*	*% contribution*	*P value*	*% contribution*	*P value*	*% contribution*
Cutting speed (m/min)	0.581	14.98	0.493	16.25	0.556	15.58	0.154	43.45
Feed (mm/rev)	0.551	16.94	0.468	17.92	0.513	18.49	0.217	28.47
Depth of cut (mm)	0.305	47.33	0.240	50.06	0.296	46.43	0.281	20.18

The reason behind selecting coated tungsten carbide tool is to achieve excellent resistance to wear and built-up edge. Thermal properties of Al/SiCp/20p/220 workpiece material and tungsten coated carbide tool is shown in Table 96.2.

Table 96.4 Thermal properties of work piece material tool insert (Dabade et al., 2009)

Thermal properties	Al/SiCp/20p/220 thermal properties	TiN Coated carbide insert thermal properties
Thermal conductivity	185	25
Heat capacity	0.837	12
Emissivity	0.672	0

Experimental details

FEA based deform 3-D simulation software has been used for experimentation. L9 orthogonal array has been applied to assess the effect of cutting parameter on the experiment layout. Accordingly, the no of experiments are reduced to 9 for effectively getting process-related information. Table 96.3 shows various parameters and the levels thereof, considered for simulation.

Table 96.5 Process parameters and their levels

Process parameters	Unit	Levels		
		1	2	3
Cutting speed	m/min	150	200	250
Feed	mm/rev	0.05	0.07	0.1
Depth of cut	mm	0.5	0.75	1

Result and discussion

All the experiments of turning operation on Al/SiCp MMC material have been performed in DEFORM 3D software. For the analysis of the performance characteristics i. e. cutting force and temperature Minitab 16 Statistical Analysis software has been used and Lower-the-better (LB) criterion has been applied. ANOVA, which is a partitioning technique is used to identify the significance of each parameter. Table 96.4 shows the results obtained for different output parameters by using Machining simulation software.

A. Analysis of simulated feed force

From Table 96.5 ANOVA results show that DOC is most significant cutting parameter for simulated feed force with 50.06% contribution followed by feed and cutting speed. From Figure 96.2 re-present calculated SN ratio for simulated feed force. It is clear that for optimum effect of feed force the values of feed, cutting speed and depth of cut have been obtained to be 0.07 mm/rev, 200 m/min and 0.75 mm respectively.

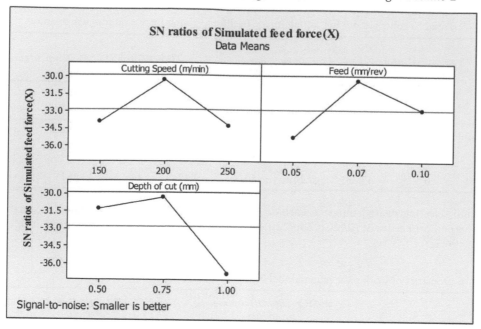

Figure 96.2 SN ratio of simulated feed force

B. Analysis of simulated cutting force

From Table 96.4 ANOVA results show that DOC is most influencing cutting parameter for simulated cutting force with 47.33% contribution followed by feed and cutting speed. From Figure 96.3, it is clear that in order to maintain the cutting force to its lowest, the values of should be set for feed, depth of cut and speed need to be set to 0.07 mm/rev, 200 m/min and 0.75 mm, respectively.

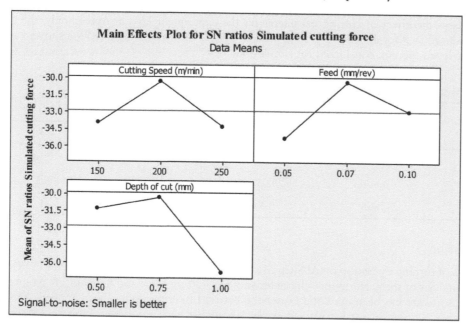

Figure 96.3 SN ratio of simulated cutting force

C. Analysis of simulated thrust force

From Table 96.4, ANOVA for simulated thrust force, it is observed that significant cutting parameter is the Depth of cut which contributes 46.43%. From Figure 96.4 which represents SN ratio for simulated thrust force, it is clear that for optimum effect of thrust force the values of feed, cutting speed and depth of cut are 0.07 mm/rev, 200 m/min and 0.75 mm respectively.

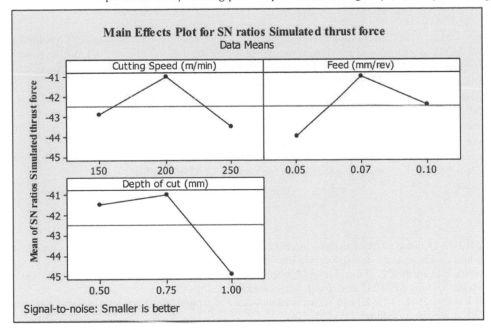

Figure 96.4 SN ratio of simulated thrust force

D. Analysis of simulated temperature of tool

From Table 96.4, ANOVA for simulated temperature of tool, shows that cutting speed is the significant factor that contributes 43.45%. From Figure 96.5, it is clear that if feed, speed and depth of cut are set to 0.1 mm/rev, 150 m/min and 0.5 mm respectively lowest tool temperature can be maintained.

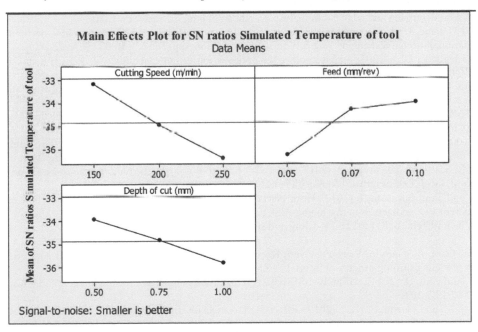

Figure 96.5 SN ratio of simulated temperature of tool

E. Confirmation experiment

The confirmation experiment is the final step to verify the improvement of cutting forces and tool temperature at optimal level of machining parameters obtained SN ratio. Anew experiment was conducted on MTAB CNC turning center. A dynamometer and optical Infrared Pyrometer was used to measure force components and insert temperature. The improvement is obtained 10% and 8% respectively in cutting forces and tool temperature.

Conclusion

- For analysis of Machining forces and tool temperature, Deform 3D software is used. Techniques like ANOVA and AOM plots are used for significance of process parameters on these machining forces and tool temperature. Taguchi methodology is used for optimisation of process parameter for machining forces and tool temperature. Based on this it concludes that,
- Cutting speed and Depth of cut have been found to be the most significant parameters for Tool temperature and Machining forces respectively.
- For Machining forces, Cutting speed of 200 m/min, feed rate of 0.07 mm/rev and 0.75 mm depth of cut have been found to be the optimum parameters for turning of AlSiCp/20p.
- For Temperature of tool, Cutting speed of 150m/min, feed rate of 0.1 mm/rev and 0.5 mm depth of cut have been found to be the optimum parameters for turning of AlSiCp/20p.

References

Al-Zkeri, I. A. (2012). 3-D Finite element analysis of effect cutting edge geometry on cutting forces, effective stress, temperature and tool wear in turning. J. Kerbala University 10(2).

Attanasio, A., Ceretti, E., Giardini, C., Filice, L., and Umbrello, D. (2008). Criterion to evaluate diffusive wear in 3D simulations when turning AISI1045 steel. Int. J. Mater. forming. 1:495–498

Bhoyar, Y. R. and Kamble, P. D. (2013). Finite element analysis on temperature distribution in turning process using deform-3d. Int. J. Res. Eng. Technol. 2(5):901–906.

Constantin, C., Bisu, C. F., Croitoru, S. M., and Constantin. G. (2010). Milling analysis by 3D FEM and experimental tests. In Annals of DAAAM for 2010 and proceedings of 21st international DAAAM symposiumm, 21.

Constantin, C., Croitoru, S.-M., Constantin, G., and Strǎjescu, E. (2012). FEM tools for cutting process modelling and simulation. U.P.B. Sci. Bull., Series D 74(4).

Dabade, U. A., and Jadhav, M. R. (2016). Experimental study of surface integrity of Al/SiC particulate metal–matrix composites in hot machining. Procedia CIRP 41:914–919.

Dabade, U. A., Dapkekar, D., and Joshi S.S. (2009). Modeling of chip-tool interface friction to predict cutting forces in machining of Al/SiCp composites. Int. J. Mach. Tools Manuf. 49:690–700.

Gardner, J. D., Vijayaraghavan, A. D., and David, A. (2005). Comparative Study of Finite Element Simulation Software. In Consortium on deburring and edge finishing, https://escholarship.org/uc/item/8cw4n2tf.

Jadhav, M. R. and Dabade, U. A. (2016a). Modelling and simulation of Al/SiCp MMCs during hot machining, In ASME international mechanical engineering congress and exposition, (pp. 50527), V002T02A023.

Jadhav, M. R. and Dabade, U. A. (2016b). Multi-objective optimisation in hot machining of Al/SiCp metal matrix composites. IOP Conf. Series: Mater. Sci. Eng. 114:1–10.

Jadhav, M. R., Patil, D. S., Mulik, P. V., Patil, S. P., and Mane, P. S. (2020). Surface roughness and tool wear analysis while turning Al/SiCp metal matrix composites. Manuf. Technol. Today 19(12):3–8.

Mohruni, A. S. Zahir, M., Yanis, M., Sharif, S., and Yani, I. (2019). Investigation of finite element modelling on thin-walled machining of Ti6Al4V using DEFORM-3D. In IOP Conf. Series: J. Phys.: Conf. Series. 1–9.

Pervaiz, S., Deiab, I., Rashid, A., and Nicolescu, M. (2014). Prediction of energy consumption and environmental implications for turning operation using finite element analysis. Proc. Inst. Mech. Eng. Part B: J. Eng. Manuf. 1–8.

Kumar, S. S., Reddy, S. P., and Murthy, A. B. (2017). Analysis of 3D tool wear for turning operation based on finite element method. Int. J. Eng. Sci. Res. Technol. 118–132.

Senthil Kumar, N. and Tamizharassan, T. (2012). Finite element analysis and optimisation of uncoated carbide cutting inserts of different tool geometries in machining AISI 1045 steel. J. Mech. Sci. 1(1):37-47.

Swamy, M., Raju, B., and Teja, B. R. (2012). Modeling and simulation of turning operation. IOSR J. Mech. Civil Eng. 3(6):19–26.

Vijayaraghavan, V., Garg, A., Gao, L., Vijayaraghavan, R., and Lu, G. (2016). A finite element based data analytics approach for modeling turning process of Inconel 718 alloys. J. Clean. Prod. 1–9.

Wanga, Z., Hu, Y., and Zhu, D. (2014). DEFORM-3D based on machining simulation during metal milling. Key Eng. Mater. 579–580:197–201.

Yanda, H., Ghani, J. A., and Haron, C. H. C. (2010). Effect of rake angle on stress, strain and temperature on the edge of carbide cutting tool in orthogonal cutting using fem simulation. J. Eng. Technol. Sci. 42(2):179–194.

Zamani, H., Hermani, J. P., Sonderegger, B., and Sommitsch, C. (2013). 3D simulation of laser assisted side milling of Ti6Al4V alloy using modified johnson-cook material model. Key Eng. Mater. 554–557.

97 A systematic review on effects of process parameters of pulsed electrodeposition on surface structure of copper thin films

Nihal Pratik Das[a], Vimal Kumar Deshmukh[b], Sanju Verma[e], Mridul Singh Rajput[b], Sanjeev Kumar[d], and H.K. Narang[f]

Department of Mechanical Engineering, National Institute of Technology Raipur, Raipur, India

Abstract

Pulsed electrodeposition technology has become an unavoidable and one of the most common electrodeposition techniques due to its capacity to deposit material to produce homogeneous, compact, and coherent surfaces with high surface polish and microhardness. In recent years, pulsed electrodeposition techniques are being used extensively in the field of electronics to manufacture flip-chip solder connections, copper interconnects and manufacturing of material needed for nanotechnology and nanobiotechnology. The prime purpose of this research is to demonstrate the current state of the art in pulsed copper electrodeposition. The review begins with a brief introduction on deposition technique followed by an introduction to electrodeposition. The research and developments in pulsed electrodeposition of copper are grouped broadly into the surface structure, surface roughness, coating/surface thickness and microhardness of copper thin films. The final section includes a summary of prospective research directions based on the review.

Keywords: Copper, electrodeposition, pulsed electrodeposition, surface roughness, surface structure, surface thickness and microhardness.

Introduction

The applications of thin-film coatings depend on their surface structure which strongly depends on deposition techniques (Jilani et al., 2017). Now-a-days, more technologies with different combinations of parameters and types of deposition methods are used to produce quality coatings at micro and nano grade. Electrodeposition offers reliability, cost and environmental advantages over other deposition technology (Society, The Electrochemical, 1999). The use of the electrodeposition technique is moving towards the production of materials needed for nanotechnology and nanobiotechnology applications (Schwarzacher, 2006). Copper electrodeposition is vastly used in the electronics field to eliminate problems of passivation (Tao and Li, 2006; Pagnanelli et al., 2015). Copper has good thermal conductivity and copper electrodeposition can be done on non-metallic materials to make them electrically conductive which is a primary concern in the electronic industry (Ibañez and Fatás, 2005).

The layout of the deposition technique is shown in below Figure 97.1.

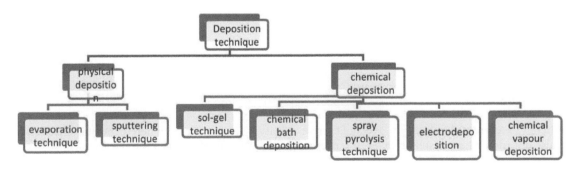

Figure 97.1 Methods of thin-film deposition

[a]nihalpratik2@gmail.com; [b]deshmukh.vimal1920@gmail.com; [c]sanjuverma.sv.sv@gmail.com; [d]msrajput.me@nitrr.ac.in; [e]skumar.me@nitrr.ac.in; [f]hnarang.me@nitrr.ac.in

Galvanostatic method of electrodeposition in the form of pulsed current are used widely in coating industry (Zhang et al., 2021; Krajaisri et al., 2022). Optimisation of deposition parameters to produce high quality coatings required innovation in methods of electrodeposition. By adding additives to electrolyte concentration reduces energy consumption (Lin et al., 2021; Xu et al., 2021). Advancement to electrodeposition technique is Pulsed current electrodeposition (PED) technique which is extensively applied in electronics field. There are numerous advantages of using pulse platting because of the reduction of porosity in films, fine grained deposits, low electrical resistance, high hardness and low surface roughness (Zhan et al., 2021; Jang et al., 2021). Three parameters can be varied independently in pulsed plating. The structure of the deposits obtained by the electrolyte supply is determined by pulse off time, pulse on time and pulse current (Devaraj and Seshadri, 1992). Nucleation and grain growth significantly impact nano grain and micrograin film in the pulsed electrodeposition process (Natter and Hempelmann, 1996). For the production of nano grained or micro grained deposits, the conditions that cause a high rate of nucleation and a slower rate of grain growth apply (Tao and Li, 2006).

This paper aims to furnish significant contributions in the areas of pulsed electrodeposition of copper. This paper will focus on optimum process parameters that must be used in pulsed electrodeposition of copper to produce homogeneous, compact and coherent films with high surface finish and desirable micro-hardness.

A brief overview of electrodeposition

When anode and cathode in a closed electric cell immersed in an electrolyte are exposed to the applied voltage, electrodeposition occurs. At the cathode's surface, a reduction process occurs, resulting in the deposition of the desired material (Mughal et al., 2014). The process is governed by Faraday's law of electrolysis. Now-a-days, the electrodeposition of metals (Ni, Co, Fe and Cu) and their alloys are used rapidly due to their unique thermophysical and magnetic properties (Oriňáková et al., 2006).In practice, there is a chance of wear out of machine components that cannot be replaced due to cost background. Hence electrodeposition technique is used to overcome such wear out parts (Venkatesh et al., 2018).

Pulsed electrodeposition technique

The surface structure of a coating can be improved to improve its qualities. As a result, the impact of process parameters on coating surface structure and mechanical characteristics must be thoroughly understood. A variety of techniques can be employed to alter a coating's surface structure. Pulse plating (PP) is one of these technologies that can be utilised to improve deposit characteristics. Pulse plating, which is essentially a discontinuous current that fluctuates in time, allows for a greater variety of parameters to be used to alter the coating structure and improve deposit quality (Leisner et al., 2007; Halmdienst et al. 2007). Time intervals with zero and non-zero current are classified as ON and OFF times (Pagnanelli et al., 2015) and denoted as t_{ON} and t_{OFF} respectively.

$$J_{av} = \frac{J_A \times t_{ON}}{t_{ON} + t_{OFF}} \tag{1}$$

$$p = \frac{t_{OFF}}{t_{ON}} \tag{2}$$

$$d = \frac{t_{ON}}{t_{ON} + t_{OFF}} \tag{3}$$

$$f = \frac{1}{t_{ON} + t_{OFF}} \tag{4}$$

Where,J_A = current density, J_{av} = average current density, t_{ON} = on period where deposition occurs, t_{OFF} = pause duration where system relaxes, p = pause-to-pulse ratio and d = duty ratio, f = frequency.

Research on pulsed electrodeposition of copper

A.   The surface structure of copper films

The crisp outcomes of various studies on the influence of various process parameters or deposition parameters on the surface structure of copper thin films during pulsed electrodeposition will be discussed in this section. The fundamentals of the effect of distinct deposition parameters on the surface structure during pulsed electrodeposition of copper on Si has been reported by Ivana and Jelena (Pc, Current, 2020). The experiment was performed with 60g/L H_2SO_4 and 240 g/L $CuSO_4 \cdot 5H_2O$ at room temperature. The findings of their study are shown in Tables 97.1–3 and is analysed graphically. It is inferred from Figure 97.2 that when frequency increases, grain size decreases and grain size is minimum at a frequency of 100 Hz. In Figure 97.3, at a frequency of 100 Hz grain structure is very fine and grain structure changes from very coarse to very fine as frequency increases. In Figure 97.4, there is no globules formation with J_A at 100 mA/ cm^2 and 120 mA/cm^2 but with J_A at 140 mA/cm^2 formation of globules occurs.

Table 97.1 Grain structure and grain size of copper films obtained by the PC regimes under current density (J_A) = 100 mA/cm^2 (Pc, current, 2020)

f (Hz)	t_{ON} (ms)	t_{OFF} (ms)	p	Samples	Grain structure	Grain size(μm)
30	5	28.3	5.66		Very coarse	10
50	5	15	3		coarse	5 to 10, Mean=7.5
80	5	7.5	1.5		fine	Less than 5, mean=2.5
100	5	5	1		Very fine	Less than 5, Mean=2.5

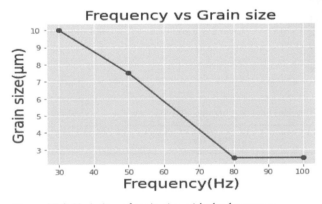

Figure 97.2 Variation of grain size with the frequency

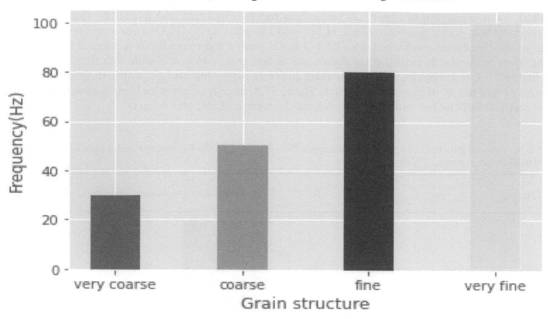

Figure 97.3 Variation of grain structure with frequency

Table 97.2 Grain structure and grain size of copper films obtained by the PC regimes under t_{ON} = 5 ms, t_{OFF} = 5 ms, frequency(f) = 100 Hz (Pc, current, 2020)

J_A in mA/cm^2	Samples	Grain structure	Grain size(μm)
100		Very fine	Less than 5, mean=2.5
120		Very fine	Less than 5, mean=2.5
140		Globules formation	Globules dia=30μm

Table 97.3 Formation of globules with frequency (f) = 100 Hz (Pc, current, 2020)

J_A in mA/cm^2	f in (Hz)	Globules formation	
100	100	No	0
120	100	No	0
140	100	yes	1

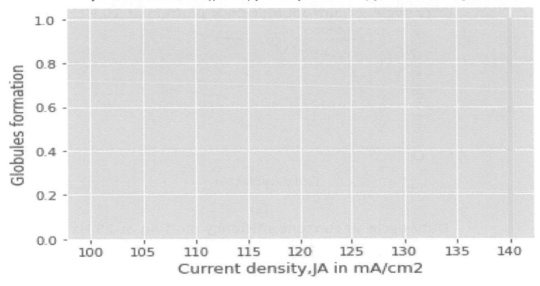

Figure 97.4 Globules formation with current density

Balasubramanian and Srikumar (Balasubramanian et al., 2009) have reported on the fundamental under-standing of the various outcomes of variable process parameters onto as-deposited surface structure during pulsed current electroplating of Cu on stainless steel. The experiment was performed with 200 g/L $CuSO_4 \cdot 5H_2O$ in 100g/L H_2SO_4. The findings of their study are analysed in Table 97.4. Hydrogen evolution occurs at higher frequencies, resulting in a reduction in current efficiency. During 80% duty cycle and pulse frequencies of 25 Hz and 50 Hz, improving ion migration enhances the rate of nucle-ation rate (Pearson and Dennis, 1990), rate of deposition (Dikusar et al., 2005) and current efficiency (Balasubramanian et al., 2009).

Table 97.4 Current efficiency and surface thickness of copper films obtained by the PC regimes under average current density (J_{av})=4 mA/cm² (Balasubramanian et al., 2009; Mohan and Raj, 2005). Where t_c= surface thickness, CE= current efficiency

f (Hz)	t_{ON} (ms)	t_{OFF} (ms)	d (%)	At 25°C		At 50 °C	
				t_c (μm)	CE	t_c (μm)	CE
10	10	90	10	6	61.25	6.876	70
	20	80	20	6.375	65	8.75	87.5
	40	60	40	7.5	75	9.125	91.25
	80	20	80	7.75	80	9.5	96.25
25	4	36	10	6	62.5	7.25	72.5
	8	32	20	6.75	68.75	8.75	88.75
	16	24	40	7.75	78.75	9.25	92.5
	32	8	80	8.5	86.25	9.75	97.5
50	2	18	10	6.25	63.75	7.5	76.25
	4	16	20	7	70	9	90
	8	12	40	8	82.5	9.375	93.75
	16	4	80	8.75	88.75	9.875	98.75
100	1	9	10	6.5	66.25	7.875	78.75
	2	8	20	7.25	72.5	8.25	82.5
	4	6	40	7	73.75	9	91.25
	8	2	80	8.25	85	8.125	81.25

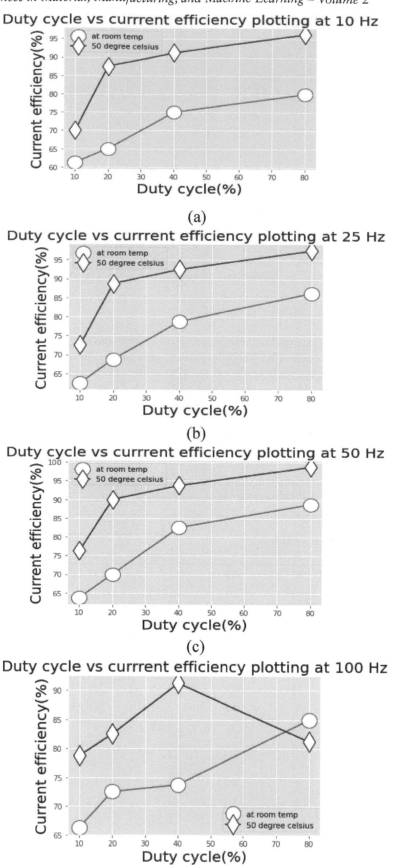

Figure 97.5 Variation of current efficiency with duty cycle at (a) 10Hz (b) 25 Hz (c) 50 Hz (d) 100 Hz

Frequency vs currrent efficiency plotting at 80% Duty cycle

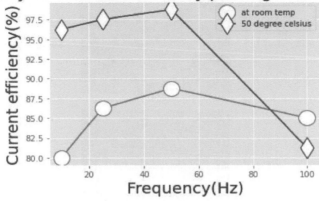

Figure 97.6 Variation of current efficiency with frequency at 80% duty cycle

The data shown in Table 97.4 is graphically presented in Figures 5 and 6. It is inferred from Figure 97.5 that current efficiency at 50°C is more than current efficiency at room temperature with varying duty cycles for all frequencies except at a frequency (f) = 100 Hz. Current efficiency (CE) value is maximum with 80% duty cycle for frequency of 10 Hz, 25 Hz,50 Hz. In Figure 97.6, at 80% duty cycle, current efficiency at 50°C is more than current efficiency at room temperature for all frequencies except at a frequency of 100 Hz.

Imaz et al. (2009) have analysed the effect of variation in process parameters onto as-deposited surface structure during pulsed electrodeposition of Cu on Ni plated steel foils. The experiment was performed with 190 g/L $CuSO_4.5H_2O$ and 43.5 mL/L H_2SO_4. The findings of their study are analysed in Table 97.5. During copper electrodeposition, chloride ions can show levelling and adsorption effects. As a result, the bath was likewise filled with a constant concentration of chloride (0.06 g/L). When 0.06 g/L of chloride and 0.2 g/L of gelatin was used as an additive, a very small size copper grain was formed as shown in Table 97.5. Hence this combination was used for the calculation of the microhardness of copper films as shown in Table 97.12.

Table 97.5 Surface structure of copper films produced with different amounts of thiourea and gelatin as additives with average current density(J_{av}) = 3.3 A/dm² and duty ratio(d) =0.33 (Imaz et al., 2009). Where, Q_c = Amount of chloride, Q_t = amount of thiourea and Q_g = amount of gelatin

Q_c (g/L)	Q_t (g/L)	Q_g (g/L)	Samples	Surface structure
0	0	0		Polyhedric, coarse columnar
0.06	0.02	0		Dendritic
0.06	0.2	0		Fibrous

| 0.06 | 0 | 0.02 | | Small, equiaxed grain size |
| 0.06 | 0 | 0.2 | | Very small grain size |

A. Surface roughness of copper films

This section will provide an overview of research on the effect of different process parameters during pulsed electrodeposition on the surface roughness of copper thin films. It was analysed that with an increase in frequency (f) average surface roughness(R_a) of copper films decreases as shown in Table 97.6. At a frequency of 100 Hz, surface roughness is 169.9 nm which is the smallest among all surface roughness measured on corresponding frequencies. So to check surface roughness value at different conditions further research work was carried out at constant frequency i.e. 100 Hz and varying current densities as shown in Table 97.7. It was found that with an increase in current densities surface roughness value increases as shown in Table 97.10. So optimum condition i.e frequency = 100 Hz and current density = 100 mA/cm² where surface roughness value was smallest has been taken for further research work on the calculation of hardness of copper thin films as shown in Table 97.8 (Pc, Current, 2020). The data shown in Tables 97.9 and 10 is graphically analysed in Figure 97.7. It is inferred from Figure 97.7(a) that R_a decrease with an increase in frequency and R_a value is minimum at the frequency of 100 Hz. In Figure 97.7(b), R_a is minimum at J_A of 100 mA/cm² and increases with an increase in J_A. In Figure 97.7(c), R_a decreases with an increase in J_{av} up to 50 mA/cm² after which R_a increases.

Table 97.9 Average surface roughness (R_a) value of copper films measured at different frequencies and J_A of 100 mA/cm². (Pc, current, 2020)

f (Hz)	J_A *(mA/cm²)*	J_{av} *(mA/cm²)*	*Average surface roughness, R_a in nm*
30	100	15	507.3
50	100	25	470.5
80	100	40	385
100	100	50	169.9

Table 97.10 Average surface roughness (R_a) value of copper films measured at frequency of 100 Hz and different current densities (Pc, current, 2020)

f (Hz)	J_A *in (mA/cm²)*	J_{av} *(mA/cm²)*	*Average surface roughness, R_a in nm*
100	100	50	169.9
100	120	60	237
100	140	70	229.1

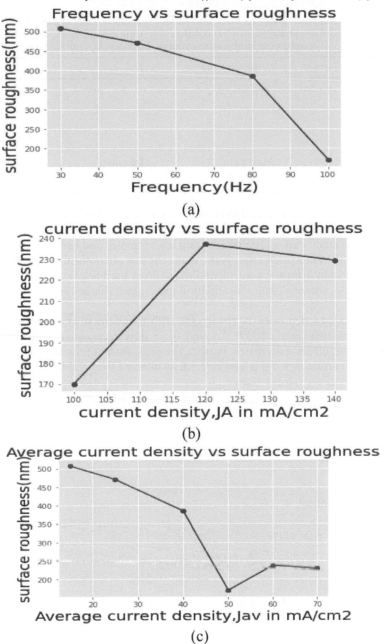

Figure 97.7 Variation of surface roughness with (a) frequency (b) current density (c) average current density

C.   Surface thickness and hardness of copper films

This section will provide an overview of research on the effect of distinct process parameters during pulsed electrodeposition on coating thickness and the hardness of copper thin films. The microhardness of a coating is highly influenced by its thickness (Ramachandramoorthy et al., 2022; Mohan and Raj, 2005). The relative indentation depth (RID=h/tc) value is used to find the hardness of composite systems (substrate and coating), where h = indentation depth and tc= coating thickness and RID value 0.1<RID<1 gives the most reliable result to measure the hardness of the composite system (Pc, Current, 2020). For calculation of values of coating microhardness, Chico-Lesage composite hardness model (C-L) has been implemented (Pc, Current, 2020). For ratios of roughly 1, where l_d is indent length measured diagonally and t_c is the coating thickness, the C–L model provides the most consistent results (Lesage et al., 2006; Lesage et al., 2006). The ratio is 1 for t_c of 40 µm in the RID range of 0.1 to 1 and $H_{composite}$ is maximum, maintaining a stable result in composite hardness values. The greatest value of $H_{composite}$ was found at the optimum deposition settings with a coating thickness of 40 m as shown in Table 97.11.

Table 97.11 Shows coating hardness value, composite hardness value and surface roughness value of copper films measured at frequencies of 100 Hz, current density of 100 mA/cm² and different coating thickness (Pc, current, 2020). Where, H_{coat} = coating hardness, $H_{composite}$ = composite hardness, t_c = coating thickness, R_a = surface roughness

Fixed parameter	Variable parameter t_c (μm)	H_{coat}	$H_{composite}$	R_a (nm)
F = 100 Hz,	10			52.42
	20			101.5
J_A = 100 mA/cm²	40	Maximum	Maximum	169.9
	60			286.3

The inverse of square root of particle size is closely related to the hardness of the as-deposited feature (Hall, 1951; Petch, 1953). As the J_A rises, the size of the grain decreases, and the hardness of the as-deposited feature increases. The J_A increases up to a certain extent after which grain size increases and hardness decreases (Ibañez and Fatás, 2005; Yang et al.). The Hall-petch effect is the name for this phenomenon. The increase in hardness occurs during mixed activation diffusion control, and the drop in hardness occurs at the start of the dominant diffusion control phase, where globules formulation takes place (Ibañez and Fatás, 2005; Fan et al., 2022).

It was found that the height of the electrodeposited layer increase with the increase in the pulsed duty cycle (Ganesan et al., 2022; Balasubramanian et al., 2009). An increase in duty cycle occurs when current on time increases or at constant on time, off-time decreases. Further, the peak current flow for a short time leads to less deposition than the peak current flow for a long interval as inferred from Table 97.4.

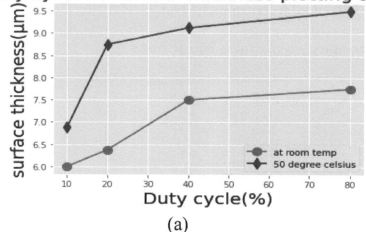

(a)

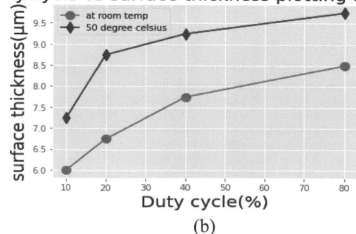

(b)

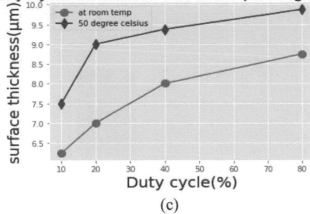

(c)

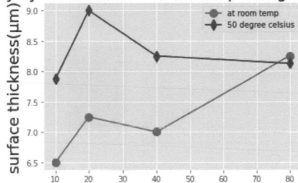

Figure 97.8 Variation of surface thickness with duty cycle at (a) 10Hz (b) 25 Hz (c) 50 Hz (d) 100 Hz

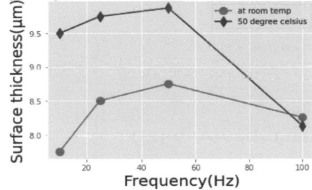

Figure 97.9 Variation of surface thickness with frequency at 80% duty cycle

The data shown in Table 97.4 is analysed graphically in Figures 8 and 9. It is inferred from Figure 97.8 that surface thickness at 50°C is more than surface thickness at room temperature with varying duty cycles for all frequencies except at a frequency of 100 Hz. Surface thickness value is maximum at 80% duty cycle for frequency of 10 Hz, 25 Hz, 50 Hz. In Figure 97.9, at 80% duty cycle, surface thickness at 50°C is more than surface thickness at room temperature for all frequencies except at a frequency of 100 Hz.

It was observed that an increase in average current density leads to an increase in hardness as shown in Table 97.9. It was also observed that a reduction in grain size occurred due to a decrease in t_{OFF}. The modification of t_{OFF} and J_A appears to be connected to the grain refining with increasing Jav (Imaz et al., 2009). The data shown in Table 97.12 is presented graphically in Figure 97.10. It is inferred from Figure 97.10 that the hardness of coating increases with an increase in average current density. The hardness value is maximum at an average current density of 10 A/dm².

Table 97.12 Hardness of copper films produced with chloride = 0.06 g/L and gelatin (GE) = 0.2 g/L as additives (Imaz et al., 2009). Where, H = Hardness

t_{ON}	t_{OFF}	d	J_A (A/dm²)	J_{av} (A/dm²)	Samples	H (GPa)
10	20	0.33	10	3.3		1.87± 0.05
10	5	0.66	10	6.7		2.38± 0.05
10	20	0.33	30	10		2.46± 0.04

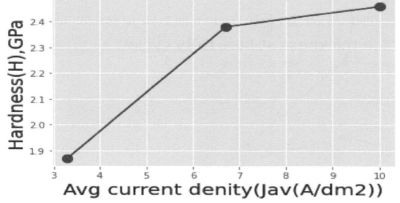

Figure 97.10 Variation of the hardness of copper films with average current density

The future direction of research

Future progress in the field of pulsed electrodeposition of copper will focus on better understanding the process' fundamental2 science and engineering, as well as expanding the technique's industrial applications. A list of future research possibilities, problems, and guidelines in the field of pulsed electrodeposition of copper is provided below:

* Fundamental understanding of the pulsed electrodeposition behaviour of copper and comparison with other materials.
* To produce compact and uniform films of acceptable microhardness, a fundamental understanding of the combined activation-diffusion control phase of the pulsed electrodeposition process is required.
* Optimisation of process parameters for pulsed electrodeposition of copper to develop the homogenous, compact and coherent surface of copper thin films.
* Optimisation of process parameters for pulsed electrodeposition of copper to develop fine grain structure, desired coating thickness and desired microhardness of copper thin films.

- Development of newer pulsed electrodeposition based techniques to improve the accuracy of copper thin films deposited by pulsed electrodeposition technique.
- Development of pulsed electrodeposition method based machine to give optimum process parameters in mixed activation-diffusion control phase to produce compact and uniform films of satisfactory microhardness.
- Selection of proper additives with specific concentrations to produce desired hardness of copper thin films.
- Modelling of pulsed electrodeposition process parameters to predict grain size, surface roughness, coating thickness and hardness of copper thin films using machine learning techniques.
- The pulsed electrodeposition technique has been automated for batch mode manufacture of deposited parts. Development of a machine tool for assessing the dimensions and precision of deposited pieces.

References

Balasubramanian, A., Srikumar, D. S., Raja, G., Saravanan, G., and Mohan, S. (2009). Effect of pulse parameter on pulsed electrode position of copper on stainless steel. Surf. Eng. 25(5):389–392. https://doi.org/10.1179/026708408X344680.

Devaraj, G. and Seshadri, S. K. (1992). Pulsed electrodeposition of copper. Plating Surf. Finish. 10:72–78.

Dikusar, A. I., Bobanova, Z. I., Yushchenko, S. P., and Yakovets, V. (2005). Throwing power of a dilute sulfuric acid copper plating electrolyte during intensive electrodeposition. Russ. J. Electrochem. 41(1):82–86. https://doi.org/10.1007/s11175-005-0009-z.

Fan, Q., Ji, X., Lan, Q., Zhang, H., Li, Q., Zhang, S., and Yang, B. (2022). An anti-icing copper-based superhydrophobic layer prepared by one-step electrodeposition in both cathode and anode. Colloids Surf. A: Physicochem. Eng. Asp. 637:128–220. https://doi.org/10.1016/j.colsurfa.2021.128220.

Ganesan, M., Liu, C. C., Pandiyarajan, S., Lee, C.T., and Chuang, H. C. (2022). Post-supercritical CO_2 electrode position approach for Ni-Cu alloy fabrication: an innovative eco-friendly strategy for high-performance corrosion resistance with durability. Appl. Surf. Sci. 577:151955. https://doi.org/10.1016/j.apsusc.2021.151955.

Hall, E. O. (1951). The deformation and ageing of mild steel: III discussion of results. Proc. Phys. Soc. Section B 64(9):747–753. https://doi.org/10.1088/0370-1301/64/9/303.

Halmdienst, M., Hansal, W. E. G., Kaltenhauser, G., and Kautek, W. (2007). Pulse plating of nickel: Influence of electrochemical parameters and composition of electrolyte. Trans. Inst. Metal Finish. 85(1):22–26. https://doi.org/10.1179/174591907X161964.

Ibañez, A. and Fatás, F. (2005). Mechanical and structural properties of electrodeposited copper and their relation with the electrodeposition parameters. Surf. Coat. Technol. 191(1):7–16. https://doi.org/10.1016/j.surfcoat.2004.05.001.

Imaz, N., García-Lecina, E., Suárez, C., Díez, J. A., Rodríguez, J., Molina, J., and García-Navas, V. (2009). Influence of additives and plating parameters on morphology and mechanical properties of copper coatings obtained by pulse electrodeposition. Trans. Inst. Metal Finish. 87(2):64–71. https://doi.org/10.1179/174591909X424807.

Jang, Y. R., Jeong, R., Kim, H. S., and Park, S. S. (2021). Fabrication of solderable intense pulsed light sintered hybrid copper for flexible conductive electrodes. Sci. Rep. 11(1):1–15. https://doi.org/10.1038/s41598-021-94024-8.

Jilani, A., Abdel-Wahab, M. S., and Hammad, A. H. (2017). Advance deposition techniques for thin film and coating. Modern technologies for creating the thin-film systems and coatings. 2(3):137–149. https://doi.org/10.5772/65702.

Krajaisri, P., Puranasiri, R., Chiyasak, P., and Rodchanarowan, A. (2022). Surface & coatings technology investigation of pulse current densities and temperatures on electrodeposition of tin-copper alloys. Surf. Coat. Technol. 435:128244. https://doi.org/10.1016/j.surfcoat.2022.128244.

Leisner, P., Møller, P., Fredenberg, M. and Belov, I. (2007). Recent progress in pulse reversal plating of copper for electronics applications. Trans. Inst. Metal Finish. 85(1):40–45. https://doi.org/10.1179/174591907X161973.

Lesage, J., Pertuz, A., Puchi-Cabrera, E. S., and Chicot, D. (2006). A model to determine the surface hardness of thin films from standard micro-indentation tests. Thin Solid Films 497(1–2):232–38. https://doi.org/10.1016/j.tsf.2005.09.194.

Lin, C., Hu, J., Zhang, J., Yang, P., Kong, X. Han, G., Li, Q., and An, M. (2021). A comparative investigation of the effects of some alcohols on copper electrodeposition from pyrophosphate bath. Surf. Interf. 22:100804. https://doi.org/10.1016/j.surfin.2020.100804.

Mohan, S., and Raj, V. (2005). The effect of additives on the pulsed electrodeposition of copper. Trans. Inst. Metal Finish. 83(4):194–198. https://doi.org/10.1179/002029605X61595.

Mughal, M. A., Newell, M. J., Vangilder, J., Thapa, S., Wood, K., Engelken, R., Carroll, B. R., and Johnson, J. B. (2014). Optimisation of the electrodeposition parameters to improve the stoichiometry of In2S3 films for solar applications using the taguchi method. J. Nanomater. 1:4–14. https://doi.org/10.1155/2014/302159.

Natter, H. and Hempelmann, R. (1996). Nanocrystalline copper by pulsed electrodeposition: the effects of organic additives, bath temperature, and pH. J Phys. Chem. 100(50):19525–19532.

Oriňáková, R., Turoňová, A., Kladeková, D., Gálová, M., and Smith, R. M. (2006). Recent developments in the electrodeposition of nickel and some nickel-based alloys. J. Appl. Electrochem. 36(9):957–972. https://doi.org/10.1007/s10800-006-9162-7.

Pagnanelli, F., Altimari, P., Bellagamba, M., Granata, G., Moscardini, E., Schiavi, P. G., and Toro, L. (2015). Pulsed electrodeposition of cobalt nanoparticles on copper: influence of the operating parameters on size distribution and morphology. Electrochim. Acta 155:228–235. https://doi.org/10.1016/j.electacta.2014.12.112.

Pc, Current (2020). Morphology , Structure and Mechanical Properties Of. 111:1–21.

Pearson, T. and Dennis, J. K. (1990). The effect of pulsed reverse current on the polarisation behaviour of acid copper plating solutions containing organic additives. J. Appl. Electrochem. 20(3):196–208. https://doi.org/10.1007/BF01076039.

Petch, N. J. (1953). The cleavage strength of polycrystals. J. Iron Steel Inst. 174:25–28.

Ramachandramoorthy, R., Kalácska, S., Poras, G., Schwiedrzik, J., Edwards, T. E. J., Maeder, X., Merle, T., Ercolano, G., Koelmans, W. W., and Michler, J. (2022). Anomalous high strain rate compressive behavior of additively manufactured copper micropillars. arXiv preprint arXiv:2201.01582.

Schwarzacher, W. (2006). Electrodeposition: A technology for the future. Electrochem. Soc. Interf. 15(1):32–33. https://doi.org/10.1149/2.f08061if.

Society , The Electrochemical (1999). Copper on-chip interconnections.

Tao, S. and Li, D. Y. (2006). Tribological, mechanical and electrochemical properties of nanocrystalline copper deposits produced by pulse electrodeposition. Nanotechnology 17(1):65–78. https://doi.org/10.1088/0957-4484/17/1/012.

Venkatesh, C., Sundara Moorthy, N., Venkatesan, R., and Aswinprasad, V. (2018). Optimisation of process parameters of pulsed electro deposition technique for nanocrystalline nickel coating using gray relational analysis (GRA). Int. J. Nanosci. 17(1–2):1–8. https://doi.org/10.1142/S0219581X17600079.

Xu, X., Zhu, Z., Xue, Z., and Zhan, X. (2021). Friction-assisted pulse electrodeposition of high-performance ultrafine-grained Cu deposits. Surf. Eng. 37(11):1414–1421.

Yang , P., Wang, N., Zhang, J., Lei, Y., and Shu, B. (2022). Investigation of the Microstructure and tribological properties of CNTs / Ni composites prepared by electrodeposition. Materials Research Express 9(3):036404.

Zhan, X., Lian, J., Li, H., Wang, X., Zhou, J., Trieu, K., and Zhang, X. (2021). Preparation of highly (111) textured nanotwinned copper by medium-frequency pulsed electrodeposition in an additive-free electrolyte. Electrochim. Acta 365(111):137391. https://doi.org/10.1016/j.electacta.2020.137391.

Zhang, W., Chen, X., Wang, X., Zhu, A., Wang, S., and Wang, Q. (2021). Pulsed electrodeposition of nanostructured polythiothene film for high-performance electrochromic devices. Sol. Energy Mater. Solar Cells 219:110775. https://doi.org/10.1016/j.solmat.2020.110775.

98 Ensemble empirical mode decomposition and Kurtosis with expert system for bearing faults diagnosis

Madhavendra Saxena[1,a] and Parag Jain[2,b]

[1]Professor, Department of Mechanical Engineering RIT Roorekee India

[2]Director RIT Roorkee India

Abstract

In this research article describe an innovative signal compression method which is based on the optimal ensemble empirical mode decomposition (EEMD) time-frequency analysis by the side of with a rule based expert system for bearing fault diagnosis. The range of bearing fault data is composed using an experimental setup which is further used for feature extraction using EEMD. Appropriate EEMD parameters are chosen for the vibration signal to be analysed so that the major feature signal can be extracted from the original signal via IMF value for the faulty bearings. The original signal of faulty bearing is decomposed by EEMD and a number of intrinsic mode functions (IMFs) are obtained. Then, the IMF with the prime Kurtosis index value is considered for additional action in terms of its make use of in an expert system. An expert system applies deterministic approach for the categorisation of bearing fault using Kurtosis values. At last, satisfactory extraction outcome are obtained for classification of bearing faults as bad, severe, and critical.

Keywords: Ensemble empirical mode decomposition, expert system, intrinsic mode functions.

Introduction

Most of the preceding work pertaining to bearing fault diagnosis was for the most part based on fast Fourier transform (FFT) investigation. Later, the time frequency analysis was dominantly applied for the same. In this work, the empirical mode decomposition (EMD) method was functional which was introduced by Huang (Tandon and Choudhury, 1999). In the action of rotating machinery, the vibration signal usually has non-stationary and nonlinear characteristics, so it is difficult to extract prominent features by using traditional Fourier transform because the collected signal is accompanied by a certain degree of noise. Therefore, EEMD is applied to decompose the vibration signal into an addition of several intrinsic mode function (IMF) components which represent the signal characteristics of different scales. The application of this tool is found in many areas such as advance digital signal processing in mechatronics systems, digital sound systems, medical equipment fault identification, etc (Srivastava and Wadhwani, 2012; Kwak et al., 2013). The present method breaks down every category of vibration signal into a small integer of IMFs (Hamadache et al, 2018; Prudhom et al, 2015). These IMFs are further used for categorisation of bearing faults using appropriate single processing and AI techniques. The common classification methods used in previous studies are decision tree, neural network, nearest neighbor algorithm, support vector machine, and so on. This article proposes a innovative method combining EEMD and expert system for classification of rolling bearing faults (Ratnam et Al, 2018; Lilo et al., 2012).

Figure 98.1 Bearing 6204 with faults to be tested

---

[a]drmadhavendra.me@ritroorkee.com; [b]director@ritroorkee.com

A. Experimental setup

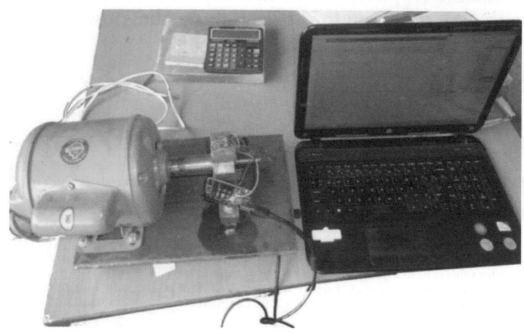

Figure 98.2 Test bearing with devices for vibration generation

Operation and analysis

A. Empirical mode decomposition

In this research work, four types of bearing faults are taken into consideration viz. inner race fault, outer race fault, ball fault, and cage fault having fault level of low, medium, and high. For each fault level, six bearings are seeded so that 18 bearings (6*3) of one type of fault are prepared and in total 72 bearings (18*4) for four types of faults are used in this work. Bearing vibration data are collected using the experimental setup for a recording duration of one second and each second consists of 1800 data points and a sampling rate of 20 kHz. The bearing fault features are extracted using the following steps as given below:

*   Break up the vibration signals into different IMFs.
*   Calculate the surface energy of each IMF.
*   Calculate the total surface energy of all IMFs.

B. Process for bearing faults analysis

Following steps are used for bearing fault analysis:

Step 1: Generate in a row, the data of bearing faults waveform (Matrix size 1x1800).
Step 2: find IMFs using the EEMD in Matlab.
Step 3: Estimate the IMFs

The IMF's are premeditated from healthy and faulty bearing vibration signals and shown in Figure 98.1.The Kurtosis of first few IMFs is extracted as fault feature index. Then expert system is in employment to the fault features as the linear transform for dimension reduction and elimination of linear dependence between the fault features. The above steps are useful to all the bearing fault data. Out of four types of faults, one set of figures are shown for low, medium and high level of inner race fault in Figure 98.2.

2.3 Kurtosis based expert system for characterisation of bearing faults

Kurtosis is a distribution function and the fourth classify moment has an important force to differentiate the condition of the bearing (Pandya et al., 2013; Lin 2010). The Kurtosis can be worn to interpret the fault in time province, frequency province, and time-frequency province. Kurtosis has different values at different points of IMF (Yang et al., 2007). Sample graph for healthy and inner race fault is shown in Figures 98.3 and 4 respectively.

Kurtosis based algorithm for characterisation of bearing faults

A. Expert system

Expert systems are computer applications which are used for solving certain types of problems because of its on-logarithmic expertise. Expert systems are widely used in numerous diagnostic applications such as playing chess, making financial planning decisions, configuring computers, monitoring real-time systems, underwriting insurance policies, and performing many services that previously required human expertise (Sawalhi et al., 2004; Harsha, 2013).

An expert system applies deterministic approaches for categorisation. The core of an expert system is a set of rules, where the 'real intelligence' from human experts in any specific area is translated into the 'artificial intelligence' in computers. The presentation of categorisation is substantially dependent on a set of IF-THEN rules and the inference engine that executes the reasoning of these rules (Harsha, 2013; Gupta et al., 2011). The most important disadvantage of the expert system is the need for a prearranged threshold value to make binary decisions, and choosing undesirable thresholds leads to less accurate categorisation (Kankar et al., 2001).

The expert systems are suitable for characterising the bearing faults because of the following reasons (Muralidharan and Sugumaran, 2013):

1. Even in absence of all the pertinent information, the expert system can decide with partial knowledge.
2. It also facilitates the expansion of the information base.

The block diagram for characterisation of bearing faults is shown in Figure 98.5. The four steps used in the block diagram for characterisation of bearing faults are explained below:

Step 1 and 2 belong to the application of the IMF of healthy and faulty bearing to calculate Kurtosis values for knowledge discovery.
Step 3 belongs to the generation of the knowledge base for the expert system.
Step 4 is the application of an expert system for characterising the test signals of bearing faults.

The knowledge base required for the expert system has been generated using the following algorithm as given below:

Step 1: interpret the time-series data of healthy and faulty bearings and find the values of Kurtosis
Step 2: Generate a knowledge base based on values of Kurtosis
Step 4: Application of knowledge base on test signals for characterisation of bearing faults

3.2 Knowledge base (rules) for expert system

The hypothetical aspects of an expert system have been discussed previous. The expert system gives output based on the knowledge base (Ratnam et al., 2018). After the comprehension discovery in the form of numeric values of Kurtosis, the following knowledge base is prepared for characterisation of bearing faults as given in Table 98.1 and 2.

Knowledge based system is used for characterisation for bearing faults according to maxima and the average value of the Kurtosis range (Liu, 2008). It is the heart of the expert system, which is more often than not composed of a set of IF-THEN or IF-THEN-ELSE rules for identifying and solving the desired problem.

Table 98.1 shows the input feature based on average value and maximum Kurtosis value range of bearing condition and justification if and then rule set up by using an expert system that is knowledge-based and shows the results in three conditions of unhealthy bearing, bad, severe, and critical.

Table 98.2 shows the bearing faults which are characterised as healthy, severe, critical and bad based on the Knowledge base for characterisation for bearing faults (Saxena et al., 2016; Saxena and Gupta, 2019).

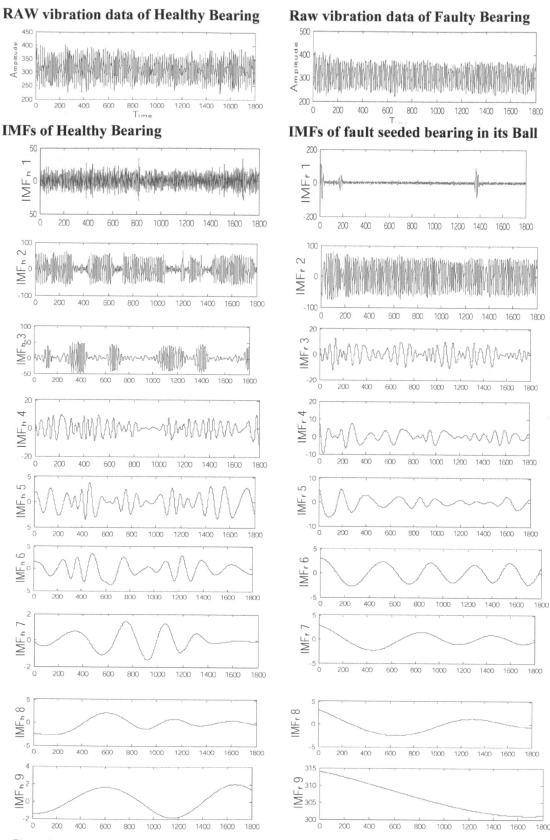

Figure 98.1 IMF of healthy and faulty bearing

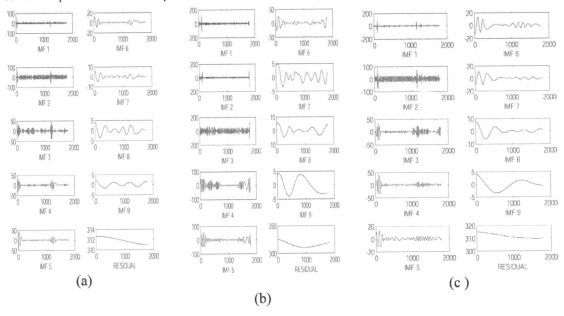

Figure 98.2 IMF of (a) Inner race rusty (b) Indentation in inner race (c) Indentation of high index

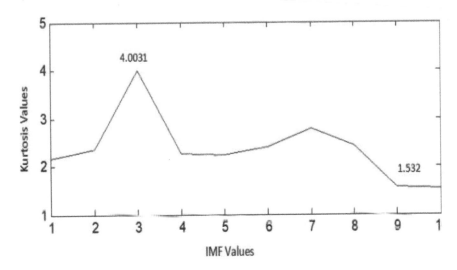

Figure 98.3 Local maxima and minima of healthy bearing

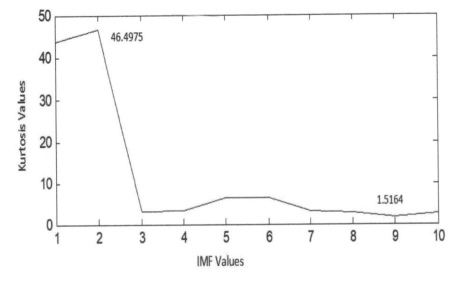

Figure 98.4 Local maxima and minima of inner race fault bearing

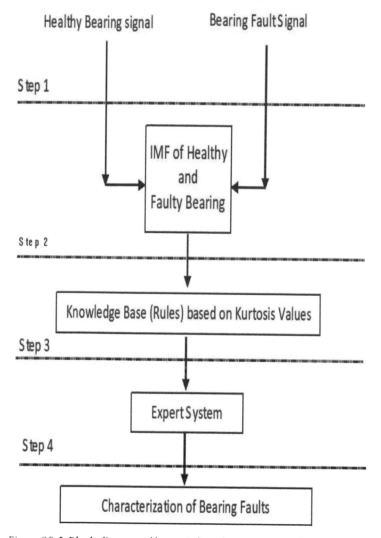

Figure 98.5 Block diagram of kurtosis based expert system for characterisation of bearing faults

Table 98.1 Rule generation for expert system

S. No.	Input Feature		Justification		Rule generation
	Average and maximum value of Kurtosis		*If k*	*And k*	*Severity level of fault*
	Average ($K_{Avg.}$)	*Maximum ($K_{Max.}$)*			
1	0–3	0–3	$K_{Avg} \geq 0$	$K_{Avg} \geq 3$	Healthy
			$K_{Max} \geq 0$	$K_{Max} \geq 3$	Healthy
2	3–9	3–15	$K_{Avg} > 3$	$K_{Avg} \geq 9$	Bad
			$K_{Max} > 3$	$K_{Max} \geq 15$	Bad
3	9–27	15–50	$K_{Avg} > 9$	$K_{Avg} \geq 27$	Severe
			$K_{Max} > 15$	$K_{Max} \geq 50$	Severe
4	27- Above	50- Above	$K_{Avg} > 27$	--	Critical
			$K_{Max} > 50$	--	Critical

Table 98.2 Knowledge base for characterisation for bearing faults

S No.	Type of Bearing Faults	Level of Fault	Kurtosis values			Severity level of fault
			Maximum	Minimum	Average	
1	Healthy	Normal	4.0031	1.532	2.767	Healthy
2	Inner race	Low	46.497	1.5164	24.006	Severe
		Medium	33.913	1.723	17.818	Severe
		High	78.449	1.801	40.125	Critical
3	Outer race	Low	6.8155	1.672	4.243	Bad
		Medium	11.980	1.646	6.813	Bad
		High	47.104	2.336	24.720	Severe
4	Ball	Low	5.523	1.415	3.469	Bad
		Medium	4.859	1.746	3.302	Bad
		High	32.089	1.6832	16.886	Severe
5	Cage	Low	19.6037	2.2025	10.9031	Severe
		Medium	7.9559	1.7471	4.8515	Bad
		High	9.0786	1.6916	5.3851	Bad

Result and discussion

The time-series data of different bearing faults are generated using the experimental setup and the above-mentioned method is applied to time series test data for the following bearing faults:

a) Inner race
b) Outer race
c) Ball
d) Cage

As per the above described method, bearing vibration data is used to find IMF using EEMD which is further used for finding the Kurtosis value. These Kurtosis values were used in a rule based expert system for characterisation of bearing faults which could successfully classify the bearing faults as healthy, bad, severe, and critical according to the condition of the bearing.

Conclusion

A bearing fault diagnosis method based on EEMD and expert system is put forward in this paper. EEMD method is suitable for analyzing complex multicomponent signals. For the fact that the vibration signal is nonlinear and unstable, EEMD method is chosen to precondition the vibration signal of the roller bearing to produce a set of IMF components. As a comparison, we replaced EEMD with wavelet and then repeated the process of feature extraction and classification. The results of the application on the real data show that the accuracy of classification based on wavelet is high, but the EEMD based method is higher; that is, EEMD method can effectively extract the signal feature effectively in the form of IMF and Kurtosis value. These Kurtosis values were used in a rule based expert system for characterisation of bearing faults which could successfully classify the bearing faults as healthy, bad, severe, and critical according to the condition of the bearing.

References

Hamadache, M., Lee, D., Mucchi, E., and Dalpiaz, G. (2018). Vibration-based bearing fault detection and diagnosis via image recognition technique under constant and variable speed conditions. Appl. Sci. 8(8):392. https://doi.org/10.3390/app8081392

Harsha, S. P. (2013). Expert Systems with Applications Fault diagnosis of rolling element bearing with intrinsic mode function of acoustic emission data using APF-KNN. Expert Systems Appl. 40(10):4137–4145.

Kwak, D. H., Lee, D. H., Ahn, J. H., and Koh, B. H. 2013. Fault detection of roller-bearings using signal processing and optimisation algorithms. Sensors. 14(1):283–298.

Lilo, M. A., Latiff, L. A., Bin, A., and Abu, H. (2012). Identify and Classify Vibration Fault Based on Artificial Intelligence Techniques. J. Theoretical Appl. Info. Technol. 94(2):464–474.

Lin, L. C. 2010. The Wavelet Tutorial. Durham: Duke University.

Liu, J. 2008. An Intelligent System for Bearing Condition Monitoring. UWSpace. http://hdl.handle.net/10012/4033

Muralidharan, V. and Sugumaran, V. (2013). Feature extraction using wavelets and classification through a decision tree algorithm for fault diagnosis of a mono-block centrifugal pump. Measurement. 46(1):353–359.

Pandya, D. H., Upadhyay, S. H., and Harsha, S. P. (2013). Expert Systems with Applications Fault diagnosis of rolling element bearing with intrinsic mode function of acoustic emission data using APF-KNN. Expert Syst. Appl. 40(10):4137–414.

Prudhom, A., Antonino-Daviu, J., Razik, H., and Alarcon, C. (2015). Time-frequency vibration analysis for the detection of motor damages caused by bearing currents. Mech. Syst. Signal Process. 1–16.

Ratnam, C., Jasmin, N. M., Rao, V. V., and Rao, K. V. (2018). A comparative experimental study on fault diagnosis of rolling element bearing using acoustic emission and soft computing techniques. Tribol. Indu. 40(3):501–513.

Sawalhi, N. and Randall, R. B. (2004). .The Application of Spectral Kurtosis to Bearing Diagnostics. Proceeding acoustic. 3–5.

Saxena, M., Bannet, O., Gupta M., and Rajoria, R.. (2016). Bearing Fault Monitoring Using CWT Based Vibration Signature. Procedia Eng. 144:234–241. doi: 10.1016/j.proeng.2016.05.029.

Saxena, M. and Gupta, M. (2019). Bearing Fault Diagnosis Using Intelligent Methods: A Review. International Conference on Recent Trends and Innovation in Engineering, Science & Technology. ICRTIEST - Poornima University.

Srivastava, A. and Wadhwani, S. (2012). Condition Monitoring for Inner Raceway Fault. Int. J. Electr. Eng. 5(3):239–244.

Tandon, N. and Choudhury, A. (1999). Review of vibration and acoustic measurement methods for the detection of defects in rolling element bearings. Tribol. Int. 32:469–480.

Yang, Y., Yu, D., and Cheng, J. (2007). A fault diagnosis approach for roller bearing based on the IMF envelope spectrum and SVM. Measurement. 40(9–10):943–950.

99 Performance of anaerobic digestion process by Utilisation of energy crop on lab scale reactor

H. M. Warade[a], Monika Tiwari[b], and Falguni Dachewar[c]

Civil Department YCCE Nagpur, India

Abstract

Storage of fossil fuel such as oil, natural gas, coal is not infinite. There is increase in demand of energy due to increase in population, therefore there is more focus towards renewable energy sources which has less cost and cannot be depleted. Biogas is generated by breakdown of organic substances without oxygen in a close reactor called as anaerobic digestion. The product we get from this process is an energy source called biogas and manure as an outlet which is further used for plant growth. The aim of our research is to check performance of Napier grass (NG) with cow dung (CD) in anaerobic digester as well as maximise the ratio of Napier grass blending with cow dung. Reactors were employed to generate the biogas production potential of organic waste substrates. The reactors are made up of either modification in water can bottles or readymade plastic cans. After several changes in components these reactors are used as a bioreactor. The study found the production of biogas was $0.43m^3$/kg VS for (NG:CD 50:50) in the average room temperature 33°C. Just as in the ratios of (NG:CD 65:35), (NG:CD 75:25), (NG:CD 85:15) and (NG:CD 95:05) same methodology was followed and activities were undertaken. In this ratios production of biogas were found 0.48, 0.40, 0.29, 0.21m3/kg VS respectively.

Keywords: Anaerobic digestion process, biogas, energy crop, methane.

Introduction

In the absence of oxygen, anaerobic digestion (AD) represents the biological breakup of organic substrate (Warade 2019). In anaerobic digestion process, its biodegradability often dictates its benefits, effects, limits and specifications, which are uniquely applicable to each substrate (Wang 1995). Napier grass and cow dung are the substrates which proves its high biodegradability in the anaerobic reactors (Rekha 2013). Several important parameters such as temperature, retention time, blending ratios, substrate-related conditions such as Organic loading Rate and pre-treatment are the control parameters in the degradation process (Lindorfer 2008). Several experimental study is generally performed in advance in order to identify the critical issues of conventional as well as modified anaerobic reactors. AD is the result of a number of metabolic relationships between different microorganism (Chen 2008). Hydrolysis, acidogenesis and methanogenesis are three phases of anaerobic digestion process (Pereira 2005). Enzymes hydrolyzing polymers into monomers such as sugar and amino acids are secreted by the first set of microorganisms (Sosnowski 2003). These are then transformed into stronger volatile fatty acids, H_2 and acetic acid, by the second set, i.e. acetogenic bacteria. Third, methanogenic, bacteria transform H_2, CO_2, and acetate to methane (Amon 2007).

Method adopted

This section elaborates the parametric study of bio methanation from lignocellulosic substrate and co-substrate. Energy crop i.e. Napier grass co digested with cattle dung or cow dung in lab scale batch mode reactors. The effect of digester temperature, atmospheric temperature, particle size, substrate to co-substrate ratios and solid and moisture content was studied in this experimental work. The feeding material used in this research is Napier Grass promoted as a main substrate and cow dung taken as a co-substrate. Anaerobic digestion is the degradation process of any organic matter without oxygen present in the reactor or digester.

[a]hmwarade@ycce.edu; [b]tiwarimonika634@gmail.com; [c]falgunidachewar96@gmail.com

A. Feedstock

• Cow dung is a disposal material of animal species also recognised as cow pats, cow pies or cow mist. Cow dung varies in colour from brownish to blackish, often darkened shortly after air contact. Cow dung that is generally dark brown colour is being used as fertiliser in agriculture farming unless the species of earthworms recycle them into the soil.
• Hybrid Napier grass is also called elephant grass owing to its tallness and energetic development. The tiller crops and a single clump can generate more than 50 tillers in favourable atmosphere and soil circumstances.
• The digested slurry was taken from a fully cow dung-based fixed dome digester of 6 cu.m size from residential house. Digested slurry was used as seeding for reactor. It collected and packed in tightly capped 20 L plastic bottles.

B. Equipment used

In this study few types of equipment are used for the preparation of substrate for mixing in the reactor. Particle size of feeding substrate is playing a very important role in degradation process. The study is based on grass bio methanation in lab scale reactor without pre-treatment of lignocellulosic material.

• Chaff cutter machine: A manufactured precision-engineered chaff cutter (table model) engine that is used for cutting of animal fodder in uniform manner also preparation of raw product in agro based industries. Chaff cutter machine are generally available in different capacity of HP but 2 HP motorised table model machines are preferable due to single phase electricity capacity. In the market chaff cutter machines are equipped with electric motor, pulley and strap.

Figure 99.1 Chaff cutter machine

The cutting level of grass in machine as shown in Figure 99.1 is about 4 to 5 mm in length pieces. These sizes of cut pieces of grass are easily movable at inlet section of the reactor. Sometimes leaves part of grass remain in long length usually 8 to 9 mm after cutting in chaff cutter due to the high sucking capacity of motor.

• Grinder: A wet grinder, which is illustrated in Figure 99.2, could be used to cut tough products or to prepare meals for paste or batter, particularly in the kitchen of India. A wet grinder for abrasive grinding utilises lubrications or heating liquid, whereas water is used to generate batter in food preparation in combinations with field grains.

Figure 99.2 Grinder

A grinder consists of revolving drum with the help of electric motor generally known as idli-dosa batter grinders. In our study this grinder is used for grinding grass material for lab scale reactor as well as pilot reactor.

C. Reactor used

The different digesters were assembled under different feedstock ratios are as follows. The substrates for this research have been gathered from the agricultural campus of the Institute.

- Reactor-I: NG:CD (50:50)

Reactor-I set up in the Plastic bottle of volume 5 lit. in which Napier grass and cow dung used as a feedstock material. The quantity of Napier Grass and cow dung has been taken as 1 kg each.

Figure 99.3 Lab scale batch reactor for ratio NG:CD-50:50

The reactor worked on water replacement method in which one fully water filled bottle was connected to reactor. A pipe attached from water bottle transferred to empty jar so that collect the water comes from bottle due to gas pressure creation in reactor (Figure 99.3). This reactor monitored continuously up to 45 to 50 days in the atmospheric temperature of 35°C.

- Reactor-II: NG:CD (65:35)

Reactor-II was also set up in the plastic bottle of volume 5 lit. in which Napier grass and cow dung used as a feedstock material (Figure 99.4). Now, in the ratio of Napier and cow dung, the quantity of Napier grass is increased by 15% making the ratio as 65% of Napier grass with 35% of cow dung. The quantity of Napier grass and cow dung has been taken as 1.3 and 0.7 kg respectively.

Figure 99.4 Lab scale batch reactor for ratio NG:CD-65:35

- Reactor-III: NG:CD (75:25)

Reactor-III was set up in the plastic bottle of volume 5 lit. in which similar methodology has adopted. Now, in the ratio of Napier and cow dung, the quantity of Napier grass is increased by 10% making the ratio as 75% of Napier grass with 25% of cow dung.

Figure 99.5 Lab scale batch reactor for ratio NG:CD-75:25

The quantity of Napier grass and cow dung has been taken as 1.5 and 0.5 kg respectively. The arrangement of this reactor was similar to previous reactors (Figure 99.5).

- Reactor-IV: NG:CD (85:15)

Reactor-IV was set up in the plastic bottle of volume 5 lit. in which the quantity of Napier grass is increased by 10% making the ratio as 85% of Napier grass with 15% of cow dung.

Figure 99.6 Lab scale batch reactor for ratio NG:CD-85:15

The quantity of Napier grass and cow dung has been taken as 1.7 and 0.3 kg respectively. The arrangement of this reactor was similar to previous reactors (Figure 99.6).

- Reactor-V: NG:CD (95:05)

Reactor-V was set up in the plastic bottle of volume 5 lit. in which the quantity of Napier grass is increased by 10% making the ratio as 95% of Napier grass with 05% of cow dung (Figure 99.7). The quantity of Napier grass and cow dung has been taken as 1.9 and 0.1 kg respectively.

Figure 99.7 Lab scale batch reactor for ratio NG:CD-95:05

After collection of feedstock materials, proper mixing ratio of substrate and co substrate were decided on the basis of volume of the respective digesters. Mixing of the materials was carried out manually and it was made sure that no impurities and lumps are present in the mixture. Around 30 to 50 days observations on each reactor were carried out. All the reactors are assembled air and water tight so that it creates pressure to replace the water due to biogas generation. Manual agitation was performed on the reactors on a daily basis in order to ensure intimate contact between the microorganisms and the substrate for effective biogas production. The gas produced by the substrates inside the anaerobic digester was passed to the water container through pipes.

D. Analytical method

Process efficiency and process consistency must be assessed to obtain an overall idea of ongoing anaerobic treatment. Process efficiency is controlled by the content of inlet as well as outlet solid substances and the production of biogas. The pH, volatile fatty acid (VFAs), total and volatile solids, and alkalinity of effluent of the reactor are shown to be stable in a given process during or at the end of the retention period. Following are the various parameters have been assessed and determined.

Results and discussion

Lab scale reactors were operated on organic substrates in which Napier grass was considered as main feeding substrates; and cow dung as a co-substrate. For the analysis of bio gas produced by anaerobic digestion, a total five PVC reactors were erected. The PVC reactors contained slurry of Napier Grass as a substrate with cow dung mixed in different ratios.

E. Performance of lab scale reactor

To find and assess the maximum capacity of the Napier grass for producing the biogas, the researcher has undertaken his research on lab scale reactor. With this intent, the first ratio tried was of Napier grass to cow dung as (NG:CD 50:50). Further ratio increased by 15% for achieving (NG:CD 65:35) and next ratios increased by 10% for continuous increment in Napier grass contribution like (NG:CD 75:25), (NG:CD 85:15) and (NG:CD 95:05) respectively.

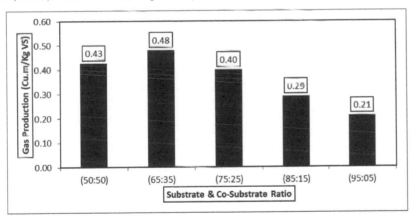

Figure 99.8 Gas production in lab scale reactor

After the setup of the digester for around 50 to 55 days, activities of monitoring the digester, proper mixing of slurry etc. were undertaken; obviously there was low gas production during these initial days. Later after few days, gas production is testified by water replacement method. We found the production of biogas was 0.43m³/kg VS for (NG:CD 50:50) in the average room temperature 33°C. Just as in the ratios of (NG:CD 65:35), (NG:CD 75:25), (NG:CD 85:15) and (NG:CD 95:05) same methodology was followed and activities were undertaken. In this ratios production of biogas were found 0.48, 0.40, 0.29, 0.21m³/kg VS respectively. On the observation of Figure 99.8, it was noticed that gas production continuously decreasing manner, when Napier grass content has been increased. It was observed that more lignocellulosic contents affect production rate of biogas, also takes more degradation time i.e., HRT in reactor also beyond a particular ratio, anaerobic digestion was not feasible.

F.   Characteristics of inlet slurry and outlet slurry

Inlet slurry contains mixture of Napier grass and cow dung. In this research, this mixture varies as per the determined ratios, since the gas production quantity and quality depend upon the inlet slurry. Outlet slurry is the homogeneous mixture of the substrates obtained from the outlet of the reactor after the extraction of gas. The outlet slurry is further used being manure or fertiliser. The characteristics of the slurry are important to study such as: Moisture content, total solids, volatile solids and fixed solids.

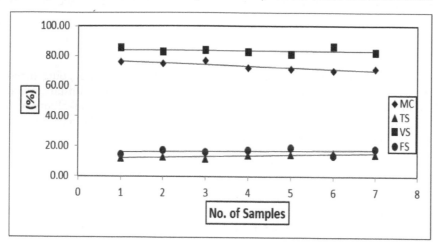

Figure 99.9 Inlet characteristics

The Figure 99.9 shows the variation in the characteristics of the substrates used in the inlet slurry. The average is taken of the overall readings of the samples wherein the average of moisture contents is 73.04%, total solids is 13.48%, volatile solids is 83.49% and fixed solids is 16.51%.

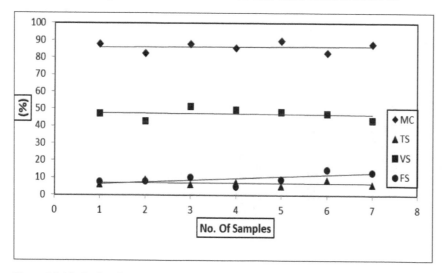

Figure 99.10 Outlet characteristics

The Figure 99.10 shows the variation in the characteristics of the outlet slurry. The average is taken of the overall readings of the samples wherein the average of moisture contents is 86.12%, total solids is 6.94%, volatile solids is 47.10% and fixed solids is 9.62%. It is obvious that in comparison with findings of the inlet slurry, apart from the moisture content, all other solid contents in outlet slurry are getting reduced.

G.   VS reduction at lab scale reactor

As it is already mentioned inevitable importance of volatile solids, in the lab scale level also it was analysed and assessed.

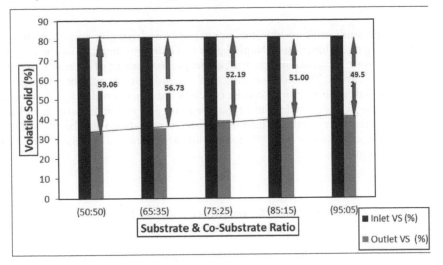

Figure 99.11 VS reduction in lab scale reactor

The methodology and activities were adopted in the lab scale reactor just as pilot scale reactor. On comparison of results (Figure 99.11) about the characteristics of inlet and outlet slurry for five ratios i.e. (NG:CD 50:50), (NG:CD 65:35), (NG:CD 75:25), (NG:CD 85:15) and (NG:CD 95:05) here, the VS reduction in Lab scale reactor was found 59.06, 56.73, 52.19, 51.00, and 49.52% respectively. It was observed that VS Reduction percentage reduced by 53.70% by increasing content from 50% to 95%. At higher ratios of Napier grass, VS reduction dropped considerably.

H. pH range at lab scale reactor

In this lab scale model, the combination of Napier grass with cow dung was taken. The similar kinds of process, methods and observations were taken as the previous.

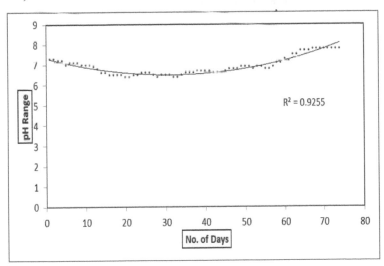

Figure 99.12 pH variation (NG:CD) in lab scale reactor

Throughout the 75 days of the observation, initially pH was 7 to 7.2. The observations were carried out continuously and after 10 days, pH was found to be less than 7. It caused decline in the gas production. The abundant rise of pH that went up to 7.8 and reached its alkaline phase. Here also, it is observed that pH value of the slurry in the reactor ranges from normal to acidic in 10 days approx.; subsequently from acidic to alkaline. The pH value recorded at an average range was 6.4 to 7.8. From the Figure 99.12 it is evident that the growth of the anaerobic microorganism is approximately normal.

Conclusion

The researcher has undertaken this study with an objective to evaluate an alternative source of energy in terms of the biogas production with a combination of Napier grass (NG) and minimum possible cow dung (CD); utilise digestate as fertiliser for cultivation and thus make proper assessment of economic merits of Napier grass cultivation and its utilization. The biogas production in combination with the various ratios of Napier grass and cow dung on lab scale reactor, it can be concluded that gas production decreased at an average rate of 19% by increase in the contribution of Napier grass and optimum ratio found to be NG:CD (65:35). On comparing outlet slurry with the inlet slurry, certain variation was observed in outlet slurry that the moisture contents is increased by 15.19%, total solids are decreased by 48.51% and fixed solids remained more or less constant.

References

Amon, T., Amon, B., Kryvoruchko, V., Zollitsch, W., Mayer, K., and Gruber, L. (2007). Biogas production from maise and diary cattle manure – influence of biomass composition on the specific methane yield. J. Agric. Ecosyst. Environ. 173–182.

Chen, Y., Cheng, J. J., and Creamer, K. S. (2008). Inhibition of anaerobic digestion process- a review. J. Bioresour. Technol. 4044–4064.

Lindorfer, H., Lopez, C. P., Resch, C., Braun, R., and Kirchmayr, R. (2008). The impact of increasing energy crop addition on process performance and residual methane potential in anaerobic digestion. J. Water Sci.Technol. 55–63.

Pereira, M. A., Pires, O. C., Mota, M., and Alves, M. M. (2005). Anaerobic biodegradation of oleic and palmitic acids: evidence of mass transfer limitations caused by long chain fatty acids accumulation onto the anaerobic sludge. J. Biochem. Microbiol. 15–23.

Rekha, B. N. and Pandit, A. (2013). Performance enhancement of batch anaerobic digestion of napier grass by alkali pre-treatment. Int. J. Chem. Technol. 558–564.

Sosnowski, P., Wieczorek, A., and Ledakowicz, S. (2003). Anaerobic co-digestion of sewage sludge and organic fraction of municipal solid wastes. J. Adv. Environemt. Res. 609–616.

Wang, X., Yang, G., Feng, Y., Ren, G. and Han, X. (1995). Optimizing feeding composition and carbon– nitrogen ratios for improved methane yield during anaerobic codigestion of dairy, chicken manure and wheat straw. J. Bioresour. Technol. 78–83.

Warade, H., Daryapurkar, R. and Nagarnaik, P. B. (2019). Review of Biogas Production from various substrates and Co-substrates through different Anaerobic Reactor. Int. J. Emerg. Technol. 235–242.

100 Application of supply chain management in iron and steel industries

Ritu Shrivastava[1,a], Shiena Shekhar[2,b], and R. B. Chadge[1,c]

[1]Mechanical Engineering BIT, Bhilai Institute of Technology, Durg, India

[2]Mechanical Engineering YCCE Yeshwant Rao Chavhan College of Engineering, Nagpur, India

Abstract

Steel and metal factories as a base enterprise have made a considerable benefaction to our nation's financial upliftment. In the previous era, metal industry regarded an experienced downfall due to the points like excess capacity and more, consequently, the regeneration Steel manufacture in Chhattisgarh has been a major focus for industrial prosperity, and steel storage has become a popular topic of research. Rational reservoir can decrease the time of the enterprise technique, make the business activity increases the response of the marketplace, advances the manufacturing and enterprise working of the company capability, decreases unimportant stockings and associated charges, and accelerate the turnover of capitals of the enterprise. Consequently the way to achieve reasonable stock through the use of superior management strategies and prototypes representing mathematically; a way to fulfil clients' service necessities through selling manufacturing capability of steel and metal industries within the enterprise must be the pursuit of the purpose. The article is written to find out the basic problems focused on the characteristics of Chhattisgarh's iron and steel manufacturing inventories, as well as the management of it 'excessive price, low income', use the ways of quick deliver chain stock control model, to resolve the troubles of the stock check.

Keywords: Inventory management, SCM system, steel industry, supply chain, stock.

Introduction

The steel industry is consistently a majority of government's public finances, making it the first to acknowledge the need to speed up manufacture. Our economy also donates a substantial amount of money through a wide range of financial mechanisms to resolve the issues of the steel sector. As the worldwide data innovation creates furthermore market contest increases, steel industry is battling lately with numerous homegrown steel ventures working execution declining, and even some seeming immense misfortunes. Revealed by Chhattisgarh steel relationship, in beyond two years, Chhattisgarh steel industry's yearly creation deals overall revenue is under 3%, well beneath the business normal benefit level which is 6% (Liu, 2012). As a result, it is crucial to examine the inventory problem in the steel enterprise production chain specifically the stock control framework under enhanced distribution networks, from the perspectives of the metal sector.

For a sensible stock control plot, firstly check the qualities of production network stock of steel manufacturing. Steel industry is a common manufacturing sector for the many part in the network of store. Manufacturing sector generally alludes to the work that accomplishes increments in the regular producing ways from raw materials to the finished goods comprises of physical or any other change Ma et al. (2001). The trademark of unrefined components as compared to handling courses and end results is quite low. As a rule, every collaboration has averaged fixed providers furthermore clients, through them around 75% of its inventory and deals are received. In the peak time of production the steel businesses those are overwhelmed through interaction, it is primarily important to set up relation of client with the executives. The metal store chain is not the same as the overall gathering venture production network, the qualities of the steel store network stock as follows:

[a]mechritu05@gmail.com; [b]shiena.shekhar@bitdurg.ac.in; [c]rbchadge@rediffmail.com

Table 100.1 Steel industry supply chain and inventory properties (Wenbo, 2013)

Manufacturing processes and time information	Data about the manufacturing process (processing time, queuing time, time of setup, in each every process involving number of machine, alternate route) Data from the calendar (information of shift, information of holiday, preventive maintenance information) Data from machines (number of machine, mean time to failure, mean time to repair, alternate data from resources, preventive maintenance time) Material structure bill
Inventory control policies information	Reorder point, and safety stock level Finished goods, raw materials, and intermediate parts inventory levels Anywhere on the shop floor, there is a stock.
Procurement and logistics information	Time it takes for a supplier to respond Size of the supply lot Purchasing capacity of suppliers Sourcing time horizon
Demand information	Priority of due date Demand for data at the start and end pattern
Policies strategies information	Policies for order control and dispatch

Assessing the steel industry's inventory issues in the supply chain

In light of the attributes in the business of steel supply chain, it will be known the absence of viable inventory network system. The end clients of steel items generally are the fabricating enterprises, and by far most of items need further profound handling. Clients have an unmistakable request to assortments, determinations, costs of steel item, what's more iron and steel undertakings transport products mostly by stream and rail route. The expansion of item stock will at last influence the steel business undertaking's business benefit. Based on broad steel item stockpiling process, we breakdown the stock issue:

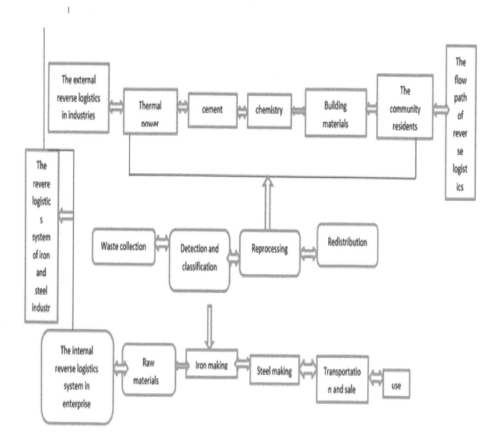

The Framework model of reverse supply chain network system in iron and steel industry

Figure 100.1 The framework model of reverse supply chain network system in iron and steel industry

A. Concentrations of steel in the industrial sector are low

There are many different ways to evaluate intensity and the associated indicators; however we adopt absolute concentration (CRn) as the index to quantify volume sales in this case. The Bhilai iron and steel group's CR4 was reconstructed in 2008, although it was still fewer than 30% when it was rebuilt in 2010. Despite the fact that the steel industry in Chhattisgarh is growing, the actual level of fixation is low (Liu and Xie, 2006). The fixation of Chhattisgarh's biggest steel companies has been increased to 43%. As per the Bain arrangement, the work in steel making sectors is going good if CR4 is 30% (Yang, 2012). Each project has a small market share, and the scope of the project isn't large enough to gain control.

B. Asymmetry of information between companies

Right now, because of certain endeavours under the steel industry store network absence of data trading furthermore sharing, or various significant data sets not executing viable connection, a few organisations can't get a handle on the genuine requirements of the downstream and the stock limit of upstream. Furthermore simultaneously production network can't acknowledge products trading and traveling, just to have enormous stock. Furthermore also stock administration of small number of little carbon enterprise is as yet in the circumstance of fake documentation and moving data, influencing numerous endeavours' data framework's similarity, with the goal that it isn't well to be coordinated (Zhong and Xu, 2004). Along these lines, how to pass on data successfully is the issue to address to work on the exhibition of store network stock the board.

C. Deficiency of SCM system

Ideas of management of inventory are reversed

Alloy making ventures are generally claimed stock, simple to cause stock intensification between the subordinate endeavours bit by bit, designing 'bullwhip impact'. To meet demand for unprocessed components on time and avoid shortages, the supply office typically prepares a good amount of security stock and implements a two stage stock management system (Wang, 2004). This will increase the project cost by forcing some materials stock to be rehashed, chime up unrefined components and assets, and face the interest burden, and so on. Furthermore paying for the capacity, authority expenses, with the misfortune, decay, also oldness hazard. Simultaneously, as the stock the board actually existing administration man-machine coinciding, some of the time with the peculiarity which can't get stock history information precisely, in the end it fabricate up the excess of stock or stock deficiency.

The inventory structure of the organisation is impractical

The design of stock association nonsensical circumstance exists in many undertakings of our nation's Iron also steel industry, area crossing, hazy obligations, and the issues existing in the different phases of big business improvement. Different divisions' capacities in big business are unique, additionally unique to the particular necessities. For the outreach group, they'd like different steel types of items, with the goal that undertaking can rapidly open the market, while the creation area desire to make single assortment yet large scale manufacturing steel item with the activity as straightforward as could be expected. Buying office trusts that request amount is the more the better, in this manner they can decrease acquirement cost and value arrangement, inspecting business, while coordination divisions trust little however many clumps appearance, so they can decrease stock, and diminish the trouble of stock the board. Clashes between offices will lead to additional decay business deals figure precision rate, to build obtainment volumes, to expand stock levels and stock expenses, to dial back coordination turnover, and at last influencing the business effectiveness furthermore the board level (Liu and Xie, 2006).

Stock forecast is deficient due to a shortage of demand

According to the data, the steel industry in Chhattisgarh has over 500 steel markets. Only 4% of the foundry industry's steel items trade for more than 1 million rupees, although these steel ventures' important parameters total more than 8000 outlets and deals rise at the same time, bringing significant deals costs. Clearly, Chhattisgarh's steel making engineering lack a comprehensive warehouses and iron and steel industries in the state depends on individual contacts in the sector to obtain market volatility internal data. Part of the venture stock interest estimation is still in the experience stage of the executives due to the lack of a clear and precise projection strategy. In this vein, once the situation has shifted, alongside the increment of stock.

Irrational inventory control

As of now, a large portion of Chhattisgarh iron and steel undertakings need of factual science sensible examination implies, regularly to make acquirement arrangement as per the deals and packages of every phase of the mixtures of iron. What's more in the capacity interaction, individual work postponed and abused what's more steel items are heaped in large quantity, to discover that the duration searching for stock products is excessive, causing the attempt to be unable to comprehend the stock control approaches of first in, first out (FIFO), resulting in a significant waste of labour, material assets, and monetary assets. The control work for stock framework to front line dealings is deteriorating these days (Liu and Xie, 2006). For example, business people only think about how much stock they have, but they don't think about whether it can be delivered to customers in the best way possible. Each provincial director from the outreach group rarely requires each part to estimate the number of assortments and requests for natural substances that the customer requires, or to effectively communicate with assistants about the stock situation of clients and ventures.t leads that venture's market data can't adequately update. Likewise, for benefit in certain offices, the sales rep will like to expand the request amount, consequently causing the general stock excessively huge.

Development of a steel enterprise inventory control model in effective supply chains

As a result of the numerous issues found in the manufacturing industry in Chhattisgarh, the inventor believes that it should lead the hypothesis of highly skilled store network into the steel business, which is said to state and examine the coordinated production network and stock control hypothesis, and then, based on theoretical examination, to ensure the issues associated with stock control of the steel industry and the inner and outer factors. It will assist businesses in examining and reviewing the critical issue of stock control that is arising in their businesses, as well as the production network where they are located, and executing comparative stock control measures in accordance with the stock control focus. Investigating the cost and season of store network multi-echelon stock control, it proposed the inter inventory control framework and the expense time improvement model, which are used to represent the sensible stock of every part effort and the ideal stock of the entire store network under the coordinated store network climate, for the current steel industry stock control giving stock model that can be made it better.

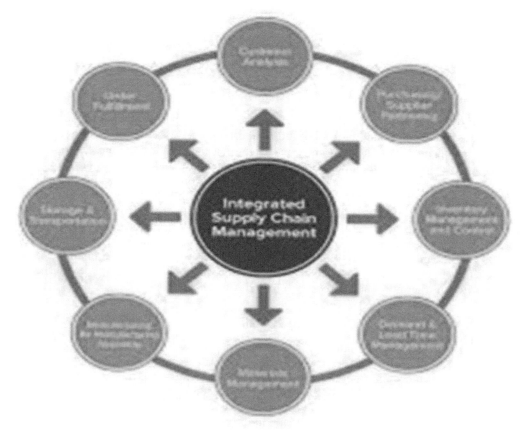

Figure 100.2 Integrated supply chain management

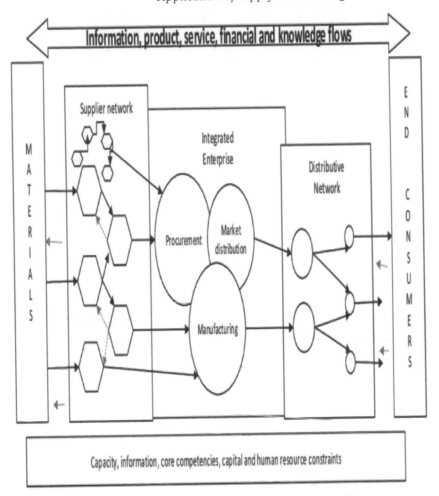

Figure 100.3 Capacity, information, core competencies, capacity and human resource constraints

Inventory control in the supply chain for the iron and steel industry

Great deal of research and evaluation of the methodology and edition of managing inventory in an effective supply chain environment, it proposes that we can obtain stock optimal control from both internal and external factors, which can be consolidated with stock popularity for comparison and collection by steel manufacturing enterprises (Cheng et al., 2003).

A. Modifying the layout of the steel production and boosting the steelmaking industry's specific focus

Steel and iron companies must consider factors metal ores, strength, water supplies, logistics, environmental proficiency, and role in the distribution, as well as make full use of international levels and various situations, based on a rise in northern regions with strict production control, transferring productivity to coastal areas. Improving economic attention must now not only be accomplished through mergers, expansions, rearrangement, and other measures to expand the dimensions of fabrication of various iron and metal company groups, but also by conquering the difficult market access to machine. Through legal and financial components, it should encourage large-scale iron and steel industries with strong capital and technology, as well as small and medium-sized businesses with room for expansion. It should be purely for the purpose of closing.

B. Establish an agile supply chain coordination mechanism for steel making companies

The integrated supply chain version finds the best cost-effective solution for us. If you want to be the most dependable scheme, you must do a lot of things. The first is the crucial criteria to develop a strong supply chain collaboration mechanism among raw cloth suppliers, steel processing companies, transport companies, iron and steel revenue corporations, and customers. Every single corporation in the logistics system contains close family members, illustrating the scenario of 'one-wing allowing, one loss,' but they may also be autonomous commercial organisations, making it critical to build an effective cooperation mechanism in such instances. The second group consists of supportive measures. Belief is the driving force behind the development of a metal enterprise with an agile supply chain, allowing the firms to gain supply coordination (Wang, 2007).

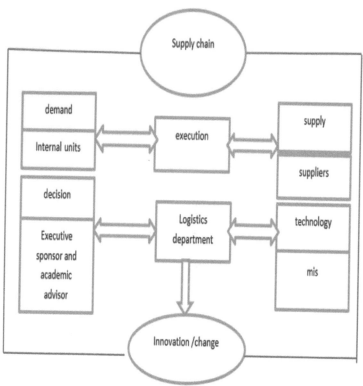

Figure 100.4 Mechanism of supply chain

Conclusion

Stock administration is a particularly significant part of business the executives in the steel and metal area that it has a nearby association with benefits. Considering the nature of Chhattisgarh's metallic area, judicious stock administration ought to assist organisations with acquiring market drive, win market endowments, and slowly move free and clear by decreasing the load of liquidity, shortening the capital cycle, and helping organisations in acquiring market drive. Because of the experienced production network and modern innovation advancement, iron and steel organisations are confronting more difficulties and conceivable outcomes in stock control. Regardless, the steelmaking business ought to zero in just on its own interesting qualities, which incorporate prior convey chain charge improvement necessities as well as an objective market response time. The inventory network centre boss stock control framework is explored in this work using the advancement of cost and time as procedures of displaying and a period cost model is proposed, joining the frameworks designing point of view with the designing technique. We created stock administration guidance for the organisation in light of the attributes of the steel business store network and veritable stock administration in Chhattisgarh. Obviously, multistage stock administration in a useful production network research is as yet in its beginning phases. It's as yet an idea with few quality applications.

References

Cheng, Z., Wang, J., and Tian, H. (2003). A research on the partner selection in agile supply chain. Machine Manufacturing. 41(2):43–45.

Li, Q. (2006). Inventory control theory and application. Beijing: Economic Science Press.

Liu, P. F. and Xie, R. H. (2006). The comparative study of modern supply chain inventory management method. Business Studies. 2:170–173.

Liu, X. L. (2012). Iron and steel industry in our country present situation analysis and countermeasures. Business Culture (5): 210–211.

Ma, S. H., Lin, Y., and Chen, Z. X. (2001). Iron and steel industry in our country present situation analysis and counter-measures. Bus. Culture. (5):210–211.

Wang, H. F. (2004). An analysis on the application of the empirical of the JIT system in our country. Dalian: Northeast University of Finance and Economics Press.

Wang, X. Y. (2007). Based on the analysis of agile supply chain logistics operation mode. Modern Manag. Sci. (5):59–61.

Wu, C. F. and Zhang, Q. P. (2007). Research on knowledge sharing in agile supply chain. J. Info. 6:35–37.

Yang, S. (2012). Based on the SCP paradigm of the organization model of iron and steel industry in China. China Collective Economy. 3:38--40.

Zhong, Z. P. and Xu., H. H. (2004). Based on the distribution enterprises of supply chain management information system. Electronic Commerce World. 7:50–52.

Wenbo, S. U. (2013). Measures to Improve the Inventory of Steel Industry in Supply Chain Environment. Manage. Sci. Eng. 7(3):90–98.

101 The prospects for low-cost and facile fabrication of zinc/graphene composite coatings and structural materials

Ayush Owhal[a], Ajay D. Pingale[b], Sachin U. Belgamwar[c], and Jitendra S. Rathore[d]

Department of Mechanical Engineering, Birla Institute of Technology and Science, Pilani, Rajasthan, India

Abstract

This work reviews a low-cost and facile method for the fabrication of graphene nanofiller reinforced zinc-based metal matrix composite (MMC). The work aims to discuss various methods for fabricating zinc/graphene composite and find the best suitable way to reinforce uniformly dispersed graphene nanofillers in a zinc matrix without damaging its structural shape. The discussions are based on the robustness of the methods to fabricate MMCs for coatings and structural materials. The work summarised that the electrochemical co-deposition is the best suitable method for mixing graphene nanofiller with zinc matrix at room temperature without any harsh processing technique, which retains the structural shape of graphene. Moreover, the in-situ electrochemical co-deposition is suitable for coating fabrication and replaces the harsh ball milling process to prepare a powder mixture of metal matrix composite (MMCs) for the powder metallurgy process. This review work may help to provide feasible solutions for challenges in fabricating zinc/graphene composites and other future MMCs.

Keywords: Ball milling, coatings, electrochemical co-deposition, graphene, metal matrix composite, powder metallurgy.

Introduction

Zinc is post-transition metal with silvery-greyish appearance. It is the twenty-fourth most abundant element in the Earth's crust. Zinc-based coatings for steel substrates are well known for sacrificial protection against corrosion and anti-bacterial protection. Apart from this, titanium-, gold-, silver-, chromium-, nickel-, and copper-based coatings have been widely tested for surface protection from corrosion and bacterial growth. But zinc-based coatings are gaining more research interest among all due to their low-cost, easy availability, high anti-corrosion and anti-bacterial properties (Owhal et al., 2021a), which makes them suitable for low-cost coating applications in public facilities and infrastructures. Zhao et al. (2017) have studied the anti-bacterial activity of coatings containing zinc and observed through in vitro bacterial experiments. They reported that the coatings could inhibit the growth of Staphylococcus aureus bacteria and had good anti-bacterial activity. Smith et al. (n.d.) have discussed the ability of zinc to form protective layers comprising basic carbonates, oxides or hydrated sulfates depending upon the nature of the environment. They reviewed the corrosion rate of zinc coatings and reported an approximately linear corrosion rate of zinc with time in most aggressive atmospheres, whereas, in milder atmospheres, the protective films formed on zinc enable a decrease in corrosion rate with time.

Moreover, recent studies have shown that zinc-based structural implant materials may biodegrade with time and resorb by the human body. Biodegradable implant materials can support the fractured tissues or bones during the healing of body fractures or injuries. After that, the implant material will degrade and absorb the body as the new tissues or bones structures reform. Zinc shows a moderate degradation rate in humans, which is neither too fast as magnesium nor too slow as iron with promising biocompatibility. This makes it one of the best metal elements for a biodegradable implant that can sustain up to the clinical role. Levy et al. reported the prospects of zinc as a structural material for biodegradable applications. They have compared the biocompatibility, corrosion behavior and mechanical properties of zinc. Figure 101.1. Some applications of zinc alloy and composite based coating and structural material. However, fabricating high strength and ductile zinc composite while retaining its homogenised properties is always one of the main goals of metallurgical engineering.

The most common technique to improve the mechanical performance of zinc is alloying. Zinc alloyed with copper in different ratios can form brass, which was first patented in 1781. Thereafter, many alloying combinations were developed with other elements such as zinc-iron, zinc-nickel, zinc-cobalt, zinc-manganese, and zinc-tin. Among those, the former three examples are the most widely reported, although zinc-manganese is probably not commercially available.

[a]ayushowhal@gmail.com; [b]ajay9028@gmail.com; [c]sachinbelgamwar@pilani.bits-pilani.ac.in; [d]jitendrarathore@pilani.bits-pilani.ac.in

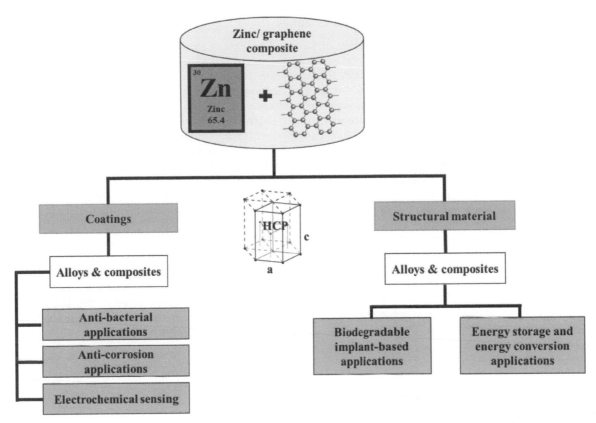

Figure 101.1 Some applications of zinc alloy and composite based coating and structural material

Recently, extensive advancement has been made to further improve the tribo-mechanical and corrosion properties by reinforcing nanofillers of ceramics (aluminium oxide, silicon oxide, and titanium oxide), carbides (silicon carbide and tungsten carbide (Hwang et al., 2018) and advance carbon-based nanofillers like graphene and MWCNTs (Owhal et al., 2022; Owhal et al., 2021b). Among all, graphene nanosheets are one of the most robust nanofillers that can enhance tribo-mechanical and anti-corrosion performance of metal matrix at very low concentrations in the nanocomposite coating the high surface to volume ratio. Graphene is a monolayer honeycomb of carbon allotrope identified in 2004 by Andre Geim and Konstantin Novoselov. Graphene has been established as advanced nanofiller in organic and inorganic binding matrices. According to literature, graphene nanosheets are highly water and oil repellent. Lin et al. reported excellent frictional properties and high wear resistance of multilayer graphene based composite materials. Kirkland et al. reported that graphene nanofillers reinforcements are an effective corrosion barrier and can protect the binding matrix from corrosion attacks. Sreevatsa et al. (2019) reported the ionic barrier ability of graphene with a single impermeable atomic layer that can improve the corrosion-resistant properties of composites. Due to the above reasons, many fabrication methods for graphene-based composites have been investigated with several metallic and non-metallic binding matrices to develop multipurpose composite materials. Sreenivasulu et al. (2018) have reported a review on graphene nanofiller reinforced polymer matrix composites, which showed various scattering strategies for graphene nanofillers and its applications. However, the polymer binding matrices are considered as inferior in mechanical properties as compared to metallic binding matrices. Graphene nanofillers are also investigated for reinforcement in copper-, nickel, zinc-, titanium-, iron-, aluminium- and tin-based single and hybrid metallic matrices to form composites, known as MMCs. The results of the reinforcement of graphene nanofillers are exciting and motivating for researchers. Many methods for the fabrication of MMCs have been developed in recent years. However, a uniform dispersion of nanofillers in a binding matrix without structural damages is still considered a significant challenge. This work discusses the prospects of various fabrication methods for zinc/graphene composites for coating and structural material.

Fabrication Methods for Zinc mmc Based Coatings

Coatings have been considered one of the best surface modification and enhancement techniques to protect base substrate material. A single layer of preventive coating material can remarkably reduce the chances of product failure during the application. Many MMC coating techniques and methods are available for different applications, providing a variety of surface modifications without affecting the cost of the product. The advanced coating methods can contribute to the enhancement of different types of surface properties such as anti-corrosion, surface wear, electro-thermal, electrochemical, toughness, microhardness, surface-topology, anti-bacterial, wettability, biocompatibility, etc (Owhal, Pingale, Khan, et al. 2021b; Pingale, Owhal, Belgamwar, et al. 2021b). However, crack, distortion, delamination, substrate contaminations, and variation in physical properties are the major challenges associated with coatings.

Some of the common coatings method for MMCs are chemical vapor deposition (CVD), physical vapor deposition (PVD) (Hussein et al., 2020), direct vapor deposition (DVD), self-propagating high-temperature method (SHS) (Yuan et al. 2011b), spraying (Aussavy et al. 2014; Sharifahmadian et al., 2013) (different conventional spray methods are as follow: (i) cold-spraying (Yandouzi et al., 2012; Hasniyati et al., 2016), (ii) thermo-plasma (Yuan et al., 2011a; Deuis et al., 1998) (iii) plasma transferred wire arc (Marantz et al., 1995; Nikiforov et al., 2016), (iv) high-velocity oxy-fuel (HVOF) (Marantz, 1991; Hanson et al., 2002; Khor, 2003), (v) detonation-gun (Poorman et al., 1955; Geetha et al., 2014), (vi) radio-frequency inductively coupled (RFIC) (Fauchais and Vardelle 2012; Fauchaise et al., 2014), (vii) D.C. blown arc (Yenni et al., 1961), laser-cladding (Liu et al. 2017) (Liu et al. 2017), electro-less deposition (Yli-Pentti, 2014), and electrochemical co-deposition (Owhal, Pingale, Khan, et al., 2021a; Pingale et al., 2020b) (see Table 101.1). Available low-cost fabrication methods suitable for co-deposition of zinc/graphene composite coatings are as follows:

Table 101.1 Various composite coating methods and their working

Coating Methods	Working	Advantages	Limitations	Reference
Spray based coatings (Thermo-plasma, HVOF, etc.)	The fine nanofillers of coating composite material are sprayed on the heated surfaces	Micro-size structures Precise control over phase structure and thickness	Oxidation, pits and voids High temperature	Nikiforov et al. (2016)
Sol-gel method	Precursors form a suspension (sol) that can gelates (gel) on surface of the substrate	Room temperature process Control over composition Organic coating and non-inorganic coatings	Only limited to the certain metal oxide types Time consuming process	Marantz et al. (1995)
Dipping, brushing, cold spray, and hand-roll painting.	Mixture of some solutions (resins composite mixture and additives) are coated as paint on the surfaces	Organic and inorganic coatings Simple and fast method Large size heavy structures can be coated easily	Solvents may catch fire Poor mechanical and tribological properties Not durable	(Popoola et al.(2016)
Vapor deposition: atomic, physical, and chemical	Evaporate or vaporise the constituents under high vacuum to condense on the surfaces	Uniformly distributed and dense compositions of coatings Organic and non-inorganic coatings	Fine cleaning of substrate is required Vaporized particles are inflammable and harmful Close chamber process High vacuum pressure Expensive	(Hussein et al. (2020)
Electrochemical co-deposition	Metal ions reduced on cathode substrate in an electrolyte bath to form coating	Low cost Thickness can be controlled Facile and scalable Varity of nanofillers can be reinforced in the metallic binding matrices	Not for non-metallic binding matrices Substrate should be metallic Certain amount of reinforcement could be added	(Owhal, Pingale, Khan, et al. 2021)

A. Hot-dip

Hot-dip can be a co-deposition technique for coating metallic substrate with zinc-based composites by dipping the substrate in a molten zinc bath with uniformly added graphene nanofillers. This batch-type technique can be used on heavy structural metallic sections. With the help hot-dip process, small components, tubes, wires, and sheets can also be coated. The hot-dip method is most common in the automotive industry for pure zinc coatings. The simplicity of this process enables a wide range of applications in the protection of steel-based automotive components. The interaction of the molten zinc and iron promotes Zn-Fe alloy formation on the surface of steel substrate, which allows proper bonding between coating and substrate (Popoola et al., 2016). The molten zinc/graphene bath and substrate surface require preparation before the co-deposition, which includes the following steps:

- Pure zinc powder mix with graphene nanofiller using ball milling to prepare uniform zinc/graphene mixture than the prepared mixture melted using furnace to prepare zinc/graphene bath.
- The substrate is cleaned with caustic soda to remove oil and grease for proper thermochemical interaction between the substrate and molten zinc/graphene bath.
- Rust and scale are removed by pickling the substrate in 10% diluted HCl (hydrochloric acid).
- The prepared substrate dipped into the molten zinc/graphene bath for coating.
- The hot-dip coated substrate can directly be further processed by centrifuging method to remove the excess coating before drying.

B. Thermal Spray

Thermal spray method is based on the melting and spray of metal alloys and composites toward a substrate using a spray gun. The zinc/graphene composite in powder form can be fed through the spray gun's nozzle for coating fabrication. The powder mixture of coating material is fed with a pressurised mixture of flammable gas and oxygen to melt the coating material, which propelled onto the surface to form a consolidated coating with an impact. The flame temperature, particles propulsion velocity and nature of material properties of the coating are three main process variables. The electric-arc and plasma spraying can also be used for thermal spraying. Prior substrate surface preparation is always required to achieve good adhesion quality of coatings. This technique allows the thermochemical interaction between substrate and deposited coating material, which results in a strong bond between them. However, these strong bonds at high temperatures fabricate permanent type coatings, making it very hard to refurbish the components after thermal spray coating.

C. Laser Cladding

Laser cladding can be another method to develop zinc/graphene composite coatings on the surface of the metallic substrate. The process involves a high-intensity laser beam to create high energy to melt metal on the surface of the metallic substrate. The melt pool of substrate material is reinforced with injected powder of nanofillers simultaneously. The composite coatings prepared by this method are dense and consistently uniform, improving wear, corrosion, and other surface-related properties such as durability and strength. This method can control coating composition using multi-feeder attachments at the focal zone of the laser, which allows required properties for engineering applications. However, the method is costly due to expensive laser equipment and high energy requirements.

D. Electrochemical Co-deposition

An electrochemical co-deposition technique is one of the low-cost and facile methods for co-deposition of composite material on conductive metallic surfaces. In this method, the metal ions and nanofiller are co-deposited from the electrolyte bath by supplying current between two electrodes (anode and cathode). The anode can be consumable and non-consumable electrodes, that provides the flow of metal ions and current. More than one substrate can be attached to the cathode to deposit an ionic cloud of metal ions and nanofillers. The process of electrochemical co-deposition starts with preparation of electrolyte bath with deagglomerated graphene, which include two steps: (i) stirring and mixing of metal salts and graphene in DI water to dilute the mixture and (ii) deagglomerations of graphene using ultrasonication, as shown in Figure 101.2 (a). Figure 101.2 (b) shows a schematic diagram of a typical electrolytic cell used for the electrochemical co-deposition for the fabrication of zinc/graphene composite. The prepared electrolyte bath is placed for the electrochemical co-deposition process with suitable process parameters such as current,

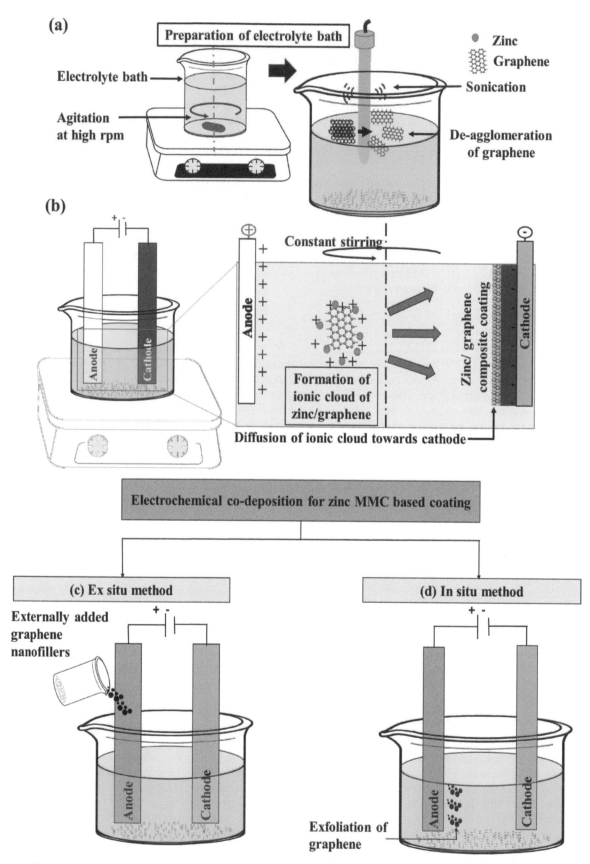

Figure101.2 Fabrication of zinc/graphene composite coating using electrochemical co-deposition method. (a) Preparation of electrolyte bath with deagglomerated graphene, (b) conventional experimental setup, (c) ex-situ method, and (d) in-situ method

temperature, pH and constant agitation. As the current supplies, the metal ions form and interact with deagglomerated graphene, which forms ionic clouds of zinc/graphene. These ionic clouds of zinc/graphene diffused on cathode substrate to form zinc/graphene composite coating.

Electrochemical co-deposition method includes following techniques:

- Ex-situ electrochemical co-deposition with nanofillers are fabricated through separate processes or procured from manufacturers (Pingale, Owhal, Belgamwar, et al., 2021a), as shown in Figure 101.2 (c).
- In-situ electrochemical co-deposition with nanofillers fabrication during the process of co-deposition (Pingale, Owhal, Katarkar, et al. 2021b), as shown in Figure 101.2 (d).

Here, in-situ methods are advantageous because of low running cost and fast fabrication due to the reduced number of steps during the process compared to ex-situ methods.

Fabrication Methods for Zinc MMC Based Structural Material

Over the years, the application of structural MMCs in different industries has gained researchers' interest. The optimisation challenge of mechanical, thermal, electrical and corrosion properties has developed the many fabrication methods of the MMCs (Pingale et al., 2021a). The selection of a suitable method depends on the desired type of nanofillers (particles, fibres, nanorods and nanosheets), distribution type (uniform and non-uniform), quantity and ratio of reinforcement in the metal matrix. MMC fabrication methods can be classified based on the processing temperature of the metal matrix during the reinforcement of nano-fillers, as shown in Figure 101.3. The three-processing temperature-based methods are (i) liquid-state, (ii) two-phase (solid-liquid) processes, and (iii) solid-state.

A. Liquid-state fabrication of MMCs

Liquid-state methods include the dispersion nanofillers into a molten metal matrix, then solidification to form MMCs. In this process, the nanofillers must be selected according to the metal matrix. To select the right nanofiller, various factors, including the melting point, strength, density, thermal expansion coefficient, the shape and size of nanofillers, thermal stability, and preparation or procuring cost, must be carefully considered in addition to the compatibility of the nanofillers with the matrix. The nanofillers and molten metal matrix is needed to be adequately bonded to achieve the required mechanical properties. Therefore, the nanofillers should get properly wet with the molten metal matrix. However, ceramics (aluminium oxide, silicon oxide, titanium oxide, etc.) are the best nanofillers for liquid-state fabrication processes of MMCs.

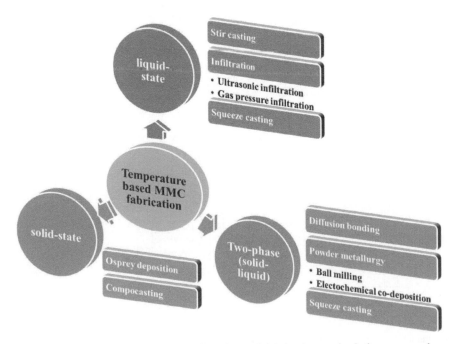

Figure 101.3 Procession temperature based MMC fabrication methods for structural material

The liquid-state fabrication can be achieved through modifications in available methods, such as stir, infiltration, and squeeze casting.

1. Stir casting method

Stir casting is established as one of the simplest and low-cost commercial MMC fabrication methods. The experimental setup for stir casting is shown in Figure 101.4(a). This method was first identified in 1968 by mechanical mixing aluminum oxide particles with molten aluminum matrix (Evans et al., 2003). However, it can be used to fabricate different MMCs as well. A mechanical stirrer is required in stir casting for the dispersion of nanofillers within the molten matrix. Achieving a uniform dispersion of nanofillers depends on the material properties and process parameters, such as the wettability between nanofillers and molten metal matrix, relative density, mixing strength, and solidification rate. Furthermore, the desired dispersion of the nanofillers in the molten metal matrix also depends on the type of the mechanical stirrer (ultrasonic, electromagnetic or centrifugal stirrer) and geometry, the stirrer placement during mixing, process temperature, and the features and specifications of the reinforced nanofillers.

2. Infiltration method

Infiltration is another liquid-state process for the fabrication of MMCs. Here, the nanofillers are uniformly dispersed to soak the metal matrix and fill the empty spaces between the nanofillers (Matsunaga et al., 2007). Different methods can be used for infiltration. Forced infiltration and spontaneous infiltration are two categories of infiltration methods with and without external pressure, respectively. The forced infiltration can be achieved by an external pressure through ultrasonic, centrifugal, electromagnetic, mechanical, or gaseous. The common forced infiltration methods are ultrasonic and gas pressure-based infiltration.

a. Ultrasonic infiltration method

Ultrasonic infiltration is the simplest method to fabricate MMCs with continuous carbon fiber reinforcement in a metal matrix (Cheng et al. 1993). Figure 101.4(b) schematically illustrates the ultrasonic infiltration process for the fabrication of MMCs. This process includes immersion of nanofillers in the molten metal matrix followed by ultrasonication process to the crucible, which transfers ultrasonic vibrations to the melt.

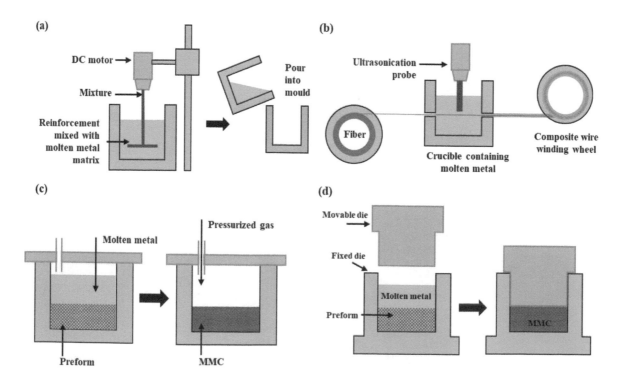

Figure 101.4 (a) Stir casting method, (b) Ultrasonic infiltration method, (c) Gas pressure infiltration method, and (d) Squeeze casting method

b. Gas pressure infiltration method

Gas pressure infiltration method is shown in Figure 101.4(c). In this process, molten metal matrix infiltrated into the nanofillers the pressure applied by an inert gas with excellent geometrical accuracy. But the damage of nanofiller structures, matrix contraction, coarse grain, a requirement of preforms, and undesirable interaction reactions are some limitations.

3. Squeeze casting method

Concept of squeeze casting dates was suggested by Chernov in 1878. In this process, the infiltration process that is carried out by applying force on molten metal using a movable die, as shown in Figure 101.4(d). This is to allow the melt to infiltrate into the dispersed nanofillers. This infiltration methods also allow reinforcement of ceramic nanofillers such as aluminum oxide, silicon carbide, and silicon oxide. A high-quality composite using this process associates with several parameters including type of nanofillers, preheating temperature of melt, tools, melting quality, external cooling, pressure, and time duration. The parts fabricated by this process have properties such as high thermal conductivity, weldability, and dimensional accuracy with smooth surface texture. The composite prepared by this technique can be applied to advance automotive parts applications.

B. Two-Phase (Solid-Liquid) Processes

Two-phase methods includes both solid and liquid phases of metal matrix during fabrication of MMCs (Mathur et al., 1991). This mean, the nanofillers are reinforced in the metal matrix in a transition region of the phase diagram where both solid and liquid phases of metal can coexist, simultaneously. Osprey deposition and compocasting are available processes that fall in the same category.

1. Osprey deposition method

Osprey deposition process is an economical method for fabrication of high quality MMCs. This process reinforces nanofillers in flowing molten matrix followed atomisation by inert gas jet to obtain final mix accumulated on a bed of reinforced matrix (Evans et al., 1985). This process is a combination of consolidation and mixture processes in powder metallurgy.

2. Compocasting method

Compocasting method is considered as an improved version of the stir-casting method (Rosso, 2006). In this method, the nanofillers are reinforced in the matrix at semi-solid processing temperature. The steps of compocasting method are given as follows:

- The semi-solid phase of the matrix stirred to crush the dendrites.
- The nanofillers are added to matrix
- The squeeze casting transforms the material into a thixotropic phase to form MMC.

C. Solid-State Fabrication of MMCs

The solid-state fabrication of MMCs involves developing solid-state bonding between the metal matrix and the nanofillers at elevated temperatures and low pressure (Kaftelen et al. 2011). These processes include diffusion bonding and powder metallurgy.

1. Diffusion bonding method

Diffusion bonding is a simple solid-state process for the fabrication of MMCs (Barabanova et al., 2010). This process includes reinforcement of nanofillers in foils and sheets of a metal matrix to fabricate MMCs, as shown in Figure 101.5. The elevated temperature at the interface of metal matrix and nanofillers allows the diffusion bond between them. The advantage of this process is its compatibility with a wide range of MMCs with precise orientation and volume fraction of na<

2. Powder metallurgy

Powder metallurgy is one of the robust and low-cost approaches for the fabrication of MMCs (Purohit et al. 2018). This process includes three basic steps. (i) preparation of a homogeneous mixture of metal matrix

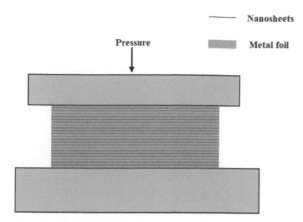

Figure 101.5 Diffusion bonding method

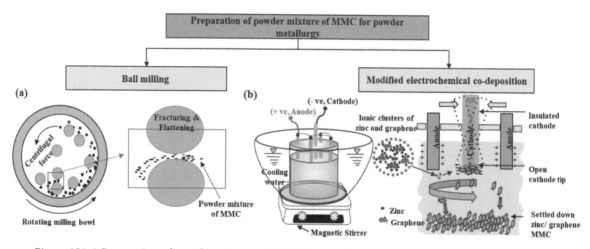

Figure 101.6 Preparation of powder mixture of MMC for powder metallurgy: (a) ball milling and (b) modified electrochemical co-deposition metallurgy

and reinforcement; (ii) Compaction of the mixture until achieving a density of approximately 75%; (iii). Sintering or heat treatment.

This process allows almost all types of nanofillers such as ceramic nanoparticles (aluminum oxide, silicon oxide, silicon oxide, titanium oxide, etc.) and carbon allotropes (graphene nanosheets, carbon nanotubes and carbon dots). There are two different approaches to prepare powder mixture of MMC are as following:

a. Ball milling

In ball milling process, the fine metal powder is precured from manufacturer and mixed with nanofillers using ball mill for several hours to obtain homogeneous mixture of MMC (Piras et al., 2019), as shown in Figure 101.6(a).

b. Modified electrochemical co-deposition

Modified electrochemical co-deposition is a facile approach to fabricating MMCs (A. Owhal et al. 2021). The experimental setup for zinc/graphene composite fabrication is illustrated in Figure 101.6(b). In this process, an electrolyte bath is prepared with diluted metal salt and uniformly dispersed nanofillers. Five electrodes setup including one cathode and four anodes can be used for co-deposition of the ionic cloud of zinc ion and graphene to form zinc/graphene MMC deposits on cathode tip, which collects on the bottom of the bath. The process completes after washing and drying the zinc/graphene composite powder. The electrochemical co-deposition method is superior to the ball milling method due to the preparation of powder mixture of metal matrix and nanofillers without harsh mixing. It avoids matrix micro welds and structural damage to nanofillers like graphene (Pingale et al., 2020a). An option of in-situ synthesis of nanofiller is also available with electrochemical co-deposition, which makes it a more facile and fast method for fabrication of MMCs.

Conclusion

In this work, various fabrication methods for zinc/graphene composite for coating and structural material application are discussed. In-situ electrochemical co-deposition method is found to be cost-effective and facile method for fabrication of zinc/graphene composite coating, which can fabricate the composite coating at room temperature. The structural material can produce by stir casting, infiltration, osprey deposition, squeeze casting, and powder metallurgy. Among all, powder metallurgy is found most suitable and low-cost method for zinc/composite. However, during powder metallurgy, the harsh mechanical processing like ball milling of zinc matrix and graphene nanofillers can damage the nanosheet type carbon structure of graphene nanofillers. The problem of the mechanical harsh mixing process can be resolved by a modified electrochemical co-deposition method that can produce a uniform zinc/graphene composite powder mixture for powder metallurgy without damaging the graphene nanosheet structure. This prospect may help in the selection of a fabrication method for the next generation of graphene based composite coatings and structural materials.

References

Aussavy, D., Costil, S., El Kedim, O., Montavon, G., and Bonnot, A. F. (2014). Metal matrix composite coatings manufactured by thermal spraying: influence of the powder preparation on the coating properties. J. Therm. Spray Technol. 23(1–2):190–196. https://doi.org/10.1007/s11666-013-9999-3.

Barabanova, O. A., Mogorychnyi, V. O., and Nabatchikov, S. V. (2010). Using diffusion bonding for producing laminated composite materials for the manufacture of highly efficient and compact heat exchange systems. Weld Int. 24(5):372–376. https://doi.org/10.1080/09507110903399240.

Cheng, H. M., Lin, Z. H., Zhou, B. L., Zhen, Z. G., Kobayashi, K., and Uchiyama, Y. (1993). Preparation of carbon fibre reinforced aluminium via ultrasonic liquid infiltration technique. Mater. Sci. Technol. 9(7):609–614. https://doi.org/10.1179/mst.1993.9.7.609.

Deuis, R. L., Yellup, J. M,. and Subramanian, C. (1998). Metal-matrix composite coatings by PTA surfacing. Compos. Sci. Technol. 58(2): 299–309. https://doi.org/10.1016/S0266-3538(97)00131-0.

Evans, A., San Marchi, C. and Mortensen, A. (2003). Metal matrix composites in industry. Boston, MA: Springer US. https://doi.org/10.1007/978-1-4615-0405-4.

Evans, R. W., Leatham, A. G., and Brooks, R. G. (1985). The osprey preform process. Powder Metall. 28(1):13–20. https://doi.org/10.1179/pom.1985.28.1.13.

Fauchais, P. and Vardelle, A. (2012). Solution and suspension plasma spraying of nanostructure coatings. In Advanced Plasma Spray Applications. InTech. https://doi.org/10.5772/34449.

Fauchais, P. L., Heberlein, J. V. R. and Boulos, M. I. (2014). Thermal spray fundamentals. thermal spray fundamentals. Boston, MA: Springer US. https://doi.org/10.1007/978-0-387-68991-3.

Geetha, M., Sathish, S., Chava, K. and Joshi, S. V. (2014). Detonation gun sprayed Al 2 O 3 –13TiO 2 coatings for biomedical applications. Surf. Eng. 30(4):229–236. https://doi.org/10.1179/1743294414Y.0000000245.

Hanson, T. C., Hackett, C. M., and Settles, G. S. (2002). Independent control of HVOF particle velocity and temperature. J. Therm. Spray Technol. 11(1):75–85. https://doi.org/10.1361/105996302770349005.

Hasniyati, M. R., Zuhailawati, H., and Ramakrishnan, S. (2016). A statistical prediction of multiple responses using overlaid contour plot on hydroxyapatite coated magnesium via cold spray deposition. Procedia Chem. 19: 181–188. https://doi.org/10.1016/j.proche.2016.03.091.

Hussein, M. A., Ankah, N. K., Kumar, A. M., Azeem, M. A., Saravanan, S., Sorour, A. A., and Al Aqeeli, N. (2020). Mechanical, biocorrosion, and antibacterial properties of nanocrystalline tin coating for orthopedic applications. Ceram. Int. 46(11):18573–18583. https://doi.org/10.1016/j.ceramint.2020.04.164.

Hwang, I., Guan, Z., and Li, X. (2018). Scalable manufacturing of zinc-tungsten carbide nanocomposites. Procedia Manuf. 26:140–145. https://doi.org/10.1016/j.promfg.2018.07.017.

Kaftelen, H., Ünlü, N., Göller, G., Lütfi Öveçoğlu, M., and Henein, H. (2011). Comparative processing-structure–property studies of Al–Cu matrix composites reinforced with TiC particulates. Compos. Part A: Appl. Sci. Manuf. 42(7):812–824. https://doi.org/10.1016/j.compositesa.2011.03.016.

Khor, K. (2003). Characterization of the bone-like apatite precipitated on high velocity oxy-fuel (HVOF) sprayed calcium phosphate deposits. Biomaterials 24(5):769–775. https://doi.org/10.1016/S0142-9612(02)00413-1.

Liu, Y,, Ding, J., Qu, W., Su, Y., and Yu, Z. (2017). Microstructure evolution of TiC particles in situ, synthesized by laser cladding. Materials 10(3):281. https://doi.org/10.3390/ma10030281.

Marantz, D. R. (1991). High-velocity flame spray apparatus and method of forming materials. US5019686A, issued 1991. https://patents.google.com/patent/US5019686A/en?oq=US+5019686+.

Marantz, D. R., David, R.. and Keith, A. 1995. High velocity electric-ARC spray apparatus and method of forming mater Als. US5442153, issued 1995. https://patents.google.com/patent/US5442153A.

Mathur, P., Apelian, D. and Lawley, E. (1991). Fundamentals of spray deposition via osprey processing. Powder Metall. 34(2):109–111. https://doi.org/10.1179/pom.1991.34.2.109.

Matsunaga, T., Matsuda, K., Hatayama, T., Shinozaki, K. and Yoshida, M. (2007). fabrication of continuous carbon fiber-reinforced aluminum–magnesium alloy composite wires using ultrasonic infiltration method. Compos. Part A: Appl. Sci. Manuf. 38(8):1902–1911. https://doi.org/10.1016/j.compositesa.2007.03.007.

Nikiforov, A., Deng, X., Xiong, Q., Cvelbar, U., Degeyter, N., Morent, R. and Leys, C. (2016). Non-thermal plasma technology for the development of antimicrobial surfaces: a review. J. Phys. D: Appl. Phys. 49(20). https://doi.org/10.1088/0022-3727/49/20/204002.

Owhal, A., Pingale, A. D., Belgamwar, S. U., and Rathore, J. S. (2021a). A brief manifestation of anti-bacterial nanofiller reinforced coatings against the microbial growth based novel engineering problems. Mater. Today: Proc. 47:3320–330. https://doi.org/10.1016/j.matpr.2021.07.151.

Owhal, A., Pingale, A. D., Belgamwar, S. U. and Rathore, J. S. (2021b). Preparation of novel Zn/Gr MMC using a modified electro-Co-deposition method: microstructural and tribo-mechanical properties. Mater. Today: Proc. 44(xxxx):222–228. https://doi.org/10.1016/j.matpr.2020.09.459.

Owhal, A., Choudhary, M., Belgamwar, S. U., Mukherjee, S., and Rathore, J. S. (2021). Co-deposited Zn-Cu/Gr nanocomposite: Corrosion behaviour and in-vitro cytotoxicity assessment. Trans. IMF 99(4):215–223. https://doi.org/10.1080/00202967.2021.1899493.

Owhal, A., Pingale, A. D., Khan, S., Belgamwar, S. U., Jha, P.N., and Rathore, J. S. (2021). Electro-codeposited γ-Zn-Ni/Gr composite coatings: Effect of graphene concentrations in the electrolyte bath on tribo-mechanical, anti-corrosion and anti-bacterial properties. Trans. IMF 1–8. https://doi.org/10.1080/00202967.2021.1979815.

Owhal, A., Pingale, J. D., Belgamwar, S. U., and Rathore, J. S. (2021). Facile and scalable co-deposition of anti-bacterial Zn-GNS nanocomposite coatings for hospital facilities: tribo-mechanical and anti-corrosion properties. JOM 73:4270–4278. https://doi.org/10.1007/s11837-021-04968-5.

Owhal, A., Pingale, A. D., Khan, S., Belgamwar, S. U., Jha, P. N., and Rathore, J. S. (2021). Facile synthesis of graphene by ultrasonic-assisted electrochemical exfoliation of graphite. Materi. Today: Proc. 44:467–472. https://doi.org/10.1016/j.matpr.2020.10.045.

Owhal, A., Pingale, A., and Belgamwar, S. (2022). Developing sustainable Zn-MWCNTs composite coatings using electrochemical co-deposition method: Tribological and surface wetting behavior. Adv. Mater. Proc. Technol. 1–14. https://doi.org/10.1080/2374068X.2022.2035968.

Pingale, A. D., Belgamwar, S. U., and Rathore, J.S. (2020a). A novel approach for facile synthesis of Cu-Ni/GNPs composites with excellent mechanical and tribological properties. Mater. Sci. Eng. B 260:114643. https://doi.org/10.1016/j.mseb.2020.114643.

Pingale, A. D., Belgamwar, S. U. and Rathore, J. S. (2020b). Synthesis and characterization of Cu–Ni/Gr nanocomposite coatings by electro-Co-deposition method: effect of current density. Bull. Mater. Sci. 43(1):66. https://doi.org/10.1007/s12034-019-2031-x.

Pingale, A. D., Owhal, A., Katarkar, A. S., Belgamwar, S. U. and Rathore, J. S. (2021a). Recent researches on Cu-Ni Alloy matrix composites through electrodeposition and powder metallurgy methods: A review. Mater. Today: Proc. 47:3301–3308. https://doi.org/10.1016/j.matpr.2021.07.145.

Pingale, A. D., Owhal, A., Belgamwar, S. U. and Rathore, J. S. (2021). Electro-codeposition and properties of Cu–Ni-MWCNTs composite coatings. Trans. IMF 99(3):126–132. https://doi.org/10.1080/00202967.2021.1861848.

Pingale, A. D., Owhal, A., Belgamwar, S. U. and Rathore, J. S. (2021). Effect of current on the characteristics of CuNi-G nanocomposite coatings developed by D.C., P.C. and PRC electrodeposition. JOM. 73(12):42994308. https://doi.org/10.1007/s11837-021-04815-7.

Piras, C. C., Fernández-Prieto, S. and De Borggraeve, W. M. (2019). Ball milling: A green technology for the preparation and functionalisation of nanocellulose derivatives. Nanoscale Adv. 1(3):937–947. https://doi.org/10.1039/C8NA00238J.

Poorman, R. M., Sargent, H. B. and Headlee, L. (1955). Method and apparatus utilizing detonation waves for spraying and other purposes. US2714563A, issued 1955. https://patents.google.com/patent/US2714563A/en.

Popoola, P.A.I., Malatji, N. and Fayomi, O.S. (2016). Fabrication and properties of zinc composite coatings for mitigation of corrosion in coastal and marine zone. In Applied Studies of Coastal and Marine Environments. InTech. https://doi.org/10.5772/62205.

Purohit, R., Dewang, Y., Rana, R. S., Koli, D. and Dwivedi, S. (2018). Fabrication of magnesium matrix composites using powder metallurgy process and testing of properties. Mater. Today: Proc. 5(2):6009–6017. https://doi.org/10.1016/j.matpr.2017.12.204.

Rosso, M. (2006). Ceramic and metal matrix composites: routes and properties. J. Mater. Process. Technol. 175 (1–3):364–375. https://doi.org/10.1016/j.jmatprotec.2005.04.038.

Sharifahmadian, O., Salimijazi, H. R., Fathi, M. H., Mostaghimi, J., and Pershin, L. (2013). Study of the antibacterial behavior of wire arc sprayed copper coatings. J. Therm. Spray Technol. 22(2–3):371–379. https://doi.org/10.1007/s11666-012-9842-2.

Smith, W. J., Goodwin, F. E. 2006. Zinc Coating. 4M 2006 - Second International Conference on Multi-Material Micro Manufacture.

Sreenivasulu, B., Ramji, B. R. and Nagaral, M. (2018). A review on graphene reinforced polymer matrix composites. Mater. Today: Proc. 5(1):2419–2428. https://doi.org/10.1016/j.matpr.2017.11.021.

Sreevatsa, S., Banerjee, A. and Haim, G. (2019). Graphene as a permeable ionic barrier. ECS Trans. 19(5):259–264. https://doi.org/10.1149/1.3119550.

Yandouzi, M., Bu, H., Brochu, M. and Jodoin, B. (2012). Nanostructured Al-Based metal matrix composite coating production by pulsed gas dynamic spraying process. J. Therm. Spray Technol. 21(3–4):609–619. https://doi.org/10.1007/s11666-011-9727-9.

Yenni, D. M., Mcgill, W. C. and Lyle, J.W. (1961). Electricarc sprayng. US2982845A. United States Patent, issued May 1961. https://dl.acm.org/doi/10.1145/178951.178972.

Yli-Pentti, A. (2014). Electroplating and Electroless Plating. In comprehensive materials processing, (pp. 277–306). Elsevier. https://doi.org/10.1016/B978-0-08-096532-1.00413-1.

Yuan, J., Zhu, Y., Zheng, X., Ji, H. and Yang, T. (2011a). Fabrication and evaluation of atmospheric plasma spraying W.C.–Co–Cu–MoS2 composite coatings. J. Alloys Compd. 509(5):2576–281. https://doi.org/10.1016/j.jallcom.2010.11.093.

Yuan, X., Liu, G., Jin, H. and Chen, K. (2011b). In situ synthesis of tic reinforced metal matrix composite (MMC) coating by self propagating high temperature synthesis (SHS). J. Alloys Compd. 509(30):L301–L303.

Zhao, Q.-M., Li, G.-Z., Zhu, H. M. and Cheng, L. (2017). Study on effects of titanium surface microporous coatings containing zinc on osteoblast adhesion and its anti-bacterial activity. Appl. Bionics and Biomech. 1–4.

102 Pool boiling heat transfer of R-600a on plain copper and Cu@GPL porous composite coating surfaces

Anil S. Katarkar[1,a], Ajay D. Pingale[2,b], Sachin U. Belgamwar[2,c], and Swapan Bhaumik[1,d]

[1]Department of Mechanical Engineering, National Institute of Technology Agartala, Tripura, India

[2]Department of Mechanical Engineering, Birla Institute of Technology and Science, Pilani, Rajasthan, India

Abstract

In this work, nucleate pool boiling experiments were performed on the plain copper and graphene nanoplatelets reinforced Cu matrix (Cu@GPL) porous composite coatings using saturated refrigerant R-600a. The Cu@GPL porous composite coatings were fabricated by a two-step electrodeposition technique. Copper sulfate pentahydrate as a source of Cu and graphene nanoplatelets (GPL) as reinforcing element were used as starting materials for fabrication of Cu@GPL porous composite coatings. The effect of coating parameters such as surface roughness, coating thickness, and porosity on the heat transfer coefficients (HTCs) and boiling characteristics of refrigerants was investigated and presented in detail. The heat transfer coefficient of Cu@GPL porous composite coating was enhanced approximately 2.36 times than that of the plain copper heating surface. The augmentation in the HTCs is primarily due to an increase in surface roughness, coating thickness, porosity and active nucleation of the Cu@GPL porous composite coatings.

Keywords: Composite, pool boiling, electrodeposition, R-600a, porosity, surface roughness.

Introduction

Nucleate pool boiling (NPB) is a very effective mechanism of the heat transfer process and it is widely employed in many industrial applications such as refrigeration, aerospace and nuclear power (Pingale et al., 2021; Pingale et al., 2020a; Pingale et al., 2021a). Mostly, the performance of a pool boiling process can be evaluated by the heat transfer coefficient (HTC) and the critical heat flux (CHF) (Walunj and Sathyabhama, 2018). To improve the performance of either HTC or CHF on a metal substrate, various types of surface treatments have been employed thus far. Among the several surface treatments, a porous coating is one of the most effective approaches to enhance the pool BHT performance of refrigerants (Jiang et al., 2021; Katarkar et al., 2021b; Dewangan et al., 2019; Katarkar et al., 2021a; Majumder et al., 2022a).

Nowadays, more researchers have focused on the porous surface because it can significantly enhance boiling heat transfer (Katarkar et al., 2020a). The porous coating is considered the most versatile and popular technique. The porous coating provides higher nucleation-site density and a large number of cavities with different sizes (Katarkar, et al., 2021). This significantly impacts bubble formation, growth and departure dynamics. The porous coated surfaces have been fabricated by passive techniques to enhance the boiling performances (Pingale et al., 2021b). Table 102.1 shows an overview of pool BHT performance on various nanostructure surfaces. Numerous techniques have been employed to fabricate uniform thin porous coating or modulated porous coating on the boiling surface such as RF sputtering, sintering, electrodeposition, electrophoretic deposition, spray coating and brazing (Karunagaran et al., 2003; Ujereh et al., 2007; Gupta and Misra, 2019; Katarkar et al., 2021b; Majumder et al., 2022b). Among them, electrodeposition is one of the appropriate porous coating techniques to enhance boiling heat transfer in pool boiling applications (Katarkar et al., 2022; Owhal et al., 2021a). Electrodeposition process is employed to form a micro/nano-porous coatings with a uniform thickness distribution on a thermally conductive surface.

Graphene nanoplatelets (GPL) are an emerging carbon material with a large specific surface area and unique 2-D honeycomb lattice structure (Pingale et al., 2020c; 2020b; Shelke et al., 2020; Owhal et al., 2022). It has higher thermal conductivity and is considered an outstanding material for thermal heat dissipation and conduction (Pingale et al., 2021b). Also, GPL has impressive physical, mechanical and electrical properties. Hence, GPL-reinforced copper matrix composites play a crucial role in boiling heat transfer applications. Higher strength and thermal conductivity make them the most suitable reinforcement material for copper matrix composites (Majumder et al., 2022b; Pingale et al., 2020d; Pingale et al., 2021b; Owhal et al., 2021b). GPL incorporated copper matrix composites can be used for heat transfer applications. In the present research work, GPL incorporated Cu matrix porous composite coatings were synthesized using a two-stage electrodeposition technique. Also, the pool BHT performance of Cu@GPL porous composite coatings with R-600a was investigated in detail.

---

[a]anil.katarkar@gmail.com; [b]ajay9028@gmail.com; [c]sachinbelgamwar@pilani.bits-pilani.ac.in; [d]drsbhaumik@gmail.com

Table 102.1 An overview of pool BHT performance on various nanostructure surfaces

Ref.	Working fluid	Coating technique	Nanostructure configuration	CHF/HTC	Comments
(Moghadasi and Saffari, 2021)	DI water	Spin coating	Fe3O4, NiO, CaTiO3	Not studied Enhanced	Highest roughness along with highest NSD
(Godinez et al., 2021)	DI water	High-temperature conductive (HTCMC)	Cu powder	Enhanced Enhanced	Reduce the contact angle and increase CHF
(Shi et al., 2015)	DI water	Electroplating process	Cu NW	Enhanced Enhanced	Increase active NSD and surface wettability
(Li et al., 2008a)	Water	PVD (Micro electro mechanical system)	Cu nanorod array	Enhanced Enhanced	Increases the bubble departure frequency
(Das and Bhaumik, 2014)	Water	e-beam evaporation system	TiO2 TF	Not studied Enhanced	Increased surface wettability and enhanced roughness
(Wu et al., 2010)	Water and FC-72	Nanofluid boiling	TiO2 and SiO2 TF	Enhanced Enhanced	Increases solid-liquid interaction and decreases hydrophilicity
(Majumder et al. 2022a)	R-134a	Powder metallurgy	Al-GNPs	Not studied Enhanced	Highest roughness along with highest NSD
(Li et al., 2008b)	R-134a	EDS and anodistion	Cu nanostructure	Not studied Enhanced	High interfacial area and high NSD
(Im et al., 2010)	Dielectric coolant PF-5060	Electrochemical deposition	Cu NW	Enhanced Enhanced	Increase NSD and decrease wall superheat.
(Hendricks et al., 2010)	Water	Micro reactor techniques	ZnO TF	Enhanced Enhanced	High NSD and high surface area
(Li et al., 2015)	Deionised water	Electrochemical deposition	Ni nano-cone array	Not studied Enhanced	Increase NSD and increase bubble departure frequency
(Saeidi and Alemrajabi, 2013)	DI water	Anodising	Nano-porous (Al alloy)	Enhanced Enhanced	Increase bubble departure frequency and high surface area
(Vemuri and Kim, 2005)	Saturated FC-72	Anodisation	Al2O3 TF	Not studied Enhanced	Increases active NSD and reduces the incipience superheat
(Kim et al., 2010)	Water	Physical vapor deposition	ZnO Nanorod	Enhanced Enhanced	Enhanced liquid spreading effect and Wettability
(Forrest et al., 2010)	Deionised water	Layer-by-layer (Lbl) assembly method	SiO2 TF	Enhanced Enhanced	Reduced hydrophilicity and Wettability
(Jo et al., 2012)	Water	Deep Reactive-Ion Etching	SiO2 TF	Enhanced Enhanced	Activated NSD and the ONB variation with Wettability

Fabrication of cu@gpl Porous Composite Coatings

The electrodeposition process was carried out in the 500 ml borosilicate glass beaker. The schematic of electrodeposition setup is shown in Figure 102.1. The bath solution was prepared using 100 g/L of copper sulfate pentahydrate (Merck, 99.99% purity), 30 g/L of sulfuric acid (Merck, 98% purity), and 100 or 200 mg/L of GPL (Alfa Aesar, 99.99% purity). The solution was prepared with analytically pure reagents and deionized water. The probe ultrasonication (20 kHz, 500 W) was used to disperse the GPL in the bath solution. Electrolyte temperature was kept at 35±1°C. A pure copper rod was used as anode and a circular copper disc (diameter = 9 mm) was used as a cathode. The substrate was mechanically polished using

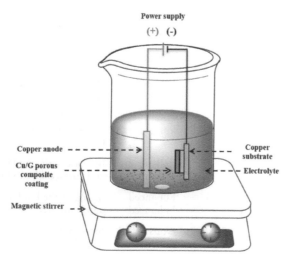

Figure 102.1 Schematic diagram of electrodeposition setup

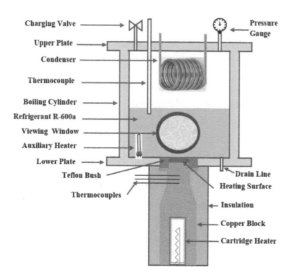

Figure 102.2 Schematic diagram of pool boiling experimental facility

emery paper (600#, 2000#, and 3000#) to achieve a mirror finish surface. The co-deposition of Cu and GPL was carried out in two stages. In the first stage, a high deposition current density of 0.4 A/cm2 was used for 90 seconds. Subsequently, in the second stage, a lower deposition current density of 0.05 A/cm2 was used for 45 minutes to fabricate microporous Cu@GPL composite coatings. Prepared coated samples were dried at 70°C in an inert atmosphere and then used for further characterisation. The morphology of the samples was examined using the high-resolution FEI-Apreo-S field emission scanning electron microscope (FESEM). Elemental analysis of samples was carried out using energy-dispersive X-ray spectroscopy (EDS) attached with FESEM. The measurement of surface roughness and porosity was conducted by Taylor Habsons 2-D profilometer and Image J software, respectively.

Experimental Apparatus and Procedure

Figure 102.2 shows the schematic of the experimental device used to study the pool boiling characteristic of R-600a. The whole experimental system includes the boiling cylinder, test section, condenser, auxiliary heater (100 W), cartridge heater (500 W), thermocouples and pressure gauge. The boiling chamber consisted of a boiling cylinder (volume 4.48 L), two stainless steel covers and two Teflon O-rings. The stainless-steel boiling cylinder was sealed with two stainless steel covers (flange), one at the upper and the other at the lower. The boiling cylinder's upper stainless-steel cover was used to fix a condenser, pressure gauge to measure the boiling pressure, and thermocouple to monitor inside temperature. A condenser was fitted at the upper stainless-steel cover to condense the refrigerant. Two viewing windows were used for the usual observation boiling process in the boiling cylinder. The working fluid (R-600a) was contained within the boiling cylinder, and its saturation condition was maintained by an auxiliary heater, which was placed inside the boiling cylinder. A 20 mm thick insulation of thermal materials was wrapped over the outside of the boiling cylinder to ensure the adiabatic condition. A Cu block was fixed at the lower stainless-steel covers of the boiling cylinder to supply heat to the test specimen. A cartridge heater was inserted from the lower end of the Cu block to supply heat to the Cu block. The three thermocouples were spaced at a vertical distance of 10 mm on the Cu block to calculate the surface temperature.

Initially, the boiling cylinder was cleaned carefully with acetone to remove the dust. The test specimen was placed on the upper surface of a Cu bar. This whole assembly of the test section (test specimen, heating surface, Teflon bush, etc.) was inserted at the lower stainless-steel covers of the boiling cylinder with the press-fit condition. After the fitting, high-pressure gas (N2) was injected into the boiling cylinder to identify the leakage. Then the refrigerant was charged in the boiling cylinder. The refrigerant was heated by the auxiliary heater for 2 hours to remove the non-condensable gases. When the system reached a steady-state, the relevant temperature data were recorded. A copper condenser was used to condense and re-circulate the refrigerant inside the boiling cylinder. After recording data, refrigerant and the test specimen were taken out from the boiling cylinder. The above entire procedure was repeated three times for each test specimen.

Data Reduction and Uncertainty

The present investigation involves the determination of HTCs (h) of the heating surface (plain Cu and Cu@ GPL porous composite coatings).

$$h = \frac{Q}{A(\Delta T)}$$

$$h = \frac{q}{(T_s - T_{sat})}$$

In addition, According to the literature (Schultz and Cole, 1979), it can be obtained that the uncertainty of heat flux and boiling heat transfer coefficient is 2.74% and 7.10%, respectively.

Results and Discussion

A. Characterisation of Cu@GPL porous composite coatings

SEM images of Cu@GPL porous composite coatings are represented in Figure 102.3(a–d). From Figure 102.3, it is clearly seen that the surface morphology becomes rougher with the increase in GPL concentration

Figure 102.3 SEM images of Cu@GPL porous composite coatings: (a, b) 100 mg/L and (c, d) 200 mg/L

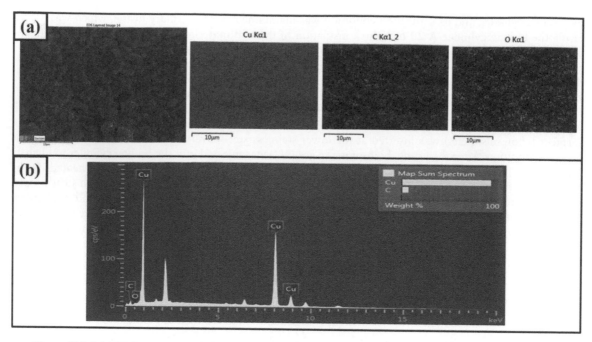

Figure 102.4 (a) EDS mapping and (b) EDS spectrum of Cu@GPL (200 mg/L) porous composite coatings

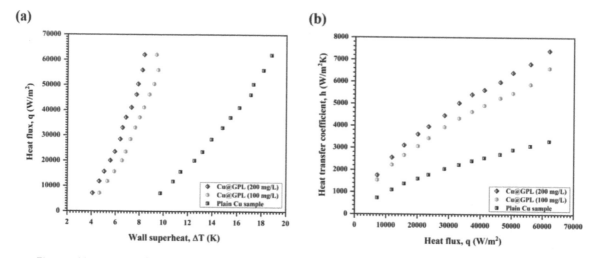

Figure 102.5 (a) Heat flux vs wall superheat and (b) HTC vs heat flux

from 100 mg/L to 200 mg/L. The measured surface roughness for Cu@GPL porous composite coatings prepared at 100 mg/L and 200 mg/L of G concentration are 0.24 ± 0.03 μm and 0.32 ± 0.03 μm, respectively. The EDS mapping and EDS spectrum of Cu@GPL (200 mg/L) porous composite coating are shown in Figure 102.4(a–b). EDS results confirm the reinforcement of G into the Cu matrix. Also, the measured porosity of Cu@GPL porous composite coatings prepared at 100 mg/L and 200 mg/L of GPL concentration are 46% and 57%, respectively.

B. Boiling Enhancement of Cu@GPL Porous Composite Coatings

To ensure the reliability of the present experiment set and procedure, a pool BHT experiment was conducted on a bare Cu test specimen with refrigerant R-600a at a saturation temperature of 10°C. The experimental results are compared with the Rohsenow correlation. The experimental results in this work are in good agreement with the Rohsenow correlation. Figure 102.5(a) shows information about the heat flux of the plain Cu and Cu@GPL porous composite coatings with respect to various wall superheats for the current investigation. It is observed that the wall superheats increase by increasing the heat flux. The behavior of all curves for the considered surfaces is nearly the same, but the Cu@GPL porous composite coatings curves are moved towards the origin. The maximum shift is for Cu@GPL (200 mg/L) porous composite coating and the minimum shift is for Cu@GPL (100 mg/L) porous composite coating concerning bare Cu surface. According to Figure 102.5(a), the value of surface wall superheat required on the Cu@ GPL porous composite coating is lesser than the bare Cu surface. Thus, the responsible parameters for such surfaces are enhanced surface coating thickness, surface roughness and high active nucleation sites (Katarkar et al., 2020b).

Figure 102.5(b) shows the variation of 'h' with 'q' for R-600a over bare Cu surface as well as Cu@ GPL porous composite coatings. In the Cu@GPL porous composite coatings, an increase in heat flux value resulted in an increase in HTC, similar to the bare Cu surface. It can be seen that the HTC of Cu@ GPL porous composite coatings is higher than that of bare Cu surface. The HTC of the Cu@GPL (200 mg/L) porous composite coating is increased by 136% over the bare Cu surface. This HTC augmentation is attributed to the increase in surface area, surface roughness, surface coating thickness and porosity of Cu@GPL porous composite coatings, and the higher thermal conductivity of GPL. The size of cavities and porosity are highly influential factors in boiling performance. Surface cavities in the surface of the porous coatings are manifested in active nucleation sites (Majumder et al., 2020; Katarkar et al., 2020a).

Conclusion

We experimentally studied the nucleate pool BHT performance of R-600a on bare Cu heating surface and Cu@GPL porous composite coatings. Pool boiling experiments were carried out at saturation temperatures of 10°C for heat fluxes ranging from 7.17 to 62.06 kW/m2. Cu@GPL porous composite coatings were successfully fabricated using a two-step electrodeposition technique on the plain Cu surface. The surface roughness, thickness, and porosity of Cu@GPL porous composite coating increased with the rise in the GPL concentration. The Cu@GPL porous composite coatings showed a higher HTC than the plain Cu surface due to the more nucleation sites, larger heat transfer area, and increased surface roughness.

Acknowledgments

The current work was performed in the PCHT Laboratory under the Mechanical Engineering Department in the NIT Agartala, Tripura, India.

References

Das, S. and Bhaumik, S. (2014). Enhancement of nucleate pool boiling heat transfer on titanium oxide thin film surface. Arab. J. Sci. Eng. 39(10):7385–7395. https://doi.org/10.1007/s13369-014-1340-z.

Dewangan, A. K., Kumar, A. and Kumar, R. (2019). Experimental study of nucleate pool boiling of r-134a and r-410a on a porous surface. Heat Transf. Eng. 40(15):1249–1258. https://doi.org/10.1080/01457632.2018.1460922.

Forrest, E., Williamson, E., Buongiorno, J., Hu, L.-W., Rubner, M. and Cohen, R. (2010). Augmentation of nucleate boiling heat transfer and critical heat flux using nanoparticle thin-film coatings. Int. J. Heat Mass Transf. 53 (1–3):58–67. https://doi.org/10.1016/j.ijheatmasstransfer.2009.10.008.

Godinez, J. C., Cho, H., Fadda, D., Lee, J., Park, S. J. and You, S. M. (2021). Effects of materials and microstructures on pool boiling of saturated water from metallic surfaces. Int. J. Therm. Sci. 165:106929. https://doi.org/10.1016/j.ijthermalsci.2021.106929.

Gupta, S. K. and Misra, R. D. (2019). An experimental investigation on pool boiling heat transfer enhancement using Cu-Al 2 O 3 nano-composite coating. Exp. Heat Transf. 32(2):133–158. https://doi.org/10.1080/08916152.2018.1485785.

Hendricks, T. J., Krishnan, S., Choi, C., Chang, C.-H. and Paul, B. (2010). Enhancement of pool-boiling heat transfer using nanostructured surfaces on aluminum and copper. Int. J. Heat Mass Transf. 53(15–16):3357–3365. https://doi.org/10.1016/j.ijheatmasstransfer.2010.02.025.

Im, Y., Joshi, Y., Dietz, C. and Lee, S. (2010). enhanced boiling of a dielectric liquid on copper nanowire surfaces. Int. J. Micro-Nano Scale Trans. 1(1):79–96. https://doi.org/10.1260/1759-3093.1.1.79.

Jiang, H., Xu, N., Wang, D., Yu, X. and Chu, H. (2021). Experimental investigation of the effect of cylindrical array structure on heat transfer performance during nucleate boiling. Int. J. Heat Mass Transf. 174:121319. https://doi.org/10.1016/j.ijheatmasstransfer.2021.121319.

Jo, H., Kim, S., Kim, H., Kim, J. and Kim, M. H. (2012). Nucleate boiling performance on nano/microstructures with different wetting surfaces. Nanoscale Res. Lett. 7:1–9. https://doi.org/10.1186/1556-276X-7-242.

Karunagaran, B., Rajendra Kumar, R. T., Senthil Kumar, V., Mangalaraj, D., Narayandass, S. K. and Mohan Rao, G. (2003). Structural characterization of DC magnetron-sputtered TiO2 thin films using XRD and raman scattering studies. Mater. Sci. Semicond. Process. 6(5–6):547–550. https://doi.org/10.1016/j.mssp.2003.05.012.

Katarkar, A., Majumder, B. and Bhaumik, S. (2020a). Effect of enhanced surfaces and materials in boiling heat transfer with HFO refrigerants: A review. Mater. Today: Proc. 26:2237–2241. https://doi.org/10.1016/j.matpr.2020.02.485.

Katarkar, A., Majumder, B. and Bhaumik, S. (2020b). Review on passive heat enhancement techniques in pool boiling heat transfer. IOP Conf. Ser. Mater. Sci. Eng. 814(1):012031. https://doi.org/10.1088/1757-899X/814/1/012031.

Katarkar, A. S., Majumder, B., Pingale, A. D. Belgamwar, S. U. and Bhaumik, S. (2021). A review on the effects of porous coating surfaces on boiling heat transfer. Mater. Today: Proc. 44:362–367. https://doi.org/10.1016/j.matpr.2020.09.744.

Katarkar, A. S., Pingale, A. D., Belgamwar, S. U. and Bhaumik, S. (2021a). Effect of GNPs concentration on the pool boiling performance of R-134a on Cu-GNPs nanocomposite coatings prepared by a two-step electrodeposition method. Int. J. Thermophys. 42(8):124. https://doi.org/10.1007/s10765-021-02876-z.

Katarkar, A. S., Pingale, A. D., Belgamwar, S. U. and Bhaumik, S. (2021b). Experimental study of pool boiling enhancement using a two-step electrodeposited Cu–GNPs nanocomposite porous surface With R-134a. J. Heat Transf. 143(12):121601. https://doi.org/10.1115/1.4052116.

Katarkar, A. S., Pingale, A. D., Belgamwar, S. U. and Bhaumik, S. (2022). Fabrication of Cu@G composite coatings and their pool boiling performance with R-134a and R-1234yf. Adv. Mater. Process. Technol. 00(00):1–13. https://doi.org/10.1080/2374068x.2022.2033046.

Kim, S., Kim, H. D., Kim, H., Ahn, H. S., Jo, H., Kim, J. and Kim, M. H. (2010). Effects of Nano-fluid and surfaces with nano structure on the increase of CHF. Exp. Therm. Fluid Sci. 34(4):487–495. https://doi.org/10.1016/j.expthermflusci.2009.05.006.

Li, C., Wang, Z., Wang, P. I., Peles, Y., Koratkar, N. and Peterson, G. P. (2008a). Nanostructured copper interfaces for enhanced boiling. Small 4(8):1084–1088. https://doi.org/10.1002/smll.200700991.

Li, S., Furberg, R., Toprak, M. S., Palm, B. and Muhammed, M. (2008b). Nature-inspired boiling enhancement by novel nanostructured macroporous surfaces. Adv. Funct. Mater. 18(15):2215–220. https://doi.org/10.1002/adfm.200701405.

Li, Y. Y., Liu, Z. H. and Zheng, B. C. (2015). Experimental study on the saturated pool boiling heat transfer on nano-scale modification surface. Int. J. Heat Mass Transf. 84:550–561. https://doi.org/10.1016/j.ijheatmasstransfer.2014.12.064.

Majumder, B., Katarkar, A. and Bhaumik, S. (2020). Effect of structured surface on contact angle using sessile droplet method. IOP Conf. Ser. Mater. Sci. Eng. 814(1):012034. https://doi.org/10.1088/1757-899X/814/1/012034.

Majumder, B., Pingale, A. D., Katarkar, A. S. Belgamwar, S. U. and Bhaumik, S. (2022a). Enhancement of pool boiling heat transfer performance of R-134a on microporous Al@GNPs composite coatings. Int. J. Thermophys. 43(4): 1–19. https://doi.org/10.1007/s10765-022-02973-7.

Majumder, B., Pingale, A., Katarkar, A., Belgamwar, S. and Bhaumik, S. (2022b). Developing Al@GNPs composite coating for pool boiling applications by combining mechanical milling, screen printing and sintering methods. Adv. Mater. Process. Technol. 00(00):1–12. https://doi.org/10.1080/2374068x.2022.2036037.

Moghadasi, H. and Saffari, H. (2021). experimental study of nucleate pool boiling heat transfer improvement utilizing micro/nanoparticles porous coating on copper surfaces. Int. J. Mech. Sci. 196:106270. https://doi.org/10.1016/j.ijmecsci.2021.106270.

Owhal, A., Pingale, A. and Belgamwar, S. (2022). Developing sustainable Zn-MWCNTs composite coatings using electrochemical co-deposition method: Tribological and surface wetting behavior. Adv. Mater. Process. Technol. 1–14. https://doi.org/10.1080/2374068X.2022.2035968.

Owhal, A., Pingale, A. D., Belgamwar, S. U. and Rathore, J. S. (2021a). A brief manifestation of anti-bacterial nanofiller reinforced coatings against the microbial growth based novel engineering problems. Mater. Today: Proc. 47:3320–330. https://doi.org/10.1016/j.matpr.2021.07.151.

Owhal, A., Pingale, A. D., Khan, S., Belgamwar, S.U., Jha, P. N. and Rathore, J.S. (2021b). Electro-codeposited γ-Zn-Ni/Gr composite coatings: effect of graphene concentrations in the electrolyte bath on tribo-mechanical, anti-corrosion and anti-bacterial properties. Trans. IMF 99(6):324–331. https://doi.org/10.1080/00202967.2021.1979815.

Pingale, A. D., Belgamwar, S. U. and Rathore, J. S. (2020a). Effect of graphene nanoplatelets addition on the mechanical, tribological and corrosion properties of Cu–Ni/Gr nanocomposite coatings by electro-Co-deposition method. Trans. Indian Inst. Met. 73(1):99–107. https://doi.org/10.1007/s12666-019-01807-9.

Pingale, A. D., Belgamwar, S. U. and Rathore, J. S. (2020b). Synthesis and characterization of Cu–Ni/Gr nanocomposite coatings by electro-Co-deposition method: effect of current density. Bull. Mater. Sci. 43(1):66. https://doi.org/10.1007/s12034-019-2031-x.

Pingale, A. D., Owhal, A., Belgamwar, S. U. and Rathore, J. S. (2021). Effect of current on the characteristics of CuNi-G nanocomposite coatings developed by DC, PC and PRC electrodeposition. JOM 73(12):4299–4308. https://doi.org/10.1007/s11837-021-04815-7.

Pingale, A. D., Owhal, A., Katarkar, A. S., Belgamwar, S. U. and Rathore, J. S. (2021a). Recent researches on Cu-Ni alloy matrix composites through electrodeposition and powder metallurgy methods: A review. Mater. Today: Proc. 47: 3301–3308. https://doi.org/10.1016/j.matpr.2021.07.145.

Pingale, A. D., Owhal, A., Katarkar, A. S., Belgamwar, S. U. and Rathore, J. S. (2021b). Facile synthesis of graphene by ultrasonic-assisted electrochemical exfoliation of graphite. Mater. Today: Proc. 44:467–472. https://doi.org/10.1016/j.matpr.2020.10.045.

Pingale, A. D., Belgamwar, S. U. and Rathore, J. S. (2020c). The influence of graphene nanoplatelets (GNPs) addition on the microstructure and mechanical properties of Cu-GNPs composites fabricated by electro-co-deposition and powder metallurgy. Mater. Today: Proc. 28:2062–2067. https://doi.org/10.1016/j.matpr.2020.02.728.

Pingale, A. D., Belgamwar, S. U. and Rathore, J. S. (2020d). A novel approach for facile synthesis of Cu-Ni/GNPs composites with excellent mechanical and tribological properties. Mater. Sci. Eng: B 260:114643. https://doi.org/10.1016/j.mseb.2020.114643.

Saeidi, D. and Alemrajabi, A. A. (2013). Experimental investigation of pool boiling heat transfer and critical heat flux of nanostructured surfaces. Int. J. Heat Mass Transf. 60(1):440–449. https://doi.org/10.1016/j.ijheatmasstransfer.2013.01.016.

Schultz, R. R. and Cole, R. (1979). Uncertainty analysis of boiling nucleation. AIChE Symp. Ser. 75:32–38. AIChE.

Shelke, A. R., Balwada, J., Sharma, S., Pingale, A. D., Belgamwar, S. U. and Rathore, J. S. (2020). Development and characterization of Cu-Gr composite coatings by electro-co-deposition technique. Mater. Today: Proc. 28:2090–2095. https://doi.org/10.1016/j.matpr.2020.03.244.

Shi, B., Wang, Y. B. and Chen, K. (2015). Pool boiling heat transfer enhancement with copper nanowire arrays. Appl. Therm. Eng. 75:115–121. https://doi.org/10.1016/j.applthermaleng.2014.09.040.

Ujereh, S., Fisher, T. and Mudawar, I. (2007). Effects of carbon nanotube arrays on nucleate pool boiling. Int. J. Heat Mass Transf. 50(19–20):4023–4038. https://doi.org/10.1016/j.ijheatmasstransfer.2007.01.030.

Vemuri, S. and Kim, K. J. (2005). Pool boiling of saturated FC-72 on nano-porous surface. Int. Commun. Heat Mass Transf. 32(1–2):27–31. https://doi.org/10.1016/j.icheatmasstransfer.2004.03.020.

Walunj, A. and Sathyabhama, A. (2018). Comparative study of pool boiling heat transfer from various microchannel geometries. Appl. Therm. Eng. 128:672–683. https://doi.org/10.1016/j.applthermaleng.2017.08.157.

Wu, W., Bostanci, H., Chow, L. C., Hong, Y., Su, M. and Kizito, J. P. (2010). Nucleate boiling heat transfer enhancement for water and FC-72 on titanium oxide and silicon oxide surfaces. Int. J. Heat Mass Transf. 53(9–10):1773–77. https://doi.org/10.1016/j.ijheatmasstransfer.2010.01.013.

103 Symmetric uncertainty based feature selection method in android malware detection

Pooja V. Agrawal[a], Deepak D. Kshirsagar[b], and Anish R. Khobragade[c]

Department of Computer Engineering and IT, College of Engineering Pune, Pune, India

Abstract

Most smartphone companies have chosen Android as their OS. Users use their mobile devices as mini-computers containing their personal information. It is necessary to detect the malware before making a serious threat to users' confidential information. It is not enough to detect android malware based on static features because it does not recognise the run-time behaviour of the application program. This study proposes a symmetric uncertainty (SU) feature selection technique with a ranker to reduce dimensionality in an android malware detection system. Based on the rank of SU, the suggested method creates feature subsets, which are then analysed using the Random Forest (RF) classifier. The proposed approach obtains a relevant feature subset for CICMalDroid2020 that includes 75 features from the original feature set and achieves a 97.3789% accuracy over the original feature set. Mutual Information (MI), Information Gain (IG), Chi-Square (CHI), and Gain Ratio (GR) are all used to compare the obtained feature subset. Compared to MI, CHI, IG, and GR, the RF classifier obtains a higher accuracy and recall of 97.3789% and 99.3981%, respectively, with SU-75 features.

Keywords: Deep learning, dynamic features, feature selection, machine learning, symmetric uncertainty.

Introduction

Mobile phone users exponentially increased worldwide for communication purposes, online shopping, banking, and social networking. As per the Statista data (Keelery, 2021), 930 million users access the internet through mobile devices, and around 1.5 billion mobile users will access the internet worldwide up to 2040. Android OS was first released in 2008 by Google has currently captured more than 70% of the market share worldwide over mobile devices (Laricchia, 2022). Malware is malicious software that targets the OS of mobile devices and desktops by finding the vulnerabilities present in that device, categorised based on its behavior (Sawaisarje et al., 2018), such as worm, virus, trojan. AV-Test Institute AV-TEST (2022) has registered 1332 million new malicious programs in Feb 2022.

Due to the open nature of Android, attackers have developed applications that entice the users, such as themes, animated wallpapers, and emoji keyboards. While installing the application from any illegitimate third-party source, malware may get installed unknowingly into the device, as it is difficult for users to differentiate between genuine and malware apps. As per the 2021 report of Kaspersky, 3464756 malicious packages were detected, out of which 17372 are mobile ransomware, and 97661 are mobile banking (Tatyana and Anton, 2022). They have detected 42.42% adware, 35.27% riskware, 2.82% banking, and 3.10% SMSware. Mobile malware attacks are carried out by phishing, installing the applications from malicious sites, downloading any email attachments from an unknown source. Once the malware gets installed in the mobile device, it tracks the user's daily activities to get confidential user information such as banking passwords and one-time-password received via SMS.

Many technologists and researchers are working to combat malware attacks; for example, Google introduced the Play Protect system by Şahin et al. (2021) in 2017 to control Play Store applications. As per McAfee's report, the Play Protect system lacks the ability to detect all types of malware, and additional detection mechanisms are essential to prevent the spread of malware. The researchers propose static, dynamic, and hybrid approaches for malware detection. The static approach (Elayan and Mustafa, 2021; Almahmoud et al., 2021; Kinkead et al., 2021; Millar et al., 2021; Zhang et al., 2021; Sahal et al., 2018; McLaughlin et al., 2017; Al Sarah et al., 2021; Mat et al., 2021; Arif et al., 2021; González et al., 2019; Wen and Yu, 2017; Mathur et al., 2021; Chakravarty, 2020; Visalakshi, 2020; Lashkari et al., 2017; Li et al., 2017; Wei et al., 2017; Sangal and Verma, 2020; Dhalaria and Gandotra, 2020) considers the application components features such as permissions and intent without executing the application over the device. In a dynamic approach (Xiao et al., 2019, Mahdavifar et al., 2020; Thangavelooa et al., 2020; Singh and Hofmann, 2017), features are extracted by running the application on a controlled virtual environment. However, static and dynamic features are combined in a hybrid approach (Mahdavifar et al., 2022; Alzaylaee et al., 2020).

[a]agrawalpv20.comp@coep.ac.in; [b]ddk.comp@coep.ac.in; [c]anishraj.comp@coep.ac.in

Deep learning (DL) and machine learning are used to detect Android malware. In the DL approach, deep neural network (DNN) (Mahdavifar et al., 2020; Mahdavifar et al., 2022; Alzaylaee et al., 2020), recurrent neural network (RNN) (Elayan and Mustafa, 2021; Almahmoud et al., 2021; Xiao et al., 2019), and convolutional neural networks (CNN) (Kinkead et al, 2021; Millar et al., 2021; Zhang et al.. 2021; Sahal et al., 2018; McLaughlin et al., 2017) are utilised, which depend on the extracted features from the different hidden layers. In the machine learning approach, different algorithms such as support vector machine (SVM), Naive Bayes (NB), RF, and decision tree are utilised by using feature selection, stacking, bagging, boosting, and ensemble learning for improving the classifier performance.

The researchers mainly carried out their research considering the static-based features rather than the dynamic-based features for detecting android malware. As shown in Table 103.1, a minimal contribution is given to dynamic analysis. Traditional research carried on dynamic features was based only on system calls; this paper considers other dynamic attributes such as binder calls used for inter-process communication and composite behaviours associated with low-level system calls.

The main contributions of the paper are as follows:

1. The paper proposes the android malware detection system using the SU method with ranker.
2. The proposed system obtains a relevant feature subset that includes top-ranked 75 features of SU on CICMalDroid2020 and achieves higher accuracy compared to the original features.

The rest of the paper is introduced as follows. Section 2 explains the recent work on detecting android malware. Further, section 3 describes the proposed system for detecting Android malware. Section 4 exhibits the system implementation and results. Finally, section 5 presents the conclusions and future scope of the proposed system.

Literature Survey

As the number of Android malware increases daily, it influences the research work to detect malware using deep and machine learning. This section reviewed recent papers that specifically detect malware from the static, dynamic features of Android mobile applications.

A. DL related recent work

In 2021, Elayan and Mustafa (2021) proposed a Gated Recurrent Unit (GRU) for malware detection on static features of the CICAndMal2017 dataset. It has achieved an accuracy of 98.2%, with three hidden layers present in GRU. Another contribution from Almahmoud et al. (2021) utilised the RNN model over the combined CICAndMal2017, CICMalDroid2020, and CICInvesAndMal2019 dataset comprising 1390 benign and 1430 malware samples. The author selected forty-five features based on frequency and cosine similarity scores from the dataset, where the model achieved 98.58% accuracy. In 2019, Xiao et al. (2019) author utilised one system call sequence and built two LSTM language models trained on the Drebin dataset. The model gained 93.7% accuracy and 96.6% recall by comparing the similarity scores calculated by these models. On the same dataset (Kinkead et al., 2021), the extracted static feature of opcode sequence represented in a vector embedding followed by a 1D convolution achieved 98% accuracy.

Millar et al. (2021) proposed a multi-view CNN model using static features such as opcodes, permissions, and AndroidAPIs of Drebin and AMD dataset for zero-day android malware detection considering all malware families, the model has achieved a 91% and 81% detection rate. In 2020, Zhang et al. (2021) proposed a text classification method. Text CNN experimented with benign apps samples from the Anzhi, and malicious applications samples from the Android Malware Genome and Contagio Community Sahal et al. (2018) furthermore generated four datasets with different benign and malware samples. TextCNN gained 96.6% higher accuracy on the fourth dataset. In 2017, McLaughlin et al. (2017) proposed a CNN model with a single convolutional layer based on opcode sequence static features. The model achieves an accuracy of 98% on the Android malware Genome project (Sahal et al., 2018).

In 2020, Mahadavifar et al. (2020) proposed a Pseudo-Label DNN (PLDNN) model based on dynamic features, namely binder and system calls, and composite behaviour of the CICMalDroid2020 dataset. With 1000 labelled samples, the model achieved an accuracy of 96.7% with five hidden layers. The same author has proposed a hybrid method Mahdavifar et al. (2022) using the semi-supervised pseudo-label stack auto-encoder (PLSAE) technique in 2021. With 100 labelled training samples, the model achieved an accuracy of 95.19% with five hidden layers on the CICMalDroid2020 dataset comprises 50621 static and 470 dynamic features. Another contribution by Alzaylaee et al., (2020) proposed a DL model which runs on real devices using the McAfee labs dataset. Multilayer perceptron (MLP) model has achieved 97.8% and 99.6% detection rate with dynamic and hybrid features with three hidden layers.

Table 103.1 Literature survey

Year	Author	Analysis Type	Dataset	Classifier	Detection Type	Accuracy
2021	Elayan and Mustafa (2021)	Static	CICAndMal2017	GRU	Binary	98.20%
2021	Almahmoud et al. (2021)	Static	CIC-[AndMal2017 MalDroid2020 InvesAndMal2019]	RNN	Binary	98.58%
2019	Xiao et al. (2019)	Dynamic	Drebin	LSTM	Binary	93.70%
2021	Kinkead et al. (2021)	Static	Drebin	CNN	Binary	98.00%
2021	Millar et al. (2021)	Static	Drebin and AMD	CNN	Malware Family	91.00% (Recall - Drebin) 81.00% (Recall - AMD)
2020	Zhang et al. (2021)	Static	Anzhi (Benign) Android Malware Genome Project and Contagio	CNN	Binary	96.60%
2017	McLaughlin et al. (2017)	Static	Android Malware Genome Project	CNN	Binary	98.00%
2020	Mahdavifar et al. (2020)	Dynamic	CICMalDroid2020	PLDNN	Malware Family	96.7%
2021	Mahdavifar et al. (2022)	Hybrid	CICMalDroid2020	PLSAE	Malware Family	95.19%
2019	Alzaylaee et al. (2020)	Hybrid	McAfee Labs	MLP	Binary	97.80% (Recall-dynamic) 99.60% (Recall-hybrid)
2021	Al Sarah et al. (2021)	Static	Drebin	LightGBM	Binary	99.50%
2021	Mat et al. (2021)	Static	Drebin AndroZoo	NB	Binary	91.10%
2021	Arif et al. (2021)	Static	Drebin AndroZoo	IG, Fuzzy	Binary	90.54%
2020	González et al. (2019)	Static	Drebin	MLP	Binary	95.82%
2017	Wen and Yu (2017)	Static	Drebin Google PlayStore	SVM	Binary	95.20%
2021	Mathur et al. (2021)	Static	AMD AndroZoo	RF	Binary	97.00%
2020	Chakravarty (2020)	Static	AMD	Randomizable filtered	Binary	93.46%
2020	Visalakshi (2020)	Static	AAGM AMD	SVM	Binary	0.748 –TPR rate
2020	Sangal and Verma (2020)	Static	CICInvesAnd Mal2019	RF	Binary	96.05%
2020	Dhalaria and Gandotra (2020)	Static	AndroMD	K-NN_RF	Binary	98.02%
2020	Thangavelooa et al. (2020)	Dynamic	APKPure Android Malware Genome Project	RF	Binary	91.70%
2017	Singh and Hofmann (2017)	Dynamic	Google PlayStore Contagio project	SVM	Binary	97.16%

B. ML related recent work

The author Al Sarah et al. (2021) uses a LightGBM model on the Drebin dataset that achieves an accuracy of 99.5% with 100 features selected by the Recursive feature elimination method. Using the Bayesian model, another contribution Mat et al. (2021) on the AndroZoo and Drebin dataset achieves 91% accuracy with 15 features selected by Chi-Square (CHI) method. In Arif et al. (2021), the dataset mentioned by Mat et al. (2021) is used to perform risk analysis associated with android applications and achieve an accuracy of 90.54% for classifying android apps into different risk levels. In the paper González et al. (2019), a genetic algorithm with an MLP model is applied on the Drebin dataset that achieves

95.82% accuracy. Author Wen and Yu (2017) proposed a feature selection method that combines principal component analysis (PCA) and ReliefF applied on Drebin, and the Google PlayStore dataset has achieved an accuracy of 95.2% with the SVM model.

Author Mathur et al. (2021) have extracted native and custom permission from Androzoo and AMD projects and achieved 97% accuracy using RF with 55 features selected by the backward elimination method. Another paper Chakravarty (2020) used the Randomizable filtered classifier on the AMD dataset comprised of 330 features has achieved 93.46% accuracy when selecting the top five features using the GR method. Visalakshi (2020) proposed KNN based relief approach for feature selection and optimised SVM on the AAGM Lashkari et al. (2017) and AMD Li et al. (2017), Wei et al. (2017) datasets. The optimised SVM with the KNN-Relief method performs better with a TPR rate of 0.748 than the traditional SVM model.

In 2020, the author Sangal and Verma (2020) used a PCA method on the CICInvesAndMal2019 dataset that achieves 96.05% accuracy with RF. Another paper Dhalaria and Gandotra (2020) proposed a CHI with a stacking approach on the AndroMD Dhalaria and Gandotra (2021) dataset. K-NN_RF has achieved 98.02% accuracy after selecting features using CHI.

Thangavelooa et al. (2020) applied a dynamic analysis method on the APKPure and Android Malware Genome Project Zhou and Jiang (2012) samples and achieved 91.7% accuracy with RF when selecting system calls, CPU usage, memory, etc., using the GR method. In 2017, Singh and Hofmann (2017), the Contagio project and GooglePlayStore samples were studied, achieving 97.16% accuracy with a 99.54% recall rate using the SVM model after selecting 31 features using a 0.2 threshold on the correlation method.

Proposed System

The proposed system comprises of data pre-processing and a filter-based feature selection method with RF for malware detection, as shown in Figure 103.1.

The android malware dataset used by the proposed work consists of different dynamic features with labels having multiple values. Dataset features are numeric, and label attribute containing a numeric value is converted into text values. Thus, the cleaned malware dataset is acquired.

Different feature selection techniques are present in machine learning to reduce the number of input features. These techniques identify the subset of features used to improve the model's performance. There are three types of feature selection methods present in ML: filter, embedded, and wrapper. The filter-based method is used in this proposed system which is faster than other methods and not dependent on any classifier. The filter methods are based on statistical measures to find the correlation between the target variable and input features. This proposed system applies SU from other filter-based feature selection methods on the cleaned malware dataset. For each feature present in the cleaned malware dataset, a score is calculated using the mathematical formula shown in (1), based on entropy and information gain.

$$SU(X,Y) = 2 * \frac{IG(X \mid Y)}{H(X) + H(Y)} \tag{1}$$

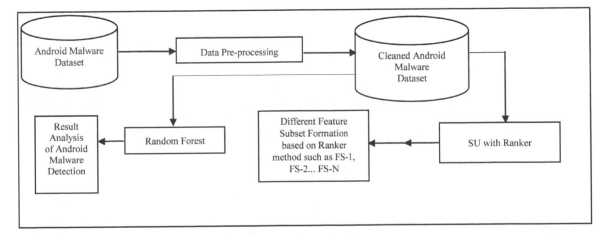

Figure 103.1 Proposed system of android malware system

where H(X) and H(Y) are the entropy of input variable X and output label Y. IG(X, Y) is information gain of X after noticing Y.

The proposed system forms different feature subsets from the original features depending on SU's rank. The system selects the first top-ranked 25 features and creates feature subset 1 (FS-1). Then, it chooses the first top-ranked 50 features and forms FS-2, and so on. In this way, different FS were created based on SU's rank. The formation of the feature subsets stops once the system achieves higher performance with minimum features using the RF classifier than the original features. In this way, the system selects a single FS as a relevant feature set and builds an RF model for detecting Android malware.

Implementation and Result Analysis

The proposed work is performed on the Windows 10 operating system having an Intel Xeon 3.6 GHz processor with 32GB RAM using WEKA Garner (1995) 3.8.3 tool. The CICMalDroid2020 dataset is downloaded from the University of New Brunswick website. This dataset contains the latest android samples from December 2017 to December 2018 collected from MalDozer, AMD, VirusTotal service, and several other sources. This dataset is generated by the author (Mahadavifar et al., 2020). The author has executed

different android applications on the CopperDroid Tam et al. (2015) virtual machine, which reconstructs high-level Android-specific and low-level operating-system-specific behaviours. The dataset comprises 470 dynamic features, including binder calls, composite behaviours, and system calls. It includes malware and benign samples from five different categories: SMSware (3904), riskware (2546), banking malware (2100), adware (1253), and benign (1795). The dataset does not contain any null and missing values. So, only numeric labels present in the dataset are modified to the text labels. Thus, for android malware detection, the cleaned malware dataset is generated.

The proposed system selects SU from the filter methods present in the WEKA tool and applies it to the cleaned malware CICMalDroid2020 dataset. The features are selected based on the top-ranked SU score of the features. The feature's score is calculated using the SU method based on IG and entropy. The score value of the feature ranges from 0 to 0.488075 in descending order. The performance of the RF classifier is evaluated using ten-fold cross-validation on the original feature set. Firstly, the system selects the top-ranked 25 features (FS-1) and evaluates the performance using an RF classifier. It has achieved lower accuracy as 96.4563%, precision as 96.9412, F1-Score as 97.9250%, recall as 98.9289% with higher FAR as 0.1705 compared to the original feature. Thus, in the next iteration, the system selects the top-ranked 50 features (FS-2), achieving lower accuracy of 96.9046%, the precision of 97.0412%, F1-Score of 98.1906%, recall of 99.3675%, and higher FAR of 0.1655 compared to the performance of the original feature. The system continues to select the top-ranked features by adding the previous 25 features to the feature set in each iteration until the highest performance is achieved. Further, the system selects the top-ranked 75 features (FS-3) and achieves higher accuracy, precision, F1-score as 97.3789%, 97.5473%, 98.4640%, and lower FAR and model built-up time as 0.1365 and 4.43 sec respectively than the original feature as shown in Table 103.2. Thus, the system selects the top-ranked 75 features achieving higher performance as summarised in algorithm 103.1. The performance of the RF classifier is evaluated based on the standard metrics as depicted in Table 103.2. Table 103.2 represents the result analysis of the RF classifier on the original features and different feature subsets. The top-ranked 75 features are listed in descending order of SU score range from 0.488075 to 0.064992 [getInstallerPackageName, getReceiverInfo, newselect, getServiceInfo, ftruncate64, FS_PIPE_ACCESS(WRITE), queryIntentServices, getApplicationRestrictions, checkPackage, resolveContentProvider, resolveIntent, getActivityInfo, isSpeakerphoneOn, CREATE_FOLDER, statfs64, FS_ACCESS(CREATE__READ__WRITE), fdatasync, getPackageInfo, isActiveNetworkMetered, remove, unlink, mprotect, FS_ACCESS(WRITE), getRingerMode, mkdir, pwrite64, FS_PIPE_ACCESS, NETWORK_ACCESS(), fchmod, getMode, FS_ACCESS(CREATE__WRITE), fchown32, NETWORK_ACCESS, access, FS_ACCESS(), getApplicationInfo, pread64, FS_ACCESS(CREATE), rename, FS_ACCESS(CREATE__READ), _llseek, getdents64, checkPermission, pipe, munmap, getStreamVolume, poll, fsync, stat64, prctl, getsockopt, FS_ACCESS, chmod, getLine1Number, sigaltstack, FS_ACCESS(READ), resolveService, nanosleep, DEVICE_ACCESS, setpgid, brk, notifyChange, getSubscriberId, mmap2, ftruncate, getAccountsAsUser, sched_yield, arm_nr_set_tls, msync, exit, TERMINATE_THREAD, writev, lstat64, clone, CREATE_THREAD].

The top-ranked 75 features are selected from other filter methods, namely MI, CHI, GR, and IG Kshirsagar and Kumar (2020), and the performance is measured with RF. Table 103.3 shows the comparative analysis of other filter-based methods with the SU method. After selecting 75 features from different filter-based methods, CHI has achieved higher accuracy, precision, recall of 97.0943%, 97.2638%, 99.3573%, and lower FAR and model built-up time of 0.1526 and 4.42 sec compared to IG, GR, MI. The SU method has achieved higher accuracy, precision, F1-score, recall of 97.3789%, 97.5473%, 98.4640%,

Table 103.2 Result analysis of RF using obtained feature subsets

Classifier	Subset	Features	Accuracy (%)	Precision (%)	F1-Score (%)	Recall (%)	FAR	Model built time (Sec)
RF	-	470	97.3357	97.4036	98.4407	99.5002	0.1448	9.69
	FS-1	25	96.4563	96.9412	97.9250	98.9289	0.1705	3.40
	FS-2	50	96.9046	97.0412	98.1906	99.3675	0.1655	3.74
	FS-3	75	97.3789	97.5473	98.4640	99.3981	0.1365	4.43

Algorithm 103.1 For getting relevant feature subset

Input: Original feature (f1... f470)
Output: Relevant feature set (f1... fn)
1. Calculate the performance of RF with the original feature set.
2. Apply SU on the original feature set.
3. Formation of different feature subsets using SU with Ranker and performance is calculated with RF classifier. While creating the feature subset, top-ranked 25 features were added in each iteration until the model achieved higher performance than the original feature.
   i.e. FS-1= {f1,f2...f25}
   FS-2= { f1,f2...f50}
   FS-3= { f1,f2...f75}

   FS-n= { f1,f2...fn}
4. If (Performance(FS-i)>= Performance(Original feature set)) then
   Relevant Feature set=FS-i
   End if

Table 103.3 Comparative analysis of filter-based methods using top-ranked 75 features with SU

Classifier	FS	Accuracy (%)	Precision (%)	F1-score (%)	Recall (%)	FAR	Model built time (Sec)
RF	Proposed Approach	97.3789	97.5473	98.4640	99.3981	0.1365	04.43
	MI	96.8615	97.0585	98.1646	99.2961	0.1643	04.91
	CHI	97.0943	97.2638	98.2994	99.3573	0.1526	04.42
	GR	95.4906	96.0135	97.3701	98.7657	0.2240	15.80
	IG	96.9564	97.1369	98.2196	99.3267	0.1599	04.67

99.3981%, lower FAR as 0.1365, and a slight increase of 0.01 sec in model built-up time as compared to the CHI method. After comparison with other methods, it is clear that the SU method is more powerful than other methods. Therefore, the proposed system has selected SU as a feature selection method for android malware detection.

Table 103.4 represents a comparison of the presented work with recent works. In the paper Chakravarty (2020), Sangal and Verma (2020), static-based feature used for malware detection. It has achieved an accuracy of 93.46% and 96.05% with a Randomizable Filtered and RF classifier, which is lower than the presented work. The paper Xiao et al. (2019), Thangavelooa et al. (2020), Singh and Hofmann (2017) is based on dynamic analysis for malware detection. It has achieved an accuracy of 93.70%, 91.70%, and 97.16%, which is less than the presented work. Our proposed work had achieved an accuracy of 97.3789% when 75 features were selected using the SU method with an RF classifier.

Conclusion and Future Scope

This study presented SU feature-selection-based method for selecting relevant features for malware detection. Here, the proposed system creates the different feature subsets and evaluates their performance until the system achieves higher performance than the original feature set using RF. After assessing the different feature subsets, top-ranked 75 features were selected with higher accuracy as 97.3789% for malware

Table 103.4 Comparison with the traditional system

Work	Dataset	Feature selection	Classifier	Accuracy (%)
Xiao et al. (2019)	Drebin	-	LSTM	93.70
Chakravarty (2020)	AMD	GR	Randomizable Filtered	93.46
Sangal and Verma (2020)	CICInvesAndMal2019	PCA	RF	96.05
Thangavelooa et al. (2020)	APKPure Android Malware Genome	GR	RF	91.70
Singh and Hofmann (2017)	Google PlayStore Contagio project	Correlation	SVM	97.16
Proposed Work	CICMalDroid2020	SU	RF	97.37

detection. This system is compared with the original feature set and traditional systems, which shows that it produces good accuracy.

In the future, we will propose a combination of evolutionary feature selection approaches with ensemble classifiers for android malware detection.

References

Almahmoud, M., Alzu'bi, D., and Yaseen, Q. (2021). ReDroidDet: android malware detection based on recurrent neural network. Procedia Comput. Sci. 184:841–846.

Al Sarah, N., Rifat, F. Y., Hossain, M. S., and Narman, H. S. (2021). An efficient android malware prediction using ensemble machine learning algorithms. Procedia Comput. Sci. 191:184–191.

Alzaylaee, M. K., Yerima, S. Y., and Sezer, S. (2020). DL-Droid: Deep learning based android malware detection using real devices. Comput. Sec. 89:101–663.

Arif, J. M., Ab Razak, M. F., Mat, S. R. T., Awang, S., Ismail, N. S. N., and Firdaus, A. (2021). Android mobile malware detection using fuzzy AHP. J. Inform. Sec. Appl. 61:102–929.

AV-TEST 2022. Malware statistics. https://www.av-test.org/en/statistics/malware/. (Accessed Feburary 21, 2022).

Chakravarty, S. 2020. Feature selection and evaluation of permission-based Android malware detection. In 2020 4th. International conference on trends in electronics and informatics (ICOEI), (48184: pp. 795–799). IEEE.

Dhalaria, M. and Gandotra, E. (2021). A hybrid approach for android malware detection and family classification. Int. J. Interact. Multimed. Artif. Intell. 6:6.

Dhalaria, M. and Gandotra, E. 2020. Android malware detection using chi-square feature selection and ensemble learning method. In 2020 Sixth international conference on parallel, distributed and grid computing (PDGC), (pp. 36–41), IEEE.

Elayan, O. N. and Mustafa, A. M. (2021). Android malware detection using deep learning. Procedia Comput. Sci. 184:847–852.

Garner, S. R. (1995). Weka: The waikato environment for knowledge analysis. In Proceedings of the New Zealand computer science research students conference, (pp. 57–64).

González, S., Herrero, Á., Sedano, J., and Corchado, E. 2019. Neuro-evolutionary feature selection to detect android malware. In International joint conference: 12th international conference on computational intelligence in security for information systems (CISIS 2019) and 10th international conference on european transnational education (ICEUTE 2019), (pp. 124–131). Springer, Cham.

Keelery, S. (2021). Number of mobile users in India. https://www.statista.com/statistics/558610/number-of-mobile-in-ternet-user-in-india (Accessed Feburary 20, 2022).

Kinkead, M., Millar, S., McLaughlin, N. and O'Kane, P. (2021). Towards explainable CNNs for android malware detection. Procedia Comput. Sci. 184:959–965.

Kshirsagar, D. and Kumar, S. 2020. Identifying reduced features based on IG-Threshold for DoS attack detection using PART. In International conference on distributed computing and internet technology, (pp. 411–419). Springer, Cham.

Laricchia, F. 2022. Worldwide market share of mobile operating systems worldwide from Jan 2012 to June 2021. https://www.statista.com/statistics/272698/global-market-share-held-by-mobile-operating-systems-since-2009/. (Accessed Feburary 20, 2022).

Lashkari, A. H., Kadir, A. F. A., Gonzalez, H., Mbah, K. F., Ghobani, A. 2017. Towards a network-based framework for android malware detection and characterisation. In 2017 15th. Annual conference on privacy, security and trust (PST), (pp. 233-23309). IEEE.

Li, Y., Jang, J., Hu, X., and Ou, X. 2017. Android malware clustering through malicious payload mining. In International symposium on research in attacks, intrusions, and defenses, (pp. 192–214), Cham: Springer.

Mahdavifar, S., Alhadidi, D., and Ghorbani, A. (2022). Effective and efficient hybrid android malware classification using pseudo-label stacked auto-encoder. J. Netw. Syst. Manag. 30(1):1–34.

Mahdavifar, S., Kadir, A. F. A., Fatemi, R., Alhadidi, D., and Ghobani, A. 2020. Dynamic android malware category classification using semi-supervised deep learning. In 2020 IEEE Intl conf on dependable, autonomic and secure computing, intl conf on pervasive intelligence and computing, intl conf on cloud and big data computing, intl conf on cyber science and technology congress (DASC/PiCom/CBDCom/CyberSciTech), (pp. 515–522). IEEE.

Mat, S. R. T., Ab Razak, M. F., Kahar, M. N. M., Arif, J. M., and Firdaus, A. 2021. A Bayesian probability model for Android malware detection. ICT Express.

Mathur, A., Podila, L. M., Kulkarni, K., Niyaz, Q. and Javaid, A. Y. (2021). NATICUSdroid: A malware detection framework for Android using native and custom permissions. J. Inform. Sec. Appl., 58: 102–696.

McLaughlin, N., del Rincon, J. M., Kang, B. 2017. Deep android malware detection. In Proceedings of the seventh ACM on conference on data and application security and privacy, (pp. 301–308).

Millar, S., McLaughlin, N., del Rincon, J. M. and Miller, P. (2021). Multi-view deep learning for zero-day android malware detection. J. Inform. Sec. Appl. 58:102–718.

Sahal, A. A., Alam, S. and Soğukpinar, I. 2018. Mining and detection of android malware based on permissions. In 2018 3rd. International conference on computer science and engineering (UBMK), (pp. 264–268). IEEE.

Şahin, D. Ö., Kural, O. E., Akleylek, S., and Kılıç, E. (2021). Permission-based android malware analysis by using dimension reduction with PCA and LDA. J. Inf. Sec. Appl. 63:102–995.

Sangal, A. and Verma, H. K. 2020. A static feature selection-based android malware detection using machine learning techniques. In 2020 International conference on smart electronics and communication (ICOSEC), (pp. 48–51). IEEE.

Sawaisarje, S. K., Pachghare, V. K., and Kshirsagar, D. D. 2018. Malware detection based on string length histogram using machine learning. In 2018 3rd. IEEE International conference on recent trends in electronics, information & communication technology (RTEICT), (pp. 1836–1841). IEEE.

Singh, L. and Hofmann, M. (2017). Dynamic behavior analysis of android applications for malware detection. In 2017 International conference on intelligent communication and computational techniques (ICCT), (pp. 1–7), IEEE.

Tam, K., Fattori, A., Khan, S., and Cavallaro, L. (2015). Copperdroid: Automatic reconstruction of android malware behaviors. In NDSS Symposium 2015, (pp. 1–15).

Tatyana, S. and Anton, K. (2022). Mobile malware evolution 2021. https://securelist.com/mobile-malware-evolution-2021/105876/ (Accessed February 22, 2022).

Thangavelooa, R., Jinga, W. W., Lenga, C. K. and Abdullaha, J. (2020). Datdroid: dynamic analysis technique in android malware detection. Int. J. Adv. Sci. Eng. Inf. Technol. 10:536–541.

Visalakshi, P. (2020). Detecting android malware using an improved filter based technique in embedded software. Microprocess. Microsyst. 76:103–115.

Wei, F., Li, Y., Roy, S., Ou, X., and Zhou, W. 2017. Deep ground truth analysis of current android malware. In International conf.erence on detect.ion of intrusions and malware, and vulnerability assessment, (pp. 252–276). Cham: Springer

Wen, L. and Yu, H. 2017. An android malware detection system based on machine learning. In AIP conference proceedings 1864(1):020136. AIP Publishing LLC.

Xiao, X., Zhang, S., Mercaldo, F., Hu, G. and Sangaiah, A. K. (2019). Android malware detection based on system call sequences and LSTM. Multimed. Tools Appl. 78(4):3979–3999.

Zhang, N., Tan, Y. A., Yang, C. and Li, Y. (2021). Deep learning feature exploration for android malware detection. Appl. Soft Comput. 102:107–169.

Zhou, Y. and Jiang, X. 2012. Dissecting android malware: Characterisation and evolution. In 2012 IEEE symposium on security and privacy, (pp. 95–109), IEEE.

104 Pool boiling of R-134a on ZnO nanostructured surfaces: experiments and heat transfer analysis

Biswajit Majumder[1,a], Anil S. Katarkar[1,b], Ajay D. Pingale[2,c], Sri Aurobindo Panda[2,d], Sachin U. Belgamwar[2,e], and Swapan Bhaumik[1,f]

[1]Department of Mechanical Engineering, National Institute of Technology Agartala, Tripura, India

[2]Department of Physics, Birla Institute of Technology and Science, Pilani, Rajasthan, India

Abstract

An experimental study was carried out to investigate the pool boiling performance of R-134a onZnO nanostructured heating surfaces. ZnO nanostructured heating surfaces were fabricated by the thermal evaporation technique followed by heat treatment. Nanostructured ZnO thin films were applied on the bare aluminum heating substrate, furthermore pool boiling performance was carried out with R-134a. Experimental data were recorded at heat fluxes ranging from 9.37 to 72.23 kW/m2k. It was found that with a rise in the thickness of nanostructured ZnO thin film, the wall superheat was decreased and the maximum decrement in wall superheat was observed for nanostructure ZnO thin film with 300 nm thickness. It was also observed that the heat transfer coefficient (HTC) of the ZnO nanostructured surfaces increased with the rise in thickness of coating. The highest value of HTC for ZnO-300 surface was61% greater than the bare Al heating surface.

Keywords: R-134a, technique for thermal evaporation, thickness of coating, ZnO.

Introduction

Among different modes of heat transfer, boiling is considered as the most efficient heat transfer mode. Due to this fact, it has wide applications in industries, especially in power plants, cooling systems, refrigeration systems, and nuclear reactors (Liang et al., 2021; Katarkar et al., 2021d; 2021c; Katarkar et al., 2021). Nukiyama (1966) first reported the boiling regimes. Researchers thereafter repeatedly reported that among different boiling regimes, the nucleate pool boiling regime is the most useful and efficient regime for transferring heat. This is because of the capability of the pool boiling regime to remove large amount of heat having small temperature differential between the heated surface and the working fluid. Due to the diversified applications in wide range of industries, boiling is the most applied phenomena by modern civilisation. Under this circumstance, it is not surprising that the pool boiling is still the field of extensive research worldwide.

Research of Jakob and Firtz (1931) established the relationship between the surface roughness and the increment in heat transfer parameter. After that, many investigations have been carried out by researchers to bolster this relationship. Later Dewan et al. (2004) categorised the surface roughness as an important passive technique to enhance heat transfer parameters. Recently, Mahmoud and Karayiannis (2021) reported that the effect of surface microstructure depends on system pressure. It was reported that in elevated operating pressure systems, like, refrigerations, power production, etc., the number of active nucleation sites increases at higher pressure without any requirement of surface modification; thus,it may not be economical. Following the same line, in case of low operating pressure system, like, electronic cooling system, etc., there would be few active nucleation sites, which need modification of surface to enhance the number of active nucleation sites. Situations like this demand modification of surfaces to enhance system efficiency.

Several studies were conducted to enhance system efficiency by modifying surfaces to enhance heat transfer efficiency. Among different surface modification techniques to enhance surface roughness, generating microporous structures (Katarkar et al., 2021b; Katarkar et al., 2020b; 2020a; Pinni et al., 2021; Katarkaret al., 2021c) in the form of coating is one of the viable solutions, which is widely accepted in industries. From literatures, it is found that for developing diverse surfaces having porous coating for boiling experiments with pool of fluids, different processes and techniques, namely, sintering, spray coating, e-beam evaporation, electrodeposition, etc., were used (Pingale et al., 2021d; Ray and Bhaumik, 2018; Memory et al., 1995; Rishi et al., 2020).

For years copper has been used extensively in heat transfer applications. Its high demand in the industry and scientific world drives the market price of copper up (Pingale et al., 2021a; Pingale et al., 2020c; Owhal et al., 2021; Pingale et al., 2020b; Owhal et al., 2021a). These issues along with the demand for lightweight material for compact technologies make researchers interested in exploring other material options to use

[a]bmajumdertit@gmail.com; [b]anil.katarkar@gmail.com; [c]ajay9028@gmail.com; [d]p20210058@pilani.bits-pilani.ac.in; [e]sachinbelgamwar@pilani.bits-pilani.ac.in; [f]drsbhaumik@gmail.com

in the heat transferring system. Aluminum due to its good thermal conductivity, greater strength and high corrosion resistance properties (Godinez et al., 2021; Shabestari et al., 2015; Ferreira et al., 2018; Haga et al., 2017), lesser weight (nearly one-third than copper) and being cost-effective (four times cheaper than copper) are increasingly getting acceptability in the heat transfer industry(Pingale et al., 2021b; Shelke et al., 2020; Pingale et al., 2020a; Pingale et al., 2021c; Belgamwar et al., 2019; Owhal et al., 2021b). To negate the deficiency of aluminium in thermal conductivity, in case of comparison made with copper, researchers are now trying to develop an aluminum composite materialas a process to enhance the roughness of the aluminum surface.

One method of surface modification is the application of additional coatings. In the literature study, it was observed that the pool boiling experiment performed with ZnO nanostructured aluminum surfaces is not very common, though it is very commonly used as paint in industries. Zinc oxide is widely used in diversified fields of industries like pigments, non-linear optics, catalysis, cosmetics, solar energy conversion, etc. But it is challenging to prepare zinc oxide nanoparticles coated with homogeneous film by traditional chemical method. In 2001, Fangli et al. (2001) reported a coating method to prepare a homogeneous coating of zinc oxide on the aluminum surface. Thereafter, Kaneva et al.(Kaneva and Dushkin, 2011) in 2011 prepared zinc oxide coating on aluminum surface by using sol-gel dip-coating method. Later, Kunkle et al. (2017) in the year 2017, prepared aluminum nanostructured surfaces by using hydrothermal synthesis technique with ZnO nanoparticles and studied the heat transfer performance of those surfaces with water; and reported that ZnO nanostructured aluminum surfaces exhibit superhydrophilicity and very high heat transfer coefficients (over 20kW/m2) which is far higher than bare aluminum surface (Owhal et al., 2022).

In the current project, ZnO nanostructured surfaces for heating were prepared using the thermal evaporation technique followed by heat treatment. Also, the boiling heat transfer (BHT) performance in a pool of R-134a on prepared ZnO nanostructured heating surfaces was investigated in detail.

Fabrication of Zno Nanostructures

The photograph of thermal evaporator facility is shown in Figure 104.1. Initially, Al substrates were mechanically polished with sandpapers to obtain a mirror finish surface. After this, Zn thin films of 100, 200 and 300 nm thickness were deposited on Al substrate (diameter = 9 mm) using a thermal evaporatorunder

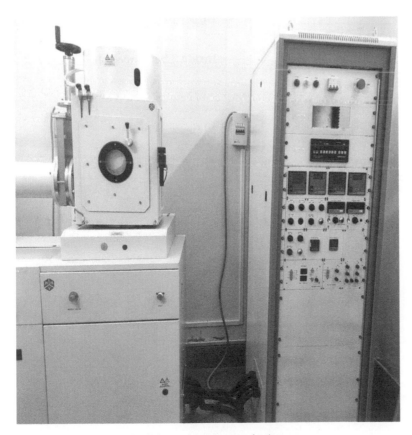

Figure 104.1 Photograph of thermal evaporator facility

a vacuum of 10–6mbar. During the evaporation of the Zn, boat made by molybdenum and pure granules of Zn (Alfa Aesar, purity 99.9%) were utilised in Thermal Evaporator. One Quartz Oscillator was utilised to monitor the Zn thin films thickness. Once the necessary thickness of Zn thin film was attained, the flap shutter was utilised to inhibit further deposition on the substrate. Finally, prepared Zn thin films were heated at 500°C in the electric furnace (Thermo Scientific) for 5 hrs to obtain ZnO nanostructure. Here, the ZnO nanostructured surfaces utilised for heating were marked as ZnO -100, ZnO-200 and ZnO-300 for ZnO nanostructured thin film having corresponding thickness of 100 nm, 200 nm and 300 nm.

The Rigaku MiniFlex-II X-ray diffractometer (XRD) was used to identify the microstructure of ZnO nanostructure films. Also, to study the surface morphology of ZnO nanostructure films, the FEI-Apreo-S scanning electron microscopy (SEM) was employed. Also, the surface roughness of ZnO nanostructure films was determined using Taylor Habsons 2D profilometer.

Experimental Apparatus and Procedure

For finding out the fact, investigations had been conducted on the research gap mentioned above. One test setup was developed to conduct this investigation. To demonstrate this test setup employed for this study, schematic representation has been presented in Figure 104.2. The most important part of this test setup is the boiling chamber. Geometrically the boiling chamber was cylindrical in shape. Other attached assemblies, shown in Figure 104.2, were supporting parts to conduct this experiment. This boiling chamber was made with SS-316 grade stainless steel, a variety of stainless steel. The boiling chamber was 315 mm long having an inner diameter of 115.4 mm and built with a 12.65 mm thick plate. There were two glass windows with a 50 mm diameter kept for viewing. Glass pieces used in this setup had a thickness of 50 mm.

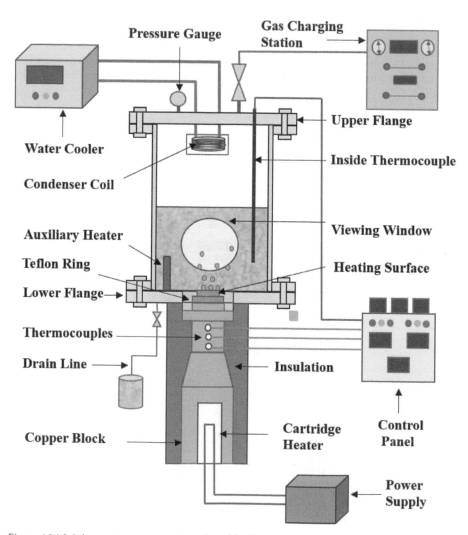

Figure 104.2 Schematic representation of pool boiling experiment setup

This provision was kept for facilitating future scope of research by accommodating high-speed camera to study specifically the bubble dynamics.

The insulation of the boiling chamber had three layers. The inner layer of this multi-layer insulation system was constituted with an insulation coating. After the inner layer, an intermediate layer of the glass-wool jacket was installed for further reinforcement; the mineral insulating material constituted the outer layer of the insulation system. This outermost insulation layer had direct contact with the environment. Any boiling chamber should be leak-proof in joints. Two flanges were used to form leak-proof removable joints with the cylindrical boiling shell for this test setup. Teflon O-rings were used at both the boiling chamber and flange connections to ensure that the vessel was hermetically sealed. An auxiliary heater was also installed within the boiling chamber, slightly above the level of the surface for testing, to heat the boiling fluid as needed.

To keep the refrigerant liquid during experimentation, vapor pressure inside the boiling chamber was kept above atmospheric pressure and bound to rise with time. To control this increasing vapour pressure, a cooling system was used. In this test setup, a coil condenser made with copper was used to cool down the generated vapor. Facility for measuring the pressure of the boiling chamber was required for controling the inside pressure of boiling chamber. To meet this demand, one pressure gauge was installed on the boiling chamber. For measuring the temperature of different parts of the test setup, K-type thermocouples were used in this experiment. The test setup as facilitated with the DC power supply to ensure the continuous flow of heat during experimentation.

On the bottom flange of the boiling chamber, a hole was retained in the central position. The aluminium heating block on which the aluminium test surface was placed, was inserted via this central hole. Top surface of the inserted aluminium-made block for heating along with aluminum-made test-surface was acted as the bottom of the pool liquid under experimentation. This aluminum-made heating block system was kept isolated from the boiling vessel or chamber by using Teflon bush arrangement. The Teflon jacket entirely insulated the aluminium heating block, with provisions for fastening it to the bottom of the heating chamber using a nut and bolt mechanism. The top and middle portion of the aluminum heating block was also housed three K-type thermocouples to measure heat flow through the heating block wall. As a source of heat, a 500 Watt cartridge heater was used and kept in hollow portion of aluminum-made heating block through the bottom end of this heating block.

Before assembling, each and every part of the test setup was gathered in a common place. Blowing air, water, and finally acetone were used to clean pieces of the boiling vessel or chamber and its accompanying parts. These parts were assembled together after drying up and then placing the test surface at the boiling chamber. After completing the assembly work, the boiling chamber was kept in that situation for a day to dry the insulation pastes used completely. Then the arrangement was vacuumed using one vacuum pump and left there for a day to check for leaks. After two days, the chamber was charged with air at 5 MPa pressure to find leakage at the chamber and joints or for any metallic faults or manufacturing defects. Simultaneously, soap water was applied to the exterior of the chamber in every junction and pipe segment. The chamber was emptied again to charge working fluids once the leak test result was satisfactory.

In this study, refrigerant R-134a was used as a working fluid. During each run of the test experiment, system pressure was kept constant and temperature of the pool was maintained at 10° C by adjusting the rate at which working liquid of the condenser flows. The refrigerant was warmed for 1 hour before the test began by one auxiliary heater to take out dissolved gas from the pool of working liquid. In a pool of 10°C, data were collected under steady-state conditions in increasing order of heat flux varied from 9.37 to 72.23 kW/m2k for R-134a. The same procedure was followed for conducting each test run, and under the same experimental conditions, three runs were conducted to complete each test experiment. This is critical for ensuring that the data collected is repeatable.

Results and Discussion

A. Characterisation of ZnO nanostructures

SEM images of prepared ZnO nanostructured thin films areshown in Figure 104.3 (a–c). From Figure 104.3, it is clearly seen that with an increase in coating thickness, the surface morphology becomesrougher.The measured surface roughness ofZnO-100, ZnO-200 and ZnO-300 is0.09±0.01 μm, 0.12±0.01 μm, and 0.15±0.01 μm, respectively. The XRDspectrum of ZnO-300 is represented in Figure 104.3 (d).XRD studies revealed that the prepared material was ZnO with wurtzite phase.

B. Analysis of pool boiling curves

This experiment was conducted to calculate the pool boiling heat transfer coefficients on a bare Al and three nanostructured ZnOheating surfaces having different thicknesses with the pool of refrigerant R-134a at 10°C saturation temperature. Figure 104.4 (a) represented the relationship between wall superheat (ΔT)

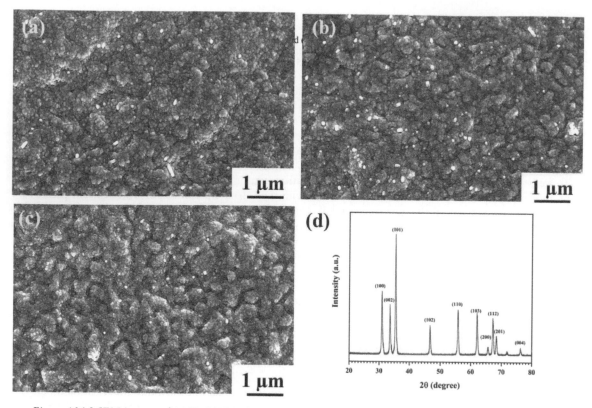

Figure 104.3 SEM images of (a) ZnO-100, (b) ZnO-200 and (c) ZnO-300, and (d) XRD spectrum of ZnO-300

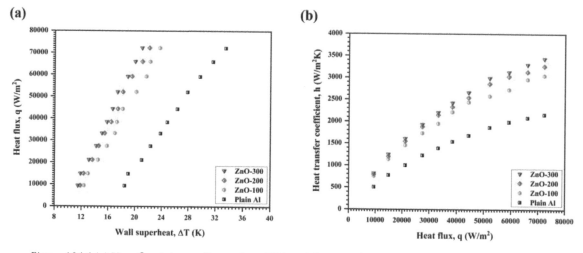

Figure 104.4 (a) Heat flux (q) vs wall-superheat (ΔT) and (b) HTC (h) vs heat flux (q)

and heat flux (q) for plain Al and ZnO nanostructured heating surfaces. It was found that the ZnO nanostructured heating surface having maximum coating thickness had lower wall superheat; i.e., wall superheat steadily increased and attained maximum value for bare Al heating surface. The onset nucleate boiling (ONB) of the ZnO nanostructured heating surfaces appeared lower than that of the bare Al heating surface. Figure 104.4 (a) pointed out that the ZnO-300 surface exhibited the smallest wall-superheat among all prepared ZnO nanostructured heating surfaces, though this surface had the highest coating thickness. Enhancement in heat transfer performance proves the rise in the supply of the volume of the cooler liquid flow on the test surface and simultaneously replacing the hotter liquid from the test surface; thus, forming a liquid cycle. The formation of such a liquid cycle can only be explained as a result of the refrigerant's capillary wicking action on the test surface due to the enhanced roughness of the test surface. As a result, the bubble departure frequency increases, limiting the production of vapour layer on the heating surface at minimum wall-superheat levels. (Katarkar et al., 2021a; Majumder et al., 2022).

Figure 104.4(b) represented the connection between heat flux (q) and HTC (h) for bare Al and ZnO nanostructured heating surfaces. According to the result, HTC of ZnO nanostructured heating surfaces rose as the coating thickness increased. HTC of ZnO nanostructured heating-surfaces appeared higher than that of the bare Al heating-surface. Maximum HTC was reported for the bare Al heating-surface is 2.16kW/m2K and the same for ZnO-100, ZnO-200 and ZnO-300 is 3.04 kW/m2K, 3.26 kW/m2K and 3.43 kW/m2K, respectively. The increment of HTCs for ZnO nanostructured heating surfaces in percentage term over bare Al test surface is calculated about 41%, 51% and 59%, respectively. So, for ZnO-300, the maximum HTC is obtained. The HTC values of ZnO nanostructured heating-surfaces are enhanced considerably because the refrigerant's capillary wicking effect is enhanced as a result of the newly formed micropores, cavities, etc. Enhancement of cavities and micropores, on the other hand, indicates the enhancement of surface roughness. This enhancement in capillary wicking effect increased the frequency of bubble departure, the main factor for the rise in HTC (Majumder et al., 2022)(Majumder et al., 2020).

Conclusion

ZnO nanostructured heating surfaces with different coating thickness has been developed over bare Al surface for experimentation by thermal evaporation technique followed by heat treatment. These developed surfaces were used for the pool boiling experiments to compare their BHT performance with bare Al surface. To get the configuration of the coating structure SEM images of ZnO nanostructured heating surfaceswere taken and found that developed structure is the configuration of micropores and microcavities. The average surface roughness values reported for all prepared ZnO-100, ZnO-200 and ZnO-300 was 0.09±0.01μm, 0.12±0.01μm, and 0.15±0.01μm, respectively. Further, it was observed that HTC of ZnO nanostructured heating-surfaces was rose when the coating thickness was increased. The highest value of HTC was observed for ZnO-300 and the recorded value was 61% higher than the bare Al heating-surface. Rapid nucleation activity is the main reason behind getting such a result. This rapid nucleation activity further enhanced the nucleation site density and is enough to indicate the enhancement of the surface roughness and porosity of the heating surface.

Acknowledgments

Current work was performed in the PCHT Laboratory under the Mechanical Engineering Department in the NIT Agartala, Tripura, India.

References

Belgamwar, S. U., Pingale, A. D., and Sharma, N.N. (2019). Investigation on electrical properties of Cu matrix composite reinforced by multi-walled carbon nanotubes. Mater. Today: Proc. 18:3201–3208. https://doi.org/10.1016/j.matpr.2019.07.196.

Dewan, A., Mahanta, P., Sumithra Raju, K., and Suresh Kumar, P. (2004). Review of passive heat transfer augmentation techniques. Proc. Inst. Mech. Eng. Part A: J. Power Energy 218(7):509–527. https://doi.org/10.1243/0957650042456953.

Fangli, Y., Shulan, H., and Jinlin, L. (2001). Preparation of zinc oxide nano-particles coated with aluminium. J. Mater. Sci. Lett. 20(16):1549–1551. https://doi.org/10.1023/A:1017959404481.

Ferreira, L.-M.-P., Bayraktar, E., Miskioglu, I., and Robert, M.-H. (2018). New magnetic aluminum matrix composites (Al-Zn-Si) reinforced with nano magnetic Fe 3 O 4 for aeronautical applications. Adv. Mater. Process. Technol. 4(3):358–369. https://doi.org/10.1080/2374068X.2018.1432940.

Godinez, J. C., Cho, H., Fadda, D., Lee, Park, S. J., and You, S. M. (2021). Effects of materials and microstructures on pool boiling of saturated water from metallic surfaces. Int. J. Therm. Sci. 165:106929. https://doi.org/10.1016/j.ijthermalsci.2021.106929.

Haga, T., Kozono, R., Nishida, S., and Watari, H. (2017). Casting of aluminum alloy clad strip by an unequal diameter twin-roll caster equipped with a scraper. Adv. Mater. Process. Technol. 3(4):511–521. https://doi.org/10.1080/2374068X.2017.1344057.

Jakob, M. and Fritz, W. (1931). Versuche über den verdampfungsvorgang. Forschung Auf Dem Gebiete Des Ingenieurwesens 2(12):435–447. https://doi.org/10.1007/BF02578808.

Kaneva, N. V. and Dushkin, C. D. (2011). Preparation of nanocrystalline thin films of ZnO by Sol-Gel dip coating. Bulg. Chem. Commun. 43(2):259–263.

Katarkar, A., Majumder, B., and Bhaumik, S. (2020a). Effect of enhanced surfaces and materials in boiling heat transfer with HFO refrigerants: A review. Mater. Today: Proc. 26:2237–2241. https://doi.org/10.1016/j.matpr.2020.02.485.

Katarkar, A., Majumder, B., and Bhaumik, S. (2020b). Review on passive heat enhancement techniques in pool boiling heat transfer. IOP Conf. Ser. Mater. Sci. Eng. 814(1):012031. https://doi.org/10.1088/1757-899X/814/1/012031.

Katarkar, A. S., Majumder, B., Pingale, A. D., Belgamwar, S., U. and Bhaumik, S. (2021). A review on the effects of porous coating surfaces on boiling heat transfer. Mater. Today: Proc. 44(November):362–367. https://doi.org/10.1016/j.matpr.2020.09.744.

Katarkar, A. S., Pingale, A. D., Belgamwar, S. U., and Bhaumik, S. (2021a). Experimental investigation of pool boiling heat transfer performance of refrigerant R-134a on differently roughened copper surfaces. Mater. Today: Proc. 47:3269–3275. https://doi.org/10.1016/j.matpr.2021.06.452.

Katarkar, A. S., Pingale, A. D., Belgamwar, S. U. and Bhaumik, S. (2021b). Experimental study of pool boiling enhancement using a two-step electrodeposited Cu-GNPs nanocomposite porous surface with R-134a. J. Heat Transf., https://doi.org/10.1115/1.4052116.

Katarkar, A. S., Pingale, A. D., Belgamwar, S. U., and Bhaumik, S. (2021c). Effect of GNPs concentration on the pool boiling performance of R-134a on Cu-GNPs nanocomposite coatings prepared by a two-step electrodeposition method. Int. J. Thermophys. 42(8):124. https://doi.org/10.1007/s10765-021-02876-z.

Katarkar, A. S., Pingale, A. D., Belgamwar, S. U. and Bhaumik, S. (2021d). Experimental study of pool boiling enhancement using a two-step electrodeposited Cu–GNPs nanocomposite porous surface with R-134a. J. Heat Transfer. 143(12):121601. https://doi.org/10.1115/1.4052116.

Kunkle, C. M., Mizerak, J. P. and Carey, V.P. (2017). The effects of wettability and surface morphology on heat transfer for zinc oxide nanostructured aluminum surfaces. In Volume 2: Heat transfer equipment; heat transfer in multi-phase systems; heat transfer under extreme conditions; nanoscale transport phenomena; theory and fundamental research in heat transfer; thermophysical properties; transport phenomena in materials (pp, 1–10). American Society of Mechanical Engineers. https://doi.org/10.1115/HT2017-4847.

Liang, G., Yang, H., Wang, J. and Shen, S. (2021). Assessment of nanofluids pool boiling critical heat flux. Int. J. Heat Mass Transfer. 164:120403. https://doi.org/10.1016/j.ijheatmasstransfer.2020.120403.

Mahmoud, M. M., and Karayiannis, T. G. (2021). Pool boiling review: Part I – fundamentals of boiling and relation to surface design. Therm. Sci. Eng.Progress. 25(May):101024. https://doi.org/10.1016/j.tsep.2021.101024.

Majumder, B., Katarkar, A. and Bhaumik, S. (2020). Effect of structured surface on contact angle using sessile droplet method." IOP Conf. Ser.: Mater. Sci. Eng. 814(1):012034. https://doi.org/10.1088/1757-899X/814/1/012034.

Majumder, B., Pingale, A., Katarkar, A., Belgamwar, S. and Bhaumik, S. (2022). Developing Al@GNPs composite coating for pool boiling applications by combining mechanical milling, screen printing and sintering methods. Adv. Mater. Process. Technol. 00(00):1–12. https://doi.org/10.1080/2374068x.2022.2036037.

Memory, S. B., Sugiyama, D. C. and Marto, P. J. (1995). Nucleate pool boiling of R-114 and R-114-Oil mixtures from smooth and enhanced surfaces—I. Single tubes. Int. J. Heat Mass Transf. 38(8):1347–1361. https://doi.org/10.1016/0017-9310(94)00263-U.

Nukiyama, S. (1966). The maximum and minimum values of the heat q transmitted from metal to boiling water under atmospheric pressure. Int. J. Heat Mass Transf. 9(12):1419–1433. https://doi.org/10.1016/0017-9310(66)90138-4.

Owhal, A., Pingale, A. and Belgamwar, S. (2022). Developing sustainable Zn-MWCNTs composite coatings using electrochemical Co-deposition method: Tribological and surface wetting behavior. Adv. Mater. Process. Technol. 1–14. https://doi.org/10.1080/2374068X.2022.2035968.

Owhal, A., Pingale, A. D., Belgamwar, S. U., and Rathore, J. S. (2021). Preparation of novel Zn/Gr MMC using a modified electro-Co-deposition method: microstructural and tribo-mechanical properties. Mater. Today: Proc. 44:222–228. https://doi.org/10.1016/j.matpr.2020.09.459.

Owhal, A., Pingale, A. D., Khan, S., Belgamwar, S. U., Jha, P. N. and Rathore, J. S. (2021a). Electro-codeposited γ-Zn-Ni/Gr composite coatings: Effect of graphene concentrations in the electrolyte bath on tribo-mechanical, anti-corrosion and anti-bacterial properties. Trans. IMF 99(6):324–331. https://doi.org/10.1080/00202967.2021.1979815.

Owhal, A., Pingale, A. D., Khan, S., Belgamwar, S. U., Jha, P. N. and Rathore, J. S. (2021b). Facile and scalable co-deposition of anti-bacterial Zn-GNS nanocomposite coatings for hospital facilities: Tribo-mechanical and anti-corrosion properties. JOM 73(12):4270–4278. https://doi.org/10.1007/s11837-021-04968-5.

Pingale, A. D., Belgamwar, S. U. and Rathore, J. S. (2020a). Effect of graphene nanoplatelets addition on the mechanical, tribological and corrosion properties of Cu–Ni/Gr nanocomposite coatings by electro-co-deposition method. Trans. Indian Inst. Met. 73(1):99–107. https://doi.org/10.1007/s12666-019-01807-9.

Pingale, A. D., Owhal, A., Belgamwar, S. U. and Rathore, J. S. (2021a). Electro-codeposition and properties of Cu–Ni-MWCNTs composite coatings. Trans. IMF 99(3):126–132. https://doi.org/10.1080/00202967.2021.186 1848.

Pingale, A. D., Owhal, A., Belgamwar, S. U. and Rathore, J. S. (2021b). Effect of current on the characteristics of CuNi-G nanocomposite coatings developed by DC, PC and PRC electrodeposition. JOM 73(12):4299–4308. https://doi.org/10.1007/s11837-021-04815-7.

Pingale, A. D., Owhal, A., Katarkar, A. S., Belgamwar, S. U. and Rathore, J. S. (2021c). Recent researches on Cu-Ni alloy matrix composites through electrodeposition and powder metallurgy methods: A review. Mater. Today: Proc. 47(July):3301–3308. https://doi.org/10.1016/j.matpr.2021.07.145.

Pingale, A. D., Owhal, A., Katarkar, A. S., Belgamwar, S. U., and Rathore, J. S. (2021d). Facile synthesis of graphene by ultrasonic-assisted electrochemical exfoliation of graphite. Materi. Today: Proc. 44:467–472. https://doi.org/10.1016/j.matpr.2020.10.045.

Pingale, A. D., Belgamwar, S. U., and Rathore, J. S. (2020b). The influence of graphene nanoplatelets (GNPs) addition on the microstructure and mechanical properties of Cu-GNPs composites fabricated by electro-Co-deposition and powder metallurgy. Mater. Today: Proc. 28:2062–2067. https://doi.org/10.1016/j.matpr.2020.02.728.

Pingale, A. D., Belgamwar, S. U., and Rathore, J. S. (2020c). A novel approach for facile synthesis of Cu-Ni/GNPs composites with excellent mechanical manufacturing and tribological properties. Mater. Sci. Eng. B 260:114643. https://doi.org/10.1016/j.mseb.2020.114643.

Pinni, K. S., Katarkar, A. S., and Bhaumik, S. (2021). A review on the heat transfer characteristics of nanomaterials suspended with refrigerants in refrigeration systems. Mater. Today: Proc. 44:1331–1335.

Ray, M. and Bhaumik, S. (2018).Structural properties of glancing angle deposited nanostructured surfaces for enhanced boiling heat transfer using refrigerant R-141b. Int. J. Refrig. 88:78–90. https://doi.org/10.1016/j.ijrefrig.2017.12.008.

Rishi, A. M., Kandlikar, S. G., Rozati, S. A., and Gupta, A. (2020). Effect of ball milled and sintered graphene nanoplatelets–copper composite coatings on bubble dynamics and pool boiling heat transfer. Adv. Eng. Mater. 22(7):1901562. https://doi.org/10.1002/adem.201901562.

Shabestari, S. G., Honarmand, M., and Saghafian, H. (2015). Microstructural evolution of A380 aluminum alloy produced by gas-induced semi-solid technique (GISS). Adv. Mater. Process. Technol. 1(1–2):155–163. https://doi.org/10.1080/2374068X.2015.1116236.

Shelke, A. R., Balwada, J., Sharma, S., Pingale, A. D., Belgamwar, S. U., and Rathore, J. S. (2020). Development and characterisation of Cu-Gr composite coatings by electro-co-deposition technique. Mater. Today: Proc. 28:2090–2095. https://doi.org/10.1016/j.matpr.2020.03.244.

105 Providing virtual aid, control and overcoming social stigma of alzheimer's disease using logistic regression and K-Nearest neighbour algorithm

Diviya M.[1,a], Hemapriya N.[2,b], Susmita Mishra[1,c], and Deivendran P.[3,d]

[1]Department of Computer Science and Engineering Rajalakshmi Engineering College Chennai, India

[2]Department of Information Technology St. Joseph's College of Engineering Chennai, India

[3]Department of Information Technology Velammal Institute of Technology Chennai, India

Abstract

In the modern era advancement in healthcare systems play a vital role. Early detection of any chronic condition is critical in preventing disease progression. The proposed work aims in one such platform using machine learning models for early prediction of Alzheimer's disease using logistic regression and k-nearest neighbour algorithm. The proposal intends to suggest a wholesome care system for the patients diagnosed with Alzheimer's disease (AD) alongside with early detection and administration of Alzheimer's and overcoming the stigma associated with it. The network of patients diagnosed with Alzheimer's could be connected on a peer to peer network to share their experiences and resources. The doctors, health care practitioners and stakeholders could be connected in the network wherein sentiment research, opinion mining, natural language processing (NLP) in stigma analysis and detection is performed, a health-wellness app which serves as an AI companion for AD patients, daily assessment tasks upon doctor's consent, virtual therapy based on AI will be implemented. The system proposes an upstanding beneficial system, initially from diagnoses, test, further-by aid, prevention, assessment and constant monitoring of the patients.

Keywords: Alzheimer's disease, artificial intelligence, AI companion, k-nearest neighbour, logistic regression, natural language processing.

Introduction

No one therapy or intervention is likely to be sufficient in treating Alzheimer's disease (AD). The current techniques are focused at helping patients retain mental control, regulate behavioural problems, and reduce symptoms of the condition. An ongoing problem for neuropsychologists has been the perception of Alzheimer's cognitive and behavioural symptoms and their relationship to underlying brain pathology, as described by Bondi et al. (2017). In a recent research, Kotecha et al. (2018) found a probable link between memory loss and scent perception. New tests for detecting AD in its early stages may be in the works. While Miguel, et al (2007) showed a possible link between certain jobs and reduced AD rates. Based on the findings, researchers believe that maintaining complex social ties, such as those connected to mentoring, teaching, or problem solving with others, may reduce the incidence of AD.

Machine learning can be used to help people with mobility issues communicate their symptoms remotely and in the comfort of their own homes.

The proposed model firmly addresses the issues in current scenario based on the advancement in machine learning and block chain technology can help bridge the gap by creating a peer-to-peer network for patients, complete with intuitive AI companions, and overcoming the social stigma that exists in the socio-cultural community by providing full-fledged virtual help and control. The proposed work may be a one-stop shop for effective assistance, assessment, and companionship, and it could be a modern value-added strategy.

Since interacting with the family (Brodaty and Donkin, 2009) individualising treatment and attending to patient comfort, quality care requires enforcing advance directives. The kith and kin are forced to struggle with the pressure of delivering treatment and providing information on facilities such as day care programs, seminars and nursing homes. There is a need for improvement in understanding the causes of the disease, early diagnosis assessment, the use of appropriate medication to postpone the worsening of symptoms, the management of Alzheimer-related behavioural problems, and the overall last stage of care are highly important to the role of care planning and caregiver support. Several treatments and medicines are now helping to alleviate behavioural effects, periodic clinical treatment, scans, tele-health technologies, helping people sustain the disease.

Two-stage strategy used by Iddi et al. (2019) addressed the progression of AD by forecasting cognition, visualising brain images and fluid biomarkers and diagnosing people with various domains simultaneously. In the first stage, mixed-effects models are employed to continually simulate various markers throughout time. Next, random forest algorithms are used to predict categorical diagnoses using predictions generated

[a]diviya.m@rajalakshmi.edu.in;[b]hemuhema2000@gmail.com;[c]susmita.mishra@rajalakshmi.edu.in;[d]deivendran1973p@gmail.com

in the first stage from continuous model-based markers (cognitively normal, mild cognitive dysfunction, or dementia). The combined impact of the two models under examination enables us to leverage their key strengths. After 2.5 years, the accuracy exceeding 80% was finally achieved (Orimaye et al., 2017; Swarbrick et al., 2019). Neurodegenerative illnesses like AD and its accompanying dementia have proven to be a difficult diagnosing challenge for doctors. Neuropsychological testing and clinical guidelines are now used to identify these conditions. The researchers employed machine learning methods to develop an automated diagnosis system based on low-level language characteristics as a result of spoken utterances. The Dementia Bank language transcript clinical dataset, which comprises of 99 patients with probable AD and 99 healthy controls, was used to train several Machine Learning models to address this problem. Even though various firms took up this research Islam et al. (2018) and Tanveer et al. (2020) stated that AD is linked to an incurable neurological brain condition. A diagnosis of Alzheimer's disease at an early stage may help prevent brain tissue damage. For AD, researchers have also suggested massive mathematical and machine learning models. In clinical research, it is standard procedure to analyse magnetic resonance imaging (MRI) for the purpose of diagnosing AD. In medical picture analysis, advanced deep learning algorithms have successfully provided results comparable to those of human (DeTure and Dickson, 2019). The authors proposed a deep convolutional neural network for the diagnosis of AD by analysing brain MRI data. For the Open Access Series of Imaging Studies dataset, they conducted enough tests to demonstrate the model's efficiency. It is impossible to prevent yet inescapable results such granulo-vacuolar degeneration and Hirano bodies, which are explained by the study of the authors. Overview of AD pathology and outlining the pathological substrates as well as associated diseases that might influence diagnosis and therapy are the primary goals of this study (Oriol et al., 2019; Ertek et al., 2014). The researchers used genetic variation data from the Alzheimer's disease neuro-imaging initiative (ADNI) cohort to compare typical machine learning models for predicting Late-Onset Alzheimer's Disease. Their experimental results demonstrate that the classification performance of the best models tested yielded ~72% of area under the ROC curve (Elahi et al., 2019). In intermittent early-onset versus late-onset Alzheimer's disease, the scientists looked at plasma proteome markers for astrocytopathy, brain degeneration, plasticity, and inflammation (EOAD and LOAD). They employed plasma analysis using ultra-sensitive immuno-based assays. The findings are evaluated using PCA, of which three key variables have been taken into account. Biomarkers of inflammation were not significantly different between groups and were only associated with age (Andrews et al., 2019).

According to the researchers' observational studies, light-to-moderate alcohol consumption lowers the risk of Alzheimer's disease, but the evidence for a causative relationship is lacking. They considered Mendelian randomisation (MR) analysis to examine is there any effect because of alcohol consumption and the test audit scores were drawn. Another experiment looked at whether age is a predictor of AD, with γ-glutamy-ltransferase levels serving as a positive control. The intake of alcohol was correlated with earlier AAOS and elevated blood concentrations of γ-glutamyltransferase. Alcohol dependence was associated with a delayed AAOS (Daunt et al., 2021). The study's goal was to apply a polygenic risk score system to identify those most at risk of cognitive decline due to Alzheimer's disease. Over the course of four years, researchers used genotyping and/or whole genome sequencing data to calculate polygenic risk scores and predict future declines in cognitive function as evaluated by the CDR-SB and the ADAS-Cog 13. AUC increased to 79.1% (CI: 75,682.6) when cognitively normal subjects were included, resulting in a prediction accuracy of 72.8% (CI: 67.977.7) for persons who would have a deterioration of at least 15 ADAS-Cog13 points over the course of four years. Using a criterion of larger than 0.6, the high genetic risk group fell, on average, by 1.4 points (CDR-SB) over the course of four years compared to the low genetic risk group (Bock et al., 2021). In order to distinguish between healthy people with approaching cognitive decline and those who aren't, a hierarchical Bayesian cognitive process is used to generate a digital cognitive biomarker. When compared to a logistic regression model, the HBCP model shows good results.

The scale of dementia is now impossible to ignore. Worldwide, there are 50 million people living with dementia, with that number estimated to triple by 2050. Studies show that this sheer number of individuals with AD cost the global economy $1 trillion in 2018. Current research gaps include evaluating the impact of newly identified genetic risk factors on AD development, gaining awareness of why neuro-nerve cells are more robust than others, and efficient clinical trials for individuals. Medical studies need to focus efforts on the most critical biological processes that drive disease. This will offer the best chance for people with AD to get about the life-changing treatments that are desperately needed. It encourages us to think carefully about how we plan our clinical trials to test potential new drugs in the smartest way. There have been many smart Alzheimer's diagnosis systems using unsupervised machine learning techniques where in brain images are studied and classified using machine learning algorithms (Razavi et al., 2019). But it isn't a full-fledged one-stop solution for assisting patients with Alzheimer's. Also, there exists a stigmatized view in the review research workforce of AD (Marjanovic et al., 2015). Even though people with dementia are among the largest users of adult social care, less than 2% of the top 200 most prolific UK dementia researchers specialize in social care and social work. The approach showed potential in predicting the underlying disease in dementia patients with clinical profiles that were unclear between Alzheimer's and vascular dementia, hinting that it could be useful in assisting doctors with diagnosis (Castellazzi et al., 2020). There exists a need to tackle the gap in developing end-of-life care and treatment for AD.

AI aims to transform the practice of medicine, but many of its practical implementations are still in their roots and need to be further explored and developed. Researchers have previously trained many models to hunt for indicators of cognitive dysfunction, including AD, utilising various sources of data such as brain scans and clinical test results. Moreover, social networks have indeed been an efficient platform for users to express their opinions among the society (Gopi et al., 2020). A team from IBM and Pfizer claims to have trained AI models by looking at linguistic trends in words from social networks to detect early indicators of AD in a new study.

Researchers say they may predict the onset of AD based on the Framingham Heart study many years before symptoms become serious enough for conventional diagnostic tools to detect them. Study participants were asked to describe a picture of a lady washing dishes, and the researchers transcribed their handwritten responses into computerised transcripts. Even without real handwriting, IBM claims its key AI model was able to identify linguistic traits that are commonly linked to early indicators of cognitive deterioration. They contained misspellings, repetitious words, and the usage of basic phrases rather than grammatically difficult statements. It gave clinicians a better understanding of how AD can affect a person's language/linguistic ability.

The current research involves in further understanding and prospective analysis of similar chronic diseases such as Parkinson's disease, Schizophrenia where neuro-imaging and analysis could be performed (Ding et al., 2019). Researches show that it is possible to diagnose early stages of AD about six years before a clinical diagnosis is made. There is no cure for AD, but in recent years, promising medications have appeared which can help stem the progression of the disorder. However, in order to do any good, these medications have to be performed early in the course of the illness. Scientists have been motivated by this race against the clock to look for ways to detect the disease sooner. machine learning models are built where PET scans of patients are scanned to help them detect the early-stage of the disease.

For the detection of AD, researchers have utilised a variety of methods, including artificial intelligence (AI). AI is being used to help diagnose AD. By combining AI with block chain technology, the system under consideration provides a one-stop solution for delivering virtual assistance, control, and reducing the social stigma associated with AD.

Methodology

It is clear that implementation phase could be divided into pre-diagnosis, post-diagnosis, stigma analysis as shown in Figure 105.1. A correct diagnosis of Alzheimer's dementia is critical for ensuring that the suspect receives the proper treatment, care, family education, and future plans. Memory impairment, such as trouble remembering things, loss of attention, problem-solving skills, reduced vocabulary in speaking or writing, poor decision-making, personality changes, depression, and a lack of social skills are all early indicators of AD (Duong et al., 2017).

The initial stage of implementation involves with a presentation prototype which provides the demonstration of the core functionality and design of the block-chain based healthcare application. The app

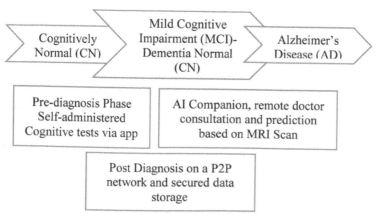

Figure 105.1 Phases involved in methodology

serves as an AI companion for AD patients, daily assessment tasks upon doctor's consent, virtual therapy based on AI and several other virtual subsidies will be implemented. The proposed model mainly involves patients and their guardians, doctors or clinicians. The application embarks a self-administered cognitive test followed by a clock drawing test which helps to detect early signs of AD. The application deploys a differential ML algorithm combined with magnetic resonance imaging (MRI) which outputs high diagnostic accuracy of neuro-generative diseases, especially AD. The patients can communicate their symptoms to the doctors in the app remotely and become a part of the network upon signing with security and privacy. The patients can login to the app with the unique ID thus all their medical data are retrieved from the blocks. They are connected to their peers and physicians to share the experiences in different stages of the disease upon diagnosis. A fine-tuned deep learning model deployed in the application will be trained to recognize the tempering in the medical images, patient speech recordings, montage of the patient behaviour, deviation in sleep, navigation thus developing a valuable diagnostic aid.

The life of a person post diagnosis of Alzheimer's becomes consequential for both the individual and the kin. Constant support, care and medication is required to improve the quality of their life. An exclusive network with doctors, patients and health care advisors integrated on a single community acts as a platform of like-minded people to share their resources and experiences. The major problem faced by any AD patient on diagnosis is social isolation. The application includes AI voice assistant which automatically triggers with remainders in defined time period, to interact with the patients, an assessment procured by the doctors with paired word to memorise. An AI (forum) chat-bot which utilize NLP to interact, understand, and respond to patient queries. Machine learning algorithms can listen to patients' speech and analyse their vocabulary and other semantic features to monitor their cognitive functional ability. The daily tasks with detailed timed schedule, provide choices, clear step-by-step instruction, napping timeout, create a safe environment, fire safety alarm, photographic memory recalling mechanism under the aegis of guardians, doctors in extreme cases.

At a time when people are most in need of care and support, a wall of stigma closes in around them (Pilozzi and Huang, 2020). Overcoming the stigma of AD explores the emotional, social and financial impact of stigma, as many of the patients conceal their diagnosis from family and friends for fear of being stigmatised. The network of patients diagnosed with Alzheimer's could be connected on a peer to peer network to share their experiences with the fellow such diagnosed people and resources and have increased the options available to physicians. It helps them overcome the feeling of dejection and loneliness and thus they could learn social interaction with people, updated with the current happenings around the world just like every other human on the planet.

The effect on the entire family can be devastating when a family member is diagnosed with AD or other dementia. It can trigger a wide variety of emotions but the patients require only utmost end-of-life care. Conflicts may arise between family members, friends or the environment in which they are surrounded. Once diagnosed with the disease, the patient's life is entirely dependent on the family members, care takers, followed by healthcare practitioners/doctors, friends, neighbours where the family members and care givers serve as the significant stakeholders. The role of care-taker plays a very important role in the patients' physical and mental health. Not only does the caregiver's job affect how the person spends their time, but it also has an impact on their overall well-being. Unless they take adequate care of their own mental and physical health, full time carers suffer from stress, depression, and other physical problems caused by excessive care commitment and tiredness. The loss of intimacy caused by the disease may be faced by a partner and a kid or child may struggle with the challenges of caring for someone who was once their biggest source of support. Because AD disrupts these strong ties, carers must seek help and support to meet their own needs as well as those of their loved ones. It is indeed vital that each of the family members must ensure to contribute their part to the betterment of the patient in every possible way. Each of them must share their responsibility, hold family meetings and gatherings, communicate or open-up share and spend time with the concerned person. It is important that the patients are not stigmatised by their own family members. If it does so, their mental state would even grim. Thus, a deeper understanding of the symptoms of dementing disorders, and socio-cultural conceptualisation of AD is of increasing importance.

A. Magnetic resonance imaging detection for AD

The prototype is developed based on the stakeholders involved. Pre-diagnosis phase involves a self-administered cognitive test which is integrated with the application. Upon diagnosis, the patient is connected to the network and gets in contact with the top rated doctors in the peer network. Initially a remote doctor consultation is scheduled via video conferencing. Upon the recommendations of the clinician, the patients are involved in getting themselves scanned. The deep fed neural network-trained algorithm is highly suggested for diagnosing and detecting the prevalence of AD, and it is fully deployed at the doctor's or clinician's end. The AI companion is trained according to the basic stages of the disease – early, mid, late.

Since several patients may encounter different symptoms, the network acts as a tool where the different symptoms, experiences are shared. NLP and sentimental analysis is performed where the unstructured data are organised in a structured format, based on which the chat-bots (using dialog flow api) and AI companion (an exclusive virtual assistance) are trained. The pre-trained AI companion serves as a saviour in the isolation of patients. Constant remainders (notifications, alerts, auto generated phone calls), assessment tasks (suggested Q and A), are set based on the wellness routine of the individual patient as prescribed by the doctors and the results and reports are hashed and updated on the individual's block. The doctors can verify, retrieve and constantly monitor the health status of an individual.

The AI companion can be viewed as a step towards human-level AI system which could be characterized as an autonomous system which helps in quasi-communication with the patients, not really on the basis of artificial intelligence, but an algorithmic structured approach which aids them as a virtual care-taker. There exist several mental health apps, systems with voice control to cope up with the progress of the mental well-being of a person. But there exists a void in an efficient virtual aid and control exclusively for AD. The AI companion focuses to achieve three key parameters such as robust reasoning, ensemble learning, improved performance, longevity and effective interaction with the patients. Analogical processing based on reasoning and learning in cognitive science (based on clinician's recommendations), a distributed peer-to-peer architecture hosted on a cluster to achieve performance and longevity, and interactivity sketching and concept maps which includes alert systems, self-administered gerocognitive exam (SAGE) tests (clock drawings, photographic recalling systems) according to different diagnosed stages. Alzheimer's disease is stigmatised due to a lack of broad knowledge and comprehension of the disease. Dementia sufferers and their families face a tremendous obstacle to their well-being and quality of life due to stigma and misinformation. Thus, an AI companion helps in progressing and maintaining the mental well-being of the individual by constantly keeping them engaged, interact and communicate to fellow patients to relieve stress and share experiences through the secured peer to peer network. The schematic views of the various technologies involved in creating the application and the corresponding results with respect to processing of individuals health records is shown in Figure 105.2.

Algorithm for the proposed work:

MR Imaging detection for AD

Step 1: Import the required data

> ***Function*** import (OASIS dataset)
> Perform data augmentation and data pre-processing
> Normalize( )
> RemoveNoise( )
> IsNull( )
> RemoveOutliers( )

Step 2: Obtain the training and target data

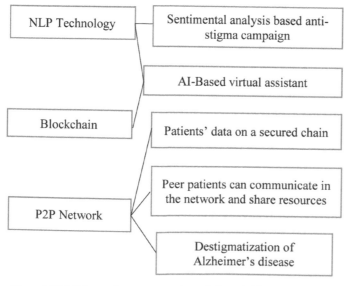

Figure 105.2 Schematic representation of virtual aid control for AD

Function GetTrainingData( )
Returns pre-processed and augmented MRI data
Function GetTargetData( )
Returns Label 1: Cognitively normal(CN)
       Label 2: Mild Cognitive Impairment(MCI)
       Label 3: Alzheimer's disease (AD)

Step 3: Perform Feature Extraction

Function GetFeatures( )
Returns features based on shape, intensity
PCAAnalysis( ) and LDAAnalysis( )

Step 4: Split the validation and training set

Function GetValidationSet( )
    Let k = 1 (Randomly divide the samples)
Perform K – fold validation
Return Kth fold as validation set
Function GetTrainingSet( )
Return the remaining folds as training data

Step 5: Select Classifier models

Function choose model ( )
Return ensemble model of Multi Nomial Logistic Regression ( ) and *k-nearest neighbours (KNN) ( )*

Step 6: Training and validation

Train the model (KNN and MultiNomialLogisticRegression) and store the outputs
Evaluate the trained outputs on the validation set and obtain the predicted class.

Step 7: when k > K, consolidate the predicted output of all folds and move to Step 8

Else perform k=k+1

Step 8: Return the confusion matrix and classification accuracy for the models.
NLP based speech analysis and AI assistant
Step 1: Model questions users based on the pre built inputs
Step 2: The inputs are stored in the database for further semantic analysis
Step 3: *GetAudioFeatures( )*

Return pitch, jitter, loudness, energy

Step 4: *SummarizeFeatures( )*

Based on the time axis, statistical features are derived followed by PCA.

Step 5: Obtain the L1*Features( )* set for each recording where L denotes the dimension of summarized features.
Step 6: *ApplyClassifier( )*

LDA( ) / SVM( )

Step 7: Perform VAD( ) (Voice activity Detector) to filter the silent part.
Step 8: Perform semantic analysis on text-data based on user-input
Step 9: Compare the true positive and false positive values
Step 10: Obtain the comparison and output features.

Removal of stigma in community among peers by semantic analysis
Step 1: Fetch the Malignant comment classification dataset
Step 2: Perform *Data analysis* – determine the skew factor of features
Step 3: *GetTargetValues( )*

Return malignant, rude, threat, abuse, loathe, others

Step 4: Data Pre-processing

RemoveStopWords( )
Lemmatize( )

Step 5: Fit the training and test data in the **LogisticRegression( )** model
Step 6: Predict the output and obtain the confusion matrix and classification report.

The pictorial representation of the various process involved in KNN algorithm has been described in Figure 105.3.

Results and Discussion

AD is a progressive brain illness. The Alzheimer's Illness Neuroimaging Initiative (ADNI) brings together experts and study data to better understand the course of the disease (AD). Prior to developing AD, persons who were previously cognitively normal (labelled as CN) begin to show signs of moderate cognitive impairment (labelled as MCI) (Label: AD). Due to the fact that each patient has many visits, the dataset is referred to as 'longitudinal.'

The process can be explained from Figure 105.3 in which the initial step is MRI data acquisition followed y various pre-processing techniques. The pre-processed data is segmented and the feature extraction phase follows. By supplying the features the data is evaluated using KNN classifier.

Features include "DXCHANGE" - diagnosis of the patient at the corresponding visits. 1 corresponds to CN, 2 is MCI and 3 is AD, "RID" and "PTID" correspond to two different identification number for each patient. As some patients have been seen multiple times, there are represented by multiple lines in the dataset, "VISCODE" corresponds to the visit number. The first one is "bl" (standing for baseline) while the other are encoded as "m#" where "#" is an number that corresponds to the number of month after the baseline. For instance, "m36" means that the visit occurred 36 months after the baseline, "EXAMDATE" corresponds to the date at the examination, "AGE" corresponds to the age AT BASELINE. The age at any visit can be computed from the AGE at baseline and EXAMDATEs, Some cofactors: 'PTGENDER' - the

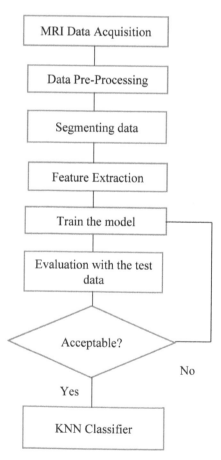

Figure 105.3 Process flow of KNN classifier

gender, 'PTEDUCAT'- Number of education in years, 'PTMARRY' - Marital status. 'APOE' - number of the APOE-epsilon4 genetic mutations which is related to Alzheimer's disease, Some cognitive tests: 'CDRSB', 'ADAS11', 'ADAS13', 'MMSE', 'RAVLT_immediate', 'RAVLT_learning', 'RAVLT_forgetting', 'RAVLT_perc_forgetting', 'FAQ', 'MOCA', Some imaging variables : 'Ventricles', 'Hippocampus', 'WholeBrain', 'Entorhinal', 'Fusiform', 'MidTemp', 'ICV' where 'ICV' is the total brain volume, Radioactive tracers : 'FDG', 'PIB', 'AV45'.

The first task consists in keeping only the first visit of each patient ("VISCODE" == "bl") and keeping only cognitively normal patient and patient with Alzheimer's disease ("DXCHANGE" == 1 and "DXCHANGE" == 3). From this subset of the initial dataset, the goal is to diagnose the patient status (1 or 3) given the other variables. All the features are taken into consideration. Since the class distribution is unbalanced in the dataset, the F1 score is found to judge the model. F1 Score of logistic regression classifier on test set: 0.965261. Correlated features can be get rid of each other. Logistic Regression is better than KNN as a good classifier.AD vs   CN vs MCI at baseline, utilizes the intermediate status subjected with Mild Cognitive Impairments (MCI). It is a diagnosis task where the "DXCHANGE" status is predicted from the other variables.

("DXCHANGE" == 1 and "DXCHANGE" == 3).

It is understood that the accuracy and F1 score corresponding to Logistic regression and KNN classifier by considering the features of cognitively normal patient and patient with Alzheimer's disease brings out the result that Logistic regression shows a better performance and confusion matrix too depicted as shown in Figures 105.4–6.

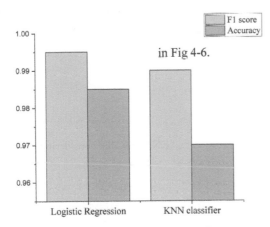

Figure 105.4 Graphical representation of accuracy and F1 score of logistic regression and KNN classifier considering cognitively normal patient and patient with AD

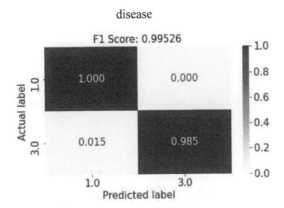

Figure 105.5 Confusion matrix for F1 score of logistic regression considering cognitively normal patient and patient with AD

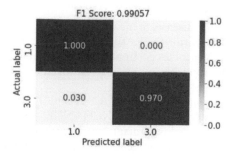

Figure 105.6 Confusion matrix for F1 score of KNN classifier considering cognitively normal patient and patient with AD

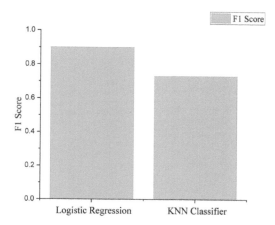

Figure 105.7 Graphical representation of F1 score of logistic regression and KNN classifier considering cognitively normal patient, patient with AD and mild cognitive impairments

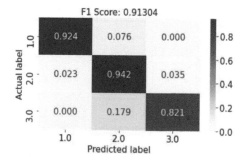

Figure 105.8 Confusion matrix for F1 score of logistic regression considering cognitively normal patient, patient with AD and mild cognitive impairments

It is clear that the accuracy and F1 score corresponding to logistic regression and KNN classifier by considering the features of cognitively normal patient, patient with AD and mild cognitive impairments brings out the result that logistic regression shows a better performance and confusion matrix too satisfies the label prediction from Figures 105.7–105.9 clearly reveals that the proposed model performs well in comparison with the other models under study.

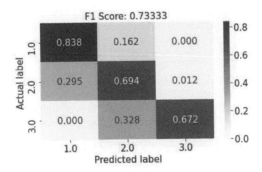

Figure 105.9 Confusion matrix for F1 score of KNN classifier considering cognitively normal patient, patient with AD and mild cognitive impairments

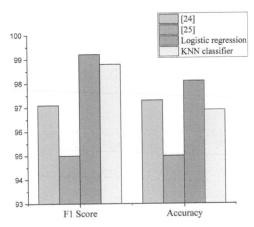

Figure 105.10 Constructive comparison between different models under study

The Figure 10 shows the constructive comparison between the proposed models as well as the models under study for comparison. It is evident that the proposed model is way better than other models taken for comparison in terms of metrics such as F1 score, accuracy.

Conclusion

The system proposes a wholesome care system, initially from diagnoses, test, aid, prevention, assessment and constant monitoring of the patients. The medical data is hashed and stored in the chain thereby retrieved based on the unique id of every patient. A healthcare assessment program can be undertaken as in where to ensure the awareness among people as suggested in (Agarwal et al., 2016).

As a result of the network, patients are able to discuss and exchange ideas and experiences. Studying people's opinions on different issues has been made easier with the use of artificial intelligence (AI) and natural language processing (NLP), both of which are used to identify the emotions and tones of text. The presence of blockchain based network helps in maintaining a decentralisation where the records and posts are not under hold of any central authority and hence the encrypted personal data is tamper-proof. The public stigma around AD can be examined using these methods and thus poses an efficient stigma analysis and removal system.

The network of patients diagnosed with Alzheimer's is connected on a peer to peer network to share their experiences and resources and have increased the options available to physicians. Sentimental analysis, natural language processing (NLP) in stigma analysis and detection, Chat-bots, video monitoring technology, therapy based on AI and virtual assistants is implemented. Block-chain is utilized to maintain information privacy, secured data storage and thus creating a decentralized P2P network.

References

Agarwal, G., Angeles, R., Pirrie, M., Marzanek, F., McLeod, B., Parascandalo, J., and Dolovich, L. (2016). Effectiveness of a community paramedic-led health assessment and education initiative in a seniors' residence building: the Community Health Assessment Program through Emergency Medical Services (CHAP-EMS). BMC Emerg. Medic. 17(1):1–8.

Andrews, S. J., Goate, A., and Anstey, K. J. 2019. Association between alcohol consumption and Alzheimer's disease: A Mendelian randomization study. Alzheimer's & Dementia. doi: https://doi.org/10.1101/190165

Bock, J. R., Hara, J., Fortier, D., Lee, M. D., Petersen, R. C., and Shankle, W. R. (2021). Application of digital cognitive biomarkers for Alzheimer's disease: identifying cognitive process changes and impending cognitive decline. J. Prevent. Alzheimer's Disease. 8(2):123–126.

Bondi, M. W., Edmonds, E. C., and Salmon, D. P. (2017). Alzheimer's disease: past, present, and future. J. Int. Neuropsychol. Soci. 23(910):818–831.

Brodaty, H. and Donkin, M. (2009). Family caregivers of people with dementia. Dialogues Clinic. Neurosci. 11(2):217.

Castellazzi, G., Cuzzoni, M. G., Ramusino, M. C., Martinelli, D., Denaro, F., Ricciardi, A., Vitali, P., Anzalone, N., Bernini, S., and Palesi, F. (2020). A machine learning approach for the differential diagnosis of alzheimer and vascular dementia fed by MRI selected features. Frontiers Neuroinformat. 14:2–5.

Daunt, P., Ballard, C. G., Creese, B., Davidson, G., Hardy, J., Oshota, O., Pither, R. J., and Gibson, A. M. (2021). Polygenic risk scoring is an effective approach to predict those individuals most likely to decline cognitively due to Alzheimer's disease. J. Prevention Alzheimer's Disease. 8(1):78–83.

DeTure, M. A. and Dickson, D. W. (2019). The neuropathological diagnosis of Alzheimer's disease. Molecul. Neurodegener. 14(1):1–18.

Ding, Y., Sohn, J. H., Kawczynski, M. G., Trivedi, H., Harnish, R., Jenkins, N. W., Lituiev, D., Copeland, T. P., Aboian, M. S., Aparici, C. M. (2019). A deep learning model to predict a diagnosis of Alzheimer disease by using 18F-FDG PET of the brain. Radiology. 290(2):456464.

Duong, S., Patel, T., and Chang, F. (2017). Dementia: What pharmacists need to know. Can. Pharm. J. 150(2):118–129.

Elahi, F. M, Casaletto, K. B., Joie, R. L., Walters, S. M., Harvey, D., Wolf, A., Edwards, L., and Cobigo, Y.. (2019). Plasma biomarkers of astrocytic and neuronal dysfunction in early-and late-onset Alzheimer's disease. Alzheimer's & Dementia. https://doi.org/10.1016/j.jalz.2019.09.004

Ertek, G., Tokdil, B., and Günaydın, I. 2014. Risk factors and identifiers for Alzheimer's disease: A data mining analysis. Industrial Conference on Data Mining.

Gopi, A., Jyothi, R., Narayana, V., and Sandeep, K. S. (2020). Classification of tweets data based on polarity using improved RBF kernel of SVM. Int. J. Informat. Technol. 1–16

Iddi, S., Li, D., Aisen, P. S., Rafii, M. S., Thompson, W. K., and Donohue, M. C. (2019). Predicting the course of Alzheimer's progression. Brain Informatics. 6(1):1–18.

Kotecha, A. M., Corrêa, A. D. C., Fisher, K. M., and Rushworth, J. V. (2018). Olfactory dysfunction as a global biomarker for sniffing out Alzheimer's disease: a meta-analysis. Biosensors. 8(2):41.

Marjanovic, S., Robin, E., Lichten, C. A., Harte, E., MacLure, C., Parks, S., Horvath, V., Côté, G., Roberge, G., and Rashid, M. 2015. A review of the dementia research landscape and workforce capacity in the United Kingdom. Cambridge: Santa Monica and Cambridge: RAND Corporation.

Miguel, S., Bolumar, F., and Garca, A. M. (2007). Occupational risk factors in Alzheimers disease: a review assessing the quality of published epidemiological studies. Occupat. Environ. Medicine. 64(11):723–732.

Orimaye, S. O., Wong, J. S. M., Golden, K. J., Wong, C. P., and Soyiri, I. N. (2017). Predicting probable Alzheimer's disease using linguistic deficits and biomarkers. BMC Bioinformat. 18(1):1–13.

Oriol, D. V. J., Vallejo, E. E., Estrada, K., and Taméz Peña, J. G. (2019). Disease Neuroimaging Initiative TA. Benchmarking machine learning models for late-onset Alzheimer's disease prediction from genomic data. BMC Bioinformat. 20(1):709.

Payan, A. and Montana, G. (2015). Predicting Alzheimer's disease: a neuroimaging study with 3D convolutional neural networks. arXiv preprint arXiv:1502.02506.

Pilozzi, A. and Huang, X. (2020). Overcoming Alzheimer's disease stigma by leveraging artificial intelligence and block-chain technologies. Brain Sci. 10(3):183.

Razavi, F., Tarokh, M. J., and Alborzi, M. (2019). An intelligent Alzheimer's disease diagnosis method using unsupervised feature learning. J. Big Data. 6:32.

Swarbrick, S., Wragg, N., Ghosh, S., and Stolzing, A. (2019). Systematic review of miRNA as biomarkers in Alzheimer's disease. Molecul. Neurobiol. 56(9):61566167.

Islam, R. 2018. Fair and Equitable Treatment (FET) Standard in International Investment Arbitration. r Singapore: Springer.

Tanveer, M., Richhariya, B., Khan, R. U., Rashid, A. H., Khanna, P., Prasad, M., and Lin, C. T. 2020. Machine learning techniques for the diagnosis of Alzheimer's disease: A review. ACM Trans. Multimedia Comput. Commun. Appl. 16(1s):1–35.

106 Review on surface characteristics of Titanium and Nickel based alloys produced by direct metal deposition

Pratheesh Kumar S. *,a, Anand K.b, Hari Chealvan S.c, and Karthikeya Muthu S.d

Department of Production Engineering PSG College of Technology, Coimbatore, India

Abstract

Direct metal deposition (DMD) is a sort of metal additive manufacturing (AM) technique which uses a layer-by-layer addition method to create objects. This paper talks about the surface characteristics of parts made by direct metal deposition. The variation of surface attributes such as roughness, finish, texture, and so on, as a function of operation parameters has been studied for a variety of materials. This study aids in determining the appropriate operating parameters, such as material feed rate, laser power, and gas flow rate, for the material chosen, so as to produce the most ideal surface characteristics. The obtained results indicate that wire feed deposition is superior to powder feed deposition. The laser power and scanning speed of the laser were discovered to be the process parameters that had the greatest influence. This research could be used to distinguish or anticipate the optimum process variables for materials that are employed across a wide range of industrial sectors.

Keywords: Additive manufacturing, direct metal deposition, surface characteristics, surface roughness, surface hardness.

I Introduction

Rapid prototyping (RP) is a phrase that is frequently utilised in a range of sectors to refer to the process of quickly establishing a replica of a genuine system or component prior to its final version or commercialization. In other ways, the focus is on rapid development of anything, with the output being a model that can be used as further models and eventually the end result might be built (Gibson, 2005; Kumar et al., 2021). The components are made using an additive process in RP technology. American Society for Testing and Materials (ASTM) International's newly formed Technical Committee agreed on the adoption of new nomenclature. The recently adopted ASTM consensus uses the term AM (Gibson et al., 2009). The fundamental premise of this technique is that a 3-D CAD model can be constructed directly without the requirement for production planning. Layering material creates parts; each layer may be a smaller cross-section of the component derived from the initial CAD data (Yan and Gu, 1996; Kumar et al., 2021).

The terms 'laser powder deposition,' 'laser metal deposition (LMD),' 'laser material deposition,' 'laser-aided direct metal deposition (DMD),' 'laser-based multi-directional metal deposition,' 'laser engineered net shaping (LENS)' or 'digital light fabrication (DLF),' and 'LMD shaping' are all used to describe the DMD process (Mahamood, 2017). It is a type of AM technology known as direct energy deposition (DED), which may be used to repair high-value objects as well as build new 3-D parts (Batut et al., 2017). The DMD technique's versatility enables the simultaneous use of a variety of materials, making it perfect for the manufacturing of functionally graded components (Mahamood et al, 2017). As a result, it offers the flexibility and potential to significantly lower the ratio of by-to-fly, particularly in the aviation sector. Additionally, this critical AM technology has been granted permission to be employed in a variety of industrial sectors, including medical, automotive, and aerospace (Pinkerton, 20210; Kumar et al., 2021).

The DMD process is implemented by continually feeding material feedstock into the molten laser focal area on substrate, resulting in the formation of melt-pool throughout the substrate, as seen in Figure 106.1 (Gu et al., 2012). The DMD approach takes the benefit of coherence and features of directionality of laser to make a melting pool on a substrate's surface as the beam hits at it. This melt-pool gets wire or powder as material input and melts it. The development of this melt-pool results in the formation of a stable track of the material deposited, which can be viewed along the laser beam. When it comes to functionally graded or composite materials, various material feedstocks are situated in multiple particle serving hoppers or wire feeders, and the materials have been supplied concurrently or repeatedly to develop the material composition of the composite selected according to its location of the component via nozzles arranged in a coaxial pattern situated next to the laser source (Mahamood, 2017; Kumar et al., 2021). The flow chart in Figure 106.2 illustrates the different stages involved in producing a part using the DMD technique.

---

aspratheeshkumarth@gmail.com; bmechanand@gmail.com; charichealvan18@gmail.com; dkarthikeral1724@gmail.com

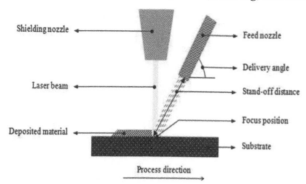

Figure 106.1 DMD process

Figure 106.2 Steps in DMD process

II Material Feeding Mechanism

DMD technology is divided into two variants that aid in the fabric feeding method used to fabricate the parts. There are both powder-fed and wire-fed systems available (Singh et al., 2020).

A. Powder feeding

In the powder feed system represented in Figure 106.1, the metallic powder is supplied or sprayed onto the laser-produced melt pool on substrate's surface. Following that, the powder melts and solidifies in the melt pool. Typically, the powder mixture is made up of a high melting point architectural metal, a low melting point binder metal, and a trace of additives such as fluxing agent or deoxidiser. This results in a clad path, which can then be utilised as a base for additional deposition, resulting in a 3-D object. Metallic powder blends have been investigated as a method of fabricating alloyed components or functionally graded structures from elemental powders. The low rate of deposition and difficulty in ensuring a high catchment efficiency, in addition to the regularly created a poor surface finish, are examples of the technique's downsides (Gu et al., 2012; Piscopo, 2019).

B. Wire feeding

Metallic powders are typically used in DMD techniques to make items with a near-net shape. In addition, wire feeding is employed in applications such as material addition and laser cladding. Both powder and wire feed laser deposition have their pros and limitations. Wire feed laser deposition often has a greater rate of deposition and material utilisation rate than powder feed laser deposition (Syed et al., 2006). In wire feed deposition, the material is generally provided from one perspective, as illustrated in Figure 106.3. There are new commercial laser cladding heads that feed the wire coaxially and split the laser into several independent beams that focus on a circular focal point. As a result, omnidirectional deposition is possible. The wire's side feeding complicates path planning, to produce the optimal outcome in porosity and surface roughness, the material should always be delivered near the leading edge of the melting pool (Akbari and Kovacevic, 2018).

DMD is an intriguing and rapidly evolving technique with several application possibilities in industries such as aerospace, medicine, and automotive. The fundamentals of the process are studied. Additionally, the effect of a variety of input variables on a variety of output variables is explored. As with any other AM technology, the surface attributes of the DMD produced component are significantly controlled by the input process variables that determine the part's surface finish. As a result, it is vital to understand how output qualities fluctuate in response to changing process parameters for various materials. The following literature review assists in gaining a better knowledge of how output process parameters vary in response to changes in process input elements.

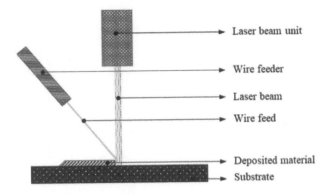

Figure 106.3 Wire feeding mechanism

Mahamood and Akinlabi (2017) investigated the consequence of processing variables on surface finish of titanium alloys formed by LMD. The laser's power has been boosted from 0.8 to 3.2 kW. The scanning velocity was set to 0.005 m/second and the rate of powder flow was maintained at 2 g/minute. Likewise, the pace of gas flow maintained at 2 l/minute. According to the findings, when the laser power was raised, the surface roughness decreased, as indicated in Table 106.1. The influence of typical powder particle size on deposit quality, powder deposition efficiency, and surface finish was examined by Kong et al. (2017), using the direct metal laser deposition (DMLD) process. The investigations demonstrated that the average size of particle influences the deposit height and efficiency in both direct and indirect ways for the Inconel 625 utilised, as smaller particle sizes are easier to concentrate, resulting in higher efficiency and larger deposits. As illustrated in Table 106.1, the study's findings will enable better powder size selection for a number of applications ranging from repair to cladding. Peyre et al. (2012) concentrated on elucidating the physical processes that contribute to poor surface finish and developing a variety of experimental strategies for resolving them. The findings indicate that non-melted or partially melted particles stick to loose surfaces, deteriorating the surface and resulting in the creation of menisci with varying curvature radii.

Gharbi et al. (2013) examined the impact of significant process variables on the surface finish of a titanium alloy (Ti6Al4V) during the DMD process. The purpose of this work was to get a better grasp of the physical causes underlying suboptimal surface finishes and to give some experimental strategies for improving them. The results indicate that surface degradation is dependent on layer thickness decrease and that increasing melt-pool volumes to favour remelting operations has a beneficial effect on roughness characteristics, as illustrated in Table 106.1. Chang et al. (2021) analysed the impurity content, mechanical characteristics, and microstructure of Ti6Al4V blocks deposited in a semi-open atmosphere utilising a powder-feed annular LMD technique. The results demonstrate that annular LMD produced Ti6Al4V can produce a gleaming silver finish in a semi-open environment and that the interstitial components' content such as nitrogen, oxygen, and hydrogen is substantially within the ASTM Grade 5 permissible limits, as shown in Table 106.1. Mahamood et al. (2017) examined the effect of rate of powder flow and laser power on the surface roughness and amount of dilution generated. The flow rate of powder was adjusted 2.88 to 5.76 g/minute, while the laser power was adjusted between 1.8 and 3.0 kW. Both the rate of gas flow and scan speed were maintained at 0.05 m/second and 4 l/minute correspondingly. According to the research, the degree of dilution increases as the laser power increases, while the Ra value decreases. As the flow velocity of powder enhanced, the dilution decreased and the Ra of the surface increased, as shown in Table 106.1.

Dadbakhsh et al. (2010) studied the use of lasers to clean the surfaces of LMD components. A series of Inconel 718 block samples were created utilising the LMD method. The samples' upper surfaces were then laser scanned with a variety of parameters. As indicated in Table 106.1, a laser may enhance the surface finish of LMD components to around 2 μm Ra, which is appropriate for a wide variety of applications in industry. Pityana et al. (2013) examined the effect of powder and gas flow rates on deposit's qualities such as physical, metallurgical, and mechanical. As shown in Table 106.1, as the powder flow rate improved, the track height, track width, and deposit weight enhanced as well, whereas the track width, deposit weight, and track height reduced. Bhardwaj et al. (2020) examined the corrosion and in-vitro bioactivity are affected by generated surface topography, with the goal of minimising surface changes after fabrication. Corrosion resistance and the impact of surface roughness is examined in-vitro, using simulated bodily fluid (SBF). According to Table 106.1, vertically generated samples have a rougher surface than horizontally produced samples, but horizontally produced samples have a corrosion resistance that is up to 75% greater.

Sadhu et al. (2020) enquired the influence the pace of cooling on fracture mitigation during multilayer DMLD of NiCrSiBC-60%WC on the substrate of Inconel 718. Within the existing testing period of 300

Table 106.1 Surface characteristics of titanium and nickel based alloy materials in DMD process

Material	Scanning speed (m/min)	Laser pulse frequency (Hz)	Laser spot size (mm)	Wire feed rate (ml/min)	Laser powder interaction distance (mm)	Powder flow rate (g/min)	Gas flow rate (l/min)	Laser power (kW)	Powder/wire Size (μm)	O2 content in shielding gas (ppm)	Surface roughness Ra (μm)	Rt (μm)	Waviness Wt (μm)	Hardness (HV)	Defects observed	Ref.
Titanium based alloy materials																
Ti6Al4V	0.1-0.4	-	1.32	-	1 - 3	23	2	0.32-3.2	45-75	20-500	0.71-3.3	5.3-165	10-8496	-	-	Mahamood, (2017); Peyre et al. (2012); Gharbi et al (2013)
	0.36	-	-	-	-	1.4	18	1	75-106	-	-	-	-	-	No gas porosity	Cheng et al. (2021)
	0.3 - 3	-	2	-	-	2.88-5.76	24	1.83	150-250	<10	13.16-21.14	-	-	318.5-345	-	Mahamood et al. (2017b); Pityana et al. (2013)
Ti-15Mo	0.3	-	2	-	-	3	10	1.9	45-105	<10	27.1 ± 4.17 (x), 52.7 ± 11.4 (y)	193 (x), 462 (y)	-	-	Less spatters (x), More spatters (y)	Bhardwaj et al. (2020)
Ti-20w/oNb	0.762	-	-	-	-	9	-	0.2	-	<10	12	-	-	-	Non-melted particles	Lewis and Schlienger (2000).
Ti6Al4V +TiC on Ti6Al4V substrate	0.2 - 0.4	-	2	-	-	3	-	0.4-0.7	45-250	-	-	-	-	304 - 631	Micro-cracks are present	Zhang et al. (2018)
Nickel based alloy materials																
Inconel 625	-	-	1	-	-	1.5	-	1	20-177	-	-	-	-	-	Porosity was present	Kpng et al. (2007)
	0.3 - 0.45	-	0.5	-	-	8 - 12	-	0.6-0.9	45-135	-	-	-	-	248 - 263	No relevant defects	Dinda et al. (2009)

(Contd...)

Material														Remarks	Reference
Inconel 718 on SS substrate	1.2	–	0.5	–	–	2.5	–	0.45	2050	9.85	–	–	–	–	Dadbakhsh et al. (2010)
Inconel 690	0.762	–	–	–	–	9	–	0.16	–	12	<10	–	–	–	Lewis and Schlienger, (2000)
Inconel 718	0.4002	–	1	–	–	6.45	–	0.65	50150	–	–	–	277 - 321	–	Zhang et al. (2011)
Ni-Co	–	–	–	–	–	–	–	5	40100	–	–	–	540	–	Gu et al. (2012)
NiCrSiBC-60%WC on Inconel 718substrate	0.3 - 0.7	–	1.6	–	–	10.8	20	0.6	45106	–	–	–	971 - 998	Multiple cracks were formed	Sadhu et al. (2020)
Colmonoy	–	–	–	–	–	–	–	–	90150	–	–	–	376	Finer grain size	Soodi et al. (2010)
Hoganas (1535-30)	0.9	–	1.5	–	–	12	–	0.85	45125	–	–	21.9	713.8	–	Gorunov et al. (2016)
Hoganas (1560-00)	0.6 -0.9	–	1.5	–	–	715	–	0.4 - 1	45125	–	–	20.5	1122	–	Gorunov et al. (2016)
Inconel 600	0.09 - 0.15	20	2.5	5.58	–	–	–	0.22	200	–	–	–	160 - 229	No cracking	Kim and Peng (2000)
Inconel 718 on Ti6Al4V substrate	0.3	20	1.59	–	–	10.7432.22	4.02 - 8.04	0.6 -1.5	53150	25 - 78	–	–	–	–	Shah et al. (2012)
Inconel 718 + SS 316L on SS 316L substrate	–	–	2.5	–	–	40.4449.92	–	0.45 - 0.75	50150	–	–	–	125 - 186	No cracks but considerable porosity was visible	Shah et al. (2012)

mm/min to 700 mm/min, cracks cannot be minimised by adjusting the rate of cooling or scan speed. As with hardness, the wear resistance of the coating increased as the pre-heating temperature of the substrate and scan speed were reduced. Soodi et al. (2010) investigated the microstructure and toughness of items manufactured using laser-assisted DMD technology. 316L stainless steel, 420 stainless steel, Stellite(R) 6, tool steel (H13), Cholmoloy (Ni-based alloy), and Aluminium Bronze were among the alloy powders studied. Microstructure and hardness values for wrought items were compared to those specified in Society of Automotive Engineers (SAE) specifications (as annealed). According to Table 106.1, laser deposited samples exhibit significantly different hardness and defect characteristics than worked samples. Dinda et al. (2009) displayed samples of Inconel 625, a nickel-based superalloy synthesised by DMD. It was revealed that the microstructure was columnar dendritic and epitaxially grew from the substrate. As demonstrated in Table 106.1, none of the samples created in this experiment exhibit significant faults such as fractures, bonding errors, and porosity, suggesting that Inconel 625 is a desirable laser deposition material.

Zhang et al. (2011) investigated the microstructure and characteristics of laser DMD zones prior to and during heat treatment. Additionally, the influence of DMD factors on rate of deposition and shape of the layer was deposited were examined. The results indicate that for rapid build-up rate of Inconel 718 alloy, a laser power of 650 W, scan speed of 5.8 mm/second, beam spot size of 1 mm, rate of powder feed of 6.45 g/minute, and corresponding specific energy of 90130 J/mm2 can be proposed. As shown in Table 106.1, following heat treatment, the laser DMD region's microhardness was much greater than the microhardness of the treatment as deposited. When introducing strength, counterface, and ductility under static loads, along with durability under cyclic bending loads, Gorunov and Gilmutdinov (2016) investigated the effect of heat treatment at 550 and 1050°C on the configuration and tribological characteristics of samples created of nickel-based metal developed by DLD material. Heat treatment has been shown to enhance the wear resistance of specimens generated by high carbide morphology emission. When heated to 550°C, the 153530 alloy sample alignment structure increases endurance when subjected to periodic bending loads, when maintaining the same wear parameters, as indicated in Table 106.1.

Kim and Peng (2000) studied the effect of location and direction of wire feeding, cladding duration and cladding rate on the laser cladding layer quality. The findings indicated that the direction and arrangement of wire feeding are crucial for wire laser deposition. If the orientation and placement of wire feeding are suitable, it can be dipped into a molten pool and dissolved by the molten metal's heat. When demonstrated in Table I, as cladding speed increases, so does the hardness of the clad layer and heat affected zone. Shah et al. (2012) reported on the results of a DMD study comparing two different aerospace alloys, Inconel 718 and Ti6Al4V. The authors evaluated the effect of laser pulse settings and powder mass flow rates on the cracking susceptibility of final deposited structures. As illustrated in Table I, substantial longitudinal tensile stresses were formed along the tracks during the deposition process, which increased in magnitude as the duty cycle increased. Gu et al. (2012) investigated the use of AM to create useful metallic parts with complicated shapes from alloys, metals, and metal matrix composites (MMCs) to fulfil the stringent criteria of the aerospace, defence, automotive, and biomedical industries. The mechanisms of densification of powder materials used in AM are reviewed, including pure metal powder, pre-alloyed powder, and multi-component metals or alloys or MMCs powder. Table I established a link for laser-based AM of metallic components, between substance, method, and metallurgical process.

Lewis and Schlienger (2000) demonstrated the feasibility of depositing any metal and many intermetallics in a single processing step into near-net form pieces using DLF and LENS process methods. As seen in Table I, the pieces are deposited with an arithmetic Ra of 10 μm, necessitating an additional finishing process for specific applications to provide high precision and a polished surface texture. Zhang et al. (2018) investigated the relationship between microstructure, Vickers hardness, and mechanical properties and process parameters and TiC concentration. As feed material, this research used powder mixtures including three distinct volume percentages of Ti6Al4V and TiC. The portion's hardness increased from 300 VH to 600 VH over its length. As shown in Table I, no significant change in hardness was seen for any of the processing settings investigated.

According to the reviewed literature, the output parameters of the parts, such as surface roughness, waviness, and hardness, fluctuate in response to variations in input elements such as scanning speed, laser power, and flow rate of powder. Additionally, the wire feed process produces items with superior surface qualities to those produced by other procedures. Steel and titanium-based alloys offer superior surface characteristics, it is stated. The research gap identified in this work is an insufficient knowledge about the effect of process factors on surface qualities of components manufactured from a variety of materials. This work addresses a research need by examining the impact of process variables on part surface quality created by powder, wire, and combination deposition techniques. The goal of this literature review is to have a greater comprehension of the relation between input and output characteristics for the two basic types of alloy materials that are usually utilised in industry. The purpose of this study is to examine the effect of process parameters such as laser power, scanning speed, powder flow rate, etc, on the surface properties of Ti and

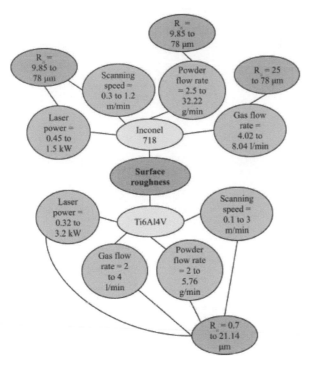

Figure 106.4 Effect of surface roughness on Ti6Al4V and Inconel 718

Ni alloys. The comprehensive method of understanding the process parameters influence on surface characteristics is enumerated in this study by systematic literature reviews. Hence, the ideal process parameters that could be used to get superior surface characteristics can be understood in this study.

III Roughness

Figure 106.4 illustrates the influence of various process factors on surface roughness of titanium and nickel-based alloy materials. For Ti6Al4V, we can deduce that increasing the laser power from 0.32 to 3.2 kW, scan speed from 0.1 to 3 m/minute, flow rate of powder from 2 to 5.76 g/minute, and flow rate of gas from 2 to 4 l/minute lead in a Ra between 0.7 to 21.14 μm (Mahamood, 2017; Peyre et al., 2012; Mahamood et al., 2017).

For Inconel 718, we can adjust the laser power between 0.45 kW and 1.5 kW, scan speed between 0.3 and 1.2 m/min, and rate of powder flow between 2.5 and 32.22 g/minute, result in a Ra range of 9.85 to 78 μm. Similarly, when the rate of gas flow is enhanced from 4.02 to 8.04 l/min, the resulting Ra value varies between 25 and 78 μm (Dadbakhsh et al., 2010; Shah, 2011; Lewis and Schlienger, 2000).

A. Topography

When Ti-15Mo is deposited vertically, the surface roughness is larger than when the material is deposited horizontally. As a result, the surface topography of vertically deposited samples is poor (Bhardwaj et al., 2020). Surface roughness is lower in titanium-based alloys such as Ti6Al4V than nickel-based alloys. As a result, the surface topography of these alloys is observed to be better.

B. Texture

The top layer of nickel-based alloy materials has a rougher surface than the rest of the product. As a result, the top layer's surface texture is rough (Dadbakhsh et al., 2010).

IV. Waviness

The effect of various process conditions on the surface waviness of titanium-based alloy material is depicted in Figure 106.5. For Ti6Al4V, we can conclude that increasing the laser power from 0.32 to 0.5 kW, powder flow rate from 2 to 3 g/minute, and scan speed from 0.1 to 0.4 m/min resulted in a Wt of 103 to 497 μm (Peyre et al., 2012; Gharbi et al., 2013).

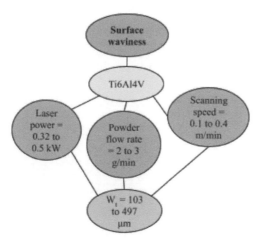

Figure 106.5 Effect of waviness on Ti6Al4V

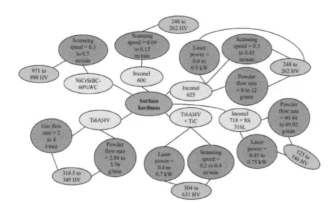

Figure 106.6 Effect of surface hardness on titanium and nickel based alloy materials

V. Hardness

The impact of various operating parameters on surface hardness of titanium and nickel-based alloy materials is depicted in Figure 106.6. We may deduce that, increasing the laser power from 0.4 to 0.7 kW, and scanning speed from 0.2 to 0.4 m/minute, the resulting Vickers hardness for the combination of Ti6Al4V and TiC material is in the range of 304 to 631 HV (Zhang et al., 2018). Variation of flow rate of powder from 2.88 to 5.76 g/min, and rate of gas flow from 2 to 4 l/min result in a surface hardness of 318.5 to 345 HV for Ti6Al4V (Pityanam et al., 2013).

For Inconel 625, we can adjust the laser power between 0.6 kW and 0.9 kW, scanning speed between 0.3 and 0.45 m/min, and powder flow rate between 8 and 12 g/minute result in a hardness range of 248 to 262 HV (Dinda et al., 2009). While the laser power is improved from 0.45 to 0.75 kW and flow rate of powder from 40.44 to 49.92 g/minute, the resultant hardness ranges between 125 and 186 HV for the combination of Inconel 718 and SS 316L (Shah, 2011). When the scanning speed of NiCrSiBC-60%WC is increased from 0.3 to 0.7 m/min, the resulting hardness ranges from 971 to 998 HV (Sadhu et al., 2002). When Inconel 600's scanning speed is increased from 0.09 to 0.15 m/min, the resulting hardness ranges from 160 to 229 HV (Kim and Peng, 2000).

Defects

The impact of numerous process settings on surface defects in nickel and titanium-based alloy materials is depicted in Figure 106.7. We may extrapolate that microcracks formed when the laser power was kept between 0.4 and 0.7 kW for the Ti6Al4V and TiC powder material combination (Zhang et al., 2018).

Similarly, pores occurred in Inconel 625 deposits if the laser power was raised from 0.6 to 0.9 kW (Dinda et al., 2009). When the laser power was kept between 0.45 and 0.75 kW, a reduction in porosity was seen in the combination of Inconel 718 and SS 316 L powder material (Shah, 2011). When the scan speed is

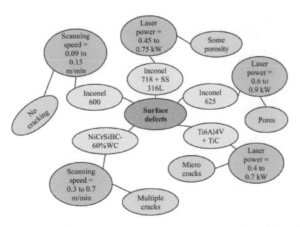

Figure 106.7 Effect of defects on titanium and nickel based alloy materials

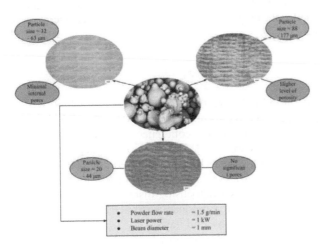

Figure 106.8 Change in microstructure at different process parameter levels on Inconel 625 sample

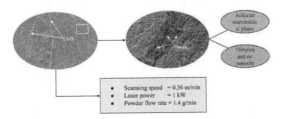

Figure 106.9 Change in microstructure at different process parameter levels on Ti6Al4V sample

held between 0.3 and 0.7 m/minute, numerous cracks appear on the surface of NiCrSiBC-60%WC (Sadhu et al., 2002). Similarly, when the scan speed was varied between 0.09 and 0.15 m/min, no cracking was identified in Inconel 600 (Kim and Peng, 2000).

A. Microstructure

The SEM image in Figure 106.8 illustrates the level of porosity identified at various process parameter settings (Kong et al., 2007). The microstructure of Ti6Al4V specimen as deposited is depicted in Figure 106.9. Numerous continuous grain boundaries (CGB) have been observed. The fracture cross-section contracts and the sample's fracture morphologies exhibit regular dimple fracture toughness characteristics. Shallow dimples cover the fracture surface (Cheng et al., 2021).

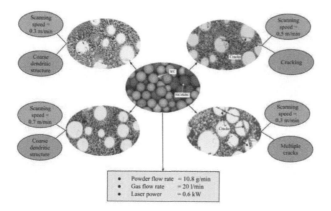

Figure 106.10 Change in microstructure at different process parameter levels on NiCrSiBC-60%WC sample

The NiCrSiBC-60%WC powder depicted in the back-scattered electrons (BSE) image above is a composite material composed of self-fluxing NiCrSiBC and casted tungsten carbide. The cross-sectional view in Figure 106.10 shows materials deposited at varying scan speeds and preheating temperatures. As the cooling rate increased, many fissures formed in the surface interface (Sadhu et al., 2002).

Summary

The ASTM consensus recently embraced the name AM. To create parts, material is stacked, with every layer holding a smaller cross-section of a component generated from initial CAD data. The DMD technique enables the concurrent use of several materials. It makes use of the coherence and a laser's directionality characteristics to produce a melt pool on the substrate's surface. DMD technology has been authorised for use in the aerospace, automotive, and healthcare industries. This procedure entails material input via a powder or a wire feed system. The attributes of the part vary according to operating factors such as scanning speed, laser power, powder defocusing distance, laser defocusing distance laser pulse frequency, laser spot diameter, powder stream rate, and gas flow rate. The component's surface characteristics include qualities such as roughness, finish, texture, waviness, topography, and flaws.

When the O2 component of the shielding gas is reduced during powder feed metal deposition, the roughness of the object reduces but the waviness increases. When scan speed and laser power both are increased, the waviness value drops. However, identical conditions result in an increase in roughness. When the separation between the powder and laser is increased, the surface waviness and roughness diminish. Thus, by increasing the distance between the laser and powder and by enhancing the laser power and scanning speed, products with less surface roughness are created. When the flow velocity of powder is raised while the laser intensity remains constant, the roughness value increases as well. The hardness of the gas rises with the flow rate. A high value of hardness can be achieved by reducing the rate of powder feed while raising the rate of gas stream. When the powder particle size is lowered, the three-dimensional surface roughness increases and the number of pores created decreases. The hardness of the part rises as the powder flow rate and scanning speed increase. When a pulsed laser beam is used instead of a continuous laser beam and rate of powder feed is increased, the Ra of the product drops while the hardness increases.

When a pulsed laser setup with decreasing frequency is utilised, the hardness value lowers first and gradually increases. To summarise, the Ra value is lower in wire feed metal deposition than in powder feed metal deposition. The surface roughness obtained with a powder and wire feed deposition procedure is identical to that obtained by a powder feed deposition process.

Conclusion

Surface characteristics are critical to part quality because they define the look and the component's surface qualities. This work discusses the surface characteristics of the DMD process component. The fluctuation of surface properties as a function of operating settings has been investigated for a wide variety of materials. This research contributes to the identification of the optimal operation parameters for achieving the best surface attributes. This research contributes to the establishment of acceptable process parameters for the fabrication of a variety of materials used in a variety of applications. This study discovered the following significant findings.

- In compared to powder feed and combined deposition techniques, the wire feed approach produces superior surface properties.
- In comparison to a continuous laser configuration, a pulsed laser configuration produces superior outcomes in both the wire and powder feed processes.
- Surface roughness is minimised when affecting attributes such as scanning speed, laser power, and laser frequency are set to high values.
- Low surface roughness results in a smooth and transparent surface finish and texture, as well as improved surface topography.
- The waviness of the surface decreases as the primary affecting factors such as laser power and scan speed rise.
- Rapid scan speed combined with low rate of powder flow and laser power results in a high surface hardness.
- When process parameters are set to low values, defects such as porosity, cracking, and so on occur infrequently compared to when they are set to high values.

References

Akbari, M. and Kovacevic, R.. (2018). An investigation on mechanical and microstructural properties of 316LSi parts fabricated by a robotized laser/wire direct metal deposition system. Addit.Manufact. 23:487–497.

Batut, B. D. L., Fergani, O., Brotan, V., Bambach, M., and Mansouri, M. E. (2017). Analytical and numerical temperature prediction in direct metal deposition of Ti6Al4V. J. Manufact. Mater. Process. 1:3.

Bhardwaj, T.,Shukla, M., Prasad, N. K., Paul, C. P., and K. S. Bindra. (2020). Direct laser deposition-additive manufacturing of Ti–15Mo alloy: Effect of build orientation induced surface topography on corrosion and bioactivity. Metal Mater. Inter. 26(7):1015–1029.

Cheng, D., Zhang, J., Shi, T., and Li, G. (2021). Microstructure and mechanical properties of additive manufactured Ti-6Al-4V components by annular laser metal deposition in a semi-open environment. Optic. Laser Technol. 135:106640.

Dadbakhsh, S., Hao, L., and Kong, C. Y. (2010). Surface finish improvement of LMD samples using laser polishing. Virtual Phys. Prototyp. 5(4):215–221.

Dinda, G. P., Dasgupta, A. K., and Mazumder, J. (2009). Laser aided direct metal deposition of Inconel 625 superalloy: Microstructural evolution and thermal stability. Mater. Sci. Eng. A: Struct. Mater 509(12):98–104.

Gharbi, M., Peyre, P., Gorny, C., and Carin, M., (2013). Influence of various process conditions on surface finishes induced by the direct metal deposition laser technique on a Ti–6Al–4V alloy. J. Mater. Proces.Technol. 213(5):791–800.

Gibson. (2005). Rapid prototyping: A review, Virtual Modelling and Rapid Manufacturing. Adv. Res. Virt. Rapid Prototyp. 7–17.

Gibson, I., Stucker, B., and Rosen, D. W. 2009. Additive manufacturing technologies: Rapid prototyping to direct digital manufacturing. New York (N Y): Springer.

Gorunov, I. and A. K. Gilmutdinov. (2016)   . Study of the effect of heat treatment on the structure and properties of the specimens obtained by the method of direct metal deposition. Int. J. Adv. Manuf. Technol. 86(912):25672574.

Gu, D. D., Meiners, W., Wissenbach, K., and Poprawe, R. (2012). Laser additive manufacturing of metallic components: materials, processes and mechanisms. Int. Mater. Rev, 57(3):133–164.

Kim, J. D. and Peng, Y. (2000). Plunging method for Nd:YAG laser cladding with wire feeding. Optic. Laser. Eng. 33(4):299–309.

Kumar, S. P., Elangovan, S., Mohanraj, R., and Ramakrishna, J. R. (2021). A review on properties of Inconel 625 and Inconel 718 fabricated using direct energy deposition. Mater. Today: Proc. 46(17):7892–7906.

Kumar, S. P., Elangovan, S., Mohanraj, R., and B. Srihari. (2021). Critical review of off-axial nozzle and coaxial nozzle for powder metal deposition. Mater. Today: Proc. 46(13):8066–8079.

Kumar, S. P., Elangovan, S., Mohanraj, R., and Ramakrishna, J. R. (2021). Review on the evolution and technology of state-of-the-art metal additive manufacturing processes. Mater. Today: Proc. 46(11):51875710.

Kumar, S. P., Elangovan, S., Mohanraj, R., and Narayanan, V. S. (2021). Significance of continuous wave and pulsed wave laser in direct metal deposition. Mater. Today: Proc.46(17):8086–8096.

Lewis, K. and E. Schlienger. (2000). Practical considerations and capabilities for laser assisted direct metal deposition. Mater. Design. 21(4):417–423.

Mahamood, R. M. (2017). Laser metal deposition process of metals, alloys, and composite materials. Cham: Springer.

Mahamood, R. M., Akinlabi, E. T., and Owolabi, M. G. 2017. Effect of laser power and powder flow rate on dilution rate and surface finish produced during laser metal deposition of titanium alloy. Mechan. Intellig. Manufact. Technol. 6(10):6–10.

Mahamood, R. M. and Akinlabi, E. T. 2017. Experimental analysis of functionally graded materials using laser metal deposition process (case study). In Functionally Graded Materials. (pp. 69–92). Cham: Springer.

Peyre, P., Gharbi, M., Gorny, C., and Carin, M.. (2012).
Surface finish issues after direct metal deposition. Mater. Sci. Forum. 706709:228–233.

Pinkerton, J. 2010. Laser direct metal deposition: theory and applications in manufacturing and maintenance. In Advances in Laser Materials Processing. (pp. 461491). Cambridge: Woodhead Publishing Limited.

Piscopo, G. Atzeni, E., and A. Salmi. 2019. A hybrid modeling of the physics-driven evolution of material addition and track generation in laser powder directed energy deposition. Mater. 12(17):2819.

Pityana, S., Mahamood, R. M., Akinlabi, E. T., and M. Shukla. (2013). Gas flow rate and powder flow rate effect on properties of laser metal deposited Ti6Al4V. Lect. Notes Eng. Comput. Sci 2203:848–851.

Sadhu, A., Choudhary, A., Sarkar, S., Nayak, P., Pawar, S. D., Nair, A. M., Pal, S. K., nath, A. K. (2020). A study on the influence of substrate pre-heating on mitigation of cracks in direct metal laser deposition of NiCrSiBC-60%WC ceramic coating on Inconel 718. Surf. Coat. Technol. 389:125–646.

Shah, K. 2011. Laser direct metal deposition of dissimilar and functionally graded alloys. PhD diss., The University of Manchester.

Singh, A., Kapil, S., and Das, M. (2020). A comprehensive review of the methods and mechanisms for powder feedstock handling in directed energy deposition. Addit. Manufact. 35:101388.

Soodi, M. Brandt, M., and S. H. Masood. 2010. A study of microstructure and surface hardness of parts fabricated by laser direct metal deposition process. Adv. Mater. Res. 129131:648–651.

Syed, W. U. H., Pinkerton, A. J., and L. Li. (2006). Simultaneous wire- and powder-feed direct metal deposition: An investigation of the process characteristics and comparison with single-feed methods. J. Laser App. 18(1):65–72.

Yan, X. and Gu, P. (1996). A review of rapid prototyping technologies and systems. Comput. Aided Desi. 28(4):307–318.

Zhang, J. Zhang, Y. Li, W. Karnati, S. Liou, F., and Newkirk, J. W. (2018). Microstructure and properties of functionally graded materials Ti6Al4V/TiC fabricated by direct laser deposition. Rapid Prototyp. J. 24(4):677–687.

Zhang, Q. L. Yao, J. H., and Mazumder, J. (2011). Laser direct metal deposition technology and microstructure and composition segregation of Inconel 718 superalloy. J. Iron Steel Res. Int. 18(4):7–378.

107 Study the effect of geometric parameters on heat transfer in metal expansion bellows using taguchi method

Sunil Wankhede[a] and Shravan H. Gawande[b]

Mechanical Engineering Department, M.E.S. College of Engineering, S. P. Pune University, Pune, India

Abstract

This paper focuses on the study of heat transfer analysis of bellows by analytical and statistical method. For analytical analysis mathematical model is proposed and implemented. For statistical analysis Taguchi design of experiments (DOE) is used. Three process parameters such as height, pitch and thickness of the convolution are selected to decide the most dominating parameter. In Taguchi DOE three parameters with three levels are taken and nine experiments are conducted to analyse the effects of geometric parameters on heat transfer. The S/N (signal to noise) ratio and analysis of variance (ANOVA) are used to regulate the process parameters. The three quality responses are extracted for three linear regressions without any interaction between parameters. The results shows good agreement between analytical and regression analysis. This study results are used as a basis for heat transfer analysis of metal bellows. Also, the result obtained by statistical and analytical methods shows a closed match for considered bellows specifications.

Keywords: Bellows, convolution, geometric parameters, metal expansion joint, thermal expansion.

Introduction

Bellows is used as the integral part of heat exchangers to provide flexibility for thermal expansion. Bellows is the most efficient energy absorbing element in many engineering system. Normally bellows is used to provide flexibility in expansion joint for shell structures of heat exchanger and piping. Bellows designs are referred from expansion joint manufacturing association (EJMA), still to select proper configuration of bellows is difficult. The main focus of the many researcher and manufacturer is on the mechanical design of bellows, still there is no measurable work on the thermal design of metal expansion bellows. Design of metal expansion joints or bellows is very critical, as the temperature variations and fluid pressure are the most variable parameters depending on individual application. Metallic bellow is the most critical and important part of the expansion joint assembly in various components.

Bellows has a function to absorb regular and irregular expansion and contraction in the various engineering system. The bellows is formed with convolutions and there are many other geometric features provided. Because of its convoluted shape, it is also known as corrugated joint. The metal bellows is widely used in air conditioning equipment, piping system, vacuum systems, aerospace equipment, oil refineries, industrial plants, chemical plants, shell and tube heat exchangers (STHE) etc. Various researchers studied the design and analytical aspect of the bellows. Bakhshi-Jooybari et al. (2010) studied, effects of pressure path on forming process for metal bellows. The pressure path affects in two stages of forming as bulge and closing. The pressure path influences the forming process which creates the bursting and wrinkling defects in bellows. Zhang et al. (2004) designed a new forming superplastic technology of Ti-6Al-4V alloy bellows, by applying gas pressure and axial compressive load. The optimum thickness obtained by specifying the three stage load route free bulging, clamping and calibrating. Li (1998) studied, effects on elliptic angles of bellows toroid enforced by internal pressure or deflection. The toroid elliptic degree should be maintained lower (lower than 15%) in manufacturing process to maintain the fatigue life and strength. Czesław and Tomasz (2014) used selective laser sintering (SLS) technology to explore the effect of specific geometric parameters on the elastic capabilities of elastic bellows. Makke et al. (2017) used a regression model to figure out the best parameter setting for the output variable. Taguchi L16 orthogonal array is utilised to cut down the number of runs and experiment time. Taguchi method is used to find the best parametric condition. Finite element model (FEM) for the bellows was constructed by Jaipurkar et al. (2017), and it was tested against experiments and published design equations.

Babin and Peterson (1990) did an experimental analysis of bellows by changing the cooling heat source. A computer based model had been developed to formulate and optimise the conceptual design of the flexible heat pipe bellows. Jaipurkar et al. (2017) did an experimental investigation for thermo-mechanical design and characterisation of flexible metallic bellows pipe. Different experiments were carried out to find the stiffness in the different movement under static load. Axial and radial deformation were measured under hydrostatic pressure of 25 bars and vibration condition. Sun and Zeng (2018) studied a heat transfer characteristics on the turbulent flow of a corrugated tube by experimentally and numerically. For the heat transfer studies, the Reynolds number (Re), Nusselt number (Nu) and Friction factor (f) were used. The corrugated tubes has better heat transfer performance than plain tubes with the same friction cost. Siginer and Akyildiz (2010) studied the effect of waviness of the corrugated tube on Nusselt number and friction factor. The velocity and temperature affect the performance, as the angle of the corrugated tube changed. Fand (1962) studied the effect of vibrations on the heat transfer from a heated surface via free convection. To improve heat transfer, they establish an oscillating relative velocity vector between the heated surface and the fluid medium. Cheng et al. (2009) studied the effect of flow-induced vibration on heat transfer for heat exchangers. The heat transfer is increased by using flow-induced vibration. Rush et al. (1999) investigate the local heat transfer and flow behaviour in sinusoidal wavy passages for laminar and transitional flow. The experimental geometry consists of a 10:1 aspect ratio channel confined by two wavy walls, with attention triggering the flow's little mixing. Microscopic mixing is associated with a significant increase in local heat transfer.

Tong et al. (2011) using a variety of sine chambers, heat transfer and flow around sinusoidal corrugated tubes were explored, and a correlation equation Nu = f (Re) was constructed for heat transfer analysis. Heidary and Kermani (2010) studied the nano-particles effect on sinusoidal wall channel under force convection. The heat flow is increased by 50% when nanoparticles are added and wavy horizontal walls are used. Mokkapati and Lin (2014) analyse the gas to liquid heat transfer performance for concentric tube heat exchanger use the twisted tape insert in corrugated tube and assess the engine performance impact. The twisted tape concentric heat exchanger in the corrugated tube saves fuel and reduces emissions. Yan et al. (2012) studied the effect of bellows construction parameter and pipe diameter on heat transfer enhancement. The average heat transfer coefficient outside and inside the bellows is 3 to 5 times greater than smooth pipes in the steady flow area, and the heat transfer enhancement is 5 to 7 times greater than smooth pipes. Rozzi et al. (2007) to increasing convective heat transfer for helical corrugated tubes, the Reynolds Numbers around 800 to transitional flow region's limit is used. Kareem et al. (2015) used numerical and experimental methods to evaluate passive heat transfer enhancement for laminar and turbulent flow zones in corrugated tubes. Kumaresan et al. (2017) use nanoparticles to increase the receiver tube heat transfer. Turbulators, nanofluid addition, and selective coating in the receiver tube of a solar parabolic trough collector are used to increase heat transfer and reduce heat loss.

Yuan et al. (2019) did the analysis of fatigue life for reinforced S shaped bellows. The different levels of plastic strain and wall thickness is considered for the hydroforming process. Ashrafi and Khalili (2015) investigated pulsing pressure hydroforming of the T-joint section of the bellows. The effects of pulsing pressure settings on part flaws and form accuracy were studied using a Taguchi design of trials. Critical process parameters impacting the final part's wrinkling, bulge height and wall thickness were identified by signal to noise ratio and analysis of variance. By using mathematical model, based on the tube upsetting technique, Rudraksha and Gawande (2017) optimise the various process parameters that impact the coefficient of friction (COF) on tube hydroforming. Friction's effect on process parameters, especially inner pressure and wall thickness, was investigated and optimised. Reddy et al. (2018) investigated the effects of process parameters such as internal pressure, axial feed and frictional coefficient on aluminium alloy by using numerical analysis and Taguchi approach for optimisation. Rudraksha and Gawande (2020) created a mathematical model based on tube upsetting technique to analyse the effect of coefficient of friction in tube hydroforming. The experiments are carried out for manufactured metal expansion bellows with varied geometrical dimensions, materials and lubricants. The influence of various lubricants on the COF was also investigated for each material. Bellows for heat transfer analysis is used and geometric parameters are varying like convolution pitch, convolution height, convolution thickness and the number of convolution. To study the effect of geometric parameters on heat transfer, the analytical analysis is done in the subsequent section.

From the literature reviews, it is observed that the most of researches focuses on forming technology, mechanical behaviour, analysis of movement test, buckling, corrosion failure, vibration and fatigue failure, but very less attention is given towards the thermal analysis of bellows. The most of the work is available on mechanical design of bellows; no considerable work is performed on design of bellows in thermal environment, in-spite of wide applications.

Analytical Analysis

This section focuses on the proposed mathematical model, to estimate and analyse the heat transfer rate through metal expansion joints or bellows. The mathematical model is developed to analyse the heat transfer rate through toroidal and U-shaped bellows. Electrical analogy technique is used for bellows analysis analogous to pipe analysis. The heat transfer equation was developed to estimate heat transfer from bellows. In the electrical analogy, the heat transfer system is represented as an equivalent electrical system. The process of conduction and convection is analogous with the process of flow of current. The heat flow problem can be solved by treating it is an equivalent electrical problem and using the empirical relations, which are developed for electrical circuits. Figure 107.1 shows the schematic diagram of bellows nomenclature and Figure 107.2 shows the thermal resistance diagram of bellows in the flow direction. Electrical analogy is used in parallel with the pipe for development of mathematical model to find heat transfer from bellows.

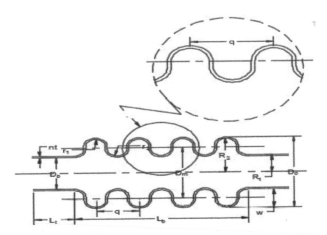

Figure 107.1 The nomenclature of bellows

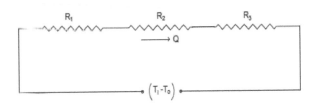

Figure 107.2 Thermal resistance of nellows (electrical analogy)

A. Heat Transfer for Semi-Toroidal and U-shape Bellows

The heat transfer for the tangent length of bellows is calculated by the Equation (1).

$$Q_{Lt} = \frac{T_i - T_o}{\dfrac{1}{2\pi R_b L_t h_i} + \dfrac{\ln(R_1 / R_b)}{2\pi L_t k} + \dfrac{1}{2\pi R_1 L_t h_o}} \tag{1}$$

The heat transfer for the crest of bellows convolution is calculated by the Equation. (2).

$$Q_c = \frac{T_i - T_o}{\dfrac{1}{2\pi R_2 L_1 h_i} + \dfrac{\ln(R_o / R_2)}{2\pi L_1 k} + \dfrac{1}{2\pi R_o L_1 h_o}} \tag{2}$$

The heat transfer for the root of bellows convolution is calculated by the Equation (3)

$$Q_r = \frac{T_i - T_o}{\dfrac{1}{2\pi R_b L_2 h_i} + \dfrac{\ln(R_1 / R_b)}{2\pi L_2 k} + \dfrac{1}{2\pi R_1 L_2 h_o}} \tag{3}$$

The heat transfer for the side wall of U-shape bellows convolution is calculated by the Equation (4).

$$Q_s = \frac{T_i - T_o}{\dfrac{1}{l \times d \times h_i} + \dfrac{n \times t}{l \times d \times k} + \dfrac{1}{l \times d \times h_0}} \tag{4}$$

The heat transfer for one convolution of bellows is presented by the Equation (5).

$$Q = Q_c + Q_r + 2 \times Q_s \tag{5}$$

The total heat transfer for the required number of bellows convolutions with tangent length is given by the Equation (6).

$$Q_t = 2 \times Q_{Lt} + n_1 \times Q_c + (n_2 + 1) \times Q_r + n_3 \times Q_s \tag{6}$$

The total heat transfer for the required number of bellows convolutions without tangent length is given by the Equation (7)

$$Q_t = n_1 \times Q_c + (n_2 + 1) \times Q_r + n_3 \times Q_s \tag{7}$$

Table 107.1 shows heat transfer analysis of bellows based on the mathematical model as per Equation 6. It is observed that as the number of convolutions increases, the heat transfer also increases. It's observed that the convolution height is the most significant geometric parameter to increase heat transfer from bellows is verify with the statistical analysis of bellows.

Statistical Analysis

This section focuses on heat transfer analysis of metal expansion bellows by statistical method. Taguchi, design of experiments, S/N ration, and ANOVA (analysis of variance) technique are used.

A. Taguchi and Design of Experiments

Design of experiments (DOE) is the technique that study any situation involves responses, it can be a function of one or more independent parameters. This technique are used to defining and investigating all probable condition involves multiple factors in experiments as follows:

- Process of planning, experiment that can analysed appropriate data through statistical methods.
- Finalise optimum or best condition for process or product
- Individual factors contribution can be estimate.

Two aspects is used to analyse experimental problem such as DOE and statistical analysis. To optimise or control the performance of any system that involves many factors then it is essential to focused on significant factors. Taguchi L9 orthogonal array are designate in the study for three levels and three parameters.

Table 107.1 Heat transfer for bellows specification

ID mm	OD mm	Convolution Height (w) mm	Convolution Pitch (q) mm	Convolution Thickness (t) mm	Heat transfer (Watt)		
					N = 7	N = 8	N = 9
88.9	108.9	10	12	0.5	14.86	16.98	19.11
88.9	108.9	10	14	0.6	15.49	17.71	19.93
88.9	108.9	10	16	0.7	16.13	18.44	20.74
88.9	112.9	12	12	0.6	17.42	19.91	22.40
88.9	112.9	12	14	0.7	18.07	20.65	23.23
88.9	112.9	12	16	0.5	18.71	21.38	24.06
88.9	116.9	14	12	0.7	20.07	22.94	25.81
88.9	116.9	14	14	0.5	20.72	23.68	26.65
88.9	116.9	14	16	0.6	21.38	24.44	27.50

Also three quality characteristics are measured for each experiment. The following objectives are attain as a results of analysed experiments.

- Optimum operating conditions are establish through signal to noise ratio analysis.
- Estimate the contribution of individual parameters by using the analysis of variance to get significant parameters.
- Perform the regression analysis of heat transfer.

B. Levels of process parameters

The method used is Taguchi, to perform the design of experiment and select suitable combination of geometric parameters for heat transfer analysis of bellows. Table 107.2 shows the parameters and its level that affect the heat transfer. Table 107.3 shows the Taguchi L9 (3^3) orthogonal array used in experimental designs for nine experiments.

C. ANOVA *and signal to noise ratio*

By considering w, q and t are the responses and heat transfer as response, the heat transfer is calculated and signal-to-noise ratio is achieved for each experiments. It has been select any one from larger is better or nominal is best or smaller is better for Taguchi method, as per objective is to maximise the heat transfer therefore 'larger is better' is selected for analysis and it is mathematically written as:

The values obtained by using the heat transfer as response for signal-to-noise ratio are shows in Table 107.4.

$$\frac{S}{N} = -10 \, log \left(\frac{1}{n} \sum_{i=1}^{n} \frac{1}{y_i^2} \right)$$

The greater delta value is observed for w, after the response obtained for S/N ration that indicates, most influencing factor in heat transfer process then q and t are influences. The rank also helps to indicate the influencing factors on heat transfer process that is rank one for w, rank two for q and rank three for t. Tables 107.5 and 6 are shows the similar responses obtained for the mean value.

Table 107.2 Factors and its level

Factors	Levels								
	Set-I			Set-II			Set-III		
	1	*2*	*3*	*1*	*2*	*3*	*1*	*2*	*3*
Convolution Height, w (mm)	10	12	14	12	14	16	10	12	14
Convolution Pitch, q (mm)	12	14	16	10	12	14	10	12	14
Convolution Thickness, t (mm)	0.5	0.6	0.7	0.5	0.6	0.7	0.5	0.6	0.7

Table 107.3 L9 Taguchi orthogonal array design

Experiment No.	Factors		
	w	*q*	*t*
1	10	12	0.5
2	10	14	0.6
3	10	16	0.7
4	12	12	0.6
5	12	14	0.7
6	12	16	0.5
7	14	12	0.7
8	14	14	0.5
9	14	16	0.6

Table 107.4 Heat transfer (HT) and signal to noise ratio of its

Experiment No	HT	HT(S/N)
1	16.9866	24.60215
2	17.7133	24.96597
3	18.4408	25.31562
4	19.9126	25.98258
5	20.6542	26.30017
6	21.3863	26.60269
7	22.9407	27.21214
8	23.6885	27.49076
9	24.4446	27.76366

Table 107.5 Response for the S/N ratios (larger is better)

Levels	w	q	t
1	24.96	25.93	26.23
2	26.30	26.25	26.24
3	27.49	26.56	26.28
Delta	2.53	0.63	0.04
Rank	1	2	3

Table 107.6 Response of means

Levels	w	q	t
1	17.71	19.95	20.69
2	20.65	20.69	20.69
3	23.69	21.42	20.68
Delta	5.98	1.48	0.01
Rank	1	2	3

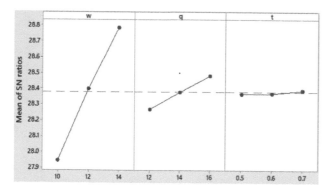

Figure 107.3 Mean effect plot for signal to noise ratio

Table 107.7 ANOVA for signal-to-noise ratio

Source	DF	Seq SS	Contribution	Adj SS	Adj MS	F-Value	P-Value
w	2	9.5930	94.12%	9.5930	4.7965	3043.2	0.000
q	2	0.5923	5.81%	0.5923	0.2961	187.91	0.005
t	2	0.0035	0.03%	0.0034	0.0017	1.10	0.476
Error	2	0.0032	0.03%	0.0031	0.0015		
Total	8	10.192	100.0%				

Table 107.8 Model summary

S	R-sq	R-sq(adj)	R-sq(pred)
0.0397001	99.97%	99.88%	99.37%

Table 107.9 ANOVA analysis for selected specification

Source	DF	Seq SS	Contribution	Adj SS	Adj MS	F-Value	P-Value
w	2	53.6046	94.24%	53.6046	26.8023	122494.14	0.000
q	2	3.2733	5.75%	3.2733	1.6367	7480.05	0.000
t	2	0.0002	0.00%	0.0002	0.0001	0.49	0.669
Error	2	0.0004	0.00%	0.0004	0.0002		
Total	8	56.8786	100.00%				

Table 107.10 Model summary

S	R-sq	R-sq(adj)	R-sq(pred)
0.0147921	100.00%	100.00%	99.98%

Results and Discussion

A. Analysis of variance for signal-to-noise ratio

For analysis considered factors are w, q and t, and the response is the signal-to-noise ratio (Figure 107.3). The results obtained from ANOVA are shown in Table 107.7. As per contribution in percentage, w is highly contributed (94.12%), whereas t has lowest contribution (0.03%) of the S/N ratio.

The S, R-sq, adjusted R and predicted R-sq, values are used effectively to select best fit model. The model summery are shown in Table 107.8 for heat transfer data as 0.0397001 is S, 99.97% is R-sq, 99.88% is R-sq(adj) and 99.37% is R-sq(pred). R-sq is used for prediction then it shows a 99.37% of variation.

Equation 2 shows the linear regression model generated by analysis of variance for signal-to-noise ratio for the selected Taguchi design.

$$HT(S/N) = 16.334 + 0.6319 \text{ w} + 0.1571 \text{ q} + 0.221 \text{ t} \tag{8}$$

B. Analysis of variance for heat transfer

The objective of this work is to select suitable specification for heat transfer analysis of bellows and manufacture the bellows for experimentation. After the concerning parameters must be selected to maximise the heat transfer performance. In the experiment w, q and t are the factors and HT is the response. Table 107.9 (HT against w, q and t in mm) shows the result of analysis of variance. In the contribution of percentage, w has highly contributed (94.24%) on the heat transfer (HT), after that the q has contributed (5.75%) on the heat transfer (HT), whereas t has negligible or no contribution (0.00%). In this analysis the effects of heat transfer interaction are assessed by w, q and t. Generally 0.05 chosen level was selected and p-value indicate the results of interaction term (0.669) is much higher than 0.05. Thus, the interaction for factor t is not significant. The p-value of w is given as 0.00, that is much less than (0.05) actual p-value. As the chosen level is 0.05, that means the effect of factor w, is significant on heat transfer. The p-value for q is also less than 0.05 and given as 0.00, that indicate the effect of q is also significant on heat transfer.

The model summery are shows in Table 107.10 for heat transfer data as, S is 0.0147921, R-sq is 100%, R-sq(adj) is 100% and R-sq(pred) is 99.98%. R-sq(pred) is indicates 99.98% variation, when it is use for prediction. Analysis of Variance generates the model of linear regression for the selected (Set-I) specification of geometric parameters shown as per Equatio 9. Similarly, for the other specification of geometric parameters (Set II and Set-III) performed the ANOVA test and developed the linear regression model as per the equations 10 and 11. The interaction plot and patterns is shown in Figure 107.4 for all combined factors along with the response that help to understand the effects of interaction on response. The blue line (w = 10mm) in row one increases as q and t increases. The strength of interaction is depends on parallelism of line. If greater the line being depart from parallel then the strength of interaction is greater; and if line is parallel then no interaction is possible. The interaction between response and factors are significant as seen in Figure 107.4.

$$HT = -2.392 + 1.49443\ w + 0.36931\ q - 0.043\ t \qquad (9)$$

$$HT = -3.058 + 1.54442\ w + 0.37660\ q - 0.171\ t \qquad (10)$$

$$HT = -2.302 + 1.48714\ w + 0.36931\ q - 0.048\ t \qquad (11)$$

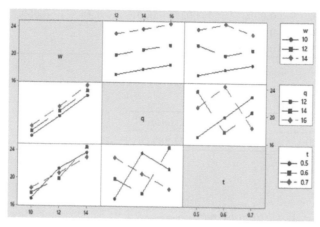

Figure 107.4 Plots of interaction

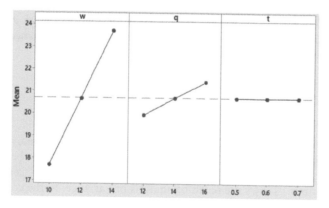

Figure 107.5 The mean effect plots for heat transfer

The significant graphs of mean effect plots are shown in Figure 107.5 and it's most useful if there are lots of factors. For every factor three levels are considered and factors w, q and t are compared for each levels. From Figure 107.5 the factors w and q mean of heat transfer values are increases as level increases and for the factor t mean of heat transfer value is almost constant or very small increase as level increase. And w, is most significant factor that shows as a levels of w increases then mean of heat transfer is also drastically increases.

Using analytical and statistical analysis, the heat transfer of set-I specification for 7, 8 and 9 convolutions is obtained and shows in Table 107.11 and comparison of results are plotted as shown in Figure 107.6. From Figure 107.6 it is found that the number of convolution increases, result in increase the heat transfer. The variation of heat transfer (ht) as a function of geometric parameters (w, q and t) for 7, 8 and 9 convolution obtained by the analytical and statistical analysis approach are seen in Figure 107.6. From Figure 107.6 and Table 107.11, the heat transfer obtained from the analytical and statistical analysis are closely match with each other for considered bellows specification.

Table 107.11 Comparison of results for analytical and regression analysis

ExpNo.	Convolution (N) = 7		Convolution (N) = 8		Convolution (N) = 9	
	Analytical Heat Transfer (Watt)	*Regression Heat Transfer (Watt)*	*Analytical Heat Transfer (Watt)*	*Regression Heat Transfer (Watt)*	*Analytical Heat Transfer (Watt)*	*Regression Heat Transfer (Watt)*
1	14.8633	14.8425	16.9866	16.9625	19.1099	19.0819
2	15.4991	15.8451	17.7132	17.6968	19.9274	19.9080
3	16.1357	16.1277	18.4408	18.4311	20.7459	20.7342
4	17.4235	17.4540	19.9126	19.9470	22.4017	22.4396
5	18.0724	18.0966	20.6542	20.6814	23.2359	23.2657
6	18.7129	18.7503	21.3862	21.4286	24.0595	24.1062
7	20.0731	20.0655	22.9407	22.9316	25.8083	25.7972
8	20.7274	20.7192	23.6885	23.6788	26.6495	26.6378
9	21.3890	21.3618	24.4446	24.4131	27.5001	27.4639

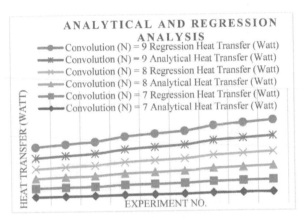

Figure 107.6 Regression and analytical analysis comparison of 7, 8 and 9 convolutions

Conclusions

To determine the heat transfer from different geometric parameters of bellows the proposed mathematical model is verified. The objective is fulfilled through the statistical and analytical analysis. From analytical and statistical analysis following conclusions are drawn.

- From the Taguchi and ANOVA analysis as per percentage contribution, w has the highest contribution (94.12%) on heat transfer (ht), after that the q has more contribution (5.81%) on heat transfer (ht) for the selected specification. Hence, w is more dominating parameters affecting the heat transfer significantly as compared to pitch and thickness of convolution.
- It is observed that the heat transfer (ht) increases with the increase in convolution height (w). Also the heat transfer obtained by the analytical and regression (statistical) analysis are shows a closed match for considered specification of bellows.
- It is absorbed that convolution height is the most significant parameter of heat transfer hence bellows with different convolution heights are selected for further work.

References

Ashrafi, A. and Khalili, K. (2015). Investigation on the effects of process parameters in pulsating hydroforming using Taguchi method. J. Eng. Manuf. Inst. Mech. Eng. 230(7):1203–1212.

Babin, B. R. and Peterson, G. P. (1990). Experimental investigation of a flexible bellows heat pipe for cooling discrete heat sources. 112(3):602–607.

Bakhshi-Jooybari, M., Elyasi, M., and Gorji, A. (2010). Numerical and experimental investigation of the effect of the pressure path on forming metallic bellows. Proc. Inst. Mech. Eng. Part B: J. Eng. Manuf. 224(1):95–101.

Cheng, L., Luan, T., Du, W., and Xu, M. (2009). Heat transfer enhancement by flow-induced vibration in heat exchangers. Int. J. Heat Mass Transf. 52(5):1053–1057.

Czesław, K. and Tomasz, K. (2014). Research of the elastic properties of bellows made in SLS technology. Adv. Mater. Res. 874:77–81.

Fand, R. M. (1962). Mechanism of Interaction between vibrations and heat transfer. J. Acoust. Soc. Am. 34(12):1887–1894.

Heidary, H. and Kermani, M. J. (2010). Effect of nano-particles on force convection in sinusoidal-wall channel. Int. Commun. Heat Mass Transf. 37:1520–1527.

Jaipurkar, T., Kant, P., Khandekar, S., Bhattacharya, B., and Paralikar, S. (2017). Thermo-mechanical design and characterisation of flexible heat pipes. Appl. Therm. Eng. 126:1199–1208.

Kareem, Z. S., Jaafar, M. N., Lazim, A. M., Abdullah, S., and Abdulwahid, A. F. (2015). Passive heat transfer enhancement review in corrugation. Exp. Therm. Fluid Sci. 68:22–38.

Kumaresan, G., Sudhakar, P., Santosh, R., and Velraj, R. (2017). Experimental and numerical studies of thermal performance enhancement in the receiver part of solar parabolic trough collectors. Renew. Sustain. Energy Rev. 77(C):1363–1374.

Li, T. (1998). Effect of the elliptic degree of Ω-shaped bellows toroid on its stresses. Int. J. Press. Vessel. Pip. 75(13):951–954.

Makke, S. K., Keste, A. A., and Gawande, S. H. (2017). Optimisation of design parameters of bellows using taguchi method. Int. Rev. Mech. Eng. (IREME) 11(10):743–747.

Mokkapati, V. and Lin, C. (2014). Numerical study of an exhaust heat recovery system using corrugated tube heat exchanger with twisted tape inserts. Int. Commun. Heat Mass Transf. 57:53–64.

Reddy, P. V., Reddy, B. V., and Rao, P. S. (2018). A numerical study on tube hydroforming process to optimise the process parameters by Taguchi method. Mater. Today: Proc. 5:25376–25381.

Rozzi, S., Massini, R., Paciello, G., Pagliarini, G., Rainieri, S., and Trifiro, A. (2007). Heat treatment of foods in a shell and tube heat exchanger: Comparison between smooth and helically corrugated wall tubes. J. Food Eng. 79(1):249–254.

Rudraksha S. P. and Gawande, S. H. (2017). Optimisation of process parameters to study the influence of the friction in tube hydroforming. J. Bio Tribo Corrosion 56:1–7.

Rudraksha, S. P. and Gawande, S. H. (2020). Influence of lubricants on coefficient of friction in tube hydroforming. J. Bio and Tribo Corrosion. 6:4.

Rush, T. A., Newell, T. A., and Jacobi, A. M. (1999). An experimental study of flow and heat transfer in sinusoidal wavy passages. Int. J. Heat Mass Transf. 52(1):1541–1553.

Siginer, D. A. and Akyildiz, F. T. (2010). Heat transfer enhancement in corrugated pipes. In 2010 14th international heat transfer conference, (pp. 853–860). American Society of Mechanical Engineers Digital Collection.

Sun, M. and Zeng, M. (2018). Investigation on turbulent flow and heat transfer characteristics and technical economy of corrugated tube. Appl. Therm. Eng. 129:1–11.

Tong, Z., Zhu, K., Lu, J., and Li, G. (2011). Numerical study of laminar flow and heat transfer characteristic in wave tubes based on sine curve. In Advance Material Research, (322, 349–352). Switzerland: Trans Tech Publication.

Yan, S., Feng, S., and Sun, Y. (2012). Experimental study of heat transfer of bellows. Adv. Mater. Res. 396–398:376–379.

Yuan, Z., Huo, S. H., and Ren, J. T. (2019). Effects of hydroforming process on fatigue life of reinforced s-shaped bellows. In Key Engineering Materials (795, pp. 296–303). Switzerland: Trans Tech Publication.

Zhang, K. F., Wang, G., Wang, G. F., Wang, C. W., and Wu, D. Z. (2004). The superplastic forming technology of Ti-6Al-4V titanium alloy bellows. In materials science forum (447, pp. 247–252). Switzerland: Trans Tech Publication.

108 Mathematical model for along-wind load on tall multistoried buildings

Suyog Dhote[1,a] and Valsson Varghese[2,b]

[1]Department of Civil Engineering St. Vincent Pallotti College of Engineering and Technology, Nagpur, India

[2]Department of Civil Engineering KDK College of Engineering, Nagpur, India

Abstract

India ranks second in the world with population density 464 per square kilometer. Continuous increase in population creates the serious problem of spaces in urban area. Lack of open spaces creates the need to construct more slender tall residential buildings. Wind forces are more predominant on tall structures, hence it is recommended to carry out dynamic analysis. The present standard includes the formulae to calculate wind induced parameters for any tall building. The computer program can be developed with the use of formulae. A dynamic analysis of residential buildings with height ranging between 80 m to 100 m with decided plan dimensions in terrain category 1 has been included. The wind induced parameters have been predicted with the help of nonlinear regression analysis. The paper describes the use of mathematical model to calculate the wind induced parameters for a building with heights 100 m, 90 m and 80 m with fixed plan dimensions. Validation of results obtained from mathematical model has been included. The technique adopted to calculate wind induced parameters on tall buildings is simple, rapid and inexpensive.

Keywords: Along wind response, IS 875 (Part 3):2015, mathematical model, regression analysis.

Introduction

The present Indian Standard gives the simplified procedure to calculate along wind response parameters for tall buildings (IS 875, Part 3). The procedure to calculate the parameters induced due to across wind on tall structures is also included (IS 875, Part 3). The procedure includes simplified formulae to calculate along as well as across wind response of reinforced concrete tall buildings. The formulae in the code make the ease in writing a computer program which again makes the analysis simple and rapid (Gu, 2009).

The main objective of this study is to develop the unique mathematical model which can be used to calculate the wind induced parameters for any tall buildings. At present, the unique mathematical model has been developed to calculate the gust factor (G) and along wind load (F) for buildings having height (H) ranges between 80 m to 100 m at each decided level with decided plan dimensions in terrain category 1. Validation of results obtained by the use of mathematical model has also been included in this paper.

II Literature Review

The paper describes the use of artificial neural network (ANN) in predicting the along-wind response of tall buildings (Nikose and Sonparote, 2019). The wind tunnel experimentation is the main source of knowledge but this method is expensive and time consuming. The paper suggested the alternative solution to calculate along-wind response of tall buildings.

Yi Li, Q. S. Li, Yong-Gui Li, 2021, The paper discusses the use of mathematical models to predict across wind response for rectangular tall buildings (Yi et al., 2021). A study of rectangular tall buildings with rounded corners has been carried out.

In this study, present IS code (IS 875 Part 3:1987) and the proposed revisions have been reviewed in comparison to the other international codes of practice ( ).

Yin Zhou, Tracy Kijewski, 2002, The study gives the comparison results for tall buildings using major international codes and standards, ASCE 7-98 (United States), AS1170.2-89 (Australia), NBC-1995 (Canada), RLB-AIJ-1993 (Japan), and Eurocode-1993 (Europe) (Yin et al., 2002).

Expected Outcomes and Need of the Study

The study of several research papers gives the analysis results of tall slender buildings using various standards. Artificial Neural Networks can be an effective tool to calculate the wind induced parameters on tall buildings. Based on the studies it is proposed to develop a new tool to carry out the analysis of tall buildings situated in Nagpur with wind speed as 44 m/s as per present Indian code of practice.

The proposed method is more rapid, easy and inexpensive.

[a]dhote.suyog7@gmail.com; [b]valsson_v@yahoo.com

101

IS 875 (Part – 3):2015

The present Indian standard deals with the forces caused by wind and their effects on the structure that should be taken into account during the design of any tall building. Wind speed varies with time and location and due to which it becomes very essential to calculate wind loads and their influence on any structure during the design. The standard specifies both the static as well as dynamic effects of tall buildings (IS 875, Part 3). The procedure involves the wind characteristics such as basic wind speed, design wind pressures, terrain categories, modification factors and force coefficients. The along wind loads on the structures can be determined using gust factor method considering effect of dynamic wind velocity. The f ormulation to calculate the basic wind speed, design wind pressures, terrain categories, modification factors and force coefficients given below.

Design hourly mean wind speed

The basic wind speed (V_b) can be determined using map of India provided in the present Indian standard. The design hourly mean wind speed at height (z) in different terrain categories ($V_{z,d}$) can be predicted considering the factors such as probability factor (k_1), Hourly mean wind speed factor (), topography factor (k_3) and importance factor (k_4). The design hourly mean wind speed can be estimated using following formula (IS 875, Part 3).

$$\bar{V}_{z,d} = V_b k_1 \bar{k}_{2,i} k_3 k_4$$

Where,

$\bar{k}_{2,i}$ -Hourly mean wind speed factor for terrain category

$$\bar{k}_{2,i} = 0.1423 \left[\ln\left(\frac{z}{z_{0,i}}\right) \right] (z_{0,i})^{0.0706}$$

A Turbulence intensity

The turbulence intensity variation with height in different terrain categories using following relations (IS 875, Part 3).
For terrain category 1:

$$I_{z,1} = 0.3507 - 0.0535 \log\left(\frac{z}{z_{0,1}}\right)$$

For terrain category 2:

$$I_{z,2} = I_{z,1} + \frac{1}{7}\left(I_{z,4} - I_{z,1}\right)$$

For terrain category 3:

$$I_{z,3} = I_{z,1} + \frac{3}{7}\left(I_{z,4} - I_{z,}\right.$$

For terrain category 4:

$$I_{z,4} = 0.466 - 0.1358 \log\left(\frac{z}{z_{0,4}}\right)$$

B Design wind pressure

The wind pressure (*pz*) at any height above mean ground level shall be obtained using the relation given below (IS 875, Part 3).

$$p_z = 0.6 V_{z,d}^2$$

C Along wind load

The force acting in the direction of wind (F) can be calculated using the following relation.

$$F = C_f A_e p_d$$

Where,

C_f– Force coefficient
A_e–Effective frontal area
p_d–Design wind pressure

III Building Parameters

A building model with elevation (H), plan dimensions (B and D) and direction of wind flow shown in Figure 108.1 below. All dimensions are considered in meters.

The height (H) of building is considered as 80 m, 90 m and 100 m with plan dimensions as breadth (B) 30 m and depth (D) 30 m in terrain category 1 with wind velocity 44 m/s. The building models have been checked for dynamic analysis as per present Indian code of practice. It is found that the dynamic analysis is required for the said buildings. A non-linear regression analysis for each building has been carried out over the results obtained from dynamic analysis. A single mathematical model has been derived which can be used for calculating wind induced parameters at each decided level for each building.

Tables from 13 show the values for gust factor (G), along wind load, shear force and bending moment values at every 10 m interval of the building having height (H) 100 m, 90 m and 80 m respectively.

Mathematical Model

After successful validation of results obtained by computer program, results are incorporated for non-linear regression analysis. The mathematical model for each building in terrain category 1 has been developed. The plan dimensions for each building considered as mentioned in previous head. Numbers of iterations are carried out to find a single equation which can be used to calculate gust factor (G) and along wind load (Fz) at each decided level for the buildings having height varies between 80 m to 100 m.

Equation 1 can be used to calculate the gust factor (G) for the buildings having height varies between 80 m to 100 m with plan dimensions mentioned in previous head at each decided level along the elevation (H) of the building in terrain category 1.

$$G = 1.669e^{0.0008(z)} \qquad\qquad (1)$$

$$F = 158ln(z) + 120.2 \qquad\qquad (2)$$

Where, z – Level at which gust factor, along wind load is to be measured

In the same manner the along wind load (FZ) can be calculated using Equation 2.

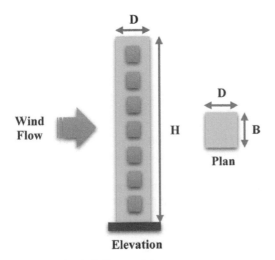

Figure 108.1 Building model

Table 108.1 Wind influenced parameters for building 1
(H = 100 m, B = 30m, D = 30m, TC = 1)

Level	G	Fz (kN)	SF (kN)	BM (kNm)
100	1.786	797.70	798	0
90	1.772	776.11	1574	7977
80	1.758	752.80	2327	23715
70	1.742	727.62	3054	46981
60	1.727	700.28	3755	77523
50	1.713	670.14	4425	115068
40	1.700	635.97	5061	159315
30	1.688	595.32	5656	209921
20	1.677	542.76	6199	266480
10	1.668	461.65	6660	328467
0	1.661	0	6660	395071

Table 108.2 Wind influenced parameters for building 2
(H = 90 m, B = 30 m, D = 30 m, TC = 1)

Level	G	Fz (kN)	SF (kN)	BM (kNm)
90	1.790	844.03	844	0
80	1.775	818.99	1663	8440
70	1.760	791.70	2455	25070
60	1.745	761.84	3217	49618
50	1.730	728.81	3945	81783
40	1.716	691.33	4637	121237
30	1.703	646.83	5284	167604
20	1.691	589.44	5873	220439
10	1.682	501.15	6374	279169
0	1.674	0.00	6374	342910

Table 108.3 Wind influenced parameters for building 3
(H = 80 m, B = 30 m, D = 30 m, TC = 1)

Level	G	Fz (kN)	SF (kN)	BM (kNm)
80	1.794	827.66	828	0
70	1.780	800.39	1628	8277
60	1.764	770.31	2398	24557
50	1.749	736.80	3135	48541
40	1.734	698.68	3834	79892
30	1.720	653.41	4487	118231
20	1.708	595.14	5082	163103
10	1.697	505.77	5588	213927
0	1.689	0	5588	269809

Validation of Results

The results obtained by mathematical models for buildings of height (H) 100 m, 90 m and 80 m with plan dimensions and other details as mentioned above are validated with the values obtained from handout on wind response of tall buildings (IS 875, Part 3).

The Table 108. 1 gives the sample validation results for GF obtained for buildings of height 100 m.

In the same manner Figure 108 3–5 show the validation of wind induced parameters for building having height (H) 90 m, breadth (B) 30 m and depth (D) 30 m in terrain category 1.

Figures 108.6–108.8 show the validation of wind induced parameters for building having height (H) 80 m, breadth (B) 30 m and depth (D) 30 m in terrain category 1.

Table 108.4 Validation for GF (H = 100 m, B = 30 m, D = 30 m, TC = 1)

Level	Target	Predicted	% Error
100	1.786	1.808	1.21
90	1.772	1.794	1.20
80	1.758	1.779	1.24
70	1.742	1.765	1.30
60	1.727	1.751	1.37
50	1.713	1.737	1.40
40	1.700	1.723	1.38
30	1.688	1.710	1.28
20	1.677	1.696	1.11
10	1.668	1.682	0.84
0	1.661	1.669	0.48

Table 108.5 Validation for along wind load (kN) (H = 100 m, B =3 0m, D = 30 m, TC = 1)

Level	Target	Predicted	% Error
100	859.06	847.82	-1.31
90	835.82	831.17	-0.56
80	810.71	812.56	0.23
70	783.59	791.46	1.00
60	754.14	767.11	1.72
50	721.69	738.30	2.30
40	684.89	703.04	2.65
30	641.11	657.59	2.57
20	584.51	593.53	1.54
10	497.17	484.01	-2.65
0	0.00	0.00	

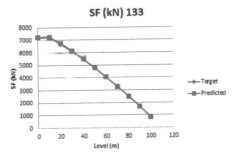

Figure 108.2 Validation for shear force (kN) (H=100m, B=30m, D=30m, TC=1)

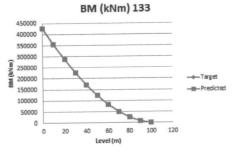

Figure 108.3 Validation for bending moment (kNm) (H=100m, B=30m, D=30m, TC=1)

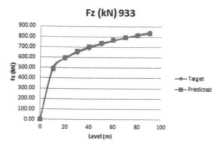

Figure 108.4 Validation for along wind load (kN)

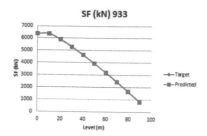

Figure 108.5 Validation for shear force (kN)

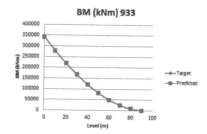

Figure 108.6 Validation for bending moment (kNm)

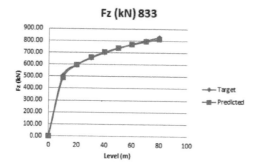

Figure 108.7 Validation for along wind load (kN)

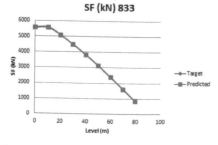

Figure 108.8 Validation for shear force (kN)

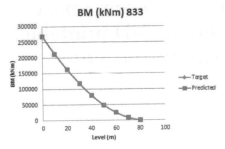

Figure 108.9 Validation for bending moment (kNm)

Conclusion

1. The present Indian standard IS 875 (Part 3):2015 describes the procedure to calculate along as well as across wind load on building with the help of formulae is simple and systematic.
2. The formulae can be incorporated in writing a computer program.
3. The values obtained for gust factor, along wind load, shear force and bending moment predicted with mathematical model and compared with target values shows percentage error up to .
4. The error values are within the permissible limit of 5%, signifies that the mathematical model can be used to calculate wind induced parameters for any building with overall height (H) varies between 80 m to 100 m with specified plan dimensions in terrain category 1.
5. The mathematical model can be useful to calculate gust factor as along wind load for any building having height varies between 80 m to 100 m.
6. The proposed method of estimation is inexpensive and rapid.
7. The several mathematical models can be developed to calculated the wind induced parameters for the buildings with other plan dimensions in different terrain categories.

References

IS 875 (Part 3); 2015, code of Practice for Design Loads (other than Earthquake) for Buildings and structures – Part 3 Wind loads, Bureau of Indian Standards, Manak Bhawan, New Delhi.

IS 875 (Part 3): 1987, code of Practice for Design Loads (other than Earthquake) for Buildings and structures – Part 3 Wind loads, Bureau of Indian Standards, Manak Bhawan, New Delhi.

An explanatory handbook on IS 875 (Part 3), Wind Loads on Buildings and Structures, IITK – GSDMA project on building codes, Department of Civil Engineering, IIT Kanpur, India.

Krishna, P., Kumar, K., Bhandari, N. M. IS: 1987 (Part 3): Wind Loads on Buildings and Structures – Proposed Draft & Commentary Document No. IITK – GSDMA – Wind 02 – V5.0, 2004, IITK – GSDMA project on building codes, Department of Civil Engineering. Kanpur: IIT Kanpur, India.

Nikose, T. J. and Sonparote R. S. (2019). Dynamic Wind response of tall buildings using artificial neural network. Cluster Comput. 22:3231–3246. https://doi.org/10.1007/s10586-018-2027-0

Chaudhary, R. and Agrawal, V. (2019). Comparative study of High Rise Building using major international codes with Indian code. J. Emerg. Technol. Innovat. Res. 299–306.

Yang, Z., Sarkar, P. P., Hu, H. (2011). An experimental study of a high-rise building model in tornado like winds. J. Fluids Struct. 27:471–486.

Gu, M. 2009. Study on wind loads and responses of tall buildings and structures. Taipei, Taiwan: The Seventh Asia-Pacific Conference on Wind Engineering.

Hajra, B. and Godbole, P. N. 2007. Computer Program for Along – Wind Response of tall Buildings. Proceedings of the Fourth National Conference on Wind Engineering, SERC, Chennai, 30th October – 1st November 2007, 259–267.

Indian standard codal 456: 2000. Indian Standard code of practice for general structural use of plain and reinforced

109 Workflow of conversion of CT scan Dicom image of reconstructed humerus fracture to 3-D printable model for preoperative usage with the help of affordable additive manufacturing process

Aniket Mandlekar[1], Atharva Wankhede[2], Jayant Giri[2,a], Atharva Chaudhari[2,b], Prajwal Gedam[2,c] , Mohanish Khotele[2,d], and Rajkumar Chadge[2,e]

[1]Concordia University, Gina Cody school of Engineering, Canada

[2]Yeshwantrao Chavan College of Engineering, India

Abstract

This paper's research work describes the application of lower-cost additive manufacturing technology for preoperative planning and the decision-making in the case of medial epicondyle fracture of the humerus. By using the computer tomography (CT) scan of fracture from which the three-dimensional (3-D) model of fractured humerus part has been reconstructed with the help of open-source software and then the 3-D model has been printed by using MakerBot entry-level printer. In which the high-temperature poly lactic acid is used to fabricate the reconstructed fractured model which could be used for surgical planning purposes. We trust that our methodology is a viable alternative to significant commercial items, as it is more cost-effective and faster in the case of manufacturing purposes. The outcome of the study highlights the process of manufacturing a less expensive 3-D printed medial epicondyle fracture within the range of 130 USD and the time required for this process is around 10 to 12 hours. It could be beneficial for the production of a similar type of fracture-related issue in which they can create customised sterilisable 3-D printed models with minimum expenditure and great adaptability.

Keywords: Additive manufacturing, computer tomography scan, preoperative planning, reconstructed 3-D model, 3-D printing.

Introduction

Currently computed tomography (CT) and magnetic resonance imaging (MRI) are widely used modalities for studying three-dimensional (3-D) information regarding the entire anatomy of internal parts of the body. Particularly MRI is used to investigate the soft tissues which is an effective method because moreover, MRI images are more detailed for the studying of small tissues which is helpful for doctors, and generally, for bone structure, CT scan is preferred as it contains a lot of information for preoperative planning of severe fractures, bone malignancies, and joint restoration. CT scan is comparatively cost-friendly and can be used in many emergencies as it requires less time for the procedure. The advancement of additive manufacturing and interactive 3-D technologies in recent years has opened new opportunities for the application of CT scans in preoperative planning purposes. The use of 3-D modelling with a combination of 3-D virtual training is redefining the revolution in the surgical medical field (Amicis et al., 2018) and providing a real hands-on application for resident doctors who can practice easily on these 3-D models before going to actual surgery. Due to advancements in the medical field, robot surgical operations are increasing day by day so in such cases, the 3-D model can create a better visualization for the doctors in which they can visualise the instrumental movements with the help of a 3-D model. From the point of view of a patient, a 3-D model can help to understand the surgical procedure directly with help of doctors, and by which surgeons can also give the full procedural knowledge to the patient's family and can explain all the complications during surgery and post-surgery by which they can mentally prepare them before original surgery could happen.

An accurate printed 3-D printed model shows the link between any tumour or fracture in bone with the normal tissue related to that part so examining it properly may be beneficial in preoperative planning to identify the safes and fast surgical strategy and also use to build the patient-specific instruments which are required for operation (Frizziero et al., 2018) or build the specific implants whichever not affordable or unavailable due to cost. Due to the combined effect generated by 3-D printing and biomaterial many new opportunities were created to fabricate the biomedical parts and scaffolds which imitate the natural growth of the tissues within the model and start a new generation in transplantation and reconstruction of fractured bones. This 3-D software which is used to generate the 3-D model is very costly and 3-D printers are also expensive. Generally, professionals are required to handle all this software. These all factors limit the use of this technology and are generally affordable by only highly specialised hospitals.

[a]jayantpgiri@gmail.com; [b]atharvachoudhari2006@gmail.com; [c]prajwalpg55@gmail.com; [d]mohanishkhotele@gmail.com; [e]rbchadge@rediffmail.com

Finally, the entire process from collecting the Digital Imaging and Communications (Dicom) images of CT scan to converting them into the proper 3-D model is generally time-consuming and inappropriate for emergencies. In this research paper, we described the conversion process of Dicom to the reconstructed 3-D printable model of medial epicondyle fracture of the humerus which can be further used for preoperative planning and bone-implant purpose for all these steps we used open-source software and the very low-cost 3-D printing procedure. The research paper aims to emphasize the possibilities, particularly for smaller institutes and especially rural hospitals which resource constraints, to develop precise 3-D printed bone replicas and customized bone models that may be used for clinical and surgical purposes.

Methodology

Our present research study explores the possibilities and implementations of the 3-D printed model of the medial epicondyle fracture of the humerus for clinical and surgical practice purposes or in preoperative planning. The methodology proposed with the use of our method is summarized in the given flowchart (Figure 109.1).

A. Image segmentation and Mesh reconstruction process

This method depends on the two types of open-source software. The first one is a 3-D slicer in which the Dicom to STL model creation has happened and the second one is Meshmixer in which the mesh reconstruction, cleaning, and smoothing of STL files have been done. In a final step, the MakerBot slicing software is used to generate the parameters for printing. The computer tomographic data of humerus fracture obtained from the internet is based on the standardisation of Dicom. Dicom is the standard used for imagining and communication in medicine related to data. For the creation of the 3-D model, the Dicom files were imported into the 3-D slicer software. By importing Dicom images into 3-D Slicer it allows visualising the internal structure with help of three views which are coronal, sagittal, and axial view. Varying the density range in a 3-D slicer can allow the creation of the 3-D model of the area of interest and one can remove unnecessary parts that are present in the 3-D model. Image segmentation divides the scanned volume into non-overlapping, linked, and homogenous areas. There are two types of segmentation procedures available one in which a manual area has been selected and manual segmentation has been done and another which is an automatic segmentation procedure in which the algorithm automatically divides the image into the selected area portion. For this research work, we adopted the semi-automatic segmentation procedure because there are a lot of Dicom images that are hard to process one by one, and the automatic segmentation is less accurate and not reliable for creating the 3-D volume (Figure 109.2).

After generating a 3-D volume model of humerus fracture, it is exported to STL file format which is a file format supported by the 3-D printer. The STL file needs more pre processing before printing. In which the fractured part of medial epicondyle bone has been repaired in the Meshmixer software. The used software aims to repair the mesh which has some errors created during the use of the 3-D slicer and reconstruct the

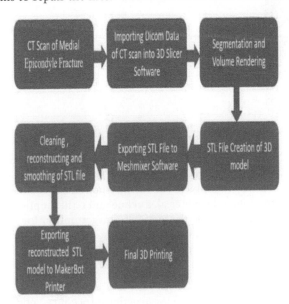

Figure 109.1 Workflow for present methodology

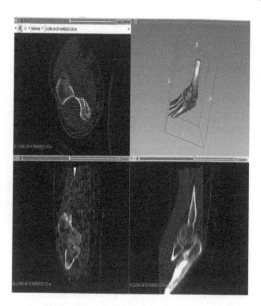

Figure 109.2 **Segmentation and STL file conversion** process in the 3-D slicer

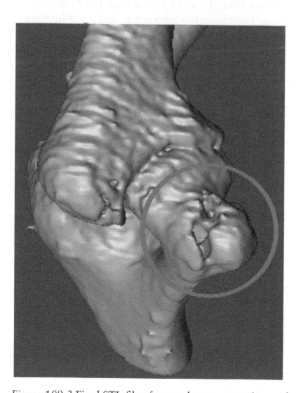

Figure 109.3 Final STL file after mesh reconstruction and error removal process in Autodesk Meshmixer

fracture area and regenerate it for 3-D printing. With help of the inspector tool in Autodesk Meshmixer, one can reduce all the defects present in the given 3-D surface of the bone and reconstruct it by choosing the close mesh option which is present in the menu bar of Meshmixer. The reconstruction procedure has been done manually to reduce errors in the STL file. After filling the material, a refine and reduce option has been selected to remove access material and at the same time enable the attract to target option for more accuracy in the whole procedure. After reducing all defects present and reconstructing the fracture area on the 3-D surface, the finalized STL file has been generated for further printing (Figure 109.3).

B. 3-D printing process

The 3-D model was ready for printing once the mesh repairing and reconstructing procedure ended. In this process, we used the MakerBot 3-D printer which is a low-cost entry-level 3-D printer and has good

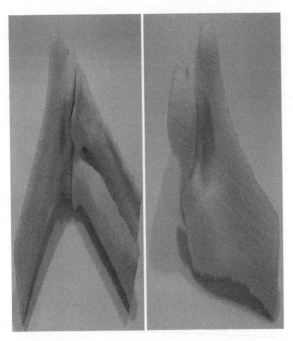

Figure 109.4 Holes in the mesh in the medial epicondyle bone area of the humerus

surface finishing due to low inertia in the moving parts. The nozzle diameter in the MakerBot printer is 0.4 mm the temperature of the extruder required is almost in the range of 180°C to 200°C. Default printing speed is range between 40 mm/s to 60 mm/second. For good accuracy in printing and smooth surface finish the layer thickness has been set to 0.25 mm. Setting up all these parameters will help to get more dimensional accuracy and precision in the printing process which may further help in surgical planning. For 3-D the high-temperature polylactic acid is used as an extruder. After this annealing procedure has been undertaken to make the 3-D printed model more usable. The stage is carried out in a laboratory oven after multiple tests to achieve full annealing using the following procedure: The 3-D printed model goes through a 30 minute heating ramp to 118°C, then a one-hour stay at 118°C, and finally a 30 minute cooling ramp. After the annealing procedure, the 3-D printed model can withstand up to almost 130°C to 140°C. After annealing the 3-D model was then sterilized in compliance with the established sterilisation (Mitsouras et al., 2015) method for medical equipment.

Results

A. Image segmentation and Mesh reconstruction process

The research study involves the CT scan of medial epicondyle fracture of the humerus which almost comprises 956 images, in which the semi-automatic procedure is for segmentation and volume rendering procedure. As the geometry-related humerus fracture is quite complex we adopted the semi-automatic approach in which for the entire 3-D reconstruction, we used a specified density value. The density value was then modified to alter the complicated geometries of the medial epicondyle part.

This semi-automatic method was necessary to facilitate the ultimate use of mesh repairing software. In actuality, gaps and segmental faults were significant in certain areas, and automated mesh repair systems were unable to accurately reconstruct the mesh (Figure.109.4). The use of a semi-automated approach for image segmentation and mesh reconstruction for mesh repairs leads to a reduction in noise and artifacts. The use of semiautomatic segmentation during CT scans reduces errors and noise significantly, according to several iterations and models. Depending on the complexity and the specifications present in patient anatomy the mesh processing for the whole model is done in about 2 to 4 hours.

B. 3-D Printing process

In this process, we observed that the printing parameters might impact the surface finishing of the printed model in such a manner that the excessive layer height present in the model can cause the direct appearance of the exterior surface of the 3-D model. Generally, the extruder flow rate also determines the surface quality of the model. On one side, the flow rate of an extruder is excess at a point then on the outer surface, an excess amount of plastic can stick to the model which impacts the dimension of the 3-D surface of a model. On the other side, if the flow rate of an extruder is less then results in poor layer adhesion, leading to layer separation, creation of gaps, and holes created in the top of the surface layer. In our study, the medial epicondyle 3-D printed model had a 0.58%–0.42% inaccuracy when compared to the digital model produced from DICOM data. In this process, the infill density is taken to 15% to avoid gaps and holes in the top layer (Table 109.1). In Figure 109.5 the printing software which is the MakerBot printer setup is generated and in Figure 109.6 final 3-D printed model is depicted. The dimension of the 3-D printed model is found to be 140 mm in length, 36 mm in height and 22 mm in width. The total material requirement is 79.12 grams which cost less than 60 USD.

C. Cost estimation

In this study FDM method of rapid prototyping is used which is a prominent and cost-effective method that consumes less time with accurate dimensions of the model. At present, the cost of the whole production process in this case study is very low as compared to other industry-level 3-D printing processes. The overall time taken to complete this process is average about 10 to 12 hours, in this almost 4 to 5 hours are required for software processing for 3-D printing preparation. The normal cost of HTPLA filament per kilogram is about 737.79 USD, with 79.12 grams required for a 3-D model of medial epicondyle bone

Table 109.1 Parameters used for 3-D printing

Sr. No	Parameters used	Value
1	Filament diameter	1.75 mm
2	Material used	HTPLA
3	Layer height	0.25 mm
4	Infill density	15%
5	Material estimated	79.12 g
6	Nozzle diameter	0.4 mm
7	Time estimated	5 hour 23 m
30	641.11	657.59
20	584.51	593.53
10	497.17	484.01
0	0.00	0.00

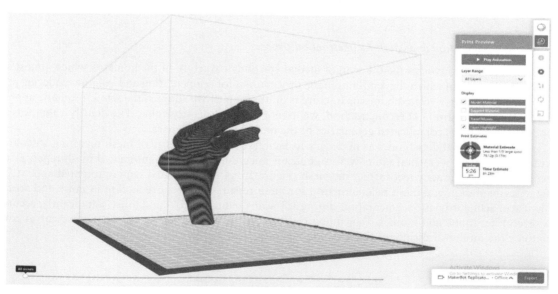

Figure 109.5 3-D Printing set-up in MakerBot printer

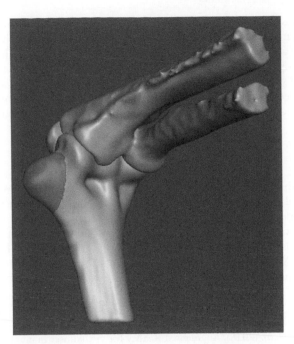

Figure 109.6 Front and back sides of 3-D printed medial epicondyle bone of humerus

which costs about 58.37 USD for a total model to print. It is evident from the research that material primarily contributes to the final additive manufacturing production cost. For printing with 15% infill density, the average production time was between 5 to 8 hours. Postproduction modifications (any cleaning or removing of support parts) and model distribution were estimated to take almost 15 to 45 minutes. And for the last sterilisation process, almost 60 to 80 minutes are required. In generally most cases the 3-D printing process has been done without the supervision of any person or it may be performed at night so the cost estimation for labour cost is calculated by 2 to 4 hours per week. The total cost estimation per bone model of medial epicondyle bone of humerus is in the range of 130 USD. We have concluded that the cost-benefit ratio in producing such a complex model is worthwhile.

Discussion

The method used in this study allows the creation of a 1:1 scale model of a bony structure which will help further for preoperative planning purposes. Resident doctors used this model for their practice purposes; it also reduces the surgical time in a complex process that requires less time. Recent research said that most of the research focused on the 3-D printing technology in the surgical field (Matsumoto et al., 2015). However, in most cases, the 3-D printing expenses are very high for the common man and the small hospitals in rural areas. In general, dental implants in India with the help of 3-D printing are almost ranging from 600 USD to 1000 USD which is very expensive for normal people to afford. The cost of 3-D virtual surgical planning in ortho in India ranges from 332 USD to 425 USD and the model preparation time takes almost 3 to 4 business days which is less effective in emergency cases (Osti et al., 2019). In our research work, almost the whole process is in a very low-cost range because we used open-source software and used the entry-level 3-D printer in which the estimated time is in the range of 6 to 10 hours, and the estimated cost per model is in the range of 140 USD. With the use of this method, many small hospitals will be benefited in the future. The traditionally manufactured models created cannot stimulate the small blood vessels and nerves which are present in the skull part (Zheng et al., 2018).

So that's the reason we used the latest 3-D printing technique which is the FDM method to print the fractured reconstructed medial epicondyle bone model. Acrylonitrile butadiene styrene (ABS) is used to print the 3-D model, which has a higher Rockwell hardness than PLA: ABS ranges from R105 to R110, whereas PLA ranges from R70 to R90 (Bizzotto et al., 2008). However, the ABS material is prohibited because of the difficulty in printing with this material. The ABS material generally requires a heated bed for printing procedures. The heating temperature for ABS material is in the range of 200 to 250°C and the heating temperature of the PLA extruder ranges from 180 to 200°C. Because of all these reasons the ABS material is more expensive than PLA. The main feature of our methodology is that we used the HTPLA extruder for printing purposes because it has many advantages over any other material. HTPLA didn't require the preheat bed for the printing procedure so that's why most of the expenses within it are reduced. The mechanical feature of HTPLA is that this material has more stiffness and brittleness. In this method, we diminished the wastage of excessive material which is created during the 3-D printing process. There are some limitations also within this study as we are printed only bone-like structures so the success rate for other anatomical structures is not investigated yet. For the checking of HTPLA material brittleness and stiffness, more 3-D printed models should be created in the future using this methodology and more clinical studies should be happening with the use of this technology in the orthopaedic department and emergency trauma cases.

Conclusion

In this study, we emphasized the user-friendly methodology which is used to create a low-cost, patient-specific, and customised sterilisable 3-D printed model of medial epicondyle bone of humerus. In which we are converting the CT scan Dicom report to a 3-D model of a reconstructed fractured bone. We trust that our methodology is a viable alternative to significant commercial items, as it is more cost-effective and faster to manufacture. This methodology is helpful for many small organisations and rural areas hospitals where they can't afford the high-end cost of production. With the help of this method, they can produce customizable sterilized 3-D printable models with minimum cost estimation.

References

Amicis, R. D., Ceruti, A., Francia, D., Frizziero, L., Simões, B. (2018). Augmented Reality for virtual user manual. Int. J. Interact. Des. Manuf. 12:689–697.

Bizzotto, N., Sandri, A., Regis, D., Romani, D., Tami, I., and Magnan, B. (2008). Three-Dimensional Printing of Bone Fractures: A New Tangible Realistic Way for Preoperative Planning and Education. J. Abbr. 10:142–149.

Frizziero, L., Francia, D., Donnici, G. Liverani, A., and Caligiana, G. (2018). Sustainable design of open molds with QFD and TRIZ combination. J. Ind. Prod. Eng. 21–31.

Matsumoto, J. S., Morris, J. M.; Foley, T. A., Williamson, E. E., Leng, S., McGee, K. P., Kuhlmann, J. L., Nesberg, L. E., and Vrtiska, T. J. (2015). Three-dimensional Physical Modeling: Applications and Experience at Mayo Clinic. RadioGraphics. 35:1989–2006.

Mitsouras, D., Liacouras, P., Imanzadeh, A., Giannopoulos, A. A., Cai, T. Kumamaru, K., George, E., Wake, N., Caterson, E. J., Pomahac, B. (2015). Medical 3D Printing for the Radiologist. Radiographics. 35:1965–1988.

Osti, F., Santi, G. M., Neri, M., Liverani, A., Frizziero, L., Stilli, S., Maredi, E.; Zarantonello, P., Gallone, G., Stallone, S., and Trisolino, G. (2019). CT Conversion Workflow for Intraoperative Usage of Bony Models: From DICOM Data to 3D Printed Models. Appl. Sci. 9:708.

Zheng, J. –P., Li, C. -Z., Chen, G. –Q., Song, G. -D., Zhang., Y. -Z. (2018). Three-dimensional printed skull base simulation for transnasal endoscopic surgical training. World Neurosurg. 111:e773–e782. doi: 10.1016/j.wneu.2017.12.169.

110 E-waste remanufacturing in Indian context

Swatantra Kumar Jaiswal[a] and Suraj Kumar Mukti[b]

Department of Mechanical Enggenering National Institute of Technology Raipur, Raipur, India

Abstract

In the last few decades, there is massive growth in the electrical and electronic industry as a result of these more products are getting obsolete, hence, there is rapid growth in the electrical and electronic waste, demand of electronic product growing enormously resulting more amount of e-waste. Remanufacturing is a method for transforming End of use product (EOU) or End of life product (EOL) to the given standards or specifications of the original product, remanufacturing is at an emerging stage in India. In this study, author aims to detect the critical factors, which will affect the e-waste remanufacturing and find out the logical relationship between them by using interpretative structural modelling (ISM). Ten factors have been taken using of literature survey and experts in the respective field. Further, these ten factors are divided into six levels based on the logical relationship among them or by using ISM model, top levels show strong dependence power and the bottom level shows strong driving power. The result shows that government support (GS), timing of return (TR) and marketing strategies (MS) are the major critical factors in the hierarchy and shows strong driving power in remanufacturing, market access (MA), product design (PD), collection strategies (CS), return intension (RI), consumption attitude (CA), assured warranty (AW) and workforce and technology (WT) shows strong dependence power as they are in top levels of ISM model. Government support (GS) are at level six of ISM model shows strong driving power it is one of the critical factor, which strongly effects the remanufacturing of e-waste, government, have to come onward and offer subsidy to both manufacturer and consumer to encourage remanufacturing. This study promotes the manufacturer for sustainable e-waste remanufacturing in India.

Keywords: e-waste; ISM, India, rmanufacturing.

Introduction

In last two decades, there is massive growth in the electronic industry as a result of these products are getting obsolete, hence, there is rapid growth in the electrical and electronic waste stream. Globally, 50 MT of e-waste is produced in a year, weighing more than all of the commercial airliners ever made (UN Report, 2019). In terms of e-waste generation India is ranked fifth in the world, behind the China, United States, Germany and Japan. India generates roughly two million tonnes of e-waste each year (TPA) (ASSOCHAM-NEC, 2018). The main causes of e-waste in India are the administration, civic and private sectors, which account for almost 70 percent of the entire waste generation, the contribution of individual homes is relatively small at about 15 per cent; the rest being contributed by the companies (Rajya Sabha Secretariat, 2011). According to GEM (Global e waste Monitor) e waste is classified into six waste categories: (a) Temperature equipment (b) Screens, monitors (c) Lamps (d) Large equipment's (e) Small equipment's (f) Small IT and telecommunications (Balde et al., 2017).

E-waste contains hazardous metals which will be dangerous for the environment and also having direct or indirect impact on human health. It contains thousands of toxic substances like barium (Ba), Antimony (Sb), Cadmium (Cd), Arsenic (As), Beryllium (Be), Polyvinyl chloride (PVC), Chlorofluorocarbons (CFCs), Mercury (Hg), Lead (Pb)and Nickel (Ni)[4]. It can be disposed of inland or incineration of the waste will be common practice in India.

Remanufacturing is a method of transforming End of use product (EOU) to the given standards or specification of the original product. In remanufacturing the first step is the collection of E-waste through various collection points then next step in remanufacturing is disassembling waste, which is carried out by a skilled workforce. In the third step disassemble waste is sorted by various labours. After sortinginspection would be done to verify either the parts are in working condition or not, then cleaning andrefurbishment of the parts would happen. In next step the refurbished part is assembled properly move towards testing, if the final product fulfils all the standards or conditions of the testing then it will be ready for the market these are the steps involved in remanufacturing as shown in figure 110. 1. After studying existing literature on e-waste it is found that very few studies have been conducted in India on e-wasteremanufacturing. In this study, author find out the potential or critical aspects of e-wasteremanufacturing in Indian context and applied interpretive structural modelling (ISM) to catch out the driving and dependency of the factors.

Literature Review

Electronic waste contain a range of toxic substances which affects social health and contaminate the surroundings if proper disposal protocol are not managed (Kiddee et al.,2013). Remanufacturing is

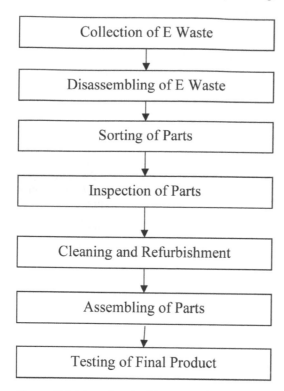

Figure 110.1 Steps involved in remanufacturing

developing a key plan which can decrease the influence on the environment (Singhal et al., 2018). In India management vision and collection centers are the key influential factors for e-waste remanufacturing (Singhal et al., 2019). Unclear government policies (Mukti and Rawani, 2016) and regulation, lack of management foresight and negligence of the environment are the main hurdles of remanufacturing (Singhal et al., 2018). The prominent external barriers of remanufacturing are the absence of collection channels for used product and consumer opposition to return the product, the casual barriers of remanufacturing are the timing of return product and uncertainty in quantity (Kumar and Dixit, 2018). In Malaysia, the potential challenges of remanufacturing are (a) Marketing and Competition (b) raw material collection (c) skill manpower and expert (d) product design (e) environmental and government (f) technology and method (Shamee,& Shamsuddin, 2019).

In India the major drivers (Animesh and Mukti, 2019) of remanufacturing can be categorised into three parts- (a) economic drivers (b) social drivers (c) environmental drivers (Mukti et al., 2014) is the potential economic driver of remanufacturing, the social driver of remanufacturing are more job opportunity and the best way of product recovery is the important environmental driver of remanufacturing (Sharma et al., 2016). Here Table 110. 1 shows the factor which affects e-waste remanufacturing in India, ten factors were taken by using literature survey and taking expert advice.

In 1973 J. N. Warfield has proposed interpretative structural modelling (ISM) which is used for finding out the circumstantial correlation among variables, it gives a basic idea about the different levels, also used for driving or dependence power among the factors. ISM is used in supplier selection (Beikkhakhian, 2015), remanufacturing (Singhal et al., 2019), knowledge management (Awan et al., 2018; Agrawal and Mukti, 2020), waste management, green manufacturing (Awan et al., 2018), manufacturing industries (Kumar et al., 2015), supply chain management (Raut et al., 2017). ISM needless expert than other Delphi or brainstorming technique.

Methodology

In this study, author has followed a series of steps, the first identification of critical factors which influence the e-waste remanufacturing then ISM model is used to find out the contextual relationship among factors.

Identification of factors

First, author have done a literature survey on the topics like e-waste management, remanufacturing, e-waste remanufacturing and find out the critical factors which effects the e-waste remanufacturing

Table 110.1 Factors affecting e-waste remanufacturing

S. No	Factors	Description	Author/Year
1	Timing of return (TR)	Uncertainty in timing of return	Singh and Srivastava (2018)
2	Assured warranty (AW)	Long term warranty and good service should be provided for the remanufactured product so that customers get attracted towards it.	Shi et al. (2019)
3	Marketing strategies (MS)	Low pricing, long term warranty, a proper advertisement should be done to promote the remanufactured product.	Singhal et al. (2019)
4	Government support (GS)	The government have to come forward and supply subsidy to both manufacturer and consumer to encourage remanufacturing.	Kumar and Dixit (2018); Kiddee et al. (2013)
5	Market access (MA)	Availability of remanufactured product to the market or the consumers	Sharma et al. (2016)
6	Workforce and technology (WT)	A skilled workforce and technology are important in every step of remanufacturing from disassembling of parts to final testing of the product.	Shamee and Shamsuddin (2019); Kumar and Dixit (2018)
7	Collection strategy (CS)	Collection of waste for remanufacturing is vital and critical because of quantity, quality and timing of return.	Singhal et al. (2019); Singh and Srivastava (2018)
8	Return intention (RI)	Consumer will to return the product or intension of consumer for returning there used product	Singhal et al. (2019); Sharma et al. (2016)
9	Consumption attitude (CA)	"one-time consumption" approach	Singh and Srivastava (2018)
10	Product design (PD)	The effectiveness and efficiency of remanufacturing depend on product design features like joining process, fastening and choice of material.	Shamee and Shamsuddin (2019)

in Indian context, then factors are sent to specialists for final assessment. Finally, ten factors have been identified which effect the remanufacturing practices in India. After factor identification, a set of questionnaires were made based on ISM model and the response of a group of experts were taken.

Interpretative structural modelling

ISM is a robust method, used for constructing logical relationships among variables to determine driving and dependence power of factor. In ISM series of steps were followed, first step is the identification of factors, structural self-interaction matrix (SSIM), reachability matrix, level partition and ISM model. In SSIM response of a group of experts were taken for developing logical relationships among factors, in this paper four symbols are used to represent the logical relationship between the factors (i and j): (a) V, if i effects j (b) A, if j effects i (c) X, if i and j both effects each other (d) O, if i and j having no effects on each other. According to responses of SSIM reachability matrix is formed by substituting binary digits in place of V, A, X and O, if the response is V then (i,j) will be 1 and (j,i) will be 0, if the response is A then (i,j) will be 0 and (j,i) will be 1, if the response is X then (i,j) will be 1 and (j,i) will be 1, if the response is O then (i,j) will be 0 and (j,i) will be 0. In level partition antecedent, reachability and intersection set for each factor are obtained from reachability matrix, factors which are common in both antecedent and reachability set lies in intersection set. If the reachability set and intersection set are the same then the factors are given top-level, successive iterations were done to get the levels of the factor, with the help of level partition construct an ISM model which shows dependence and driving power of factors.

Result and discussion

In this study, author identify the factors which effect the e-waste remanufacturing practices in India, ten factors are identified through previously published articles and expert's opinion etc. as shown in Table 110.1, after factor identification, logical relationship were developed by the experts using symbols as shown in Table 110.2. Thereafter researcher prepared a reachability matrix by replacing the symbol with the binary codes (1 and 0) as per the rules defined shown inTable 110. 3. Further, factors are divided into levels as shown in Table 110.4, finally, ISM based model (Figure 110. 2) is prepared by using levels of Table 110.4. ISM model divides the factors into six levels, top levels show strong dependence power and the bottom

Table 110.2 SSIM

Factors	TR	AW	MS	GS	MA	WT	CS	RI	CA	PD
TR		V	V	A	V	X	V	X	O	V
AW			X	A	V	X	O	O	V	X
MS				A	V	V	X	X	V	V
GS					V	X	V	V	V	V
MA						X	X	X	A	X
WT							V	O	V	V
CS								X	X	V
RI									X	V
CA										X
PD										

Table 110.3 Reachability matrix

Factors	TR	AW	MS	GS	MA	WT	CS	RI	CA	PD	DRP
TR	1	1	1	0	1	1	1	1	0	1	7
AW	0	1	1	0	1	1	0	0	1	1	6
MS	0	1	1	0	1	1	1	1	1	1	8
GS	1	1	1	1	1	1	1	1	1	1	10
MA	0	0	0	0	1	1	1	0	0	1	4
WT	0	1	0	1	1	1	1	0	1	1	7
CS	0	0	1	0	1	0	1	1	1	1	6
RI	1	0	1	0	1	0	1	1	1	1	7
CA	0	0	0	0	1	1	1	1	1	1	6
PD	0	1	0	0	1	0	0	0	1	1	4
DEP	3	6	6	2	10	7	8	6	8	10	

Table 110.4 Level partition

Factors	Reachability set	Antecedents sets	Intersection set	Level
TR	1,2,3,5,6,7,8,10	1,4,8	1,8	V
AW	2,3,5,6,9,10	1,2,3,4,6,10	2,3,6,10	III
MS	2,3,5,6,7,8,9,10	1,2,3,4,7,8	2,3,7,8	IV
GS	1,2,3,4,5,6,7,8,9,10	4,6	4,6	VI
MA	5,6,7,10	1,2,3,4,5,6,7,8,9,10	5,6,7,10	I
WT	2,4,5,6,7,8,10	1,2,3,4,5,6,9	2,4,5,6,9	III
CS	3,5,7,8,9,10	1,3,4,5,6,7,8,9	3,5,7,8,9	II
RI	1,3,5,7,8,9,10	1,3,4,7,8,9	1,3,7,8,9	II
CA	5,6,7,8,9,10	2,3,4,6,7,8,9,10	6,7,8,9,10	II
PD	2,5,9,10	1,2,3,4,5,6,7,8,9,10	2,5,9,10	I

level shows strong driving power. Government support (GS), timing of return (TR) and marketing strategies (MS) shows strong driving power. Market access (MA), product design (PD), collection strategies (CS), return intension (RI), consumption attitude (CA), assured warranty (AW) and workforce and technology (WT) shows strong dependence power as they are in top levels of ISM model. Government support (GS) is at level six of ISM model which shows it is one of the critical factors which strongly effects the e-waste remanufacturing , the government have to come onward and offer subsidy to both manufacturer and consumer to encourage remanufacturing.

   Further, level five comprises timing of return (TR) which shows strong driving power there is uncertainty in the timing of return of used product, which directly affect the remanufacturing. Marketing strategies are at level four which is in at the bottom level and shows driving power for that low pricing, long term warranty and proper advertisement should be done to promote the remanufactured product. At level two and three, assured warranty (AW), workforce technology (WT), collection strategy (CS), return intention (RI) and consumption attitude (CA) are having strong dependence power, this factor depends on the other

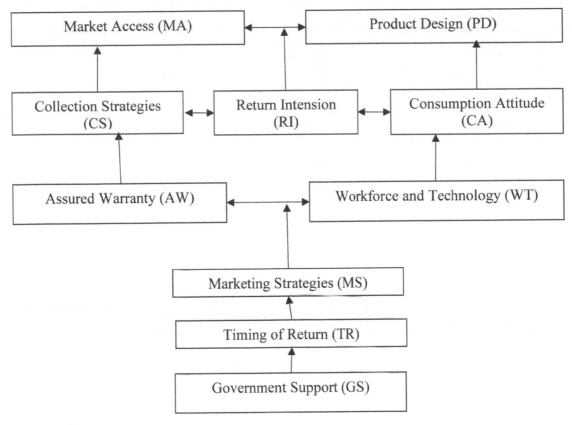

Figure 110.2 ISM model

factors which are at bottom of the model. Market access (MS) and product design (PD) are at level one in the model having a strong dependency on the other factor. Market access would be eliminated by the availability of remanufactured products to the market or the consumers more easily.

Conclusion

Remanufacturing is a method of transforming End of use product (EOU) or End of life product (EOL) to the given standards or specifications of the original product. This research aims to identify and analyse the critical factors affecting e-waste remanufacturing in India and find out the logical relationship between them by using interpretative structural modelling (ISM). Ten factors have been taken from the literature and experts in the respective field. Further, these ten factors are divided into six levels based on logical relationships among them or by using ISM model. The result shows that government support (GS), timing of return (TR) and marketing strategies (MS) are the major critical factors in the hierarchy that have strong driving power or these factors influences the other factor, market access (MA) and product design (PD) are at level one which means these factors depends on the all other factors. This study promotes the manufacturer for remanufacturing practices in India.

References

Agrawal, A. and Mukti, S. K. (2020). Knowledge Management & It's Origin, Success Factors, Planning, Tools, Applications, Barriers and Enablers: A Review. Int. J. Knowl. Manag. 16(1):43–82.

Animesh, A. and Mukti, S. K. 2019. Case Study of Critical Success Factors Affecting Knowledge Management in Small-and Medium-Sized Enterprises in Developing State: Steel Sector. In Advances in Industrial and Production Engineering. (pp. 825–831). Springer.

ASSOCHAM-NEC. 2018. India among the top five countries in e-waste generation : ASSOCHAM-NEC study. pp. 1–4.

Awan, U., Kraslawski, A., and Huiskonen, J. (2018). Understanding influential factors on implementing social sustainability practices in Manufacturing Firms: An interpretive structural modelling (ISM) analysis. Procedia Manuf. 17:1039–1048. doi: 10.1016/j.promfg.2018.10.082.

Balde, C. P., Forti, V., Gray, V., Kuehr, R., and Stegmann, P. 2017. The global e-waste monitor 2017: Quantities, Flows and Resources. Bonn, Geneva, and Vienna: United Nations University, International Telecommunication Union, and International Solid Waste Association.

Beikkhakhian, Y., Javanmardi, M., Karbasian, M., and Khayambashi, B. (2015). Expert Systems with Applications The application of ISM model in evaluating agile suppliers selection criteria and ranking suppliers using fuzzy TOPSIS-AHP methods. Expert Syst. Appl. 42(15–16):6224–6236. doi: 10.1016/j.eswa.2015.02.035.

Kiddee, P., Naidu, R., and Wong, M. H. 2013. Electronic waste management approaches : An overview. (33, pp. 1237–1250). doi: 10.1016/j.wasman.2013.01.006.

Kumar, A. and Dixit, G. (2018). An analysis of barriers affecting the implementation of e-waste management practices in India : A novel ISM-DEMATEL approach. Sustain. Prod. Consum. 14:36–52. doi: 10.1016/j.spc.2018.01.002.

Kumar, D., Agrawal, R., and Sharma, V. (2015). Enablers for Competitiveness of Indian Manufacturing Sector : An ISM-Fuzzy MICMAC Analysis,. Procedia - Soc. Behav. Sci. 189:416–432. doi: 10.1016/j.sbspro.2015.03.200.

Mukti, S. K., Tripathi, P., and Rawani, A. M. (2014). Identification of factors and indicators for success measurement of ERP system. Int. Proc. Econ. Dev. Res. 75:117.

Mukti, S. K. and Rawani, A. M. (2016). ERP system success models: A literature review. ARPN J. Eng. Appl. Sci. 11(3):1861–1875.

Rajya Sabha Secretariat. 2011. E-Waste in India. Res. Unit Rajya Sabha Secr. 24(6):293–297. doi: 10.2524/jtappij.24.6_293.

Raut, R. D., Narkhede, B., and Gardas, B. B. (2017). To identify the critical success factors of sustainable supply chain management practices in the context of oil and gas industries: ISM approach. Renew. Sustain. Energy Rev. 68:33–47. doi: 10.1016/j.rser.2016.09.067.

Shamee, A. and Shamsuddin, A. 2019. End-of-life Electrical and Electronic Equipment Remanufacturing Prospects in Malaysia. IOP Conf. Ser.: Mater. Sci. Eng. 530:012033. doi: 10.1088/1757-899X/530/1/012033.

Sharma V., Garg, S. K., and Sharma, P. B. (2016). Identi fi cation of major drivers and roadblocks for remanufacturing in India. J. Clean. Prod. 112:1882–1892. doi: 10.1016/j.jclepro.2014.11.082.

Shi, J., Zhou, J. and Zhu, Q. (2019). Barriers of a closed-loop cartridge remanufacturing supply chain for urban waste recovery governance in China. J. Clean. Prod. 212:1544–1553. doi: 10.1016/j.jclepro.2018.12.114.

Singhal, D., Tripathy, S., and Kumar, S. (2018). ScienceDirect ScienceDirect ScienceDirect ScienceDirect School for Costing models capacity optimization between capacity and operational efficiency. Procedia Manuf. 20:452–457. doi: 10.1016/j.promfg.2018.02.066.

Singhal, D., Tripathy, S. and Kumar, S. (2019). Sustainability through remanufacturing of e-waste : Examination of critical factors in the Indian context. Sustain. Prod. Consum. 20:128–139. doi: 10.1016/j.spc.2019.06.001.

Singh M. and Srivastava, R. K. (2018). Resources , Conservation & Recycling Analysis of external barriers to remanufacturing using grey-DEMATEL approach : An Indian perspective. Resour. Conserv. Recycl. 136:79–87. doi: 10.1016/j.resconrec.2018.03.021.

111 Diabetes prediction using deep learning and machine learning approach

S. W. Shende[a], Aniket Patil[b], Abiturab Vora[c], and Amisha Sherekar[d]

Information Technology Department Yeshwantrao Chavan College of Engineering (YCCE), Nagpur, India

Abstract

As the prevalence of diabetes is rising, and an increasing number of people are being afflicted on a daily basis. People must discover a remedy for this awful sickness. There are several technologies accessible in this generation for checking diabetes in patients, such as a glucometer. The patient's blood sugar level is checked with a glucometer, which determines if the sugar level is high or low. However, it is not only hazardous to the blood, but it also causes a variety of diseases such as blindness, renal disease, kidney trouble, heart disease, and other conditions that result in thousands of deaths each year. As a result, it is critical that it develops a system that can accurately diagnose diabetes patients based on their medical records. We offer an approach for the diagnosis of diabetes based on both machine learning and deep neural networks. UC Irvine's machine learning repository database contains the Pima Indian Diabetes (PID) data set, which was retrieved from there. The results on the PID dataset suggest that using a deep learning technique, an auspicious system for diabetes prediction can be designed, with a prediction accuracy of 85 percent, according to the findings.

Keywords: Deep neural network, machine learning, random forest.

I Introduction

Diabetic is a condition that affects people for a long time, and it's difficult to get rid of it after you've been diagnosed. Diabetes is caused by a person consuming an excessive amount of sugar. However, if a person's health is poor or they are over 40, you cannot assume that they will be harmed by diabetes. As people get older, their chances of developing diabetes increase. Diabetic issues affect the majority of adults over the age of 50. According to past research, diabetes will affect roughly 50 million individuals worldwide by 2040, making the new generation weaker than the previous generation. There is no healthy eating plan for the new generation. New varieties of junk food are now available on the market, influencing people's health and, as a result, increasing the diabetes patient's capacity. As a result, humans must exercise control over these (Santiago et al, 1978; Perveen et al., 2018).

The PIMA dataset, also known as the Pima Indian dataset, was used to train and test the model. This dataset is publicly available online. Both machine learning and deep learning methods are used to create the separate model a binary classifier is used in the machine learning approach to predict whether or not the output will be diabetic. In order to make accurate predictions, we employed the random forest classifier, which combines decision trees and gives us the results in the correct order. Predicted 86% and 78% accuracy levels were met. In deep learning approach feed forward neural network architecture is used.

II Related Studies

There are a number of new methods that have been developed in recent years. With the advancement of modern technology, a variety of new methods have been developed for the diagnosis of diabetes mellitus. The following is a brief summary of the work done in this field. For the prediction of diabetes mellitus, a machine learning system was examined by the author in (Islam et al., 2017). K-nearest neighbor (K-NN), artificial neural network (ANN), support vector machine (SVM), classification tree, and logistic regression are the algorithms employed by the systems. The following measures are employed in order to assess the overall performance of the system: accuracy, sensitivity, preciseness, specificity, true positive rate (TPR), false positive rate (FPR), rate of misclassification, F1 measure, and receiver operating characteristic (ROC) curve. The logistic regression algorithm has a maximum accuracy of 78% and an error rate of 0.22%. Precision and negative predictive values are 82% and 73%, respectively, when Nave Bayes and logistic regression are used. The dataset is cross-validated ten times. A study by Heydari et al. (2015) compared various diabetes categorisation methods. AI, Bayesian networks, decision trees and SVM, as well as K-NN have all been incorporated into this system. The system that performed the best and used an artificial neural network was able to achieve an accuracy rate of 97.44%. SVM, 5-nearest neighbors, decision trees, and

[a]swshende@ycce.edu; [b]aniket.patil@gmail.com; [c]voraabiturab786@gmail.com; [d]aameesha17feb@gmail.com

Bayesian networks had accuracy rates of 81.19%, 90.85%, 95.03%, and 91.60%, respectively. WEKA was employed as a simulation tool by the authors. By adopting the dropout method, Ashiquzzaman et al. (2017) suggested a deep learning-based framework for diabetes mellitus prediction. A dropout layer separates each of the two completely connected layers. A single node in the output layer determines the outcome. It achieved a maximum accuracy of 88.41% using the Pima Indian Diabetes (PID) dataset to test the system. Zhu et al. (2015) suggested an approach that uses many classifiers to improve the predictive accuracy of complicated diseases such as diabetes, by applying machine learning techniques. A dynamic weighted voting technique was developed for use in that system by the researchers. A T2DM data collection as well as a PID dataset are used to evaluate the system. On the PID dataset, the system achieved the greatest accuracy of 93.45% by employing MFWC with k = 10 and the MFWC algorithm.

III Methodology

In this section we discuss about the two different methodologies used to address the problem. First one is the machine learning approach where the feature extraction is required to carry out explicitly. We used random forest classifier for it. Secondly we build the series of deep neural networks to identify which combination gives us the best accuracy in terms of training and validation.

A. Machine learning approach

As depicted in Figure 111.1, initially the data-set is pre-processed. Feature selection is carried through the different nine parameters in the existing database and the extraction gets carried out. The data-set gets divided in 70:30 ratios for training and testing respectively.

For model preparation the various classification methods like KNN, Random forest classification, Gaussian Naïve Bayes classifiers being used in the past (Ayon and Islam, 2019; Sisodia and Sisodia, 2018; Annamalai and Nedunchelian, 2021; Zhou et al., 2020; Huang et al., 2007). In our model we choose a random forest classifier (Choubey and Paul, 2017; Çalişir and Doğantekin, 2011). The adoption of the random forest classifier is mostly due to the fact that it employs a variety of decision trees. Because it incorporates a variety of decision trees and calculates the average of all decision tree subsets, the accuracy of the model is improved.

B. Deep learning approach

A deep neural network (DNN) (Islam et al., 2017) is a complex neural network structure that consists of a neural network with several hidden layers between the input and output layers, as well as an input layer and an output layer. A DNN is designed for the purpose of forecasting results and detecting patterns and relationships within a data collection. In this case, multiple sorts of learning algorithms are employed in order to arrive at the final outcome. Similarly to the human neuron, the elements or nodes of neural networks are interconnected with one another. The precision of the output is dependent on the strength of the inter-unit connections. There are several hidden layers in a deep neural network, and there are numerous neurons in each hidden layer in a deep neural network. Figure 111.2 illustrates a straightforward deep neural network layout. The output of each layer of nodes is dependent on the output of the layer before it. It is common in neural networks for the output layer neurons to lack an activation function since the class labels are typically represented by the last output layer, which is why this is true.

In the operation of deep neural networks, the activation function is crucial. It is necessary for activation functions to incorporate intricate mapping characteristics they must be non-linear in order to incorporate

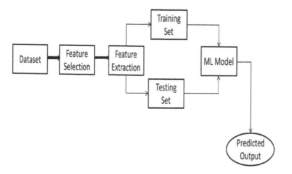

Figure 111.1 Machine learning based prediction model

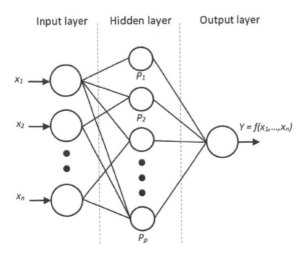

Figure 111.2 Densely connected two layer deep learning architecture

the much-needed non-linearity attribute that enables them to mimic any function, which is not always possible. Additionally, activation functions are critical in the suppression of the infinite linearly weighted sum from neurons, which is otherwise impossible. It is also vital to avoid large amounts of data accumulating at the top of the processing hierarchy.

$$Y = \Sigma(input * weight) + bias \tag{1}$$

We use the total six hidden layer of the DNN and each layer consist of 16, 12, 16, 12, 10, 2 number of neuron for this system. For diabetes prediction, we use numerous hidden layers and distinct neurons in separate layers. With the total six hidden layer and the number of neurons in each hidden layer is 16, 12, 16, 12, 10 and 8, we get the best results. The input layer is number eight, while the output layer is number one. In a neural network, neurons calculate the weighted sum of their inputs, add bias, and then decide whether or not they should be 'fired.'

Results and Discussion

A. Experimental setup

We employed the input layer, hidden layer, and output layer in a deep neural network to predict diabetes mellitus. For processing, we used a machine with an Intel Core i3 processor and 8 GB of RAM. We used Keras, an open-source machine learning toolkit written in the Python programming language, as well as the Google Colab.

B. Dataset

UC Irvine's machine learning repository database (Pima Indian Diabetes Data Set, 2018)is used to obtain the PID dataset. In the datasets, there are a number of features that acts as independent variables, along with one target (dependent) variable, which is the outcome variable. User's BMI, sugar level, and other factors such as their age and previous pregnancies are all considered independent variables in the analysis. The dataset's objective is to use diagnostic measures included in the collection to determine whether a patient has diabetes. Numerous constraints applied to the selection of these examples from a larger database. All patients are Indian women over the age of 21.We have used a total of four PID datasets according to the year format from 2010 – 2013. The dataset has a total of 768 records and a total of 8 feature parameter on which diabetes is predicted. Some of the important parameter which predicts diabetic disease is Pregnancy, Glucose, Blood Pressure, Skin Thickness, Insulin, BMI, Diabetes Pedigree Function, and Age. is given below and statistics of the same shown in Figure 111.3.

C. Result analysis

The training models were compared based on the performance of each of the classifier methods after executing trials on the generated training set. The following is a comparison of the previously specified

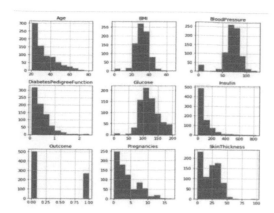

Figure 111.3 Statistics of feature set

Table 111.1 Classification results

ML classifier	Accuracy
Gaussian Naïve Bayes	0.78
KNN	0.77
Decision Tree	0.70
Random forest DT	0.86

Table 111.2 Deep neural network configuration and metrics

Model	Dense Layers-configuration(# Neurons)	Trainable parameters	Training accuracy %	Validation accuracy %
1	DL-1(12), DL-2(12), DL-3(16), DL-4(16),DL-5(14), DL-6(2)	1012	0.78	0.79
2	DL-1(16), DL-2(12), DL-3(16), DL-4(12),DL-5(10), DL-6(2)	912	0.80	0.86
3	DL-1(18), DL-2(16), DL-3(16), DL-4(12),DL-5(10), DL-6(2)	1094	0.79	0.77
4	DL-1(64), DL-2(128), DL-3(64), DL-4(32),DL-5(16), DL-6(8), DL-7(2)	19914	0.96	0.70
5	DL-1(12), DL-2(16), DL-3(16), DL-4(14),DL-5(2)	856	0.78	0.78

accuracy measures that were finally acquired. We achieved the maximum accuracy of 86% and minimum up to 80% using Random forest decision tree.

In deep learning approach we used conventional deep neural network with different number of fully connected dense layers with specific number of neurons in each layer. The number of epoch is constant = 200 for all our 5 experimentation as shown in Table 111.2. For the experimentation, the various parameters used are as follows: the Adams optimizer, learning rate = 0.0001, categorical cross entropy loss, and metric used is accuracy for both validation and training. The results are shown in Figure 111.4.

Model #   Training Validation Accuracy   Training Validation Loss

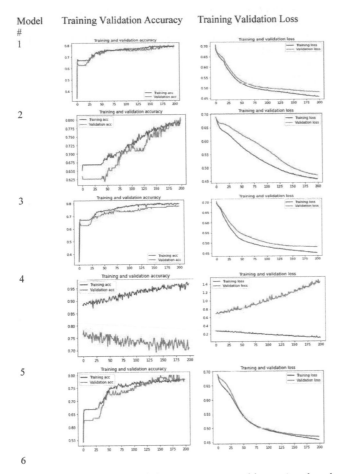

Figure 111.4 Training-validation accuracy and loss using deep learning model

Conclusion

In machine learning approach the adoption of the random forest classifier that employs a variety of decision trees, improves the result of classification. As it consist of various decision trees and takes the average of all decision tree subsets, resulting in increased accuracy. Moreover in deep learning approach we got the accuracy more than 85%. We observed that for highly dense hidden layers with large number of neurons the testing accuracy reached up to 0.96%, (model #4) with large number of learnable hyper parameters. We tested in total five different configuration of deep neural networks, and founds to be the optimal one with configuration of six hidden layer and the number of neurons in each hidden layer is 16, 12, 16, 12, 10 and 8 (model #2). By adjusting the different hyper parameter like number of epochs, learnable rate, optimiser and batch normalisation the accuracy rate can be improved further.

References

Annamalai, R. and R. Nedunchelian. (2021). Diabetes Mellitus Prediction and Severity Level Estimation Using OWDANN Algorithm. Computat. Intellig. Neurosci. 2021. https://doi.org/10.1155/2021/5573179

Ashiquzzaman, A., Tushar, A. K., Islam, M., Kim, J.-M. (2017). Reduction of overfitting in diabetes prediction using deep learning neural network. arXiv preprint arXiv:1707.08386.

Ayon, S. I. and Islam, M. (2019). Diabetes Prediction: A Deep Learning Approach. Int. J. Info. Eng. Electronic Bus. 11:2.

Çalişir, D. and Doğantekin, E. (2011). An automatic diabetes diagnosis system based on LDA-Wavelet Support Vector Machine Classifier. Expert Syst. Appl. 38(7):8311–8315.

Choubey, D. K. and Paul, S. (2017). GA_RBF NN: a classification system for diabetes. Int. J. Biomed. Eng. Technol. 23(1):71–93.

Dwivedi, A. K. (2017) Analysis of computational intelligence techniques for diabetes mellitus prediction. Neural Comput. Appl. 13(3):1–9.

Heydari M., Teimouri, M., Heshmati, Z., and Alavinia, S. M. (2015). Comparison of various classification algorithms in the diagnosis of type diabetes in Iran. Int. J. Diabetes Develop. Countries. 17.

Huang, Y., McCullagh, P., Black, N., and Harper, R. (2007). Feature selection and classification model construction on type 2 diabetic patients' data. Artif. Intell. Med. 41(3):251–262.

Islam, M. M., Iqbal, H., Haque, M. R., and Hasan, M. K. (2017). Prediction of Breast Cancer using Support Vector Machine and K-Nearest Neighbors. In Proc. IEEE Region 10 Humanitarian Technology Conference (R10- HTC), (pp 226–229). Dhaka.

Perveen, S., Shahbaz, M., Keshavjee, K., Guergachi, A. (2018). Metabolic syndrome and development of diabetes mellitus: Predictive modeling based on machine learning techniques. IEEE Access. 7:1365–1375.

Pima Indian Diabetes Data Set, 2018. Available: https://archive.ics.uci.edu/ml/datasets/Pima+Indians+Diabetes. (Accessed May 1, 2018).

Santiago, J. V., Davis, J. E., and FISHER, F. (1978). Hemoglobin A1c levels in a diabetes detection program. J. Clinic. Endocrinol. Metabol. 47(3):578–580.

Sisodia, D. and Sisodia, D. S. (2018). Prediction of diabetes using classification algorithms. Proce. Comput. Sci. 132:1578–1585.

Zhou, H., Myrzashova, R. and Zheng, R. (2020). Diabetes prediction model based on an enhanced deep neural network. Eurasip J. Wirel. Commun. Netw. 2020:113.

Zhu, J., Xie, Q., and Zheng, K. (2015). An Improved Early Detection Method of Type-2 Diabetes Mellitus Using Multiple Classifier Systems. Info. Sci. 29:114.

112 Performance evaluation of different hole profiles using 2-D CFD model

Ashish Pawar[1], Dinesh Kamble[2], and D.B. Jadhav[3]

[1]Mechanical Engineering Department, VIIT, Pune, India

[2]Professor, Mechanical Engineering Department, VIIT, Pune, India

[3]Asst. Prof. Bharati Vidyapeeth Deemed to be University College of Engineering, Pune, India

Abstract

The impact of turbine inlet temperature on overall turbine performance is critical. The increase in turbine inlet temperature leads to an increase in the Turbine's overall output. As a result, complex ways for increasing the turbine inlet temperature must be devised. However, due to the current material property restriction, increasing the turbine blade temperature to a higher level without cooling is not conceivable. Hence, the appropriate cooling solutions must be devised to enhance the turbine inlet temperature. As a result, one of the techniques is to use perforations to cool the blades. The current paper examines the effects of varied hole profiles on turbine blade cooling using CFD. In this research, computational and theoretical methods were used to analyze geometries with four different types of hole profiles (planar, semi-circular, square, and triangular). According to the results of the experiment, square hole profiles were the most effective, followed by triangular hole profiles. The square hole profile has the highest heat transfer coefficient and performance efficiency factor. The aircraft's operating efficiency can be improved by boosting thermal efficiency or propulsive efficiency. Thermal efficiency can be improved by employing high strength temperature resistant lightweight materials with a thermal barrier coating and the best cooling holes. The inlet temperature was discovered to have a significant impact on the overall performance of the turbine. Providing the best cooling holes is one of the most effective ways to operate the turbine at high temperatures. The hole profile was discovered to be an important factor in turbine cooling.

Keywords: CFD, efficiency factor, performance, thermal efficiency, turbine blade cooling.

Introduction

A gas turbine is a form of internal combustion (IC) engine in which an air-fuel mixture is burned under specific conditions, resulting in hot gases that provide power. The compressor, combustion chamber, and turbine are the three parts of a gas turbine engine. When high-temperature, high-pressure exhaust gases move through a turbine stage, energy is extracted from the flow, and the temperature and pressure of the gas are reduced (Han, 2004). The blade's design is critical because it is the component that is subjected to the most stress in the system. Any item in close proximity to a moving fluid generates lift, which is an aerodynamic force perpendicular to the flow (Dyson et al., 2014; Brahmaiah and Kumar, 2014). The GT blade has an aerofoil form because aerofoils are more effective lifting geometries that can generate greater lift while reducing drag. With the surge in power consumption, enhancing turbine efficiency is becoming increasingly important. As a result, gas turbine blade cooling is required to reduce corrosion, fuel consumption, and weight while also increasing thrust and efficiency (Singh and Shukla, 2015, Soghe and Andreini, 2013). A temperature increase of 16°C above the critical temperature of turbine blade material could reduce the life to half of its original value. Gas turbines are also subjected to high mechanical loads, such as centrifugal and gas pressure forces (Gao et al., 2008; Liu et al., 2010; Kohli and Thole, 1997). Improved designs include film cooling, internal cooling, impingement flow thermal barrier coating, or a mix of all to overcome this and extend the turbine's life. Film cooling allows the turbine to operate above the melting point of the material (Shen et al., 1996; Tartinville and Hirsch, 2009; Terrell et al., 2008; Lu et al, 2007). A tiny layer of air is formed on the blade's external surface, reducing heat transfer (Wang et al., 2019). Hence, the provision of cooling holes is necessary for cooling the turbine to extend its life and reduce corrosion (Easterby et al., 2021; Hosseini, 2021; Pawar et al., 2021). With the advent of advanced manufacturing, intricate shapes can be created. Electrochemical Machining (EMG) is one such technique which can be used to manufactures holes of different shapes in super-alloys used in turbines (Sharma, 2017). Advanced computational techniques can be used to predict the effectiveness of these micro holes (Oliver et al., 2019; Bhattacharyya, et al., 2002). The heat transfer rates vary depending upon the type and shape of these cooling holes. In this study, the CFD analysis of the plane, triangular, square and semi-circular holes is performed to determine the best hole shape in terms of heat dissipation. The heat transmission coefficient (h) and Nusselt Number (Nu) for each hole profile are determined in this study using Ansys Fluent.

I Numerical Approach

A. Preparation of computational domain

The computational domains of the hole profiles were designed on the ANSYS 18 design modeler tool. Figure 112.1 shows the dimensions of the hole profiles. On the other hand, Figure 112.2 (a), (b), (c), (d) depict the computational domain geometries of the plain, triangular, square and semi-circular hole profiles respectively. As shown, the hole diameter is 2 mm with the height as 38 mm. The hole contours have a maximum diameter of 3 mm. These dimensions were selected based on the range of dimensions that can be manufactured using the EMG process (Sharma, 2017). The simulations were performed on a workstation computer with six CPU cores.

The computational domain was designed in such a way that the hole profiles were assigned the solid material present in the study and the hole portion was assigned fluid properties for the flow of the cooling fluid. Appropriate geometry contacts were generated in order to efficiently simulate the convection.

B. Meshing

Simulation results mostly dependent on the meshing pattern and quality. In this context the 2-D elements are selected with suitable edge length in such a way that the plots of stress gradients can be precisely obtained. The mesh was generated using the Automatic method in ANSYS 18 which blends quadrilateral and triangular 2-D elements according to the spatial orientation of the geometry. Figure 112.3 shows the meshed geometry of the Semi-circular hole profile.

Table 112.1 shows the mesh metrics for the above meshed model. Face size is varied in the range of 0.08 to 0.16 with the increment of 0.02 and element quality has been checked. Aspect ratio and the skewness criteria of element quality check was highly satisfied with face size of 0.08. The number of nodes and elements obtained with the 0.08 face size are 32781 and 31028 respectively. Increase in the number of nodes and elements leads to higher processing time.

C. Grid independence study

Grid independence study is accomplished to select the best maximum face size with exceptional mesh quality and minimum number of nodes and elements. Figure 112.4 shows the plot between the average

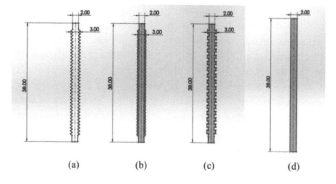

Figure 112.1 Dimensions of hole profiles (a) triangular, (b) semi-circular (c) square, (d) plane

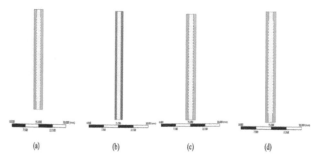

Figure 112.2 Boundary conditions applied to (a) triangular, (b) plane (c) square, (d) semi-circular

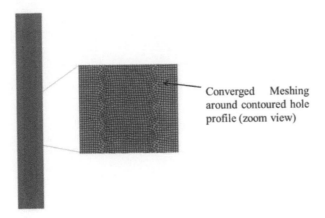

Converged Meshing around contoured hole profile (zoom view)

Figure 112.3 2-D meshing of semi-circular hole geometry

Table 112.1 Mesh metrics for plane hole meshed geometry

Max. Face Size		*0.08*	*0.1*	*0.12*	*0.14*	*0.16*
Elemental Quality	min	0.35025	0.30238	0.34337	0.40547	0.29445
	max	0.99972	0.99989	0.99979	0.99935	0.9996
	average	0.90821	0.89125	0.89597	0.88295	0.89221
Aspect Ratio	min	1	1	1	1.0001	1
	max	4.4239	2.421	2.3825	2.399	2.3652
	average	1.0958	1.1067	1.1111	1.1193	1.1054
Skewness	min	1.31E-10	1.31E-10	5.24E-10	1.34E-04	4.77E 04
	max	0.64383	0.7077	0.67134	0.61165	0.71677
	average	0.10382	0.12249	0.11833	0.13254	0.12323
No. of Elements		31028	19945	14310	10253	7501
No. of Nodes		32781	21283	15469	11232	8317

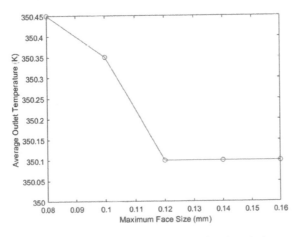

Figure 112.4 Grid independence study for plane hole geometry

outlet temperature and the maximum face size. It is evident from the graph that value of average outlet temperature 350.35 K is constant during 0.08 – 0.12 mm range. Thereafter, it reduces with increase in face size. Hence, the best face size value is 0.12 mm as it provides accurate results with a comparatively smaller number of nodes and elements. The meshing method is software-based automatic method which is an amalgamation of quadrilateral and triangular elements. This meshing method is appropriate for this study due to the simplicity of the geometry.

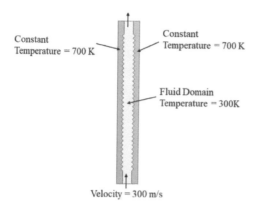

Figure 112.5 Loads and boundary conditions applied to turbine blade

D. Design of computational domain and boundary conditions

Figure 112.5 shows the boundary conditions that are applied to the computational domain. Air with 300m/s velocity and 300 K temperature is taken at the inlet. The outer solid body temperature is kept constant as 700 k. The k-ε turbulence model is employed with second order upwind spatial method, the k-ε turbulence model was chosen due to its suitability with analysis consisting of free surface flow region (Bhattacharyya et al., 2002; Zhu et al., 2019). The results were extracted using the ANSYS post-processing module.

Results and Discussion

The cases were run on a convergence criterion of residual count of less than 0.001. The cases converged after 3000 iterations and were transferred to the ANSYS post processing tool to analyse the results. As shown in Figure 112.6 (a), the temperature contours for the plane hole profile, the average outlet temperature upon probing is 350.46 K. Similarly, for the triangular, square and semi-circular profiles, the average outlet temperatures are 367.076 K, 365.267 K and 370.4 K respectively.

(i)   It can be seen that the semi-circular profile has better heat dissipation than other hole profiles as it is able to expel more heat to the cooling fluid. This is due to the larger surface area exposed to the cooling fluid. The plain hole profile has the least average outlet temperature due to the absence of intricate surface features which enhance the heat dissipation. The triangular and the square hole profiles are equally effective in the heat transfer process. Qualitative analysis is performed by determining the heat transfer coefficient and Nusselt number. This calculation is essential to determine the efficiency of heat transfer for the individual hole profiles.

The determination of the heat transfer coefficient ad the Nusselt Number of the hole profiles is essential to determine the thermal dissipation of the individual hole profiles. As discussed before, the cooling of the turbine blade is essential to improve the life and performance of the turbine. The heat transfer coefficient of the hole profiles was determined using the formulas given below.

$$Q = mCp\Delta T \ ... \ eq. \ (i)$$

$$Q = hA\Delta T \ .... \ eq. \ (ii)$$

where, $\Delta T1$ = (Tout – Tin), the difference between the outlet and the inlet temperature of the cooling fluid and $\Delta T2 = (Tsur - Tb)$, which is the difference between the surface temperature (700 K) and the average of the outlet and the inlet temperature, i.e. Tb.

$$Nu = ... \ eq. \ (iii)$$

The Nusselt Number (Nu) is calculated using the formula given below.

After probing these values from the post processing module, the values of the heat transfer coefficient were determined and are tabulated in the Figure 112.7. It is evident that the square hole profile has the

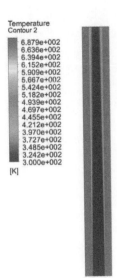

(a) Plane Hole Profile

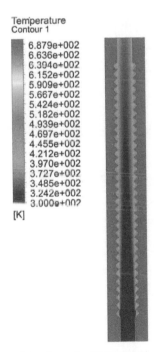

(b) Triangular Hole Profile

Figure 112.6 A and B temperature contours for hole profiles

maximum heat transfer coefficient of 459.25 (W/m2K). The triangular and semi-circular hole profiles have heat transfer coefficients of 344 and 313 (W/m2K), respectively. The plain hole profile has the least heat transfer coefficient of 258 (W/m2K). The Nusselt Number was calculated for every hole profile and the square profile had the maximum Nusselt Number of 87.258 followed by triangular, semi-circular and plane hole profiles with Nusselt Numbers of 65.38, 59.622 and 49.0651, respectively. This data is shown in Figure 112.8. It is clear that the square profile has the best heat transfer properties as compared to other hole profiles.

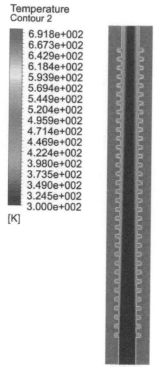

(c)Square Hole Profiles

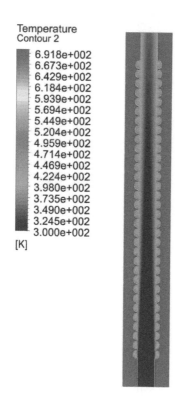

(d) Semi-circular Hole Profiles

Figure 112.6 C and D Temperature contours for hole profiles

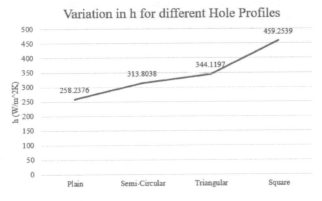

Figure 112.7 Variation in h for different hole profiles

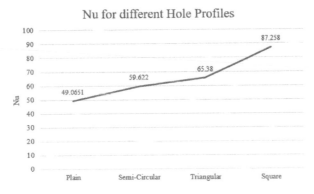

Figure 112.8 Variation in Nu for different hole profiles

Conclusion

In this study, the hole profiles for turbine cooling are analysed using CFD. The square, triangular, semi-circular and plane hole profiles are designed and computational models are created. Later, grid independence study is conducted to find the best element size for the meshing. The boundary conditions are applied and the k-ε turbulence model is used to solve the cases. After studying the post-processing results, it is evident that the semi-circular hole profile has good heat dissipation property as the outlet temperature of the cooling fluid is 370 K, due to a large surface area. Upon calculation of the heat transfer coefficient and the Nusselt Number, the square hole profile is the best hole profile amongst the others as it has the largest heat transfer coefficient of 459.25 and the largest Nusselt Number of 87.258. The triangular hole profile is the second-best hole profile for such an application. The plane hole profile not as efficient in terms of thermal properties as compared to the other hole profiles. This shows that the square profile holes could be an effective solution to enhance the performance of the turbine blades. This can be implemented in turbines using the EMG technique.

References

Bhattacharyya, B., Mitra, S., Boro, A. K. (2002). Electrochemical machining: new possibilities for micromachining. Robotics Comput. Integ. Manufact. 18(3–4):283289.

Brahmaiah, K. H. and Kumar, M. L. (2014). Heat Transfer Analysis of Gas Turbine Blade Through Cooling Holes. Int. J. Comput. Eng. Res. 4(7):2250–3005.

Dyson, T. E., Bogard, D. G., and Bradshaw, S. D. (2014). Evaluation of CFD simulations of film cooling performance on a turbine vane including conjugate heat transfer effects. Int. J. Heat Fluid Flow. 50:279–286. doi: 10.1016/j.ijheatfluidflow.2014.08.010.http://www.ijceronline.com.

Easterby, C. C., Moore, J. D., and Bogard, D. G. (2021). CFD Evaluation of Internal Flow Effects on Turbine Blade Leading-Edge Film Cooling With Shaped Hole Geometries. https://doi.org/10.1115/GT2021-59780

Gao, Z., Narzary, D. P., Han, J. C. (2008). Film cooling on a gas turbine blade pressure side or suction side with axial shaped holes. Int. J. Heat Mass Transfer. 51(9–10):21392152

Han, J. C. (2004). Recent Studies in Turbine Blade Cooling. Int. J. Rotating Machin. 10. https://doi.org/10.1155/S1023621X04000442

Hosseini, E. (2021). Optimization of cooling structures in gas turbines: A review. J. Mech. Eng. Sci. 35(6):1846.

Kohli, A. and Thole, K. A. (1997). A CFD investigation on the effects of entrance crossflow directions to film-cooling holes. American Society of Mechanical Engineers, Heat Transfer Division. 350:223232.

Liu, J. S., Malak, M. F., Tapia, L. A., Crites, D. C., Ramachandran, D., Srinivasan, B., Muthiah, G., and Venkataramanan, J. (2010). Enhanced Film Cool. Effectiveness New Shaped Holes. 15171527.

Lu, Y., Allison, D., Ekkad, S. V. (2007). Turbine blade showerhead film cooling: Influence of hole angle and shaping. Int. J. Heat and Fluid Flow. 28(5):922931.

Oliver, T. A., Bogard, D. G., Robert, Moser, D. (2019). Large eddy simulation of compressible, shaped-hole film cooling. Int. J. Heat Mass Transfer. 140:498–517.

Pawar, A., Kamble, D., and Ghorpade, R. R. (2021). Overview on electro-chemical machining of super alloys. Mater. Today Proc. 46(40):696–700. doi: 10.1016/j.matpr.2020.12.017.

Sharma, A. 2017. Introduction to Computational Fluid Dynamics Development, Application and Analysis. John & Wiley sons. doi:10.1002/9781119369189

Shen, J. R., Wang, Z., Ireland, P. T., Jones, T. V., Byerley, A. R. (1996). Heat Transfer Enhancement Within a Turbine Blade Cooling Passage Using Ribs and Combinations of Ribs With Film Cooling Holes. J. Turbomach. 118(3):428434.

Singh, P. and Shukla, V. (2015). Heat Transfer Analysis of Gas Turbine Rotor Blade Cooling Through Staggered Holes using CFD. Int. J. Eng. Res. V4(7):538–545. doi: 10.17577/ijertv4is070261.

Soghe, R. D. and Andreini, A. (2013). Numerical characterization of pressure drop across the manifold of turbine casing cooling system. J. Turbomach. 135(3):1–9. doi: 10.1115/1.4007506.

Tartinville, B. and Hirsch, C. (2009). Modelling of Film Cooling for Turbine Blade Design. 22192228. https://doi.org/10.1115/GT2008-50316

Terrell, E. J., Mouzon, B. D., and Bogard, D. G. (2008). Convective Heat Transfer Through Film Cooling Holes of a Gas Turbine Blade Leading Edge. 833844. https://doi.org/10.1115/GT2005-69003

Wang, W., Pu, J., and Wang, J.- H. (2019). An experimental investigation on cooling characteristics of a vane laminated end-wall with axial-row layout of film-holes. Appl. Thermal Eng. 148:953962. https://doi.org/10.1016/j.applthermaleng.2018.11.104

Zhu, X. D., Zhang, J. Z., Tan, X. M. (2019). Numerical assessment of round-to-slot film cooling performances on a turbine blade under engine representative conditions. Int. Commun. Heat Mass Transfer. 100:98110.

113 Prediction of MRR in EDM process using ANN model

Rohit Paliwal[a], Sanchit Dhasmana[b], Rishabh Singhal[c], Sandeep Kumar Singh[d], and Navriti Gupta[e]

MED, Delhi Technological University, Delhi, India

Abstract

Electric discharge machining is known for its outstanding potential to cut hard materials, economic good surface finish which is the limitation of many conventional machining techniques. But excessive tool wear and low material removal rate (MRR) are the challenges in electric discharge machining. Its prediction is possible by using Artificial neural network (ANNs) to achieve better results. The software used to predict MRR was MATLAB using NNTOOL. A feed-forward backdrop neural network with a back-propagation algorithm is used in this study to predict the MRR in the EDM process. The dataset we used was L24 and the parameters used as input variables were Current, Spark on Time, Spark off Time, and MRR as the output variable. To prevent resubstituting errors, the data was split into two datasets labeled as a training and a testing dataset. The splitting method we used to spilt our data in training, testing and validation is hold-out validation technique. The model has the highest accuracy of 99.23% was achieved at the 18th epoch. The training function we used was TRAINBR and the adaptive learning function is learngd. The activation function of our model is Tansigmoid and the performance function used was mean squared error.

Keywords: Artificial neural network, electric discharge machining, feedforward back-propagation, material removal rate.

Introduction

Electric discharge machining is a non-traditional, non-conventional machining technique for cutting extremely hard materials. This method is also popular with other names like spark Machining, Wire erosion, wire burning, spark eroding and a few more (Ho and Newman, 2003; Marafona and Chousal, 2006; Abbas et al., 2007; Pérez et al., 2000; Rao et al., 2021). The metal from the workpiece is removed utilizing a high-frequency electrical spark discharge from graphite in this procedure however, suitable only for a good conductor or metallic workpiece.

Electric discharge machining works on the principle of spark discharge between workpiece and tool. An electric discharge occurs when two current-carrying wires, which function as anode and cathode, are placed closer enough, i.e. workpiece and tool, causing a little quantity of metal to be eroded at the point of contact, resulting in small cracks. This spark produces a tremendous amount of heat, causing the metallic part inside the sparking region to melt and evaporate (Abu Qudeiri et al., 2018; Natarajan et al., 2016; Gaikwad and Jatti, 2018; Bhatt and Goyal, 2019). If both electrodes are composed of the same material, the electrode linked to the positive pole erodes quicker than the electrode negative electrode terminal. The tool is shaped similarly to the imprint that must be formed on the work. An EDM system consists of a power generator (to convert alternating current to direct current), a tool, a servo motor to regulate tool feed, and dielectric fluid (Abu Qudeiri et al., 2018).

Few applications of EDM are Micro-hole drilling in nozzles, gear wheel manufacturing, rotary form cutting, fine holes or slots in the hard material used for turbine and compressor blades. A few of the advantages of the EDM process over conventional machining techniques are that complex shapes can be produced, high tolerance, an excellent surface polish may be obtained at a low cost, and there is no deformation or disturbance of the tool or work, among other benefits. But few cons are also there which include high power consumption, excessive tool wear, low material removal rate (MRR) (Mahardika et al., 2008; Dave et al., 2012; Somashekhar et al., 2010).

Artificial neural network (ANNs) is the study that gives computers the ability without being explicitly programmed. Thus, the capacity to enhance behavior based on experience is the application of ANN. It explores algorithms that learn from given data, fit a model onto a training set of data that minimises the error on the training set and these could be utilised for a variety of activities such as decision making, prediction, problem-solving, and so on (Arunadevi and Prakash, 2021).

ANN works with the models and given a new problem or task it turn up with the problem's solution along with a function (f). ANN is being used in big data cloud, self-driving cars, autonomous robot control, natural language processing, and many more (Naveen Babu et al., 2019; Wang et al., 2019; Ganapathy et al., 2019; Singh et al., 2020).

Using ANN an accurately predictive model can be built that helps us to predict the variables in EDM namely MRR, electrode wear rate (EWR), surface quality (SQ) (Naveen Babu et al., 2019; Singh et al., 2020). Heat affected zone, dimension accuracy, total machining time, micro crack density, etc. different optimisation techniques such as the Taguchi method, genetic algorithm technique, etc. helped for the prediction of performance parameters of the EDM process follow.

Literature Review

Fenggou and Dayong (2004) describe a technique as a method for automatically determining and optimizing processing parameters in the EDM sinking process using ANN in their research paper. They determined that the genetic technique and back propagation methods may efficiently train neural networks and that the node deletion algorithm could scientifically determine and optimise the topology of an ANN. Their study demonstrates that using an ANN to automatically determine and optimise EDM sinking process parameters, using the current peak value as the core, is useful.

Gao et al. (2008) presented a method for optimising EDM process parameters that combine the Levenberg-Marquardt algorithm and genetic algorithms. To depict the link between MRR and input parameters, they created an Artificial Neural Network model using this method as well as a 3-26-1 network design. It demonstrated that the net performs better in terms of generalisation and convergence speed. They concluded that the parameters that were optimised using the genetic algorithm improved MRR.

Panda and Bhoi (2005) determined the Material Removal Rate utilizing an artificial feed-forward neural network. Their study was grounded on the Levenberg-Marquardt backpropagation approach with the suitable design of the logistic sigmoid activation function (MRR).

Padhee et al. (2012) studies the impact of EDM on the SR and MRR of an Al/SiC metal matrix composite. The authors investigated the link among four input variables: pulse-on duration, pulse peak current, mean gap voltage, SiC percent volume fraction, and process outputs. Researchers also discovered the best circumstances for NSGA-II outputs. For such SR and MRR, practical models were created using machining characteristics such as pulse-on time, pulse-off time, and discharge current.

Golshan et al. (2012) employed the NSGA-II multi-objective optimisation approach to get feasible solutions. The NSGA-II had also been utilised to improve the outputs of the EDM technique while utilising a powder-mixed dielectric. Investigators employed numerical models to forecast MRR and SR based on four process variables: discharge current, powder concentration (silicon) inside the dielectric liquid, pulse-on duration, and duty cycle, with EN-31 tool steel as that of the workpiece. The RSM has been used to investigate the influence of the control variable on answers and to create predictive models.

Choudhary et al. (2014) draws comparison on the work of various authors utilizing the latest optimisation techniques and different process and performance parameters in EDM for obtaining cost-effective manufacturing methodology for the industries.

Joshi and Pande (2010) developed an integrated(FEM-ANN-GA) model to select optimum EDM process parameters and found the approach very efficient as the optimum process parameters suggested were giving an expected performance of the EDM.

Tsai and Wang (2001) created an ANN model of the EDM process and utilised it for the optimisation of the input parameters by using GA, also worked on a methodology for optimisation of the EDM process for various workpiece and tool material pairs.

Wang et al. (2003) examined several ANN training techniques for MRR and surface roughness prediction. The authors observed the results as the adaptive-network-based fuzzy inference system (ANFIS) using bell-shaped membership function performed much better for MRR predictions however ANFIS, radial basis function neural network (RBFE) and tangent multi-layered perceptrons seem to be additional accurate in roughness prediction.

Su et al. (2004) created a mixed version of the EDM process that predicts MRR and surface roughness using ANN and GA. The synaptic weightages were optimised using GA.

Methodology

ANN has been applied to simulate the performance parameters of the EDM process on several occasions. As of now, the ANN parameters like- the learning and the training algorithm, amount of hidden layer neurons, and so on have been randomly adjusted to produce an effective ANN model equation. As a result, an attempt has been made in this paper to construct the complete factorial design to acquire the best ANN process parameter values.

An ANN model's model equation (for a single hidden layer model) may be stated using a weight and bias matrix and is expressed as:

$$a^2 = f^2(W^2 f^1(W^1 p + b^1) + b^2) \qquad (1)$$

Where a 2 is the second layer's output vector, f 1 is the transfer function, W1 and W2 are the weight matrices of the hidden and output layers, respectively, p is the input vector, and b1 and b2 are the bias vectors of the first and second layers, respectively.

A. Variations in the setting of ANN parameters

The parameters of the procedure in numerous studies, ANN topologies, learning/training techniques, and the number of hidden neurons have been modified to achieve an efficient ANN model. The importance and significance of all these process parameters to the ANN model, though, have not even been adequately proven. To evaluate the effectiveness and importance of ANN topologies, learning/training techniques, and the number of hidden neurons on an ANN model, these process parameters were chosen as process parameters in a complete factorial design method. We used hit and trail method to set the various parameters for training the model and it was observed that number of neurons were effecting the no. of epochs in training the data model and to increase the no. of epoch we took no. of neurons in hidden layer as 10 in count.

B. Training, validation, and testing data

The entire factorial design process was used for a total of 27 experimental runs. Training and testing datasets were created from the 24 sets of experimental data, all these 24 rows data was used in training as well as testing of model using nntool in Matlab and remaining three rows of data from the L27 data set were used for validation of our created model. This training data set is utilised to approximate that function among input and output parameters. This ANN training was terminated based on the early stopping criterion determined by validator data. This test dataset would be the unseen data to the trained model that was used to evaluate a properly trained model's performance and generalisation error. Generalisation refers to how well the trained model approximates an unknown data set, and early stopping conditions are critical for lowering generalisation error.

ANNs use backward propagation of error learn from their mistakes and give optimum output. This error is calculated in a supervised training phase where the actual output and predicted output are compared. The weight of the attributes is adjusted so that this difference is reduced using backpropagation.

We will make use of an activation function which is used to calculate the weighted sum of input and biases using the gradient descent method. The sigmoid function is used as the activation function, it is defined as:

$$A = \frac{1}{1 + e^{-x}} \qquad (2)$$

Validation technique

To prevent resubstituting errors, the data was split into two datasets labeled as a training and a testing dataset. A total of 89% of the entire data i.e. 24 rows were used for training and testing and the remaining 11% of the data set i.e. three rows were used for validation of model. The training and testing of the data was performed on L24 data out of L27. The total validation checks were ten as shown in Figure 113.3. This splitting method is called the hold-out validation technique. In this method, there was a possibility that uneven distribution of data could be present in the training and test dataset. To fix this problem, the data was distributed equally in the training and test dataset. This process is called stratification.

Data set

The given parameters varied between-

- Current - [8 , 16] A
- Spark on Time - [200 ,400] microseconds
- Spark Off Time - [1600 , 1800] microseconds
- MRR - [0.0888, 0.51667] gm/min

Table 113.1 MRR for different current, spark on time, spark off time for training and testing model (Das et al., 2013)

Current (ampere)	Spark on time (microsec)	Spark off time(microsec)	MRR(gm/min)
8	200	1800	0.14187
12	200	1800	0.19778
16	200	1800	0.21207
8	200	1700	0.1537
12	200	1700	0.20833
16	200	1700	0.38095
8	200	1600	0.15353
12	200	1600	0.38889
16	200	1600	0.40667
8	300	1800	0.1037
12	300	1800	0.1625
16	300	1800	0.29305
8	300	1700	0.12273
12	300	1700	0.31759
16	300	1700	0.32667
8	300	1600	0.22578
12	300	1600	0.31
16	300	1600	0.34167
8	400	1800	0.0888
12	400	1800	0.23985
16	400	1800	0.26538
8	400	1700	0.17778
12	400	1700	0.25909
16	400	1700	0.29063

The parameters taken while constructing the neural network-

- Network type-feedforward backdrop
- Input data set - 24
- Target data set - 24
- Training function-TRAINBR
- Adaptive learning function-LEARNGD
- Performance function-MSE (mean square error)
- Number of layers   - 2
- Number of hidden layers - 1
- Number of input neuron - 3
- Hidden layer neuron - 10
- Output neuron - 1
- Activation Function-Tansigmoid

Learngd function was used as an adaptive learning function as it calculates the weight change dW for every given neuron from its input value and given error.

We have used Trainbr's LM algorithms as a training function and the hyperparameter is mu

Initial mu - The default value is 0.005.

The decreasing factor for mu - The default value is 0.1.

Increase factor for mu -The default value is 100.

Maximum value for mu - The default value is 10e11.

The parameters taken for testing were-

- Epochs- 10000
- Time - Infinite
- Minimum gradient- 0.0000001

- Max-fail- 1000
- Mu- 0.005
- mu-dec- 0.1
- mu-inc- 100
- mu-max- 100000000000

Numerous parameters are defined in ANN modeling, some of which are maintained fixed whereas others are modified to build a robust ANN model. Three input parameters and one output parameter are reflected in ANN as three input and one output neuron. Per Fausett, the backpropagation design with one hidden layer is enough for the majority of applications.

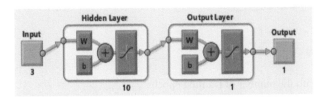

Figure 113.1 Neural network created for training data set

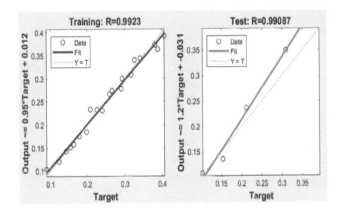

Figure 113.2 Mean squared error on training and validation datasets

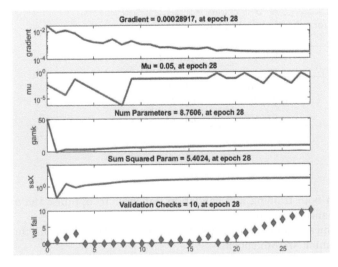

Figure 113.3 Variation of different model parameters with each iteration. Here the optimal solution was reached at 18th epoch and 10 more epochs were used for validation

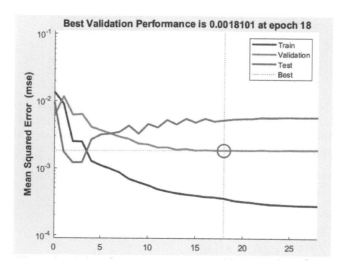

Figure 113.4 Variation of mean squared error with the number of training epochs

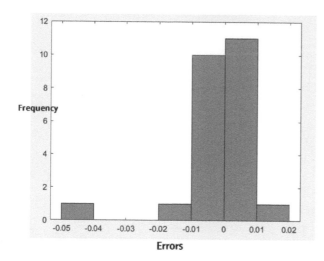

Figure 113.5 Error histogram of the model

Table 113.2 Values of MRR for the given for validation dataset with the prepared model (Das et al., 2013)

Current (Ampere)	Spark On Time(microsec)	Spark Off Time(microsec)	MRR	Predicted MRR(gm/min)
8	400	1600	0.17536	0.1596
12	400	1600	0.25278	0.2679
16	400	1600	0.51667	0.4598

Result and Discussion

The model converges to high accuracy on both training and validation datasets. To optimise the model mean squared error was used as a performance function.

The above figure shows the variation of mean squared error with the number of training epochs, thus we find that the model reaches the global optimum at 18th epoch.

The training R-value is **0.9923**

The mean squared error is '0.001237'.

The mean average percentage error is '8.65%'.

The ANN predicted results are very closer to 1, in a way this signifies the accuracy of the process. The error histogram signifies that most of the errors are within +- 1 percent limit which shows the high accuracy of the model.

Conclusions

In this paper ANN-based model is being used to predict the experimental output values. The experimental results obtained are very close to the ANN predicted results. This clearly signifies the high accuracy of the process.

ANN is used for prediction and other optimisation techniques such as GA, PSO is being used in synchronisation with the conventional techniques of optimisation such as Taguchi, RSM, and others.

Future Scope

The ANN-based optimisation is a very accurate method of predicting the experimental results. In Industry 4.0, there is an emphasis on the digitisation of manufacturing practices. ANN is an AI/ML-based approach to predicting the parameters. Also Genetic Algorithm, PSO(Particle Swarm Optimisation) are being used for the optimisation purpose. The application of ANN is widely practiced across the manufacturing sector, with more and more researchers applying AI-based optimisation techniques.

References

Abbas, N. M., Solomon, D. G., and Bahari, M. F. (2007). A review on current research trends in electrical discharge machining (EDM). Int. J. Mach. Tools Manuf. 47(7-8):1214–1228. doi:10.1016/j.ijmachtools.2006.08.026.

Abu Qudeiri, J. E., Mourad, A. H. I., Ziout, A.,Abidi, M. H., and Elkaseer, A. (2018). Electric discharge machining of titanium and its alloys: review. Int. J. Adv. Manuf. Technol. 96:1319–1339. https://doi.org/10.1007/s00170-018-1574-0.

Arunadevi, M. and Prakash, C. P. S. (2021). Predictive analysis and multi objective optimisation of wire-EDM process using ANN. Mater. Today: Proc. 46:6012–6016. doi:10.1016/j.matpr.2020.12.830.

Bhatt, D. and Goyal, A. (2019). Multi-objective optimisation of machining parameters in wire EDM for AISI-304. Mater. Today: Proc. 18:4227–4242. doi:10.1016/j.matpr.2019.07.381.

Choudhary, S. K., Jadoun, R. S. Kumar, A., and Ajay (2014). Latest research trend of optimisation techniques in electric discharge machining (EDM). IJREAT Int. J. Res. Eng. Adv. Technol. 2(3).

Das, M. K., Kumar, K., Barman, T. K., and Sahoo, P. (2013). Optimisation of surface roughness and MRR in EDM using WPCA. Procedia Eng. 64:446–455. Doi:10.1016/j.proeng.2013.09.118.

Dave, H. K., Desai, K. P., and Raval, H. K. (2012). Modelling and analysis of material removal rate during electro discharge machining of Inconel 718 under orbital tool movement. Int. J. Manuf. Syst. 2(1):12–20.

Fenggou, C. and Dayong, Y. (2004). The study of high efficiency and intelligent optimisation system in EDM sinking process. J. mater. Process. 149(13):83–87. doi:10.1016/j.jmatprotec.2003.10.059.

Gaikwad, V. S. and Jatti, V. S. (2018). Optimisation of material removal rate during electrical discharge machining of cryo-treated NiTi alloys using Taguchi's method. J. King Saud University - Eng. Sci. 30(3):266–272. doi:10.1016/j.jksues.2016.04.003.

Ganapathy, S., Balasubramanian, P., Vasanth, B., and Thulasiraman, S. (2019). Comparative investigation of artificial neural network (ANN) and response surface methodology (RSM) expectation in EDM parameters. Mater. Today: Proc. doi: 10.1016/j.matpr.2020.05.499.

Gao, Q., Zhang, Q.-H., Su, S.-P., and Zhang, J.-H. (2008). Parameter optimisation model in electrical discharge machining process. J. Zhejiang University-Sci. A: Appl. Phys. Eng. 9(1):104–108. doi:10.1631/jzus.a071242.

Golshan, A., Gohari, S., and Ayob, A. (2012). Multi-objective optimisation of electrical discharge machining of metal matrix composite Al/SiC using non-dominated sorting genetic algorithm. Int. J. Mechatron. Manuf. Syst. 5(5/6):385.

Ho, K. and Newman, S. (2003). State of the art electrical discharge machining (EDM). Int. J. Mach. Tools Manuf. 43(13):1287–1300. doi:10.1016/s0890-6955(03)00162-7.

Joshi, S. N. and Pande, S. S. (2010). Intelligent process modelling and optimisation of die-sinking electric discharge machining. Appl. Soft Comput. 11(2011):2743–2755.

Mahardika, M., Tsujimoto, T., and Mitsui, K. (2008). A new approach on the determination of ease of machining by EDM processes. Int. J. Mach. Tools Manuf. 48:746–760. doi:10.1016/j.ijmachtools.2007.12.012.

Marafona, J. and Chousal, J. (2006). A finite element model of EDM based on the Joule effect. Int. J. Mach. Tools Manuf. 46(6):595–602. doi:10.1016/j.ijmachtools.2005.07.017.

Natarajan, U., Hyacinth Suganthi, X., and Periyanan, P. R. (2016). Modeling and Multiresponse optimisation of quality characteristics for the Micro-EDM drilling process. Trans. Indian Inst. Met. 69(9):1675–1686. doi:10.1007/s12666-016-0828-5.

Naveen Babu, K., Karthikeyan, R., and Punitha, A. (2019). An integrated ANN ā PSO approach to optimise the material removal rate and surface roughness of wire cut EDM on INCONEL 750. Mater. Today: Proc. 19:501–505. Doi: 10.1016/j.matpr.2019.07.643.

Padhee, S., Nayak, N., Panda, S., Dhal, P., and Mahapatra, S. (2012). Multi-objective parametric optimisation of powder mixed electro-discharge machining using response surface methodology and non-dominated sorting genetic algorithm. Sadhana 37(2):223–240.

Panda, D. and Bhoi, R. (2005). Artificial neural network prediction of material removal rate in electro discharge machining. Mater. Manuf. Process. 20(4):645–672. doi: 10.1081/amp-200055033.

Pérez, J., Llorente, J., and Sanchez, J. (2000). Advanced cutting conditions for the milling of aeronautical alloys. J. Mater. Process. Technol. 100(1-3):1–11. doi:10.1016/s0924-0136(99)00372-6.

Rao, P. S., Dora, S. P., and Purnima, N. S. (2021). Influence of WC/Co powder metallurgy electrodes made by micron and nano particles on EDM performance. Met. Powder Rep. 76(6):52–58. doi: 10.1016/S0026-0657(21)00304-0.

Sahua, S. N., Murmub, S. K., and Nayakc, N. C. (2019). Multiobjective optimisation of EDM process with performance appraisal of GA based algorithms in neural network environment, Mater. Today: Proc. 18:3982–3997. doi: 10.1016/j.matpr.2019.07.340.

Singh, N. K., Singh, Y., Kumar, S., and Sharma, A. (2020). Predictive analysis of surface roughness in EDM using semi-empirical, ANN and ANFIS techniques: A comparative study. Mater. Today: Proc. 25:735–741. doi: 10.1016/j.matpr.2019.08.234.

Somashekhar, K. P., Ramachandran, N., and Mathew, J. (2010). Optimisation of material removal rate in micro-EDM using artificial neural network and genetic algorithms. Mater. Manuf. Process. 25:467–475.

Su, J. C., Kao, J. Y., and Tarng, Y. S. (2004). Optimisation of the electrical discharge machining process using a GA-based neural network, Int. J. Adv. Manufact. Technol. 24:81–90.

Tsai, K. M. and Wang, P. J. (2001). Comparison of neural network models on material removal rate in EDM. J. Mater. Process. Technol. 117:111–124.

Wang, J., Sánchez, J. A., Iturrioz, J. A., and Ayesta, I. (2019). Artificial intelligence for advanced non-conventional machining processes. Procedia Manuf. 41:453–459. Doi: 10.1016/j.promfg.2019.09.032.

Wang, K., Gelgele, H. L., Wang, Y., Yuan, Q., and Fang, M. (2003). A hybrid intelligent method for modeling the EDM process. Int. J. Mach. Tool Manufact. 43:995–999.

114 CFD Simulation forfluid flow through a circular chamber by using ANSYS

Aamir M. Shaikh[a] and Dayanand A. Ghatge[b]

Department of Mechanical Engineering, Karmaveer Bhaurao Patil College of Engineering, Satara, India

Abstract

Fluid particles flow through a conduit so that it fills the entire portion, and as much it enters one end of the inlet section as it leaves the other end of the pipe at the outlet section at the same time, the flow is steady rate condition. The liquid flow does not change with time at any point of the pipe. Such fluid transportation systems are used in the chemical industry, oil industry, etc. In order to determine the effect of fluid flow flowing through the circular chamber, this paper adopted ANSYS/Fluent software to analyse the responses like pressure, velocity. We consider the flow velocity is 7m/s here, and the Reynolds number is more than 4000. This paper examines the two-dimensional flow behavior of a turbulent flow type to find pressure, velocity, density using the software. Also, see the effect of the speed of particles at the outlet for a circular chamber.

Keywords: circular chamber, fluid, conduit, steady rate.

Introduction

Liquid, while it flows through a pipe or any other closed section, it covers the entire portion of the tube. The liquid flow is said to be steady when it enters one end of the pipe as it exits the other end of the pipe simultaneously. In the circular chamber also, the water flows in this condition. The liquid flow does not change concerning the time at any point of the pipe. The path of any liquid particle as it moves through the pipe is called a streamline (Scmat and Katz, 1958).

The authors found the location of vena-contracta using CFD simulations in terms of pressure and velocity profiles for a jet of water from the narrowing end of the orifice meter. He also performed Computational Fluid Dynamics (CFD) of fluid flow from the orifice meter at different conditions (Hoskote, 2021).

Several engineering applications can be modeled as cylinders, including offshore structures, bridge piers, and pipelines. The author performed the study of the flow field around cylinders for identical flow conditions ranging from laminar (Reynolds number [Re] = 2) for the same conditions or characteristics (Yuce and Kareem, 2016).

The author performed the study to obtain the numerical simulation for various flow properties and turbulence characteristics of various cross-sections like a circular, square, and rectangular at subsonic Mach number. Using CFD software ANSYS fluent, the author validates the result for the nozzle. The study is for high-speed jets, and the small vortices remain stable. The equations for the nozzle are formulated theoretically (Rakesh, 2019).

Hirani (2013) used CFD simulation to investigate the resistance coefficient for Y-shape fitting in the plumbing system. For Y-shape, analysis is performed for a different angle. The author observed the effect of angle corresponds to the resistance coefficient. Again he found pressure drops at a different bent angle for water flow (Hirani, 2013).

Bejena et al. (2021) focused on CFD simulation of fluid flow in a pipe having pipe length was 1.5 m and 0.075 m inner diameter with the effect of fluid viscosity for high viscous fluid at laminar flow to determine the velocity profile, pressure drop, and velocity vector at various fluid viscosities. We observed this hydrodynamic behavior of fluid concerning viscosity (Bejena et al., 2021).

Ferede, studied fluid flow in the T-junction for horizontal pipe having 100 mm diameter and the vertical branch with 10 mm diameter. Velocity distribution and average velocity profile with numerical analysis for temperature pressure and velocity are carried out (Ferede, 2019).

E. Shashi Menon performed a study on fluid flow in pipes. Minor losses in pipelines from the valve, fitting, pipe enlargement, pipe contractions, frictional head loss are analysed by calculating the pressure drop in gas (Menon, 2015).

By considering the above papers, the above authors studied the different cases like closed sections, orifice meters, cylinders for identical flow conditions, high-speed jets nozzles, Y-shape fitting, etc., so there is scope to perform the analysis of a circular chamber to visualise the flow behavior. The shape of this chamber is not having a straight and uniform cross-section. Here for curved shapes, we can find the effect of two halves portions using ANSYS Fluent, irrespective of other forms used by authors.

---

[a]amir.shaikh@kbpcoes.edu.in; [b]dayanand.ghatge@kbpcoes.edu.in

Problem definition

Figure 114.1 shows the entire conduit with a circular section of with an inner diameter of 50 units and an outer diameter of 100 units. The total length is 200 units. Inlet and outlet sections are of uniform diameter having 30 units. It has been noticed that most fluids, especially liquids, are transported in circular pipes. This is because pipes with a circular cross-section can withstand large pressure differences between the inside and the outside without undergoing significant distortion. Other shapes are usually used in different applications where the pressure difference is relatively small. In Tables 114. 1 and 114. 2, the various parameters and properties of the fluid are mentioned, respectively.

II. PROBLEM DEFINITION

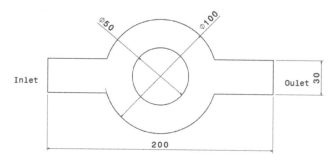

Figure 114.1 Geometry

Table 114.1 Parameters

Sr. No.	Parameter	Dimensions
1	Inner diameter (D1)	50 mm
2	Outer diameter (D2)	100 mm
3	Inlet & outlet pipe diameter (d)	30 mm
4	Total length (L)	200 mm
5	Inlet Velocity (V1)	7 m/s

Table 114.2 Property of fluid

Sr. No.	Property of fluid used (water)	Magnitude
1	Chemical Formula	H2O
2	Boiling point	1000C
3	Density	998.2 Kg/m³
4	Specific Heat	4182 J/kg-k
5	Thermal Conductivity	0.6 w/m-k
6	Viscosity	-s

Governing equations

Steady flow is described as that type of flow in which the fluid qualities like velocity, pressure, density, etc., at a point do not change with time. Thus, for a steady flow, mathematically, we have partial differential equations. The Continuity equation for 2D flow is written as below

$$\left(\frac{\partial u}{\partial t}\right)_{x0,\ y0,} = 0;$$

$$\left(\frac{\partial v}{\partial t}\right)_{x0,\ y0,} = 0;$$

$$\left(\frac{\partial p}{\partial t}\right)_{x0,\ y0,} = 0;$$

$$\left(\frac{\partial \rho}{\partial t}\right)_{x0,\ y0,} = 0$$

where () is a fixed point in a fluid field where these variables are being measured w.r.t. time (Khan et al., 2021).

It is assumed that no-slip boundary conditions at all the walls, flow is turbulent, fluid is incompressible, steady, uniform, and irrational.

Fluid flow is based on the law of conservation of momentum or the momentum principle. The total force on fluid equals the change in flow momentum per unit of time.

Euler's Equation of Motion: This equation of motion gives the forces due to gravity and pressure are taken into consideration by the movement of a fluid element along a streamline.

$$\frac{dp}{\rho} + g.\,dz + v.\,dv = 0$$

The above equation is as per the Euler's equation of motion.

Ansys cfd solver settings and boundary conditions:

Solver type: Pressure based
Velocity Formulation: Absolute
Time: steady
Energy equation: ON
Model: K-Epsilon
Material: Water liquid
Cell zone conditions: fluid is used for part body
Boundary conditions: Inlet Velocity = 7m/s (normal to boundary)
Solution Methods:
Pressure velocity coupling: Simple
Gradient: Least squares cell-based
Pressure: Second order
Momentum: second order-upwind
Solution initialisation: hybrid
Calculation activities: Autosave every (iterations) is 20

Ansys meshing

In CFD meshing, our domain of the circular chamber is divided into subdomains which helps in a numerical grid to a fluid body and boundary, analogous to meshing in finite element simulations. Meshing algorithms are used to produce clusters of grid points, determining the accurateness of a CFD simulation. The mesh influences the accuracy, convergence, and speed of the solution (Vogel, 1929, Atmaca et al., 2021). These small parts of the circular chamber are called elements. Two-dimensional elements do the meshing of an entire region of the circular section. The fine meshing option is chosen to get a more accurate result. Initially, the analysis is performed for the coarse mesh option. Finally, the total number of elements are 1207, and the total number of nodes are 2618.

Result and discussion

Consider that the flow velocity at the inlet is 7m/s here and that the Reynolds number will be calculated using the values of density, velocity, diameter, and viscosity. We examine the two-dimensional flow behavior by considering the Reynolds number.

In this section, ANSYS Fluent results discussed with contour plots

When the inlet velocity V1 is equal to 7m/s, partial streamlines are shown in Fig. below

Convergence

After dividing the domain into several small parts and applying boundary conditions, the solution gets converged after some iterations—the scaled residual plays a vital role in linking the solution.

Solution is converged at 7.3792e-04, 2.4454e-05

Adaption

The position value in between -0.1 m to +0.1 m (-100 mm to +100 mm)
Figure 114.4 shows adaption function plot of position on x axis and adaption function on y axis. For wall body the highest value is 1.50e-08 at position of -0.0375m (-37.5 mm)

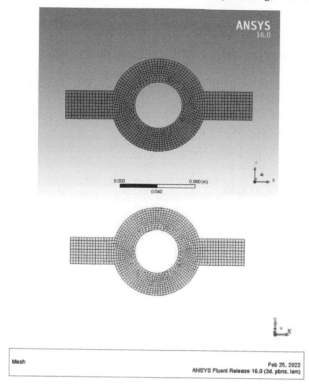

Figure 114.2 Mesh

Table 114.3 Nodes and elements

Domain	Nodes	Elements
Part body	2618	1207
Iterations: 500		

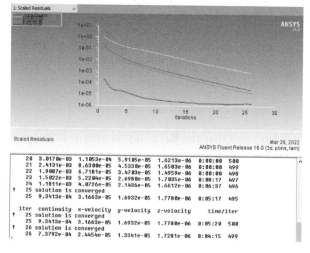

Figure 114.3 Scaled residual

Figure 114.5 shows strain rate highest values is 7.20e-06 at position of -0.05m (-50 mm) and 8.00e-06 at position of +0.05m (+50 mm).

Residuals Mass Imbalance & Static Pressure

In mass imbalance of the residuals, the difference between the mass entering and leaving the cell is considered indicated in Figure 114.6.

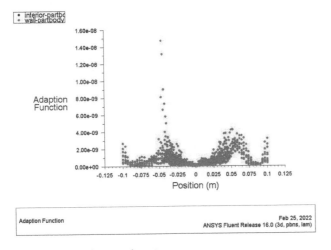

Figure 114.4 Adaption function

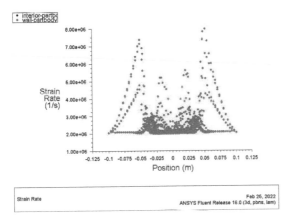

Figure 114.5 Derivatives strain rate

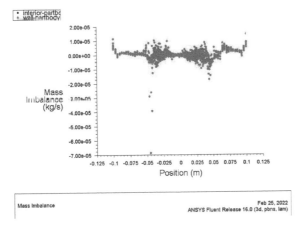

Figure 114.6 Residuals mass imbalance

Figure 114.7 indicates static pressure in pascals (Pa). The static pressure is 1.80e+12 Pa at -0.1 m position and 1.70e+12 Pa at +0.1 m position.

Pathline

Tracing the path of a particular particle is obtained here, which gives the pathline of that particle, and the stream is not considered.

In Figure 114.8, it has been observed that the particles are crossed with each other.

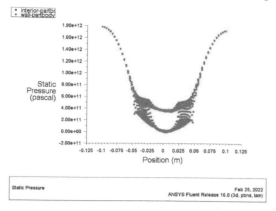

Figure 114.7 Static pressure

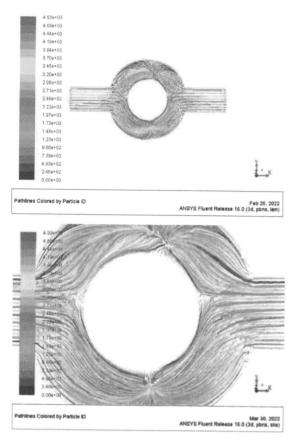

Figure 114.8 Pathline

E. Velocity vector

In the case of a uniform cross-section, an average velocity is constant in an incompressible type of flow. The fluid velocity in the chamber changes from point to point. The velocity is zero at the boundary of the wall body, and it increases gradually towards the center of the domain. So we consider an average velocity Vavg value for our study purpose (Çengel and Cimbala, (2021).????).

As inlet velocity is 7 m/s, fluid particles enter the lower section faster than the upper chamber because of gravity.

From the Figures 114.10–114.12, velocity reduces from 7 m/s to 4 m/s

To check the flow type, we can calculate Re from obtained value.

$$\therefore \text{Reynolds Number (Re)} = \frac{\rho V D}{\mu}$$

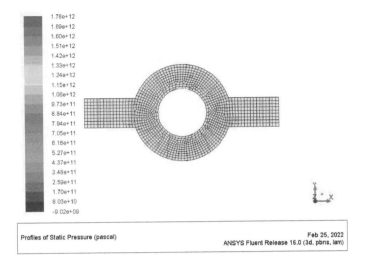

Profiles of Static Pressure (pascal)

Feb 25, 2022
ANSYS Fluent Release 16.0 (3d, pbns, lam)

Figure 114.9 Static pressure

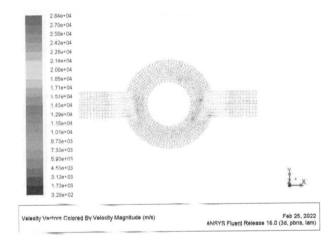

Velocity Vectors Colored By Velocity Magnitude (m/s)

Feb 25, 2022
ANSYS Fluent Release 16.0 (3d, pbns, lam)

Figure 114.10 Velocity vector

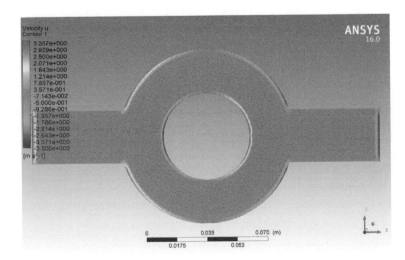

Figure 114.11 Velocity u contour

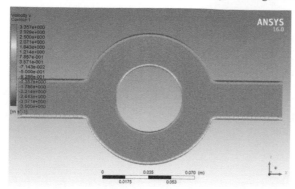

Figure 114.12 Velocity v contour

Where,
ρ = Density of fluid=998.2 Kg/m3
V= Velocity= 4m/s (software value)
D= Diameter=30mm= 0.03m
µ=Viscosity of fluid=0.001003 kg/m-s
By putting above values we get,
    Here Re is more than 4000, so flow is turbulent.

$$\therefore \text{Reynolds Number (Re)} = \frac{998.2 \times 4 \times 0.03}{0.001003}$$

$$\therefore Re = 119784$$

Pressure change:

In ANSYS FLUENT, there is a facility to extract pressure contour lines or profiles. In our problem of the circular chamber, we obtain these plots to know the values of various characteristics. The pressure value is mentioned in Pascals (Pa) as per the software. The pressure drop might still be decreased without affecting the flow, which will be discussed below.

Here Figure 114.13 indicates a density contour in which density remains constant. While in Figure 114.14, variation pressure is recorded.

The maximum pressure is 1.740e+012 Pa. On the lower side of the circular chamber, the pressure value is 2.096e+011 Pa. Similarly, on the upper side of the circular chamber, the pressure value is 7.762e+011 Pa.

The shape of the circular chamber initially has a uniform cross-section up to 50 mm from the inlet, then splits into two circular parts, upper and lower. After passing the fluid through these two circular halves, it again passes through one uniform cross-section at the outlet. Due to this shape, the pressure values are different in the upper and lower halves.

Results obtained from software:

The values obtained from the software are shown in Table 114.4. The mass flow rate of 125 kg/s is possible in this chamber at a water particle velocity equal to 7 m/s.

Conclusion

The results are obtained for an inlet velocity 7 m/s. From the above results, we conclude that the flow is possible in the chamber. Flow behavior is observed particularly in this case where the outer diameter is two times that of the inner diameter. Here velocity reduces from 7 m/s to 4 m/s. An approximate 42% loss will be there due to the shape of the chamber. These simulations indicate that these studies can be performed easily with CFD. Fluid splits into two curved halves then obtains an outlet flow, but there are losses due to circular obstruction. Loss due to obstruction and friction is also needed to concentrate. One other reason for energy loss is eddies formation in a chamber.

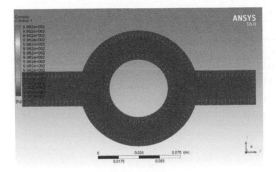

Figure 114.13 Density contour

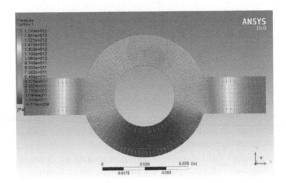

Figure 114.14 Pressure contour

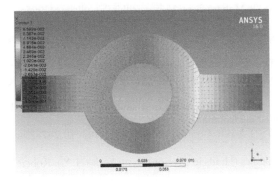

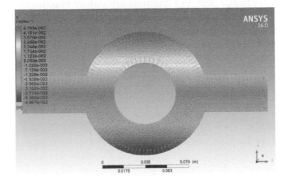

Figure 114.15 X and Y contour

Table 114.4 Mass flow rate, area and mass-weighted average.

Mass flow rate	
Mass flow rate	(kg/s)
interior-part body	181.08032
wall-part body	124.94417
Net	124.94417
Area of surfaces (0 1) projected onto plane (1, 0, 0): 0	
Area	(m2)
interior-part body	6.2227264e-05
wall-part body	0.017881337
Net	0.017943564
Mass-Weighted Average	
Static pressure	(pascal)
Part body	6.9026113e+11
Net	6.9026113e+11

References

Çengel, Y. A. and Cimbala, J. M (2021). Fluid mechanics: Fundamentals and applications. eds. Y. A. Cengel and J. M. Cimbala. (1st ed. p. cm). (McGraw-Hill series in mechanical engineering).

Atmaca, M., Çetin, B., Ezgi, C. and Kosa, E. (2021). CFD analysis of jet flows ejected from different nozzles. Int. J. Low-Carbon Technol. 16(3):940–945. https://doi.org/10.1093/ijlct/ctab022.

Bejena, B., Prabhu, S. V. and Gundaboina, S. (2021). Computational fluid dynamics simulation and analysis of fluid flow in pipe: Effect of fluid viscosity. J. Comput. Theor. Nanosci. 18:805–810. 10.1166/jctn.2021.9680.

Ferede, N. A. (2019). Numerical analysis and CFD simulation of fluid flow in T-Junction pipe by using ANSYS CFX 1. Int. J. Scient. Eng. Res. 10(10). doi: 10.13140/RG.2.2.35259.23845/1.

Hirani, A. (2013). CFD simulation and analysis of fluid flow parameters within a y-shaped branched pipe. IOSR J. Mech. Civ. Eng. 10:31–34. 10.9790/1684-1013134.

Hoskote, S. (2021). CFD analysis of an orifice meter using ANSYS. Maharashtra: Vidyavardhini's College of Engineering And Technology.

Khan, S., Ibrahim, O. and Aabid, A. (2021). CFD analysis of compressible flows in a convergent-divergent nozzle. Mater. Today: Proc. 46. 10.1016/j.matpr.2021.03.074.

Marchisio, D. L. and Barbato, M. C. (2021). A detailed CFD analysis of flow patterns and single-phase velocity variations in spiral jet mills affected by caking phenomena. Chem. Eng. Res. Des. 174:234–253. ISSN 0263-8762. https://doi.org/10.1016/j.cherd.2021.07.031.

Menon, E. S. (2015). Fluid flow in pipes. In Transmission pipeline calculations and simulations manual. ed. E. S. Menon, (pp. 149–234). Kidlingron, Oxford : Gulf Publishing Company.

Rakesh, N. (2019). Analysis of flow of nozzle by using ANSYS. J. Mech. Continua Math. Sci. 1. doi: 10.26782/jmcms. spl.2019.08.00088.

Semat, H. and Katz, R. (1958). Physics, chapter 9: Hydrodynamics (Fluids in Motion). New York (N. Y.): Robert Katz Publications.

Vogel, G. (1929). Experiments to determine the loss at right angle pipe tees. In Hydraulic laboratory practice ed. J. R. Freeman, (pp. 470–472), New York (N. Y.): American Society of Mechanical Engineers.

Yuce, M. and Kareem, D. (2016). A numerical analysis of fluid flow around circular and square cylinders. J. Am. Water Works Assoc. 108:E546–E554. 10.5942/jawwa.2016.108.0141.

115 Prediction of dynamic strain ageing at high temperatures for Ti-6Al-4V alloy by developing processing maps using different instability criteria

Mohd Abdul Wahed[a]

Department of Mechanical & Production Engg.Deccan College of Engineering & Technology, Hyderabad, Telangana, India

Abstract

Processing maps play a key role in signifying stable and unstable regions for a manufacturing process carried out at high temperatures. It also represents the behaviour of a particular material, by presenting modifications in the microstructural growth via high temperatures. In this research work, processing maps are generated based on the true stress vs. true strain information of Ti-6Al-4V alloy at 10^{-2} /s–10^{-4} /s and from 700°C–900°C, to predict dynamic strain ageing. The true stress - true strain information is obtained based on the temperature, strain as well as strain rate by carrying out high temperature uniaxial tensile tests. Depending on this, initially, strain rate sensitivity is calculated from log (true stress) and log (strain rate) curves and then strain rate sensitivity contour maps are developed for different strains. Secondly, efficiency maps are developed by considering strain rate sensitivity, to signify the efficiency that is associated with the quantity of internal entropy developed. Further, instability maps are developed based on Prasad's and Murty's criteria, in order to accurately predict dynamic strain ageing that is to be eluded while working at high temperatures. Finally, processing maps are generated by overlaying instability maps over efficiency maps. The results show that the Murthy's criterion predicts dynamic strain ageing more accurately when compared to Prasad's criterion for Ti-6Al-4V alloy at high temperatures.

Keywords: Dynamic strain ageing, iinstability criteria, rocessing maps, strain rate sensitivity, Ti-6Al-4V alloy.

Introduction

Titanium is a unique metal, as it is found in abundance from earth's crust. Among different titanium alloys, Grade V/ Ti-6Al-4V alloy is foremost, as it contributes 50% of the total manufactured components and is considered as the forefront in the titanium manufacturing (Leyens and Peters, 2003). Ti-6Al-4V alloy includes incomparable properties for instance high strength and melting point, low density as well as good corrosion and erosion resistance. It has number of industrial applications i.e. aerospace, biomedical, defence, marine, etc. Ti-6Al-4V alloy is regarded as a polycrystalline material, as it enjoys high temperature deformation. It is achieved with equiaxed as well as bimodal microstructures (Gillo Giuliano, 2011). Alabort et al. (2015) studied hot deformation of Ti-6Al-4V alloy via 1D tensile tests over 700°C950°C and at 10^{-5} /s - 10^{-2} /s, from material characterization, FEA as well as applications aspects. A processing map is an important tool in order to get optimized parameters in case of high temperature/ hot deformations. Sun et al. (2018) examined strain rate sensitivity (m) factor, efficiency (η) factor as well as instability (ξ) factor to develop 3-D processing maps depending on the isothermal compression tests performed over 410°C490°C and at 10-3 /s - 1/ s and then studied the hot deformation performance of Al alloy (6A02). Similarly, for any individual material, processing map responds in terms of different microstructural mechanisms. Ghasemi et al. (2017) carried out 1-D compression tests on Titanium alloy (BT9) over 1000°C1100°C and at 10^{-3} /s - 10^{-1} /s to inspect the flow stress as well as strain hardening performance. Finally, generated processing maps have been used to relate microstructural development with hot working behaviour. Processing map also envisage whether a particular region is safe or unsafe while producing critical components, particularly at high temperatures. Meng et al. (2018) performed compression tests on Titanium alloy (ATI 425) from 700°C900°C and at 10^{-3} /s - 1/ s to generate the processing maps. In developed processing maps, they observed that the higher value of efficiency means high microstructural growth, while the lower value of efficiency relates to low microstructural growth. Likewise, Zhe et al. (20170 carried out compression tests on Titanium alloy (TB17) over 775°C - 905°C and at 10^{-3} /s – 10 /s and then generated processing maps in order to characterize the high temperature behaviour. Depending on this, they envisaged the hot deformation parameters as well as validated with microstructural growth. Park et al. (2002) studied the isothermal compressed flow performance of Ti-6Al-4V alloy in hot deformation conditions and then generated processing maps depending on dynamic material model (DMM) to get a mutual relationship between hot deformation, constitutive modelling and microstructural development. Additionally, the obtained processing maps have been applied in FEA, to calculate the flow as well as microstructure

[a]wahedmohdabdul@gmail.com

reliability in the forging method. Seshacharyulu et al. (2000; 2002) developed/generated processing maps in order to acquire optimal parameters as well as microstructure view/ control in the hot deformation of Ti-6Al-4V alloy over 750°C1100°C and at 3 x 10⁻⁴ /s - 10 /s. Depending on the stress vs. strain data obtained from the uniaxial compression tests, Cai et al. (2016) generated processing maps for Ti-6Al-4V alloy centred on different instability criteria as well as corroborated with the microstructural development. Luo et al. (2009) conducted compression tests to study the effect of strain on the microstructure, efficiency as well as instability factors by generating processing maps for three diverse compositions of Titanium alloys. The distinction of microstructure on the modification of Boron on Ti-6Al-4V alloy deformed over 750°C1000°C and at 10⁻³ /s - 10 /s for various strains have been investigated by developing processing maps (Sen et al., 2010). Prasad et al. (2015) generated processing maps for diverse metallurgical mechanisms using optimal settings for diverse materials over diverse temperatures and at diverse strain rates, to study the efficiency as well as flow instabilities by carrying out compression tests. Li et al. (2009) performed compression tests for hydrogenated Ti-6Al-4V alloy from 760°C920°C and at 10⁻² /s - 10 /s through maximum reduction in height. They generated processing maps to recognize the effect of hydrogen on volume fraction as well as grain size. Xing et al. (2019) performed tensile tests for steel alloy at room temperature (RT) from 10⁻⁴ /s - 10² /s, in order to calculate the mechanical properties. They reported that strain rate sensitivity (m) becomes negative when strain rate has been greater than 100 /s and this has been recognised due to dynamic strain ageing (DSA). Lukaszek-Solek and Krawczyk (2015) combined high temperature behaviour as well as compression testing by applying a material model. Furthermore, evaluated microstructural growth based on processing maps and observed that the generated processing maps have been found to be safe in the stable area. Wang et al. (2016) performed compression tests for Tungsten alloy over 1250°C1550°C and at 10⁻³ /s - 1 /s to examine the high temperature behaviour, microstructural development, stable as well as unstable regions for which processing maps have been generated. Results show that the stable area in processing maps may be classified into multiple areas, while unstable areas have been related to individual microstructure. However, the above literature completely lacks the generation of processing maps based on different instability criteria at high temperatures to predict DSA with respect to uniaxial tensile tests for Ti-6Al-4V alloy. Hence, in this work, the tensile tests were carried out from 700°C900°C and at 10⁻² /s - 10⁻⁴ /s, then the strain rate sensitivity maps, efficiency maps and instability maps based on different instability criteria have been developed. Lastly, processing maps have been developed to predict the DSA based on different instability criteria for Ti-6Al-4V alloy.

Material and test details

Ti-6Al-4V alloy of 1.3 mm thickness comprising of fine equiaxed microstructure with (α + β) phases has been received as presented in Figure 1 and chemical composition has been presented in Table 115.1.

The uniaxial tensile test samples have been prepared as per sub-sized ASTM E8 standard as presented in Figure 2. Delta glaze liquid coating has been applied on the Ti-6Al-4V alloy specimens, in order to avoid

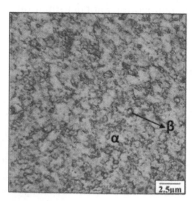

Figure 115.1 Optical image of Ti-6Al-4V alloy

Table 115.1 Chemical composition of Ti-6Al-4V alloy (%wt)

Al	V	Fe	O	C	Ti
5.98	4.07	0.22	0.12	0.02	Balance

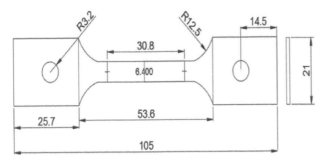

Figure 115.2 Geometry of tensile test sample (mm)

(a)   (b)

Figure 115.3 (a) Computer control UTM of 50 kN and (b) high temperature split furnace

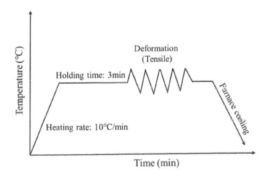

Figure 115.4 Heating - tensile deformation - cooling of specimens

the oxidation effect at high temperatures. The tensile tests have been performed over 700°C–900°C at a gap of 50°C and at 10^{-2}/s–10^{-4}/s by using 50kN UTM as presented in Figure 115.3.

The specimens have been heated at a constant rate of 10°C/min and then after reaching the preferred temperature held for three to five minutes to steady the temperature. This has been followed by uniaxial tensile test till specimen breaks and subsequent furnace cooling as shown in Figure 115.4. Each tensile test has been carried out three times in order to confirm results repeatability.

Results and discussion

Depending on the uniaxial tensile tests data, the true stress - true strain curves were developed over 700°C900°C and at 10^{-2} /s - 10^{-4} /s as presented in Figure 5 (a), (b) as well as (c) respectively. Using this data, the strain rate sensitivity (m) maps, efficiency (η) maps as well as instability (ξ) maps have been developed, which were finally used to generate processing maps.

Flow stress characterisation

Figures 5 (a), (b) and (c) shows flow stress is noticeable at low temperatures and higher strain rates, while flow stress decreases with the rise of temperature and at lower strain rates. From this, it has been observed

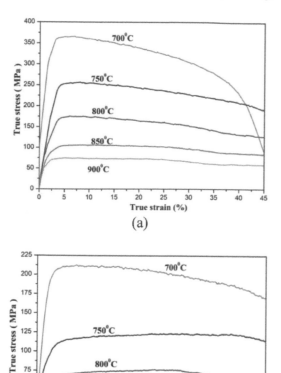

Figure 115.5 True stress vs true strain curves developed over diverse temperatures and strain rates (a) 10^{-2} /s, (b) 10^{-3} /s as well as (c) 10^{-4} /s respectively

that apart from 700°C and at 10^{-2} /s curve, other true stress vs. true strain curves are steady. Consequently, most of the curves start to soften.

The existence of flow softening at high temperatures $(\alpha+\beta)$ deformation has been due to adiabatic heating as well as dynamic recovery, similarly reported by Alabort et al. (2015). The adiabatic heat produced rises the tensile temperature of the specimens. Generally, at all strain rates, serrations have been observed in the true stress vs. true strain curves signifying the competing nature of work hardening as well as flow.

Strain rate sensitivity (m)

Strain rate sensitivity is an important factor in describing high temperature deformation. It responds to strain rate as well as defined by a correlation among equivalent/ flow stress (σ) and strain rate () (Leyens and Peters, 2003) as shown in '(1)',

$$\sigma = K\dot{\varepsilon}^m \qquad (1)$$

where K is a constant as well as m is strain rate sensitivity.
With the application of log function on both the sides of '(1)', we obtain,

$$m = \frac{\Delta \log \sigma}{\Delta \log \dot{\varepsilon}} \qquad (2)$$

m has been calculated by means of '(2)'. Usually, m ranges from (0.10.8) at high temperatures (Gillo Giuliano, 2011). In this research work, m has been obtained by fitting lines in between log (true stress) and log (strain rate) at high temperatures for different strain values as presented in Figure 6 for 0.35 strain, similarly considered by Sun et al. (2018) and Wang et al. (2016). Thus, m has been developed for all the strains in the form of contour maps i.e., 0.15, 0.25, 0.35 and 0.45 as presented in Figure 7. It has been observed that m varies as the strain value varies and maximum positive value of m = 0.79 has been attained at the 0.45 strain as shown in Figure 7 (d) and the minimum positive value of m = 0.17 has been attained at the 0.15 strain as shown in Figure 7 (a). Positive m value increases as the strain value increases steadily over 700°C - 900°C and 10^{-2} /s - 10^{-4} /s. However, m shows negative sign at 0.35 strain over 700°C725°C and at 10^{-2} /s - 10^{-4} /s representing dynamic strain ageing, similarly reported by Xing et al. (2019).

In the development of a processing map, two factors i.e., efficiency as well as instability are functions of m (Sun et al., 2018) indicating the significance of m. These two factors are discussed further in detail.

Efficiency maps (η)

For generating processing maps, DMM i.e. Dynamic material model (Ghasemia et al., 2017; Meng et al., 2018) has been applied consisting of a dimensionless factor known as efficiency (η). It refers to the energy dissipation over the microstructural evolution at varied temperatures. Though, working in a high temperature range, it has been developed between temperature vs. strain rate known as an efficiency map (Zhe et al., 2017). It consists of diverse regions which can be related to different microstructural developments such as dynamic recovery (DRV) and recrystallization (DRX), high temperature deformation, etc. Efficiency (η) is described by '(3)' Meng et al. (2018),

Efficiency (η) plays a key part in describing the possibility of a specific method (Park et al., 2002). The efficiency (η) maps at different strains 0.15, 0.25, 0.35 and 0.45 over diverse temperatures as well as strain rates are presented in Figure 8.

From the presented figures, it has been observed that as the strain value increases, the efficiency also increases i.e. continuously increases from 0.15 to 0.45 strain. A maximum efficiency of 84% has been attained at 0.45 strain, it characterizes that greater is the efficiency, superior is the workability, as reported by Lukaszek-Solek and Kraczyk (2015).

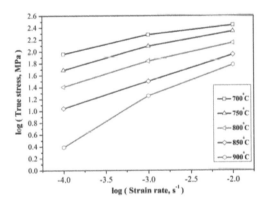

Figure 115.6 Log-log curves of true stress-strain rate for high temperatures at 0.35 strain

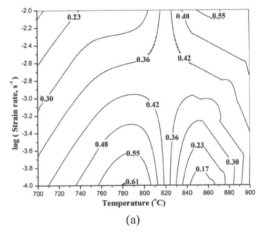

(a)

Figure 115.7 Strain rate sensitivity contour maps for Ti-6Al-4V alloy developed at diverse strains (a) 0.15, (b) 0.25, (c) 0.35, as well as (d) 0.45 (Contour numbers in the maps represent strain rate sensitivity)   (*continues*)

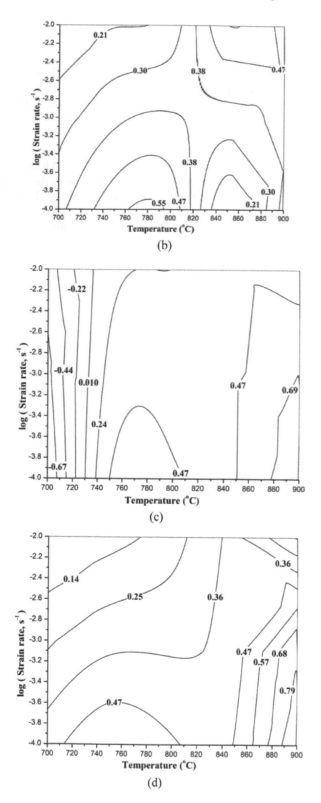

(b)

(c)

(d)

Figure 115.7 Continued

Instability maps (ξ)

An instability factor analogous to extreme principles of irreversible thermodynamics (Seshacharyulu et al., 2000) has been applied to discover the instability that took place at high temperature tensile deformation i.e. dynamic strain ageing (DSA), adiabatic bands, flow instability, etc.

In this work, two instability criteria have been used i.e. Prasad's and Murthy's. Prasad's instability criterion is presented by '(4)' (Seshacharyulu et al., 2002) and Murthy's instability criterion is presented by '(5)' (Cai et al., 2016). Instability maps have been developed using these two criteria at different strains 0.15, 0.25, 0.35 and 0.45 over diverse high temperatures as well as strain rates as presented in Figure 9 and Figure. 10 respectively. Contour numbers with negative sign as well as shaded area denote DSA.

$$\xi\,(\dot{\varepsilon}) = \frac{\partial \ln\left(\frac{m}{m+1}\right)}{\partial \ln \dot{\varepsilon}} + m \ < 0 \qquad (4)$$

$$\xi\,(\dot{\varepsilon}) = \frac{2m}{\eta} - 1 < 0 \qquad (5)$$

Statistically, DSA means $\xi\,() < 0$, i.e. negative value; whereas non DSA means $\xi\,() > 0$, i.e. positive value (Cai et al., 2016). Instability factor identifies whether a particular method is stable or not under diverse working environments (Luo et al., 2009).

In an unstable area, negative (-ve) value is related to instability viz. greater is the -ve value, greater the instability, while smaller is the negative value, smaller is the instability. Thus, a positive value is related to stability viz. higher is the positive value, greater are the chances of taking place a stable deformation, as reported by Wang et al. (2016).

Development of processing maps

Processing maps as presented in Figure 11 have been developed by superimposing instability map centered on Prasad's criterion as presented in Figure 9 and instability map centered on Murthy's criterion as presented in Figure 10 on efficiency maps as presented in Figure 8. In order to predict the dynamic strain ageing (DSA), these processing maps have been developed at 0.35 strain only, as m has been found to be negative at this strain.

It has been observed from Figure 11 (a and b) that Prasad's criterion doesn't predict any DSA, while Murthy's criterion predicts the DSA accurately for the considered high temperatures and strain rates. From Murthy's criterion, DSA occurs over 700°C725°C and at 10-2 /s - 10-4 /s which is denoted by a shaded area as presented in Figure 11 (b). Further, the percentage of efficiency in the DSA region is low (≤ 30%), while in the non DSA region, a maximum efficiency of 80% has been obtained. So, DSA region must be eluded while working at high temperatures as well as at different strain rates. Hence, for ensuring the safety

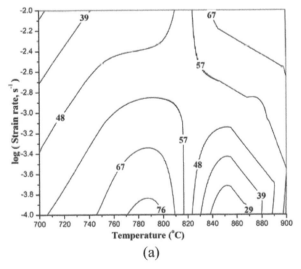

(a)

Figure 115.8 Efficiency maps for Ti-6Al-4V alloy developed at diverse strains (a) 0.15, (b) 0.25, (c) 0.35, as well as (d) 0.45 (Contour numbers in the maps represent efficiency)   (*continues*)

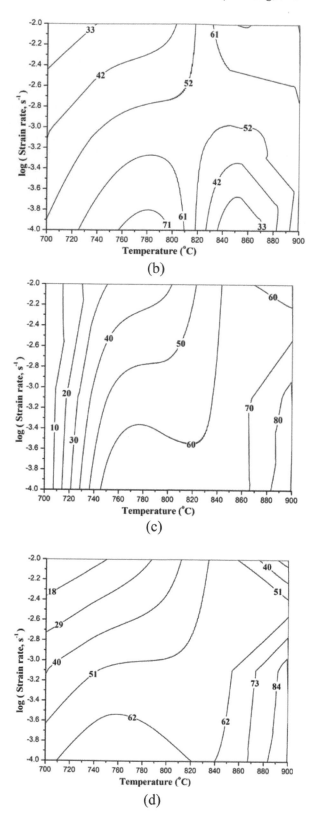

Figure 115.8 Continued

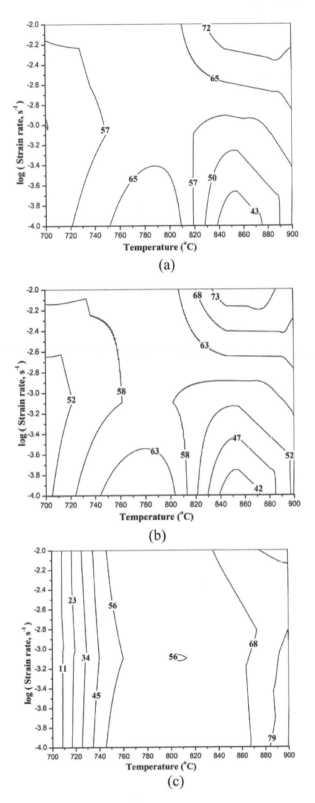

Figure 115.9 Instability maps for Ti-6Al-4V alloy based on Prasad's criterion developed at diverse strains (a) 0.15, (b) 0.25, (c) 0.35, as well as (d) 0.45 (Contour numbers with negative sign signify instability)   (*continues*)

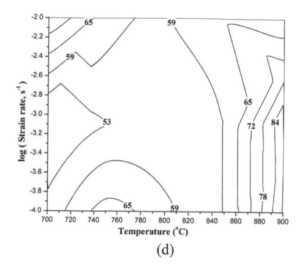

(d)

Figure 115.9 Continued

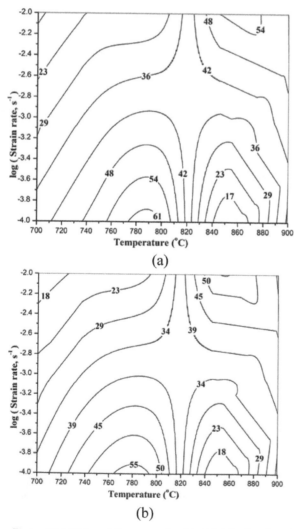

(a)

(b)

Figure 115.10 Instability maps for Ti-6Al-4V alloy based on Murthy's criterion developed at diverse strains (a) 0.15, (b) 0.25, (c) 0.35, as well as (d) 0.45 (Contour numbers with negative sign as well as shaded area represents instability)

(*continues*)

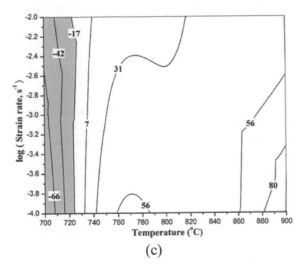

(c)

Figure 115.10 Continued

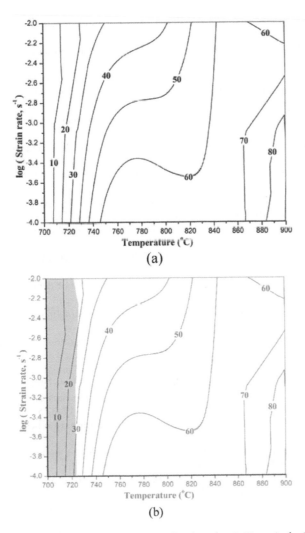

(a)

(b)

Figure 115.11 Processing maps developed at 0.35 strain for Ti-6Al-4V alloy based on (a) Prasad's and (b) Murthy's criterion (Contour numbers represent efficiency, while shaded area denotes instability)

during hot tensile deformation, Murthy's criterion can be considered for accurate prediction of DSA region in Ti-6Al-4V alloy at high temperatures.

Conclusions

This research work can be concluded as follows:

- In true stress - true strain curves, serrations have been observed signifying the competing nature of work hardening as well as flow softening leading to the hot deformation of Ti-6Al-4V alloy.
- Strain rate sensitivity (m) is a key factor that depends on temperature as well as strain rate in predicting DSA. Also, it is applied in developing efficiency as well as instability maps.
- Processing maps have been generated at 0.35 strain (as m has been found to be negative) based on Prasad's and Murthy's instability criteria. Prasad's criterion doesn't predict DSA, while Murthy's criterion predicts DSA.
- From Murthy's criterion, DSA occurs over 700°C725°C and at 10-2 /s - 10-4 /s, it is denoted with shaded area and low efficiency (≤ 30%), whereas in the non DSA region, a maximum efficiency of 80% has been achieved.
- Therefore, in order to accurately predict DSA for Ti-6Al-4V alloy at high temperatures and for ensuring the safety during hot tensile deformation, Murthy's criterion can be considered.

References

Alabort E, Putman, D. and Reed, R. C. (2015). Super plasticity in Ti-6Al-4V: Characterisation, modelling and applications. Acta Materialia. 95:428–442.

Cai, j., Zhang, X., Wang, K., Wang, Q., and Wang, W. (2016). Development and Validation of Processing Maps for Ti-6Al-4V Alloy Using Various Flow Instability Criteria. J. Mater. Eng. Perfor. 25:4750–4756.

Ghasemia, E., Zarei-Hanzakia, A., Farabia, E., Tesarb, K., Jagerb, A., and Rezaee, M. (2017). Flow softening and dynamic recrystallization behaviour of BT9 titanium alloy: A study using processing map development. J. Alloy. Comp. 695:1706–1718.

Gillo Giuliano, G. 2011. Superplastic forming of advanced metallic materials methods and applications. Sawston: Wood head Publishing Limited.

Leyens C and Peters M. 2003. Titanium and Titanium alloys fundamentals and applications. Wiley-VCH Verlag GmbH & Co.

Li, M. Q. and Zhang, W. F. (2009). Effect of hydrogen on processing maps in isothermal compression of Ti-6Al-4V titanium alloy. Mater. Sci. Eng. 502:32–37.

Lukaszek-Solek, A. and Krawczyk, J. (2015). The analysis of the hot deformation behaviour of the Ti–3Al–8V–6Cr–4Zr–4Mo alloy, using processing maps, a map of microstructure and of hardness. Mater. Desig. 65:165–173.

Luo, J., Li, M., Yu, W. (2009). Effect of the strain on processing maps of titanium alloys in isothermal compression. Mater. Sci. Eng. 504:9098.

Meng, Q. G., Bai, C., and Xu, D. (2018). Flow behaviour and processing map for hot deformation of ATI425 titanium alloy. J. Mater. Sci. Technol. 34:679–688.

Park, N. K, Yeom, J. T, and Na, Y. S. (2002). Characterization of deformation stability in hot forging of conventional Ti-6Al-4V using processing maps. J. Mater. Proc. Technol. 130131:540–545.

Prasad, Y. V. R. K, Rao, K. P., and Sasidhara, S. 2015. Hot working guide: a compendium of processing maps. United Satetes: ASM international.

Sen, I., Kottada, R. S., and Ramamurty, U. (2010). High temperature deformation processing maps for boron modified Ti-6Al-4V alloys. Mater. Sci. Eng. 527:6157–6165.

Seshacharyulu, T., Medeiros, S. C., Frazier, W. G., and Prasad, Y. V. R. K. (2000). Hot working of commercial Ti-6Al-4V with an equiaxed α-β microstructure: materials modelling considerations. Mater. Sci. Eng. 284:184–194.

Seshacharyulu, T, Medeiros, S. C, Frazier, W. G, Prasad, Y. V. R. K. (2002). Microstructural mechanisms during hot working of commercial grade Ti-6Al-4V with lamellar starting structure. Mater. Sci. Eng. 325:112–125.

Sun, Y., Cao, Z., Wan Z., Hu, L., Ye, W., Li, N., and Fan, C. (2018). 3D processing map and hot deformation behaviour of 6A02 aluminium alloy. J. Alloy. Comp.

Wang, J., Zhao, G., and Li, M. (2016). Establishment of processing map and analysis of microstructure on multi-crystalline tungsten plastic deformation process at elevated temperature. Mater. Desig. 103:268–277.

Xing, J., Hou, L., Du, H., Liu, B., and Wei, Y. (2019). A New Explanation for the Effect of Dynamic Strain Aging on Negative Strain Rate Sensitivity in Fe–30Mn–9Al–1C Steel. Mater. 12:3426.

Zhe, W., Xinnan, W., and Zhishou, Z. (2017). Characterization of high-temperature deformation behaviour and processing map of TB17 titanium alloy. J. Alloy. Comp. 692:149–154.

116 Development of low cost deburring setup with feedback capability

S. T. Bagde[1,a], A. V. Kale[2,b], N. V. Lotia[3,c], and G. H. Waghmare[1,d]

[1]Mechanical Engineering Yeshwantrao Chavan College of Engineering, Wanadongri Nagpur, India

[2]P. R. Pote Patil College of Engineering and Management, Amravati, India

[3]Mechanical Engineering Anjuman College of Engineering and Technology, Sadar, Nagpur, India

Abstract

Commercially available deburring setups are costly and complicated because of the instrumentation and it is needed to select specific setups for specific applications. This research work addresses combination of operations like deburring and inspection on same system. The aim was to simplify the deburring setup to minimum essential components for the construction, instrumentation and to obtain feedback of deburred edge profile from it. Thus a simple deburring set up is developed and manufactured. Important aspect in this deburring setup is that the instrumentation and sensors are not used for correcting the path and maintaining uniform force but for giving feedback about the profile in terms of deflection and change of speed. The deburred part edges are compared with those of master part using image processing and acceptance or rejection is decided accordingly. Developed Deburring setup is versatile, use minimum and simple instrumentation for obtaining feedback about the profile being deburred, thus making it cost effective. Experimentation is carried out on this setup on eight samples each of four materials. Data is recorded in terms of deflection and change of speed indicating presence of burr. These data is analysed, using models developed by earlier researchers, for deburring force. It is also calculated from the grinding wheel geometry. When the values of deburring forces obtained from experimental results and those from the wheel geometry are compared, it is observed that the percentage difference is within 10% which means this setup is capable of deburring like existing set ups and has feedback capability.

Keywords: Automated deburring setup, deburring, robotic deburring, image processing.

Introduction

A burr is a small flake of metal attached on the surface or edges of a work piece. Most machining processes like milling, drilling, turning, and broaching generates burr. Deburring is removal of burrs using simple hand deburring to Robotic deburring. Deburring is necessary to avoid injuries to workers, to facilitate assemblies avoiding clearance restrictions caused by burrs, to reduce stress concentration at sharp corners causing part failures and to enhance aesthetics of work piece or assembly.

Literature review:

There are five types of deburring operations viz. manual, mechanical, thermal, chemical and electrical deburring. The deburring setup will also act as an on-line inspection station. Aurich et al (2009) discussed the issue of burr formation versus burr removal. Recent research focuses on burr control rather than burr avoidance Benati et al. (1999) focuses on the burr measurement methods. Burrs detection and measuring its size are very complicated matter as intrinsic variability in shape and dimension. There are no standards available for their acceptable dimensions and measurement techniques especially for sheet steel products. Fredrick Procter and Karl Murphy (1989) mentioned about the need for improved deburring technology for burr removal, its measurement and modelling of burr formation; they suggest the use of sensors for feedback, robot control and the automated process planning. Karpat and Ozel (2008) presented investigation of the effects of cutting conditions, heat generation, and resultant temperature distributions at the tool and work piece interface of high-speed machining with chamfered tools through analytical and thermal modelling. Pagilla and Yu (2001) presented complete dynamic model describing the dynamic behaviour of the robot for surface finishing processes like deburring, grinding, chamfering, and polishing. Kazerooni and Guo (1987) developed direct-drive, active compliant end-effectors (active RCC). The design and construction of a fast, light-weight, active end-effectors which can be attached to the end-point of a commercial robot manipulator are presented. Kazerooni (1986) theoretically and experimentally investigated the deburring process of manufactured parts as a frequency domain control problem especially if done by industrial robot manipulators. Robot oscillations and small uncertainties in the part location with respect to the robot need compensation for precision deburring. Norberto et al. (1987) developed an indirect force control strategy designed to operate with industrial robotic deburring applications. The system is developed to deburr high-quality knives. Bin and Masyoshi (1996) suggested a controller design for adaptive control of robot manipulator in constrained motion. For development of this adaptive motion and force

---

[a]asana123_in@yahoo.com; [b]bsvssngp@gmail.com; [c]cn_lotia@yahoo.com; [d]dgwaghmare@gmail.com

control of manipulators in constrained motion, parametric uncertainties both in the robot and contact surface are considered. Ling and Li-Chen (1998) proposed robotic deburring with controlled conditions. The robotic deburring removes the burrs at a constant tangential cutting force. A constant contact normal force and tangential cutting force is to be maintained to achieve smooth removal of burr.

Duelen et al. (1992) proposed industrial robots with hybrid position/force controller for stable metal removal. The hybrid control problem is solved at both servo and motion planning control levels. The main feature of the developed control scheme is its reliability and robustness. Feng-Yi and Li-Chen (1998) suggested a new hybrid position/force control of robot manipulators through adaptive fuzzy control approach to overcome manipulator control problems. Giacomo et al. (2005) proposed robotic deburring of planar work pieces with an unknown shape using mechatronics. He used suitable design of the deburring tool with hybrid force/velocity control law. Bone and Elbestawi (1994) developed automated robotic edge deburring using sensing and control elements of a system. The accurate sensing and control of the deburring chamfer depth was achieved through combination of force and vision information during the deburring pass and a new form of adaptive generalized predictive control (GPC) combined with learning control, termed GPC with Learning (GPCL).

Shimokura and Liu (1994) developed a lead through programming for teaching a deburring robot based on human skilful motion. The tool feed rate can be adjusted in accordance with the varying burr characteristics, such as burr size and material properties through robot programming. Liao et al. (2008) presented a dual-purpose compliant tool head for modelling and control of an automated polishing/deburring process. Tool compliance can be provided through pneumatic spindle that can be extended and retracted by three pneumatic actuators. This tool head can be used for polishing and deburring by integrating a pressure sensor and a linear encoder. Lee et al. (1993) presented a force model for deburring found suitable for up cut and down cut grinding. An algorithm of 2-D vision system for burr detection is developed using this model. The algorithm uses the frequency, cross section area and height of burr which are the functions of burr contour. A burr contour tracking method (BCTM) is developed to obtain burr data.

The paper addresses combination of operations like deburring and inspection on same system.

Research gap and development of setup:

Major deterrents in use of robots for deburring are the cost, complexity of tooling and difficulty in programming the robot to follow complex contours to be de-burred The aim was to simplify the deburring setup with minimum essential components for the construction/instrumentation and to obtain feedback of de-burred edge profile from it. Thus a simple deburring set up is developed and manufactured. A new setup having bidirectional compliance is developed with instrumentation for feedback capability. It has a table which can move in x-y direction along with a nice fitted over this table for work piece clamping. Deburring compliant tool mounted on z axis is as shown in Plate 116.1.

Experimentation is carried out on this setup with four materials and observed that this setup is capable of deburring with feedback capability.

Experimentation

The experimentation was carried out for four materials (mild steel, stainless steel, copper and aluminium). Eight samples of each material having rollover burr are used.

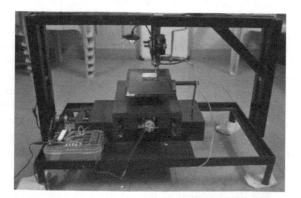

Plate 116.1 Deburring setup Triangular element of the grit

Methodology:

The setup is operated with two speeds of tool alternatively. Two speeds (5000 rpm and 10000 rpm) are used for deburring. During operation, inspection is carried out in terms of deflection and variation of speed of the deburring tool. Kiel software is used for path generation and controlling deburring parameters like feed/plunging depth of deburring tool. Work piece is clamped in a vice which is fixed on the table. The movement of table in x and y directions is achieved through two separate DC motors. A cone shaped grinding wheel (A3) is used for deburring which is mounted in the collate fitted on motor shaft with axis vertically downward. The material of the grinding wheel is Zirconium alumina.

The motors of x-y table are actuated through controller hardware. The deburring path is obtained using Keil programming. This path is fed to the controller and deburring operation is performed. The deflection of tool and change in speed is recorded in database. Deburring and inspection is performed at same station. Initially, the image of work piece with the burr is stored. This image is used for the comparison with the image of deburred work piece. The feedback received from the instrument is cross checked with the stored image. After the deburring is over, the image of finished work piece is obtained. This image is compared with master image using image processing. Automatic inspection is done through image processing using MATLAB software. graphic user interface (GUI) is developed using VB for inspection unit.

Experimental results:

Eight sets of experiments for each material i.e. four experiments for first level of speed (5000 rpm) and remaining four experiments for second level of speed (10000 rpm) were conducted. The data generated is voluminous, sample data is shown in Table 116.1.

Figures 116.1 and 116.2 shows the sample graphs for deflection and change in speed plotted with respect to time intervals. These graphs are compared with image of work piece to check the correctness of output parameters through the occurrences of burr

Table 116.1 Experimental results for one edge (with 10000 RPM)

Sr No	Time	Actual RPM of grinding wheel(N)	Deflection of grinding wheel*10000(δ)
1	0.50	9197	9385
2	1.00	9322	9512
3	2.00	9225	9413
4	3.00	9124	9310
5	4.00	9342	9533
6	5.00	9257	9446
48	46.00	9270	9459
49	48.50	9223	9411
50	50.00	9290	9480

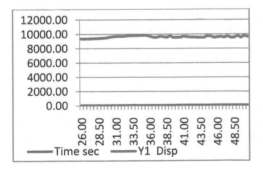

Figure 116.1 Graph of time vs deflection

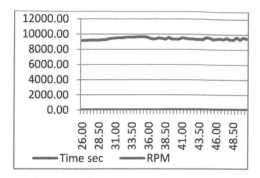

Figure 116.2 Graph of time vs speed

Deburring Force Calculations

Using experimental results

Rubenstein (1972) divided the cutting force in grinding into the:

- force for chip formation
- force arising from the finite radius of curvature of the cutting edge,
- the friction force between the flank wear land and the work piece,
- cutting force for the grains
- force for the grains to plough the work piece and
- Friction force between the wheel bond and the work piece material.

The ploughing force can be neglected as it is very small.

The cutting force can be simply represented by the friction force and the chip formation force. The grinding process becomes a pure friction process with very small normal force when there is no chip formation. When there is either no friction force or it is much smaller than the chip formation force the grinding process becomes a pure chip formation process.

Hence, the force model for deburring processes is given by Lee et al (1993) as:

$$F_n = \frac{2k_c}{D}(\frac{Vw}{Vs})A_{work} + 2k_t\, a_{root} L \qquad (1)$$

$$F_t = \frac{2\phi k_c}{D}(\frac{Vw}{Vs})A_{work} + 2\mu k_t\, a_{root} L \qquad (2)$$

and

$$\text{Total force F} = \sqrt{F_t^{\,2} + F_n^{\,2}} \qquad (3)$$

In experimentation, four materials are used i. e. mild steel, stainless steel, aluminium and copper. The properties of these materials and values of constants are given in Table 116.2.

Eight samples of each material are used for deburring with roll over burr (Pate 2). This type of burr is obtained on four edges of these samples by milling. Each edge is divided into fifty equal parts by the software on time scale. Thus 200 readings of speed and deflection are obtained per sample. The values given in Table 116.3 are average values of the averages of eight samples for each material.

Figure 116.3 shows enlarged view of tool contact, h, d and l are contact dimensions in various planes with respect to work piece edge. From geometry of Figure 116.3, h = d tan θ: d= plunging depth of tool = 0.05 mm, (adjusted in the program): θ = 600 for grinding wheel. Therefore, h = 0.087 mm and l = 0.1 mm. Values of the terms in Equations 1 and 2 with values of deburring force from Equation 3 are as shown in Table 116.3.

Table 116.2 Values of strengths, K_c and K_t

Sr. No	Material	S_{ut}	S_{yt}	S_{ys}	BHN	K_c	K_t
1	Mild steel	379	218	140	110	273	17.5
2	Stainless steel	610	240	135	135	300	16.9
3	Aluminium	225	112	65	60	140	8.30
4	Copper	310	170	104	90	213	13.0

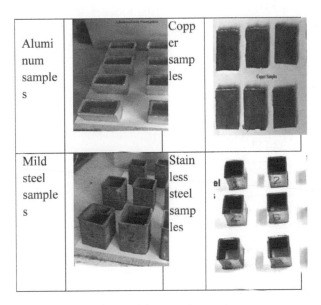

Plate 116.2 Samples used for experimentation

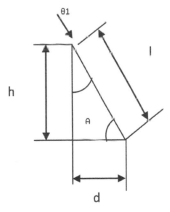

Figure 116.3 Enlarged view of tool contact with work piece

Based on wheel geometry

A force model for analysis of deburring force is developed. The single grit of conical shape is considered for this model. The origin of this coordinate system is fixed at the point O. Each grit has its local rotating coordinate system. The local coordinate system 'tya' is fixed to the individual grits as shown in Figure 116.4. The origin of this coordinate system is fixed at the tip of individual grits. The direction 'y' of the rotating coordinate system is the same as the direction 'y' of the fixed reference system as given by Ghosh (1993).

Only half of the individual grit is involved in the cutting operation at any instance i.e. only half portion of the grit is inside the work.

Table 116.3 Average values of parameters

S.No	1	2	3	4
Material	Mild Steel	Stainless Steel	Aluminum	Copper
$V_{w, mm/min}$	120	120	120	120
RPM	8785	8403	8691	8686
Vs	413774	395761	409365	409114
δ	8964	9263	9582	9215
a_{root}	0.8964	0.9263	0.9582	0.9215
a	0.05	0.05	0.05	0.05
w_{root}	17.928	18.526	19.163	18.431
A_{burr}	8.036	8.581	9.181	8.5
$A_{chamfer}$	0.002	0.002	0.002	0.002
A_{work}	8.039	8.583	9.183	8.503
F_n	3.222	1.065	1.607	2.559
F_t	0.9	1.207	0.452	0.717
Ftotal N	3.345	4.529	1.67	2.658

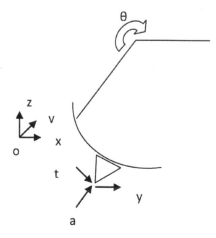

Figure 116.4 Enlarged view of the work piece-grinding wheel contact region

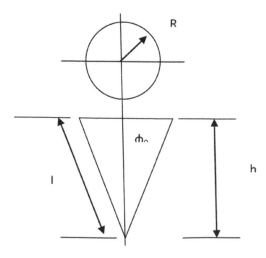

Figure 116.5 Grit geometry

From Figure 116.5,

$$R = h \tan \phi_0 \qquad (4)$$

$$l = \frac{R}{\sin \phi_0} \qquad (5)$$

For mathematical treatment, grit is shown as triangular element. The triangular element will undergo the similar cutting action as of in single point cutting tool. The mathematical model can be represented by summation of all forces on this element with reference to Figure 116.6.

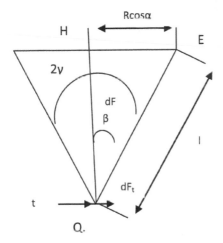

Figure 116.6 Triangular element of the grit

$$F_{total} = F_{xtotal} + F_{ztotal} \qquad (6)$$

$$F_{xtotal} = \sum_{i=1}^{N} F_{x\,grit_i} \qquad (7)$$

$$F_{ztotal} = \sum_{i=1}^{N} F_{z\,grit_i} \qquad (8)$$

Where, N is the total number of grits engaged with work.
It can be calculated from the relation
N = contact length / size of a single grit
N = 0.1 / 0.024 = 4

$$N = 0.1 / 0.024 = 4$$

$$F_{x\,grit} = -2\sin\theta \int_{\alpha=0}^{\alpha=\pi/2} dFt - 2\cos\theta \int_{\alpha=0}^{\alpha=\pi/2} dF\alpha$$

$$F_{z\,grit} = 2\cos\theta \int_{\alpha=0}^{\alpha=\pi/2} dFt - 2\sin\theta \int_{\alpha=0}^{\alpha=\pi/2} dF\alpha$$

$$F_{y\,grit} = 0$$

Where,

$$\int_{\alpha=0}^{\alpha=\pi/2} dF_t = k_a \int_{\alpha=0}^{\alpha=\pi/2} \sin\beta\, dA$$

Putting the value

$$\int_{\alpha=0}^{\alpha=\pi/2} dF_\alpha = k_a \int_{\alpha=0}^{\alpha=\pi/2} \frac{\cos\beta dA}{\sqrt{1 + \tan^2\phi_0 \sin^2\phi}}$$

$$\int_{\alpha=0}^{\alpha=\pi/2} dF_t = k_a \int_{\alpha=0}^{\alpha=\pi/2} \sin\beta lRd\alpha/2 \text{ and,}$$

$$\int_{\alpha=0}^{\alpha=\pi/2} dF_\alpha = k_a \int_{\alpha=0}^{\alpha=\pi/2} \frac{\cos\beta lRd\alpha/2}{\sqrt{1 + \tan^2\phi_0 \sin^2\phi}}$$

For the wheel of Zirconium Alumina, values are as follows:
ϕ_0- semi cone angle= 300
H – Height of grit inside the work = 0.012 mm
R – Radius of cone at height h = 0.014 mm
l – Slant height of cone = 0.028 mm
D – Diameter of tool = 15 mm
θ – Angular position of grit = 2700
β is the angle from centre line at which the resultant cutting force will act = 600
α = 450
ψ = 600
N = 4
k_a = unit cutting force = 2.5 * Sut
Substituting the values of Sut and k_a from Table 116.4, total force is calculated from Equation 6. The values of forces for four materials are as shown in Table 116.5.

Table 116.4 Values of ultimate tensile strength and k_a

Sr. No.	Material	Sut	k_a
1	Mild steel	379	947.5
2	Stainless steel	610	1525
3	Aluminium	225	562.5
4	Copper	310	775

Table 116.5 Total deburring force values for different materials (analytical)

Sr No	Material	F_{xgrit}	F_{zgrit}	F_{xtotal}	F_{ztotal}	F_{total}
1	Mild steel	0.504 N	0.356 N	2.016 N	1.424 N	3.44 N
2	Stainless steel	0.812 N	0.402 N	3.248 N	1.608 N	4.86 N
3	Aluminium	0.3 N	0.148 N	1.2 N	0.532 N	1.73 N
4	Copper	0.408 N	0.204 N	1.82 N	0.816 N	2.64 N

Table 116.6 Comparison of the values of deburring forces

Sr No.	Material	Experimental $F_{total, N}$	Analytical $F_{total, N}$	% Difference
1	Mild steel	3.35	3.44	2.6
2	Stainless steel	4.52	4.86	9.4
3	Aluminium	1.56	1.73	4.04
4	Copper	2.51	2.64	4.9

Results and Discussion

The force values are computed with the help of experimental results and analytical models for four different materials. Considering four edges for deburring, there are 200 readings of experimental values for a single sample, average values of the averages of eight samples are presented in Table 116.6.

From Table 116.6, it is observed that the percentage difference between the experimental and analytical results is within 10%, which indicates that the low cost deburring setup can be used efficiently and economically as compared to the costly robotic setups. The expressions used for calculating force values are pertaining to existing deburring setups. If commercialized on large scale, this setup may prove to be very useful and economical for all types of industries. In that case, the deburring forces will vary according to the wheel geometry and its type. Performance of deburring setup can be checked and corresponding corrections could be made for actual deburring forces with the help of instrumentation.

Conclusions

Following conclusions can be drawn:

1. Developed set up is simple, easy to manufacture with less instrumentation.
2. This method of deburring eliminates need of path correction. Deburring is possible without instrumentation. Instrumentation is used only to obtain feedback about the profile. It is not used for path correction.
3. Deburring setup can be used as a feedback device to obtain information about the manufactured profile subjected for deburring.
4. It is possible to identify the presence of burr while carrying out the deburring operation by image processing
5. Feedback while deburring can open out new methods of manufacturing which eliminates need of an independent workstation for quality control. Such workstation can ensure 100% quality checking for parameters.
6. Simplification of deburring process and value addition to it is successfully demonstrated in this work.
7. The low cost deburring setup can be used efficiently and economically as compared to the costly robotic setups.
8. If commercialized on large scale, this setup may prove to be very useful and economical for all types of industries.

Future Scope

The set up presented here is capable of carrying out the finishing (deburring) and inspection simultaneously, but there is always a scope for development. Following are some of the suggestions for future work:

1. The setup is capable of deburring a rectangular or square shaped geometry. This is to be modified for any shape by adding interpolation facility.
2. The instrumentation for feedback is done for receiving data about the deflection and speed of the tool. In future, this is to be modified for receiving more data like force acting on tool / work piece etc.
3. The data received is stored into database and after completion of a cycle, it is transferred for analysis. In future, a Graphic User Interface to be developed so that the data can be displayed directly on screen during the processing of work piece, like medical equipment.
4. The present set up can be programmed for rectangular/square shaped geometry. In future the CAD based set up is to be developed where the machine will trace the path as per the CAD drawing on screen. In that case the tool path will be generated through CAD model.

Limitations

The methodology used has certain limitations as listed below:

1. The deburring tool used in this application is tapered, whenever there is a change in dimension of the work piece, the conical tool contacts the edge to be deburred at varying radii. This introduces error in displacement readings.
2. The controller of the system is designed to have linear interpolation in X and Y directions only.
3. The recirculation ball screws can be used for reducing friction.

References

Aurich, J. C., Dornfeld, D., Arrazola, P. J., Franke, V., and Leitz, L. (2009). Burrs—Analysis, Control and Removal. CIRP Ann. Manuf. Technol. 58:519–542.

Benati, F., Butler, C., Gatti, S., Sacerdotti, F., andYang, Q. (1999). A robot-based burr measurement system for the automotive industry. IMTC/99. Proceedings of the 16th IEEE Instrumentation and Measurement Technology Conference. IEEE. doi: 10.1109/IMTC.1999.776120

Bin, Y. and Massayoshi, T. (1996). Adaptive control of robot manipulator in constrained motion - controller design. J. Dyn. Syst. Meas. Control Trans. ASME. 117.

Bone, G. M. and Elbestawi, M. A. (1994). Sensing and Control for Automated Robotic Edge Deburring. IEEE Trans. Ind. Electron. 41(2).

Duelen G., Munch H., Surdilovic D., and Tim, J. (1992). Automated force control schemes for robotics & burring: development and experimental evaluation.

Feng-Yi, H. and Li-Chen, F (2000). Intelligent robot deburring using adaptive fuzzy hybrid position/force control. IEEE Trans. Robot. Automation. 16(4).

Ghosh S. (1993). Modelling and analysis of grinding process. PhD diss., The University of Maryland.

Giacomo, Z., Giovanni, L., and Antonio, V. (2005). A mechatronic design for robotic deburring. Dubrovnik, Croatia: IEEE ISIE.

Karpat, Y. and Ozel, T. (2008). Analytical and thermal modelling of high speed machining with chamfered tools. J. Manuf. Sci. Eng. 130.

Kazerooni, H. and Guo, J. (1987). Direct-drive, active compliant end-effector (active RCC). Proceedings. 1987 IEEE International Conference on Robotics and Automation. IEEE. doi: 10.1109/ROBOT.1987.1087870.

Kazerooni, H., Bausch, J. J., and Kramer, B. M. (1986). An approach to automated deburring by robot manipulator. J. Dyn. Syst. Meas. Control Trans. ASME. 108:359.

Lee, K. C., Huang, H. P., and Lu, S. S. (1993). Burr detection by using vision image. Int. J. Adv. Manuf. Technol. 8:275–284.

Liao, L., Xi, F., and Liu, K. (2008). Modeling and control of automated polishing/deburring process using a dual-purpose compliant toolhead. Int. J. Mach. Tools Manuf. 48:1454–1463.

Ling, I. C. and Li-Chen, F. (1998). Manipulator for automated deburring with on-line adaptive hybrid force/position control of a flexible cutting trajectory modification. Proceedings of the 1998 IEEE international conterence on robotics. 1.S:Au:otnatloi, Leuven, Belgium.

Norberto, P. J., Gabriel, A., and Nelson, E. (2007). Force control experiments for industrial applications: a test case using an industrial deburring example. Assem. Autom. 27(2):148–156.

Pagilla, P. R. and Yu, B. (2001). Robotic surface finishing processes: modeling, control, and experiments. J. Dyn. Sys. Measur. Control. 123.

Proctor Frederick, M. and Murphy Karl, N. (1989). Advanced Deburring System Technology. Mechanics of Deburring and Surface Finishing Process. ASME PED. 38:115.

Rubenstein C. (1972). The mechanics of grinding. Int. J. Mach. Tool Des. Res. 12:127–139.

Shimokura, K. I. and Liu, S. (1994). Programming deburring robots based on human demonstration with direct burr size measurement. IEEE.

117 Multi-response optimisation using hybrid AHP and GRA technique in face milling of EN-31 steel

Vijay Kumar Sharma[a], Talvinder Singh[c], and Kamaljeet Singh[b]

Department of Mechanical Engineering, Chitkara University Institute of Engineering and Technology, Chitkara University, Punjab, India

Abstract

In the multi-response optimisation technique, the determination of the weight percentage of each output response is indeed an essential task. In this study Analytical Hierarchy Process (AHP) is used to determine the weight fraction of output response and it is coupled with the GRA method to form a hybrid system of multi-criteria decision making while machining EN-31 steel under minimum quantity lubrication technique. The machining parameters are coupled with the MQL variables as per Taguchi-based orthogonal array L27. Then optimisation of the calculated multi-response factor is carried out by employing Taguchi based analysis of means (ANOM) and analysis of variance (ANOVA) to simultaneously optimise the tool flank wear, surface roughness, and surface microhardness. The optimised machining conditions are obtained as Vc = 110 m/min, f = 60 mm/min, d = 0.2 mm, LC = 60% and LFR = 150 ml/hr and this ideal parametric combination is further authenticated using confirmation experiments. An in-depth analysis of worn-out edges of the tool inserts revealed that adhesion is the main cause of flank wear. However, the formation of built-up edge and micro-chippings are also noticed at higher cutting speeds especially when the lubricant flow rate is low.

Keywords: Minimum Quantity Lubrication, Grey Relational Analysis, Analytical Hierarchy Process, Surface Roughness, Tool Flank Wear.

Introduction

A high amount of heat generation while machining the hard carbon steel such as EN-31, is the foremost reason for spoiling the machining performance criteria such as tool behaviour, surface integrity, and productivity. The unmatchable trend of competition forces the manufacturing industries to produce the parts at a high rate. Thus, machining at high speed is preferred to meet the requirements of higher production rates. However, employing high-speed machining invites high-temperature developments and leads to deterioration of the quality of produced parts (Weinert et al., 2004). Therefore, the usage of the lubricant is compulsory for lowering the temperature in machining. Now, the limitation with the usage of lubricant is its harmful effects on the worker's health and on the environment (Schwarz et al., 2015). Other than these issues, the surplus usage of lubricant is avoided due to higher expenses related to it. minimum quantity lubrication (MQL) presents itself as a solution to this problem MQL is considered a possible alternative to other cooling strategies in which a very low amount of cutting fluid (50500 ml/hr) is delivered to the cutting area (Babu et al., 2019; Singh et al., 2018). The aerosol mixture entering the machining area helps in the formation of a protective lubrication layer between the chip-tool and work-tool interfaces which leads to improved machining performance (Liao and Lin, 2007; Liew, 2010). Initially, the MQL technique was limited to the machining of relatively softer materials like aluminium alloy. Kishawy et al. (2005) observed lesser flank wear in MQL assisted machining of aluminium alloy as compared to machining under conventional cooling techniques. It was found that MQL is effective in high-speed machining (Kishawy et al., 2005). De Lacalle et al. (2006) also found lower flank wear under MQL as compared to flood cooling during milling of aluminium alloy Similar outcomes of surface integrity for experimental investigation of Conger et al. (2009) while milling Aluminium alloy AI-6061. MQL performed better than dry cutting. Yazid et al. (2019) also found the MQL as a valuable approach in milling. The improved surface finish was detected with lower lubricant discharge when optimisation of MQL parameters along with machining parameters is done in the milling of aluminium alloy. Tosun et al. (2010) evaluated the surface roughness of aluminium alloy during milling operation with the MQL technique. Different milling and MQL parameters were selected to check their response. Observations showed improved surface finish for MQL based machining than conventional wet cooling. In recent times, many researchers have employed the MQL technique in the machining of hard to metals such as steel alloys, titanium alloy, etc. Sharma et al. (2021) used the MQL system while milling EN-31 steel and compared the outcomes of tool wear and surface properties with flood cooling. Statistical results revealed that the response of MQL matches the performance of flood cooling.

The MQL system has certain input parameters which can be optimised or can be coupled with machining parameters to obtain an ideal combination for overall improvement in machinability. Lubricant flow rate (LFR) is the chief input variable associated with the MQL and a lot of researchers optimised the quantity of lubricant supplied to the cutting zone. Sales et al. (2009) found that with the increase in LFR the tool wear

[a]vijayk.sharma@chitkara.edu.in; [b]talvinder.singh@chitkara.edu.in; [c]kamaljeet.singh@chitkara.edu.in

reduces. Appropriate penetration of cutting fluid in the cutting area was observed by Cai. et al. (2012) at a higher lubricant supply rate during MQL assisted end milling of titanium alloy. Thus, resulting reduction in the surface roughness and cutting forces. Hassanpour et al. (2016) and Singh et al (2021) also found the identical effects of LFR on surface roughness in MQL based milling of alloy steel. Apart from lubricant flow rate, Liu et al. (2011) worked on the optimisation of other MQL variables like spraying distance, air pressure, and spraying angle for lowering the cutting temperature. The optimum parametric setting of the MQL parameters for lower cutting temperature during milling on titanium alloy is obtained as 25 mm spray distance 0.6 MPa air pressure and 135o spraying angle. During the milling of carbon steel, Rooprai et al. (2021) reported an enhancement in surface finish by optimising MQL parameters by combining them with machining parameters. Yan et al. (2012) worked on the optimisation of the nozzle position to enhance the lubricating effects of the MQL system while machining steel alloys. Ideal position of the nozzle position with 120o nozzle-work angle, 60o Nozzle-spindle axis angle, and 20 mm spraying distance. Improvement in the surface qualities is also detected by Rana et al. (2021) by optimising the MQL nozzle position at 45o nozzle elevation angle 40 mm spraying distance in milling of AISI 52100.

The main purpose of this is to concurrently optimise the output responses Surface Roughness (Ra), Flank wear (VB), and Surface microhardness (Hv) in MQL based face milling of EN-31 steel employing multi-response optimisation techniques Grey relational analysis (GRA). All three output responses are assigned the weights by employing Analytical Hierarchy Process (AHP) method. The objective of parametric optimisation is accomplished by exploiting Taguchi-proposed Analysis of Means (ANOM) and Analysis of Variance (ANOVA) which are the statistical tools that are utilised by many researchers to discover the degree of influence and significance of input variables on output characteristics (Rooprai et al., 2021; Yan et al., 2012; Rana et al., 2021). SEM analysis is carried out for the detailed analysis of tool wear patterns along with wear mechanisms and their effects on the surface finish.

Material and Methods

Specimen preparation and experimental setup

EN-31 steel is one of the difficult to cut alloy steel having high compressive strength, is chosen as specimen material. The chemical composition of EN-31 is as 0.978% carbon, 0.353% Manganese, 0.215% Silicon, 1.01% Chromium, 0.089% Sulphur and 0.086% Phosphorous. It has applications in the making of die parts, injection molding parts, bearings, several automotive parts, etc (Sharma et al., 2022). Grinding and milling operations are performed to convert raw strips of EN-31 steel into specimens of sise 75 mm × 75 mm × 16 mm. The experimentation is accomplished on the HURCO-VM10 vertical milling centre upon which additional MQL equipment is fitted. The lubricant flow rate is controlled by the air and oil-controlling valves. The mixture of water and Soluble oil (SAFCO RUBRIC) is chosen as a lubricant. Three triangular carbide inserts (HRC: 81) are mounted on Widia face milling cutter having BT-40 type shank and of 50 mm diameter.

For the current work, the parameters associated with the MQL system are coupled with machining parameters. For determination of the range and level of input parameters the pilot experimentation is executed employing one parameter at a time approach and published literature is also utilised (Liao and Lin, 2007; Sales et al., 2009; Rooprai et al., 2021; Yan et al., 2012; Rana et al., 2021). On the basis of both and because of a certain design of experiment-related limitations, some parameters are kept on a constant value. These are: Spraying distance = 50 mm; Nozzle angle = 45o and machining length = 450 mm (Sales et al., 2009; Sales et al., 2009; Sharma et al., 2022). The different variables of experimentation are tabulated in Table 117.1. Taguchi's Proposed design of the experiment is utilised to arrange input parameters and L27 orthogonal array is formed. Tool flank wear was measured by removing inserts after each experiment using a Sipcon Vision Measurement system (Model SVI-3D-CNC). A lot of researchers proposed that the

Table 117.1 Experimental conditions

Cutting speed, Vc	110, 165. 220 (m/min)
Feed rate. f	60, 120, 180 (mm/min)
Depth of cut, d	0.2, 0.4, 0.6 (mm)
Cutting condition	MQL, air pressure: 7 bar
Lubricant flow rate, LFR	50, 100, 150 (ml/hr)
Lubricant concentration, LC	50, 60, 70 (%)

flank wear is observed to check the performance of the tool, and also, its development disturbs the finish of machined surface (Sharma et al., 2021; Sharma et al., 2022) The average values of Surface roughness (Ra) are measured at three points on the specimen using Mitutoyo surface roughness tester. Surface microhardness is determined using Vicker hardness tester.

Analytical hierarchy process: procedure and implementation

The AHP method is developed by Saaty and Vargas, (1991) which is utilised to ease the decision-making that involves multiple criteria. In the multi-response optimisation techniques determination of the weight percentage of each output response is indeed an essential task. The limitation of conventional optimisation techniques is the individual optimisation of the output response. The optimal setting of one output response may not give the ideal result for other responses. AHP is a subjective method of weight determination which caught the attention of the researchers for determining the weight fraction of output responses. The relative importance depends on the requirements of the researcher which are further derived by the industrial requirements (Saaty and Vargas, 1991; Marinoni, 2004). Table 117.2 presents the measure of comparisons which quantifies the relative priority of the output responses. Scale values 1, 3, 5, 7, and 9 determines the extent of priority of one response in comparison with other, and Scale value 2, 4, 6, and 8 are allocated to in-between value if needed. For the current study, tool wear is considered moderately important over surface roughness and strongly important over surface microhardness. Similarly, surface roughness is considered moderately important over surface microhardness. Thus, scales are selected accordingly.

Once the scaling is accomplished, the pairwise assessment matrix [P] is formulated. Designated scale values are put in the cells of matrix row-wise showing the pair of responses and in transpose cell, reciprocal value of scale is placed (column-wise) as shown in Table 117.3. Thereafter, the normalisation of values of the pairwise assessment matrix is calculated by dividing the value of each cell by summation of all column values. Then, the weight of every output response is calculated employing Equation 1 and the weight matrix [W] is formed. The normalised value and the response's weight calculations are tabulated in Table 117.4. These calculated weights are further utilised in the multi-response optimisation technique GRA for concurrent optimisation.

Table 117.2 Measure of comparisons (Saaty and Vargas, 1991)

Relative importance of factors	Scale of importance
Both factors are identically important	1
One factor moderately important as compared to another	3
One factor strongly important as compared to another	5
One factor very strongly important as compared to another	7
One factor very extremely important as compared to another	9

Table 117.3 Pairwise assessment matrix [P]

Factors	R_a	V_B	H_v
R_a	1	3	5
V_B	1/3 (0.33)	1	3
H_v	1/5 (0.2)	1/3 (0.33)	1
	1.53	$\sum_{i=1}^{i=n} a_{ij}$ 4	9

$$\text{Weight of response } (W_i) = \frac{\sum_{j=1}^{j=n} b_{ij}}{no.of\ factors}, \qquad (1)$$

Where, i = 1,2,3 and $\Sigma_{j=1}^{j=n} b_{ij}$ is the sum of normalised values in ith row.

Table 117.4 Normalised matrix and responses' weight

Factor	R_a	V_B	H_v		Responses' Weight $\sum_{j=1}^{j=n} b_{ij}$ [W]
R_a	0.6535	0.6928	0.5555	1.9018	0.6339
V_B	0.2156	0.2309	0.3333	0.7798	0.2599
H_v	0.1307	0.0762	0.1111	0.318	0.106

As the AHP method is entirely based on human judgement and scaling of relative priority, inconsistency may exist. Thus, is become mandatory to check the consistency of calculated weight. The process of consistency check starts by multiplying the pairwise comparison matrix [P] with weight matrix [W] to form a matrix [C] as given in Equation 2. Then, [C] matrix is divided by weight matrix [W] to obtain the values of the consistency vector, and the consistency vector-matrix [Cv] is generated as shown by Equation 3.

$$[P]^* [W] = [C] \tag{2}$$

$$\begin{bmatrix} 1 & 3 & 5 \\ 0.33 & 1 & 3 \\ 0.2 & 0.33 & 1 \end{bmatrix} \times \begin{bmatrix} 0.6339 \\ 0.2599 \\ 0.106 \end{bmatrix} = \begin{bmatrix} 1.94 \\ 0.787 \\ 0.318 \end{bmatrix}$$

$$[C] \div [W] = [C_v] \tag{3}$$

$$\begin{bmatrix} 0.6339 \\ 0.2599 \\ 0.106 \end{bmatrix} \div \begin{bmatrix} 1.94 \\ 0.787 \\ 0.318 \end{bmatrix} = \begin{bmatrix} 3.06 \\ 3.02 \\ 3.01 \end{bmatrix}$$

In the next step of consistency checking, the maximum eigenvalue of the comparison matrix, λmax is calculated by using Equation 4. As the value of λmax, consistency index (CI) is calculated by using Equation 5. After that, consistency ratio (CR) is by dividing consistency index by random index (RI) as mentioned in Equation 6. The value of RI is taken as 0.58, from a random consistency table formulated by Saaty and Vargas (1991), where the 1st row (n) specifies the no. of output responses and the 2nd row is the Random consistency index. If CR ≤ 0.1, the inconsistency is within satisfactory limits. Otherwise, the inconsistency is large and the decision-maker is advised to rethink the relative importance of the responses.

$$\lambda_{\max} = \frac{\Sigma c_V}{number\ of\ factors} \tag{4}$$

$$CI = \frac{\lambda_{\max} - n}{n-1} \tag{5}$$

$$CR = \frac{CI}{RI} \tag{6}$$

The value of λmax, CI, and CR is calculated as 3.03, 0.016, and 0.0286 respectively. The value of RI is taken as 0.58 as there are three output responses (n = 3). As the CR value is less than 0.1, which signifies that calculated weight fractions are consistent and pairwise comparison is accurately done.

Grey relational analysis

In current work, GRA is exploited to simultaneously optimise VB, Ra, and Hv by converting them into a single response called GRG. The various steps of GRA are as given below (Singh et al., 2021; Saaty and Vargas, 1991).

Step 1: Grey relational generation for normalisation of data

The first step is to normalise the calculated response data and is depends upon the type of output response. If output response is to be minimised, then the normalisation is done employing 'smaller-the-better' approach using Equation 7. If higher value of output response is desired, then the 'larger-the-better' approach is used (Equation 8).

$$x_{is}^*(k) = \frac{\max x_i(k) - x_i(k)}{\max x_i(k) - \min x_i(k)} \quad (7)$$

$$x_{il}^*(k) = \frac{\max x_i(k) - x_i(k)}{\max x_i(k) - \min x_i(k)} \quad (8)$$

Here, i = 1, 2...., m & k = 1, 2...., n. 'm' is the no. of experiments and 'n' is the total no. of responses. xi*(k) represents the normalised value for the ith outcome. xi(k) represents the original sequence, max xi(k) and min xi(k) represent the highest and lowest values of the xi(k).

Step 2: Calculation of the grey relational coefficient (GRC)

In this step, the value of GRC is calculated using a formula as given as eq. (9);

$$\text{GRC}, \xi_i(k) = \frac{\Delta_{min} + \xi\Delta_{max}}{\Delta_{oi}(k) + \xi\Delta_{max}} \quad (9)$$

Where Δoi(k) is the deviation sequence of the reference sequence and the comparability sequence and is calculated by the formula given as Eq. 10;

$$\Delta_{oi}(k) = \|x_o(k) - x_i(k)\| \quad (10)$$

The absolute differences of comparing sequences are represented by Δmin and Δmax, where Δmin & Δmax show the lowest & highest difference values. xo(k) is the reference sequence and xi(k) is the comparable sequence. ξ is known as the identification coefficient. The value of ξ varies from 0 to 1 (commonly taken as 0.5).

Step 3: Calculation of the Grey Relational Grade (GRG)

Formula is used to determine the value of GRG considering weight fraction 'W' of each factor as calculated through AHP method, is presented as Equation 11. The higher value of GRG in the table indicates the better results of multi-responses in combination.

$$\text{GRG}, \gamma_i = \sum_{k=1}^{n} \xi_i(W_i k) \quad (11)$$

Results and Discussions

Face milling experimentation is executed as per L27 orthogonal array without any interruption for a machining length of 450 mm (6 passes of 75 mm). Every trial is done thrice to eliminate the irregularity in the output of responses during machining and further analysis is done considering the mean of output response values of every experiment. After each experiment, the inserts are changed with fresh ones for independent measurement of wear land on cutting edge. The measured values of average Ra, VB, and Hv for all experiments are given in Table 117.5. After that, all steps of GRA were implemented to convert the all three responses into one factor GRG and presented in Table 117.6.

Analysis of variance (ANOVA) is executed with 95% confidence level to know the significance of input parameters on GRG. Various statistics of ANOVA are calculated by using MINITAB 19 software and given in Table 117.7. At 95% confidence level of ANOVA, the variables with the P-value below 0.05, is termed as significant one. The P-value of LC is more than 0.05 which makes it non-significant for output response. Also, the F-ratio value of LC (0.35) is less than the critical value of F-ratio which is 3.63 for a factor having a degree of freedom of two along with the degree of freedom of error as sixteen. All other parameters are significant ones as they have P-values below 0.05 and F-ratio values are greater than 3.63. ANOM is used for obtaining the optimal parametric setting for GRG. The mean GRG is calculated for all input parameters in Minitab 19 software and given in Table 117.8. The graphs showing the influence of input parameters

Table 117.5 Taguchi proposed L27 array and output measurements

Exp. No	VC	f	d	LC	LFR	Ra (μm)	VB (μm)	Hv (HRV)
1	110	60	0.2	50	50	0.34	107	327
2	110	60	0.4	60	100	0.47	109	324.2
3	110	60	0.6	70	150	0.65	129	330.1
4	110	120	0.2	60	150	0.46	98	331
5	110	120	0.4	70	50	0.68	147	344.5
6	110	120	0.6	50	100	0.65	144	336.3
7	110	180	0.2	70	100	0.61	135	347.9
8	110	180	0.4	50	150	0.67	156	341
9	110	180	0.6	60	50	1.02	228	351.6
10	165	60	0.2	60	150	0.29	132	314.2
11	165	60	0.4	70	50	0.48	168	324
12	165	60	0.6	50	100	0.59	192	319.6
13	165	120	0.2	70	100	0.44	132	329.6
14	165	120	0.4	50	150	0.54	141	325.8
15	165	120	0.6	60	50	0.87	194	355.2
16	165	180	0.2	50	50	0.64	211	348.6
17	165	180	0.4	60	100	0.63	189	342.6
18	165	180	0.6	70	150	0.93	222	343.1
19	220	60	0.2	70	100	0.39	159	332.2
20	220	60	0.4	50	150	0.53	161	329.4
21	220	60	0.6	60	50	0.81	224	343.3
22	220	120	0.2	50	50	0.64	227	344.2
23	220	120	0.4	60	100	0.71	172	352.4
24	220	120	0.6	70	150	0.83	196	347.8
25	220	180	0.2	60	150	0.65	235	338.7
26	220	180	0.4	70	50	0.91	247	349.6
27	220	180	0.6	50	100	1.13	238	357.9

on GRG are also generated in Minitab 19 application and are presented in Figure 117.1. From plots, it is observed that the GRG is reduced rapidly as Vc rises from 110 m/minute to 220 m/minute. The same pattern is observed for the GRG with a change in the feed rate and cutting depth when their values change from 60 mm/min to 180 mm/min and 0.2 mm to 0.6 mm respectively. The value of GRG increases rapidly with the growth in LFR from 50 ml/hour to 100 ml/hr. A further upsurge of LFR didn't give promising outcomes in the GRG value is nearly constant at LFR of 100 ml/hour and 150 ml/hour. Thus, the minimum quantity of lubricant that is sufficient for lubrication and cooling effect at the cutting zone can be selected as 100 ml/hr. This suggests that the mist solution effectively penetrates the cutting zone as the discharge rate of MQL is increased and provides adequate lubrication between the tool, workpiece, and chip surface. This also encourages effective cooling in the cutting zone and therefore, results in lesser tool wear and also, drops the probability of worsening the surface finish of the part.

From Figure 117.1, it is detected that the value of GRG hasn't changed too much with the lubricant concentration as slops sticks near the mean line of GRG. Still, the level with the best value is selected as the optimal level. Ideal parametric set for GRG is noted as Vc = 110 m/min, f = 60 mm/min, d = 0.2 mm, LC = 60% and LFR = 150 ml/hour i.e., Vc1-f1-d1-LC2-LFR3. The confirmation runs are executed to confirm the authentications of parametric optimisation. The face milling with three repetitions is carried at optimum parametric setting and VB, Ra, and Hv are measured. The output value of the previous best parametric

Table 117.6 Calculation table of GRA

Exp. No.	Normalised value			Deviation sequence			GRC			GRG
	Ra	VB	Hv	Ra	VB	Hv	Ra	VB	Hv	
1	0.940	0.940	0.293	0.060	0.060	0.707	0.894	0.892	0.414	0.842
2	0.786	0.926	0.229	0.214	0.074	0.771	0.700	0.871	0.393	0.712
3	0.571	0.792	0.364	0.429	0.208	0.636	0.538	0.706	0.440	0.572
4	0.798	1.000	0.384	0.202	0.000	0.616	0.712	1.000	0.448	0.759
5	0.536	0.671	0.693	0.464	0.329	0.307	0.519	0.603	0.620	0.551
6	0.571	0.691	0.506	0.429	0.309	0.494	0.538	0.618	0.503	0.555
7	0.619	0.752	0.771	0.381	0.248	0.229	0.568	0.668	0.686	0.606
8	0.548	0.611	0.613	0.452	0.389	0.387	0.525	0.562	0.564	0.539
9	0.131	0.128	0.856	0.869	0.872	0.144	0.365	0.364	0.776	0.408
10	1.000	0.772	0.000	0.000	0.228	1.000	1.000	0.687	0.333	0.848
11	0.774	0.530	0.224	0.226	0.470	0.776	0.689	0.516	0.392	0.612
12	0.643	0.369	0.124	0.357	0.631	0.876	0.583	0.442	0.363	0.523
13	0.821	0.772	0.352	0.179	0.228	0.648	0.737	0.687	0.436	0.692
14	0.702	0.711	0.265	0.298	0.289	0.735	0.627	0.634	0.405	0.605
15	0.310	0.356	0.938	0.690	0.644	0.062	0.420	0.437	0.890	0.474
16	0.583	0.242	0.787	0.417	0.758	0.213	0.545	0.397	0.701	0.523
17	0.595	0.389	0.650	0.405	0.611	0.350	0.553	0.450	0.588	0.530
18	0.238	0.168	0.661	0.762	0.832	0.339	0.396	0.375	0.596	0.412
19	0.881	0.591	0.412	0.119	0.409	0.588	0.808	0.550	0.460	0.704
20	0.714	0.577	0.348	0.286	0.423	0.652	0.636	0.542	0.434	0.590
21	0.381	0.154	0.666	0.619	0.846	0.334	0.447	0.372	0.599	0.443
22	0.583	0.134	0.686	0.417	0.866	0.314	0.545	0.366	0.615	0.506
23	0.500	0.503	0.874	0.500	0.497	0.126	0.500	0.502	0.799	0.532
24	0.357	0.342	0.769	0.643	0.658	0.231	0.438	0.432	0.684	0.462
25	0.571	0.081	0.561	0.429	0.919	0.439	0.538	0.352	0.532	0.489
26	0.262	0.000	0.810	0.738	1.000	0.190	0.404	0.333	0.725	0.419
27	0.000	0.060	1.000	1.000	0.940	0.000	0.333	0.347	1.000	0.408

Table 117.7 ANOVA table for GRG

Source	DF	SS	MSS	F-Value	P-Value	Significance
VC	2	0.0566	0.0283	12.89	0.0005	Significant
f	2	0.1270	0.0635	28.91	0.000	Significant
d	2	0.1627	0.0813	37.04	0.000	Significant
LC	2	0.0015	0.0007	0.35	0.7073	Insignificant
LFR	2	0.0176	0.0088	4.02	0.0386	Significant
Error	16	0.0351	0.0021			
Total	26	0.4007				

Table 117.8 Response table for mean GRG

Level	VC	f	d	LC	LFR
1	0.6160	0.6495	0.6632	0.5658	0.5311
2	0.5799	0.5707	0.5656	0.5772	0.5846
3	0.5060	0.4816	0.4731	0.5588	0.5861
Delta	0.1101	0.1679	0.1901	0.0184	0.0550
Rank	3	2	1	5	4

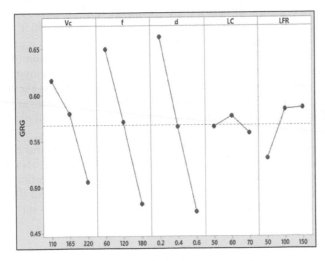

Figure 117.1 Main effects plots of GRG

Table 117.9 Results of confirmation experiments

Parametric setting	GRG
Initial best parametric setting (Vc2-f1-d1-LC2-LFR3)	0.848
Optimum parametric setting (Vc1-f1-d1-LC2-LFR3)	0.916
Improvement in GRG	**8.01%**

combination (Experiment -1) in Table 117.6 is replaced with the output values of the confirmation run and the value of GRG is determined. The outcomes of the confirmation run and calculations are given in Table 117.9 and it is noted that GRG is improved by 8.01%. The mean values of VB, Ra, and Hv noted at the optimal setting are 105 µm, 0.31 µm, and 331 HRV respectively.

After completion of the experimentation, the flank faces of the tool inserts are investigated in the metallurgic microscope with 200 x magnification. Also, the SEM images of the flank face are captured and are presented in Figure 117.2.

Tool Wear Analysis

During the measurement of tool wear, dissimilar kinds of wear modes are detected. SEM image of the minimum flank wear is presented as Figure 117.2(a) which occurred for Experiment-4 performed with lesser cutting speed (110 m/min) and with LFR is 150 ml/hor. It is clear from the SEM image that the maximum width of wear land is observed at the tooltip and the wear land of almost constant width extended over the whole length of the active cutting edge. The wear at low cutting speed is mainly caused by adhesion and clear marks of abrasion are visible in Figure 117.2(a). Also at lower cutting speed, the non-appearance of the BUE resulted in enhanced surface integrity. Figure 117.2(b) indicated that the prime cause of wear is started with the chipping caused by the adhesion of specimen material to cutting edge when MQL with 50 ml/hour is supplied at cutting speed of 165 m/min. At high LFR notching may absent at low cutting speed, frictional heat at the tool-work interface causes the adhesive wear (Yan et al., 2012; Yıldırım et al., 2019). However, as the LFR is taken at lower level, BUE at various portions of cutting edge is observed which are presented in Figure 117.2(b).

The maximum flank wear is observed during replication of Experiment-26 and is shown in Figure 117.2(c). Exaggerated wear is noticed at localised points of flanks indicating the reduction of effectiveness of MQL at high speed and lower LFR. The grouping of higher cutting speed and lower LFR gave rise to extreme temperature developments. Development of the built-up edge is noticeable on the cutting edge which disturbs the finish of the machined surface. Also, this built-up edge breaks, as the cutting force increases especially at the point of re-entering the tool which cause excessive chipping. At high speed, wear is mainly caused by adhesion and the degree of adhesion diminishes with the growth of lubricant supply (Liew, 2010; Sharma et al, 2021). While cutting hard steels, the combination of the built-up edge (BUE) and the stress concentration can lead to deformation of the cutting edge and instigates the chipping of material (Liew, 2010; Sharma et al, 2021; Yıldırım et al., 2019). A higher LFR facilitates effective dispersion of the lubricant at the cutting area. This results in sufficient cooling of the cutting area and hence developments of the built-up edge are comparatively low. However, repeated cooling and heating can progress the high temperature stresses on the cutting edge (Sharma et al, 2021; Yıldırım et al., 2019), that can instigate the development of micro-cracks on the exposed tool edge. Clear evidence of micro-cracks is detected in SEM images also for current study.

The SEM image of the tool rake for Experiment-26 is presented in Figure 117.3. Wear on the rake face is primarily due to the abrasion which is caused by friction of sliding chip on the rake face. The breaking of the built-up edge also breaks the material from the rake face of the tool and metal layer separation can be seen in SEM images (Figure 117.3). The combination of the stress developments in the chip and sliding action results in excessive temperature developments which cause diffusion of chip particles on the rake

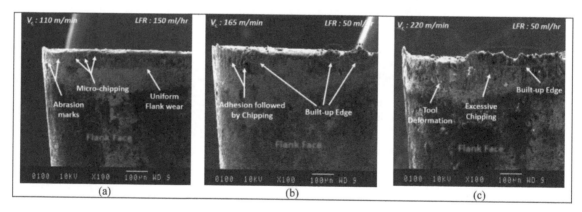

Figure 117.2 SEM images of tool flank part

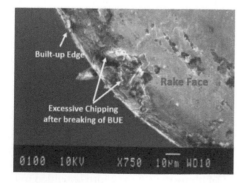

Figure 117.3 SEM image of the tool rake face

face. Although the diffusion is at a satisfactory distance from the main cutting edge, yet it can help to determine the degree of heat generated during machining (Sharma et al, 2021; Yıldırım et al., 2019). The conclusion is drawn while observing the flank and rake face, the key mechanism responsible for the tool wear is adhesion. However, as the tool slides over the surface, friction between the rubbing surface tends to produce heat and also causes abrasion (Yan et al., 2012). The marking pattern due to surface rubbing in all SEM images in Figure 117.2 clearly illustrates the signs of abrasion in the flank face. No evidence of diffusion is noticed on the tool flank.

Conclusions

- The main aim of this study is to concurrently optimise VB, Ra, and Hv using hybrid AHP-GRA methodology. The key conclusions of the research are:
- As per outcomes of ANOVA, Depth of cut is found most significant parameter for multi-responses followed by feed rate. Lubricant concentration hasn't shown much significance for the multi-response.
- As per outcomes of ANOM, the optimal parameter setting is discovered as Vc = 110 m/min, f = 60 mm/min, d = 0.2 mm, LC = 60% and LFR = 150 ml/hr. It is noted that the value of GRG is improved by 8.01 % at the optimal parametric setting at which the mean values of VB, Ra, and Hv are measured as 105 μm, 0.31 μm, and 331 HRV respectively. This hybrid system of AHP-GRA can be employed in other machining process-related researches in manufacturing industries.

The in-depth analysis of worn-out edges revealed that adhesion is the main cause of flank wear. Also, there are clear signs of abrasion in both the rake and flank face of the tool. The formation of built-up edge and micro-chippings are also noticed at higher cutting speeds especially with lower lubricant flow rates.

References

Babu, M. N., Anandan, V., Muthukrishnan, N., and Santhanakumar, M. (2019). End milling of AISI 304 steel using minimum quantity lubrication. Measurement. 138:681–689.

Cai, X. J., Liu, Z. Q., Chen, M., and An, Q. L. (2012). An experimental investigation on effects of minimum quantity lubrication oil supply rate in high-speed end milling of Ti–6Al–4V. Proc. Ins. Mech. Eng., B: J. Eng. Manuf. 226(11):1784–1792.

Conger, D. B., Emiroglu, U., and Altan, E. (2019). An experimental study on cutting forces and surface roughness in MQL milling of aluminum 6061. Machines. Technol. Mater. 13(2):86–89.

De Lacalle, L. L., Angulo, C., Lamikiz, A., and Sanchez, J. A. (2006). Experimental and numerical investigation of the effect of spray cutting fluids in high speed milling. J. Mater. Process. Technol. 172(1):11–15.

Hassanpour, H., Sadeghi, M. H., Rasti, A., and Shajari, S. (2016). Investigation of surface roughness, microhardness and white layer thickness in hard milling of AISI 4340 using minimum quantity lubrication. J. Clean. Produc. 120:124–134.

Kishawy, H. A., Dumitrescu, M., Ng, E. G., and Elbestawi, M. A. (2005). Effect of coolant strategy on tool performance, chip morphology and surface quality during high-speed machining of A356 aluminum alloy. Int. J. Machine Tools Manuf. 45(2):219–227.

Liao, Y. S. and Lin, H. M. 2007. Mechanism of minimum quantity lubrication in high-speed milling of hardened steel. Int. J. Machine Tools Manuf. 47(11):1660–1666.

Liew, W. Y. H. (2010). Low-speed milling of stainless steel with TiAlN single-layer and TiAlN/AlCrN nano-multilayer coated carbide tools under different lubrication conditions. Wear. 269(78):617–31.

Liu, Z. Q., Cai, X. J., Chen, M., and An, Q. L. (2011). Investigation of cutting force and temperature of end-milling Ti–6Al–4V with different minimum quantity lubrication (MQL) parameters. Proc. Ins. Mech. Eng. B: J. Eng. Manuf. 225(8):1273–1279.

Marinoni, O. (2004). Implementation of the analytical hierarchy process with VBA in ArcGIS. Comput. Geosc. 30(6):637–646.

Rana, M., Singh, T., Saini, A., Singh, J., Sharma, V. K., Singh, M., and Rooprai, R. S. (2021). Multi response optimization of nozzle process parameters in MQL assisted face milling of AISI 52,100 alloy steel using TGRA. Mater. Today: Proc. 44:31773182.

Rooprai, R. S., Singh, T., Singh, M., Rana, M., Sharma, V. K., and Sharma, S. (2021). Multi-variable optimization for surface roughness and micro-hardness in MQL assisted face milling of EN31 steel using Taguchi based grey relational analysis. Mater Today: Proc. 43:31443147.

Saaty, T. L. and Vargas, L. G. 1991. Prediction, projection and forecasting: applications of the analytic hierarchy process in economics, finance, politics, games and sports, (pp. 1131). Boston: Kluwer Academic Publishers.

Sales, W., Becker, M., Barcellos, C. S., Landre, J., Bonney, J., and Ezugwu, E. O. (2009). Tribological behaviour when face milling AISI 4140 steel with minimum quantity fluid application. Industrial Lubrication and Tribology.

Schwarz, M., Dado, M., Hnilica, R., and Veverková, D. (2015). Environmental and Health Aspects of Metalworking Fluid Use. Pol. J. Environ. Stud. 24(1).

Singh, G., Gupta, M. K., Mia, M., and Sharma, V. S. (2018). Modeling and optimization of tool wear in MQL-assisted milling of Inconel 718 superalloy using evolutionary techniques. Int. J. Adv. Manuf. Technol. 97(1):481494.

Singh, T., Sharma, V. K., Rana, M., Saini, A., Rooprai, R. S., and Singh, M. (2021). Multi response optimization of process variables in MQL assisted face milling of EN31 alloy steel using grey relational analysis. Materi. Today: Proc. 47:40624066.

Sharma, V. K., Rana, M., Singh, T., Singh, A. K., and Chattopadhyay, K. (2021). Multi-response optimization of process parameters using Desirability Function Analysis during machining of EN31 steel under different machining environments. Mater. Today: Proc. 44:31213126.

Sharma, V. K., Singh, T., Rana, M., Singh, A. K., and Chattopadhyay, K. (2022). Experimental investigation of tool wear in face milling of EN-31 steel under different machining environments. Mater. Today: Proc. 50:21352142.

Tosun, N., and Huseyinoglu, M. (2010). Effect of MQL on surface roughness in milling of AA7075-T6. Mater. Manuf. Proces. 25(8):793798.

Weinert, K., Inasaki, I., Sutherland, J. W., and Wakabayashi, T. (2004). Dry machining and minimum quantity lubrication. CIRP Annals. 53(2):511537.

Yan, L., Yuan, S., and Liu, Q. (2012). Influence of minimum quantity lubrication parameters on tool wear and surface roughness in milling of forged steel. Chin. J. Mech. Eng. 25(3):419429.

Yazid, M. Z. A., Zainol, A., and Mustapaha, A. M. (2019). Effect of Machining Parameters in Milling Aluminium Alloy 7075-T6 under MQL Condition. Int. J. Eng. Adv. Technol. 9(2):109113.

Yıldırım, Ç. V., Sarıkaya, M., Kıvak, T., and Şirin, Ş. (2019). The effect of addition of hBN nanoparticles to nanofluid-MQL on tool wear patterns, tool life, roughness and temperature in turning of Ni-based Inconel 625. Tribol. Int. 134:443-456.

118 An autonomous misplaced trolley arrangement system using RFID

Meenakshi Prabhakar[a], Abdul Wahaab Samsu Gani[b], Mohamed Hisham Mohamed Ibrahim[c], Valenteena Paulraj[d], Joshuva Arockia Dhanraj[e], and Seenu N.[f]

Centre for Automation & Robotics (ANRO), Department of Mechanical Engineering, Hindustan Institute of Technology and Science, Chennai, Tamil Nadu, India

Abstract

People sometimes arrange trolleys in a trolley depot but most of the time people tend to leave the trolley somewhere far from the trolley depot. This causes frustration among the people who are working there and they had to search and find the trolley and arrange it in the trolley depot. So, to overcome this problem, a system has been developed which makes the trolley reach the trolley depot automatically without any manpower. An Arduino UNO, RF reader, RF tag, and an ultrasonic sensor are used in this setup to overcome the problem. RF tags are placed all over the supermarket once the pathway is known. RF tags are placed either on the floor or ceiling of the supermarket which has a command programmed so that trolley reads the tag using an RF reader and gets the direction thus helping the trolley to reach the trolley depot. In the future, the project can be further improved by implementing an automatic billing system, load cell for detecting goods in the trolley, and SLAM navigation. So, the project mainly focuses on rearranging the trolley to its destination place thus reducing the work and time.

Keywords: RFID tags, autonomous parking systems, misplaced trolleys, arrangement systems.

Introduction

Robots are widely being used in and around industries for increasing production. One of how robots become more advanced is being autonomous vehicles. These autonomous vehicles can be used in many applications. Shopping malls are becoming the hotspot of the city to hang out for youngsters. Supermarkets in malls are always overcrowded. This project aims to develop a smart trolley that can navigate itself.

Shopping malls and supermarkets are overcrowded these days. Customers buy things and load them in their vehicles from the trolley. Those trolleys are left in the parking making the area more congested. Also, the trolleys become a shortage for the customers inside the market since others leave them in the parking area itself. Thus we come up with a solution called a smart trolley. These smart trolleys can navigate themselves to the parking lot given separately the trolleys, so it will be easy for the customers to access the carts for shopping.

The main objective is to develop a trolley, attached to a device. The trolley will be capable of avoiding dynamic obstacles. It catches the signal from the nearest RIFD tags and finds the shortest distance for reaching its destination. The trolleys can navigate themselves and reach the parking lot.

After doing some groundwork, the smart trolley is designed and developed. The circuit design is checked and the connections are done. After that, the gyroscope is installed.

In this article, a smart trolley model for splendid shopping is built. Supermarkets are locations that are usually busy on Saturdays and Sundays. The major issue that individuals experience when they go shopping is getting stuck in a long line of store clerks and becoming unhappy (Figure 118.1). The study offers a smart trolley system to solve this problem. A microcontroller, an ARM CPU, a bar code scanner, ultrasonic sensors, and an Android phone are all included in this setup. For presenting information about discount coupons, costs, and the total revised invoice, a mobile phone with a Wi-Fi connection is necessary says Arathi and Shona (2017).

Peradath et al. (2017) address the concept of a supermarket automation trolley based on an RFID reader (Figure 118. 2). The trolley is equipped with an RFID reader and a specific hardware and software system to make the transaction more convenient. The RFID card, which is fixed to the reader, is used to identify those goods that are above a certain amount. When the item is placed in front of the reader, the amount of the items is added to the customer's buying bill and displayed on the display screen. It also has a function that allows you to remove items from the trolley and deduct the cost from the total. The cost is computed once the products have been added, and when it exceeds a certain level, an alarm signal is delivered to warn the customer that their budget has been exceeded. All of this data is sent to a computer for billing updates over a wireless network termed the Zigbee network, which simplifies the payment system even faster and saves time.

[a]vijiprabha2000@gmail.com; [b]abdulwahaab.dme@gmail.com; [c]hishamsteel2000@gmail.com; [d]valenteena0701@gmail.com; [e]joshuva1991@gmail.com; [f]ail2seenu.n@gmail.com

Figure 118.1 Unarranged trolleys

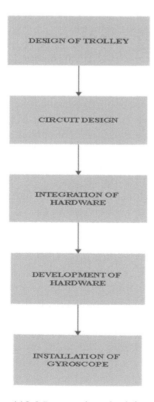

Figure 118.2 Proposed methodology

When shopping at supermarkets, a shopping cart is a must-have item. However, there were left through-out the supermarket after they had been utilised. There were also concerns about shopping cart safety, such as falling down an escalator. Customers who are in a hurry to find desired items at a supermarket, on the

Figure 118.3 3-D model of the prototype

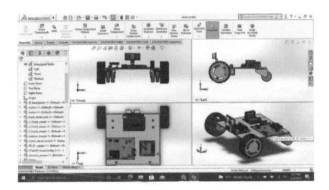

Figure 118.4 Multiple views of the design

other hand, will find it inconvenient and time-consuming. To tackle these issues, an advanced human following and line-following shopping cart with a smart shopping system are created. For obstacle avoidance, the shopping cart was integrated with an ultrasonic sensor. Customers can also simply trace the things they have purchased which was experimented with by Nayak and Kamath (2015).

The adoption of a smart trolley with innovative RFID technology is recommended to make shopping easier for customers. It has a lot of benefits, including allowing users to quickly search for commodities without relying on promoters, notifying customers of the number of products purchased, and helping customers to keep a better watch on their financial budget when buying. The major purpose of this research is to identify the results of a market survey on how people utilise a smart trolley with new capabilities called RFID. The scope of the research is conducted among Giant Hypermarket personnel, customers, and suppliers in Bandar Seri Manjung, Perak. Customers, employees, and suppliers received a total of 250 surveys. A total of 200 surveys were gathered, including copies from 140 consumers, 50 giant hypermarket employees, and ten vendors. The study looked at three different aspects of time management, budgeting, and product layout. Customers picked these characteristics as indicators to emphasise the necessity of a smart trolley rollout. The most significant part of customer input, according to the mean, is budget control. This is because, according to Ishak et al. (2006)this new smart trolley may assist customers with their shopping expenses, hence increasing their cash flow.

Yogalakshmi and Maik (2020)mean that innovation has changed dramatically across the world, making life easier for everyone. Humans have always attempted to develop an efficient approach to satisfy the purpose of having or displaying a severe avaricious inclination towards time spent unnecessarily. The Internet of Items (IoT) is the simplest method to connect things and solve issues, making life easier. We have created a trolley that is attached to an RFID reader and eliminates counter systems in retail places in our suggested system. It also assists in the delivery of things ordered from a store to the customer's preferred shopping centre location. The QR lock guarantees that the trolley's journey is safe and secure until it gets to its destination. A theft alert is placed over the seal if the lock that has been sealed with a QR code is attempted to be opened manually without the necessary QR code.

The goal of this article is to create a user-friendly, dependable, and automated parking system. Even if the location is known, numerous cars may seek the few available parking places, causing significant traffic congestion. In this article, we develop and build a smart parking system prototype. The automatic vehicle parking system is completely automated, with each user receiving a unique ID that corresponds to the trolley that has been assigned to him or her. This type of technology is beneficial for addressing the issue of limited parking space in congested areas. The ratio of individuals in India who own vehicles and motorbikes has lately grown due to an increase in economic activity and an improvement in living quality, boosting metropolitan traffic. As a result, parking difficulties will be a significant barrier in facilitating transportation networks and ensuring the quality of urban life. It is difficult for vehicles to find a parking place in most urban locations, especially during rush hour. The problem stems from the fact that you don't know where the open spaces are at the moment says Sharanya et al. (2017).

Design

The robot is designed in SolidWorks. The dimension of the bot is 330 mm x 200 mm x 85 mm. The back wheels have two geared DC motors. It is equipped with Arduino UNO, an RFID reader, an ultrasonic sensor, and a motor driver on the top.

Circuit and software requirements

The hardware components required are Arduino UNO, RFID tags, RFID reader, Ultrasonic sensor, DC motors, motor driver, and battery.

The circuit design contains a microcontroller which is the working brain of the system. The RFID reader is attached to Arduino UNO (Saeliw et al., 2019; Bairagi, 2018). The ultrasonic sensor is connected to the board for the collision avoidance system. Two DC motors are used for each motor and it is connected to the motor driver. The circuit design was built in Proteus Suite.

In the project, the software being used is Arduino IDE, Proteus design suite, and SolidWorks. The Arduino IDE is free software that allows to creation and upload of code to the Arduino board. It is the key software that is being used in programming the controller i.e. Arduino UNO. Arduino UNO is programmed in such a way that it meets the application of reading the RF tag by an RF reader and then processing it to control the motor (Premananthan et al., 2020). The concept of obstacle avoidance is also programmed in the controller which helps to detect, stop and avoid the obstacle (Figure 118.5).

Proteus design suite is software primarily used for simulating electrical design. The program that is being created, is simulated in the Proteus design suite to ensure the working of the designed circuit. The software helps to understand the working of the circuit by uploading the program into the controller and thus helps in reducing failure on the hardware side. Proteus design suite consists of many components that can be placed and simulated in software (Premananthan et al., 2020). In our project, Arduino UNO is programmed using Arduino IDE, and other components (Ultrasonic sensor, RF reader, RF tag, motor driver, etc) are connected to Arduino UNO using the software component library.

SolidWorks is a cutting-edge 3-D design software and product design platform.it is used to design the body of the trolley. When all the programming and electronic part is done, the outer body is designed to protect the components (Das et al., 2020; Sherlie et al., 2021). Measurements are taken physically and a rough sketch diagram is drawn to get an idea. Once the final diagram is done, the 3-D model design of the robot is created in solid works to get the virtual look of the body of the robot. At last, the 3-D model design is printed using a 3-D printer to get the real-world body of the robot (Figures 118.3 and 118.4).

Experimental setup

The trolley is manually driven by a human when the load is added to the trolley. When the customer unloads the goods from the trolley, the trolley is sometimes left behind due to some human laziness or work. The trolley when unloaded triggers the autonomous mode and it tries to find the destination. The

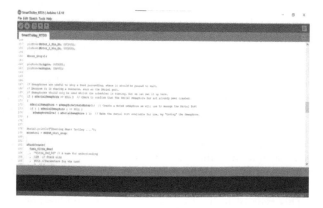

Figure 118 5 RTOS programming

Figure 118.6 Front view of the prototype

Figure 118 7 Top view of the prototype

trolley is driven automatically by reading RF tags from the ceiling using an RF reader. RF tags are placed all over the supermarket and are named 1, 2, 3,.... for our convenience which helps us to code the RF tags easily (Chatterjee et al., 2019; Arpita et al., 2021). When the command is received from the RF tag (for example forward, backward, left, right) the Arduino UNO process the data and gives the output to the motor. The trolley moves accordingly reading all the RF tags on its way and reaching the destination (Figures 6 and 7). When there are obstacles (human or random object) in-between the path, the trolley stops and waits until the object is moved away from the allotted path. For this obstacle avoidance is used and is achieved by using an ultrasonic sensor (Shailesh et al., 2021).

Results and Discussion

The final system demonstration is done and output is achieved successfully. The robot starts from the place it is left and moves towards the parking lot on the north side. Thus, the robot turns north from its leftover position and travels to the parking lot by catching the nearest RFID tags. The robot does not change even when it comes in contact with an obstacle. The robot stops once it detects obstacles and again moves when its way is clear.

Conclusions and Future Scopes

In this study, a smart trolley has been designed and developed. The trolley is capable of navigating itself to the given parking lot. Supermarkets and shopping malls are being more crowded these days. This smart trolley with a built-in collision avoidance system can be a good solution for modern shopping. The developed prototype is capable of catching the nearest RFID signal and thus finding the shortest route to reach the parking lot. Science and technology are growing too fast as the days pass by. More innovations can make our life simple and convenient. This proposed system is very much effective in all aspects. The Challenges that the trolley can face include, delay in receiving the RFID signals, deviating from its path, etc. Considering these drawbacks we can use AI and image processing for making this system more efficient. The trolley can even be programmed with an automatic billing system and SLAM navigation which makes our shopping even more convenient. Developing a dynamic collision avoidance system will also be a good ideology for the smart trolley. Artificial intelligence is one step away from providing automated shopping for us. Many more innovations can be done using this concept as a base.

References

Arpita, R., Shashidhara, K. S., and Dakulagi, V. (2021). Smart Trolley with Automatic Master Follower and Billing System. In Techno-Societal 2020, 163–170. Cham: Springer.

Arathi, B. N. and Shona M. (2017). An elegant shopping using Smart Trolley. Indian J. Sci. Technol. 10(3):36.

Bairagi, P. (2018). Compact smart parking system based on autonomous trolley and RFID system. Int. J. Adv. Sci. Technol. 4(9).

Chatterjee, A., Manna, S., Rahaman, A., Sarkar, A. R., Ghosh, Ansari, A. (2019). An automated RFID based car parking system. 2019 International Conference on Opto-Electronics and Applied Optics (Optronix). 13.

Das, T. K., Tripathy, A. K., AND Srinivasan, K. (2020). A Smart Trolley for Smart Shopping. 2020 International Conference on System, Computation, Automation, and Networking (ICSCAN). IEEE. 15.

Ishak, I. C., Muslim, M. M., Ismail, S. B., and Mohd, M. A. 2006. A smart trolley with RFID implementation: A Survey Among Customers. Lumut, Malaysia:Malaysia University Kuala Lumpur.

Nayak, M. and Kamath, K. (2015). Fabrication of an automated electronic trolley. IOSR J. Mech. Civil Eng. 12(3):7284.

Peradath, A. N., Purushothaman, A. N., Gopinath, A. N., Km, A. N., and Joe, N. (2017). Rfid based smart trolley for supermarket automation. Int. Res. J. Eng. Technol. 4(7):197580.

Premananthan, G., Nagaraj, B., and Divya, N. Sensor-based integrated smart trolley system using Zigbee-experimental analysis. Mater. Today: Proc. 12.

Saeliw, A., Hualkasin, W., Puttinaovarat, S., and Khaimook, K. (2019). Smart Car Parking Mobile Application based on RFID and IoT. Int. J. Interac. Mobile Technol. 13(4). doi: 10.3991/ijim.v13i05.10096.

Shailesh, S., Deb, P. S., Chauhan. R, and Tyagi, V. (2021). Smart Trolley. 2021 International Conference on Advanced Computing and Innovative Technologies in Engineering (ICACITE). IEEE. 242245).

Sharanya, B. and Harshitha, S. (2017). Automatic Smart Car Parking System Using RFID. Int. J. Adv. Res. Innov. Ide. 3(2).

Sherlie, A., Akhila, P., Muhafeez, G. M., Prateek, R., and Chandra, D. (2021). Smart Trolley with social distance monitoring. New arch-Int. J. Contemp. Archi. 11:8(1s):240–245.

Yogalakshmi, C. N. and Maik, V. (2020). Innovative automated shopping trolley with RFID and IoT technologies. In Artificial Intelligence and Evolutionary Computations in Engineering Systems, 461–471. Singapore: Springer.

119 High frequency dielectric response of rhombohedral lead zirconate titanate

Balgovind Tiwari[1,a], Babu Thodeti[1,b], and R.N.P. Choudhary[2,c]

[1]Physics Department, IIIT RKV, AP-RGUKT, Vempalli, Kadapa, India

[2]Physics Department, ITER, SOA University, Bhubaneswar, Orissa, India

Abstract

In this paper, an attempt has been made to investigate the dielectric properties of pure and manganese modified lead zirconate tianate (PZT) ceramics, at higher frequencies such as 50 kHz, 500 kHz and 1 MHz. PZT bulk ceramics of rhombohedral crystal symmetry, have been processed at high temperatures through solid solution synthesis. Manganese has been substituted at the zirconium site of PZT, in various amounts. The electrical characteristics of the samples have been recorded by LCR meter at various frequencies and temperatures. The nature of dielectric constant and tangent loss of the characterised samples, has been investigated as a function of both frequency and temperature. The ferroelectric nature of samples has been confirmed by the presence of dielectric peaks. The amount of diffusiveness in the compounds has been evaluated through Curie-Weiss law.

Keywords: Bulk ceramics, dielectric response, ferroelectric, PZT.

Introduction

Lead zirconate titana (PZT), an exceptional smart material, is being used in many industrial applications such as photovoltaics (Su et al., 2019) sensors (Li et al., 2003), memories (Melnick et al.,1992), etc. This is due to its outstanding smart material response like photoferroelectricity, piezoelectricity, ferroelectricity, pyroelectricity, etc. Ever since its discovery, PZT has been mostly used and/or employed in piezoelectric applications. Because the material has capability to generate the charge carriers across the boundaries when an external stimulus in the form of pressure/stress is applied. This capability is usually called as piezoelectric effect. The advantage of PZT is that, it exhibit the converse/reverse piezoelectric effect i.e. PZT undergoes structural deformation when the external stimulus in the form electric/magnetic field is applied. But later, it has been discovered that the PZT also exhibit ferroelectricity in it, owing to its non-centrosymmetric crystal symmetry. Due to the ferroelectric nature of PZT, it has been widely employed in several applications like ferroelectric memories, ferroelectric transistors, ferroelectric sensors, ultrasonic transducers, etc.

PZT is a ferroelectric material of solid solution type that belongs to the ABO_3 perovskite structural family. The lead ions are situated at A-site. Zirconium and titanium ions are situated together at B-site. Oxygen ions form octahedron in each crystal where these atoms are usually occupying the face centre positions of ABO_3. One can have a look at the diagram provided in recently published paper (Tiwari et al., 2020) for the schematic representation of PZT ABO_3 structure.

Since PZT is a well-established ferroelectric material, the presence of ferroelectricity in the broad range of both temperature domain and frequency, domain decides the applicability in applications. This can be achieved through the dielectric characterisation of PZT. This is what has been done in this work i.e. PZT and the Mn modified samples have been dielectrically characterised to study/investigate existence of ferroelectricity with respect to temperature and frequency. Also, this study majorly helps to identify the dielectric loss the material which reduces the performance of material.

Strictly, lots of work (Jaffe et al., 1955; Heartling et al., 1971; Toacsan et al., 2007; Shaw et al., 2002; Zhang et al., 2004; Tiwari and Choudhary, 2008; Das, 2007; Kuma et al. ,2006) has been covered on the dielectric properties of PZT but very less on Mn modified PZT. In this regard, we have prepared pure PZT and modified it with Mn in various amounts such as 2%, 6%, and 10%. The structural analysis has already been reported in the earlier part of this work (Zhang and Whatmore, 2004). It has been identified that PZT and Mn modified PZT samples exhibit rhombohedral crystal symmetry. Due to Mn ions substitution, not much change in the structure has been observed. The existence of rhombohedral structure in PZT is due to the presence of large amount zirconium in the material. In this paper, we report detailed investigation on the dielectric properties of as prepared compounds at very high frequencies i.e. 50 kHz, 500 kHz and 1000 kHZ.

Method of synthesis

The preparation of compounds has been done, as it is reported in various works of the existing literature (Tiwari et al., 2020). Also, the process parameters at which the samples have been processed through solid

[a]balgovindtiwari@rguktrkv.ac.in; [b]r131786@rguktrkv.ac.in; [c]ramchoudhary@soa.ac.in

state mixing of powders, are reported in the earlier part of the work (Zhang and Whatmore, 2004). But to assist the readers, the experimental constraints/conditions are given below:

1. Dry grinding – 1 hour
2. Wet grinding – 1 hour (in methanol)
3. Final calcination – 1000°C for 3 hours
4. Sintering - 1100°C for 2 hours

It is worth to note that the grinding (both wet and dry methods) of samples has been repeated multiple times. The electrical properties such as parallel capacitance, complex impedance, phase, dielectric loss, dielectric permittivity, etc. have been recorded as functions of both temperature and frequency. Main reasons to prepare samples through this method is as follows: easy preparation and cost effective. For more in-depth understanding and analysis, one can refer the bibliography.

Results and Discussion

This section gives some basic understanding on the dielectric response of Mn modified PZT ferroelectrics, at 50 kHz, 500 kHz and 1 MHz i.e. higher frequencies.

Figure 119.1 shows the fluctuation of relative dielectric constant (εr) of compounds, as a function of temperature. The magnitude of dielectric constant, of every compound, is increased with increment in temperature. But the increase has been observed for only upto certain temperature at which the maximum of relative dielectric constant can be obtained. The temperature corresponding to the maximum dielectric constant (ε_{max}) of material renders the value of Curie temperature (T_c). The value of T_c gives an idea of phase during which the material can be ferroelectric or non-ferroelectric (i.e. paraelectric) (Tiwari et al., 2020). So, T_c is considered as a specific temperature at which the phase transition of a material takes place, from ferroelectric phase to paraelectric phase. One can infer from the figure that the T_c of PZT has been enhanced very large in magnitude. For instance, the T_c of PZT at 1 MHz frequency is 373°C. But with the substitution of 2%, 6% and 10% of Mn, T_c values has been increased to 427.4°C, 436.9 °C, and 436.9°C respectively.

Hence, these materials exhibit ferroelectric nature at sufficiently high temperatures. But the magnitude of permittivity (ε_r) has been slightly enhanced and reduced with the substitution of 2%, and 6%, 10% of Mn

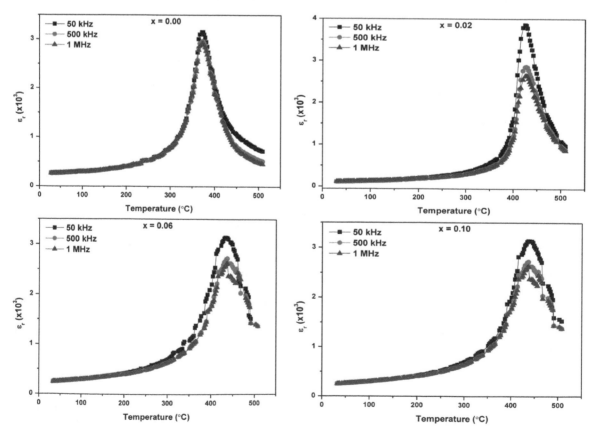

Figure 119 1 Fluctuation of relative dielectric constant of PZT and Mn modified PZT compounds with temperature

respectively. Therefore introduction of Mn ions at B-site in PZT, altered the dielectric properties such as T_c and ε_{max}, which could be useful for electronic applications. Further, the presence of peaks in plots indicate the evidence of ferroelectricity i.e. the samples are ferroelectric in nature (Tiwari et al., 2020).

Shift of these peaks towards the high temperature region indicate the increase in T_c values. The change in width of the peaks with change in Mn concentration indicate the existence of diffused phase transition in the compounds. Regarding the diffusivity, the values and analysis has been made the following sections.

Figure 119.2 represent the fluctuation of dielectric loss of as prepared compounds, as a function of temperature. Dielectric loss (tan δ) is the measure of loss of energy in the form of heat. At high temperature and high frequency phases, the samples lose their energy in the form of heat called as dielectric loss and/ or tangent loss. One can have a look at Figure 119.2, and could grasp that the magnitude of tan δ of all compounds is almost very much negligible. So, PZT and Mn modified PZT's show very less at higher frequencies and temperatures. At each respective frequency, the value of tan δ of PZT has shown continuous decrement, with increase in the amount of Mn substitution. This shows that the presence of Mn ions in PZT could reduce the dielectric loss of bulk PZT ceramic. The appearance of nearly asymmetrical peaks indicate the relaxation of dipoles of the samples.

Figure 119.3 represent the plots from which the amount of diffusiveness in the compounds can be obtained. Figure 119.3 shows the fluctuation of $\ln(1/\varepsilon-1/\varepsilon_{max})$ with respect to $\ln(T-Tc)$. The magnitude of diffusivity in the compounds has been obtained from the slopes of curves. Based on the well-known Curie Weiss law and/or expression, one can quantify the value of diffusivity from the following expression: $\ln(1/\varepsilon-1/\varepsilon_{max})$ $\ln(T-T_c)\gamma$, where γ is the magnitude of diffuseness (Tiwari et al., 2020). One can identify from the figure that the value of diffusivity is changing with change in concentration. The values of diffusivity are mentioned in Table 1.

Figure 119.4 represent the nature of ac conductivity of as prepared compounds, as a function of inverse absolute of temperature. The conductivity of ferroelectric material can be investigated based on its response to external ac electric signal. The conduction that takes place in ferroelectrics is mainly due to the charge carriers that are weakly bounded. When an electric field is applied, the ordered motion of charged particles results in electrical conduction. Since ferroelectrics are assumed not to have free charge carriers, the

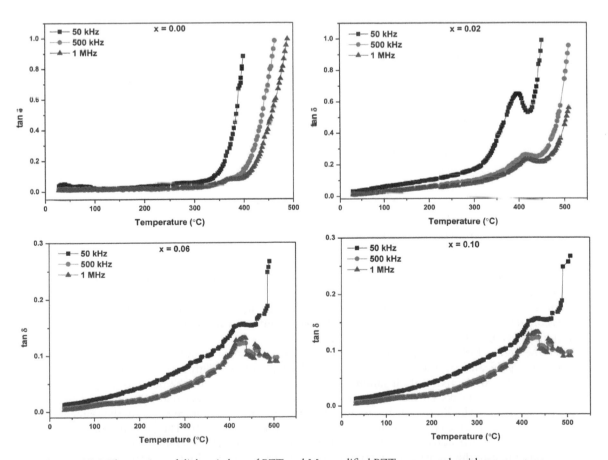

Figure 119.2 Fluctuation of dielectric loss of PZT and Mn modified PZT compounds with temperature

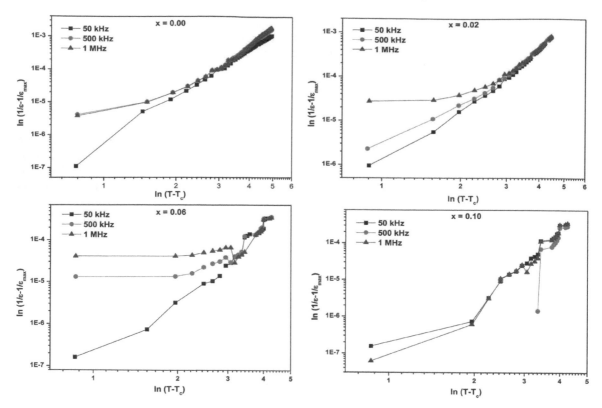

Figure 119.3 Fluctuation of $\ln(1/\varepsilon - 1/\varepsilon_{max})$ versus $\ln(T-T_c)$

Table 119.1 Diffusivity values of samples

Sample	Frequency	γ
x = 0.00	50 kHz	1.24
	500 kHz	1.33
	1 MHz	1.30
x = 0.00	50 kHz	1.25
	500 kHz	1.16
	1 MHz	1.26
x = 0.00	50 kHz	1.76
	500 kHz	1.04
	1 MHz	1.57
x = 0.00	50 kHz	1.04
	500 kHz	1.26
	1 MHz	1.12

existence of conductivity is may be due to low mobile ionic charges. Hence, the conduction in ferroelectrics takes place through hopping of charge carriers.

Using parameters of dielectric information, the conductivity has been calculated from the following expression: $\sigma_{ac} = \omega \varepsilon r \varepsilon 0 \tan\delta$ where $\varepsilon 0$ is termed as the permittivity value in the free space, and ω is regarded as value of the angular frequency (Tiwari et al., 2020).

From the Figure 119.4, it can be noticed from the figure that the magnitude ac conductivity of all samples shows significant variation over wide range of temperatures. With increment in the magnitude of temperature, the conductivity of all samples is increasing accounting for the presence of peaks at higher temperatures. These peaks correspond to the dielectric relaxation peaks that can observed near Curie temperature region. The increment in conductivity with rise in temperature suggests the semiconducting nature i.e. negative coefficient of resistance behaviour (Tiwari et al., 2020). The existence of conduction in the materials can be attributed to the hopping charge carriers. Hence, the increment in ac conductivity with temperature indicate the increase in rate of hopping of charge carriers with temperature. Hence, the nature of graphs suggests that the transport properties of materials are thermally activated obeying Arrhenius law.

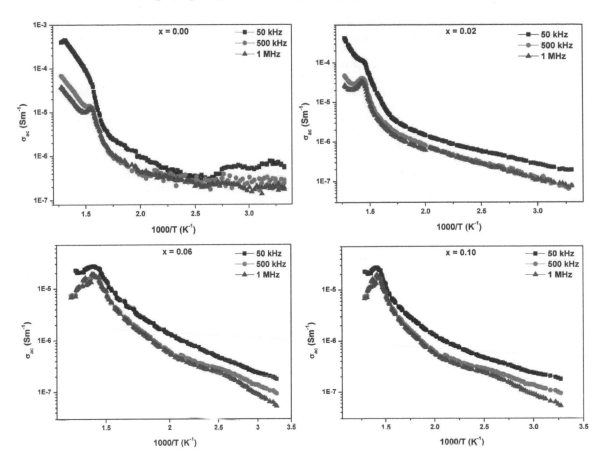

Figure 119.4 Fluctuation of ac conductivity of PZT and Mn modified PZT, with temperature

When the temperature increases, the polarons become active since they get the required thermal energy and jump over the grain boundaries. Different slopes of the graphs indicate different activation energies of the compounds (Tiwari et al., 2020).

Conclusions and Recommendations

The samples have been synthesized by calcining and sintering the powders at 1000°C and 1100°C respectively. Introducing Mn ions in PZT, enhances the Curie temperature showing a maximum of 436.9°C. The compounds became more diffusive because of the presence of Mn ions. The ac conductivity nature of the samples is found to obey the Arrhenius law. The conduction in samples is mainly due to the hopping charge carriers. The rate of hooping of charge carriers has been increased with increment in temperature. The dielectric loss of modified samples has been reduced very much. It is recommended to carry out further investigation photovoltaic properties Mn modified PZT based nanomaterials and/or composites.

References

Das, P. R. (2007). Investigations of structural, dielectric, electrical properties of some Tungsten Bronze Ferroelectric Vanadates. Ph.D Thesis, IIT-Kharagpur: India.

Heartling, G. H. and Land, C. E. (1971). Hot-Pressed (Pb,La)(Zr,Ti)O$_3$ Ferroelectric Ceramics for Electrooptic Applications. J. Am. Ceramic Soci. 54:111.

Jaffe, B., Roth, R. S., and Marzullo, S. (1955). Properties of Piezoelectric Ceramics in the Solid-Solution Series Lead Titanate-Lead Zirconate-Lead Oxide: Tin Oxide and Lead Titanate-Lead Hafnate. J. Res. Natl. Bur. Stand. 55:239–248.

Kumar, A., Choudhary, R. N. P., Singh, B. P., and Thakur, A. (2006). Effect of strontium concentration on electrical conduction properties of Sr-modified BaSnO$_3$. Ceram. Int. 32:73–83.

Li, Y., Horowitz, R., and Evans, R. (2003). Vibration control of a PZT actuated suspension dual-stage servo system using a PZT sensor. IEEE Trans. Magn. 39:932–937.

Melnick, B. M., Scott, J. F., and Paz De Araujo C. A. (1992). Thickness dependence of D.C. leakage current in lead zirconate-titanate (PZT) memories. Ferroelectrics. 135:163–168.

Shaw, C. P., Gupta, S., and Stringfellow, S. B., Navarro, A., Alcock, J. R., and Whatmore, R. W. (2002). Pyroelectric properties of Mn-doped lead zirconate–lead titanate–lead magnesium niobate ceramics. J. Eur. Ceram. Soc. 22:2123–2132.

Su, R., Zhang, D., Wu, M. (2019). Plasmonic-enhanced ferroelectric photovoltaic effect in 0-3 type BaTiO 3 -Au ceramics. J. Alloy. Comp. 785:584–589.

Tiwari, B. and Choudhary, R. N. P. (2008). Complex Impedance Spectroscopic Analysis of Mn modified $Pb(Zr_{0.65-x}Mn_xTi_{0.35})O_3$ Electroceramics. J. Phys. Chem. Solids. 69:2852–2857.

Tiwari, B. and Choudhary, R. N. P. (2009). Effect of Mn-substitution on structural and dielectric properties of $Pb(Zr_{0.65-x}Mn_xTi_{0.35})O_3$ ceramics. Solid State Sci. 11:21922.

Tiwari, B., Babu, T., and Choudhary R. N. P. (2020). Dielectric, impedance and modulus spectroscopy of $Pb(Zr_{0.52-x}Ce_xTi_{0.48})O_3$ (x = 0.00, 0.10) ferroelectric ceramics. Physica. Scripta. 95:115806.

Tiwari, B., Babu, T., and Choudhary, R. N. P. (2020). AC Impedance and Modulus Spectroscopic Studies of $Pb(Zr_{0.35-x}CexTi_{0.65})O_3$ (x = 0.00, 0.05, 0.10, 0.15) Ferroelectric Ceramics. Mater. Chem. Phys. 256:123655.

Tiwari, B., Babu, T., and Choudhary, R. N. P. (2020). Effect of manganese doping on dielectric characteristics of lead zirconate titanate of different zirconium/titanium ratios. IOP Con. Series: Mater. Sci. Eng. 764:012–029.

Tiwari, B., Babu, T. and Choudhary, R. N. P. (2020). Synthesis of $Pb(Zr_{0.35-x}Mn_xTi_{0.65})O_3$, x = 0.00, 0.02, 0.06, 0.10 ceramics and their structural, dielectric characteristics. Mater. Res. Expr. 7:055701.

Toacsan, M. I., Ioachim, A., Nedelcu, L., and Alexandrub, H. V. (2007). Accelerate ageing of PZT-type ceramics. Prog. Solid State Chem. 35:531–537.

Zhang, Q. and Whatmore, R. R. (2004). Hysteretic properties of Mn-doped $Pb(Zr,Ti)O_3$ thin films. J. Eur. Ceram. Soc. 24:277–282.

120 Component fault detection and isolation for PEMFC systems using unknown input observers

Vikash Sinha[a], Pankaj Bhokare[b], Ashish Shastri[d], and Rupesh Patil[c]

Department of Mechanical Engineering Navsahyadri Education Society's Group of Institutions, Pune, India

Abstract

An unknown input observer (UIO) based component fault detection and isolation (CFDI) technique for proton exchange membrane fuel cell system (PEMFCS) is presented. The algorithm is derived by assuming single component fault at any time instant and devised with the assumption that sensors and actuators are fault free. First, a number of UIOs are designed for PEMFCS in such a way that one particular observer will be sensitive to a particular component and residuals are determined. Then, a residual based CFDI algorithm is formulated which is not incorporated in previous works on fault diagnosis of such systems. The effectiveness of the algorithm is shown with simulation results of the cathode pressure model of PEMFCS. The results show novel UIO based component fault diagnosis of the system using decision table.

Keyword: CFDI, linear and non-linear systems, PEMFCSs, UIOs.

Introduction

For mobile and stationary power applications, fuel cell systems (FCSs) are undergoing extensive development. Proton exchange membrane fuel cell system (PEMFCSs) are particularly beneficial in ground vehicle and stationary power applications. These systems are used in a wide range of power generation applications, including distributed and backup power generation, as well as automotive applications. PEMFCSs have become more appealing for various applications due to the rising demand for low-emission power generation equipment and cars. These generate almost no emissions and have a high power density and quick start (Pukrushpan et al., 2004). Despite the fact that there have been several studies on fuel cell modelling, only a few studies have been used to control and observe FCSs. The model includes transient phenomena such as compressor inertia and flow dynamics, manifold filling dynamics, and membrane humidification. FCS stack voltage, efficiency, and power, are affected by several variables. (Pukrushpan et al., 2004) describes a 2-D heat and mass transfer model for a PEMFCS with channels.

FCSs are complicated and severely nonlinear, making it challenging to discern altering parameters even in static operation modes. Following Luenberger's (1971) pioneering work on estimation problems, various types of estimators for linear and non-linear systems have been developed (Rajamani and Ganguli, 2004; Yang et al., 2010; Wang et al., 2006; Darouach et al., 1994; de Lira et al., 2012; Mondal et al., 2008). It is difficult to measure all input signals in a system at one time. An observer with the ability to estimate states using unknown inputs is highly valuable in fault diagnosis and robust control. For ideal linear and non-linear systems, researchers have developed many forms of unknown input observer (UIOs) (Darouach et al., 1994; Mondal et al., 2008). Darouach et al. (1994) proposed a full-order Luenberger type observer. De Lira et al. (2012) developed a linear parameter varying fault diagnosis technique for PEMFCS. The model-based fault diagnosis compares the monitored system's real behaviour to a predicted behaviour generated using a mathematical model in real time. The presence of a defect is assumed if a considerable disparity (residual) is found between the model outputs and the measurement signals (Mondal et al., 2008).

A UIO-based component fault detection and isolation (CFDI) technique for PEMFCS is proposed in this paper. First, a number of UIOs for the system are designed, and residuals are calculated. The residuals are then used to formulate CFDI algorithm. There are two steps for implementing CFDI technique. The defect is detected in the first step, and the faulty zone is isolated. The erroneous parameter is isolated in the second phase. The key advantage of this CFDI technique is that the second phase is only performed when one of the components fails. In comparison to typical parameter identification-based CFDI techniques, the complexity of fault isolation is greatly decreased. Simulation results may easily indicate the usefulness of CFDI technique and its on-line implementation when compared to those available in the literature (Sinha and Mondal, 2018; Kim et al., 2007; Sinha and Mondal, 2020; 2021; 20022). If a set of measurements is given, a collection of residuals with varied sensitivities that correlate to possible defects can be generated. In most circumstances, isolating the fault and determining its magnitude may be accomplished by evaluating how the faults affect the residuals in real time. The high interdependence of complex system parameters in PEMFCS makes their stabilization very sensitive and a suitable technique is still lacking for any post diagnostic adjustment of the control parameters. In this work, a novel UIO based CFDI algorithm is formulated which is not incorporated in previous works on fault diagnosis of PEMFCSs.

[a]vikash.me15@nitp.ac.in; [b]pankaj.bhokare@rediffmail.com; [c]ashishshastri.mech@gmail.com; [d]principalnavs@gmail.com

A general PEMFCS model is described in Section 2. Section 3 describes the CFDI algorithm for a single component fault at any time instant. In Section 4, the results and discussion are presented, followed by conclusions in Section 5.

PEMFCS Model

Thermodynamic, electro-chemical, electrical, and fluid mechanics principles are all used in the modelling of FCS. Here, we focus on some of the most often utilized PEMFCS. The supply and return manifolds, air compressor, and hydrogen tank are among the auxiliary components. The FCS is a volumetric capacitor with two electrodes, the cathode and anode, sandwiching an electrolyte inside a membrane with bipolar plates called MEA depicted in Figure 120.1 (Pukrushpan et al., 2004).

With the help of a catalyst, hydrogen gas travels over the anode and splits into electrons and hydrogen protons. The protons flow to the cathode and electrons pass externally for power generation. Meanwhile, water is produced when hydrogen protons and electrons react with oxygen flowing through the cathode. In operational circumstances, the temperature is 80°C. There are some aspects to the model built specifically for control and diagnosis research.

Pukrushpan et al. (2004) constructed and validated a mathematical model that contains crucial aspects such as dynamics (transient) impacts but ignores parameter spatial fluctuation. The anode and cathode volumes of several FCSs are assumed to form one stack. The sub-models of PEMFCS are discussed as follows:

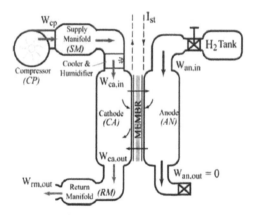

Figure 120.1 Scheme of PEMFCS model (Pukrushpan et al., 2004)

Cathode pressure model

This model describes the behaviour of air flow in the cathode. It works on the theory of mass conservation, and the psychrometric and thermodynamic properties of air (Pukrushpan et al., 2004). There are a number of assumptions which are as follows:

- All gases obey the law of ideal gas.
- Stack temperature is equal to the cathode air temperature.
- The flow properties outside the cathode are assumed to be the same as those within the cathode.
- The vapour condenses into liquid form when the gas relative humidity exceeds 100%. The water present in the stack either evaporates or accumulates in the cathode.
- The cathode layer and flow channel are combined to single volume by ignoring spatial variations.

As shown in Figure 120.1, the air compressor, the supply manifold, and the cathode dynamics are included in this model. The following equations are used to describe the cathode pressure model. Equation (1) represents a lumped parameter model used for dynamic behaviour of the compressor. Equation (2) represents the supply manifold model including pipe and stack manifold. Equations (35) give the states as oxygen (O_2), nitrogen (N_2), and vapour partial pressures of the cathode pressure model respectively.

$$\frac{d\omega_{cp}}{dt} = -\eta_{cm}\frac{k_t k_v}{R_{cm}J_{cp}}\omega_{cp} + \eta_{cm}\frac{k_t}{R_{cm}J_{cp}}v_{cm} - \frac{c_p T_{atm} k_{cp}}{\eta_{cp}}\left|\left(\frac{p_{sm}}{p_{atm}}\right)^{\frac{\gamma-1}{\gamma}} - 1\right|\phi \qquad (1)$$

$$\frac{dp_{sm}}{dt} = -\frac{\gamma R_a T_{sm} k_{sm,out}}{V_{sm}}p_{sm} + \frac{\gamma R_a T_{cp}}{V_{sm}}k_{cp}\phi\omega_{cp} + \frac{\gamma R_a T_{sm} k_{sm,out}}{V_{sm}}p_{ca} \qquad (2)$$

$$\frac{dp_{O_2}}{dt} = -\frac{R_{O_2}T_{ca}}{V_{ca}}\left(\frac{x_{O_2,in}k_{ca,in}}{1+\omega_{ca,in}} + \frac{x_{O_2,out}k_{ca,out}}{1+\omega_{ca,out}}\right)\left(p_{O_2} + p_{N_2} + p_{v,ca}\right)$$

$$+\frac{R_{O_2}T_{ca}}{V_{ca}}\frac{x_{O_2,in}k_{ca,in}}{1+\omega_{ca,in}}p_{sm} + \frac{R_{O_2}T_{ca}}{V_{ca}}\frac{x_{O_2,out}k_{ca,out}}{1+\omega_{ca,out}}P_{rm} - \frac{R_{O_2}T_{ca}}{V_{ca}}M_{O_2}\frac{n}{4F}I_{st} \qquad (3)$$

$$\frac{dp_{N_2}}{dt} = -\frac{R_{N_2}T_{ca}}{V_{ca}}\left(\frac{\left(1-x_{O_2,in}\right)k_{ca,1}}{1+\omega_{ca,in}} + \frac{\left(1-x_{O_2,out}\right)k_{ca,out}}{1+\omega_{ca,out}}\right)\left(p_{O_2} + p_{N_2} + p_{v,ca}\right)$$

$$+\frac{R_{N_2}T_{ca}}{V_{ca}}\frac{\left(1-x_{O_2,in}\right)k_{ca,1}}{1+\omega_{ca,in}}p_{sm} + \frac{R_{N_2}T_{ca}}{V_{ca}}\frac{\left(1-x_{O_2,out}\right)k_{ca,out}}{1+\omega_{ca,out}}P_{rm} \qquad (4)$$

$$\frac{dp_{v,ca}}{dt} = -\frac{R_{v,ca}T_{ca}}{V_{ca}}\left(\frac{\omega_{ca,in}k_{ca,1}}{1+\omega_{ca,in}} + \frac{\omega_{ca,out}k_{ca,out}}{1+\omega_{ca,out}}\right)\left(p_{O_2} + p_{N_2} + p_{v,ca}\right)$$

$$+\frac{R_{v,ca}T_{ca}}{V_{ca}}\frac{\omega_{ca,in}k_{ca,1}}{1+\omega_{ca,in}}p_{sm} + \frac{R_{v,ca}T_{ca}}{V_{ca}}\frac{\omega_{ca,out}k_{ca,out}}{1+\omega_{ca,out}}P_{rm} + \frac{R_{v,ca}T_{ca}}{V_{cu}}\frac{M_v n\left(1+2A_{fc}n_d\right)}{2F}I_{st}$$

$$-\frac{R_{v,ca}T_{ca}}{V_{cu}}\frac{M_v n A_{fc}D_w}{t_m}\left[f\left(p_{v,ca}\right)p_{v,ca} - f\left(p_{v,an}\right)p_{v,an}\right] \qquad (5)$$

Anode pressure model

The anode pressure model is very similar to the cathode pressure model (Pukrushpan et al., 2004). The anode is assumed to be supplied with pure hydrogen gas from a hydrogen tank in this model. The following equations are used to describe the anode pressure model. Equation (6) represents the anode supply model. Equations (7) and (8) denote the states of the hydrogen (H2) and vapour partial pressure. Equation (9) shows the return manifold model.

$$\frac{dp_{sm,an}}{dt} = -\frac{R_{H_2}T_{sm,an}k_{sm,an,out}}{V_{sm,an}}p_{sm,an} + \frac{R_{H_2}T_{sm,an}k_{sm,an,out}}{V_{sm,an}}p_{an} \qquad (6)$$

$$\frac{dp_{H_2}}{dt} = -\left(\frac{k_1}{1+\omega_{an,in}} + k_{H_2,out}\right)\frac{R_{H_2}T_{an}}{V_{an}}p_{H_2} - \left(\frac{k_1}{1+\omega_{an,in}} + k_{H_2,out}\right)\frac{R_{H_2}T_{an}}{V_{an}}p_{v,an}$$

$$+k_{H_2,out}\frac{R_{H_2}T_{an}}{V_{an}}\left(p_{O_2} + p_{N_2} + p_{v,ca}\right) - \frac{R_{H_2}T_{an}}{V_{an}}M_{H_2}\frac{n}{2F}I_{st} + \frac{k_1}{1+\omega_{ca,in}}\frac{R_{H_2}T_{an}}{V_{an}}p_{sm,an} \qquad (7)$$

$$\frac{dp_{v,an}}{dt} = -\left(\frac{\omega_{an,in}k_1}{1+\omega_{an,in}} + k_{v,an,out}\right)\frac{R_{v,an}T_{an}}{V_{an}}p_{v,an} - \left(\frac{\omega_{an,in}k_1}{1+\omega_{an,in}} + k_{v,an,out}\right)\frac{R_{v,an}T_{an}}{V_{an}}p_{H_2}$$

$$+k_{v,an,out}\frac{R_{v,an}T_{an}}{V_{an}}\left(p_{O_2}+p_{N_2}+p_{v,ca}\right)+\frac{\omega_{an,in}k_1}{1+\omega_{an,in}}\frac{R_{v,an}T_{an}}{V_{an}}p_{sm,an}$$

$$+\frac{R_{v,an}T_{an}}{V_{an}}\frac{M_vA_{fc}nD_w}{t_m}\left[f\left(p_{v,ca}\right)p_{v,ca}-f\left(p_{v,an}\right)p_{v,an}\right]-\frac{R_{v,an}T_{an}}{V_{an}}\frac{M_vA_{fc}n_dn}{F}I_{st} \qquad (8)$$

$$\frac{dp_{rm}}{dt} = -\left(\frac{R_aT_{rm}k_{ca,out}}{V_{rm}}+\frac{C_{D,rm}A_{T,rm}\varsigma_{rm}}{\sqrt{\overline{RT}_{rm}}}\right)p_{rm}+\frac{R_aT_{rm}k_{ca,out}}{V_{rm}}\left(p_{O_2}+p_{N_2}+p_{v,ca}\right) \qquad (9)$$

where the symbols and notations are having standard meaning as used in (Pukrushpan et al., 2004; Kim et al., 2011).

CFDI Algorithm

The CFDI algorithm assuming single component fault at any time instant is presented. This technique is developed assuming fault-free actuators and sensors. Consider a PEMFCS (Mondal et al., 2008) as.

$$\dot{x}(t) = (A+\Delta A)x(t)+(B+\Delta B)u(t)+gx(t)+Gv(t) \qquad (10)$$

$$y(t) = Cx(t)+Dv(t) \qquad (11)$$

$$\dot{x}(t) = \left(A+\Delta A+\Delta A_f\right)x(t)+\left(B+\Delta B+\Delta B_f\right)u(t)$$
$$+gx(t)+\Delta g_f x(t)+Gv(t) \qquad (12)$$

$$\dot{x}(t) = (A+\Delta A)x(t)+(B+\Delta B)u(t)+Ed(t)$$
$$+gx(t)+Gv(t) \qquad (13)$$

$$\text{where, } Ed(t) = \Delta A_f x(t)+\Delta B_f u(t)+\Delta g_f x(t) \qquad (14)$$

The system has now been separated into N subsystems, each with a small number of parameters. Assuming ith subsystem to be malfunctioning, the equations for the system are given by

$$\dot{x}_{(i)}(t) = (A+\Delta A)x_{(i)}(t)+(B+\Delta B)u(t)+E_{(i)}d_i(t)$$
$$+g\left(x_i(t)\right)+Gv(t) \qquad (15)$$

$$y_i(t) = C_{(i)}x_{(i)}(t)+D_{(i)}v(t) \qquad (16)$$

The system residuals are obtained by

$$r_i(t) = y_i(t)-C_{(i)}\hat{x}_i(t) \qquad (17)$$

In this way 'm' number of UIO is designed for which 'm-1' number of residuals are determined. Now, observing the behaviour of residual with respect to corresponding threshold $\varepsilon_{(i)}$ values, following decision Table 120.1 is formed (Mondal et al., 2008; Sinha, 2021). The occurrence of fault is proven since $r_{(1)}$ lies below $\varepsilon_{(1)}$ but $r_{(2)}$ and $r_{(3)}$ crosses above $\varepsilon_{(1)}$. Table 120.1 can be used to isolate the faulty component. A correctly built UIO estimates states when unknown inputs are also present. When fault persists in the ith component or fault-free case, the residual within threshold values. Otherwise, this value exceeds the threshold. Noise, uncertainties, and inputs all influence the magnitude of threshold values. As a result, N number of UIOs for N number of components can be designed. When N > 2, (N-1) UIOs are adequate to isolate a faulty component because the remaining component is identified as the faulty when (N-1) components are proven fault-free.

Results and Discussion

Using [11], the system and input matrices are given by

$$A = \begin{bmatrix} -22.9610 & -22.9610 & -22.9610 \\ -46.4930 & -46.4930 & -46.4930 \\ -0.3295 & -0.3295 & -0.3295 \end{bmatrix}$$

$$B = \begin{bmatrix} 367.5 & 7.7390 & -0.2296 \\ 367.5 & 29.104 & -0.4649 \\ 367.5 & 0.1180 & -942.225 \end{bmatrix}$$

Now, a fault signal is given in component 1 at fifty seconds and CFDI technique is now implemented. Three observers are developed for component 1 by using

$$E_{(1)} = \begin{bmatrix} 1 & 0 & 0 \end{bmatrix}^T, \; E_2 = \begin{bmatrix} 0 & 1 & 0 \end{bmatrix}^T, \; E_3 = \begin{bmatrix} 0 & 0 & 1 \end{bmatrix}^T \text{ and}$$

$$C_{(1)} = \begin{bmatrix} 1 & 0 & 0 \\ 0 & 0 & 1 \end{bmatrix}^T, \; C_2 = \begin{bmatrix} 1 & 0 & 1 \\ 0 & 1 & 0 \end{bmatrix}^T, \; C_3 = \begin{bmatrix} 1 & 0 & 0 \\ 0 & 0 & 1 \end{bmatrix}^T.$$

These observers are designed to estimate the states. The states and residuals of the cathode pressure model of PEMFCS are plotted in Figs. 120.2–120.4. Suitable threshold values $\varepsilon_{(1)}$ = [0.09 0.010]^1 are chosen to find out the occurrence of fault. As $r_{(1)}$ lies below $\varepsilon_{(1)}$ while $r_{(2)}$ and $r_{(3)}$ crosses above $\varepsilon_{(1)}$, indicates the occurrence of fault. Table 120.1 is used to identify faulty component 2. As sinusoidal input is applied to the system, the states are following similar trends. The residual plots in Figures 120.3–120.4 show fault-free and faulty scenarios of the system respectively.

Table 120.1 Decision table (Sinha, 2021)

r(t)	Previous Obs.	r(i) > ε(i)	Decisions	Remarks
r(1)		No	Component 1 may be faulty	Check r(2) , r(3)
r(2)	r(1)< ε(1)	No	No fault	Go back to check r(1) again
	r(1)< ε(1)	Yes	Component 1 is faulty	Fault is detected and isolated
r(3)	r(1)> ε(1)	No	Component 2 is faulty	Do
	r(1)> ε(1)	Yes	Component 3 is faulty	Do

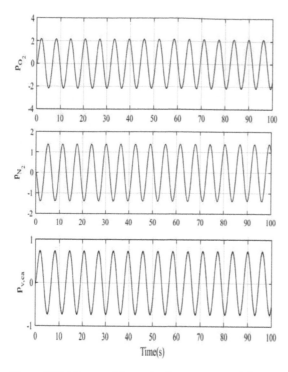

Figure 120.2 States of the cathode pressure model x(t)

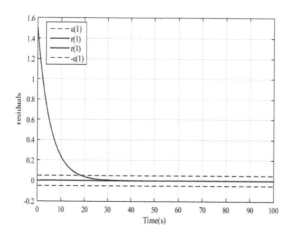

Figure 120.3 Residual r1(t)

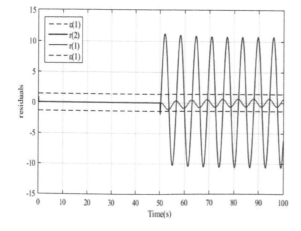

Figure 120.4 Residual r2(t)

Conclusions

A UIO based technique for PEMFCSs is presented. The CFDI algorithm is derived by single component fault at any time instant. Several UIOs are developed to obtain residuals. Thus, N number of UIOs can be designed for N number of components. When N > 2, (N-1) UIOs are adequate to identify defective component because the rest component is the faulty one when (N-1) components becomes fault-free. The residuals are then used to formulate a CFDI algorithm. Simulation results are plotted in application to the cathode pressure model of PEMFCS show the effect of residual based CFDI algorithm. The significance of present work is development of novel UIO based CFDI algorithm to detect component fault of the system. This work can be further extended for designing a unified approach to detect and diagnose all types of PEMFCS faults, namely, actuator fault, sensor fault, and component fault simultaneously.

References

Darouach, M., Zasadzinski, M., and Xu, S. J. (1994). Full order observers for linear systems with unknown inputs. IEEE Transactions on Automatic Control 39:606–609.

De Lira, S., Vicenc P., and Quevedo J. (2012). Fault detection and isolation of a real PEM fuel cell using interval LPV Observers. 8th IFAC Symposium on Fault Detection, Supervision and Safety of Technical Processes (SAFEPROCESS), Mexico City, Mexico. 90–95.

Kim, E-S., Kim, C-J., and Eom, K-S. (2007). Nonlinear observer design for PEM fuel cell systems. Proceedings of International Conference on Electrical Machines and Systems. Seoul, Korea. 1835–1839.

Luenberger, D. G. (1971). An introduction to observers. IEEE Trans. Automat. Contr. 16:596–602.

Mondal, S., Chakraborty, G., and Bhattacharyya, K., (2008). Robust unknown input observer for nonlinear systems and its application in fault detection and isolation. J. Dyn. Syst. Meas. Control T ASME. 130:0445031 0445035.

Pukrushpan, J. T., Peng, H., and Stefanopoulou, A. G. (2004). Control-oriented modelling and analysis for automotive fuel cell systems. J. Dyn. Syst. Meas. Control T ASME. 126:14–25.

Rajamani, R. and Ganguli, A., (2004). Sensor fault diagnosis for a class of non-linear systems using linear matrix inequalities. Int. J. Control. 77:920–930.

Sinha, V. 2021. Observer based fault diagnosis and fault tolerant control of proton exchange membrane fuel cell systems. PhD Thesis, NIT Patna, India.

Sinha, V. and Mondal, S., (2018). Recent development on performance modelling and fault diagnosis of fuel cell systems. Int. J. Dyn. Control. 6:511–528.

Sinha, V., and Mondal, S. (2020). Sensor FDD based FTC design for time-delay PEMFC systems, Int. J. Recent Technol. Eng. 9:348–353.

Sinha, V., and Mondal, S. (2021). Adaptive unknown input observer approach for multi-fault diagnosis of PEM fuel cell system with time-delays. J. Control Decision. 8:222–232.

Sinha, V., and Mondal, S., (2022). Robust adaptive fault estimation observer-based FTC design for time- delay PEMFC systems. J. Ins. Eng.: Series B. doi: 10.1007/s40031-022-00723-9.

Wang, C., Nehrir, M. H., and Gao I I. (2006). Control of PEM fuel cell distributed generation systems. IEEE Transactions on Energy Conversion 21:586-595. 6] Frank, P.M., 1994. On-line fault detection in uncertain nonlinear systems using diagnosis observers: A Survey. Int. J. Syst. Sci. 25:2155–2166.

Yang, Q., Autocue, A., and Bouamama B. O. (2010). Model based fault detection and isolation of PEM fuel cell. Conference on Control and Fault Tolerant Systems, Nice, France. 825–830.

121 Combination of transfer learning and incremental learning for anomalous activity detection in videos

Yash Rathore[a], Vandana Japtap[b], Parth Gawande[c], Sarthak Oke[d], and Aditya Rao[e]

School of Computer Engineering & Technology, MIT-WPU, Pune, India

Abstract

Human action recognition has turned into an excellent exploration region in computer vision issues. There are numerous applications where human activity detection can be consolidated, for example, analysing the video information which will assist with observing information produced by CCTV cameras.

The main drawback of training video data is that assuming new classes should be instructed to the framework, the system should be retrained without any training, and all classes retaught to the system. This particular problem has many difficulties, for example, putting away and holding information and responding training costs hence we use incremental learning to avoid catastrophic forgetting which is learning new examples without compromising previously learned knowledge.

There is a requirement for transfer learning algorithms because existing models are reused to take care of another test or issue and to save time and resources from being required to prepare different machine learning models to complete similar tasks while at the equivalent time upgrading their overall performance.

In this paper, a system is proposed which uses transfer learning along with incremental learning to detect different types of human activities in videos. The proposed system is evaluated on a benchmark dataset, KTH, UCF101 (UCF101 human actions dataset. experiments performed on complicated video datasets demonstrate the ability of the proposed method for low complexity incremental learning while achieving significantly better accuracy than existing such models.

Keywords: Computer vision, convolutional neural network, human action recognition, incremental learning, machine learning.

Introduction

These days, because of video surveillance, the world is being observed much more than it was in a couple of years. A report from IHS Technologies anticipated that before the finish of 2019, consistent video observation will create over 2500 PBs which is far in excess of 566 PBs of information produced in 2015. They have additionally predicted that by the end of 2021, this information will shoot up to 3.3 trillion video hours of information caught through video surveillance cameras done worldwide. Breaking down such a colossal measure of information is a need of 60 minutes. One of the reviews that can be performed on this sort of information is human action identification. Human activity detection assumes an imperative part in giving data about human-to-human interaction, their character, their relational connection, or their mental state. A portion of the issues which make human activity detection a difficult errand in videos or still pictures are listed below: encompassing variations like changing background, occlusions, and various perspectives. Every individual has their own way of completing an activity as a part of variations of entertainer's development. . A few sorts of activities like signals, gestures, collaborations, and joint activities by the group are some of the examples. The insufficient measure of training video recordings is also a factor to include. Two critical conceptions in human activity detection are activity identification and activity localisation. Activity recognition or identification is a process that lets us know what actions are happening in the video and activity localisation in videos provides the location of action performed.

Video surveillance fills in as the remote eyes of the executives and the security forces. It furnishes the security workforce with early notification of breaches in security, including hostile and terrorist acts, and is a piece of the arrangement to safeguard personnel and resources. It is a basic subsystem for an extensive security plan.

There's a bewildering exhibit of factors in video content; long structure, short structure, resource, speed, lighting, and sound quality. These factors, combined with a reasonable problem statement and use case, intensely impact the video analysis methods expected to extricate and expand its worth. All around, video analytics throughout the course of recent years have progressed extraordinarily in capacities permitting clients to catch, index, and search faces, colours, and objects.

There are various procedures that assist the analysis of videos without any problem. In various applications like the railway, traffic state detection, sports video analysis, flames, and smoke detection there

[a]yashrathore02061999@gmail.com; [b]vandana.jagtap@mitwpu.edu.in; [c]parthgawande2000@gmail.com; [d]sarthakoke@gmail.com; [e]aditya.rao1890@gmail.com

are various strategies and procedures to examine the videos. Semantic event extraction and slow motion extraction is an essential research work to analyse sports related videos where accurate decisions have to be made in very small time.

Classifiers are used in activity recognition to identify the class of activity with spatial temporal features. There are two different approaches in activity recognition: local and global approach. Local approach uses a local descriptor that describes the characteristic of a point. These characteristics are local features which describe the image patch. This image patch is a small cluster of pixels. 'Scale invariant feature transform (SIFT) and space-time interest points (STIP)' are local descriptors generally used in videos (Willems et al., 2008).

Global approach uses the entire image instead of a small image patch. In this approach a single vector is used to generalise an entire image.

Machine learning technologies help us to infer a lot of knowledge from digital data. In machine learning the data is given before training which makes its underlying structure more static. In contrast to the traditional approaches of machine learning, incremental learning continuously trains the model based on constantly arriving data streams.

Incremental learning approach learns data which data arrives over time with limited memory resources without compromising the model's accuracy. The motive of incremental learning is to avoid catastrophic forgetting which is learning new examples without compromising previously learned knowledge. There are few criteria which let us know whether a learning process is incremental learning or not. They are:

- Whenever a new example arrives the model should learn the new examples.
- There is no need for the presence of original data, whenever new data arrives.
- Whenever new data arrives, the system should be able to retain the previously learned knowledge.
- The model should have a room to add newly learned classes.

Transfer learning for machine learning is when existing models are reused to take care of another test or issue. Transfer learning is definitely not a distinct kind of ML algorithm, rather it's a procedure or technique utilised while training models. The information created from past preparation is reused to assist in performing a new task. The new undertaking will be connected somehow or another to the recently prepared task, which could be to classify objects in a particular file type. The original trained model usually requires a high level of generalisation to adapt to the new unseen data.

Transfer learning is generally used:

- To save time and resources from being required to prepare different AI models without any preparation to complete comparable tasks
- As productivity saving in areas of ML that require high measures of resources, for example, picture categorisation or natural language processing
- To discredit an absence of labelled data held by an association, by utilising pre-trained models.
- The main benefits of transfer learning for machine learning include:
- Eliminating the requirement for an enormous arrangement of labelled training data for each new model.
- Working on the productivity of machine learning advancement and deployment for multiple models.
- A more summed up way to deal with machine critical thinking, utilising various calculations to tackle new difficulties.
- Models can be prepared inside reproductions rather than genuine conditions.

The paper is divided into three sections. Section 2 contains a description of other work that has been done in human action recognition and incremental learning. Section 3 comprises a comprehensive description of the proposed system and Section 4 contains the evaluation of the approach and its comparison with other available methods. Conclusions about the study are stated in Section 5.

Related work

In literature surveys about the work done for the human activity detection from video and different methods used to find and analyse those activities are mentioned.

Michau and Fink (2021) proposed unsupervised transfer learning for anomaly detection. This methodology finds abnormalities in operational circumstances that only other units in a fleet have encountered. They suggest using adversarial deep learning to ensure that the distributions of the different units are aligned, and we introduce a novel loss, inspired by a dimensionality reduction tool, to ensure that the intrinsic variability of each dataset is conserved. To detect the abnormalities, we apply a state-of-the-art once-class technique. A new framework for performing unsupervised transfer learning (UTL) for one-class classification problems is proposed in this approach. As a result, it differs from other UTL applications in the literature, which typically aim to find a common structure between datasets in order to perform clustering or dimensionality reduction.

Hu et al. (2016) proposed Video anomaly detection using deep incremental slow feature analysis networks. It can detect global anomalies such as crowd panic with pinpoint accuracy. A set of anomaly maps generated from the network at various scales is used to detect local anomalies. The proposed method is universal and practical, performing well in a variety of scenarios with little human intervention and low memory and computational requirements. Extensive experiments on different challenge datasets are used to validate the benefits.

Wu et al. (2021) proposed an explainable and efficient deep learning framework for video anomaly detection. It uses pre-trained deep models to extract high-level concept and context features for training the denoising autoencoder (DAE), which requires low training time while attaining equivalent detection performance to the leading approaches.

Ade and Deshmukh (2013) proposed methods for incremental learning: a survey. In this approach they discuss the present methods of incremental learning that are accessible. This study provides an overview of current incremental learning research that will be useful to investigate scalars. There are various applications for incremental learning methods, such as online incremental learning or batch incremental learning. Incremental learning has a wide range of uses in a variety of fields.

Hacene et al. (2018) proposed transfer incremental learning using data augmentation (TILDA). It uses pre-trained DNNs as feature extractors, a nearest-class-mean-based approach for robust feature vector selection in subspaces, majority voting, and data augmentation at both the training and prediction stages. Experiments on difficult vision datasets show that the suggested technique can perform low-complexity incremental learning with much higher accuracy than current incremental methods.

Chengwu Liang et al. (2016) proposed a system in which RGB-D cameras were used, which captured both RGB and depth data. In this approach a feature coding strategy called LASC is used along with a local action descriptor to improve action recognition's performance. An affine subspace dictionary and statistical information of high order is used in LASC strategy which helps it acquire more important information than LLC and SD for classification.

Jagadeesh and Patil (2016) proposed a system for detecting human activities in videos by using optical flow and SVM classifier. In this approach, at first, optical flow between 100 frames which are extracted from each video is computed. After computing the optical flow, a binary image is generated. Then the feature vector will get taken out from the binary images using histogram of oriented gradient (HOG) descriptor. The model is then trained by using the training feature provided to classifier, support vector machine (SVM) classifier used for classification. This model is used to identify the movements like walking, jogging, running, boxing, hand waving, and hand clapping from the video. This approach was very efficient under a controlled environment but the accuracy of the model decreased when used on real-time videos.

Varol et al. (2018) proposed a system which used long-term temporal convolutions (LTC) to identify human actions. In this paper an artificial neural network with LTC with increased temporal extends is proposed which improves the accuracy of human action identification. It also showed the importance of high-quality optical flow for improving the accuracy of learning action models. In this paper they have used the HMDB51 dataset. The major limitation was, the dataset they used had limited training data due to which they were not able acquire high accuracy.

Liu et al. (2018) suggested a system that used a universal context-aware LSTM network for skeleton-based human action recognition. In this approach, a universal context-aware attention long-short term memory (LSTM) encoder is projected which uses a global context memory cell.

In each frame, this new class of LSTM is skilled in selectively aiming on informative joints. In this approach they also introduced coarse-grained and fine-grained attentions.

Chuankun Li et al. (2019) suggested a system which incorporates multiple networks to acquire structures for 3-D human action identification. In this method temporal information is exploited by using three views created in the spatial domain which are given to three convolutional neural networks. These views are generated using joint trajectory maps. The model was tested on NTU RGB+D dataset. The model was inefficient to recognise some actions like reading, writing. The model was also not able to recognise multi-person interaction.

Meng et al. (2019) proposed an end-to-end data augmentation network called sample fusion network (SFN)for skeleton-based human action recognition . SFN generates new samples with the help of a LSTM. In this approach they have also used adaptive weighting strategy to increase the performance of human action identification method during testing. They have used the NUCLA dataset. This model was not able to produce more diverse samples as the dataset contained fewer classes than other dataset. Other network architectures such as DenseNet can be used with sample fusion networks to increase the accuracy of human action recognition. Efficiency of sample fusion network approach can also be improved by using other modalities such as depth, RGB etc.

Zhang et al. (2019) proposed a novel adaptive scheme for skeleton-based human action recognition. In this approach two view adaptive neural networks are used. One of the two networks is VA-RNN, which is built with the use of recurrent neural networks with LSTM and the other one is VA-CNN, which is built using convolutional neural networks. In each network the most suitable observation viewpoint is learned and determined and then the skeletons are transformed into those viewpoints for recognition. When different initialisation seeds are utilised, this approach shows performance fluctuation.

Polikar et al. (2002) introduced an algorithm that can incorporate and learn newly arrived data. The algorithm is called Learn++. In this algorithm there is no need for the presence of original data to learn new examples. The algorithm uses ensemble learning and several hypotheses are produced by using training datasets. Other incremental learning algorithms can be evaluated and their accuracies and performances can be compared such as RBF networks, Fuzzy ARTMAP, TopoART, and IGNG.

This paper works in the situation where classes are gradually appearing and disappearing (Sun et al., 2016). In this methodology base learner is upheld for each class. These learners are updated when new data arrives. This is known as class-based ensemble learning. The limitation of this approach is that its ability waning after use on non-evolved classes.

Rosenfeld and Tsotsos (2018) contemplated a deep adaptation network. The network uses a linear combination of already learned examples to learn the new examples that arrive. This network can handle the tasks performed by multiple domains. Swapping between several learned examples can be controlled by the new architecture that is learned which will allow a single network to process tasks of various domains. The network works efficiently when tasks are highly correlated to each other.

Yang et al. (2019) established an adaptive approach. The system mainly has two special goals, the first goal is to provide flexibility and speed to the model and the second goal is to have capacity sustainability. The first goal is achieved by scaling the streaming data. As stream data changes continuously, the second is also achieved.

Didwania and Jagtap (2020), provided a good survey of different approaches used for anomaly detection from the videos, where they have mentioned various incremental and transfer learning approaches. Findings of the paper are that incremental learning is used very rarely for video-based anomalous activity detection. In this survey various research gaps of multiple methods are provided.

Proposed method

In this research KTH dataset is used, it comprises Six various types of human actions performed in video. The human actions that are performed are, boxing, walking, jogging, running, hand waving, and hand clapping) shown in Figure 121.1. Second data set used is the action recognition data set UCF101, it contains multiple different types of human activities that are demonstrated in video (jump rope, swing, wall pushups, skiing, drafting, diving…etc). All these actions are accomplished by 25 different subjects in four diverse scenarios. These four diverse scenarios are: outdoors s1, outdoors with scale variation s2, outdoors with different clothes s3 and indoors s4 as shown in Figure 121.2. This dataset contains 2391 sequences which are captured using static cameras with 25 fps rate. Each sequence has a length of four seconds on average. The sequences are down sampled to 160*120 pixels. Totally the sequences are deposited using AVI file format. For each mishmash of six actions, four scenarios and 25 subjects, there are in total 25 x 6 x 4 = 600 video files generated.

System architecture

In this section the proposed classification model is explained in detail. The proposed system is divided into three modules: Data collection and pre-processing part, model engine and performance tuning part and finally model output part.

Data collection and pre-processing

In this piece of the proposed framework, the information in the form of videos is being gathered from the source. The video information ought to be of extension .avi. When the video is caught, it is then converted into frames. When the recordings are changed into frames, the pre-processing of the frames is performed. Subsequent to pre-processing those edges are likewise added to the frame array. This frame array is then given as an input to our three layered convolutional neural network for the training process.

Figure 121.1 KTH Dataset

Figure 121.2 UCF101 - Action recognition data set

Model engine and performance tuning

The feature extraction is finished by the 3-D CNN itself. After that, the VGG19 pre-trained model is used for classification. In the proposed framework, to execute incremental learning, the designated checkpoints are determined. These checkpoints will be utilised when new models show up. These designated spots are determined in the way that iteration is considered as a checkpoint, assuming the precision at that specific cycle is more prominent than the past one. At the point when we are utilising the last checkpoint to retrain the model, the recently previously learned knowledge stays intact, which prompts settling the issue of catastrophic forgetting. The outcome is then given to the model output.

Model output

In this part of the system, the model will classify the actions in the films into categories like walking, jogging, running, boxing, hand waving, hand clapping, and jump rope, swing, wall push-ups, skiing, drafting and diving. Here users can see the result of classification through a user interface.

B. Network architecture

The 3-D-convolutional neural network along with incremental learning and transfer learning is used in the proposed system. One of the problems with the 3-D CNN training is that they require a sufficiently big number of labelled examples. In comparison to 2-D, 3-D networks have a lot more parameters, which results in over fitting on small datasets. 3-D networks can be trained to remember certain examples, but they don't make generalisations for all datasets, therefore, lots of actions end up in the wrong classes. In order to avoid this problem we can choose the right parameters for initialisation strategy and just "fine-tune" our network on the available labelled dataset. We used a pre-trained VGG19 for fine tuning and classification.

The network architecture is shown in Figure 121.3. Here we are using a 3-D convolution neural network with eight layers. This neural network contains four convolutional layers, two max-pooling layers and two fully-connected layers along with three dropout layers and one flatten layer. Kernel size utilised are 3*3*3 and three 32 filters are also used. Three convolutional layers are used for feature extraction. The max-pooling size used is 3*3*3 and dropouts used are 0.25, 0.25 and 0.5 in three layers. Flatten layer is used sandwiched between the convolutional layer and fully connected layer.

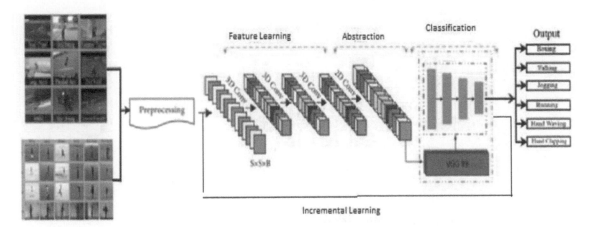

Figure 121.3 Network architecture

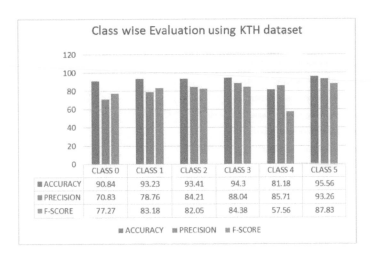

Figure 121.4 Class wise evaluation using KTH

The main role of a flatten layer is to transform a feature matrix into a feature vector that can be fed to a fully connected layer. In the first dense layer 512 hidden units are used and in the final dense layer hidden units are equivalent to the total number of classes i.e. six. The final dense layer provides us the probability of each class using which we classify whether the action in video is among any of the six classes.

Experimental results and analysis

The KTH dataset and the UCF 101 dataset, which contained six classes, were used to train and evaluate the proposed model. The assessment metrics used are accuracy, f-score and precision. Formulae used for accuracy, f-score and precision are given as follows:

Accuracy: The accurateness can be defined as the percentage of correctly classified instances.

Accuracy = (TP + TN)/(TP + TN + FP + FN)

Where, number of true positives, true negatives, false negatives and false positives are used respectively.

Precision: The precision is the percentage of relevant instances among the retrieved instances.

Precision = TP/TP + FP

F-score: It is a metric used to estimate a test's accuracy. It includes both precision and recall. It is a harmonic mean of precision and recall. The F-score does not take the true negatives into interpretation.

F-score = 2*TP /(2*TP + FP + FN)

The class-wise metrics evaluation was conducted. In the KTH Dataset, all six classes have an average accuracy of greater than 91.42%, with the exception of class four (running), which has an average accuracy of 81.18%. Figure 121.4 shows the evaluation metrics for each class. The figure provides a lot of information and it includes values of all the three measures for all the six classes. Class 0, Class 1, Class 2, Class 3, Class 4, Class 5 are boxing, hand clapping, hand waving, jogging, running and walking respectively.

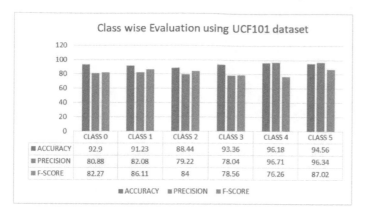

Figure 121.5 Class wise evaluation using UCF101

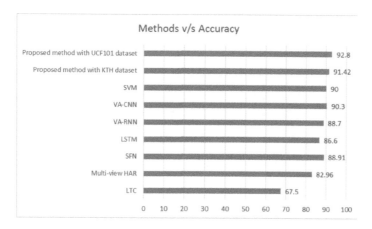

Figure 121.6 Comparatively analysis

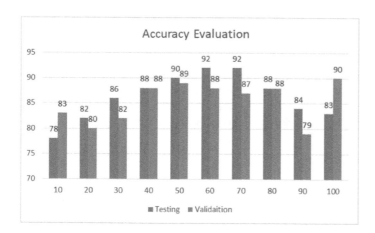

Figure 121.7 Comparatively analysis

In the UCF 101 dataset, all six classes have an average accuracy of greater than 92.8%. Figure 121.5 shows the evaluation metrics for each class. The figure provides a lot of information. It includes values of all the three measures for all the six classes. Class 0, Class1, Class 2, Class 3, Class 4, Class 5 are jump rope, swing, wall pushups, skiing, drafting and diving respectively.

The proposed method was also compared to my state-of-art method listed in related work as shown in Figure 121.5. The proposed system has outperformed all the other methods mentioned in the earlier.

The suggested system has a 91.42% accuracy on the KTH dataset and a 92.8% accuracy on the UCF 101 dataset, and the accuracy of the other models are 90% for SVM, 90.3% for VA-CNN, 88.7% for VA-RNN, 88.6% for LSTM, 88.91 for % for SFN, 82.6% for Multi View HAR, and 67.5% for LTC. As a result, from Figures 121.4–121.6, it can be concluded that the proposed methodology used for both the datasets works best and outperforms other existing classification models. So overall the proposed system has outperformed all the mentioned methods.

Testing and validation accuracies of the proposed model for 100 epochs with the intervals of 10 epochs is shown Figure 121.7. The KTH dataset model works best for 60 epochs and UCF101 dataset model works best for 70 epochs. Hence, the final model is shown for the mean which is 65 epochs.

Conclusion

Nowadays there is a huge advancement in information delivered by videos, as there is video surveillance done all over the place. Human activity identification is one of the noticeable areas of research in computer vision. In this paper, a system is recommended that can distinguish and perceive human activities in videos utilising incremental learning. The proposed system was effectively ready to recognise various human activities, for example, running, boxing, strolling, running, hand waving, and hand applauding to list a few from the KTH and UCF datasets. This proposed system outperforms many existing methods in terms of complexity and accuracy.

References

Ade, R. R. and Deshmukh, P. R. (2013). Methods for incremental learning: a survey. Int. J. Data Min. Knowl. Manag. Process (IJDKP) 3(4).

Didwania, P. and Jagtap, V. (2020). Anomalous activity detection in videos using increment learning: A survey. Eur. J. Eng. Technol. Res. 5(3):297–300. doi: https://doi.org/10.24018/ejeng.2020.5.3.1803.

Hacene, G. B., Gripon, V., Farrugia, N., Arzel, M., and Jezequel, M. (2018). Transfer incremental learning using data augmentation. MDPI Academic Open Access Publishing.

Hu, X., Hu, S., Huang, Y., Zhang, H., and Wu, H. (2016). Video anomaly detection using deep incremental slow feature analysis network. Inst. Eng. Technol.

Jagadeesh, B. and Patil, C. M. (2016). Video based action detection and recognition human using optical flow and SVM classifier. In IEEE international conference on recent trends in electronics, information & communication technology (RTEICT).

Li, C., Hou, Y., Wang, P., and Li, W. (2019). Multiview-based 3-D action recognition using deep networks. IEEE Trans. Hum. mach. Syst. 49(1).

Liang, C., Qi, L., He, Y., and Guan, L. (2016). 3D human action recognition using a single depth feature and locality-constrain affine subspace coding. IEEE Trans. Circuits Syst. Video Technol. 28(10):1051–8215.

Liu, J., Wang, G., Duan, L. Y., Abdiyeva, K., and Kot, A. C. (2018). Skeleton-based human action recognition with global context-aware attention LSTM networks. IEEE Trans. Image Process. 27(4).

Meng, F., Liu, H., Liang, Y., Tu, J., and Liu, M. (2019). Sample fusion network: An end-to-end data augmentation network for skeleton- based human action recognition. IEEE Trans. Image Process. 28 (11).

Michau, G. and Fink, O. (2021). Unsupervised transfer learning for anomaly detection: Application to complementary operating condition transfer. Knowl. Based Syst. 216:106816.

Polikar, R., Udpa, L., and Honavar, V. (2002). Learn++: An incremental learning algorithm for supervised neural networks. IEEE Trans. Syst. Man Cybern. 31(4):497508. doi:10.1109/5326.983933C.

Rosenfeld, A and Tsotsos, J. K. (2018). Incremental learning through deep adaptation. IEEE Trans. Pattern Anal. Mach. Intell. doi: 10.1109/TPAMI.2018.2884162.

Sun, Y., Tang, K., Minku, L. L., Wang, S., and Yao, X. (2016). Online ensemble learning of data streams with gradually evolved classes. IEEE Trans. Knowl. Data Eng. 28(6). doi: 10.1109/TKDE.2016.252667.

Varol, G., Laptev, I., and Schmid, C. (2018). Long- term temporal convolutions for action recognition. IEEE Trans. Pattern Anal. Mach. Intell. 40(6).

Willems, G., Tuytelaars, T., and Van Gool, L. (2008). An efficient dense and scale-invariant spatio-temporal interest point detector. In: Computer vision – ECCV 2008. ECCV 2008: Lecture notes in computer science, ed. D. Forsyth, P. Torr and A. Zisserman, (5303). Berlin, Heidelberg: Springer. https://doi.org/10.1007/978-3-540-88688- 4_48.

Wu, C., Shao, S., Tunc, C., Satam, P., and Hariri, S. (2021). An explainable and efficient deep learning framework for video anomaly detection. Cluster Comput. https://doi.org/10.1007/s10586-021-03439-5.

Yang, Y., Zhou, D. W., Zhan, D. C., and Xiong, H. (2019). Adaptive deep models for incremental learning: considering capacity scalability and sustainability. ACM. doi: 10.1145/1122445.1122456.

Zhang, P., Lan, C., Xing, J., Zeng, W., Xueand, J., and Zheng, N. (2019). View adaptive neural networks for high performance skeleton-based human action recognition. IEEE Trans. Pattern Anal. Mach. Intell. 41(8).

122 AHP approach for measuring service quality of passenger car

Mangesh D. Jadhao[1,a], Arun P. Kedar[2,b], and Ramesh R. Lakhe[3,c]

[1]Department of Mechanical Engineering, G.H. Raisoni Institute of Engineeriring and Technology, Nagpur, India

[2]Department of Mechanical Enineering, Yashwantrao Chauhan College of engineering, Nagpur, India

[3]Director, Shreyas Quality Management System, Nagpur, India

Abstract

Choosing the best service provider is an important goal for everyone. Managers of service stations are also interested in learning about areas for improvement and how to address them. To date, the majority of the work has sought to measure service quality using the SERVQUAL technique. First, the idea and determinants of service quality are investigated in this study. The proposed service quality framework is then evaluated using an analytic hierarchy process. To elucidate the methodology, a case study in the vehicle servicing industry is presented. The research indicated that five characteristics of service quality (tangible, reliability, responsiveness assurance, and empathy) were effective for this study. The created model is useful to assess the quality of service offered by auto repair shops.

Keywords: AHP, service quality, SERVQUAL.

Introduction

Because of the increasing number of automobiles on the road and people's changing lifestyles, India's automobile industry is growing at a rapid pace. India is the world's largest automaker and Asia's fourth-largest auto market. The automobile industry is the most competitive due to globalisation. In this highly competitive environment, everyone is attempting to stay afloat. Every business wants to expand its network across the country in order to attract more consumers and provide them with door-to-door service. However, the service sector of this profession is a significant area where businesses can provide services following product sales and track their clients over time. Client happiness has an impact on a company's market share and profitability, hence customer retention is a crucial concern for all businesses (Anderson et al., 1994). If the service quality is inadequate, customers have the option to switch to another brand. Providing higher service quality for long-term benefits is now a primary challenge for everyone. In order to keep the vehicle's quality, it's critical to meet customers' expectations in the competition. That is to say, service quality should be measured in order to benefit both customers and the firm. Service and its associated standards of service were chosen for investigation and review in this study.

Furthermore, the quality of service in the vehicle service sector, particularly for passenger cars, is somewhat diverse. Because the passenger vehicle service sector's service quality is more crucial than ever due to the country's growing population and the number of passenger cars on the road, this sector was chosen for implementation in an underdeveloped country like India. Although there are several ways for measuring service quality, SERVQUAL (service quality) methodology (Parasuraman et al., 1985) was chosen for this study since it is the most generally used and preferred methodology. SERVQUAL has a large number of research articles in a variety of fields.

The analytic hierarchy process (AHP) is perhaps the most well-known and widely utilised method in multiple-criteria decision making (Saaty, 2005). The AHP considers both qualitative and quantitative components of a problem (Salgado et al., 2012). AHP is now widely employed (Schmidt et al., 2015; Yildiz and Yildiz, 2015; Jayamani et al., 2017; Nosal and Solecka, 2014; Bartuskova and Kresta, 2015; Emrouznejad and Marra, 2017, Li et al., 2018).

The automobile industry's service quality is complex and varies across the board. Multiple factors are involved, as well as a degree of judgment uncertainty. In the subject of service quality, Manny's ideas and models are applied. However, because there is a scarcity of study in this subject, assessing the service is challenging. To deal with this ambiguity, you'll need a powerful approach. AHP is one of the most effective strategies for addressing the problem of service quality assessment.

[a]mangesh.jadhav@raisoni.net; [b]arunkedar64@gmail.com; [c]rameshlakhe786@gmail.com

Though the objectives are stated above, this research comprises of

(1) A literature review that summarises the service idea, service quality concept, and car service quality and critique of SERVQUAL.
(2) The research design is then used to define the service quality dimensions that will be employed in this study.
(3) The AHP is then used in the approach. Develop a hierarchical approach for evaluating service quality as well.
(4) Finally, the findings are assessed and the research is completed.

Literature Review

A. Service concept

Many countries' economic growth is based on the service sector, and the service sector still dominates the world today. Services are available in a variety of settings, including healthcare, restaurants, hotels, banks, transportation, and insurance. Many experts have attempted to define the services, but there is no single description that is universally acknowledged. Kotler (1994) provides the most thorough definition of service, defining it as 'an act or performance that one party can supply to another that is basically intangible and does not result in the ownership of anything.' Its manufacturing could be tested as a physical product or not 'services' are defined as 'actions, processes, and results' by Zeithaml (2000). Actions refer to the service provider's actions, processes to the procedures involved in providing the service, and results to how the consumer perceives the service's delivery.

It's more difficult to define and measure service quality. It's a way of thinking. Many researchers Ghobadian et al., (1994), Buzzell et al. (1987), Zeithaml (2000) have attempted to define service quality and have concluded that it is an important strategy for attracting and retaining clients. Goetschand and Davis (1998) state that service is to do labour for someone else. It's a description of what you will do for the consumer and how you will do it (Edvardsson, 1996). According to Bouman and Van der Wiele (1992), customer expectations are met when service meets them. To determine service quality, a comparison of expectations and performance is made (Bolton and Drew, 1991; Parasuraman et al., 1988).

It's more difficult to assess service quality than it is to assess product quality. Service performance is more difficult to assess than production performance. Since 1970, many scholars have researched the measuring of service quality. Parasuraman et al. (1985) were the first to create the SERVQUAL model. Many academics have used this model to evaluate services in a variety of industries, including hospitals, restaurants, education, health services, banking, tourism, vehicle servicing, and schools.

Tangibles, dependability, responsiveness, access, competence, courtesy, credibility, communication, security, and comprehension were the first ten service quality aspects established by Parasuraman et al. (1985). These ten dimensions were then whittled down to five: Tangible, Reliability, Responsiveness, Assurance, and Empathy.

Automobile service quality

Emerging lifestyle patterns, rising motor vehicle technical sophistication, and the corporate sector's expanding prominence in economic transformation processes all promise well for India's automobile services. We foresee a significant growth in the number of different types of motor vehicle users in the coming years as a result of the creation of a new culture that priorities conveniences and luxury. Accurate assessment is critical to the success of passenger car service quality. Valid measurement is critical for developing effective tactics. SERVQUAL is the most comprehensive methodology for assessing service quality. In the vehicle industry, the SERVAQUAL technique is also used. In the realm of automotive service, there has been very little research. As a result, the automobile industry's service quality is mostly unknown (Berndt, 2009; Izogo and Ogba, 2015; Bouman and Van der Wicle, 1992; Keshavaraz et al., 2009). Because of SERVQUAL's limitations, researchers had to resort to other models. The quality of service was assessed using a neural network (Behara et al., 2002).

Mersha and Adalakha (1992) used the Delphi approach to determine the most essential dimension. A performance-only model (SERVPERF) is used to evaluate attributes (Corrêa et al., 2007). Various authors use the literature survey to derive five service quality dimensions for evaluating passenger automobiles. The research gap observed through the literature review is that the majority of work is carried out by SERVAQUAL methodology only. Therefore, since service quality measurement is a multi-criteria decision making process, there is scope for research work carried out by the analytic hierarchy process of the MCDM method. The previous study did not include the sub-criteria of car cleaning after service or pickup and drop facility. Measuring the service quality of Nagpur city service stations was not included in the previous study.

Research Design

According to literature surveys, the dimensions for evaluating the service quality of passenger cars are determined as tangible, reliability, responsiveness, assurance, and empathy Because these dimensions were not previously considered in the measurement of service quality through the analytic hierarchy process. In evaluating the Passenger car service quality, tangibles mean the appearance of physical facilities, equipment, and personnel (Bouman and Van der Wiele, 1992, Berndt, 2009, Brito, 2007). Within this study, it comprises modern technical equipment. It is important to diagnose the problem with the vehicle. The dress code of employees shows the dignity they maintain in the dealership. Lastly, the reception and workshop areas are clean, which is essential for a healthy environment.

The term 'reliability' refers to the capacity to guarantee that. At a certain time, the vehicle will be ready for delivery (Berndt, 2009). The customer should be able to schedule an appointment at a time and date that is convenient for him or her (Bouman and Van der Wiele, 1992). Repairs are error-free, which means that the appropriate repair should be done the first time (Mersha and Adlakha, 1992). Finally, the service stations bill should be clear and exact, and service staff should fully explain the work performed (Behara et al., 2002).

The factor of responsiveness refers to a company's readiness to assist consumers and deliver accurate and timely service. It includes immediate attention from service workers when customers arrive at the service station (Berndt, 2009). It is the responsibility of service employees to guarantee that spare parts are available for vehicle repair and maintenance (Behara et al., 2002). Customers should constantly be helped and their doubts should be cleared by service station workers (Corrêa et al., 2007).

The assurance dimension defines the ability of service professionals to inspire trust and confidence in customers while interacting. The technical competence of service employees boosts their confidence (Liou and Chen, 2006). Customers always hope that the service station would charge them a reasonable service fee (Liou and Chen, 2006). Customers should have faith in mechanics' work if they are trustworthy (Parameshwaran, 2010).

Empathy refers to a person's ability to care for and understand their consumers. It is critical to treat clients with respect and to understand their demands in order to provide high-quality service (Behara et al., 2002; Parameshwaran, 2010). Staff should think about the customer's needs and work from the heart (Berndt, 2009; Bouman and Van der Wiele, 1992). Customers should be able to visit service station at times that suit them (Behara, 2002). Due to his busy schedule, the customer did not have time to service his vehicle. The service station can pick up the vehicle from the owner's house or office and drop it off whenever it is convenient for him finally, they return the clean car to the customer after repairs and upkeep the evaluation criteria and sub-criterion groups are listed in Table 122.1.

Methodology

Analytic hierarchy process

AHP is the most extensively used MCDM method for decision making in service quality assessment. AHP is a quantitative technique for constructing a multi-attribute problem that is difficult. The key benefit of this strategy is that it is simple to use when dealing with various criteria. It is simpler to comprehend, and it can efficiently manage both qualitative and quantitative data. AHP is frequently employed in a variety of fields, including social, political, economic, and management science, to solve unstructured problems. Some results in the literature have been reported as an application of the analytic hierarchy approach (Zahedi, 1986; Podvezka, 2009; Emrouznejad and Marra, 2017; Dos Santos et al., 2019).

The AHP includes the following steps:

- Create a decision-making issue hierarchy based on the primary purpose.
- Evaluate the overall rating of alternatives, weighing the Ratings with relative criteria;
- Determine the relative priorities of Criteria on a pair-by-pair basis, reflecting their respective relevance concerning the element at the higher level; Priorities for sub-criteria are also important.
- Double-check the pairwise comparison's consistency.

Data collection

The questionnaire was broken into three part to match the conceptual setting indicated above. The first section provides background information on the customer and their car. The questionnaire's second section provides five assertions of key service quality criteria for customers to rate the importance of service criteria at the vehicle service centre. The respondents were given definitions of each service dimension to reduce bias in interpretation. A nine-point significance scale was used to grade the responses (Table 122.2). The

Table 122.1 Criterion and sub criterion

Criterion	Sub-Criterion	Definition
Tangible (C1)	Modern equipment (ME)	Up to date technical equipment is available.
	Dress code(DC)	Staff wearing proper dress code with a visible identity card.
	Reception and workshop (RW)	Reception and workshop area was clean.
Reliability (C2)	Promise time (PT)	Vehicle ready at-promise time.
	Appointment (AP)	Got the appointment as per desired date and time.
	Error-free repair (EFR)	Repairs carried out error-free.
	Bill (B)	Clear and correct bill provided
Responsiveness (C3)	Instant attention (IA)	Received instant attention on arrival
	Spare parts (SP)	Spare parts immediately available.
	Willing to help (WH)	Staff always helped willingly
Assurance (C4)	Knowledge (KN)	Staffs were knowledgeable and skillful to perform their task
	Service charged (SC)	Reasonable service charged.
	Trust (TR)	Mechanics trustworthiness
Empathy (C5)	Operating hour (OH)	Operating Hours were convenient.
	Friendliness (FR)	Friendliness of service advisor
	Understanding (UN)	Understand the customer needs
	Interest (IN)	The staff has the best interest at heart.
	Pick up and Drop (PD)	Pick up and drop facility for service/repair
	Car Clean(CC)	Car clean after servicing

Table 122.2 Importance scale (Saaty, 2012)

Intensity of importance	Definition	Explanation
1	Equal importance	An activity is equally favoured with each other.
3	Moderate importance	Experience and judgment slightly favour one activity over another.
5	Strong Importance	Experience and judgment strongly favour one activity over another.
7	Very strong importance	An activity is favoured very strongly over another.
9	Extreme importance	The evidence that favours one action over another is of the greatest possible quality.
2,4,6,8	Values in the middle	The difference in scale values between consecutive scale values.
1/9,1/8.....1	Reciprocals of above values	Used for inverse comparison.

questionnaire's third section has sixteen assertions of sub-criteria related to each criterion. The responses were graded on a nine-point significance scale once again.

The questionnaire was distributed to consumers of three top car servicing outlets in Nagpur, India, over the course of four months. The owners who had done paid service and had a 1200 CC car or above were considered for this research work. The replies were first checked to see if they had visited all three service facilities (Alternatives 1, 2, and 3) during the previous four months. Customers were requested to complete the survey based on prior experiences with the service centre. A total of 28 customers took part in the poll.

Three respondents were eliminated after inconsistencies were discovered, resulting in a response rate of 90%.

Microsoft Excel was used to record and compute the raw data from customers. The proposed effort uses the AHP methodology to assess the quality of passenger transportation service. The priority weight of criterion, sub-criteria, and alternatives was computed in this study using Saaty's Eigenvector technique (2012).

Case Study

Constructing hierarchy structure

The problem of service quality was broken down into three levels of hierarchy: major criteria, sub-criteria, and alternatives (Figure 122.1). The first level considers the relative importance of the primary service quality dimensions' main criteria in relation to the aim, i.e., the best service provider. Customers were asked to priorities the goal's most important criterion. The relative importance of sub-criteria within the primary criteria is the second level. Customers were asked to rank sub-criteria in relation to the primary criteria. At the third level, they compared service providers' performance in terms of sub-criteria of service dimensions. On a nine-point preference scale, clients were asked to indicate their preference for the service provider.

B. Pairwise comparison

As previously stated, a questionnaire was utilised to collect the respondents' pairwise compared evaluations for the three levels of the hierarchy (Figure 122.1). The pairwise comparison matrix was constructed using these as inputs. The values range from 1 to 9, and the pairwise comparison matrix assigns their reciprocal values. The values of the comparison are supplied in the transverse cell, and the factors in a row are compared to the factors in a column. When comparing the crossing cell and its corresponding cell, the crossing cell and its matching cell are stronger if the factor in the row is stronger (more significant) than the factor in the column. Table 122.3 depicts this.

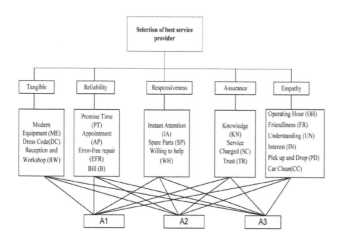

Figure 122.1 A framework for evaluating a service quality model

Table 122.3 Pairwise comparison of main criteria concerning the goal

Criteria	C1	C2	C3	C4	C5
C1	1	¼	1/2	1/6	½
C2	4	1	2	1/2	2
C3	2	½	1	1/4	1
C4	6	2	4	1	4
C5	2	½	1	¼	1
Column total	15	4.25	8.5	2.16	8.5

The Eigenvector method is used to analyse the comparison matrix at the first level.

Normalise matrix

Divide each item in each column by the total of the elements in that column to normalise the matrix above. Table 122.4 shows the normalised matrix. Divide the relevant cell value by the column total in Table 122.3 to get the cell entry in Table 122.4. The relative weight of the traits is represented by the last column in Table 122.4, which is the average of the row entries. (Winston, 1993) has demonstrated that this approach can accurately approximate the precise weights.

Table 122.4 Normalise matrix

Criteria	C1	C2	C3	C4	C5	Priority weight
C1	0.066	0.058	0.058	0.076	0.058	0.064
C2	0.266	0.235	0.235	0.230	0.235	0.241
C3	0.133	0.117	0.117	0.115	0.117	0.120
C4	0.4	0.470	0.470	0.461	0.470	0.455
C5	0.133	0.117	0.117	0.115	0.117	0.120

i. Eigen vector: Eigen vector (Priority weight) is an average of row in normalise matrix (Table 122.4)

$$W_c = [0.064, 0.241, 0.120, 0.455, 0.125]T$$

ii. Largest Eigen Value: Largest Eigen value is calculated by the following equation

$$\lambda_{max} = \Sigma W \times Column\ Total$$
$$\lambda_{max} = [(0.064 \text{x} 15) + (0.241 \text{x} 4.25) + (0.120 \text{x} 8.5) + (0.455 \text{x} 2.166) + (0.125 \text{x} 8.5)]$$
$$\lambda_{max} = 5.013$$

D. Consistency ratio

The consistency ratio (CR) is calculated using the formula CR = CI/RI, where the consistency index (CI) is calculated using the method:

$$Matrix\ order = 5$$

$$Consistancy\ index\ (CI) = (\lambda_{max} - n)/(n\text{-}1)$$
$$= (5.013\text{-}5)/\ 5\text{-}1) = 0.0325$$

$$Consistency\ ratio\ (CR) = CI/RI$$
$$= 0.0325/1.12$$
$$= 0.029$$

Where RI stands for Random Consistency Index.

Table 122.5 generates the RI value, which is related to the Matrix dimension. Saaty (2012) revealed that starting the AHP analysis with a consistency ratio (CR) of 0.10 or less is appropriate. If the consistency ratio exceeds 0.10, the judgments must be changed in order to identify and correct the inconsistency's source.

We can presume that our matrix of judgments is reasonably consistent because the proportion of inconsistency CR is less than 0.10. We can continue the decision-making process using AHP because the value of 0.029 for the proportion of inconsistency CR is less than 0.10.

Table 122.5 Random consistency index (Saaty, 2008)

n	1	2	3	4	5	6	7	8	9	10
RI	0	0	0.58	0.90	1.12	1.24	1.32	1.41	1.45	1.49

Table 122.6 Overall priority weights of alternatives

	C1	C2	C3	C4	C5	Priority weight	Ranking
Weight	0.06	0.24	0.12	0.45	0.12		
A1	0.34	0.22	0.34	0.37	0.33	0.32	2
A2	0.23	0.28	0.25	0.26	0.22	0.26	3
A3	0.43	0.50	0.41	0.37	0.45	0.42	1

Similarly, all pairwise comparisons of sub-criterion and alternatives for main criterion and sub-criteria were produced using the same technique as described above for determining Sub-criterion and alternate weightings are given a higher importance.

Derive overall priority

By multiplying and adding the priority weights of main criteria and options while considering sub-criteria. You may get the overall priority weight of each alternative. Table 122.6 shows the results.

The following equation is used to generate overall priority weights for alternatives.

$$W_o = \sum w_c \times w_a$$

Where = Overall priority weight of alternative
W_a = Main criteria weight
W_c = Sub criteria weight with respect to alternative
Alternative 1= 0.06(0.34) + 0.24(0.22) + 0.12(0.34) + 0.45(0.37) + 0.12(0.33) = 0.32
Alternative 2= 0.06(0.23) + 0.24(0.28) + 0.12(0.25) + 0.45(0.26) + 0.12(0.22) = 0.26
Alternative 3= 0.06(0.43) + 0.24(0.50) + 0.12(0.41) + 0.45(0.37) + 0.12(0.45) = 0.42

The order of the service dimensions is shown in Table 122.4. When evaluating the 'service station,' customers ranked 'assurance' as the highest priority with a weight of 45%. 'reliability' is given second priority with a weight of 24%. From the customer's perspective, 'responsiveness' and 'empathy' are both equally crucial, with 12%. The term 'tangible' is the least priority, with 6%.

AHP was used to compare service stations in this study. The service stations performance is ranked as shown in Table 122.6. All estimates put Alternative 3 (A3) at 42%, Alternative 1 (A1) at 32%, and Alternative 2 (A2) at 26%. Among the alternatives, Alternative A3 has the best service quality performance. It also does not imply that this service station provides excellent service. In fact, in light of the other options, service station should improve their service quality by taking into account the factors that determine a client's value.

Findings and Discussion

The goal of this study was to create a model that could assess perceived service quality in the automotive service sector as well as the performance of top passenger car service stations in Nagpur. As a result, data is collected from service station clients in order to assess the performance of alternative service stations. The findings revealed that in order to provide satisfying and quality service, service stations should focus on tangibles, responsiveness, and empathy. The findings of this study give managers an important insight into the factors that influence consumers' perceptions of the quality of passenger car service. Service stations can improve their quality and give better service to clients by addressing their particular weaknesses.

Managerial Implications and Recommendations

Using the AHP approach explained in this paper, managers can review business measures and quantify a different estimate in a way that gives them a strategic edge in managing service levels.

The disparity in satisfaction reveals which component of the business will be improved in order to strengthen its competitive position. It allows a corporation to get a strategic advantage over its competitors. Customers, on the other hand, have many options and may not be able to maintain a competitive advantage. Consumers are now satisfied with a company's services. They may realise that a competitor's superior offerings satisfy them even more. All strategies aimed at gaining a competitive advantage must be market-driven and market-aware. This is only possible if client perceptions of competitors are known. As a result, the AHP methodology outlined in this study aids management in developing and sustaining a relevant, competitive plan for improving service quality.

Limitations and Research Directions

Because the clients in our sample had visited all three passenger car service stations in the previous four months, one limitation of this study is that it did not look into respondents' switching behavior. This issue could be addressed in the future. Another drawback of the existing technique is that it lacks clear guidance on how to proceed. While it identifies which areas of service need to be changed, it does not provide guidance on how to implement a proper action plan to address the current framework's flaws. Future research in this area could broaden the framework. The third constraint is that the current framework only takes into account SERVQUAL's service dimensions. Future research could look towards expanding the framework given here by adding other dimensions.

In terms of generalisability, the authors suggest that, while the model was developed for a passenger car service station in this study, it could be used to a range of service industries to compare service performance to that of competitors.

References

Adeniran, A. O. and Fadare, S. O. (2018). Assessment of passengers' satisfaction and service quality in murtala muhammed airport (MMA2), Lagos, Nigeria: Application of SERVQUAL model. J. Hotel Bus. Manag. 7(188):2169–0286.

Ahmad, I. and Verma, M. K. (2018). Application of analytic hierarchy process in water resources planning: A GIS based approach in the identification of suitable site for water storage. Water Resour. Manag. 32(15):5093–5114.

Ambekar, S. S. (2013). Service quality gap analysis of automobile service centers. Indian J. Res. Manag. Bus. Soc. Sci. 1(1):38–41.

Anderson, E. W., Fornell, C., and Lehmann, D. R. (1994). Customer satisfaction, market share, and profitability: Findings from Sweden. J. Mark. 58(3):53–66.

Baki, N. A., Norddin, N. I., and Azaman, W. A. W. (2017). Application of analytic hierarchy process for selecting best student. J. Appl. Environ. Biol. Sci. 7(1):69–63.

Bartusková, T. and Kresta, A. (2015). Application of AHP method in external strategic analysis of the selected organisation. Procedia Econom. Finance 30:146–154.

Behara, R. S., Fisher, W. W., and Lemmink, J. G. (2002). Modelling and evaluating service quality measurement using neural networks. Int. J. Oper. Prod. Manag. 22(10):1162–118.

Berndt, A. (2009). Investigating service quality dimensions in South African motor vehicle servicing. Afr. J. Mark. Manag. 1(1):001–009.

Bolton, R. N. and Drew, J. H. (1991). A longitudinal analysis of the impact of service changes on customer attitudes. J. Mark. 55(1):1–9.

Bouman, M. and Van der Wiele, T. (1992). Measuring service quality in the car service industry: Building and testing an instrument. Int. J. Ser. Ind. Manag. 3(4):0–0.

Büyüközkan, G., Çifçi, G. and Güleryüz, S. (2011). Strategic analysis of healthcare service quality using fuzzy AHP methodology. Expert Syst. Appl. 38(8):9407–9424.

Buzzell, R. D., Gale, B. T., and Gale, B. T. (1987). The PIMS principles: Linking strategy to performance. New York (N Y): Free Press.

Chen, K. K., Chang, C. T., and Lai, C. S. (2009). Service quality gaps of business customers in the shipping industry. Transp. Res. Part E: Logist. Transp. Rev. 45(1):222–237.

Corrêa, H., Brito, E. P. Z., Aguilar, R. L. B. and Brito, L. A. L. (2007). Customer choice of a car maintenance service provider. Int. J. Oper. Prod. Manag.

Dos Santos, P. H., Neves, S. M., Sant'Anna, D. O., de Oliveira, C. H., and Carvalho, H. D. (2019). The analytic hierarchy process supporting decision making for sustainable development: An overview of applications. J. Clean. Prod. 212:119–138.

Edvardsson, B. (1996). Making service-quality improvement work. Manag. Serv. Qual.: Int. J. 6(1):49–52.

Emrouznejad, A. and Marra, M. (2017). The state of the art development of AHP (1979–2017): A literature review with a social network analysis. Int. J. Prod. Res. 55(22):6653–6675.

Galeeva, R. B. (2016). SERVQUAL application and adaptation for educational service quality assessments in Russian higher education. Qual. Assur. Educ.

Ghobadian, A., Speller, S. and Jones, M. (1994). Service quality. Int. J. Qual. Reliab. Manag. 11(9):43–66.

Goetsch, D. L. and Davis, S. (1998). Understanding and implementing ISO 9000 and ISO standards. New Jersey: Prentice Hall.

Haming, M., Murdifin, I., Syaiful, A. Z. and Putra, A. H. P. K. (2019). The Application of SERVQUAL distribution in measuring customer satisfaction of retails company. 17(2):25–31.

Jadhao, M. D., Kedar A. P. and Lakhe R. R. (2018). Evaluation of passenger car service quality through fuzzy AHP. Int. J. Eng. Sci. Invention. 7(4):30–35.

Jayamani, E., Perera, D. S., Soon, K. H. and Bakri, M. K. B. (2017). Application of analytic hierarchy process (AHP) in the analysis of the fuel efficiency in the automobile industry with the utilisation of Natural Fiber Polymer Composites (NFPC). In IOP conference series: materials science and engineering (Vol. 191, No. 1, p. 012004). IOP Publishing.

Katarne, R., Sharma, S. and Negi, J. (2010). Measurement of service quality of an automobile service centre. In Proceedings of the 2010 international conference on industrial engineering and operations management, (pp. 286–291), Dhaka, Bangladesh.

Keshavarz, S., Yazdi, S. M., Hashemian, K. and Meimandipour, A. (2009). Measuring service quality in the car service agencies. J. Appl. Sci. 9(24):4258–4262.

Kotler, P. (1994). Marketing management, analysis, planning, implementation, and control. London: Prentice-Hall International.

Li, H., Ni, F., Dong, Q. and Zhu, Y. (2018). Application of analytic hierarchy process in network level pavement maintenance decision-making. Int. J. Pavement Res. Technol. 11(4):345–354.

Liou, T. S. and Chen, C. W. (2006). Subjective appraisal of service quality using fuzzy linguistic assessment. Int. J. Qual. Reliab. Manag.

Longaray, A. A., Gois, J. D. D. R., and Da Silva Munhoz, P. R. (2015). Proposal for using AHP method to evaluate the quality of services provided by outsourced companies. Procedia Comput. Sci. 55:715–724.

Lotayif, M. S. (2017). Empirical assessment and application of SERVQUAL Evidence from UAE Banks. Int. J. Bus. Manag. 12(4).

Mersha, T. and Adlakha, V. (1992). Attributes of service quality: the consumers' perspective. Int. J. Serv. Ind. Manag. 3(3):0-0.

Narayan, B., Rajendran, C., Sai, L. P. and Gopalan, R. (2009). Dimensions of service quality in tourism–an Indian perspective. Total Qual. Manag. 20(1):61–89.

Nikfalazar, S. (2016). Statistical process control application on service quality using SERVQUAL and QFD with a case study in trains' services. TQM J. 28(2):195–215.

Nikkhah, A., Firouzi, S., Assad, M. E. H. and Ghnimi, S. (2019). Application of analytic hierarchy process to develop a weighting scheme for life cycle assessment of agricultural production. Sci. Total Environ. 665:538–545.

Nosal, K. and Solecka, K. (2014). Application of AHP method for multi-criteria evaluation of variants of the integration of urban public transport. Transp. Res. Procedia 3:269–278.

Parasuraman, A., Zeithaml, V. A., and Berry, L. L. (1985). A conceptual model of service quality and its implications for future research. J. Mark. 49(4):41–50.

Singh, R., Shankar, R., Kumar, P. and Singh, R. K. (2012). A fuzzy AHP and TOPSIS methodology to evaluate 3PL in a supply chain. J. Model. Manag. 7(3):287303.

Winston, W. L. (1993). Operations research: Applications and algorithms. Duxbury Press.

Yildiz, S. and Yildiz, E. (2015). Service quality evaluation of restaurants using the AHP and TOPSIS method. J. Soc. Adm. Sci. 2(2):53–61.

Zahedi, F. (1986). The analytic hierarchy process—a survey of the method and its applications. Interfaces 16(4):96–108.

Zeithaml, V. A. (2000). Service quality, profitability, and the economic worth of customers: what we know and what we need to learn. J. Acad. Mark. Sci. 28(1):67–85.

123 Image analysis for crop acreage estimation using remote sensing for digraj region

Amoli Belsare[1,a], Priyanka Khadgi[1,b], and Sanjay Balamwar[2,c]

[1]Electronics and Telecommunication Department, Yeshwantrao Chavan College of Engineering, Nagpur, India

[2]Maharashtra Remote Sensing, Application Centre (MRSAC), Nagpur, India

Abstract

Agriculture plays a crucial role in Indian economy which includes cultivation of soil, growing crops and uplifting livestock. Prediction of the crop production, crop yield and determining the harvest of the crop are few important applications of remote sensing in agriculture. In this paper, a crop classification and acreage estimation were carried out using geospatial technology. The selected study area is located near Digraj, Maharashtra, India. The multi-temporal, Quickbird Satellite images of September 2013, January 2015, January 2017, March 2019 and December 2020 were used for determination of crop acreage estimation in the proposed work. The algorithm evaluation is carried out with comparison of output from QGIS and ERDAS IMAGINE. Classification accuracy was calculated using the error matrix and results from each software were compared against one another. The classification accuracies were in the range of 8495% for QGIS and 9398% for ERDAS IMAGINE. On comparison ERDAS IMAGINE showed better results.

Keywords: Crop acreage estimation, crop yield estimation, remote sensing, GIS, time series.

Introduction

Agriculture in developing countries will face some major challenges in coming years such as increasing food demand from rapidly growing population as well as economic growth, decreasing productivity in high productivity region, increasing vulnerability of agriculture to potential climate change, etc. Due to the multiple needs for agricultural commodities, estimation of crop yield and acreage is of utmost importance. Remote sensing (RS) is used to analyse and interpret objects on earth surface systematically and observe change in physical object over the period of time (Fichera et al., 2012; Yedage et al., 2013; Zhang et al., 2017; Kussul et al., 2017; Quadir et al., 2019; Pimpale et al., 2019). RS image analysis mostly used to collect physically inaccessible earth surface data and analysis it for various application areas. Main focus of current study in remote sensing is based on agriculture which would help in increasing crop yield estimation and monitoring. These applications mainly include: soil sensing, crop classification, cop condition assessment, crop land estimation, far mapping, mapping of land management practices and compliance monitoring (Karale et al., 2014; Su and Zhang, 2020; Singh and Kumar, 2020). In developing countries agriculture will face some major challenges, increasing food demand being the basic reason due to increase in the population growth.

Some of the other major reasons are climate change, decreasing productivity, and increasing economic growth. Thus, there is a need for crop yield estimation as it can be used for determination of the crop growth in advance. This would in turn help to determine appropriate cost of all the agricultural commodities as well as provide assurance to the food security. This can be done through remote sensing and image processing can improve decision making for vegetation measurement. Such analysis uses multispectral remote sensing which involves the visible, near infrared, and short-wave infrared images bands captured in a number of wavelength bands. Abundance of land cover maps observed from earth observation through RS sensors could provide a satisfactory efficiency in analysing using image processing approaches. One of the most important tasks in image processing is image classification which is used to analyse land use and land cover classes. It can be done through supervised or unsupervised classification. The supervised classification, classifier training is performed using prior known sample maps from captured image to identify different spectral classes present in image. This approach identifies objects on earth surface captured according to respective spectral signatures. In this method, a predefined region of interest from image captured is to be selected by user as training site. Further, these training sites are used as references for the supervised classification of all other pixels in the image. The Semi-Automatic Classification Plugin (SCP) approach is used to classify the image in QGIS. The images are preprocessed for band selection and post-processing of the images after classification improves the results. An automatic workflow to examine land cover classification includes use of set of for raster image processing tools. ERDAS Imagine is a raster-based software package and is an extremely effective way to process and analyse raster data. It is designed specifically to extract information from imagery. In this paper we have used two different software to compare crop classification results.

This paper is composed of four sections; introduction, literature review, methodology, and conclusion. The introduction section will set the context of the study. The second section, the literature review focuses

---

[a]adbelsare@ycce.edu; [b]priyanka.khadgi@gmail.com; [c]sanjaybalamwar@gmail.com

on the related work done by various researchers and concludes their findings. The third section of methodology is further divided into four subsections- study area and data collection, classification, accuracy assessment and crop acreage estimation. The final section illustrates the results obtained from the proposed study and discusses possible limitations of the approach and implications of the results.

Related Work

Many techniques and model have been proposed by researchers regarding crop classification and prediction of crop yield before actual harvesting (Prasad et al., 2006; Maurya et al., 2011; Karale et al., 2014; Song et al., 2019; Su and Zhang, 2020). In literature one can also find several supervised-unsupervised or hybrid approaches on the crop type identification and crop acreage estimation for different geographical areas (Jovanović et al., 2014; Bhuyar et al., 2020; Wang et al., 2019; Desai et al., 2020). In recent work rice area extraction on spatial-temporal data using regional parametric syntheses approach has been discussed (Cao et al., 2021). This paper also focuses on importance of cultivation area selection in remote sensing application development. Similarly, mapping of betel leaf acreage analysis is presented in (Hudait and Patel, 2022). A machine learning based approach is implemented in this work for yield estimation. A summary of few literatures work on crop acreage estimation and image classification is given in Table 123.1. It consists of the methodologies used, image details, evaluation parameters and remarks. In most of the paper's accuracy is the general parameter used for algorithm evaluation. Most of the studies have been conducted on any one type of crop like rice, wheat, sugarcane, corn and soybean etc. The crop estimation mainly dependent on time series data and is the backbone for complete yield prediction in corresponding area. The literature shows application of ML based algorithm could improves estimation of crop.

Methodology

In proposed work supervised classification of images was performed using two software's- QGIS and ERDAS IMAGINE. Entire workflow followed is given in Figure 123.1. After acquiring the required images, classification was performed. Accuracy assessment was done and various parameters were calculated followed by crop acreage estimation. All results from both the software's were compared against one another. Detailed steps for crop image classification are depicted in Figure 123.2. Georeferencing, band set selection and processing, training sample selections, classification and accuracy assessment are the important steps in RS image analysis. Following section describes the detailed process flow for crop yield estimation for Digras area.

A. Study area and data collection

The study area falls near Digraj, Maharashtra, India. The multispectral images were taken from Quickbird satellite which involves images of different months of four years. The time series data were as follows - September 2013, January 2015, January 2017, March 2019 and December 2020. The images were in JPG format having resolution of 531m and the bands used for classification of image were red, green and blue. A multi-spectral monochrome image of the same scene, where each of them taken with a different sensorsensitive to a different wavelength are used in the analysis. Each image is referred to as a band. This multibands image data consists of Red, Green, Blue (RGB) color image bands.

Classification

After loading the image in QGIS, georeferencing was the first step followed. All the images were georeferenced using co-ordinates from Google Earth. These co-ordinates were fed to georeferencer in QGIS and georeferenced image was obtained. Band set was defined which was the input image for SCP and training input file was created for collecting training areas and calculation of respective spectral signatures. These training areas were taken in the form of region of interests (ROIs) which were manually created. Supervised classification was performed for classification of images in both QGIS and ERDAS IMAGINE. Classification outputs obtained from QGIS and ERDAS IMAGINE are shown in Table 123.2. The obtained output using supervised classifier gives more accurate qualitative output for December 2020 in QGIS. The obtained results were verified by using standard procedure in ERDAS IMAGINE. Georeferencing of images using Google Earth was done before classification in QGIS whereas no georeferencing was required before classification in ERDAS IMAGINE. The co-ordinate system was set to WGS 84: EPSG 4326.

In QGIS software SCP plug-in was used for setting band-set and also for performing classification. Several regions of interests were created. The image was classified into three classes – crops, land and

Table 123.1 Summary of work done on crop yield estimation techniques

Paper ref no.	Methodology	Image details	Evaluation parameters	Findings
Prasad et al. (2006)	NDVI, non-linear Quasi-Newton multi-variate optimisation method	Study area-Iowa, US Crops-soybean, corn	R2 -values: Corn: 0.78 Soybean: 0.86	Proposed Quasi-Newton method helps the countries for better crop production where it is weather dependent.
Maurya et al. (2011)	Hybrid classification techniques, unsupervised, supervised and time series based	Satellite-MODIS (TERRA) Study area-11 districts of Madhya Pradesh Climatic requirements-warm and moist Time series-June 1st week to October 2nd week MODIS Images Crop-Soybean Soil-Well drained and fertile loam soils	Accuracy: Acreage area and production: 80.72%	The proposed classification technique helps in accurate estimation of acreage and production of soybean crop in parts of Madhya Pradesh
Yedage et al. (2013)	Digital image processing (DIP), GIS based analysis, Crop concentration index, supervised classification (MXL, Minimum Distance, Mahalonobis)	Study Area- Solapur District Satellite used-IRS P6 LISS-III Crop-sugarcane	Accuracy: Mahalonobis classification: 95% MXL accuracy: 86.1%	When compared with traditional algorithms ANN and J48 algorithm gave better results.
Karale et al. (2014)	Supervised classification, Maximum likelihood algorithm, Euclidean distance	Satellite used-Rapideye Crops-Corn, sugar, beet, soybean Software used-ERDAS Imagine	Accuracy: RapidEye image processing: 90%	For feature detection and land cover mapping of agriculture landscapes RapidEye imagery and all other satellite images with high spatial and spectral resolution are suitable.
Zhang et al. (2017)	NDVI,Pareto boundary method	Satellite-Landsat-5, Landsat-8, MODIS Study area-North Korea Crops-Rice, Maize, Soybeans Total no of images taken- 32 (MODIS time series data), 3 (Landsat images)	Accuracy: Classification of Landsat images (2008): 74.4% Classification of Landsat images (2013): 69.8% Classification of Landsat images (2014): 73.1%	Major crops were characterised from the featured derived from NDVI profiles. Even for limited ground truth data, the proposed approaches are used for crop mapping and acreage estimation.
Kussul et al. (2017)	Multilevel DL architecture, unsupervised neural network (NN), supervised NNs (MLP, Random forest, CNN)	Satellite used-Landsat-8, Sentinel-1A RS Satellites Study Area-Kyiv region of Ukraine Image Data-19 Multitemporal scenes, 4 Landsat-8 images, 15 Sentinel-1 images Software-TensorFlow Study season- 2015 vegetation season	Accuracy: RF: 88.7% ENN: 92.7% Ensemble of 1D: 93.5% Ensemble of 2D: 94.6%	Certain summer crop can be better discriminated as the ensemble of CNN exceeds the ensemble of MLPs.

(Contd...)

Table 123.1 (Contd...)

Paper ref no.	Methodology	Image details	Evaluation parameters	Findings
Quadir et al. (2019)	Integrated two-stage ISODATA clustering, Hybrid Technique, NDVI	Satellite-IRS-P6 Sensor-AWiFS Study area-5 districts of Maharashtra Crops-Wheat, rabi sorghum, sugarcane Software-ERDAS, Imagine, ArcGIS, LOCATE	Accuracy: Pure-class pixels: 93-98% Mixed-class-pixel: 2-7% Wheat classification: 95%	As compared to hierarchical classification technique, the hybrid technique is simple, time saving, less subjective and requires less expertise.
Song et al. (2019)	Supervised classification, NDVI, Correlation coefficient method, DVI	Satellite-Landsat 5 Study area-Bulandshahr District Soil-Alluvial Crops-Kharif Software-ERDAS imagine Image Data-8 October 2000 and 21 September 2014	Accuracy: Supervised classification (2000): 97.9% Supervised classification (2014): 99.5%	The observed relationship between the declination of natural vegetation land area and increment in crop production and crop land area was good.
Su and Zhang (2020)	Normalised Difference Vegetation Index (NDVI), Leaf Area Index (LAI), Support Vector Machine, Decision Tree, Neural Network	Study area-Ottawa, China, Haryana (India), Ukraine, USA, Satellite- MODIS, Landsat, RapidEye Crops-corn, rice, wheat	Weather-based prediction system, Accuracy-93.64% Sensitivity-89.36% Specificity-91.72%	Machine learning techniques outperform the conventional regression based analysis for prediction.

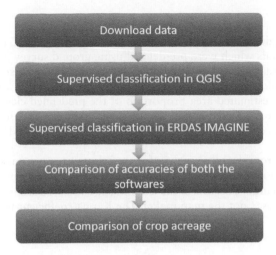

Figure 123.1 Workflow followed

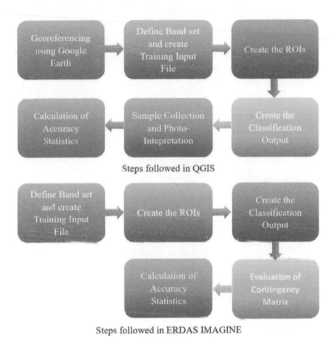

Figure 123.2 Methodology followed in QGIS and ERDAS IMAGINE

built-up. In ERDAS IMAGINE several AOIs were created in signature editor and were saved in a single AOI layer. Minimum distance algorithm was selected for classification in both the software's.

This classier accepts the Euclidean distance d(x, y) calculated between spectral signatures of image and training spectral signatures as per equation (1):

$$d(x, y) = \sqrt{\sum_{i=1}^{n}(x_i - y_i)^2} \qquad (1)$$

Where: x and y are spectral signature vectors of an input and training area from captured image;
n = number of image bands

Thus, the distance is calculated for every pixel in the image, assigning the class of the spectral signature that is closer, according to the discriminant function.

Accuracy assessment

This is an essential step in algorithm in order to evaluate the algorithm performance and minimise the errors. ROIs were created using stratified random points which were pre-defined using ground survey and used as reference for the accuracy assessment. In literature, one can find many statistical parameters such as overall accuracy (TA), user's accuracy (UA), producer's accuracy (PA), and Kappa coefficient (KC) to estimate the algorithm performance. To represent the estimated area proportion of the predefined classes in images, the corresponding evaluation parameters are calculated based on area-based error matrix. These parameters can be In QGIS random points were created using SCP plug-in whereas in ERDAS IMAGINE contingency matrix was used to calculate accuracy.

These performance analysis parameters estimated from the error matrix for the selected remote sensed image. Overall Accuracy depicts correct proportion of mapping of all of the reference sites with classification output. Producer's accuracy describes the map accuracy in terms of the map maker and user's accuracy and is the accuracy from the point of view of a map user, not the map maker. The Kappa coefficient tests the accuracy of a classification results.

The formulae used for calculation of all the parameters are as follows:

$$UA = \frac{p}{q} * 100 \qquad (2)$$

$$PA = \frac{p}{r} * 100 \qquad (3)$$

$$TA = \frac{p}{s} * 100 \qquad (4)$$

$$KC = \frac{(TC * TCS) - \sum r * q}{TS^2 - \sum r * q} \qquad (5)$$

Where, : p Total number of correctly classified pixels in each category (diagonal total)
q :Total no of classified pixels in that category (row total)
r : Total no of reference pixels in that category (column total)
s : Total no of reference pixels

Table 123.3 lists all the above evaluation parameters for the time series data analysis carried. This accuracy assessment analysis shows that, PA for land and crop field is well classified with the reference classification pixels as compared to built-up area. Whereas TA and KC evaluation parameters depicts that due to minimum error there is scope for improvement in proposed supervised classification results in QGIS and can be improved for more accurate crop acreage estimation.

Crop acreage estimation

For estimating crop acreage, the pixel sum of each obtained class from classifier output is then multiplied with the resolution. Crop acreage for different years was calculated using the following formula:

$$Crop\ Acerage = Total\ number\ of\ Pixels * \text{Re} solution$$

The area obtained was in meters. Table 123.4 depicts the area of captured time series data for the area of study. From this table, it can be seen that the acreage area obtained using maximum likelihood classifier in QGIS is similar to the area obtained in ERDAS IMAGINE. Thus, it can be concluded that supervised

Table 123.2 Classification outputs of QGIS and ERDAS IMAGINE

Month, Year	Input Image	QGIS Classification Output	ERDAS IMAGINE Classification Output
September, 2013			
January, 2015			
January, 2017			
March, 2019			
December, 2020			

Table 123.3 User's accuracy, producer's accuracy, overall accuracy and kappa coefficient

Date	QGIS					ERDAS				
Month, Year	Class	UA	PA	TA	KC	Class	UA	PA	TA	KC
September, 2013	Crop	100	99.3889	95.7906	0.7639	Crop	99.6974	9.3450	96.8568	0.9122
	Land	83.33	100			Land	85.9060	100		
	Built-up	100	12.247			Built-up	91.8367	95.7446		
January, 2015	Crop	93.7500	95.5138	90.2802	0.8091	Crop	100	98.9130	98.6441	0.9682
	Land	91.6667	88.5712			Land	95.7672	97.8378		
	Built-up	100	100			Built-up	93.2203	98.2142		
January, 2017	Crop	100	99.6393	94.0097	0.8740	Crop	100	99.7512	98.4822	0.968
	Land	100	100			Land	93.5483	96.6667		
	Built-up	77.7778	100			Built-up	96.9697	95.0495		
March, 2019	Crop	100	75.0176	84.6669	0.7129	Crop	94.6488	98.6062	93.7735	0.8880
	Land	88.8889	97.5976			Land	96.7032	86.69995		
	Built-up	50	100			Built-up	77.5510	95		
December, 2020	Crop	84.6154	95.0571	82.8122	0.6980	Crop	98.7805	100	98.9313	0.9810
	Land	100	78.5849			Land	99.0591	98.3978		
	Builtup	37.5	100			Builtup	99.2424	94.9275		

Table 123.4 Comparison of crop acreage

Date	QGIS (m²)	ERDAS IMAGINE(m²)
September, 2013	528628.6431	537710.1305
January, 2015	345575.1624	381087.6271
January, 2017	427856.5878	434882.4671
March, 2019	159341.1137	236207.6904
December, 2020	349115.6511	372972.8002

classifier shows good results as compared to standard approach from ERDAS. Accuracies measurement for the selected time series data were calculated on the basis of error matrix due to unavailability of ground truth data. Classification results obtained were good but have scope to improve with ML approach.

Conclusion

Performance evaluation of proposed approach includes kappa coefficients generated from contingency matrices with error of omission (Type II error), where pixels are not reported as crop when they should be; and error of commission (Type I error), where pixels are reported into crop image when they should not be; are taken into account for analyses. The accuracies of the time series data as listed in Table 123.3 ranges from 8495% for QGIS and 9398% for ERDAS IMAGINE and values of kappa coefficients ranges from 0.710.87 in QGIS and 0.88SC0.98. It can be observed that for most of the images ERDAS IMAGINE provided better results than QGIS which is also evident from the values of kappa coefficients. Also, on comparison it was found that both gave satisfactory results. From Table 123.3, it can be seen that the acreage area obtained are near about same in both the cases. Thus, it can be concluded that ERDAS IMAGINE gave better classification results.

Minimum errors were found and mixing of classes was observed due to similar spectral signatures. This happened as the Earth's natural phenomenon cannot be captured using mathematical formulae only. Thus, for accurate results the proposed work can be extended to machine learning algorithms and implemented through AI based solutions.

References

Bhuyar, N., Acharya, S., and Theng, D. (2020). Crop classification with multi-temporal satellite image data. Int. J. Eng. Res. Technol. (IJERT) (9/06):221–225. ISSN: 2278-0181.

Cao, D., Feng, J. Z., Bai, L. Y., Xun, L., Jing, H., Sun, J., and Zhang, J. (2021). Delineating the rice crop activities in Northeast China through regional parametric synthesis using satellite remote sensing timeseries data from 2000 to 2015. J. Integr. Agric. 20(2):424–437.

Desai, K., Rajesh N. L., Shanwad, U. K., Ananda, N., Koppalkar, B. G., Desai, B. K., Rajesh, V. Kumara, K., and Chandralekha (2020). Geospatial techniques for paddy crop acreage and yield estimation. Curr. J. Appl. Sci. Technol. 39(14):71–79. doi: 10.9734/CJAST/2020/v39i1430704.

Fichera, C. R., Modica, G., and Pollino, M. (2012). Land cover classification and change-detection analysis using multi-temporal remote sensed imagery and landscape metrics. Eur. J. Remote Sens. 45(1):1–18. doi: 10.5721/EuJRS20124501.

Hudait, M. and Patel, P. P. (2022). Crop-type mapping and acreage estimation in smallholding plots using Sentinel-2 images and machine learning algorithms: Some comparisons. Egypt. J. Remote Sens. Space Sci. 25:147–156.

Jovanović, D., Govedarica, M., and Rašić, D. (2014). Remote sensing as a trend in agriculture. Res. J. Agric. Sci. 46(3):32–37.

Karale, Y., Mohite, J., and Jagyasi, B. (2014). Crop classification based on multitemporal satellite remote sensing data for agro-advisory services. In Proc. of SPIE (9260):926-004–1, doi: 10.1117/12.2069278.

Kussul, N., Lavreniuk, M., Skakun, S., and Shelestov, A. (2017). Deep learning classification of land cover and crop types using remote sensing data. IEEE Geosci. Remote Sensing Lett. 14(5):778–782. doi: 10.1109/LGRS.2017.2681128.

Maurya, A., Tripathi, S., Soni, S., and Soni, P. (2011). Estimation of acreage & crop production through remote sensing & GIS technique. Geospatial world forum.

Pimpale, A. R., Rajankar, P. B., Wadatkar, S. B., and Ramteke, I. K. (2019). Application of remote sensing and GIS for acreage estimation of wheat. Int. J. Eng. Bus. Enterp. Appl. (IJEBEA) 12(2):167–171.

Prasad, A. K., Chai, L., Singh, R. P., and Kafatos, M. (2006). Crop yield estimation model for Iowa using remote sensing and surface parameters. Int. J. Appl. Earth Obs. Geoinf. 8:26–33. DOI:10.1016/j.jag.2005.06.002.

Quadir, A., Abir, I. A., Nawaz, S., Khan, N., Mohammad, A. A., Olukunle, R. K., Akhtar, N., Anees, M. T., Hossain, K., and Ahmad, A. (2019). Crop acreage and crop yield estimation using remote sensing and GIS techniques Bulandshar District. Indian J. Ecol. 46(3):470–474.

Singh, K. and Kumar, S. S. (2020). Crop yield prediction techniques using remote sensing data. Int. J. Eng. Adv. Technol. (IJEAT) (9/3):3683–3689. doi: 10.35940/ijeat.C6217.

Song, J., Xing, M., Ma, Y., Wang, L., Luo, K., and Quan, X. (2019). Crop classification using multitemporal landsat 8 images. Int. Geosci. Remote Sens. Symp. 2407–2410. doi: 10.1109/IGARSS.2019.8899274.

Su, T. and Zhang, S. (2020). Object-based crop classification in Hetao plain using random forest. Earth Sci. Inform. 14:119–131. doi: 10.1007/s12145-020-00531-z.

Wang, L., Dong, Q., Yang, L., Gao, J., and Liu, J. (2019). Crop classification based on a novel feature filtering and enhancement method. Remote Sens. 11:1–18. doi: 10.3390/re110404455.

Yedage, A. S., Gavali, R. S., and Patil, R. R. (2013). Remote sensing and gis base cropacreage estimation of the sugarcane for solapur district, Maharashtra. Gold. Res. Thoughts. (2):1–12.

Zhang, L, H., Liu, Q., Shang, J., Du, X., and Zhao, L., Wang, N. and Dong, T. (2017). Crop classification and acreage estimation in North Korea using phenology features. GISci. Remote Sens. 1–26. doi:10.1080/15481603.2016.1276255.

124 Latch circuit based portable site-specific landslide early warning system

Prashant B. Pande[a], Jayant M. Raut[b], Shantanu R. khandeshwar[c], Rahul G. Sangole[d], Harshal G. Sangole[e], and Durga Thakre[f]

Civil Engineering, Yeshwantrao Chavan College of Engineering, Nagpur, India

Abstract

This paper aims to present a low cost landslide early warning system based on latch circuit. Landslide has a very destructive past in India, not just in India but also everywhere in the world. There is certain method to detect and forecast it namely as landslide early warning systems (EWSs). These systems have a very great role in alerting and evacuating peoples when landslide occurs. Because of such systems the management, evacuation and alerting process becomes rapid. These systems are deployed in the ground zero or can be monitored by various satellite monitoring. Now a day's IoT based monitoring systems have been also introduced. The effectiveness of such system depends upon the detection of the displacement of the ground. But there was always a need of such systems which is robust, accurate enough and should be available at low cost at same time.

In this paper a simple but effective device to detect the displacement of the soil is introduced. Our study area is under the umbrella of the development of early warning system device which is based on latch circuit principal and its use in geotechnical engineering specifically as detection of movement of ground. In our review, we tend to find such devices are really less or negligible which are based on the principle of latch circuit to detect ground movement, and which can deploy according to the site specific requirement. This device can be categorised under the self-maintaining circuits. The developed instrument can be employed in various soil and landslide zone which can be assembled in grid pattern. The methodology of this device is under transitional phase for this kind of ground movement detection.

Together with its benefits and limitations are illustrated in this paper, motivating that the presented solutions delivered the planned aim. The system has been tested in outdoors in natural slopes and in various subsurface soil conditions on different soil. The obtained results are extremely satisfactory.

Keywords: Early warning system, landslides, latch circuits.

Introduction

Natural disasters are very frequent and uncommon in nature one can predict the zone but not the exact location, the time span but not exactly when it is going to happen. Landslides are one of the very common phenomenons in natural disaster events and there is an urge to rectify, monitor and combat it (Guzzetti et al., 2020).

India, a country which is well known for its various facts, acclaimed for its rich geography and topography, but it has a vulnerable past to landslide. Over 14% of Indian landmass is susceptible to landslides. India has the longest mountainous chain on the planet earth, the Himalayas formed due to the collision of the Gondwana and Eurasian plate. At this time also, these plates are moving and colliding but in very low pace resulting the earthquake and landslides. After observing the landslides and factors responsible for it we tend to know that it is happening in an exceeding frequency in a span of few months (Dikshit and Satyam, 2018; Casagli, et al., 2017) and makes it more difficult for us to predict and manage this alarming situation as all its nature is unexpected, Now a day 5G system has also been studied to combat this (Li et al., 2021).

In India the geological survey of India (GSI), National landslide status map (NLSM) categorised the whole landmass of India into various states, districts and divisions which further classify into high, modest, and low zones defining the status of landslides on grounds (gsi.gov.in). Moreover, the challenge in front of us is likely to be, anyone cannot predict when and wherever the landslide may occurs. By using many high-end technologies, we were just able to find out the areas which are more prone to it. This develops the need of site-specific warning system. The landslide doesn't begin initially but gives a significant sign of failing by dispersion and sliding of rock fall (Lacasse and Nadeem, 2009). We used the temperamental signs of landslides to design our early warning system which can maintain the advantages of currently used early warning systems but overcome the limitations of it such as cost, portability, robustness etc. and such devices to be deployed on the ground (Pradhan et al., 2005; Kumara et al., 2019). There is always need of alerting devices which detect ground movements and here as comes our solution of latch circuit-based ground movement detection device.

For the detection of ground movements there are several kinds of devices available, but we thought of development of latch circuit based EWSs. All this kind of early warning system usually need that's the

[a]prashantbpande21@gmail.com; [b]jmrv100@gmail.com; [c]khandeshwar333@yahoo.com; [d]rahulsangole2409@gmail.com; [e]harshalsangole1312@gmail.com; [f]durga.thakre2002@gmail.com

sensors to be buried underground for subsurface detection (Boonyang and Biansoongnern, 2016; Ju et al., 2015). By investigating furthermore in depth and by experimentation the grasp of concept makes additionally clears. In field we reciprocate the essential trials on ground and on different soil materials on different slope conditions. That clears our idea practically. The proposed device is currently under transitional phase and the preliminary experimental results are extremely encouraging.

In this paper, the introduction to latch circuit, methodology, design overview, implementation strategy, installation process, field trials, its advantages and limitations, flexibility and practical challenges are discussed.

Study Area

This study is revolving around latch circuit, it may be categorised under short circuit which may occur in the complementary circuit or in integrated circuit (IC). More precisely or specifically it is circuit path, which is unintended or of low impendence, which triggers the path when it gets overload or circuit gets breaks that's disrupts the local settle connection and makes the circuit to get break. If alarming device is connected to it such as buzzer it will starts to alarm in form of beep or buzz sound. Similar distributed sensor network solution has also been presented by (Chu and Patton, 2021) to monitor and study landslide.

These circuits can be used in various forms in electronics applications in a view of their low cost, ease in manufacturing, convenience on ground, sturdy, and versatile, at the same time precise and multi-practicability (Intrieri et al., 2012; Ramesh, 2012). Considering all its benefits and limitations of assorted EWSs we thought to develop landslide early warning system using latch-based principle. By reviewing various literatures we get to know the present fundamental situation that, the most EWSs offered as a solution are very costly and require further optimisation discussed by Thuro and Wunderlich (2014) and Costanzo and di Massa (2015). Some of the current available solutions shares are mentioned in the Figure 124.1. The main motive to settle on latch circuit is, very fewer devices are based on it, which are employed and being used in the field of geotechnical engineering.

Design Overview

The designed displacement detecting system is modular, comprised of segments of smooth 10 cm plates and latch circuit which is installed in little depth in ground site, like the presently existing systems (Figure 124.2). This system is based on the short of circuit in the continuous latch grid due to the movements between two successive joints placed in ground. The system for connecting the joints, over which the local displacement measurement system can be inserted, has been designed and enforced. The local displacement measurement system has been chosen to provide precise measures of linear displacements with limited costs.

Methodology

The working principle in this circuit of the sensor is based on latch and connective latch grid. There are many types of sensors available in market. The pros to choose this sensor are enlisted above. For selecting this type of circuit is, it's construction is very simple and economical it can also be made work on the AC supply when needed by making minor changes which includes connecting the 12v relay to it and components are easily available in the electronic Market (Figure 124.2).

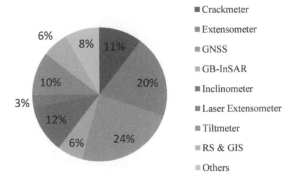

Figure 124.1 EWSs available in field and frequency of usage

In this sensor the power supply which is provided is 3.7v DC and the components used are transistor bc547 resistance 680 ohms and 2.2k respectively an indicator led a Switch and rechargeable lithium-ion battery of 3.7v a ring magnet and some metal plates and pipe All the connections are made according to the above provided circuit diagram (Figure 124.3). In this case we have used the bc547 transistor which has three terminals collector base emitter. The base and the emitter pin are short Circuited this implies the current flows from VCC to the ground through 2.2k resistor any current will not flow in the transistors due to which the circuit will remain open, and the buzzer will not work (Figure 124.2). While in another case when the base and the emitter is open Circuited then the current flows from VCC to 2.2 k resistor this because of this the transistor will work and flow of the current starts from the collector pin to emitter pin which leads in flow of current through the buzzers which will makes an alarming sound in order to warn us.

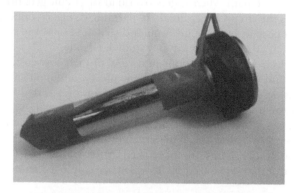

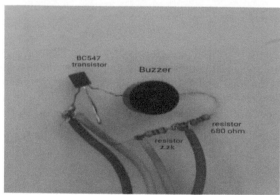

Figure 124.2 Construction details of device

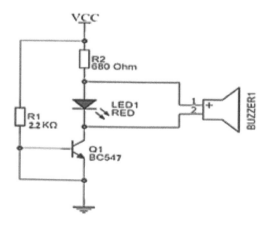

Figure 124.3 The circuit diagram of latch EWS device

Implementation Strategy

Concept of this sensor is when any activity occurs on the steep slope the loose rocks and soil eventually tends to fall downwards when loose rocks and any such thing meets our sensors within fraction the circuit gets breaks, and it indicates that there is such movement going in the land slide prone area. Another placement is a small grip made up of flexible copper wire which relate to each other with the help of small button magnets in the

junction and mounted externally on the site on which the system is to be install. Whenever any displacement is going on the land, they placed magnet gets disconnected from each other and hence results in breaking the circuit. It is a clear indication that the landslide may happen. In region many such small grids can be made and placed one after the other to monitor larger areas. This system is flexible enough for further development that it can also be connected to the sirens to warn the area and can be monitored remotely.

Device shown (Figure 124.4) is a newly developed latch circuit-based sensor which helps in detecting the rock fall, soil displacement, movement of ground in internal layer of soil placed on the mountain and the hilly areas which is a detachable circuit which beeps when the circuit breaks in any pattern weather it is due to rock fall or due to the crack in the land.

For constructing this device, we have used a stainless-steel pipe and a small metal cap for the pipe and a 25 mm magnet. The battery and the circuit are mounted in the pipe housing and the buzzer is placed beside the magnet and for the crack detecting device we are using button magnet and flexible copper wire which are connected in a checkers pattern grid which whenever the magnet gets disconnected from the connection, it beeps and indicates that there is some displacement in the soil and ground. The above image shows the actual depiction of the device explaining the design concept more effectively.

Installation Process

The installation Process, for both the system of sensor is very easy and requires no skilled technicians for the sensor we just must drill a hole on the bottom of the slope which is chosen to be monitored and insert the epoxy adhesive in it and then insert the device gently in it.

After that to install the grid system we must make small mesh like structure of conducting wires of low tensile strength those can detect the minor deflection in the soil strata and the slope this will be mounted on the areas which can develop the cracks while landslides.

Field Trials

Before beginning of taking actual measurement on ground we first check the accuracy of our device in open air and set it as per the site and our testing requirement. The preliminary test initiative was in open atmosphere where we check the basic wiring and connection were properly connected, to ensure the proper functioning of the device. After that the said device was deployed in the experimental ground (Figure 124.1) as per the installation procedure mention below.

The trials were taken on steep slope and different type of soil and ground conditions in a very rugged style. In the test, there were different phases in which the instrument is mounted in various surface and subsurface condition on ground by varying depth. The Six styles of testing were adopted on ground, first

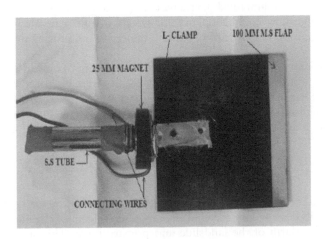

Figure 124.4 Actual image of device

with the position of magnet open above surface on ground and second covering the entire device in soil primarily at 5, 10, 15, 20 and 25 centimeters. The instrument setup was employed in one of these conditions through the testing. While performing tests the current is supplied through DC powered batteries with help of connected wires. The warnings were successfully received after every trial in form of the detachment of the circuit or in form of the detachment of button magnet. After the first phase of test conducted device was covered under the ground and a substantial amount of soil superimposed upon it. The same test was performed on it and it was showing great performance under various testing conditions whenever any rock or soil lump gets disturb its circuit connection, leaving every time the buzzer alarming and hence alerting.

Further testing was conducted on different topographical nature primarily at the clayey soil, sandy soil, rocky soil, gravels, and debris of earth materials. In this ground testing the device was subjected to the vertical as well as inclined direction spilling the ground to push in two different directions by means of the rolling debris particle/rubble over it and observed surface area coverage. (Figure 124.2).

By such manner it is attainable to simulate the condition of operation during which the instrument would possibly operate in actual condition. In next phase of testing the connected latch grid was placed by varying the spacing specifically at 25, 50, 75, 100 cm between two latch grid to obtain the optimum spacing and surface area coverage. Throughout the experiment a variety of different measurement were taken in different soil and site conditions whereas simulating different operation condition.

Key Feature

Other sensor which works on the complicated circuits principle are hard to maintain and are way too expensive than the solution. We have provided a better solution to the foreseen problem. Efficiency of the sensor totally depends up on the location of installation of the sensor. The built-up cost of the sensor is about 100 INR. If multiple sensors relate to each other preparing a large chain more area can be covered with the help of this systems 10 sensors can cover up to 400 square feet of area can be installed in grid pattern to monitor a larger surface area in low cost.

The sensor in terms of convince is a bit tougher than the satellite monitoring of the slope but is accurate and provides the status of actual on ground situation and is economical for the landslide prone area. The Sustainability of this device is that it works in all weather conditions on change in altitude and the variation in temperature doesn't have much harm to it. It does not require much electricity; it can further be run on the solar panel making it itself sustainable monitoring system of its own kind.

Flexibility

The sensor has very diversified numbers of flexibilities some of which are as follows: -

* Sensor when collaborated with the moisture and vibration sensor can also monitor the soil moisture and vibration and can be upgraded into multipurpose sensor.
* It can also work on AC as well as DC current supply with minor changes.
* As, it works on both the types of current it can also be made portable which can be connected to electrical grid and monitoring systems.
* If the system is powered by the solar panel, then it becomes more sustainable, as many of the sensors are placed in isolated location so, it will become self- maintained device which operate in low cost.
* If we connected this sensor to an Arduino board, we could configure this system to auto generate warning SMS for the locals of the area, creating an option in front of it that connect to monitoring devices and generate real time warnings.
* With the help of the internet we can monitor all the sensors form anywhere at any point of time in a day, if necessary, changes have installed. The same entity of development and flexibility were discussed while developing an EWS's (Pandey et al., 2007; Sharma and Mahajan, 2018).
* It can also generate a quick warning system thorough this by the help of sirens of long ranges which will warn the peoples and vehicles. It will make a most effective way of saving human life especially for the roads in hilly regions.

Practical Challenges

The practical challenges can be encountered in the form of maintenance and life span of the power source (if connected only to the batteries power supply). Secondary the challenge may be in form of the alerting to people, that how early people interpret the alarm of the landslide and prepare for it. This situation completely comes under the management authority, how eager and accurate they react (Chua et al., 2021).

Third the challenge may encounter as the harsh weather and site conditions e.g., Suffering from very heavy rainfall, ice or snow cover and other human activities such as mining etc which causes heavy vibration in ground (Van Westen et al., 2005).

It is very important to have appropriate knowledge about the upcoming instabilities and risk. Therefore, to combat this situation pre investigations and constantly site update and maintenance should perform. The management & maintenance operations are important key asset also mentioned (Shano et al., 2020). If we connect this sensor to Arduino board and make it run on computer programming then it's efficiency and accuracy can be improve but at the same time when computer programming enters in the picture the sensor no more is economical, but with the help of computer programming and making it digital it will no longer have any kind of limitations and the wholesome system can be monitored remotely from anywhere at any place or time.

$$\text{EFFICIENCY} = \frac{\text{NUMBER OF TIMES DEVICE WORK EFFECTIVELY}}{\text{NUMBER OF TEST PERFORMED}} *100$$

Result and Discussion

In our field trials on different type of soil and steep slope, the device was able to successfully to detect the displacement and movement of soil. The trials were extremely satisfactory, and the device was working properly as it was designed. The condition of testing on actual landslide ground was beyond our reach. But we were able to create maximum possible simulative situation to test as per on real land sliding ground discussed in the paragraph of Field trials. We find out, it will be able to maintain the purpose for which it has been deployed. Below (Figure 124.1 and 124.2) shows the effective depth at which the device should be deployed in various subsurface and surface soil conditions.

We conducted our test in five different soil conditions with variable depth specifically at, the magnet and the M.S flap placed above the ground surface, 5,10,15,20 and 25cms below ground surface. It is found that the maximum efficiency was in the rocky soil, with the magnet placed above the ground surface. The same results were obtained on gravels and boulders and in earth fill having considerable amount of moisture content, getting maximum efficiency at above ground surface condition (Figure 124.2). But in the case of clayey soil and the sandy soil the maximum efficiency was getting when the device was placed 15cms and 10cms respectively below ground surface (Figure 124.1).

In all test conditions when the device was placed 25 cms below the ground level it was very less efficient to detect the ground movement, making it less ideal position to place the device. Later the test was conducted on connected grid circuit. The connected grid was developed by connecting two devices to one grid, confirming the surface area coverage with respect to varying spacing and optimum efficiency.

For clayey soil the optimum efficiency was obtained when the device was placed at 75 cms spacing covering the 13 sq.m surface area maintain 87% efficiency (Figure 124.6.3).

For Sandy soil the optimum efficiency was obtained when the device was placed at 50 cms spacing covering the 10 sq.m surface area maintain 92% efficiency (Figure 124.4).

For Rocky soil the optimum efficiency was obtained when the device was placed at 100 cms spacing covering the 18 sq.m surface area maintain 95% efficiency (Figure 124.5). For Gravel & Boulders the optimum efficiency was obtained when the device was placed at 75 cms spacing covering the 15 sq.m surface area maintain 92% efficiency (Figure 124.6).

For Earth fill (having moisture) the optimum efficiency was obtained when the device was placed at 75 cms spacing covering the 7 sq.m surface area maintain 87% efficiency (Figure 124.7).

Figure 124.5.1 Installed device before landslide

Figure 124.5.2 Installed device after landslide

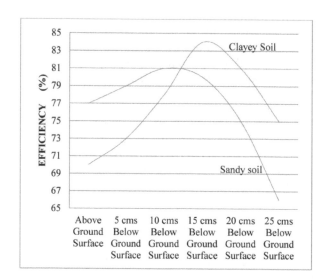

Figure 124.6.1 Performance on varying depth (Clayey and Sandy Soil)

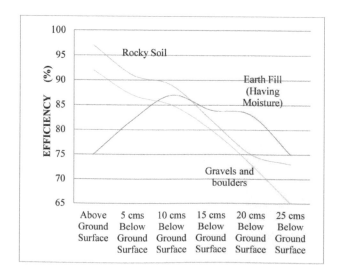

Figure 124.6.2 Performance on varying depth (Earth Fill, Gravels and boulders and Rocky Soil)

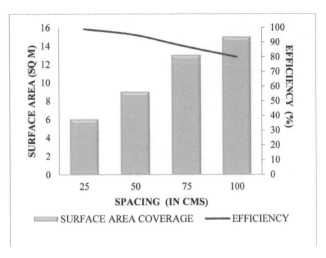

Figure 124.6.3 Performance on varying spacing for clayey soil

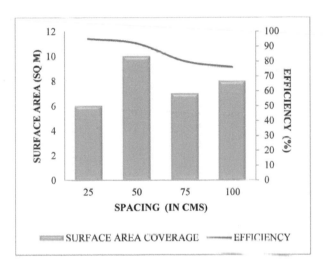

Figure 124.6.4 Performance on varying spacing for sandy soil

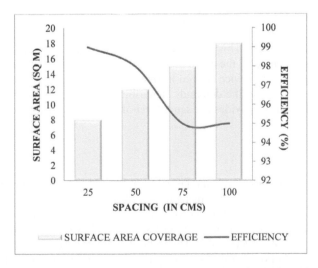

Figure 124.6.5 Performance on varying spacing for rocky soil

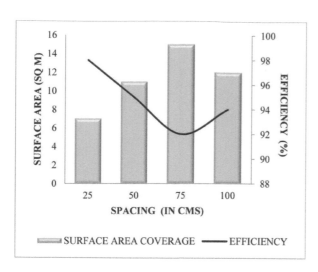

Figure 124.6.6 Performance on varying spacing for gravel and boulders

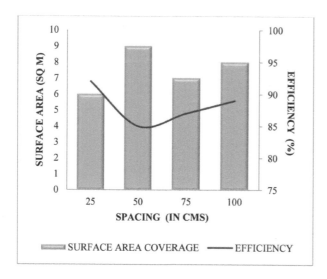

Figure 124.6.7 Performance on varying spacing for earth fill (having moisture)

The final test was to identify the typical surface coverage area range at optimum efficiency with respect to spacing provided. The results were based on grid surface area coverage. The maximum coverage was getting in the rocky soil of nearly 18Sq.m with 95% efficiency. This is a tentative range of covered surface area. The basic principle is the more the device connected in grid, the more the coverage area will be covered but should be offered optimum efficiency at same time. The device is been in transitional phase so as it offers many flexibilities to modify as per the ground or site specific demands. The obtained results have shown that the instrument is working accurately and basically provides the same level of precision as provided by the same sensors of these categories.

Conclusion

This paper aims to present a solution on implementing site-specific landslide EWSs which overcomes the current limitation of available landslide early warning system. Focusing on specific requirements and practical issues based on current ongoing experiences. The importance of early warning system (EWS), in geotechnical engineering is identified and solution in terms of latch circuit based EWS has been developed.

From our results and rigorous testing, it can be concluded as this device can be deployed in various soil types. The ideal range to position the device will be above ground surface position to 15cms below ground surface. From the performance perspective it should be install in connected grid circuit to get optimum results. The typical surface area can be increase by deploying more devices in connected grid. Very few researchers have used Latch circuit as a EWS for detecting the displacement and movement of ground. We provided a low-cost solution to it which can be easily upgradable as per ground requirement by installing minimal changes. The device has been in transitional phase so as it offers many flexibilities to modify as per the ground or site-specific demands.

We provided solution to specific problem, and it was able to fulfill the requirement. Moreover, it can upgrade to meet more technical or IoT based device solution as further research may continue for this device to provide more optimistic solution.

References

Boonyang, P. and Biansoongnern, S. (2016). Development of low cost vibration sensors network for early warning systems of landslides. Elsevier.

Casagli, N., Frodella, W., Morelli, S., Tofani, V., Ciampalini, A., Intrieri, E., Raspini, F., Rossi, G., Tanteri, L., and Lu, P. (2017). Spaceborne, UAV and ground-based remote sensing techniques for landslide mapping, monitoring and early warning. Geoenviron. Disasters. 4(9).

Chua, M., Pattonb, A., Roering, J., Siebert, C., Selkera, J., Waltera, C., Udell, C. (2021). Sitka Net: A low – cost, distribution sensor network for landslide monitoring and study. doi:10.1016/j.ohx.2021.e00191

Costanzo, S. and di Massa, G. (2015). Low cost radars integrated into landslide early warning system. Advances in Intelligent Systems and Computing. AISC. 354, 1119.

Dikshit, A. and Satyam, D. N. (2018). Early warning system using tilt sensors in chibo, Kalimpong, Darjeeling Himalayas, India. Nat. Hazards. 94:727–741.

Guzzetti, F., Gariano, L. S., Peruccaccimaria, S., Teresa, B. M., Ivan, M., Mauro, R., and Massimo, M. (2020). Geographical landslide early warning system. Earth Sci. Rev. 200. doi: 10.1016/j.earscirev.2019.102973

https://www.gsi.gov.in/webcenter/portal/OCBIS/pageQuickLinks/pageNLSM?_adf.ctrl-state=19ckvxzdx6_5&_afrLoop=37033554801878384.

Intrieri, E., Gigle, G., Mugnai, F., Fanti, R., and Casagli, N. (2012). Design and implementation of a landslide early warning system. doi: 10.1016/j.enggeo.2012.07.017.

Ju, N. P., Huang, J., He, C. et al. (2015). A real time monitoring and early warning system for landslides in Southwest China. J. Mountain Sci. 12:1219–1228.

Kumara, V., Gupta, V., Jamira, I., and Chattoraj, S. L. (2019). Evaluation of potential landslide damming: case study of Urni landslide, Kinnaur, Satluj valley. Elsevier.

Lacasse, S. and Nadeem, F. (2009). Landslide risk assessment and mitigation strategy. Berlin Heidelberg: Springer.

Li, Z., Fang, L., Sun, X., and Peng, W. (2021). 5G IoT–BASED geohazard Monitoring and early warning system and its application. Eurasip. J. Wirel. Commun. Netw.(160).

Pandey, A., Dabral, P. P., Chowdary, V. M., and Yadav, N. K. (2007). Landslide hazard zonation using remote sensing and GIS: A case study of Dikrong river basin, Arunachal Pradesh, India. Earth Sci. Rev. 54:1517–1529.

Pradhan, B., Singh, R. P. and Buchroithner, M. F. (2005). Estimation of stress and its use in evaluation of landslide prone region using remote sensing data. Adv. Space Res. 70(3).

Ramesh, M. V. (2012). Design, development and deployment of wireless sensor network for detection of landslides. Ad Hoc Networks. 13:2 18.

Shano, L., Raghuvanshi, T. K., and Meten, M. (2020). Landslide susceptibility evaluation and hazard zonation techniques – a review. Geoenviron. Disaster. 7(18).

Sharma, S. and Mahajan, A. K. (2018). Comparative evaluation of GIS-based landslide susceptibility mapping using statistical and heuristic approach for Dharamshala region of Kangra Valley. India. Geoenviron. Disaster. 5(4).

Thuro, K., and Wunderlich, T. (2014). Low cost 3D early warning system foe alpine instable slopes: The aggenalm landslide monitering syestem. Springer.

Van Westen, C. J., van Asch, T. W. J. and Soeters, R. (2005). Landslide hazard and risk zonation -Why is it still so difficult?. Springer.

125 Investigation of effect of addition of alumina nano particles in carbon fibre reinforced polymers composite

Ashish Mogra[1,a], Tiju Thomas[2,b], and Rakesh Chaudhari[1,c]

[1]Mechanical Engineering Department SVKM's, NMIMS, MPSTME HMT Shirpur, India

[2]R & D, Mechanical Engineering HMT Banglore, India

Abstract

Carbon-reinforced fibre composites are popularly used in number of application replacing metals to reduce the overall weight of the products. The present work shows a characterization of hybrid carbon fibre reinforced polymer mixed with alumina nanoparticles. The mechanical properties were measured using uniaxial tensile test. The yield strength, ultimate tensile strength, % elongation and reduction in cross-section area were measured of polymers composites with addition of different amount of alumina nano particles. The microstructures were observed to study behaviour of polymer under at mechanical and thermal loading conditions. Differential Scanning Calorimetry was used to determine crystallinity of different composites. thermo-gravimetric analysis (TGA) was also carried out to investigate thermal degradation of the composites with temperatures.

Keywords: Alumina nano particle, composites, carbon fibre, crystallinity, polymer.

Introduction

The very essential criteria for a material is to provide strength, durability, lightweight, economic effectiveness and safety. It becomes difficult to impart all properties in single metals required in specific applications. This makes researchers to work on finding of new composites fulfilling the requirements of these properties (Declan et al., 2017).

Composites are made by mixing of two or more materials which possess considerably high strength, high modulus, high corrosion, low density, low coefficient of thermal expansion (CTE), high wear resistance excellent resistance to fatigue and creep (Weise et al., 2019; Bing et al., 2020). The composites are mostly classified on the basis of reinforcement used to fix the matrix and it providing the strength to composites. Polymers matrix composites, metal matrix composites, ceramic matrix composites and carbon matrix composites are the popular categories used to replace the metals4. Fibre-reinforced plastics (FRP's) find important applications replacing the use of costly metals as these provide better mechanical properties than many metals (Fuda et al., 2015). FRP are made combining polymer matrix with fibres that helps to improve the strength and other mechanical properties. Glass, basalt, wood, asbestos and carbon are the common fibres used in forming the composites whereas polyester, epoxy and other thermosetting plastics are usually used as core polymers (Thomas et al., 2017; Dickson et al., 2017).

Number of natural polymers and fibres are used in composites to enhance specific properties. The strength of fibre material is always greater than matrix material which makes them load-carrying element in the composites (Yao et al., 2018). Glass fibres (GFs) are very durable and exhibit outstanding strength, thermal stability, wear and chemical resistance. However the machining of these fibres is comparatively slow and challenging. The despoil is a major challenge of these fibres at the end of the service life (Mao et al., 2021). Basalt fibre exhibits enhanced properties compare to glass fibre and it is also cheaper than other types of fibres. But these fibres have some concern about static strength at elevated temperatures (Chacón at el., 2019). A Graphene carbonaceous fibre displays excellent strength and electrical conductivity than other fibres increasing their application potential in conductive cables, supercapacitors, actuators, etc. (Ding et al., 2021; Okayasu and Tsuchiya, 2019). The biodegradable composites are also finding new space for certain applications with existing commercial composites (Kanak et al., 2015).

In metal matrix composites different metals or alloys are added in the form of particle or fibre with the ceramics or organic compounds. Aluminium, magnesium and titanium are commonly added as matrix materials to provide support to reinforcement. Powder blending and consolidation, foil diffusion bonding, stir and squeeze casting, pressure infiltration and spray deposition are used for manufacturing metal matrix composites. These matrix exhibits excellent strength-to-density ratios, significant fatigue resistance and other high-temperature properties (Zakaria et al., 2019). Carbon-fibre-reinforced polymers (CFRP) has huge potential in designing lightweight structures. Carbon fibre reinforced plastics are popularly used in automobile, ships, aerospace, sports equipment's etc. due to their high strength to weight ratio and excellent rigidity (Maqsood and Rimašauskas, 2021).

Carbon fibre is of extremely small size of about 5 to 10 micrometers and is made of thin, strong crystalline filament of carbon. The use of carbon fibres offers different directional properties as compared to number of other metals those are mostly used in automobile and aerospace applications. In addition, it helps to reduce the overall weight significantly which is not possible in metals. As CFRP are composed artificially the properties of these composites can be tailored with variation in length, direction, quality of reinforced filer and type of polymer matrix (Jakubczak et al., 2020). It exhibits number of advantages such as significant stiffness, high tensile strength, high chemical resistance, high-temperature tolerance and low weight, low thermal expansion. Due to these properties carbon fibre finds significant applications various fields in replacement of steel and other light weight alloys especially in aircrafts, racing cars and sport equipment's These composites are almost five times stronger compare to different steels(Higuchi et al., 2019).

Different techniques such as open and close mould processes, pultrusion are used to develop stable, reliable and robust forms of CFRP composites. The machining of CFRP composite is very challenging due to requirement of accuracy and dimensional stability. The important consideration is needed on machinability, cutting force, geometries of cutting tools, wear rate, delamination, coating etc. for effective and economical production of quality CFRP composites (Hamed, 2021). A novel additive manufacturing successfully used for forming carbon fibre reinforced plastic (CFRP) composites. This method is very instrumental in enhancement of intrinsic mechanical properties of thermoplastic material used in additive manufacturing. The carbon fibre reinforcement into plastic is mostly used to form CFRP composites yielding significant properties with light weight. The composition of reinforced carbon fibre, internal structure and bonding with metal matrix greatly affect the properties of the composite (Zhao et al., 2020). It has been investigated that mixing of nanoparticles in carbon fibre significantly improves the tensile strength and other properties of composite. It also helped to enhance the wear resistance at high pressure wet friction condition. The size, shape as well as distribution of the particles in the matrix have very substantial influence on performance of composite (Hu et al., 2020; Xu and Gao, 2015). Carbon, ceramic or mineral based nanoparticles has found very effective in enhancing the fatigue resistance fracture tightness and delamination in composites. Addition of alumina nanoparticles offers excellent fracture resistance are minimum energy absorption in case of impact loading which is very advantageous for high speed vehicles. The investigation on adding of alumna nanoparticles in carbon fibre reinforced polymers has resulted in improvement in tensile and flexural strength with uniformity in composition of the composites (Naureen, 2005).

The present research investigates the performance of the fibre-reinforced polymer (FRP), composites with addition of Al_2O_3 nanoparticles. It also studied the effect of amount of nanoparticles on mechanical properties and metallurgy of the composites.

Materials and Methodology

In present work the carbon fibre composite with alumina nano-particles were fabricated by using conventional hand lay-up and open molding technique. A carbon fibre cloth was cut into desired dimension as per the size of available mould. The resin and hardener were mixed proportionally in the ratio of 5:4 to obtain translucent coloured solution. Carbon-fibre sheets were adhered together with the help of prepared solution. The carbon fibre composites of 20 mm x 20 mm size and 5 mm thickness are fabricated using composites molding machine (CMM) as shown in Figure 125.1. Temperature variation is attained with the help of heating coils attached between lower-fixed and upper-moving plate. Piston attached to upper plate is used to create pressure using 2 HP motor. Three types of carbon fibre with nanoparticle fabricated using epoxy resin as a binder with different proportion of alumna particles. One fibre sample is prepared without adding alumna whereas other two are formed by adding 2 gms (4% of total weight) and 6 gms (12% of total weight) alumina nanoparticles. The rage of alumna added in composite is varied as is gives good mechanical properties especially strength, light weight and corrosion resistant etc. (Manjunath and Purohit, 2013).

Further carbon fibre clothes were stacked up over one another with addition of binder till the weight reached to 50 gms. The compression pressure was applied for 24 hours to bind the composite specimens. The details of the fabricated specimen are as given in Table 125.1.

Figure 125.1 shows compression molding machine used for fabrication of composite carbon fibre. Figure 125.2 shows the prepared composite fibre.

A uniaxial tensile test was conducted on all three samples to determine yield strength, ultimate tensile strength, % elongation, and reduction in cross section area of carbon fibre composite. The test was performed to the tensile direction of specimens extracted from the fabricated composites is kept in the direction perpendicular to direction of the fibres. The tests were performed at room temperature as per ASTM

Figure 125.1 Compression molding machine

Table 125. 1 Details of specimen

Sr No.	Types of Specimen	Alumina Nanoparticles weigh in (gms)	Nomenclature
1	Carbon fibre with 0% Al_2O_3	0	CF-0A
2	Carbon fibre with 2% Al_2O_3	2	CF-2A
3	Carbon Fibre with 6% Al_2O_3	6	CF-6A

Figure 125.2 Fabricated samples (a) CFRP (CF-0A), (b) CFRP (CF-2A), (c) CFRP (CF 6A)

D3039 standards using high capacity AIM UT-40 C Universal testing machine of 800 N capacity. The standard gauge length of the specimen was 25 mm and the cross sectional area was 75 mm². The test was carried out at low strain rate of 0.001 to 0.01 S⁻¹.

Thermo gravimetric analysis (TGA) and differential scanning calorimetry (DSC) were performed for thermal analysis of the composites. The characterisation and morphology of the composite fibre was analysed by scanning electron microscopy (SEM).

Result and Discussion

The carbon fibre composite specimens are tested to understand behaviour under mechanical and thermal loading conditions. Uniaxial tensile test is conducted to measure different strength of the composites. The results of the test help to understand the structural deformations depending on the amount of nanoparticles in the composites.

Tensile test

Three sets of samples are cut to bone shaped specimens and subjected to tensile test using universal testing machine. The stress-strain diagrams are plotted for all the samples. Different mechanical properties such as ultimate tensile strength, % elongation and % reduction are evaluated. The results of tensile test conducted on three composites are tabulated in Table 125.2.

It is observed that maximum ultimate strength is exhibited by carbon fibre composite with 6% Al_2O_3 as compare to others two composites. The increase in strength can be attributed higher elastic modulus

Table 125.2 Tensile test results

Properties	CF-0A	CF-2A	CF-6A
Max. force (N)	19870	22769	24401
Max. displacement (mm)	14	16	12
Ultimate tensile strength (N/mm2)	413	549	1070
% Reduction in area	51	63	56
% Elongation in length	1.667	1.547	1.111

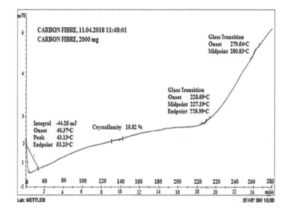

Figure 125.3 Heat flow vs temperature curve (CF-0A)

formed due to mixing of more amounts of alumina particles in fibre composite. All the samples revealed very little or no ductility and resulted in failure at low strain. The catastrophic failure is observed in all samples with no plastic deformation before breaking. Almost similar reduction of 55% in cross sectional area is observed all samples. The maximum strength in carbon fibre composite samples with no alumina particles and 2% alumina particles is very low as compare to sample with 6% alumina particles. The strength of fibres is showing extreme variation as failure stress is very much related to existence of defects and cracks. Moreover voids and cracks in fibres also affect in strength. It is found that the extremely small cracks on the surface of carbon fibre due to collimation in bundles reduce the fracture strength more than 50% (Sharma and Patnaik, 2018).

Differential scanning calorimetry

The graphs are between heat energy and temperature with respect to time is plated for thermal analyses using Differential Scanning Calorimetry. The graphs are used to calculate melting enthalpies required for the phase transitions (Figure 125.3).

The crystallinity of 15.82% is measured for each carbon fibre specimen (CF-0A) and it is observed that glass transition took place at two points i.e.: one with onset temperature 228°C and another at onset temperature 279°C. The crystallinity of the carbon fibre specimen (CF-2A) was obtained as 10.91% and glass transition took place at two points one with onset temperature 167°C and the other point was at onset temperature 217°C.

The crystallinity of the carbon fibre specimen (CF-6A) was obtained as 9.91% and glass transition took place at two points one with onset temperature 170°C and the other point was at onset temperature 270°C.

The existence of two glass transition temperature clearly indicate the presence of two mixture in different forms revalinging presence of two or more materials in composites (Table 125.3). Furthermore, it can be said that the carbon atoms tends to move from rigid state to flexible state at the glass transition temperatures. The lowest crystallinity is observed in CF-6Â specimen indicates the improvement in conductivity with increase in alumna nanoparticles. The glass transition temperature in all specimens is above the room temperature making it hard and brittle. In addition higher glass transition temperature can be attributed to stiffening of chain due to tight packing of atoms, segmental rotations and lower molecular weight (Choudhary et al., 2019).

Table 125.3 DSC test results

Properties	CF-0A	CF-2A	CF-6A
Crystallinity (%)	15.82	10.91	9.91
First glass transition temperature (°C)	228	167	170
Second glass transition temperature (°C)	279	217	270

Table 125.4 Thermo gravimetric analysis test results

Properties	CF-0A	CF-2A	CF-6A
Temp [10% weight loss] (°C)	700	740	920
Temp [50% weight loss (°C)	900	790	980

Table 125.5 Specification of DSC and TGA

S. No.	Particular	Values/Ranges
1	Temperature range	RT to 1100°C
2	Heating rate	0.02 to 250 K/min
3	Crucible volume	Up to 100 μL
4	Weighing accuracy	0.005%
5	Weighing precision	0.0025%

Thermo-gravimetric analysis test

Thermo-gravimetric analysis (TGA) is carried out by measuring a mass of sample by gradual change in temperature with respect to time. This analysis was carried out to estimate thermal degradation behaviour of composites. The temperature at which 10% and 50% loss in sample occurred is measured for all samples (Table 125.4).

The two-stage degradation is observed in all carbon fibre composite samples. The first stage of 10% degradation occurred at 700°C in CF-0A and CF-2A samples whereas at 920°C in CF-6A samples. Similar trends are observed in second stage of degradation in all samples. The delay in degradation in CF-6A can be attributed to presence of larger amount of alumina particles (Table 125.5).

Scanning electron microscopy test

A scanning electron microscope (SEM) is carried out to understand the fibre orientation, reinforcement in matrix and dispersion of nanoparticles. It is observed that all the fibres show uniform reinforcement of fibres in the matrix in CF-0A samples. The agglomeration of fibres is one location builds strong attractive fibre van der Waals forces resulted in stress concentration decreasing strength of the composites. From the micrograph of CF-0A (2 gms alumina mixed in polymers) shows excellent dispersion of nano-particles in fibres. Some of the broken fibres are observed near the surface. It is also observed that the tensile fracture surface does not show pulling out of the fibres from the matrix. This also supports the negligible amount of elongation during the tensile test along leading to catastrophic fracture. The small amount of zinc nano-particles is observed to be adhered to fibres with thin layer of epoxy resins ensuring the strengthening nature of binding matrix with nano filler. Moreover the accumulation of nano particles at one location with attachment of few fibres is revealed in CF-6A composite polymers. It is supported by significant strength in CF-6A sample due to decrease in stress concentration (Figure 125.4).

Conclusion

Hybrid composite of carbon fibre and varying amount of alumina nanoparticles was prepare and characterized to investigate the effect of alumina particles at high temperature. It has been found that the increase in amount nanoparticle resulted in the enhanced tensile strength due to improved interaction between the constituents. Similar trends have been observed for the yield strength. The decrease in the % elongation is observed in composites with addition of alumna nanoparticles. The composite without addition of alumina particles showed higher crystallinity. All samples revealed two sets of glass transition temperatures. The increase in 10% degradation temperature as well as 50% degradation temperature is measured specimens with addition of nano particles. The matrix showed good binding nature with the fibres ensuring the enhanced strength.

Figure 125.4 Scanning electron microscopy test (a) Plain carbon fibre (CF-0A), (b) Carbon fibre with 2 gms alumina particles (CF-2A) (c) Carbon fibre with 2 gms alumina particles (CF-6A)

Industrial applicability and future scope

Carbon fibre reinforced polymer (CFRP) is a type of composite material used in repairing and strengthening of reinforced concrete structure. It will be very useful as substitute to different steel in construction industry. Also the light weight of these composites providing same strength and other mechanical properties finds effective use in automobile sector. As per the future direction is concerned, research can be done by varying the different types of nanoparticle as well as their concentrations.

References

Carolan, D., Ivankovic, A., Kinloch, A. J., Stephan Sprenger, S., and Taylor, A. C. (2017). Toughened carbon fibre-reinforced polymer composites with nanoparticle-modified epoxy matrices. J. Mater. Sci. 52(3):1767–1788.

Chacón, J., Caminero, M. A., Núñez, P. J., García-Plaza, E., García-Moreno, I., and Reverte, J. M. (2019). Additive manufacturing of continuous fibre reinforced thermoplastic composites using fused deposition modelling: Effect of process parameters on mechanical properties. Comp. Sci. Technol. 181:107688.

Choudhary, M., Singh, T., Sharma, A., Dwivedi, M., and Patnaik, A. (2019). Evaluation of some mechanical characterization and optimization of waste marble dust filled glass fiber reinforced polymer composite. Mater. Res. Exp. 6(10):105702.

Dickson, A. N., Barry, J. N. McDonnell, K. A., and Dowling, D. P. (2017). Fabrication of continuous carbon, glass and Kevlar fibre reinforced polymer composites using additive manufacturing. Add. Manuf. 16:146–152.

Ding, L., Liu, L., Wang, X., Shen, H., and Wu, Z. (2021). Effects of connecting materials on the static and fatigue behavior of pultruded basalt fiber-reinforced polymer bolted joints. Construc. Build. Mater. 273:121683.

Ebrahimi, S. H. (2021). Singularity analysis of cracks in hybrid CNT reinforced carbon fiber composites using finite element asymptotic expansion and XFEM. Int J. Sol Struct. 214:1–17.

Higuchi, R., Yokozeki, T., Nagashima, T., and Aoki, T. (2019). Evaluation of mechanical properties of noncircular carbon fiber reinforced plastics by using XFEM-based computational micromechanics. Compos. A: Appl. Sci. Manuf. 126:105556.

Hofstätter, T., Pedersen, D. B., Tosello, G., and Hans N. Hansen, H. N. (2017) Applications of fiber-reinforced polymers in additive manufacturing. Procedia CIRP. 66:312–316.

Hu, Y., Cheng, F., Bingyan, Yi Ji, Yuan, Z., and Xiaozhi Hu, X. (2020). Effect of aramid pulp on low temperature flexural properties of carbon fibre reinforced plastics. Compos. Sci. Technol. 192:108095.

Jakubczak, P., Bieniaś, J., and Droździel, M. (2020). The collation of impact behaviour of titanium/carbon, aluminum/carbon and conventional carbon fibres laminates. Thin-Walled Struct.155:106952.

Kalita, K., Chaudhari, R., and Ramachandran, M. (2015). Mechanical characterization and finite element investigation on properties of PLA-jute composite. Int J. Compos. App. 123(13).

Manjunath, C. M. and Purohit, D. (2013). A Study of Microstructure and Mechanical Properties of Aluminium Silicon Carbide Metal Matrix Composites (MMC's). Int. J. Eng. Res. Technol. 2:690–700.

Mao, J. J., Lai, S. K. Zhang, W., and Liu, Y. Z. (2021). Comparisons of nonlinear vibrations among pure polymer plate and graphene platelet reinforced composite plates under combined transverse and parametric excitations. Compos. Struct. 265:113767.

Maqsood, N. and Rimašauskas, M. (2021). Characterization of carbon fiber reinforced PLA composites manufactured by fused deposition modeling. Compos. C. 4:100112.

Ning, F., Cong, W., Wei, J. Wang, S., and Zhang, M. (2015). Additive manufacturing of CFRP composites using fused deposition modeling: effects of carbon fiber content and length. Proceedings of the ASME 2015 International Manufacturing Science and Engineering Conference MSEC2015. doi: doi:10.1115/MSEC2015-9436

Okayasu, M. and Tsuchiya, Y. (2019). Mechanical and fatigue properties of long carbon fiber reinforced plastics at low temperature. J Sci Adv. Mater. Dev. 4(4):577–583.

Pereszlai, C., Geier, N., Poór, D. I., Balázs, B. Z., and Póka, G. (2021). Drilling fibre reinforced polymer composites (CFRP and GFRP): An analysis of the cutting force of the tilted helical milling process. Compos. Struct. 262:113646.

Qin, B., Li, B., Zhang, J., Xie, X., and Li, W. (2020). Highly sensitive strain sensor based on stretchable sandwich-type composite of carbon nanotube and poly (styrene–butadiene–styrene). Sens. Actuator Phys. 315:112357.

Shahid, N., Villate, R. G., and Barron, A. R. (2005). Chemically functionalized alumina nanoparticle effect on carbon fiber/epoxy composites. Composites Science and Technology. 65(14):2250–2258.

Sharma, A. and Patnaik, A. (2018). Experimental investigation on mechanical and thermal properties of marble dust particulate-filled needle-punched nonwoven jute fiber/epoxy composite. J. Manag. 70(7):1284–1288.

Weise, B. A., Wirth, K. G., Völkel, L., Morgenstern, M., and Seide, G. (2019). Pilot-scale fabrication and analysis of graphene-nanocomposite fibers. Carbon. 144:351–361.

Xu, Z. and Gao, C. (2015). Graphene fiber: a new trend in carbon fibers. Mater. Today. 18(9):480–492.

Yao, H., Zhou, G., Wang, W., and Peng, M. (2018). Silica nanoparticle-decorated alumina rough platelets for effective reinforcement of epoxy and hierarchical carbon fiber/epoxy composites. Compos.: Appl. Sci. Manuf. 110:53–61.

Zakaria, M., Hazizan R., Akil, Md., Helmi M. Kudus, A., Ullah, F., Javed, F. and Nosbi, N.. (2019) Hybrid carbon fiber-carbon nanotubes reinforced polymer composites: A review. Compos. B: Eng. 176:107313.

Zhao, X., Wang, X., Wu, Z., and Wu, J. (2020) Experimental study on effect of resin matrix in basalt fiber reinforced polymer composites under static and fatigue loading. Construc. Build. Mater. 242:118–121.

126 A review on friction stir welding of aluminium

Gopi Krishnan P.[1,a]*, Suresh Babu B.*[2,b]*, Akash M.*[2]*, Kowsikraj K.*[2]*, Kishore S.*[2]*, and Noel Benitto S.*[2]

[1]Department of Mechanical Engineering, Dr. N.G.P. Institute of Technology, Coimbatore, Tamilnadu, India

[2]Department of Mechanical Engineering, Sri Krishna College of Technology, Coimbatore, Tamilnadu, India

Abstract

Friction stir welding (FSW) is stated as a greener joining process that utilises frictional force to join materials without the need of shield gas, flux, electrode and filler material. The applications that do not necessarily require the above accessories find FSW as a potential route for joining. FSW has already emerged as a viable route for joining light metals such as Aluminium, Titanium and so on. The research study in FSW is focussed in the development of better joints in harder materials such as steel, Copper etc,. Many suggestions are given by the researchers in obtaining the better FSW weldment. Welding between the dissimilar materials, change of tilt angle, orientation of advancing and retreating sides, tool geometry, tool material, optimisation of FSW process parameters, double pass welding, post and pre-weld treatment are the different variants of FSW used in the literature. Apart from the above, new variants have been evolving over time. The extensive research in the area of current trends of friction stir welding is given in this article.

Keywords: Friction stir welding, macrostructure and microstructure, mechanical properties, optimisation, tool wear,

Introduction

Friction stir welding (FSW) is a new joining process that has attracted the interest of researchers worldwide. FSW was invented by The Welding Institute (TWI) in Cambridge, UK in 1991. Friction stir welding (FSW) is a type of solid-state welding processes in which a rotating tool rotates, plunges, and then traverses through the edges of two workpieces to form a weld. The combined action of rotation of tool and friction between the tool and metal generates the heat energy along the faying surfaces. The generated heat is dissipated towards the non-faying surfaces of workpieces. A required amount of force is supplied perpendicular to the movement of tool and along the axis of the tool, which mechanically deforms the shape of the workpieces. The passing of tool towards the edges of workpieces is known as the tool traverse, which is used to distribute the heat energy uniformly to the edges. Therefore, the tool rotation, axial force and tool traverse are the essential process parameters for FSW process. The success of the welding in terms of the mechanical strength mainly depends on these three process parameters. The working procedure of the joining process is shown in Figure 126.1. A rotating pin is the one that stirs the plasticised material to move from the advancing side to the retreating side.

FSW is being used in many applications from household to aerospace. The process has many advantages compared to the conventional welding processes such as arc welding, gas welding etc,. The mechanical properties of the metals which undergo fusion welding seem to degrade owing to the change of microstructure. But, the welds carried out using FSW have better mechanical properties due to the fact that the weld temperature has not gone above 80% of its melting point. The metalworking at sub-melting point temperature does not degrade the properties of the parent metal. Another important aspect of FSW is the absence of fumes and metal spatter during welding. The process is customised to the applications in recent times, such that it can be used to weld in different orientations such as horizontal, vertical and inclined positions too. Any type of joints such as butt-joint, lap joint, T-joint etc., can be welded using this technology. There is no need of consumables, filler metals and shielding gas for FSW, which makes it an attractive option of joining the metals. Due to these advantages, FSW is termed as the green joining process, as it does not produce health hazardous gases which may affect the environment and people. Most of the applications do not require the cleaning of metal surface.

Despite the advantages, few limitations are also seen in FSW. The exit hole is created at the end of the welding which needs attention. As the workpiece undergoes tremendous force due to tool rotation, tool traverse and axial load, heavy-duty clamping setup is required for arresting the degrees of freedom of the workpiece. Large axial force is required to insert the tool probe inside the workpiece. But, other than the limitations, most applications are suitable to accommodate friction stir welding.

[a]gopikrishnan4682@gmail.com; [b]b.sureshbabu@skct.edu.in

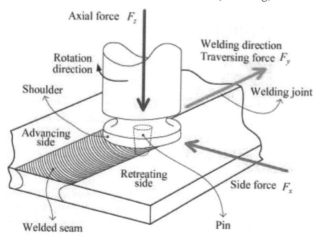

Figure 126.1 Friction stir welding process

Literature Review

In the earlier years after the invention of FSW process, the compatibility of the joining operation was examined with the light weight alloys. FSW was used in the fabrication work of components in aviation, transportation, aerospace and propulsion systems (Thomas and Nicholas, 1997). The attractive feature of sub-melting point temperature was utilised in joining Aluminium components. Application oriented Friction stir welding the temperature does not reach the melting point. Trials undertaken up to the present time show that a number of lightweight materials suitable for the automotive, rail, marine, and aerospace transportation industries can be fabricated by FSW. Friction stir welding is primarily used for joining the lightweight metals such as aluminium, titanium, magnesium and so on. Later the process is tried out for the denser materials such as steel, copper etc. Friction stir welding is also used to join the composite materials made out of light metal matrices and ceramic reinforcements. The characteristics of the weldments with two different metals are also studied in the literature. The concept of dissimilar welding comes to existence after this. This dissimilar welding comprising of two different metals provides better welds due to their sub-melting point temperature. The mechanical working with the axial force also helped to consolidate the required shape of the joint. The testing of mechanical and tribological properties is also studied with the process. The flow pattern of the plasticised metal is analysed using the finite element modelling using many software tools. Thermal flow analysis is also carried out to understand the welding process to obtain the sound welds. Many innovative works are carried out in this welding process nowadays worldwide.

FSW is used as a viable route for the fabrication of components in vehicles and propulsion systems due to its better adaptability (Bhat et al., 1997). The fabrications in cryotanks and primary structures of propulsion systems are carried out. Braun et al. (2000) compared the tensile strengths of the weldments carried out by laser beam welding and friction stir welding of 1.6 mm aluminium alloy sheet 6013-T6. Pores were observed in the fusion welds made by laser beam welding, whereas the joints made by FSW were free of porosity. The tensile strength also approached 80% of the base metal. Fatigue testing of FSW weldments were studied by (Jata et al., 2000). Aluminium alloy AA7050-T7451 was welded using FSW and the welds of as-weld and post-treatment conditions were analysed to study the fatigue behaviour of the alloy. Post-weld heat treatment had inferior properties as compared to the as-weld alloy, thereby claiming the flexibility and adaptability of FSW process. Rhodes et al. (1997) stated that there is no evidence of complete melting of AA7075 in his work. The authors suggested that the aluminium alloy which was diffi-cult to weld by fusion processes could be welded satisfactorily by friction stir welding. The microstructural characteristics of Aluminium alloy 6063 after welding with FSW was studied by (Sato et al., 1999). The results show that the precipitates obtained were responsible for the hardness of weldments rather than the grain size. The precipitates lose their density and were dissolved above the temperature of 675 K. After the establishment of FSW, different tool profiles were also tried out to improve the joining efficiency. Colligan (1999) stated that the friction stir welding process uses stirring action to initiate the joining of metals and much of the welding is carried out by extrusion.

Thomas et al. (1999) experimented the process with 12% chromium alloy and low carbon steel and conducted the tensile and bend test. The authors confirmed that the weld metal was no different from the parent metal. The properties were found similar and a suggestion to improve the tool material in view of the work material was given. Thomas et al. (1999) confirmed the existence of dynamic continuous recrystallisation microstructure in welds of 6061-T6 aluminium alloy. The grains are sheared and their size also got reduced to 10 μm in weld nugget as compared to 100 μm for parent metal. The hardness at the welds also reduced to 55–65 VHN as compared to that of parent metal with 85–100 VHN. Tang et al. (1998) worked on the study of thermal behaviours of FSW welding on Aluminium and concluded that the peak temperature in the weld seam was well below 0.8 times its melting point. They also proved that the temperature distribution was nearly isothermal under the pin shoulder and so it plays important role in welding process. Liu et al. (1997) experimented the dissimilar welding between Aluminium alloys 2024 and 6061 and studied the flow patterns and strength of weld nugget. They found significant reduction in the microhardness values of both parent metals to a range of 40–50%.

Perspective on the Trends in FSW

A. Microstructure

The problem with the most fusion welding processes, which is the microstructure changes, is completely addressed well in the friction stir welding process. The critical issue of the change of the properties using fusion welding, though the good weld is done, lies in the change of the microstructure. Microstructure changes lead to the change in the mechanical properties of the weldment and subsequently to the intended application. In the early days, FSW is considered as the alternative to the riveting process in the aerospace applications. The weight of the steel rivets was crucial in an aircraft and so, as the alternative, FSW was used which provided less weight comparatively with a sound weld. Later, due to the advancements in FSW, the process is extended to other light materials such as Magnesium, Titanium etc., as an exclusive process for joining similar to fusion welding.

Liu, G., et al investigated the microstructure of the welding of AA6061 with different rotation speed and translational speed. The results showed that there is a substantial change in hardness values of the welded specimen with the parent material. This is regarded to the change in the size of the grains from 100 microns to 10 microns due to the dynamic recrystallisation of particles (Liu, G., et al). Nunes, A. (Nunes, A.) performed the experiments in the same alloy with different tool rotation speeds (300 to 1000 rpm) and same travel speed (2mm/s). Heat input and temperature distribution are examined in the perpendicular directions of the welding. The highest temperature recorded was 80% of the melting point of the parent material and the temperature distribution is also found to be isothermal under the pin shoulder. Sato et al. (1999) studied the microstructure of the age-hardenable AA6063 alloy. Hardness of the weld region is characterised by the refinement of grains mainly due to the precipitate distribution and slightly on the grain size.

This research finding is strengthened by the investigation by Lee et al. (2003). This work also revealed the different shapes of the regions of weldment such as thermo-mechanically affected zone, heat affected zone and stir zone. Every zone had the unique microstructure due to the dynamic recovery of grains after welding. Rotational speed is the major parameter contributing to the microstructure of the welded alloy (Lee et al., 2003). Peel et al. (2003) also reported the change of the microstructures in the zones due to the stir action of the tool along with the frictional force over the specimen. The thermal input to the zones has more correlation with the mechanical input during the welding of AA5083 alloy plates. Figure 126.2 also shows the microstructure of the various zones (Peel et al., 2003).

Dehghani et al. (2015) joined the AA7075 alloys using different rotation speeds and travel speeds. The results showed the 15 % increase in the hardness of the weld zone as compared to the base metal due to the precipitates distribution with relatively less ductility. The bending tests showed the decreased strength due to the microstructure evolution (Dehghani et al., 2015).

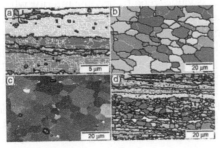

Figure 126.2 Figure showing the microstructure of a) Parent material b) stir zone c) SEM image of stir zone d) Partially recrystallised zone (Peel et al., 2003)

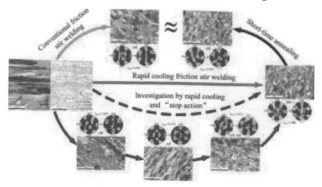

Figure 126.3 Different routes of heat treatment after welding by FSW (Liu et al., 2015)

Microstructural evolution of the grains through heat treatment is also tried out by the researchers. Liu et al verified the effect of the heat treatment in three ways: a) conventional FS welding b) Rapid cooling after FS welding and c) Intermittent Rapid cooling after FS welding as shown in Figure 126.3. Rapid cooling is carried out through liquid CO_2. Surprisingly the grains evolved in a different manner for the intermittent phases but the final microstructure after the cooling process resembles close to that obtained using conventional FS welding (Liu et al).

B. Properties testing

The joining of the metal and alloys always modify the physical properties of the base materials. Though friction stir welding uses the sub-melting point temperatures for joining materials, due to the change of microstructure, the mechanical properties also behave in a different manner. It is clearly seen in the strength testing of the welded specimens. Literature shows that all the mechanical testing procedures are tried out to verify the strength of Aluminium alloys. In the initial years of the development of friction stir welding, pure Aluminium and its alloys are welded and their properties are tested. The improvisation of the property is done from the tool point of view, specimen point of view and the process parameter point of view. Tool rotation speed, weld travel speed and axial force are the process parameters generally considered for the friction stir welding of aluminium. The change in the range of these parameters gets reflected in the strength obtained in the weldment. Later, aluminium composite materials are also welded and tested for mechanical and tribological properties. Almost all testing methods such as tensile, hardness, compression, fracture toughness, fatigue, impact, corrosion, shearing, bending, etc., in the welded aluminium plates and sheets are investigated in the literature. The right mix of the parameter setting and welding conditions are required for the better weld joints of specimens.

Boz and Kurt (2004) presented a study of Aluminium 1080 alloy welded with the stirrers of different shapes such as square and cylindrical cross-sections. The better welding is obtained with the stirrer containing low screw threaded cylindrical profile, suggesting that the tensile property gets improved with the good stirring action (Boz and Kurt, 2004). Zhao et al. (2005) confirmed the change of mechanical properties with the change in the stir pins in the tool (Zhao et al., 2005). The work investigated by Scialpi et al. (2007) used different designs of shoulder geometry for joining of Al6082. It showed the chage in the bending strength of the weldment substantially (Scialpi et al., 2007). Lee et al. (2004) conducted the experiments to verify the hardness levels of the weld zone with that of the other zones in age hardenable 6005 Al alloy in the experiments. The results showed that the change of the hardness values are not significant between the parent metal and heat affected zone (Lee et al., 2004). The harness values tend to become inferior when the thickness of the specimen increased drastically in few cases (Rao et al., 2010).

Rao et al. (2013) reported the variation of the mechanical properties with respect to the advancing and retreating sides due to the asymmetric material flow during welding (Rao et al., 2013). The tensile strengths of the specimen are verified at various locations through local stress-strain curves. The material strength of the welded specimens is also reportedly varying with the change of the process parameters (Azimzadegan and Serajzadeh, 2010; Bahrami et al., 2014). The design of the tool geometry also resulted in the improvisation of mechanical strengths in most cases (Wan et al., 2014). Few other articles suggested the effectiveness of bobbin tools in the improvements of the mechanical strength of the weldments (Lafly et al., 2006; Wang et al., 2015; Thomas et al., 2009).

C. Optimisation Problems

The modification of mechanical properties by the change of the tool geometry or the process parameters have led to the introduction of optimisation in the weld process. Optimisation procedures are employed for

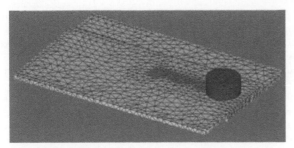

Figure 126.4 Finite element model of the FSW process (Fratini et al., 2009)

getting the optimum level of process parameters or tool geometry or the combination of both. Conventionally Taguchi technique is widely used by many researchers for the optimised process parameter identification (Lakshminarayanan and Balasubramanian, 2008; Koilraj et al., 2012; Vijayan et al., 2010) (Gopi et al). The demand for the more sophisticated set of optimised parameters led to the use of evolutionary algorithms and nature inspired algorithms in friction stir welding.

Fernandez and Murr (2004) reported the decline in the tool wear of the friction stir welding of Aluminium matrix composite (Al359 + 20 % SiC) through the decreasing of rotation speed and increasing of weld speed (Fernandez and Murr, 2004). The same trend is observed in the similar works for welding of the Aluminium composite (Prado et al., 2003; Shindo et al., 2002). Tansel et al. (2009) optimised the operating conditions for the welding with the use of genetically modified neural network systems (Tansel et al., 2009).

Tutum et al. (2010) investigated the optimum welding conditions for the materials based on the evolutionary methods. NSGA-II is used to find the local optimum points of FS welding with the input of process parameters and tool geometry. Later the local points are transported to the design space to discover the global optimum point (Tutum et al., 2010). Fratini et al. (2009) employed the finite element method to discretise the welded regions of the materials (Figure 126.4). Neural based search algorithm is used to converge on the optimum values. The average grain size is predicted from the FEM analysis and the confirmatory tests are conducted to check the adequacy of the prediction (Fratini et al., 2009).

Single objective optimisation is used in the earlier times and then, the multi objective optimisation is tried out for obtaining the right mix of the required conditions of the weld process. Shojaeefard et al. (2014) obtained the FEM model from the FS welding of AA5083 alloy (Shojaeefard et al., 2014).

Conclusion

Friction stir welding is considered as the green welding process due to its inherent properties unlike the fusion welding processes. It is used for the joining of Aluminium in the earlier days and then extended to other light materials such as Titanium, Magnesium etc., The versatility of the process encourages the researchers to shift their focus towards the hard materials such as copper, steel etc., Satisfactory weldments are nowadays obtained by the selection of the optimal range of process parameters. The working conditions of the welding process have been refined through many trials thereby obtaining the weldments with much improved strength. Microstructure changes due to the dynamic recrystallisation during welding lead to the increase of the strength of the weldment. Aluminium alloys and composites are used in all engineering applications in which the friction stir welding process finds extensive usage for obtaining the better weld with the optimum use of resources.

References

Azimzadegan, T. and Serajzadeh, S. (2010). An investigation into microstructures and mechanical properties of AA7075-T6 during friction stir welding at relatively high rotational speeds. J. Mater. Eng. Perform. 19:1256–126.

Bahrami, M., Besharati Givi, M. K., Dehghani, K. and Parvin, N. (2014). On the role of pin geometry in microstructure and mechanical properties of AA7075/SiC nano-composite fabricated by friction stir welding technique. Mater. Des. 53:519–527.

Bhat, B. N., Ledbetter, F. E. and Marshall, G. C. (1997). Materials and processing technologies for highly reusable vehicles and propulsion systems. In 33rd Jt. Propuls. Conf. Exhib., American Institute of Aeronautics and Astronautics. doi: 10.2514/6.19972857.

Boz, M. and Kurt, A. (2004). The influence of stirrer geometry on bonding and mechanical properties in friction stir welding process. Mater. Des. 25(4):343–347.

Braun, R., Donne, C. D. and Staniek, G. (2000). Laser beam welding and friction stir welding of 6013-T6 aluminium alloy sheet. Materwiss. Werksttech. 31:1017–1026.

Colligan, K. (1999). Material flow behavior during friction stir welding of aluminum. Weld. J. (Miami, Fla) 78:229-s.

Dehghani, K., Ghorbani, R. and Soltanipoor, A. R. (2015). Microstructural evolution and mechanical properties during the friction stir welding of 7075-O aluminum alloy. Int. J. Adv. Manuf. Technol. 77:1671–1679.

Fernandez, G. J., and Murr, L. E. (2004). Characterisation of tool wear and weld optimisation in the friction-stir welding of cast aluminum 359+ 20% SiC metal-matrix composite. Mater. Charact. 52(1):65–75.

Fratini, L., Buffa, G. and Palmeri, D. (2009). Using a neural network for predicting the average grain size in friction stir welding processes. Comput. Struct. 87(17–18):1166–1174.

Jata, K. V., Sankaran, K. K. and Ruschau, J. J. (2000). Friction-stir welding effects on microstructure and fatigue of aluminum alloy 7050-T7451. Metall. Mater. Trans. A, 31(9):2181–2192.

Koilraj, M., Sundareswaran, V., Vijayan, S. and Koteswara Rao, S. (2012). Friction stir welding of dissimilar aluminum alloys AA2219 to AA5083 – Optimisation of process parameters using Taguchi technique. Mater. Des. 42:1–7.

Krishnan, P. G. and Siva, K. (2016). Taguchi optimisation of process parameters in friction stir welding of the AA7010–SiC–Al$_2$O$_3$ hybrid composite. High Temp. Mater. Process. An Int. Q. High-Technol. Plasma Process. 20(3):185–196.

Lafly, A. L., Alléhaux, D., Marie, F., Donne, D. and Biallas, G. (2006). Microstructure and mechanical properties of the aluminium alloy 6056 welded by friction stir welding techniques. Weld World 50:98–106.

Lakshminarayanan, A. K. and Balasubramanian, V. (2008). Process parameters optimisation for friction stir welding of RDE-40 aluminium alloy using Taguchi technique. Trans. Nonferrous Met. Soc. China 18(3):548–554.

Lee, W. B., Yeon, Y. M. and Jung, S. B. (2003). Evaluation of the microstructure and mechanical properties of friction stir welded 6005 aluminum alloy, Mater. Sci. Technol. 19(11):1513–1518.

Lee, W. B., Yeon, Y. M. and Jung, S. B. (2004). Mechanical properties related to microstructural variation of 6061 Al alloy joints by friction stir welding. Mater. Trans. 45(5):1700–1705.

Liu, G., Murr, L. E., Niou, C. S., McClure, J. C. and Vega, F. R. (1997). Microstructural aspects of the friction-stir welding of 6061-T6 aluminum. Scr. Mater. 37:355–361.

Liu, X., Sun, Y. and Fujii, H. (2017). Clarification of microstructure evolution of aluminum during friction stir welding using liquid CO2 rapid cooling. Mater. Des. 129:151–163.

Peel, M., Steuwer, A., Preuss, M. and Withers, P. (2003). Microstructure, mechanical properties and residual stresses as a function of welding speed in aluminium AA5083 friction stir welds. Acta Mater. 51(16):4791–4801.

Prado, R., Murr, L., Soto, K. and McClure, J. (2003). Self-optimisation in tool wear for friction-stir welding of Al 6061+20% Al$_2$O$_3$ MMC. Mater. Sci. Eng. A 349(1–2):156–165.

Rao, D., Heerens, J., Pinheiro, A., Dos Santos, J. and Huber, N. (2010). On characterisation of local stress–strain properties in friction stir welded aluminium AA5083 sheets using micro-tensile specimen testing and instrumented indentation technique. Mater. Sci. Eng. A 527(18–19):5018–5025.

Rao, D., Huber, K., Heerens, J., dos Santos, J. and Huber, N. (2013). Asymmetric mechanical properties and tensile behaviour prediction of aluminium alloy 5083 friction stir welding joints. Mater. Sci. Eng. A 565:44–50.

Rhodes, C., Mahoney, M., Bingel, W., Spurling, R. and Bampton, C. (1997). Effects of friction stir welding on microstructure of 7075 aluminum. Scr. Mater. 36(1):69–75.

Sato, Y. S., Kokawa, H., Enomoto, M. and Jogan, S. (1999). Microstructural evolution of 6063 aluminum during friction-stir welding. Metall. Mater. Trans. A 30(9):2429–2437.

Scialpi, A., De Filippis, L. and Cavaliere, P. (2007). Influence of shoulder geometry on microstructure and mechanical properties of friction stir welded 6082 aluminium alloy. Mater. Des. 28(4):1124–1129.

Shindo, D. J., Rivera, A. R. and Murr, L. E. (2002). Shape optimisation for tool wear in the friction-stir welding of cast AI359-20% SiC MMC. J. Mater. Sci. 37(23):4999–5005.

Shojaeefard, M. H., Akbari, M. and Asadi, P. (2014). Multi objective optimisation of friction stir welding parameters using FEM and neural network. Int. J. Precis. Eng. Manuf. 15(11):2351–2356.

Tang, W., Guo, X., McClure, J. C., Murr, L. E. and Nunes, A. (1998). Heat input and temperature distribution in friction stir welding. J. Mater. Process. Manuf. Sci. 7:163–172.

Tansel, I. N., Demetgul, M., Okuyucu, H. and Yapici, A. (2009). Optimisations of friction stir welding of aluminum alloy by using genetically optimised neural network. Int. J. Adv. Manuf. Technol. 48(1–4):95–101.

Thomas, W. M. and Nicholas, E. D. (1997). Friction stir welding for the transportation industries. Mater. Des. 18:269–273.

Thomas, W. M., Threadgill, P. L. and Nicholas, E. D. (1999). Feasibility of friction stir welding steel. Sci. Technol. Weld. Join. 4:365–372.

Thomas, W. M., Wiesner, C. S., Marks, D. J. and Staines, D. G. (2009). Conventional and bobbin friction stir welding of 12% chromium alloy steel using composite refractory tool materials. Sci. Technol. Weld. Join. 14(3):247–253.

Tutum, C. C., Deb, K. and Hattel, J. (2010). Hybrid search for faster production and safer process conditions in friction stir welding. In Asia-Pacific Conference on Simulated Evolution and Learning, (pp. 603–612). Springer, Berlin, Heidelberg.

Vijayan, S., Raju, R. and Rao, S. R. K. (2010). Multiobjective optimisation of friction stir welding process parameters on aluminum alloy AA 5083 using Taguchi-based grey relation analysis. Mater. Manuf. Process. 25(11):1206–1212.

Wan, L., Huang, Y., Guo, W., Lv, S. and Feng, J. (2014). Mechanical properties and microstructure of 6082-T6 aluminum alloy joints by self-support friction stir welding. J. Mater. Sci. Technol. 30(12):1243–1250.

Wang, F., Li, W., Shen, J., Hu, S. and dos Santos, J. (2015). Effect of tool rotational speed on the microstructure and mechanical properties of bobbin tool friction stir welding of Al–Li alloy. Mater. Des. 86:933–940.

Zhao, Y. H., Lin, S. B., Wu, L. and Qu, F. X. (2005). The influence of pin geometry on bonding and mechanical properties in friction stir weld 2014 Al alloy. Mater. Lett. 59(23):2948–2952.

127 Bond strength of waste-create bricks

Sanjay P. Raut[a] and Uday Singh Patil[b]

Department of Civil Engineering, Yeshwantrao Chavan College of Engineering, Nagpur, India

Abstract

The bond strength and other features of waste-create bricks made of recycled paper mill sludge (RPMS) and cement were measured in this research for various compositions in order to expand their uses in building structures. The modified bond wrench test as per ASTM C-1072 and shear bond strength test setup on brick triplets were used to conduct flexural and shear bond strength tests on waste-create brick samples joined with various materials. To compare the results, bond strength was also carried out on the conventional bricks i.e. burnt clay bricks (BCB) and fly-ash bricks (FAB). The outcome of the investigation shows the higher compressive strength for RPMS and cement brick over the conventional bricks and the bond strength of waste-create bricks containing paper pulp waste shows the significant bond strength as compared to the conventional bricks.

Keywords: Bond strength, recycled paper mill sludge, rice husk ash, waste-create bricks.

Introduction

Bond strength determines the strength, durability, and application of brickwork. As a result, having a good bond strength is important since a poor bond might reduce the wall's compressive strength, tensile strength, or shear strength, allowing water to enter and cause damage. A perfect bond also ensures the resistance of stresses caused by various types of loading conditions. The structural behavior is also influenced by the bond strength of mortar and masonry. It is evident that masonry is strong in compression but weak in flexural tension because of a bond interface between mortar and masonry. Thus, allowable compressive stresses are larger than flexural tension stresses in a building, limiting the usage of masonry as a structural element. Several hypotheses for brick bond strength have been presented in past, all of which assume that the bond of brick with mortar remains intact after the brick or mortar fails. According to the experiment on stack bonded prisms, when the brick-mortar bond strength is poor, the prism failure is accompanied by a failure of the brick mortar bond. Factors such as the surface features of the brick, which may or may not have any bearing on the deformation of the brick or mortar, can have a significant impact on the bond (Sarangapani et al., 2005). Furthermore, the shear bond strength determines the modulus of masonry that influences the strength (Reddy and Vyas, 2008). The investigation of bond strength of natural hydraulic lime (NHL) mortar suggests that NHL mortar has high water retention, which allows for a strong bond that compares well to Portland cement and cement/lime mortars (Pavia and Hanley, 2009). Experiments on bricks made by sintering dried water treatment plant sludge and agricultural wastes revealed that developed bricks were light and had high compressive strength (Chiang et al., 2006). When sewage sludge ash was used to make bricks, it was discovered that increased fire temperature and decreased amount of ash resulted in a reduction in water absorption (Lin and Weng, 2001). The investigation into the potential use of cigarette butts for brick manufacture reveals a decrease in the density of bricks and an improvement in thermal conductivity performance (Kadir et al., 2010). An experiment with Fal-G bricks and hollow blocks found that hot water curing causes more hardening and resulted in increased strength in the bricks and blocks (Kumar, 2002). The usage of fiber-reinforced mudbrick has been found to increase the compressive strength of interface layers of fibrous materials when used as a building material (Binici et al., 2004). The results of using cotton and limestone powder waste in the development of bricks showed that it has a high energy absorption capacity and a low density (Algin and Turgut, 2007). The linear shrinkage and density of clay-sand-rice husk ash mixed bricks are both reduced as the RHA content is increased (Rahman, 1987). Experiments with recycled paper processing residue in the production of bricks had revealed increased compressive strength and improvement in thermal conductivity (Sutcu and Akkurt, 2009). A review of textile sludge incorporated bricks had been found to satisfy Indian standards (Patil et al., 2021). The compressive strength of concrete developed by fly ash & RHA was found to be greater than the control concrete (Padole et al., 2019). The present study thus compares the performance of recycled paper pulp-cement bricks with burnt clay bricks (BCB) and fly ash bricks (FAB) in terms of flexure and shear bond strength.

---

[a]sprce22@gmail.com; [b]patil.udaysingh4@gmail.com

Scope of the present study

The flexure bond strength and shear bond strength of masonry for various compositions were investigated in this experimental study. A modified bond wrench test is used to measure flexure bond strength, whereas a test on brick triplets is used to measure shear bond strength. The bond between the bricks was made by using cement mortar, fevicol bond, and cement mortar with chicken mesh in between bricks.

Materials Used in an Experimental Program

Burnt clay brick

Brick is typically constructed of clay and sand, which are mixed and moulded in various ways before being fired in kilns. Clays have provided the primary material of construction for ages, and their qualities vary depending on the purpose for which they are designed. Clay bricks are utilised in a wide range of structures, from homes to factories, as well as tunnels, waterways, bridges, etc. IS 1077:1992 was referred for the specification of the burnt clay bricks IS 1077 (1992) and IS 5454:1978 was referred for the properties like dimensional characteristics, and physical characteristics of the bricks (IS 5454, 1978). Locally available BCB was investigated in the experiments. Table 127.2 shows the mean of three specimens for each of the tests listed. The codal provisions of IS 3495 (1992) were implemented for the tests (IS 3495 (Part 1 to 4) 1992). The compressive strength was found to be 3.1 MPa, which was lower than the Indian Standard. The water absorption value was 14.12% which was in the range as mentioned in the standard. For the conventional brick, the density of the brick was normal, and the efflorescence test shows that there was no perceptible deposit of salt.

Fly ash bricks

India's rapid expansion of thermal power generation capacity has resulted in a massive amount of fly ash production (50 MT/year). The current disposal methods pollute the ecosystem and cause ecological imbalance. Fly ash disposal is wasting vast swaths of land that could be used for housing, agriculture, or other beneficial reasons. In the construction industry, fly ash bricks are a helpful and environmentally favorable product. The raw ingredients are blended in a pan mixer to produce a homogeneous semi-dry mix. The amount of water to use for this mixing is not predetermined and is determined by the volume to be blended. This semi-dry mixture is poured into the hydraulic/vibratory press's moulds. The moulded bricks are then air-dried (2 to 3 days) depending on weather conditions, before being cured for 14–28 days. The developed bricks are sound, compact, and uniform in size and shape. The specification of IS 13757:1993 was referred for fly ash bricks (IS 13757, 1993). The compressive strength for the brick was 3.12 MPa and water absorption was 14.64% by weight and the density was 1750 kg/m^3 which were considered to be normal and these show the NIL efflorescence.

Recycled paper mill sludge

Recycled paper mills produce around 15% sludge of their actual production and its disposal is a major concern for the industry. Sludge from recycled paper mills is a material that can be used in a variety of industrial applications, although it is rarely used. Due to its low utility, it is simply thrown away thereby posing disposal problems. Recycled paper mill sludge thus generated is usually land-filled, although incineration is becoming increasingly widespread. Various attempts have been made to increase the number of sludge management alternatives. Solid waste management could be solved by using recycled paper mill sludge to create value-added products. The creation of novel brick compositions is necessary to improve the performance of traditional bricks in terms of environmental protection, durability, and energy efficiency, etc. In the process of bringing some innovations to the brick industry, the recycled paper pulp waste was used as another source of manufacturing brick.

Table 127.1 shows the various compositions of bricks which are made up of RPMS, cement, and RHA. Table 127.2 shows that the modified brick's compressive strength was much higher compared to traditional bricks. However, the water absorption exceeds the limit by a much higher percentage and is not in the range as mentioned by the Indian Standard.

Rice husk ash

RHA waste forms when the outer coating of rice grains accumulates while milling operation; it accounts for around 20% of the world's annual rice production of 300 million metric tonnes. Rice husk has proven to be challenging to work with and dispose of. Its use in construction, on the other hand, is environmentally friendly. The whitish-gray ash from rice husks was used to make bricks. RHA was employed after it was sieved at 600 microns.

Characterisation of RPMS and RHA

According to the XRF analysis (Table 127.3), RPMS is predominantly composed of silica (Si) and calcium (Ca) (Ca).The proximate analysis is presented in Table 127.4, and the ultimate analysis is presented in Table 127.5.

From Figure 127.1 of TG curves of RPMS samples, it can be seen that 45% mass loss was observed between 290 and 300°C. The three distinct mass losses were seen in the curve. The first occurred between

Table 127.1 Different compositions of brick

S.N.	Sample Name	Material		
		RPMS	*Cement*	*RHA*
1	A	80%	10%	10%
2	B	75%	10%	15%
3	C	70%	10%	20%
4	D	90%	10%	--
5	E	85%	15%	--
6	F	80%	20%	--

Table 127.2 Properties of brick

S.N.	Brick Type	Comp. Strength (MPa)	Water Absorption (%)	Density (kg/m³)	Efflorescence
1	BCB	3.10	14.12	1695	NIL
2	FAB	3.12	14.64	1750	NIL
3	A	15.00	100.5	588	NIL
4	B	14.77	85.4	560	NIL
5	C	11.90	62.5	540	NIL
6	D	9.94	100	670	NIL
7	E	9.62	92.8	702	NIL
8	F	10.68	83.3	765	NIL

Table 127.3 Elemental analysis

	O %	Ca %	Si %	Al %	Mg %	S %	Ti %	K %	Fe %	Na %	Cu %	P %	Cl %	S %	Mn %	Sr %
RPMS	15.83	14.94	60.57	2.06	3.59	1.07	0.15	0.16	0.92	0.22	0.05	0.03	0.41			
RHA	46.36	1.44	34.46	3.31	0.37	0.67	0.32	2.44	1.69	0.13		0.6		0.67	0.12	0.1

Table 127.4 Proximate analysis of RPMS

S. N.	Wt. gms	Moist %	Ash %	Volatile Materials %	Free Carbon %	GCV Kcal/kg
1.	420	5.8	40.6	44.7	8.9	2372

Table 127.5 Ultimate analysis of RPMS

S. N.	Wt. gm	C %	H %	N %	S %	O %
1.	420	22.7	2.5	0.3	0.4	23.6

30 and 280°C i.e 7.5% is due to the removal of water from pores. The material is thermally deteriorated and sintered after a second mass loss. As a result, RPMS bricks can withstand temperatures up to 300°C.

Figure 127.2 shows the XRD results of virgin and binder mixed RPMS. The samples were shown to have amorphous patterns between 25 and 30 (2θ peak), which are typical of commercial cement of 43 grade. Even after varying degrees of cement addition to RPMS (5–20%wt), the composition of the materials has not changed.

XRD tests were used to determine if RHA silica was amorphous or crystalline. Figure 127.3, the broad peaks in RHA samples indicate an amorphous form of silica.

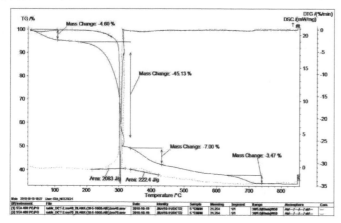

Figure 127.1 TG-DTA of RPMS

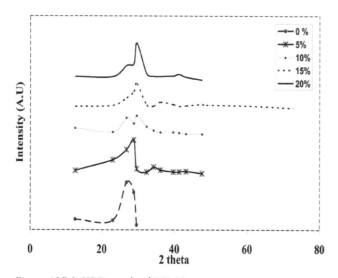

Figure 127.2 XRD result of RPMS-cement

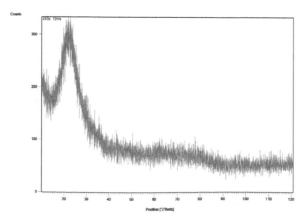

Figure 127.3 Broad peak at 2θ = 22°C, RHA

The presence of pores with fibrous character is readily seen in SEM images of RPMS (Figure 127.4). As a result, the moisture is trapped in these pores and the fibrous envelopes prevent moisture from reaching the surface. Because of its fibrous character, it has a high energy absorbing capacity and thus possesses a good compressive strength.

Figure 127.5 shows an SEM image of RHA samples. A porous silica structure of active nature was observed. Regular spherical structures of varying size between 35–50 µm were observed in parallel rows. SEM revealed that the porous and amorphous structure of RHA has a significant quantity of active silica.

Mortar

A cement mortar ratio of 1:4 was prepared using Ordinary Portland cement (OPC- Grade 43) and sand passed through a 2.36 mm IS sieve. The water-cement ratio was kept between 0.55 to 0.60 to maintain workability. The types of mortars which were used are listed below:

CM 1: Cement Mortar of 1:4
CM 2: Cement Mortar of 1:4 with Chicken mesh
CM 3: Fevicol Bond

Experimental program

The bond's strength was determined by measuring the flexure or shear bond strength on the various types of bricks using the three types of bonding criteria stated before. The flexure bond strength was performed on conventional type brick and modified brick on 5 brick high bonded prism stacks that had been cured for around 28 days. The shear bond strength was carried out on the brick triplets for various combinations.

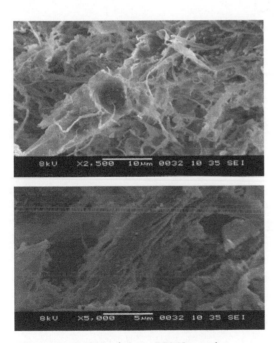

Figure 127.4 SEM of virgin RPMS sample

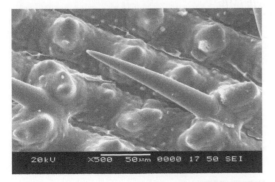

Figure 127.5 SEM image of RHA sample

Flexural bond strength: modified bond wrench test

Using a modified bond wrench test setup illustrated in Figure 127.6, the test was done on prism stacks with a height to thickness ratio of 4. The prism's bottommost brick was clamped, and a pulley mechanism was employed to apply the load to the prism's top-most brick, resulting in a moment, which caused flexure failure amongst the masonry unit and the mortar. This method is utilised because joints near to the grip may not be strained equally when the specimen is gripped at a few discrete points. As a result, a single joint failure far away from the grips is ideal.

Shear bond strength test

This test was evaluated for bricks of varying proportions bonded with cement mortar and fevicol bond. The test configuration is depicted in Figure 127.7, and the horizontal mobility of the top and bottom bricks were restricted, however, the horizontal movement of central brick was allowed. The hydraulic jack was used to apply the horizontal shear load until the bond gets ruptured and then the strength is determined.

Results and Discussion

Shear bond strength test

The results obtained for all compositions with a combination of using the chicken mesh in cement- mortar in between are presented in Table 127.6 and the respective graph (Figure 127.8) is also plotted for values of shear bond test versus types of brick for different combinations of mortars. However, when compared

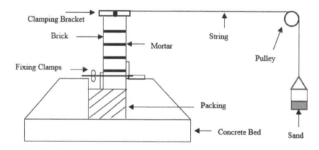

Figure 127.6 Modified bond wrench test

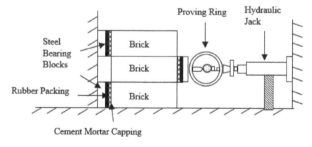

Figure 127.7 Shear bond test setup

Table 127.6 Shear bond strength

SN	Type of brick	Bond type and shear bond strength (MPa)		
		CM 1	CM 2	CM 3
1	BCB	0.101	—	—
2	FAB	0.195	—	—
3	A	0.043	0.062	0.0124
4	B	0.041	0.058	0.0119
5	C	0.042	0.058	0.0119
6	D	0.038	0.041	0.0083
7	E	0.038	0.041	0.0083
8	F	0.040	0.043	0.0090

to the bond comprising only cement mortar and fevicol, the samples incorporating chicken mesh (CM 2) indicate an improvement in bond strength.

Flexure bond strength test

Table 127.7 presents the flexural bond strength values for all discrete compositions. It is seen that only the brick-and-mortar interface failed in the test, indicating bond failure. Thus it can be inferred that, If the bonding at this interface is good, either brick failure or a combination of brick and bond failure will occur. The brick-and-mortar interface, on the other hand, will fail if the bond strength is weak. As the test results show, a high-strength cement mortar with a rich mix also forms a strong bond with the brick. The use of chicken mesh improves bond strength even further. As seen from the table, for conventional bricks using the cement mortar of 1:4 the bond strength is high as compared to the other various modified bricks.

Figure 127.9, reveals that modified bricks of mortar type CM 1 showed less flexure strength value as compared to the mortar type CM 2. Figure 127.10 shows the results of the compression test carried out on both conventional as well as modified bricks. The graph indicates the higher compressive strength for bricks A, B, and C which is made using RHA, and bricks D, E, and F also shows the compressive strength much higher than the BCB and FAB.

The water absorption and density of several bricks for varied compositions are shown in Figure 127.11. It is seen that with the percentage decrease of waste in bricks the value of water absorption also decreases. However, the brick types D, E, and F show that as the percentage of waste decreases and the amount of cement increases the density also increases. The typical observation is that density and water absorption have a significant degree of inverse association. The water absorption of these bricks decreases as their density rises. Regardless of the mix fraction, the brick's compressive strength increases as the density increases.

(Table 127.2) The density of BCB and FAB ranges from 1600 to 1800 kg/m^3, whereas, the density of modified bricks was ranged from 540 to 800 kg/m^3. This demonstrates that adopting modified brick will significantly lower the structure's weight and will provide working comfort and handling.

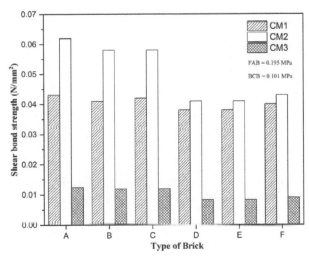

Figure 127.8 Graph showing result of shear bond strength

Table 127.7 Flexural bond strength

SN	Brick type	Bond type and flexural bond strength (MPa)		
		CM 1	CM 2	CM 3
1	BCB	0.210	—	—
2	FAB	0.281	—	—
3	A	0.103	0.147	0.027
4	B	0.101	0.143	0.027
5	C	0.098	0.138	0.027
6	D	0.057	0.068	0.025
7	E	0.060	0.070	0.025
8	F	0.062	0.072	0.025

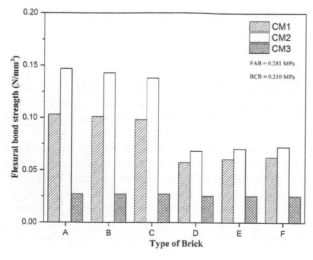

Figure 127.9 Graph showing results of flexural bond strength test

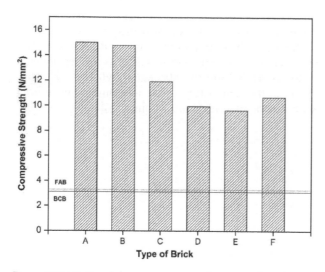

Figure 127.10 Graph between compressive strength and types of brick

The flexure bond strength was seen to be greater than the shear bond strength for all compositions (Figure 127.12). For conventional bricks, the use of rich mortar appears to be sufficient for achieving a good bond of 0.21 MPa to 0.281 MPa. The bond strength drops to 0.062 MPa when applied to the modified bricks.

In comparison to the preceding mortar, every attempt at bond augmentation yields somewhat better outcomes. The cement mortar with chicken mesh in between (CM 2) gives similar results but marginally higher than the CM 1 mortar.

Table 127.7 shows that as compared to cement mortar CM1, the CM 2 mortars have a superior bonding with the brick. It also demonstrates that higher compressive strength bricks inevitably result in better shear or flexure strength. As evidenced in the Figure 127.13 above, the shear bond strength is nearly 60% of the flexural bond strength.

Conclusion

The following broad conclusion emerges after the exhaustive investigations:

- The use of a 1:4 cement-mortar ratio resulted in flexural bond strengths of around 0.10 MPa for brick types A, B, and C, but less than 0.10 MPa for brick types D, E, and F.
- The bond strength can be improved by using chicken mesh in between the brick and mortar. As a result, rich mortars with additives, such as 1:4, produce stronger bonds.

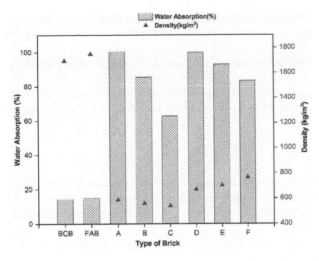

Figure 127.11 Graph of % water absorption vs density (kg/m³)

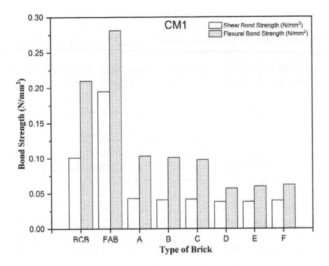

Figure 127.12 Relation between bond strength and type of brick for CM 1

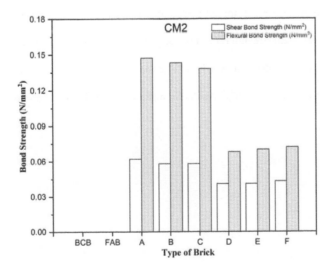

Figure 127.13 Bond strength vs type of brick for CM 2 mortar

- The compressive strength of RPMS and RHA incorporated bricks is greater than that of conventional bricks, and the lightweight bricks shrunk but did not shatter under compressive load due to the fibrous nature of RPMS.
- The high voidage and cellulosic nature of RPMS causes an increase in the water absorption value of bricks by almost 100%.
- The amount of sludge added to the mixture has an inverse relationship with the density of the brick. The pore size of the brick grows as the mixture absorbs more water, resulting in a lower density than conventional bricks.

Future Scope

From the test results, we can say that this type of brick can be used in masonry structures. Using the modified bricks and by building a model house we can check the parameters like humidity or by placing it on a shake table, we can find out the seismic resistance capacity of the house. And also, if we manufacture a wall panel, we can check the wind effect on the panel.

References

Algin, H. M. and Turgut, P. (2007). Cotton and limestone powder wastes as brick material. J. Constr. Build. Mater. 1074–1080.

Binici, H., Aksogan, O., and Shah, T. (2004). Investigation of fiber-reinforced mud brick as a building material. J. Constr. Build. Mater. 313–318.

Chiang, K. Y., Chou, P. H., and Chien, K. L. (2006). Novel lightweight building brick manufactured from water treatment plant sludge and agricultural waste. J. Waste Manag. 26:189–194.

IS 5454 (1978). Indian standard: Method For sampling of clay building bricks, BIS, New Delhi, 1979(First Revision)

IS 1077 (1992). Indian standard: common burnt clay building bricks-specifications. BIS, New Delhi, 1992 (Fifth Revision)

IS 3495 (Part 1 to 4): (1992). Indian standard: Methods of tests of burnt clay building bricks. BIS, New Delhi, 1992 (Third Revision).

IS 13757 (1993). Indian standard: Burnt clay fly ash building bricks- specification, BIS, New Delhi.

Kadir, A. A., Mohajerani, A., Roddick, F., and Buckeridge, J. (2010). Density, strength, thermal conductivity and leachate characteristics of light-weight fired clay bricks incorporating cigarette butts. Int. J. Environ. Sci. Eng. 179–184.

Kumar, S. (2002). A perspective study on fly ash- lime- gypsum bricks and hollow blocks for low-cost housing development. J Constr. Build. Mater. 519–525.

Lin, D. F. and Weng, C. H. (2001). Use of sewage sludge ash as brick material. J. Environ. Eng. 922–927.

Padole, D., Patil, U., and Padade, A. (2019). Partial replacement of cement by fly ash & rha and natural sand by quarry sand in concrete (May 4, 2019). In Proceedings of sustainable infrastructure development & management (SIDM), Available at SSRN: https://ssrn.com/abstract=3382786 or http://dx.doi.org/10.2139/ssrn.3382786.

Patil, U., Raut, S. P., Ralegaonkar, R. V., and Madurwar M. V. (2021). Sustainable building materials using textile effluent treatment plant sludge: A review. Green Mater. 1–15. https://doi.org/10.1680/jgrma.21.00027.

Pavia, S. and Hanley, R. (2009). Flexural bond strength of natural hydraulic lime mortar and clay brick. J. Mater. Struct. 913–922.

Rahman, M. A. (1987). Properties of clay- sand- rice husk ash mixed bricks. Int. J. Cem. Compos. Lightweight Concr. 9(2).

Reddy, B. V. and Vyas, V. U. (2008). Influence of shear bond strength on compressive strength and stress-strain characteristics of masonry. J. Mater. Struct. 1697–1712.

Sarangapani, G., Reddy, B. V., and Jagadish, K. S. (2005). Brick-mortar bond and masonry compressive strength. J. Mater. Civil Eng. ASCE 17(2).

Sutcu, M. and Akkurt, S. (2009). The use of recycled paper processing residue in making porous bricks with reduced thermal conductivity. J. Ceram. Int. 2625–2631.

128 A systematic review on advancement in Modi Handwritten Character Recognition

Samrudhi Bhalerao[a] and H.D.Gadade[b]

Department of Computer Engineering and IT, College of Engineering, Pune, India

Abstract

This review paper is based on MODI handwritten characters recognition. Handwritten characters recognition work is now a challenging task for the last few years because different writing styles for each person are different. Character recognition is nothing but turning human handwritten language into machine language. Many works have been done on other Indian languages until the present moment. Many Indian historical documents are written in many Indian script languages. But not so much work is done for the MODI script. So that it is now necessary to preserve such documents for the future, several recognition systems have been created for multiple languages, and research on Indian scripts is ongoing. In recent decades, the different writing patterns of every handwritten optical character recognition (HOCR) have been a severe problem. The world recognised India because of its unity in variety. Due to the country's geographical and cultural diversity, many spoken languages and written scripts were formed and used for regular contact. Several brave efforts were made towards HOCR for different Indian hands. Because of the variety and structure of Indian characters, creating the character recognition system was deemed an active topic of study. This paper investigates the previous research done on MODI handwritten character recognition. Most of the work has been done in English or regional languages, but there is a need to work in ancient languages like MODI.

Keywords: Character recognition, classification, MODI script, recognition, segmentation.

Introduction

According to research in India, India speaks around 780 languages. Out of that, 220 languages are vanishing. MODI script was an old script that lapsed from proper use in the 19th century before Devanagari was officially accepted (Kulkarni et al., 2015). MODI was used for written communication at various phases for around 600 years. MODI became popular and was regularly used to write Marathi, as shown in Figure 128.1. History researcher Chandorkar's MODI script emerged from the Ashoka period's 'Mouryi' (Bramhi) script. According to another research, the Modi letter was created during the reigns of 'Peshwai' (Pune) and 'Chatrapati Shivaji Maharaj,' it was widely employed for writing in the 17th century. Shalgoankar found that the earliest MODI document is stored in the 'Bharat Itihas Sanshodhan Mandal (BISM)' in Pune between 1429 and 1389 A.D. The MODI script was also called 'Adyakalin' in the 12th century, 'Yadavakalin' in the 13th century, 'Bahamanikalin' in the 14th to 16th centuries, and 'Shivakalin' in the 17th century, and 'Chitnisi' as in 18th century. Figure 128.1 shows MODI characters with consonants and vowels.

Many academics were drawn to handwritten character identification as pattern recognition, image processing, natural language processing, and document analysis are the most promising fields. Today's environment is extremely fast-paced and highly mechanised. Automation has become associated with technology. Everything is because we, as people, want to do our tasks as quickly as possible. As a result, the further we organise, the simpler and quicker our work gets. Digitalisation is the next massive thing in today's fast-paced society. Because we live in the computer age, we want all accessible information to be digitised and kept in computers, which have quicker computational capabilities.

However, the challenge with translating facts into the digital world is to teach the computer the real-world facts in question. MODI is an ancient language built in the 12th century created by 'Hemadpant.' It is used for writing purposes only. These characters are generally cursive, as shown in Figure 128.1, so it isn't easy to recognise each character written by different writers. Most research now moves towards handwritten characters recognition which is most important for image processing, natural language processing, and pattern recognition.

There are two modes of Handwritten Characters Recognition.

- Online characters recognition mode
- Offline characters recognition mode

---

[a]bhaleraosp20comp@coep.ac.in; [b]hdg.comp@coep.ac.in

In online mode, a gadget wherein popularity is achieved when characters are below creation. In offline mode, it may be a framework in which first written by hand documents are generated, scanned, put away within the computer, and after that, they are recognised. For offline mode, the input picture from the scanner is acquired for the offline character recognition system. Following its acquisition of the scanned digital print, the second stage is pre-processing, as shown in Figure 128.2. Pre-processing improves the image and makes it acceptable for segmentation. The procedures include erosion, dilation, opening, closing, and smoothing operations. After this pre-processed image is segmented using the binarisation method. The binarisation method uses global thresholding to transform a greyscale image into binary. The procedures done in the last two phases to generate the pre-processed picture appropriate for segmentation include detecting lines in the binary image using the Sobel approach, dilating the image, and filling the holes.

The primary purpose of this research is to offer a comprehensive review of state of the art in offline handwriting recognition systems for MODI characters. We looked at the research publications on handwritten characters' recognition based on information collected. Figure 128.2 explain the basic blocks of a block diagram of a character recognition system.

MODI Character Recognition Systems

Machine learning-based methods for feature extraction and classification

Extraction of features is the initial stage in the data pre-processing method, which separates and minimises an originating organising original data into more manageable groups. Consequently, processing will be simplified; the essential element of these massive data sets is many variables. These variables demand a significant amount of processing power to process. As a result, feature extraction assists in locating the ideal feature from giant data sets by selecting and integrating elements from components, resulting in a reduction in the amount of data. Feature extraction assists in finding the perfect feature from giant data sets by

Figure 128.1 Basic MODI characters set

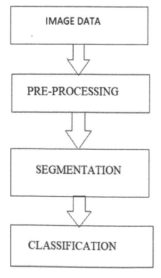

Figure 128.2 Basic character recognition diagram

selecting and integrating elements from components, resulting in a reduction in the amount of data. Table 128.1 shows some classification and features extraction techniques used in machine learning. Classification is classifying a set of data classes; it may be used to organise unstructured data. The procedure begins with forecasting the trajectory of provided data points. Classification predictive analysis emulates the mapping function between categorical input and discrete output variables. The primary purpose is to determine which category the new data will belong to.

Singh and Budhiraja (2011) introduced zoning, projection histogram, distance profile feature as extraction methods and K nearest neighbour (KNN), probabilistic neural network (PNN), and support vector machine (SVM) methods as classification techniques for Gurumukiscript. With the help of these methods, they got an accuracy rate of KNN 72.54% and SVM 73.02%. Affine and Moment Invariant Moment Invariant Gharde and Ramteke (2016) is one of two approaches used to feature extracted from handwritten separated data.

Machine learning is a subfield of artificial intelligence inspired by psychology and biology that works on understanding from such a collection of data. It may be used to address a wide range of issues. A machine learning formula is presented data examples unique to a given problem and a solution that solves the issue for each case. Once training is finished, the system can achieve higher responses.

Gharde and Ramteke (2016) introduce different machine learning. As a classifier in machine learning approaches, the support vector machine is utilised. When doing classification, this support vector machine employs the linear kernel function. Using SVM, three datasets are combined, and this system has a high recognition accuracy for handwritten MODI script data.

For MODI handwritten character recognition, Chandure and Vandana (2016) introduced Deep Convolutional Neural Network (DCNN), KNN, and SVM. They analysed Devanagari and MODI Vowels using an algorithm with the chain code feature extraction method. According to this study, the recognition rate for Devnagari is more than the MODI vowels.

Aydin (2021) proposed two different methods for the recognition of handwritten documents as well as printing document scans through a scanner or digital camera with optical character recognition. He used the supervised method is used. One of those is machine learning-based OCR, while the other is Microsoft office document management library. Text classification is done with the Naive Bayes algorithm. This optical character recognition (OCR) is getting high accuracy for computer-based documents. It is getting low in handwritten characters. It is around 4560%, and phase accuracy was 53% using naive Bayes.

Sachdeva and Mittal (2021) suggested various machine learning techniques for handwritten characters Devnagari recognition. He works on compound surfaces. Features are applied to the SVM, sequential minimal optimisation (SMO), multi-layer perceptron (MLP), and simple logistic recognition. This classification model achieved a recognition rate: SVM achieved 99.88%, SMO achieved 99.72%, simple logistic regression achieved 99.04%, and MLP achieved 97.7%.

Ramteke and Katkar (2012) researched the structural similarity approach, and with that recognition, the rate was satisfactory, as shown in Table 128.1. Besekar (2012) chain code, an image-centroid method of extraction of features, and a neural network model for classification. (Joseph and George 2020). Suggested SVM for classification with his model accuracy rate was 99.3% which are most satisfactory for characters recognition (Chandure and Inamdar, 2021). Also introduced is the SVM model for classification, and the result is most acceptable.

Table 128.1 summarises approaches for machine learning algorithms for MODI character recognition with methodology, results, and future scope.

Deep learning-based methods for classification

Deep learning methods are used for solving complex problems that require finding hidden data for the extensive dataset. It is nothing but a deep understanding of intricate relationships between many interdependent variables. Table 128.2 shows MODI handwritten character recognition employs deep learning approaches for categorisation.

Joseph and George (2020) For the MODI character recognition, as a feature extractor, a CNN autoencoder has been installed is suggested. The feature set size was reduced from 3600 to 300 using the Convolutional Neural Network (CNN) autoencoder. Following that, the retrieved characteristics were classified using SVM. Character recognition accuracy was 99.3%, which is higher than other recorded accuracy of MODI character recognition accuracy. This work's essential contribution is attaining a high accuracy rate in character recognition of MODI script.

Anam and Gupta (2015) looked into 22 MODI scripts with different vowels and consonants with numerals. They developed a recognition system that used Otsu's binarisation algorithm and Kohonen neural network method got a lower recognition rate for similar characters. Still, the overall recognition rate

Table 128.1 Studies based on machine learning-based methods for classification and feature extraction

Project title	Year of publication	Methodology	Result	Future scope
Recognition of characters in Indian MODI Script. Gharde and Ramteke (2016)	2016	SVM	Accuracy rate dataset 1 89.71% Dataset 2 90.14% Dataset 3 89.33%	Increasing samples in datasets and adding samples in datasets check the recognition rate for using different kernel functions in the future.
Performance analysis of handwritten Devnagari and MODI Character recognition system. Chandure and Vandana (2016)	2021	BPM, SVM, KNN	Accuracy rate BPM 37.5%, KNN 60%, SVM 65% using chain code as feature extraction	Use different classification algorithms for improving accuracy, such as the neural network method.
Classification of Documents from Images with Optical character Recognition Methods. Aydin (2021)	2016	Naive Bayes	The accuracy rate was 53%	Taking significant dataset recognition rate needs to improve.
Handwritten Offline Devanagari compound character recognition using machine learning. Sachdeva and Mittal (2021)	2021	SVM, SMO, M L.P., simple logistic recognition	Accuracy rate was SVM 99.88%, SMO 99.72 %, simplelogistic 99.04 %, and mutilayerperception 97.7%.	Take a large dataset for more accurate rate recognition. And apply more machine learning techniques to it.
Recognition of Offline MODI . Ramteke and Katkar G.S (2012)	2012	A measured structural similarity approach is used for structural handwritten characters recognition.	The performance rate was 9197%	Take more text set datasets and compare alphabets that compare handwritten characters using A measured structural similarity approach.
Special Approach for Recognition of Handwritten MODI script's vowels Besekar (2012)	2012	For feature extraction Chain code, image, centroid method used, with classification neural network model used.	Additional features like centroid are used for testing so that accuracy is improved.	In the future, take many samples and extend the work using all characters.
Handwritten Character Recognition of MODI Script using Convolutional convolutional neural network based Feature extraction method and support vector machine classifier Joseph and George (2020).	2020	SVM used for classification	99.3% accuracy rate was reported.	Modi script segmentation.
Handwritten MODI character recognition using transfer learning with discriminant feature analysis. Chandure and Inamdar (2021).	2021	SVM is used to obtain a classifier model.	A satisfactory accuracy rate was obtained.	Improve inter-class misclassification for the respective dataset as well as CNN invariance properties.

was 72.6%, satisfactory for the handwritten characters dataset—Chandure and Vandana (2021) Research on MODI and Devnagari handwritten scripts. Comparers self-created handwritten dataset compares and development is done with the support of supervised transfer learning method on a framework. It utilises a deep convolutional neural network (DCNN). They achieved 92.32% and 97.25% recognition rates for MODI and Devanagari scripts. As shown in Table 128.2 Dixit et al. (2021) CNN classification models were

Table 128.2 Deep-learning-based method for classification

Project title	Year of publication	Methodology	Result	Future Scope
Handwritten Character Recognition of MODI Script using Convolutional Neural Network-Based Feature Extraction Method and Support Vector Machine Classifier. Joseph and George (2020)	2020	CNN autoencoder	The accuracy rate was 99.3%	Take more concentration on Modi segmentation in the future.
An Approach for Recognizing Modi Lipi using Otsu's Binarisation Algorithm and Kohonen Neural Network. Anam and Gupta (2015)	2011	otsu's binarisation algorithm and Kohonen neural network method	72.6% accuracy rate for handwritten characters dataset	In the future, they will extend the scope of work with different Indian languages script available and improve the recognition rate.
Handwritten MODI Character Recognition Using Transfer Learning with Discriminant Feature Analysis. Chandure and Vandana (2021)	2021	DCNN	Modi script accuracy was 92.32%. Devnagari script accuracy was 97.25%	It will encourage examining the technique using an additional dataset, emphasizing improving inter-class misclassification and the CNN's invariance qualities.
Handwritten Digit Recognition using machine and Deep Learning Algorithms. Dixit et al. (2021)	2021	Developed SVM(support vector machine), CNN(Convolutional neural network), MLP (Multilayer Perceptron)	Compare the three models and conclude that CNN has the best accuracy rate of the other two models for image character recognition.	Use applications in government sectors like medical for treatment of patients, in national forces for equipment checking in national parties, and fingerprint detection and surveillance system for detection of the suspected person.
An Ancient Indian Handwritten Script Character Recognition by Using Deep Learning Algorithm. Mahajan and Tajne (2021)	2021	Deep learning model name as AlexNet	The accuracy rate was 89.72%	Improve accuracy rate using different deep learning models.

getting more accuracy than other machine learning-based models. Mahajan and Tajne (2021) AlexNet is used as a classification model that achieves a satisfactory recognition rate.

Table 128.2 summarises approaches for deep learning algorithms for MODI character recognition with methodology, results, and future scope.

Discussion

We observed and analysed some points listed below.

* Need to generate a Standard dataset for handwritten MODI script
* Need more research on old Indian handwritten languages
* The handwritten style for the same person may be changed, which affects for accurate recognition of handwritten characters
* In machine learning-based methods, the accuracy rate is minimum, and machine learning-based methods contain feature extraction methods that may be time-consuming.
* Technology can analyse enormous amounts of data and identify precise trends and patterns that humans may miss.
* ML algorithms could handle many-dimensional and complex information in unexpected environments.
* Data gathering is one of the most challenging aspects of Machine Learning. Furthermore, data collection has a cost. Again, gathering data through surveys may contain a considerable amount of fake and erroneous data. We frequently encounter situations where we discover an unbalance in data, resulting in poor prediction accuracy.

- To solve a Machine Learning challenge, several algorithms may be used. They run models with several algorithms and determine the best accurate method based on the findings.
- Machine learning algorithms are sometimes time-consuming, and sometimes it takes more CPU power to process a large number of datasets.
- In deep learning-based methods, the accuracy rate was satisfactory.
- In deep learning, Properties are quickly inferred and adequately modified to get the required outcome. At that point, it will not be essential to extract characteristics. This reduces the need for time-consuming machine learning techniques.
- The same Neural Network-based methods are applicable for any other applications. In deep learning methods, differences in data are learned automatically.
- Mainly parallel computation is performed, so machine performance is much satisfaction when a large dataset is used. Deep learning architecture is more flexible and will adopt new problems in the future.
- It is not a specific model to aid you in selecting the correct deep learning resources since it necessitates an understanding of topology, training technique, and other characteristics. As a result, it is harder to adopt by less competent individuals.

Conclusion

The overall paper is based on previous studies on MODI handwritten character recognition. According to the survey, we state that there is a need for a lot of work to recognise characters to generate one standard dataset. Future work will be to maximise the accuracy of the published dataset. Researchers are researching this field by applying Deep learning and machine learning methods. Not much research was done for MODI handwritten characters script, so it is now to improve handwritten character recognition research work. But so far 100% accuracy rate was not noted, so in the future need, research increasing the accuracy rate. According to observations, MODI character identification is still in its early stages. MODI is an old character. It is not on the list of approved Indian language hands, which adds that it has received less study attention than other Indian writings. Pattern identification in MODI language is problematic for various reasons, such as letters from similarity and irregularity in writing style in contrast to other scripts. So far, just a few papers have been published, and all of them have been studied in this paper.

References

Anam, S. and Gupta, S. (2015). An Approach for Recognizing Modi Lipi using Otsu's binarisation algorithm and kohonen neural network. Int. J. Comput. Appl. 111(2). (0975–8887).

Aydin, O. (2021) Images with optical character recognition methods. Anatol. J. Comput. Sci. https://doi.org/10.48550/arXiv.2106.11125

Besekar, D. N. (2012). Special approach for recognition of handwritten MODI Script's Vowels. Int. J. Comput. Appl. 48–52.

Chandure, S. and Inamdar, V. (2021). Handwritten Modi character recognition using transfer learning with discriminant feature analysis. IETE J. Res. 1–11.

Chandure, S. L. R., and Inamdar, V. (2016). Performance analysis of handwritten devnagari and MODI character recognition system. In 2016 International conference on computing, analytics and security trends (CAST) college of engineering, Pune, India.

Dixit, R., Kushwah, R., and Pashine, S. (2021). Handwritten digit recognition using machine and deep learning algorithms. Comput. Vis. Pattern Recognit. 176. https://doi.org/10.48550/arXiv.2106.12614

Gharde, S. S. and Ramteke, R. J. (2016). Indian recognition of characters in modi script. In 2016 International conference on global trends in signal processing, information computing and communication, 978-1-5090-0467-6/16/$31.00 ©2016. IEEE.

Joseph, S. and George, J. (2020). Handwritten character recognition of MODI script using convolutional neural network based feature extraction method and support vector machine classifier. In 2020 IEEE 5th International conference on signal and image processing (ICSIP), (pp. 32–36).

Kulkarni, S., Borde, P., Manza, R., and Yannawar, P. (2015). Review on recent advances in automatic handwritten MODI script recognition. Int. J. Comput. Appl. 115(19):975–8887.

Mahajan, K. and Tajne, N. (2021). An ancient indian handwritten script character recognition by using deep learning algorithm. In 1st International conference on emerging scientific applications in engineering and technology, pp 59–67.

Ramteke, A. S. and Katkar, G. S. (2012). Recognition of offline MODI script. Int. J. Res. Eng. I.T. Soc. Sci. (IJREISS). 2(11):102–109. ISSN 2250-0588.

Sachdeva, J. and Mittal, S. (2021). Handwritten offline devanagari compound character recognition using machine learning. In ACI'21: Workshop on advances in computational intelligence ISIC 2021. Delhi, India.

Singh, P. and Budhiraja, S. (2011). Feature extraction and classification techniques in OCR systems for handwritten Gurmukhi Script–a survey. Int. J. Eng. Res. Appl. 1(4):1736–1739.

129 Composting of domestic solid waste by effective microorganism coupled with vermicomposting

Khalid Ansari[1,a], Tejaswani Niwal[1,b], Shantanu Khandeshwar[1,c], and Tripti Gupta[2,d]

[1]Department of Civil Engineering, Yashwantrao Chavan College of Engineering, Nagpur, India

[2]Department of Civil Engineering, Shri Ramdeobaba College of Engineering & Management, Nagpur, India

Abstract

Solid waste is unreliable, undesirable, and dangerous to both the environment and humans. Waste is a type of material that is discarded after it has served its purpose. This useless waste is produced due to citizens' daily routines and activities. Municipal solid waste management is becoming increasingly important in developed countries. The current paper is focused on an analysis of the best approach for solid waste management using rapid composting methods. In the composting process, natural matter from solid waste decomposes and is maintained by microorganisms with or without oxygen. The composting process is time-consuming. To reduce the period of composting process, the three different rapid composting methods are performed with effective microorganisms traditional composting (C1), effective microorganisms composting (C2), and effective microorganisms with vermicomposting (C3). The results were analysed by using physicochemical parameters and the FTIR method.

Keywords: Composting, effective microorganisms, FTIR, vermicomposting.

Introduction

Solid waste management is carried out through systematic generation, accumulation, storage, transportation, handling, and recovery, becoming a significant problem worldwide because of the rapidly increasing population, industrialisation, and other business activities (Kale, 2012; Allen et al., 1994). The environmental impact and dumping of wastes, which are vermicomposting, reduce soil vulnerability, such as eroding, as well as humic acid and heavy metal accumulation in polluted nourish nutrients and improves the physical and chemical properties of fertiliser (absorption, strength, bulk density, electrical conductivity, microorganism population, enzymes (Bakari et al., 2016; Berthomieu and Hienerwadel, 2009). Additionally, it is used as a biofertiliser or biopesticide since it contains bacteria not accessible by vermicomposting and can inhibit diseases. According to the researcher's investigation Chang and Hudson (1967), chemical fertilisers are being used in excess, significantly degrading soil conditions. As a result, the researchers advocate vermicomposting to increase soil productivity. The drawback of this technique is expanse requirements with careful management (Coury and Dillner, 2008; Dasho, 2007; Diaz et al., 1993). Appropriately, researchers developed efficient microbe technology for economic and time savings. Effective microorganisms (EM) technology is being utilised to alleviate problems associated with solid waste management. This low-cost method was developed by Duygu et al. (2009) using microorganisms such as actinomyces, photosynthetic bacteria, lactic acid bacteria, and others (Fogarty and Tuovinen, 1991). These microorganisms cooperated and synchronised in fluid cultures. The procedure of Effective Microorganisms was designed to be primarily integrated with the composting method (Gomez-Brandon et al., 2012; Guo et al., 2012).

Researchers devised the Bokashi technique, which enhances soil texture without reducing nitrogen, achieves an optimum C: N ratio (10:1), and is also proven to be superior to regular composting (Haug, 1980; Jiménez and Garcia, 1989; Khalil et al., 2013). This technique can be used for aerobic and anaerobic processes, recognising beneficial bacteria for use in important microorganism vaccination. EM not only improves the evaluation of organic matter decomposition, but it also improves the soil's macronutrients such as nitrogen (N), phosphorous (P), and potassium (K) (Kale, 2012; Lokman and Jusoh, 2013). Furthermore, because of their separation from degrading debris, active bacteria release foul-smelling gases such as ammonia (NH_3) and hydrogen sulphide (H_2S), which act as a disinfectant in landfills and the surrounding environment (Manyuchi and Phiri, 2013). The goal of this research is to look at the effects of incorporating effective microorganisms into composting procedures for organic waste containing cow dung, wet soil, and activated EM solution, as well as EM with vermicomposting methods using the same

[a]khalidshamim86@rediffmail.com

materials (organic waste, cow dung, wet soil, and activated EM solution), and to come up with estimated physicochemical parameters for traditional composting, EM composting, and EM with vermicomposting.

Methodology

A. Composting sample and process

For the composting process, samples like vegetable leaves, fruit peels, leftover food, eggshells, and garden waste are collected from Kalamna market, Nagpur, India. The composting approach was made in a plastic bin associated with aerobic conditions with small holes of 2 mm diameter placed at 1.5 cm equal distance to all sides of the container and at the bottom to allow excess water to drain as shown in Figures 129.1– 129.3. In addition, fine sand and gravel beds are kept for 50 mm to maintain the compost's moisture. In this research, three bins experimental setup is developed for compost treatment such as traditional composting (C1), EM composting (C2) and EM with vermicomposting (C3), with cotton cloth cover at the top to avoid the excessive loss of heat and system handle manually for analysis. The EM solution product named Maple EM.1 was purchased from Maple Orgtech Pvt. Ltd, Maharashtra and for activation of the solution, the materials are taken in quantity as 20 ml of (EM.1) solution, 400 ml of water and 400 gm of jaggery were mixed well and kept away from sunlight for ten days. The pH of the activated EM solution is around 2.95~3. For traditional composting (C1), the slurry of cow dung and wet soil is mixed with solid waste. For EM composting (C2), the mixture includes a slurry of cow dung and moist soils with solid waste and

1) Solid waste- 332gm.
2) Cow dung with water- 300gm with 800ml.
3) Wet soil- 800gm.

Figure 129.1 Traditional composting (C1)

1) Solid waste- 1095gm.
2) Cow dung with water- 300gm with 800ml.
3) Wet soil- 1300gm.
4) EM solution

Figure 129.2 EM composting (C2)

1) Solid waste- 640gm.
2) Cow dung with water- 300 gm with 800ml.
3) Wet soil- 800gm.
4) EM solution
5) Earthworms- 55gm

Figure 129.3 EM with vermibed composting (C3)

allowed to semi decompose and then activated EM solution is sprayed over it and mixed well. The exact process is done for EM with vermicomposting (C3); just the addition of cultured earthworms of size (3~5) cm is added to make the process vermicomposting. All bins mixtures were turned after every 23 days until the composting process was completed; it was done to maintain the porosity and moisture of composting by spraying water at regular intervals.

B.   Calculation of carbon to nitrogen ratio

The potential C/N ratio of compost mixture was determined, subsequent chart (Table 129.1) below is used to calculate the estimated percentage of carbon and nitrogen in an element. As reactors contain fresh vegetable waste and cow dung, the total carbon and nitrogen were calculated by using the following formula: Total carbon value = vegetable waste (lbs.) × 10% of carbon + Cow dung (lbs.) × 20% of carbon Total nitrogen value = vegetable waste (lbs.) × 1% of nitrogen + Cow dung (lbs.) × 1% of nitrogen The C/N ratio was calculated by using the following formula: C/N = carbon content/nitrogen content

C.   Fourier transform infrared spectroscopy

The Fourier transform infrared spectrum is a vibrational spectroscopy system that depicts microscopic structures exposed to ultraviolet radiation. The absorption of this light causes movement by depositing energy into vibrational modes (hinga and parr, 1994; popovicheva et al., 2015; roosmalen and langerijt, 1989; ruggeri and takahama, 2016; shalaby, 2011; venkatesan et al., 2012). When subjected to ir radiation from a light source (a source of ir energy), a molecule absorbs only the frequencies corresponding to its molecular modes of vibration (venkatesan et al., 2012). Each band's vibratory spectrum exhibits variations in frequency and amplitude that are unique to that band. Distinct vibrational frequencies are associated with different bonds (o-h, c-o, c=c, c-c). FTIR mentioned the bonding nature of materials which highlight the connection of link is strong; organic matter can be identified as an absorption band in the infrared spectrum, indicating that it exists (Venkatesan et al., 2012)

Data Analysis

The monitoring of composting was done every two to three days a week for different composting containers. As a result, the pH, temperature, odour, colour, carbon to nitrogen ratio, chemical and humic acid changes were observed with the initials raw materials in composting treatment with and without effective microorganisms.

Results and Discussion

In India, enormous amounts of solid garbage are generated daily due to growing industrialisation and the migration of humans from small towns to metropolises, which are critical issues with solid waste management and disposal. Moreover, due to recent lousy management, solid waste treatment facilities must improve and shift to new technology. The two primary components of municipal solid waste are accumulation and separation, which requires prior observation to ensure that people know proper waste handling and management. In today's world, where daily waste is generated in massive quantities, effective municipal solid waste management is critical.

A.   Change in odour

The comparison of odour in three different profiles of composting is shown in Figure 129.4. It was found that the unpleasant odour of all existing composting methods decreases with time (Haug et al., 1987).

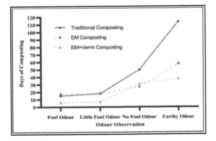

Figure 129.4 Variation of odour in different composting methods

For example, the unpleasant odour of C2 and C3 were rapidly reduced as compared to the C1 composting technique; C1 (traditional composting) took 114 days, C2 (EM cComposting) took 58 days, and C3 (EM and vermicomposting) took 38 days for diverting unpleasant odour to earthy odour.

B. Change in colour

Figure 129.5 depicts three different composting methods using colour profiling. Compared to C1 and C2 composting methods, the colour of C3 composting changes quickly; nevertheless, there was no change in colour for some initials days in C1, C2, and C3, and changes occur during C1–15 days, C2–8 days, and C3–7 days. C1 changed colour to brown after 27 days, and C3 changed colour to brown after 18 days. The dark brown colour developed after 63 days in C1, 29 days in C2, and 33 days in C3. The dark black colour appeared in C3 at 40 days. The obtained results were concurred with those reported in the previous analysis, mentioned that developed compost must be brownish-black.

C. Change in temperature

Composting is a biological reaction that requires the cooperation of microorganisms and their enzymes. The change in composting temperature indicates the transformation of microorganisms within the compost matrix; this is the analysis caused by the collection of microorganisms' metabolic warmth within the compost matrix; thus, it indicates the metabolic intensity and organic waste change rate of microorganisms within the matrix (Popovicheva et al., 2015). On day 1, the temperature of various composting techniques is C1, C2 and C3: 27°C, 26°C and 27°C, respectively. Between 10–25 days, the temperature of three composting profiles rises and shows 42°C, indicating that all composting treatment processes reached the thermophilic phase (>40°C). As per observation, the different composting profiles reached the maturation phase in different stages: C1–114 days, C2–60 days, and C3–45 days.

D. Change in pH values

The interchange in pH value during the three different composting profiles are shown in Figure 129.6. The pH alters in all the treatments in a range of 6–10 (Roosmalen and Langerijt, 1989). The pH value recorded from the C1 composting method was 8.2, C2 7.5, and C3 7.6. The starting values for C1, C2, and C3 composting procedures were reduced in two weeks and moderately rise and reached the highest point, the maximum peak of the C1 was 9.3 after 68 days, C2 was around 8.4 on day 40 and 7.3 on day 45 for C3. The results indicated as per researchers and recommend a scale for good quality compost is 6–8.5 (Popovicheva et al., 2015; Roosmalen and Langerijt, 1989; Ruggeri and Takahama, 2016; Shalaby, 2011; Venkatesan et al., 2012).

Table 129.1 Carbon and nitrogen percentage

Material	%Nitrogen	%Carbon
Fruit Wastes	8	0.5
Fallen leaves	20–35	0.4–1.0
Grass Cuttings, fresh	10–15	1–2
Kitchen Scraps	10–20	1–2
Vegetable wastes, starchy	15	1.0
Manure, cow	12–20	0.6–1.0
Vegetable wastes, fresh, leafy	10	1.0

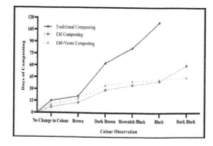

Figure 129.5 Variation of colour in different composting methods

E. Carbon to nitrogen ratio

The carbon-to-nitrogen ratio for three different composting procedures is shown in Figure 129.7. According to research, the carbon-to-nitrogen ratio was reduced in composters because of inorganic substances that precipitate organic mixtures. The beginning values of the C/N ratio were as follows: C1 represented a C/N ratio of 34, C2 represented a C/N ratio of 36, and C3 represented a C/N ratio of 35. After the composting procedures were completed, the end C/N ratio for C1 was 11.11, C2 was 11.5, and C3 was 9.5. These results compare approving with researchers. According to researchers, the C/N ratio must be slighter than 15~20 for use without any restriction and considered mature compost; as per analysis, the ratio above 20 is not regarded as unmatured compost because carbon is not in the available form (Shalaby, 2011).

F. Outcome of Metals

During the degradation of organic solid waste, heavy metals such as iron (Fe) and copper (Cu) increase. The level of heavy metals in composting processes C1, C2, and C3 rises with time, except for Cu, which decreased proportionally over time. According to experts, the accumulation of nutrients and heavy metals grows rapidly during the composting process, indicating the compost's development. Heavy metals such as Fe and Cu effectively identify components involved in plant development (Ruggeri and Takahama, 2016). Figures 129.8 and 129.10 illustrate the various shapes of heavy metals. The time required to observe the effects of heavy metals were 114 days for C1, 60 days for C2, and 45 days for C3. Cu concentrations in the final samples were 0.44 mg/kg for C1, 1.20 mg/kg for C2, and 0.68 mg/kg for C3. Fe final concentrations in C1 were 4.44 mg/kg, 8.64 mg/kg in C2, and 6.08 mg/kg in C3. Mn concentrations were finalised at 3.00 mg/kg for C1, 6.80 mg/kg for C2, and 2.28 mg/kg for C3. Zn final concentrations were 2.32 mg/kg for C1, 3.64 mg/kg for C2, and 3.16 mg/kg for C3. Manganese (Mn) and Zinc (Zn) concentrations were significantly higher in C2 composting than in C1 or C3 composting, as illustrated in Figure 129.10 should be executed for both rural and urban areas. The amount of waste remains after the treatment, it is necessary to dispose of that waste in secure landfills. The administration has grasped various decisions and enterprises to enhance the waste handling system, but the path is too far away from achieving the goals of effectual solid waste management. Various scientific methods have been developed for the handling of waste; according to research composting methods are virtuous for solid waste treatment. The present study is based on the rapid composting techniques with effective microorganisms; this process is simplest for \converting the organic waste into the useful by- product sand used for the improvement of soil texture.

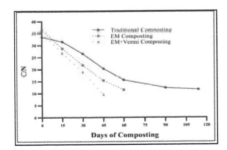

Figure 129.6 Variation of pH in different composting method

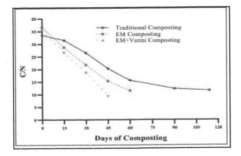

Figure 129.7 Variation of C/N ratio in different composting method

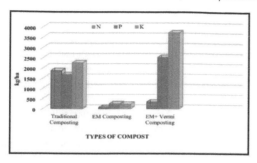

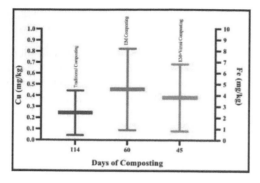

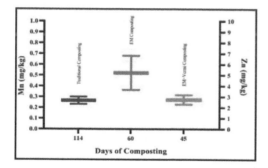

Figure 129.8 Variation of NPK, Fe, Cu, Mn and Zn in different days

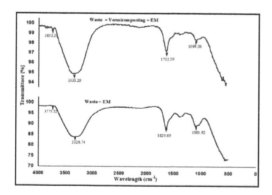

Figure 129.9 FTIR Spectra for different composting

Conclusion

It is invigorating to study the development in composting processing, evaluation yield by composting performance comprises the comparatively secure and economical for controlling the solid wastes and productive for soil reformation. The research associated with rapid composting methods terminated that using effective microorganisms' technique reduces the time duration and processing can be easily operated from the results and analysis, it can be terminated that similar physical and chemical parameters are shown by all rapid composting methods. The decrease in the C/N ratio and total organic carbon means that a microbe has absorbed an organic composite individual. The implementation of effective microorganisms in fertiliser expands the micro-nutrient content. The effective microorganism with vermicomposting is a more beneficial composting process than traditional composting as per the results of parameters and the heavy metals were under the standard limitation.

References

Allen, D. T., Palen, E. J., Haimov, M. I., Hering, S. V., and Young, J. R. (1994). Fourier transform infrared spectroscopy of aerosol collected in a low pressure impactor (LPI / FTIR): Method development and field calibration. Aerosol Sci. Technol. 21(4):325–342.

Bakari, S. S., Moh'd, L. M., Maalim, M. K., Aboubakari, Z. M., Salim, L. A., and Ali, H. R. (2016). Characterisation of household solid waste compost inoculated with effective microorganisms. Mod. Environ. Sci. Eng. 2(3):194–200.

Berthomieu, C. and Hienerwadel, R. (2009). Fourier transform infrared (FTIR) spectroscopy. Photosynth. Res. 101(2):157–170.

Chang, Y. and Hudson, H. J. (1967). The fungi of wheat straw compost. Transact. British Mycol. Soc. 50(4):649–666.

Coury, C. and Dillner, A. M. (2008). A method to quantify organic functional groups and inorganic compounds in ambient aerosols using attenuated total reflectance FTIR spectroscopy and multivariate chemometric techniques. Atmos. Environ. 42(23):5923–5932.

Dasho, S. T. (2007). Application of the technology of effective microorga nisms in Bhutan, ministry of agriculture royal government of Bhutan. Kitanakagusuku, Kishaba: EM research organisation.

Diaz, L. F., Savage, G. M., Eggerth, L. L., and Golueke, C. G. (1993). Composting and recycling municipal solid waste. Hercules, California, USA: Lewis Publishers, Cal Recovery Inc.

Duygu, D., Yildiz, K., Baykala, T. and Acikgoz, I. (2009). Fourier Transform Infrared (FT-IR) Spectroscopy for Biological Studies. Gazi Univ. J. Sci. 22(3):117–121.

Fogarty, A. and Tuovinen, O. (1991). Microbiological degradation of pesticides in yard waste composting. Microbiol. Rev. 55(2):225–233.

Gomez-Brandon, M., Lores, M., and Dominguez, J. (2012). Species- specific effects of epigeic earthworms on microbial community structure during first stages of decomposition of organic matter. PLoS One. 7:1–8 (e31895).

Guo, X., Gu, J., Gao, H., Qin, Q., Chen, Z., Shao, L., Chen, L., Li, H., Zhang, W., Chen, S., and Liu, J. (2012). Effects of Cu on metabolisms and enzyme activities of microbial communities in the process of composting. Bioresour. Technol. 108:140–148.

Haug, R. T. (1980). Composting engineering principles and practice. Ann Arbor (Mich.): An Arbor Science.

Hinga, T. and Parr, J. (1994). Beneficial and effective microorganisms for a sustainable agriculture and environment. Atami, Japan: International Nature Frming Research Center.

Jiménez, E. I. and Garcia, V. P. (1989). Evaluation of city refuse compost maturity: A review. Biolog. Wastes. 27(2):115–142.

Jusoh, M. L. C., Manaf, L. A., and Latiff, P. A. (2013). Composting of rice straw with effective microorganisms (EM) and its influence on compost quality. Iranian J. Environ. Health Sci. Eng. 10(1):1–9.

Kale, D. K. (2012). Solid waste management by use of Effective microorganisms technology. Asian J. Exp. Sci. 26 (1):5–10.

Khalil, A. I., Hassouna, M. S., Shahee, M. M. and Bark, M. A. A. (2013). Evaluation of the composting process through the changes in physical, chemical, microbial and enzymatic parameters. Asian J. Microbiol. Biotechnol. Environ. Sci. 15(1):25–42.

Kumar, M., Ou, Y. -L. and Lin, J.-G. (2010). Co-composting of green waste and food waste at low C/N ratio. Waste Manag. 30(4):602–609.

Manyuchi, M. and Phiri, A. (2013). Vermicomposting in solid waste management: A review. Int. J. Sci. Eng. Technol. 2.1234–1242.

Popovicheva, O. B., Kireeva, E. D., Shonija, N. K., Vojtisek-Lom, M. and Schwarz, J. (2015).

Roosmalen, G. R. V. and Langerijt, J. V. D. (1989). Green waste composting in the Netherlands. Biocycle. 30(4):32–35.

Ruggeri, G. and Takahama, S. (2016).

Shalaby, E. A. (2011). Prospects of effective microorganisms technology in wastes treatment in Egypt. Asian Pac. J. Trop. Biomed. 1(3):243–248.

Venkatesan, S., Pugazhendy, K., Sangeetha, D., Vasantharaja, C., Prabakaran, S., and Meenambal, M. (2012). Fourier transform infrared (FT-IR) spectoroscopic analysis of spirulina. Tamil Nadu, India.

Van Fan, Y. (2017). Evaluation of effective microorganisms on home scale organic waste composting. J. Environ. Manage. 15(216):41–48.

Van Fan, Y., Lee, C. T., Klemeš, J. J., Chua, L. S., Sarmidi, M. R., and Leow, C. W. (2018). Evaluation of Effective Microorganisms on home scale organic waste composting. J. Environ. Manag. 216:41–48.

130 Evaluation of image segmentation on various color spaces in view of image processing on mobile devices

Hemantkumar R. Turkar[a] and Lalit B. Damahe[b]

Computer Science & Engineering, Rajiv Gandhi College of Engineering & Reasearch Nagpur, India

Abstract

Recent advances in imaging and mobile phone industries, smart phones can offer an attractive medium for delivering image-based services at a low cost with high memory and speed. Segmentation in image processing is the area concerned with the analysis and evaluation of the digital images based on the problem domain and relevant knowledge base. Image pre-processing and image post-processing are two important processes in digital image processing. In this paper relative analysis of existing clustering approaches (KM), (MKM), (EMKM) and proposed optimised algorithm is performed. Qualitative and quantitative analysis is done on the different evaluation parameters namely mean square error (MSE), mean absolute error (MAE) and peak signal-to-noise ratio (PSNR). After experimentation it observed that the proposed optimise algorithm is preferred method for image segmentation.

Keywords: Colour space, HSV, image segmentation, YCbCr.

Introduction

No widespread and standard description of image segmentation evaluation parameters for evaluators. As many of these performance measures are reliant on the applications (Gonzalez and Woods, 2005). The inexperienced evaluators which may use the wrong evaluation parameters for the wrong application gives improper result. Typical performance indicators include accuracy, robustness, sensitivity, adaptability, reliability and efficiency. Various researchers used different evaluation parameters to quantitatively evaluate image segmentation algorithms. Many of the performance evaluation parameters applied to a large set of images including synthetic image data set and standard image data set demonstrate that some of the evaluation parameters prefer under-segmentation preference whereas other prefers over segmentation preference. The various evaluation parameters used in segmentation of image processing are random index, mean absolute error, boundary detection error, mean square error (MSE), peak signal-to-noise ratio (PSNR) etc. The estimation techniques are dependent on the evaluation parameters like average colour between regions, texture, entropy etc.

Techniques for image segmentation

Image segmentation, is principally the method used to identify the similarities and dissimilarities in the given image (Khaire and Thakur, 2012). This leads to form the various regions, that is, partitioning the given image based on the certain parameters, for example, pixel intensity; colour; texture; pixel connectivity; etc. Generally, the input to the image segmentation process is the image and output of this process is, mostly, the attributes related to the concerned image which later can be used for the desired purpose.

Grey or colour image are generally used in image segmentation. Various colour spaces are utilised to process the given colour image. Image segmentation method is easy technique used in various image processing problem areas such as pattern, object and face recognition etc. The various methods of image segmentation are classified based on similarity and discontinuity properties (Saini and Arora, 2014). Discontinuity property method is also known as boundary based method similarity property method is known as region based methods. Edge detection is used for identifying boundary of a region in discontinuity method. Partitioning an image in region based methods into sections, which is related to some predefined condition. Image segmentation methods classified as edge, region, thresholding and clustering based techniques.

---

[a]hemantturkar@rediffmail.com; [b]damahe_l@rediffmail.com

Edge based technique

In image processing for image partitioning the edge based techniques relates a huge set of approaches made on data about edges in the image. There are many edge detecting operators like Sobel for detecting edges. Gaps in grey level, size etc. are detected by these boundaries. The fundamental step includes smoothing of image and Edge localisation (Ravi and Khan, 2012).

Region based technique

The aim of image segmentation is to divide an image into segments. The region based approach is used for determining the region directly. The regions are formed on the basis of likenesses in an image. The principle of region based image segmentation technique is making large homogeneous region (Khan and Ravi, 2013).

i)   *Region growing method:* It also referred as pixel based image segmentation technique because it uses early seed point. The seed point is an initial pixel from where we initialise the process of region growing. Grow regions by recursively containing the adjacent pixels which are similar and connected to the seed pixel.

ii) *Region splitting and merging method:* Given method separates image into likely number of regions based on similarity measure. Then combine the segments on the basis of measure (Peng et al., 2013). This method is also known as quad tree method. The first step of quad tree is to set up some criteria such as texture, variance, mean etc. Then the whole image is split into four quadrants, check each quadrant for matching. If the quadrants are not uniform then divide into four new quadrants. The last step is to compare the adjacent regions and merge if uniform according to the similarity measure. The working of quad tree is shown in Figure 130.1.

Thresholding based technique

Picture thresholding has a focal situation in utilisations of picture division in light of its natural properties and effortlessness of execution. This technique is oldest and frequently used technique of image segmentation. Thresholding the various techniques are broadly divided into global thresholding technique and local thresholding technique (Sharma et al., 2012). The initial and most crucial step is selection of threshold value (T). Threshold value can be selected using several methods such as optimal thresholding and multispectral thresholding. Only one threshold value which performs as a cut off value is used in global thresholding technique whereas the multiple threshold values are used in local thresholding technique. The advantage of this method is its simplicity while the disadvantage of this method is noise sensitivity and over or under segmentation (Zuva et al., 2011). Image segmentation based on Otsu method is proposed Panchbhai et al. (2012) for Blood cell images.

Clustering based technique

Bunching is a course of making gatherings of groups of same items though various articles are isolated in various bunches. Clustering technique Bora and Gupta (2014) is essentially utilised in face acknowledgment, finger impression acknowledgment, object acknowledgment, optical person acknowledgment and

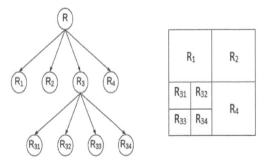

Figure 130.1 Region splitting and merging method

factual information investigation and so on. Presently a day's grouping calculations are fundamentally utilised in the clinical picture handling space for picture division (Christ et al., 2014). Grouping is partitioned into two primary sorts specifically, hard and soft clustering. Hard grouping has a place with a solitary bunch while soft grouping has a place with various bunches.

Colour Space

An identifying colour precisely is important in commercial fields for to know many colour names. For this purpose colour spaces are broadly classified into linear and non-linear colour spaces (Chen et al., 2008). Most popular colour spaces are discussed in the next section.

RGB colour space

RGB is a truncation for red–green–blue and this shading space depends on the RGB shading model. A tone is acquired by blending the three base tones. On the off chance that you utilise none of the three base tones the outcome is dark. By mixing all three base colours with full intensity, you'll get a white colour. RGB is a most widely used colour model for computer graphics due to its similarity to human visual perception.

HSV colour space

It is a shading model that portrays colours (hue or tint) as far as their shade (saturation or amount of grey) and their brilliance (value or luminance).

YCbCr colour space

YCbCr is one of the most popular colour spaces in image processing. The first luminance component of YCbCr is sensitive to human eye and analogous to grey scale image. The chrominance component contains the colour information and human eye is less sensitive for this two planes. It represents colours in terms of luma (Y) and two chroma (Cb and Cr). YCbCr colour space efficiently represents the images in image processing. So, YCbCr is commonly used colour space in digital video domain.

Performance Evaluation Parameters

There is no far reaching and standard depiction for picture division assessment boundaries that can be utilised by the evaluators. Since the greater part of these presentation measures are application reliant, unpracticed evaluators can utilise the erroneous assessment boundaries for the inaccurate application in this way giving incorrect result. Ordinary execution markers incorporate an exactness, power, affectability, flexibility, unwavering quality and productivity. Different specialists have been utilising distinctive assessment boundaries to quantitatively assess picture division calculations (Zhang and Gerbrands, 1994). A significant number of the presentation measurements applied to sufficient arrangement of pictures incorporates synthetic image data set and standard image data set shows that a portion of the assessment boundaries like under-division inclination where as others like over division inclination (Jada et al., 2015). The different assessment boundaries utilised in division of picture handling are random index, mean absolute error, boundary detection error, MSE, PSNR etc. These assessment strategies depend on boundaries, for example, texture, entropy, average colour between regions, root mean square etc. The outcomes are assessed on three boundaries in particular MSE, PSNR and mean square error (MAE).

Mean square error

Quantitative analysis is a mathematical situated technique to discover the exhibition of the calculations with next to no human mistake (Siddiqui and Isa, 2011). It ascertains the square mistake between the pixels of unique and coming about picture. Quantitative analysis is an assessed boundary to ascertain the exhibition of the calculations as far as MSE. MSE is assessed by utilising condition (1).

For calculation of MSE value V_i is the pixel which is part of jth cluster and N is an image.

$$MSE = \frac{1}{N}\sum_{j=1}^{k}\sum_{i \in c_j}\left\|v_i - c_j\right\|^2 \tag{1}$$

Peak signal-to-noise ratio

It is a proportion between the greatest possible power of a signal and the power of corrupting noise. PSNR is most handily characterised through the mean squared error (MSE). From above condition the PSNR for a 8 bit grey scale picture is given by the situation (2).

$$PSNR = 10\log_{10}\left(R^2\big/MSE\right) \tag{2}$$

Where R is the greatest worth in input picture information. Assuming pixels of advanced picture are addressed utilising n bits, $R = 2^n - 1$.

Mean absolute error

It is a proportion of distinction between two neighbouring factors as shown in the following equations (3).

$$MAE = \frac{\sum_{i=1}^{n}|e_i|}{n} \quad \text{and} \quad MAE = \frac{\sum_{i=1}^{n}|y_i - x_i|}{n} \tag{3}$$

It is feasible to communicate MAE as the amount of two parts i.e. amount conflict and allotment conflict. Amount conflict is the outright worth of the Mean Error. Allotment conflict is MAE shorts the amount conflict.

Experimental Results

Different investigations are completed on cell phones. It utilises various upsides of k (number of bunches) for HSV and YCbCr shading space. Subjective examination is regularly utilised cycle in picture handling. In this cycle, the probabilistic assertion about the calculations strength and shortcoming are handled. It is done based on human visual insight. The original images to carry out the experimental analysis are given in Figure 130.2.

Qualitative Analysis on HSV and YCbCr colour space

Figures 130.3 and 130.4 illustrates the partitioning outcome of HSV and YCbCr colour space images for KM, MKM, EMKM and the proposed optimised algorithm (PA). All the pictures are partitioned for k = 3.

The original YCbCr images are presented in the first row, Second row images are segmented images by using K-Means clustering algorithm, third row images are segmented images by using MKM clustering algorithm, fourth row images are segmented images by using EMKM, and fifth row images are segmented images by using our Proposed approach(PA). In all these algorithms, The YCbCr colour space as shown in Figure 130.4 it provide better segmentation result as compare to HSV colour space presented in Figure 130.3.

Figure 130.2 Original images

Quantitative analysis on HSV and YCbCr colour space

Quantitative analysis is a numerically oriented procedure to figure out the performance of algorithms without any human error. Performance analysis helps to seek out the most efficient technique/method used for segmenting an image by thoroughly analysing the used parameters values. Various types of quality parameters/metrics can be used for the sole purpose of evaluating performance analysis of different type of method used for segmenting an image. One of the parameter mean square errors (MSE) is calculated on segmentation of four selected images.

The examination of executed calculations is done as far as MSE, peak signal-to-noise proportion (PSNR) and mean absolute error (MAE). Every boundary is determined for KM, MKM, EMKM and proposed approach for HSV pictures with the bunch count of k, where k = 3, 4 and 5 separately. Table 130.1 shows the rundown of MSE, PSNR and MAE assessment for the HSV shading space.

The examination of executed calculations is done as far as MSE, Peak signal-to-noise proportion (PSNR) and MAE. Every boundary was determined for KM, MKM, EMKM and proposed approach for YCbCr pictures with the bunch count of k, where k = 3, 4 and 5 separately. Table 130.2 shows the rundown of MSE, PSNR and MAE assessment for the YCbCr shading space.

Conclusion and Future Scope

The proposed approach is evaluated on the cell phone with existing algorithms. The proposed algorithm can be compare with other clustering algorithms with other color model (Turkar and Thakur, 2019; Turkar and Sakarkar, 2021) so that the efficiency of the algorithm can be measured. The results confirm that the

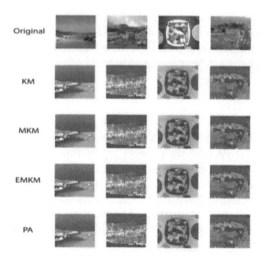

Figure 130.3 Segmentation of HSV images for K = 3

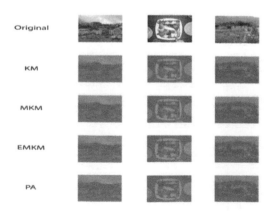

Figure 130.4 Segmentation of YCbCr images for K = 3

Table 130.1 Evaluation of approaches for selected HSV images using MSE, PSNR and MAE

HSV Color-Space	No. of cluster	Mean Square Error (MSE)			PSNR			MAE		
		k=3	k=4	k=5	k=3	k=4	k=5	k=3	k=4	k=5
KM	Image-1	32401.79	32347.84	4938.49	3.02	3.03	11.19	0.56	0.56	0.08
	Image-2	34693.24	12155.82	14830.5	2.72	7.28	6.41	0.6	0.21	0.25
	Image-3	3642.16	14436.37	36220.04	12.51	6.53	2.54	0.06	0.25	0.62
	Image-4	26093.2	2402.32	20912.03	3.96	14.32	4.92	0.45	0.04	0.36
MKM	Image-1	31904.79	32347.84	4938.49	3.09	3.03	11.19	0.55	0.56	0.08
	Image-2	34306.94	12155.82	14830.5	2.77	7.28	6.41	0.59	0.21	0.25
	Image-3	3273.87	14436.37	36220.04	12.98	6.53	2.54	0.05	0.25	0.62
	Image-4	25811.95	2402.32	20912.03	4.01	14.32	4.92	0.44	0.04	0.36
EMKM	Image-1	31235.5	32347.84	4938.49	3.18	3.03	11.19	0.54	0.56	0.08
	Image-2	33490.41	12155.82	14830.5	2.88	7.28	6.41	0.58	0.21	0.25
	Image-3	2552.58	14436.37	36220.04	14.06	6.53	2.54	0.01	0.25	0.62
	Image-4	25009.98	2402.32	20912.03	4.14	14.32	4.92	0.43	0.04	0.36
PA	Image-1	30609.58	32347.84	4933.49	3.27	3.03	11.19	0.53	0.56	0.08
	Image-2	33125.62	12155.82	14830.5	2.92	7.28	6.41	0.57	0.21	0.25
	Image-3	1929.14	14436.37	35220.04	15.27	6.53	2.54	0.03	0.25	0.62
	Image-4	24536.16	2402.32	20912.03	4.23	14.32	4.92	0.42	0.04	0.36

Table 130.2 Evaluation of approaches for selected YCbCr images using MSE

YCbCr colour-space	No. of Cluster	Mean square error (MSE)			PSNR			MAE		
		k=3	k=4	k=5	k=3	k=4	k=5	k=3	k=4	k=5
KM	Image-2	28224.28	265.32	33550.84	3.62	23.89	2.87	0.49	0.01	0.58
	Image-3	2437.01	19942.15	36095.16	14.26	5.13	2.55	0.04	0.34	0.62
	Image-4	2676.88	5135.13	19036.07	13.85	11.02	5.33	0.04	0.08	0.33
MKM	Image-2	27431.68	265.32	33550.84	3.74	23.89	2.87	0.47	0.01	0.58
	Image-3	2436.8	19942.15	36095.16	14.26	5.13	2.55	0.04	0.34	0.62
	Image-4	1818.9	5135.13	19036.07	15.53	11.02	5.33	0.03	0.08	0.33
EMKM	Image-2	27161.26	265.32	33550.84	3.79	23.89	2.87	0.47	0.01	0.58
	Image-3	1122.41	19942.15	36095.16	17.62	5.13	2.55	0.01	0.34	0.62
	Image-4	1315.8	5135.13	19036.07	16.93	11.02	5.33	0.02	0.08	0.33
PA	Image-2	26706.44	265.32	33550.84	3.86	23.89	2.87	0.46	0.01	0.58
	Image-3	643.81	19942.15	36095.16	20.04	5.13	2.55	0.01	0.34	0.62
	Image-4	823.48	5135.13	19036.07	18.97	11.02	5.33	0.01	0.08	0.33

proposed approach produces the lowest MSE values for almost all images. These findings strongly support the quantitative analysis, which shows that the proposed approach has successfully outperformed the conventional clustering algorithms. The proposed approach has been proven to produce a better segmentation as compared to other conventional algorithms regardless of the number of clusters used.

References

Bora, D. J. and Gupta, A. K. (2014). A novel approach towards clustering based image segmentation. Int. J. Emerg. Sci. Eng. (IJESE) 2(11). ISSN: 2319–6378.

Chen, T., Chen, Y. and Chien, S. (2008). Fast image segmentation based on K-Means clustering with histograms in HSV color space. IEEE (MMSP) 322–325.

Christ, M. C. J., Sivagowri, S. and Babu, P. G. (2014). Segmentation of brain tumors using meta heuristic algorithms. Open J. Commun. Softw. 1(1).

Gonzalez, R. C. and Woods, R. E. (2005). Digital image processing. (2nd ed.), Pearson Education Asia.

Jada, F. B., Aibinu, A. M. and Onumanyi, A. J. (2015). Performance metrics for image segmentation techniques: A Review. researchgate. Conference: International Engineering Conference IEC 2015.

Khaire, P. A. and Thakur, N. V. (2012). An Overview of image segmentation algorithms. Int. J. Image Process. Vis. Sci. 1(2):62–68.

Khan, A. M. and Ravi, S. (2013). Image segmentation methods: A comparative study. Int. J. Soft Comput. Eng. (IJSCE) 3(4). ISSN: 2231–2307.

Panchbhai, V. V., Damahe, L. B., Nagpure, A. V. and Chopkar, P. N. (2012). RBCs and parasites segmentation from thin smear blood cell images. Int. J. Image Graph. Signal Process. 10:54–60. (Published Online September 2012 in MECS) (http://www.mecs-press.org/) DOI: 10.5815/ijigsp.2012.10.08.

Peng, B., Zhang, L. and Zhang, D. (2013). A survey of graph theoretical approaches to image segmentation. ACM Digital Library.

Ravi, S. and Khan, A. M. (2012). Operators used in edge detection: A case study. Int. J. Appl. Eng. Res. 7(11). ISSN 0973-4562.

Saini, S. and Arora, K. (2014). A study analysis on the different image segmentation techniques. Int. J. Inf. Comput. Technol. 4(14). ISSN 0974-2239.

Sharma, N., Mishra, M. and Shrivastava, M. (2012). Colour image segmentation techniques and issues: An approach. Int. J. Sci. Technol. Res. 1(4).

Siddiqui, F. U. and Isa, N. A. M. (2011). Enhanced moving k- means (EMKM) algorithm for image segmentation. IEEE Trans. Consum. Electron. 57(2).

Turkar, H. K. R. and Sakarkar, G. (2021). An optimised approach for image segmentation on mobile devices. Turk. J. Comput. Math. Educ. (TURCOMAT) 12(2):3018–3024.

Turkar, H. K. R. and Thakur, N. S. V. (2019). Performance comparison of clustering algorithms based image segmentation on mobile devices. In Proceeding of CISC 2017, cognitive informatics and soft computing (pp. 581–591), DOI:10.1007/978-981-13-0617-4_56.

Zhang, Y. J. and Gerbrands, J. J. (1994). Objective and quantitative segmentation evaluation and comparison. Signal Process. 39(1-2):43–54.

Zuva, T., Olugbara, O. O., Ojo, S. O. and Ngwira, S. M. (2011). Image segmentation, available techniques, developments and open issues. Can. J. Image Process. Comput. Vis. 2(3).

131 Data sanitisation techniques for transactional datasets using association rule hiding techniques

Nitin Jagtap[a] and Krishankant P. Adhiya[b]

Department *of Computer Engineering,* SSBT College of Engginerring and Technology, Bambhori, Jalgaon, India

Abstract

The data mining and rule generation techniques are most popular due to the high utilisation of transactional data in e-commerce websites. Mining services require balanced data to generate accurate results, yet privacy concerns may lead consumers to supply false information. Several strategies based on random disturbance of database files have recently been presented to safeguard client privacy in data analysis. Every day, online businesses deal with thousands of transactions, leading to privacy concerns. Clustering algorithm concealing is a privacy approach that focuses on hiding confidential material produced by online department's shops, Face book information, and other sources. These strategies are used to detect sensitive rules and offer privacy to sensitive rules, resulting in Lost and Ghost rules. Techniques created so far have failed to provide improved results. This paper proposed a rule hiding technique for hiding sensitive information from generated rules by hash base apriori. First, we implement hash-based apriori to generate the association rules and various feature selection techniques has been used for hiding the rules. The inexpensive experimental analysis evaluates both rules from DB' rules and DB rules and evaluates the data quality and data loss. The proposed system can reduce 0.12% data loss on an adult dataset.

Keywords: Association rule mining, Apriori algorithm, hash apriori, rule generation, rule hiding techniques, transactional dataset.

Introduction

The technique of automatically extracting information from data is known as 'data mining.' Data analysis is an excellent method to turn data into insights while more data is collected, with the volume of data growing every five years. It's widely employed in various fields, including marketing, fraud detection, and scientific research. Data gathering can be used on any extensive data collection, and while it can discover hidden correlations, it can't find patterns that aren't already there in the data set. Mining is the technique of extracting new and usable insights from data, and it has become a helpful tool for corporate analysis and decision-making. Data sharing may benefit research and commercial cooperation in a variety of ways. On the other hand, large data repositories include sensitive regulations and personal data that must be maintained before being disseminated. Private information data mining has become a popular research area in the information retrieval and information security sectors, driven by the numerous competing demands of information sharing, confidentiality, and knowledge extraction.

The methodology optimised for knowledge discovery and the theory of association processing is to describe the consistency contained in an extensive database. This procedure can allow an individual or an organisation to re-identify information that is confidential. Studying in this area uses numerous methods to approach the issue differently. The methods used in the study are focused on covering strategy, database recovery strategy, iteration-hidden rules, and computational nature. Most of the techniques lead to affecting their use of data, breaching privacy and having numerous adverse effects. The withdrawal symptoms involve covering non-sensitive rules incorrectly and mistakenly applying fake rules.

In addition, the rest of the work is presented under the following headings: Section II examines existing methods to association data mining and data hiding techniques with privacy preservation techniques created by previous researchers, while Section 3 depicts the materials and techniques employed in the creation of the future system, and Section 4 provides the implementing algorithm explanation. In Section 5, we discuss the experimental strategy for implementing the business work and conclusions obtained with our technology, as well as a comparison with numerous state-of-the-art techniques. Section 6discusses the result as well as its future potential.

[a]ntnjagtap@gmail.com; [b]adhiyakp@gmail.com

Literature Survey

The apriori Mining Algorithm is improved by the Linked List Structured Hash Table (Mar and Oo, 2020). Several industrial firms gained helpful information from client transactions made on various occasions. The Apriori algorithm is one of the most important data mining techniques for locating frequent item groups in large datasets and uncovering information using the association rule. When dealing with many transactions, the Apriori technique loses time traversing the whole database, searching for frequently occurring item sets. Its memory needs are likewise inefficient. The modified apriori approach may increase the cost by adding and calculating the third threshold.

So the suggested study uses linked-list and hash tables to harvest common item sets from a big transaction database. The database is searched using Improved apriori, and hash tables are employed to count common 1-item groups. In the database, the following standard item sets were recorded using the linked list: performance algorithms for fuzzy association rule mining (Rahman et al., 2019). The performance of fuzzy association rule mining is compared to that of conventional rule mining. Performance is evaluated using the apriori algorithm, fuzzy apriori, and evolutionary genetic fuzzy apriori-DC. Data mining is one of the newest and fastest developing machine learning fields today. Data mining uses a variety of approaches to find common patterns in large or high-dimensional datasets.

Improved Apriori data mining algorithm (Cong, 2020). The traditional apriori method is described, and then the modified approach is evaluated. In the experiments, the modified apriori algorithm outperforms the regular apriori algorithm. Preventing Maverick purchasing utilising Association Rule Mining and FP-Growth (Isa et al., 2021), NE makes choices, particularly when acquiring products. The association rules mining approach links materials and projects. This study's framework was the cross-industry standard process for data mining (CRISP-DM). As a consequence, each project has its own set of supplies. This information will aid NE in making buying decisions. Association methods for software metrics, Nave Bayes, and rule mining (Tua and Sunindyo, 2019). Adding a feature selection process using classic feature selection methods like subset selection and software defect prediction results in a worse distribution but greater accuracy. It has low recall performance, resulting in poor results. To increase the method's performance using software metrics, the ARM feature selection process should be integrated into the software prediction process. Software metrics are interconnected and cannot be ignored.

A genetic algorithm for mining fuzzy association rules (Kar and Kabir, 2019), the genetic cooperative-competitive learning algorithm (GFS.GCCL) and the structural learning algorithm on vague environment are compared (SLAVE). Genetic algorithms provide fuzzy rules for pattern classification in GFS.GCCL. The slave technique learns fuzzy association rules iteratively. This experiment utilised the Iris and Wine datasets. On real-world datasets, the SLAVE technique beats the GFS. GCCL algorithm. E-commerce prospect client identification. The apriori association rule algorithm Zheng (2020) enhances database search. The new approach simplifies the algorithm.

In ecommerce, the Apriori algorithm proposes links. Ecommerce is becoming a bigger part of the company. With problematic items, how can e-commerce platforms propose goods to customers and attract new users? The store's business platform weighs in. Based on association criteria, apriori generates a technique with strong association. The e-commerce platform may propose a product to a customer based on their consumption habits and degree of product connection, saving them time when browsing and improving their shopping experience. The Apriori algorithm and UML are used to design correct object-oriented database association rules (Angulakshmi et al., 2020). A social network may be used to build a realistic scenario of the Electricity Bill Deposit System (EBDS) and user-interested commodities. In both cases, an Apriori algorithmic rule is used to locate repeated information sets using correct association rules. The rules of the association are defined in the unified modeling language. The present master's project uses an Apriori algorithm to analyse EBDS data.

IAR: A New FP-Growth Method (Kreesuradej and Thurachon, 2019). Uncovering incremental association rules with FP-Growth We also add an FPISC tree on top of the FUFP tree. For incremental association rule discovery, the FPISC-tree beats the FUFP-tree. Instead of reprocessing the original pathways and using them to find frequent item sets from FPISC-tree, this technique retrieves frequent item sets and uses their support count to update the new support count of the incremental database. The RCFP-growth algorithm is used to mine association rules in silk relics (Dong et al., 2019). The auxiliary pattern design system describes an RCFP-growth technique to extract association rules from the silk relic's database, which features imbalance, enormous data, and a user-specified mining aim. It saves memory and time compared to Apriori and frequent pattern growth (FP-growth) approaches. The RCFP-growth algorithm creates certain pattern designs—sequential association rule mining for Autonomous Hierarchical Task Extraction (Ghazanfari et al., 2020).

For both Markov decision processes (MDPs) and factored MDPs, SARM-HSTRL is a unique approach based on sequential association rule mining (SARM). After discovering causal and temporal linkages between states in different trajectories, the proposed technique extracts a task hierarchy that encodes these ties as termination conditions of discrete sub-tasks. We show that the extracted hierarchical policy is optimum in MDPs and factored MDPs. SARM-HSTRL extracts this hierarchical optimum policy without dynamic Bayesian networks in single and multiple task trajectories. The generated hierarchical task structure is also compatible with trajectories and offers the most efficient, reliable, and compact structure. Many test beds' numerical results compare the proposed SARM-HSTRL technique's performance to current HRL algorithms in terms of sub-goal accuracy, extracted hierarchies' validity, and learning time. The linked list based hash table improves Apriori mining (Mar and Oo, 2020). With the Apriori method, you can find common item groupings in huge datasets and discover information using the association rule. When dealing with large databases, the Apriori method wastes time looking for frequently recurring item sets. Its memory demands are inefficient. The third criteria may raise the cost of the modified Apriori technique.

So the suggested study uses linked list and hash tables to harvest common itemsets from a big transaction database. The database is searched using Improved Apriori, and hash tables are employed to count common 1-item groups. The database was then searched for the following standard item sets.

Extraction of Frequent Itemsets using Apriori (Shayegan and Namin, 2021), This is a major problem for scientists. In an effort to discover a solution, researchers have proposed many solutions. Association Law One proposed schema is mining, which looks for links in transactional data. Locate frequent item sets first to find such linkages.

To identify frequent item sets, this study uses the Apriori technique and the Apache Spark distributed platform. We now offer an improved version of Apriori that prioritises the detection of MFIs. In dense datasets, the suggested approach beats algorithms like YAFIM, HFIM, and Apriori by an average of 38%. Make use of the Apriori algorithm (Tang et al., 2020). Apriori was used to find strict association rules, while support degree and confidence were used to find consumer purchase rules. A hidden association rule may be used to improve product layout by extracting important data and quantifying the degree of relationship between pieces. It may employ the degree of support rule to encourage purchases, the degree of solid rule to encourage impulse purchases, the degree of promotion rule to promote bundle sales, and so on. This research may assist shops enhance their company and service.

Intrusion detection with Hadoop Map Reduce Using a Priori Hashing Algorithm (Azeez et al., 2019), the intrusion detection solution uses the Hadoop Map Reduce architecture to discover and detect network intrusions using association rules. The KDD dataset was used to assess the solution's efficacy and dependability. Our findings show that our method detects network intrusions reliably and effectively. Frequent patterns in variable-bandwidth networks: distributional mining (Lin et al., 2019), a method for quickly mining common patterns across a large network. The proposed method may also discover the best number of computer nodes to efficiently use computing resources and load balance. The suggested method's execution efficiency and load balancing performance have been experimentally proven.

Rapid exploration of high-utility association rules (Mai et al., 2020), The mining of non-redundant high-utility association rules (NR-HARs). We search within a semi-lattice of mined high-utility itemsets for closed and generator itemsets. The constructed lattice is then used to generate rules efficiently. This novel approach outperformed previous approaches in terms of speed and memory use. The suggested method works well with external systems like the Internet of Things (IoT) and distributed computer systems. Many firms use IoT and other computer technologies to monitor data and make decisions. On-going data input through IoT or any other information system. The ability to foresee customer needs and make timely business choices allows management to anticipate consumer needs. The study of crypto currency behaviour in Hard times using association rules (Hernández et al., 2021), Time series were turned into transaction matrices and the Apriori algorithm was used to discover the association rules between various currencies, choosing whether the rules were comprised of the currencies' price or volume. It was split into two subgroups. We found that before the Bitcoin crash, prices dominated in determining the association rules and that after the crash, transaction volumes dominated. This is a neutrosophic association rule mining algorithm (Abdel-Basset et al., 2018). It analyses object membership, indeterminacy, and non-membership functions, resulting in an efficient decision-making system that considers all ambiguous association rules. To validate the approach, we compare it to neutrosophic mining. Using the provided technique increases the number of developed association rules.

Method for mining geographic points of interest (Lian et al., 2018). The rule rankings are generated immediately after finding the rule proportions. This article provides a new approach for ranking rules and estimating the percentage of items for each rule called Mean-Product of Probabilities (MPP). MPP selects the optimum rule based on the rankings generated by DBSCAN (Density-Based Scanning Algorithm with Noise). A visual representation of the rules is then employed to assess whether the two preceding conditions were met—evolutionary computation for mining association rules (Telikani et al., 2020). Traditional ARM's long processing times have been addressed by evolutionary computing-based ARM. Despite multiple studies, no complete evaluation of current evolutionary ARM techniques exists. We review recent ARM evolutionary computation research. This paper investigates the application of ARM methods in evolutionary computations. Evolutionary ARM algorithms were grouped into four types: swarm intelligence-based, physics-inspired, and hybrid approaches.

On Spark, there's Adaptive-Miner (Rathee and Kashyap, 2018). Hadoop is a well-known open-source software system that uses Map Reduce to store and analyse huge datasets across cheap hardware clusters. However, hadoop is inefficient because a highly iterative algorithm like Apriori requires a lot of disc I/O. These days, several map reduce-based parallel computing solutions are available. Spark, a framework for distributed computations, has piqued their curiosity.

As a result, we created adaptive-miner, a distributed association rule mining solution based on spark that employs an adaptive strategy to detect common patterns. Adaptive-miner adapts partial dataset processing. To reduce time and space complexity, adaptive-miner prepares execution plans before each iteration. Adaptive-miner adjusts to the dataset's characteristics. As a result, it outperforms current static association rule mining techniques. Association rule mining and social network analysis to identify technology influencers (Ampornphan and Tongngam, 2020), improved productivity, quality, cheaper industrial production costs, or higher-value commodities are all benefits of new inventions. Patent papers are considered as significant sources of information that lead to breakthroughs. Technological trends are predicted by experts' subjective experience. However, the enormous volume of text data in the obtained patent documents makes it difficult for those specialists to get reliable information. As a consequence, objective technical trend forecasting is becoming increasingly realistic. These include technological overview, investment volume, and technology life cycle. Also, technical attributes may be utilised to categorise patent documents to facilitate business decision-making. Patent data mining and social network analysis are the study's primary contributions.

Association rule mining for distribution terminal unit fault analysis (Zhang et al., 2021), an association rule mining method was utilised to analyse DTU faults. First, we looked at common DTU error causes. The Eclat algorithm's performance was compared to the FP-growth and Apriori algorithms utilising several DTU fault databases. Then an Eclat algorithm-based DTU fault analysis was proposed. The approach worked with a real DTU fault database. Finally, the method's effectiveness was shown. Gradual Association Rule Mining under Dynamic Threshold Aqra et al. (2019) is an approach that allows recovery of information that satisfies numerous thresholds without learning the technique from scratch. In tests, the proposed technique exhibited high accuracy and a processing time reduction of almost two-thirds.

Numerical association rules mining (Jaramillo et al., 2021) from a defined schema the variable mesh optimisation (VMO) meta-heuristic is a unique optimisation technique for numerical association rule mining. This work can categorise categorical data given a defined rule structure. Unlike earlier strategies that discredited continuous variables, ours optimises continuous variable intervals. The quality of the rules developed was compared to four population-based algorithms using an actual dataset. Uncertainty unlabelled landslide inventory reduction Using AI t-SNE Apriori association rules data mining and clustering (Althuwaynee et al., 2021). The three-phase approach was chosen. After examining spatial cluster patterns for landslide, no landslide, vulnerable slopes, and unlabelled features, the dimension was reduced using t-distributed stochastic neighbour embedding (t-SNE). Second, the Apriori technique based on association rule mining was used to find standard links in the inventory using landslide antecedent factors (derived from topography and land cover maps). Finally, images from Landsat TM (Thematic mapped) and ETM+ (Enhanced thematic map per) were utilised to verify the findings. These approaches may be used to classify and categories missing or outdated geographic information (reduced dependency on paid remote sensing sensors and field surveys).

Creating a multiple-criteria decision-making model for stock investing (Cheng et al., 2021), the data is collected and processed to produce decision trees and association rules. The analytical findings are used to build an investment decision model for investors who seek to lower risk while improving profits. This stock investing judgment strategy uses numerous factors. This study will assist investors in three ways. 1. Apriori and decision tree algorithms are used to investigate relevant data to discover implicit investment knowledge. 2. A stock selection model based on an effective investment decision model is constructed. 3. Implications of association rules for investment decision-making are improved. To predicting mid-long-term landslide movement which is used to the map reduce architecture to improve the speed of the Apriori algorithm (Guo et al., 2021). Comparing the computational efficiencies of the I Apriori MR algorithm to the original Apriori MR technique, the IApriori MR algorithm was shown to be more efficient in processing large-scale data than the original Apriori MR method. A mid-long-term early warning study of landslides in the three parallel rivers validated the method's viability. This means that IApriori MR can anticipate landslides more accurately than the FP-growth approach under the same conditions. This method may help avoid and control landslides technologically. Ahead of capturing dynamic regional traffic congestion (Xie et al., 2019). Exploring the spatiotemporal linkages of regional traffic congestion IntraT-ST-Apriori addresses the static elements of regional traffic congestion by adding both time and spatial parameters. However, the InterT-ST-Apriori method solves the dynamic nature of regional traffic congestion by using both time and spatial features. Case studies for the Tianjin metropolitan road network are based on actual data. Apriori technique may reveal underlying linkages that underlie regional traffic congestion. This approach also shows the propagation patterns of congestion utilising internet-ST-Apriori.

Lin et al. (2016) created PSO2DT, a PSO-based algorithm for hiding sensitive item sets and reducing the sanitisation process's side effects. Each particle in the algorithm represents a group of unwanted transactions that have been later removed. To estimate the particles and reduce side effects, the algorithm created a fitness function. Using state-of-the-art GA-based approaches, it was able to effectively mask sensitive item sets with the discovered undesirable transactions. As compared to GA-based solutions, PSO2DT can provide better results for the sanitising issue if it examines a few more metrics. The method, on the other hand, treats optimisation as a single-objective issue.

With GA, Afshari et al. (2016) developed a multiple objective technique for hiding sensitive association rules. The goal of this approach was to support database security while increasing the value and confidence of the rules mined. A selection of association rules obtained from specific datasets were considered sensitive rules in this research. The method employed a binary transactional dataset as the input, with only minor modifications made to disguise the sensitive rules. Despite the fact that the modification has minor side effects, the end result is inaccurate.

In 2014, Khan et al. (2014) proposed an algorithm named improved GA with a new fitness function for hiding association rules. Foe result Khan et al. used two databases. Results showed their work has less information loss, lost rules and ghost rules.

In 2014 and 2015 Lin et al. (2014) proposed the cpGA2DT and sGA2DT, pGA2DT algorithms for hiding the sensitive itemsets by removing the victim transactions based on Gas by setting fitness and cross over and mutation function.

Lin et al. (2018) proposed a multi-objective algorithm for hiding the sensitive itemsets with strategy transaction delete. CS algorithm is another widely used metaheuristic search algorithm. Be inspired by necessitate offspring parasitism of some cuckoo species, in 2009; Yang and Deb (2009) proposed a CS algorithm to effectively solve optimisation problems. Afshari et al. (2016) presented cuckoo optimisation algorithm for association rule hiding (COA4ARH). In this paper, association rule hiding was achieved by the distortion technique. In addition, three fitness functions were defined, which made it possible to achieve a solution with the fewest side effects. A lot of experiments were conducted on the different database and the experimental results showed that COA4ARH hid all association rules and it had fewer side effects, such as lost rules (LR) and ghost rules (GR).

Research Gap

The following are the privacy preservation difficulties that existing algorithms face, which the proposed methodology aims to address.

* The most serious problem here is the occurrence of side effects during the sanitisation process: i) Misses cost ii) Ghost cost.

Proposed System Design

Figure 131.1 describe the proposed system architecture for the generation of hiding rules on extensive transactional dataset. In the execution in transactional dataset sometimes contains miss-classified install as well as null values; due to this problem, system not able to generate effective rules. To eliminate this problem, it needs to pre-process and normalise the entire dataset before generating the classification rules. The hash base Apriori classification algorithm has been used to generate the frequent itemset. Those generated rules are stored in a local repository and applied hide rule algorithm on sensitive rules. Before applying this algorithm, we first need to identify those rules which contain sensitive information or sensitive attributes. After the execution of the rule hiding algorithm, we get rules that known as Hide rules.

Nevertheless, in other hand the initial transactional dataset was processed with hiding sensitive rules according to identification from rule hiding algorithm. The sensitive hide rules are considered a sanitised database, and generate FIM by using hash-based approach. The generated rules from the hash-based approach and initially generated hide rules have been validated consequently. After validation, few rules are created, such as ghost and lost rules. According to both rule sets, we can identify the data loss and data quality on both datasets and evaluate the entire execution's performance analysis.

This function is used for calculating the fitness for generation of sensitive rules, using optimisation algorithm. The fitness function is having the ability to find actually similarity using evolutionary approach. The algorithm 1 executes this procedure frequent classification rules and generate sensitive rules.

The advantages of proposed system are as follows:

Fitness function is used for calculating the fitness score for all sensitive rules. Then according to the fitness score rule hiding algorithm hide the item to convert original datasets into sanitised dataset.

A. Algorithm 1 Fitness function for detection of sensitive rules

Input: Instances ARM [rule]={G1,G2.......Gn} with set of attributes, List FR rules

Output: Fitness score F(x).

Calculate each dbc fitness with FR rules using below function

$$f(x) = \sum_{i=0}^{n} dbc[i...n] \sum_{j=0}^{n} (FR[j])$$

Return f(x)

Among the records of ARM[rule], we are calculating here fitness score of each ARM record first, then sort all records base on fitness score. Finally create a hide rule are created with selective information.

B. Algorithm2. Hiding rules generation using Fitness function

Input: Database Db, FR Rules, Alfa, DI{A1, A2....An}
Output: DB' generated dataset
1: for (r' = A,B → C from FR != Null)
2: Update FR ← FR − r'
3: create DBc ← where -A, B, → −C
4: for each (Read dbc to DBc != Null)
5: Compute fitness using below formula
FF(dbc) = FF(dbc,FR)
  End for // Finish 4 for
6: display DBc in order set
7: while (Conf (r') > = Alfa * Conf(B→C))
  FirstRow= DBc[0] // Select first row from DBc
8: Update the each DB record using discriminatory itemset as dbc from −A to A.
9: Re-compute Conf(r')
  End while // Finish 7 while
  End for // Finish 1 for
10 : DB' == DB
11 : Return DB'

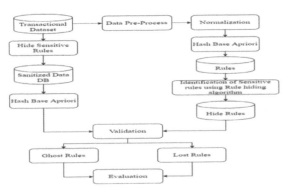

Figure 131.1 Proposed system of association rule hiding technique design

Dataset Information

The first experiment has used adult dataset which is taken from UCI Machine learning laboratory. data set having 48832 rows and for the train portion 32551 and for the test portion 16281 rows respectively. This dataset having 15 attributes with class attributes .class attributes is decision making attributes. Numerical attributes converted into categorical one so age above 30 and below or equal to 30 ((????) https://archive.ics.uci.edu/ml/datasets/adult.).

Result and Discussion

The quantitative assessment of the describes a data sanitisation technique using hash base apriori and fitness based rule hiding techniques and algorithms are presented We utilised the proposed algorithms Mar and Oo (2020) to get frequent classification rules, a standard approach for extracting frequent rules. The Java (JDK 1.7) software platform implemented all computations and utility measurements. The tests were carried out on a 2.7 GHz Intel Core i3 computer with 4 GB of RAM and Windows 7 installed. First, we went over the data sets that we used in our tests. The selected dataset was then pre-processed to reduce noise and random error in order to obtain an accurate result. The rules were then extracted using the Hash Based Apriori Algorithm and then sanitised dataset using the Rule Hiding Algorithm. Finally, we offer the outcomes of the various approaches' evaluations and comparison amongst them.

Misses cost (MC).The proportion of rules that can be retrieved from the information set but not from the converted data set is quantified by this metric. The following formula was used by MC to arrive at his conclusion.

$$MC = |FR| - |FR \cap FR'| / |FR|$$

Ghost cost (GC).This metric estimates the proportion of instructions that could be extracted from the converted data set and not from the actual database. GC determined using the formula provided.

$$GC = |FR| - |FR \cap FR'| / |FR'|$$

MC and GC are both essentially 0%. However, as a result of the process of change, MC as well as GC may not be 0%.

Figure 131.2 shows the information loss (MC and GC) with different alfa values. The six different experiments show the better identification alfa. Here for given input to this system as alfa=1.2, p=0.9 and DIs= {Sex = female, Age =Young} for given adult dataset. The y shows the percentage of information loss in whole data. DIs is itemset which is basically used for selection of sensitive attributes.

Conclusion

This work proposes new algorithms that inherit the qualities of hash apriori and a concealing rule generation approach algorithm that performs better at hiding rules. This program separates binary transactions information into blocks in order to create a sanitised dataset. The blocking strategy performs better when it comes to concealing rules. The alpha and beta algorithms additionally include additional parameters. Work to identify additional parameters and complexity of these methods in future studies. The experimental

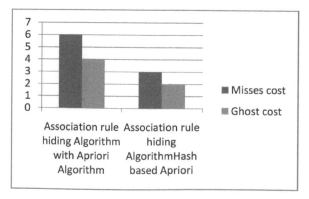

Figure 131.2 Information loss data sanitisation methods when alfa = {1.1 to 1.6}

research revealed that the system outperformed the state-of-the-art execution speed, utility, and correctness. In future studies, we hope to refine the suggested system's conceptualisation using other optimisation approaches, such as a parallel genetic algorithm, to investigate concealing a group of instructions in one optimizing run rather than one rule in each run.

References

Abdel-Basset, M., Mohamed, M., Smarandache, F., and Chang, V. (2018). Neutrosophic association rule mining algorithm for big data analysis. Symmetry 10(4):106.

Afshari, M. H., Dehkordi, M. N. and Akbari, M. (2016). Association rule hiding using cuckoo optimisation algorithm. Expert Syst. Appl. 64:340–351.

Althuwaynee, O. F., Park, Y., Aydda, A., Hwang, I. T., Lee, Y. K., Kim, S. W., and Lee, M. S (2021). Uncertainty reduction of unlabeled features in landslide inventory using machine learning t-SNE clustering and data mining apriori association rule algorithms. Appl. Sci. 11(2):556.

Ampornphan, P. and Tongngam, S. (2020). Exploring technology influencers from patent data using association rule mining and social network analysis. Information 11(6):333.

Angulakshmi, M. (2020). Association rule modeling using uml and apriori algorithm. In International conference on emerging trends in information technology and engineering (ic-ETITE), IEEE.

Aqra, I. (2019). Incremental algorithm for association rule mining under dynamic threshold. Appl. Sci. 9(24):5398.

Azeez, N. A., Misra, S., Ayemobola, T. J., and Damaševičius, R. (2019). Network intrusion detection with a hashing based Apriori algorithm using Hadoop MapReduce. Computers 8(4):86.

Cheng, K. C., Wang, K. H., and Fu, C. K. (2021). Establishing a multiple-criteria decision-making model for stock investment decisions using data mining techniques. Sustainability 13(6):3100.

Cong, Y. (2020). Research on data association rules mining method based on improved apriori algorithm. In International conference on big data & artificial intelligence & software engineering (ICBASE), IEEE.

Dong, S., Liu, M., and Zhang, S. (2019). Association rules mining of silk relics database with the RCFP-growth algorithm. In Chinese control conference (CCC), IEEE.

Ghazanfari, B., Afghah, F. and Taylor, M. E. (2020). Sequential association rule mining for autonomously extracting hierarchical task structures in reinforcement learning. IEEE Access 8:11782–11799.

Guo, W. (2021). Method for mid-long-term prediction of landslides movements based on optimised apriori algorithm. Appl. Sci. 9(18):3819.

Hernández, C., Benito, J., García-Medina, A. and Porro, M. A. V. (2021). Study of the behavior of cryptocurrencies in turbulent times using association rules. Mathematics 9(14):1620.

Isa, N., Neddy, S. K. and Mohamed, N. (2021). Association rule mining using FP-growth algorithm to prevent maverick buying. In IEEE 11th IEEE symposium on computer applications & industrial electronics (ISCAIE), IEEE.

Jaramillo, I. F., Garzás, J. and Redchuk, A. (2021). Numerical association rule mining from a defined schema using the vmo algorithm. Appl. Sci. 11(13):6154.

Kar, S. and Kabir, M. M. J. (2019). Comparative analysis of mining fuzzy association rule using genetic algorithm, In 2019 International conference on electrical, computer and communication engineering (ECCE), IEEE.

Khan, A., Qureshi, M. S. and Hussain, A. (2014). Improved genetic algorithm approach for sensitive association rules hiding. World Appl. Sci. J. 31(12):2087–2092.

Kreesuradej, W. and Thurachon, W. (2019). Discovery of incremental association rules based on a new FP-growth algorithm. In IEEE 4th International conference on computer and communication systems (ICCCS), IEEE.

Lian, S., Gao, J. and Li, H. (2018). A method of mining association rules for geographical points of interest, ISPRS Int. J. Geo-Inf. 7(4):146.

Lin, C. C. (2019). A distributed algorithm for fast mining frequent patterns in limited and varying network bandwidth environments. Appl. Sci. 9(9):1859.

Lin, C. W., Hong, T. P., Yang, K. T., and Wang, S. L. (2015). The GA-based algorithms for optimizing hiding sensitive itemsets through transaction deletion Eng. Appl. Artif. Intell. 42(2): 210–230.

Lin, C. W., Zhang, B., Yang, K. T., and Hong, T. P. (2014). Efficiently hiding sensitive itemsets with transaction deletion based on genetic algorithms. Sci. World J. 2014:1–13.

Lin, C. W., Zhang, Y., Fournier-Viger, P., Djenouri, Y., and Zhang, J. (2018). A metaheuristic algorithm for hiding sensitive itemsets, in proceeding of international conference on database and expert systems applications, Regensburg, Germany 492–498.

Lin, J. C. W., Wu, T. Y., Viger, P. F., and Lin, G. (2016). High-utility itemsets in privacy-preserving utility mining. Eng. Appl. Artif. Intell. 55:269–284.

Mai, T., Nguyen, L. T. T., Vo, B., and Yun, U. (2020). Efficient algorithm for mining non-redundant high-utility association rules, Sensors 20(4):1078.

Mar, Z. and Oo, K. K. (2020). An improvement of apriori mining algorithm using linked list based hash table. In International conference on advanced information technologies (ICAIT), IEEE.

Rahman, T., Kabir, M. M. J. and Kabir, M. (2019). Performance evaluation of fuzzy association rule mining algorithms. In 4th International conference on electrical information and communication technology (EICT), IEEE.

Rathee, S. and Kashyap, A. (2018). Adaptive-Miner: An efficient distributed association rule mining algorithm on Spark. J. Big Data 5(1):1–17.

Shayegan, M. J. and Namin, P. A. (2021). An approach to improve apriori algorithm for extraction of frequent itemsets. In 7th International conference on web research (ICWR), IEEE.

Tang, K. T., Sun, Y., and Lee, P. H. (2020). Apply apriori algorithm in supermarket layout research. In International conference on modern education and information management (ICMEIM), IEEE.

Telikani, A., Gandomi, A. H. and Shahbahrami, A. (2020). A survey of evolutionary computation for association rule mining. Inf. Sci. 524:318–352.

Tua, F. M. and Sunindyo, W. D. (2019). Software defect prediction using software metrics with naive bayes and rule mining association methods. In 2019 5th International conference on science and technology (ICST), (Vol. 1). IEEE.

Xie, D. F., Wang, M. H., and Zhao, X. M. (2019). A spatiotemporal Apriori approach to capture dynamic associations of regional traffic congestion. IEEE Access 8:3695–3709.

Yang, X. S. and Deb, S. (2009). Cuckoo search via Levy flights. In Proceeding of world congress on nature & biologically inspired computing (NaBIC), (pp. 210–214), Coimbatore, India.

Zhang, X., Tang, Y., Liu, Q., Liu, G., and Ning, X. (2021). A Fault analysis method based on association rule mining for distribution terminal unit. Appl. Sci. 11(11):5221.

Zheng, L. (2020). Research on E-commerce potential client mining applied to apriori association rule algorithm. In International conference on intelligent transportation, big data & smart city (ICITBS), IEEE.

132 Irrelevance of dimensional approach as applied to formulate equations in convective heat transfer

Narendra J. Giradkar[a], Jayant P. Giri[b], Vivek M. Korde[c], and Rajkumar B. Chadge[d]

Department of Mechanical Engineering, Yeshwantrao Chavan College of Engineering, Nagpur, India

Abstract

Investigators working in the field of heat transfer by forced convection had correlated their results in terms of dimensionless number. These correlations are unable to explain the phenomenon of heat transfer by forced convection; and were formulated earlier, due to non-availability of computing facilities. Since then, it has become a trend to present correlations in the form of dimensionless numbers. In this investigation we have shown that, how the above correlations had failed to explain the phenomenon of heat transfer by forced convection and gave erratic results when applied for same conditions. The data available in research paper were processed to form empirical correlations to determine heat transfer coefficient in the form of independent variables or parameters. For formulating the correlations in the form of independent variables, we had given different correlations for gasses and liquids as their microscopic structure is totally different. These correlations based on independent variables represent the phenomenon of heat transfer and may be used for designing heat exchangers, for developing the algorithm of software related to design of heat exchangers and in academics also.

Keywords: Dimensionless Number, Heat transfer, Heat transfer coefficient, Forced Convection.

Introduction

Dimensionless numbers are being used extensively in correlating experimental results on momentum, heat and mass transfer in pipe flow, mixing, fixed and fluidised beds and in many more operations. However, the investigators in this field did not realise that the indiscriminate use of dimensionless numbers can be conceptually wrong and may result in the wrong interpretation of experimental facts.

To explain the irrelevance of Reynolds and other dimensionless numbers in heat transfer studies and also to present a conceptual mechanism of forced convection in tubes, let us consider the Dittus Boelter equation

$$\left(\frac{hD}{k}\right) = 0.023 \left(\frac{Du\rho}{\mu}\right)^{0.8} \left(\frac{Cp\mu}{k}\right)^{b} \tag{1}$$

where 'h' is heat transfer coefficient, 'D' is diameter of pipe or tube through which fluid is flowing, 'k' is thermal conductivity of fluid, 'u' is the velocity of fluid, 'ρ' is mass density of fluid, 'μ' is dynamic viscosity of fluid and 'Cp' is specific heat capacity of fluid

Equation 1 is extensively used in designing heat exchangers for gases as well as for liquids. The value of exponent b is taken as 0.4 and 0.3 for heating and cooling fluids inside tubes respectively. The above equation, for heating fluid, can be written as,

$$h = 0.023 \left(\frac{Cp^{0.4} k^{0.6} u^{0.8} \rho^{0.8}}{D^{0.2} \mu^{0.4}}\right) \tag{2}$$

In these equations Cp, ρ, and k represent the thermal behaviour of the fluid whereas parameters D, u, ρ and μ control the turbulence in pipe.

---

[a]njgiradkar@gmail.com; [b]jayajtpgiri@gmail.com; [c]vmkorde@gmail.com; [d]rbchadge@rediffmail.com

If we critically examine these equations, they create confusion and not fit well in to well established concepts. For example,

(i)   In transfer processes thermal diffusivity defines the heat transfer capacity of the system. In pipe flow, the heat transfer process is expressed in terms of various independent variables as,

$$u\left(\frac{\partial T}{\partial x}\right) = \frac{k}{\rho Cp}\left[\frac{\partial^2 T}{\partial r^2} + \frac{1}{r}\left(\frac{\partial T}{\partial r}\right)\right] \tag{3}$$

Analytical solutions of equation 3 for streamline flow assuming parabolic velocity profile are available in literature. Because of complex nature of turbulent flow, such analytical solutions are not possible and we have to depend upon empirical relations. But in doing so we have to remember that the thermal behaviour of the system is controlled by thermal diffusivity $\left(\frac{k}{\rho Cp}\right)$. If this is true then the heat transfer coefficient should be directly proportional to thermal conductivity and inversely proportional to specific heat. Equations (1 or 2) does not agree with these basic concepts where exponent of specific heat capacity is positive.

(ii) Reynolds number $\left(\frac{Du\rho}{\mu}\right)$ is used as an index or measure of turbulence in pipe flow, for more than a century. It states that if we increase D, u, ρ; or decrease μ the turbulence will increase which in turn should increase heat transfer coefficient. Eagle and Ferguson (1930) even called it as a 'coefficient of turbulence'. But contrary to this understanding, if Reynolds number is increased by increasing diameter heat transfer coefficient decreases. Thus, from Table 132.1, correlations based on dimensional analysis or analogies shows decrease in heat transfer coefficient, if Reynolds number is increased from 5000 to 20000, by increasing tube diameter from 5 mm to 20 mm.

Table 132.1 Effect of pipe diameter on heat transfer coefficient

$$\left[\begin{array}{l} Calculated\ by\ correlations\ for\ conditions:\ G = 1000\frac{kg}{s-m^2},\ Cp = 4000\frac{J}{kg-K}, \\[2mm] k = 0.4\frac{W}{m-K},\ \mu = 1\times10^{-3}\frac{N-s}{m^2} \end{array}\right]$$

S/No.	Author	Correlation	Heat transfer coefficient $\left(\frac{W}{m^2-K}\right)$	
			D = 0.01 m, Re = 10,000	D = 0.02 m, Re = 20,000
01	Dittus-Boelter (1930)	$h = 0.023\dfrac{G^{0.8}k^{0.6}Cp^{0.4}}{\mu^{0.4}D^{0.2}}$	3662	3189
02	Gnielinsky (1976)	$h = \dfrac{3\ Cpf/8\left(1-1000/Re()\right)}{1+12.7\sqrt{f/8}\left(Pr^{0.67}-1\right)f/8}$	3602	3382
03	Martinelli analogy (1942)	$h = \dfrac{G\ Cp\sqrt{f/2}\left(Tw\text{-}Tc\right)/(Tw-Tb)}{5\left[Pr+\ln\left(1+5Pr\right)+0.5\ln\dfrac{Re}{60\sqrt{f/2}}\right]}$	3346	3128
04	Petukhov (1970)	$h = \dfrac{G\ Cp\ f/8}{K_1 + K_2\left(Pr^{2/3}-1\right)\sqrt{f/8}}$ where $f = \sqrt{1.82\times\ln_{10}}$, $K_1 = 1+3.4f,\ k_2 = 11.7+1.8\times Pr^{-1.3}$	3961	3532

There are three possibilities in the present situation

(a) Equations in Table 132.1 may not be representing the experimental facts.
(b) Reynolds number may not define the degree of turbulence. Or
(c) Dittus and Boelter (1930) and other equations may not represent experimental findings and also the Reynolds number may not have definite relation with turbulence.

Irrelevence of Dimensional Approach in Pipe Flow

Forced convection in tubes involve many independent variables like D, u, ρ, μ, Cp, k; many of which cannot be manipulated individually without disturbing others while conducting experiments. To tackle this complex situation, investigators in this field have taken the help of a mathematical tool called the dimensional theorem, which reduced the above mentioned six variables into two. To show that an ideal mathematical theorem when applied to a complex irreversible process like that of heat transfer may result into the wrong interpretation of experimental facts, we will discuss this approach in short.

Heat transfer coefficient is a function of turbulence defining parameters such as velocity, density, viscosity, diameter and thermal properties such as specific heat capacity and thermal conductivity.

$$h = f(D, u, \rho, \mu, Cp, k) \tag{4}$$

which can be expressed as

$$h = a\, D^b\, u^c\, \rho^d \mu^e\, Cp^f k^g \tag{5}$$

And for a dimensionally homogeneous equation i.e. the correlation coefficient 'a' to be dimensionless, with the help of dimensional analysis, the Equation (5) can be expressed as

$$h = a\, D^{c-1}\, u^c\, \rho^c\, \mu^{f-c}\, Cp^f\, k^{1-f} \tag{6}$$

or

$$\left(\frac{h\,D}{k}\right) = a\left(\frac{D\,u\,\rho}{\mu}\right)^c \left(\frac{Cp\,\mu}{k}\right)^f \tag{7}$$

The users of this technique were happy that they have converted Equation (5) with six independent variables into Equations (6 or 7) with only two independent variables viz. the Reynolds number and Prandtl number. But while doing so, they did not realise, that they have transformed four basic independent variables into dependent, which is a fundamental mistake. Because of this conversion of independent variables into dependent, it became necessary to study the effect of only two variables since the effect of remaining four was then fixed automatically without experimental verification.

Irrelevance of Reynolds Number in Fluidised Beds

The particle Reynolds number $\left(\dfrac{D_p\, u \rho_g}{\mu}\right)$ is used for defining dynamic state of the fluidised bed while correlating experimental results on fluid bed particle heat transfer. One such relation as suggested by Kunni and Levenspiel (1969) for diatomic gases is

$$\frac{h}{D_p\, k_g} = 0.03\left(\frac{D_p\, u\rho_g}{\mu}\right)^{1.3} \tag{8}$$

The minimum fluidisation velocity for fine particles when viscous forces are operating can be calculated by using Wen and Yu (1966) relation,

$$u_{\mathrm{mf}} = \frac{D_p^2\left(\rho_s - \rho_g\right)g}{1650\,\mu} \tag{9}$$

and is thus a strong function of solid density. Since the particle Reynolds number $\left(\dfrac{D_p u\rho_g}{\mu}\right)$ does not include the effect of solid density, two beds operating at the same value of particle Reynolds Number but having different solid densities, cannot have the same degree of turbulence. Thus the bed with lighter particles may fluidise vigorously whereas the other one with heavier particles may not fluidise at all even though the particle Reynolds Number is the same in two beds.

Another situation where the particle Reynolds number fails to define the dynamic state of the bed can be that of fluidised bed boilers operated under pressure. Since the basic intensive property of fluidisation is 'minimum fluidisation velocity', if we are interested in keeping the similar dynamic conditions in two beds operating at different pressures, the same gas velocity must be maintained in both the cases and not the same particle Reynolds number.

It can be concluded that the heat transfer results expressed in terms of Nusselt, Reynolds, Prandtl numbers in case of pipe flow as well as fluidized beds do not explain the heat transfer mechanism satisfactorily. As a matter fact the empirical correlations available in the literature are misleading and create confusion.

Heat Transfer Mechanism

When a fluid is heated inside the tube, the tube wall acts as a source of energy and the bulk of fluid the sink. The laminar boundary layer acts as a transfer medium. Naturally, in studying heat transfer mechanism we have to consider dynamic and thermal behaviour of both, the sublayer and the turbulent core, at various operating conditions.

In Figure 132.1, the flow structure at turbulent conditions consisting of laminar sublayer and turbulent core along with buffer layer is shown; and the overall turbulence depends upon the liquid particle mobility and eddy turbulence in sublayer and turbulent core respectively. The buffer layer is normally unstable and its behaviour unpredictable. We shall try to evaluate overall turbulence by analysing liquid particle mobility in sublayer and eddy turbulence in core separately.

Fluid particle mobility in laminar flow

When the Reynolds number is increased in streamline region, the fluid particle mobility increases and ultimately attains a mixed state at a critical value of 2100. Thus, in laminar flow, the fluid particle mobility is completely defined by Reynolds number and can be expressed by a relation:

$$\text{Fluid particle mobility} = f_1\left(\frac{D\,u\,\rho}{\mu}\right) \tag{10}$$

Another important parameter associated with turbulence in tubes is energy consumption or loss of potential energy in terms of pressure drop in Joules per kg per meter length of tube and is given by:

$$\frac{\Delta P_f}{\rho} = 32\frac{u\,\mu}{\rho D^2} \tag{11}$$

It is evident from Equation (10) that when the viscosity is increased, the fluid particle mobility decreases whereas the Equation (11) shows the increase in energy consumption with increase in viscosity. Therefore, the energy consumption cannot be taken as a measure of fluid particle mobility at streamline conditions.

The heat transfer coefficient in this region will increase with increase in mobility of fluid particles and thus will depend upon Reynolds number. Analytical solutions of Equation (03) given in literature assuming conduction as a mode of heat transfer with parabolic velocity profile express heat transfer coefficients in terms of Graetz number or Peclet number which do not include the effect of viscosity. However, since the fluid particle mobility decreases with increase in viscosity, the heat transfer coefficient should also show the similar behaviour.

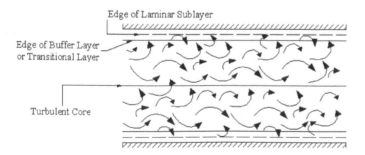

Figure 132.1 Flow structure of fluid in turbulent flow through a pipe

Eddy turbulence in turbulent core

At streamline conditions, when the viscous forces are operating, the fluid particle mobility in terms of Reynolds number is a measure of dynamic state of flow. At very high turbulence, when the effect of viscous forces is negligible, the inertial forces, which are independent of fluid viscosity, are predominant and thus the friction factor remains almost constant showing negligible variation with Reynolds number. Energy supplied for maintaining the flow is consumed by eddies. Therefore, the energy consumed by eddies in J/kg-m can be taken as a measure of eddy turbulence in the turbulent core.

The friction factor is function of Reynolds number and is given in numerous forms. However, the relation proposed by Drew et al. (1932)

$$f = 0.0014 + 0.125 \, \text{Re}^{-0.32} \tag{12}$$

is more conceptual since it indicates the effect of viscous and inertial forces distinctly. As per this equation at very high values of Reynolds number, the friction factor becomes constant and its value is 0.0014.

The pressure drop in pipe flow is given by the relation:

$$\Delta H_f = \frac{4f \, l \, u^2}{2g \, D} \tag{13}$$

Since we are interested in the core turbulence, the friction factor can be taken as 0.0014 and then the energy consumed per meter length of tube in J/kg is given by the equation:

$$\frac{\Delta P_f}{\rho} = 0.0028 \frac{u^2}{D} \tag{14}$$

The eddy turbulence, therefore can be expressed in terms of energy consumption by relation:

$$\text{Eddy turbulence} = f_2 \left(\frac{u^2}{D} \right) \tag{15}$$

It must be remembered that the eddy turbulence has no relevance to Reynolds number.

Overall turbulence

We have seen that the Reynolds number defines fluid particle mobility in laminar flow, whereas the eddy turbulence in the turbulent core depends upon the energy consumption and vice versa is not true. In most of the engineering problems, the heat exchangers are designed for high turbulence, but there is limit to the same because of excessive pressure drop. The flow structure at normal working conditions consists of laminar sublayer, buffer layer and turbulent core. The nature of the buffer layer is not well defined and therefore the behaviour of sublayer and core at various conditions is discussed for qualitative evaluation of overall turbulence in pipe flow.

The flow in laminar sublayer is laminar and therefore the mobility of fluid particles in it will depend upon Reynolds number or in other words will be directly proportional to tube diameter and fluid velocity and inversely proportional to viscosity.

Similarly, as discussed earlier, the core turbulence which depends upon energy consumption (Equation 15) will be directly proportional to fluid velocity and inversely proportional to tube diameter.

The controlling resistance in case of liquids lies in the viscous sublayer and therefore, to enhance the rate of heat transfer, the sublayer resistance has to be decreased which is effected by increasing eddy turbulence in the core or by increasing sublayer mobility. The eddies shown by circular arrows in Figure 132.1 strike the sublayer and thus create disturbance in it by giving part of their energy to the latter. Some of the eddies may even penetrate the sublayer and get absorbed in it allowing the equivalent amount of liquid to leave the sublayer and join mainstream. The eddies mix the warmer and cooler fluids so effectively that heat is transferred very rapidly between the edge of viscous sublayer and turbulent bulk of fluid. It is thus apparent that the most of the temperature drop between tube wall and bulk of liquid occurs in the sublayer. Naturally the fluid particle mobility in the sublayer, which is controlling the rate of heat transfer, will also depend upon the eddy formation in turbulent core. The overall turbulence in the pipe is thus the combined effect of fluid particle mobility in the sublayer and the eddy turbulence in the turbulent core and can be qualitatively represented by combining equations (10) and (15) as

$$\text{Overall turbulence} = f_3 \left(\frac{D \, u \, \rho}{\mu} \right) + f_4 \left(\frac{u^2}{D} \right) \tag{16}$$

from where it is evident that the effect of fluid velocity and viscosity on overall turbulence and consequently on heat transfer coefficient is normal. However, the tube diameter has an opposing effect on liquid particle mobility and eddy turbulence. As the tube diameter increases, the sublayer mobility increases and the eddy turbulence decreases. Thus, the overall turbulence can be affected when the tube diameter is increased in one of the following manners:

i.   Overall turbulence increases, if the increase in fluid particle mobility in sublayer is more than the decrease in eddy turbulence,

ii. overall turbulence remains constant, if the increase in fluid particle mobility in sublayer is equal to the decrease in eddy turbulence and,

iii overall turbulence decreases, if the fluid particle mobility in sublayer is less than the decrease in eddy turbulence

Experimental Results

When the literature on heat transfer in pipe flow and fluidized systems was searched it was observed that not a single author has studied the effect of basic parameters like diameter, viscosity, specific heat or conductivity systematically. All have expressed their results in terms of dimensionless numbers. Therefore, the effect of these fundamental variables was studied individually and that too for liquids and gases separately. Only the conclusions are given here.

(i)   Liquids in tubes (14)

    a)   Tube diameter range 9 mm – 20 mm

$$h = 287.5 \frac{G^{0.74} \ k^{0.9} \ D^{0.35}}{\mu^{0.3} Cp^{0.3}} \tag{17}$$

    b)   Tube diameter range 20 mm – 50 mm

$$h = 23.02 \frac{G^{0.58} \ k^{0.9}}{\mu^{0.3} Cp^{0.3} D^{0.6}} \tag{18}$$

(ii) Gases in tubes (07)

    a)   Tube diameter range 3 mm – 5 mm

$$h = 7.281 \frac{G^{0.76} D^{0.84} k^{1.34}}{\mu^{1.29} Cp^{0.63}} \tag{19}$$

    b)   Tube diameter range 5 mm – 76 mm

$$h = 0.0188 \frac{G^{0.75} k^{1.34}}{\mu^{1.29} Cp^{0.63} \ D^{0.3}} \tag{20}$$

As per Equations (17 and 19) in small diameter tubes, the heat transfer coefficient increases with tube diameter and thus viscous sublayer controls the heat transfer process. In larger tubes (Equation 18 and 20), the heat transfer coefficient decreases with increase in tube diameter and thus, as per Equation (16), the heat transfer is controlled by eddy turbulence.

(iii) Fluid – particle heat transfer (Giradkar, 1994)

    (a)   Fluidised systems

$$h = 1.702 \frac{k_g^{1.34} u_{mf}^{0.95} \rho_g^{0.95} D_p^{0.13} \left[1 + 2.25 \left(u - u_{mf}\right)\right]}{\mu^{1.29} Cp_g^{0.63}} \tag{21}$$

where 'kg' is thermal conductivity of gas, 'u_{mf}' is minimum fluidisation velocity of gas, 'ρ_g' is density of gas, 'D_p' is diameter of solid particle and 'Cp_g' is Specific Heat Capacity of gas

    (b)   Fixed beds

$$h_f = 1.702 \frac{k_g^{1.34} u_f^{0.95} \rho_g^{0.95} D_p^{0.13}}{\mu^{1.29} Cp_g^{0.63}} \tag{22}$$

The through discussion of these results is given in part (2) and (3) for liquids and gases respectively and in part (4) for fluidised and fixed beds.

Conclusions

i.   Reynolds number can be taken as a measure of fluid particle mobility at only stream line conditions.

ii. When flow is turbulent, overall turbulence depends upon sublayer mobility and core turbulence as given by Equation (16).

iii. Dimensional approach converts independent variables into dependent and hence use of dimensionless numbers in correlating experimental results must be avoided as it may lead to wrong interpretation as happened in case of convective heat transfer.

iv. Viscous sublayer acts as a medium of heat transfer between the bulk of fluid and solid wall and thus the heat transfer coefficient depends upon its thermal properties and dynamic state whereas the pressure drop is mainly dependent upon the turbulence in the core. Therefore, the use of momentum heat transfer analogies is illogical.

v.   The use of particle Reynolds number in fluidized system is conceptually wrong.

vi. The experimental results for liquids and gases are represented by Equations (17) through (20) indicate that for small diameters heat transfer coefficient increases with increase in diameter and for larger diameters it decreases with increase in diameter. Similarly specific heat has a negative effect on heat transfer coefficient. When we compare these results with correlations based on dimensional approach or momentum-heat transfer analogies (Table 132.1) where the heat transfer coefficient is shown to be a inverse function of diameter and positive function of specific heat, it can be concluded that the dimensional analysis or analogies failed to represent experimental facts and therefore their use in heat transfer is irrational.

References

Dittus, F. W. and Boelter, L. M. K. (1930). Heat transfer in automobile radiators of the tubular type. Berkeley, California: University of California Press.

Drew, T. B., Koo, E.C., and Mcadams, W. H. (1932). The friction factor for clear round pipes. Trans AIChE. 28:56.

Eagle A. and Ferguson R. M. (1930). The coefficents of heat transfer in tube for water. Proc. Ins. Mech. Eng. 2:9851035.

Giradkar N. J. 1994. Fluid particle heat transfer in fluidised beds as applied to drying of groundnut pods. PhD diss, Nagpur University, Nagpur.

Gnielinski V. (1976). New equations for heat and mass transfer in turbulent pipe and channel flow. Int. Chem. Engg. 16:339368.

Korde V. M. 2000. Effect of tube diameter and l/d ratio on heat transfer coefficient in forced convection. PhD diss, Nagpur University, Nagpur.

Kripalani V. M. 1997. Forced convection in tubes for gases. PhD diss, Nagpur University, Nagpur.

Kunni D. and Levenspiel O. 1969. Fluidization Engineering. New York (N Y):Wiley.

Martinelli R. C., Southwell C. J., Alves G., Graig H. L., Weinberg E. B., Lansing N. F., and Boelter L. N. K. (1942). Heat transfer and pressure drop for a liquid flowing in the viscous region through vertical tube. Trans. Am. Inst. Chem. Eng. 38:493530.

McAdams, W. H. 1954. Heat transmission. New York (N Y). McGraw Hill.

Patil P. D. 1989. Mechanism of forced convection in pipes. PhD diss, Nagpur University, Nagpur.

Petukhov B. S. (1970). Heat transfer and fluid friction in turbulent pipe flow with variable physical properties. Adv. Heat Transf. 6:503264.

Wen C. Y., and Yu Y. H. (1966). AIChE. J. 12:610.

Korde, V., Giradkar, N., and Giri, J. (2021). Formulation of empirical correlation for internal forced convection. Mater. Today: Proc. doi:10.1016/J.MATPR.2021.04.569

133 Parameters of microstructural investigations for novel materials

Bazani Shaik[a], P. Siddik Ali, and M. Muralidhara Rao

Professor, Mechanical Engineering, Ramachandra College of engineering, Vatluru, Eluru, Andhra Pradesh, India

Abstract

By its significant benefits over the traditional fusion welding technique of aluminium alloys, the solid-state Friction stir welding technology has enormous promise in the automotive, aerospace, and construction sectors. An experimental analysis of 'FSP of different alloys is addressed in this paper. To estimate the tensile properties, impact resistance, elasticity, and micro-hardness of different FSW joints, mathematical modelling equations are established. The process parameters are adjusted to provide the highest tensile strength and hardness characteristics possible. Post-weld heat treatment is performed, and the metallurgical characteristics of the FS welded are reported for various tool rotational speeds combinations. Alumina and its alloys are commonly utilised as nonferrous alloys for a wide range of industrial applications. Aluminium combines superior mechanical strength with lightweight properties, and as a result, it is progressively overtaking metal in industrial uses where the strength-to-weight ratio is crucial. Aluminium welding is primarily connected with a higher thermal expansion, solidification shrinkage, and the dissolving of hazardous gases in the molten metal during welding in traditional process of welding'.

Keywords: Environment-friendliness, industrial applications, new and novel materials.

Introduction

Similar joining of brass material in FSW Taguchi L_9 orthogonal array of rotational speed 2000 rpm, tensile strength 101 MPa and pressure upset is 151 MPa and parameter of weld analysis used GRA, ANOVA, S/N ratio on outputs prediction observed 0.814 of grade value 0.5% and improved strength and quality of joint weld (Shaik et al., 2019; Padhy et al., 2018). Dissimilar friction stir weld joints of AA6016T4 and ADC12 decrease fatigue strength and observed Microstructural structures at stir zone changes drastic and tensile strength decreased on offset tool change value (Uematsu et al., 2017; Guo et al., 2018). Al-Si-Mg alloy of high rotational speed, low welding speed temperature increases on weld line at 0.3 to 2.3 C/mm increased forces of plunge on welding speed and stress flow is higher and heat conduction loss is increased and variation forces of tool geometry and clamping system on forces of temperature mainly influenced quality of joints.AA6061T651 and AA5083H321 joining of FSW on fatigue test and defects of kissing bond 0.3 to 1.0 mm and defect toe flash and lap joint has low fatigue. Meng et al. (2018), Kawashima et al. (2018) investigated toe-flash defect is less for the effect of fatigue performance. As well as the lap joint did not perform the butt joint in detail (Ma et al., 2018; Wan and Huang, 2018). Dissimilar joining of 2060T8 and 2099T83 alloys of lap joints on weld speed 200 mm/min free defect joints on 50 to 400 mm/min and maximum tensile strength 495.4 Mpa and elongation 10%, welding speed 200 mm/min and rotational velocity 800 rpm with depth of plunge is 0.1 mm and similar on 93.1% and 80% of 2060 alloy of the cold lap, hook formed at nugget zone (Arcade et al., 2018; Liu et al., 2018). Studied on review of titanium material of strength is higher on weight ratio. The corrosive resistance is higher and aerospace industry on ability improves on manufacturability of temperature is high directly focuses on titanium sheets on FSW of both similar and dissimilar weld joints on selection of tool position, weld nugget cooling, composition material, flow material complex on tool travels, etc and focuses on properties of subsurface on mechanical properties and Microstructural evolution texture field of research recent development on current challenges (Ugender, 2018; Singh et al., 2018). Improved properties of fatigue on AA7075T73 alloy of specimens weld 106 cycles with tensile strength is 15 Mpa and stress amplitude is 200 Mpa and 105 cycles increased to 50 Mpa base metal due to heat effects on thermomechanical affected zone and heat-affected zone is decreased and laser peening of residual stresses is 330 Mpa and hardness is 24 HV and compressive residual stress of maximum is 310 Mpa at surface depth of 4 μm and regions compressed down depth is 50 μm surface and

[a]drbazanishaik@rcee.ac.in

properties of fatigue is improved for the prevention of fatigue cracks at surface (Choi et al., 2018; Shaik, 2022). AA6061T6 and Ti6Al4V alloys of plunge depth is small due to low input of heat and bonding are observed without compounds of layer is intermetallic at interface joint and improved ultrasonic on average grain size is decreased on fragmentize of titanium alloy of maximum tensile strength of joint is 236 Mpa and reached 85 % of AA6061T6 alloy at typical 54 Mpa and ductile fracture (Jagathesh et al., 2016; Yu et al., 2017) has at stir zone. Dissimilar joining of aluminum and steels has low welding heat and minimize formation extent of intermetallic compounds arc mainly to increase weld strength and a layer of transition, coating intermedia for improvement of good strength, the structure of composite quality is high for the success of a commercial. AA5052H32 alloy of joint for to improve quality and stress residual is low of rotational speed 500 rpm of tensile strength is maximum 208.9 Mpa and rotational speed 700 rpm, travel speed 65 mm/min of output response of tensile strength is (Kundu and Singh, 2016; Winarto et al., 2019) maximum 200 Mpa and axial load 7 KN, tilt angle 1.5 and equiaxed finer grains on stir zone and efficiency of joint is low and deformation of plastic is failed at joint (Chularis et al.., 2020; Bora et al., 2021). Joining of stainless steel lap welded sheet 3.5 % of weight NaCl of grain size 3 μm free defect lap joint and carbon steel of (Bhushan and Sharma, 2021) grain size.

Experimental Work

Computer numerically controlled (CNC) Friction Stir Welding is a special purpose machine with different tools used after studies of literature survey as depth on investigations used tools are taper threaded tool, straight cylindrical threaded tool using M2 grade super high-speed steel tool shoulder diameter of 18 mm and probe length is 6 mm. The base material chemical compositions for the elements of alloy measured on the American society of metals. The direction of a spindle rotation with the same direction of tool weld.

The AA7075T651 aluminium alloys with 6 mm thickness of plates are placed at the advanced side and AA6082T651 is at the retreating side for the improvement of mechanical properties in a welded joint. The Sampling plates are cutter machines for work pieces and shaping machines for samples that produce plates with uniquely flattering surfaces as a result of changing a mechanism of two paired edges. A butt weld is formed on holding materials with fixtures placed on the retreating side, with parameters of rotating speed, welding speed, tilt angle, and axial force examined. Testing are produced in accordance with American Society of Testing Materials (ASTM) standards, as specified in the central composite design matrix, and the specimens for testing are prepared in accordance with the ASTM standards.

Microstructural Investigations on Aluminium Alloys of on FSW by Using Three Different Tools

Figure 133.1 shows a De Wintor Inverting Trinocular Metallurgical Microscope being used to conduct microstructural studies on the aluminium alloys AA7075T651, AA6082T651. The specifications of a magnifying instrument are shown in Table 133.1. The common macrostructure of several aluminium alloys friction stir welded in different zones is seen in Figure 133.2.

Sample-micro-structural investigations on FSP

Figure 133.3a shows microstructures at 100 x magnification and etchant. The Keller's Reagent solvent is used. It has an annealed AA6082 parent metal microstructure with precipitate Mg2Si particles in a basic aluminium alpha solid solution. Eutectics granules roll in a clockwise direction. The main grains get longer.

Figure 133.3b shoulder zone of AA6082 dominant taper threaded tool components The sample illustrates the process parameter shoulder zone with microscopic fragmented AA6082 elements. The fusing of AA7075 materials is seen in the top layer.

Figure 133.1 Inverted Trinocular Metallurgical De Wintor microscope

Table 133.1 Specifications of metallurgical microscope

S.No	Name	Specifications
1	Magnification	50 X to 1000 X
2	Eye piece	Paired at 15 X and 10 X
3	Objective	5X, 10X, 20X, 25X, 45X, 50X ,100X
4	Power	12 V, 50 w halogen lamp
5	Shaft	X-Y direction with 360 degrees rotation
6	Polarisation	Polarizer prism
7	Microscale	Attached with ocular scale of 0.01 mm

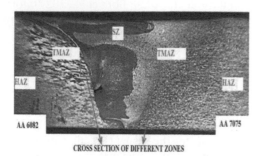

Figure 133.2 Shows a typical macrostructure of a distinct Al alloys FSW joint with different zones

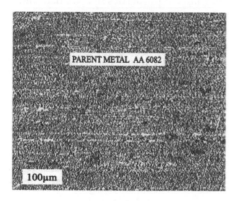

Figure 133.3a Annealed parent metal AA6082

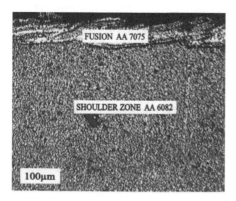

Figure 133.3b The shoulder zone predominant constituents

Figure 133.3c shows a TMT zone on the receding side of AA6082 for a TTT sample. The AA6082 model presented contains a thermomechanical transition zone containing directed flow.

Figure 133.3d displays AA6082, a nugget with an interface zone. The parent metal displays grain movement vertically, whereas the weld nugget on the left represents broken grains.

Figure 133.3e depicts the weld nugget region with fragmented and dynamic recrystallisation. Both metals have evolved a different layer. With grain sizes of 10 microns, re-crystallisation. Components have fragmented and dynamically recrystallised.

Figure 133.3f shows an alternative layering of alloy components with grain size measurements show that both components have fusion and homogenous re-crystallisation distribution.

Figure 133.3g shows the AA7075 interaction zone just on left as well as the nugget region on the right hand side of the advanced side process, as well as the morphology occurred. The advanced side of FSW process with of AA7075 parent metal is seen in Figure 133.3g. The source metal has a micro-structure in rolling temper. The sheet was cold processed by roll, and the primary particles of alpha aluminium extended in parallel with the orientation forming. Cu-Al2, Mg2-Si, Zn-Al2, and Mg-Al2 components expanded in lengths with the direction perpendicular of precipitation.

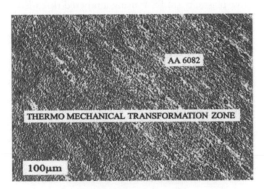

Figure 133.3c Thermo mechanical transformation zone TTT for samples

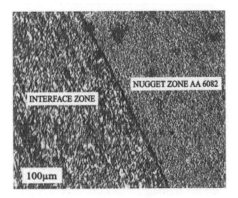

Figure 133.3d Samples of the interface region

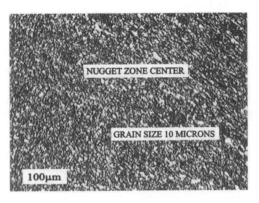

Figure 133.3e Using a tapering threaded tool to centre a nugget zone on FSW

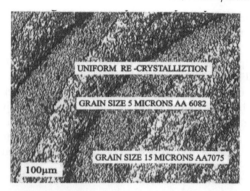

Figure 133.3f Samples of both components exhibiting effective plasticity on FSW using a tapered threaded tool

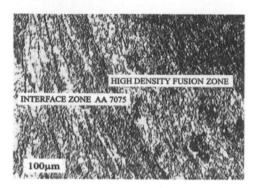

Figure 133.3g Diagram using a taper threaded tool, sample the interaction zone for AA7075 on FSW

Conclusions

This paper concludes the microstructures investigations to improve strength at different zones. Eutectics granules roll in a clockwise direction. The main grains get longer, and the eutectics form. Both metals have evolved a different layer. The top zone of the FSW is characterised by shattered grains and re-crystallisation, with grain sizes ranging from 10 to 15 microns. Fragmentation and dynamic recrystallisation have occurred in the components of two alloys. The Applications are very helpful for many industries benefit to the society resolve problems easily.

References

Arcade, G. R., Shukla, S., Liu, K., and Mishra, R. S. (2018). Friction stir lap welding of stainless steel and plain carbon steel to enhance corrosion properties. J. Mater. Process. Technol. 259:259–269. https://doi.org/10.1016/j.jmatprotec.2018.04.048.

Bhushan, R. K. and Sharma, D. (2021). Optimization of Friction Stir Welding parameters to maximize hardness of AA6082/Si 3 N 4 and AA6082/SiC composites joints. Silicon. 12:1–19.

Bora, B., Kumar, R., Chattopadhyaya, S., and Borucki, S. (2021). Analysis of variance of dissimilar Cu-Al alloy Friction Stir Welded joints with different offset conditions. Appl. Sci. 11(10):4604.

Choi, W., Morrow, J. D., Pfefferkorn, F. E., and Zinn, M. R. (2018). Welding parameter maps to help select power and energy consumption of Friction Stir Welding. J. Manuf. Process. 33:35–42.

Chularis, A. A., Rzaev, R. A., and Syndetov, M. K. (2020). Friction Stir Welding of aluminium and copper alloys. Weld. Int. 34(4–6):230–241.

Guo, S., Shah, L., Ranjan, R., Walbridge, S., and Gerlich, A. (2018). Effect of quality control parameter variations on the fatigue performance of aluminum Friction Stir Welded joints. Int. J. Fatigue. doi: https://doi.org/10.1016/j.ijfatigue.2018.09.004.

Jagathesh, K., Jenarthanan, M. P., Babu, P. D., and Chanakyan, C. (2016). Analysis of factors influencing tensile strength in dissimilar welds of AA2024 and AA6061 produced by Friction Stir Welding (FSW). Aust. J. Mech. Eng. OI:10.1080/14484846.2015.1093229.

Kawashima, T., Sano, T., Hirose, A., and Tsutsumi, S. (2018). Femtosecond laser peening of Friction Stir Welded 7075-T73 Aluminum alloys, J. Mater. Process. Technol. 262:111–122.

Kundu, J. and Singh, H. (2016). Friction Stir Welding of AA5083 aluminum alloy: Multi-response optimization using Taguchi-based grey relational analysis. Adv. Mech. Eng. 8(11):1–10. DOI: 10.1177/1687814016679277.

Liu, Z., Ji, S., and Meng, X. (2018). Joining of magnesium and aluminum alloys via ultrasonic-assisted Friction Stir Welding at low temperature, Springer-Verlag London Ltd., part of Springer Nature. Int. J. Adv. Manuf. Technol. https://doi.org/10.1007/s00170-018-2255-8.

Ma, Z., Jin, Y., Ji, S, Meng, X., Ma, L., and Li Q. (2018). A general strategy for the reliable joining of Al/Ti dissimilar alloys via ultrasonic-assisted Friction Stir Welding. J. Mater. Sci. Technol. https://doi.org/10.1016/j.jmst.2018.09.022.

Meng, X., Xu, Z., Huang, Y., Xie, Y., Wang, Y., Wan, L., Lu, Z., and Cao, J. (2018). Interface characteristic and tensile property of friction stir lap welding of dissimilar aircraft 2060-T8 and 2099-T83 Al-Li alloys. Int. J. Adv. Manuf. Technol. 94:1253–1261, doi 10.1007/s00170-017-0996-4.

Padhy, G. K., Wu, C. S., and Gao, S. (2018). Friction stir based welding and processing technologies - processes, parameters, microstructures, and applications: A review. J. Mater. Sci. Technol. 34:1–38.

Shaik, B. (2022). Investigations on microstructures by using Friction Stir processing, intelligent manufacturing and energy sustainability. Smart Innovation, Syst. Technol. 265. https://doi.org/10.1007/978-981-16-6482-3_53.

Shaik, B., Gowd, G. H., and Prasad, B. D. (2019). Experimental and parametric studies with friction stir welding on aluminium alloys. Mater. Today: Proc. 19:372–379. ttps://doi.org/10.1016/j.matpr.201907.615,

Singh, K., Singh, G., and Singh, H. (2018). Review on friction stir welding of magnesium alloys. J. Magnes. Alloy. 000:1–18.

Uematsu, Y., Kakiuchi, T., Mizutani, Y., Ishida, Y., and Fukunaga, K. (2017). Fatigue behavior of dissimilar Friction Stir welds between wrought and cast aluminum alloys. Sci. Technol. Weld. Joining. DOI:10.1080/13621718.2017.1361669.

Ugender, S. (2018). Influence of tool pin profile and rotational speed on the formation of friction stir welding zone in AZ31 magnesium alloy. J. Magnes. Alloy. 6:205–213.

Wan, L. and Huang, Y. (2018). Friction stir welding of dissimilar aluminum alloys and steels: A review, Springer-Verlag London Ltd., part of Springer Nature. Int. J. Adv. Manuf. Technol. https://doi.org/10.1007/s00170-018-2601-x.

Winarto, W., Anis, M., and Eka, F. B. (2019). Mechanical and microstructural properties of Friction Stir Welded dissimilar aluminum alloys and pure copper joints. MATEC Web of Conferences IIW 2018. Bali: Indonesia.

Yu, H., Zheng, B., and Lai, X. (2017). A modeling study of welding stress induced by friction stir welding. J. Mater. Process. Technol. doi:https://doi.org/10.1016/j.jmatprotec.2017.11.022.

134 Performance and emission characteristics of 4-stroke diesel engine with the influence of swirler using plastic pyrolysis oil as fuel

Sundarraj Moorthi, Meikandan Megaraj, and Raja Thandavamoorthi

Department of Mechanical Engineering, Veltech Rangarajan Dr. Sagunthala R & D Institute of Science and Technology, Chennai, India

Abstract

Energy creation and energy recovery from waste is an indigenous process. Plastic waste is an ideal energy resource because of its heating value, and it can be converted into oil through the pyrolysis method. The present investigation is to study the performance and emission characteristics of 4-stroke diesel engines powered by blended pyrolysis oil extracted from waste. Low-density polyethylene plastic wastes with various blends ratio of petroleum and pyrolysis oil like 50:50, 70:30, and 80:20. The characteristic was analysed and compared with standard diesel fuel operation by introducing a swirler in the engine's intake manifold and the testings were carried out by inducing and without inducing swirler. The emission parameters NO_x, CO, O_2, and CO_2, has been reduced to 4 to 5% in the engine when it is operated with standard diesel fuel with induced in swirler, and the emission parameters of NO_x, CO, O_2, and CO_2 is tested in the engine when it is operated with blended fuel showed near emission results of CO, O_2, CO_2 of engine operated in standard diesel fuel and 20 to 25% reduced NO_x Emission.

Keywords: Engine, emission, pyrolysis, swirler, waste plastics.

Introduction

The extreme usage of plastic material leads to the growth of plastic production and creates pollution to the environment. Several researchers carried out extensive research to identify the methods for reducing plastic pollution and recycling of waste plastic materials into a valuable product. Most of the researchers discussed the pyrolysis method and its advancement in recycling plastic material. Pyrolysis is the best method suitable for recycling a massive amount of plastic material. However, the resulting outcome from the pyrolysis process is pyrolysed oil, char, and gaseous products. Mangesh et al. (2020) discussed the suitability of polypropylene, low-density polyethylene, and high-density polyethylene oil extracted from the pyrolysis process (Mangesh et al., 2020). Kalargaris et al. (2017a) identified diesel engine power's combustion and performance characteristics with polypropylene and polyvinyl chloride plastics-based pyrolysed oil. The results showed 60–70% of pyrolysed oil blends would not affect the engines brake thermal efficiency. Still, it increases NO_x, UHC, CO, and CO_2 emissions (Kalargaris et al., 2017a). Kalargaris et al. (2017a) discussed the plastic pyrolyzed oil extracted from the same set of plastic at different temperatures. The result showed a higher amount of plastic oil blend leads to ignition delay and 34% of reduced brake thermal efficiency and higher emissions (Kalargaris et al., 2017b). Sarker et al. (2011) reported that catalysed distillation process would reduce the sulfur content in the pyrolysed oil. Astrup et al. (2009) evaluated significant greenhouse gases and emissions during recycling plastic waste materials (Astrup et al., 2009). Bajpai et al. (2009) discussed about Karanja based biodiesel blend showed reduced carbon monoxide emission level and decreases the engines brake thermal efficiency. Sundarraj and Meikandan discussed waste plastic-based pyrolysed oil into fuel for engine to reduce natural energy resources and reported that about 80% of waste plastics can be converted into useful petroleum products (Moorthi and Megaraj, 2021).

Bridjesh et al. (2018) experimented to identify engine performance and emission characteristics of plastic-based oil mixed with the additives of diethyl ether and 2-methoxy ethyl acetate. The resulting outcome showed NO_x emission is lower by 5.5%, and HC and CO emissions decreased by 5.9% (Bridjesh et al., 2018). Sukjit et al. (2017) studied butanol and DEE effects on diesel engines and showed lower HC, CO, NO_x emissions (Sukjit et al., 2017). Yilmaz (2013) studied swirl number and its effects on combustion characteristics, and it was observed that increasing swirl number causes increasing velocity, temperature, and improved gas concentration. Kaplan (2019) presented increasing swirl number in the combustion chamber causes turbulence intensity that will raise heat transfer near the unburned gas, leading to improved combustion performance. None of the work is discussed about improving engine performance by increasing air fuel proportion and fuel combustion. Samiran et al. (2019) experimented swirl characteristics of syngas combustion using the computational fluid dynamics (CFD) simulation method. The resulting outcome showed that increasing swirl number causing lower NO_x and CO emissions (Samiran et al., 2019). Also, a lower swirl number causing higher temperature leads to preignition and higher NO_x and CO emissions. Xiangrong Li et al., 2010 proposed a double swirl chamber to improve the engine's combustion performance. The outcome result showed an improved mixing level of air with fuel (Li et al., 2010). Giani and Dunn-Rankin, 2013 developed swirl vanes to improve the tangential velocity of flame in the combustion burner. The examined results showed the effective size of swirl vanes improves thermal transport properties and creates high tangential velocity near the wall surfaces promotes proper mixing of fuels (Giani and Dunn-Rankin, 2013). Reang et al. (2022) investigated the effect of biodiesel blended with rice wine alcohol, the experimentation revealed that the mixing of rice wine alcohol changes the physical properties of fuels like reduced flash and fire points. Dey et al. (2020) studied the performance and emission characateristics of an engine operated with palm oil and ethanol based diesel fuel. From the report it was identified that the increased hydrocarbon emission for higher loading conditions due to improper burn. Hence the engine should operate at an optimum speed for reducing nitrogen oxides formation (Dey et al., 2020).

Reang et al. (2020) experimented with the four stroke engine fuelled with neem methyl ester blended with rice wine alcohol and diesel blend. The outcome result revealed that reduced carbon monoxide and carbon dioxide level with improved nitrogen oxides and higer heat release rate (Reang et al., 2020) The engine operated with linseed methyl ester and diethyl ether mixed with diesel fuel and its consequence were identified by Reang et al. the performance test were carriedout on an constant engine speed it was identified that the blended fuel increases 4.48% brake thermal efficiency of engine with improved NOx percentage (Reang et al., 2019).

In this present work, the performance characteristics and emission analysis of direct ignition diesel engine were tested by using low-density polyethylene plastics based pyrolyzed oil (LDPEPO) blended with diesel like 20% LDPEPO + 80% diesel (LDPEPO80), 30% LDPEP0 + 70% diesel (LDPEPO70), 50% LDPEP0 + 50% diesel (LDPEPO50) were used as fuel. A swirler is induced in the engine's intake manifold to make the turbulence flow of air inside in engine, leading to a homogeneous mixture of air fuel.

Materials and Methods

A. Low-density polyethylene pyrolysed oil extraction

The low-density polyethylene plastic wastes are collected from the local municipal solid waste storage area. Those plastic materials are washed, dried, and shredded through a mechanical shredder machine in a particle size ranging from 2 mm² to 5 mm². The shredded plastic materials are utilised as feedstock material for the pyrolysis process. In the pyrolysis process, collected plastic materials (3 Kg) are heated up to 450°C and maintained temperature for 3050 minutes to get oil from the pyrolysis reactor. In addition, to improve the pyrolytic oil yield, the zeolite catalyst is added to the pyrolysis reactor after the reactor temperature reaches above 40°C. The collected LDPE oil (1.97 Kg) is filtered to remove ash and dust particles then the properties of extracted oil are analyzed. Table 134.1. shows the properties of standard diesel fuel and waste LDPE plastics based pyrolysed oil.

Table 134.1 Properties of diesel fuel and waste LDPE pyrolytic oil

Properties	Standard	Diesel	Waste LDPE
Density @30°C gm/ml	IS1448, P32	860	803
Kinematic viscosity @40°(cSt)	IS1448, P25	2.107	2.38
Flashpoint °C by Pensky-Martens Closed Cup methods	IS1448, P21	50	25
Fire point °C by Pensky-Martens Closed Cup methods	IS1448, P21	56	31.5
Cetane number	ASTM D976 - 91	50	67.64
Gross calorific value Kcals/kg	IS1448, P6	10150	9833

B. Development of swirler

The designed swirler was manufacture with the help of a 3-D printing machine, and acrylonitrile butadiene styrene (ABS) is used as filament material. ABS is a thermoset type polymer that will not get affected by the temperature presents in the intake manifold of the diesel engine. Before manufacturing, the swirler design was subjected to CFD analysis to find the feasibility and velocity outcome air from the designed swirler, leading to avoiding the failure of the designed model and wastage of the designed model manufacturing cost. Based on the outcome results from CFD, the swirler are manufactured, and the images are shown in Figure 134.1.

C. Experimental setup

The experimentation was carried out on a three-cylinder four-stroke direct injection diesel engine having a rated power capacity of 150 KW and an operating speed range of 29008000 rpm. The experimental setup specifications are listed in Table 134.2. It consists of a three-cylinder engine, Eddy current dynamometer, Engine Data Acquisition & control software (E-DACS), Throttle controller, fuel consumption meter, four gas analyser, and smoke meter. The engine was allowed to run for 5 minute before supplying each fuel blend so the engine can attain its perfect operating conditions.

Result and Discussion

A. Brake thermal efficiency

The brake thermal efficiency (BTE) of tested fuel with and without inducing swirler in the experimental setup is presented in Figure 134.2. The brake thermal efficiency of all fuel blends is comparatively increased about 4% when inducing a swirler in the engine's intake manifold due to increased mixing level of air with fuel, leading to improved combustion.

Preheating of fuel reduces viscosity, enhances air-fuel mixing, and provides complete combustion. Compared with the other fuel, the BTE of diesel is high because of the higher calorific value and lower cetane number (Venkata Ramanan and Yuvarajan, 2015). Enrichment of oxygen in the engine to improve its combustion performance, and its test result shows 48% increased brake thermal efficiency (Baskar and Senthilkumar, 2016). The BTE of pure LDPEPO100 is shown nearer value to the diesel value for all loading conditions.

B. Specific fuel consumption

Specific fuel consumption (SFC) defines the ratio between fuel consumption per hour to the brake power. The differences in specific fuel consumption with loads for fuel blends are shown in Figure 134.3. The SFC

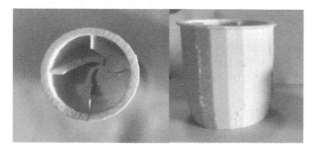

Figure 134.1 Developed swirler models

Table 134.2 Specification of the experimental setup

Engine type	:	*Three cylinder, four stroke, water cooled*
Bore diameter X stroke length	:	108 mm × 120 mm
Fuel tank capacity	:	3.3 Litre
Engine power	:	50 HP @ 2150 RPM
Torque of the engine	:	180 Nm
Firing order	:	1-2-3
Compression ratio	:	18.5 →: 1

of LDPEPO100, LDPEPO80, LDPEPO70, and LDPEPO50 was higher than standard diesel fuel at all loading conditions with the absence of swirler and the lower calorific value, which results in increased fuel consumption rate to maintain constant power output. After inducing a swirler in the engine's intake manifold, the fuel consumption rate for all fuel blends is comparatively lowered due to the increased swirl number.

C. Carbon monoxide (CO) emission

The carbon monoxide (CO) emission concerning different loading conditions is presented in Figure 134.4. The CO emission of engine induced with swirler is comparatively reduced 1.5% for all fuel blends. Due to increased combustion rate at a lower speed, the CO emission is lower when the engine's speed increases, the CO emission increases. Also, increasing the engine brake power leads to an excessive quantity of intake fuel and causing higher CO emissions. CO emission value of a high speed engine is higher when the load is also increased. The increased blending ratio of pyrolysed oil with diesel causes higher CO emissions (Verma et al., 2018). In the present study, it was showed lower values.

D. Hydrocarbon emission (HC)

The hydrocarbon emission concerning different loading conditions is presented in Figure 134.5. The outcome results for all the fuel blends are similar to the carbon monoxide emissions results of engine operated in standard diesel fuel. At lower speed and load, the hydrocarbon formation is low due to the lean mixture of fuel. Increasing speed and load on the engine amplifies rich mixture and causes lowered combustion performance and higher hydrocarbon emissions (Caliskan and Mori, 2017). The hydrocarbon emission level is comparatively low when the engine is induced with a swirler because of the improved combustion rate. The unburned fuel causes HC emissions which occur nearer to the cylinder wall because of insufficient temperature in this area, and the air-fuel mixture is lower than the center of the combustion chamber.

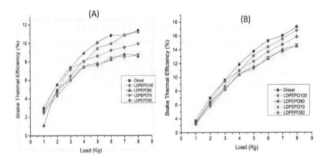

Figure 134.2 Brake thermal efficiency engine (A) without swirler (B) with swirler

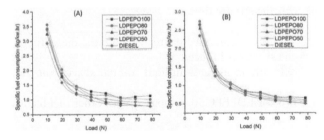

Figure 134.3 SFC (A) without swirler (B) with swirler

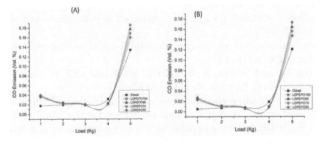

Figure 134.4 Carbon monoxide emission of the engine (A) without swirler (B) with swirler

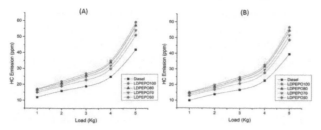

Figure 134.5 Hydrocarbon emission of the engine (A) without swirler (B) with swirler

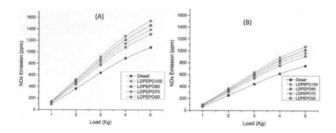

Figure 134.6 NO$_x$ emission of the engine (A) without swirler (B) with swirler

E.   Oxides of nitrogen (NOx) emissions

The NO$_x$ emission concerning different loading conditions is presented in Figure 134.6. It was witnessed that the NO$_x$ emission increases with loads on all fuel blends. The emission generated by the LDPEPO100 blend was higher compared with all other fuel blends. It happens because the temperature increases in the cylinder during the combustion process causing preignition of air fuel in the combustion chamber. Addition of silver oxide nanoparticles to the fuel will shorter ignition period also decreases NO$_x$ emission (Devarajan et al., 2018). The split injection of fuel to the engine significantly reduces NO$_x$ generation (Sindhu et al., 2018). In the present case inducing swirler in the intake manifold causes the homogeneous mixture of air-fuel, increasing the effective combustion rate. It is well known that the molecular weight of oxygen is higher than nitrogen. Hence, oxygen has a higher concentration on the perimeter of the boundary than that of the nitrogen, reducing NO$_x$ formation.

Conclusion

The present work aims at experimenting the performance and emission characteristics of low density pyrolysed oil with different ratios of diesel blend by inducing swirler in the intake manifold of diesel engine. From the experimentation results, the following conclusions are obtained.

- The brake thermal efficiency value of LDPEPO100 is maintained nearer value to the diesel fuel for all loading conditions, which is increased when the swirler is induced.
- The specific fuel consumption value is lowered with an increase in load compared with the value of SFC without inducing in swirler.
- The hydrocarbon, CO, and NO$_x$ emission is low for LDPEPO50 compared with other fuel blends.

References

Astrup, T., Fruergaard, T., and Christensen, T. H. (2009). Recycling of plastic: Accounting of greenhouse gases and global warming contributions. Waste Manage. Res. 27(8):763–772.

Bajpai, S., Sahoo, P. K., and Das, L. M. (2009). Feasibility of blending karanja vegetable oil in petro-diesel and utilization in a direct injection diesel engine. Fuel 88(4):705–711.

Baskar, P. and Senthilkumar, A. (2016). Effects of oxygen enriched combustion on pollution and performance characteristics of a diesel engine. Eng. Sci. Technol. Int. J. 19(1):438–443.

Bridjesh, P., Periyasamy, P., Vijayarao, A., Chaitanya, K., and Geetha, N. K. (2018). MEA and DEE as additives on diesel engine using waste plastic oil diesel blends. Sustain. Environ. Res. 28(3):142–147.

Caliskan, H. and Mori, K. (2017). Thermodynamic, environmental and economic effects of diesel and biodiesel fuels on exhaust emissions and nano-particles of a diesel engine. Transp. Res. Part D: Transp. Environ. 56:203–221.

Devarajan, Y., Munuswamy, D. B., and Mahalingam, A. (2018). Influence of nano-additive on performance and emission characteristics of a diesel engine running on neat neem oil biodiesel. Environ. Sci. Pollut. Res. 25(26):26167–26172.

Dey, S., Reang, N. M., Deb, M., and Das, P. K. (2020). Study on performance-emission trade-off and multi-objective optimization of diesel-ethanol-palm biodiesel in a single cylinder CI engine: A Taguchi-fuzzy approach. Energy Sources, Part A: Recovery, Utilization, Environ. Effects, 1–21.

Giani, C., and Dunn-Rankin, D. (2013). Miniature fuel film combustor: Swirl vane design and combustor characterization. Combust. Sci. Technol. 185(10):1464–1481.

Kalargaris, I., Tian, G., and Gu, S. (2017a). Combustion, performance and emission analysis of a DI diesel engine using plastic pyrolysis oil. Fuel Process. Technol. 157:108–115.

Kalargaris, I., Tian, G., and Gu, S. (2017b). The utilisation of oils produced from plastic waste at different pyrolysis temperatures in a DI diesel engine. Energy 131:179–185.

Kaplan, M. (2019). Influence of swirl, tumble and squish flows on combustion characteristics and emissions in internal combustion engine-review. Int. J. Automot. Eng. Technol. 8(2):83–102.

Li, X., Sun, Z., Du, W., and Wei, R. (2010). Research and development of double swirl combustion system for a DI diesel engine. Combust. Sci. Technol. 182(8):1029–1049.

Mangesh, V. L., Padmanabhan, S. Tamizhdurai, P. and Ramesh, A. (2020). Experimental investigation to identify the type of waste plastic pyrolysis oil suitable for conversion to diesel engine fuel. J. Clean. Prod. 246:119066.

Moorthi, S. and Megaraj, M. (2021). Indigenous development of single screw conveying machine for pyrolysis of waste plastics using nano zeolite particles in fixed bed reactor. Adv. Mater. Process. Technol. 1–13.

Reang, N. M., Dey, S., Deb, M., and Barma, J. D. (2022). Effect of diesel – biodiesel – alcohol blends on combustion, performance, and emission characteristics of a single cylinder compression ignition engine. Environ. Prog. Sustain. Energy 41(2):e13752.

Reang, N. M., Dey, S., Debbarma, B., Deb, M. and Debbarma, J. (2020). Experimental investigation on combustion, performance and emission analysis of 4-stroke single cylinder diesel engine fuelled with neem methyl ester-rice wine alcohol-diesel blend. Fuel. 271:117602.

Reang, N. M., Dey, S., Debbarma, J., and Deb, M. (2019). Effect of linseed methyl ester and diethyl ether on the performance–emission analysis of a CI engine based on Taguchi-Fuzzy optimisation. Int. J. Ambient Energy 1–15.

Samiran, N. A., Chong, C. T., Ng, J.-H., Tran, M.-V., Ong, H. C., Valera-Medina, A., Chong, W. W. F., and Jaafar, M. N. M. (2019). Experimental and numerical studies on the premixed syngas swirl flames in a model combustor. Int. J. Hydrog. Energy 44(44):24126–24139.

Sarker, M., Rashid, M. M., and Molla, M. (2011). Waste plastic conversion into hydrocarbon fuel like low sulfur diesel. J. Environ. Sci. Eng. 5(4).

Sindhu, R., Rao, G. A. P., and Murthy, K. M. (2018). Effective reduction of NOx emissions from diesel engine using split injections. Alex. Eng. J. 57(3):1379–1392.

Sukjit, E., Liplap, P., Maithomklang, S., and Arjharn, W. (2017). Experimental investigation on a DI diesel engine using waste plastic oil blended with oxygenated fuels. SAE Technical Paper.

Venkata Ramanan, M. and Yuvarajan, D. (2015). Performance study of preheated mustard oil methyl ester on naturally aspirated CI engine. Appl. Mech. Mater. 787:761765.

Verma, P., Zare, A., Jafari, M., Bodisco, T. A., Rainey, T., Ristovski, Z. D., and Brown, R. J. (2018). Diesel engine performance and emissions with fuels derived from waste tyres. Sci. Rep. 8(1):1–13.

Yilmaz, I. (2013). Effect of swirl number on combustion characteristics in a natural gas diffusion flame. J. Energy Res. Technol. 135(4).

135 Development of IoT enabled mini tiller for house gardening application

Rajesh, R.[1,a], Pratheesh Kumar, S.[1,b], Vignesh Kumar, N. R.[2,c], Jayasuriya, J.[3,d], Santhosh, B.[3,e], and Mathan Kumar, S.[3,f]

[1]Department of Production Engineering PSG College of Technology Coimbatore, India

[2]ME Product Design and Commerce Department of Production Engineering PSG College of Technology Coimbatore, India

[3]ME Manufacturing Engineering Department of Production Engineering PSG College of Technology Coimbatore, India

Abstract

In recent times, many agricultural industries have started using Internet of Things (IoT) technology to improve productivity in farming and for other reasons such as less human intervention, time and cost. Today a driving force behind increase in agriculture production is low cost IoT. The adoption of IoT solutions for agriculture is growing constantly and Business The mini tiller in the current market is powered by fuel engines or electricity. The main problem in the electric powered tiller is expensive and not affordable by farmers. Hence in this work, the design and development of IoT enabled mini tiller has been undertaken with the goal of commercializing technology at a low cost. The activities include development of concepts followed by selection of concept using a weighted matrix multi-criteria decision-making method. Finally, an effort is made to develop a prototype of the IoT enabled mini tiller machine. In this work power management data, humidity data are accessible from the sensors and it is stored in Google cloud data storage. These functions are controlled by the Arduino Uno board along with Arduino programming. Mobile access application is developed to control and monitor the IoT enabled mini tiller machine.

Keywords: Arduino, cost, IoT, mini tiller, sensors, weighted matrix multi- criteria decision making.

Introduction

Agriculture is the foundation of Indian economy. India's growth of agricultural products-based industries is of prime importance to the national economy. Much of the Indian population relies on the farming and agro-based industries. Lack of mechanisation is one of the main road blocks to increase agricultural productivity. Weeds are one of the key reasons to lose agricultural productivity. Weed is a routine term usually used to identify a plant that is deemed unwanted. The word weed is widely used in human-controlled settings for unwanted plants in fields and gardens. The advancement in the technology ensures that the sensors are getting smaller, sophisticated and more economic. The networks are also easily accessible globally for smart farming. Smart farming is the answer to the problems that the industry is currently facing. Smart farming is done using smart phones and Internet of Things (IoT) devices. The required information is collected by farmers from agricultural farms. The toothed form of mini tiller resembles the shape of chisel plower, but it serves a different purpose. The cultivators with minimum tooth are handled by a single person to take care of the small size gardens. The rotary tillers of similar size combine the functions of harrow and cultivator into a multipurpose machine. In most of the occasion, mini tiller is self-propelled or it is attached to the tractor with two wheels or tractor with four wheels. In case of tractors with two wheels, it is rigidly fixed and powered through couplings to the tractor's transmission. In four-wheel tractors, it is attached through a three-point hitch and power take-off supports in driving the vehicle. Power tiller is a prime mover in which the operator walking behind it takes the direction of travel and its regulation for field service. Power tiller is a walking tractor often used in puddle soil for rotating cultivation and can more efficiently replace animal power and help to increase employment in humans. The small and marginal farmers were the major users of custom power hiring tiller. A power tiller is the need of the hour for all farming operations such as initiation of tillage for small land holding farmers. The machine is the source self-employment opportunity for youth in rural areas. There is also an optional riding facility in some power tiller models. In small or medium-sized farms where four-wheel tractors application is limited there is an opportunity to

[a]rrh.prod@psgtech.ac.in; [b]spk.prod@psgtech.ac.in; [c]vigneshkumar845@gmail.com; [d]21mp01@psgtech.ac.in; [e]21mp02@psgtech.ac.in; [f]21mp31@psgtech.ac.in

use these mini tillers. The power tiller used to prepare seed beds in the low-land paddy fields. Power tiller is an important power source for different agricultural operations like preparing seed beds, sowing and fertiliser application. In situation of broad spaced row crops, tillers are also useful in intercropping and also useful in harvesting the cereal crops in upland conditions. Highest quantity of power tillers, used in India, are fitted with a rotary unit fitted at the rear end for forward movement and tillage operation. The protection of these tillers needs special attention. Power tillers that are referred to as walking tractors and alternate names such as single axle tractors, hand tractors were conceived as equipment for the preparation of seedbeds with rotary tillers and for transport. The tiller with two wheels has different attachments that is used to perform several kinds of farm labour such as harvesting, planting tillage and transport. As the tillage machine is mounted on a tractor with two wheels, it is referred as power-tiller. The tiller with two wheels is classified as tractors used for professional farming and gardening purpose. Small tractors used in rural areas are well suited for the gardening applications. Small tractor structure is simple and this enables easy service, maintenance and repair.

Literature Review

Shabbir et al. (2018) applied new technology to avoid problems faced in conventional time-consuming, hard-working and costly farming process. It was found that the machines used for farming in India are meant for high level farming purpose. The machines used in farms are costlier and not affordable to farmers. In this work, an effort has been made to overcome this issue. The farms which used this mini cultivator working model were found achieving the main aim of the work. Swapnil et al. (2015) developed a machine that would completely remove weeds and unnecessary crops that affect the yielding in the field. The wear of the power tiller is studied and service life improvement suggestions are provided in this journal. The power transmission to the wheels is through chain mechanism. Naque et al. (2013) designed a machine significantly to reduce the time and effort needed for development. Furthermore, this machine manufacturing cost is less than other machines available in market. In this work, new machine reduced the labour cost involved in conventional process for small-scale farmers. Thakur and Jagadale (2018) identified that multipurpose tool carrier with matching tillage tools is a cost-effective method for small farmers to plough their land. The reason for promoting this tool carrier is that it reduces the mechanisation cost involved in farming. It's a big time and money saver for field operations. Manivelprabhu et al. (2015) studied the existing model of the rotavator blade in the hyper mesh software and identified that shear, principal and Von misses stress are high in old design. The redesign changes reduced the above stress values compared to the older design. Rahman et al. (2020) found a new approach to reduce the wear in the worm gear of the rotary hoe transmission box. The minimum wear was observed in the gear bronze. Mandloi et al. (2017) found that addition of multipurpose tillage tool to the mini tractor supports soil bed preparation in the middle portion of Gujarat State. Dwivedi et al. (2018) focused on the different weeding techniques used for agricultural implants which will have a scope for future use. There are some side effects to the implants used in traditional methods. The successful properties of the solar-powered weeding machine are ideal for weeding purposes and have no side effects. Narang and Tiwari (2005) identified the power tiller's output factor. The demand for light weight power tiller in the market was considered as an important reason for this work. The parameters such as the fuel efficiency and field power are crucial in determining performance. Ademiluyi and Oladele (2008) discussed the VST Shakti owner tiller's field success on Sawah rice plots in Nigeria and Ghana. These study sites are located approximately 40 Km northwest of Kumasi on the Kumasi-Sunyani main road in the district of Ahafo Ano South, in the northwest part of the Ashanti region. The state of the art indicated that there is no work carried out on the IoT enabled tiller hence this work has been initiated as a scope of applying emerging technologies for agriculture problem.

Problem Definition and Objectives

The consumers using electric powered mini tiller face difficulty in accessing the electricity. The mobility of the vehicle cannot be controlled. The electric powered mini tillers available in the market are manually operated and costly. The cost of mini tiller in the market is expensive and not affordable by farmers. The objectives of the work are mentioned as follows.

1. To develop concepts of mini tiller with objective of cost minimisation.
2. To select best concept using weighted matrix method.
3. To understand and apply the IoT flow diagram for developing mini tiller prototype.

Methodology

The methodology of the work consists of activities involved in design and development of IoT enabled mini tiller machine. In this design and development process comprises steps from customer analysis to prototype development. Initially existing product data and customer requirements are collected. Thereby the concepts are developed based on the requirements and existing product. These concepts are modelled with the help of Creo design software. The selection of concepts is done using multi criteria decision making process that is weighted sum method. In this selection based on attributes of the products. The product next step is prototype development. In this prototype development IoT section is designed based on Arduino programming. In this section application is developed for controlling the machine. Data collected from this machine is stored in Google cloud storage and these data are accessed by using mobile phone or personal computer through programming as shown in Figure 135.1.

Concept Generation and Selection

Conceptual design is an important stage in the development of this mini tiller. This stage focuses on the development of model concepts for a tiller machine using various concept development methods (Arunachalam et al., 2014). Weighted matrix method is used to select the best generated concept based on a set of criteria. The morphological analysis methodology is used to merge the selected design concepts from a range of solutions obtained for individual components of the product through a structured search process.

In this activity, the solutions for the given problem are modelled using PTC Creo. It is found that CAD systems offer better visualisation characteristics compared to a 2-D sketch. The free hand sketches roughly provide an outlook of the concept solution for the given problem. The pencil is used mostly in the concept generation stage whereas CAD application is used in detail design such as assigning form and dimensions for parts. The better flexibility is obtained in pencil work but rather modelling supports better visualisation.

The concept 1 and 2 indicate the solutions modelled through a CAD package. The solution 1 shows the tiller blade position aligned in front of the CAD model. In the CAD model 2 the tiller blade is positioned at the centre of the mini tiller. In the concept 1, rubber tires are used as the source of movement whereas in the concept 2 metal wheels with serrations are used for source of movement. The motor positioning and frame design is different in two concepts. The initial rough sketching was done as a reference for modelling. The sketching in the product development process is considered as a link between generated idea and model developed. It provides better understanding of the solution for a stated problem. CAD system has such constraints which make them unable to be used during the conceptual design stage. In this current study as many solutions as possible was developed by means of rough sketching.

The CAD packages are the fast means to quickly develop the ideas and also support data retrieval at an early stage of design. In this work, CAD model developed is evaluated using concept selection tools to select the best concept. In this work, two concepts are developed as shown the Figure 135.2. For the

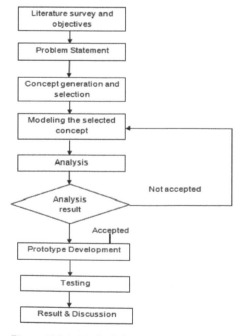

Figure 135.1 Methodology

a) Concept

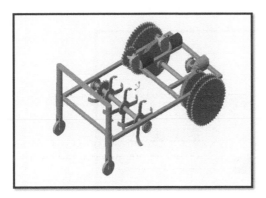

b) Concept

Figure 135.2 Concept models

concept selection process multi criteria decision making process is used. Pugh chart and weight matrix are two of the significant methods that is used to evaluate concepts in concepts selection process. In weighted sum method there will be multiple alternatives based on the multiple criteria. The below equation is called as weighted sum method. In this concept evaluation process, the models are ranked to identify the most effective concept based on a set of criteria. Otto and Wood (2012) considered evaluation criteria for concept decision mainly depends on the type of product to be designed. Hence differ from product to product.

$$A_i^{WSMscore} = \sum_{j=1}^{n} w_j a_{ji} \; for \; i = 1, 2, 3...m$$

The concept 2 has been selected based on the portability constraints and ease of manufacturing and assembly. The rigidity of the design also is better in concept 2.

Internet of Things

IoT is the means of communicating the physical objects that is provided with an IP address through internet connection to the main server for storage of data either in main server or in cloud. It is also called as IoT and Machine-to-Machine. IoT consist of three components. They are sensors, connectivity, peoples and processes. Internet of Things is used to connect devices embedded in various systems to the internet.

When devices can represent themselves digitally, they can be controlled from anywhere. The connectivity then helps to capture more data from many places ensure more ways to increasing efficiency. There are many challenges faced nowadays that is scalability, security, technological standardisation, software complexity and technical requirements. In this work IoT enabled mini tiller is developed with the help of sensors and Arduino programming. The flow diagram of IoT is shown in Figure 135.3.

Arduino UNO, Ultrasonic sensors, humidity sensor and soil moisture sensors are used to build an IoT enabled mini tiller machine. Data from these sensors are stored in Google cloud storage and these data are accessible from any time. For this mini tiller application is developed for controlling the machine with the help of MIT inverter. MIT Application Inventor is an intuitive, visual programming environment that allows everyone – even children – to build fully functional apps for smartphones and tablets. MIT application interface shown in Figure 135.4.

Embodiment design is a stage in the design process, where the design is developed from concept of a technical product in accordance with the economic and technical requirements where the further information will lead directly to the manufacture of parts in the product. The materials where selected based on the properties for the fabrication of IoT enabled tiller is shown in Figure 135.6. Finite element analysis was performed to ensure fail safe design for various structural elements in the mini tiller.

Fabrication

Mini tiller DC Motor is placed at the top of model and with the help of DC Motor chain drive pulley is rotated and which rotates the chain drive. The wheel shaft is rotated with the help of chain and wheel base to which blades are attached. Blades start its rolling motion, due to tractive effect, plough are moving forward in linear direction. The final CAD assembly with BOM is shown in Figure 135.5.

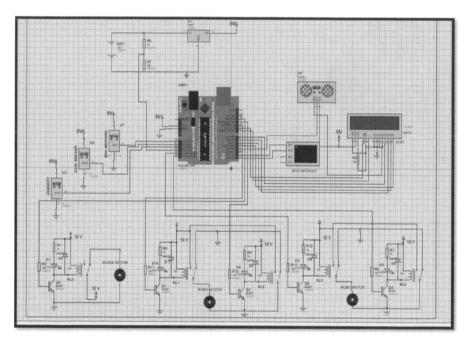

Figure 135.3 Schematic diagram for IoT enabled mini tiller

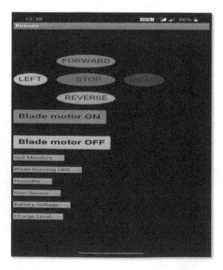

Figure 135.4 MIT application interface

The programming of Arduino UNO is done using C language where in the Arduino Software Integrated Development Environment (IDE) is used. The ATmega328 on the Arduino UNO incorporates pre burnt boot loader that provides access to upload new code. It is the use of an external hardware programmer. It is used to bypass the boot loader and program the microcontroller through the ICSP (In Circuit Serial Programming) header. Arduino IDE software is available for both Windows and Linux operating system. The functioning of the IoT enabled mini tiller is shown in the Figure 135.7.

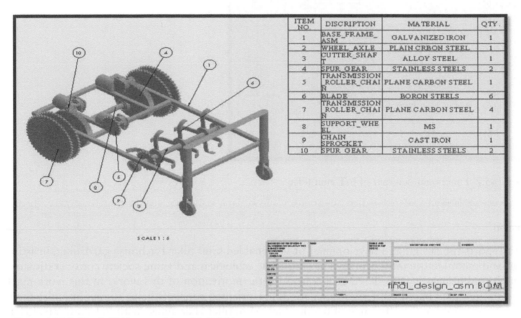

ITEM NO.	DISCRIPTION	MATERIAL	QTY.
1	BASE_FRAME_ASM	GALVANIZED IRON	1
2	WHEEL_AXLE	PLAIN CRBON STEEL	1
3	CUTTER_SHAFT	ALLOY STEEL	1
4	SPUR_GEAR	STAINLESS STEELS	2
5	TRANSMISSION_ROLLER_CHAIN	PLANE CARBON STEEL	1
6	BLADE	BORON STEELS	6
7	TRANSMISSION_ROLLER_CHAIN	PLANE CARBON STEEL	4
8	SUPPORT_WHEEL	MS	1
9	CHAIN_SPROCKET	CAST IRON	1
10	SPUR_GEAR	STAINLESS STEELS	2

Figure 135.5 Final CAD assembly with BOM

Table 135.1 Technical specification of mini tiller

Weight	20 kg
Input	12v
Torque	10 N-m
Drive shaft Diameter	9 mm
Rpm	1000 (rpm)

Figure 135.6 IoT enabled mini tiller

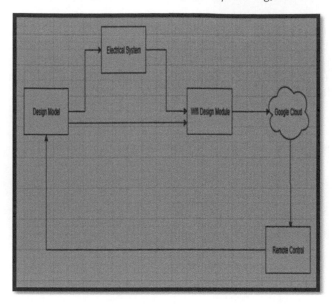

Figure 135.7 Functional diagram of IoT mini tiller

Conclusion

The work reports the design and development of IoT enabled mini tiller for house gardening, insisting on the design and manufacture until prototyping stage. The ecological and aging society consideration of various agriculture business is considered as a reason for the motivation of this work. In this work machine structures are designed based on design calculation. IoT oriented concepts helps to revolutionise the farm equipment's and improve the efficiency of the product. Mini tiller is designed with IoT concepts. IoT technology machine is operated by remote control-based application in mobile and also equipped with sensing various data like humidity, soil moisture, battery capacity from the sensors. The technical specification of minitiller is mentioned in Table 135.1. These data are stored in Google cloud with the help of Arduino programming. These data are retrieved and accessed at any time with the help of internet. This IoT enabled power tiller prototype is capable of primary and secondary tillage operation and is most suitable for house gardening. The developed prototype is tested and slight refinement of the equipment will support the customers in market.

IX.ANNEXURE

$$\text{Design load} = P_t \left(\text{Tagential force}\right) \times K_s$$

$$P_t = \frac{1020N}{V}$$

$$P_t = \frac{1020 \times 1.2}{10.58}$$

$$P_t = 115.68 \text{ Kgf}$$

$$\text{Design load} = 115.68 \times 3.5$$

$$\text{Design load} = 404.88 \text{ kgf}$$

$$\text{FOS} = \frac{\text{Breaking load}}{\text{Design load}}$$

$$= \frac{1000}{404.88}$$

$$\text{FOS} = 2.46 \cong 3 \left(\text{Design safe}\right)$$

Acknowledgment

The authors are thankful to the PSG Management and Department of Production Engineering for providing the facilities in the open laboratory for fabrication of prototype.

References

Ademiluyi, S. Y. and Oladele, O. I. (2008). Field performance of vstshakti power tiller Onsawah rice plots in Nigeria and Ghana. Bulg. J. Agric. Sci. 14:517–522.

Arunachalam, M., Prakash, A. R., and Rajesh, R. (2014). Foldable bicycle: Evaluation of existing design and novel design proposals. APRN J. 9:706–710.

Dwivedi, A., Doltade, A., Lahane, S. and Bhagat, A. (2018). design and development of solar power weeding machine. Int. J. Sci. Eng. Res. 9:235–237.

Mandloi, K., Swarnkar, R., Yoganandi, Y. C.. Patel, P. and Dabhi, K. L. (2017). Dabhi development and evaluation of a multipurpose tool bar for mini tractor suitable for the cropping pattern of middle Gujarat region. Int. J. Agric. Eng. 10:450–456.

Manivelprabhu, C., Sangeetha, N. and Ramganesh, T. (2015). Design modification and structural analysis of rotavator blade by using hyperworks 12.0. Altair Technology Conference.

Naque, M. A., Rizvi, A. A., Tijare, A. V. and Tupkar, A. B. (2013). Design, development and fabrication of soil tiller and weeder.Int. J. Innov. Eng. Technol. 2:367–370.

Narang, S. and Tiwari, P. S. (2005). Krishi Vidyapeeth 2004. Field Performance evaluation of a lightweight power tillers. J. Agric. Eng. 42.

Otto, K. and Wood, K. (2012). Product design techniques in reverse engineering and new product development. South Asia: Pearson Education.

Rahman, M.ur., Mahmood, Z., Ali, T., Mufti, M. A. I., Khan J. S. and Ahmad, M. (2020). Effect of gear composition and soil interaction on surface wear of worm gear of self propelled rotary hoe. Sarhad J. Agric. 31:22–29.

Shabbir, J. K., Kurkute, C. L., Saurabh, S., Lad, S. S., Jadhav, H. K. and Daithankar, T. R. (2018). Design, development and fabrication of mini cultivator and tiller. Int. J. Sci. Eng. 3:135–138.

Swapnil, K. L., Kadam, G. B., Jadhav, K. P., Gawade, V. S. Garje, A. and Gosavi, A. (2015). Design, development and operation of 3.5HP Power tiller. Int. J. Recent Res. Civil Mech. Eng. 2:149–154.

Thakur, N. and Jagadale, M. (2018). Design and performance evaluation of a low cost muti-purpose tool carrier with matching tillage tools. Int. J. Curr. Microbiol. App. Sci.

136 Post-earthquake condition survey, damage evaluation and rehabilitation of buildings: Case study of masjid al-rahman, ranau

A. M. Sarman[1,a], S.R.M. Tanin[2,b], E. M. Mazlan[3,c], and S.F. Suhadi[2,d]

[1]Faculty of Engineering, Universiti Malaysia Sabah, Green Materials and Advanced Construction Technology (GMACT) Research Unit, Universiti Malaysia Sabah, 88400 Kota Kinabalu, Sabah, Malaysia

[2]Faculty of Engineering, Universiti Malaysia Sabah, 88400 Kota Kinabalu, Sabah, Malaysia

[3]Development and Maintenance Department, Universiti Malaysia Sabah, 88400 Kota Kinabalu, Sabah, Malaysia

Abstract

Safety assessment or evaluation of damaged structures is a crucial part in the recovery process of a post-earthquake event specifically in the Ranau earthquake events. This research is done with the purpose of assessing a building affected by the Ranau earthquakes through the employment of visual condition survey, evaluating the damages sustained by the structural members of the building by using CSP 1 Matrix, JKR BARIS and CIDB QLASSIC, and suggesting rehabilitation strategies based on the data collected from the condition survey and damage evaluation. In this case study, three types of building condition assessment (BCA) – namely Condition Survey Protocol (CSP) 1 Matrix, JKR Building Assessment Rating System (BARIS) and CIDB Quality Assessment System in Construction (QLASSIC) – was employed to assess the damages sustained by Masjid Al-Rahman, Ranau due to the magnitude 6.0 and 5.2 earthquakes that hit the district in 2015 and 2018. A comparison of all three types of BCA showed that both CSP 1 Matrix and JKR BARIS which possess almost similar evaluation methods is the most suitable to be used in this case study as they enable for fast and thorough assessment. Although the CIDB QLASSIC provides a very detailed evaluation with regards to more diverse building types, it requires more time and skilled assessors. It is also more suited for the quality assessment of a newly completed development rather than a damaged premise. From the data obtained, the possible reasons for damage and failure of structure were discussed and possible methods for retrofitting and improving future constructions have been recommended.

Keywords: Building condition assessment, CIDB QLASSIC, CSP 1 matrix, JKR BARIS, post-earthquake, rehabilitation.

Introduction

Malaysia is situated entirely on the Sunda tectonic plate, between two major boundaries of the Australian Plate and Eurasian Plate in the west of Peninsular Malaysia, and the Philippine Sea Plate and Eurasian Plate at Borneo Malaysia (Hutchison, 2005). For a long time, Malaysia has been perceived as a country that is safely tucked away from major seismic activity.

For the most part, Malaysia is generally seismically stable with very few modern histories of volcanic activity. Some sleeping fault lines and semi-active fault lines in Malaysia which are located in the West and East of the country had been triggered from frequent earthquakes in neighboring countries such as Sumatera and the Philippines (Hamid and Mohamad, 2013).

The safety of building structures in Malaysia subjected under the effects of seismic loads exuding from earthquakes has perpetually been a concern to the general public. The matter in question has been brought back into the centre of attention ever since the 26 December 2004 earthquake in Banda Aceh which generated a series of large tsunamis up to a height of 30 metres. With an estimated death toll of 227, 898 (Szakács, 2021) across 14 countries in the Indian Ocean, the 2004 incident was subsequently named the 10th deadliest natural disaster in history.

Despite its proximity to the epicentre of the 2004 earthquake, Malaysia was spared the kind of damage that struck countries hundreds of miles farther away. The reported casualties were only limited to open coastal areas and there were no other fatalities among building occupants. However, there were a few reports of building swaying including structural and non-structural defects in the form of cracks.

[a]sr.asmawan@gmail.com; [b]saidatulradziah@ums.edu.my; [c]elis@ums.edu.my; [d]samirahnana99@gmail.com

This occurrence has increased fear of safety among occupants, especially residents of high-rise buildings. According to Abdul Hamid and Mohamad (2013), reinforced concrete (RC) structures in West Malaysia were hugely affected by past earthquakes varying from 4.4 to 9.4 Richter scale that occurred in Banda Acheh, Pulau Nias and Padang, Sumatera.

Furthermore, studies show that most construction industry stakeholders waste a significant amount of money searching for, validating, and recreating facility information that should be readily available. This lack of information must be filled in order to increase stakeholder satisfaction, and one possible way to improve asset knowledge is provided by Building Condition Assessment (BCA) which is a technical assessment by a professional assessor to evaluate the physical status of building elements and services and to determine facility maintenance needs is an example of this (Sarman et al., 2018). The condition survey protocol (CSP) 1 matrix is a method for assessing the condition of a building by analysing the flaws of a structure. JKR BARIS is almost same to CSP 1 Matrix, however the assessment is a little more precise than CSP 1 Matrix, which explains in particular evaluation. In order to address substandard workmanship quality in construction, the Malaysian government, through one of its bodies, the Construction Industry Development Board (CIDB), created the QLASSIC sometime in the late 1990s, which resulted in the establishment of the QLASSIC guideline (Sholichah et al., 2018).

Only until three years ago, after the 5 June 2015 6.0 Richter scale earthquake struck a small town in Ranau, Sabah around 7.15 am has the awareness of seismic effect on structural members of buildings really heightened (Cheng, 2016). Dr Kerry Sieh, Director of the Earth Observatory of Singapore (EOS), at Nanyang Technological University (NTU) said the earthquake that lasted for about 30 minutes was registered as the strongest to hit Malaysia since 1976 with 18 reported fatalities. With the epicentre located 16 km Northwest of Ranau, the tremors were experienced by the residence in Kundasang, Tuaran, Tambunan, Kota Belud and Kota Kinabalu. The Malaysian Meteorological Department (MMD) confirmed 100 aftershocks following the incident.

These series of earthquake occurrences, either originated locally or through tremors felt from quakes in neighboring countries, have shattered the common perception shared by the society globally that Malaysia is a zone free of seismic crisis. Other than casualties resulting from these incidents, evaluation of structural and non-structural damages sustained by buildings especially those located close to the epicenter is of utmost importance (Hasgur, 2012).

More recently on 28 September 2018, a shallow, large earthquake struck in the neck of the Minahasa Peninsula, Indonesia, with its epicentre located in the mountainous Donggala Regency, Central Sulawesi. The 7.5 Richter scale quake was located 77 km away from the provincial capital Palu and was felt as far away as Samarinda on East Kalimantan and also in Tawau. The disaster also highlighted the risk of tsunami waves that may reach Sabah between 40 to 120 minutes with waves of one to five metre heights, as mentioned by Professor Dr Felix Tongkul et al. (2020). Dr Felix also mentioned that the threat of tsunami in Tawau is great as most of its topography is low and the district is exposed to tsunami waves from the Sulawesi Sea, which is also located close to the source of the tsunami.

The presented case study concentrates on the post-earthquake condition survey, damage evaluation and rehabilitation of Masjid Al-Rahman in Ranau, Sabah which suffered substantial damage due to the 2015 and 2018 earthquakes. Currently, the project of reconstructing the mosque has already started and parts of the structures have been demolished but there is a request to salvage parts of the failed structure for sentimental value (Mazlan et al., 2015). This shall be the purpose behind this particular study and the possible ramifications shall these structures be maintained or be part of the reconstruction.

This study is conducted to fulfil the objectives as portrayed as follows.

a) To assess a building affected by the Ranau earthquakes trough the employment of visual condition survey.
b) To evaluate the damages sustained by the structural members of the buildings.
c) To determine the rehabilitation efforts based on the data collected from the condition survey and damage evaluation.

A. Seismicity of Sabah

Sabah's seismicity is known nationwide where there are 95 earthquakes recorded between 1897 and 2011 with 50 earthquakes recorded to have occurred since the 1960s. During that period, the earthquakes in 1966 and 1991 have caused substantial damage. The recent most destructive earthquake is the magnitude 6.0 which occurred at 7.15 am on 5 June 2015. There was more than a total of 100 aftershocks with a staggering 33 aftershocks in the period of two days after the major one was triggered.

There are three seismic zones in Sabah which are Central-North Zone (West Coast of Sabah, Ranau, Keningau, Kudat and Telupid), Labuk-Bay – Sandakan Basin Zone (Sandakan, Kinabatangan and Beluran) and Dent-Semporna Peninsular Zone (Lahad Datu, Semporna and Tawau). These zones are as illustrated in Figure 136.1.

In general, the North Borneo region has a moderate rate of earthquakes influenced by the local tectonics with the largest reported earthquake occurred in 1923 with a magnitude of 6.9 (Simons et al., 2007). According to Khalil et al. (2017), Sabah, unlike most other Malaysian states, is characterised by a common seismological activity; generally a moderate magnitude earthquake is experienced at a roughly 20 years' interval originating from two major sources, either local (e.g. Ranau and Lahad Datu) or regional (e.g. Kalimantan and South Philippine subductions).

B. Masjid Al-Rahman, Ranau

The proposed premise, Masjid Al-Rahman, Ranau before the 2015 earthquake shown in Figure 136.2 has a long history of 40 years where it was constructed and designed as a mosque for the small town of Ranau which was able to cater up to 1,500 congregates at a time. According to Noor et al. (2019), Building owners and design team should consider environment condition throughout the year during the design and construction stage. Thus, the premise has gone through two major renovations and extension works which include new structures during the construction stage. However, there is no original drawing to indicate the renovation and expansion additional physical works.

The major earthquake of magnitude 6.0 that hit Ranau on 5 June 2015 had the mosque badly destroyed in certain areas and succumbed to deteriorations in other areas. The prominent visual evidences acquired by bystanders proved that the major deterioration is on the roof and its supporting structures (Sarman et al., 2015). Figure 136.3 shows the condition of the premise as of March 2018.

The new mosque reconstruction was launched in October 2017 by the former Chief Minister of Sabah, Tan Sri Musa Aman which would cater up to a congregation of 2,500 Muslims. Currently, the project has kicked off and parts of the structures have been demolished but there is a request to salvage parts of the failed structure for sentimental value. This study will determine the possible ramifications should these structures be maintained or be part of the reconstruction.

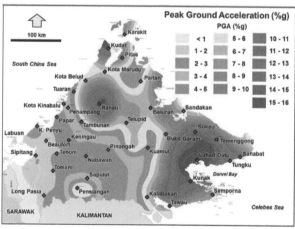

Figure 136.1 Peak ground acceleration (PGA) contour map of sabah

Figure 136.2 Masjid Al-Rahman, Ranau prior to the 2015 earthquake

Figure 136.3 Condition of premise as of March 2018

The proposed premise for the investigation is currently situated at 20 km away from the magnitude 6.0 earthquake epicentre on 5 June 2015, 25 km away from the recent magnitude 5.2 earthquake epicentre on 8 March 2018, 25 km away from concentrated zone of earthquake aftershocks and within the 49 km radius from most aftershocks following the 5 June 2015 earthquake.

Methodology

A. Desktop Study

Desktop study covered all three elements presented by the research title which are earthquake, condition survey and damage evaluation as well as rehabilitation of a building affected by earthquake (Mazlan et al., 2017). In regard to the aforementioned elements, the desk study included a detailed study on the geological features and the seismic history of Sabah in general and Ranau specifically. This included the causes of earthquakes in general, the geologic formations on which the location is founded upon, the fault lines nearby the proposed location and also a timeline of earthquake occurrences from as early as the recorded time.

A detailed study on the methods of building condition assessment is also crucial for this study so that a thorough job will be done when the time comes for the on-site inspection (Wahida et al., 2012). The condition survey protocol (CSP) 1 matrix is an assessment method for evaluating building condition. These protocols serve as a framework for Building Surveyors to analyse any building fault depending on priority and condition. This matrix has its own grading system to assist the examiner in thoroughly assessing the condition of the school facility (Sarman et al., 2018). Moreover, the Malaysian government, through one of its authorities, Construction Industry Development Board (CIDB) has created the QLASSIC to address substandard workmanship quality in construction. Physical evaluations will be carried out by the appointed assessors using relevant building inspection techniques (Bin Sulaiman et al., 2019). An in-depth understanding of the subject is needed where in this case the Condition Survey Protocol (CSP) 1 Matrix, JKR Building Assessment Rating System (BARIS) and CIDB Quality Assessment System in Construction (QLASSIC) were chosen as the condition assessment modules. The desktop study also encompassed rehabilitation strategies which include building pathology and structural integrity and failure. This information is useful in order to determine and suggest rehabilitation approaches based on the data collected from the building condition assessment.

B. On-Site Inspection

The condition assessment of the investigated building structures was conducted through on-site inspections. Inspections were made on the components which were requested to be maintained and be used with the newly constructed structures (Hairudin et al., 2020). Photographs of these components were taken as reference to the ratings given. The visual inspections were carried out according to the methods as presented by the CSP 1 Matrix, JKR BARIS and CIDB QLASSIC.

C. Data Analysis

Condition ratings obtained from the on-site inspection were computed using the CSP1 Matrix, JKR BARIS and CIDB QLASSIC ratings. These ratings will be used to suggest the suitable approaches for repair, rehabilitation of retrofitting of the structures. The data obtained has used to decide the ramifications should the proposed structures be maintained or demolished and be part of the reconstruction of the other parts of the structure.

Result and Discussion

A. CSP 1 Matrix Assessment

The information obtained from the assessment can be summarised in the CSP 1 Matrix schedule of building condition as presented in Table 136.1. From here it can be seen that the defects sustained by the building mostly fall into the yellow colour in the matrix which indicate the need for condition monitoring. Four assessed elements fall into the danger zone which is indicated by the red colour in the matrix that means serious attention is needed where some form of rectification should be done to avoid serious injuries (Yacob et al., 2016).

From Table 136.1, the overall building rating is obtained where the score 13.4 immediately places the premise under the red colour in the matrix which indicates the building is dilapidated and is in serious need of rehabilitation.

The overall summary of the assessment can be found in the Executive Summary which will follow the building condition schedule. The recommendation is decided based on the overall score of the building.

B. JKR BARIS Assessment

The data obtained from the assessment is summarised in Table 136.2 which shows the building assessment schedule for JKR BARIS assessment. From the table, it can be observed that the assessed elements fall into the average (grey) to very poor (yellow and red) assessment range in the matrix. Only two are observed to be in acceptable condition (blue) in the matrix.

The assessment method of JKR BARIS is almost similar to CSP 1 Matrix, with JKR BARIS having a five-point scale with a total score of 25 rather than a four-point scale with a total score of 20. This means that the JKR BARIS assessment is a little bit more detailed in its analysis as compared to CSP 1 Matrix, which explains why in this particular assessment there are two elements that fall into the blue colour matrix.

Table 136.1 Schedule of building condition (CSP 1 Matrix)

No.	Area	Defects	Condition Survey Protocol (CSP) 1				
			Condition Assessment [a]	Priority Assessment [b]	Matrix Analysis [c] (a × b)	Photo No./ (Sketch No.)	Defe Plan T
A	EXTERIOR	**Wudhu Area** 1 Failed column face	5	4	20	1	1
		Minaret					
		2 Failed dome supports	4	3	12	2	2
		3 Failed column face	5	3	15	3	3
		Main Foyer					
		4 Exposed rebar on column	4	3	12	4	4
		External Walkway					
		5 Failed rood beam	4	4	16	5	5
B	INTERIOR	**Main Area** 6 Failed ceiling	5	4	20	6	6
		Interior Columns					
		7 Honeycombing	3	3	9	7	7
		8 Major cracking	3	2	6	8	8
		9 Galvanization on dowels	4	3	12	9	9
		Roof (Interior Main Area)					
		10 Shear cracking on roof beam	4	3	12	10	10
		Total marks [d] (Σ of c)	134				
		Number of defects [e]	10				
		Total score (d/e)	13.4				
		Overall building rating	Dilapidated				

Table 136.2 Schedule of building condition (JKR BARIS)

No.	Area	Defects	Buliding Assessment Rating System (BARIS)				
			Condition Assessment [a]	Priority Assessment [b]	Matrix Analysis [c] (a x b)	Photo No./ (Sketch No.)	Defe Plan T
A	EXTERIOR	**Wudhu Area** 1 Failed column face	5	5	25	1	1
		Minaret 2 Failed dome supports	4	3	12	2	2
		3 Failed column face	5	5	25	3	3
		Main Foyer 4 Exposed rebar on column	4	4	16	4	4
		External Walkway 5 Failed rood beam	5	4	20	5	5
B	INTERIOR	**Main Area** 6 Failed ceiling	5	4	20	6	6
		Interior Columns 7 Honeycombing	3	4	12	7	7
		8 Major cracking	3	3	9	8	8
		9 Galvanization on dowels	3	3	9	9	9
		Roof (Interior Main Area) 10 Shear cracking on roof beam	4	3	12	10	10
		Total marks [d] (Σ of c)			160		
		Number of defects [e]			10		
		Total score (d/e)			16		
		Overall building rating			Poor		

The overall building rating shown in Table 136.2 indicates that the premise is considered poor in the matrix with the colour yellow and a total score of 16. The rating means that the premise is in need of rehabilitation to avoid further damages that may pose harm to anyone in the vicinity.

CIDB QLASSIC assessment

Only consider 56% of total architectural works (excluding roof, external finishing and external works). Pro-rate the percentage to find the breakdown weightage.

Internal finishes							
Floor	=	18	=	0/56 × 82%	=	0	
Internal wall	=	18	=	0/56 × 82%	=	0	
Ceiling	=	8	=	8/56 × 82%	=	11.71	
Door	=	8	=	0/56 × 82%	=	0	
Window	=	8	=				
Fixtures (internal)	=	8	=				
Roof	=	10	=	10/56 × 82%	=	14.64	
External wall	=	10	=	0/56 × 82%	=	0	
Apron and perimeter drain	=	4	=	0/56 × 82%	=	0	
Material and functional tests	=	20	=				
Total (Considered)		46/100				26.36	

From the analysis, it can be seen that the total considered score for the premise out of 100 in total is only 26.36%. This value is considered quite low in the QLASSIC standard in terms of quality of workmanship. However, this method is not done accurately since the assessment calls for a skilled assessor who has undergone a QLASSIC assessment training course with CIDB. Moreover, it should be kept in mind that QLASSIC is not the suitable method of assessment especially for the nature of the assessed premise (Sholichah et al., 2018). QLASSIC was developed with the intentions of being an independent quality assessment system to assess the quality of workmanship for building works (Norizam and Malek, 2013). In other words, this assessment method is most suited for newly completed developments where quality assessment is needed in detail for the purpose of obtaining occupational certificate (OC) and not so much for assessing premises that are affected by seismic loadings. Nevertheless, this does not mean that the score obtained is invalidated. The low score supports the observation where construction deficiencies are seen as the leading cause of most of the structural degradation that were observed. This indicates that workmanship for the mosque is not up to the design standards, let alone comply with seismic considerations (Che-Ani et al., 2014). Other than that, it can also be concluded that poor quality of workmanship can pose danger to a construction and the effect may only be detected once major deformities are shown.

Conclusion and Recommendations

- After assessing the damages sustained by the premise, it can be concluded that the premise has reached its ultimate limit states (ULS) and has experienced structural failure due to poor workmanship, poor construction joints and materials, construction was not according to current design criteria and/or specifications, and partly due to the premise being over 40 years of age.
- A comparison of all three types of building condition assessment revealed that the CSP 1 Matrix and JKR BARIS, have nearly identical evaluation methods which are the best fit for this case study because they allow for quick and thorough assessment. Although the CIDB QLASSIC provides a more detailed evaluation for a wider range of building types, it takes more time and skilled assessors. It is also better suited to evaluating the quality of a newly completed development rather than a damaged premise.
- Based on the information gathered, physical deformities show that the structural strength of the concrete is not sufficient to resist moments or forces.
- Furthermore, shear links were absent in most of the column reinforcements as well as the existence of inappropriate distance between nominal links.
- Adding to that, the premise was not designed to withstand seismic forces.
- Although some retrofitting works can be devised and may increase stiffness or rigidity of concrete components, such practice may unnecessarily add potential for structural failure despite periodical maintenance and changes (Samah et al., 2014).
- Such practice will only promote existing structures as elastic hinges which will act as a trigger for structural failure when integrated into a newer design which is mainly made up of highly ductile structural members.

All in all, it is strongly recommended that the existing structures not to be included in the newer design for it may cause structural collapse or reach the ULS envelope.

References

bin Sulaiman, S., Jusoh, A., Ying, K. S., and Soheilirad, S. (2019). Customer satisfaction in conquas and qlassic certified housing projects. J. Public Value and Adm. Insights. 2(1), 1017. https://doi.org/10.31580/jpvai.v2i1.478.

Che-Ani, A. I., Norngainy, M. T., Suhana, J., Mohd Zulhanif, A. R., and Hafsah, Y. (2014). Jurnal teknologi full paper building condition assessment for new houses : A case study in Terrace. J. Teknol. 70(1):43–50.

Cheng, K.-H. (2016). Plate tectonics and seismic activities in sabah area. Trans. Sci. Technol. 3(1):47–58. Retrieved from http://transectscience.org/

Hairudin, A. R., Che-Ani, A. I., Hussain, A. H., Sarman, A. M., and Mazlan, E. M. (2020). The development of Visual Basic.NET application utilizing condition survey protocol matrix. Test Eng. Manag. 15053–15064. Hamid, N. H. A. and Mohamad, N. M. (2013). Seismic assessment of a full-scale double-storey residential house using fragility curve. Procedia Eng. 54:207–221.

Hasgur, Z. (2012). A novel post-earthquake damage survey sheet: Part I- RC buildings. In 15th world conference on earthquake engineering (15WCEE).

Hutchison, C. S. (2005). Geology of North-West Borneo. https://doi.org/10.1016/B978-044451998-6/50023-5.

Khalil, A. E., Abdallah, N. M., Bashandy, G. M., and Kaddah, T. A. H. (2017). Ultrasound-guided serratus anterior plane block versus thoracic epidural analgesia for thoracotomy pain. J. Cardiothorac. Vasc. Anesth. 31(1):152–158.

Mazlan, E. M., Che-Ani, A. I., Sarman, A. M., Mydin, M. A. O., and Usman, I. M. S. (2015). Common defects on floating mosque: Tengku tengah zaharah mosque, kuala ibai, kuala terengganu. Appl. Mech. Mater. 359–362.

Mazlan, E. M., Che-Ani, A. I., Sarman, A. M., and Tawil, N. M. (2017). Analisa faktor kecacatan pada usia bangunan masjid terapung: Aplikasi matriks CSP1. J. Des. Built. 1(17):9–17.

Noor, J. M., Sarman, A. M., Che-Ani, A. I., Latiff, R. A., and Wahi, W. (2019). Identifying the critical components to extend concrete flat roof service life in equatorial climates: A review. Int. J. Recent Technol. Eng. 7(6):281–284.

Norizam, A. and Malek, M. A. (2013). Perception on quality assessment system in construction (QLASSIC) implementation in Malaysia. Malays. Constr. Res. J. 13(2).

Samah, A. H., Tawil, N. M., Mahli, M., Che-Am, A. I., and Abd-Razak, M. Z. (2014). Building condition assessment using condition survey protocol matrix: A case of school building. Res. J. Appl. Sci. https://doi.org/10.3923/rjasci.2014.565.572.

Sarman, A. M., Che-Ani, A. I., Mazlan, E. M., and Yahya, F. (2018). dome structural in coastal area condition assessment using csp1 matric application. Int. J. Adv. Mech. Civil Eng. 5(2):84–88. [ISSN 2394-2827] – Non-ISI/Non-SCOPUS.

Sarman, A. M., Mohd. Nawi, M. N. Che-Ani, A. I., and Mazlan, E. M. (2015). Concrete flat roof defects in equatorial climates. Int. J. Appl. Eng. Res. 7319–7324.

Sholichah, E., Purwono, B. and Nugroho, P. (2018). Exploring the potential of integration quality assessment system in construction (QLASSIC) with ISO 9001 quality management system (QMS). Int. J. Qual. Res. https://doi.org/10.1088/1755-1315.

Simons, W. J. F., Socquet, A., Vigny, C., Ambrosius, B. A. C., Haji Abu S., Promthong C., Subarya, C., Sarsito, D. A., Matheussen, S., Morgan, P., and Spakman, W. (2007). A decade of GPS in Southeast Asia: Resolving Sundaland motion and boundaries. J. Geophys. Res. 112:B06420. doi:10.1029/2005JB003868.

Szakács, A. (2021). Precursor-based earthquake prediction research: proposal for a paradigm-shifting strategy. Front. Earth Sci. 670.

Tongkul, F., Roslee, R., and Daud, A. K. T. M. (2020). Assessment of tsunami hazard in Sabah--Level of threat, constraints and future work. Bull. Geol. Soc. of Malays. 70.

Wahida, R. N., Milton, G., Hamadan, N., Lah, N. M. I. B. N., and Mohammed, A. H. (2012). Building condition assessment imperative and process. Procedia - Soc. Behav. Sci. 65(ICIBSoS):775–780. https://doi.org/10.1016/j.sbspro.2012.11.198.

Yacob, S., Ali, A. S. and Peng, A.-Y. C. (2016). Building condition assessment: lesson learnt from pilot projects. In MATEC Web of Conferences. https://doi.org/10.1051/matecconf/20166600072.

137 Vision-based methodology for monitoring anomalies in cast components using vision system and to control feed cycle in robotic deburring process

S. Kishorekumar[a], G. Rajesh[b], R.M. Kuppan Chetty[c], Joshuva Arockia Dhanraj[d], and D.Dinakaran[e]

Centre of Automation and Robotics (ANRO), Department of Mechatronics and Engineering, Hindustan Institute of Technology and Science, Chennai, Tamilnadu, India

Abstract

Sharp edges, burr, and fins in cast components are common difficulties in the casting industries due to poor deburring quality induced by positional mistakes and profile abnormalities and also robotic machining processes are affected by workpiece material and its physical properties like homogeneity. A vision-based monitoring system will give better results in the deburring process, thereby reducing the positional errors and accommodating profile anomalies to enhance the process control and at present the MATLAB is used for comparing master image with the defect image. Comparing to MATLAB Omron FHV7 is easy to interface with real tie applications and also it is industrial AI smart camera. The proposed system is capable of predicting the orientation and burr size of cast components through a template matching algorithm, where an Omron FHV7 is used and capture the image of the component of interest and compared with the reference image of the same from this information the number of feed cycles required for the Yaskawa MH5 6 axes robot is obtained. The novelty of this work is to interface vision sensor to deburring tool using PLC control system and layout matching technique utilises FH\FHV simulation software to find the abnormality in the cast component.

Keywords: Deburring, feed cycle, image processing, PLC control system, vision sensor.

Introduction

The utilization of robots for deburring in casting components had been restricted due to the formation of burr irregularity, a lack of real-time trajectory programming, the control system applied on the robot, the tolerances of the components, and inaccuracies in the movement of the robot. As a result, most of the deburring operations are carried out manually with low usage of CNC machinery or robots. The manual method is time utilizing, expensive, and requires a very high level of ability and knowledge to maintain uniformity (Kim et al., 2008). The majority of these manual processes utilise handheld rotary tools that cause more vibration which leads to health and safety issues and also may cause damage to costly components. Because of rigidity and accuracy, CNC machinery is often preferred (Minu and Shetty, 2015). However, the price of these machines is expensive for deburring cells, and they are rigid because of the fewer axes and range of motion. In contrast, robots are less rigid and precise but offer a greater workload and higher controlled axes with lower costs than CNC machinery (Kim and Chung, 2006; Tania and Rowaida, 2016). The regular deburring operation depends on the pre-programmed path. The major disadvantage of this approach is that it assumes the part is located correctly, located at a known position, and the robot proceeds along a hard programmed path; this technique shows the irregularity in deburring due to the incorrect position between the tool holding device and the predetermined path which is followed by the robot. This may result in leaving burrs at the component's edges or the cutting tool tip can be damaged. For complicated geometries, like splines and arcs, many points must be instructed besides the base for the robot to carry out a precise trajectory (Maini and Aggarwal, 2009).

As a result of using standard teach programming to automate the robotic deburring process is unsuitable when parts components get more complicated, and the system must be able to build the necessary robot trajectories using CAD models even though, the CAD prototype has no details about the irregularities, the path may not perfectly match the shape on the contour of the deburred surface due to changes in the part of the difference between the part edge and the exact path the robot has interpolated. The control of both motion and force has been established to avoid inaccuracies in the robotic position to assure that the

[a]Kishore9445138717@gmail.com; [b]rajeshthiru3@gmail.com; [c]kuppanc@hindustanuniv.ac.in; [d]Joshuva1991@gmail.com; [e]dinakaran@hindustanuniv.ac.in

cutting tool maintains touch with the workpiece at any time (Shrivakshan and Chandrasekar, 2012). Most of the studies in this area of the control methods could be classified into two categories: Hybrid position/frce control and impedance control (Chiou and Li, 2009). These methods demand precise details of the force interaction that connects the manipulators and the environment, and it is hard to achieve on ordinary industrial manipulators developed for position control (Guo et al., 2016; Matuszak and Zaleski, 2014).

The current force control technique for deburring has the inbuilt feature of leaving the deburred surface with an impression of the original and is unable to differentiate the position deflection of a large number of burrs and the end effector to find a solution to the problem.

Baskoro et al. (2008) suggested an impedance control method and a flexible algorithm find the cavities and the burrs on the components in the deburring process. The velocity can be altered when the deburring tool hits a large number of burrs and cavity defects, and the impedance parameters can be changed using a fuzzy inference system to enhance force control performance. Schimmels (Gu et al., 2013; Yang et al., 2019) has proposed a technique for enhancing the positioning capability and to improve the effective stiffness through multi-directional compliance and constraint of a robotic manipulator when the robot comes into contact with the workpiece for excessive deflection. To ensure compliance and to eliminate chatter caused by air compressibility, Xu et al. (2012) designed a new active pneumatic tool based on a single pneumatic actuator with a passive chamber. The actuators are directed to increase torque in the opposite direction of the deflection in this active control technique. Dewangan and Sahu (2020) developed a new method for modeling and controlling an automated deburring process that uses a dual-purpose complaint tool head with a pneumatic spindle that can be extended and retracted by three pneumatic actuators to enable tool compliance. To reduce random vibration generated by the contact of the force end-effector and workpiece during the deburring process, Daniali and Vassoughi designed an adaptive critic-based neuro-fuzzy controller (Lee et al., 2016). This method used here by nirosh jayaweera is to measure various points on the deburred portion and utilise these to build the robot path locally. Even if there are process errors such as robot inaccuracies, part geometry deviations from nominal, the robot can maintain an accurate edge contour (Wang et al., 2006).

The novelty of this work describes about detecting edges of the defective components or burr is by FH/FHV7 simulation software by uploading the master image or reference image to the vision system. It starts to compare the master image with the defect workpiece by using a template matching algorithm and the orientation, burr size, depth of cut, and no. of cycle is obtained. The trajectory plan for no. of feed, cycle is pre-programmed by the PLC control system it works based on the depth of cut and burr size. This analysis is displayed by Omron FH\FHV7 software.

The paper is categorised as follows Section 2 describes about the experimental setup of the proposed system while Section 3 describes about the methodology of deburring process and burr identification. This is followed by a discussion of the result and conclusion.

Experimental Setup

In this proposed system, the component is delivered through a conveyor, then picked up by a six-axis robot, which is then displayed to the vision sensor. The workpiece image is taken using the vision sensor, and the captured images are transferred to the Personal Computer for data acquisition and image processing are shown in Figure 137.1. In general, the vision sensor captures the state of the cast component. The

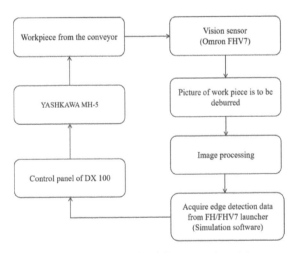

Figure 137.1 Process of the deburring using vision system

component image is acquired by the vision sensor and the images are delivered to the personal computer for data acquisition and image processing. The acquired images are compared to the master image or reference images in this process i.e, The camera has previously read the master image or reference image to validate parameters such as positional invariant.

The Yaskawa MH5 robot has a payload of 5 kg and it is a material handling robot it has both the controllers of Yaskawa DX200 and Motoman DX100. The MH5 can perform Tig and Mig welding. For the robotic application, the Motoman MH5 is perfect for automation and the robot's lightweight and compact base design is ideal for ceiling or wall mount installations, tabletop, and saving valuable floor space. This robot was mainly used for the deburring process in the project; it is used for pick and place. The YASKAWA MH5 robot as shown in Figure 137.2 is articulated in structure and has a maximum reach of 706 mm (27.79 inches) it consists of electrical service at the forearm and it is a six-axis articulated robot with repeatability of ± 0.02 mm.

The robot is controlled by a DX100 control system, and the data is communicated through PLC programming language and the PC is used for handling software. And also it has the main switch and the door lock which was located in the front and the programming pendant is to be hung by the hook below the button and in the upper right corner, an emergency button is installed. Below Figure 137.3 shows the front view of DX 100.

The programming pendant is equipped with the buttons and keys, used to conduct the manipulator teaching operations and then to edit the jobs. Below Figure 137.4 shows the Programming teach pendant.

The control group is used to control a group of axes at a time it is classified into three different categories, Robot is used as a manipulator, The coordinate systems are used to operate the manipulator, the coordinate system used here are explained as follow, in the joint coordinate system each axis of the manipulator moves independently, in Cartesian coordinate system manipulator of the tooltip moves parallel to the X, Y and Z axes. In the user coordinate system at any angle and the point, XYZ-Cartesian coordinates are defined. In a cylindrical coordinate system Θ axis is moved around the S-axis and the R-axis is to be moved parallel to the L-axis arm and the tooltip of the manipulator moves parallel to the Z-axis of the vertical motion, In the tool coordinate system the tool mounted in the effective direction of the wrist flange of the manipulator is defined as Z-axis this controls the endpoint of the coordinates of the tool.

Figure 137.2 Robotic arm with a deburring tool

Figure 137.3 Front view of DX100 control system

The vision sensor OMRON FHV7 100F smart camera as shown in Figure 137.5 is used to capture the defect components image and it has advanced inspection in the compact housing, it has expanded functionality, performance, the software tools, camera, communication, and more. The features of, the image processor have been built inside the camera unit in this series, In this camera, the sensor has the lightning of high power which is capable of lighting evenly across the view of a wide field. And it provides sufficient lighting even though a polarizing filter (closed) is used. lens focus can be adjusted to take the real image for the view of a specific field and need of installation distance I/O power supply connector is used to obtain inspection results in which the external output line changes the setup for the input line and the power supply line all to be combined to the one connector. In order to transfer the data, an Ethernet connector is provided, the input command from the PLC to control the FHV7 and the measurement result and the inspection result can be the output from the FHV7 to PLC. And then we can transfer the image to the computer and the camera is water-resistant. We can also use it in wet conditions. All the cables from the camera are flexible. And this allows sensors to be safely used on the moving parts. Smart click connectors are used to make connections easy and quick with the definitive clear click into the place mechanism.

The high power lighting is built-in LED and we have the perfect compact camera with the most challenging detection applications. The tool of color data in the camera can look at the color differences between the workpiece and the registered image of a good product. There are two different levels of judgment factors that can be helped to determine an overall pass or fail of a product. The first factor is a color deviation which is defined as the difference in the density between the workpiece break and the reference color. And the second factor is color difference which is defined as the difference between the workpiece color and the reference color. And the other features in the smart camera include 11 different images to refine the image for quicker more stable inspection. We can also save the multiple programs in the camera and recall them when the camera is changing, so we don't need the other camera to reteach the sensor with each variation. The camera is fixed on one side, and it scans the workpiece for the detection of burr. And after detecting the burr, it will send the signal to the PC, and using the FH\FHV7 simulator it will send another signal to the controller of the robot. And it will take a certain cycle of functions to remove the burr. The vision sensor with 2 megapixels enables high precision inspections for wider areas. It reduces the necessity of various cameras to capture different reference points. The autofocus lens has a focal length range of 59 mm to 2000 mm. Even when producing components are of various in size, the focusing length can be modified by

Figure 137.4 Programming pendant

Figure 137.5 OMRON FHV7 smart camera with backlight

parameters. This technique reduces the need for mechanical operation during product replacement, results in more effective system. Vision sensors may varies based on lens size. Schedule various heights of different components and switch between them during changeover. The FHV7 smart camera gives the preferable solution for the problems like increased cycle time caused by inspection point in additional to production quality. It doesn't need to split the angle of vision into various sections and configured to different cameras or high speed vision system can be installed.

Methodology

This approach describes about detecting edges of the defective components or burr is by FH/FHV7 simulation software by uploading the master image or reference image to the vision system. It starts to compare the master image with the defect workpiece by using a template matching algorithm and the orientation, burr size, depth of cut, and no. of cycle is obtained. The trajectory plan for no. of feed, cycle is pre-programmed by the PLC control system it works based on the depth of cut and burr size. This analysis is displayed by Omron FH\FHV7 software.The process of the robotic deburring is shown in Figure 137.6.

The process of identification of burr on the components will have several steps to follow as shown in Figure 137.7. This whole process is divided into two modules of discrete units of functionality each one of them is depending upon others. The first segment of the process is picking up the workpiece from the conveyor and correcting the orientation error next part is the identification of burrs, fins, and sharp edges. To pick the workpiece from the conveyor, the robot must know two variables, position, and orientation.

- The XYZ coordinates the position of the workpiece.
- The angle of orientation of the workpiece.

The above two variables will be found by observing the conveyor through the vision system and the image which is taken as input by the vision system will be processed through the Omron FH/FHV simulation software image processing to find the exact value of the coordinates as well as angle. These values will be given to the robot's control panel as the feedback. After picking up the workpiece from the conveyor, it will be shown to another vision system for further process. The image of the workpiece will be taken as input and processed by the Omron FH/FHV simulation software for burr identification.

Image acquisition is also known as Digital imaging. It is a digital encoded creation of the representation of the visual characteristics of the object such as structure (interior) or the physical scene of an object. The term is often assumed to include or imply the storage, processing, printing, display, and compression of each image. The main advantage of digital image over the analog image is such as film photography, it is

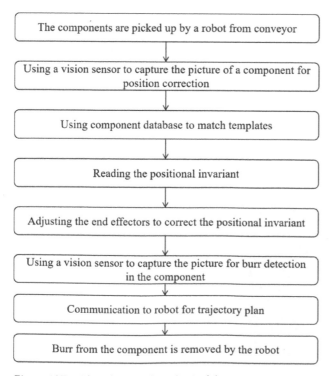

Figure 137.6 Flowchart of the robotic deburring process

the ability to make copies digitally indefinitely without loss of any image quality. The image acquisition is to be classified by the radiation of electromagnetic or other waves of attenuation of the variable, as they reflect off or pass through the objects. And it conveys the information of the image that constitutes. In all the classes of digital image acquisition, the given information is to be converted into the digital image signals by the image sensor that has been processed by the computer and output will be gained as a light image (visible).

Result and Discussion

In general, the vision sensor captures the state of the cast component. The component image is acquired by the vision sensor and the images are delivered to the personal computer for data acquisition and image processing. Below Figure 137.8 shows the defect and non-defect cast components.

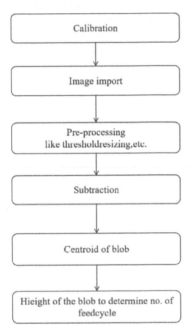

Calibration

Image import

Pre-processing like thresholdresizing,etc.

Subtraction

Centroid of blob

Hieight of the blob to determine no. of feedcycle

Figure 137.7 Process of burr identification

Figure 137.8 (a) Non-defect component (b) defect component

The acquired images are compared to the master image or reference images in this process. Below Figure 137.9 is a captured reference component in FH\FHV7 simulation software.

After feeding the master image in personal computer orientation process starts to acquire angle data as shown in the Figure 137.10 and Table 137.1.

The below mentioned Table 137.2 provide us the information about angle measured for orientation by guardian angle and also by vision sensor.

The camera has previously read the master image or reference image to validate parameters such as positional invariant. If the images aren't matched, the current image coordinates (X1, Y1) will be communicated to the robotic controller through the FH\FHV7 simulation software in the personal computer The robot's position will then change accordingly with the master image's orientation, and the robot will again display the component to the vision sensor to complete the orientation check and in order to detect the burr values B0, B1. Figure 137.11 and Table 137.3 shows the burr detection of component and burr coordinates.

When a burr is spotted, the coordinates of the burr are communicated to the robot controller through FH\FHV7 simulation software, As shown in Figures 137.12 and 137.13 which will initiate the robot's trajectory plan that is pre-programmed.. After completion of this process, it moves on to the deburring operation. The deburring tool is installed near the robot which rotates at 1000 rpm speed.

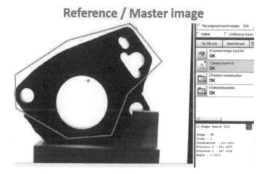

Figure 137.9 Master/reference image

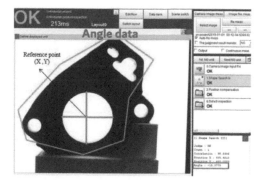

Figure 137.10 Orientation of cast component

Table 137.1 Orientation angle data

Shape search 3	
Judge	OK
Count	1
Correlation	90.6406
Position X	605.4219
Position Y	460.0509
Angle	−15.0779

Table 137.2 Angles measured for orientation

Sl.no	*Orientation*	
	Angle measured by guardian angle in degree	*Angle measured by vision sensor in degree*
1	10°	10.0584°
2	11°	11.0724°
3	12°	12.0687°
4	13°	13.0789°
5	14°	14.0249°
6	15°	15.0459°
7	–10°	– 10.0229°
8	–11°	– 11.0557°
9	–12°	– 12.0884°
10	–13°	– 13.0669°
11	–14°	– 14.0682°
12	–15°	– 15.0779°

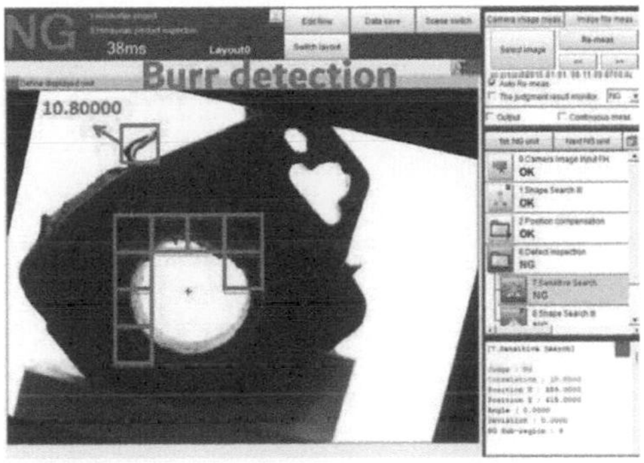

Figure 137.11 Burr detection of defect component

Table 137.3 Burr coordinates position in X and Y value

Sensitive search	
Judge	*NG*
Correlation	10.0000
Position X	889.0000
Position Y	615.0000
Angle	0.0000
Deviation	0.0000
NG Sub region	8

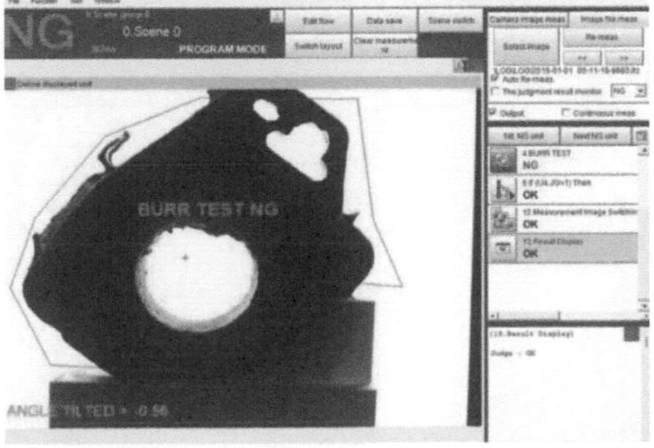

Figure 137.12 Burr testing result for defect material

Implementation of Robotic Deburring Cycle

Feed cycle for robotic deburring process (Figure 137.14) is done by feeding five different programs which was pre-programmed based on burr size to PLC control system so that the feed cycle of the robotic arm may change accordingly and number of feed cycle is obtained.

Below mentioned Equation (1) and Table 137.4 indicates the number of feed cycle which is obtained by measured burr and depth of cut.

$$K = \frac{MB}{D} \tag{1}$$

Where,
K = No. of feed cycle
MB = Burr measured using vision sensor
D = Depth of the cut

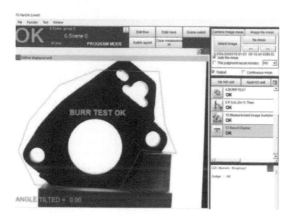

Figure 137.13 Burr testing result for non-defect material

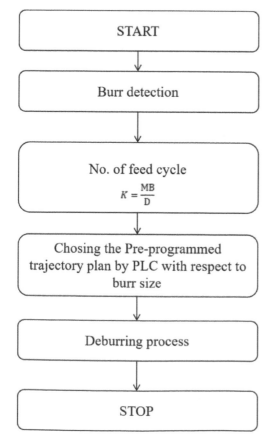

Figure 137.14 Implementation of robot deburring cycle

Table 137.4 Burr detected by vision sensor

Sl.no	Burr detection		
	Burr measured by vernier caliper in mm	Burr measured by vision sensor in mm	No. of feed cycle in mm
1	2 .4	2.2896	4.5792
2	2.8	2.7789	5.5578
3	3.5	3.1445	6.2867
4	4.2	3.9784	7.9568
5	4.6	4.2924	8.5848
6	5.8	5.4547	10.9094
7	6.2	6.0206	12.0412
8	7.8	7.7056	15.4112
9	8.5	8.2056	16.4112
10	10	9.8000	19.6

Conclusion

The robotic deburring process could be a revolutionary technology in the casting industry. Deburring results in advantages like quality consistency, high production, and safety in the workplace. Iron foundries will use this technology in the casting industry by taking bound precautions to confirm success. This develops the casting industry and fixes the new standards for workplace practices. This paper, presents a methodology to find the problem in the number of feed cycle, which is interfaced with PLC and vision system. The burr data measured by the vision sensor is uploaded to the FH\FHV7 software. To acquire edge detection, shape search, orientation, and burr detection is done, these data is sent to the PLC control system using Ethernet communication, where trajectory plan is pre-programmed based on the depth of cast component with different burr size, the program is generated as per the burr data. This technique is proposed and verified through the test using the robotic system. The robot will remove the burr based on the number of feed cycles which is programmed into the PLC control system. This technique gives a compact and cheap finishing robotic system that minimises the programming and the cast component fixture time on the worker in the foundry industries.

References

Baskoro, A. S., Kabutomori, M., and Suga, Y. (2008). Automatic welding system of aluminum pipe by monitoring backside image of molten pool using vision sensor. J Solid Mech. Mater. Eng. 2(5):582–592.

Chiou, Y. C. and Li, W. C. (2009). Flaw detection of cylindrical surfaces in PU-packing by using machine vision technique. Measurement 42(7):989–1000.

Dewangan, D. K. and Sahu, S. P. (2020). Driving behavior analysis of intelligent vehicle system for lane detection using vision-sensor. IEEE Sens. J. 21(5):6367–6375.

Gu, W. P., Xiong, Z. Y., and Wan, W. (2013). Autonomous seam acquisition and tracking system for multi-pass welding based on vision sensor. Int. J. Adv. Manuf. Technol. 69(1):451–460.

Guo, B., Shi, Y., Yu, G., Liang, B., and Wang, K. (2016). Weld deviation detection based on wide dynamic range vision sensor in MAG welding process. Int. J. Adv. Manuf. Technol. 87(9):3397–3410.

Kim, C. H. and Chung, J. H. (2006). Robust coordination control of a pneumatic deburring tool. J. Robot. Syst. 22(S1):S1–S13.

Kim, C., Chung, J. H., and Hong, D. (2008). Coordination control of an active pneumatic deburring tool. Robot. Comput. Integr. Manuf. 24(3):462–471.

Lee, T. J., Yi, D. H., and Cho, D. I. (2016). A monocular vision sensor-based obstacle detection algorithm for autonomous robots. Sensors 16(3):311.

Maini, R. and Aggarwal, H. (2009). Study and comparison of various image edge detection techniques. Int. J. Image Process. (IJIP) 3(1):1–11.

Matuszak, J. and Zaleski, K. (2014). Edge states after wire brushing of magnesium alloys. Aircr. Eng. Aerosp. Technol: Int. J.

Minu, S. and Shetty, A. (2015). A comparative study of image change detection algorithms in MATLAB. Aquat. Procedia 4:1366–1373.

Shrivakshan, G. T. and Chandrasekar, C. (2012). A comparison of various edge detection techniques used in image processing. International J. Comput. Sci. Issues (IJCSI) 9(5):269.

Tania, S. and Rowaida, R. (2016). A comparative study of various image filtering techniques for removing various noisy pixels in aerial image. Int. J. Signal Proc. Image Process. Pattern Recognit. 9(3):113–124.

Wang, M. L., Huang, C. C., and Lin, H. Y. (2006). An intelligent surveillance system based on an omnidirectional vision sensor. In 2006 IEEE conference on cybernetics and intelligent systems. 1–6.

Xu, Y., Yu, H., Zhong, J., Lin, T., and Chen, S. (2012). Real-time seam tracking control technology during welding robot GTAW process based on passive vision sensor. J. Mater. Process. Technol. 212(8):1654–1662.

Yang, Y., Sang, X., Yang, S., Hou, X., and Huang, Y. (2019). High-precision vision sensor method for dam surface displacement measurement. IEEE Sens. J. 19(24):12475–12481.

138 Monitoring of depth of cut using tri-axial forces in robotic deburring process

K. Ganesh Ram[a], G. Rajesh[b], D. Dinakaran[c], R. M. Kuppan Chetty[d], and Joshuva Arockia Dhanraj[e]

Centre of Automation and Robotics (ANRO), School of Mechanical Science, Hindustan Institute of Technology and Science, Chennai, Tamilnadu, India

Abstract

Most of the industries are moving towards robotic automation for all sorts of micro finishing operations. The robotic deburring process is one of the most complicated machining operation. Robotic arms are implemented for the removal of excess burrs and fins to attain a higher surface finish. The irregular force in the robotic arm leads to positional errors, workpiece damage, and higher cycle time. A lot of research is carried out to compensate the trajectorial errors and enable them for deburring operations. The constant contact force control can only lead to attaining better outputs. To obtain this, an experimental study has been developed using L27 Array-Based Taguchi method which aims to ensure the precise deburring of the component without damaging the tool or the workpiece, by preventing it from the excess force. A tri axial force sensor is also used in this study to measure the machining forces in all three directions. This paper focuses on developing an optimisation model which analyses the tri-axial force and depth of cut values required to determine number of feed cycles to be given for the robotic deburring operation of an aluminium 6061 component.

Keywords: Accuracy optimisation, sensor, tri axial force.

Introduction

Industrial revolution 4.0 implements robotic automation as a replacement for all sorts of manual work. The industries implemented industrial robotic arms in various applications like milling, grinding and assembly, etc (Cheng et al., 2016). Most of the engineering metal part-production industries come across a common problem after machining the workpieces is the undesirable formation of excess raised small edges which remain attached with the workpiece called burrs. These burrs are formed in most of the machining processes at the surface or at the edges of the workpieces. Deburring is a mandatory finishing process because these excess burrs result in the improper quality of components, dimensional errors, and injure the human workers (Kuss et al., 2016). There are various methods to reduce the burr formation but it cannot be eliminated entirely, therefore deburring is the only essential process for burr removal after machining (Jin et al., 2020). These workpieces with burrs are considered as defective components and scraped or reworked once again manually (Rahul et al., 2021). Most industries do not implement robots in this deburring process because of a lack of flexibility compared to human workers. Industries consider that the technologies to be developed to satisfy industrial needs and most of the robotic setup are unable to adapt to the required changes therefore, the manual deburring process is performed. In this manual deburring process to obtain the standard quality of surface finish, heavy manual work and more skilled labour are required. Most of the manual work time is spent on re-measuring the burring and processing in slow flow mode (Aurich et al., 2009). The current development features of the robot will increase high performance and accuracy. Implementing actuators, sensors and controllers is used to equip the robot to perform adaptively to the requirements (Lakshminarayanan et al., 2021). In the past decade, the industrial revolution consider that the manual deburring process is not an economical process in industries, because employing more workers for deburring is more expensive than making a component. This manual process takes more duration to complete the task, which will also influence the factor of production and work efficiency. These issues are faced by the industries on behalf of the manual deburring process. Therefore, semi-automated CNC systems were operated for machining processes. CNC systems have more work efficiency but high investments & maintenance, high power consumption, and highly skilled labourers were required to obtain the output (Schreck et al., 2014). To overcome these issues, robotic automation is adapted for deburring processes rather than manual work to upgrade the industry to the latest industrial standards (Pappachan

[a]shivkarthick64@gmail.com; [b]rajeshthiru3@gmail.com; [c]dinakaran@hindustanuniv.ac.in; [d]kuppanac@hindustanuniv.ac.in; [e]Joushuva1991@gmail.com

et al., 2017). Most industrial robots were invented and introduced into industries for robotic machining such as paint spraying, materials handling, and spot welding. In some cases, industrial robots are implemented only for assembling components to make a final product (e.g. like a car) in most of the industrial sectors (Onstein et al., 2020). The application of industrial robots in many operations like advanced deburring is still limited due to the lack of knowledge and complexity of industrial robots compared to CNC machines. The wide range of working capacity and flexibility in all the 6-axis of the industrial robot which is provided at economical cost leads to more implementation of robots in the industry (Solvang et al., 2007). For applications that require low cutting forces implementation of robotic arm is the affordable alternative for the CNC system, which has great high potential, especially in deburring process (Klimchuk et al., 2017). Most of the research confirmed that robots are the replacement of CNC machinery which provides essential benefits for the industry. On this basis, this paper proposes an industrial-oriented technique of adaptive deburring using an industrial robotic arm, which is used as the base content for the related tri axial force analysis study. Industrial robots have the capacity to process the job in a compact space which provides a large free workspace in the industry. The addition of excess axes can modify the extendable workspace instantly (Kabit et al., 2020). Industrial robots are capable of deburring large-sized complex-shaped parts and multi-faces machining in a single setup. The robotic automation can rapidly retool as required in their manufacturing lines into robot-integrated systems. Industrial robots can machine hard materials economically. Robots develop the latest trajectory generation which can be controlled using a control pendant simply, compared to the Cartesian machines because the high nonlinearity arises between actuated space and the operational space at mapping process (Brunette et al., 2018). The improvised position and force control of the industrial robot is used to machine hard metal tools also on a limited scale (Xiong et al., 2018). Most industries consider deburring a challenging task by using an industrial robot because robotic arm movements are determined only to reach the point-to-point programmed coordinates repeatedly. The robotic arm moves in line with programmatic movements and it cannot be adjusted according to the tool wear after programming. The robots cannot understand the contact force or pressure required for deburring a component, like human labour to who is trained to adjust the force as required. To compensate for these issues, feedback force control system devices are designed and implemented into robotic systems. When setting up a robot for the deburring process, we want to note several factors like cutting force, feed force, stiffness of the robotic arm and other parameters for developing a program. Most industries consider the industrial robots to be low profile in stiffness and to be reworked again and again to get the accuracy as compared to CNC systems (Dong et al., 2018). The large-sized burrs are used to stop or damage the tool during the continuous robotic deburring process. There are various novel concepts proposed for robotic deburring, the researchers Tao et al. (2016) and Bottin et al. (2021) acknowledge that high stiffness is must require for the perfect precision of the cut and to obtain force control over the robot Pappachan et al. (2017) has explained about the radial function (RBF) in neural network determined on sliding mode control method. This control method is used to learn about the conventional SMC scheme used to track joint positions of the robot manipulator and uncertain control actions at operations. To estimate the burr dimensions, Kakoi et al. (2020) prefer a sensor-less force controller which is actuated only from integrated motor sensors to estimate dimensions of burrs. The sensors are also implemented to estimate the burr profile (Kakoi et al., 2020; Hu et al., 2020), which leads to low accuracy and makes the environment dirt. The clean environment makes the ambient air clean which is able to use the sensors effectively (El Naser et al., 2020). Imbalance stiffness of the robotic arm was a major issue for the low usage of industrial robots. As a result of this issue, the robotic arm has been developed precisely, for deburring by (Guo et al., 2019; Vuong et al., 2017). This robotic arm has features similar to the CNC machine. Equally, another researcher has designed an approach to increase stiffness, he proposed to implement an external manipulator to be connected with an industrial robot. Matric chain to one of the parallel robots by which the high stiffness can be obtained. However, these approaches require complex external expensive equipment and high time-consuming for industrial applications. The stiffness is adjusted to reduce vibrations and to improve the surface finishing. To adjust the stiffness according to the operational task, variable stiffness actuators (VAS) mechanisms are implemented which proposed the tuned mechanism to change its stiffness in a wide range, from zero to (theoretically) infinite by has been developed by (Kishore et al., 2015). The L27 Taguchi orthogonal array method is adopted in this paper to find the average cutting force values using 27 experiments with different factors and levels respectively. This application can minimise cost, number of experiments and non-productive data (Hassan et al., 2012). The cutting speed and federate are the important cutting parameters taken Özdemir (2019) in which the authors explain about the chemical property and mechanical property of the workpiece. The various researches also explain about the cutting tool and its dimensional analysis. The researchers also state about the importance of cutting parameters and angle of cut (Akhtar et al., 2021). Similarly, this paper also adopts the cutting parameters of tool rotation speed, feed and pressure given to the tool to run. Mostly Taguchi L27 orthogonal array methodology is used in CNC machining techniques. Mostly the alloy metals are only used force analysis (Rivière-Lorphèvre et al., 2019). The depth of cut is an important factor in deburring process which states that, the rate of metal removed with respect to cutting motion of the tool till the specific depth. To obtain depth of cut various cutting force models are listed in the literature (Rivière-Lorphèvre et al., 2019). This paper calculated the depth of cut using the force output.

Methodology

A. Process model

This paper focuses on the force analysis and depth of cut to be given by the robotic arm in deburring process using a fixed deburring tool. Machining parameters, tool features and workpiece material are the important factors for metal cutting force analysis. By optimal selection the machining parameters such as feed rate, depth of cut and cutting speed which all are the factors that influences, to attain the result. The concept of the fixed tool and moving workpiece is implemented for this process which is used for positioning and orientation of the workpiece with respect to deburr.

B. Workpiece Analysis

The aluminium 6061 of ø75mm with 1.5 mm of burr is the workpiece material used for this experiment. The aluminium 6061 metal is used in manufacturing of ABS, receiver tanks and propeller tanks in automobile industries. It is corrosive resistant and it has the ability to withstand high loads. Aluminium has self-oxide layer which is corrosive resistance. The below Tables 138.1 and 138.2 explains about the chemical and mechanical properties of the workpiece Aluminium 6061.

The workpiece is picked by the gripper at the end effector. The robotic path planning has been made for obtaining instant accurate results. In deburring process, to obtain maximum force accuracy, the triaxial force, namely the radial thrust force (Fx), the tangential cutting force (Fy), and the feed force (Fz), were monitored and depth of cut has been calculated. The methodology is explained in Figure 138.2.

Table 138.1 Mechanical properties of aluminium 6061

Young's modulus (E)	68 Gpa (9900) ksi
Tensile strength (σt)	124–290 Mpa
Elongation (ε) at break	12–25%
Poisson's ratio (ρ)	0.3
Density (ρ)	273g/Cm

Table 138.2 Chemical composition of aluminium 6061

Al	Mg	Si	Fe	Cu	Zn	Ti	Mn
95.85	0.8	0.6	0.6	0.20	0.5	0.15	0.15

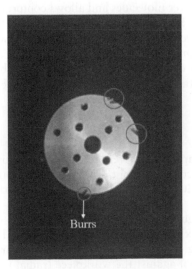

Figure 138.1 Workpiece before deburring

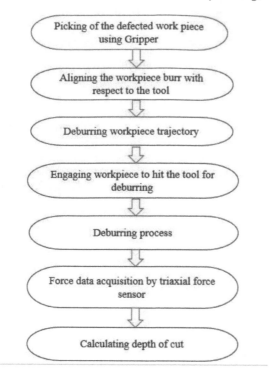

Figure 138.2 Process model

Experimetal Setup

L27 orthogonal array

This paper adapts the orthogonal Array analysis which consist of 5 columns and 27 rows. An orthogonal array (L27) which can be used to allocate the test factors and their interaction. By using three factors - three levels setup, the total number of experiments conducted is 27. However, as a few more factors are added for further study with the same type of material, therefore, it was decided to utilise the L27 array.

Experimental setup

The experimental setup consists of a 6-axis Yaskawa MH5 Robot. It has the capacity of 5 kg of the payload and 6 degrees of freedom which is controlled by the motoman DX100 controller. The Yaskawa robot has been adaptively utilised for this deburring process, which is fixed with a one-foot depth foundation according to the industrial standard for vibration reduction. The deburring workpiece is fixed at the end effector therefore the trajectory planning can be simpler for deburring. The tool position deflection is caused by irregular spindle force therefore, the spindle force of the tool is also monitored. Atlas Copco GX5 roto xtend fluid compressor connected with air dryer which eliminates water molecules and allows compressed 6 bar of airflow to the robot through the pneumatic hose. Figure 138.1 represents the Workpiece before deburring. The Direction Control Valve (DCV) of the robot is connected with Kobe's pneumatic gun with a deburring tool as shown in Figure 138.3 which is generated according to the program command. The cylinder end radius of the deburring tool 10 mm × 20 mm × 70 mm which is made up of tungsten carbide which rotates at 1250 rpm. The Figure 138.3 shows about the deburring tool setup which is connected with a pneumatic gun.

Tri-axial force sensor

To calibrate the deflection, position errors and force of the robotic arm a sensitive force sensor is required which should have high sensitivity in all three perpendicular directions without any influence of other direction forces respectively. To acquire all these parameters the industrial standard Kistler Triaxial force sensor Type 9251A is utilised as shown in Figure 138.4 below.

This Triaxial force sensor has high rigidity and a measuring range with low crosstalks. Crosstalks is a phenomenon of the transmission of signals influencing from one axis to another which is minimal. This sensor is utilised for measuring the triaxial components of cutting forces specifically, feed force, thrust force, and tangential forces. The force required to withstand the tool against the workpiece (radial force

Fx), the relative internal centroid force of the workpiece with spindle axis feed force (Fz), and tangential force (Fy) readings are monitored. A tri-axial force sensor is fixed between the end effector and the gripper with a work piece. This triaxial sensor is fixed closer to the workpiece to acquire three orthogonal forces precisely acting in arbitrary directions. A sensor fixture has been developed which is made especially for the sensor to fix in between the robot end-effector and the gripper with a workpiece.

The fixture is designed for mounting of the tri axial force sensor. This fixture consists of Ø 75 mm of outer diameter. The upper casing consists of six holes with Ø 9mm diameter and a middle hole with Ø18 mm as shown in Figures 138.5 and 138.6. These holes are made for the screw holdings. These figures also explain about the lower casing which also consists of the same dimensions as the upper casing but it has a raised shaft of 7mm in height.

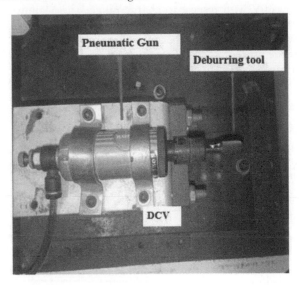

Figure 138.3 Deburring tool

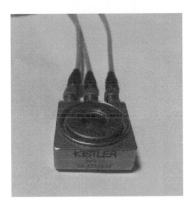

Figure 138.4 Triaxial force sensor

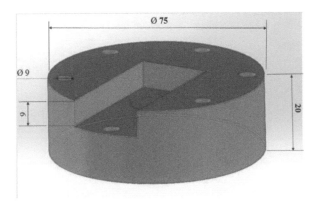

Figure 138.5 Upper casing for the force sensor

The sensor is fixed in between the fixture jaws and screwed with the robot end effector. The force sensor is fixed between these two casing as shown in the Figure 138.7. The Kistler triaxial force contains 3 pair quartz sensor which can able to measure three orthogonal component forces directly. This sensor has a measuring range of +/– 100 to 10000 pc, frequency range of 0–20 kHz with high repeatability. The below Figure 138.8 explains about a piezoelectric industrial charge amplifier type 5073 which is mounted on the top of the robotic arm which converts electrical charge signals into a low impedance voltage range of approximately 4 to 16 ohms which is exactly proportional to the three components Fx, Fy, and Fz of the cutting force. This amplifier is specialised to protect charge input against electrostatic discharge. It consists of a high-frequency range of 0–20 kHz.

The amplifier measures the electrical charge relatively to the level of zero. At the activation of digital input measurement, the output signal always shows 0V with respect to charge range. The Manuware is interfaced for output data acquisition. The charge amplifier consists of three sensor digital input ports through which the sensor data are imported as signals. The signal out port discharges the sensor data. The tri-axial force values are transmitted as signals at the process of edge finishing and excess metal removal. The values are denoted in various symbols which has been explained in nomenclature (Table 138.4).

Total cutting force modelling and experimental test

The 27 experimental values are obtained with respect to 3 factors and 3 levels of cutting parameters. The Table 138.3 shows the L27 array analysis to obtain force values at deburring process.

The tri-axial force sensor is interfaced with Kistler Manuware software to obtain the force values. NI DAC is also utilised for acquiring force values. The total cutting force (Ft) for each parameter is calculated using equation (1).

$$F_t = \sqrt{\left(F_x^2\right) + \left(F_y^2\right) + \left(F_z^2\right)} \tag{1}$$

By substituting the triaxial values in the above equation (1) the total force value has been obtained. The sensitivity of the sensor is also required to calculate depth of cut. To obtain the sensitivity of the sensor a graph has been prepared as shown in Figure 138.9 below.

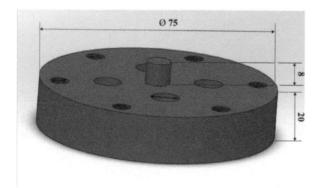

Figure 138.6 Lower casing for the force sensor

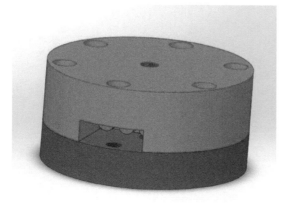

Figure 138.7 The whole fixture model

Table 138.3 Total force analysis using L27 model

S.N	Cutting parameters			Force Sensor values in Voltage			
	P in bar	Ts in rpm	f in mm	Fx in N	Fy in N	Fz in N	Total cutting force in N
1	6	800	0.2	0.4	0.2	0.2	0.489
2	6	800	0.2	0.4	0.1	0.3	0.509
3	6	800	0.2	0.4	0	0.2	0.447
4	6	800	0.2	0.4	0.2	0.3	0.538
5	6	800	0.2	0.4	0.2	0.3	0.538
6	6	800	0.2	0.4	0.1	0.3	0.509
7	6	800	0.2	0.4	0	0.2	0.447
8	6	800	0.2	0.4	0	0.3	0.500
9	6	800	0.2	0.4	0.2	0.2	0.489
10	8	1000	0.3	0.6	0.2	0.2	0.663
11	8	1000	0.3	0.6	0.2	0.2	0.663
12	8	1000	0.3	0.6	0.3	0.3	0.734
13	8	1000	0.3	0.6	0.2	0.3	0.701
14	8	1000	0.3	0.6	0.2	0.3	0.700
15	8	1000	0.3	0.6	0.3	0.2	0.7
16	8	1000	0.3	0.6	0.3	0.1	0.678
17	8	1000	0.3	0.6	0.1	0.3	0.678
18	8	1000	0.3	0.6	0.5	0.4	1.024
19	10	1200	0.4	0.8	0.5	0.3	0.989
20	10	1200	0.4	0.8	0.3	0.4	0.943
21	10	1200	0.4	0.8	0.3	0.2	0.877
22	10	1200	0.4	0.8	0.3	0.4	0.943
23	10	1200	0.4	0.8	0.4	0.4	0.979
24	10	1200	0.4	0.8	0.4	0.3	0.943
25	10	1200	0.4	0.8	0.2	0.4	0.916
26	10	1200	0.4	0.8	0.4	0.2	0.916
27	10	1200	0.4	0.8	0.2	0.2	0.489

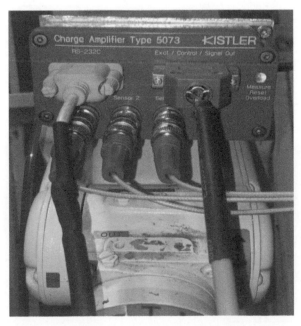

Figure 138.8 Kistler charge amplifier type 5073

This Figure 138.9 represents the average sensitivity of tri axial force sensor which is acquired at the deburring process with respect to the bar pressure given to the tool. The series of orange line represent the accurate sensitivity of tri- axial force sensor values. Using these graphical method force values are acquired instantly. Here, the voltage of 0.3 outputs at 5.8 bar pressure is the average value which is calculated as sensitivity value of the sensor at deburring process. The 5.8 bar pressure is equal to 0.58. The tolerance of the sensor sensitivity is given as ± 1 to be added with the sensitivity value acquired at deburring process therefore, let us take the sensitivity of the sensor (k) = 1.58. There is no S.I units required for force sensor sensitivity so it is neglected.

Result and Discussion

The Figure 138.10 explains the first 9 experimental result average of total force values obtain at the first factor of cutting parameters with six Bar of Pressure, tool rotating speed of 800rpm and feed rate (f) = 0.2.

The Figure 138.11 explains the 918 experimental result average of total force values with second factor of cutting parameters with eight Bar of Pressure, tool rotating speed of 1000 rpm with feed rate of 0.3.

The final factors of cutting parameters with ten Bar of Pressure, tool rotating speed of 1200rpm with feed rate of 0.4 has been tested and the result average value of last nine experiments are shown in the Figure 138.12

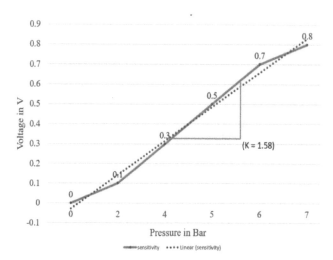

Figure 138.9 Sensitivity calibration

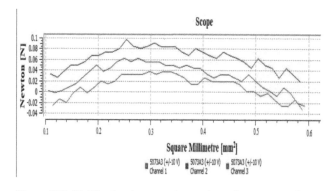

Figure 138.10 The 1ˢᵗ nine experimental result average value

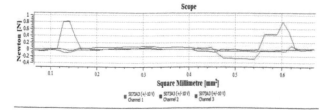

Figure 138.11 The 2ⁿᵈ nine experimental result average value

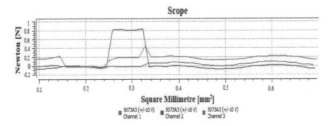

Figure 138.12 The 3rd nine experimental result average value

The average of total force values are acquired by equating the triaxial force value averages at equation (1) and tabulated in Table 138.1. After formulating the triaxial force values in this equation (1), the average value of total cutting force is valuated (Ft) = 0.538 N/mm². To obtain depth of cut various cutting force models are listed in the literature. The following equation (2) is the formula to obtain depth of cut using cutting force by sensor sensitivity.

$$F_t = k \times h \times d_c \tag{2}$$

As we all know the cutting force depend on the sharp edges and burr chip thickness and depth of cut which is proportional to sensor sensitivity or cutting force co-efficient. The burr thickness values h = 1.21 mm are taken using digital Vernier caliper. The sensitivity of the sensor is 1.58 N/mm². To obtain the depth of cut value the formula is reformed as follows.

$$d_c = \left(\frac{F_t}{k \times h} \right) \tag{3}$$

By substituting the values in Equation 3, depth of cut value has been obtained (dc) = 0.305 mm.
This model is used for analysing of the depth of cut required for deburring an aluminium 6061 metal.

Conclusion and Future Scope

The total cutting force is analysed sucessfully, using triaxial force components at three different cutting parametres using Taguchi L27 orthogonal array. The result of the experiments of cutting forces are gradually increased with depth of cut. The depth of cut value is determined and estimation of number of cycles required for deburring process is successfully derived. The required number of deburring cycles can be determined by multiplying the depth of cut value with 'n' number of cycles. This calculation value has to be similar to the height of the burr. Here, the height of the burr is 1.5 mm and depth of cut is 0.3 therefore, five cycles are taken by the robotic arm to remove the 1.5 mm of burr in aluminum 6061 metal.

Acknowledgment

I sincerely thank Hindustan institute of technology and science for providing me this necessary facilities and support. I show my gratitude to my guides for their support and motivation.

Table 138.4 Nomenclature

Ft	Total cutting force
Fx	Radial force
Fy	Tangential force
Fz	Feed force
K	Sensitivity of the force sensor
H	Burr thickness
Dc	Depth of cut
F	feed
Ts	Tool speed
P	Pressure for rotating the tool
mm	millimetre

References

Akhtar, M. N., Sathish, T., Mohanavel, V., Afzal, A., Arul, K., Ravichandran, M., Rahim, I. A., Alhady, S. S., Bakar, E. A., and Saleh, B. (2021). Optimisation of process parameters in CNC turning of aluminium 7075 alloy using L27 array-based taguchi method. Materials 14(16):4470.

Aurich, J. C., Dornfeld, D., Arrazola, P. J., Franke, V., Leitz, L., and Min, S. (2009). Burrs—analysis, control and removal. CIRP Ann. 58(2):519–542.

Bottin, M., Cocuzza, S., and Massaro, M. (2021). Variable stiffness mechanism for the reduction of cutting forces in robotic deburring. Appl. Sci. 11(6):2883.

Brunette, A., Gambao, E., Koskinen, J., Heikkilä, T., Kaldestad, K. B., Tyapin, I., Hovland, G., Surdilovic, D., Hernando, M., Bottero, A., and Anton, S. (2018). Hard material small-batch industrial machining robot. Robot. Comput. Integr. Manuf. 54:185–199.

Cheng, G. J., Liu, L. T., Qiang, X. J., and Liu, Y. (2016). Industry 4.0 development and application of intelligent manufacturing. In 2016 International conference on information system and artificial intelligence (ISAI), (pp. 407–410), IEEE.

Dong, C., Liu, H., Yue, W., and Huang, T. (2018). Stiffness modeling and analysis of a novel 5-DOF hybrid robot. Mech. Mach. Theory 125:80–93.

El Naser, Y. H., Atali, G., Karayel, D., and Özkan S. S. (2020). Prototyping an industrial robot arm for deburring in machining. Academic Platform J. Eng. Sci. 8(2):304–309.

Guo, W., Li, R., Zhu, Y., Yang, T., Qin, R., and Hu, Z. (2019). A robotic deburring methodology for tool path planning and process parameter control of a five-degree-of-freedom robot manipulator. Appl. Sci. 9(10):2033.

Hassan, K., Kumar, A., and Garg, M. P. (2012). Experimental investigation of material removal rate in CNC turning using Taguchi method. Int. J. Eng. Res. Appl. 2(2):1581–1590.

Hu, J., Kabir, A. M., Hartford, S. M., Gupta, S. K., and Pagilla, P. R. (2020). Robotic deburring and chamfering of complex geometries in high-mix/low-volume production applications. In 2020 IEEE 16th International Conference on Automation Science and Engineering (CASE), (pp. 1155–1160). IEEE.

Jin, S. Y., Pramanik, A., Basak, A. K., Prakash, C., Shankar, S., and Debnath, S. (2020). Burr formation and its treatments—a review. Int. J. Adv. Manuf. Technol. 107(5):2189–2210.

Kabit, A. M., Hartford, S. M., Gupta, S. K., and Pagilla, P. R. (2020). Robotic deburring and chamfering of complex geometries in high-mix/low-volume production applications. In 2020 IEEE 16th international conference on automation science and engineering (CASE), (pp. 1155–1160). IEEE.

Kakoi, H., Yanagihara, K., Akashi, K., and Tsuchiya, K. (2020). Development of vertical articulated robot deburring system by using sensor feedback. In IOP Conference Series: Materials Science and Engineering (vol. 886, no. 1, p. 012035). IOP Publishing.

Kishore, D. S., Rao, K. P., and Ramesh, A. (2015). Optimisation of machining parameters for improving cutting force and surface roughness in turning of Al6061-TiC in-situ metal matrix composites by using Taguchi method. Mater. Today: Proc. 2(4-5):3075–3083.

Klimchuk, A., Ambiehl, A., Garnier, S., Furet, B., and Pashkevich, A. (2017). Comparison study of industrial robots for high-speed machining. In Mechatronics and robotics engineering for advanced and intelligent manufacturing, (pp. 135–149).

Kuss, A., Drust, M., and Verl, A. (2016). Detection of workpiece shape deviations for tool path adaptation in robotic deburring systems. Procedia CIRP. 57:545–550.

Lakshminarayanan, S., Kana, S., Mohan, D. M., Manyar, O. M., Then, D. and Campolo, D. (2021). An adaptive framework for robotic polishing based on impedance control. Int. J. Adv. Manuf. Technol. 112(1):401–417.

Onstein, I. F., Semeniuta, O. and Bjerkeng, M. (2020). Deburring using robot manipulators: A review. In 2020 3rd international symposium on small-scale intelligent manufacturing systems (SIMS), (pp. 1–7). IEEE.

Özdemir, M. (2019). Optimisation with Taguchi method of influences on surface roughness of cutting parameters in CNC turning processing. Mechanics 25(5):397–405.

Pappachan, B. K., Caesarendra, W., Tjahjowidodo, T. and Wijaya, T. (2017). Frequency domain analysis of sensor data for event classification in real-time robot assisted deburring. Sensors 17(6):1247.

Rahul, M. R., Bhute, R. Y., Chiddarwar, S. S., Dalvi, M., and Sahoo, S. R. (2021). Burr registration and trajectory planning of 3D workpiece using computer vision. In Advances in mechanical engineering (pp. 107–113). Singapore: Springer. IntechOpen.

Rivière-Lorphèvre, E., Huynh, H. N., Ducobu, F. and Vermined, O. (2019). Cutting force prediction in robotic machining. Procedia CIRP 82:509–514.

Schreck, G., Surdilovic, D., and Krueger, J. (2014). Hephestos: Hard material small-batch industrial machining robot. In ISR/Robotik 2014; 41st international symposium on robotics, (pp. 1–6).

Solvang, B., Korondi, P., Sziebig, G., and Ando, N. (2007). Sapir: Supervised and adaptive programming of industrial robots. In 2007 11th international conference on intelligent engineering systems 2007, (pp. 281–286). IEEE.

Tao, Y., Zheng, J. and Lin, Y. (2016). A sliding mode control-based on a RBF neural network for deburring industry robotic systems. Int. J. Adv. Robot. Syst. 13(1):8.

Vuong, N. D., Li, R., Chew, C. M., Jafari, A., and Polden, J. (2017). A novel variable stiffness mechanism with linear spring characteristic for machining operations. Robotica 35(7):1627–1637.

Xiong, R., Lai, Z., Guan, Y., Yang, Y., and Cai, C. (2018). Local deformable template matching in robotic deburring. In 2018 IEEE international conference on robotics and biomimetics (ROBIO) (pp. 401–407). IEEE.

139 Al6061 Alloy based particulate reinforced composites produced by stir casting-A review

Hemanth Raju T.[1,a], Mohammed Nazmul Shaikb[1,b], Mohammed Inshal Ashfar[1,c], Manoj M. R.[1,d], Tammann Gouda Patil[1,e], and Udayashankar S.[2,f]

[1]Mechanical Engineering, New Horizon College of Enginering Bengaluru, India

[2]Mechanical Engineering, Visvesvaraya Technogical University, Belgaum, India

Abstract

Numerous Aluminium alloys and composites have been produced in recent years to improve material performance. Al6061 is a high-strength aluminium alloy with a wide range of applications as a result of remarkable material qualities it demonstrates. It is a widely used matrix material for the fabrication of aluminium matrix composites, and it has a number of advantages (AMCs). Al6061 aluminium MMCs made using the stir-casting technique are the subject of this study, which has been designed to provide an overview. It describes the typical stir-casting manufacturing technique, as well as the variables, the different reinforcements used, and the mechanical characteristics of Al6061 composites. The stir-casting technology has been proven in numerous research studies to be widely used and acceptable for the production of Al6061 composites reinforced with SiC, B4C, Al_2O_3, TiC, and other inorganic, organic, hybrid, and nonmaterial materials, among other materials. The bulk of the studies found that increasing the amount of reinforcing fibre in the composites improved the mechanical and tribological characteristics of the composites. Also, as compared to mono-reinforced composites, hybrid composites demonstrated superior material properties. Aside from that, it is believed that industrial and agricultural waste products will be utilised in hybrid composites. Future study could concentrate on the production and characteristics of Al6061 composites produced by stir casting, as these materials have significant potential as advanced materials in their own right.

Keywords: Al6061 alloy, metal matrix composites, microstructure, stir casing.

Introduction

Metal matrix composites (MMCs) are becoming an increasingly popular material for complex aerospace applications due to their flexibility to be customized through the incorporation of appropriate reinforcements (Prashant et al., 2012). A wide range of uses for MMCs are found in a variety of industries, most notably the automobile industry, where the demand for lightweight materials has increased as a result of environmental concerns. Manufacturing of diesel engine pistons and connecting rods has been done using MMCs (Suresh et al., 2013; Rajesh et al., 2016; Adediran et al., 2021). Also demonstrated is the fact that these materials have a substantial potential for use in braking discs for railway brake equipment (Bharath et al., 2014; Ranganathaiah et al., 2014; Pazhuhanfar and Eghbali, 2021; Bhat et al., 2021), as well as in other applications. Recent decades have seen a substantial increase in the interest in metal matrix composites (MMCs) based on aluminium (Mohal et al., 2019; Bhasha and Balamurugan, 2021). A composite material is formed by the incorporation of a ceramic substance into a metal matrix, and this material is attractive due to its interesting combination of physical and mechanical qualities (Balaji et al., 2021; Rajasekhar and Yedula, 2017; Mohamed et al., 2018; Maurya et al., 2019). Carbon, SiC, TiC, tungsten, boron, Al_2O_3, fly ash and zirconium are just a few of the reinforcements that have been studied in the production of aluminium MMCs (Chammala et al., 2019; Raju et al., 2021b; Raju et al., 2021). In addition to being harder than steel, carbon fibre composites offer higher strength and wear resistance than steel (Chandra et al., 2021; Adhikary et al., 2018; Adhikary et al., 2021). The low density of aluminium alloys, which allows them to be suitable for a wide variety of applications while staying lightweight, has continued to be the focus of substantial research (Adhikary et al., 2017; Nagendra et al., 2021; Nagendra and Prasad, 2020; Srinath et al., 2021). In the manufacture of wear-resistant components, these alloys are increasingly being used in place of cast iron and bronze. In an earlier work, it was discovered that particle reinforcement

[a]hemanth.bhadravathi@gmail.com; [b]nazmulshaik143@gmail.com; [c]inshu8118@gmail.com; [d]manojmr77@gmail.com; [e]basupatil9291@gmail.com; [f]udaya_creative@yahoo.com

improved the mechanical parameters of Al-matrix composites (Srinath and Prasad, 2020; Nagabhushana et al., 2020b; Puneeth and Prasad, 2019; Bopanna and Prasad, 2020). The particulate reinforced MMCs are widely employed since the particles are easily available and the processing technology used to make the particulate reinforced MMCs is both efficient and cost-effective to manufacture (Bellie et al., 2021b; Jayanth et al., 2021). MMCs are commonly reinforced with a range of fibres, particles, and whiskers, with aluminium alloys acting as the predominant base metal (Subbian et al., 2021; Rajesh et al., 2020; Jose et al., 2017b; Bellie et al., 2021a). Particle-reinforced MMCs, and in particular aluminium based composites, had already established themselves as feasible options for industrial applications among the many types of recently produced composites (Shanmugam et al., 2021; Jose et al., 2020; Madhusudan et al., 2020), despite the fact that the technology is still in its early stages.

There are three different ways to make aluminium metal matrix composites. The first is to make them in a liquid condition, the second is to make them semisolid, and the third is to make them powder (Padmini et al., 2019; Nagendra et al., 2018; Nagabhushana et al., 2020a). Metal matrix composites (MMCs) are generated by incorporating ceramic particles into a molten metallic matrix through the use of liquid state processing techniques. Stir casting is used to create composite materials in the liquid state (Nagabhushana et al., 2019; Adhikary et al., 2016; Nagendra et al., 2019; Srinath and Prasad, 2018). This process involves mechanically mixing a ceramic particulate in addition to a molten matrix metal during the casting process, which results in a composite material in the liquid state. In order to make the final product, the liquid composite material is produced using typical casting processes (Sathish et al., 2020; Adhikary et al., 2015b; Adhikary et al., 2015c; Adhikary et al., 2015d). Al 6061 is a widely used aluminium alloy that has a low density and excellent heat conductivity, yet it is fragile due to its high brittleness. It is used in aerospace and defense applications. Al6061 Alloy is an aluminium alloy that was used to fabricate composites (Adhikary et al., 2015a; Adhikary et al., 2014). It is mostly composed of silicon and magnesium alloying elements in significant amounts. It is used in a variety of applications that demand high strength and resistance to fatigue (Adhikary et al., 2013b; Adhikary et al., 2013a). The stir casting equipment used for the development of MMCs is shown in the below Figure 139.1.

Alloy 6061 offers a medium to high strength, increased formability, machinability, and surface quality, superior corrosion resistance, better weldability, and workability, and it is readily accessible (Sathish et al., 2021). It is also relatively inexpensive (Bellie et al., 2021b; Jayanth et al., 2021). Al6061 alloy is widely employed in a variety of industries, including automotive, shipping, sports, aerospace, defense, and structural applications, among others (Gopal et al., 2018; Damodharan et al., 2019). The most difficult step in developing composites is selecting the matrix type and ensuring that it is compatible with the reinforcement; thus, careful matrix selection is required. The tensile strength of a material is increased when it is reinforced (Adhinarayanan et al., 2020; Nagabhushana et al., 2020). Increasing the load bearing capability of the reinforcement, in addition shrinking the grain size of the matrix material, both help to improve the mechanical characteristics of aluminium 6061 composites (Srinath and Prasad, 2019a). The particle distribution of the base metal was found to be homogeneous, according to the results of the analysis. Due to the precise refining of the matrix grains, a strong link can be formed between the reinforcement and matrix (Srinath and Prasad, 2019b; Jose et al., 2017a). Due to the absence of reaction products, cavities, and pores, the interface between the particles and the aluminium matrix was left pristine. The particles aided in increasing the composite's hardness and UTS (Damodharan et al., 2021; De Poures et al., 2020; Raju et al., 2021a).

Al6061's chemical composition is listed in Table 139.1.

The Al6061 alloy based particulate reinforced metal matrix composites are extensively used in automobile applications. In automobile components they are used in parts such as connecting rods, pistons, piston

Figure 139.1 Stir casting equipment

Table 139.1 Composition of Al6061 aluminium alloy

Element	Al	Mg	Si	Fe	Cu	Zn	Ti	Mn	Cr
Quantity (wt.%)	Balance	0.90	0.50	0.50	0.30	0.20	0.10	0.10	0.25

rings, cylinder liners, bearings and brake drums. They are also used in aircraft, helicopters and spacecrafts. In spacecraft they are used in components such as tubes, plates, and panels. In aircraft they are used in fuselage, wing, and support structures.

Literature Review

Al6061 aluminium alloy based composites literature review has been done over the last decades and are as discussed as below.

Prashant et al. (2012), studied the fabrication and characterization of Al6061-Graphite Particulate Composites and discovered that composites comprising Al6061 at concentrations of 6, 9, and 12 wt. % graphite. The stir casting method was successfully used to create graphite particles in three phases of mixing. Stir cast composite optical micrographs reveal a rather homogeneous dispersion of graphite particles within the Al6061 metal matrix. Given the presence of porosity in all of the composites, it was revealed that the experimental densities were lesser than the theoretical densities in all of the situations studied. It was discovered that the hardness of composite samples decreases in direct proportion to the percentage of particles in the sample. The use of graphite enhanced the ultimate tensile strength of the material. The tensile strength of reinforcement is proportional to its volume fraction. Tensile strength of the composite improves as the volume fraction of graphite particulate increases. Incorporating graphite, on the other hand, resulted in a greater improvement in tensile properties. It was discovered that the wear rate of the Al6061-Graphite particulate composite lowers up to 6 %, but then begins to rise again after that.

Suresh et al. (2013), studied the microstructure and wear characteristics of TiB2 on Al6061 MMCs and discovered that TiB2 reinforcement improved the aluminium's hardness property. Aluminium's micro-hardness value is increased by adding reinforcement. As a consequence, it is apparent that increasing the amount of TiB2 in the aluminium will surely enhance its strength. The microstructure images clearly demonstrated a rise in the composition of TiB2. The samples were subjected to wear tests, and the results were used to determine the wear resistance and COF of the samples. The graph demonstrates that increasing the TiB2 content of the aluminium composite increases its wear resistance properties. This illustrates that the strengthening will increase the aluminium's wear resistance. The sample was subjected to a variety of loads, and it was discovered that the weight loss rose in direct proportion to the rise in load. Even when applied with considerable pressures, TiB2 demonstrates less weight loss than the other composition materials when compared to their respective counterpart materials. The findings of this article demonstrate that adding TiB2 reinforcement enhances the mechanical characteristics of aluminium composite.

Rajesh et al. (2016), investigated the mechanical characteristics of an Al6061 MMC Reinforced With Fused Zirconia Alumina and reported that mechanical qualities such as ultimate tensile strength, BHN, and impact strength varied by means of composite combinations. As per testing data, the optimal matrix and reinforcement weight percentages for mechanical qualities are 90 and 10, correspondingly.

Adediran et al. (2021), investigated the characteristics of Al6061 alloy reinforced with particulate waste glass and found that when comparing the WGP/Al6061 composites to the control samples, the stir casting method utilized to create the WGP/Al6061 composites resulted in a homogeneous distribution of waste glass particulate at an 8 % weight fraction. When the composite's weight percentage was increased, the composite's hardness and tensile strength improved, while wear rate and porosity decreased. Microstructural investigations show that the waste glass particles were homogeneously distributed at the optimal WGP addition rate of 8%.

Bharath et al. (2014), reported that stir casting was used to produce Al6061-Al$_2$O$_3$ MMCs, and the mechanical and wear characteristics of the materials were investigated. They discovered that using a melt stirring procedure that includes three phases mixing as well as reinforcing particle preheating, composites including Al6061 with 6, 9, and 12 wt.% Al$_2$O$_3$ particulates could be successfully made. Optical micrographs acquired during the stir cast composite process reveal a rather homogeneous dispersion of Al$_2$O$_3$ particles within the Al6061 metal matrix, which is consistent with previous findings. It was discovered that primary-Al dendrites and eutectic silicon were present in the composites' microstructure. Separation of Al$_2$O$_3$ particles happened in interdendritic zones as well as in eutectic silicon during the experiment. As the weight % of Al$_2$O$_3$ particles in the composites increases, the composites get tougher. Comparing manufactured composites to cast Al6061, their tensile and yield strengths were higher, but their ductility was lower.

Aside from that, the ultimate tensile strength of Al_2O_3 increases with an increase in the wt.% of Al_2O_3. As cast Al6061 alloy displayed the largest weight reduction when subjected to a constant load of 19.62 N and a constant rotational speed of 300 rpm, whereas Al6061+12% Al_2O_3 composites exhibited the least amount of weight loss. Additionally, as cast 6061Al alloy exhibited a higher wear rate than Al6061-Al_2O_3 composites, and as cast Al6061 alloy exhibited the greatest weight loss.

Ranganathaiah et al. (2014), studied the fabrication and characteristics of Al6061 Fiber MMCs reinforced with e-glass fibre stir casting technique was used to show that e-glass fibres may be successfully incorporated into the Al6061 alloy matrix to create the composites. As per the microstructure analysis, the manufactured composite material has a relatively uniform dispersion of reinforcements. When contrasted to an unreinforced matrix, Al6061's ultimate tensile strength is dramatically increased when e-glass fibre is added. Furthermore, the ultimate tensile strength of e-glass fibre begins to deteriorate at a concentration of 6wt%. When contrasted to an unreinforced matrix, the insertion of e-glass fibre increases the compressive strength of Al6061 substantially or greatly. However, after 8wt. %, the compressive strength of e-glass fibre begins to deteriorate. The samples hardness improved as the composite's reinforcing content increased.

Pazhuhanfar and Eghbali (2021), worked on the Production and characteristics of Al6061-TiB2 composites were explored, and it was discovered that the particle dispersion is not homogeneous in the as-cast composite. Certain particle clusters and particle-free zones have an effect on the casted composite's mechanical characteristics. Roll-bonded composites have a microstructure that is devoid of particle clusters and agglomerates. In proportion to the increase in the number of deformation cycles, particle-free zones began to disappear, resulting in a homogenous dispersion of reinforcing particles throughout the matrix structure. According to this conclusion, greater work hardening causes a reduction in the ductility of the aluminium matrix. Hardness, ultimate tensile strength and elongation to failure of all developed composites improved following the application of post-deformation ageing. It is possible to ascribe this discovery to two factors: the occurrence of static recrystallisation while solutionising, and the homogeneous dispersion of magnesium silicate precipitates in the microstructure; Microvoids form at the particle/matrix interfaces when TiB2 reinforcement particles are used as reinforcement. As a result of the proliferation and coalescence of these voids under the influence of local tensile strains, the final fracture occurs. Because of the preferential nucleation of voids on these particles, the quantity of ductile dimples grows when the number of ARB cycles increases, and the fracture transforms into a ductile rupture as the number of cycles increases.

Bhat et al. (2021), reported that when contrasted to Al6061, the recently fabricated Al6061 and 5% Al_2O_3 MMCs exhibit superior wear properties. The hardness value of the novel composite material, as determined by the Rockwell number, has also been greatly increased. The knowledge base is expanding as speed and load exert a bigger impact on the wear rate. Under constant load circumstances, wear increases with increasing speed. When the speed remains constant while the load increases, wear occurs. When the speed remains constant as the load increases, Al6061-Al_2O_3 exhibits higher wear than Al6061-Al_2O_3. Additionally, it was revealed that Al6061 wears faster than Al6061-Al_2O_3 wear under continuous load and rising speed.

Mohal et al. (2019), reported that the stir casting technique was successfully employed to create a MMC based on aluminium and reinforced with Al_2O_3-SiC. The percentage elongation of the Al6061/Al_2O_3-SiC MMC in its as-synthesized state drops dramatically as the Al_2O_3 content increases. The composite's decreased elongation could be explained by an increase in brittleness and the addition of tougher alumina particles. Additionally, the micro-hardness of the as-developed Al6061/Al_2O_3-SiC MMC increases with increasing Al_2O_3 percentage up to 6%, and then decreases with increasing Al_2O_3 percentage. This is related to the increased porosity in reinforced composites and may also be attributable to a high percentage of clustered reinforcing particles.

Bhasha and Chinnamahammad (2021), studied the mechanical characteristics of an Al6061/RHC/TiC hybrid composite were examined, and it was observed that the occurrence of more than 9% TiC results in heterogeneous dispersion and clustering, as proven by SEM with an EDS pattern. The chemical process initiated by the insertion of reinforcing lead leads to the formation of the intermetallic element $Al_{12}Mg_{17}$, as well as the main constituents Al, SiO_2, and TiC. By adding TiC, the chemical process is further slowed and thermal stability is increased. Sample S2 demonstrated a 46.71% increase in ultimate strength, a 36.72% increase in compression strength, and a 48.85% increase in flexural strength when contrasted to the embryonic specimen. When contrasted to the base alloy, the presence of reinforcement particulates resulted in a 22.92% increase in the microhardness of specimen S2, showing a considerable improvement in performance. In addition to improving mechanical properties, decreasing grain slippage during loading has the added benefit of enhancing the ability for the grain to act as a lubricant throughout the whole manufacturing process. Because of the inclusion of foreign particulates into the composite, the modulus of elasticity and Poisson's ratio of the material both increase. The impact strength and elongation properties of the material diminish as the RHC and TiC levels rise. Sample S4 demonstrates the greatest reduction in impact energy at 52.17% and the greatest reduction in elongation property at 73.50%, respectively. Particle de-boning and dimples and ridges were found on the fracture surfaces of the impact test surfaces, indicating a heterogeneous failure mode.

Balaji et al. (2021), studied the synthesis and Investigation of Al6061/Al$_2$O$_3$/TiC Hybrid MMC, reported that the hardness value of stir produced aluminium alloy 6061 with Al$_2$O$_3$ and TiC reinforced composites is considerably higher than that of base Al6061 alloy. The incorporation of Al$_2$O$_3$ and TiC particles within an aluminium matrix material results in a rise in the BHN of the Al6061 matrix material, which is a desirable property. By employing the response surface methodology, we were able to accurately calculate both the material removal rate and the surface roughness of aluminium MMCs throughout the turning process. The following were the conclusions reached by the previous investigation: Cutting speed, depth of cut, and feed rate all rise as a result of this, and the total amount of material removed per minute increases from 0.48 cubic metres per minute to 3.60 cubic metres per minute. In example, increasing the depth of cut improves the chip thickness and consequently the rate of material removal. Surface roughness is lowered from 0.181 m to 0.038 m by decreasing feed and depth of cut from 0.297 mm/rev, 0.45 mm to 0.1 mm/rev,0.3mm. These results were achieved using the composition Al606185p/Al$_2$O$_3$ 10p/TiC5p. The surface roughness of pure cast aluminium 6061 is 0.031 m when cut at the same speed, depth of cut and feed as other aluminium alloys. When the surface roughness values of Al606185p/Al$_2$O$_3$ 10p/TiC5p and AA6061 are assessed, the Al6061 has the best value. The fundamental reason for this is that when the volume percentages of Al$_2$O$_3$ and Tic grow, the work piece's hardness increases, resulting in reduced surface roughness. The values for surface roughness also grow in direct proportion to the depth of cut as the depth of cut increases. In accordance with the evidence at hand, increasing the depth of cut causes a corresponding rise in cutting force and vibration, which leads to a subsequent increase in surface roughness. Response surface approach is used to determine the rate of material removal and surface roughness in turning of aluminium MMCs and Al6061. R2>0.95 indicates that the fitted value is extremely near to the experimental value.

Rajasekhar and Yedula (2017), carried out the experimental study on the wear behaviour of Al6061/SiC/Zr hybrid MMCs and demonstrated the successful production of Al6061/SiC/Zr hybrid MMCs. When Zr particles are included in the composite, they operate as a load-bearing element, increasing the composite wear rate. The testing results indicated that raising the SiC reinforcement from 10% to 20%, combined with 2% Zr, considerably enhanced the wear rate for applied loads of 10N, 20N, and 30N.

Mohamed et al. (2018), examined the mechanical characteristics of Al6061-SiC-Gr MMCs and discovered that Al6061 hybrid composites with homogeneous SiCp and Gr particle distribution were successfully manufactured by liquid metallurgy. The inclusion of SiCp greatly enhanced the hardness of composites, with the greatest hardness obtained at 15% SiCp. When a small amount of SiCp is introduced to Al6061 at a low wt. %, the tensile strength increases while the percentage elongation reduces. The Al6061-SiC composite wear rate reduced as the SiC content increased, whereas the Al6061-Graphite composite first declined by up to 5% before gradually increasing.

Maurya et al. (2019), studied the impact of SiC Particles on the Production of Aluminium-Based MMCs was researched, and it was found that SiC particles dispersed in the Al6061 matrix in a homogenous and uniform manner. The density of the composite grew in proportion to the silicon carbide content. By introducing 8% silicon carbide into the composite, the density was raised by 1.11%. The maximum hardness determined by the Rockwell hardness test was 51 HRB. If Al6061/8wt percent SiC is compared to the base alloy Al6061, the hardness of Al6061/8wt percent SiC is increased by 27.5%. Comparing the maximum tensile strength of the Al6061 alloy (276 MPa) to that of the Al 6061/8wt percent SiC composite (298 MPa), the Al6061/8wt. % SiC composite has a higher maximum tensile strength. Because of the incorporation of strong SiC particles, the tensile strength of the fabricated composite was significantly enhanced. The use of SiC particles helped to improve the UTS. The Al6061/8wt% SiC composite has a maximum UTS of 324.7 MPa when compared to the Al661 alloy. Due to the higher UTS, the composite's ductility was diminished.

Fazludheen et al. (2019), investigated the Mechanical Characteristics and Microstructure study of the Al6061/SiC/TiB2 MMCs and it was revealed that stir casting process was effectively used to fabricate Al-SiC-TiB2 composites of varied concentrations. By increasing the reinforcing content, the density of the composites was decreased. As a result, Al-SiC-Gr composites were discovered to outperform Al-SiC-Gr composites. As a consequence, these composites are suitable for applications requiring large weight reductions. The wear and hardness of the test materials were evaluated. Greater reinforcing area percentage results in greater tensile strength and hardness of the underlying matrix structure. The inclusion of Al-SIC-TIB2 at a greater concentration greatly decreases the rate of elongation of the hybrid MMCs. We may deduce from the data that Al-SiC-TiB2 composites, instead of Al-SiC-TiB2 composites, may be regarded an extraordinary material in areas needing lightweight and increased mechanical characteristics.

Most recent research work on Al6061 based particulate reinforced composites

The most recent research work on Al6061 based particulate reinforced composites is shown in the below Table 139.2.

From the above literature survey it is noted that Al6061 alloy can be widely used as a matrix material and it can be combined with different reinforcement particles to form a composite material. The mechanical and wear properties of Al6061 alloy can be increased with the addition of reinforcement particles.

Conclusion

In this review, the stir-casting technique was utilized to synthesize Al6061 MMCs. Among the many manufacturing methods, stir casting has been selected as the most economical and widely employed. The following points summarize the outcomes from the reviews.

- Al6061 is a widely used aluminium alloy with a wide range of uses. Several Al6061 composites were produced using stir-casting technique by reinforcing them with a variety of organic and inorganic elements.
- In terms of characteristics, the manufactured composites performed better than the base alloy. Stir-casting method process variables such as stirrer speed and time, blade design, size of reinforcement, and melting temperature are all important in determining the final properties of Al6061 composites.
- The properties of Al6061 composites are highly impacted by variables in the stir-casting process. An optimization technique may be employed in order to acquire the desired parameters.
- It has been observed that adding reinforcement helps the molten composite solidify, leading in grain refinement.
- Increasing the reinforcing particle weight fraction had a substantial impact on the mechanical and tribological features of the Al6061 composites.
- It is possible that increasing the wt. % of reinforcing particles in the composite will enhance the mechanical and tribological behaviour of the composite. More importantly, increasing the amount of reinforcement introduced beyond a certain point results in the formation of pores and agglomeration, which degrades the characteristics of the material during the process.

Acknowledgements

The authors would like to convey their appreciation to the Management and Principal of NHCE, Bengaluru-560103, for their support in conducting the research.

Table 139.2 Most recent research work on Al6061 based particulate reinforced composites

Author/Year	Matrix and reinforcement	Volume fraction	Particle size	Process
Prashant S N/ 2012	Al 6061+ Graphite	6%, 9%, 12%	125μm	Stir casting
S. Suresh/ 2013	Al 6061+TiB$_{24}$	0%, 4%, 8%, 12%	40 μm	Stir casting
P.V.Rajesh /2016	Al 6061 + ZA40	5%, 10%, 15%	30μm	Stir casting
Adeolu Adesoji Adediran/2021	Al 6061 + Waste Glass	0%, 2%, 4%, 6%, 8%, 10%,	23 μm	Stir casting
Bharath V/2014	Al 6061 + Al$_2$O$_{34}$	6%, 9 %, 12%	125μm	Stir casting
Ranganathaiah C. K/2014	Al 6061+ E-Glass Fiber	0%, 2%, 4%, 6%, 8%, 10%,	30μm	Stir casting
Y. Pazhuhanfar/2021	Al 6061 + TiB$_2$	3%, 6%, 9%	20 μm	Stir casting
Avinash Bhat /2021	Al 6061+ Al$_2$O$_{34}$	5%,	50μm	Sir casting
Sachin Mohal /2019	Al 6061+ Al$_2$O$_3$+ SiC$_4$	3%, 6%, 9%	45 μm	Stir casting
A. Chinnamahammad Bhasha/2021	Al 6061+RHC+TiC	0,%, 3% , 6%, 9%, 12%	55μm	Stir casting
N. Balaji/2021	Al 6061+ Al$_2$O$_3$+ TiC$_4$	5%, 10%,	25 μm	Stir casting
Rajasekhar Sivapuram/2017	Al 6061+SiC+ Zr$_4$	10%, 15%, 20%	50 μm	Stir casting
S.S.Mohamed/2018	Al 6061 + SiC+ Gr	5%, 10%, 15%	45 μm	Stir casting

Declaration of competing interest

The authors say that they are unaware of any conflicting financial interests or personal ties that would appear to have affected the work presented in this research.

References

Adediran, A. A., Akinwande, A. A., Balogun, O. A., Adesinaa, O. S., Olayanjuc, A. and Mojisola, T. (2021). Evaluation of the properties of Al-6061 alloy reinforced with particulate waste glass. Sci. Afr. 12(2):1–9.

Adhikary, P., Bandyopadhyay, S., and Kundu, S. (2017). Application of artificial intelligence in energy efficient H.V.A.C. system design: a case study. ARPN J. Eng. Appl. Sci. 12.

Adhikary, P., Bandyopadhyay, S., and Mazumdar, A. (2018). C.F.D analysis of air-cooled HVAC chiller compressors. ARPN J. Eng. Appl. Sci. 13.

Adhikary, P., Kundu, S. and Mazumdar, A. (2021). CFD analysis of small hydro plant turbines: case studies. ARPN J. Eng. Appl. Sci. 16.

Adhikary, P., Kundu, S., Roy, P. K., and Mazumdar, A. (2013a). Optimum selection of hydraulic turbine manufacturer for SHP: MCDA or MCDM tools. World Appl. Sci. J. 28.

Adhikary, P., Roy, P. K., and Mazumdar, A. (2013b). Fuzzy logic based optimum penstock design: Elastic water column theory approach. ARPN J. Eng. Appl. Sci. 8.

Adhikary, P., Roy, P. K., and Mazumdar, A. (2014). Multi-dimensional feasibility analysis of small hydropower project in india: A case study. ARPN J. Eng. Appl. Sci. 9.

Adhikary, P., Roy, P. K., and Mazumdar, A. (2015a). Maintenance contractor selection for small hydropower project: A fuzzy multi-criteria optimization technique approach. Int. Rev. Mech. Eng. (I.RE.M.E.) 9.

Adhikary, P., Roy, P. K., and Mazumdar, A. (2015b). optimal renewable energy project selection: A multi-criteria optimization technique approach. Glob. J. Pure Appl. Math. 11.

Adhikary, P., Roy, P. K., and Mazumdar, A. (2015c). Selection of small hydropower project site: A multicriteria optimization technique approach, ARPN J. Eng. Appl. Sci. 10.

Adhikary, P., Roy, P. K., and Mazumdar, A. (2015d). Turbine supplier selection for small hydro project: Application of multi-criteria optimization technique. Int. J. Appl. Eng. Res. 10.

Adhikary, P., Roy, P. K., and Mazumdar, A. (2016). C.F.D. Analysis of micro hydro turbine unit: A case study. ARPN J. Eng. Appl. Sci. 11.

Adhinarayanan, R., Ramakrishnan, A., Kaliyaperumal G., De Poures, M., Babu, R. K., and Dillikannan, D. (2020). Comparative analysis on the effect of 1-decanol and di-n-butyl ether as additive with diesel/LDPE blends in compression ignition engine. Energy Sources Part A: Recovery Util. Environ. Eff.

Balaji, N., Balasubramani, S., and Pandiaraj, V. (2021). Fabrication and analysis of Al6061/Al$_2$O$_3$/TiC hybrid metal matrix composite, Paideuma J. XIV(3):24–35. ISSN NO: 0090-5674.

Bellie, V., Gokulraju, R., Rajasekar, C., Vinoth, S., Mohankumar, V., and Gunapriya, B. (2021a). Laser induced breakdown spectroscopy for new product development in mining industry. Mater. Today: Proc. 45.

Bellie, V., Gokulraju, R., Rajasekar, C., Vinoth, S., Mohankumar, V., and Gunapriya, B. (2021b). Laser induced breakdown spectroscopy for new product development in mining industry. ARPN J. Eng. Appl. Sci. 16.

Bharath, V., Nagaral, M., Auradi, V, and Kori, S. A. (2014). Preparation of 6061Al-Al$_2$O$_3$ MMC's by Stir casting and evaluation of mechanical and wear properties. In 3rd International conference on materials processing and characterization (ICMPC 2014). Procedia Mater. Sci. 6:1658–1667.

Bhasha, A. C. and Balamurugan, K. (2021). Studies on mechanical properties of Al6061/RHC/TiC hybrid composite. Int. J. Lightweight Mater. Manuf. 4.

Bhat, A., Kakandikar, G., Deshpande, A., Kulkarni, A., and Thakur, D. (2021). Characterization of Al$_2$O$_3$ reinforced Al6061 metal matrix composite. Mater. Sci. Eng. Appl. 1(1):11–20.

Bopanna, K. D. and Prasad, M. S. G. (2020). Thermal characterization of aluminium-based composite structures using laser flash analysis. J. Inst. Eng. (India): Series C 101.

Chammala, F., Ansari, A., Deepak, T. P., Mohammed Hashir, A. M., and Mohammed Najeeb, P. T. (2019). Study of mechanical properties and micro structure of the composition Al6061/SiC/TiB2 metal matrix composite. Int. J. Eng. Sci. Comput. 22935–22937.

Chandra, R. P., Halemani, B. S., Chandrasekhar, K. M., Ravitej, Y. P., Raju, T. H., and Udayshankar, S. (2021). Investigation and analysis for mechanical properties of banana and e glass fiber reinforced hybrid epoxy composites. 47.

Damodharan, D., Gopal, K., Sathiyagnanam, A. P., Kumar, B. R., De Poures, M. V., and Mukilarasan, N. (2021). Performance and emission study of a single cylinder diesel engine fuelled with n-octanol/WPO with some modifications. Int. J. Ambient Energy 42:779–788.

Damodharan, D., Kumar, B. R., Gopal, K., De Poures, M. V., and Sethuramasamyraja, B. (2019). Utilization of waste plastic oil in diesel engines: A review. Rev. Environ. Sci. Bio/Technol. 18: 681–697.

De Poures, M. V., Gopal, K., Sathiyagnanam, A. P., Kumar, B. R., Rana, D., Saravanan, S., and Damodharan, D. (2020). Comparative account of the effects of two high carbon alcohols (C5 & C6) on combustion, performance and emission characteristics of a DI diesel engine. Energy Sour. Part A: Recovery Util. Environ. Eff. 42:1772–1784.

Gopal, K., Sathiyagnanam, A. P., Kumar, B. R., Saravanan, S., Rana, D., and Sethuramasamyraja, B. (2018). Prediction of emissions and performance of a diesel engine fueled with n-octanol/diesel blends using response surface methodology. J. Clean. Prod. 184: 423–439.

Jayanth, B. V., Depoures, M. V., Kaliyaperumal, G., Dillikannan, D., Jawahar, D., Palani, K., and Shivappa G. P. M. (2021). Laser induced breakdown spectroscopy for new product development in mining industry. Energy Sources, Part A: Recovery Util. Environ. Eff.

Jose, A. S., Athijayamani, A., and Jani, S. P. (2020). A review on the mechanical properties of bio waste particulate reinforced polymer composites. Mater. Today: Proc. 37.

Jose, A. S., Athijayamani, A., Ramanathan, K., and Sidhardhan, S. (2017a). Effects of an addition of coir-pith particles on the mechanical properties and erosive-wear behavior of a wood-dust-particle-reinforced phenol formaldehyde composite. Mater. Tehnol. 51.

Jose, S., Athijayamani, A., Ramanathan, K., and Sidhardhan, S. (2017b). Effects of aspect ratio and loading on the mechanical properties of prosopis juliflora fibre-reinforced phenol formaldehyde composites. Fibres Text East. Eur. 25.

Madhusudan, M., Kumar, S., Kurse, S., Shanmuganatan, S. P., John, J., and Haseebuddin M. R. (2020). Behavioral studies of process parameters and transient numerical analysis on friction stir welded dissimilar alloys. Mater. Today: Proc. 37.

Maurya, M., Maurya, N. K., and Bajpai, V. (2019). Effect of SiC reinforced particle parameters in the development of aluminium based metal matrix composite. Evergr. Joint J. Nov. Carbon Resour. Sci. Green Asia Strat. 6(3):200–206.

Mohal, S., Sharma, A., Sharma, R., Kumar, V., and Goyal, R. R. (2019). Investigation of mechanical properties of Al6061/Al$_2$O$_3$-SiC composites fabricated by stir Casting. J. Compos. Theory. 12(7):785–791.

Mohamed, S. S., Abdallah, S. A., and Alazemi, H. K. H. (2018). An investigation on the mechanical and physical properties of Al6061-SiC-Gr metal matrix composites. Int. J. Eng. Appl. Sci. 5(3):1–5.

Nagabhushana, N., Rajanna, S., and Ramesh, M. R. (2020a). Erosion studies of plasma-sprayed NiCrBSi, Mo and flyash cenosphere coating. Mater. Manuf. IOP Conf. Series: Mater. Sci. Eng. 925.

Nagabhushana, N., Rajanna, S., Ramesh, M. R. and Pushpa, N. (2020b). Influence of temperature on friction and wear behavior of aps sprayed nicrbsi/flyash and nicrbsi/flyash/tio2 coatings. J. Green Eng. 10.

Nagabhushana, N., Rajanna, S., Mathapati, M., Ramesh, M. R., Koppad, P. G., and Reddy N. C. (2019). Microstructure and tribological characteristics of APS sprayed NiCrBSi/flyash cenosphere/Cr2O3 and NiCrBSi/flyash cenosphere/Mo composite coatings at elevated temperatures. Mater. Manuf. Mater. Res. Express 6.

Nagendra, J. and Prasad, M. S. G. (2020). FDM process parameter optimization by taguchi technique for augmenting the mechanical properties of nylon-aramid composite used as filament material. J. Inst. Eng. (India): Series C 101.

Nagendra, J., Prasad, M. S. G., Shashank, S., and Ali S. M. (2018). Comparison of tribological behavior of nylon aramid polymer composite fabricated by fused deposition modeling and injection molding process. Mater. Manuf. Int. J. Mech. Eng. Technol. 9.

Nagendra, J., Prasad, M. S. G., Shashank, S., Vijay, N., Ali, S. M., and Suresh, V. (2019). Nylon-aramid polymer composite as sliding liner for lube-less sliding bearing by fused deposition modeling. In AIP Conference Proceedings. 2057.

Nagendra, J., Srinath, M. K., Sujeeth, S., Naresh K. S., and Prasad M. S. G. (2021). Optimization of process parameters and evaluation of surface roughness for 3D printed nylon-aramid composite. Mater. Today: Proc. 44.

Padmini, B. V., Niranjan, H. B., Kumar, R., Padmavathi, G., Nagabhushana, N., and Mohan, N. (2019). Influence of substrate roughness on the wear behaviour of kinetic spray coating. Mater. Today: Proc. 27.

Pazhuhanfar, Y. and Eghbali, B. (2021). Processing and characterization of the microstructure and mechanical properties of Al6061-TiB$_2$ composite. Int. J. Miner. Metall. Mater. 28(6):1080–1089.

Prashant, S. N., Nagaral, M. and Auradi, V. (2012). Preparation and evaluation of mechanical and wear properties of 6061Al reinforced with graphite particulate metal matrix composite. Int. J. Metall. Mater. Sci. Eng. (IJMMSE) 2(3):85–95. ISSN 2278-2516.

Puneeth, H. V. and Prasad, M. S. G. (2019). Biological factors influencing the degradation of water-soluble metal working fluid. Sustain. Water Resour. Manag. 5.

Rajasekhar, S. and Yedula, H. R. (2017). Experimental Investigation on Wear Rate of Al6061/SiC/Zr hybrid metal matrix composite. Int. J. Eng. Res. Mech. Civil Eng. (IJERMCE) 2(9):117–119. ISSN (Online) 2456-1290.

Rajesh, A., Gopal, K., De Poures, M. V., Rajesh Kumar, B., Sathiyagnanam, A. P. and Damodharan, D. (2020). Effect of anisole addition to waste cooking oil methyl ester on combustion, emission and performance characteristics of a DI diesel engine without any modifications. Fuel 278.

Rajesh, P. V., Roseline, S., and Paramasivam, V. (2016). Evaluation of mechanical properties of Al6061 Metal Matrix Composite Reinforced With Fused Zirconia Alumina. Int. J. Adv. Eng. Res. 11(1):1–9.

Raju, T. H., Chandan, B. B., Vinaya, B., Udayashankar, S., Jagadeesha, T., and Gajakosh, A. K. (2021a). Influence of zircon particles on the characterization of Al7050-Zircon composites. Mater. Today: Proc. 47:2241–2246.

Raju, T. H., Kumar, R. S., Udayashankar, S., and Gajakosh, A. (2021b). Influence of dual reinforcement on mechanical characteristics of hot rolled AA7075/Si3N4/Graphite MMCs. J. Inst. Eng. (India): Series D. 102.

Ranganathaiah, C., Sanjeevamurthy, K., Prasad, R., Harish, S,. and Chandra, B. T. (2014). Preparation and evaluation of mechanical properties of Al6061 reinforced with E-Glass fiber metal matrix composites. Int. J. Mech. Prod. Eng. Res. Dev. 4(2):43–48.

Sathish, T., Arul, S. J. Gopal, K., Velmurugan, G., and Nanthakumar, P. (2021). Comparison of yield strength, ultimate tensile strength and shear strength on the annealed and heat-treated composites of stainless steel with fly ash and ZnO. Mater. Today: Proc. 46:3165–4348.

Sathish, T., Kaliyaperumal, G., Velmurugan, G., Arul S. J., De Poures, M. V., and Nanthakumar, P. (2020). Investigation on augmentation of mechanical properties of AA6262 aluminium alloy composite with magnesium oxide and silicon carbide. Mater. Today: Proc. 46.

Shanmugam, R., Dillikannan, D., Kaliyaperumal, G., De Poures, M. V., and Babu, R. K. (2021). A comprehensive study on the effects of 1-decanol, compression ratio and exhaust gas recirculation on diesel engine characteristics powered with low density polyethylene oil. Energy Sources Part A: Recovery Util. Environ. Eff. 43.

Srinath, M. K. and Prasad, M. S. G. (2018). Numerical analysis of heat treatment of TiCN coated AA7075 aluminium alloy. In AIP Conference Proceedings. 1943.

Srinath, M. K. and Prasad, M. S. G. (2019a). Corrosion analysis of TiCN coated Al-7075 alloy for marine applications: A case study. J. Inst. Eng. (India): Series C. 100.

Srinath, M. K. and Prasad, M. S. G. (2019b). Surface morphology and hardness analysis of TiCN coated AA7075 aluminium alloy. J. Inst. Eng. (India): Series C 100.

Srinath, M. K. and Prasad, M. S. G. (2020). Mathematical modeling on the residual stresses in coatings due to heat treatments. Mater. Sci. Forum 978 MSF.

Srinath, M. K., Nagendra, J., Puneeth, H. V., and Prasad M. S. G. (2021). Micro-structural, physical and tribological properties of HVOF spayed (TiC + Cr2O3) composite coatings. Mater. Today: Proc. 44.

Subbian, V., Siva Kumar, S., Chaithanya, K., Arul, S. J., Kaliyaperumal, G., and Adam, K. M. (2021). Optimization of solar tunnel dryer for mango slice using response surface methodology. Mater. Today Proc. 46.

Suresh, S., Shenbaga, N., and Moorthi, V. (2013). Process development in stir casting and investigation on microstructures and wear behavior of TiB$_2$ on Al6061 MMC. Procedia Eng. 64:1183–1190.

140 Experimental study and analysis of vibratory stress relief system for hard turned EN 31 component

Pankaj Bhokare[a] and K. V. Ramana[b]
Department of Mechanical Engineering, K. L. University, Vaddeshwaram, India

Abstract

When the cutting forces are removed from the workpiece in a single-point cutting tool machining operation, stresses are created in the body. External charges have no effect on these pressures. To eliminate residual stresses in turned components, VSR methods can be applied. The previous works using VSR methods are mainly performed on welded structures while this work aims to relieve the residual stresses of smaller hard turned EN 31 machined components. The primary purpose of this research is to get a better knowledge of the process for reducing residual stresses in materials after they have been machined. Experiments have demonstrated that VSR can minimise residual tensions significantly. It is critical to comprehend the factors decreasing residual stresses. The experiments are conducted on the hardened steel EN 31. Time, natural frequency, and amplitude are VSR parameters considered for experiments. Dia. 28 turned cylindrical components are used to reduce the residual stress. Taguchi L16 orthogonal array is used to prepare a mathematical model. Frequency plays important role in the VSR for reducing residual stress. The experimental results indicate that predicted residual stress values (56 and 137.8) are very close to actual residual stress values (61.6 and 148.4). Thus, the results are validated by experiments and mathematical models respectively.

Keywords: amplitude, EN31, natural frequency, residual stresses, Taguchi OA, turned component, VSR.

Introduction

Residual stress generated in different production process forecasts has been an area of research. There are many causes for residual stresses to be observed. Residual stresses in the components are caused by all manufacturing and joining processes, like forging, casting, turning, grinding, soldering, heating, and non-conventional machining. These also are caused by confining the material to a small area due to differences in plastic flow. Different Destructive and Non-Destructive methods are used for the measurement of residual stresses. Several NDT instruments were developed and used successfully for the measurement of residual stresses like ultrasonic, X-ray diffraction, eddy current, and Barkhausen sound method. Various destructive methods are also available for measuring residual stress. X-ray diffraction has been a well and precise technique for deciding levels of residual stress on solid material (Dawson and Moffat, 1980). XRD for residual stress measurements is very cost-efficient and easy to access with a compact and automated diffractometer. Bhokare and Ramana (2019) has determined the mathematical model to calculate residual stress for various machining parameters. The residual stress measurement was using the X-ray diffraction technique.

Mechanical, thermal, and electromagnetic approaches are used to relieve residual stress. Hammering or vibrations are used in the mechanical method. Vibrations are used to introduce energy into the part in the vibration method. For the strained atomic structure, the energy generated by heat and the energy generated by vibrations is the same. The energy helps to restructure the crystalline structure, which reduces strain and stabilises the thing without distorting it. For a fixed period of time, the VSR process introduces low frequency and high amplitude vibrations into a component based on its weight. Without deformation or changing of tensile strength, fatigue resistance, and yield point, static equilibrium is restored and residual stress is released. The most effective vibrations are resonant vibrations because stress is better distributed in resonance frequency vibrations. The most common pieces of equipment employed are a robust variable-speed vibrator mounted to the object and electronic control panels. Both of these products are kept in a small cabinet that can be moved about. There are two simple criteria to follow for all operations:

* Provide rigid support for the object by isolation that allows it for free vibration.
* In order to transfer the maximum vibrating power produced, the vibrator should be directly linked to the item.

[a]bhokarepankaj79@gmail.com; [b]proframana@kluniversity.in

Vibration or cyclic loading is used to release residual stress in produced components in a technique known as vibratory stress alleviation. A cantilevered beam was investigated in order to construct a model that might predict residual stress (Hassan, 2014). Various VSR devices have been developed in recent years to minimise residual stress. An experiment is conducted to reduce residual stress (Vardanjani et al., 2016). The beam is clamped by a tool holder, and the chuck connects it to the spindle with an eccentric cam-like element. The revolving spindle, which is applied to the beam, forces the cam to provide a cycle load. In Wang et al. (2015), a shaker platform is created to introduce the right amount of dynamic stress for reducing residual stress. Small workpieces cannot be dealt with directly by the eccentric mass motor as they are too small to clamp. The eccentric weight motors were physically fastened to the work by (Jurčius et al., 2010). The most serious issue is that the frequency introduced by modern VSR machines is too lower than the natural frequency of many workpieces. The workpiece is unable to create adequate dynamic stress to induce resonance.

Because of its affordability and outstanding results, the VSR treatment is frequently used in welding, foundry, and industries. Few investigations is performed on aluminum parts and hard-turned EN 31 machined components. Rao et al. (2005) Used a fatigue testing system to investigate and express the stress fluctuation in stainless elements. Mayer et al. (2005) Discovered that 319-T7 has a durability limit of 107 cycles and that stress measurements can predict material life. In the reduction of residual stress is investigated and the composition development of magnesium alloy using the cantilever system (Wang et al., 2015). Saurav et al. (2017) used electromagnetic wave VSR system and reduced the residual stress of AISI 316 welded component. Rao et al. (2007) studied the cyclic stress-strain of 304L specimens.

Important details concerning VSR theory should be mentioned. Dawson (1975)stated that vibrations are necessary for stress release due to the *production* of strain in major stress regions. Lattice slip, in other words, can lessen strain. According to the standard model presented by Klotzbucher and Kraft (1987), when residual stress along with vibrating stress surpasses the material's yield strength, the stress is alleviated. The premise is that the subsequent plastic flow will allow the previously stressed area to restore to a low level of residual stress after the vibrational amplitude is reduced. This is due to the fact that metals' plastic deformation processes are suggestive of dislocation movement. Internal pressures and thermally generated lattice vibrations are thought to cause dislocation segments to relocate towards a more stable configuration after being liberated from weakly pinning point defects. Dislocation mushrooming occurred as the vibration increases, and the mushrooming ceased at the conclusion of the vibration, according to Walker et al. (1995). This demonstrated that as the dislocation increased, the strain decreased. These observations suggest that the strain energy was released during vibration as a result of dislocation proliferation.

VSR technology, which uses electric motors with eccentric mass to excite treated parts to vibrate near their resonance frequencies, is a more assuring way to reduce the residual stress than thermal aging. VSR is a low-cost, high-efficiency, and environmentally benign technology (Sun et al., 2004; Gao et al., 2018; Yang, 2009), however, it is ineffective for parts whose excitation frequency is lower than their natural frequency. Furthermore, the injected vibratory stress distribution is unequal as the excitation frequency and extended wavelength are low, resulting in uneven residual stress release across the portion.

The ultrasonic VSR (UVSR) approach was used by Shalvandi et al. (2013) to investigate the residual stress alleviation of tiny Almen strips. In this investigation, it has been seen that the percentage reduction of residual stress is approximately similar to thermal aging. The stress reduction is affected by vibrational amplitude and treatment duration. Du et al. (2010), Du et al. (2011) found that stress is decreased for the photoresist SU-8 layer. The vibrational effects on cross-linked networks like defect removal, homogeneity, and local damage may be considered internal stress reduction. UVSR technology can be used on nonmetallic items, according to the study.

Zhang and Wang (2017), Wang and Wang (2015), Wang et al. (2013) studied UVSR for a rectangular cross-section slender rod, and the studies revealed considerable impacts on residual stress alleviation. They came to the conclusion that the important condition for UVSR technology is given by the addition of initial and extra-vibratory stresses. Although a few UVSR applications have been presented in recent years, the majority of the work has been presented on Almen samples or thin rods. Only a few cases of use on machined parts and welded structures is described. Furthermore, the majority of the current work consisted of feasibility studies that lacked an in-depth examination of the characteristics and operating mechanisms.

In previous works, the VSR system for relieving the residual stress is applied to welded structures on a wide scale but their application to small machined components is still lacking in the literature. The concept of VSR treatment is considered using a computational model (Yang, 2009). The two main parameters in VSR are considered as frequency and amplitude of vibration and both the resonant and non-resonant vibration can reduce the magnitude of residual stresses conditionally. The resonant vibrations are most effective in VSR treatment. Due to their high cost and feasibility issues, this work considers the usage of sub-resonant vibration. The natural frequency of the machine component is very high so we used sub-resonant VSR to reduce the residual stress. The work discussed in this study aims to enhance the stability of EN 31 components that have been hard turned. VSR is found to disperse induced residual stresses and decrease workpiece deformation (Guo, 2010). Experimental validation is carried out based on this concept. By investigating the influence of VSR on the dimensional stability of hard-turned EN 31 machine components, a method to improve their stability is developed.

The rest of the paper is as follows. Section 2 discusses the experimentation on EN 31 workpiece material. The results and discussions are illustrated in Section 3 with conclusions in Section 4.

Experimentation

Test specimen

For this work, the selected workpiece material is EN31 steel cylindrical shape having a diameter of 28 mm. EN31 is a high-tensile steel alloy with good ductility and wear resistance. Due to these properties, EN31 is widely used in the manufacturing of mechanic opponent sent. It contains nickel, chromium, and molybdenum.

Experiment details

The turning is done on BFW make CNC machine using different machining parameters. Residual stresses generated due to turning operation are measured by the X-ray diffraction method. In this work, the Taguchi method was employed to formulate the experimentation. By Taguchi technique, process parameters are compared and the total number of experiments is reduced. The process parameters examine included Amplitude, Frequency, and Time.

VSR Setup

The experimental VSR setup for the current study is depicted in Figure 140.1. The workpiece material used in the study is the bar of EN 31. Further, the workpiece hardened to 55 Rockwell Hardness. To reduce discrepancies in conducted trials, the test pieces were machined on a CNC turning center. Figure 140.2 shows the actual setup used for VSR. The pieces after turning are shown in Figure 140.3. The International Standards nominated Cemented carbide turning tool inserts are CNMG coated.

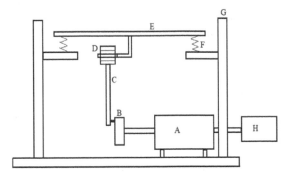

Figure 140.1 VSR setup A = Motor, B = Eccentric disc, C = Connecting rod, D = Bearing, E = Exciter table, F = Spring, G = Support, H = DC Supply

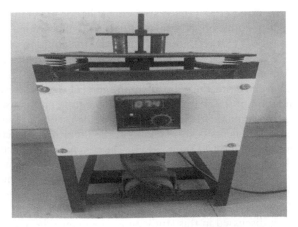

Figure 140.2 The actual setup used for VSR

VSR is used to provide controlled and monitored vibration to the test specimens. For the experiment of mechanically actuated vibration, a shaker is used. Vibration frequency is controlled using a frequency controller. To vary the amplitude of vibration, the eccentric disc has holes with known distances. The eccentric disc connected to the electric motor further translates vibrations through the connecting rod. A Special purpose fixture created to hold the specimen with a shaking tabletop. The frequency is steadily increased up to the required level and kept constant throughout the experiment.

Selection of the orthogonal array

The test was conducted at various frequencies and amplitudes. Table 140.1 shows the factor information. Taguchi L16 orthogonal array is used to conduct the experimentation and record the test results. Amplitude, frequency, and time are considered primary factors for investigation while the secondary factors are not considered. We used these parameters to estimate the values of residual stress and as a criterion for comparing workpieces, before and after VSR treatment. This method is mentioned in some corporations and papers as well as in the research done by (Bhokare and Ramana, 2019; Teggari et al., 1982), as a good method of measuring and comparing residual stress, when other methods are hard to perform or are impossible. Furthermore, this method uses sub-resonant VSR that minimises the total cost incurred on making the experimental setup. In this study, the L16 orthogonal array of the Taguchi experiment is selected for three parameters with four levels as shown in Table 140.2.

Results and Discussion

Sixteen experiments are conducted for the above-mentioned sets of parameters. And the result recorded as per the Table 140.3.

In addition, a numerical model was developed by a Minitab, and the analysis of variance (ANOVA) by the Minitab is shown in Figure 140.4. The model's quality of fit is shown by the R^2 coefficient. After taking into account the important variables, the coefficient R^2 suggests that the model can fit 84.4% of the overall variability.

Main effect plot

There is a considerable influence when different levels of a component have diverse effects on the reaction. In a main effects plot, a line graphs the response mean for each factor level. The main effect plot for the above experiment is shown in Figure 140.5. This plot clearly indicates that the frequency of the vibration is having significant effect in the process.

Figure 140.3 Turned test piece

Table 140.1 Factor information

Symbol	Factors	Levels			
		1	2	3	4
X1	Amplitude (mm)	3	5	6	7
X2	Frequency (Hz)	750	850	950	1050
X3	Time (sec)	3	5	6	7

Table 140.2 Taguchi orthogonal array

Amplitude	Frequency	Time	Residual stress after VSR in N/mm²	Residual stress before VSR in N/mm²
3	750	3	60.3	67
3	850	4	7.5	8.5
3	950	5	89.0	99
3	1050	6	156.5	174.3
5	750	4	13.9	15.3
5	850	3	252.8	280.9
5	950	6	194.5	216
5	1050	5	267.4	297.1
6	750	5	237.5	263.9
6	850	6	287.4	319.3
6	950	3	544.0	604.5
6	1050	4	493.6	548.5
7	750	6	192.0	213.2
7	850	5	62.0	68.7
7	950	4	66.5	74
7	1050	3	485.0	539

Table 140.3 ANOVA

Factor	DF	Adj SS	Adj MS	F-Value	P-Value
Amplitude	1	82031	82031	4.43	0.057
Frequency	1	111057	111057	6.00	0.031
Time	1	26671	26671	1.44	0.253
Error	12	222005	18500		
Total	15	441763			

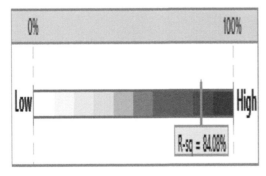

Figure 140.4 Percentage of variation

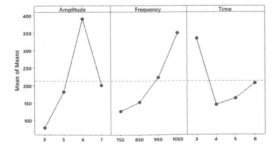

Figure 140.5 Main effects plot for means

Mathematical Model

An optimisation plot is a tool in the Response optimiser that shows how different experimental settings affect the projected responses for a stored model. The study's optimisation graphic is shown in Figure 140.6.

Equation 1 shows the regression equation for prediction through random variable values. Using the regression equation one can easily predict the residual stress after the VSR process.

$$-3237 + 367X_1 + 4.35X_2 + 159X_3 - 37.7X_1^2 + 58.4X_3^2 - 0.802X_2 X_3 \tag{1}$$

Validation

Two validation experiments were performed by taking random to verify residual stress and results are recorded in the Table 140.4 below.

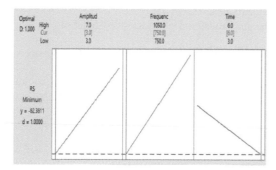

Figure 140.6 Optimisation plot

Table 140.4 Validation experimentation

Expt.No.	Factors used	Actual residual stress	Predicted residual stress	Error
1	7, 1050, 6	61.6	56	9.09 %
2	4, 750, 5	148.4	137.8	

Conclusion

This paper proposed regression technique for predicting the residual stress after VSR responses. An experimental investigation of the relation between VSR parameters and residual stresses is discussed in this work. Following conclusions are derived.

- Residual stress in manufacturing components can be reduced via cyclic loading. This method can be applied for operations that do not require a change in the part's crystalline lattice.
- Mathematical model has been developed based on control factors like amplitude, frequency and time.
- From the analytical observations it is clear that the frequency of vibration plays important role in the reducing the residual stress.
- Using the regression equation one can further predict the residual stress value with the factors mentioned in this study.
- It is clear from the experimental validation results that the predicted residual stress values (56 and 137.8) are close to actual residual stress values (61.6 and 148.4). Thus, the results are validated by experiments and the mathematical model respectively.

References

Bhokare, P. and Ramana, V. (2019). Optimisation of process parameters for residual stress in hard turning of EN 31 using Taguchi OA. Int. J. Recent Technol. Eng. 8(1):2597–2600.

Dawson, R. and Moffat, D. G. (1980) Vibratory stress relief: A fundamental study of its effectiveness. J. Eng. Mater. Technol. 102(2):169–176.

Dawson, R. (1975). Residual stress relief by vibration. Ph.D. thesis, Liverpool.

Du, L. Q., Wang, Q. J., and Zhang, X. L. (2010). Reduction of internal stress in SU 8 photoresist layer by ultrasonic treatment. Sci. China Ser. E: Technol. Sci. 53:3006–3013.

Du, L. Q., Wang, Q. J., and Zhang, X. L. (2011). Application of ultrasonic stress relief in the fabrication of SU-8 Micro Structure. Key Eng. Mater. 483:3–8.

Gao, H., Zhang, Y., Wu, Q., Song, J., and Wen, K. (2018). Fatigue life of 7075-T651 aluminium alloy treated with vibratory stress relief, Int. J. Fatigue 108:62–67.

Guo, J. K. (2010). Study on micro yield mechanism of aluminium alloy thick plate subjected to vibratory stress relief and experimental study. Master's Thesis, Central South University, Changsha, China.

Hassan, A. H. (2014). Vibratory stress relief analysis. Asian J. Appl. Sci. 7(5):273–293. 10.3923/ajaps.2014.273.293.

Jurčius, A., Valiulis, A. V., Černašejus O., Kurzydłowski, K. J., Jaskiewicz, A., and Lech-Grega, M. (2010). Influence of vibratory stress relief on residual stresses in weldments and mechanical properties of structural steel joint. J. Vibroengineering 12(1):133–141.

Klotzbucher, E. and Kraft, H. (1987). Residual stresses in science and technology, (p. 959), USA: DGM Metallurgy Information.

Mayer, H., Ede, C., and Allison, J. E. (2005). Influence of cyclic loads below endurance limit or threshold stress intensity on fatigue damage in cast aluminium alloy 319-T7. Int. J. Fatigue 27:129–141.

Rao, D. L., Chen, L. G., Ni, C. Z., and Zhu, Z. Q. (2005). The mechanism for vibratory stress relief of stainless steel. Trans. China Weld Inst. 26:58–60.

Rao, D., Wang, D., Chen, L. and Ni, C. (2007). The effectiveness evaluation of 314L stainless steel vibratory stress relief by dynamic stress. Int. J. Fatigue 29:192–196.

Saurav, M., Arivarasub, M., Arivazhagana, N., and PhaniPrabhacark, K. V. (2017). The residual stress distribution of CO_2 laser beam welded AISI 316 austenitic stainless steel and the effect of vibratory stress relief. Mater. Sci. Eng. A 703:227–235.

Shalvandi, M., Hojjat, Y., Abdullah, A., and Asadi, H. (2013). Influence of ultrasonic stress relief on stainless steel 316 specimens: A comparison with thermal stress relief, Mater. Des. 46:713–723.

Sun, M. C., Sun, Y. H. and Wang, R. K. (2004). The vibratory stress relief of a marine shifting of 35 bar steel, Mater. Lett. 58.

Teggari, R., Merchant, H. C., and Bodre, R. A. (1982). Vibration stress relief in single crystals. In Proceedings of ASME conference on productive applications of mechanical vibrations 3, (November 14–19, 1982, pp. 59–74), Phoenix, AZ.

Vardanjani, M. J., Ghayour, M., and Homami, R. M. (2016). Analysis of the vibrational stress relief for reducing the residual stresses caused by machining. Exp. Tech. 40(2):705–713.

Walker, C. A., Waddell, A. J., and Johnston, D. J. (1995). Vibratory stress relief—an investigation of the underlying processes. Proc. Inst. Mech. Eng. Part E: J. Proc. Mech. Eng. 209(1):51–58. 10.1243.

Wang, J. S., Hsieh, C. C., Lin, C. M., Kuo, C. W., and Wu, W. (2013). Texture evolution and residual stress relaxation in a cold rolled Al-Mg-Si-Cu alloy using vibratory stress relief technique. Metallur. Mater. Trans. A 44:806–818.

Wang, J. S., Heish, C. C., Lai, H. H., Kuo, C. W., Wu, T. Y., and Wu, W. (2015). The relationships between residual stress relaxation and texture development in AZ31 Mg alloys via the vibratory stress relief techniques. Mater. Charact. 99:248–253.

Wang, R. Y. and Wang, S. Y. (2015). Mechanism analysis and experimental research of the ultrasonic vibration ageing. J. Chin. Agric. Mechaniz 36:59–62.

Yang, Y. P. (2009). Understanding of vibration stress relief with computation modeling. J. Mater. Eng. Performance 18:856–862.

Zhang, L. and Wang, S. Y. (2017). Feasibility analysis and experiment research of ultrasonic vibratory stress relief. Mach. Des. Manuf. 7:140–142.

141 A Comprehensive study on mechanical properties of Ti alloys produced by casting, milling and direct metal deposition

Pratheesh Kumar S.[a], Anand K.[b], Mohanraj R.[c], and Karthikeyan R.[d]

Department of Production Engineering, PSG College of Technology, Coimbatore, India

Abstract

A cast part is made by pouring hot molten metal or plastic into a mould and letting it harden inside the form. Whereas, as the cutter advances into the workpiece, milling removes material. On the other hand, in direct metal deposition, objects are created by layering metal powders. The mechanical properties of parts produced by casting, milling, and direct metal deposition (DMD) methods are discussed in this study. The variation of mechanical properties such as yield strength, Ultimate Tensile strength, elongation, hardness, ductility, and so on as a function of influencing operating parameters has been studied for titanium materials. This study helps to determine the suitable process parameters, such as laser power, scanning speed and rotation speed. The observed result establishes that DMD is preferable than casting and milling. In comparison to other process parameters, the laser power, scanning speed, and gas flow rate were determined to have the highest influence. To better understand titanium materials, have the optimum mechanical qualities for diverse industrial uses, this investigation might be put to use.

Keywords: Casting, direct metal deposition, gas flow rate, laser power, milling, scanning speed.

Introduction

Casting is an ancient manufacturing technique that dates all the way back to 6000 years ago. Even now, it is primarily used. Then, in the late 18th century, the industrial revolution altered manufacturing processes in both the United States and Europe. Milling came into play and was primarily used. Machines were widely used to manufacture products. Then, in the early 1900's, mass production was introduced following factory electrification. Then, approximately 40 years ago, additive manufacturing became available. They were initially prohibitively expensive but have since been commercialised and made widely available. The solidified component, referred to as casting, is ejected or broken free from the mould to complete the process. A pattern is used to create a mould in sand during the casting process. A gating system is incorporated, and the mould cavity is filled with molten metal. After cooling, the metal is removed from the casting (Campbell, 2011). Milling is one of the most frequently used processes for precision machining of custom parts. It can be classified into two types: up milling and down milling. Up milling is accomplished by feeding in the opposite direction of the cutter rotation at the point of engagement. Down milling occurs with the feed direction parallel to the cutter rotation direction at the point of disengagement (Maekawa et al., 2013).

Additive manufacturing is a form of 3-D printing. This process constructs parts layer by layer in order to deposit the material in accordance with the design data. It is a computer controlled process that deposits materials to create three-dimensional objects (Pfeiffer et al., 2021; Laguna et al., 2021).

Three broad categories have been established for additive manufacturing technology. The second technology completely melts the material, including direct laser metal sintering. The third category is stereolithography, which employs a process called photo polymerisation. Materials such as biochemical, ceramics, metals, and thermoplastics can be used in additive manufacturing (Pérez et al., 2020). The advantages of additive manufacturing include shape freedom, lightweight design, and the potential for tooling elimination. The disadvantages include high manufacturing costs, restricted component sizes, and a discontinuous manufacturing process (Ford and Despeisse, 2016).

Casting

Casting is a manufacturing process in which a molten metal, such as metal or plastic, is introduced into a mould, allowed to solidify, and then ejected or broken out to create a fabricated part as shown in

---

[a]spratheeshkumarth@gmail.com; [b]mechanand@gmail.com; [c]mohanraj839@gmail.com; [d]karthiramaraj1999@gmail.com; [e]njgiradkar@gmail.com

Figure 141.1. Casting is used to create complex-shaped components that would be difficult or uneconomical to manufacture using other methods, such as cutting from solid material (Campbell, 2015).

Part building terminology

The two major terms that have been processed during the casting process are solidification and dendrite formation. Solidification occurs at the metal-metal interface. The grain structure of pure metals is determined by heat transfer into the mould and the metal's thermal properties, as illustrated in Figure 141.2, as well as grain flow during the casting process, as illustrated in Figure 141.3. Solidification continues towards the mould centre, where the initial skin formed near the mould wall has undergone rapid heat removal, resulting in the formation of randomly oriented grains (Campbell, 2015). Dendrite formation occurs during the casting process; nucleation occurs at the metal interface; nuclei are randomly oriented and grow to form dendrites. Dendrite growth occurs when dendrites are slowly cooled, allowing them to grow long, whereas rapid cooling causes dendrites to grow short. Because dendrites become grains, slow cooling results in a grain structure with large grains and rapid cooling results in small grains in the solidified metal. As the solidification process progresses, additional arms and dendrites form until the melt crystallises during the casting process (Swift and Booker, 2013).

Milling

Milling is a type of machining that utilises rotary cutters to remove material from a workpiece by advancing the cutter into it. This has been accomplished by using a cutter with numerous teeth and spinning it rapidly or by advancing the material slowly through the cutter; most frequently (Louhenkilpi, 2014).

Part building terminology

Milling is a subtractive process in which excess material other than the part required is removed using cutting tools, as illustrated in Figure 141.4. Machining processes remove materials by generating chips or

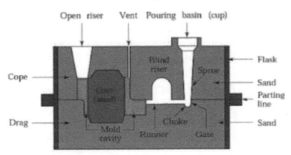

Figure 141.1 Schematic representation of casting process

Figure 141.2 Grain structure in casting

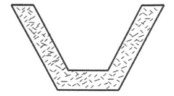

Figure 141.3 Grain flow in casting

physically separating them. Low resistance to plastic deformation materials form a continuous chip. Due to their high resistance deformation, brittle materials form small, easily fractured, or segmented chips (Sahoo and Sahu, 2014). Two technical events occur concurrently during the chip removal process.

Direct Metal Deposition

Direct metal deposition (DMD) is an additive manufacturing technique that utilises a powder jet to repair, hard face, and repurpose forging dies. DMD is a process in which heat is generated using a concentrated heat source of various types, sufficient to melt the substrate's surface and form a small melt pool as shown in Figure 141.5. Laser pulse parameters were also shown to have a significant effect on residual stress fields generated during DMD (Xu, 2021).

Part building terminology

DMD manufactures parts in a layer-by-layer fashion. High cooling rates and a distinct thermal history result in the formation of features such as grains that are elongated in the build direction, dendritic solidification structures, and prominent crystallographic textures, as illustrated in Figure 141.6. DMD deposits material at a faster rate and is not constrained by a build chamber, allowing for the production of larger parts in less time. Adjusting each powder feeding system layer by layer enables the fabrication of parts with chemical gradients. DMD is an enticing additive manufacturing technique for the fabrication of complex material systems (Pinkerton, 2010; Kumar et al., 2021).

Casting and milling are both traditional manufacturing processes used in plant machinery and transmission housings. DMD is a rapidly growing and innovative manufacturing technique with numerous application possibilities in industries such as aerospace and automotive. The principle of the process is investigated, as well as the mechanical properties. Additionally, the effect of a variety of input variables on a variety of output variables is examined. Mechanical properties of cast, milled, and DMD components are strongly influenced by the input parameters that define the part's quality. As a result, the fluctuation of output parameters as a result of changing process parameters is critical for titanium material. The following literature review assists in gaining a better understanding of how output process parameters vary in response to changes in process input factors.

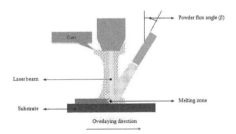

Figure 141.4 Grain flow in milling

Figure 141.5 Schematic representation of DMD

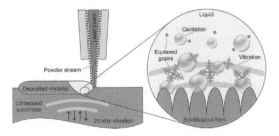

Figure 141.6 Grain structure control in DMD

Cha et al. (2021) examined the mechanical properties of a high-speed titanium alloy using a variety of casting processes. The rotation speed is set to 180 revolutions per minute, the gas flow rate is set to 6 litres, the UTS value is 250 MPa, and the hardness value is determined by the findings. Sankaranarayanan et al. (2013) used the twin roll casting process to investigate the microstructure and mechanical properties of titanium matrix composites. According to the findings, the casting speed is 400 rpm, the grain size is 20, the hardness value is 71 MPa, and the UTS value decreases as the gas flow rate decreases.

Zhang et al. Gupta et al. (2020) examined the thermal exposure and mechanical properties of titanium alloys produced using various casting technologies. The rotation speed is set to 140 revolutions per minute and the gas flow rate is set to 12; the elongation value obtained is 1.17; as the rotation speed increases, the mechanical properties value decreases. Li et al. (2021) used stir casting to investigate the microstructure and mechanical properties of titanium material. The casting speed is set to 800 rpm and the stirring temperature to 8, and the UTS value obtained is 300MPa, which decreases with increasing casting speed. Yang et al. (2022) investigated the mechanical properties and wear resistance of titanium alloys using the centrifugal casting process. The grain size is 15 mm, the rotation speed is 170 rpm, and the vibration time is also calculated. The mechanical properties values vary and have been ranged and the mechanical properties values increase as a result of the findings.

Zhang et al. (2020) used mechanical milling to investigate the mechanical properties of a novel titanium alloy. The milling speed was set to 120 rpm, the milling time was set to 50 seconds, and the feed rate was set to 16mm/sec. As expected, the UTS, hardness, and ductility values decreased as the milling speed increased. Yao et al. (2018) used wet mechanical milling to investigate the mechanical properties of titanium alloy and hydriding. The milling time is 70 seconds, the standoff distance is 1, the UTS value is 923 MPa, and the hardness is 321.4 HV, with the value decreasing as the feed rate is increased.

Camara et al. (2021) examined the morphological and mechanical properties of titanium materials prepared using a high energy ball milling process. The milling speed was set to 180 rpm and the feed rate was set to 18mm/sec. The elongation value was 4.1 and the ductility value was 19. Both values decreased as the standoff distance decreased, as observed. Zhu et al. (2010) synthesised and mechanical milled titanium storage materials. The milling time is 80 seconds and the feed rate is 18 mm/second. The UTS value obtained is 490 MPa and the elongation is 4, which will gradually increase as the standoff distance increases. Clinktan et al. (2019) examined the mechanical properties of titanium composites and the milling process. The milling speed was set to 140 rpm and the milling time was 90 seconds. The mechanical properties, such as ductility, were determined to be 18 and decreased as the feed rate increased.

Dong et al. (2017) used the milling process to investigate the interface strength and mechanical properties. The milling speed was set to 120 rpm, the standoff distance was set to 4, the yield strength was 520 MPa, and the ductility was set to 12. As the feed rate decreased, the UTS increased as a result of the findings. Cai et al. (2021) investigated the microstructure and mechanical properties of titanium-based materials. The milling time is 80 seconds, the hardness value is 238 HV, and the young's modulus value is 45 MPa, which increases as the hardness value decreases.

Liu and Liu (2019) investigated the magnetic properties of titanium magnetic SMA using direct laser metal deposition and heat treatment. The laser power used is 800 W, the powder flow rate is 18 g/minute, and the mechanical properties such as ultimate tensile strength are 490 MPa, and the value of UTS increases as the laser power is increased.

Rabiey et al. (2020) used a laser deposition process to modify the mechanical and wear properties of low stacking titanium materials. The scanning speed is set to 500 g/minute, the powder flow rate is 8g/min, the UTS value is 300 MPa, and the properties values obtained can be decreased as the input parameter powder flow rate is also decreased. Nong et al. (2021) used direct laser metal deposition to investigate the microstructure and mechanical properties of titanium. The laser power was set to 1500 W, the powder flow rate was 16 g/minute, the scanning speed was 500 mm/min, the elongation value was 19.76%, and the UTS value increased as the scanning speed decreased.

According to the reviewed literature, the output parameters of the parts, such as yield strength, ultimate tensile strength, elongation, hardness, and ductility, vary as a function of input factors such as casting speed, rotation speed, milling speed, laser power, scanning speed, and powder flow rate. Additionally, the mechanical properties of parts produced by DMD are superior to those produced by casting and milling. The research gap identified in this work is a lack of adequate on the mechanical properties of parts manufactured from titanium based materials. This work addresses a research need by examining the parts produced by casting, milling, and DMD processes. The purpose of this literature review is to gain a better understanding of the relationship between input and output parameters for the titanium-based alloys used in industry. This research could be used to establish process parameters for fabricating components for titanium based materials. The mechanical properties of titanium based alloys produced by casting, milling and DMD are presented in Tables 141.1, 141.2 and 141.3 respectively.

Table 141.1 Mechanical properties of titanium based alloy materials in casting

S. No.	Material	Casting speed (rpm)	Grain size (mm)	Rotation speed (rpm)	Gas Flow rate (l/min)	Stirring End temperature (°C)	Vibration time (sec)	Pressure (MPa)	Yield Strength (MPa)	Ultimate Tensile strength (MPa)	Elongation	Hardness (HV)	Ref.
1	Ti-45Al	400	35	160	8	575	160	-	-	400.12	0.92	-	Cha et al. (2021)
2	TiAl	900	40	160	8	540	150	-	-	593.74	1.45	-	Sankaranarayanan et al. (2013)
3	TiB3	900	35	160	10	550	200	-	120.65	407.38	20.74	-	Gupta et al. (2020)
									222.94	453.65	13.69	-	
4	Ti6Al4V	800	40	140	6	500	180	-	0	570	0.8	-	Li et al. (2021)
									900	950	1.5	-	
									925	900	2.5	-	
5	Ti-25Nb	800	20	170	4	500	140	-	563	665	13.8	-	Yang et al. (2022)
6	Ti-35Nb	1000	30	190	8	350	160	-	636	750	2.2	85	Zhang et al. (2020)

Table 141.2 Mechanical properties of titanium based alloy materials in milling

S. No.	Material	Milling time (hour)	Standoff distance (mm)	Feed rate (mm/s)	BPR	Yield strength (MPa)	Ultimate tensile strength (MPa)	Elongation	Hardness (HV)	Young's modulus (GPa)	Milling speed (rpm)	Ductility (%)	Ref.
1	TiC	70	2	16	5	499.05	582.39	26.37	205.07	110.75	100	-	Yao et al. (2018)
						761.77	945.99	4.75	275.06	127.69	100	-	
						744.66	834.14	1.19	320.18	114.42	100	-	
						-	627.96	0.56	343.08	96.1	100	-	
						-	-	-	424.83	-	100	-	
2	Ti-35Nb	70	2	16	5	355	412	7	-	-	140	-	Câmara et al. (2021)
3	Ti-45Al	70	2	14	8	-	400.12	0.92	-	-	-	20	Zhu et al. (2010)
4	TiAl	60	4	-	4	-	593.74	1.45	-	-	200	-	Clinktan et al. (2019)
5	Ti6Al4V	80	-	-	6	1062	1108	7.18	-	-	160	-	Dong et al. (2017)
6	Ti6Al4V gr	70	8	18	-	-	-	-	472	55.6	-	-	Cai et al. (2021)
7	TiB2-Tic	100	4	-	5	-	-	-	17.6	392	120	10	Liu and Liu (2019)

Table 141.3 Mechanical properties of titanium based alloy materials in DMD

S. No.	Material	Laser Power (W)	Scanning Speed (mm/min)	Powder Flow Rate (g/min)	Yield Strength (MPa)	Ultimate Tensile Strength (MPa)	Elongation	Ref.
1	Ti-6Al-4V	100	-	-	250	400	0.8	Rabiey et al. (2020)
		150	-	-	150	700	1	
		200	-	-	900	900	1.75	
		250	-	-	850	950	2.5	
		300	-	-	800	950	3	
		350	-	-	800	900	3.5	
		400	-	-	850	950	3	
2	TiAl	1000	400	14	-	593.74	1.45	Nong et al. (2021)

Mechanical Properties

Yield strength

Figure 141.7 discusses the effect of yield strength on titanium alloy based material. We can deduce from the casting process for the material Ti6Al4V that the casting speed is 800 rpm, the grain size is 40 mm, the rotation speed is 140 rpm, the stirring end temperature is 500 degrees Fahrenheit, the vibration time is 180 seconds, and the yield strength is 900925 MPa. The milling process for the material Ti6Al4V uses a speed of 160 rpm and a time of 80 seconds, yielding yield strength of 1062 MPa. The laser power used to deposit the material was varied between 100 and 400 kW, and the yield strength was obtained between 250 and 850 MPa. DMD is the best of the three processes because it requires fewer input process parameters to achieve the desired mechanical properties than the other two.

Ultimate Tensile Strength

Figure 141.8 discusses the effect of yield strength on titanium alloy based material. We can infer from the casting process for Ti6Al4V that the casting speed is 800 rpm, the grain size is 40 mm, the rotation speed is 140 rpm, the stirring end temperature is 500 degrees, the vibration time is 180 seconds, and the UTS is 570900 MPa. The milling speed is 160 rpm and the milling time is 80 seconds for the material Ti6Al4V. The UTS value obtained is 1108 MPa. The laser power used to deposit the material is varied between 100

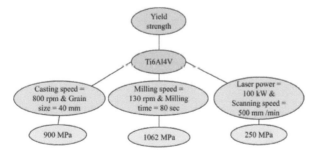

Figure 141.7 Effect of yield strength on Ti6Al4V

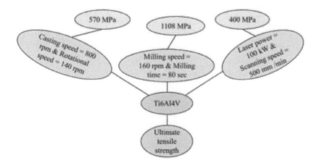

Figure 141.8 Effect of ultimate tensile strength on Ti6Al4V

and 400 kW, and the UTS value obtained ranges between 400 and 950 MPa. DMD is the best of the three processes because it requires fewer input process parameters to achieve the desired mechanical properties than the other two.

Elongation

Figure 141.9 discusses the effect of elongation on titanium alloy based material. We can infer from the casting process for the material Ti6Al4V that the casting speed is 800 rpm, the grain size is 40 mm, the rotation speed is 140 rpm, the stirring end temperature is 500 degrees Fahrenheit, the vibration time is 180 seconds, and the elongation value is 0.81.5%. The milling speed is 160 rpm and the milling time is 80 seconds for the material Ti6Al4V. The elongation value obtained is 7.18%. The laser power used to deposit the material is varied between 100 and 400 kW, and the elongation value obtained is 0.83%. DMD is the best of the three processes because it requires fewer input process parameters to achieve the desired mechanical properties than the other two.

Hardness

Figure 141.10 discusses the effect of hardness on titanium alloy based material. We can infer from the casting process for the material Ti6Al4V that the speed of casting is 800 rpm, the grain size is 40 mm, the rotation speed is 140 rpm, the stirring end temperature is 500, the vibration time is 180 seconds, and the hardness obtained is 5265 HV. The milling speed is 160 rpm and the milling time is 80 seconds for the material Ti6Al4V. The hardness obtained is 760 HV. The laser power used to deposit the material was varied between 100 and 400 kW, and the hardness was determined to be 332 HV. DMD is the best of the three processes because it requires fewer input process parameters to achieve the desired mechanical properties than the other two.

Strength

Figure 141.11 discusses the effect of strength on titanium alloy based material. We can infer from the casting process for the material Ti6Al4V that the speed of casting is 800 rpm, the grain size is 40 mm, the rotation speed is 140 rpm, the stirring end temperature is 500, the vibration time is 180 seconds, and the strength obtained is 200 MPa. The laser power used to deposit the material was varied between 100 and 400 kW, and a strength of 125 MPa was obtained. DMD can be chosen over the other two processes because it requires fewer input process parameters to achieve the desired mechanical properties.

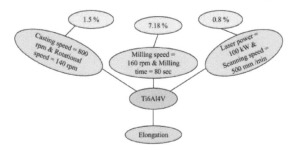

Figure 141.9 Effect of elongation on Ti6Al4V

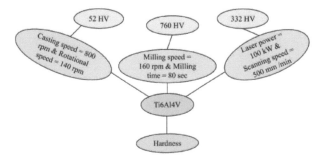

Figure 141.10 Effect of hardness on Ti6Al4V

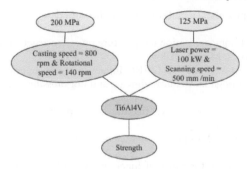

Figure 141.11 Effect of strength on Ti6Al4V

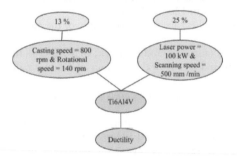

Figure 141.12 Effect of ductility on Ti6Al4V

Ductility

Figure 141.12 discusses the effect of ductility on titanium alloy based material. The milling speed is 160 rpm and the milling time is 80 seconds for the material Ti6Al4V. The ductility obtained is 13%. The laser power used to deposit the material is varied between 100 and 400 kW, and the ductility is increased by 25%. DMD can be chosen over the other two processes because it requires fewer input process parameters to achieve the desired mechanical properties.

Conclusion

Mechanical properties of a component are critical in determining the range of usefulness of a material and the expected service life of the component. This article discusses the mechanical properties of casting, milling, and DMD process components. This research contributes to the identification of the most efficient process parameters for achieving the desired mechanical properties. This study could be used to predict the optimal process parameters for titanium materials. This study discovered the following significant findings.

- The casting process increases ultimate tensile strength by adjusting input parameters such as casting speed, rotation speed, and major mechanical properties such as yield strength.
- The milling process increases the major mechanical properties such as ultimate tensile strength and young's modulus by adjusting the input parameters such as milling speed and feed rate.
- By adjusting the input parameters such as laser power and scanning speed, the mechanical properties such as yield strength and UTS are increased.
- In comparison to casting and milling, the DMD process results in superior mechanical properties.
- As a result, for all three major manufacturing processes, a high value for the input parameters results in a high value for the mechanical properties.

References

Alam et al. (2022). Influence of aluminum addition on the mechanical properties of brass/Al composites fabricated by stir casting. Mater. Today: Proc. 48(1):811–814.

Cai, X., Ren, X., Sang, C., Zhu, L., Li, Z., and Feng, P. (2021). Dissimilar joining mechanism, microstructure and properties of Ti to 316 stainless steel via Ti-Al thermal explosion reaction. Mater. Sci. Eng. A Struct. Mater. 807(140868):140868.

Campbell, J. 2011. Complete casting handbook. Oxford: Butterworth-Heinemann. doi: 10.1016/c2011-0-04123-6.

Câmara, N. T. (2021). Impact of the SiC addition on the morphological, structural and mechanical properties of Titanium composite powders prepared by high energy milling. Adv. Powder Technol. 32(8):2950–2961.

Campbell, J. (2015). Complete casting handbook: Metal casting processes, metallurgy, techniques and design. Oxford: Butterworth-Heinemann.

Cha, J. W., Jin, S.-C., Jung, J.-G., and Park, S. H. (2021). Effects of homogenisation temperature on microstructure and mechanical properties of high-speed-extruded Titanium alloy. J. Magnes. Alloy.

Clinktan, R., Senthil, V., Ramkumar, K. R., Sivasankaran, S., and Al-Mufadi, F. A. (2019). Influence of B4C nanoparticles on mechanical behaviour of Titanium nanocomposite through mechanical alloying and hot pressing. Ceram. Int. 45(15):18691–18700.

Ding et al. (2020). Processing, microstructure and mechanical properties of a novel mg matrix composites reinforced with urchin-like CNTs@ SiCp. Diam. relat. mater. 109(1):87–108.

Dong, L., Chen, W., Zheng, C., and Deng, N. (2017). Microstructure and properties characterisation of titanium materials doped with grapheme. J. Alloys Compd. 695:1637–1646.

Fathi et al. (2020). Investigation on mechanical properties and wear performance of functionally graded AZ91-SiCp composites via centrifugal casting. Mater. Today Commun. 24(1):101–169.

Filippov et al. (2018). Microstructural, mechanical and acoustic emission-assisted wear characterization of equal channel angular pressed (ECAP) low stacking fault energy brass. Tribol. Int. 123(1):273–285.

Ford, S. and Despeisse, M. (2016). Additive manufacturing and sustainability: an exploratory study of the advantages and challenges. J. Clean. Prod. 137:1573–1587.

Gupta, M. K., Singla, A. K., Ji, H., Song, Q., Liu, Z., Cai, W., Mia, M., Khanna, N., and Krolczyk, G. M. (2020). Impact of layer rotation on micro-structure, grain size, surface integrity and mechanical behaviour of Titanium alloy. J. Jpn. Res. Inst. Adv. Copper-Base Mater. Technol. 9(5):9506–9522.

Jia et al. (2021). Microstructure and properties of Ni-Co-Mn-Al magnetic shape memory alloy prepared by direct laser deposition and heat treatment. Opt. Laser Technol. 141(1):10–19.

Kumar, S. P., Elangovan, S., Mohanraj, R., and Ramakrishna, J. R. (2021). A review on properties of Inconel 625 and Inconel 718 fabricated using direct energy deposition. Mater. Today. 46(5). doi:10.1016/j.matpr.2021.02.566

Laguna, O. H., Lietor, P. F., Godino, F. J. I., and Corpas-Iglesias, F. A. (2021). A review on additive manufacturing and materials for catalytic applications: Milestones, key concepts, advances and perspectives. Mater. Des. 208(109927):109927.

Li, P., Yang, H., and Gao, M. (2021). Microstructure and mechanical properties of multi-scale in-situ Ti and CNTs hybrid reinforced composites. J. Jpn. Res. Inst. Adv. Copper-Base Mater. Technol. 14:2471–2485.

Liu, S. and Liu, D. (2019). Effect of hard phase content on the mechanical properties of TiC-316 L stainless steel cermets. Int. J. Refract. Hard Met. 82:273–278.

Louhenkilpi, S. (2014). Continuous milling of steel: Treatise on process metallurgy, (pp. 373–434). Elsevier.

Maekawa, K., Obikawa, T., Yamane, Y., and Childs, T. H. C. (2013). Metal machining: theory and applications. Oxford: Butterworth-Heinemann.

Nong, X. D., Zhou, X. L., Wang, Y. D., Yu, L., and Li, J. H. (2021). Effects of geometry, location, and direction on microstructure and mechanical properties of Ti fabricated by directed energy deposition. Mater. Sci. Eng. A Struct. Mater. 821(141587):141587.

Pérez, M., Carou, D., Rubio, E. M., and Teti, R. (2020). Current advances in additive manufacturing. Procedia CIRP 88:439–444.

Pfeiffer, S., Florio, K., and Puccio, D.. (2021). Direct laser additive manufacturing of high performance oxide ceramics: A state-of-the-art review, J. Eur. Ceram. Soc.

Pinkerton, A. J. (2010). Laser direct metal deposition: theory and applications in manufacturing and maintenance. Adv. Laser Mater. Process. 461–491.

Rabiey, M., Schiesser, P., and Maerchy, P. (2020). Direct metal deposition (DMD) for tooling repair of Titanium. Procedia CIRP 95:23–28.

Sahoo, M. and Sahu, S. (2014). Principles of Milling, (3rd ed.). McGraw-Hill Education.

Sankaranarayanan, S., Sabat, R. K., Jayalakshmi, S., Suwas, S., and Gupta, M. (2013). Effect of hybridizing micronsized Ti with nano-sized SiC on the microstructural evolution and mechanical response of Mg–5.6Ti composite. J. Alloys Compd. 575:207–217.

Soffel et al. (2021). Interface strength and mechanical properties of Inconel 718 processed sequentially by casting, milling, and direct metal deposition. Journal of Mater. Proc. Technol. 291(1):11–21.

Swift, K. G. and Booker, J. D. (2013). Casting processes. Manufacturing process selection handbook, (pp. 61–91). Oxford: Butterworth-Heinemann

Xu, W. (2021). Direct additive manufacturing techniques for metal parts: SLM, EBM, laser metal deposition. In Reference module in materials science and materials engineering.

Yang, Q., Lea, Y., Liu, X. H., Zhang, T. T., Cao, Y., Jiao, C. R., and Li, Y. H. (2022). Microstructure and mechanical properties of Titanium alloy plate produced by HCCM horizontal continuous casting. J. Alloys Compd. 893(162302):162302.

Yao, L., Han, H., Liu, Y., Zhu, Y., Zhang, Y., and Li, L. (2018). Improved dehydriding property of polyvinylpyrrolidone coated Ti hydrogen storage nano-composite prepared by hydriding combustion synthesis and wet mechanical milling. Prog. Nat. Sci. 28(1):7–14.

Zhang, J.-Y., Li-jie, Z., Jian, F., and Bing, J. (2020). Effect of thermal exposure on microstructure and mechanical properties of Ti alloy produced by different casting technologies. Trans. Nonferrous Met. Soc. China 30(7):1717–1730.

Zhu, Y., Liu, Z., Yang, Y., Gu, H., Li, L., and Cai, M. (2010). Hydrogen storage properties of Ti–C system hydrogen storage materials prepared by hydriding combustion synthesis and mechanical milling. Int. J. Hydrogen Energy 35(12):6350–6355.

142 Protection of power system in the presence of distributed generation by estimating phase angle

P. S. Patil[a], R. M. Moharil[b], and R. M. Ingle[c]

Department of Electrical Engineering, YCCE, Nagpur, India

Abstract

In traditional grid power flow is usually radial with non-directional overcurrent relay. Owing to inclusion of distributed generation power flow becomes bi-directional for which the system requires a directional overcurrent relay. This paper presents a method where the fault direction is estimating by using 3 point method. An algorithm is used to find the direction of fault. For a froward fault the angle estimated in the range of $\pi/4$ rad to $3\pi/4$. The algorithm is tested for different fault condition and found to be working satisfactorily.

Keywords: Directional relay, distributed generation, overcurrent, protection.

Introduction

With the alarming situation such as in increased alertness of conventional energy resources i.e. depletion of fossil fuels, high cost to construct new large power plants, global warming, availability of land, limitations of transmitting the power over a long distance and the revolution in the use of solar energy has exploited the use traditional energy sources. The use of dispersed energy i.e. distributed generation (DG) sources attracted the researchers toward the use of renewable energy sources (Patil and Ramteke, 2015; Patil, et al., 2021). In due course of time the distributed generation (<500kW) and micro grid has been very popular.

Issue with Distibuted Generation

As every coin has two sides, although DG grid has certain advantages but also has certain limitations to put be implemented directly. such as:

- Protection
- Islanding
- Stability
- Optimal placement

Protection Issue

In this paper address the issue of protection issue in the distributed network due to use of distributed generation. It has been widely known that the during fault the fault current changes a lot under the fact that the value of fault impedance is very less, but in case of high resistance fault the change is similar to the load change taking place in the system. At the same time In the traditional grid the power flow is unidirectional but with the introduction of distributed generation the power flow is bidirectional (Patil and Ramteke, 2015) the conventional grid is designed to for unidirectional power flow i.e. from source towards the load Figure 142.1. This becomes a major challenge for the electrical engineer to handle this situation. The present article proposes a method where the conventional 51 (non-directional) overcurrent relays (OCR) are to be simply equipped with the proposed algorithm. Where the fault location is detected with the aid of the only parameter that is the current. To equip the directional feature both voltage and current sensors are required, at the same time the directional relays (67) are to be present at both the ends of the bus bar Figure 142.2.

To mitigate the protection issue in DG system, several researchers have come with their own protection tactics to deal the problem of protection in dc network (Saleh, 2017).

For recognition of faults in LV-DC grids K. A. Saleh et al. has used Hybrid passive overcurrent relay (Saleh, 2017) whereas F. Ponci et al. (Monti, and Ponci, 2014). have used wavelets and artificial neural

---

[a]psp4india16@gmail.com; [b]337@ycce.in; [c]Rohan.ingle@gmail.com

Figure 142.1 Traditional distributed network (all non-directional relay)

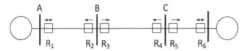

Figure 142.2 Distributed network with bidirectional power flow (directional relay)

networks to address the issue of overcurrent protection in dc microgrid protection. These protection methods rapidly detected the fault, but these methods were implemented on low voltage and short length applications like shipboards, due which basic characteristic of protective relay was found missing, such as coordination and selectivity. Their methods were unable to differentiate through faults.

Meghwani et al. (2017) have proposed a non-unit type protection technique, where the current derivatives were used for detecting the fault in dc network. Whereas in (Patil and Ramteke, 2015) has also used derivative method but for AC network. In case of high-resistance fault, the change in current is less as a result the derivative of current is found to be less, same is the case for large load change where the proposed method fails to detect the normal conditions

Bayati (2021) has used the slope of the fault current to detect the fault in DC network which was a non-unit protection scheme. Patil and Ramteke (2015) and Fletcher (2014) proposed that the current differential relaying scheme is used to mitigate the current issues, whereas this system requires a large data handling, ultimately causing additional data handling and transferring architecture. This technique too mal operates for a high-resistance fault. Hwang et al. (2013) has proposed to use fault current limiters at bus end to limit the value of fault current.

Algorithm for the Proposed Scheme

In the proposed method the fault is determined with the 51 which is non directional overcurrent relay. If the faut has occurred in the system then the proposed algorithm comes into the picture. Once the fault is detected, the proposed algorithm becomes active which helps in determine the direction of fault. For which the change in current is estimated by determining the derivative of the current. In this paper the three-point method is used to determine fwd (I'_f) and rev (I'_r) derivative are obtained. The phase angels of the obtained results are compared with that of the reference phase angle, through which fault direction is estimated.

$$\angle Iref = \frac{\angle(+I') + \angle(-I')}{2} \tag{1}$$

The source angle is associated with the estimated phase angle of faut current.

$$\alpha = \angle(Iref) - \angle(Ia) \tag{2}$$

Here the locations is kept at bus B. Figure 142.3 shows the vector diagram for the upstream and fault directions.

Where

V_A is the bus voltage of Bus A

V_C is the bus voltage of Bus C

δ is the angle between bus voltage Bus A and C

I_{F1} is the current when the fault is in the downstream zone

I_{F2} is the current when the fault is in the upstream zone

If the change is stuck between "+3 $\pi/4$ > $\angle$ α > $\pi/4$" then the fault is downstream hence a trip indication will be created, otherwise there will be no trip indication.

If $3\pi/4$ > $\angle$ α > $\pi/4$ then downstream fault /trip

or

If $-3\pi/4$ > $\angle$ α > $-3\pi/4$ then upstream fault /no trip

The detailed algorithm is shown in Figure 142.3.

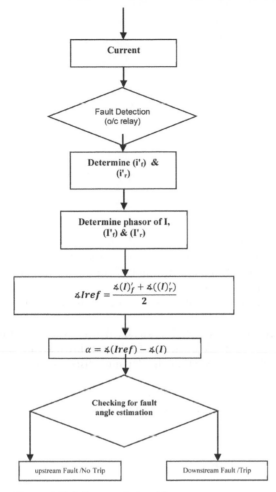

Figure 142.3 Proposed algorithm

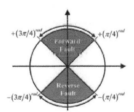

Figure 142.4 Fault estimation characteristic

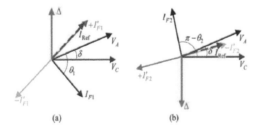

Figure 142.5 Phasor diagram a) downstream fault b) upstream fault

System under consideration

To prove above algorithm under high resistance fault or dead short circuit, system shown if Figure 142.6 is considered.

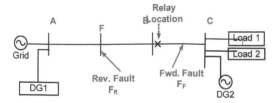

Figure 142.6A System under consideration

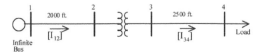

Figure 142.6B IEEE 4 bus system

As discussed in the section here a 4-bus system is considered, to cope with the future work, and measuring purpose an additional bus is considered. Starting from synchronous motor at start and loads are connected at far end of the feeder. Two DG sources are introduced one at point of common coupling and one at the relay location. To test the validity of the system the most severe fault i.e. three phase fault f_m and least sever fault i.e single phase to ground f_L are created. The fault was also created for upstream and downstream fault.

To achieve the dynamic changes in the system, induction generator (IG) is switched at t = 3 second faults are created upstream and downstream of the location of relay. A Upstream fault and downstream fault both f_m and f_L is made. The performance of proposed scheme is tested for the dynamic disruptions appearing in the system under consideration with respect to load angle,

The Ratings of various parameters are taken from (Patil and Ramteke, 2015).

F_m and F_r are created by varying the angles of sending end occurrence of upstream fault 0.2

- 0.3
- occurrence of downstream 0.5.6
- variation in sending voltage is made as $-\pi/4$, $+\pi/4$, $-\pi/6$, $+\pi/6$

Results

Figures 142.7 and 142.8 shows the generation of trip signal only for a downstream fault whereas Table 142.1. shows the from case studies result for two bus system.

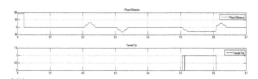

Figure 142.7 Fr sending end voltage angle: $-\pi/4$

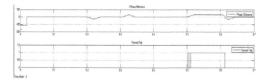

Figure 142.8 Fr sending end voltage angle: $-\pi/6$

Table 142.1 Phase angle detected

S.N.	Fault	Angle measured	Conclusion
1)	F_r	$-7\pi/18$	Upstream
2)	F_f	$7\pi/18$	Downstream
3)	F_r	$-46\pi/18$	Upstream
4)	F_f	$11\pi/18$	Downstream

Figure 142.9–142.11 show the load bus voltage, current and load angle. There is disturbance for a short duration during transient period due to DG.

Figures 142.12–142.14 show the detection of fault for f_m and f_L and trip is generated only for dwonstream fault only.

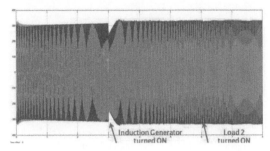

Figure 142.9 Voltage profile vs t (without fault)

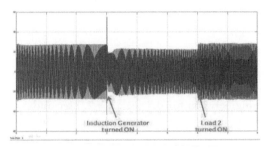

Figure 142.10 Current vs t (without fault)

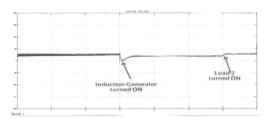

Figure 142.11 Load angle vs t

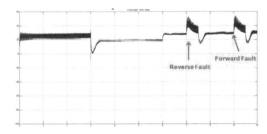

Figure 142.12 Load angle vs t for f_L

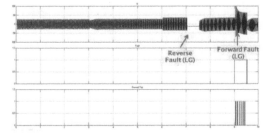

Figure 142.13 Current for downstream fault and trip signal

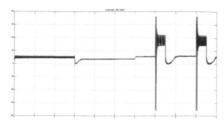

Figure 142.14 Load angle vs t (f_m)

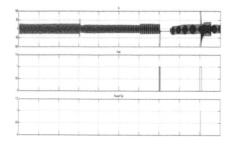

Figure 142.15 Current and trip signal f_L

To in cooperate the effect of f_m and f_L are created for 0.5 sec at t = 7 sec. Figure 142.13 shows the deviation of load angle when a upstream f_m fault has occurring between for 0.5 sec a t = 7.0 sec and a downstream f_L fault occurring again for 0.5 sec at t = 9.0 second. For same condition Figure 142.11 show the variation of current (a), fault detection (b) and trip command(c). The fault is detected only in case if fault is downstream only.

Figure 142.15 show the variation of fault current and issue of trip command for a downstream fault

Conclussion

The data from Table 142.1 concludes that the projected method works satisfactorily, this method works regardless direction of power flow. Direction of the fault is projected accurately for the downstream as well as for upstream fault. When the fault is downstream the projected phase angle lye in the range of $\pi/4$ rad to $3\pi/4$ rad for which trip indication is generated. In case of for upstream the phase position expected, lye in the range of $-\pi/4$ to $-3\pi/4$ and so no trip signal is generated. Hence it is stated that the proposed method is correct. However, further examination of this method on dynamic system. It was found that the trip signal is generated only in case of downstream fault and not for upstream fault.

Hence it can be concluded that the proposed method is valid in dynamic system also.

References

Asmus, P. (2010) Microgrid virtual power plants and our distributed energy future. The Elect J. 23(10):72–82.

Bayati, N., Baghaee, H. R., Hajizadeh, A., and Soltani, M. (2021). Localized protection of radial DC microgrids with high penetration of constant power loads. IEEE Syst. J. doi: 10.1109/JSYST.2020.2998059.

Bhargav, R., Bhalja, B. R., and Gupta, C. P. (2019) Algorithm for fault detection and localisation in a mesh-type bipolar DC microgrid network. IET Gener. Transmiss. Distrib. 13(15):3311–3322.

Chakraborty A. (2011). Advancements in power electronics and drives in interface with growing renewable energy resources. Renewable Sustain. Energy Rev. 15(4):1816–27.

Dhar, S., Patnaik, R. K., and Dash, P. K. (2018). Fault detection and location of photovoltaic based DC microgrid using differential protection strategy. IEEE Trans. Smart Grid. 9(5):4303–4312.

Dragi˘cevi´c, T., Lu, X., Vasquez, J. C. and Guerrero, J. M. (2016). DC microgrids— Part II: A review of power architectures, applications, and standardization issues. IEEE Trans. Power Electron. 31(5):3528–3549.

Fletcher, S. D. A., Norman, P. J., Fong, K., Galloway, S. J., and Burt, G. M. (2014). High-speed differential protection for smart DC distribution systems. IEEE Trans. Smart Grid. 5(5):2610–2617.

Hwang J. S., Khan, U. A., Shin, W., Seong, J., Lee, J. G., Kim, Y. H., Lee, B. W. (2013). Validity analysis on the positioning of superconducting fault current limiter in neighboring AC and DC microgrid. IEEE Trans. Appl. Supercond. 23(3).

Kakigano, H., Nomura, M. and Ise, T. (2010). Loss evaluation of DC distribution for residential houses compared with AC system. in Proc. Int. Power Electron. Conf. (IPEC)-ECCE Asia, Sapporo, Japan. pp. 480–486.

Meghwani, A., Srivastava, S. C., and Chakrabarti, S. (2017). A non-unit protection scheme for DC microgrid based on local measurements. IEEE Trans. Power Del. 32(1):172–181.

Monti, W., Li, A., and Ponci, F. (2014). Fault detection and classification in medium voltage DC shipboard power systems with wavelets and artificial neural networks. IEEE Trans. Instrum. Meas. 63(11):2651–2665.

Park, J. Do., Candelaria, J., Ma, L.and Dunn, K. (2013). DC ring-bus microgrid fault protection and identification of fault location. IEEE Trans. Power Del. 28(4):2574–2584.

Patil, P. and Ramteke, M. R. (2015). Impact of Distributed Generation on power distribution system: Over-current protection by phase angle estimation. IEEE Power, Communication and Information Technology Conference (PCITC) pp. 844849.

Patil, P. S., Badar, A., and Hinge, T. P. (2021). Application of Linear Programming for Overcurrent Protection. International Conference on Intelligent Technologies (CONIT). pp. 1–4.

Saleh, K. A., Hooshyar, A., and El-Saadany, E. F. (2017). Hybrid passive over current relay for detection of faults in low-voltage DC grids. IEEE Trans. Smart Grid. 8(3):1129–1138.

Yeap, Y. M., Geddada, N., Satpathi, K., and Ukil, A. Time- and frequency domain fault detection in a VSC-interfaced experimental DC test system. IEEE Transact. Ind. Info. 14(10): 4353–4364.

143 Durability of mortar with nano silica and carbon nano-tubes in sulphate environment

Mohd Moonis Zaheer[a], Varisha[b], and Syed Danish Hasan[c]

Department of Civil Engineering, Zakir Hussain College of Engineering and Technology, Aligarh Muslim University, Aligarh, India

Abstract

Attack of sulphate is a progressively-acting degradation process that can lead to cracking, increased permeability, and ultimately loss of composites' strength. In this paper, mortar specimens containing 1.0 wt.% nano silica (NS) and 0.3 wt.% carbon nano-tubes (CNTs) were formed for analysing the mechanical and durability properties. CNTs of four different types, untreated and treated with a COOH group with different outside diameters (10–20 nm and 30–50 nm), were used to prepare mixes. Six different mortar mix combinations were prepared for the present study and tested at 28, 56, 90, and 120 days. For comparative purposes, the mortar specimens for all ages were cured under both potable and magnesium sulphate solution. Durability studies were carried out by investigating the coefficient of water absorption and sorptivity measurements. Scanning electron microscope (SEM) and energy-dispersive X-ray spectroscopy (EDS) analysis were done for the products formed as a result of hydration to study the microstructure. In the present work, an effort has been made to understand the synergistic behaviour of mortar admixed with NS and CNTs in terms of its mechanical and durability properties, which has scarcity in previous literature. The research could suggest that NS and CNTs are good reinforcing materials for improving the cement mortar's mechanical and durability properties. Among all the tested samples for strength and durability, samples mixed with 1.0 wt% NS and 0.3 wt% treated CNTs with the COOH group performed better.

Keywords: Carbon nano-tubes (CNTs), durability property, mechanical property, nanosilica (NS), sulphate attack, sorptivity.

Introduction

Of the different sulphates such as magnesium, sodium, ammonium, and calcium that lead to degradation in cementitious composites, the starting two being more common, with the magnesium sulphate causing more deterioration compared to others (Bonen and Menashi, 1992). The resistance to sulphates in mortar or concrete can be obtained by using ordinary portland cement (OPC) low in its tricalcium aluminate (C_3A) content or sulphate resisting cement. The conditions in which sulphate resistant Portland cement is not readily available, OPC can be used along with nanomaterials like nano silica or carbon nano-tubes at appropriate addition levels to minimize deterioration to sulphate attack (Al-Akhras, 2006). However, the resistance to sulphates of the blended mixtures depends majorly on the chemical composition of the nanomaterial used and the amount to be added. In particular, the decrement in the permeability of the cementitious composites on adding nanomaterial is important for improving the resistance against sulphate attack (Sidney et al., 2003). External sulphate attack would not occur without water's presence to dissolve the sulphate ions from their natural state and bring them into contact with mortar or concrete's surface (Batilov, 2016). The sulphate attack is because of the compounds formed as a result of a chemical reaction between sulphate ions and the tricalcium aluminate (C_3A), calcium silicate hydrate (CSH), and calcium hydroxide (CH) present in the cement (Grengg, 2017). These products are ettringite (AFt), gypsum, thaumasite, and brucite, responsible for expansions and cracks in mortar or concrete. Consequently, it causes reduced strength and durability of structures (Kishar et al., 2013).

The most efficient way to prevent cementitious materials from the adverse effects of sulphate is to reduce the permeability or use various mineral additives. Of the various mineral additives that have been used in cementitious composites and have a fairly good effect on sulphate resistance are carbon nano-tubes (CNTs) (Vu, 2002). The ability of CNTs in augmenting sulphate resistance comes from their crack arresting mechanism due to its high aspect ratio and extremely high strength (Hou et al., 2014). The effect of CNTs for preventing the sulphate attack of mortars or concrete is in two ways: the pozzolanic and the micro filler effect. The introduction of this additive in cement is responsible for reducing the micropores by filling them and improving the strength of the mortar or concrete (Imbabi et al., 2013). Another widely used mineral additive

[a]mohdmooniszaheer@gmail.com; [b]rizwanvarisha@gmail.com; [c]sdhasan.amu@gmail.com

for mitigating the effects of sulphate on mortar or concrete is nanosilica (NS), which expedites the hydration reactions of cement (Atahan and Arslan, 2016). Since products of hydration start to form among cement particles due to NS inclusion, a denser and compact microstructure could be obtained compared to the control mixture (Naskar and Chakraborty, 2016). Thus, the resulting cementitious materials are more durable.

It is reported that the application of nanomaterials in cementitious materials by adding NS significantly improved the performance of mortar and concrete (Han et al., 2017). A state-of-the-art development on recent advances in nanosilica application in concrete was presented by (Barbhuiya et al., 2020). A nanosilica substitution of 2% was found to be sufficient for obtaining mixtures with significant sulphate resistance against external sulphate attacks (Arel and Thomas 2017). The study showed that nanosilica replacement in mortar and concrete production results in a positive performance in terms of strength, impermeability, and durability. Recently, (Mohsen et al., 2017) studied the ideal dosage of CNTs for improving cementitious composites' strength and reported that 0.1–0.3% CNTs content gave optimal results. A nanomaterial may also be used in combination with other nanomaterials to improve durability further. The study of mortar's durability admixed with CNTs and NS was checked by (Lee et al., 2018) and observed compressive strength enhancement by 12–76% in addition to improved corrosion resistance. In one of the studies, two types of MWCNTs were used with different concentrations (0.1–0.5 wt.%), and their effect on mechanical properties was investigated and found 0.3 wt.% as the optimum content (Zaheer 2019). Furthermore, (Han et al., 2013) found that multi-walled carbon nano-tubes (MWCNTs) could decrease the absorption and permeability of water and air permeability of cementitious materials.

Previous literature shows a scarcity of investigations on the integrated impacts of NS and CNTs on the durability properties of mortar. Therefore, to fill this gap, the effect of adding NS with untreated and treated CNTs in mortar on its strength and durability properties was studied under sulphate environment. The microstructure was also investigated through SEM and EDS analysis to justify the experimental findings.

Experimental Work

In the current work, the performance of mortar prepared with and without NS and CNTs are compared for mechanical and durability properties. The materials used and the methodology adopted are described in the following subsections.

Materials

Ordinary portland cement (OPC) from Ultra-Tech Company was used to prepare mixes. The grade of cement taken was 43. The specific gravity and soundness of the cement are found to be 3.15 and 1.20 mm, respectively.

The fine aggregates consisted of Indian Standard sand obtained from Bhagirath Construction Company, Bangalore. The equal amount by weight of grades 1, 2, and 3 were mixed to obtain Indian Standard sand. The sand conforms to zone 3 with a specific gravity of 2.62. The fineness modulus was found to be 2.58. The bulk density in the loose state was obtained as 1660 kg/m^3.

High range water reducing admixture (HRWRA) of polycarboxylates was used in the study. The relative density and pH of the admixture were 1.105 and 5.75, respectively.

Four industrial-grade MWCNTs (Type I, II, III, and IV) were used in the current study. Type I and III are untreated, while type II and IV are functionalized with –COOH group. Type I and II CNTs have average diameters and lengths of 10–20 nm and 1–5 μm. The specific surface area for Type I and II CNTs is 370 m^2/g. Whereas these specifications for Type II and IV are 30–50 nm, 10–20 μm, and 400 m^2/g, respectively. All CNTs grades had 99 + % purity.

Mortar mix design

Mortar mix has been formed with and without the inclusion of nanomaterials to compare mortar's mechanical and durability behaviour. The water-to-cement ratio and cement to sand proportion were kept equal to 0.55 and 1:3, respectively. The amount of various materials used in different mix types are shown in Table 143.1. A 0.4% HRWRA by cement weight was used for all the samples. Multi-walled carbon nano-tubes (MWCNTs) and nano silica (NS) were added at dosages of 0.3% and 1.0% by cement weight.

Specimen preparation

Prismatic flexural specimens of size 40 mm × 40 mm × 160 mm were formed. An electromagnetic vibrating table was used to ensure good compaction after pouring the mixture into molds conforming to IS: 10078 (1982). All specimens were kept in a potable water for 28 days before sulphate exposure. After 28 days,

Table 143.1 Mix proportions of mortar with various wt% of NS and CNTs

Systems	Mixes	Cement (g)	Sand (g)	NS/CNT (wt.%)
CS	OPC	140	420	0.0+0.0
NS	OPC + 1% NS	140	420	1.0+0.0
U1	OPC + 1% NS + 0.3% untreated CNT	140	420	1.0+0.3
T1	OPC + 1% NS + 0.3% treated CNT-COOH	140	420	1.0+0.3
U2	OPC + 1% NS + 0.3% untreated CNT	140	420	1.0+0.3
T2	OPC + 1% NS + 0.3% treated CNT-COOH	140	420	1.0+0.3

CS: Control specimen with OPC only; NS: Specimen with NS + OPC; U1: Specimen with OPC+ NS+ CNT (type 1); T1: Specimen with OPC + NS+ CNT functionalized with COOH (type 2); U2: Specimen with OPC + NS + CNT (type 3); T2: Specimen with OPC + NS + CNT functionalized with COOH (type 4); Water/cement raio: 0.55; Sand/cement ratio: 1:3.

half of the specimens were transferred to a plastic curing tank having a 10% magnesium sulphate solution. The remaining samples were cured in the potable water tank and tested in flexure at the same age (56, 90, and 120 days) of samples immersed in sulphate solution. Different stages of specimen making are shown in Figure 143.1.

Sulphate solution

The 10% $MgSO_4$ solution was made for curing the specimens. Appropriate amount of sulphate solution was prepared for the curing tank to satisfy the recommended minimum solution-to-mortar volume ratio of 4. The pH of the solution was manually adjusted in the range of 67 weekly with 0.5 N H_2SO_4 for the entire curing period of 120 days to ensure the sulphate ion concentration in the medium.

Testing Procedures

Flexural strength

Flexural test was done on a three-point loading frame. It consists of a load and displacement measuring attachments. The specimen was tested under center point loading over a simply supported span of 100 mm, as shown in Figure 143.2(a). The loading rate was maintained as 50 ± 10 N/s during testing. A flexural specimen after failure is shown in Figure 143.2(b). The flexural strength is determined as per IS: 4031 (Part 8) – 1988.

Compressive strength

The compressive strength test was done on a 40mm sized cube as per IS: 4031 (Part 8) – 1988. Each prism was tested in a compression testing machine of 45 kN capacity. The rate of loading was 200 kg/cm²/minute.

Figure 143.1 Dispersion process of nanomaterials and mortar sample preparation

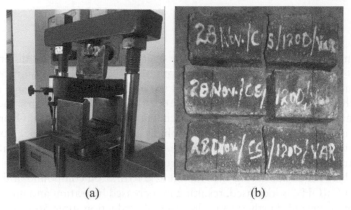

Figure 143.2 Flexural test under center point loading (a); Specimens showing major failure location (b)

Coefficient of water absorption

This test was done in accordance with ASTM C 642 (1997). It measures the amount of water absorbed by the specimen in 60 min. It gives us the rate of absorption of water by oven-dried samples for 60 min. and is obtained by:

$$K_a = [Q/A]^2 \times 1/t \tag{1}$$

Where, K_a = coefficient of water absorption; Q = quantity of water absorbed by an oven-dried sample in time t and A = total surface area of the specimen through which water penetrates in time t = 60 min.

Sorptivity

This test is performed as per ASTM C 642 (1997). For sorptivity measurements, a cubical specimen of 40 mm × 40 mm × 40 mm of different mix types was used (CS, NS, U1, T1, U2, and T2). The cubes were waxed from four sides to restrict evaporation. Sorptivity determination requires a scale, a stopwatch, and a wide shallow pan containing water. In this process, mass is measured initially and overtime at an interval of 5 minute, for 60 minutes duration. viz. 5, 10, 15, 20, 25, 30, 35, 40, 45, 50, 55, and 60 min. by keeping samples immersed only to a depth of 5mm in water. The various steps in sorptivity measurement are shown in Figure 143.3. Specimens were taken out from the water, and excess water was soaked off with an absorbent towel and weighed again after every interval. Further, the samples are immersed in water after each interval, and the stopwatch is started again for time measurement. The graph between increase in mass by area over the density of water and the square root of the time elapsed was drawn. The tangent of the line by plotting the best fit of these points is termed sorptivity. Soptivity is expressed as:

$$S = \frac{I}{\sqrt{T}} \tag{2}$$

where, I = cumulative water absorption per unit area of inflow surface; $\sqrt{T}$ = time elapsed (s) and S = sorptivity (m/s$^{1/2}$).

SEM and EDS analysis

The changes in the mortar microstructure because of the inclusion of NS and CNTs compared to CS were studied by SEM images taken at University Sophisticated Instruments Facility (USIF) Centre, AMU, Aligarh. Also, the elemental composition of different materials in the cementitious matrix was investigated by EDS analysis. After the destructive tests, a small piece of mortar was taken from different samples and stored in ethanol to prevent it from hydration. Later on, the pieces were coated with a gold layer of approximately 20 using an automated sequence sputter coating machine. The magnification range varies from × 1500 to × 3000.

Results and Discussion

Flexural strength

The flexural strength of the nano admixed specimens should increase over time, but a slightly different result was seen here because of sulphate water curing. The flexural strength of all the samples was improved till 56 days, followed by a reduction at 90 and 120 days. The order of strength improvement at each age is shown in Figure 143.4. Mix T2 exhibited maximum values of flexural strength than other specimens. Increments of about 53% and 39% are seen for T2 than control specimens at 28 and 56 days. Overall, T2 shows a higher flexural strength at 120 days, even after the sulphate attack. This behaviour could be attributed to the addition of nanomaterials, which improves the mechanical properties of the matrix because these nanoparticles act as catalysts and fillers both, thus increasing the hydration reaction and consequently densifying the matrix. CNTs act as nucleating sites for the C_3S hydration compound and get coated with CSH gel. The formation of $Ca(OH)_2$ is increased, resulting in increased hydration and improving the composites' strength in the initial period. Furthermore, the strength reduction after 56 days can be related to two reasons. Firstly, ettringite (expansive) might be produced due to calcium aluminate and calcium sulphate reaction. The availability of sulphates also leads to the increased formation of ettringite (Taylor, Famy, and Scrivener 2001), causing a reduction in strength. Secondly, due to magnesium sulphate's presence considerable damage and debonding of the CSH gel happens, thus causing a reduction in flexural strength at 90 and 120 days.

Compressive strength

A similar pattern as observed for flexure strength was seen for compressive strength also. Initially, the compressive strength increases from 28 to 56 days, and then a reduction in the values is obtained, as seen in Figure 143.5. Maximum compressive strength at 28 and 56 days is for T2, which is 19% and 16% more

Figure 143.3 Sorptivity measurements

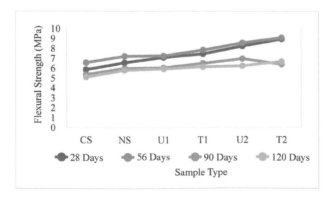

Figure 143.4 Flexural strength of mortar at different ages under sulphate water

than CS, respectively. The decrement in the values is seen with the curing age because of sulphate attack. Even in sulphate solution, mix T2 performed better than CS and other mixes at 90 and 120 days. T2 exhibited a compressive strength value of 32 MPa at 120 days, which is 16% more than the CS value at the same curing period. The increase in strength by adding NS and CNTs is due to these nanomaterials' pore filling ability and pozzolanic property. In the beginning, the cement paste's hydration process increases due to the addition of nanoparticles, thus forming more CSH gel, ettringite, and gypsum (Mittermayr et al., 2015). After 28 days, even transferring the specimens into sulphate solution, the value gets increased because of the expansive product's (ettringite) compaction effect. Ettringite fills the pore spaces and compacts the matrix leading to increased strength during the initial period of sulphate exposure (Najjar et al., 2017). The stress development forms micro-cracks in the structure, leading to reduced strength at later ages.

Coefficient of water absorption

Table 143.2 shows the coefficient of water absorption values obtained for different cubical specimens. Results showed that the CS mix has a higher percentage of coefficient of water absorption than nano-modified mixes. It is seen that up to 56 days, the coefficient of water absorption decreases followed by its increase. The reason may be attributed to magnesium sulphate's presence which causes degradation and debonding of the CSH gel leading to increased water absorption. Again, the outcomes found that mix T2 is more effective with lesser coefficient of water absorption.

Sorptivity

The results of sorptivity of mortar specimens are represented in Figure 143.6. The sorptivity values were observed to be increased for CS compared to the other mixes. This may be due to the porous nature of the mortar mix (Najigivi et al., 2012). According to the results, by adding NS and CNTs in the mix, the concrete specimens' capillary porosity is decreased upto 56 days. Consequently, the sorptivity of samples is decreased. Furthermore, beyond 56 days, the sorptivity is increased. For instance, in control specimens, the sorptivity of specimens is decreased by 2.4%, from 28 to 56 days. After that, at 90 and 120 days, the sorptivity increases by 0.6% and 1.8%, respectively. For different mixes, a falling trend in sorptivity is

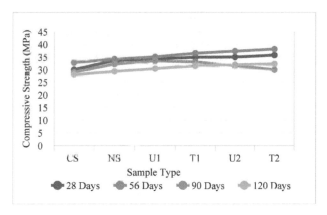

Figure 143.5 Compressive strength of mortar at different ages under sulphate water

Table 143.2 Average coefficient of water absorption value of different mixes

Sample type	Age (days)			
	28	*56*	*90*	*120*
CS	4.0826×10^{-4}	2.8561×10^{-4}	4.6895×10^{-4}	4.7635×10^{-4}
NS	1.5926×10^{-4}	1.1546×10^{-4}	1.6928×10^{-4}	1.9186×10^{-4}
U1	1.2315×10^{-4}	1.2621×10^{-4}	1.7173×10^{-4}	1.7537×10^{-4}
T1	6.4896×10^{-5}	5.5427×10^{-5}	6.7439×10^{-5}	7.0176×10^{-5}
U2	5.5067×10^{-5}	4.9251×10^{-5}	5.6701×10^{-5}	6.1710×10^{-5}
T2	4.3256×10^{-5}	3.5634×10^{-5}	4.7817×10^{-5}	5.1563×10^{-5}

observed at all curing ages. This may be attributed to the pore filling property of the nanomaterials that improved the pozzolanic behaviour, thereby forming a compact and denser microstructure (Jo et al., 2007). The sorptivity was found to be the least for the T2 mix.

SEM and EDS analysis

A microscopic investigation of the mortar specimen was done to support the results obtained from various tests. Figure 143.7 (a) represented the SEM image of the CS specimen at 28 days. It showed the formation

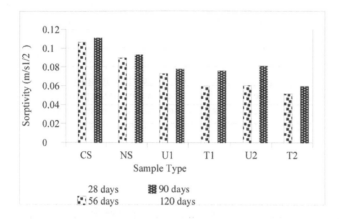

Figure 143.6 Sorptivity of mortar samples at different ages under sulphate water

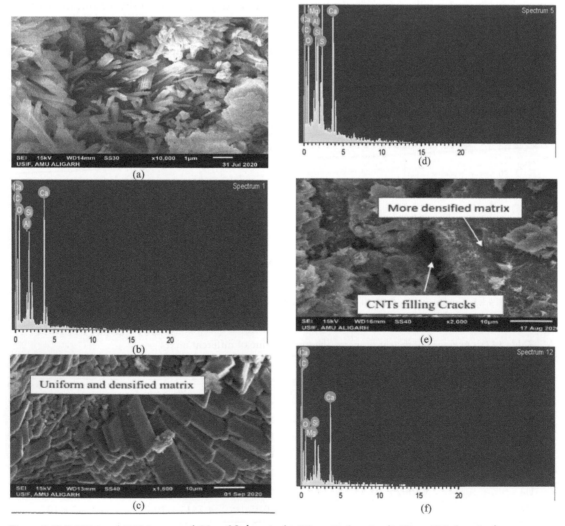

Figure 143.7 SEM and EDS images of CS at 28 days (a, b); T2 at 56 days (c, d); T2 at 120 days (e, f)

of characteristic products viz. CSH gel and CH (Portlandite) in the shape of cylindrical needles which are responsible for attaining the strength of mortar. EDS analysis was implemented to authenticate the strengths, locate the CNTs in the matrix, and determine the neighboring materials' composition. Figure 143.7 (b) illustrates the EDS analysis results of CS specimens at 28 days near a CNT location. It could be seen that calcium is the primary element. Silica was present as an outcome of the cement hydration in the matrix. The presence of gold is due to the etching of the sample by a thin gold layer applied before testing. Other elements in the EDS analysis might be related to various impurities. SEM image of the T2 specimen at 56 days is shown in Figure 143.7 (c), which shows a uniform and densified matrix. The obtained SEM image is strong evidence for the improved experimental values shown in Figure 143.5. Figure 143.7 (d) represented the EDS result of the T2 sample at 56 days. A higher percentage of silica, calcium, and carbon could be seen due to the bonding between silica and CNTs, thus representing a strong bond with the hydration products. The microstructure changes of the T2 specimen at 120 days are presented in Figure 143.7 (e). It showed a more densified matrix due to the bridging of fine cracks by nanoparticles. It became proof of the highest strength of T2 than other mortar mixes at 120 days, even after the sulphate attack. Similarly, Figure 143.7 (f) showed the EDS analysis of the T2 sample after 120 days of curing. It showed carbon in a considerably high amount, indicating the presence of CNTs in the matrix. Sulphur was also seen due to sulphate water curing.

Conclusion

From the present work, the following main conclusions are drawn:

- The inclusion of NS and CNTs improved various mechanical properties. NS improved the pozzolanic activity and formed a compact CSH gel, whereas CNT worked as a reinforcing fibre between the cement paste and fine aggregate, which caused the strength enhancement. The flexural strength of mixes with NS and CNTs is much better than control ones. Further, specimens with treated CNTs performed better than untreated ones. A maximum of 39% increase was observed for T2 at 56 days. After 56 days, the values of all the specimens start degrading due to sulphate attack. A similar pattern was seen for compressive strength measurements.
- The absorption coefficient of water and sorptivity of various nano admixed mortars were initially decreased because of the pore filling effect. Beyond 56 days, an increasing trend is observed due to sulphate's presence which causes deterioration and debonding of the CSH gel, leading to increased water absorption and sorptivity.
- SEM and EDS analysis presented that the nano admixed mortar has shown a denser and compact microstructure due to the exhaustion of CH and added CSH gel formation with nanosilica. CNTs confined the crystalline growth of the matrix, which is responsible for improved mechanical strength.
- This research suggested that NS and CNTs are good reinforcing materials for improving the cement mortar's mechanical and durability properties. Also, treated CNTs performed relatively better than the untreated CNTs, especially T2.

Acknowledgment

The experiments of this work were carried out in the Materials Laboratory, Civil Engineering Department, ZHCET, AMU, Aligarh. The authors want to praise each person with the help of whom this work gets completed properly.

References

Arel, H. Ş. and Thomas, B. S. (2017). The effects of nano-and micro-particle additives on the durability and mechanical properties of mortars exposed to internal and external sulfate attacks. Results Phys.7:843–851.

ASTM C 642 (1997). Standard test method for density, absorption, and voids in hardened concrete. ASTM International. United States: Annual Book of ASTM Standards.

Atahan, H. N. and Arslan, K. M. (2016). Improved durability of cement mortars exposed to external sulfate attack: The role of nano & micro additives. Sustainable Cities Soci. 22:40–48.

Barbhuiya, H., Moiz, M. A., Hasan, S. D., and Zaheer, M. M. (2020). Effects of the nanosilica addition on cement concrete: A review. Mater. Today: Proceedings. 32:560–566.

Batilov, I. B. (2016). Sulfate resistance of nanosilica contained Portland cement mortars. PhD diss. University of Nevada, Las Vegas.

Bonen, D. and Menashi, D. C. (1992). Magnesium sulfate attack on portland cement paste-I. Microstructural analysis Cement and concrete research. 22(1):169–180.

Grengg, C. (2018). Microbial induced acid corrosion in sewer environments.

Han, B., Zhang, L., Zeng, S. (2017). Nano-core effect in nano-engineered cementitious composites. Compos. A: Appl. Sci. Manufact. 95:100–109.

Han, B., Yang, Z., Shi, X., and Yu, X. (2013). Transport properties of carbon-nanotube/cement composites. J. Mater. Eng. Perform. 22, no. 1: 184–189.

Hou, P., Cheng, X., Qian, J., Zhang, R., Cao, W., and Shah, S. P. (2015). Characteristics of surface-treatment of nano-SiO2 on the transport properties of hardened cement pastes with different water-to-cement ratios. Cement Concrete Compos. 55:26–33.

Imbabi, M. S., Collette, C. and Sean, M. (2012). Trends and developments in green cement and concrete technology. Int. J. Sustain. Built. Environ. 1(2):194–216.

IS 10078 (1982). Specification for Jolting Apparatus Used for Testing Cement. Bureau of Indian Standards (BIS), New Delhi.

IS 4031-Part 8 (1988). Methods of Physical Tests for Hydraulic Cement (Part 8 - Determination of Transverse and Compressive Strength of Plastic Mortar Using Prism. Bureau of Indian Standards (BIS), New Delhi.

Jo, B. W., Kim, C., H., Tae, G., H., and Park, J., B. (2007). Characteristics of cement mortar with Nano-SiO$_2$ particles. Construct. Build. Mater. 21(6):1351–1355. https://doi.org/10.1016/j.conbuildmat.2005.12.020.

Kishar, E. A., Doaa, A. Ahmed, M. R. M. and Rehab, N. (2013). Effect of calcium chloride on the hydration characteristics of ground clay bricks cement pastes. Ben. Uni. J. Basic Appl. Sci. 2(():20–30.

Lee, B. Y., and Kurtis, K. E. (2017). Effect of pore structure on salt crystallization damage of cement-based materials: Consideration of w/b and nanoparticle use. Cement Concrete Res. 98:61–70.

Lee, H. S., Balasubramanian, B., Gopalakrishna, G. V. T. Kwon, S-J, Karthick, S. P., and Saraswathy, V. (2018). Durability performance of CNT and nanosilica admixed cement mortar. Construct. Build. Mater. 159:463–472.

Mittermayr, F., Rezvani, M., and Baldermann, A., (2015). Sulfate resistance of cement-reduced eco-friendly concretes. Cement Concrete Compos. 55:364–373.

Mohsen, M. O., Taha, R., Taqa, A., A., and Shaat, A. (2017). Optimum carbon nanotubes' content for improving flexural and compressive strength of cement paste. Construct. Build.Mater. 150:395–403.

Nabil, M. Al-A. (2005). Investigation of the effect of metakaolin (MK) replacement of cement on the durability of concrete to sulfate attack. Cem. Concr. Res. 36(9):1727–1734.

Najigivi, A., Abdul, R.Suraya, A., Aziz, F., N. Mohd, S., and Mohamad, A. (2012). Water absorption control of ternary blended concrete with nano-SiO2 in presence of rice husk ash. Mater. and struct. 45(7):1007–1017.

Najjar, M. F., Nehdi, M. L., Soliman, A. M., and Azabi, T. M. 2017. Damage mechanisms of two-stage concrete exposed to chemical and physical sulfate attack. Construct. Build. Mater. 137:141–152.

Naskar, S. and Chakraborty, A. K. (2016). Effect of nano materials in geopolymer concrete. Perspectives Sci. 8:273275.

Sidney, J., Francis, Y., and David, D. (2003). Concrete. New Jersey (United States): Pearson Education.

Taylor, H. F. W., Famy, C., and Scrivener, K. L. (2001). Delayed ettringite formation. Cement Concrete Res. 31(5):683693.

Vu, Dinh Dau. (2004). Strength properties of metakaolin-blended paste, mortar and concrete. 0222-0222.

Zaheer, M. M. (2019). Experimental Study of Multi-Walled Carbon Nano-tubes in Cement Mortar for Structural Use. Jordan J. Civil Eng. 13(4).

144 Development and analysis of Paperless SOPs and Travelcard for the Manufacturing Industry

Rohit Patil[1,a], Prakash Pantawane[2,b], Ashutosh Ramtirthkar[3,c], and Sachin Ambade[4,d]

[1]Department of Manufacturing Engineering and Automation College of Engineering, Pune, India

[2]Department of Production Engineering College of Engineering, Pune, India

[3]Department of Mechanical Engineering, Yashwantrao Chavan College of Engineering, Nagpur, India

[4]Cummins Power Generation Business Unit, Phaltan, India

Abstract

In today's era of industrialisation, the big concern that everyone should look upon is the protection of our mother nature from pollution and degradation. The paper industry depletes the forest cover and causes harm to the environment by polluting air, water, soil, etc. Various alternatives are available to avoid the use of papers in the manufacturing industry, which needs contemplation. One of the initiatives that the organisations are thinking about is 'Green Manufacturing'. So, under this initiative organisations are more focused to reduce the use of paper in the industry. In this work, an attempt has been made to have paperless 'Standard Operating Processes' and 'Travel Cards' for the manufacturing industries. This paper addresses the various alternate solutions that organizations can employ to reach the goal of the paperless industry. Furthermore, an attempt has been made to outline possible modifications or automation for the processes, where papers are used traditionally.

Keywords: Paperless, software, standard operating procedure, travel card.

Introduction

It is very difficult to find the correct figure but still according to the U.S. Environmental Protection Agency, in their report have mentioned that the world produces 300 million tons of paper every year (World Meteorological Organization Climate Report Met). This is an alarming condition considering today's technological development. By using appropriate ways for their application one can reduce the use of paper. The idea of the paperless document was introduced way back in the 1960s for deposited securities in the USA. Further in the 1970s Non-certified (Undocumented) Securities were introduced as Paperless Securities (Pleshkevich, 2006). Now in the current scenario, it became essential to reduce the use of paper so, many organizations are trying to reduce the use of paper as much as possible. Today's technology is one of the nourishing parameters for achieving paperless work. There does not have fixed way to reduce the use of paper. Anyone by anyhow can achieve their need. This will also reduce global warming. Which is one of the worst cases emerging these days. This is mainly due to the accumulation of carbon dioxide in the atmosphere. If we go deeper into this, we will find that this accumulation is happening is just because there is no way or method available to use it. Only trees can take carbon dioxide inside.

Trees are one of the important parts of our environment as they are the most valuable life source on earth. They benefit every life form directly or indirectly and the Earth is connected to them to maintain a natural balance. Thus, concluding will be deforestation is one that is causing global warming, so the least use of paper can reduce global warming miserably (Nayyar and Arora, 2019). So, to reduce the use of paper to save trees, this study has been carried out a study in an MNC company.

Doing all the corporate work without the use of paper is such a complicated task, as the corporate industry involves lots of documents which are needed to be saved in hard copy. But to save paper to help environmentally, the company wanted to do all the work in paperless form. So, this paper defines the methodology used to reduce the use of paper. This problem is with the Standard Operating Procedure document and Travel Card document.

---

[a]rrpatil1998@gmail.com; [b]pdpantawane@gmail.com; [c]iw776@cummin.com; [d]sachinamb2@rediffmail.com

Literature Review

The points such as What is Paperless society? Why go paperless? How to go for paperless? was cleared from the case study also the ways such as document management, audio, video format, etc. were discussed in the paper which was helpful for our purpose (Gupta, 2015). Discussion over the points, of how paperless technology is advantageous and why it should be implemented, the focusing point was how the use of paper can be decreased also the major issues such as global warming and way to reduce it were discussed in the paper. How the traditional way of using paper is disadvantageous and how technology can overcome them by using information and Communication Technology, e-administration (e-Government), Information Society. The paper discusses the way to improve the quality and accessibility of services (Orantes-Jimenez et al., 2015). Ways that can be accompanied which will reduce the use of paper and how it will be beneficial to the concerned field as well as to the environment in an informative way. Four fields where they have put their ideas to go paperless. Also, the advantages and challenges in actual implementation (Prasetyo et al., 2020). The information on electronic form technology, google forms, the use of add-ons, and form publishers will be more beneficial tools for industrial purposes (Nemec, 2019). The utilisation of the software would be an effective tool for paperless students' assignments. The paper discusses papered examination over paperless examination and assignments along with their advantages and limitations. In their report, they critically explained assignment marking with help of computer-based technology (Plimmer and Apperley, 2007). The concept of green computing is the use of technology to reduce energy consumption. This concept is discussed in the paper to improve the environmental condition (Qadri et al., 2013). The paper focuses on green computing technology that is of using computing resources effectively. Paper concludes that there should be e-document for government purposes, publications of e-books rather than printed books also newspapers in an e-paper way. Also, it has been stated that every secondary level education should be taken through e-books (Agarwal et al., 2014). A case study was carried out in a school which was attempted to go for a paperless environment. For this purpose, the author took a survey from school, and with a questionnaire, he has found that a paperless environment is more advantageous than the traditional way of taking exams (Carr, 2005). Cloud storage on paperless thesis examination discusses the advantages of cloud storing the data and design considerations to achieve such a system to eliminate hard copy prints (Rahman, 2018). An empirical approach to go for a paperless office. This report covers every aspect such as challenges, data analysis, relative advantages, the relation between relative advantages, and the adoption of a paperless office (Dorji, 2018).

Current Industrial Practises

Standard operating procedure

A Standard operating procedure is a document containing detailed information about the operations. It includes information such as steps involved, machine used, equipment used, and other useful information in proper sequence. So, these are set of instructions compiled by an organization to help workers to carry out their tasks. The main objective to have a standard operating procedure (SOP) at workstations is to make work easier for employees, improve quality, and efficiency, ensure employee safety, etc. (Roughton and Crutchfield, 2008; Schmidt and Pierce, 2016).

Travel card

A travel card is a document that is used in industries to record/ track/and mark the work done on a particular station. This document contains every operation carried out on each station and this card goes along with the product from the very start to end, that is dispatch.

Earlier these SOPs are in printed format which was printed in A3 size paper and on the other side travel cards were also in printed format but on A4 size paper.

Paper Consumption and Wastage

In these years various things are evolved in the fields of manufacturing and technology. People are thrived towards it leaving behind the thinking of other problems. Nowadays we see the concept of Industry 4.0 which is the Fourth Industrial Revolution in the field of manufacturing making it Smart Manufacturing which deals with IoT-related terms to achieve a smart factory (Rai et al., 2021). These are the best parts happening around us but the problem related to our environment is needed to be focused on in this booming era. So, sustainability in the manufacturing system needed to be addressed wherever required. In the manufacturing industry, we observe three major fields that are Products, Processes, and Systems. In these different levels, we further observe the different areas, the one which is useful for our study is the 3R concept that is 'Reduce', 'Reuse' and 'Recycle'. This is the concept of green technology for sustainability (Houshyar et al., 2014).

The problem in the company is the use of paper needed for SOP document and travel card document. Considering the SOP document, if there are any changes in the process then that should be changed in SOP document. Also, this document is not an easy task as it again involves approval, signature, and all. And if again found an error, then one has to write it again, correct it, and print it again. So, ultimately this incurs more time, more cost, and wastage of paper.

Taking travel cards into consideration, a travelcard is a document that is used by line employees to mark their completed as well as incomplete work. This travel card is required for each product. So, if a line such as an assembly line, manufacturing line which consists of several stations and also if various products are coming out of line each day, we need travel card for each of them. This is a very big issue company has faced because this is a document that consists of 1215 pages printed on both sides. So, the company has been spending a huge amount on this. Now under the Green Industry company wanted it to reduce to zero. As this document is only important till the dispatch of the product. Once the product is dispatched then this document is discarded.

So, in the company, there we needed to fix both problems. That is to make SOPs online as well as cater to costs involved in the printing travel card. Further, waste in the form of papers is mentioned in the Figures and Tables section.

Paperless Process Methodology

These two problems were resolved by two different approaches. Though the objective is the same for both the requirements were different. As in the concerned company, SOPs were needed on the assembly station because new manpower joins every year. So, to understand the process, it is required to understand SOP first and if it is with video, it will become easier as visuals make things more understandable. Here we are going to explain the way for both paperless projects separately.

Paperless standard operating procedure implementation and analysis

As mentioned in the abstract it is a document having all information related to the process and steps involved to complete a task in a particular way and an operator has to follow the steps as well as the sequence of operation to avoid errors in the work. To date, the company used to update it on a computer in MS Office and later used to print it on paper and keep this on the station. Now to make this digital and to avoid the use of paper we made use of tablet PCs everywhere in the plant. This made work easier. So, the updated pdf can be inserted into the PCs through the common folder. Again, automation in this is making videos of particular operations done on each station. We took shootings of each step, and process on each station, and added SOP to it as subtitles(de Giorgio, et. al. 2021). Also, have added voice to depict whatever is written as subtitles. So, for this purpose, we used the 'Wondershare Filmora 9.0' application to edit videos and add subtitles to them. Also taking line operators into account and their knowledge we found it difficult for them to understand English. So, the need arises that the video should be understandable to all. So, we carried a translation of subtitles from English to Hindi by using 'Google Translate' and to get a voice out of it. We used paid online website named 'voicemaker' which converts text to speech. The need to go for this is if anyone knowing English can directly go through the subtitles but when one has a problem with it then he can go for audio. The online software voice maker also allows us to download voice in .mp3 format so it became easier to add to videos through video editing software. Figure 144.1.

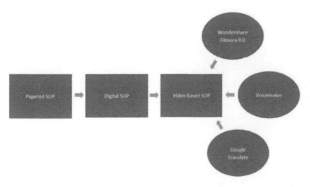

Figure 144.1 Flow diagram for SOP

represents the whole methodology discussed so far. This shows the flow from papered SOP to digital SOP, digital SOP to vdeo-based SOP. And finally, software, applications, web-application used to achieve so are directed towards video-based SOP.

Paperless travel cardimplementation and analysis

This is one of the critical tasks which was difficult to implement in the plant. This travel card has to travel through nine different stations and lastly should merge all station's work information on the 9th station. As 9th station is a quality check station where the whole product is checked to its specifications, parts, fittings, etc. The quality person on station nine had to check every work, station by station and if there is an error in any work, he has to inform to line coordinator who again calls the line operator to fix the problem. So, he (quality person) needs every station's worksheet with him to check if any work is incomplete/ wrongly done according to the customer's requirement. Again, this travel card has to go to the testing area where the product is checked for its performance under different loading and operating conditions. After this travel card has to go to the ATA area. For this purpose, the company uses a tugger so now, to date they using stickers that they were pasting directly on papered travel cards, and by viewing this the tugger person would get to know that which part is to be moved to the ATA area. Finally, this document is needed to be checked in the pre-dispatch area to whole quality check well before to assure quality product to the customer.

This is the final stage, now this travel card has to be saved at this location to address further queries/ problems. So, to make this whole possible we firstly used 'Google Spreadsheet' but due to the company's policy, this did not work. Again, this is done through the help of 'portable document format (PDF)'. We found this to be very easy as we can simply tick on the operations carried and also could save the name of the operator by just scanning his identity card and save it, and transferring it to the next station through the common folder. This is one of the options given by us. Again, there were some problems related to PDF such as it is not traceable that is if a non-working person comes and he signed in the place of an actual worker, it would be difficult for regular workers. Also, there we do not have options and tools as in MS Excel. So, we are searching for better options so that we could display each station's work on each display particularly.

Figure 144.2 is the flow diagram for the methodology used to make travel cards digital. So, it all started from the concept to go paperless, now to make it digital there was a wide range of applications available which were again suitable for some of the requirements and also lagging in some functions. Thus, the Digital Travel card arrows to three elliptical texts (Google Spreadsheet, Portable Document Format (PDF), Software-Based (Future Scope)).

Table 144.1 shows the paper required in every area of the plant. There are two categories, one is fixed which is the paper required on an assembly line. The above table shows figures for fixed paper consumption over a year. Another category is flexible which is when process changes then we need to change it in the SOP. This is taken under the flexible category because in any company process does not change frequently. There are rare changes in the process so when it changes, we need to change it in the SOP.

Cost saving for paperless SOP

Table 144.2 represents the total cost saving achieved in terms of cost saved for pages and cost saved for printing as well as labour cost. Total cost saving as mentioned in the table is huge just to print papers and cost of papers. So, this initiative somehow helped economically as well as environmentally.

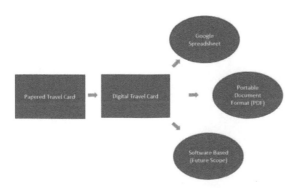

Figure 144.2 Flow diagram for travel card

Cost-saving of paperless travel card

Statistical data of cost-saving achieved from the paperless travelcard is quoted in Table 144.3. This was a huge paged document so cost saving is also more as compared to cost saving for paperless SOP. The table depicts figures for pages, printing and stationery cost, and total savings.

Table 144.1 Paper consumed in every area of the plant

Paper consumed	
Description	Quantity (Nos.)
Pages required on assembly lines	
Line-1	90
Line-2	90
Line-3	90
Pages required in other AREAS	
Panel	50
Test c	90
Canopy	40
ATA	20
Nigeria kitting	10
Normal kitting	30
Dispatch	20
Noise pad	10
Quality gauge room	20
Total Pages	**560**

Table 144.2 Cost saving for paperless SOP

Cost saving	
Cost for pages	
Pages need to be changed/year (Flexible) (Nos.)	5000
Total pages required (Nos.)	5560
Pages in one rim (Nos.)	500
Total Rim Consumed/year (Nos.)	10
Cost of one rim (A3 Size)	₹ 605.00
Total cost for rims consumed	₹ 6,050.00
Printing and stationery cost	
Colour printing cost for A3 size Paper	₹ 10.00
Total printing quantity	₹ 11,120.00
Total printing cost	₹ 1,11,200.00
Total cost	₹ 1,17,250.00
Labour cost	₹ 50,000.00
Total savings	**₹ 1,67,250.00**

Table 144.3 Statistical data of cost saving

Material	Cost/Qty.	Qty.	Total Cost
Monitors	₹ 28,500	16	₹ 4,56,000
MS stands	₹ 11,450	16	₹ 1,83,200
IT material	₹ 5,000	16	₹ 80,000
Electrical material and Protection cover	₹ 7,000	16	₹ 1,12,000
Total	**₹ 51,950**	**16**	**₹ 8,31,200**

Table 144.4 Total investment of company

Description	Total
Pages required for one set of travelcard (Nos.)	6
Total printing quantity (Nos.)	12
Cost for pages	
Total product plan per year (Approx.)	11000
Total pages required (Nos.)	66000
Total printing quantity (Nos.)	132000
Pages in one rim (Nos.)	500
Total rim required/year (Nos.)	132
Cost of one rim	₹ 300.00
Total cost for rims	₹ 39,600.00
Printing and stationery cost	
Printing cost/page (A4 size)	₹ 1.50
Total printing cost	₹ 1,98,000.00
Total cost savings	
Total cost saving	₹ 2,37,600.00
Labour cost/year	₹ 1,00,000.00
	₹ 3,37,600.00

The above Table 144.4 shows the total investment the company did for the whole digitalisation project. Though the figures in the table are more than the savings, this is a setup cost that the company has to bear for one time. The statistical data tables show the results of going paperless and yearly cost savings due to that. Besides savings, it is a great step to help the environment.

Conclusion

Looking at today's environmental condition it is much needed that everyone should give attention to our nature/environment. Day by day it is worsening and it is essential to focus on the problems that are causing these issues. So, the company took a great step to move forward as a green industry. Also, this initiative will save the companies nearly 600000 per year. It is beneficial to use computer technology and there is much software available which makes things easier. But paperless technology cannot be standardized, as there are no standards that define the way to achieve paperless technology. So, it whole depends on the individual.

References

Agarwal, S. and Nath, A. (2014). Green Computing and Sustainable Environment – Introduction of E-documents and Replacements of Printed Stationeries. Int. J. Innov. Res. Info. Secur. 1(5).
Carr, M. R. (2005). An analysis of the feasibility of a paperless environment – The case of the Mona School of Business. Manag. Environ. Quality Int. J. 16(4):286–290. doi: 10.1108/14777830510601172
de Giorgio, A., Roci, M., and Maffei, A. (2021). Measuring the effect of automatically authored video aid on assembly time for procedural knowledge transfer among operators in adaptive assembly stations. Int. J. Product. Res. doi: 10.1080/00207543.2021.1970850
Dorji, T. (2018). Going Paperless Office. Royal Institute of Management, Simtolha, India.
Gupta, S. (2015). Paperless Society – From Vision to Fulfillment. Glob. J. Enter. Info. Sys. 7(1):45. doi: 10.18311/gjeis/2015/3034
Houshyar, A. N., Hoshyar, A. N., and Sulaiman, R. B., (2014). Review paper on sustainability in manufacturing system. J. Appl. Environ. Biol. Sci. 4(4)7 11.
Nayyar, N. and Arora, S.. (2019). Paperless Technology – A Solution to Global Warming. International Conference on Power Energy, Environment and Intelligent Control (PEEIC). doi:10.1109/PEEIC47157.2019.8976599
Nemec, K. D. (2019). The Application of Paperless Processes to Improve Data Management within Small to Medium Businesses. University Honors Program pHd. diss, Honors College Theses. 438.
Orantes-Jimenez, S-D., Zavala-Galindo, A., and Vazquez-Alvarez, G. (2015). Paperless Office: a new proposal for organization. Systmics Cybernetics Informatics. 13(3).
Pleshkevich, E. A. (2007). The Paperless Document: The Beginning of a New Stage of Communications. Scient. Technic. Info. Proce. 34:24–26. doi:10.3103/S0147688207010042

Plimmer, B. and Apperley, M. (2007). Proceedings of the 8th ACM SIGCHI New Zealand chapter's international conference on Computer-human interaction: design centered HCI. 1 8. doi: 10.1145/1278960.1278961

Prasetyo, S. E. and Kusumawardani, S. S. (2020). A review of the challenges of paperless concept in the society 5.0. Int. J. Ind. Eng. Manag. 2(1):15 24. doi: 10.24002/ijieem.v2i1.3755

Qadri, S. F. (2013). Green computing and energy consumption issues in the modern age. IOSR J. Comp. Eng. 12(6): 91 98. doi: 10.9790/0661-1269198

Rahman, F. N. A., Ferdiana, E., and Kusumawardani, S. S. (2018). Integrated cloud storage on paperless thesis examination. 4 th International Conference on Science and Technology (ICST). doi:10.1109/ICSTC.2018.8528573

Rai, R., Tiwari, M. K. Ivanov, D., and Dolgui, A. (2021). Machine learning in manufacturing and industry 4.0 applications. Int. J. Product. Res. 4773 4778. Doi: 10.1080/00207543.2021.1956675

Roughton, J. E. and Crutchfield, N. 2008. 11th Standard or Safe Operating Procedures (SOP), Job Hazard Analysis. Oxford, U K: Butterworth-Heinemann.

Schmidt, R. H. and Pierce, P. D. (2016). The use of standard operating procedures (SOPs). In Handbook of Hygiene Control in the Food Industry. Cambridge: Woodhead Publishing.

World Meteorological Organization Climate Report Met (2021, October, 31). https://public.wmo.int/en/media/press-release/state-of-climate-2021-extreme-events-and-major-impact

145 Utilization of cupola slag blended with coal ash for the development of sustainable brick

Sangita Meshram[1,a], S. P. Raut[1,b], and M. V. Madurwar[2,c]

[1]Department of Civil Engineering, Yeshwantrao Chavan College of Engineering, Nagpur, India

[2]Department of Civil Engineering, Visvesvaraya National Institute of Technology, Nagpur, India

Abstract

The accumulation of mismanaged industrial burnt residue has caused an increase in environmental concern, especially in developing nations. Recycling these wastes into a long-term construction material appears to be a viable option to both the pollution problem and the cost of green building design. Industrial and burned surplus materials are being used to generate sustainable construction materials. The purpose of this study is to learn more about the possibility of using industrial waste coal ash and cupola Slag to replace natural soil in brick manufacture. From the five different mix proportions M1 has 75% CS, 15% CA, 10% cement gives highest value of compressive strength 4.36 N/mm^2 and lowest water absorption value 7.133%. When it is compared to fly ash bricks and burnt clay brick which are commercially available has compressive strength of 5.8 and 3.5 N/mm^2. Water Absorption in case of burnt clay and fly ash bricks are 18.43% and 27.11% which is more than as compared to cupola slag bricks. Density for M5 proportion is lowest among all the mix proportions which is 1698.06 kg/m^3. This research is focused on various combinations of mixes for manufacturing blocks while maintaining 10% cement content. This research shows that combining cupola slag and coal ash to make sustainable bricks yields satisfactory results.

Keywords: Cement, coal ash, cupola slag, physico-mechanical properties, sustainable brick.

Introduction

Brick is broadly utilised manufacturing and structural constituent around the globe. Brick is playing an important part in structure and production for number of years because of its extraordinary properties such as low costs, more strength, and great durability (Zipeng et al., 1018). As the world's population continues to rise, so does the demand for structure, which drives up the price of construction resources. Application of various waste materials to develop blocks reduces the demand for natural available construction material, which leads to be a sustainable solution for construction industry (Raut et al., 2011). Nowadays, the waste materials are blended with raw materials which lead to expansion of new material, used to develop sustainable bricks having more competitive improved physio-mechanical properties (Cecile et al., 2014). Various agro-industrial wastes are being employed to create sustainable building blocks (Pappu et al., 2007; Raut et al., 2022). Cupola slag is a by-product obtained from cast iron industry when liquefied steel is purified from impurities in cupola furnaces (Balaraman and Ligoria, 2015). Approximately 57% of waste is generated during the manufacturing of cast iron in cupola furnaces. Industry produces between 50 and 3000 tonnes of C. I. (Mistry et al., 2016). Cupola slag is utilized as a partial replacement of cement in concrete (Alabi and Joseph , 2013). Cupola slag has also been used in concrete to test its performance as a partial replacement for fine and coarse aggregate mixtures (Balaraman and Elangoval, 2018). Cupola slag used as a supplementary cementitious material based on its chemical composition (Meshram. and Raut, 2021). For the manufacturing of concrete, pavement, and mortar, cupola slag has been utilised as an alternate material for fine, coarse aggregate and cement. Furthermore, no research on the utilisation of cupola slag in bricks has been conducted. It could be useful in the production of building components. Cupola slag used to develop bricks reduces environmental pollution and improves construction efficiency, making the materials more sustainable and cost-effective.

Materials and Methods

Raw materials

Cupola slag and coal ash were used as raw material in manufacturing of bricks. Coal ash utilised in this research was procured from Purti power plant, Khursapur, Nagpur, as per Figure 145.1. Cupola slag (CS) was obtained from CP foundry works, Kamptee road, Nagpur, Maharashtra as presented in Figure 145.2.

[a]sangitameshram3@gmail.com; [b]sprce22@gmail.com; [c]mangesh_bits@yahoo.com

Cupola slag is a steel industry waste product. Also 53 grade OPC cement is used as a binding material. It was used for all proportion with a consistent proportion. A total of 100 brick specimens in varied concentrations were prepared.

Manufacturing of brick specimens

Cupola slag is a dense material with high iron content. Slag must be sieved before being used in bricks. A 1.18mm IS sieve was used for sieving. Cupola slag, coal ash, and cement are used in various proportions to make bricks. The dimensions of the mould (190 × 90 × 90 mm) were used for the preparation of bricks. All of the raw materials' densities were first determined. Table 145.1 shows the quantities of cupola slag, coal ash, and cement that were dry mixed together. Hand mixing was used to incorporate the water into the dry mixture, and mixing was continued till a homogeneous consistency was achieved. A frog having measurements of 140 × 60 × 10 mm was made for the manufacturing of the brick. The proper moulds were used to make the bricks (Figure 145.3). The bricks are then sun-dried in a rain-protected region for seven days.

Test methodology

The elemental composition of CS was studied using an XRF spectrometer (Parth metallurgical services, Nagpur). The specific gravity of materials was calculated using IS 2386 (Part III):1963. The grain size analysis of the cupola was also investigated. IS 2185 (Part 1): 2005 is used to calculate block densities. Compressive

Figure 145.1 Coal ash

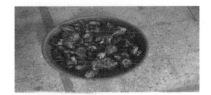

Figure 145.2 Cupola slag

Figure 145.3 Brick specimen

Table 145.1 Mix proportions

Mix	No of specimen	Cupola slag (%)	Coal ash (%)	Cement (%)
M1	20	75	15	10
M2	20	70	20	10
M3	20	65	25	10
M4	20	60	30	10
M5	20	55	35	10

strength was conducted using IS 3495 (Part 1): 1992. Until failure, all created bricks were subjected to at a steady rate per minute of 14 N/mm² compressive load is applied. The weight of oven dry bricks and the weight of bricks immersed in water for 24 hours were used to calculate the water absorption of bricks.

Results and Discussion

Chemical properties of raw materials

The chemical properties of raw materials are represented in Table 145.2. Cupola slag consist of more amount of lime, silica and alumina, thus it's perfect for creating bricks. Calcium oxide, iron oxides and magnesium oxide, were also found in significant amounts in the cupola slag. The presence of considerable amounts of silica increases the binding properties of the bricks. The concentration of silica in coal ash is likewise higher. Cupola slag shows higher calcium oxide content than coal ash that may help with pozzolanic reaction. Cupola slag shows higher percentage of SiO_2 and CaO. More cementitious hydration products were generated when Al_2O_3 and MgO were present. The existence of considerable amounts of silicon, aluminium, calcium oxides in the cupola slag suggests a high reactivity potential for the material. Cupola slag contains higher proportion of lime, which helps to improve binding qualities. Loss of ignition was observed in coal ash.

Physical properties of raw materials

Specific gravity and density Table 145.3 shows the specific gravity of cupola slag and coal ash. Cupola slag has a higher specific gravity and density than coal ash, according to research. Coal ash has a specific gravity

Table 145.2 Chemical properties

Chemical composition (%)	Cupola slag (CS)	Coal ash (CA) [11]
CaO	23.6	9.51
SiO_2	45.6	58.73
AL_2O_3	14.5	20.07
Fe_2O_3	10.6	6.22
MgO	0.872	1.64
SO_3	0.496	0.42
Na_2O	0.872	0.12
K_2O	0.747	0.97
P_2O_5	0.591	-
BaO	0.079	-
CrO_3	0.104	-
TiO_2	1.24	-
NiO	0.008	-
CuO	0.007	-
Mno	1.51	-
ZnO	0.01	-
ZrO_2	0.04	-
V_2O_5	0.591	-
MoO_3	0.003	-
LOI	-	0.79

Table 145.3 Physical properties

S.N	Tests conducted	Test results		
		Cement	Coal ash	Cupola slag
1	Specific gravity	3.15	1.87	2.917
2	Density (kg/m³)	1440	940	1796
3	Standard consistency (%)	31	-	-
4	Fineness (%)	4	-	-
5	Water absorption (%)	-	10	0.26

of 1.87 and a density of 940 kg/m3, whereas cupola slag has a specific gravity of 2.917 and a density of 1796 kg/m3. The use of coal ash and cupola slag in combination may help to reduce weight of bricks.

Particle size distribution

Grain size analysis of cupola slag is shown in Figure 145.4. Gradation curve of raw material shows varied range of particles. It was revealed that sand made nearly 96 % of the cupola slag particles (Table 145.4). It indicated that the cupola slag may be utilised to produce bricks. A total of 4% of the particles were in the category of silt. Table 145.4 shows that in the gravel and clay categories, there are no particles. The grain size distribution of raw materials has a considerable impact on the performance of bricks.

Mechanical properties

Bulk density

Figure 145.5 shows the bulk density results for brick specimens incorporating cupola slag. After introducing CS in bricks, an increase in bulk density was found. As shown in table 145.5, once the proportion of CS was reduced from 80% to 60%, resulting in an 18% decrease in dry density that is from 1999.13 kg/m^3 to 1698.06 kg/m^3 (Table 145.5). The specific gravity and density of raw materials are essential factors in calculating the weight of clay bricks. For M5 brick mix density is less when it compared fly ash and burnt clay bricks which are commercially available as per Table 145.9.

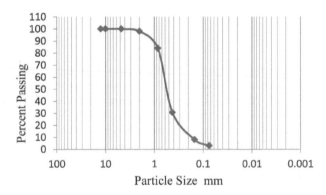

Figure 145.4 Particle size distribution curve

Table 145.4 Particle size distribution

Percentage Distribution	Size Specifications (in mm)	Cupola Slag
Clay	<0.002	0
Silt	0.002–0.075	4
Sand	0.075–2	96
Gravel	>2	0

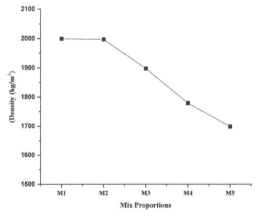

Figure 145.5 Density for brick specimens

Table 145.5 Density of cupola slag bricks

Mix proportion	Density (kg/m³)
M1	1999.13
M2	1996.96
M3	1897.33
M4	1778.87
M5	1698.06

Water absorption

The strength and quality of bricks are both affected by water absorption. The water absorption results for brick samples including CS were shown in Figure 145.6. Water absorption increases from 7.1330.07% when cupola slag in the brick mixes decrease from 75%55% (Table 145.6). Higher brick classes can be utilized for water absorption less than 15%, as specified by IS 1077:1992, and for brick category up to 12.5, values less than 20% can be utilised. As a result, adding CS to masonry bricks can help make them more durable, cost-effective, and ecologically beneficial. When related to burnt clay and fly ash bricks, cupola slag bricks have lower water absorption values for all mix proportions (Table 145.9). Water absorption was less than 22% in all of the brick specimens, indicating that it can be utilised as a moderately weather resistant bricks. Bricks incorporating cupola slag along with coal ash can be efficiently used as a sustainable construction material.

Compressive strength

The compressive strength for all mix proportions is depicted in Figure 145.7 Compressive strength decreases as the percentage of cupola slag in the various mix decreases. Figure 145.8 shows the compressive testing machine used for the testing of bricks. Figure 145.9 indicates the compressive testing conducted on bricks. As shown in Table 145.7, the compressive strength of M1 to M5 mixtures drops from 4.36 N/mm² to 2.9 N/mm². M1 (70 CS: 20 CA: 10 cement) mix shows highest compressive strength. This is due to chemical composition present in cupola slag. According to IS 1077:1992, the lowest compressive strength should be 3.5 N/mm². The IS requirements for compressive strength were met by all mix proportions bricks. It also meets the 4 N/mm² criteria for class 1 bricks. Consequently, the produced blocks can be used to construct

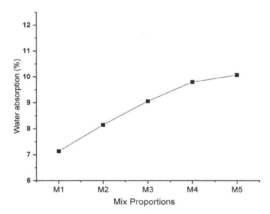

Figure 145.6 Water absorption for cupola slag bricks

Table 145.6 Water absorption

Mix proportion	Water absorption (%)
M1	7.133
M2	8.15
M3	9.06
M4	9.8
M5	10.07

sustainable structures. Fly ash bricks have the lowest compressive strength of 5.4 N/mm², while burnt clay bricks have the lowest compressive strength of 3.5 N/mm². Cupola slag has comparable value of compressive strengths for all mix proportions when related to burnt clay and fly ash bricks as given in Table 145.9.

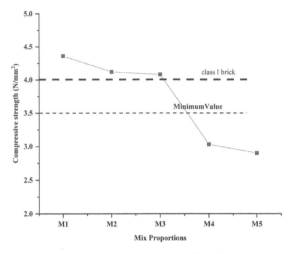

Figure 145.7 Compressive strength of brick specimens

Figure 145.8 Compression test on brick

Figure 145.9 Compression test

Table 145.7 Compressive strength

Mix proportion	Compressive strength (N/mm²)
M1	4.36
M2	4.12
M3	4.08
M4	3.03
M5	2.9

Table 145.8 Efflorescence of brick specimens

Mix Proportion	Efflorescence
M1	Nil
M2	Nil
M3	Nil
M4	Slight
M5	Slight

Figure 145.10 Efflorescence of brick

Table 145.9 Comparison between cupola slag bricks, burnt clay bricks, fly ash bricks

Types of bricks	Density (kg/m³)	Compressive strength (N/mm²)	Water absorption (%)
Burnt Clay Bricks	1650	3.5	27.11
Fly ash bricks	1700	5.8	18.43
Cupola Slag bricks	1698.06	4.36	7.133

Efflorescence

Brick efflorescence is caused by soluble salts in the bricks. On the surfaces of bricks, efflorescence is in the form of crystalline, salty coating. According to IS 3495: 1992, efflorescence is 'Nil' when salt deposits are negligible. A 'slight' is defined as a thin layer of salts covering less than 10% of the visible surface area of the brick. Efflorescence is nil in case of mix proportions from M1 to M3 (Table 145.8). There are slight deposits of salts are visible in case of M4 and M5 mix proportions as shown in Figure 145.10.

Conclusions

Cupola slag is dumped on land and then disposed of; it pollutes the soil, which has a straight effect on the environment. Hence, its effective use in the building of bricks provides a partial solution to these problems. Cupola slag, a cast iron industry by-product, was blended with coal ash to make bricks with a consistent cement ratio. Significant amount of alumina and silica present in cupola slag make it suitable to develop sustainable bricks. From all the different mix proportions M1 has 75% CS, 15% CA, 10% cement gives highest value of compressive strength 4.36 N/mm² and lowest water absorption value 7.133%. When it is compared to fly ash and burnt clay brick which are commercially available has compressive strength of 5.8 and 3.5 N/mm². Water absorption for fly ash and burnt clay bricks were 18.43 % and 27.11% which is more than the cupola slag bricks for all the mix proportions. Density for M5 proportion is lowest among all the proportions is 1698.06 kg/m³. Therefore, it is comparable with commercially available bricks. It is useful in making light weight bricks. All of the mixes had a water absorption value of less than 20%, suggesting that they complied with the standards. Since, cupola slag is a by-product obtained from cast iron manufacturing industry. Cupola slag disposed in open land which caused land pollution. It can be used as a construction material to develop sustainable bricks. Results indicated that the cupola slag can be effectively utilised in making bricks. Hence results in environmentally friendly and long-term construction material.

References

Alabi, A. S. and Joseph, O. A. (2013). Investigation on the potentials of cupola furnace slag in concrete. Int. J. Integrat. Eng. 5(2).

Balaraman, R. and Elangoval, N. S. (2018). Behaviour of cuplola slag in concrete as partial replacement with combination of fine and coarse aggregate. Mater. Sci. 14.

Balaraman, R. and Ligoria, S. A. (2015). Utilization of cupola slag in concrete as fine and coarse aggregate. Int. J. Civil Eng. Technol. 8(6):6–14.

Cecile, B., Borredon, M. E. Vedrenne, E., and Vilarem, G. (2014). Development of eco-friendly porous fired clay bricks using pore-forming agents: A review. J. Environ. Manag. 143:186–196.

IS 2386 (Part 3):1963 – Methods of tests for aggregate of concrete.

IS 2185(Part 1): 2005 – Concrete Masonry Units- Specification.

IS 3495 (Part 1): 1992 – Methods of tests of Burnt Clay Building bricks.

IS 1077:1992 - Common Burnt Clay Building Bricks – Specification.

Meshram, S. S. and Raut, S. P. 2021. Utilization of cupola slag as a sustainable construction material. In Advances in Civil Engineering and Infrastructural Development, pp. 43–49. Singapore: Springer.

Mistry, V. K., Patel, B. R., and Varia, D. J. (2016). Suitability of concrete using cupola slag as replacement of coarse aggregate. Int. J. Sci. Eng. Res. 7(2).

Oruji, S., Nicholas, A., Brake, Ramesh, K., Guduru, L. N., Gunaydın-Şen, O., Kharel, K., Rabbanifar, S., Hosseini, S., and Ingram, E. (2019). Mitigation of ASR expansion in concrete using ultra-fine coal bottom ash. Construct. Build. Mater. 202:814–824.

Pappu, A., Saxena, M., and Asolekar, S. R. (2007). Solid wastes generation in India and their recycling potential in building materials. Building and environment. 42(6):2311–2320.

Raut, S. P., Patil, U. S., and Madurwar, M. V. (2022). Utilization of phosphogypsum and rice husk to develop sustainable bricks. Mater. Today: Proc.

Raut, S. P., Ralegaonkar, R. V., and Mandavgane, S. A (2011). Development of sustainable construction material using industrial and agricultural solid waste: A review of waste-create bricks. Construct. Build. Mater. 25(10):4037–4042.

Zipeng, Z., Wong, Y. C., Arulrajah, A., and Horpibulsuk. S., (2018). A review of studies on bricks using alternative materials and approaches. Construct. Build. Mater. 188:1101–1118.

146 Critical study of analysis and design parameters considering ductile detailing of reinforced concrete structure

Monali Wagh[1,a], Anshul Nikhade[2], and Ashutosh Bagde[3]

[1]Department of Civil Engineering, Yeshwantrao Chavan College of Engineering, Nagpur, India

[2]Department of Civil Engineering, KDK College of Engineering, Nagpur, India

[3]Jawaharlal Nehru Medical College, Datta Meghe Institute of Medical Sciences, Sawangi, Wardha, India

Abstract

The ductility factor is important in making a structure more earthquake-resistant. When ductile members are used to construct a structure, it can withstand massive deformations before failing. When it comes to overloading, ductility factor is important. Before failing, the structure will undergo huge deformation, giving the occupants ample notice. The STAAD-Pro was used to analyse a G+10 storied building. During the analysis, it was discovered that the deflections at the upper levels were significantly higher. As a result, the shear walls were used to reduce the deflections. In a ductile detailed building, the majority of the parameters listed above are found to be on the lower side. This project also includes a comparison of ordinary moment resisting frame (OMRF) and special moment resisting frame (SMRF) detailing.

Keywords: Deformation, ductility, earthquake resistant, modelling.

Introduction

Ductility is defined as a material's capacity to withstand significant deformations without rupturing before failure. The importance of ductility in earthquake-resistant structures cannot be overstated. When the ductility factor is used in a structure, it can withstand huge deformations without failing. It is advantageous in the case of overloading. The structure can deform huge before failing, which serves as a warning to the occupants. This will reduce the loss of life (Wagh et al., 2014). The goal of this project is to conduct a critical analysis of the IS 13920 (1993) standard, analyse the structure with and without the ductility factor, and investigate the importance of ductile detailing in steel quantities. IS 13920 (1993) provisions were critically examined in this project. Lateral reinforcement is vital for ductility because it eliminates precipitate shear failures and reduces the compression zone, which increases the deformation capability of a reinforced concrete beam. Concrete members in bridges with inadequate flexural ductility, low shear strength, and insufficient lap length for starter bars were constructed prior to the application of the new seismic design standards, all of which contribute to earthquake failure (Saadatmanesh et al., 1994). Shear walls are designed to withstand seismic and wind stresses, and they may even be built on soils with poor foundations using different ground improvement techniques (Amar et al., 2021).

Literature Review

Zhanga et al. (2022) evaluated the performance of G+2 story frame against cyclic lateral loading condition by using ultra-high strength (UHS) steel bars with shear wall. Concealed bracing reinforced by UHS steel bars can effectively reduce shear strength degradation, regulate shear wall damage, and increase the structure's reparability. Based on the failure process, an ideal load-deformation response was created to reflect the damage situation.

Requena-Garcia-Cruz et al. (2021) investigate the ductile response behaviour of RC framed buildings using different non-invasive retrofitting techniques. The addition of steel braces provided the greatest benefit, according to the findings. Single braces and steel jackets, on the other hand, have proven to be the best solutions in terms of benefit and cost. Due to the critical role that joints play in the resistant capacity of RC structures, solutions that increase joint stiffness have shown to have a higher improvement. It was also discovered that by increasing the stiffness of the system, the values of the fundamental periods were reduced by up to 30% when retrofitting elements and materials were added.

---

[a]wagh.monali04@gmail.com

Ozdemir A et al. (2021) analysed the impacts of a steel strip retrofitting approach on the performance of damaged reinforced concrete shear walls. The steel strip retrofitting method for rebuilding reinforced concrete shear walls with and without apertures that had been damaged at heavy and moderate levels was very successful in restoring the shear walls to their pre-damage performance levels.

Omar El-Kashif and Ahmed Abdalla (2019) investigated the response of FRP-retrofitted RC shear walls subjected to lateral loads, using the general-purpose finite element code ANSYS. The wall capacity was improved by 9.5%, 18.15%, 27.32%, and 32.14%, respectively, by using one vertical layer of CFRP and raising the concrete strength from 25 MPa to 30 MPa, 35 MPa, 40 MPa, and 45 MPa.

Ugaldea and Lopez-Garcia (2017) observed that, all residential building in Chile that is more than five stories tall is almost entirely made of reinforced concrete shear walls. Despite the fact that numerous buildings were subjected to ground accelerations larger than those authorised by the Chilean seismic design code, just 2% of the residential building inventory in Chile was badly damaged by the 2010 earthquake (Mw 8.8).

Structure Modeling

Nagpur's Impressa Classic structure

In STAAD-Pro, the modelling of the G+10 storied RCC building at Koradi Road, Nagpur, was completed in the same manner as the Impressa Classic building.

Building specifications:

1) Building length - 56.78m
2) Building width - 20.42 m
3) The building's height is 33m
4) Column size (0.500 m × 0.450 m) and (0.450 m × 0.50 m)
5) Beam dimensions: 0.450 m × 0.450 m
6) 0.1 m slab thickness
7) Self-weight (DL) – 14KN/m for 0.23 m thick wall
8) Self-weight (DL) – 7 KN/m for 0.15 m thick wall
9) Maximum live load – 2.5 KN/m²
10) Earthquake load a) X-directional earthquake load, b) Z-directional earthquake load
11) Thickness of the shear wall – 0.2 m
12) Response reduction factor [R]: a) 5 for SMRF, b) 3 for OMRF

STAAD-pro is used to analyse and design this G+10 storied building.
This G+10 storied building is analysed for zone 2, zone 3, zone 4, zone 5.

Structural Analysis and Design

Though ductile detailing is recommended for high-rise structures in zone 4 and 5, according to IS 13920(1993). Following are the analysis parameters used to compare ordinary moment resisting frame (OMRF) and special moment resisting frame (SMRF), with ductile detailing.

1) Deflection
2) Shear in the X-axis (Fx)
3) Moment along the Z-axis (Mz)

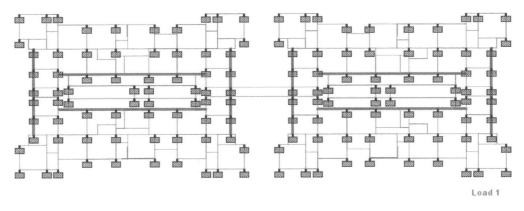

Figure 146.1 Plan of structure and position of shear wall

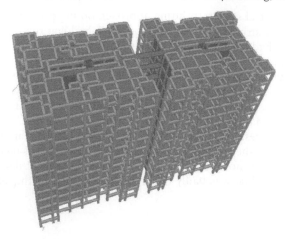

Figure 146.2 3-D view of the structure

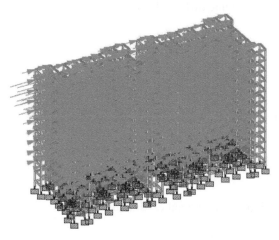

Figure 146.3 Earthquake load in X-direction

We chose the side column for comparison.

Exclusive analysis data for these columns is exported to Excel for further comparison and analysis. Graphs and mathematical models are created based on the data.

However, two load combinations have been discovered to be critical for columns. Depending on the column orientation, these are 1.5 (DL+EQX) or 1.5(DL+EQZ).

Table 146.1 Comparison of deflections for all zones (s.de column)

L/C	Floor Level	With shear wall								Without shear wall							
		OMRF				SMRF				OMRF				SMRF			
		II	III	IV	V	II	III	IV	V	II	III	IV	V	II	III	IV	V
1.5(DL+EQX)	0	8.16	8.23	8.33	8.49	8.10	8.14	8.21	8.30	54.60	55.18	55.95	57.10	54.22	54.56	55.03	55.72
1.5(DL+EQX)	1	14.06	14.17	14.36	14.66	13.92	14.01	14.13	14.30	108.33	109.53	111.13	113.52	107.54	108.25	109.21	110.65
1.5(DL+EQX)	2	19.79	19.93	20.22	20.66	19.56	19.69	19.87	20.13	163.35	165.24	167.75	171.51	162.10	163.23	164.73	166.99
1.5(DL+EQX)	3	25.51	25.67	26.07	26.66	25.18	25.36	25.60	25.95	216.76	219.36	222.82	228.02	215.03	216.59	218.67	221.78
1.5(DL+EQX)	4	31.12	31.31	31.81	32.56	30.68	30.91	31.21	31.66	266.49	269.80	274.21	280.83	264.28	266.27	268.92	272.89
1.5(DL+EQX)	5	36.52	36.74	37.35	38.26	35.98	36.25	36.62	37.17	311.30	315.29	320.62	328.61	308.63	311.03	314.23	319.02
1.5(DL+EQX)	6	41.64	41.89	42.61	43.68	40.99	41.32	41.75	42.39	350.54	355.17	361.34	370.60	347.45	350.23	353.93	359.49
1.5(DL+EQX)	7	46.42	46.70	47.52	48.76	45.68	46.05	46.54	47.28	383.97	389.16	396.09	406.48	380.50	383.62	387.78	394.01
1.5(DL+EQX)	8	50.84	51.15	52.07	53.44	50.01	50.42	50.97	51.79	411.63	417.31	424.87	436.22	407.85	411.25	415.79	422.60
1.5(DL+EQX)	9	54.85	55.19	56.19	57.69	53.94	54.39	54.99	55.89	433.75	439.80	447.87	459.98	429.71	433.34	438.19	445.45
1.5(DL+EQX)	10	57.99	58.34	59.40	60.98	57.02	57.49	58.13	59.08	449.87	456.18	464.60	477.22	445.67	449.45	454.50	462.07
1.5(DL+EQZ)	0	11.92	11.98	12.13	12.37	11.79	11.86	11.95	12.09	71.79	72.53	73.52	75.00	71.29	71.74	72.33	73.22
1.5(DL+EQZ)	1	22.39	22.52	22.84	23.33	22.12	22.26	22.46	22.75	134.96	136.42	138.36	141.27	133.99	134.86	136.03	137.78
1.5(DL+EQZ)	2	33.64	33.84	34.37	35.15	33.19	33.42	33.74	34.21	194.71	196.90	199.83	204.22	193.24	194.56	196.32	198.95
1.5(DL+EQZ)	3	45.66	45.94	46.69	47.81	45.00	45.34	45.79	46.46	250.35	253.29	257.22	263.12	248.38	250.15	252.51	256.04
1.5(DL+EQZ)	4	58.11	58.48	59.47	60.97	57.23	57.68	58.28	59.18	301.45	305.14	310.07	317.45	298.99	301.21	304.16	308.59
1.5(DL+EQZ)	5	70.66	71.12	72.37	74.26	69.55	70.11	70.87	72.00	347.60	352.02	357.90	366.72	344.66	347.31	350.84	356.13
1.5(DL+EQZ)	6	83.04	83.58	85.10	87.38	81.68	82.37	83.28	84.65	388.38	393.47	400.25	410.42	384.99	388.04	392.11	398.21
1.5(DL+EQZ)	7	95.03	95.66	97.44	100.11	93.44	94.24	95.30	96.91	423.36	429.05	436.63	448.01	419.57	422.98	427.53	434.35
1.5(DL+EQZ)	8	106.46	107.18	109.21	112.25	104.64	105.55	106.77	108.60	452.12	458.31	466.57	478.95	448.00	451.71	456.66	464.09
1.5(DL+EQZ)	9	117.02	117.79	120.04	123.42	114.98	115.99	117.34	119.37	474.28	480.84	489.59	502.72	469.90	473.84	479.09	486.97
1.5(DL+EQZ)	10	125.37	126.19	128.60	132.22	123.17	124.25	125.70	127.88	489.86	496.66	505.72	519.31	485.33	489.41	494.85	503.00

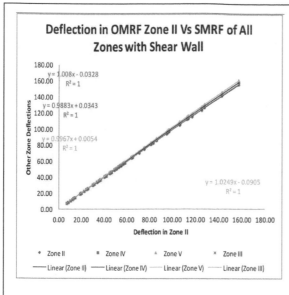

Graph 1

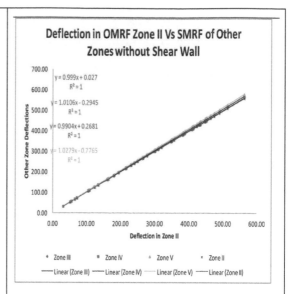

Graph 2

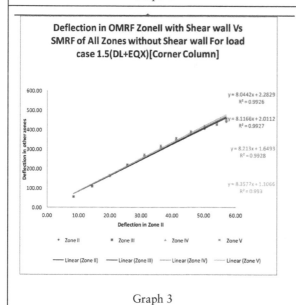

Graph 3

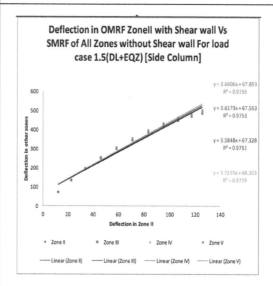

Graph 4

Table 146.2 Comparison of F_x of OMRF and SMRF for all zones (side column)

Load condition	Floor Level	With shear wall								Without shear wall							
		II	III	IV	V	II	III	IV	V	II	III	IV	V	II	III	IV	V
		OMRF	OMRF	OMRF	OMRF	SMRF	SMRF	SMRF	SMRF	OMRF	OMRF	OMRF	OMRF	SMRF	SMRF	SMRF	SMRF
		Fx	Fx	Fx	Fx	Fx	Fx	Fx	Fx	Fx	Fx	Fx	Fx	Fx	Fx	Fx	Fx
1.5(DL+EQX)	0	2770.71	2780.95	2794.60	2815.08	2763.88	2770.03	2778.22	2790.51	3221.28	3230.99	3243.94	3263.36	3214.80	3220.63	3228.40	3240.05
1.5(DL+EQX)	1	2441.22	2450.29	2462.38	2480.52	2435.17	2440.61	2447.87	2458.75	2640.83	2647.13	2655.54	2668.15	2636.62	2640.41	2645.45	2653.02
1.5(DL+EQX)	2	2133.05	2141.03	2151.68	2167.64	2127.73	2132.52	2138.91	2148.49	2127.36	2130.74	2135.25	2142.01	2125.11	2127.14	2129.84	2133.90
1.5(DL+EQX)	3	1835.79	1842.66	1851.81	1865.54	1831.22	1835.33	1840.83	1849.07	1680.14	1681.07	1682.30	1684.15	1679.52	1680.08	1680.82	1681.93
1.5(DL+EQX)	4	1550.74	1556.50	1564.17	1575.67	1546.91	1550.36	1554.96	1561.87	1296.57	1295.52	1294.13	1292.03	1297.27	1296.64	1295.80	1294.55
1.5(DL+EQX)	5	1278.19	1282.86	1289.09	1298.43	1275.08	1277.88	1281.62	1287.22	968.85	966.28	962.86	957.72	970.57	969.02	966.97	963.89
1.5(DL+EQX)	6	1018.06	1021.70	1026.55	1033.83	1015.63	1017.81	1020.73	1025.10	688.18	684.52	679.65	672.33	690.62	688.42	685.50	681.11
1.5(DL+EQX)	7	769.76	772.45	776.03	781.41	767.97	769.58	771.73	774.96	447.09	442.80	437.07	428.47	449.96	447.38	443.94	438.79
1.5(DL+EQX)	8	532.26	534.09	536.53	540.19	531.04	532.14	533.60	535.80	240.52	236.08	230.17	221.30	243.47	240.81	237.26	231.94
1.5(DL+EQX)	9	303.71	304.78	306.20	308.34	302.93	303.64	304.49	305.77	65.25	61.27	55.97	48.02	67.90	65.52	62.33	57.56
1.5(DL+EQX)	10	85.40	85.86	86.49	87.42	85.08	85.37	85.74	86.30	-73.79	-76.48	-80.06	-85.44	-71.99	-73.61	-75.76	-78.98
1.5(DL+EQZ)	0	2668.46	2677.83	2690.33	2709.08	2662.21	2667.84	2675.33	2686.58	2993.87	3002.87	3014.87	3032.86	2987.87	2993.27	3000.47	3011.27
1.5(DL+EQZ)	1	2418.20	2427.28	2439.38	2457.54	2412.15	2417.59	2424.86	2435.75	2790.93	2800.19	2812.54	2831.07	2784.75	2790.31	2797.72	2808.84
1.5(DL+EQZ)	2	2167.55	2176.17	2187.67	2204.91	2161.80	2166.97	2173.87	2184.22	2556.54	2565.71	2577.94	2596.28	2550.43	2555.93	2563.27	2574.27
1.5(DL+EQZ)	3	1914.54	1922.54	1933.22	1949.23	1909.20	1914.00	1920.41	1930.01	2300.49	2309.30	2321.05	2338.68	2294.62	2299.90	2306.95	2317.53
1.5(DL+EQZ)	4	1659.40	1666.65	1676.31	1690.79	1654.57	1658.92	1664.71	1673.41	2026.83	2035.05	2046.02	2062.48	2021.34	2026.28	2032.86	2042.73
1.5(DL+EQZ)	5	1402.71	1409.07	1417.56	1430.28	1398.47	1402.29	1407.38	1415.01	1739.79	1747.24	1757.19	1772.10	1734.81	1739.29	1745.25	1754.20
1.5(DL+EQZ)	6	1145.10	1150.48	1157.66	1168.43	1141.51	1144.74	1149.05	1155.51	1442.63	1449.16	1457.88	1470.95	1438.27	1442.19	1447.42	1455.26
1.5(DL+EQZ)	7	887.04	891.37	897.14	905.80	884.15	886.75	890.21	895.41	1137.69	1143.18	1150.49	1161.45	1134.04	1137.33	1141.71	1148.29
1.5(DL+EQZ)	8	628.85	632.06	636.35	642.78	626.70	628.63	631.20	635.06	826.33	830.63	836.36	844.96	823.46	826.04	829.48	834.64
1.5(DL+EQZ)	9	370.42	372.48	375.23	379.34	369.05	370.29	371.93	374.40	509.42	512.06	516.38	522.35	507.43	509.22	511.60	515.19
1.5(DL+EQZ)	10	112.82	113.69	114.84	116.57	112.24	112.76	113.45	114.49	185.30	186.76	188.72	191.65	184.32	185.20	186.37	188.13

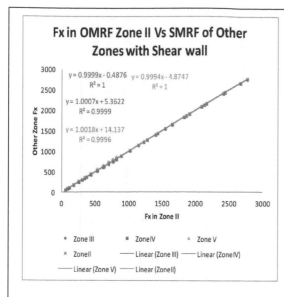

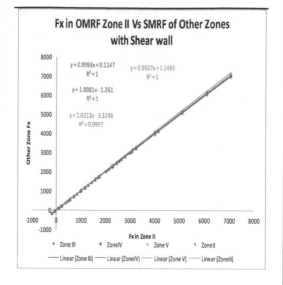

Graph 5

Graph 6

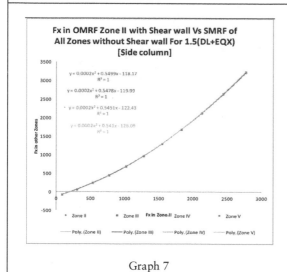

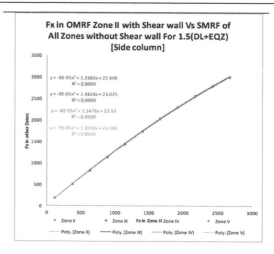

Graph 7

Graph 8

Table 146.3 Comparison of M_z of OMRF and SMRF for all zones (side column)

		With shear wall								Without shear wall							
		OMRF	OMRF	OMRF	OMRF	SMRF	SMRF	SMRF	SMRF	OMRF	OMRF	OMRF	OMRF	SMRF	SMRF	SMRF	SMRF
		II	III	IV	V	II	III	IV	V	II	III	IV	V	II	III	IV	V
L/C	Floor Level	Mz	Mz	Mz	Mz	Mz	Mz	Mz	Mz	Mz	Mz	Mz	Mz	Mz	Mz	Mz	Mz
1.5(DL+EQX)	0	131.8	133.2	135.3	138.5	130.6	131.5	132.8	134.7	1040.5	1051.7	1066.7	1089.1	1033.0	1039.7	1048.7	1062.2
1.5(DL+EQX)	1	58.1	58.4	59.6	61.5	56.8	57.4	58.1	59.3	855.7	866.2	880.3	901.3	848.6	855.0	863.4	876.0
1.5(DL+EQX)	2	73.5	73.9	75.4	77.8	71.9	72.6	73.6	75.0	860.8	872.2	887.3	910.1	853.2	860.0	869.1	882.8
1.5(DL+EQX)	3	73.0	73.4	75.1	77.5	71.4	72.1	73.1	74.6	813.6	825.1	840.6	863.7	805.9	812.8	822.1	835.9
1.5(DL+EQX)	4	70.1	70.6	72.3	74.8	68.5	69.2	70.2	71.7	740.8	752.1	767.1	789.8	733.2	740.0	749.1	762.6
1.5(DL+EQX)	5	65.9	66.3	68.0	70.5	64.3	65.0	66.0	67.5	654.7	665.3	679.6	700.9	647.5	653.9	662.5	675.3
1.5(DL+EQX)	6	60.7	61.2	62.8	65.2	59.2	59.9	60.8	62.3	562.6	572.3	585.2	604.7	556.1	561.9	569.7	581.3
1.5(DL+EQX)	7	54.8	55.3	56.8	59.1	53.5	54.1	55.0	56.4	469.3	477.8	489.0	506.0	463.7	468.7	475.5	485.7
1.5(DL+EQX)	8	48.7	49.2	50.6	52.6	47.5	48.1	48.9	50.1	379.6	386.6	395.9	409.9	375.0	379.2	384.8	393.1
1.5(DL+EQX)	9	42.6	42.9	44.1	45.9	41.5	42.0	42.7	43.8	293.7	298.9	306.0	316.5	290.2	293.3	297.5	303.8
1.5(DL+EQX)	10	31.6	31.6	32.1	32.8	30.9	31.2	31.5	31.9	223.5	227.0	231.7	238.6	221.2	223.3	226.1	230.3
1.5(DL+EQZ)	0	-40.3	-40.5	-40.6	-40.8	-40.3	-40.3	-40.4	-40.6	-73.9	-74.5	-75.4	-76.8	-73.4	-73.8	-74.4	-75.2
1.5(DL+EQZ)	1	-29.3	-29.5	-29.6	-29.8	-29.3	-29.4	-29.4	-29.6	-101.4	-102.3	-103.5	-105.3	-100.8	-101.4	-102.1	-103.1
1.5(DL+EQZ)	2	-19.0	-19.1	-19.1	-19.2	-19.0	-19.1	-19.1	-19.1	-88.3	-89.2	-90.3	-91.9	-87.8	-88.3	-88.9	-89.9
1.5(DL+EQZ)	3	-19.5	-19.6	-19.7	-19.7	-19.6	-19.6	-19.6	-19.7	-87.0	-87.9	-89.0	-90.8	-86.4	-86.9	-87.6	-88.7
1.5(DL+EQZ)	4	-17.4	-17.5	-17.5	-17.6	-17.4	-17.4	-17.5	-17.5	-81.7	-82.6	-83.7	-85.5	-81.2	-81.7	-82.4	-83.4
1.5(DL+EQZ)	5	-14.9	-15.0	-15.0	-15.0	-15.0	-15.0	-15.0	-15.0	-75.8	-76.7	-77.8	-79.4	-75.3	-75.8	-76.4	-77.4
1.5(DL+EQZ)	6	-12.3	-12.5	-12.5	-12.4	-12.5	-12.5	-12.5	-12.5	-69.4	-70.1	-71.1	-72.7	-68.8	-69.3	-69.9	-70.8
1.5(DL+EQZ)	7	-9.9	-10.0	-10.0	-9.9	-10.0	-10.0	-10.0	-10.0	-62.6	-63.3	-64.2	-65.6	-62.2	-62.6	-63.1	-63.9
1.5(DL+EQZ)	8	-7.3	-7.3	-7.2	-7.1	-7.5	-7.4	-7.4	-7.3	-55.9	-56.5	-57.3	-58.4	-55.5	-55.9	-56.3	-57.0
1.5(DL+EQZ)	9	-1.3	-1.2	-0.9	-0.3	-1.6	-1.5	-1.3	-1.0	-49.0	-49.4	-50.0	-50.9	-48.7	-49.0	-49.3	-49.8
1.5(DL+EQZ)	10	15.7	16.0	16.8	18.1	15.0	15.3	15.8	16.6	-50.4	-50.7	-51.2	-51.9	-50.1	-50.3	-50.6	-51.0

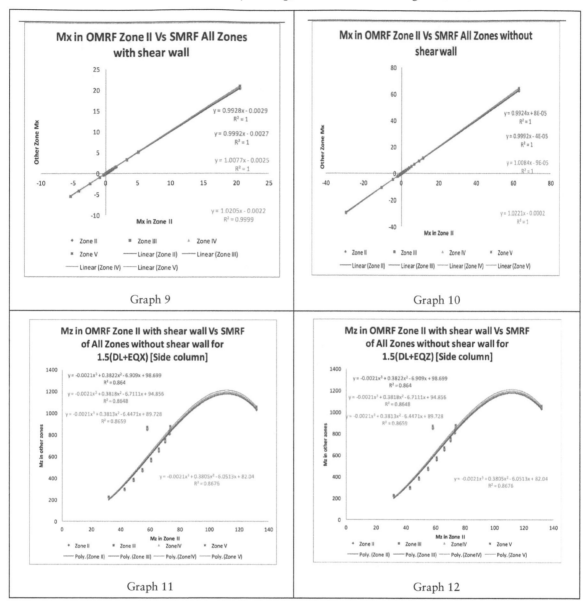

Graph 9

Graph 10

Graph 11

Graph 12

Results and Discussions

Table 146.1 shows the comparison of deflections of OMRF and SMRF for all zones for side column. Graph 1, 2, 3 and 4 shows the deflection of G+10 storied multistoried building analyse in STAAD-pro with and without shear wall in ordinary and special moment resisting frame in zone 2, 3, 4, and 5. With the consideration of special moment resisting frame, deflection of the G+10 storied building with shear wall in side column decreases with 2.283.08% in zone 5. With the implementation of shear wall in Building, deflection reduced with 7080% in all ordinary and moment resisting frame. Deflection of G+10 storied building increases with floor height and zone.

Table 146.2 shows the comparison of shear in X direction (Fx) of OMRF and SMRF for all zones of side column. Graph 5, 6, 7 and 8 shows the shear in X direction of G+10 storied multistoried building analyse in STAAD-pro with and without shear wall in ordinary and special moment resisting frame in zone 2, 3, 4, and 5. Shear in X direction of the G+10 storied structure with shear wall in side column falls by 12% in zone 5 when special moment resistant frame is taken into account. Shear in X direction in all conventional and moment resistant frames was decreased by 2025% after shear walls were installed in the building. With floor height, the shear in X direction of a G+10 storied building decreases. With the rises of zone, the shear in X direction of a G+10 storied building rises.

Table 146.3 shows the comparison of moment in the Z direction (Mz) of OMRF & SMRF for all zones (side column). Graphs 9, 10, 11, and 12 depict Mz of a G+10 storied multistory building analysed in STAAD-pro with and without shear wall in ordinary and special moment resisting frames in zones 2, 3, 4, and 5. When special moment resisting frame is taken into account, moment in the Z direction of the G+10 storied structure with shear wall in side column falls by 23.5% in zone 5. Moment in the Z direction was reduced by 8590% in all conventional and moment resistant frames after shear walls were put in the structure for 1.5 DL+1.5 EQx. Moment in the Z direction was reduced by 4570% in all conventional and moment resistant frames after shear walls were put in the structure for 1.5 DL+1.5 EQz. The moment in the Z direction of a G+10 storied structure reduces as floor height increases. The moment in the Z direction of a G+10 storied structure increases as the zone rises.

After analysis of G+10 storied building, design is carried for beams and columns in zone V because deflection, shear, and moment, all these parameters are maximum in case of zone V. Design all columns and beams in the STAAD-Pro for the V Zone, the results of the quantities of steel in OMRF and SMRF shown in Table 146.4 are as follows-

Table 146.4 Comparison of quantities of steel in OMRF and SMRF for V zone

Components	Type of building	Weight of Bar Dia [N]					Total weight [N]	Increase in % of steel
		8 mm	10 mm	12 mm	16 mm	20mm		
Columns	OMRF	84036.38	0	199864.39	303917.66	443982.38	1031800.81	1.727742649
	SMRF	87256.17	0	212115.17	309562.28	441007.47	1049941.09	
Beams	OMRF	165912.42	345921.1	348885.56	398982.97	364697	1624399.07	1.374988513
	SMRF	175840.77	337970.1	351479.12	401105.53	380650.25	1647045.76	

Comparison of quantities of steel indicates that about 1 to 2% increments with shear wall in Special Moment Resisting Frame than OMRF

Table 146.5 Reduction in displacement, Fx and Mz

Column type	L/C	% Reduction in displacements	% Reduction in Fx	% Reduction in Mz
Interior column	1.5(DL+EQ$_X$)	0.753.2	0.241.3	0.32.3
	1.5(DL+EQ$_Z$)	0.63.1	0.231.8	0.94.8
Side column	1.5(DL+EQ$_X$)	0.753.2	0.241.3	0.95
	1.5(DL+EQ$_Z$)	1.13.3	0.231.9	0.28

Conclusion

1) By providing a shear wall, displacements at various levels were reduced. Displacements were reduced by 70–80%.
2) The 1.5 (DL+EQX) or 1.5 (DL+EQZ) load combinations were found to be critical.
3) The ductility factor was used to detect reductions in forces and moments, as shown in the Tables 146.5 above-
4) Although the reduction in deflections is not significant, the structure can withstand more displacement due to ductile detailing of joints, reducing the risk of failure.
5) Steel quantities of Special Moment Resisting Frame with shear wall increase about 12% higher than ordinary moment resisting frame.
6) When detailing of members and joints is done according to IS 13920, extra bar lengths must be calculated carefully, and beams, columns, and joints must be properly detailed.
7) Ductile detailing on the jobsite necessitates close supervision.
8) Though this is outside the scope of the project, it has been realised that a methodology for estimating the maximum deflection or displacements of a ductile joint can take before failure is needed. With this estimate, we may be able to provide thinner sections as a safety precaution and lowering the structure's cost.
9) A comparison of IS 13920, (1993) and STAAD-Pro design reinforcement detailing of beam-column joints reveals that careful detailing of members and joints is required.

References

Amar, G., Sanjay Gokul, V., Vamsi, K. K., and Rakesh, D. Analysis and design of reinforced concrete rectangular shear Wall. Int. J. Innov. Res. Technol. 2(12):360–366.

IS 13920, (1993). Ductile detailing of reinforced concrete structures subjected to seismic forces.

Ozdemir, A., Kopraman, Y., and Ozgür, A. (2021). Hysteretic behavior of retrofitted RC shear wall with different damage levels by using steel strips. J. Build. Eng. 44:103394.

Requena-Garcia-Cruz, M., Morales-Esteban, A., and Durand-Neyra, P. (2021). Optimal ductility enhancement of RC framed buildings considering different non-invasive retrofitting techniques. Eng. Struct.242:112572.

Saadatmanesh, H., Ehsani, M. R., and Li, M. W. (1994). Strength and ductility of concrete columns externally reinforced with fiber composite straps. ACI Struct. J. 434–447.

Ugaldea, D. and Lopez-Garcia, D. (2017). Behavior of reinforced concrete shear wall buildings subjected to large earthquakes. Procedia Eng. 199:3582–3587

Ugaldea, D., and Lopez-Garcia, D. (2019). Finite element modeling of RC shear walls strengthened with CFRP subjected to cyclic loading. Alexandria Eng. J. 58:189–205.

Wagh, M. R., Nikhade, A. R., and Nikhade, H. R. (2014). Analysis and design considering ductile detailing of reinforced concrete structure. Int. J. Sci. Eng. Technol. Res. 3(11):2987–2991.

Zameeruddin, M. and Sangle. K. (2021). Performance-based seismic assessment of reinforced concrete moment resisting frame. J. King Saud Univ. Eng. Sci. 33:153–165.

Zhanga, J., Liua, J., , D., and Huang, X. (2022). Hysteretic behavior of high-performance frame-shear wall composite structure with high-strength steel bars. J. Build. Eng. 45:103416.

147 Fluid-particle heat transfer in fluidised and fixed beds

Narendra J. Giradkar[a], Jayant P. Giri[b], Vivek M. Korde[c], and Rajkumar B. Chadge[d]

Department of Mechanical Engineering, Yeshwantrao Chavan College of Engineering, Nagpur, India

Abstract

The particle Reynolds number $Re_p \left(\dfrac{D_p u \rho_g}{\mu} \right)$, where 'Dp' is fluid particle diameter, 'u' is velocity of fluid, 'ρ_g' is density of gas and 'μ' is dynamic viscosity of fluid; is extensively used in design and operation of fluidized systems to define bed dynamics. However, the bed turbulence greatly depends upon solid density ρ_s and since the Re_p does not include the effect of ρ_s, Re_p is unable to define bed dynamics. Moreover, dimensional approach converts independent variables into dependent and hence the heat transfer correlations in terms of Re_p, Nusselt Number 'Nu' and Prandtl Number 'Pr' do not represent the actual experimental facts. Experimental results, therefore, should be presented in terms of measured parameters and the use of dimensionless numbers should be avoided.

Processing of experimental results from many investigations is analysed for.

i) Heat transfer between the particles and the gas entering the bed near the grid in bubbling bed in situations like heat recovery systems,

ii) Heat transfer in particulate fluidization or in the emulsion phase of the bubbling bed in situations like endothermic/exothermic catalytic reactions.

iii) In fixed beds like catalytic reactors or dehydration of solids

The heat transfer mechanism is presented.

Keywords: Heat Transfer, Fluidized Bed, Fixed Bed, Dimensional Aproach, Heat transfer Mechanism in Fluidized Bed, Heat transfer Mechanism in Fixed Bed.

Introduction

The knowledge of fluid-particle heat transfer is necessary in many processes; viz the catalytic reactors in petrochemical industries, fluidised bed combustion, dehydration of foods, mineral processing and heat recovery systems. The comprehensive survey of literature is given by many authors including Zenz (1978), Kunni and Levenspiel (1969), Barker (1965), Botterill (1975), Gelperin and Einstein (1975), Gutfinger and Abuaf (1974), Frantz (1961) and Deshmukh and Giradkar (1985). It is observed that numerous correlations are given in the literature to predict heat transfer between gas and particles; but with little agreement. Almost all investigators have correlated their results in terms of Nusselt number, Reynolds number and Prandtl number as given by,

$$\left(\frac{h\, D_p}{k_g} \right) = a \left(\frac{D_p u \rho_g}{\mu} \right)^b \left(\frac{Cp_g \mu}{k_g} \right)^c \tag{1}$$

Where 'h' is heat transfer coefficient, 'k_g' is thermal conductivity of gas, 'Cp_g' is Specific heat capacity of gas

The value of exponent of Prandtl number is normally taken as 0.33 and for diatomic gas like air it becomes constant and thus its effect has been included in correlation coefficient by many authors. The Reynolds number is $\left(\dfrac{D_p u \rho_g}{\mu} \right)$ taken as an index of turbulence by almost all of the investigators. However, there is gross confusion as far as the value of its exponent 'b' is concerned. Many authors (Frantz, 1961; Gutfinger and Abuaf, 1974; Kettenring et al., 1950; Kunni and Levenspiel, 1969; Walton et al., 1952) have shown heat transfer coefficient to be a positive function of Reynolds number assigning various values to its exponent ranging from 0.67 to 1.7. Chang and Wen (1966) and Bhattacharya and Pei (1974) did not observe any effect of gas velocity on heat transfer coefficient in fluidized state. While Mann and Feng (1968) have even shown heat transfer coefficient to be an inverse function of Reynolds number giving its

[a]njgiradkar@gmail.com; [b]jayajtpgiri@gmail.com; [c]vmkorde@gmail.com; [d]rbchadge@rediffmail.com

exponent 'b' as -6.1...! Commenting on these results, Kunni and Levenspiel (1969) states that the variation in heat transfer coefficient calculated by proposed correlations by various authors be two thousand-fold! The improper experimentation is one of the major reasons for these discrepancies, which has been discussed by Deshmukh and Giradkar (1985) in details and need not be repeated here. Another reason behind the general disagreement between various authors is the indiscriminate use of dimensionless numbers.

Irrelevance of Dimensionless Numbers

Investigators in this field have taken particle Reynolds number as an index of turbulence while correlating their results. By index of turbulence, we mean that if two beds have the same value of Reynolds number, then their dynamic condition will be identical. However, particle Reynolds number cannot define bed turbulence because of the following reasons.

Effect of solid density

The free fall velocity of a small particle, also known as terminal velocity, is calculated by considering drag force ($F_d = 3\pi\mu u D_p$) and is given by the relation:

$$u_t = \frac{D_p^2(\rho_s - \rho_g)g}{18\mu} \tag{2}$$

Similarly, on the basis of Ergun's investigations [06], Wen and Yu [19] have derived a relation for minimum fluidization velocity, u_{mf}, by considering the viscous and inertial forces.

The relation proposed by them is,

$$24u_{mf}^2 + 1650\frac{\mu u_{mf}}{D_p\rho_g} = \frac{D_p(\rho_s - \rho_g)g}{\rho_g} \tag{3}$$

For small light particles, when the effect of inertia is negligible, the equation (3) becomes,

$$u_{mf} = \frac{D_p^2(\rho_s - \rho_g)g}{1650\mu} \tag{4}$$

It is clear from these relations that the fluidized state is greatly influenced by particle size as well as its density; and since the effect of solid density is not included in particle Reynolds number $\left(\frac{D_p u \rho_g}{\mu}\right)$; it will not be able to define dynamic state of the bed. For instance, if we have two beds containing particles of the same size but different densities, their minimum fluidisation velocity will be different; and if we pass gas through these beds at the same velocity chosen to be in between the minimum fluidisation velocities of the two types of particles; even though the Reynolds number $\left(\frac{D_p u \rho_g}{\mu}\right)$ is the same in two beds, the bed with lighter particles will be in fluid state whereas the other one with heavier particles will remain in fixed bed condition. While drying chillies in fluidised bed Laul and Giradkar (1979), noticed that when hot air was passed through a bed of wet chillies the material remained in the fixed bed condition when the gas velocity was not enough to fluidise it. But after some time, when the material was partially dried and became lighter, it started fluidising with the same gas velocity. When the material was further dried, the fluidisation was quite vigorous with the same gas velocity. And thus, because the particle Reynolds number $\left(\frac{D_p u \rho_g}{\mu}\right)$ was not able to define the condition of the bed they used reduced gas velocity (u/u_{mf}) to maintain constant drying condition instead of linear gas velocity.

Effect of gas density

Since the particle Reynolds number is of no use in fluidised state because it does not include the effect of solid density, and since the bed consists of emulsion at minimum fluidisation velocity and gas bubbles, Deshmukh and Giradkar (1985) correlated the available data on gas particle heat transfer by the following relation,

$$Nu = 0.027\left(Re_{mf}\right)^{1.2}\left(\frac{u}{u_{mf}}\right) \tag{5}$$

Here Re_{mf} (Reynolds number based on minimum fluidised bed velocity) was supposed to define the dynamic state of emulsion at minimum fluidisation condition and the reduced gas velocity (u/u_{mf}), the effect of excess gas. Now, consider two beds of the same particle size and solid density; one operating at atmospheric pressure and the other at ten atmospheres. The minimum fluidisation velocities in two beds will be almost identical (equation (4)) on the other hand, the value of the $Re_{mf}\left(\dfrac{D_p u_{mf}\rho_g}{\mu}\right)$ at ten atmospheres will be ten times that of bed operating at atmospheric pressure due to difference in gas density. It is thus evident that we cannot take Re_{mf} as a parameter to define the dynamic state at incipient fluidisation; and therefore, to avoid confusion, the minimum fluidisation velocity u_{mf} must be taken as an intensive property of the system at all conditions.

Reynolds number in fixed beds

As mentioned earlier Ergun [06] considered viscous and inertia forces for flow of fluid through channels in fixed bed with voidage 'ε_f' and developed a correlation for pressure drop as given below,

$$-\Delta P_f = \frac{\rho_g L (1-\varepsilon_f)}{D_p \varepsilon_f^3}\left[\frac{150\,\mu\,u_f(1-\varepsilon_f)}{D_p \rho_g} + 1.75 u_f^2\right] \tag{6}$$

where the two terms on RHS of Equation (6) correspond to viscous and inertia forces respectively. Defining friction factor f_f for fixed beds as

$$f_f = \frac{-\Delta P_f\; D_p\; \varepsilon_f^3}{\rho_g L\; u_f^2 (1-\varepsilon_f)} \tag{7}$$

and the modified Reynolds number Re' as

$$Re' = \frac{D_p u_f \rho_g}{(1-\varepsilon_f)\mu} \tag{8}$$

the equation (6) gives

$$f_f = \frac{150}{Re'} + 1.75 \tag{9}$$

The use of particle Reynolds number in fixed beds is thus theoretically rational, since that the solid density does not affect the flow condition dynamics in fixed beds. Without giving any thought to the difference between the dynamics of fixed and fluidized beds, the users of fluidized bed technique also used the particle Reynolds number in their investigations and have gone wrong because the solid density does affect the fluidized state as discussed earlier in details.

Apart from these conceptual limitations, dimensional approach has one more theoretical draw back. It converts independent variables into dependent as is evident from equation (1). A system, in which the study of six independent variables viz Dp, u, ρ_g, μ, Cpg and kg is necessary to define behaviour of heat transfer coefficient, has been represented by this equation in which the study of only two parameters will be sufficient to correlate the results; thus converting four independent variables into dependent.

It is thus evident that to study the conceptual mechanism of heat transfer, the available data should be processed in terms of actual measured parameters and the use of dimensionless numbers like Nu, Re or Pr should be totally avoided.

Heat Transfer in Fluidized Beds

When a gas at velocity 'u' is passed through a bubbling bed, the part of the gas equivalent to umf keeps the particles in suspension and the remaining excess gas equivalent to (u- umf) form small bubbles at the grid. After release from the gas distributor, these tiny bubbles travel up and while travelling, they coalesce and grow and thus reduce in number. This growth of bubbles occurs in the bottom most section of the column 10 cm to 15 cm in depth after which they attain a maximum size and then travel in the bed with constant velocity. Thus, the bubbling bed consists of two phases viz (i) the emulsion phase or continuous phase containing solid particles at minimum fluidization condition and (ii) the gas bubbles.

The hot gas passing through the emulsion transfers its heat directly to particles through the boundary layer surrounding them. The heat transfer coefficient in the emulsion phase thus depends upon the thermal and dynamic characteristics of the boundary layer. The mode of heat transfer through this gas film is by conduction and convection, which in turn will depend upon the gas velocity (which is umf), particle size, and gas properties. In the emulsion phase, therefore

$$h_{mf} = f(Cp_g, k_g, \rho_g, \mu_g, D_p, u_{mf})$$
(10)

The excess gas passing in the form of bubbles will also contribute to heat transfer and including its effect the heat transfer coefficient can be expressed by a relation,

$$h = f\,[Cp_g, k_g, \rho_g, \mu_g, D_p, u_{mf}\,(u - u_{mf})]$$
(11)

Because of large surface of solid particles and growing gas bubbles, even though the heat transfer coefficients are low, the heat transfer rates near the grid are very high. It is observed that, depending upon the particle size, the hot gas entering the column may transfer whole of heat energy within a bed depth of one centimetre or even less near the gas distributor (Deshmukh and Giradkar, 1985; Gelperin and Einstein, 1975; Kettenring et al., 1950; Kunni, and Levenspiel, 1969; Walton et al., 1952).

The data we are going to consider is for air and that too at ambient conditions and therefore the gas properties can be taken constant. The equation (11) then can be written as,

$$h = p\ u_{mf}^c\ D_p^d\left[1 + q\left(u - u_{mf}\right)\right]$$
(12)

where the constant 'p' includes the effect of gas properties and constant 'q' is invariant with particle size and its density.

For a particular material with constant solid density and particle size, the minimum fluidisation velocity is fixed and, therefore, Equation (12) becomes

$$h = h_{mf}\left[1 + q\left(u - u_{mf}\right)\right]$$
(13)

where

$$h_{mf} = p\ u_{mf}^c\ D_p^d$$
(14)

The Equation (14) gives the value of heat transfer coefficient for gas-particle heat transfer in the emulsion phase; whereas the bracket in the Equation (12) includes the effect of growing bubbles near the grid.

Generalised correlations

As mentioned earlier and discussed in details by Deshmukh and Giradkar (1985), in many investigations proper care was not taken while conducting experiments and therefore, such data gave inconsistent results. After critical examination, the data obtained at controlled conditions by Heertjes and McKibbins (1956), Chang and Wen (1966), Bhattacharya and Pei (1974) and Deshmukh and Giradkar (1985) were selected to study the heat transfer mechanism.

The results of these four investigations were combined together and were processed for the variation in particle size, minimum fluidisation velocity and excess gas. The particle size affects porosity or the channel size whereas the minimum fluidisation velocity depends upon particle size and also upon solid density. Therefore, particle size (D_p) and minimum fluidisation velocity (u_{mf}) were considered as independent variables along with excess gas (u-u_{mf}).

The heat transfer coefficient at incipient fluidisation was obtained by processing the data of individual investigation for a particular particle size by using Equation (13) considering (u-umf) as an independent variable. The summary of these results is given in Table 147.1. After obtaining hmf in this manner, the Equation (14) was then used to determine the effect of Dp and umf; the expression obtained was,

$$h_{mf} = 226.41\ u_{mf}^{0.95}\ D_p^{0.13}$$
(15)

These results are represented in Figure 147.1.
The generalised correlation obtained in the form of equation (12) was,

$$h = 226.41\ u_{mf}^{0.95}\ D_p^{0.13}\left[1 + 2.25\left(u - u_{mf}\right)\right]$$
(16)

The Equation (16) considers the heat transfer in the entry region during formation and growth of bubbles. The heat transfer in the upper section of the bubbling bed where the heat transfer occurs in the emulsion phase and also in the particulately fluidised bed can be obtained by using Equation (15).

Heat transfer near the grid

Investigations of Heertjes and McKibbins (1956) and Deshmukh and Giradkar (1985) give heat transfer coefficients in the entry region. They have passed air over wet silica gel and monitored temperature gradient in the column by using suction thermocouple. As given in Table 147.1 Heertjes had used 50 mm column and since such narrow beds are sluggish in nature (Rowe and Everett, 1972), Deshmukh used 138 mm diameter column. If we observe hmf values of Heertjes given in column (7) of Table 147.1 their increase with particle diameter does not seem to be consistent. For example the hmf obtained for 0.925 mm particles is lower than 0.725 mm diameter particles. The reason behind unsatisfactory quality of experimental results is the nylon grid distributor they have used where the actual gas velocity almost doubles when it enters the bed and thus creates disturbances in the entry region resulting in the inconsistent results with high values of heat transfer coefficients as is evident from Figure 147.1. For smooth entry of the gas, Deshmukh had used packed bed air distributor, which had improved the quality of fluidisation giving consistent values of heat transfer coefficients, which are reported in column (7) of Table 147.1. It may be mentioned that nylon grid air distributor gives lower umf and higher hmf values than the packed bed air distributor as reported in Table 147.1. Equation (16) should be used to calculate heat transfer coefficients in the entry region of such beds.

Heat transfer in emulsion phase

Bhattacharya and Pei (1974) used unsteady state technique to measure heat transfer coefficients. They fluidised 3 mm and 5 mm iron oxide pellets in a 50 mm column by using air. The particle temperature was measured by inserting thermocouples inside some of the Fe_2O_3 pellets, which were fluidising in the upper section of the bed. A covered thermocouple was used to measure the gas temperature. The particles were heated indirectly by using microwave technique and change in temperature was recorded as a function of time after putting off the power supply to microwave generator. The summary of their results is given in Table 147.1 and represented in Figure 147.1. They were puzzled to note as to why the heat transfer coefficients remained constant when the gas velocity was increased beyond the minimum fluidisation value. The experimental technique adopted by these authors was such that the tagged particles, over which the heat transfer data was taken, were surrounded by the gas at incipient fluidisation in the emulsion phase and naturally the heat transfer coefficient, which depended solely on minimum fluidisation velocity, did not increase with superficial gas velocity. The gas bubbles while travelling in the bed, create momentary displacement or disturbance in the continuous or emulsion phase; but it seems that, that does not affect the overall turbulence in the emulsion confirming our assumption that the minimum fluidisation velocity defines turbulence in the particle boundary layer. The Equation (15) gives heat transfer coefficients in the emulsion phase.

Heat transfer in particulate fluidisation

For efficient operation of the fluidised bed there has to be thorough contact of gas with solids. Gas bubbles reduce the probability of gas solid contact and thus the efficiency of the operation is also affected. To

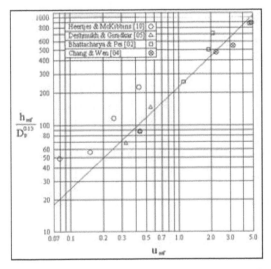

Figure 147.1 Effect of particle size and minimum fluidisation velocity on heat transfer coefficient

increase the efficiency of gas fluidised beds, the bubble formation is prevented by using internal baffles, which break the bubbles. When the gas velocity is increased beyond the incipient value, the particle to particle distance increases in such beds and thus there is uniform expansion of the bed without gas bubbles. Such beds are called the particulately fluidised beds. It may be mentioned that ideal particulate fluidisation is observed in case of liquid fluidised beds.

The gas solid heat transfer data for particulate fluidization available in the literature is that of Chang and Wen (1966) only and is given in Table 147.1. They used 50 mm steel column to fluidise aluminium, brass and stainless-steel balls of 4.75 mm and 6.35 mm diameter by air. The particulate conditions were maintained in the bed by using internal baffles. The solid temperature was measured by inserting thermocouple inside a particle and a suction thermocouple was used to record gas temperature at unsteady state conditions. Initially, the bed was heated by passing fluidizing gas at higher temperature; then the hot gas supply was discontinued and cold gas was passed. The temperature gradient in the bed was observed up to a height of 50 mm above which the temperature of the bed was constant. Hence to measure heat transfer from the hot tagged particle to the surrounding gas at uniform bed temperature, the tagged particle was placed at a distance of 88 mm from the grid. The summary of their results is given in Table 147.1 and represented in Figure 147.1. Equation (15) should be used to calculate heat transfer coefficient in such beds. They were also puzzle to observe that the heat transfer coefficient did not change when the gas velocity was increased beyond the minimum fluidization value. This strange behaviour of particulately fluidised beds will be explained in details in the discussion.

Heat transfer in fixed beds

Kunni and Levenspiel (1969) have discussed fluid particle heat transfer in fixed beds in details. It is rather difficult to determine heat transfer coefficients in fixed beds due to reasons cited earlier (Deshmukh and Giradkar, 1985), in case of fluidissed beds. However, since the incipient fluidisation represents the end point of fixed bed and starting point for fluidised state, the conditions valid at incipient fluidisation can be extended to fixed or static bed also. Thus, the Equation (15), which is valid at incipient fluidisation, will give the heat transfer coefficients in fixed bed also. Thus, the expression for fixed bed becomes,

$$h_f = 226.41 \; u_f^{0.95} \; D_p^{0.13} \qquad\qquad (17)$$

where 'h_f and 'u_f' represent heat transfer coefficient and superficial gas velocity in the fixed bed respectively.

Table 147.1 Experimental results for fluidised beds

(01)	(02)	(03)	(04)	(05)	(06)	(07)
S/No	Author	Dt (mm)	Material	Dp (mm)	Umf (m/s)	Correlation h = p[1+ q (u-umf)]
01)	Heertjes and McKibbins (1956)	50	Silica Gel	0.360	0.0763	h = 17.51 [1+3.90(u- umf)]
02)	Heertjes and McKibbins (1956)	50	Silica Gel	0.510	0.1466	h = 21.01 [1+5.43(u- umf)]
03)	Heertjes and McKibbins (1956)	50	Silica Gel	0.725	0.2526	h = 45.50 [1+2.75(u- umf)]
04)	Heertjes and McKibbins (1956)	50	Silica Gel	0.925	0.3415	h = 35.41 [1+0.68(u- umf)]
05)	Heertjes and McKibbins (1956)	50	Silica Gel	1.100	0.4162	h = 93.18 [1+2.19(u- umf)]
06)	Deshmukh and Giradkar (1985)	138	Silica Gel	0.725	0.3247	h = 26.75 [1+2.29(u- umf)]
07)	Deshmukh and Giradkar (1985)	138	Silica Gel	0.926	0.4388	h = 35.41 [1+2.23(u- umf)]
08)	Deshmukh and Giradkar (1985)	138	Silica Gel	1.204	0.5529	h = 61.17 [1+2.24(u- umf)]
09)	Bhattacharya and Pei (1974)	50	Fe2O3	3.000	1.1065	h =118.56[1+0.18(u- umf)]
10)	Bhattacharya and Pei (1974)	50	Fe2O3	5.080	1.8981	h =252.05[1+0.03(u- umf)]
11)	Bhattacharya and Pei (1974)	50	Fe2O3	5.080	2.0848	h =356.92[1-0.06(u- umf)]
12)	Chang and Wen (1966)	50	Aluminium	4.750	2.236	h = 238.5
13)	Chang and Wen (1966)	50	Aluminium	6.350	3.226	h = 283.9
14)	Chang and Wen (1966)	50	Brass naval	4.750	4.635	h = 437.2
15)	Chang & Wen (1966)	50	Stainless steel	4.750	4.830	h = 442.9

Discussion

The heat transfer coefficient depends upon thermal properties of the fluid and the overall turbulence in the bed. In this section we will study the effect of turbulence. Normally the velocity increases turbulence and hence the heat transfer coefficient; but Chang and Wen found heat transfer coefficient to be independent of velocity. We have to explain this strange behaviour. We will first study turbulence in the fixed bed with the help of Ergun's (1952) work.

Fluid mobility in viscous sublayer

While developing a correlation for friction factor in fixed bed (Equation 9), Ergun assumed pressure drop due to viscous and inertia forces for flow through the channels in between the particles as given by Equation (6). The flow structure in the channel will be a viscous sublayer adjacent to the particle surface and the turbulent eddies in between the sublayers of two adjacent particles.

The first term on the RHS of Ergun correlation (Equation 9) represents the contribution of viscous sublayer and second term the contribution due to eddies in the friction factor. When the modified Reynolds number $\left(\dfrac{D_p u_f \rho_g}{(1-\varepsilon_f) \mu}\right)$ is less than about unity, the fluid passes through the channels at streamline condition.

At higher values of modified Reynolds number, may be more than 1000 or so, the friction factor is almost constant and pressure drop is due to turbulent eddies only. When the flow through fixed bed is started and the modified Reynolds number is gradually increased in stream line region, the mobility of fluid molecules also increases and at the critical value the streamline structure is smashed due to onset of mixing. The mobility of fluid molecules at stream line conditions, thus depends upon the value of modified Reynolds number; and qualitatively represented by expression,

$$\text{Fluid mobility in stream line flow} = f1\left(\frac{D_p u_f \rho_g}{(1-\varepsilon_f) \mu}\right) \tag{18}$$

Eddy turbulence

The inertia forces control the dynamic conditions at high turbulence. The friction factor (Equation 9) becomes independent of Reynolds number and, therefore, the Reynolds number loses its significance. Power consumption in Joules per kilogram of fluid per meter bed height, as per Equation (6) then becomes,

$$\frac{-\Delta P_f}{\rho_g L} = 1.75 \frac{(1-\varepsilon_f) u_f^2}{\varepsilon_f^3 D_p} \tag{20}$$

Whatever power is supplied is consumed by eddies and, therefore, power consumption can be taken as an index of eddy turbulence and can be expressed qualitatively as,

$$\text{Eddy Turbulence} = f2\left[\frac{(1-\varepsilon_f) u_f^2}{\varepsilon_f^3 \rightleftarrows D_p}\right] \tag{21}$$

It is thus clear that the modified Reynolds number and power consumption define fluid particle mobility in the sublayer and eddy turbulence in the core and the vice versa is not true.

Overall Turbulence

In practice very high fluid velocities are not preferred due to excessive pressure drop, and, as stated earlier, the channel flow structure consists of viscous sublayer over the particle surface and eddies in the core. When the heat transfer occurs from solid particles to gas, the solids act as a source and channel core as sink, which is at uniform temperature due to turbulent eddies. The viscous sublayer adjacent to particle surface functions as a medium for the energy transfer, which occurs by conduction, aided by convection currents. The eddies strike the surface of sublayer, receives energy and transfer it effectively to the turbulent core. Some of the eddies may even penetrate the sublayer, get lost in it and transfer back the equivalent amount of hot fluid into the core. Eddies, thus, not only help in transferring energy from sublayer but also create disturbance in it and thus enhance its convective power. It is thus clear that the heat transfer rate will depend upon the thermal properties of the sublayer, its dynamic condition and also upon the eddy turbulence. Since the sublayer is in the laminar condition, its dynamic state is defined by its mobility as given by equation (18) in terms of modified Reynolds number. Similarly, the power consumption can be taken as an index of eddy turbulence as expressed by equation (21). And then the overall turbulence, which will affect the heat transfer coefficient, will thus depend upon both, the sublayer mobility as well as eddy turbulence and can be expressed as,

$$\text{Overall turbulence in the fixed bed} = f1\left[\frac{D_p u_f \, \rho_g}{\left(1-\varepsilon_f\right)\mu}\right] + f2\left[\frac{\left(1-\varepsilon_f\right)u_f^2}{\varepsilon_f^3 \, D_p}\right] \tag{22}$$

It may be mentioned that when particle size is changed in fixed bed, the voidage is not substantially affected; only the void size is changed.

Equation (22) gives qualitative nature of overall turbulence in the fixed bed. However, since the incipient fluidization represents the end point of fixed bed condition or the starting point of the fluidized state, a similar equation will give the condition of overall turbulence in the emulsion phase of the fluidised bed also. Therefore

$$\text{Overall turbulence in the emulsion phase} = f3\left[\frac{D_p u_{mf} \, \rho_g}{\left(1-\varepsilon_{mf}\right)\mu}\right] + f4\left[\frac{\left(1-\varepsilon_{mf}\right)u_{mf}^2}{\varepsilon_{mf}^3 D_p}\right] \tag{23}$$

Effect of particle size

As per Equation (22 and 23), the effect of gas velocity and viscosity on overall turbulence or heat transfer coefficient is normal. However, since the particle diameter is in the numerator of the first term and in the denominator of the second, of these equations, the overall turbulence may increase or may remain unchanged or may even decrease when the particle size is increased, depending upon the change in magnitude of sublayer mobility and eddy turbulence. When the particle size is increased, the heat transfer coefficients at incipient fluidization and in fixed beds as given by Equations (15) and (17) respectively increase because the sublayer mobility increases but the eddy turbulence decreases to the lesser extent showing a net increase in overall turbulence.

Effect of gas velocity in particulate fluidisation

Channel flow in particulately fluidised bed is similar to that in fixed bed and therefore Equation (22), which represents overall turbulence in fixed bed, will be valid for particulate fluidisation also. Therefore, the Equation (22) can be written in the form

$$\text{Overall turbulence in particulate fluidization} = f5\left[\frac{D_p}{\mu}\frac{u}{\varepsilon}\frac{\rho_g\varepsilon}{\left(1-\varepsilon\right)}\right] + f6\left[\frac{u^2}{D_p\varepsilon^2}\frac{1-\varepsilon}{\varepsilon}\right] \tag{24}$$

When superficial gas velocity is increased beyond the minimum fluidization value, there is uniform expansion of the bed and thus voidage increases. However, due to increase in voidage, the actual gas velocity in the bed (u/ε) remains unaltered. It is thus clear from Equation (24), that the sublayer mobility increases but eddy turbulence decreases when superficial gas velocity is increased. In the investigations of Chang and Wen (1966), the heat transfer coefficient did not change with gas velocity because the effect of sublayer mobility and eddy turbulence on overall turbulence must be cancelling each other keeping the overall turbulence unchanged.

Effect of gas properties

The effect of gas properties on heat transfer coefficient could not be established due to lack of data. The Equations (15, 16 and 17) are valid for air at ambient conditions. However the channel flow in fixed beds and flow of fluids in tubes are very similar i.e. both having laminar sublayer and turbulent eddies. Therefore the data of Pickett et. al. (1979) available in a 3.12 mm tube for helium–argon mixtures could be used to study the effect of fluid properties. The details of the analysis are given by Kripalani (1997) who studied heat transfer mechanism in tubes for gases. He observed that the heat transfer coefficient increased with tube diameter up to 5 mm and then decreased as shown by the following relations:

Tube size 3–5 mm

$$h = 7.281 \frac{G^{0.76} D^{0.84} k^{1.34}}{\mu^{1.29} \, Cp^{0.63}} \tag{25}$$

Tube size 6 mm to 76 mm

$$h = 0.0188 \frac{G^{0.75} k^{1.34}}{\mu^{1.29} \, Cp^{0.63} \, D^{0.3}} \tag{26}$$

It is worth noting that as far as the effect of particle size in fluidised and fixed beds (Equations 15 and 17) and tube diameter in pipe flow, equation (25) on heat transfer coefficient is concerned, all show a very similar behaviour.

By considering the effect of fluid properties for forced convection in tubes and in fluid particle heat transfer in fluidised and fixed beds to be similar, the Equation (16) and Equation (17) can be written as,

$$h = 1.702 \frac{k_g^{1.34} u_{mf}^{0.95} \rho_g^{0.95} D_p^{0.13} [1 + 2.25(u - u_{mf})]}{\mu^{1.29} Cp_g^{0.63}} \tag{27}$$

and

$$h_f = 1.702 \frac{k_g^{1.34} u_f^{0.95} \rho_g^{0.95} D_p^{0.13}}{\mu^{1.29} Cp_g^{0.63}} \tag{28}$$

Conclusions

1. The use of particle Reynolds number as an index of turbulence in fluidised state is irrational.
2. The overall turbulence in the fixed bed, at incipient fluidisation and in the particulately fluidised bed depends upon the fluid mobility in the laminar sublayer and eddy turbulence in the channels and is qualitatively expressed in the form of Equation (22) through (24).
3. For heat transfer between the gas entering a bubbling bed and particles near the grid, Equation (27) should be used to calculate heat transfer coefficient
4. For fluid particle heat transfer in the emulsion phase of a bubbling bed and in a particularly fluidised bed, Equation (27) should be used by replacing superficial gas velocity u by u_{mf}.
5. Equation (28) should be used for heat transfer in fixed beds.
6. For air at ambient conditions, Equations (15, 16 and 17) can be used for appropriate conditions.

References

Barker, J. J. (1965). Heat transfer in packed beds Ind. Eng. Chem. 52:33.

Bhattacharya, D. and Pei, D. G. T. (1974). Heat Transfer in Fixed Beds. Ind. Engg. Chem. (Fund.) 13:199.

Botterill J. S. M. 1975. Fluidized Bed Heat Transfer. New York (N Y): Academic Press, New York.

Chang T. M. and Wen C. Y. (1966). Mechanics of fluidization. Chem. Engg. Prog. Symp. Series. 62(67):111.

Deshmukh S. T. and Giradkar J. R. (1985). Role of Emulsion and Bubbles in Fluid Bed Gas Solid Heat Transfer. Indian J. Technol. 23:441.

Ergun, S. 1952. Fluid Flow through Packed Columns. Chem. Engg. Prog. 48:89.

Frantz J. F. 1961. Fluid-to-Particle Heat Transfer in Fluidized Beds. Chem. Engg. Prog. 52:358.

Gelperin N. I. and Einstein V. G. (1975). Heat Transfer in Fluidized Beds. In Davidson and Harrison. Fluidization. Academic.

Gutfinger C. and Abuaf, N. 1974. Advances in heat transfer. New York (N Y): Academic Press.

Heertjes P. M. and McKibbins S. W. (1956). Heated Turbulent Flow of Helium Argon Mixtures in Tube. Chem. Engg. Sci. 5:161.

Kettenring, K. N., Nanderfield, H. L., and Smith, J. N. (1950). Heat and Mass Transfer in Fluidized Systems. Chem. Eng. Prog. 46:139.

Kripalani V. M. (1997). Forced convection in tubes for gases. PhD Diss. Nagpur University, Nagpur.

Kunni, D. and Levenspiel, O. 1969. Fluidization Engineering. New York (N Y): Wiley.

Laul M. S. and Giradkar J. R. 1979. Developments in Drying. Princeton (N. J): Science Press.

Mann S. and Feng, L.C.C. 1(968). Gas-Solid Heat Transfer in Fluidized Bed. Ind. Eng. Chem. Porc. Des. And Dev. 327.

Pickett P. E., Taylor M. F., and McElgott D. M. 1979. Heated turbulent flow of helium argon mixtures in tubes. Int. J. Heat Mass Transf. 22:705719

Rowe P. N. and Everett 1972. D. J. Trans. Fluidized bed Bubbles Viewed by X-rays. Inst. Chem. Eng. 50:42.

Walton, J. S., Olson R. L., and Levenspiel O. 1952. Ind. Eng. Chem. 44:1474.

Wen, C. Y. and Yu, Y. H. (1966). AIChE. 29. 174.

Zenz, F. A. 1978. Encyclopaedia of chemical technology. John Wiley & Sons.

148 Optimum Sprue design and analysis using InspireCast for performance improvement in casting process

S. S. Chaudhari[a], Pratik Hingnekar[b], Rohit Rayanutala[c], and Shivam Kawale[d]

Department of Mechanical Engineering, Yeshwantrao Chavan College of Engineering, Wanadongri, Nagpur, India

Abstract

Proper sprue design is an essential part of the gating system in the casting process as it regulates the flow of molten metal into the mold cavity which directly affects the quality of cast product. This project aims to propose an optimum sprue design for sand casting by performing simulations on InspireCast so that the performance of the casting process can be improved. In this project, three cases are considered viz. two tapered sprue, two straight sprues, and single taper sprue whose 3-D CAD models are designed, and then simulations are performed by using the boundary conditions based on observations from the foundry. The results are compared for all three cases by simulating both filling and solidification process which shows various defects like Air Entrapment, Mould Erosion, Cold Shuts, etc. The conclusions drawn from the simulations point out that the optimal sprue design is one with two tapered sprues. Furthermore, certain changes such as increasing the mass flow rate from 3.4 kg/s to 6 kg/s, increasing pouring temperature from 1300°C to 1400°C, and changing the dimensions of the tapered sprue lead to better quality casting product as suggested by the simulation.

Keywords: Casting defects, numerical simulation, sand casting, sprue design.

Introduction

Casting is one of the most primitive and widely utilised manufacturing processes. Bulky and complex parts can be easily made by using this process. Implementation of the casting process can also be done for mass production purposes in a relatively cheaper way. However, casting offers some demerits as well which are not limited to internal defects only, but also uneven surface finish and poor dimensional accuracy. Sprue is an important part of the gating system because during filling molten metal flows through the sprue into the mold cavity, hence acting as a regulator for metal flow. Many casting defects are directly related to the sprue design such as air entrapment, cold shuts, mould erosion, microporosity, misruns, etc. One way to counter these defects is to use optimum parameters for the designing of gating system which may include proper sprue size and shape, addition of risers, proper ingate dimensioning, etc.

Determination of properties and the nature of the molten metal flow is difficult when done through physical experimentation methods as the complete casting process is carried out at an elevated range of temperature i.e. around 1300–1400°C. Moreover, in local foundries, conventional hit and trial methods are used for designing the gating system and for finding out optimum parameters affecting the quality of cast product. In order to bridge the gap between theoretical conclusions and experimental observations, simulation analysis can be implemented in the Standard Operating Procedure (SOP) followed. Numerical simulations are one of the best methods to tackle this problem and to produce required casting without any defects. This project uses InspireCast® for performing simulations, thereby proposing optimum sprue design and parameters required to produce sound casting.

Rajkumar et al. (2021) proposed an improved Multi-gate design by comparing different gating connections in series and parallel combinations. It is found that the centre sprue runner extension parallel connection (CS-RE-PC) was the most suitable of all which gives the best flow rate and almost defect-free casting. Location of sprue(s) in multi-gate design is crucial for the determination of optimum parameters required for producing the desired casting. Kermanpur et al. (2008) studied the effect of different casting variables including gating design, mould surface roughness, pouring time for automotive cast parts by using numerical simulations. They used brake disc and flywheel casting for performing comprehensive experimentation

[a]sschaudharipatil@rediffmail.com; [b]pratikashokhingnekar@gmail.com; [c]rohitrvr13@gmail.com; [d]shivmkawale@gmail.com

along with simulation on FLOW-3D®. The turbulent flow was simulated by using the two-equation k-ε turbulence model. Apart from the gating system, the soundness of casting depends on mould design as well. Iqbal et al. (2012) conducted simulations by modifying three different mould designs and studied the effect of mould parameters on the final cast product. Moreover, the effect of changes in the gating system on the casting is also studied by comparing the results with experimental data. It is found that design optimization in the gating system results in smooth filling of mould without any defects. Effects of using different sprue bases have been investigated by Baghani et al. (2013) and the possibility of the formation of vortex flow in different sprue designs is also studied. They have reached the conclusion that the volumetric flow rate of molten metal is approximately 1.7 times that of water. Increasing well size produced a vortex flow at the bottom of the sprue base, which increased the surface velocity of liquid metal in the runner. Using a rather big sprue well could eliminate vena contracta but ingate velocity was observed to be independent of well size. Using a curved sprue base could remove vortex flow at the bottom of the sprue while keeping a nearly full contact between liquid metal and the runner wall (Baghani et al., 2015).

Choudhari et al. (2013) identified hots spots by using solidification simulation facilitating optimised placement and design of feeders which eventually increases the yield percentage while ensuring sound casting. It is observed that 50 mm riser diameter with 5mm sleeve thickness shows similar result as actual trials performed at foundry. Using sleeve as feed aid helps in reducing riser dimensions from 50 mm to 60 mm, thereby increasing casting yield. On performing intensive literature review it is found that available simulation studies don't cover all conditions and parameters required for sprue design optimisation. Further, the effect of sprue design on solidification and filling defects is not studied.

This project compares three different arrangements of sprue(s) in the gating system in order to determine the arrangement which leads to the desired outcomes in the final cast product. Filling and solidification in the three arrangements were studied and results regarding the velocity of molten metal flow, solidification rate and defects arising due to different sprue designs and arrangements are also compared.

Theoretical background

The casting which is used for this project is the side frame of a ginning machine that is used for the separation of cotton fibres from cotton seeds or lint. Figure 148.1 represents the 3-D model of side frame casting. The part is manufactured by following the standard Gravity sand mould casting process. The material used for the casting is Standard Flake FG200 Cast Iron which has certain machining properties as final machining is needed before assembly. The properties of the material are given in Table 148.1. Sand preparation plays an important role in obtaining a sound casting. Additives like Bentonite, Lustrous Carbon additives (LCA) along with moisture are added in the moulding sand for the acquisition of proper binding property in moulding sand. The sand which is used for the manufacturing of side frame casting is green sand. The various properties of green sand along with its composition is shown in Table 148.2.

Figure 148.1 3-D model of side frame of ginning machine

Dimensions of the components of gating system:
Sprue length: 150 mm
Taper sprue upper diameter: 63 mm
Taper sprue lower diameter: 45 mm
Straight sprue diameter: 45 mm
Ingate width: 32 mm
Gate thickness: 15 mm
Riser length: 80 mm
Riser diameter: 30 mm

Methodology Adopted

The major defects observed in the casting product studied were cold shuts, microporosity, and air entrapment. Shrinkages can be reduced by the addition of risers in the gating system. Risers act as reservoirs for extra molten metal, this molten metal flows out and fills the gaps whenever there is shrinkage. Porosity and air entrapment arise due to the nature of the molten metal flow. The flow can be controlled to a certain extent by optimising the gating system.

Firstly, the casting process for the side frame was studied at the foundry. Notable observations include the gating system design, required pouring time, material properties, sand properties and composition, tapping and pouring temperature. Secondly, after the gathering of required data, a solid model of the casting product was created using Solidworks®. The complete analysis and simulation is performed on InspireCast®. InspireCast® by Altair Engineering Inc. is free software, especially used for small casting which is easy to use and gives highly accurate results. For obtaining three cases, modifications are done in the gating system and sprue geometry. The models are then imported into the simulation software, where initial conditions like pouring temperature, material grade, sand type, flow rate, risers, pouring points etc. are specified. After importing the model, meshing is done and filling and solidification sequence is performed.

Finally, the results obtained for each of the models were compiled and compared to determine the optimum design of the sprue which led to minimising the defects arising due to the molten metal flow.

Figure 148.2 Cope and drag

Table 148.1 Green sand properties with composition

Sr. No	Properties	Values
1	Silica	90% and above
2	Moisture	3.84%
3	Total clay (TC)	12–13%
4	Active clay (AC)	9–9.5%
5	Loss in ignition (LOI)	5.5–6.5%
6	Grain fineness number	50–55
7	Green compression strength	900–1500 KPa
8	Compactibility	40–48%
9	Permeability	130–180

Table 148.2 FG 200 cast iron properties

Sr. No	Properties	Values
1	Density	7150 kg/m^3
2	Elastic modulus	92.4 GPa
3	Brinell's hardness	187–241 BHN
4	Poisson's ratio	0.21
5	Tensile strength	207 MPa
6	Coefficient of thermal expansion	$(1.1–1.5) \times 10^{-5}$ 1/K
7	Melting point	1127–1204°C
8	Specific heat capacity	460 J/kg-K
9	Thermal conductivity	20–80 W/m-K

Simulation

The complete simulation is carried out for three different variations based on the number of sprues and sprue design (Figures 148.3–148.5).

a.   Gating system with 2 tapered sprues.
b.   Gating system with 2 straight sprues.
c.   Gating system with only one tapered sprue.

   Apart from these three cases sprue design changes are also incorporated in first case for the reduction in microporosity. Meshing is done in the background automatically using Altair Hypermesh and Simlab which are submodules of InspireCast. A powerful algorithm is used to calculate optimum mesh size which gives the best results. Formulation of finite element based model is done for solving the governing Navier-Stokes equation for filling and solidification process. By using a biphasic air-model for computation, the effect of air on casting is effectively predicted. The boundary conditions consists of initial pouring temperature and mass flow rate. The pouring temperature of molten metal at the time of pouring is taken as 1300°C. For the first two cases, when pouring is done from two points, the mass flow rate is taken as 3.4 kg/s and for the last case, the mass flow rate is taken as 6 kg/s. The mass flow rate is considered as per the time required for the filling of the casting that is observed in local foundries.

Table 148.3 Filling and solidification time

Case No	Filling time	Solidification time
1	23.48 seconds	1092.95 second
2	23.12 second	981.24 second
3	12.89 second	851.22 second

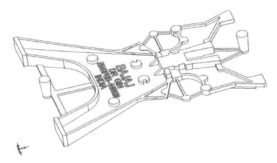

Figure 148.3 Gating system with two tapered sprues

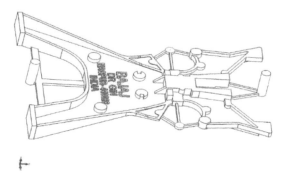

Figure 148.4 Gating system with two straight sprues

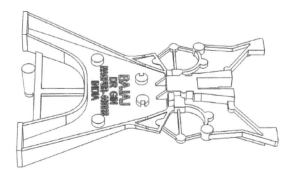

Figure 148.5 Gating system with only one tapered sprue

Results and Discussion

Air entrapment

Case 1 < Case 2 < Case 3

The reason for air entrapment is the nature of the flow. The more turbulent the flow, the more is the air trapped in it.

Case-1 The turbulences in the flow of molten metal are decreasing throughout the cross-section of the sprue because the smaller diameter of the tapered sprue at the bottom end acts as a choke which helps to keep the sprue full of molten metal and eliminate any unwanted air present inside it.

Case-2 The turbulence as the molten metal flows down does not decrease which leads to more air bubbles being trapped in the flow and hence increasing the air entrapment.

Case-3 Even though the sprue is kept tapered to minimise the turbulence, a high mass flow rate through the sprue causes the highest turbulence among all three cases and hence higher air entrapment.

The blue region shows the area with air entrapment fraction 0.00 and the red region shows the area with air entrapment fraction 0.90.

Average air entrapment fraction

case 1 0.50

case **2 0.67**

case 3 0.80

Air entrapment fraction is the amount of air entrapped in specific regions. Air entrapment of 1.0 implies that the entire region is filled with air and 0.00 implies that the entire region is filled with metal.

Higher flow rate in case-3 results in higher air entrapment.

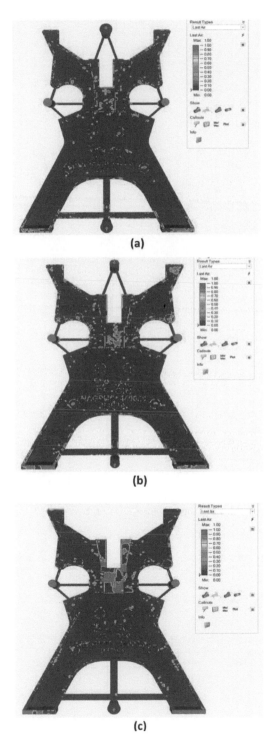

(a)

(b)

(c)

Figure 148.6 Air entrapment at the end of filling for all three cases

Microporosity

Case 3 < Case 2 < Case 1

Microporosity appears more prominently in the thinner regions of the casting in all three cases although the concentration of microporosity varies in all three.

Maximum microporosity

Case 1 - 2.39%

Case 2 - 2.2%

Case 3 - 2.01%

Microporosity in case 1 is 17.84% more than case 2 and 33.33% more than case 3.

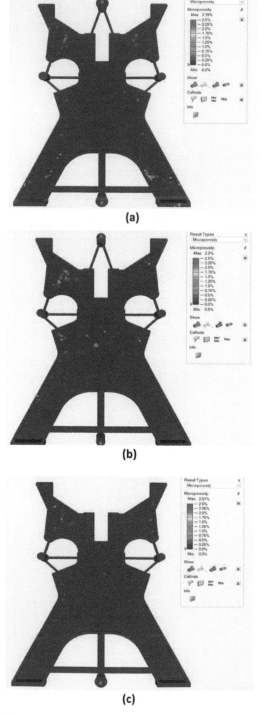

Figure 148.7 Microporosity observed for all three cases

Coldshuts

Case 1 < Case 3 < Case 2

Cold shuts occur when the molten metal starts solidifying before filling the cavity. A proper mass flow rate and thickness of ingates are important for countering the cold shut defect.

Case-1 Filling starts at both ends. The turbulence of the flow also decreases as the molten metal flows down the sprue. Given these points, solidification does not begin in most of the cavity until filling is completed, hence leading to a minimum cold shut(s).

Case-2 The turbulence of the flow does not decrease hence increasing cold shuts.

Case-3 Molten metal is being poured into a single tapered sprue therefore filling starts only at one end. The mass flow rate in case 3 is higher compared to case 1 and 2 but unlike in case 2 the tapered sprue decreases the turbulence in the stream leading to lower cold shuts.

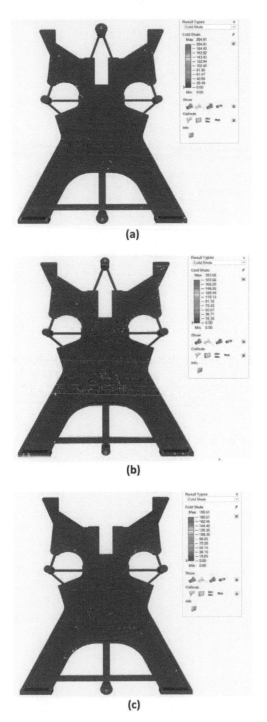

Figure 148.8 Coldshuts observed for all three cases

Mould erosion

Case 1 < Case 2 < **Case** 3

In general mould erosion occurs due to insufficient binding of sand particles in the mould. But mould erosion also depends upon the velocity of the metal at the time of filling.

Case-1 Flow rate is 3.4 kg/s and the sprue is tapered hence the turbulence decreases as the fluid flows through the sprue leading to minimum mould erosion.

Case-2 Flow rate is 3.4 kg/s and the sprue is straight hence the turbulence barely decreases, leading to high turbulence entry of the fluid at the ingates resulting in comparatively higher mould erosion.

Case-3 Flow rate is 6 kg/s and the sprue is tapered. Since only a single sprue is used in this case higher flow rate was necessary in-order to fill the mould completely before solidification could set in. The geometry of the sprue helps in decreasing.

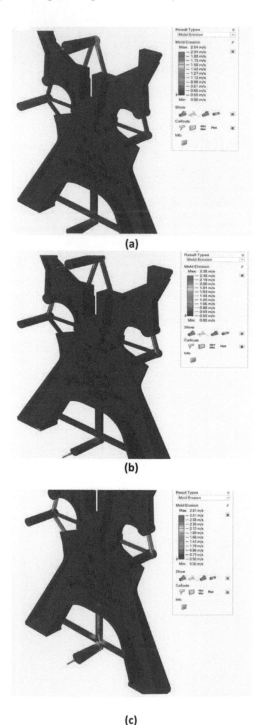

Figure 148.9 Mould erosion observed for all three cases

Proposed sprue design:

Based on the simulation results it has been concluded that among the three cases studied case-1 leads to better quality casting product as compared to the other two. The drawback in case 1 is that compared to the other two cases higher microporosity levels have been observed. To tackle this problem, changes in the sprue design as well as pouring temperature and mass flow rate were incorporated in case 1.

The sprue top diameter was increase to 75 mm from 63 mm. The pouring temperature was increased to 1400°C from 1300°C and the flow rate was increased from 3.46 kg/second. Simulation carried out using these parameters shows that microporosity level in the proposed case is 20% less than in case 1.

Conclusions

From the above discussion, we can see that each of the three variations come with their own set of advantages and disadvantages. By conducting simulation and experimentations for the filling and solidification process following conclusions are drawn:

• Sprue design in case 1 is optimal as it leads to the casting product with the least number of defects among the three cases studied.

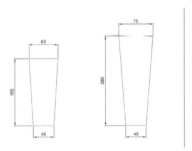

Figure 148.10 Proposed sprue design for minimising microporosity

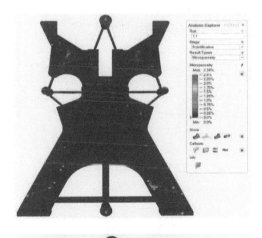

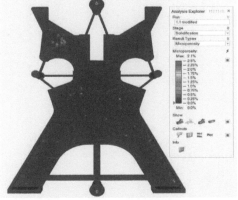

Figure 148.11 Microporosity before and after sprue design changes

- Yield Percentage is :
  Case -1 – 89.99%
  Case -2 – 91.2%
  Case -3 – 93%
- Even though the yield percentage in case 3 is the highest, the defects observed in this case can be seen in the results.
- From the conclusions drawn from the simulation higher mass flow rate leads to lower microporosity. But higher mass flow rate also leads to increased air entrapment and mould erosion.
- The gating system used in the foundry is case 1. The casting process can be further improved by increasing the pouring temperature to 1400°C and increasing the mass flow rate to 6 kg/second. Implementing the above parameters into the existing casting process can decrease the microporosity by 20% leading to better quality casting.
- Higher pouring temperature is recommended so that the molten metal fluidity can be increased and defects such as air entrapment and microporosity, which generate cavities in the casting, are eliminated.

References

Baghani, A., Bahmani, A., Davami, P., Varahram, N., and Shabani, M. O. (2015). Application of computational fluid dynamics to study the effects of sprue base geometry on the surface and internal turbulence in gravity casting. Proc. Ins. Mech. Eng. L: J. Mater.: Des. App. 229(2):106–116.

Choudhari, C. M., Narkhede, B. E., and Mahajan, S. K. 2013. Optimum design and analysis of riser for sand casting. IEEE International Conference on Industrial Engineering and Engineering Management. 1151–1155.

Iqbal, H., Sheikh, A. K., Al-Yousef, A., and Younas, M. (2012). Mold design optimization for sand casting of complex geometries using advance simulation tools. Mater. Manufact. Proces. 27(7):775–785.

Kambadahalli, H. R., and Chheda, U., and Bhallamudi, R. (2013). Flow Through Multi-Gate Gating System: Experimental and Simulation Studies. Adva. Manufact. 2B.

Kermanpur, A., Mahmoudi, S., and Hajipour, A. (2008). Numerical simulation of metal flow and solidification in the multi-cavity casting moulds of automotive components. J. Mater. Proc. Technol. 206(1–3):62–68.

Rajiv, N., Umar, P., and Mohamed, U. (2019). CFD analysis of fluid flow in sand casting. Int. J. Trend Scient. Res. Dev. 3.

Rajkumar, I., Rajini, N., Alavudeen, A., Ram Prabhu, T., Ismail, S. O., Mohammad, F., Al-Lohedan, H. A. (2021). Experimental and simulation analysis on multi-gate variants in the sand casting process. J. Manufact. Proces. 62:119–131.

149 Energy audit of textile industry

Sujata Bose[1,a], Shital Bharaskar[2], Swapnil Khubalkar[3,b], and Yogesh Pahariya[4,c]

[1]Department of Electrical Engineering, G H Raisoni College of Engineering, Nagpur, India

[2]Department of Electrical Engineering, Sandip University, Nashik, India

Abstract

The textile sector has the lowest energy utilisation efficiency and is one of the most energy-intensive industries. The power loom sector is a key contributor to the Indian textile industry, accounting for 62% of all textile production in the country and hence it is a major contributor to our export earnings. The Indian textile industry's lifeline is the decentralised power loom sector. Bhiwandi, known for its power looms is a key textile centre in western India. As we all know, energy is one of the most important sources of a country's economic growth. In the case of emerging countries, the energy sector takes on a key role, as ever-increasing energy demands necessitate massive investment to supply them. In comparison to industrialized nations, India's energy intensity is quite high, thus it's important to balance total energy inputs with total energy consumption. It also helps to identify all energy streams in the system and quantify energy usage using a discrete function. Lower efficiency and waste are the causes of increased energy intensity. To reduce specific energy consumption for sustainable development in a cost-effective way continuous energy monitoring and process tracking are required. The industrial audit has grown in importance over the last several decades as the drive to reduce ever-increasing energy costs and move toward a more sustainable future has pushed it to the forefront. In this paper, we selected an industry (Chur textile) and carried out the survey, collected required data and did an analysis on how to save energy by reducing energy consumption areas including financial areas as well.

Keywords: Decatizing. efficiency, energy audit, mercerization, textile.

Introduction

As per Indian Energy Conservation Act 2001, energy audit is defined as the verification, monitoring and analysis of the use of energy including the submission of a technical report containing recommendations for improving energy efficiency with cost-benefit analysis and an action plan to reduce energy consumption (https://beeindia.gov.in). Energy (both electrical and thermal), labour, and materials are frequently determined to be the top three operational expenses in every industry (Khude, 2017). If the manageability of the cost or potential cost savings in each of the above components were compared, energy would invariably come out on top, and so the energy management function is a critical area for cost reduction. An energy audit will assist in better understanding how energy and fuel are used in any industry, as well as identifying areas where waste can occur and where improvements can be made (Nadimuthu, 2021; Rajput and Singh, 2016; Nagaveni et al., 2019; Jadhav et al., 2017. In general, an energy audit is the process of turning conservation concepts into reality by providing technically possible solutions while also taking into account economic and other organisational factors within a set time limit.

ISO50001 is a standard for power management in the international commerce sector, with an impact on power consumption that can be monitored and modified by the company. The ISO 50001 Energy Management System (EnMS) standard is designed to assist businesses in implementing the systems and procedures necessary to improve energy efficiency (Pahariya, 2001; 2004). A related study for the industry is given in (Mahalaxme et al., 2000; Shende et al., 2001). This international standard aims to reduce greenhouse gas emissions and other environmental implications, as well as energy expenditures, via effective energy management. The objective of the energy audit is to find methods to lower the amount of energy used per unit of product, determine where energy savings may be made in a facility, and increase the efficiency of a plant or company. The primary objectives of the energy audit are shown in Figure 149.1. The scope is limited to studying energy metering, monitoring and control and recommending suitable system for future monitoring, to establish the scope for optimization of load factors through a detailed load management study. Not only saves money or energy for an Industry but it also allows the country to optimise its energy resources.

[a]sujata.bose.mtechps@ghrce.raisoni.net; [b]swapnil.khubalkar@raisoni.net

Details of energy audit

The sort of energy audit to be conducted is determined by the following factors (https://beeindia.gov.in): the function and type of industry, the depth to which the final audit is required, and the potential and size of cost savings expected. As a result, energy audits can be divided into the following categories as shown in Figure 149.2.

Preliminary audit

A preliminary energy audit is simply a data collection activity that tries to establish a knowledge of how energy is utilized in industry and lay the groundwork for a more complete energy audit. The first step in a preliminary auditing procedure is to interview important personnel such as the factory manager and owner to obtain information about the plant's schematic design, production program and capacity, yearly production, raw material usage, and monthly and annual fuel consumption statistics.

Detailed audit

A comprehensive audit provides a thorough energy project implementation plan for an area/facility, since it evaluates all major energy-using systems. This audit provides the most precise assessment of energy savings and costs. It takes into consideration the interplay of all projects, accounts for the energy consumption of all important equipment, and contains precise energy cost savings projections as well as project costs.

A thorough energy audit seeks to determine the real energy efficiency of certain end-users and operations. Based on energy conservation potential discovered during the preliminary assessment. Detailed energy audit (DEA) is a long-term inspection procedure that incorporates cost-benefit analysis and encompasses all forms of data such as production processes, equipment efficiency and performance, energy consumption, and economic and financial assessment of energy performance improvement strategies. The audit findings must be presented in a report, which must also include an action plan outlining the project's goals for implementation. Phase 1, 2, and 3 are the three stages of detailed energy audits: pre-audit phase, audit phase, post-audit phase.

Methodology

In the industry, auditing refers to determining the efficiency of production based on energy use. This is the procedure to be followed-

- Survey
- Data collection

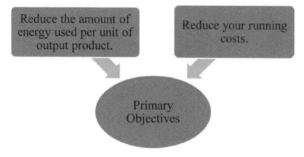

Figure 149.1 Primary objectives of energy audit

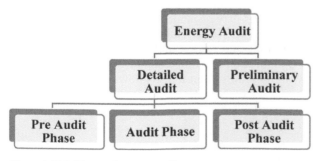

Figure 149.2 Types of energy audit

- Measurement and testing on-site
- Data analysis
- Recommendations

For efficient functioning, a defined technique for conducting an energy audit is required. An initial site assessment is usually recommended since the preparation of the processes required for an audit is crucial. Energy auditing procedure is shown in Figure 149.3.

Company profile

Here, is a fabric manufacturing plant with value added processing facility is considered which has total combined capacity 250000 meters per day.

- Type: Textile industry
- Industry sector: Dying and printing mill
- Account class: HT1A

Process of industry

Process flowchart is explained in Figure 149.4.

Electricity consumption

The Supply is 22 KV. The company has a transformer that reduces the voltage to 415 volts by stepping down. Electricity use per year is 3300000 KWH i.e. annual units and the bill for electricity is Rs.22800000. The connected load is 696 KW. Contract demand is 600 KVA. The minimum billing demand is 360 KVA.

Energy Distribution in Textile Industry

Energy audit had been conducted in the textile industry. Required data is collected by visiting and communicating with industrial persons or the auditing team. Installed load details had been gathered and plotted in Table 149.1. The power distribution of the textile industry is shown in Figure 149.5.

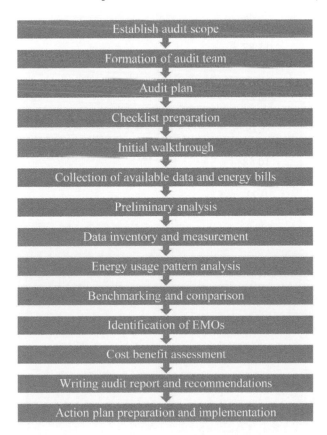

Figure 149.3 Flowchart for energy audit procedure

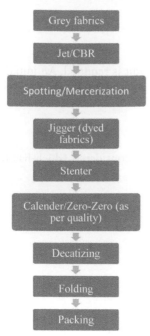

Figure 149.4 Process flowchart

Table 149.1 Power distribution in textile industry

Description		kW	% Loading
Lighting		5.52	0.79
Machinery	Jigger	203.00	29.03
	Jet	19.80	2.83
	CBR	55.95	8.00
	Mercerizer	60.00	8.58
	Stenter	97.00	13.87
	Calender	29.00	4.15
	Decatizer	22.00	3.15
	CSR (Zero Zero)	59.00	8.44
	Folding	6.60	0.94
	Singeing	37.00	5.29
	Total	589.35	84.28
Compressor		74.00	10.58
Office		5.40	0.77
Sewage plant		5.00	0.72
water pump		20.00	2.86
Total power		699.27	100.00

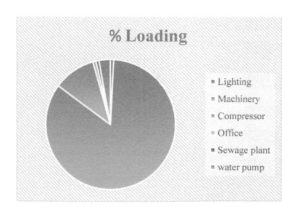

Figure 149.5 Power distribution of textile industry

Energy Conservation Opportunities in Systems Driven by Motor

To achieve greater performance from motor it is required to run all motors- within a 10% tolerance of the rated voltage, Operation from a sinusoidal voltage supply, Frequency within a 5% tolerance, Operation with a voltage imbalance of less than or equal to 1%.

Motor loading analysis

Readings got using power analyser at each motor in on load condition are shown in Tables 149.2 and 149.3.

Conclusion of motor input power quality analysis

- There is scope for optimisation of RMS voltage.
- Average value of % voltage unbalance factor is less than 1% at majority of motor terminals and a few places it is slightly above 1% recommended by NEMA MG standard. There is no need to carry out de-rating of motors.

Table 149.2 Power quality analysis at motor terminals

Name of machinery (motor)	Average RMS voltage (V)			Average RMS current (A)			Active power (KW)		
	Max.	Avg.	Min.	Max.	Avg.	Min.	Max.	Avg.	Min.
Jigger 10(5.5 kW)	407.65	405.11	402.52	10.85	10.73	10.62	4.94	4.92	4.91
Jigger 8(5.5 kW)	398.94	395.35	392.31	4.60	3.83	3.69	2.12	1.09	1.01
Jigger 26(4 kW)	215.49	214.25	212.53	0.47	0.46	0.46	0.11	0.10	0.10
Jet 4(2.2 kW)	434.6	432.8	430.7	3.18	3.14	3.09	0.91	0.9	0.89
Jet 5(2.2 kW)	216	215.8	215.7	5.75	5.74	5.73	1.11	1.1	1.09
CBR 55.95 kW	326.36	325.86	325.32	48.64	48.46	48.30	22.07	21.99	21.92
Mercerizer 60 kW	266.63	193.19	7.53	82.52	62.36	0.00	14.58	3.75	3.37
Stenter 97 kW	135.62	134.83	133.58	119.08	96.44	95.47	23.25	6.24	3.49
Calender 29 kW	127.62	126.49	125.31	37.82	27.25	24.13	0.54	0.48	0.00
Decatizer 22 kW	225.17	224.36	223.12	15.04	11.05	9.20	6.17	4.42	3.59
CSR 37 kW	153.45	153.15	152.71	43.50	31.05	27.38	13.76	6.47	3.69
CSR 22 kW	225.17	224.36	223.12	15.04	11.05	9.20	6.17	4.42	3.59
Folding 3(2.2) kW	215.98	215.88	215.78	7.90	7.62	7.31	1.66	1.49	1.32
Singeing 37 Kw	197.1	196.7	196.2	37.29	27.8	26.2	14.6	5.49	3.14
Compressor 2(37 Kw)	163.1	162.3	161.7	37.12	28.4	26.9	11.3	5.18	3.83

Table 149.3 Power quality analysis at motor terminals

Name of machinery (motor)	Reactive power (KVAR)			Apparent power (KVA)			P.F.		
	Max.	Avg.	Min.	Max.	Avg.	Min.	Max.	Avg.	Min.
Jigger 10(5.5 kW)	5.86	5.7	5.54	7.66	7.53	7.4	0.664	0.654	0.644
Jigger 8(5.5 kW)	2.49	2.38	2.27	3.12	2.62	2.5	0.684	0.417	0.387
Jigger 26(4 kW)	1	0.9	0.78	0.18	0.17	0.17	0.618	0.603	0.518
Jet 4(2.2 kW)	2.12	2.08	2.03	2.39	2.35	2.3	0.496	0.486	0.478
Jet 5(2.2 kW)	1.84	1.84	1.83	2.15	2.15	2.14	0.516	0.514	0.511
CBR 55.95 kW	16.32	16.26	16.19	27.44	27.4	27.3	0.81	0.8	0.8
Mercerizer 60 kW	29.03	19.92	10.43	30.87	21.8	0	0.44	0.43	0.42
Stenter 97 kW	41.09	38.58	36.07	47.82	39	38.6	0.49	0.48	0.46
Calender 29 kW	5.62	5.93	0.87	14.26	10.3	9.1	0.68	0.42	0.2
Decatizer 22 kW	7.97	5.98	5.11	10.08	7.44	6.24	0.61	0.59	0.57
CSR 37 kW	14.41	12.54	12.08	20.01	14.3	12.7	0.69	0.44	0.29
CSR 22 kW	7.97	5.98	5.11	10.08	7.44	6.24	0.61	0.59	0.57
Folding 3(2.2) kW	2.46	2.42	2.4	2.95	2.85	2.73	0.565	0.523	0.48
Singeing 37 Kw	16.69	15.29	15.08	22.03	16.4	15.5	0.66	0.33	0.2
Compressor 2(37 Kw)	13.92	12.71	12.48	18.03	13.8	13.1	0.63	0.37	0.29

Table 149.4 Percentage motor loading assessments

Name of machinery (motor)	Specification	Rated current (I)	Actual RMS current (A)	% Motor loading
Jigger 10(5.5 kW)	KW: 5.5	11.25	10.73	95.377
Jigger 8(5.5 kW)	5.5 KW	11.25	3.83	34.04
Jigger 26(4 kW)	KW: 4	7.5	0.46	6.133
Jet 4(2.2 kW)	2.2 KW	4.5	3.14	69.77
Jet 5(2.2 kW)	KW: 2.2	4.5	5.74	127.55
CBR 55.95 kW	KW:55, HP:75, EFF.: 94%, N:2965	112.5	48.46	43
Mercerizer 60 kW	Vtg: 400V, I: 120A, KW: 60, F: 59Hz, PF: 0.78, RPM: 1750	120	62.36	51.96
Stenter 97 kW	KW: 97	180	96.44	54
Calender 29 kW	Spindle Motor: V: 398V, I: 56A, KW: 29, RPM: 2300, F: 77.84Hz	56	27.25	48.66
Decatizer 22 kW	KW: 22, 3phase I.M.	45	11.05	24.55
CSR 37 kW	V:415V, 68A, 37KW, 2000RPM	68	31.05	45.66
CSR 22 kW	KW: 22, 3phase I.M.	45	11.05	24.55
Folding 3(2.2) kW	KW: 2.2	4.5	7.62	169.33
Singeing 37 Kw	KW: 37, I: 68A, V: 415V, RPM: 2000	68	27.76	40.823
Compressor 2(37 Kw)	V:415V, 68A, 37KW, 2000RPM	68	28.38	41.735

Table 149.5 List of motors loaded below 50%

Name of machinery (motor)	Specification	Rated current (I)	Actual RMS current (A)	% Motor loading	Actual power input (KW)
Jigger 8(5.5 kW)	5.5 KW	11.25	3.83	34.04	2.12
CBR 55.95 kW	KW:55, HP:75, EFF.: 94%, N:2965	112.5	48.46	43	22.07
Decatizer 22 kW	KW: 22, 3phase I.M.	45	11.05	24.55	6.17
CSR 37 kW	V:415V, 68A, 37KW, 2000RPM	68	31.05	45.66	13.76
CSR 22 kW	KW: 22, 3phase I.M.	45	11.05	24.55	6.17
Singeing 37 Kw	KW: 37, I: 68A, V: 415V, RPM: 2000	68	27.76	40.823	14.58
Total actual input KW					64.87

Table 149.6 List of over loaded motors

Name of machinery (motor)	Specification	Rated current (I)	Actual RMS current (A)	% Motor loading
Jet 5(2.2 kW)	KW: 2.2	4.5	5.74	127.55
folding 3(2.2) kW	KW: 2.2	4.5	7.62	169.33

- Average value of % Total voltage harmonics at motor terminal is below 5% limit at majority of motor terminals and few places is above 5% recommended by IEEE 519-1992 Standard.
- Majority of motors are correctly loaded except few motors which are oversised and under sized.

Motor loading assessment

Energy saving potential by conversion of lightly loaded motor below 50% on permanent basis from permanent delta connection to star connection

Energy saving potential by conversion of lightly loaded motors from Delta connection to star connection

Out of various motors few motors are loaded below 50%. All such motors are undersized. Motors are designed to deliver maximum efficiency at 75% load. Recommended to convert connection of mentioned motor from delta connection to permanent star connection or by augmenting such motors with overloaded motor according to their required rating.

Actual Power consumption of lightly loaded motor = 64.87KW

Energy saving potential by conversion of lightly loaded motor connection from delta connection to star connection per year = 64.87 × 0.10 × 22 × 350 = 49,950 KWH

Energy cost saving potential considering working hours.22, days 350 and energy cost of Rs.6.96 per kWh = Rs.6.96x 49,950.00 = Rs.3,47,452 annually.

While conversion of lightly loaded motor connection from delta connection to star connection. Recommended to record motor current in delta connection as well as star connection. If current in star connection is lower than Delta only such motor loading is below 50% and should be converted to star connection. If current in star connection is higher than delta such a motor connection should be transferred to delta and used in delta.

List of over loaded motors

Motors listed in Table 149.7 are over loaded motors based on loading condition and measurements taken during Energy Audit. Recommended to cross check loading and replace over loaded motors with energy efficient motors to avoid frequent failures if any.

Conclusion

An energy audit is a valuable tool for achieving significant energy savings. Reduced energy costs are a crucial factor in improving a company's competitiveness; as a result, conducting an energy audit of an industrial site is not only a particular responsibility mandated by European Directives but also a real potential for businesses. The investigation revealed that a variety of energy-saving techniques may be identified. The corporation used the findings of this energy audit to develop its long-term energy-saving plan. There are several textile industries in India. As a result, by using the aforesaid approach in all textile mills, a significant amount of energy may be saved. The main expectation of this paper was to build on the previous energy efficiency work that had been accomplished and further reduce consumption and costs, as such it can be seen as very successful: it accomplished discovery measures that reduce the consumption by around 49950 kWh per year, which translates in around Rs 3,47,452 per year, with nil investment. From this perspective, it can be replicated in the remaining factories.

References

Chaudhari, M., Babu, K., Khubalkar, S. W., and Daigavane, P., (2019). Off-grid hybrid solar power conditioning unit for critical and non-critical loads. In 2019 International Conference on Intelligent Computing and Control Systems (ICCS). 969–974.

Chaudhari, M., Babu, K., Khubalkar, S. W., and Talokar, S. (2019). Off-grid hybrid online solar power conditioning unit for domestic purposes. In 2019 International Conference on Computing, Power and Communication Technologies (GUCON). 121–126.

https://beeindia.gov.in/

Jadhav, V., Jadhav, R., Magar, P., Kharat, S., and Bagwan, S. U. (2017). May. Energy conservation through energy audit. In 2017 International Conference on Trends in Electronics and Informatics (ICEI). 481–485.

Khude, P. (2017). A review on energy management in textile industry. Innov Ener Res, 6(169):2.

Mahalaxme, S., Khubalkar, S., and Bharadwaj, S. (2000). Low voltage distribution box monitoring-new way to monitor power in industry. 2020 4th International Conference on Trends in Electronics and Informatics (ICOEI). 48184:214–216.

Nadimuthu, L. P. R., Victor, K., Basha, C. H., Mariprasath, T., Dhanamjayulu, C., Padmanaban, S., and Khan, B. (2021). Energy conservation approach for continuous power quality improvement: A case study. IEEE Access. 9:146959-146969.

Nagaveni, P., Kumar, M. S., Nivetha, M., Amudha, A., and Emayavaramban, G. (2019). Electrical energy audit–an experience in a small scale textile mill. Int. J. Innov. Technol. Explor. Eng. 8(10):4102–4107.

Pahariya, Y. (2001). Economical effect of using oversize induction motor. IEEMA J. Mumbai. 21(7):24–32.

Pahariya, Y. (2004). Energy conservation in compressed air system. IEEMA J. Mumbai. 24(11):27–29.

Patil, M. D., Vadirajacharya, K., and Khubalkar, S. W. (2021). Design and tuning of digital fractional-order PID controller for permanent magnet DC motor. IETE J. Res. 1–11.

Rajput, S. K. and Singh, O. (2016). October. Energy audit in textile industry: a study with ring frame motor. In 2016 International Conference on Control, Computing, Communication and Materials (ICCCCM). 1–4.

Shende, A., Khubalkar, S. W., and Vaidya, P. (2021). Hardware implementation of automatic power factor correction unit for Industry. J. Phys.: Conference Ser. 2089(1):012032.

150 Are students using the time effectively during the COVID-19 pandemic?

Mohamed Ifham[1,a], Senthan Prasanth[2,b], Kuhaneswaran Banujan[1,c], and B.T.G.S. Kumara[1,d]

[1]Department of Computing and Information Systems, Sabaragamuwa University of Sri Lanka, Belihuloya, Sri Lanka

[2]Department of Physical Sciences and Technology, Sabaragamuwa University of Sri Lanka, Belihuloya, Sri Lanka

Abstract

The spread of the COVID-19 pandemic impacts everyone's lives somehow. The adverse impact on educational systems in all regions has a wide-ranging extension effect. A total lockdown creates new challenges for students to study and teachers to properly direct the class, transforming such a physical classroom system into a virtual classroom. Despite the widespread use of online learning in schools throughout the COVID-19 epidemic, there are inadequate facts about the elements that influence student time utilisation with this innovative learning environment in a crisis. This study focuses on many aspects a student goes through during the pandemic. It's vital to understand whether students utilise their time efficiently during the pandemic, especially virtual learning systems. In this study, two machine learning models, namely decision tree and artificial neural network, were created to predict whether a student has utilised their time properly or not. The data set for the model was gathered via online questionnaires, which consist of 19 questions based on various parameters. Then the dataset is preprocessed using different mechanisms before feeding the data to the models. The final results of ANN show better accuracy of 65.91 and a lesser error rate of 34.08 after the comparison between the two models.

Keywords: artificial neural network, covid19 impact, decision tree, machine learning, virtual learning.

Introduction

The World Health Organization (WHO) has declared COVID-19 as a pandemic in early March of 2020. In order to reduce the infection rate, countries all around the world had to implement several preventative steps. Colleges and universities decide to employ online streaming tools like Zoom for teaching to minimise the consequences of a lockdown. All public venues, markets and restaurants were shut down during the lockdown. Consequently, students' mental health suffered due to the government's self-quarantine recommendations. Mental disorders, restlessness, and depression symptoms were all linked to several of them (Ron and Cuéllar-Flores, 2020). Humans' perceptions of psychological processes change to assist them in surviving in a dynamic environment (Ron and Cuéllar-Flores, 2020). Like the shifting stimulus, the lockdown caused a considerable increase in difficulty retaining track of time, with people unsure of which day and time they are spending. Time distend also associated with a greater sense of boredom throughout lockdown (Cellini et al., 2020). As a result of the shutdown and lockdown situations worldwide, social media was utilized to keep people updated on COVID-19 pandemic information. Other than that, during the outbreak, it has frequently been used as a medium to meet individuals' desires for human relationships. As a result, many students watched more multimedia content to pass the time (Stockdale, L. A. and Coyne, 2020), and learn more about COVID-19 (Bao et al., 2020). The psychological requirement became a compulsive practice with negative psychological health consequences.

COVID-19 pandemic seems to have a significantly larger influence on education during the academic year, continuing to be in the coming years. School systems, institutions, or universities no longer offer physical training. Students' college academic lives have been disturbed by a decline in household stipends, lack of access to internet supplies, and high network access rates. Furthermore, 1.5 billion pupils worldwide are deprived of fundamental education (Lee, 2020)., resulting in major psychological consequences for their health. Changes in students' daily schedule, such as disturbed sleeping patterns, lack of outdoor activity, and social quarantine, have all influenced their mental health. It is expected that the continuous virus

[a]ifham547@gmail.com; [b]sprasanth@appsc.sab.ac.lk; [c]bhakuha@appsc.sab.ac.lk; [d]btgsk2000@gmail.com

expansion, travel limitation, and the shutdown of educational institutions throughout the world are forecasted to have a major effect on students' education, mental health and social lives (Odriozola-González et al., 2000).

There are several activities computers could do better than humans in many domains nowadays. However, on the other hand, our amazing brains still have the advantage of rational thinking, inspiration, and imagination. It's important to utilize a machine learning model which can interpret the human's way of thinking in this study. Artificial neural networks, which are motivated by the composition of the brain, are indeed the solution to making computer systems more human alike and assisting machines in reasoning like people.

Another widely used machine learning model is the decision tree (DT). They can be applied to classification and regression cases while supporting linear and non-linear data samplesDT are being used to make a series of decisions based on facts and performance, as the names indicate. The linearity of the data determines its effectiveness because it does not use a linear classifier. Therefore, we have utilized the above two machine learning models in this study and compared the results to get the optimum solution.

This paper researched and analysed the probable repercussions of the COVID-19 pandemic on students' lifestyles. Our study examines whether a student has used his time wisely during the covid19 pandemic and built a machine learning model to predict the time utilisation based on many parameters. Furthermore, our research aims to analyse dependent between the mental state of students of all ages using various characteristics such as sleep pattern style, daily fitness practice, and social support concerning time utilisation. We also look at the time utilising those students employ to cope with the current scenario.

Literature Review and Related Works

Quarantine has been linked to a significant effect on mental health issues such as stagnation, stress, and anxiety (Salari et al., 2020). The shift in the personal evaluation of time flow, increasing multimedia usage and sleep patterns have also been documented (Cellini et al., 2020). These changes were explored, and these parameters were linked to students' psychological well-being and routine activities between 14 and 24. Our research found a substantial difference in the amount of time spent studying following the initial COVID-19 outbreak, which could be attributable to the change to virtual classrooms.

The authors Ahuja and Banga (2019) analysed datasets that included the implementation of two testing's. Afterwards, depending on the outcomes, the authors evaluated the sufficient stress at the start and the end of the evaluation employing a decision tree model (Adnan et a., 2021) that assist in identifying stress gauge a student suffer. Another study by Slavich (2019) is concerned about one's own life based on recent occurrences and shortage of time to perform responsibilities.

The paper by Gaikwad and Paithane (2017) contains EEG data collected with headgears, defined anxiety levels, and improved classification performance. A new study on machine learning algorithms is utilised regarding the COVID-19 pandemic behaviour (Alballa and Al-Turaiki, 2021), emphasising the effectiveness of the machine learning model towards two essential components. COVID-19 pandemic observation uses commonly accessible clinical and biochemical information (Bokam et al., 2021). There's debate on different methodologies, how to train enormous datasets, and how to extract features. Many machine learning approaches used in each procedure were supervised learning classifiers. Machine learning algorithms identify predictive and therapeutic qualities consistent with current research findings (Parthiban et al., 2021). The GAD-7 scaling mechanism is used by Cao et al. (2020) to diagnose anxiety and depression, panic diagnosis disorders, and social phobia (Ye et al., 2020). It also looks at how resilience, coping, and social support play a part in dealing with psychological disorders. Plenty of research on students' virtual learning in education has enriched both the knowledge and social consequences of adopting various forms of virtual learning (Valverde-Berrocoso et al., 2020). Student satisfaction is a significant performance metric when measuring the success of online learning in schools (Shim, T. E. and Lee, 2020; Aristovnik et al., 2020).

Furthermore, most individuals who didn't think online classes were effective said they weren't concentrating on hobbies or talents and were exhausted and depressed (Ali et al., 2021). The most likely explanation is that these individual students' difficulty comprehending and adjusting to online learning resulted in a lack of interest in other external activities, which indicates mental health problems (Yousef et al., 2017; Kendler, 2016).

A total of 89.3% of participants who took classes online during the shutdown said they were useless (Ali et al., 2021). Most of the students may not believe virtual classes are effective because of internet inaccessibility in some regions due to weak technical infrastructure, an absence of interaction between teachers and students, and different technological struggles faced by both students and teachers, such as not recognizing how to explore the e-learning application (Adnan and Anwar, 2020).

During the quarantine, the majority of the students who did not feel sleepless were studying a different new skill based on their interests. This finding is consistent with a prior study that links sufficient sleep and cognition (Carskadon, 2011). More research has shown how a freshly learned skill is changed from a transitory to a permanent condition while sleeping, a process known as post-training consolidation (Christova et al., 2018).

The findings also imply that students who could keep themselves occupied by employing on a project felt as though time was passing them by faster. According to an ongoing study, emotions and feelings might influence one's experience of time (Van Wassenhove et al., 2008; Eagleman, 2008). As a result, the students could stay focused by keeping themselves occupied with their academic work and hobbies. And also, they are compelled to be bored and not even have to keep track of time throughout the day. Our research focuses on finding whether the students utilize their time efficiently during the COVID-19 pandemic and the dependent parameters.

Methodology

In this research, we intend to find the effect of the COVID-19 pandemic on students' time utilisation of different ages by considering several parameters. After carefully investigating the above attributes, we created two machine learning models to evaluate the time utilisation of each student concerning the COVID-19 pandemic. Figure 150.1 shows the high-level methodology we are intended to build our prediction models.

Data collection

Data analysis would be a huge step forward in analyzing student expectations for a new or changed behavior. Before creating any machine learning model, pinpointing the primary requirements of the particular analysis is essential.

The dataset for this study was derived from (Chaturvedi et al., 2020). It consists of a 19-questions of an online questionnaire, with a mix of multichoice questions and the option for responders to submit any opinion relative to the target attribute. The data was collected through Google application forms, which requested students to be registered to an electronic mail address to partake.

The question set was given out using social media networks, electronic mail, and networking apps. One thousand one hundred eighty-two responses of the student's dataset were evaluated in this study. In order to be precise on our dataset attributes relevant to the target attribute, we discarded five attributes from the 18 attributes, excluding the target variable 'time utilisation'. Table 150.1 shows a brief sample data set attributes and relative data types of the dataset we have used in our study.

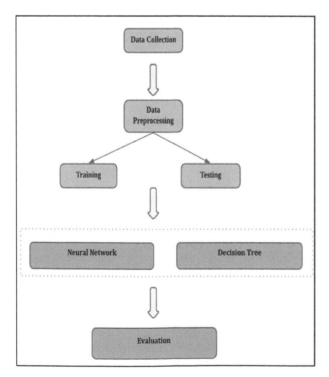

Figure 150.1 An overview of the methodology

Table 150.1 Data set attributes and relative data types of the dataset

No	Attribute	Datatypes
1	Region of residence	String
2	Age of student	Integer
3	Time devoted to online classes	Integer
4	Experience with online classes (rating)	String
5	Time devoted on self study	Integer
6	Time devoted on fitness	Integer
7	Time devoted on sleep	Integer
8	Time devoted on social media	Integer
9	Time devoted on TV	Integer
10	Each day's meal count	Integer
11	Weight fluctuations	String
12	Any health problems during lockdown	Boolean
13	Do you feel a stronger bond with your family, close friends, and relatives?	Boolean
14	Target: Time utilised	Boolean

Data preprocessing

This is a process responsible for performing data pre-processing and removing incomplete, noisy, or untrustworthy information from the system. This step is crucial in developing a forecasting system to determine students' time utilisation during the COVID-19 pandemic. As part of this procedure, the Pandas library was used to help achieve the goal mentioned above.

Under the above circumstances, the tasks below were fulfilled during the preprocessing stage.

- We saved the original information as a comma separated values (CSV) file. Originally, the necessary dataset was taken in an excel file. The collected information has been transformed to a CSV format to make the rest of the operations easier.
- We used pandas to load the CSV.
- We determined the number of rows and columns.
- Removed columns that are no longer needed.
- We checked for redundant and null values.
- We made all variables of the same type.
  Integer data types have been used to convey eight attributes, as shown in Table 150.1. Meanwhile, three attributes used Boolean data type, and the remaining three attributes of the other characteristics are displayed in a string format. Finally, the features of the string and Boolean data types have been transferred to the integer data type.
- Identification and substitution of outliers
- Normalised the data set.

Feature engineering

The dataset was input into the selected machine learning algorithms after the data preprocessing stage, but slight model overfitting was discovered, as shown in Figure 150.2. Following some study, the next step was to determine the best selection of qualities that may make a meaningful impact on the data set. For the above goal, the Pearson correlation analysis approach was used. Highly associated qualities towards the target attribute were deleted using the above technique. Since the strongly correlated variables would not assist the models any further, we were opted to remove those. The dataset was then rebuilt based on the important attributes and supplied into the supervised algorithms of choice. This approach concludes with the construction of the forecasting models, which is the next stage in determining the most suited algorithm with higher accuracy.

Classification Model

Artificial neural network

An artificial neural network (ANN) is a machine learning technique that simulates the way the nervous system of the human brain work. It uses supervised learning to make adjustments and learn on its own as

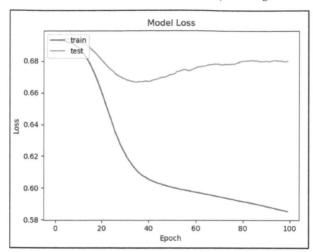

Figure 150.2 Model overfitting

new information is received. As a result, they're a great resource for non-linear statistical data processing. Input, hidden, and output neurons are the three neurons that make up an ANN. Neurons are the nodes of an artificial neural network, which are best described as a weighted directed graph. The Artificial Neural Network receives an input message in a pattern from an external source.

Decision tree

A decision tree (DT) is a mechanism for precisely categorizing many classes. It asks a series of questions regarding the dataset's quality and properties. The decision tree classification approach can be represented as a binary tree. Each of the internal nodes is presented with a query, and the data on that node is further separated into individual statistics with various attributes. The tree's leaves represent the classes from which the dataset is divided.

Results and Discussion

A detailed study was conducted during this phase to determine the relationship between the prevalent traits and the target column attribute.

Once the most significant attributes were discovered, the dataset was split into two non-equal parts. Training and testing are the two parts our data set are divided. The data is then fed into supervised learning algorithms (ANN and decision tree) to determine the optimum for creating the final prediction models.

The possible outcomes of predicting the models' performances were thoroughly assessed. Precision, accuracy, F1-score, recall, mean squared error, and confusion matrix was employed to achieve this goal. The K-fold cross-validation process was the major method used in this study to describe the best model.

Valuation metric

One thousand one hundred eighty-two student details with 13 question features and one target value were integrated for deep analysis, as previously stated in this study. Seventy percent of the datasets were used for training, while thirty percent were used for testing. The results of ANN and DT "Evaluation Metric" and "Confusion Metric" are reported in Tables 150.2 and 150.3.

The evaluation metrics are used to evaluate the models to determine their efficacy. This generally comprises training a machine learning algorithm on a dataset, then utilising the prediction model to make predictions on the dataset that was not used during training and comparing the forecasts to the current dataset's projected values. To assess the effectiveness of the suggested model, we compute accuracy, recall, precision, and F1-score.

The K-fold cross-validation approach was used with five folds to discover the optimal technique for dealing with any percentage of training and testing data and the outcome obtained from the tables as mentioned earlier.

The accuracy produced by the machine learning algorithms for each fold is shown in Table 150.4 the mean accuracy. Overall, accuracy decreases after the second fold and continues to drop until the fourth fold, whereas ANN techniques indicate an optimum level of accuracy. Even though DT shows a little gain in accuracy at the fifth fold, ANN has the highest mean accuracy.

In conclusion, it was noticed and ensured that ANN had surpassed other procedures with higher accuracy, 65.95%, with a low error rate of 34.08%, based on all types of analyses done thus far.

A training accuracy curve is created from training data samples that show how well the model is training. The testing curve created from testing data samples illustrates how well the models' algorithm generalizes. According to the training and testing curve in Figure 150.3, extending the training epochs could improve the ANN model accuracy.

However, no progress in testing accuracy was detected as the number of epochs was extended, although training accuracy rose. The model is expected to be a decent learning curve when utilising the same epochs. However, utilizing the hyperparameter to alter the curves, there is still plenty of space to improve the accuracy of the training and testing curves. Our training and testing accuracy appear to be well-weighted, and future studies could improve this by expanding the training datasets. The resulting prediction model was created using ANN. It is adequate to forecast whether students have optimised their time during the pandemic and whether the questions asked from the survey are sufficient to have a final model.

Table 150.2 Evaluation metrics without hyper-parameter tuning for machine learning model

Classifier	Accuracy	Precision	Recall	F1- Score	MSE
DT	55.49	51.21	51.85	51.53	44.50
ANN	65.91	65.14	65.51	65.32	34.08

Table 150.3 Confusion metrics obtained for the classifiers

Classifier	Confusion matrix			
	TN	FP	FN	TP
DT	113	80	78	84
ANN	120	61	60	114

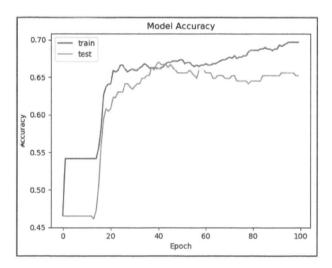

Figure 150.3 Model accuracy vs epoch values

Table 150.4 Performance of individual classifiers after cross-validation with five-folds

Classification	Accuracies measured with five folds of K-fold cross-validation					
	first-fold	second-fold	third-fold	fourth-fold	fifth-fold	Mean accuracy
DT	57.8	56.9	57.6	55.9	57.2	56.77
ANN	65.5	65.6	61.0	59.6	61.1	62.63

Conclusions and Future Works

The emergence of the COVID-19 pandemic has thrown each individual's life into chaos. The closing of schools and higher education institutions was one of the first adjustments made to slow the expansion of the virus within the students. New teaching approaches for online education delivery were created to prevent further interruptions of studies [29]. However, these actions may have long-term effects on students' lifestyles. As a result, there is a great requirement to document and research the implementation's impact. Our goal in this research is to examine the COVID-19 pandemic on students' academic, health, and social lives and highlight the time utilisation for different activities during the pandemic. We created two machine learning models, ANN and DT, to compare and find whether the students' questions were relevant to each student's time utilisation during the pandemic.

Subsequently of the circumstances in the COVID-19 epidemic, many students were likely to face tension, worry, and despair. As a result, providing emotional support is crucial to students. Most of the students have said they have not utilised the time properly, and our prediction model ANN shows a better accuracy concerning time utilisation.

Our study has certain substantial limitations that should be addressed. The first constraint is the sampling approach utilized in this study. It is built on internet technology and voluntary involvement, leading to selection bias. Technologically disadvantaged students were unable to participate because of the enforced travel restrictions. Another limitation is that the study was conducted cross-sectionally with no follow-up time for the participants. Future studies in this domain could look into the limitations while collecting the data and be precise on the validity of the data set.

References

Adnan, M. and Anwar, K. (2020). Online learning amid the COVID-19 pandemic: students' perspectives. J. Pedagogical Sociol. Psychol. 2:45–51.

Adnan, N., Murat, Z. H., Kadir, R. S. S. A., and Yunos, N. H. M. (2021). University students stress level and brainwave balancing index: comparison between early and end of study semester. 2012 IEEE Student Conference on Research and Development (SCOReD). 42–47.

Ahuja, R. and Banga, A. (2019). Mental stress detection in University Students using machine learning algorithms. Procedia Comput. Sci. 152:349–353.

Alballa, N. and Al-Turaiki, I. (2021). Machine learning approaches in COVID-19 diagnosis, mortality, and severity risk prediction: A review. Info. Medicine Unlock. 24:100564.

Ali, A., Siddiqui, A. A., Arshad, M. S., Iqbal, F., and Arif, T. B. (2021). Effects of COVID-19 pandemic and lockdown on lifestyle and mental health of students: A retrospective study from Karachi, Pakistan. Ann Med Psychol. 180(6):S29S37.

Aristovnik, A., Keržič, D., Ravšelj, D., Tomaževič, N., and Umek, L. (2020). Impacts of the COVID-19 pandemic on life of higher education students: A global perspective. Sustainability. 12:8438.

Bao, Y., Sun, Y., Meng, S., Shi, J., and Lu, L. (2020). 2019-nCoV Epidemic: Address Mental Health Care to Empower Society. Lancet. 395:e37–e38.

Bokam, Y., Guntupalli, C., Gudhanti, S., Manne, R., Alavala, R., and Alla, N. (2021). Importance of pharmacists as a front line warrior in improving medication compliance in Covid-19 patients. Indian J. Pharma. Sci. 83:398–401.

Cao, W., Fang, Z., Hou, G., Han, M., Xu, X., and Dong, J. (2020). The psychological impact of the COVID-19 Epidemic on college students in China. Psychiatry Res. 287:112934.

Carskadon, M. A. (2011). Sleep's effects on cognition and learning in adolescence. Progress Brain Res. 190:137143.

Cellini, N., Canale, N., Mioni, G., and Costa, S. (2020). Changes in sleep pattern, sense of time and digital media use during COVID-19 lockdown in Italy. J. Sleep Res. 29:e13074.

Chaturvedi, K., Vishwakarma, D. K., and Singh, N. (2020). COVID-19 and its impact on education, social life and mental health of students: A survey. Children Youth Serv. Rev. 121:105866.

Christova, M., Aftenberger, H., Nardone, R., and Gallasch, E. (2018). Adult gross motor learning and sleep: Is there a mutual benefit? Neural Plasticity. https://doi.org/10.1155/2018/3076986

Eagleman, D. M. (2008). Human Time perception and its illusions. Curr. Opin. Neurobiol. 18:131–136.

Gaikwad, P. and Paithane, A. (2017). Novel Approach for Stress Recognition Using EEG Signal by SVM Classifier. 2017 International Conference on Computing Methodologies and Communication (ICCMC). 967–971.

Johnson, N., Veletsianos, G., and Seaman, J. (2020). US Faculty and Administrators' Experiences and Approaches in the Early Weeks of the COVID-19 Pandemic. Online Learn. J. 24:6–21.

Kendler, K. S. (2016). The phenomenology of major depression and the representativeness and nature of DSM criteria. Am. J. Psychiatry. 173:771–780.

Lee, J. (2020). Mental Health effects of school closures during COVID-19. Lancet Child Adolescent Health. 4:421.

Odriozola-González, P., Planchuelo-Gómez, A., Irurtia, M. J., and de Luis-García, R. (2000). Psychological effects of the COVID-19 outbreak and lockdown among students and workers of a Spanish University. Psychiatry Res. 290:113108.

Parthiban, K., Pandey, D., and Pandey, B. K. (2021). Impact of SARS-CoV-2 in online education, predicting and contrasting mental stress of young students: A machine learning approach. Augmented Human Res. 6:1–7.

Ron, A. G. and Cuéllar-Flores, I. (2020). Psychological impact of lockdown (confinement) on young children and how to mitigate its effects: Rapid review of the evidence. Anales de pediatria. 57.

Stockdale, L. A. and Coyne, S. M. (2020). Bored and online: Reasons for using social media, problematic social networking site use, and behavioral outcomes across the transition from adolescence to emerging adulthood. J. Adolescence. 79:73–183.

Salari, N., Hosseinian-Far,A., Jalali, R. Vaisi-Raygani, A., Rasoulpoor, S., and Mohammadi, M. (2020). Prevalence of stress, anxiety, depression among the general population during the COVID-19 pandemic: A systematic review and meta-analysis. Global. Health. 16:1–11.

Shim, T. E. and Lee, S. Y. (2020). College students' experience of emergency remote teaching due to COVID-19. Children Youth Serv. Rev. 119:105578.

Slavich, G. M. (2019). Stressnology: the Primitive (and problematic) study of life stress exposure and pressing need for better measurement. Brain, Behavior, Immunity. 75:3–5.

Valverde-Berrocoso, J., Garrido-Arroyo, M. D. C., Burgos-Videla, C., and Morales-Cevallos, M. B. (2020). Trends in educational research about e-learning: a systematic literature review (2009–2018). Sustainability. 12:5153.

Van Wassenhove, V., Buonomano, D. V. Shimojo, S., and Shams, L. (2008). Distortions of subjective time perception within and across senses. PloS One. 3:e1437.

Ye, Z., Yang, X., Zeng, C., Wang, Y., Shen, Z., and Li, X. (2020). Resilience, social support, and coping as mediators between Covid-19-related stressful experiences and acute stress disorder among college students in China. Appl. Psychol.: Health Well-Being. 12:1074–1094.

Yousef, S., Athamneh, M., Masuadi, E., Ahmad, H., Loney, T., and Moselhy, H. F. (2017). Association between depression and factors affecting career choice among jordanian nursing students. Rontier. Public Health. 5:311.

151 Thyristor controlled LC compensator for current harmonic reduction

Sandeep R. Gaigowal[1,a], Sandeep Bhongade[b], and Sachin Ambade[2,c]

[1]Department of Electrical Engg.,Yeshwantrao Chavan College of Engineering, Nagpur, India

[2]Department of Electrical Engg., Shri G S Institute of Technology, Indore, India

Abstract

Modern power system requires control on all the power system parameters like bus volage, power angle and line reactance. Flexible alternating current transmission system (FACTS) controllers provide control over active and reactive power control in the grid. Static VAR (SVC) is shunt connected FACTS device which mainly provides reactive power control in the power system. The improvement in design of SVC is Thyristor controlled LC (TCLC) which is shunt compensator and it is also taken into consideration as it has many advantages over the Fixed capacitor Thyristor controlled reactor (FC-TCR) like cost reduction and less THD. The TCLC circuit is made up of coupling inductor in series with the FC-TCR circuit. In this paper, TCLC is implemented in power distribution system to regulate point of common coupling (PCC) voltage with current harmonics reduction and its performance analysis is compared with FC-TCR compensator. Delta connection of TCLC circuit is simulated in MATLAB Simulink software and harmonic spectrum is analyzed. Delta connected TCLC is used for dynamic reactive power control which also regulates the voltage at load end in power system with current harmonics reduction. Using delta connected TCLC compensator, low order harmonics are significantly reduced as compared to FC-TCR compensator.

Keywords: Static VAR compensator, Thyristor controlled LC, total harmonic distortion.

Introduction

Electric power demand is increasing day by day. Along with thermal power generation renewable power is adding more power in the existing grid which results in overloading of existing transmission network. New transmission network is not expanded in the proportion of generation. Mainly environmental issues, land acquisition problems and long building cycles makes hurdles in the expansion of transmission network. Flexible alternating current transmission system (FACTS) controllers can be used to utilize existing network upto its full thermal limits. FACTS are capable of altering power system parameters like active and reactive power flows, bus voltage, power angle and impedance of transmission lines. It plays very important role in today's power system operation and control. It can be categorized as conventional FACTS controllers and VSC based FACTS controllers (Hingorani and Gyugyi, 2001; Mathur and Varma, 2002). Conventional FACTS controllers use Thyristor or GTO to control current and voltage across the circuit. Main role is to reactive power exchange with the system. VSC based FACTS controllers requires high rated and high costly power semiconductor devices. Static VAR compensator (SVC), one of the FACTS devices, is used for the reactive power compensation in the transmission line. SVC is shunt compensator and it mainly constitute TCR and fixed capacitor (FC) conveniently used for absorbing and injecting reactive power respectively. TCR circuit comprises reactor whose current is controlled by an antiparallel thyristor circuit. The reactive power to be absorbed by the reactor can be controlled with the firing angle of thyristors. As the firing angle increases from $090°$ the current flowing through reactor decreases and hence the reactive power absorbed by the reactor can be controlled. Whereas capacitor injects fixed amount of reactive power depending upon the capacitance of the capacitor. Traditionally SVC was used in the transmission system as it is cheap. However, operation of SVC injects lower order harmonics in the current entering the system and it can make the system unstable (Joshi et al., 2016; Patil. and Baviskar, 2015; Gyugyi, 1988). Voltage source inverter (VSC) based STATCOM compensator can be used for reactive power control with the faster response. But STATCOM is costlier (about 30%) than SVC. Hence modification in reducing the harmonic injection by SVC were investigated further. Resulting parallel combination of SVC with passive filters were suggested. But the cost of this combined system becomes very high. A combined system of SVC and STATCOM can eliminate Hamonic injection and it can control reactive power in no linear loads. Some research papers proposed combined compensating circuit comprises of thyristor controlled reactor (TCR) and TSC to reduce harmonics. Cost effective reactive compensation can be easily provided by SVC.

---

[a]sandeep_rg5@rediffmail.com; [b]bhongadesandeep@gmail.com; [c]sachinamb2@rediffmail.com

FC-TCR controls reactive component of current and it regulates the bus voltage. But it will remain harmonics in the compensating current (Soares and Verdelho, 2000; Gyugyi, 1988; Lee and Kim, 2007; Gyugyi et al., 1978; 1976; 1993). It can be reduced by TCLC. If TCLC is connected in delta connection, it further reduces harmonics.

Table 151.1 gives comparison between SVC, STATCOM and TCLC compensator. SVC mainly consists of TCR and TSC circuits which uses inductor and capacitor mainly and it is controlled by thyristor switching. SVC can provide reactive power compensation with the advantage of low cost, low losses, simple control scheme. STATCOM uses high rating power electronics switches which increases its cost of installation. It gives wide range of control as well as faster control. But disadvantage is its high cost. As compared to SVC and STATCOM, TCLC compensator provide reactive power compensation with low cost of installation and wide range reactive power control. It gives very less harmonics injection in the circuit. Shaded portion in the Table 151.1 shows undesirable characteristics.

Literature cited above found that shunt compensator like FC-TCR provide reactive power control but injects harmonics in the current. VSC based FACTS device like STATCOM provide wide range of control in active and reactive power at PCC. But it consists of high rated power semiconductor switches which increases its cost of installation and operation increased. In this paper TCLC is proposed a reliable solution for reactive power control with low THD. Thyristor controlled LC (TCLC) which is cheap and injects less harmonics. In this paper, TCLC is demonstrated for shunt compensation with harmonics reduction. TCLC consists of a reactor in series with the FC-TCR. This reactor is used to limit the surge current as well as to reduce the harmonic injection (Wang et al., 2016). Paper starts with introduction section and then TCLC is presented. After that, system studies in simulation results and discussion are given. Simulation results are shown in MATLAB software. At the end of the paper, work is concluded in conclusion section.

Thyristor Controlled LC Compensator

The traditional SVC made up of FC-TCR can compensate reactive power by controlling the firing angles of the anti-parallel pair of thyristors. However, in FC-TCR, lower order (3rd, 5th, 7th, 9th) harmonic are generated in current, which then affects the system performances. To overcome this problem the STATCOMs, SVC-STATCOM combined, SVC-PPF were studied for reactive power control with faster response and harmonics in currents are reduced. But the initial cost of all these systems is so high which is not economical. PV-STATCOM is the FACTS controller which can be connected in shunt. Solar inverter in PV system can be utilized as STATCOM. In day time it can inject active power in the grid and in the night time, solar is disconnected and solar inverter can be worked as STATCOM to control bus voltage. Since it is VSC based system, it is costly (Rode et al., 2018; Azharuddin, M. and Gaigowal, 2017; Sandeep et al., 2020; Patil and Kadwane, 2018). In previous research work, Current harmonics are analyzed by implementation of Multi-Quasi-Proportional-Resonant Controller for Thyristor-Controlled LC-Coupling Hybrid Active Power Filter (TCLC-HAPF) (Sou et al., 2022). Power quality issues are mitigated using Hybrid compensator (Chakraborty et al., 2022).

TCLC compensator can provide reactive power control. TCLC circuit is made up of series connected inductor L_C with the FC-TCR circuit which consist of capacitor C_{PF} connected in parallel with Inductor L_{PF}. The FC-TCR is used for the reactive support in the system and the coupling inductor is used to remove the harmonics injected by the FC-TCR. Compensator SVC alike a fixed capacitor-thyristor controlled reactor (FC-TCR) generates harmonics in the current, the proposed TCLC can significantly reduce the harmonics in current (Wang et al., 2014; 2016; IEEE Standard 519, 2014). The contribution of this paper is as follows.

1) The TCLC is studied for reactive power control and hence bus voltage is regulated.
2) Harmonic content in compensating current is analysed in FC-TCR and TCLC circuit. The harmonics in the current through TCLC is reduced to a great extent.

Table 151.1 Comparison SVC, STATCOM and TCLC (Wang et al., 2017)

	SVC	*STATCOM*	*TCLC*
Power Electronics switches uses	*	* * *	*
Harmonic injection	* * *	*	*
Cost of compensator	*	* * *	*
Loss	*	* *	*
Reactive power control	* * *	* * * *	* * *
Control scheme complexity	*	* *	*

Increasing * mark means increasing level of characteristics and shaded portion indicated unitrable part

TCLC circuit connected in each phase can reduce harmonics in the compensating current. In the proposed work, TCLC circuit is connected in delta connection to reduce harmonics further.

Simulation Results

Reactive power compensation is demonstrated using FC-TCR and TCLC compensator circuit. In this paper, Simulation in MATLAB software is presented. Main objective of the FC-TCR or TCLC compensator is to provide reactive power compensation with the aim that harmonics content will be reduced and point of common coupling (PCC) voltage is regulated.

(A) Thyristor controlled reactor (TCR) is shunt connected FACTS controller and it can be used in power system for controlling reactive power in the system. TCR circuit mainly consists of inductor L in series with antiparallel thyristors connected as shown in Figure 151.1 Current flowing inductor L can be controlled by controlling firing instants of thyristors. Current expression is given by,

$$IL(\alpha) = \frac{V}{\omega L}\left(1 - \frac{2\alpha}{\pi} - \frac{\sin 2\alpha}{\pi}\right) \tag{1}$$

Equation (1) shows the TCR current variation with changing value of firing angle α.

For different firing angles, current waveform is drawn. It is shown in Figure 151.2

To provide dynamic reactive power support Fixed capacitor – TCR (FC-TCR) compensator can be used. A fixed capacitor is connected across the TCR. It provides inductive or capacitive VAR as per requirement.

To minimise harmonics, FC-TCR circuit can be connected in delta connection. A significant harmonic can be reduced.

(B) Traditional SVC contains harmonics of high magnitude in current. To reduce these harmonics one inductor L_C is connected in series with the FC-TCR which is called as the coupling inductor. This circuit as a whole is referred as Thyristor Controlled LC i. e. TCLC. The TCLC circuit shown in Figure 151.5 connected in each phase of the system having the parameters shown in figure.

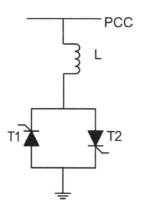

Figure 151.1 Thyristor controlled reactor

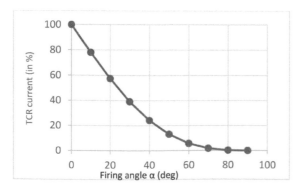

Figure 151.2 TCR current in % with respect to different angle α

The simulation results as well as the harmonic analysis is shown below. Figure 151.6 shows resulting current through TCR of TCLC circuit and Figure 151.7 shows resulting current through TCLC circuit. Figure 151.8 shows regulated PCC voltage. Voltage at PCC is regulated at 110 V. Harmonic analysis is shown in Figure 151.9 It is observed that the harmonics of lower order are injected in the system.

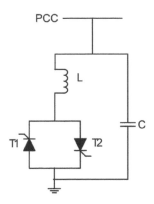

Figure 151.3 FC-TCR circuit

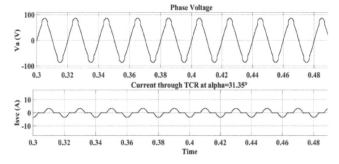

Figure 151.4 Delta connected FC-TCR circuit

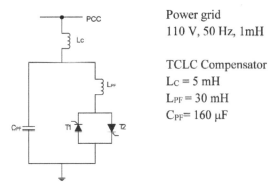

Power grid
110 V, 50 Hz, 1mH

TCLC Compensator
L_C = 5 mH
L_{PF} = 30 mH
C_{PF} = 160 μF

Figure 151.5 Thyristor controlled LC compensator

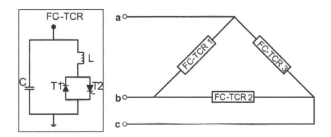

Figure 151.6 Current through TCR of TCLC circuit

(C) The TCLC in per phase scheme will have harmonics in its current. To overcome this problem of harmonics the TCLC is connected in delta. The TCLC connected in delta as shown in Figure 151.10 having the parameters shown in Figure 151.10. Three TCLC circuits TCLC1, TCLC2 and TLLC3 are connected in delta connection. Here along with the 3rd order harmonics the THD is reduced significantly as compared to all other connection schemes. The simulation results as well as the harmonic

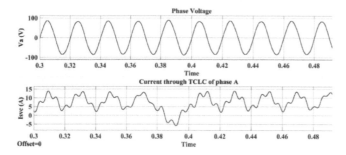

Figure 151.7 Current through TCLC circuit

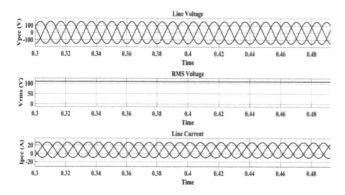

Figure 151.8 PCC voltage and current

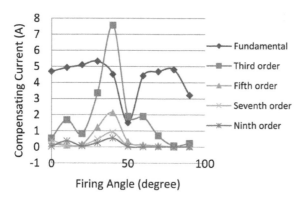

Figure 151.9 Harmonic analysis at different firing angles

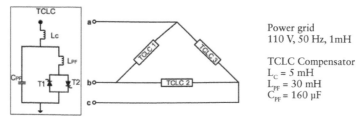

Figure 151.10 Delta connection thyristor controlled LC compensator

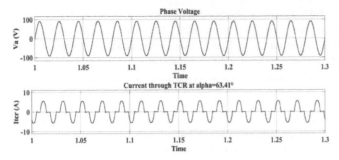

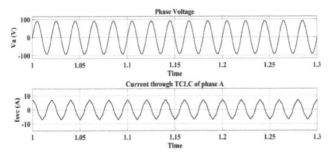

Figure 151.11 Current through TCR of TCLC circuit

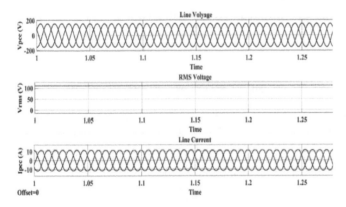

Figure 151.12 Current through TCLC circuit

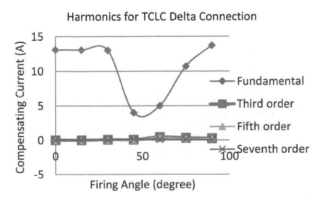

Figure 151.13 Voltage and current at PCC

Figure 151.14 Harmonic analysis with different firing angles

analysis is also shown below. Figure 151.11 shows resulting current through TCR of TCLC circuit and Figure 151.12 shows resulting current through delta connected TCLC circuit. Figure 151.12 shows regulated PCC voltage. Voltage at PCC is regulated at 110 V. Harmonic analysis is shown in Figure 151.14. It is observed that harmonics of lower order are almost reduced.

Discussion

In this paper, current harmonic spectrum of FC-TCR circuit, TCLC compensator and delta connected TCLC compensator are analyzed. Simply FCTCR circuit injects current harmonics in the system. Specifically low order harmonics are observed in the harmonic spectrum. To reduce low order harmonics, TCLC is proposed. Figure 151.9 shows harmonic spectrum of TCLC compensator. Harmonics are found reduced but low order harmonics 3rd, 5th are not found reduced considerably. TCLC circuit is connected in delta connection to reduce low order harmonics. Figure 151.11 shows current harmonic spectrum of delta connected TCLC. It is found that harmonics are reduced. Low order harmonics 3rd and 5th are considerably reduced. PCC voltage in Delta connected TCLC compensator is shown in Figure 151.13 and it is found regulated at 110 V.

Conclusion

Reactive power compensation is provided by inductor and capacitor circuits to regulate bus voltage. SVC consists of L and C circuits controlled by Thyristor. SVC provides reactive power compensation for load compensation. FC-TCR is SVC circuit which controls dynamic reactive power requirements. The improvement in traditional FC-TCR design is introduced as Thyristor-controlled LC (TCLC) which is cheap as compared to other SVC compensators. In this paper, TCLC circuit is analyzed for bus voltage regulation by controlling reactive power. MATLAB simulation results show that PCC voltage is regulated to the reference value and THD is found reduced using TCLC circuit. To reduce low order harmonics i. e. 3rd and 5th, TCLC circuit is connected in delta. MATLAB simulation results of TCLC current is analyzed and it is found that THD in TCLC current is reduced considerably. Simulation results conclude that delta connected TCLC compensator provide reactive power compensation with reduced low order harmonics and PCC voltage is regulated at reference PCC voltage.

References

Azharuddin, M. and Gaigowal, S. R. (2017). Voltage regulation by grid connected PV-STATCOM. International Conference on Power and Embedded Drive Control (ICPEDC), IEEE.

Chakraborty, S., Mukhopadhyay, S., and Biswas, S. K. (2022). A hybrid compensator for mitigation of power quality issues in distribution systems. IEEE International Conference on Power Electronics, Smart Grid, and Renewable Energy (PESGRE). doi: 10.1109/PESGRE52268.2022.9715834

Gyugyi, L. 1988. Power electronics in electric utilities: Static var compensators. Proceeding of IEEE. 76(4):31–35.

Gyugyi, L. (1988). Power electronics in electric utilities: static var compensators. Proceedings of the IEEE. 76(4):483494.

Gyugyi, L., Otto, R. A., and Putman, T. H. (1978). Principles and applications of static, thyristor-controlled shunt compensators. IEEE Transaction on Power Apparatus and Systems. PAS-97(5):19351945.

Gyugyi, L. (1976). Reactive power generation and control by thyristor circuits. Paper SPCC 77-29 presented at Power Electronics Specialist Conference. 1520.

Gyugyi, L. (1993). Dynamic compensation of AC transmission lines by solid-state synchronous voltage sources. IEEE/PES Summer Meeting. 2227.

Hingorani, N. g. and Gyugyi, L. 2001. Understanding facts: concept and technology of facts. Wiley-IEEE Press.

IEEE Standard 519. 2014. IEEE recommended practices and requirements for harmonic control in electrical power systems. doi: 10.1109/IEEESTD.2014.6826459

Joshi, B. S., Mahela, O. P., and Ola, S. R. 2016. Reactive power flow control using static VAR compensator to improve voltage stability in transmission system. IEEE International Conference on Recent Advances & Innovation in Engineering (ICRAIE).

Lee, W. and Kim, T. (2007). Control of the Thyristor controlled reactor for reactive power compensation and load balancing. Second IEEE Conference on Industrial Electronics and Applications. 201206.

Mathur, R. M. and Varma, R. K 2002. Thyristor-based facts controllers for electrical transmission system. Wiley-IEEE Press.

Patil, A. P. and Baviskar, P. V. (2015). Implementation of static VAR compensator (SVC) for power factor improvement. Int. J. Emerging Technol. Adv. Eng. 4:2529.

Patil, S. and Kadwane, S. G. (2018). Hybrid optimization algorithm applied for selective harmonic elimination in multilevel inverter with reduced switch topology. Springer. 24(2):17.

Rode, N. R., Gaigowal, S. R., and Patil, P. S. (2018). Cascaded H-bridge inverter based PV-STATCOM. International Conference on Smart Electric Drives and Power System (ICSEDPS), IEEE.

Sandeep, R., Gaigowal, Renge, M. M. (2020). DSSC to Improve power system loadability index. Adv. Electric. Comp. Technol. 10751083.

Soares, V. and Verdelho, P. (2000). An instantaneous active and reactive current component method for active filters. IEEE Trans. Power Electron. 15(4):660–669.

Sou, W. K., Chan, P. I., Gong, C., and Lam, C. S. (2022). Finite-set model predictive control for hybrid active power filter. IEEE Transactions on Industrial Electronics IEEE.

Sou, W. K., Chao, C. W., Gong, C., Lam, C. S., and Wong, C. K. (2022). Analysis, design, and implementation of multi-quasi-proportional-resonant controller for thyristor-controlled LC-coupling hybrid active power filter (TCLC-HAPF). IEEE Transact. Ind. Electron. 69(1).

Wang, L., Lam, C. S., and Wong, M. C. (2016). A hybrid-STATCOM with wide compensation range and low dc-link voltage. IEEE Trans. Ind. Electron. 63(6):3333_3343.

Wang, L., Lam, C. S., and Wong, M. C. (2014). An adaptive hysteresis band controller for LC-coupling hybrid active power filter with approximate constant switching frequency. IEEE PES Asia-Pacific Power and Energy Engineering Conference (APPEEC). 15.

Wang, L., Lam, C. S., Wong, and M. C. (2017). Design of a thyristor controlled LC compensator for dynamic reactive power compensator in smart grid. IEEE Transaction on SMART GRID. 5462

Wang, L., Lam, C. S., and Wong, M. C. 2016. Modeling and parameter design of thyristor controlled LC-coupled hybrid active power filter (TCLC-HAPF) for unbalanced compensation. IEEE Transaction on Industrial Electronics.

152 Solve selective harmonic elimination problem with a new metaheuristic optimisation algorithm

Sarika D. Patil[1,a], Akshay D. Kadu[1,b], and Pratik Dhabe[2,c]

[1]Department of Electrical Engineering, Yeshwantrao Chavan College of Engineering, Nagpur, India

[2]Department of Electrical Engineering, Shri Sant Gajanan Maharaj College of Engineering, Shegaon, India

Abstract

In many applications, power conversion is conveniently done through power electronic devices in power system. Mostly in high power conversions, multilevel inverter plays a main role. Different modulation techniques used so far for switching purpose. But high switching frequency indicates high switching loss thereby affecting system's efficiency. Hence fundamental switching frequency modulation technique come into existence and becomes popular to ensures the system stability. Low switching frequency modulation scheme is implemented for multilevel inverter successfully by many researchers. The switching angles are calculated by different algorithms and provided for switching of multilevel inverter. This research paper proposes a new metaheuristic algorithm for calculation of the offline firing angles of inverter. It is a combined approach of both ACO optimization algorithm and Newton-Raphson method. This new algorithm is implemented for 1-phase, 9-level inverter to get the expected fundamental voltage with harmonics reduction. Here, the initial guess in Newton-Raphson algorithm is provided by the output of ACO algorithm which eliminates the drawback of assuming initial guess in N-R method in this new algorithm. Hence this proposed algorithm is simple in design, with less computational efforts, and it will converge very fast as compared to other conventional methods. The effectiveness of the algorithm is confirmed through all the simulated and the experimental results. All the results are simulated in MATLAB environment.

Keywords: Multilevel Inverter, N-R method, ACO algorithm, SHE Equation, Odd harmonics component.

Introduction

Multilevel inverters play major role for medium voltage as well as high voltage applications. Hence they are proved to be the best choice in power conversion system (Rodriguez et al., 2007). Applications like traction drives, hybrid vehicles, power quality issues multilevel inverters are mostly used (Rodrigues et al., 2002; Kouro et al., 2010). It has different topologies like FC- Flying capacitor multilevel inverter (Huang and Corzine, 2006), CHB- Cascaded H Bridge multilevel inverter (Corzine and Familiant, 2002), DC-Diode clamped multilevel inverter (Rodrigues et al., 2002), Various switching methodologies have been implemented in multilevel for enhancement of power quality improvement. These methods are SPWM, SHEPWM, SVPWM, OHSW method (Edpuganti, and Rathore, 2005; Sirisukprasert, 1999), and OTHD method (Dahidha et al., 2015). With the help of these techniques, desired output can be obtained. But output waveform of multilevel inverter includes some harmonics. Removal of higher order harmonics is taken care by filters. But minimisation of lower order harmonics cannot be easily achieved. The output equations of multilevel inverter are nonlinear transcendental equations. Finding solution to these equations is not easy and straightforward. It requires some iterative methods or optimisation algorithms to get the exact solutions. The iterative algorithm like Newton–Raphson (N–R) algorithm is implemented for getting the best solutions to these equations (Kumar et al., 2008). But it requires correct initial guess for achieving desired solution. Due to this drawback, it is not used so far. The resultant polynomial theory is implemented in (Chiasson et al., 2003). Using this algorithm, finding solution with different modulation indices is possible. But this algorithm becomes complicated when variable dc input system is implemented. To overcome these drawbacks of iterative methods, recently optimization methods come into existence. These methods are very fast to respond, require less computational time and doesn't require any initial guess also. Secondly the convergence rate is very fast for these algorithms. Namely these algorithm includes, Differential evolution (DE) algorithm (Aghdam et al., 2007), Genetic algorithm (GA) (Baskaran et al., 2012), Particle swarm optimisation (PSO) algorithm (Taghizadeh and Tarafdar, 22010). Dorigo et al first developed a metaheuristic algorithm called Ant colony optimisation algorithm for resolving the salesman's problem. This algorithm is a probabilistic approach for solving complex optimisation problems. The Ant Colony algorithm is generally used in many

---

[a]sdpatil79@gmail.com; [b]Akshaykadu001@gmail.com; [c]pratikdhabe5052@gmail.com

applications where solution to the constrained problems is a main concern. It is an optimisation algorithm used to solve the nonlinear equations with less complexity. In this algorithm, the ants travelled the distance followed by each other by laying down the substance known as pheromone. So in this way they choose the shortest path to reach upto the food source. Hence maximum amount of pheromone deposition path becomes the most preferred path for each individual ant. ACO algorithm proved its effectiveness over the other algorithms in many applications. It has better performance over GA, DE, and PSO (Sundareswaran et al., 2007). Also, all optimum solutions can be found out even if the population size is less.

A new metaheuristic algorithm is presented in this paper. It includes combination of two algorithms i.e. optimisation algorithm, ACO and Newton Raphson method. This new algorithm not only used to find the optimum solution for solving nonlinear transcendental equations but also used to reduce harmonics and THD. Also the fundamental component is preserved at the chosen level. It's very easy to find the global minimum solution with this new proposed algorithm. Computation time is also very less as compared to other algorithms.

She Problem Formulation

Selective harmonic equations

The inverter output is always a stepped output shown in Figure 152.1 for the levels it is designed for. It is observed that the waveform has only odd harmonics. If waveform has half wave symmetry then even order harmonics are absent. Equation (1) below shows the expression for output voltage waveform using Fourier series expansion.

$$v_0(t) = a_v + \sum_{n=1,3,\ldots}^{\infty} a_n \cos(n\omega_0 t) + b_n \sin(n\omega_0 t) \tag{1}$$

The count of harmonics which is to be reduced depends upon the number of levels of inverter and all the firing angles obtained should be less than 90° or $\pi/2$ radians. The expansion of Equation (1) for seven-level multilevel inverter will be the three nonlinear transcendental equations can be given as under. Where Equation (2) is the fundamental equation to be satisfied at chosen level. m_a indicates the modulation index. The Equation (3), Equation (4) and (5) represent the third, fifth, and seventh harmonic equations.

$$\cos(\alpha_1)+,\ldots\ldots\ldots\ldots\ldots,+\cos(\alpha_4) = 3 \times m_a \tag{2}$$

$$\cos(3\alpha_1)+,\ldots\ldots\ldots\ldots\ldots,+\cos(3\alpha_4) = 0 \tag{3}$$

$$\cos(5\alpha_1)+,\ldots\ldots\ldots\ldots\ldots,+\cos(5\alpha_4) = 0 \tag{4}$$

$$\cos(7\alpha_1)+,\ldots\ldots\ldots\ldots\ldots,+\cos(7\alpha_4) = 0 \tag{5}$$

The Total Harmonic Distortion is the ratio of summation of harmonic component's power to the power of the fundamental frequency represented by Equation (6)

$$THD = \frac{\sqrt{\sum_{n=3,5,7,\ldots\ldots}^{\infty} (V_n)^2}}{V_1} \tag{6}$$

For solving selective harmonic elimination problem firing angles need to be calculated offline through numerical methods. So many conventional methods implemented for this purpose. Here lower order harmonics are not completely removed but can be minimised to a great extent. So problem of minimising harmonics and keeping the fundamental component at chosen level is considered as a optimisation problem which requires a systematic approach.

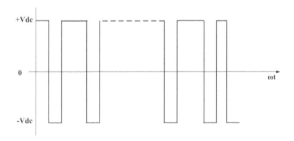

Figure 152.1 Inverter output voltage

Proposed Metaheuristic Algorithm

The objective function to be minimised is given by equation (7) with the constraints for firing angles

$$f = \min\left\{\left(100\frac{V_1^* - V_1}{V_1^*}\right)^4\right\} + \sum_{s=2}^{S}\frac{1}{h_s}\left(50\frac{V_h}{V_1}\right)^2 \tag{7}$$

By applying proposed metaheuristic algorithm desired solution can be obtained. The steps for proposed algorithm are as follows.

Stage 1: Construct objective function with constraints

Stage 2: Execute Ant Colony algorithm to get function minimum value.

Stage 3: Run Newton Raphson algorithm with output from ACO algorithm as a initial guess.

Stage 4: Print best value after algorithm converges.

Stage 5: Stop the programme.

In new proposed metaheuristic algorithm, the combination of two algorithms viz ACO along with N-R algorithm is presented. Hence only dependency of finding solution on iterative method is less. Also assumption of initial guess for N-R method is neglected. In new algorithm, output of ACO algorithm will be the input to N-R algorithm and treated as a initial guess. In ACO algorithm, ants follow the path by depositing a substance called as pheromone on the edges they travelled. The follower ants when travelled, follow the path with extreme deposition amount. Ultimately all the ants will choose the shortest path to reach the food source. The concept of this algorithm is implemented to obtain the optimized values for firing angles of inverter. Figure 152.2 indicates the flowchart for metaheuristic algorithm. The benefit of this algorithm is

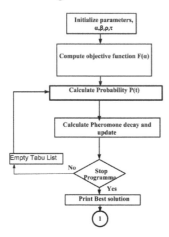

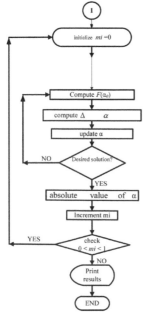

Figure 152.2 Flowchart of metaheuristic algorithm

that it converges very fast i.e. within five to six cycles it converges, when the population size is less. But the time for convergence is uncertain when the population size is more. Hence conventional iterative method is somewhat flexible to use for such condition. But assumption of initial guess close to exact solution is a major concern in Newton Raphson algorithm. So the combination of both the algorithms i.e. ACO algorithm and N-R method is proposed in this new metaheuristic algorithm. The beauty of this new algorithm is that it doesn't depend on the initial guess as output of ACO algorithm will be the input to N-R method and the algorithm converges very fast with less computational time. After sufficient iterations, when minimum value of objective function is attained, the programme terminated.

After calculating offline optimised switching angle values for firing the switches of 1-phase, 9-level inverter. Here cascaded H- Bridge inverter topology having equal DC input is used for nine-level inverter shown in the Figure 152.3. The phase voltage output can be measured through the load. Here only resistive load is used. The parameters considered for Ant Colony algorithm is shown in Table 152.1.

Different set of switching angles is obtained in accordance with modulation index for different levels of multilevel inverter. Only one set of switching angle is selected for firing the switches of inverter. This selection should be very accurate which will take care of reduction in the harmonics and obtaining the expected chosen fundamental component. Also THD should be very less. Different switching angles with different multilevel topologies is shown in Table 152.2.

Table 152.1 ACO parameters

Particulars	Amount
Population	100
Evaporation -ρ	0.85
Constant -α	1.00
Pheromone – τ	1.00
Iteration count	200

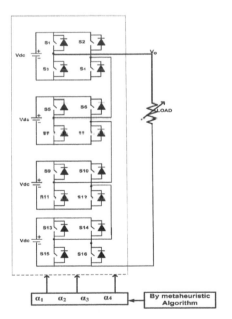

Figure 152.3 Implementation for first phase, nine-level MLI

Table 152.2 Switching angles

MLI	α_1	α_2	α_3	α_4	α_5
Five level	24.75°	57.22°	-----	----	----
Seven level	12.52°	38.81°	72.33°	----	----
Nine level	11.55°	23.10°	33.27°	42.93°	----
Eleven level	16.31°	25.48°	34.68°	57.85°	7.3°

Simulation Results

The new proposed metaheuristic algorithm is implemented for first phasenine-level inverter. The firing angles obtained offline through algorithm with respect to modulation index m = 0.8 are $\alpha_1 = 11.55°$ $\alpha_2 = 23.10°$, $\alpha_3 = 33.27°$, $\alpha_4 = 42.93°$. This set of firing angles is selected and found to be most suitable choice for among all obtained sets because it will minimise the harmonics as well as it will maintain the desired fundamental component. Also THD is found to be very less i.e.1.2% only as shown in simulated results.

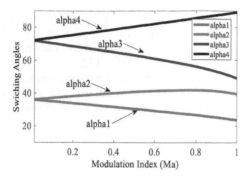

Figure 152.4 Firing angles for nine level MLI

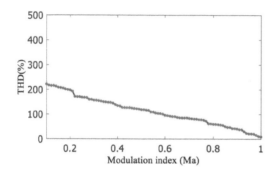

Figure 152.5 % THD

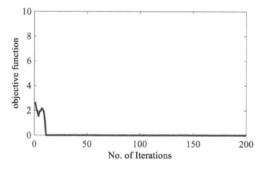

Figure 152.6 Objective function

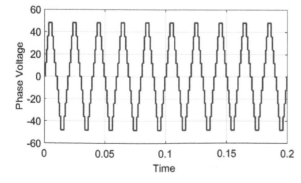

Figure 152.7 Phase voltage

Hardware Results

The complete analysis of 1-phase nine level MLI is done successfully with the help of selected switching angles. All the experimental results proved the usefulness of new proposed metaheuristic algorithm for reducing harmonics. Figure 152.9 shows the block diagram of test setup for experimental purpose. And Figure 152.10 shows the complete hardware setup. Here equal DC input of 12V is used for all H-Bridges of inverter. The fundamental frequency and switching frequency of the inverter is 50 Hz. Here only resistive load is used. Figure 152.11 shows the experimental results obtained for single phase output voltage and corresponding FFT analysis.

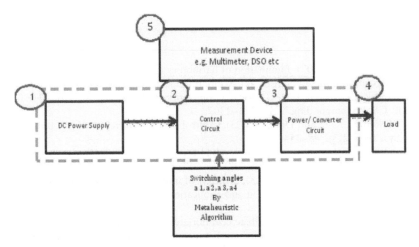

Figure 152.8 FFT analysis

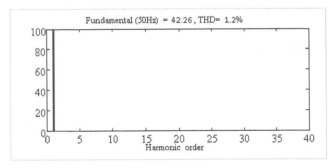

Figure 152.9 Block diagram of test setup

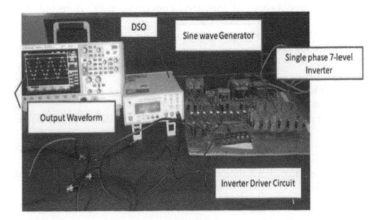

Figure 152.10 Hardware setup

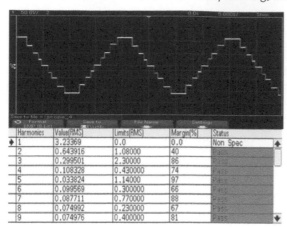

Harmonics	Value(RMS)	Limits(RMS)	Margin(%)	Status
1	3.23369	0.0	0.0	Non Spec
2	0.643916	1.08000	40	Pass
3	0.299501	2.30000	86	Pass
4	0.108328	0.430000	74	Pass
5	0.033824	1.14000	97	Pass
6	0.099569	0.300000	66	Pass
7	0.087711	0.770000	88	Pass
8	0.074992	0.230000	67	Pass
9	0.074976	0.400000	81	Pass

Figure 152.11 Experimental results

Conclusion

The new proposed algorithm is implemented successfully for calculation of offline firing angles for first phase nine-level cascaded multilevel inverter. All possible solution sets can be obtained through this algorithm. This proposed algorithm proved its feasibility over the other conventional algorithms in terms of its speed and accuracy and ensures its effectiveness. This algorithm can be effectively implemented for any level of MLI for obtaining the desired switching angles. Also the simulated results and experimental results show achievement of desired expected results.

References

Aghdam, M., Fathi, S., and Gharehpetian, G. (2007). Elimination of harmonics in a multi-level inverter with unequal dc sources using homotopy algorithm. Proceeding IEEE Int. Symp. Ind. Electron. 578–583.

Baskaran, J., Thamizharasan, S., and Rajtilak, R. (2012). GA based optimization and critical evaluation SHE methods for three level inverters. Int. J. Soft Comput. Eng. 2(3):2231–2307.

Chiasson, J. N., Tolbert L. M., McKenzie, K. J., and Du, Z. (2003). Control of a multilevel converter using resultant theory. IEEE Trans. Cont. Sys. Tech. 2(3):345–354.

Corzine, K. and Familiant, Y. (2002). A new cascaded multilevel H-bridge drive. IEEE Trans Power Electron. 17(1):125–131.

Dahidha, M. S. A., Konstantinou, G., and Agelidis, V. G. (2015). A review of multilevel selective harmonic elimination PWM: Formulations, solving algorithms, implementation and applications. IEEE Trans. Power. Electron. 30(8). 4091–4106.

Dorigo, M. Ant colony optimization web page. http://iridia.ulb.ac.be/mdorigo/ACO/ACO.html.

Edpuganti, A. and Rathore. A. (2015). A Survey of Low Switching Frequency Modulation Techniques for Medium-Voltage Multilevel converters. IEEE Trans. Ind. Electron. Soc. 5I(5):4212–4228.

Huang, J. and Corzine. K. A. (2006). Extended Operation of Flying Capacitor Multilevel Inverters. IEEE Trans Ind. Electron. 21(1):140–147.

Kouro, S., Malinowski, M., Gopakumar, K., Pou, J., Franquelo, L. G., Wu, B., Rodriguez, J., Perez, M. A., and Leon, J. I. (2010). Recent advances and industrial applications of multilevel converters. IEEE Trans.Ind. Electron. 57(8):2553–2580.

Kumar, J., Das, B., and Agarwal, P. (2008). Selective Harmonic Elimination Technique for a Multilevel Inverter. Fifteenth National Power Systems Conference (NPSC). IIT Bombay.

Rodriguez, J., Lai, J. S., and Peng, F. Z. (2002). Multilevel inverters: A Survey of topologies, controls , and applications. IEEE Trans Ind. Electron. 49(4):724–738.

Rodriguez, J., Bernet, S., Wu, B., Pontt, J. O., and Kouro, S. (2007). Multilevel voltage-source-converter topologies for industrial medium-voltage drives. IEEE Trans. Ind. Electron. 54(6):2930–2945.

Sirisukprasert, S. (1999). Optimized harmonic stepped-waveform for multilevel inverter. Phd diss., Virginia Polytechnic Institute and State University.

Sundareswaran, K., Jayant, K., and Shanavas, T. N. (2007). Inverter harmonic elimination through a colony of continuously exploring Ants. IEEE Trans. Ind. Electron. 54(5):2558–2565.

Taghizadeh, H. and Tarafdar Hagh, M. (2010). Harmonic elimination of cascade multilevel inverters with nonequal dc sources using particle swarm optimization. IEEE Trans Ind. Electron. 57(11):3678–3684.

153 Design and analysis of single-phase liquid immersion cooling (SLIC) for motor controller of a super-mileage vehicle

Vinayak Khatawate[a], Aditya Devkar[b], Aditya Jain[c], Bhavik Panchal[d], and Kshitij Jaiswal[e]

Dwarkadas J. Sanghvi College of Eng., Mumbai, India

Abstract

The traditionally used cooling methods like forced air cooling and indirect liquid cooling technologies are not very efficient for enclosed compact electronics and printed circuit boards (PCB). Single-phase liquid immersion cooling employs a dielectric liquid circulation mechanism where the coolant flows directly over the hot electronic components and uses a heat exchanger unit to cool the heated coolant. In this paper, the concept of Single-phase Liquid Immersion Cooling of motor controllers for a super-mileage electric vehicle has been analysed. The motor controller PCB is completely submerged in a dielectric coolant mitigating the risk of failure and effectively cooling them for better performance, thereby increasing their life. This research includes calculations, designs and simulations of the controller casing using CFD and thermal analyses.

Keywords: Electric vehicle, CFD, liquid immersion cooling, super-mileage vehicle.

Introduction

A great amount of energy is lost in the form of heat such as the brakes, motors, batteries, etc. Single-phase liquid immersion cooling system for the battery and motor controller minimises these losses. Traditionally, batteries and controllers have been cooled with forced air convection in Shell-Eco Marathon Prototype vehicles. Thermal management is not only important to improve the efficiency of these components but also to mitigate the risk of fire (Pajari, 1985).

Research done by Sundin and Sponholtz (2020) explained the use of Single-phase Liquid Immersion coling in battery thermal management system and a similar approach can also be used for other heating electronic components like the motor controller. Some researchers provided details concerning some thermal test methods, in several surveys and documents such as Trinamic (2013) and Texas Instruments (2012). Among these methods, thermal characterization of electronic devices has been addressed by Rogie et al. (2017) with the finite volume method by means of the Flowtherm code. It was found that the calculations of power generation by electronic package for various parameters is insufficient (Baïri, 2016).

Air has a very low heat capacity, so a large volume of air is required and the difference in temperature between the inlet and outlet air temperature is low. Another challenge is the surface of these electronic components needs to be exposed to air restricting the physical location of the components on the vehicle. This also requires filtering and conditioning of air to prevent corrosion (Bereke et al., 2013).

Single-phase Liquid Immersion Cooling involves entirely immersing electronic components in a dielectric heat transfer fluid that is cycled by a pump (Incropera, 2021). The flow causes the heat to be transferred by direct conduction between the coolant and the heating component. This is subsequently passed via heat transmission equipment like radiators or heat exchangers (Incropera, 2021). Because liquid coolants have a greater thermal conductivity and heat density than air, they function exceptionally well as a cooling medium (Sundin and Sponholtz, 2020). Direct immersion in liquid single-phase dielectric coolants (Single-phase liquid immersion cooling) would theoretically provide the optimum performance for maintaining the right temperature range for the controllers, with the least amount of temperature volatility, least expense and complexity for the system (Sundin and Sponholtz, 2020).

This paper shows a possible method to improve the efficiency of an electric vehicle. An attempt was made to do so by implementing a direct liquid cooling technology on the electrical components of the vehicle. With effective heat dissipation, the controller and the electronics can run safely reducing the risk

[a]Vinayak.Khatawate@djsce.ac.in; [b]adityadevkar91@gmail.com; [c]adityajain050300@gmail.com; [d]panchalbhavik155@gmail.com; [e]kshitijjaiswal1407@gmail.com

of overloading of these components. The entire system including the dielectric fluid, tubes, pumps and the radiator were carefully selected for an efficient overall system.

Cooling Load Determination

Cooling load required for the vehicle was determined. It is the amount of heat that needs to be dissipated by the cooling system.

The current battery used in the vehicle is a 48V 15Ah Li-ion battery. This gives us an energy of 0.72 kWh. Thus, the effective power of the system is 720W. Whereas, the motor being used in the vehicle has the power rating of 180W. The motor manufacturer mentioned the motor efficiency to be 90% due to transmission losses, we considered the effective efficiency of the transmission system to be 80% as a safety concern. The energy lost in heat would be close to 144 W. This loss is majorly due to the motor and the controller losses (Incropera, 2021). Since the hub motor rated efficiency was 90%, therefore cooling system was not required. Thus, the losses need to be subtracted for a 180W motor which comes close to 19W and thus the power lost to heat is 126 W (according to the Equation (2)) (Incropera, 2021).

$$\dot{Q} = (1 - \eta)\, P_{max} = (1 - 0.8)720W = 144W \tag{1}$$

$$Q = 144W - P_{motor}(1 - \eta) = 144W - 180W(1 - 0.9) = 126W \tag{2}$$

Here, Equation (1) is used to calculate the total energy that was lost in the form of heat. Equation (2) is used to calculate the total power lost due to heat by the system which is given by the difference of the total heat energy lost and the product of power rating of the motor and unity minus the motor rated efficiency.

To accurately verify this analysis, OptimumG's vehicle dynamics simulation software Optimum Lap was used which is an open source software. The important data like vehicle weight, motor specifications, tire data etc. were added, and a performance curve was obtained. The track details of Shell Bangalore, India, (Shell-Eco Marathon Track, 2022) (Figure 153.1) were used and obtained from Google Maps to accurately model the losses.

The torque-elapsed distance (Figure 153.2), speed-elapsed distance (Figure 153.3) and the time-current (Figure 153.4) graphs were plotted and studied for taking a sharp turn and for a long straight path. The motor current was determined for these instances. In the calculation it was assumed that voltage remains constant at 48 V according to the ideal conditions of the battery.

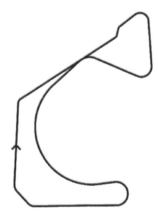

Figure 153.1 Shell Bangalore, India

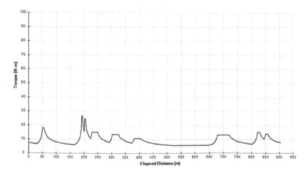

Figure 153.2 Torque-elapsed distance

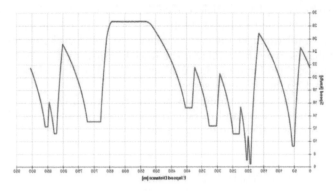

Figure 153.3 Speed-elapsed distance

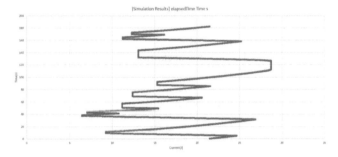

Figure 153.4 Time-current

$$P = VI \tag{3}$$

Using Equation (3) the current was mapped, and this was used to calculate the heat generated in the cells (Chen, 2013). Using this current the resistive heating in the battery was mapped.

$$H = I^2 R \ joules \tag{4}$$

$$Power = \frac{H}{time} W \tag{5}$$

From Equations (4) and (5), the resistive losses in the batteries were found to be 25.42 W.

The heat Generation in cells at different temperatures is tabulated in Table 153.1. The batteries perform efficiently when the temperature of its surrounding is close to 21°C (Chen, 2013). The lithium-ion batteries tend to heat up while charging and discharging at a higher C rating. This data is for a single cell and at 0.5°C discharge rate where initial battery temperature is 30°C. The battery in our vehicle contains 65 cells and this amounts to 143W. Thus, the coolant must be able to dissipate 143W of heat.

Design and Calculations

An airtight container was designed to fill the PCB of the motor controller with the dielectric coolant. To go ahead with the calculations, it was important to decide the fluid to be used as that would play a pivotal

Table 153.1 Heat Generation in cells at different temperatures (Chen, 2013)

Battery Temp	Discharge Rate				
	0.25C	*0.5C*	*1C*	*2C*	*3C*
−10°C	0–2.09W	0–5.29W	0–10.82W	0–24.71W	-
0°C	−0.33–2.05W	0–5.20W	0–10.21W	0–19.52W	0–29.93W
10°C	−0.24–1.43W	0–4.37W	0–8.87W	0–16.72W	0–24.79W
20°C	−0.44–0.87W	0–3.32W	0–4.92W	0–13.78W	0–4.92W
30°C	−0.46–0.85W	−0.33–1.86W	0–4.56W	0–10.39W	0–16.48W
40°C	−0.43–0.71W	−0.44–1.62W	0–3.70W	0–7.88W	0–14.21W

role in the design of the various factors (Saylor et al., 1988). Dielectric coolant (EC-110) from Engineered Fluids would suit the purpose (ElectroCool Dielectric Coolants). Table 153.2 mentions the properties of this fluid coolant which was further used in simulations. Figure 153.5 explains the entire cycle the coolant must go through. The coolant is stored in the reservoir, from there it is allowed to flow to the motor controller where the heat transfer takes place as shown in Figure 153.6. Warmer coolant travels to the radiator to be cooled and with the help of a pump back to the coolant reservoir.

A box was designed to fit the PCB in position and let the fluid flow around the heated PCB in a closed chamber. This was made airtight by using rubber gaskets and machined aluminium (Al6061). Two holes were drilled on the sides of this case for the coolant to flow and they were sealed with gaskets and adhesives to make it watertight.

Table 153.2 Characteristic of the coolant used (ElectroCool Dielectric Coolants)

ElectroCool Dielectric Coolants	
Characteristic of electrocool Dielectric Coolants	
Product ID	*EC-110*
Typical application	Outdoor and sealed system electronics cooling and insulation
appearance	clear
Fluid behaviour	Non-Compressible, Isotropic, Newtonian
Dielectric strength	>60kV
Resistivity(ohm-cm)	$>1 \times 10^{14}$
Dielectric constant	2.08
Refractive index	1.441
Pour point (°C)	−57
Flash point(°C)	193
ISO 4460 particle Cnt.	10/10/2012
Total sulphur(ppm)	0
Density. g/cc @16 °C	0.82
Coefficient of thermal expansion, volume/°C	0.00067
Kinematic 0°C	43.1
Viscosity 40°C	8.11
cSt 100°C	2.22
Thermal 0°C	43.1
Conductivity 40°C	0.1359
(W/m °C) 100°C	0.1325
Specific 0°C	2.0608
Heat 40°C	2.2121
(kJ/kg °C) 100°C	2.439
Global warming potential	0
Biodegradability	>95%
Materials compatibility warranty	Yes
Shelf life (Yrs)	25

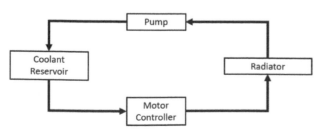

Figure 153.5 Coolant flow diagram

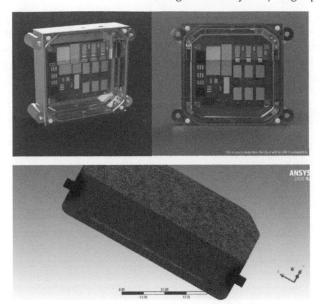

Figure 153.6 Rendered image of the PCB with coolant

It was recommended by the manufacturer to have a flow rate of 2 L/minute for 1 kW of power (ElectroCool Dielectric Coolants). Therefore, a flow rate of 0.252 L/minute of flow and similarly 0.286 L/minute for the battery is required to dissipate 126 W (from Equation).

$$Flow\,rate : 0.126 kW \times \frac{2L\,/\,min}{1 kW} = 0.252\,L\,/\,minute \tag{6}$$

A hand calculation was done to obtain preliminary results. A flat square copper plate was considered at 70°C. Equation (7) is the heat transfer equation for a flat plate. Equation (8) was required to calculate heat transfer coefficient. Equation (9) is the Nusselt number equation for the given problem. Equation (10) was used to calculate the thermal diffusivity. Equation (11) was used to calculate the fluid temperature T_f at the outlet.

$$q = hA(T_s - T_L) \tag{7}$$

$$h = \frac{N_u k}{L} \tag{8}$$

$$N_u = p_r^{0.33}(0 \cdot 664) R_e^{0.5} \tag{9}$$

$$T_d = \frac{k}{\rho C_p} \tag{10}$$

$$Q = \dot{m} C_P (T_f - T_i) \tag{11}$$

Plugging in the values gives the outlet temperature to be 44°C. The 3-D design was simulated on a CFD software to understand the working and obtain much more realistic values. The result obtained from the

CFD analysis was close to the hand calculations (45°C) as seen in Figures 153.7, 153.8 and 153.10. The fluid was not very comfortable in moving below the PCB and was getting accumulated as seen in Figures 153.9 and 153.10. Thus, to ensure better circulation the height of the PCB from the base was increased. In Figure 153.11 it can be observed that the PCB is cooled and the recirculation of the coolant from the sidewalls causes even better cooling due to turbulence.

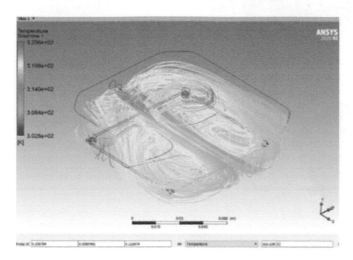

Figure 153.7 Temperature streamlines (ISO view)

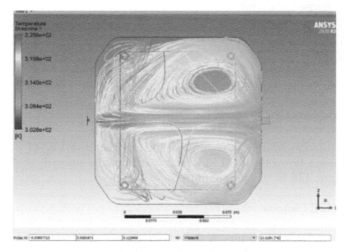

Figure 153.8 Temperature streamlines across the PCB (top view)

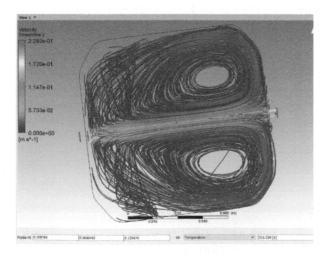

Figure 153.9 Velocity streamline of the fluid

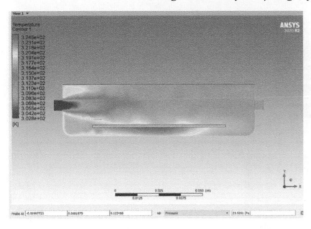

Figure 153.10 Temperature contour of the fluid

Radiator Design and Study

The heat absorbed by the fluid must be transferred to the surrounding. Initially, the fluid coolant was to be cooled by forced air convection by placing copper tubing on the rear side of the vehicle. The hand calculation shows that a large area is required to achieve desired results. To verify this, a simulation was done on ANSYS which gave similar results as seen in Figures 153.12 and 153.13. Copper tubes increased the weight drastically and hence it was decided that 120 mm standard radiator (Figure 153.14) will serve the required purpose. This will be placed on the body of the chassis at the rear end. When the vehicle is moving, the air around the body is at a lower pressure and thus will allow the air inside at higher pressure to escape and radiate heat efficiently.

For the outlet temperature to be 30°C when the inlet is at 45°C (ambient air temperature is 28°C), the volume flow rate of the radiator is 0.25 l/min and the heat to be dissipated is 126 W (from Equation 6).

For clean un-fined heat exchanger, the overall heat transfer coefficient can be determined as:

$$UA = \frac{1}{R_{total}} = \frac{1}{\dfrac{1}{h_i A_i} + R_{wall} + \dfrac{1}{h_0 A_0}} \tag{12}$$

$$Q = UA\Delta T_m \tag{13}$$

In Equations (12) and (13), U is the overall heat transfer coefficient and A is the area of the wall. Their product is the inverse of the total thermal resistance offered. The total thermal resistance is the sum of the resistance to convective heat transfer at both the inlet and the outlet, and the resistance offered by the wall thickness. Since the wall thickness is very small, the wall thermal resistance can be neglected.

Like the equations used in the previous calculation at the controller chamber; Reynold's number, Nusselt number and Prandtl number were calculated, and these were used to calculate convective heat transfer coefficients (h).

Using the Equations (12) and (13), the overall heat transfer coefficient U = *42.5 W/ m^2.K* and $\Delta T_m = 7°C$. Thus, area of the wall A = *0.40 m^2*.

Pump Calculations

As the volume flow rate suggested by the manufacturer was not very large the pressure drop observed was also minimal close to 50 Pa. A small 3W electrical pump with maximum flow rate of 1 L/minute was selected. This was to be operated at a lower current to obtain required flow.

Due to lower speeds and laminar flow the pressure drop is not very high. To maintain the suggested volume flow, rate a proper tubing was to be selected. An online calculator helped us map the required values as shown in Figure 153.15 and Figure 153.16 (Copely Developments Ltd, 2021). This suggested that for a hose length of 2m the quantity of flow would be 0.417L/minute (FOS=2). Also, an appropriate diameter for the same was 5.5 mm.

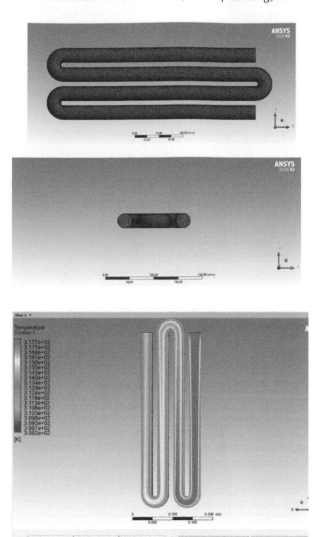

Figure 153.12 Radiator temperature gradient

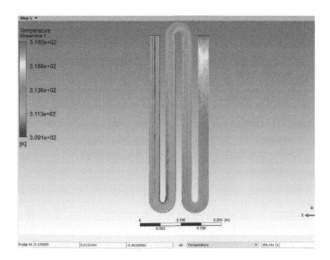

Figure 153.13 Radiator temperature streamline

Figure 153.14 120 mm radiator CAD

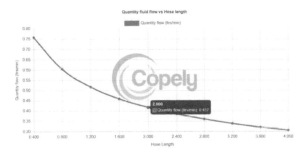

Figure 153.15 Quantity flow (lit/minute) – hose length (m) (Copely Developments Ltd, 2021)

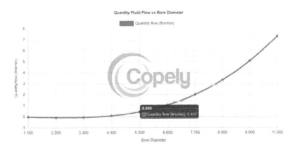

Figure 153.16 Quantity flow (lit/minute) – bore diameter (mm) (Copely Developments Ltd, 2021)

Conclusion

The following paper was written after an attempt to increase the efficiency of the motor controller while building an electric vehicle. The method aims to maximise efficiency for electric vehicles in order to aid the shift from IC engine vehicles. The limitation of this method is that the actual results also depend on the ambient temperature which has not been considered here. The above cooling system also reduces the chances of electrical failure. Single phase liquid immersion cooling can be easily implemented for commercial electric vehicles.

C_p = Specific heat

References

Baïri, A. (2016). Free convective overall heat transfer coefficient on inclined electronic assembly with active QFN16 package. Int. J. Numer. Methods Heat Fluid Flow. 26(5):1446–1459. doi: 10.1108/hff-04-2015-0142

Bereket, J., Hermann, G., Bazzi, A. 2013. Copper corrosion by atmospheric pollutants in the electronics industry. doi: https://doi.org/10.1155/2013/846405

Chen, K. 2013. Heat generation measurements of prismatic lithium-ion batteries. http://hdl.handle.net/10012/7936

Copely Developments Ltd (2021). Thurmaston Lane, Leicester, LE4 9HU. https://www.copely.com/tools/flow-rate-calculator/

ElectroCool Dielectric Coolants. Engineered for single-phase, liquid immersion cooling of electronics. https://www.engineeredfluids.com/electrocool

Incropera, F. 2021. Liquid Immersion Cooling of Electronic Components. In Fundamentals of Heat and Mass Transfer. Hoboken, N J: Wiley.

Pajari, J. (1985) Safety measures for prevention of pcb accidents. Environ. Health Perspect. 60:347350.

Rogie, E., Bissuel, B., Laraqi, V., Daniel, N., Kotelon, O., and Cécile, M. (2017). State of the art of thermal characterization of electronic components using computational fluid dynamic tools. Int. J. Numer. Methods Heat Fluid Flow. 27. doi: 10.1108/HFF-10-2016-0380.

Saylor, J. R., Lee, T. Y., Simon, T. W., Tong, W., Wu, P. S., and Bar-Cohen, A. (1988). Fluid Selection and Property Effects in Single-and Two-Phase Immersion Cooling. IEEE Transactions on Components, Hybrids, and Manufacturing Technology. 11(4):557565. https://doi.org/10.1109/33.16697

Shell-Eco Marathon Track. (2022). https://www.google.com/maps/place/Shell+Eco-Marathon+Track/@13.1558226,77.6997271,440m/data=!3m1!1e3!4m5!3m4!1s0x3bae1d68b7c18dcb:0x69610d2c4859c20a!8m2!3d13.1546081!4d77.6995581

Sundin, D. W. and Sponholtz, S. (2020). Thermal management of li-ion batteries with single-phase liquid immersion cooling. IEEE Open J. Vehicular Technol. 1:8292. doi: 10.1109/OJVT.2020.2972541

Texas Instruments, 2012. Semiconductor and IC package thermal metrics.

Trinamic Motion Control 2013. Trinamic application note 005, Rev. 1.01. www.trinamic.com

154 Seismic analysis of multi-storied buildings with plan irregularity in different seismic zones

Deenay Ambade[a] and Vaishali Mendhe[b]

Department of Civil Engineering, Yeshwantrao Chavan College of Engineering, Nagpur, India

Abstract

Seismic forces are unpredictable and irregular; so, seismic analysis is required to design earthquake resistance structures to assure safety against seismic forces. The major cause of building failure during an earthquake is irregularities in the buildings. The seismic analysis for the G+7 building is performed using STAAD. Pro software for the various seismic zones and the different plan shapes (i.e., Rectangular shape, C-shape, L-shape, and T-shape). The structures are designed to resist earthquake forces in accordance with IS 1893 (Part 1): 2016. The primary goal of this study is to examine the responses of regular and irregular buildings subjected to seismic load in different seismic zones, such as base shear, storey drift, and storey displacement.

Keywords: Regular and irregular, seismic analysis, seismic zones, STAAD.Pro, storey drift.

Introduction

General

In recent years, the infrastructure and construction industries have seen rapid growth. People are currently dealing with issues such as land shortages and high land prices. The population growth and the onset of the industrial revolution resulted in the movement of people from rural to urban regions, implying that the construction of multi-storey structures has become necessary. The multi-storied buildings are improperly designed for resisting lateral force. It might lead to the structure's total collapse (Kumar et al., 2014). The structural irregularity is seen in buildings because of the architectural and service necessity. Irregularities are inevitable in the construction of buildings. A detailed analysis of the structural behaviour of structures with irregularities is required for earthquake design (Shahare and Mohod, 2021; Cotipalli et al., 2021). In this, the structural behaviour of multistoried regular and irregular structures subjected to seismic load is studied for the different seismic zones. India's seismic zones are classified into four seismic zones according to IS 1893 (part 1): 2016 (i.e., zone 2, 3, 4, and 5). According to IS 1893 (part 1): 2016, zone 5 is expected to have the highest level of earthquake intensity, while zone 2 is expected to have the lowest level of earthquake intensity (Panchal and Dwivedi, 2017). This study deals with the comparative analysis of results obtained from the analysis of G+7 storey buildings situated in different seismic zones with plan irregularity using staad.pro software.

Background

Kumar et al. (2014) performed a case study on a multi-storey residential building's seismic analysis. The seismic study considers a G+15 storey structure in zone 2. The seismic behaviour of irregular buildings was explored (Shahare and Mohod, 2021). The goal of this study is to examine the behaviour of irregularities (plane and shape) on structures under seismic effects using the equivalent static method and response spectrum methods. Cotipalli et al. (2021) studied earthquake analysis of different structures in seismic zone 5. In seismic zone 5 studies lateral forces at all levels of building and in all soil types. Buildings of various heights with both regular and irregular plan shapes are analysed and designed in various seismic zones for the G+6 building. The variation in steel %, Max SF, Max BM, and Max deflection are all compared in each seismic zone. Lingeshwaran et al. (2021) performed a comparative investigation on symmetrical and asymmetrical constructions subjected to seismic stresses. Buildings with symmetric and asymmetric G+ 9 storeys are modelled using the etabs programme. The storey drift and displacement variations in the various structures are compared. Varma and Kumar (2021) etabs was used to study the dynamic analysis of G+20 residential buildings in zones 2 and 5. The effects of lateral loads on base shear and storey drift on the buildings are studied, and the results from zones 2 and 5 are compared.

---

[a]ambade.deenay3@gmail.com; [b]vaishalimendhe@gmail.com

Objectives

1. The main objective is to examine the behaviour of a multi-storied building due to seismic load.
2. To compare the responses of regular and irregular structures subjected to seismic load for different zones.
3. To calculate the storey drift and storey displacement at each storey of the multi-storied building for the different seismic zones.

Methodology

General

In this paper, an equivalent static analysis method with STAAD pro software is used to perform the seismic evaluation for the G+7 school building for different seismic zones with plan irregularity. According to the specifications of IS 875 (Part 1): 1987 and IS 875 (Part 2): 1987, the structure is subjected to dead and live load values. IS 1893(Part 1): 2016 is used to calculate the seismic load.

Modelling

Details of the structure

Table 154.1 Details of the structure

Parameters	Description
Height	29 m
Number of floors	G+7
Floor to floor height	3.625 m
Dimensions of column	0.35 m × 0.65 m
Dimensions of beam	0.23 m × 0.55 m
Slab thickness	0.15 m
Wall thickness	0.23 m
Density of concrete	25 KN/m^3
Density of brickwork	20 KN/m^3
Supports	Fixed

Seismic parameters

Table 154.2 Detail of the seismic parameters

Parameters	Description
Seismic zones	2, 3, 4, 5
Response reduction factor	5 (SMRF)
Importance factor	1.5
Type of soil	Medium
Damping ratio	5%
Earthquake direction	X and Z direction

Model 1 (Rectangular shape)

As shown in Figure 154.1, the G+7 building of a Rectangular shape was modelled using staad.pro.

Model 2 (C shape)

As shown in Figure 154.2, the G+7 building of a C shape was modelled using staad.pro.

Model 3 (T shape)

As shown in Figure 154.3, the G+7 building of a T shape was modelled using STAAD.pro.

Model 4 (L shape)

As shown in Figure 154.4, the G+7 building of a L shape was modelled using STAAD.pro.

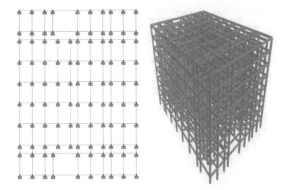

Figure 154.1 Shows a plan and 3-D representation of a rectangular-shaped building

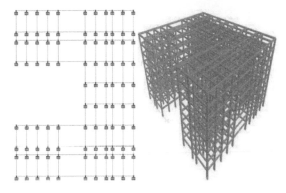

Figure 154.2 Shows a plan and 3-D representation of a C-shaped building

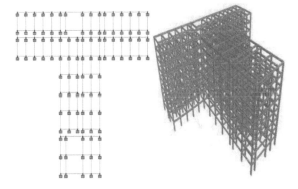

Figure 154.3 Shows a plan and 3-D representation of a T-shaped building

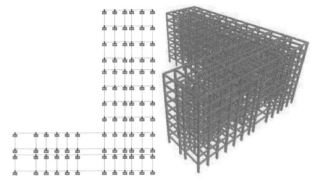

Figure 154.4 Shows a plan and 3-D representation of a L-shaped building

Results and Discussions

Results of base shear

From Table 154.3, base shear increases in zones 3, 4, and 5 when compared to zone 2, in X-direction and Z-direction.

Table 154.3 Base shear in KN

Shapes	Zone II	Zone III	Zone IV	Zone V
Rectangular shape	8372.83	13396.53	20094.79	30142.19
C shape	8050.94	12881.51	19322.66	28983.39
T shape	8140.91	13025.45	19538.17	29307.26
L shape	7949.15	12718.64	19077.96	28616.95

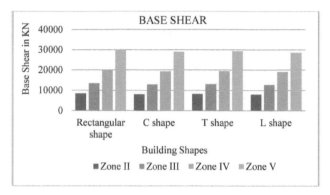

Figure 154.5 Base shear in KN

Results for storey displacement

Table 154.4 Storey displacement for zone 2 in mm

Storey Height (m)	Rectangular shape		C-shape		T-shape		L-shape	
	Directions		Directions		Directions		Directions	
	X	Z	X	Z	X	Z	X	Z
3.625	9.2	5.1	9.5	5.2	9.3	5.4	10.0	5.7
7.25	21.4	14.1	22.2	14.4	22.0	15.2	23.5	15.8
10.875	33.6	23.8	34.9	24.5	34.8	25.9	37.1	26.8
14.50	45.4	33.4	47.1	34.4	47.2	36.5	50.2	37.6
18.125	56.2	42.3	58.4	43.7	58.7	46.4	62.2	47.7
21.75	65.6	50.1	68.0	51.9	68.6	54.9	72.6	56.3
25.375	72.7	56.1	75.5	58.4	76.3	61.6	80.6	63.0
29.00	77.0	60.1	79.9	62.9	81.0	66.1	85.4	67.4

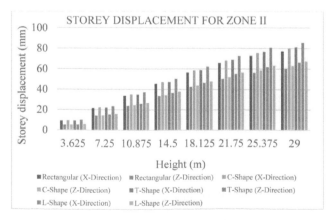

Figure 154.6 Storey displacement for zone 2 in mm

Table 154.5 Storey displacement for zone 3 in mm

Storey height (m)	Rectangular shape		C-shape		T-shape		L-shape	
	Directions		Directions		Directions		Directions	
	X	Z	X	Z	X	Z	X	Z
3.625	14.7	8.2	15.2	8.4	14.9	8.7	16.0	9.1
7.25	34.2	22.5	35.4	23.1	35.2	24.3	37.6	25.2
10.875	53.8	38.1	55.8	39.2	55.7	41.5	59.3	42.9
14.50	72.6	53.5	75.4	55.1	75.5	58.4	80.3	60.2
18.125	89.9	67.7	93.4	70.0	93.9	74.2	99.6	76.2
21.75	105	80.1	109	83.1	110	87.8	116	90.4
25.375	116	89.8	121	93.5	122	98.5	129	101
29.00	123	96.2	128	101	130	106	137	108

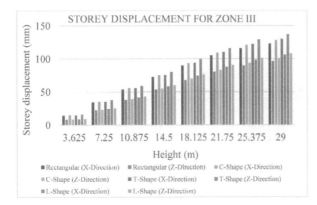

Figure 154.7 Storey displacement for zone 3 in mm

Table 154.6 Storey displacement for zone 4 in mm

Storey height (m)	Rectangular shape		C-shape		T-shape		L-shape	
	Directions		Directions		Directions		Directions	
	X	Z	X	Z	X	Z	X	Z
3.625	22.1	12.3	22.8	12.5	22.4	13.0	24.0	13.6
7.25	51.3	33.8	53.2	34.6	52.9	36.5	56.4	37.9
10.875	80.6	57.2	83.7	58.8	83.6	62.2	89.0	64.3
14.50	109	80.2	113	82.7	113	87.7	120	90.3
18.125	135	102	140	105	141	111	149	114
21.75	157	120	163	125	165	132	174	135
25.375	175	135	181	140	183	148	193	151
29.00	185	144	192	151	194	159	205	162

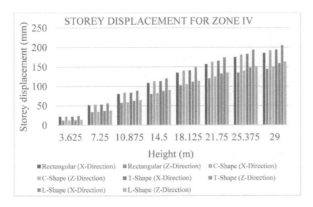

Figure 154.8 Storey displacement for zone 4 in mm

Table 154.7 Storey displacement for zone 5 in mm

Storey height (m)	Rectangular shape Directions		C-shape Directions		T-shape Directions		L-shape Directions	
	X	Z	X	Z	X	Z	X	Z
3.625	33.1	18.4	34.2	18.8	33.6	19.5	35.9	20.4
7.25	76.9	50.7	79.7	51.9	79.2	54.7	84.6	56.8
10.875	121	85.8	126	88.1	125	93.3	133	96.5
14.50	163	120	170	124	170	131	181	136
18.125	202	152	210	158	211	167	224	172
21.75	236	180	245	187	247	198	261	203
25.375	262	202	272	210	275	222	290	227
29.00	277	216	288	226	292	238	308	243

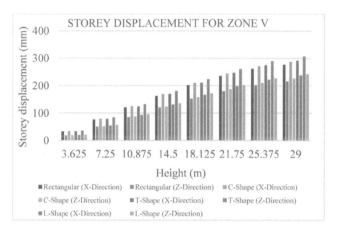

Figure 154.9 Storey displacement for zone 5 in mm

From the results of storey displacement, it is observed that storey displacement goes on increasing from ground floor to top floor in the X direction and Z direction. Table 154.4 to table 154.7 indicates the effects of storey displacement in both the X and Z directions for various plan shapes of building in various zones. Figures 154.6–154.9 show the same results. From these data, we may conclude that storey displacement is higher in L shape structures than in other buildings.

Results for storey drift

Table 154.8 Storey drift for zone 2 in mm

Storey height (m)	Rectangular shape Directions		C-shape Directions		T-shape Directions		L-shape Directions	
	X	Z	X	Z	X	Z	X	Z
3.625	9.2	5.1	9.5	5.2	9.3	5.4	10.0	5.7
7.25	12.2	9.0	12.6	9.2	12.7	9.8	13.5	10.1
10.875	12.2	9.7	12.7	10.1	12.8	10.7	13.6	11.0
14.50	11.8	9.6	12.2	10.0	12.4	10.6	13.1	10.8
18.125	10.8	8.9	11.3	9.3	11.5	9.8	12.1	10.0
21.75	9.3	7.7	9.7	8.2	9.9	8.5	10.4	8.6
25.375	7.2	6.1	7.4	6.5	7.7	6.7	8.0	6.7
29.00	4.3	4.0	4.4	4.5	4.7	4.5	4.8	4.4

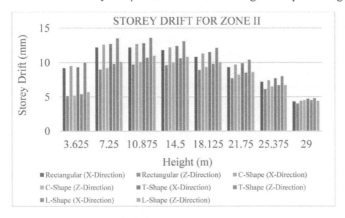

Figure 154.10 Storey drift for zone 2 in mm

Table 154.9 Storey drift for zone 3 in mm

Storey height (m)	Rectangular shape Directions		C-shape Directions		T-shape Directions		L-shape Directions	
	X	Z	X	Z	X	Z	X	Z
3.625	14.7	8.2	15.2	8.4	14.9	8.7	16.0	9.1
7.25	19.5	14.4	20.3	14.7	20.3	15.6	21.6	16.2
10.875	19.6	15.6	20.3	16.1	20.5	17.2	21.7	17.7
14.50	18.8	15.3	19.6	15.9	19.8	17.0	20.9	17.3
18.125	17.3	14.3	18.0	14.9	18.3	15.7	19.3	16.0
21.75	14.9	12.4	15.5	13.1	15.9	13.7	16.6	13.8
25.375	11.5	9.7	11.9	10.4	12.3	10.7	12.8	10.7
29.00	6.9	6.4	7.1	7.1	7.6	7.2	7.7	7.0

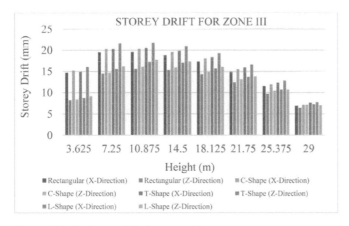

Figure 154.11 Storey drift for zone 3 in mm

Storey drift increased with floor height up to the third storey, reaching a maximum value at the third storey and then start to decrease up to the top storey. Tables 154.8–154.11 indicate the effects of storey drift in both the X and Z directions for various plan shapes of building in various zones. Figures 154.10–154.13 show the same results. From these data, we may conclude that storey drift is greater in L-shaped structures, but rectangular buildings have less storey drift than other types of buildings.

Table 154.9 Storey drift for zone 4 in mm

Storey height (m)	Rectangular shape Directions		C-shape Directions		T-shape Directions		L-shape Directions	
	X	Z	X	Z	X	Z	X	Z
3.625	22.1	12.3	22.8	12.5	22.4	13.0	24.0	13.6
7.25	29.2	21.5	30.4	22.1	30.4	23.5	32.4	24.3
10.875	29.3	23.4	30.5	24.1	30.7	25.8	32.6	26.5
14.50	28.2	23.0	29.3	23.9	29.7	25.4	31.4	26.0
18.125	26.0	21.4	27.0	22.4	27.5	23.6	29.0	24.0
21.75	22.4	18.6	23.3	19.6	23.8	20.5	25.0	20.7
25.375	17.2	14.5	17.8	15.6	18.5	16.0	19.2	16.1
29.00	10.3	9.7	10.6	10.7	11.3	10.8	11.5	10.6

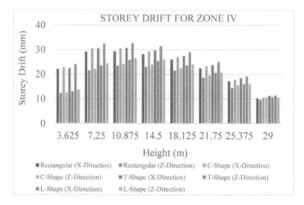

Figure 154.12 Storey drift for zone 4 in mm

Table 154.10 Storey drift for zone 5 in mm

Storey height (m)	Rectangular shape Directions		C-shape Directions		T-shape Directions		L-shape Directions	
	X	Z	X	Z	X	Z	X	Z
3.625	33.1	18.4	34.2	18.8	33.6	19.5	35.9	20.4
7.25	43.8	32.3	45.6	33.1	45.6	35.2	48.6	36.4
10.875	44.0	35.1	45.8	36.2	46.1	38.6	48.9	39.7
14.50	42.4	34.5	44.0	35.9	44.6	38.1	47.1	39.0
18.125	39.0	32.1	40.5	33.6	41.3	35.4	43.4	36.0
21.75	33.6	27.8	34.9	29.4	35.8	30.7	37.5	31.1
25.375	25.8	21.8	26.7	23.4	27.7	24.1	28.8	24.1
29.00	15.4	14.5	15.9	16.0	17.0	16.2	17.3	15.9

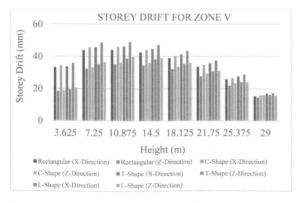

Figure 154.13 Storey drift for zone 5 in mm

Conclusions

Many studies have been conducted related to earthquakes, considering zones, shapes, vertical irregularities, and the type of soil. But no specific study has been conducted considering their combined effect. The purpose of this research is to understand the behaviour of a structure considering the combined effect of plan irregularities and seismic zones. Also, it is important because an earthquake leads to damage to property and the loss of lives. Hence, understanding the structural performance under seismic load is essential before construction and to make the building safe and stable. This study could be helpful for the future construction of buildings. In this, the analysis of multistorey buildings with plan irregularities subjected to earthquake load by using STAAD pro software was carried out. From the results, the following conclusions can be drawn:

Base shear depends on factors like soil conditions at the site, weight of the structure, etc. Base shear is minimum in zone 2 as compared to the other zones and maximum in zone 5 for all the plan shapes of the buildings. There is an increase in base shear by 37.5%, 58.4%, and 72.3% in zones 3, 4, and 5 respectively compared to zone 2 for all plan shapes of buildings.

The earthquake loads acting on a building cause some movement of the structure. The maximum storey drift for G+7-storey buildings occurs on the third floor. The storey displacement in the X direction is reduced by 10%, 6.4%, and 5.2% in the rectangular shape, the C shape, and the T shape, compared to the L shape. Similarly, in the Z direction, the reduction of storey displacement is 11%, 6.6%, and 1.9% in rectangular, C, and T shapes compared to the L shape.

According to the results, the storey drift and storey displacement are less for rectangular-shaped buildings than for C-shaped, T-shaped, and L-shaped buildings. In all seismic zones, L-shaped buildings are more likely to get damage than other buildings. All the shapes of buildings considered for analysis are safe in displacement and drift as the obtained values are less than the permissible limits.

In the highly seismically active zones, the effect of the earthquake is greater. As a result, the storey drift and displacement of the building are also higher. To reduce the effect of the earthquake in these zones, providing lateral load resisting systems such as shear walls, braced frames, and moment frames and also increasing the size of the column and beam can be useful to reduce the effect of the earthquake.

From this, we can conclude that buildings are more likely to get damaged in zone 5 and need to be designed properly. Also, it is observed that the regular building performs better than the irregular building.

References

Cotipalli, V., Varma, V. N. K., and Kumar, U. P. (2021). Earthquake analysis of regular and irregular structures for all the soil types in seismic zone V. Mater. Today: Proc. doi:10.1016/J.MATPR.2020.11.932

IS 1893 (Part 1): 2016. Criteria for Earthquake Resistant Design of Structures Part 1: General Provisional and Buildings. Bureau of Indian Standard.

IS 456: 2000. Indian Standard Criteria for Earthquake Resistant Design of Structures Part 1: General Provisions and Buildings. Fifth Revision 2002.

Kumar, E. P., Naresh, A., Nagajyothi, M., and Rajasekhar, M. (2014). Earthquake analysis of multi storied residential building - a case study. Int. J. Eng. Res. Appl. 4(11).

Lingeshwaran, N., Koushik, S., Reddy, T.M. K., and Preethi, P. (2021). Comparative analysis on asymmetrical and symmetrical structures subjected to seismic load. Mater. Toda: Proc. 45(7):6471–6475.

Panchal, A. and Dwivedi, R. (2017). Analysis and Design of G+6 Building in different seismic zones of India. Int. J. Innov. Res. Sci. Eng. Technol. 6(7).

Shahare, J. J. and Mohod, M. V. (2021). Seismic Response of Irregular Structures. Int. J. Adv. Res. Idea. Innov. Technol. 7(3):1765.

Varma, V. N. K. and Kumar, U. P. (2021). Seismic response on multi-storied building having shear walls with and without openings. Mater. Toda: Proc. 37.

155 Machine learning based credit card fraud detection

Shakti Kinger[a] and Varsha Powar[b]

School of Computer Engineering and Technology, Dr. Vishwananth Karad MIT-WPU, Paud Road, Kothrud, Pune, India

Abstract

The COVID-19 pandemic has brought in new challenges for businesses, including both offline and online businesses. The restrictions to people's movement to some level have made it impossible to buy products and services offline, resulting in sharp growth of reliance on online business. At the same time, online transactions have significantly increased credit cards fraud, impacting consumers' confidence in online commerce. As a result, there is a critical need to build the finest machine learning approach feasible to prevent practically all fraudulent credit card transactions. Our study is to recognise 100% of wrong fraud/scam transactions because credit card fraud is an average example in the grouping. So, in this paper, we have cantered pre-processing data set collections and trained them on Catboost and Random Forest.

Keywords: CatBoost, credit card fraud, fraud detection, machine learning, random forest.

Introduction

As the world moves toward a cashless future, online transactions will become increasingly important. Modern fraud does not necessitate the presence of the perpetrators in the crime scene. They can carry out their evil deeds in the privacy of their own houses, concealing their identity in a variety of ways. Identity concealment strategies include the usage of a VPN, diverting communication through the layers of the dark network (e. g. Tor), and so on, and tracing them back is difficult.

It is impossible to estimate the consequences of financial losses that are occurred due to online fraud. Once the card information is stolen, it can either be used by the fraudsters themselves or the information can be sold to others — for example in India, it's estimated that the card information of roughly 70 million people is sold in the black market (Dubey et al., 2020). The event occurred in the recent past, the mid-2000s when a group of fraudsters collaborated across multiple nations that resulted in the theft of over 32,000 credit cards details (Martin, 2022). This is thought to be the largest credit card scam in history. As a result of the lack of efficient security solutions, credit card companies and consumers have lost billions of dollars (Zhang et al., 2019).

The massive losses suffered by banks and other financial organisations sparked a surge in interest in studies aimed at preventing and mitigating fraud. The privacy constraints imposed by the corporations due to the severe competition in the area have restricted sharing of these processes with the general public. The other is to ensure that fraudsters are unable to enter the field. There was no standard dataset for the same reasons. Until 2015, when the deception was exposed by researchers dataset for detecting (Pozzolo and Bontempi, 2015).

The topic of detecting credit card fraud is seen to be one of the best for testing computational intelligence algorithms (Pozzolo et al., 2017). The highly unequal distribution of classes as a result of the small proportion of fraud transactions relative to the entire number of transactions (no more than 0.1%) is one of many concerns in this situation Pozzolo and Bontempi (2015). Another is the concept of drift, which may be explained mathematically as an altered distribution across time difficulty due to the tide relation with human progress over time. The major cause of imbalanced data is an unbalanced distribution (Daumé, 2012). It is the distribution from which the data is drawn that is unbalanced, not the data itself. The remainder of the paper is organised in the following manner. In Section 2, a quick summary of the existing methodologies usually employed to handle this problem, with a focus on machine learning. After explaining the dataset in detail in Sections 3 and 4, we will describe the metrics we used to compare it. The results of testing common classifiers (RF Shirodkar et al. (2020), and CatBoost Prokhorenkova et al. (2018)) using traditional measures will be presented, followed by a comparison of them customer designed saving measure in two scenarios viz. one with SMOTE Chawla et al. (2002) over-sampling and other without SMOTE. The comparison is done using two versions of SMOTE, viz with sensitive versions with and without using a manually designed Savings measure.

[a]shakti.kinger@mitwpu.edu.in; [b]varsha.powar@mitwpu.edu.in

Related Work

In Puh and Brkić (2019) this paper researchers, with professional consultation, 331 expert variables are established and 30 are chosen to reduce data dimensionality. The training set is used to build and fit several models, such as logistic regression and decision trees. Logistic Regression, Decision Tree, Naive Bayes, Random Forest, Boosted Tree, AdaBoost, Neural Network SVN, KNN, and Random Under Sampling and SMOTE were the algorithms compared. DT is one of the strategies. It is simple to construct, but each transaction must be checked individually (Delamaire et al., 2004). With an uneven European credit card fraud detection (ECCFD) dataset, Khatri et al. (2020) investigated various models. Meanwhile, in Mohammed et al. (2018), researchers investigated fraud detection using three classifiers: random forest, balanced Bagging Ensemble, and Gaussian Naive Bayes and concluded that balanced bagging ensemble produces better prediction results while the random forest is most adapted to large data sets. In Dhankhad et al. (2018), researchers proposed a classifier that combines multiple classifiers and compares it to RF and XG- boost using traditional measures. In Awoyemi et al. (2017), Sailusha et al. (2020), Varmedja et al. (2019) the binary classification of imbalanced credit card fraud has compared the performance of Naive Bayes, K-nearest neighbour, and Logistic regression models. Based on all of the evaluation measures, KNN has surpassed the competition. Another approach is to use LightGBM. Taha and Malebary (2020) used LightGBM to do their experiment on two datasets. The ECCFD dataset is the first, and the UCSD-FICO Data Mining Contest 2009 dataset is the second. Despite the fact that Bahnsen et al. (2013) [used BMR wrapping and classifiers like LR, RF, and DT, they did not use SMOTE to rebalance the data, instead opting for under- sampling. Dhankhad et al. (2018), Dornadula and Geetha (2019) Random forest trains the behavioural aspects of conventional and nonstandard transactions using random tree-based and CART-based approaches. While Random forest can get results even with a tiny dataset, its results are still face a challenge with unbalanced data. As part of future work, we would be focusing on rectifying these uneven datasets.

Proposed Work

We found that tree-based algorithms worked better in fraud detection, than other logistics and probabilistic algorithms. Among those we found that random forest and Catboost algorithms gave the best results but Catboost was not so widely used. We also found that the lack of a proper dataset was another issue faced by researchers and students.

The motivation of this project was to compare Catboost and random forest and to check the effect of artificial random over sampling to get better train models.

The proposed method follows a procedure outlined below in three steps.

Step 1: Get the dataset from the repository.

Step 2: Balance the dataset by deploying an appropriate sampling technique.

Step 3: Machine learning models development

The approach is depicted in the flowchart in the Figure 155.1

Simulation environment

We simulate the environment using components as follows:

1) *Software:* For our work, we have used a 64-bit Windows 10 laptop, Jupyter Notebook running with Python.

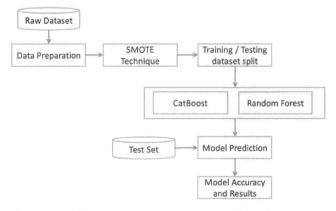

Figure 155.1 System architecture

The python libraries used in the environment are Numpy, Pandas, Scikit-Learn for data analysis, for data visualisation we have used plotting libraries Matplotlib and Seaborn, and libraries for machine learning classifiers.

2) *Hardware:* The hardware used in the environment consists of a CPU with an Intel(R) Xeon(R) processor @ 2.20 GHz and 7.00 GB of RAM.

Dataset

The dataset used in our work is taken from the open source data science community, Kaggle Pozzolo et al. (2015b). The dataset covers real world transactions done by European cardholders during two days in September 2013, as released by Universite Libre de Bruxelles (ULB). Out of 284,807 transactions, there are 492 cases of fraud using 31 different features that included 'time', 'amount', and 'class'. This dataset, which can be accessible in the associated work section, is extensively utilised by many researchers and practitioners; thus, it was chosen for our study to do a comparative evaluation of some of the metric values of our proposed model with other work already done in this field.

A closer analysis of the dataset reveals that it's skewed heavily. The reason for this is that the dataset favours the genuine class. We can see this because just 492 transactions out of 284315 are not authentic. When compared to the total number of transactions, there are just 0.172% fraudulent transactions Figure 155.2.

From the dataset beloare are the features used:

- Time delta between transactions and the first transaction in the collection.
- Amount refers to the total amount of the transaction (used for example-dependent cost-sensitive learning).
- The output variable is the class with value '1' resembling to fraud transaction and '0' to a normal transaction.

We also searched for other datasets that provide essential information and do not hide it but the datasets available were not up to the mark or were artificial which really didn't matter but the original one was better.

Data pre-processing: The dataset didn't have any missing values and all features except class were numerical float64 and V1- V28 PCA transformed. However, as can be seen from Figure 155.3 the distribution of raw data across the entire sample shows that except for a few transactions, the majority of the rest of the transactions are of lower value (<5000). The same information (data is right skewed) is seen from a histogram shown in Figure 155.4. Further, we logarithmically scaled the data to normalise this skew — Figures 150.5 and 150.6 show data and histogram after logarithmic scaling.

We also scaled time to 0–1 linearly, although, it didn't really make much difference as tree-based algorithms are not sensitive towards data variance. We also plotted genuine vs time and fraud vs time graphs to look for time-based trends Figure 155.7 and correlation matrix to understand the dependence of unknown features towards the target class Figures 150.8 and 150.9.

Algorithms

1) *Random forest (RF):* Random forest is an ensemble model and a supervised learning technique. The problem in Decision Trees has been overcome by the Random Forest method that's made up of a series of decision trees (Shalev-Shwartz and Ben-David, 2014). The method creates a 'forest' ouonsemble of decision trees, where each tree functions as a weak learner and further uses the 'bagging' method for combining these decision trees to form a robust learner (Shirodkar et al., 2020). Figure 155.10

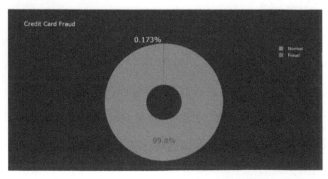

Figure 155.2 Fraud distribution

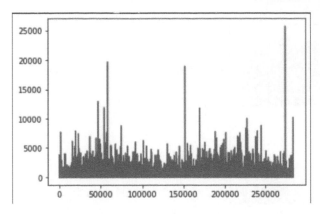

Figure 155.3 Amount before logarithmic scaling

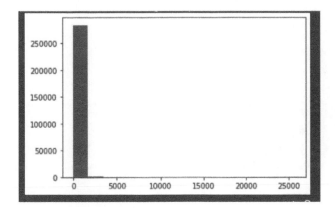

Figure 155.4 Histogram representation of amount before logarithmic scaling

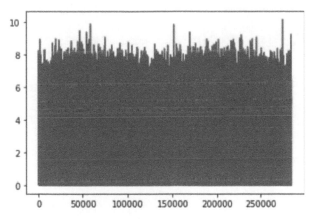

Figure 155.5 Amount after logarithmic scaling

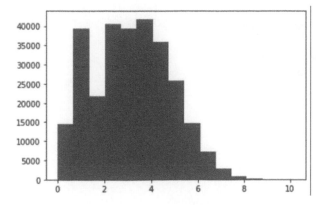

Figure 155.6 Histogram representation of amount after logarithmic scaling

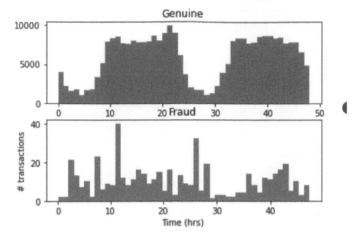

Figure 155.7 Transactions vs time(hours)

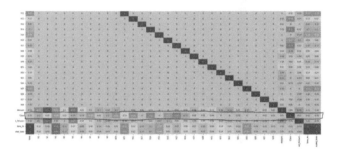

Figure 155.8 Correlation matrix

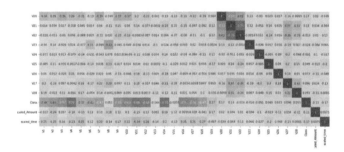

Figure 155.9 Randomised under sampled dataset with equal fraud and genuine transactions

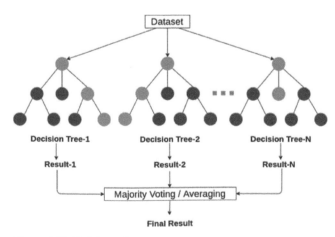

Figure 155.10 RF working

depicts this method with multiple decision trees with averaging the results across these decision trees. The primary objective of the bagging method is to combine several learning models to improve the final output. The random forest prediction is determined by a majority vote over the individual trees' projections. To create a random forest, we must first define the method that each tree employs, as well as the distribution of its independent and identically distributed random variables. Random forest are effective and faster when dealing with the imbalanced dataset that is skewed even with large number of features (Seeja and Zareapoor, 2014).

2) *CatBoost:* The word 'CatBoost' is derived from two disagreements: 'Category' and 'Boosting.' The word "boost" comes from a gradient boost. Gradient boosting library functions are useful in machine learning. CatBoost is a simple- to-implement machine learning method with a lot of power. It is widely used in a variety of professional tasks such as fraud detection, recommendation items, and forecasting, and it performs admirably. It produces excellent results and works quickly on a small amount of data. It can return an outstanding result with relatively fewer data. Unlike other machine learning algorithms that only perform well after learning from extensive data.

Prokhorenkova et al. (2018) provided a new gradient boosting toolbox, and this technique competes in terms of quality with other current boosting implementations. According to studies, CatBoost makes two advancements:

• The use of ordered boosting as an alternative in comparison to the traditional algorithm
• Categorical feature processing algorithm.

These two breakthroughs will eliminate target leakage in implementations of gradient boosting methods that are currently available.

Figure 155.11 shows single tree development stages for CatBoost

1) Using some measures split calculation is done
2) Transformation of categorical/text features to numerical features (optional)
3) Bootstrap options are used for selecting the tree structure.
4) Finally leave values are calculated.

Resampling technique

To address the imbalance class issue in a dataset, resampling approaches are widely utilised (Scikit-Learn-Contrib, 2022). The overall number of genuine instances in the dataset used here is 284,315, with 492 records for fraudulent transactions. The data set is undeniably imbalanced with 99.827% valid data and merely 0.173% fraudulent data. As the performance of the algorithms is linked to the imbalance class issue, resampling strategies come into action (He and Garcia, 2009). Undersampling, oversampling, and a combination of both undersampling and oversampling are the three basic types of resampling techniques.

1) *Undersampling:* Often there is a need to compact a dataset and make it balanced. Techniques like undersampling are deployed to achieve this objective. Undersampling adds an advantage of lowering the learning phase's cost (Pozzolo et al., 2015a). One of the drawbacks of undersampling strategies is that they remove a big portion of the training set, predominantly in the situations where a vast majority of the class instances are relatively huge, resulting in the loss of significant examples, which can make classification and prediction difficult.

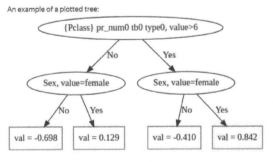

Figure 155.11 CatBoost tree dig

2) *Oversampling:* In contrast to undersampling, oversampling strategies try to conserve the majority class instances while replicating the minority class instances thus balancing an uneven training set. The problem with this method is that it can lead to poor model performance in certain circumstances where generating minority data in the training set is a challenge (García et al., 2008; Cieslak and Chawla, 2008).

3) *SMOTE:* 'The Synthetic Minority Over-sampling Technique' Chawla et al. (2002) is an algorithm that 'over-samples the minority class by taking each minority class sample and injecting synthetic examples along the line segments joining any/all of the 'k' minority class nearest neighbours.' The objective is to interpolate across samples of the same class to create new minority examples. This results in clusters forming around each minority observation. The classifier creates larger decision areas with surrounding instances from the minority class by constructing synthetic observations. We have used SMOTE technique to balance the dataset for fraud detection Figure 155.12.

Evaluation metrics

For evaluating the two modules we have used following metrics accuracy, precision, recall, F1-sdcore and receiver operating characteristic curve (AUC). In one way or another, all of the evaluation criteria employed in the proposed technique are dependent on a confusion matrix Hanley and Mcneil (1982) A confusion matrix, also known as an error matrix Stehman (1997), is a common method for evaluating the performance of machine learning algorithm. A confusion matrix yields four important results values viz. true positive (TP), true negative (TN), false positive (FP), and false negative (FN) (Fawcett, 2006; Chicco et al., 2021).

One method for assessing how often the classifier properly identifies a data point is to look at the accuracy, also known as the error rate (Deepai, 2022). Equation (1) Guido and Müller (2021) depicts accuracy as the proportion of correctly classified instances (TP (fraud) and TN (non-fraud)) to the total number of instances.

$$Accuracy = TP + T\ N/TP + FN + T\ N + FP \tag{1}$$

Precision and recall are two evaluation criteria that produce different results in different ways. Precision and recall are frequently put to the test. When Precision increases, Recall decreases, and when Precision decreases, Recall increases (A. I. C3, 2022). As illustrated in Equation (2), precision, also known as the positive predictive value, assesses the right prediction of positive cases out of a total of positive cases:

$$Precison = Positive\ predicted\ value = TP\ /TP + FP \tag{2}$$

Another important evaluation criteria that are utilised in detecting fraudulent credit card transactions are recall (also known as true positive rate, TPR) and sensitivity (Masís 2021). Its value is based on its capacity to detect affirmative cases. The greater the recall number, the more likely it is that fraudulent conduct will be detected. As a result, it's critical to get the highest recall value to prevent missing any occurrences of fraud. Equation (3) expresses recall as follows:

$$Recall = Sensitivity = True\ positive\ total\ positives = TP\ /TP + FN \tag{3}$$

Precision is essential, but it pales in comparison to recall. Precision is concerned about all anticipated positive cases' real TP cases. The number of FP cases is considered to be reasonable and the model is still considered to be better performant, as long as there is no impact on the FN cases. As a result, in credit card fraud detection, Recall is more crucial than Precision. F1-Score, on the other hand, evaluates the model's performance by combining the Precision and Recall values. The F1-Score is evaluated among other assessment

Figure 155.12 Dataset balancing after applying SMOTE

metrics when comparing two or more models Prusti and Rath (2019), hence the classifier with the highest F1-Score must be picked, as mentioned in (Daumé, 2012). F1-Score is expressed as Equation (4):

$$F1 = 2 \times Precision \times Recall /Precision + Recall \qquad (4)$$

Results and Discussion

Various standards for calculating correlation have been used to determine which algorithm is best for recognizing fraud transactions. The most often used criteria for determining the outcomes of machine learning algorithms are accuracy, recall, and precision. To determine the completeness of the referenced measurements, a contingency table can be employed. These measures were used to evaluate the model's effectiveness. On SMOTE data, the models were evaluated, the predicted outcome was verified, and the ROC Curve/ the AUC values were analysed.

The results from the two methods viz. random forest and Catboost, used in this study are summarised in Figure 155.13, Classification report of RF Figure 155.14 and CatBoost Figure 155.16 are showing different metrics for both algorithms. From the confusion matrix for RF Figure 155.15 and CatBoost Figure 155.17

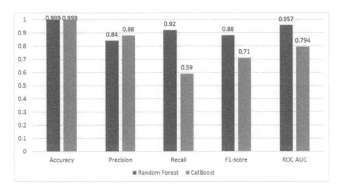

Figure 155.13 Evaluation metrics

```
              precision    recall  f1-score   support

           0       1.00      1.00      1.00     71095
           1       0.84      0.92      0.88       107

    accuracy                           1.00     71202
   macro avg       0.92      0.96      0.94     71202
weighted avg       1.00      1.00      1.00     71202

Accuracy: 0.9996207971686188
ROC AUC Score for Random Forest Classifier: 0.9578173340528303
```

Figure 155.14 Classification report for RF

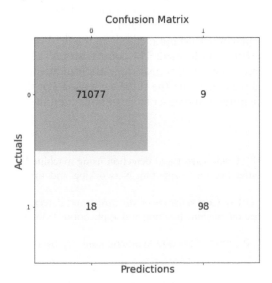

Figure 155.15 Confusion matrix RF

```
            precision   recall  f1-score   support

        0       1.00     1.00      1.00      71029
        1       0.88     0.59      0.71        173

  accuracy                         1.00      71202
 macro avg       0.94     0.79      0.85      71202
weighted avg      1.00     1.00      1.00      71202
```

```
Accuracy: 0.998806213308615
ROC AUC Score for Cat Boost Classifier: 0.7946991365653221
```

Figure 155.16 Classification report for CatBoost

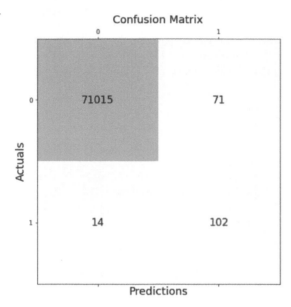

Figure 155.17 Confusion matrix CatBoost

it can be seen that the true positive and true negatives are better in the case of RF and hence its accuracy is better than compared to CatBoost. False positive is less in case of CatBoost and hence its having a better precision. Whereas False negatives are less in case of RF and hence its Recall is better than CatBoost.

Conclusion

Credit card fraud is a big source of concern for businesses. Companies and individuals have incurred significant financial losses due to these scams. As a result, companies are constantly investing in the development of innovative approaches and ways to aid in the detection and prevention of fraud.

The major goal of this study was to examine different machine learning methods in the field of fraudulent transactions detection. As a result, there was a differentiation formed. The comparison revealed that the Catboost algorithm produces the best results, i.e. correctly detects if transactions are fraudulent or not. To determine this, usage of accuracy, recall, precision, and the area under the curve were used. For this type of challenge, a high recall value is essential. Selection of features and dataset have an impact on the results.

References

Awoyemi, J. O., Adetunmbi, A. O., and Oluwadare, S. A. (2017). Credit card fraud detection using machine learning techniques: A comparative analysis. In International Conference on Computing Networking and Informatics (ICCNI). https://doi.org/10.1109/iccni.2017.8123782

Bahnsen, A. C., Stojanovic, A., Aouada, D., and Ottersten, B. (2013). Cost sensitive credit card fraud detection using Bayes minimum risk. In 2013 12th international conference on machine learning and application. 1:333–338.

C3.ai (2022). https://c3.ai/glossary/machine-learning/precision/

Chawla, N. V., Bowyer, K. W., Hall, L. O., and Kegelmeyer, W. P. (2002). SMOTE: Synthetic minority over-sampling technique. J. Artif. Intell. 16:321–357.

Chicco, D., Tötsch, N., and Jurman G. (2021). The Matthews correlation coefficient (MCC) is more reliable than balanced accuracy, bookmaker informedness, and markedness in two-class confusion matrix evaluation. BioData Mining. 14(13).

Cieslak, D. A. and Chawla, N. V. (2008) Start globally, optimise locally, predict globally: Improving performance on imbalanced data. In Proceedings of the 2008 Eighth IEEE International Conference on Data mining. 143–152.

Daumé, I. (2012). A course in machine learning. CIML Inf. 5–73.

DeepAI, (2022). https://deepai.org/machine-learning-glossary-and-terms/accuracy-error-rate.

Delamaire, L., Abdou, H., and Pointon, J. (2004). 61–61.

Dhankhad, S., Mohammed, E., and Far, B. (2018). Supervised machine learning algorithms for credit card fraudulent transaction detection: A comparative study. In 2018 IEEE international conference on information reuse and integration (IRI). 122–125.

Dornadula, V. N. and Geetha, S. (2019). Credit card fraud detection using machine learning algorithms. Procedia Comput. Sci. 165.

Dubey, S. C., Mundhe, K. S., and Kadam, A. A. (2020). Credit card fraud detection using artificial neural network and backpropagation. In Proceedings of the 2020 4th international conference on intelligent computing and control systems (ICICCS). 268–273).

Fawcett, T. (2006). An introduction to ROC analysis. Pattern Recognit. Lett. 27.

García, V., Mollineda, R. A., and Sánchez, J. S. (2008). On the k-NN performance in a challenging scenario of imbalance and overlapping. Pattern Anal. Appl. 11269–11280.

Guido, S. and Müller, A. C. (2021). Introduction to machine learning with python a guide for data scientists. Sebastopol, CA: O, Reilly.

Hanley, J. A. and Mcneil, B. J. (1982). The meaning and use of the area under a receiver operating characteristic (ROC) curve. Radiology. 143.

He, H. and Garcia, E. A. (2009). Learning from imbalanced data. IEEE Trans. Knowl. Data Eng. 21:1263–1284.

Khatri, S., Arora, A., and Agrawal, A. P. (2020). Supervised machine learning algorithms for credit card fraud detection: A comparison. In Proceedings of the 2020 10th International Conference on Cloud Computing. 680–683.

Martin, T. (2022). https://www.uswitch.com/credit-cards/guides/credit-card-fraud-the-biggest-card-frauds-in-history/

Masís, S. (2021). Interpretable machine learning with python: learn to build interpretable high-performance models with hands-on real-world xamples. Birmingham, UK: Packt Publishing Ltd.

Mohammed, R. A., Wong, K. W., Shiratuddin, M. F., and Wang, X. (2018). Scalable machine learning techniques for highly imbalanced credit card fraud detection: a comparative study. In Pacifc rim international conference on artifcial intelligence, pp. 237–246.

Pozzolo, A. D. and Bontempi, G. (2015). Adaptive machine learning for credit card fraud detection. PhD diss., Université Libre de Bruxelles.

Pozzolo, A. D., Boracchi, G., Caelen, O., Alippi, C. and Bontempi, G. (2017). Credit card fraud detection: a realistic modeling and a novel learning strategy. IEEE Trans. Neural Netw. Learn. Syst. 29(8):3784–3797.

Pozzolo, A. D., Caelen, O., and Bontempi, G. (2015a). When is undersampling effective in unbalanced classification tasks. In Proceedings of the Joint European Conference on Machine Learning and Knowledge Discovery in Databases. 200–215.

Pozzolo, A. D., Caelen, O., Johnson, R. A., and Bontempi, G. (2015b). Calibrating probability with undersampling for unbalanced classification. In Proceedings of the 2015 IEEE Symposium Series on Computational Intelligence. 159–166.

Prokhorenkova, L., Gusev, G., Vorobev, A., Dorogush, A. V., and Gulin, A. (2018). 6638–6648.

Prusti, D. and Rath, S. K. (2019). Fraudulent transaction detection in credit card by applying ensemble machine learning techniques. In Proceedings of the 2019 10th international conference on computing, communication and networking technologies (ICCCNT). 1–6.

Puh, M. and Brkić, L. (2019). Detecting credit card fraud using selected machine learning algorithms. In Proceedings of the 2019 42nd international convention on information and communication technology, electronics and microelectronics (MIPRO). 1250–1255.

Sailusha, R., Gnaneswar, V., Ramesh, R., and Rao, G. R. (2020). Credit card fraud detection using machine learning. In 4th International Conference on Intelligent Computing and Control Systems. doi:10.1109/ICICCS48265.2020.9121114

Scikit-Learn-Contrib (2022). imbalanced-learn 0.9.1. 22–22.

Seeja, K. and Zareapoor, M. (2014). Fraudminer: A novel credit card fraud detection model based on frequent itemset mining. Sci. World J. 1–10.

Shalev-Shwartz, S. and Ben-David, S. (2014). From theory to algorithms. Understanding machine learning. Cambridge: Cambridge University Press.

Shirodkar, N., Mandrekar, P., Mandrekar, R. S., Sakhalkar, R., Kumar, K. C., and Aswale, S. (2020). Credit card fraud detection techniques-A survey. In Proceedings of the 2020 International Conference on Emerging Trends in Information Technology and Engineering. 1–7.

Stehman, S. V. (1997). Selecting and interpreting measures of thematic classification accuracy. Pattern Recognit. Lett. 62.

Taha, A. A. and Malebary, S. J. (2020). An intelligent approach to credit card fraud detection using an optimised light gradient boosting machine. IEEE Access. 8:25579–25587.

Varmedja, D., Karanovic, M., Sladojevic, S., Arsenovic, M., and Anderla, A. (2019). Credit card fraud detection - machine learning methods. In 18th International Symposium INFOTEH-JAHORINA (INFOTEH).

Zhang, X., Han, Y., Xu, W., and Wang, Q. (2019). HOBA: A novel feature engineering methodology for credit card fraud detection with a deep learning architecture. Inf. Sci. 557:302–316.

156 Characterisation and electrochemical corrosion behaviour of electrodeposited Ni-Al$_2$O$_3$ nanocomposite coatings on steel

Ziouche Aicha[1,a], Mokhtari Majda[2,b], Amirouche Hammouda[3], Baraa Hafez[4,c], Hicham Elmsellem[4,d], Zoubiri Nabila[1,e], and Allou Djillali[1,f]

[1]Research Centre in Industrial Technologies CRTI, Cheraga, Algiers, Algeria

[2]Laboratory of Valorisation and Technology of Sahara Resources (VTRS), Univeristy of EChahid Hamma / Kakhder, 39000 Eloued Algeria 2'University larbi Ben M'hidi, Oum ElBouaghi, Algeria

[3]Department of Pharmaceutical Sciences, College of Pharmacy, and Health Sciences, Ajman University, Ajman, UAE

[4]Laboratory of Analytical Chemistry Materials and Environment (LC2AME) Faculty of Sciences University of Mohammed Premier, Oujda, Morocco

Abstract

In the present work, a Ni-Al$_2$O$_3$ nanocomposite coating was successfully elaborated by electrodeposition technique using direct current. The alumina nanoparticles were incorporated in nickel matrix to create the Ni-Al$_2$O$_3$ nano-coating. The effect of the alumina amount variation from 520 g/l in the Ni nanocomposite coating on the microstructure, mechanical and corrosion behaviour has been evaluated. The analysis of X-rays diffraction showed that the deposed layers of aluminium oxide is α-Al$_2$O$_3$ type with the cell parameters D (A°) = 1.1472. It was reported that the hardness and corrosion behaviour of the Ni- Al$_2$O$_3$nanocomposite coating is influenced by the amount of Al$_2$O$_3$nanoparticles. Nano indentation hardness value, thickness measurement and corrosion potential values were improved with the amount of Al$_2$O$_3$ nanoparticles in the composite coating.

Keywords: Al$_2$O$_3$nanoparticles, electrodeposition coating, nickel matrix, corrosion resistance.

Introduction

Metal and ceramic composite deposition coatings have been widely explored in recent years for a different application, including corrosion protection coatings (Owczarek and Adamczyk, 2016).

Electrodeposition process is one of the most common methods for creating metallic coatings with superior corrosion resistance. Incorporating inert insoluble micro or nanoparticles like TiO$_2$, Al$_2$O$_3$, or ZrO$_2$ into metal matrixes is an effective way to increase anti-corrosion protection (Nazarov and Thierry, 2007; Lekbir et al., 2201).

Electrodeposition coating of metals and alloys is an important procedure for obtaining a good solution for metal corrosion phenomenon that is also a low-cost process. Chemical and electrochemical reactions are at the heart of electrodeposition coating, which is used to deposit metals, alloys, and metal matrix composite materials (Owczarek and Adamczyk, 2016).

Electrodeposition process is based to the phenomena of the reduction of metal ions with an impressed current. In contrast, conversion method, sol gel coating and electroless processes do not need an imposed electric current.

The physical properties of nanomaterials it is so different from bulk materials having the same composition because the nonmetric nature of the structure imposed.

In the case the of incorporation of metallic oxides in a metal matrix by electrodeposition process, the sizes of the particles take values from nano-meter to micro-meter.

However, the electrodeposition process presents a good way to make a coating or thin film with a crystalline structure (Nazarov and Thierry, 2007; Lekbir et al., 2201; Ziouche et al., 2021).

The composition of the coating and structure properties are optimized by electrodeposition parameters such as current density, particles nature, pH, temperature, agitation and particles concentration (Yan et al., 2010; Douche et al., 2010). Many research focused on nickel based composite coatings due to its superior

[a]Aicha_ziouche@yahoo.fr; [b]madjda.mokhtari@yahoo.fr; [c]Baraahafez@msn.com; [d]h.elmsellem@gmail.com; [e]n.zoubiri@crti.dz; [f]d.allou@crti.dz

anticorrosion performance, high hardness, oxidation resistant and good thermal stability (Yan et al., 2010; Allou et al., 2021). Electrochemical deposition has chooses not only as a cost- effective alternative to physical methods (PVD, CVD) to fabricate thin films, but also because it is a simple and an advantageous method. Various methods of characterisation have been used to evaluate the anti-corrosion performance of metallic coatings, such as corrosion potential and current measurements, corrosion rate and mechanical and structural (Shipley and Feijó, 1999; Chkirate et al., 2021; Ziouche et al., 2018).

The objective of this research work is to study the effect of the Al_2O_3 amount on corrosion resistance, mechanical, microstructure and morphology of electrodeposition Ni- Al_2O_3 nanocomposite coatings (Aljourani et al., 2010; Hegazy et al., 2011; Elmsellem et al., 2015).

Experimental Procedure

Materials and test solution

The metal used in electrodeposition process was mild steel cuted from a pipeline used in petroleum industry.

Before depositing the coating, a homogenisation treatment was applied to the steel samples. The chemical composition of the steel substrate is presented in Table 156.1 and the metallographic structure in the Figure 156.1, it is composed of ferrite, pearlite and a minor amount of bainite.

Before the deposition the steel specimens were policed mechanically with different grades papers, pickled and activated by dipping in diluted HCl for few seconds followed by washing with deionized water. After the electrodeposition, the surfaces were washed with distilled water (Aljourani et al., 2010; Hegazy et al.,2011; Elmsellem et al., 2015).

A Ni-Al_2O_3 alloy bath was prepared using a different chemical reagents and deionized water. The chemical composition of the used is: $NiCl_2$ $6H_2O$ 24 g/L, $NiSO_4$ $6H_2O$, 21.4 g/L; $H_3BO_3$18.54 g/L Na_3 $C_6H_5O_7$ $2H_2O$, 5.85 g/L. the active metallic surface area for the all depositions is about 1 cm². The composite coatings of Ni-Al_2O_3 were electrodeposited by adding a fixed quantity (5, 10; 15 and 20 g/L) of commercial grade Al2O3 nanopowder (Sigma Aldrich).for the homogenization of the bath deposition, Ni-Al_2O_3 bath is stirring with alumina powder for two hours and was used for electroplating. During the experience of the deposition, the conditions of agitation were imposed in order to avoid settling down of the composite particles due to gravity (Chkirate et al., 2021). The electrodeposition process was carried out under potentiostatic conditions in a conventional three-electrode cell using potentiostat/galvanostat controlled by nova software. The counter electrode and the reference electrode are a platinum electrode and a saturated calomel electrode (SCE).

Caracterisation methods

The surface morphology and the thickness of the Ni-Al2O3 nanocomposite coatings were measured with scanning electron microscopy (SEM) (ZEISS-Gemini SEM 300). The structure was performed with X ray diffraction and the mechanical characterization was performed with Nano indentation techniques.

Figure 156.1 Microstructure of mild steel used in electrodeposition (Magnification X500)

Table 156.1 Chemical composition of the mild steel used in electrodeposition process

e	C	Mo	Cu	Ni	Mn	Cr	P	Si	V
Amount W%	0.09	0.008	0.194	0.073	1.46	<0.055	0.031	0.052	0.0043

Corrosion behaviour

In the study of corrosion for coating steel, using the electrochemical method can determine the stability of the potential according to time, current and the corrosion rate by extrapolating the straight TAFEL from polarization curve (Elmsellem et al., 2015; Ziouche et al., 2017; Kourim et al., 2021). The information that we can extract from electrochemical method, can give us great benefit concerning corrosion rate and corrosion resistance (Chetouani et al., 2005).

Results and Discussion

Microstructural and mechanicscharacterisation

The structure of the nanocomposite coating is presented on the Figure 156.2. It is observed that nickel morphology obtained in the composite coatings in the presence of aluminium oxide changes as with the incorporation of the nanoparticles in the electrodeposited coating (O'M et al., 1977; Hamed, 2010; Lebrini et al., 2007).

The thickness of the coating increases when the amount on Al_2O_3 nanoparticles added increase, this is due to the rise in the rate of Al_2O_3 in the coating of Ni- Al_2O_3.

The analysis of X-ray diffraction spectra showed that the aluminium oxide formed in the coating is of α- Al_2O_3 type with the cell parameters D (A°) = 1.1472.

The following table exposes the diffractions parameters of our coatings.

The pure nickel deposit that growth of the crystals in the presence of Al_2O_3, one notices also the formation of the small masses with an intermediate size of the particles of nonuniform Al_2O_3 on the whole of the coatings (Krishnaveni et al., 2009).

To obtain the indices of Miller of the plans of our layers, we used software EVA 4 Bruker.

The Figure 156.2a shows the X-ray diffraction spectra obtained starting from a pure nickel coating and the preferential orientations of the peaks are (111), (200).

The Figure 156.3b, represents the coatings of Ni-Al_2O_3 with a percentage of alumina 20%, In these spectra the presence of Al_2O_3 in the coatings, the preferential orientation of the orientation crystals changes, the peak of diffraction (200) of Nickel decreases, while the peak of diffraction (111) increases in intensity compared to the Ni deposit.

The hardness measurements of the coatings were determined by the nano-indentation technique. The results are presented in the following table.

These results show that, the elaborate coatings have a good hardness for micrometric layers, the Ni Al_2O_3 coating presented a good result of the hardness raised compared to that of Ni, it is can be explained the presence to the variation of the concentration of the Al_2O_3 particles in the nickel matrix. It reaches a maximum value for a concentration of 20 g/l seen the nature of our composite coatings (metal-ceramics) and consequently it improves the mechanical properties of the coatings (Nasirpouri et al., 2014).

Figure 156.2 Morphology of the various elaborate deposits a) Ni-Al_2O_3 5 g/l; b) Ni-Al_2O_3 10g/l, c) Ni- Al_2O_3 15g/l, d) Ni- Al_2O_3 20 g/l

Table 156.2 Thickness measurement of Ni and Ni- Al_2O_3 coating

Sample	Thickness (µm)
Ni Pure	2.7
Ni + 5 g/l Al_2O_3	3.6
Ni + 10 g/l Al_2O_3	4.9
Ni + 15 g/l Al_2O_3	7.6
Ni + 20 g/l Al_2O_3	10.3

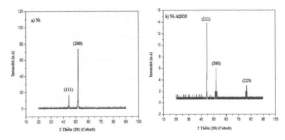

Figure 156.3 X-ray diffraction spectra of a) Ni pure; b) Ni- Al$_2$O$_3$

Table 156.3 Crystallographic orientations of the various coatings

(Hkl)		(111)		(200)		(223)	
Sample	2θ°	Intensity (μ.a)	2θ°	Intensity (μ.a)	2θ°	Intensity (μ.a)	
Ni pur	44.46	17	51.94	79	/	/	
Ni-Al$_2$O$_3$	44.6	14	51.59	7	76.48	3	

Table 156.4 The hardness of the various coatings according to the concentration variation of Al$_2$O$_3$ particles

Code	Hardness (Hv)
Ni Pure	11.49
Ni + 5 g/l Al$_2$O$_3$	17.83
Ni + 10 g/l Al$_2$O$_3$	22.92
Ni + 15 g/l Al$_2$O$_3$	31.07
Ni + 20 g/l Al$_2$O$_3$	37.82

The polarisation curve of coating steel, in 3.5 mol/l of NaCL is shown in Figure 156.4.

We observe when the percentage of Al$_2$O$_3$ increase, the corrosion potential takes a positive value, which can have explained by the amelioration of corrosion proprieties given by the coating to this steel.

The Tafel extrapolation method was used to calculate different parameters including corrosion potential (Ecorr), corrosion current density (icorr), and corrosion rate. The results are listed in Table 156.5.

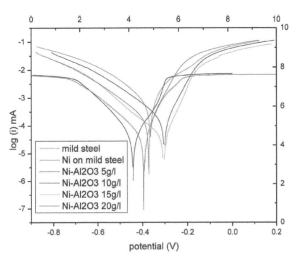

Figure 156.4 Polarisation curves for the deposition of pure Ni and Ni- Al$_2$O$_3$ at different concentration of Al$_2$O$_3$ nanoparticles in 3.5 wt% NaCl solution

Table 156.5 Electrochemical corrosion data extracted from polarisation curves of coatings containing different concentrations of Al_2O_3 nanoparticles

Sample	Ecorr (mV)	Icorr (µA m−2)	Vcorr (mmpy)	Rp (ohm)
Mild steel	−561.4	11.689	0.9167	324.51
Ni coating	−529.248	10.952	0.8652	369.79
Ni- Al2O3 5g/l	−397.651	9.952	0.6835	384.67
Ni -Al2O3 10g/l	−386.056	9.689	0.6619	408.64
Ni -Al2O3 15g/l	−304.896	9.375	0.5467	423.82
Ni -Al2O3 20g/l	−301.293	9.236	0.4903	429.61

It can be clearly seen that the corrosion resistances of all composite coatings are higher than that of pure nickel coating.

According to the results shown in Table 156.5, the corrosion current decreases and the corrosion potential shifts more positive values with the increase of Al_2O_3 nanoparticle concentration in the plating bath. The coating produced by the amount adding of 20 g/L Al_2O_3 to the bath deposition showed the highest corrosion potential (−301, 293 mV) compared to the other coatings. It is also confirmed for the corrosion rate with the lower values of corrosion rate (0,4903 mmpy). This reveals can prove that there is significant enhancement in corrosion resistance with the incorporation of Al_2O_3 nanoparticle into Ni matrix. It is seen that the highest polarisation resistance, (Rp) indicating the best corrosion protection, was observed in the case of composite coating produced by added 20 g/L Al_2O_3 in the electrolyte bath.

Conclusion

In the summery of the results obtained and discussion from this work following conclusion can be regrouped:

- Ni-Al_2O_3 nanocomposite coatings were successfully formed by electrodeposition technique on steel substrate surfaces.
- The thicknesses of the elaborate coatings are of the micrometric order.
- The test of quality of deposit carried out for all the electrodeposits leads us to conclude that these coatings have a very good adherence.
- The diffraction of x-rays to show that, the structure is quite crystalline. The peaks are very well solved. The pure nickel deposit shows the preferential orientations of the peaks (111) and (200), and for the peaks of Al_2O_3 an orientation (223) is of α type.
- Hardness by the test of nano-indentation, on the various coatings, showed that the latter have a good hardness, which increases with the addition of Al_2O_3 particles dispersed in Ni matrix, this increase is due primarily to the increase in the concentration of the solid particles Al_2O_3, and which causes to improve the mechanical properties of the deposits.
- The tests of corrosion revealed that the Ni coatings, Ni- Al_2O_3 have a good corrosion resistance considering good resistance to recorded polarization and the decrease the corrosion rate has fur with measurement with the dialogs of Al_2O_3 in the various coatings.

References

Aljourani, J., Golozar, M. A., and Raeissi, K. (2010). The inhibition of carbon steel corrosion in hydrochloric and sulfuric acid media using some benzimidazole derivatives. Mater. Chem. Phys. 12:320–325.

Allou, D., Ould Brahim, I., Cheniti, B., Miroud, D., and Ziouche, A. (2021). Effect of post weld heat treatment on microstructure and mechanical behaviors of weld overlay Inconel 182 on 4130 steel substrate using SMAW process. Metallog. Microstruct. Analysis. 10(5):567–578.

Chetouani, A., Hammouti, B., Benhadda, T., and Daoudi, M. (2005). Inhibitive action of bipyrazolic type organic compounds towards corrosion of pure iron in acidic media. Appl. Surf. Sci. 249:375385.

Chkirate, K., Azgaou, K., Elmsellem, H., El Hajjaji, S., and Essassi, E. M. (2021). Corrosion inhibition potential of 2-[(5-methylpyrazol-3-yl)methyl]benzimidazole against carbon steel corrosion in 1 M HCl solution: Combining experimental and theoretical studies. J. Molecul. Liquid. 321:114750.

Douche, D., Elmsellem, H., Guo, L., Hafez, B., Tüzün, B., Louzi, A. E., Bougrin, K, and Himmi, B. (2020). Anti-corrosion performance of 8-hydroxyquinoline derivatives for mild steel in acidic medium: Gravimetric, electro-chemical, DFT and molecular dynamics simulation investigations. J. Molecul. Liquid. 308:113042.

Elmsellem, H., Harit, T., Aouniti, A., Malek, F., Riahi, A., and Chetouani, A. (2015). Adsorption properties and inhibition of mild steel corrosion in 1 M HCl solution by some bipyrazolic derivatives: experimental and theoretical investigations. Protect. Metal. Phy. Chem. Surf. 51(5):873–884.

Hamed, E. (2010). Studies of the corrosion inhibition of copper in Na2SO4 solution using polarization and electrochemical impedance spectroscopy. Mater. Chem. Phys. 121:70–76.

Hegazy, M. A., Ahmed, H. M., and El-Tabei, A. S. (2011). Investigation of the inhibitive effect of p-Substituted 4-(N,N,N- dimethyldodecyl Ammonium bromide) benzylidene-Benzene 2-yl-amine on corrosion of carbon steel pipelines in acidic medium. Corros. Sci. 53:671–678.

Kourim, A., Malouki, M. A., Ziouche, A., Boulahbal, M., and Mokhtari, M. (2021). Tamanrasset's clay characterization and use as low cost, ecofriendly and sustainable material for water treatment: Progress and challenge in copper Cu (II). Defect Diffu. Forum. 406:457–472

Krishnaveni, K., Narayanan, T. S. N. S., and Seshadri, S. K. (2009). Corrosion resistance of electrodeposited Ni–B and Ni–B–Si3N4 composite coatings. J. Alloys Compd. 480:765–770.

Lebrini, M., Lagrenée, M., Vezin, H., Traisnel, M., and Bentiss, F. (2007). Experimental and theoretical study for corrosion inhibition of mild steel in normal hydrochloric acid solution by some new macrocyclic polyether compounds. Corros. Sci. 49(5):2254–2269.

Lekbir, C., Dahoun, N., Guetitech, A., Ouaad, K., and Djadoun, A. (2017). Effect of immersion time and cooling mode on the electrochemical behavior of hot-dip galvanized steel in sulfuric acid medium. J. Mater. Eng. Perform. 26(6):2502–2511.

Nasirpouri, F., Sanaeian, M. R., and Samardak, A. S. (2014). An investigation on the effect of surface morphology and crystalline texture on corrosion behavior, structural and magnetic properties of electrodeposited nanocrystalline nickel films. Appl Surf. Sci. 292:795–805.

Nazarov, A. and Thierry, D. (2007). Application of volta potential mapping to determine metal surface defects. Electrochem. Acta. 52:7689–7696.

O'M, J., Bochris, A. K., and Reddy, N. 1977. Modern Electrochemistry. New York (N Y): Plenum Press.

Owczarek, E. and Adamczyk, L. (2016). Electrochemical and anticorrosion properties of bilayer polyrhodanine/isobutyltrie thoxysilane coatings. J Appl. Electrochem. 46:635–643

Shipley, A. M. and Feijó, J. A. (1999). The use of vibrating probe technique to study steady state extracellular currents during pollen germination and tube growth. In fertilization in higher plants. molecular and citological aspects. Verlag: Springer.

Yan, M., Gelling, V. J., Hinderliter, B. R., Battocchi, D., Tallman, D. E., and Bierwagen, G. P. (2010). SVET method for characterizing anti-corrosion performance of metalrich coatings. Corrosion Sci. 52(8):2636–2642.

Ziouche A., Zergoug M. A, Boucherrou N. A, Boudjellal H. C, Mokhtari M. A., and Abaidia, S. (2017). Pulsed eddy current signal analysis of ferrous and non-ferrous metals under thermal and corrosion solicitations. Russ. J. Nondestruct. Test. 53(9):652–659.

Ziouche, A., Haddad, A., Badji, R., .Bedjaoui, W., and Abaidia, S. (2018). Microstructure, corrosion and magnetic behavior of an aged dual-phase stainless steel. J. Mater. Eng. Perform. 27(3):1249–1256.

Ziouche, A., Hammouda, A., Boucherou, N., Elmsellem, H., and Abaidia, S. (2021). Corrosion protection enhancement on aluminum alloy and magnesium alloy by mo-ceo2 conversion coating. Moroccan J. Chem. 9(3):386–393.

157 An efficient technique for optimal sitting and sizing of capacitors in reconfigured network

B. Y. Bagde[a], Javed Shaikh[b], and Ameya Saonerkar[c]

Department of Electrical Engineering Yeshwantrao Chavan College of Engineering Nagpur, India

Abstract

Capacitors are placed optimally in a reconfigured distribution system to reduce power loss and maintain the voltage profile of buses within the acceptable limits subject to various system constraints. This primary objective of reduction in power losses is realized using network reconfiguration along with optimum capacitor placement. Since the objective has number of solutions, a new heuristic-based optimisation technique, Teaching-Learning based optimization (TLBO) is used to achieve the objective. Network reconfiguration is the method of modifying the topological assembly of feeders by altering the close or open position of sectionalizing switches and tie lines. TLBO is an optimision technique based on population consist of two parts; teacher and learner phase. In this technique a cluster of learners is treated as population and the subjects offered to these learners are considered as control variables. Result of any learner is considered as degree of fitness of given objective function and the best solution obtained is declared as the teacher. Since majority of the distribution systems are radial in nature hence radial distribution system is selected as a test case. It is observed that the suggested approach improves the quality of solution compared to the other approaches given in literature. Solution to the problem of reduction in loss and improvement in voltage profile is obtained using MATLAB (R2020) on IEEE 33-bus distribution test case.

Keywords: Optimisation, power loss, reconfiguration, TLBO, voltage profile.

Introduction

Radial distribution systems are very common due to its commercial viability and simple design. There is high power loss and reduction of voltages in the distribution system. Power utilities aims at reducing these losses at distribution side to improve the overall efficiency of power transfer. Shunt capacitors when installed in the distribution network ensures reduction in power losses, improves bus voltages and thus makes system stable. Though the capacitors provide necessary reactive power to the system, the capacitor allocation must be done optimally so that desired performance of the distribution system is achieved with minimum resources.

Deregulation of electricity across globe has given rise to new market place based on electricity transactions competitively. This competition exists in main activities of any power system namely generation, distribution and transmission. Since the system has become more complex owing to restructuring more advanced techniques are needed to deal with electricity which is being traded in competitive market. The traditional optimisation techniques have many disadvantages like its handling capacity of algebraic functions is limited; most of the techniques find it difficult to incorporate dynamic characteristics. These techniques are extremely sensitive to initialisation problem and choice of objective functions as well as constraints is limited. These methods are time consuming hence unsuitable for real time operation of power system. These problems of traditional optimisation techniques are overcome by usage of heuristic optimisation techniques like evolutionary algorithms (EAs). The proposed work suggests a methodology to decrease the loss in power and improvement in voltage profile with simultaneous implementation of capacitor placement and distribution system reconfiguration in optimal way. Previous works on similar lines include capacitor allocation and network reconfiguration in two consecutive steps where branch exchange technique is applied for reconfiguration of the network (Peponis et al., 1996). In Simulated annealing (SA) is implemented to get best placement of capacitors in reconfigured network (Dan and Baldick, 1996). Optimal solution is obtained with simultaneous approach to these two problems with the main objective of minimisation of active power loss in (Chang, 2008),and the Ant colony search algorithm (ACSA) gives superior results in terms of optimal capacitor placement and these solutions are then compared with SA and Genetic algorithm (GA). A modified PSO algorithm is applied in (Pooya et al., 2011) to obtain the optimised placement of capacitors in reconfigured system with objective of cost reduction and power loss reduction. In a heuristic-based technique is implemented to assess reliability of the system with different

---

[a]by.bagde@rediffmail.com; [b]jawedshaikh01@yahoo.com; [c]10471@ycce.in

schemes of protection and separation nodes in the distribution system (Montoya and Ramirez, 2012). Harmony search algorithm (HSA) deals with optimal placement of capacitors under reconfiguration. In a combination of classical method and heuristic method uses deterministic approach especially for system reconfiguration and optimum placement of capacitors using a heuristic technique (Montoya and Ramirez, 2012). Another such combination is of minimum spanning tree (MST) algorithm and GA. Here MST used for deciding reconfigured network while GA is for capacitor allocation.

In this work, system performance is investigated under reconfigured network and optimum capacitor placement based on TLBO algorithm. This algorithm attempts to implement network reconfiguration along with capacitor placement simultaneously to achieve maximum decrease in active power loss and enhancement of bus voltages. Simulation results are then compared with the results reported in literature.

Problem Formulation

Statement of problem

The proposed work attempts to find the optimum size of capacitors to be used in distribution network which gives minimum power loss while satisfying operating limits. The objective function of real power loss minimization is,

$$minimize \ f = \min\left(P_{loss}\right) \tag{1}$$

Mathematically, this is expanded as,

$$min\left(P_{loss}\right) = \sum_{i=2}^{n_n}\left(P_{gni} - P_{dni} - V_{mi}V_{ni}Y_{mni}\cos\left(\partial_{mi} - \partial_{ni} + \theta_{ni}\right)\right) \tag{2}$$

where P_{loss} is the active power loss, is the real power generated at bus ni, is the real power demanded at bus ni, V_{mi} is bus mi voltage, V_{ni} is voltage of bus represented as ni, Y_{mni} is considered as admittance between bus ni and mi, voltage phase angle at bus mi is taken as ∂_{mi}, voltage phase angle at bus ni is ∂_{ni} and θ_{mi} is taken as admittance angle. The objective function (2) must satisfy below given limitations.

1.   Equation of power flow must be maintained

$$P_g = P_D + P_{loss} \tag{3}$$

2.   Bus voltage magnitudes should be within limits

$$V_{min} \le \left|V_k\right| \le V_{max} \tag{4}$$

3.   Distribution system configuration is always radial
4.   Load bus must.be intact

Radial distribution network

To retain radial arrangement of distribution network, an approach of branch and bus incidence matrix has been used. Let A be the branch to node matrix in which each row has entries of branch and each column j has entries of node. These entries are made as per following rule,

   aij = 0 if there is no linking in between branch i and node
   designated as j
   aij = −1 if direction is from branch i to the node j aij = 1 if direction is assumed to be from node j to branch i

The matrix is formed excluding reference node. If determinant of A turns out to be either 1 or −1, the system can be declared as radial in nature. If determinant comes out to be zero then system is not radial and it can be discarded (Figure 157.1).

Problem formulation

$$P_{k+1} = P_k - P_{Loss,k} - P_{Lk+1}$$

$$P_{k+1} = P_k - \frac{R_k}{\left|V_k\right|^2}\left\{P_k^2 + \left(Q_k + Y_k\left|V_k\right|^2\right)^2\right\} - P_{Lk+1} \tag{5}$$

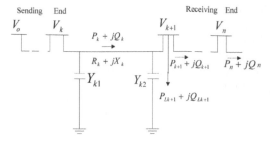

Figure 157.1 Section of distribution system

$$Q_{k+1} = Q_k - \frac{X_k}{|V_k|^2}\left\{P_k^2 + \left(Q_k + Y_k|V_k|^2\right)^2\right\} - Y_{k1}|V_k|^2 - Y_{k2}|V_{k+1}|^2 - Q_{Lk+1} \tag{6}$$

$$|V_{k+1}|^2 = |V_k|^2 + \frac{R_k^2 + X_k^2}{|V_k|^2}\left(P_k^2 + Q_k^2\right) - 2\left(R_k P_k + X_k Q_k'\right)$$

$$|V_{k+1}|^2 = |V_k|^2 + \frac{R_k^2 + X_k^2}{|V_k|^2}\left(P_k^2 + \left(Q_k + Y_k|V_k|^2\right)^2\right) - 2\left(R_k P_k + X_k(Q_k + Y_k|V_k|^2)\right) \tag{7}$$

The line losses occurring between k and $k + 1$ are calculated as,

$$P_{Loss}(k, k+1) = R_k \frac{\left(P_k^2 + Q_k^2\right)}{|V_k|^2} \tag{8}$$

The feeder power loss is given as,

$$P_{Loss} = \sum_{k=1}^{n} P_{Loss}(k, k+1) \tag{9}$$

The total power loss reduction in the distribution network is calculated based on above equations.

Power flow

For load-flow computation, backward forward sweep methodology has been applied. In this method, in every computational iteration two steps are performed. In case of forward sweep, voltage drops are calculated with updates in current or power flow. Thus, voltages of nodes are updated beginning from the source node and in the forward direction till end node of distribution network while maintaining substation voltage to actual value. In this step, power calculated in each branch is same as power calculated during backward step.

In the backward sweep power flow or current is obtained with updating of voltages. In this step power flow is calculated for branches beginning from the last branch to the starting node in reverse or backward direction. The power flows in every branch are computed using node voltages of preceding iteration. Thus, voltages obtained in the forward direction are assumed constant in backward sweep and power flows in branches are updated. The iterations get converged if voltage difference is below the tolerance otherwise power flows are updated in each branch in backward direction with current values of voltages. The process is repeated until solution gets converged.

Teaching-learning based optimisation

Teaching-learning-based optimisation (TLBO) algorithm was proposed based on teaching learning process which evaluates how a teacher impacted result of learners in given class. This algorithm works through two components of teaching learning process i.e., teacher phase and learner phase and assesses teaching learning ability of a classroom (Baran and Wu, 1989). In teachers' phase, teacher is the one who shares knowledge and inspires the learners for improvement in the results. Similarly in learners' phase, learners gain knowledge through discussions with peers which also refines their results.

TLBO is a population-based optimisation technique where number of learners can be equated with population and courses offered to learners can be assumed to be variables of the problem. Learners' grades can be compared to fitness of the given objective and the best value amongst the population can be determined.

Teacher phase

In this phase teacher is responsible for the learning. A teacher put in efforts to enhance performance of the entire class in the course taught by him. At given iteration i, let the number of courses is m which are analogous to number of variables of given objective function, total number of learners in given classroom or size of population is n and $M_{j,i}$ can be said to be the average result of the class in a specific course j. If all the courses are considered for the entire population of learners, then the best learner k_{best} will have best result $X_{total} - k_{best,i}$. This best learner can be considered as the teacher. The mismatch of the current average result of each course and the teacher (k_{best}) for each course then can be stated as,

$$\text{difference\_mean}_{j,k,i} = r_i \left(X_{j,kbest,i} - T_F M_{j,i} \right) \tag{10}$$

Where, best learner has result of $X_{j,kbest,i}$ for the course j. T_F is a teaching factor which modifies the class average and r_i is any arbitrary number in [0, 1] range. T_F is given by,

$$T_F = \text{roundup}\left[1 + \text{random}(0,1)\{2-1\} \right] \tag{11}$$

Based on on difference_mean$_{j,k,i}$, the current solution is then updated in the teacher stage,

$$X'_{j,k,i} = X_{j,k,i} + \text{difference\_mean}_{j,k,i} \tag{12}$$

Where, $X'_{j,k,i}$ is the modified and final solution if the fitness value is the best compared to $X_{j,k,i}$. Once teacher phase is over, all the best fitness values are now input to the learners stage.

Learner phase

In this phase of algorithm, the students further improve their knowledge by interacting with their fellow students. Every learner gains new facts from other randomly selected better informed learner. Let for population size of n, a learners phase with two learners P and Q can be specified as,

$$X'_{total-Q,i} \neq X'_{total-P,i} \tag{13}$$

Where, $X'_{total-P,i}$ and $X'_{total-Q,i}$ are the new modified values of $X_{total-P,i}$ and $X_{total-Q,i}$ of P and Q at the close of teacher phase.

$$X''_{j,P,i} = X'_{j,P,i} + r_i \left(X'_{j,P,i} - X'_{j,Q,i} \right) \text{ if } X'_{total-Q,i} > X'_{total-P,i} \tag{14}$$

$$X''_{j,P,i} = X'_{j,P,i} + r_i \left(X'_{i,Q,i} - X'_{j,P,i} \right) \text{ if } X'_{total-P,i} > X'_{total-Q,i} \tag{15}$$

$X''_{j,P,i}$ is now a new solution if it has better fitness value.

Mulation Results and Discussions

The test case is IEEE 33 bus radial distribution system and consists of five tie lines and 32 sectionalising switches as shown in Figure 157.2. The test case data has been taken from (Venkata and Waghmare, 2014).

The generator rating and system voltage are 100 MVA and 12.66 kV respectively. The loads connected to this test system are 3.715 MW and 2.3 MVar. The capacitor has different sizes ranging from 200 KVar to 1200 KVar with step size of 2 KVar. Switches 33 to 37 are normally open switches, and 1 to 32 are normally closed switches. The main objective of reduction in loss by capacitor placement is applied to test system using MATLAB programming environment 2019a. voltage bounds are 0.9 minimum and 1.00 pu maximum. The TLBO variables used are population size of 50 and number of iterations as 50 based on the studies carried out in standard literature (Duong et al., 2019).

Base case of test system is identified as the system without configuration or capacitor placement. When simulations are performed on base case power loss of 202.49 kW is obtained. If the system is reconfigured, best solution is obtained for open switches 7, 9, 14, 32, 37 and power loss reduces to 125.93 KW. With network reconfiguration and optimal placement of capacitors, the switch position is unaltered but due to capacitors at select buses reduces power loss further to 97.47 KW as shown in Table 157.1.

When the results obtained by TLBO is compared with the results of optimisation techniques reported in literature as shown in Table 157.2, for reconfigured case the power loss is improved to 125.93 KW compared to 126.34 KW obtained using modified imperialist competitive algorithm (MIC) (Prakash, D. B. and Lakshminarayana, 2017). Similarly for reconfigured system with capacitor placement power loss has improved from 97.63 KW in MIC algorithm to 97.47 KW. The minimum voltages at the buses of the system are within the bounds.

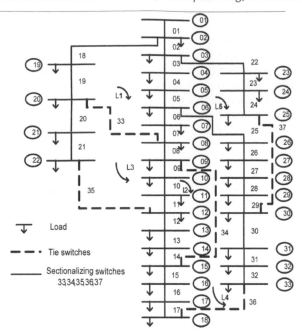

Figure 157.2 33 bus distribution system

Table 157.1 Power loss in 33-bus system

	Base Case	*Reconfigured Case*	*Reconfigured and capacitor placed system*
Tie switches	33,34,35, 36, 37	7, 9, 14, 32, 37	7, 9, 14, 32, 37
Power loss (KW)	202.49	125.93	97.47
Voltage (min pu)	0.91	0.93	0.95
Optimal locations (bus-size of capacitor) (KVar)	--	--	6 - 680, 11 - 340, 24 - 360, 29 - 620, 32 - 300

Table 157.2 Comparison of methods of optimisation

Algorithm		*Power losses (KW)*	*Minimum voltage (pu)*
TLBO	Reconfiguration	125.93	0.94
MIC (Prakash and Lakshminarayana, 2017)		126.34	0.95
TLBO	Reconfiguration and capacitor placement	97.47	0.95
MIC (Prakash. and Lakshminarayana, 2017)		97.63	0.95

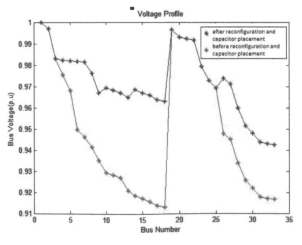

Figure 157.3 Voltage profiles at buses

Figure 157.3 plots the improved voltage profiles after the network reconfiguration and optimised capacitor placement in the network. Bus number 18 which was at lowest voltage prior to reconfiguration and capacitor placement stands improved at 0.965 pu. Overall voltage profile at the buses is improved.

Conclusion

TLBO algorithm has been effectively applied to obtain optimal solution of only reconfiguration of the system and reconfiguration with capacitor placement in 33 bus radial distribution system. Compared with the results obtained using MIC Algorithm in the literature (Prakash, D. B. and Lakshminarayana, 2017) the power loss has reduced however with no significant changes in bus voltages. TLBO algorithm has the potential of application to other power system areas of optimisation for different objective functions such as reduction in reactive power loss, improvement in active power flow, enhancement of system stability etc.

References

Abd El-salam, M. F., Beshr, E., and Eteiba, M. B. (2018). A new hybrid technique for minimizing power losses in a distribution system by optimal sizing and siting of distributed generators with network reconfiguration. Energies. 11(12):3351.

Abril, I. P. (2017). Algorithm of inclusion and interchange of variables for capacitors placement. Electr. Power Syst. Res. 148:117–126.

Bagde, B. Y., Umre, B. S., and Narde, N. (2018). Power Loss optimisation using distribution system reconfiguration in presence of DG. J. Eng. Appl. Sci. 13(1).

Bagde, B. Y., Umre, B. S., Bele, R. D., and Gomase, H. (2016). Optimal network reconfiguration of a distribution system using biogeography-based optimization. IEEE 6th International Conference on Power Systems New Delhi, India, pp. 1–6.

Baran, M. E. and Wu, F. F. (1989). Network reconfiguration in distribution system for loss reduction and load balancing. IEEE Trans. Power Del. 4(2):1401–1407.

Chang, C. F. (2008). Reconfiguration and capacitor placement for loss reduction of distribution systems by ant colony search algorithm. IEEE Trans. Power Syst. 23(4):1747–1755.

Pooya, R., Mehdi V. (2010). Distribution system efficiency improvement by reconfiguration and capacitor placement using a modified particle swarm optimization algorithm. IEEE Electrical Power and Energy Conference, pp. 1–6.

Dan, J. and Baldick, R. (1996). Optimal electric distribution system switch reconfiguration and capacitor control. IEEE Trans. Power Syst. 11(2):890–897.

Devabalaji, K. R., Yuvaraj, T., and Ravi, K. (2018). An efficient method for solving the optimal sitting and sizing problem of capacitor banks based on cuckoo search algorithm. Ain Shams Eng. J. 9(4):589–597.

Duong, M. Q., Pham, T. D., Nguyen, T. T., Doan, A. T., and Tran, H. V. (2019). Determination of optimal location and sizing of solar photovoltaic distribution generation units in radial distribution systems. Energies. 12(1):174.

El-Fergany, A. A. and Abdelaziz, A. Y. (2014). Artificial bee colony algorithm to allocate fixed and switched static shunt capacitors in radial distribution networks. Electric Power Component. Syst. 42(5):427–438.

Hussain, A. N., Al-Jubori, W. K. S., and Kadom, H. F. (2019). Hybrid Design of Optimal Capacitor Placement and Reconfiguration for Performance Improvement in a Radial Distribution System. Hind. J. Eng. https://doi.org/10.1155/2019/1696347 Mishra, S., Das, D., and Paul, S. (2017). A comprehensive review on power distribution network reconfiguration. Energy Syst. 8(2):227–284.

Montoya, D. P. and Ramirez, J. M. (2012). Reconfiguration and optimal capacitor placement for losses reduction. IEEE/PES, transmission and distribution: Latin America Conference and Exposition, pp. 1–6.

Montoya, D. P. and Ramirez, J. M. 2012. Reconfiguration and optimal capacitor placement for losses reduction. IEEE Conference.

Peponis, G. J., Papadopulos, M. P., and Hatziargyriou, N. D. (1996). Optimal operation of distribution networks. IEEE Trans. Power Syst. 11(1):59–67.

Narde, N. and Bagde, B. Y. (2017). Optimal placement of DG in distribution system using evolutionary algorithm. Int. J. Eng. Technol. Manag. Appl. Sci. 5(5):889–895.

Pooya, R., Mehdi, V., and Hajipour, E. (2011). Reconfiguration and capacitor placement in radial distribution systems for loss reduction and reliability enhancement. 116th Internatonal Conference on Intelligent system application to power systems (ISAP), pp. 1–6.

Prakash, D. B. and Lakshminarayana, C. (2017). Optimal siting of capacitors in radial distribution network using whale optimization algorithm. Alexandr. Eng. J. 56(4):499–509.

Tamilselvan, V., Jayabarathi, T., Raghunathan, T., and Yang, X.-S. (2018). Optimal capacitor placement in radial distribution systems using flower pollination algorithm. Alexandr. Eng. J. 57(4):2775–2786.

Venkata R. and Patel, V. (2013). An improved teaching-learning-based optimization algorithm for solving unconstrained optimization problems. Scientia Iranica. 20(3):710–720.

Venkata, R. and Waghmare, G. G. (2014). A comparative study of a teaching–learning-based optimization algorithm on multi-objective unconstrained and constrained functions. J. King Saud University Comp. Info. Sci. 26(3):332346

Zou, F., Chen, D., and Xu, Q. (2019). A survey of teaching–learning-based optimization. Neurocomput. 335:366383.

158 A comparative study of a plane NACA2412 airfoil with a NACA2412 airfoil attached with gurney flap using CFD

Sahil Waquar Ahmad Khan[1], Abhiram Dapke[2], Jayant Giri[3], and Neeraj Sunheriya[3]

[1]Department of Mechanical Engineering, Yeshwantrao Chavan College of Engineering, Nagpur, India

[2]University of Maryland, United States

[3]Department of Mechanical Engineering, Yeshwantrao Chavan College of Engineering, Nagpur, India

Abstract

In this paper, the study aims to demonstrate how a gurney flap improves the efficiency of the airfoil, by using Autodesk CFD software. The purpose of a gurney flap is to enhance the aerodynamic features of an airfoil shape, like increasing lift force with a lesser increase of drag force, so that it can resist the weight of an aerodynamic device in the air by providing more thrust. In this research, National Advisory Committee for Aeronautics (NACA)2412 airfoil is selected to test the impact of a gurney flap. First, the coordinates of the airfoil are imported using the airfoil website database into Fusion 360 CAD software, and then two 3-D wings were created from 2-D airfoil cross-sections, one without a gurney flap and another with a gurney flap that is 2% of the airfoil chord length in height and 0.5% in width. CFD simulation was carried out on both 3-D models by using Autodesk CFD software, at 0°, 4°, 8°, 12°, 16°, and 20° angles of attack. The change in aerodynamic properties of airfoil due to gurney flap are plotted into graphs by using the values of lift and drag forces for both airfoils in different scenarios. This experiment demonstrates that the gurney flap enhanced the maximum lift force by 28% as compared to a plane airfoil at the same wind speed and angle of attack and a 22.05% rise in the lift to drag ratio is observed.

Keywords: Airfoil, autodesk CFD, CFD analysis, gurney flap, NACA2412, simulation

Introduction

An airfoil is a cross-section that is used in an aircraft's wings or tail, racing cars, or turbine blades to produce lift during its relative motion along with a gas or a fluid. When the airfoil travels through a fluid region, it induces an aerodynamic force on the airfoil surface. The perpendicular component of this force is known as lift force, whereas the force parallel to it gives drag force. There are many types of airfoils, used in different vehicles to improve their aerodynamic features. Airfoils are generally round and have a larger radius at the leading edge in case of subsonic application, and for supersonic use, airfoils have a sharper leading edge. All airfoil has sharper trailing edges at the end.

National Advisory Committee for Aeronautics (NACA) system is one of the most used airfoil classification systems to define airfoil geometry. NACA, today known as the National Aeronautics and Space Administration (NASA), created the NACA system in 1930. Due to recent progress in the aerodynamic field and to meet higher demands for efficient airfoils, complex geometry airfoils are preferred over NACA airfoils. In the NACA system, an airfoil is represented by a series of digits after the word NACA. These numbers represent the values of airfoil dimensions which, when used in their respective equations, will generate the profile or cross-section of an airfoil.

There are different types of NACA system series to design the NACA airfoil which are four- digit, five-digit, six- digit, modifications of four- and five-digit series, six series, seven series, eight series. The slope of an airfoil known as the camber line, and the thickness distribution on both sides of the camber line, are two key airfoil geometry variables in all NACA series methods.

Among these methods, the NACA four-digit series airfoil is used in this experiment to test the working efficiency of gurney flaps with NACA 2412.

In the NACA four-digit series method, geometry or shape of wing cross-section is defined by four digits, in which the digit at the first position represents the value of maximum camber expressed in terms of percentage of chord width, and the second digit divided by 10 gives the location of this camber from leading or front side of an airfoil in chord length percent. Finally, the remaining numbers in the series indicate the maximum thickness value.

In NACA 2412, the first digit shows the maximum camber in the airfoil is 2% of chord length. The second digit is the distance of this camber which is 40% of the chord length of the airfoil from the leading edge. Finally, the last two numbers indicate that the value of the maximum thickness of the airfoil is 12% of chord length.

NACA 2412 had been a part of airplanes like Cessna 175, Avia B-534, Baumann BT-120 Mercury, Baumann B-100 Mercury, and Bell 65 ATV. From the airfoil tools website database, the coefficient of lift for NACA 2412 is maximum when the angle of attack is in the range 13–16°. After that, the stall stage occurs, and the coefficient of lift decreases.

A Gurney gurney flap called a wicker bill is a flat and tiny plate provided at the right angle on the trailing edge end at the lower surface of the airfoil. This device, invented by the renowned race car driver and team manager Daniel Gurney in 1971, increased the downward force in racing cars for better handling, especially during the cornering at high velocity. The length of this strip ranges from 0.5–2% of the airfoil chord. This small strip increases the aerodynamic properties of an airfoil and serves the same purpose as an airfoil design which gives higher pressure at the pressure surface of the airfoils to increase lift force. The Gurney flap keeps the boundary layer during airflow linked to the trailing edge at the suction surface of the airfoil. This minor modification can lead to an increase in lift coefficient up to 30%, as per an experiment at (Maughmer and Bramesfeld, 2008), conducted on an airfoil with gurney flaps in a subsonic wind tunnel, at 10^6 Reynold's number.

This method of adding a gurney flap to increase the airfoil lift property has been an area of research for decades. Many research and experiments were performed regarding the optimum gurney flap parameters or dimensions because along with improving lift force on an airfoil by the raising pressure below the airfoil, the gurney flap also increases drag force if proper height and thickness are not assured.

Yoo (2000) described research to understand the importance of gurney flap modification on NACA23012 airfoil. In this experiment, the RAMPANT code was used to conduct CFD analysis for 2D flow on an airfoil with turbulence model as standard k-ε. Conclusions derived from this experiment were that (1). Gurney flap increases the load on airfoil length, particularly on the trailing edge, (2). Same lift coefficients can be achieved for lower drag coefficients as compared to plane airfoils, (3). A higher ratio of lift to drag is obtained at high lift coefficients with a gurney flap.

At (Chen and Bo, 2022), the gurney flap was studied for tilt rotor-wing applications. In this research, they carried out an analysis on an A821201 airfoil attached with a gurney flap for heights 1%c, 2%c, and 4%c, with a 0.2 Mach number. Here, c denotes the chordal length of the airfoil. The results show an increase in camber during airflow along the airfoil surface which increased the lift coefficient. Along with the lift coefficient, an increase in drag coefficient value was also observed. In this analysis, results show the highest lift to drag coefficient ratio with a minimum stall angle of 22°, when gurney flap height is 2%c.

Li-Shu and Gao, (2019analysed a set of cases where an S809 airfoil was tested to study how a gurney flap affects airfoil performance. For this, first gurney flaps with a common height of 2%c and thickness of 0.2%c, 0.6%c, and 1.0%c were used for investigation including a baseline airfoil also. It was observed that the effect of gurney flap width/thickness is small as compared to its height and the flap increased the camber of the airfoil by adding a delay before fluid flow converges at end of the trailing edge. Also, another objective of this research was to determine whether a rectangular or triangular-shaped gurney flap is more efficient. This study concluded that the maximum value of lift coefficient was increased due to the gurney flap by 20.65% when the thickness value was 0.2%c. Also, a triangular-shaped gurney flap was found to be more efficient than a rectangular-shaped gurney flap.

Fadl et al. (2018) performed a 2-D investigation on S822 and S823 airfoils by replacing the 4% chord length trailing edge airfoil portion with a gurney flap. This study was performed by using seven values of the Reynolds number. The heights of gurney flaps that were used for simulations are 1%, 2%, and 3% of airfoil chord length. The authors concluded that a gurney flap at 2%c height gave a maximum lift to drag ratio and was most efficient because a 3%c gurney flap increased drag coefficient and hence deteriorating the airfoil performance.

Numerical analysis done by Aramendia et al. (2018) on a DU91W250 airfoil, describes the process of selecting an optimized gurney flap configuration to get an improved lift to drag ratio with a suitable value of angle of attack (AOA). At Yang et al. (2020), detailed research was performed to determine the gurney flap significance on airfoil subjected to turbulent flow. The data after the study was completed, says that height is a very important dimension for the gurney flap than its thickness, and a 14.35% rise in the lift and drag ratio was achieved by implementing different heights and keeping thickness as 0.75%c. Another important result was that the gurney flap was effective for airfoil only for low turbulent flow conditions.

This phenomenon is also described by Graham et al. (2017), who experimented on an airfoil wing inside a wind tunnel at 27 m/s free stream velocity with a Reynolds number of 1.5×10^6 on an SD7062 airfoil cross-section wing of 0.303 m span. In this experiment, three gurney flaps of height 1%, 2%, and 4% of chord length were used. Along with this, for each flap height, 1%, 2%, 4%, and 6% of chord length as gurney flap thickness were tested. From the test conducted inside the wind tunnel, it was found that lift augmentation occurred for all airfoils wing with a gurney flap than baseline configuration. 1%c flap thickness is found to give maximum lift coefficient. This research concluded that the lift of the airfoil is directly proportional to the gurney flap height and inversely proportional to flap thickness.

A similar concept of employing suitable dimensions for a gurney flap is discussed in research done at (Saha et al., 2017), which used CFD software to track the flow of air along an airfoil surface with a gurney flap. Just like the experiment at [8], the investigation was done at subsonic flow condition at 20 m/s free stream velocity inside a wind tunnel. One of the objectives of this CFD analysis was to determine the proper flap height by testing flap heights from 2 to 5% of the chord length of the airfoil. Same as the Research work described above, the results suggested that 2% of chord length is a suitable height for a gurney flap.

Another detailed research was conducted on the Gurney flap subjected to high turbulent flow conditions at (Kumar et al., 2021). In this experiment, the same idea of using different gurney flap height airfoils for analysis was applied, but here along with height, the flap thickness was also considered for testing, which was 0.25%, 0.5%, and 0.75% of chord length. The research concludes that the height of the flap has more influence on airfoil efficiency than the thickness of the flap and suggests a suitable range of 0.25% to 0.75% of chord length that can be used as flap thickness.

It can be observed that the above research and experiments describe the significance of a gurney flap, and its parameters. This helps us to decide the range of gurney flap parameters like height, thickness, and shape. It was observed that much of the research and numerical analysis work is mostly done on two-dimensional airfoil profiles. Therefore, the objective of this research is to determine the gurney flap effects on a three-dimensional wing having NACA2412 airfoil, with a span of 3 m and chord width of 1 m using Computational Fluid Dynamics. This method for conducting CFD analysis and study is referenced by the research process of Chakraborty et al. (2021). By analysing the results of the authors from (Maughmer and Bramesfeld, 2008; Yoo, 2000; Chen and Bo, 2022; Li-Shu and Gao, 2019; Fadl et al., 2018; Aramendia et al., 2018; Yang et al., 2020; Graham et al.,2017; Saha et al., 2017; Kumar et al., 2021), it is clear that the gurney flap height of 2%c gives a better lift to drag ratio in many scenarios and thickness should be lower than 1%c. Thus, 2%c height and thickness of 0.5%c for the gurney flap were considered during CFD simulations.

Research Approach

To perform the simulations and compare the results of the analysis, the software used is Autodesk CFD 2021. The analysis and simulation have been performed in the steps given below.

Geometry creation

The coordinates to create the 2-D geometry of the NACA 2412 airfoil were taken from the airfoil tools website's database. These coordinates were used in Autodesk fusion 360 software to create a 2-D sketch profile as shown in Figure 158.1 and Figure 158.2.

2-D sketch to 3-D CAD model

Once the sketch was created from the coordinates, extrude command was used to convert this sketch into a 3-D Cad Model of a Wing. 1 m and 3 m were selected as Chord length of airfoil and span of wing respectively. After this, the CAD model was imported as a STEP file to perform CFD analysis on it, using autodesk CFD software.

Setup of CAD model and wind tunnel creation

The STEP file of the CAD model was imported in Autodesk CFD software, and then a rectangular volume or domain was created around the wing with distance in X-direction (along the chord of the airfoil) being 200%c, the distance between upper and lower boundary or height of domain is 110%c and length of the domain along Y- direction is 450%c.

Defining the boundary conditions

In this Rectangular Volume, the surface of the rectangle facing toward the leading edge or front side of the wing acts as the inlet side of the wind tunnel, and the surface towards the trailing edge as the exit of the wind tunnel. The inlet side of the domain as shown in Figure 158.6 with a green arrow, is assigned with a Normal Velocity condition with a magnitude of 125 m/sec, whereas at the outlet side of the wind tunnel, Gauge Pressure is set to 0 Pascal. The rest of the faces of the wind tunnel are side, top, and bottom surfaces assigned Slip/Symmetry conditions. This Slip/Symmetry condition is provided to consider viscous effects as negligible and avoid boundary layer effects.

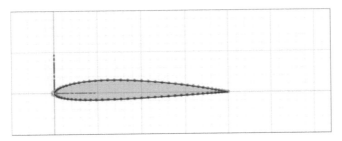

Figure 158.1 Geometry of NACA 2412 at 0°

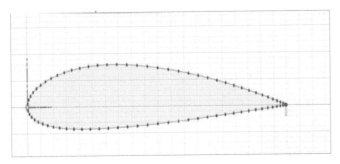

Figure 158.2 Geometry of NACA 2412 at 0° with Gurney flap

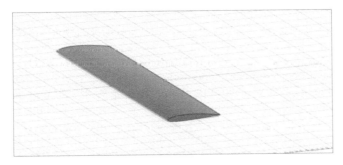

Figure 158.3 3-D Wing of NACA 2412 airfoil

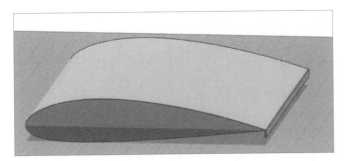

Figure 158.4 NACA 2412 airfoil wing with Gurney flap at trailing edge

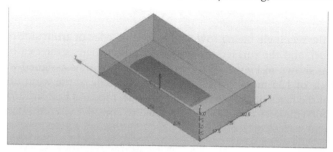

Figure 158.5 Wind tunnel around wing

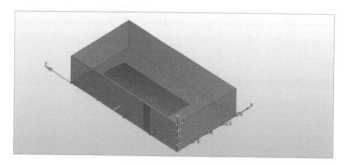

Figure 158.6 Boundary conditions for wing tunnel

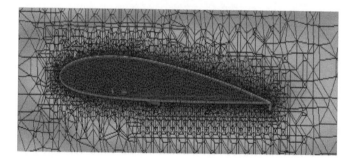

Figure 158.7 Mesh generation around airfoil

Meshing process

Once the wind Tunnel was created with the proper boundary conditions, meshing was performed on the model. For the model to analyse the flow and pressure distributions along the wing surface, it is necessary to provide a meshing having a suitable element size. For this purpose, while performing meshing for all conditions, a minimum of 31,000 elements and 10 layers around the edges of the airfoil are provided to obtain a proper meshed model for simulation. During the meshing of the airfoil wing with a gurney flap, a special or manual meshing was performed at the flap area to capture accurate results.

Setting initial conditions for analysis

Table 158.1 Parameters of CFD analysis

Model used for turbulence	Standard k-epsilon
Air density (ρ)	1.269 kg/m³
Number of iterations	100
Angle of attacks	0°, 4°,8°,12°,16°,20°
Gurney flap height	0.02 m (2% of chord Length)
Dynamic viscosity	0.00001817 kg/(m-second)
Gurney flap thickness	0.005 m (0.5% of chord length
Inlet velocity for wind tunnel	125 m/second
Boundary layers during meshing	5–10

Simulations results

CFD Analysis and simulation results of NACA 2412 (Mark and Götz., 2008). Airfoil wing without Gurney flap: -

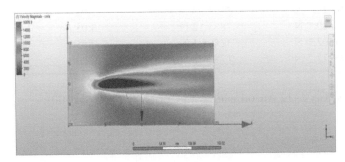

Figure 158.8 Velocity-magnitude contour

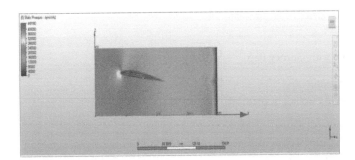

Figure 158.9 Static pressure contour

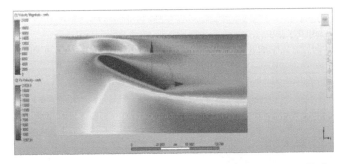

Figure 158.10 Vx-velocity contour (velocity contour across X-direction)

Figure 158.11 Vz-velocity contour (velocity contour across Z-direction) with Gurney flap

Table 158.2 Analysis results for NACA 2412 using standard k-epsilon

Sr. No	Angle of attack (α)	Lift (In Newtons)	Drag (In Newtons)	Lift to drag ratio (L/D)
1	0°	1469.25 N	3613.52 N	0.41
2	4°	8200.38 N	3958.79 N	2.07
3	8°	16026.5 N	5504.94 N	2.91
4	12°	23620.7 N	7835.12 N	3.01
5	16°	11502.7 N	32585.1 N	0.35
6	20°	38377.7 N	17177.8 N	2.23

CFD Analysis and simulation results of NACA 2412 Airfoil wing with Gurney flap (Yoo, 2000): -

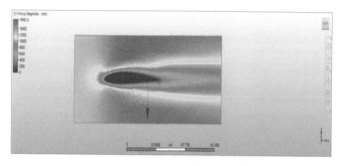

Figure 158.12 Velocity-magnitude contour with Gurney flap

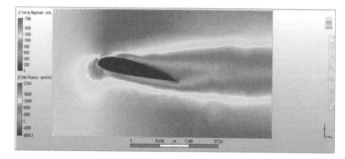

Figure 158.13 Static pressure contour with Gurney flap

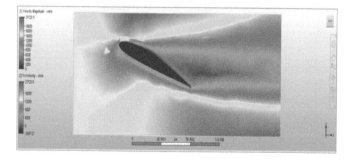

Figure 158.14 Vx-velocity contour (velocity contour across X-direction) with Gurney flap

Figure 158.15 Vz-velocity contour (velocity contour across Z-direction) with Gurney flap

Table 158.3 Simulation results for NACA 2412 with gurney flap using standard k-epsilon

Sr. No	Angle of attack (α)	Lift (In Newtons)	Drag (In Newtons)	Lift to drag ratio (L/D)
1	0°	3539.2 N	3593.78 N	0.98
2	4°	11635.4 N	15716.2 N	0.74
3	8°	23913.6 N	6473.31 N	3.69
4	12°	33123.3 N	9849.23 N	3.36
5	16°	42067.6 N	14957.9 N	2.81
6	20°	49334.1 N	20894.8 N	2.36

Result

(Maughmer and Bramesfeld, 2008). In a plane NACA2412, the results obtained are described in Figures 158.16–158.18. A steady increase in lift and drag force values can be observed until stalling begins after 12°. Before stalling, lift force values were higher than drag force, but after stalling drag force values began to rise, decreasing lift to drag ratio. This stage occurs as airflow was higher on the top surface (suction side) of the airfoil than on the bottom surface (pressure side) before stalling, which means suction side pressure is less than the lower surface pressure of the airfoil which provides lift to the wing. When the wing exceeds the critical angle of attack, airflow at the pressure side becomes higher than on the suction side, thus pressure

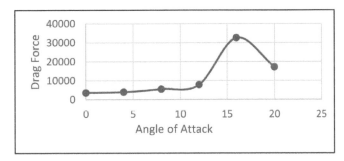

Figure 158.16 Lift vs attack angle

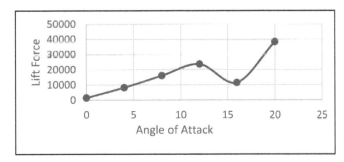

Figure 158.17 Drag vs attack angle

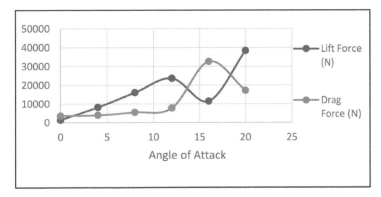

Figure 158.18 Lift and drag force comparison

difference between both sides decreases, and the wing drag force increases. This process before stall angle is seen in Figure 158.9, where plane airfoil at 8° by is represented by pressure plane, in which suction side pressure represented by cyan colour is less than the pressure at pressure side shown in light green colour.

(Yoo, 2000). For NACA2412 with Gurney flap at the trailing edge, results obtained from simulations are shown in Figures 158.19–158.21. It can be observed that values of lift force in Figure 158.19 for a gurney flap airfoil are greater as compared to a plane airfoil in Figure 158.16. Along with this improvisation in lift force, as we have observed in previous research work, the drag force also increases, but this increment is very low as compared to the lift force value. Therefore, by comparing Figures 158.18 and 158.21, it shows that using the Gurney flap we can achieve a higher lift and drag ratio for the same velocity and angle of attack as compared to plane airfoils, without any external force or energy application.

In Figure 158.22, the lift and drag ratio curve for both the airfoils at a different angle of attack is plotted for comparison of results for plane and gurney flap airfoils. We can see that initially, the Gurney flap causes an increase in the lift to drag ratio but as we raise the Attack angle, the lift to drag ratio value increases and improves the aerodynamic performance of the airfoil. By comparing the highest values of lift to drag ratio for both airfoils, the airfoil with a gurney flap has a 22.5% greater lift to drag ratio value than the plane airfoil.

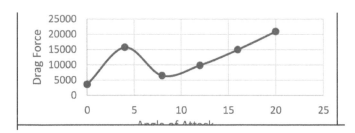

Figure 158.19 Lift vs attack angle

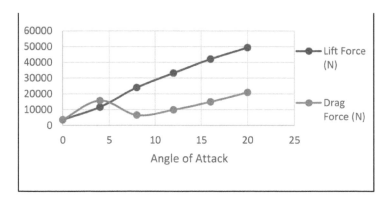

Figure 158.20 Drag vs attack angle

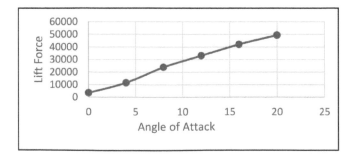

Figure 158.21 Lift and drag force comparison for Gurney flap airfoil wing

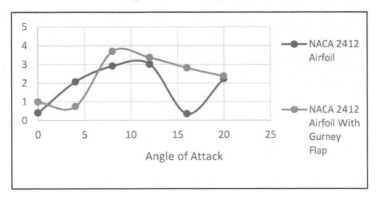

Figure 158.22 L/D comparison of plane NACA2412 airfoil and NACA2412 airfoil with Gurney flap

Conclusion

In this article, after performing CFD Analysis on plane and gurney flap attached airfoil at 0°, 4°, 8°, 12°, 16°, and 20° angle of attack, we obtained the values of lift and drag forces for every condition and these results after simulations were used for giving following conclusions: -

1. The gurney flap wing gives a higher lift force at the same velocity and angle of attack than the normal wing.
2. Gurney flap extends the camber of the airfoil, resulting in better airflow along the flap surface and reducing premature boundary layer separation which increases the lift properties of the airfoil.
3. Height of 2% chord length is suitable and is the optimum value for the gurney flap to get the benefit of lift augmentation from the airfoil before stalling occurs. Even though the thickness of the flap has less impact on the gurney flap, gurney flap thickness should be less than 1%c between 0.250.75% as it can increase load and drag force of airfoil, resulting in degradation of airfoil characteristics.
4. The maximum lift force on a gurney flap attached airfoil, at a 20° attack angle is about 28% more as compared to a plane airfoil at the same velocity.
5. In Figure 158.22, the lift and drag ratio of the normal airfoil compared with the gurney flap airfoil, shows that the gurney flap has significant importance in improving the aerodynamic features of an airfoil. Gurney flap airfoils have a higher lift to drag ratio than the typical NACA airfoils, indicating that the lift force increases with a smaller increase in drag force. In this analysis, the gurney flap increased the lift and drag ratio by 22.05%.

Thus, using a Gurney flap improves overall the efficiency of the airfoil without any complex mechanism or device.

Acknowledgment

We are grateful to Siemens centre of excellence, Department of Mechanical Engineering, Yeshwantrao Chavan College of Engineering, Nagpur for providing valuable guidance and insightful comments wherever necessary to complete this research work.

References

Aramendia, I., Saenz-Aguirre, A., Fernandez-Gamiz, U., Zulueta, E., Lopez-Guede, J. M., Boyano, A., and Sancho, J. (2018). Gurney Flap Implementation on a DU91W250 Airfoil. Proceedings 2(23):1448. https://doi.org/10.3390/proceedings2231448

Chakraborty, S., Bhattacharjee, S., Karmakar, S., Sadhukhan, S., Mukherjee, A., and Kumar, R. (2021). A Comparative Study of Boeing 737 and NACA 2412 Airfoils using CFD. Int. J. Eng. Res. Technol. 9(11).

Chen, H. and Bo, C. 2022. Lift Enhancement of Tiltrotor Wing Using a Gurney Flap. Int. J. Aerospace Eng. Hindawi. https://www.hindawi.com/journals/ijae/2022/1245484/

Fadl, A., Abbasher, M., Taha, O. Y., and Seory, A. E. (2018). A numerical investigation of the performance of wind turbine airfoils with gurney flaps and airfoilshape alteration. J. Eng. Sci. Technol. 16.

Graham, M., Muradian, A. and Traub, L. W. (2017). Experimental study on the effect of gurney flap thickness on airfoil performance. J. Aircraft. 55:16. 10.2514/1.C034547.

Kumar, A., Chaubdar, P., Sinha, G. S., and Harichandan, A. B. (2021). Performance analysis of NACA4412 airfoil with gurney flap. Proceedings of International Conference on Thermofluids, pp 167–176.

Li-Shu, H. and Gao, Y. W. 2019. Effect of Gurney Flap Geometry on a S809 Airfoil. Int. J. Aerospace Eng. Hindawi. https://www.hindawi.com/journals/ijae/2019/9875968

Maughmer, M. D. and Bramesfeld, G. (2008). Experimental Investigation of Gurney Flaps. J. Aircraft. 45:2062–2067.

Saha, S. K., Alam, M. M., and Hasan, A. B. M. (2017). Numerical investigation of gurney flap aerodynamics over a NACA 2412 airfoil. AIP Conf. Proc. 1980.

Yang, J., Yang, H., Zhu, W., Li, N., and Yuan, Y. (2020). Experimental study on aerodynamic characteristics of a gurney flap on a wind turbine airfoil under high turbulent flow condition. Appl. Sci. 10(20):7258. https://doi.org/10.3390/app10207258

Yoo, N. S. (2000). Effect of the Gurney flap on a NACA 23012 airfoil. KSME Int. J. 14:1013–1019. https://doi.org/10.1007/BF03185804

159 Improve Stabilisation of black soil by Bagasse ash, coir fibre, and plastic strips

Prashant B. Pande[a], Jayant M. Raut[b], Shantanu R. khandeshwar[c], Kartik C. Mankar[d], Anikesh K. Rangari[e], and Vaibhav R. Suryawanshi[f]

Department of Civil Engineering, Yeshwantrao Chavan College of Engineering, Nagpur, India

Abstract

Black cotton soil is located over many regions of India, exclusively in Maharashtra, Gujarat, Madhya Pradesh, and some regions of Karnataka. Black cotton soil is the clay-rich soil. The key constituent of this soil is montmorillonite which regulates the water holding capacity. Thus, the swelling and shrinking behaviour is often noticed during wet and dry condition. Owing to high swelling and shrinkage characteristic of this soil, the foundation and other soil structures encounter with the failure. To overcome this problematic conduct of black soil, stabilization of soil is necessary to improve the engineering characteristic of soil and make it suitable as a foundation as well as construction material. Stabilisation of soil refers to improve strength of soil which is sufficient research work was conducted out by the other researchers. Solid waste can be used as a reinforcing material to reduce the probability of failure accordingly to improve the strength of soil. In this research work is attempted to enhance properties and hence stability of by using solid waste materials which is one of the prominent alternatives without altering other useful engineering properties of soil. Also waste products are cheap, easily available, and environmentally friendly.

Bagasse ash, coil fibre, plastic is used as an admixture in this research work. The liquid limit, permeability, and the strength test i.e. direct shear and unconfined compression were performed on the soil with and without admixture. The significant enhancement in soil properties as well as in the strength is observed. The liquid limit of soil decreased from 11 to 5. The shear strength is increased from 11 to 22.

Keywords: Black cotton soil, shear strength, coir and plastic fibre, soil stabilisation.

Introduction

Soil stabilisation is nothing but treatment of mechanical, physical and chemical characteristic of soil by adding some materials which the characteristics properties of soil. Thus, on transformation of this engineering characteristic of soil, the overall shear strength and bearing capacity is improved. Physical properties of soil get change which results in improvement permanently by doing stabilisation of soil. Properties consist of bearing capacity, mechanical strength, shear strength, permeability, compressibility, durability and plasticity. Soil stabilisation can be achieved mechanically by adding cementing material, pozzolanic materials, fibres materials admixture as bagasse ash, lime, fly ash, rice husk, coconut husk, coir fibre, plastic and other waste products, etc. chemically by adding chemicals which effectively enhance the characteristic engineering properties of soil and also stabilisation is done physically by compaction. Soil stabilisation is done to avoid crack and failure in foundation and to withstand structural load, Impact loads, Seismic load and all other loads and transmission of force by construction work. Soil stabilisation by using waste product is ongoing trend and necessity in all over the world. By using waste product like plastic, polymer, fly ash, coconut husk, rice husk on one hand there is stabilisation of soil and on other hand solve the problem of excessive production and deposition of this waste byproduct and these materials are also hazardous to nature.

Black cotton soil is very expansive in nature and can be found in various region of India such as in Maharashtra, Tamil Nadu, Uttar Pradesh, Madhya Pradesh, etc. It contains montmorillonite mineral; due to this it has swelling and shrinkage property when there is extremely change in moisture content. Various elements are present in this soil like lime, magnesium, iron in adequate quantity. It also includes nitrogen, phosphorous and organic matter in low proportion. Hence this soil is more productive in lowlands rather than on uplands. Beside this, it has low bearing capacity and are highly compressible. Due to presence of swelling and shrinkage tendency, it causes a variation in the volume of soil that leads to the formation of cracks in the foundation or in the construction of building structural system. Due to the formation of cracks, durability and strength of structure gets greatly reduced causing the failure of structure. The load

[a]prashantbpande21@gmail.com; [b]jmrv100@gmail.com; [c]khandeshwar333@yahoo.com; [d]kartikcmankar@gmail.com; [e]13aniketrangari@gmail.com; [f]vaibhavsuryawanshi015@gmail.com

bearing capacity is excessively low causing several types of failure like general failure, punching failure, etc. therefore soil stabilization of expansive soil is required before construction work.

Literature Review

Trivedi et. al (2013) investigates the index properties by adding pre-determined fraction of Fly Ash in soil. Soil is greatly affects by addition of smallest fraction of Fly Ash and also all index properties change with the adding Fly Ash in soil. The sample is collected from site near the Godhra (Gujarat), and Fly Ash is added (w/w) in proportion of 10, 20, 30, 40 % and experiments is done in laboratory. For 10% adding Fly Ash the value of OMC gives maximum value 29.27% from 21.38%. Adding 20% of fly ash experimental laboratory values of all index properties gives its maximum value, excluding OMC. Adding 20% fly ash experimental laboratory values of MDD remains same at 1.41gm/cc. The UCS value changes from 0.148 to 0.152 (N/mm^2) by adding 20% fly ash content. In adding 20% fly Ash value of CBC is 20.53% from value 5.64%. Further for the range of 20 to 30% index properties decreases rapidly and for range of 30 to 40% the index properties remain stable. Therefore, observed that for adding 20% of fly ash content best results of soil stabilisation is obtained.

Kodicherla and Nandyala (2019) investigate the stabilization of soil by randomly mixing of fly ash and Coir Fibre and experimental values of laboratory test on soil are carried out and observation is done. By addition of percentage (%) of fly ash plastic index decreases by adding 0-30% of fly ash the MDD value decrease from the [17.5] to [15.8] MDD value [KN/m^3]. The UCS value of soil increased by up to 50% after adding 20% fly ash. By adding 20% fly ash, CBR test value is increased by 6% further addition of 1% of coir fibre to this mixture the CBR test value increased by 9%.

Badiger (2019) studied the strength of soil by adding coir fibre. Shear capacity of soil is increase by % addition of coir fibre. By adding 2% of coir fibre the MDD value decreases from 18.13 to 13.29 in standard proctor test. By adding 0-2% of coir fibre the maximum shear strength is improve from 0.537 to 1.2 in Vane shear test. The compression strength is increase from 3.25 to 19.92 by adding 0 to 2% coir fibre in unconfined compression test.

Kumar et al (2017) wanted to upgrade the strength of clay soil so he prepared soil-bagasse ash-lime mixture. He prepared nineteen specimens to examine the properties of soil with the addition of 5%, 10%, 15%, 20%, 25% and 30% of Bagasse ash along with 0%, 5%, and 10% lime of the above specimens. He conducted UCS test, Standard Proctor test to inspect the OMC, MDD and compressive strength of soil mixture. He determined that in standard proctor test, dry density increases by increment bagasse ash percentage upto 20% and therefore it started decreasing. He also concluded that there was a decrease in OMC with increasing bagasse ash percentage and also marginal increase in MDD with increase in lime percentage.

In bagasse ash potential ability. In their study Hydrated lime, bagasse ash is mixed with the black cotton soil. They prepared samples using different proportions of bagasse and hydrated lime with [0%], [6%], [10%],[18%],and 25% by the dry mass at a proportion of 3:1, respectively. They conducted (FSR) Free Swell Ratio, Unconfined compression test and California bearing ratio test for the samples by curing period of (3), (7), and (28) days. The result obtained explained that stabilization of black cotton soil by using bagasse ash, hydrated lime enhances the strength and facilitates subsist environmental problems caused by sugar industry waste materials.

Rao and Chittaranjan (2011) used various agricultural waste as ash of rice husk, bagasse, sugarcane, groundnut shell in order to stabilise expansion of soil. He treated subgrade soil with these waste materials with percentage as 0%, 3%, 6%, 9%, 12%, 15%. He conducted CBR test for each percent. The result obtained shows great improvement with the increment in CBR value upto certain optimum content.

Kiran R.G. et.al (2013) add different percentage of bagasse ash (4%, 8% and 12%) and cement in expansive soil. He tested various strength parameter like UCS test, CBR test. He observed that MDD value rise in by 1.516 [g/cc] to [1.65] g/cc with the incorporation of 8% bagasse ash by [8%] cement content. It also shows increase in CBR value [2.12] to 5.43 with incorporation of 4% bagasse ash and with 8% cement content. The UCS value had changes from [84.92] [KN/m^2] to [174.91] KN/m^2 with addition of 8% bagasse ash with 8% cement.

In 2018, Akansha and Mittal, attempted the enhancement in properties of black soil by utilising Baggase ash and coir fibre. They performed California Bearing Ratio (CBR) test by utilising various rates of Bagasse ash of virgin soil. It is observed that addition of 10% debris and 4% coir fibre in natural soil gave the maximum bearing capacity and strength in the specimen. The percentage increase in the new CBR value for 2.5 mm penetration in submerged and unsubmerged is recorded as 52.42 % and 159.65%.

Arthi, et .al (2017) studied the enhancement in the properties of black cotton soil by incorporation coir pith in various quantities. She conducted various test like OMC, MDD, Atterberg limits and UCS and the results were computed. The result obtained showed that 2 to 3% addition of coir pith has reduced the plasticity index as well as there is effective increase in the UCS, MDD value of black soil. It also increases CBR strength with increment in coir pith percentage.

Sayali et al. (2017) studied experimental programs managed for stabilization of expansive nature of soil using plastic waste in Amravati, Andhra Pradesh. Conducted sequence of CBR test to find the required optimum amount of plastic content to maximise the value CBR. They concluded that by the adding 4% plastic content in soil CBR percentage increases and then it started decreasing with further increment. Thus, they proved that by using plastic as soil stabiliser is economical, convenient can also reduce the risk of increasing pollution.

Earlier in Chitranjan (2012) has worked on soil stabilisation by using the contents as sugarcane baggass ash for stabilising the sub grade soil. He treated subgrade of soil the baggass ash individually [0%], [,6%], and [12%]. Different tests were conducted on different soil by various proportion of baggass ash and result has shown great rise in CBR value along with rise in ash proportion upto certain optimum proportion.

In 2013 B.M. Patil et al. investigate the impact of Baggassash, run batted in grade 81 considering characteristics by base course and subgrade of soil for versatile pathway. In general, in the market many took clay soil material as subgrade and grades 3 material as base course. But he took completely different proportions for getting optimum mixture. Thus, soaked cosmic radiation value for subgrade clay soil and grade 3 material results were seen very much changed due to baggass ash and run batted in grade 81. Thus, optimum combined he get for subgrade soil is 76:20.04 along with 77:20:03 for base course. In 2013 he checked hydrocarbon mixture victimisation Ash being Substitution to rock powder was acting like filler. Rutting test was carried out for the impact of extreme temperature stability of Bituminous concrete. Low atomic number 20 used fly ash for Analysis. He also tried six different check samples of 5%, 6% and 7% of filling material. Optimum hydrocarbon material has changed his mind to think by Marshall method. Raised temperature solidity had been derived as a live of dynamic solidity. An increase in dynamic solidity observed 27.4% to 30% as amount of ash accumulated by (5% to 7%).

Jianhing et al. (2013) examined bituminous blend taking Fly ash as an alternative for lime-stone dust was acting as filling material. Considering impact of excessive temperature balance of Asphalt concrete become analysed with the aid of using rutting check. Less calcium fly ash turned into material being used. Six check samples had been organised with 5%, 6% and 7% of filling content. Optimum asphalt Content material become decided with the aid of using Marshall way. High temperature balance was derived as a degree of dynamical balance. A boom in dynamical balance become stated from 27.4% to 30% along with amount of the ash become improved 5% to 7% accordingly.

Materials

Expansive soil

The expansive soil was used procured from the Nagpur region which was air dried and then pulverized. The soil used for experimentation was sieved with 4.75 mm. The index properties of soil comprising particle size distribution, consistency limits and the specific gravity were performed as per IS. It was found as clay with high plasticity i.e. (CH). The results of sieve analysis are as shown in Figure 159.1. The various soil parameters regarding index properties of the expansive soil are enlisted in Table 159.1.

Table 159.1 Soil index properties

Soil properties	Values
Liquid limit (w_L)	50.00%
Plastic limit (w_p)	24.1%
Plasticity index (I_p)	25.9%
specific gravity (G)	2.64
Maximum dry unit weight (Υ_{dmax})	14.4 kN/m^3
Optimum moisture content (w_{opt})	25%
Gravel	0%
Sand	9.47%
Silt	47.23%
Clay	43.30%
USCS (IS) Classification	CH

Figure 159.1 shows a graph between number of blows and the water content.

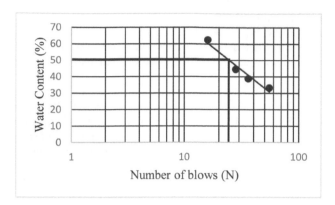

Figure 159.1 Liquid limit

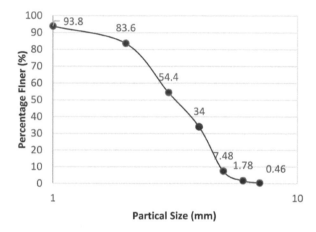

Figure 159.2 Sieve analysis results

The above figure 159.2 shows a particle size distribution curve.

Bagasse ash

The bagasse ash of sugarcane was procured from a sugar factory near to Nagpur. According to Cordeiro et al. (2008) the basic characteristics of ash relied on particles size of bagasse ash. The 425 μ sieve was used for sieving the ash. The bagasse ash was (Figure 159.3) having specific gravity of 2.36 and non-plastic behaviour was observed after classification.

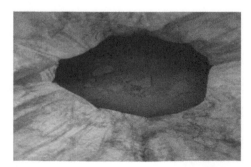

Figure 159.3 Bagasse ash

Experimental Work

Liquid limit

Soil may appear in one of four states; solid, semi-solid, plastic and liquid which rely on water content. In each state, the consistency and behaviour of a soil is different and as a result its engineering properties too. Thus, the change in the soil's behaviour can decide the boundaries between each state of the soil. Liquid limit is one of the very important property of cohesive soil and it gives information regarding the state of consistency of soil on the site. Liquid limit can also help in determining the consolidation properties of soil while calculating permissible bearing capacity and settlement of foundation. Liquid limit is determined by using Cassagrande's apparatus. Water content at which 25 blows were essential to cause the groove to close gives the liquid limit value.

First liquid limit of normal black cotton soil is determined by Cassagrande's apparatus (Figure 159.4). Then liquid limit of (soil + 15% bagasse ash) mixture is determine. Similarly liquid limit of (soil + 30% bagasse ash) mixture determined. Variation of liquid limit is compared.

Permeability test

Permeability is a measure of how easily the liquid can flow through porous medium. It gives idea about the stability of foundations, seepage through embankments and helps in solving the problems relating yield of water strata, stability of earthen dams' seepage through earthen dams, and embankments of canal bank exaggerated by seepage, settlement etc. In general, there are two methods of determining permeability of soil i, e, constant head permeability method and other one is falling head permeability method. For more permeable soils or soil with more discharge, constant head permeability method is used and for less permeable soils or soil with low discharge falling head method is applicable. The coefficient of permeability is determined by using Jodhpur Permeameter apparatus.

First coefficient of permeability is determined for normal black cotton soil and then coefficient of permeability for mixture of soil + 30% bagasse ash is found and result is observed.

Direct shear test

Direct shear test is a used to measure the shear strength properties of the soil. The cohesive soils have some problems in controlling the strain rates in either of the drained or undrained loading. In granular soil the loading is always drained.in many engineering problems like design of foundation, retaining walls, slab bridges, pipes, sheet pilling, the value of angle of internal friction and cohesion of the soil are needed for the design. A specimen is poured in the shear box and a confining stress is applied vertically to the specimen and the upper ring is pulled laterally until the sample fails. A failure envelope of the soil can be obtained from the result when compares between normal and shear stress and will show us the cohesive intercept. The value of the graph which is the shear strength parameters is necessary for the further determination of the results.

Direct shear test is carried on normal soil with 15% water content also test is carried on mixture of soil with 30% bagasse ash + 1% plastic fibre + 0.2% coir fibre at 15% water content.

Test is carried on normal soil with 25% water content also test is carried on mixture of soil with 30% bagasse ash + 1% Plastic fibre + 0.2% coir fibre at 25% water content and result is discussed.

Figure 159.4 Determination of liquid limit using Casagrande apparatus

UCS test

The unconfined compression test (UCS) is used for determining the unconfined compressive strength of soil. It is primarily useful for saturated, cohesive soils recovered from thin sampling tubes and not applicable for cohesionless or coarse-grained soils. This test is strained controlled. When the sample is loaded on the equipment, the pore pressure (water within the soil) undergoes changes that do not have enough time to dissipate. UCS refers or represents the actual max uniaxial compression that a sample may withstand without being deformed (without failure).

Similarly, UCS test is carry on normal soil with 15% water content also test is carried on mixture of soil with 30% bagasse ash + 1% plastic fibre + 0.2% coir fibre at 15% water content.

UCS test is carry on normal soil with 25% water content also test is carried on mixture of soil with 30% Bagasse ash + 1% plastic fibre + 0.2% coir fibre at 25% water content and result is discussed.

Result and Discussion

Effect of bagasse ash on liquid limit of soil

Table 159.2 give the values of liquid limit with different proportions of soil and bagasse ash. It is found that liquid limit goes on decreasing with increase in the proportion of bagasse ash in the soil.

Liquid limit of normal sample soil i. e soil without any admixture is found to be 50%. The liquid limit for the soil mixed with 15% bagasse ash is found to be 33.35% as shown in Figure 159.5. The liquid limit for the soil mixed with 30% bagasse ash is found to be 28.90% as shown in Figure 159.6. There is decrease

Table 159.2 Liquid limit with different proportions of soil and bagasse ash

Bagasse ash (%) mixed with soil	Liquid limit (%)
0	50
15	33.35
30	28.90
40	26.36

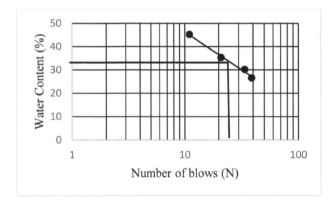

Figure 159.5 Liquid limit for admixture of 15% bagasse ash

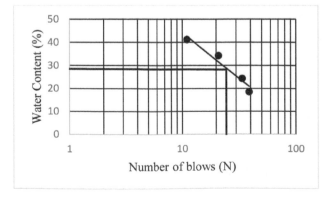

Figure 159.6 Liquid limit for admixture of 15% bagasse ash

in liquid limit by adding bagasse ash. The liquid limit for the soil mixed with 40% bagasse ash is found to be 26.36% as shown in Figure 159.7.

Liquid limit decreases with increase in percentage of bagasse ash as presented in Figure 159.7. This indicates that the plastic property of soil can be reduced the above graph shows the decreasing nature of the liquid limit with the in the percentage of bagasse ash with the soil.

Initially when 15% bagasse was added the liquid limit decreases to significant decrease in liquid limit by 27% with respect to normal soil. Although when 15% bagasse was added the liquid limit decreases, the rate of increase was dropped to 13.34%. In case of 40% bagasse ash the liquid limit reduced by only 8.78% which is very much lower than the falling percentage observed initially.

Permeability test

Table 159.3 shows the coefficient of permeability (K) values for normal soil and with mix soil (soil + bagasse ash).it is found that permeability decreases as we increase the percentage of bagasse ash. The rate of decrease in coefficient of permeability (K) is observed to be more as compared to the further increase in percentage of bagasse ash as shown in Figure 159.8.

The permeability test was carried out to understand the behaviour of soil for various percentage of bagasse ash and percent reduction in permeability was found to be 27.06% (Figure 159.9).

Coefficient of permeability of normal sample soil i. e soil without any admixture is found to be 8.371×10^{-3} cm/s. When the soil mixed with 15% bagasse ash the coefficient of permeability obtained as 6.105×10^{-3} cm/s. The coefficient of permeability for the soil mixed with 30% bagasse ash is found to be 5.638×10^{-3} cm/s. The liquid limit for the soil mixed with 40% bagasse ash is found to be 5.179×10^{-3} cm/s.

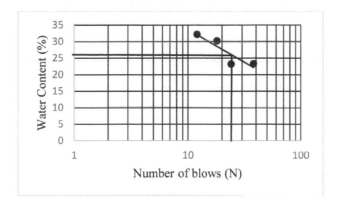

Figure 159.7 Liquid limit for admixture of 40% bagasse ash

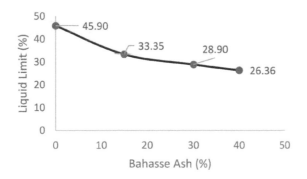

Figure 159.8 Variation of liquid limit with bagasse ash (%)

Table 159.3   Values of K for a different percentage of bagasse ash

Bagasse ash (%)	K (cm/s)
0	8.371×10^{-3}
15	6.105×10^{-3}
30	5.638×10^{-3}
40	5.179×10^{-3}

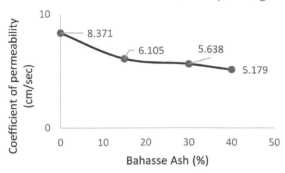

Figure 159.9 Variation of coefficient of permeability with bagasse ash

These results show that the bagasse ash reduces the permeability. It is found that increasing the percentage of bagasse ash to the soil decreases the permeability to a considerable extent initially i.e., approximately 30%. The significant reduction is observed in the addition of 15% later it goes slightly lower reduction in the values of coefficient of permeability. It is revealed that the effect of bagasse ash on the coefficient of permeability is limited 15 to 30%.

Unconfined compression test

In this test, two samples were prepared. First sample is normal soil and other is normal soil with 30% bagasse ash along with 1% plastic fibre and 0.2% coir fibres for the water content of 25% and 15% respectively and the compressive strength is computed.

(i)   For WC 15%
      Table 159.4 represents the values of unconfined compressive strength for normal soil and mix soil. For water content of 15% the unconfined compressive strength shows a considerable rise with the addition of bagasse ash, coir fibre, and plastic fibre respectively.
(ii) For WC 25%
      Table 159.5 represents the values of unconfined compressive strength for normal soil and mix soil. For water content of 25% unconfined compressive strength shows a great increment and it rises to 187 KN/m².

As this test is conducted to determine the compressive strength of the soil, addition of bagasse ash shows a great increment in the compressive strength to a great extent approximately about 11 to 22% respectively (Figures 159.10 and 159.11).

Direct shear test

In this test, normal stress and shear stress are determined for both normal soil and mix soil (bagasse ash +1% plastic fibre + 0, 2% coir fibre). Then after determining this stresses graph is plotted between these two stresses in order to determine C and φ.

Table 159.4 Unconfined compressive strength values for the samples at WC of 15%

Description of material	Unconfined compressive strength (KN/m²)
Normal soil	139
Mix soil with 30% bagasse ash + 1% plastic fibre+0.2% coir fibre	155

Table 159.5 Unconfined compressive strength values for the samples at WC of 25%

Description of material	Unconfined compressive strength (KN/m²)
Normal soil	153
Mix soil with 30% bagasse ash + 1% plastic fibre+0.2 % coir fibre	187

The above Table 159.6 represents shear strength for particular load and proportion. For a water content of 15% the percent increase in shear failure increases gradually to a considerable load but it started decreasing as the load is increased.

Table 159.7 represent shear strength for load and proportion. For a water content of 25% the percent increase in shear failure increases gradually to a considerable load but it started decreasing as the load is increased.

The direct shear strength test results shown significant improvement in the shear strength of about 35 to 45% approximately (Figure 159.12).

Figure 159.10 UCS specimen of normal soil

Figure 159.11 UCS specimen of mix soil

Table 159.6 Shear strength for 15% water content for given loads and proportion

Contents	Loads (Kg)			
	1.416	2.832	4.248	5.664
Normal soil	*% increase in shear at failure*			
Mix soil with 30% bagasse ash + 1% plastic fibre + 0.2% coir fibre	47.313	39.786	34.562	30.692

Table 159.7 Shear strength for 25% water content for given loads and proportion

Contents	Loads(kg)			
	1.416	2.832	4.248	5.664
Normal soil	*% Increase in shear at failure*			
Mix soil with 30% bagasse ash + 1% Plastic fibre+0.2% coir fibre	45.273	33.267	24.675	18.167

Figure 159.12 Shear specimen after testing and dial gauge reading

Conclusion

The intention of proposed research work to is fulfilled as the findings shows the improvement in engineering characteristics of black cotton soil by use of bagasse ash, coir fibre and plastic strips. It is also concluded that strength of black cotton soil can be effectively improved by treating soil with bagasse ash, plastic, and coir fibres. Use of black cotton soil with bagasse ash, plastic and coir fibres reveal a considerable rise in shear strength and decrease liquid limit. Inclusion of bagasse ash has a slight effect on liquid limit. With the addition of 15% bagasse ash liquid limit decreases by about 10.55% and with further addition in bagasse ash i.e. by 30% it decreases by about 4.99%. As bagasse ash is a fine particulate material, it fills up the voids which results in decrease in permeability. The decrease in permeability was found to be 27.06% respectively. The increase in compressive strength at 15% water content was found to be 11.51% whereas at 25% water content it was found to be 22.22% respectively by addition of coir fibre. The shear strength of the soil also increases significantly to about 35% to 45% by use of plastic strips.

References

Ahmed, A. G. A. (2014). Fly ash Utilization in Soil Stabilization. International Conference on Civil Biological and Environmental Engineering, CBEE. 76–78.

Anas, A., (2011). Soil Stabilization Using Raw Plastic Bottles. Proceedings of Indian Geotechnical Conference. 15–17.

Arora, K. R., (2004). Soil Mechanics and Foundation Engineering. Delhi: Standard Publishers Distributors.

Arthi, P. D., *Ahire, A. R., More, G. S., and Nikam, P. A.* (2017). Stabilization of black cotton soil using coir pith. Int. *Res. J. Eng. T*echnol. 2.

Chamberlin, K. S. (2014). Stabilization of Soil by Using Plastic Wastes. IJETED. 204–218.

Choudhary, A. K., Jha, J. N., and Gill, K. S. (2010). A study on CBR behavior of waste plastic fibre reinforced soil. Eurasian J. Educ. Res. 15(1).

Dia, J. and Liu, Z (2012). Influence of Fly Ash Substitution forMineral Powder on High Temperature Stability of Bituminous Mixture. Energy Procedia. 16:91–96.

Dutta, M. (1997). Waste disposal in engineered landfills. London: Narosa Publishing House.

Gautam, A. and Mittal, S. K. (2018). Stabilazion of black cotton soil using bagasse ash and coir fiber. Int. J. Adv. Res. Ideas Innov. Technol. 4(5):724–727.

Hassan, H. J. A., Rasul, J., and Samin, M. (2021). Effects of plastic waste materials on geotechnical properties of clayey soil. Transp. Infrastruct. Geotechnol. 8:390–413.

IS 2720 (Part 2) -1973, for moisture content.

IS: 2720 (Part 7) – 1980, for light compaction.

IS: 2720 (Part 16) – 1987 (Re-affirmed 2002), for UCS.

Kiran R. G. and Kiran L. Analysis of strength characteristics of black cotton soil using bagasse ash and additives as stabilizer. Int. J. Adv. Res. Sci. Eng. Technol. 7.

Kumar, V. K. M and Parkash, V. (2021). Soil Stabilization of Clayey Soil Using Jute. Int. J. Res. Technol. Stud. 4(10):15513–15519.

Madavi, S. D., Patel, D., and Burike, M. 2017. Soil stabilization using plastic waste. Nagpur, India: Department of Civil Engineering, SRPCE.

Manikandan, A. T., Ibrahim, Y., Thiyaneswaran, M. P., Dheebikhaa, B., and Raja, K. (2017). A study on effect of bottom ash and coconut shell powder on the properties of clay soil. Int. J. Adv. Res. Sci. Eng. Technol. 4(2):550–553.

Manuel, M and Joseph, S. (2014). Stability analysis of kuttanad clay reinforced with PET bottle fibers. Int. J. Adv. Res. Sci. Eng. Technol. doi: 10.17577/IJERTV3IS110331

Nagle R. and Jain. R. (2014). Comparative study of UCS of soil, reinforced with natural waste plastic material. Int. J. Sci. Eng. Res. 4(6):304–308.

Patil, B. M., Patil, K. A. (2013). Effect of Pond Ash and RBI Grade 81 onProperties of Subgrade Soil and Base Course of Flexible Pavement. Int. J. Civil. Arch. Struct. Construc. Eng. 7(12).

Raj, P. P. 2005. Soil Mechanics and Foundation Engineering. India: Pearson Education.

Rao, A. V. N and Chittaranjan, M. (2011). Applications of agricultural and domestic wastes in Geotechnical Applications. J. Environment. Res. Dev. 5(3).

Raut, J. M., Khadeshwar, S. R., Bajad, S. P., and Kadu, M. S. 2014. Simplified design method for piled raft foundations. In Advances in Soil Dynamics and Foundation Engineering, ASCE, ed. R. Y. Linag, J. Qian, and J. Tao, 462–471. Shanghai, China: National Academics of Sciences, Engineering, and Medicine.

Saini, H., Khatti, J., and Acharya B. (2019). Stabilization of black cotton soil by using sugarcane bagasse ash. Int. J. Scientif. Res. Rev. 7(1):109–116.

Trivedi, J. S., Nair, S., Iyyunni, C. (2013). Optimum Utilization of Fly Ash for Stabilization of Sub-Grade Soil using Genetic Algorithm. Procedia. Eng. 51:250–258.

160 A literature review on pavement friction measurement methods on different pavement conditions

Yogesh Kherde[1], Saloni Patne[1], Shubhangi Golhar[1], Aniket Pathade[2,a], Mayur Banarse[3], and Sneha Atram[1]

[1]Civil engineering department, YCCE, Nagpur, India

[2]Jawaharlal Nehruh, Medical College, Datta Meghe Institute of Medical Sciences, Sawangi (M) Wardha, India

[3]Civil engineering department, Ram Meghe Institute of Technology and Research, Badnera, Amravati, India

Abstract

Driving safety improvement has become the most focused area in transportation systems. The friction between pavement surface and wheel plays an important role in maintaining safety and sufficient braking distance. Therefore, improvement in pavement friction is one of the significant parameters to maintain adequate skid resistance in reduction of road accidents. Several factors such as material characteristics, construction methods, climatic conditions, surface texture and moisture condition creates an unique impact on skid resistance of various types of pavement further it aggravates due to progressive wearing and polishing pavement surface, hence it is recommended to adopt specific friction coefficient to enhance driving performance and road safety, which can be evaluated by adopting appropriate techniques and methodologies for measuring frictional properties and skid resistance of the pavement surface. Much research has been carried out in this area and various measurement methods and different equipment's with test procedures have been developed to measure friction factor and slip resistance worldwide.

This paper summarise the key research studies on the pavement friction measurement and describe various methods for measurement and evaluation of skid resistance to provide a comprehensive overview of the pavement surface friction measurement techniques. This paper also attempts to frame guidelines for improvement and development of new methods for friction resistance measurement.

Keywords: Friction factor, pavement, measurement devices, skid resistance.

Introduction

Road engineers around the world are engaged with problems faced by road friction and accident risks associated. As it is the force which opposes the motion between wheels and surface of motion. Many research provide the limiting value of friction coefficient (Wallman and Åström, 2001; Yan et al., 2020). This limiting value decides various parameters necessary for pavement construction. These values may also vary according to temperature, pavement texture variation, climatic variation (Solminihac et al., 2008). Additionally, Pavement friction developed at contact of tyre-pavement, majorly depends upon texture of pavement (Yan et al., 2020). The microtexture and macrotexture of pavement surface is responsible for skid resistance, rolling resistance, tyre wear and visibility loss due to accumulated water. These properties can be adjusted by pavement design. Texture of pavement is affected by factors like aggregate size and type, texture orientation. Deviation in the layer of surface from true planar surface results in pavement texture. Pavement texture is categorised into a range of amplitudes (A) and wavelength (λ) by world road association as given in Table 160.1 (Kouchaki et al., 2018).

Nowadays where the whole lot is developing and updating, automotive provides higher speed opportunities to users (Mataei et al., 2016). Manoeuvres of vehicles such as cornering and braking require sufficient friction between tyre-pavement surface to maintain stability of the vehicle. These manoeuvres also cause wear-tear, these leads towards skid resistance and tear wear capacity of textured pavement also affects surface drainage (Gunaratne et al., 200). The pavement surface should be designed in such a way that it provides comfortable and safe condition to the users. Skid resistance affects the driving safety on road surface and pavement and has hilarious effects on accidents occurring, mainly on wet pavement which results in low skid resistance value. Hence it is obvious to emphasize research of characteristics of pavement surface

---

[a]aniketpathade@gmail.com

Table 160.1 Wavelength and amplitude of pavement texture

Texture type	Wavelength (λ)	Amplitude (A)
Micro texture	$\lambda < 0.5$mm	A = 1–500µm
Macrotexture	0.5mm $< \lambda <$50mm	A = 0.1–20mm
Mega texture	50mm $< \lambda <$ 500mm	A = 0.1–50mm

(Mataei et al., 2016). Previously conventional tests were performed for texture property and skid resistance determination like drainage tests, sand patch test, grip tester, accelerated polishing, locked wheel skid tester, sideways-force coefficient, overflow meter, circular track meter (Yan et al., 2020; Solminihac et al., 2008). Tests mentioned are generally performed on the field after construction of pavement. And various lab tests such as British pendulum (BPT) is used all over the world but is found lacking over some areas such as data fluctuation on coarse surfaces. Secondly Dynamic Friction Tester is high speed testing of pavement friction properties, but some special hand-held compactor and special mold is required (Yu et al., 2010). Several researchers are working on different devices and methods for fast and economic and more accurate analysis of pavement surface (Mataei et al., 2016).

This paper presents a research study on the portrayal of friction characteristics on pavement surfaces, as well as a discussion of the various methodologies and technologies utilized for evaluating and quantifying pavement characteristics. Firstly, this paper summarises the understanding of pavement friction, and detailed study on depiction of the surface of pavement and its frictional properties. Then, friction mechanics and factors affecting the properties of friction of micro texture and macro texture surfaces and other parameters affecting skid resistance and friction coefficient. After that, the discussion on advantages and disadvantages of methods and devices currently used for friction coefficient and properties measurement was elaborated. Lastly concluding solutions and ideas are proposed for better results and reduction in current defects.

Pavement friction overview

Defining Pavement friction, vertical force related to horizontal force developed while tyre-pavement interaction as tyre moves along surface. To the driver friction is a measure of time required to stop a moving vehicle or can say safety measures whereas adequate friction reduces the accident rate. According to ASTM E 867 (2006), specification friction resistance 'prevention of loss of traction by surface travelled' (Mataei et al., 2016; Kowalski et al., 2010). Minor or major accidents caused are directly or indirectly related to pavement friction. hence, detailed study on pavement parameters and tire pavement relations is necessary (Mataei et al., 2016).

1) Friction coefficient is dimensionless coefficient (μ) defined as normalised friction, i.e., frictional reaction divided by normal reaction.
2) Skidding is movement of locked wheel/tyre on pavement.
3) Skid resistance is a friction reaction developed at interaction of tire and surface.
4) Relative velocity between wheel and surface of motion at Centre of interaction area is known as slip speed.
5) The resulting quotient when slip speed is divided by operating speed is called longitudinal slip

Friction level varies by pavement characteristics like texture and vehicle tire. Improving friction between tire surface and pavement depends upon texture, speed, water presence on surface (Masad et al., 209). Also, construction materials, weathering influence, construction techniques (Dewey, 2001).

Various authors have given their theories of friction and its characteristics. Moore and (1972) and Do (2000) have explained friction on hysteresis friction and characteristics. Elastomers of tyre while tyre surface interaction, the adhesion and hysteresis friction, and its consequences are explained by Moore (1972) Figure 160.1. The stopping distance on wet pavement is affected by interaction of elastomers, hysteresis and adhesion forces, adherence or intermolecular binding force at surface level creates adhesive components. Vander Waals force also known as dipole forces between surface irregularities exposed to one another provides attraction between each other and does not allow any further movement (Mataei et al., 2016; Persson, 2000).

Adhesion force is dominant up to the critical slip, the true contact area on interaction of tyre and surface. Actual contact intermediate of tyre and surface of pavement shows shear strength and adhesion. Adhesion friction is major until critical slip is reached. Two third of friction force is included under adhesion friction on wet pavement while speed driving (Mataei et al., 2016). Hysteresis friction at rough surface shows energy loss, as the elastomers undergo deformation alternately (i.e. the vehicle tyre undergoes compression and returns to its normal state) (Lindner, 2004).

Yandell related texture scales and hysteresis friction (Don et al., 2000). Kummer (1966) hysteresis friction is maximum when sliding speedily. And adhesion is maximum when sliding with slow sp eed. Factors affecting friction are listed in Table 160.2 as presented by Wallman and Åström (2001).

Pavement surface and aggregates affected by surface type, surface age, texture of surface, aggregates and its geological properties, then vehicle factors like speed, angle made by tyre in the wheel motion direction, and slip ratio, and loading factors including geometry of road, environment and flow of traffic and temperature (Wallman and Åström, 2001).

Characteristics of pavement surface

Macro texture refers to the irregularity of the rough texture of the paved surface. This coarse aberration arises from the voids between the aggregates. In asphalt pavement the extent of those parts depends on the properties of the mixture (aggregate form, distribution of coarse combination and size) utilised in the pavement construction. The friction physical phenomenon element is laid low with the macro texture (Noyce et al., 2005; Kouchaki et al., 2018; Masad et al., 2008).

The tyre treads depth and macro-texture influenced the hydroplaning in two ways. Ibrahim (2007) observed that a 0.5–3 mm variation in surface roughness causes a 16 km/hour (10 mph) difference in hydroplaning onset. To prevent hydroplaning, raise the micro texture of the pavement between 0.2 and 0.5 mm and increase the hydroplaning speed by up to 20% (Ong et al., 2005).

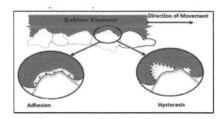

Figure 160.1 Schematic plot of tyre pavement friction

Table 160.2 Parameters affecting pavement friction

Pavement surface characteristics	• Microtexture • Macrotexture • Megatexture • Unevenness • Material properties • Temperature

Vehicle operating parameters	• **Slip speed**
	  – Vehicle speed
	  – Braking speed
	• **Driving maneuver**
	  – Turning
	  – Overtaking

Tire properties	• Foot print • Tread design and condition • Rubber compression and hardness • Inflation pressure • Load • Temperature

Environment	• **Climate**
	  – Wind
	  – Temperature
	  – Water (rainfall, condensation)
	  – Snow and ice
	• **Contaminants**
	  – Debris, dirt, mud
	• Materials (salt, sand) anti-skid

Note: Critical parameters are shown in bold

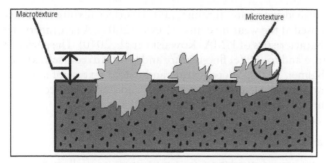

Figure 160.2 Pavement surface micro and macro texture

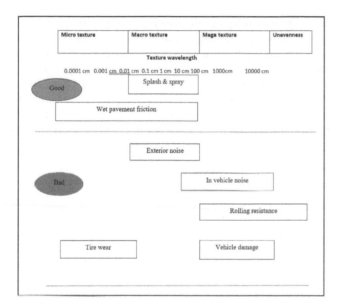

Figure 160.3 Texture wavelength influence on tire-pavement interaction

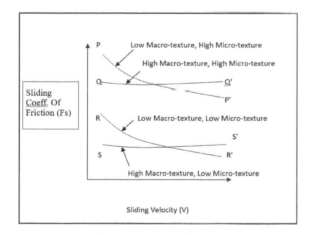

Figure 160.4 Effect of pavement surface microtexture and macrotexture on pavement

There are few direct methods, such as profilometers and indirect methods, such as the sand patch, grease patch, and British pendulum test for determining pavement texture. The sand patch and grease patch methods are used to indirectly measure macro texture, whereas the British pendulum technique is used to directly measure macro texture. They discovered a number of connections between wear rate and pavement friction of tyre on concrete pavement, which they were able to derive using real texture attributes (Gunaratne et al., 2000).

Micro texture reduction occurred to the polished aggregate (Ibrahim, 2007). Surface texture is influenced by the characteristics of coarse aggregate exposed at the wearing course (Dewey, 2001). A laboratory device is developed to evaluate pavement friction characteristics of HMA (Kowalski et al., 2010). The surface texture is classified into (a) smooth, (b) fine texture and round (c) fine-texture and gritty, (d) coarse-texture and rounded, and (e) coarse-texture and gritty (Balmer, 1978). Micro texture and macro texture of a pavement surface could be classified into (a) smooth and polished surface, (b) smooth and harsh surface, (c) rough and polished surface, (d) rough and harsh surface (Kokkalis and Panagouli, 1998).

Environmental factors impact on skid resistance

The key environmental elements that affect tyre pavement friction are temperature, rainfall, and pollution.

Influence of temperature

Seasonal variations in pavement skid resistance are a significant component, and previous experimental investigations have investigated the topic over the last few decades. Summer wet road surface skid resistance is lower than winter wet road surface skid resistance. Because the lowest values of skid resistance are attained during the summer, design is essential. This explains why, as temperature rises, hysteresis losses fall and rubber resilience rises. Both variables interact to lower the value of skid resistance as there is rise in temperatures (Kennedy et al., 1990), resulting in decreased microtexture and macrotexture (Henry, 2000). Evidence from in-service micrographs shows that due to natural aging, the aggregate surface micro roughness increases in winter and aggregates microroughness decreases in summer due to increased traffic polishing (Smith, 2008).

Influence of rainfall

Water in the form of rain or condensation causes short-term changes in skid resistance. The slip resistance measured immediately after the rain may show higher values than in periods of drought. This shows that there is a relationship with the impact of pollution from vehicle oils and their pollutants (Henry, (2000). Water mixed with dust, mud, and oil in the low bed during a dry time minimizes friction on the road surface. When measurements are obtained right after a wet season, this impact is diminished. Because rainfall cleans the road of contaminants. It has been demonstrated that when there is a drought, the number of landslides falls and reaches a minimum after seven days without rain (Hill and Henry, 1981). The link between temperature and precipitation affects is not well understood. Temperature variation impacts on microtexture, while rain and freeze-thaw cycles affect macrotexture (Ongel and Harvey, 2009).

Contaminants

Dirt, sand, oil, water, snow, and ice are the most common contaminants found on the road surface. Contamination's impact on surface aggregates were explored. When contaminants combine with various aggregates, a substantial change in behaviour is noticed, which speeds up polishing. The addition of fine but hard impurities significantly reduces the measured slip resistance and helps to polish the surface of the aggregate. While the addition of coarse but harsh impurities significantly improves the measured slip resistance and the surface scratches and rubs (Wilson, 2013).

Friction measurement methods and devices

There are several methods and equipment can be adopted to measure the slip resistance of road surfaces. Each device has a specific function and each device measures slightly different parameters. The slip resistance measurement can be divided into two methods: 1. Field measurement 2. Laboratory and portable tester

The force that happens when the tyres are constricted sliding on the road surface is used to determine field slip resistance. To reflect the true state of the field, these field measuring methods must be exact, highly repeatable, and reproducible.

Device based on locked wheel principle

The locked wheel device named ASTM E-274 is used to evaluate the pavement friction in the U.S. For testing smooth tyre (ASTM E-524) and ribbed tyre (ASTM E-501) are used (Mataei et al., 2016; Henry, 2000).

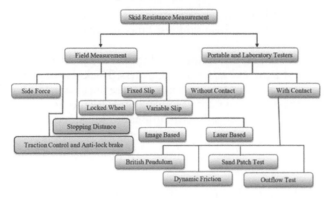

Figure 160.5 Classifications of friction measurement methods

The gadget is mounted on a trailer and pulled behind the vehicle that is being measured (Leu and Henry, 1978). In this method, the operator applies a sudden brake and computes the frictional properties of pavement surface without using anti-lock brakes. These measurements are done at an operational speed of 40 m/hour and (64 km/hour).

The various disadvantages of locked wheel devices are listed (Saito et al., 1996). The price of a locked wheel device is higher than other field devices (Kummer and Meyer, 1963). In a locked wheel device slip speed will be 100%.

Device based on side force principle

The lateral force device monitors the amount of road friction in proportion to the vehicle's travel direction. In addition, the lateral force measurement device determines the lateral guide force perpendicular to the movement direction. When cornering, the lateral force device reflects the vehicle's capacity to maintain control (Hall et al., 2009). The continuous friction measurement on the test track is a benefit of the lateral force measuring equipment (Hall et al., 2009). The MuMeter and the Lateral Force Coefficient Road Inventory Machine (SCRIM) are two measuring devices of the lateral force device. The yaw angle of Mu-meter (ASTM E-670) is 75 degrees and SCRIM is 20 degrees. The Side-force device is sensitive to micro textures and not to macro textures (Kowalski et al., 2010; McCarthy et al., 2018). Mu-Meter (ASTM E670) is widely used on airport runway (Yan et al., 2020). The operational speed of the vehicle is 40 m/hour and 64 km/hour. These devices can also perform the measurements at high speed.

Device based on fixed slip principle principle

These devices measure pavement friction at constant slip between 1020% slip speed (Mataei et al., 2016). These devices used anti- lock braking conditions to measure pavement friction. The various fixed slip devices that are widely used are runway friction tester (RFT), airport surface friction tester (ASFT), norse meter (ROAR), saab friction tester (SFT) and grip tester (GT). De Solminihac et al. (2008)studied the step-by-step process, harmonise and analyse the skid resistance values measured with GT. The grip test measured skid resistance values are expressed in grip number (GN) (Wallman and Åström, 2001; De Solminihac et al., 2008). Device based on stopping distance principle.

To determine pavement friction stopping distance measuring is the easiest method. When the driving vehicle reaches its desired speed, it locks the wheels and measures the speed of the vehicle up stopping point from the start point. ASTM E445, a method for determining stopping distance on the surface with a passenger wheel and tyre, is described. The coefficient of friction is as follows:

$$\mu = v^2/2 \times g \times d$$

Where,
μ = coefficient of friction
d = stopping distance
v = vehicle brake application speed
g = acceleration due to gravity

Stopping distance is important because with time skid resistance of a pavement deteriorates under traffic and actual stopping distance of pavement will vary with age. Also skid resistance of pavement surface will differ with wet-weather.

In case of car ASTM smooth tyres sliding on smooth pavement, stopping distance is dependent on factors such as tyre type, loading, inflation pressure, thread path and water film thickness. Thickness of water film increases at a decreasing rate as on wet pavement stopping distance increases, because thickness of water film increases at a decreasing rate as skid resistance on wet pavement decreases. Stopping distance increases with vehicle speed on wet pavement, because square of speed is directly proportional to the stopping distance. Hence, control on water film thickness is important to decrease stopping distance (Ong and Fwa, 2010).

Device based on variable slip principle

Friction estimation on tyre road can be approached by variable slip (Gustafsson, 1993). Variable slip works on the function of slip of tyre and pavement surface. Slip is relative motion of tyre and pavement surface. Device variable slip gives initial slip curve variation according to tyre material and pavement surface effects on friction value after the peak value of the curve. Some of the devices are French IMAG, ROAR, SALTAR system.

Device based on traction control and anti-lock brake principle

Traction control system is vehicle which manages force applied on pavement surface by vehicle tyre. This system includes roll stability, electronic stability, traction control, etc. (Falconer, 2020). Anti-lock brake systems and traction control systems are used to collect information about friction on pavement. Studies state that both the systems are reasonable.

Laboratory skid resistance measurement methods

The dynamic friction tester (DFT) and the British Pendulum Tester (BPT) are two portable testers used to analyse road surface friction (AASHTO T 278 or ASTM E 303). (ASTME 1911). At low speeds, the equipment may also be used in the laboratory and in the field. Frictional properties are determined by the kinetic energy lost by the revolving disc or pendulum when it touches the road surface. The loss of kinetic energy in the friction force is used to assess road friction. The major benefit of DFT is its capacity to quantify friction at various speeds. At each measurement rate, the coefficient of friction evaluated with DFT and BPT showed a strong correlation. Both testers are very easy to use and portable (Mataei et al., 2016).

The volumetric technique is frequently used to determine the macro texture of pavement. This process involves evenly scattering a known amount of homogeneous sand, glass, beads, and crystals on the pavement surface in a circular pattern. The mean texture depth was calculated using the circle's diameter (MTD) (Balmer, 1978). Macro structure measurement can be evaluated by Outflow Meter Test (OFT) (Wallman and Åström, 2001; Balmer, 1978). Flow meter is used to measure pavement water drainage capacity (Aktaş et al., 2011).

The SPM equipment includes a wide screen, container, brush and disk (Hall et al., 2009). A total of 25 ml sand which is lower than 6.33 mm grains are used for making slurry and chip seal and also 50ml sand which is bigger than 6.33 mm are used for making chip seal. The operator pores a known amount of sand into a cleansed surface in a circular. The diameter of the circle is then determined, and the mean of four to five measurements is utilised to compute mean texture depth (MPD).

Recent technological advancements have resulted in the development of speedier and more reliable measuring methods. Some of the novel data collecting methods include interferometry, structured light, numerous 2-D profiling techniques, and the laser scanning position sensor (SLPS). Figure 160.8 depicts a variety of topographic data collecting approaches that work at or near the target size and may be used to road pavements (Johnsen, 1997).

Interferometry and the stylus profiling techniques are the two different methods used for the measuring topographic data at scales that cover a section of the target scales for determining pavement texture (Johnsen, 1997).

Structured light and the SLPS are the two new methods of acquiring surface topography. These methods proved that there is some limitation while measuring surface the surface asperities in the full range of different surface elevations. The SLPS is a specific technique designed specifically for acquiring topographic data from pavement surfaces. This device can be easily used for in-situ measurement (Johnsen, 1997).

Stereo photography is a qualitative historical tool for visual inspection of surface characteristics. Visual inspection requires a unique focusing tool and a pair of images (stereo pair), each taken at a specific angle perpendicular to the inspected surface. This technique has the potential to measure the topographic features of the surface. The pictures taken from the surface texture can be analysed by digital scanning systems and computer algorithms (Johnsen, 1997).

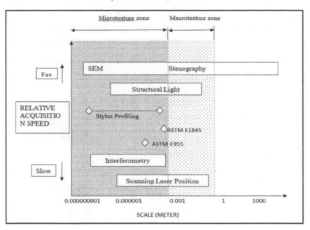

Figure 160.6 Different data acquisition methods

Conclusion

Increasing problems due to friction measurement have led towards various inventions and theories for friction measurement and texture characteristics. Still all of them have few disadvantages like working speed, accuracy, economic, etc. which makes these inventions and methods unpopular. Different methods are discussed in this paper giving their positive and negative points. Some aspects for reducing current limitations in future development. Providing a guide for deep examination of these methods.

Major disadvantage of these devices is consuming time for performance. The sand patch test, British pendulum, outflow test, circular test meter, dynamic friction are slow tests. The circular track meter, sand patch test, outflow test only used in small areas. Traffic control requires, and lane closure required for all above devices.

Some of the mentioned devices cannot work in some places or with different conditions. Talking about lock wheel testing devices, works only on straight segments of surface no curves, roundabouts, or T-sections can be included. Also, it can avoid slippery spots.

The fixed slip device is not accurate. It works on specific slip speed, which generally varies with critical slip speed value. Thus, variation is particularly seen on ice- and snow-covered areas. Huge quantity of water is required in the process and a skilled person for data reduction. Side force testing devices are reactive on irregular surfaces. cracks and potholes can even destroy tires. British pendulum test device shows undependable behaviour on coarse textured surfaces. Also overflow test works only on non-porous pavement.

The locked wheel tester, British pendulum tester, dynamic friction test devices collect spot measurement hence cannot use for continuous friction evaluation. It is more beneficial to use devices having low cost. Variable slip test and locked wheel test are costly methods as they require more operating cost. To determine speed dependency of skid resistance by locked wheel tester several repetitions must be done which is more time consuming.

Using suitable devices for network evaluation is more beneficial, British pendulum tester, dynamic friction test devices collect spot measurement hence cannot be used for network evaluation. The outflow test, sand patch, circular texture meter is also unable for network evaluation.

The condition of the surface of pavement impacts on skid resistance of pavement. The skid resistance varies in dry and wet pavement condition. The method which is able to measure pavement friction and surface texture in wet conditions is termed as a standard method. The dynamic friction, British pendulum test sand patch cannot measure skid resistance in wet conditions.

Till now friction coefficient measurement and resistance characteristics are examined only in direction of tire motion and normal to it, like in British pendulum test. But according to the critical other directions of resistance measurement, even diagonal directions of motion are responsible. Hence it becomes mandatory to develop a device and method which considers the diagonal directions for friction and resistance to skid measurement.

The measurement methods which are not affected by operator procedure and climatic conditions are more suitable to use. The operator procedure and wind affect British pendulum testers. The sand patch test is also affected by operator procedure. The methods which need complicated data processing are disadvantageous to use as it requires more time. Data processing is complicated in variable slip method.

Coming across all these flaws in methods and devices mentioned in Table 160.3, for friction characteristics used till date, innovation of some developed innovative methods and devices become mandatory.

Table 160.3 Overview of friction measurement devices

Test method	Associated Standard	Description	Equipment
Locked wheel (Wallman, and Åström, 2001; Choubane et al., 2004)	ASTM E-274	Low vehicle and locked skid trailer are required for testing. Water 0.05mm is placed on front test tire and the tire is lowered as needed and locks the tire by braking system. The resistive drag force and wheel load applied to the pavement is used to calculate coefficient of force. The friction is reported as friction number (FN) or skid number (SN).	
Side-force (Kowalski et al., 2010; McCarthy et al., 2018)	ASTM E-670	The average side force perpendicular to the direction of travel of the tire is measured to computed Mu number (MuN) or Sideway force coefficient (SFC)	
Fixed slip (Wallman and Åström, 2001; Yan et al., 2020) De Solminihac et al. (2008)	ASTM E-274	The device is mounted on a trailer or vehicle. Water 0.5 mm is placed on the test tire and tire rotation is prohibited up to certain percentage of vehicle speed by the anti-locking braking system. The friction is reported as friction number (FN).	
Variable slip (Gustafsson, 1993)	ASTM E -1859	Variable slip (0100%) works on the function of slip of tire and pavement surface. Wet pavement is prepared of water thickness (0.02 in {0.5 mm}) on which the wheel is rotated freely. tire speed is reduced and speed, rotation speed, travel distance, frictional force is noted at max 2.5 mm intervals.	
Stopping distance measurement (Yan et al., 2020; Ong, and Fwa, 2010)	ASTM E- 445	Water is sprayed on the pavement surface. A vehicle drives at the same speed (64 km/hour). The wheels of the vehicle are locked, and the travelling distance is measured until the vehicle stops. Otherwise, anti- lock braking system and different speeds have been utilised	
Deacceleration rate measurement (Mataei et al, 2016; Kowalski et al., 2010).	ASTM E- 2101	Testing is done in the winter season. Brakes are applied on a vehicle to lock wheels which travel at standard speed, up to the rates of deceleration is measured. For friction computation rate of deceleration is noted.	
Outflow Meter (Wallman and Åström, 2001; Balmer. 1978)	ASTM E-2380	This a simple volumetric test method used to measure water drainage ability through surface texture and interior voids. Its results are an indication of hydroplaning potential of a wet surface	
Circular texture meter (Balmer, 1978)	ASTM E-2157	Surface texture is measured by non-contact laser device in diameter 11.25 in (286 mm) circular profile of pavement surface at 0.868 mm interval. It rotates at 6 m/minute and produces mean profile depths and profile traces for pavement surface.	
British Pendulum Tester (Mataei et al., 2016)	ASTM E-303	The BPT is easy to use. It generates low-speed sliding contact between pavement surface and skidding tyre. The pendulum swing height gives the information about friction properties. Five readings are recorded, and data is collected	

| Dynamic friction test (Wallman and Åström, 2001; Yan et al., 2020; Persson, 2000) | ASTM E-1911 | Three rubber sliders are mounted on the lower surface of a disk that rotates with its plane parallel to the test surface. The torque required to rotate these rubber sliders is measured by dynamic friction tester. Watering is done automatically using a water tank at rate 3.6 L/minutes. Rotational torque and downward load are measured. The velocity is also measured to indicate the relationship between coefficient of friction and speed. | |
| Sand patch method (SPM) (Yan et al., 2020; De Solminihac et al., 2008) | ASTM E 965, ISO 10844 | Sand patch method is based on volumetric methods. Glass beads of known volume are spread over the circle on the surface and the mean texture depth is calculated. | |

References

Aktaş, B., Gransberg, D. D., Riemer, C., and Pittenger, D. 2011. Comparative Analysis of Macrotexture

Balmer, G. G. (1978). Pavement Texture: Its Significance and Development. Transp. Res. Record. 666:1–6.

Balmer, G. G. (1978). Transportation Research Record, (1978).

Choubane, B., Holzschuher, C., and Gokhale, S. (2004) Precision of locked-wheel testers for measurement of roadway surface friction characteristics. Transp. Res. Record: J. Transp. Res. Board. 1869:145–151.

De Solminihac, H., Chamorro, A., and Echaveguren, T. 2008. A friction management method to assess paved road networks. 7th International Conference on Managing Pavement Assests, TRB Committee AFD10.

Dewey, G. R. (2001) Aggregate Wear and Pavement Friction. Transportation Research Board, Annual Meeting, Washington DC, pp. 17.

Do, M.-T., Zahouani, H., and Vargiolu, R. (2000). Angular parameter for characterizing road surface microtexture. J. Transpor. Res. Board. 1723:66–72.

Falconer, H. 2020. United States Patent (10).

Gunaratne, M., Bandara, N., Medzorian, J., Chawla, M., and Ulrich, P. (2000). correlation of tire wear and friction to texture of concrete pavements. J. Mater. Civil Eng. 12(1):46–54.

Gustafsson, F. 1993. Slip-based Tire-Road Friction. Linkoping, Sweden: Linkoping University.

Hall, J. W., Smith, K. L., Titus-Glover, Wambold, J. C., Yager, T, J., and rado, J. 2009. Guide for Pavement Friction. Transportation Research Board of the National Academie.

Henry, J. J. (2000). Evaluation of pavement friction characteristics. Transport. Res. Board. 291.

Hill, B. J. and Henry, J. J. (1981). Short-term, weather-related skid resistance variations. Transp. Res. Rec. 836:76–81.

Ibrahim, M. A. (2007). Evaluation of the safety of flexible pavement using skid resistance measurements. Build. Environ. 42(1):325–329.

Johnsen, W. A. 1997. Advances in the Design of Pavement Surfaces. Phd, diis, Worcester Polytechnic Institute.

Kennedy, C. K., Young, A. E., and Butler, I. C. (1990). Measurement of skidding resistance and surface texture and the use of results in the United Kingdom, in: Surface Characteristics of Roadways. International Research and Technologies: 1st Symposium on Surface Characteristics, State College, Pennsylvania, USA. 1031:87–102.

Kokkalis, A. G. and Panagouli, O. K. (1998). Fractal evaluation of pavement skid resistance variations. I: surface wetting. Chaos, Solitons Fractals. 9(11):18751890. http://dx.doi.org/10.1016/S0960-0779(97)00138-0.

Kouchaki, S., Roshani, H., Prozzi, J. A., Garcia, N. G., Hernandez, J. B.(2018). Field investigation of relationship between pavement surface texture and friction. Transport. Res. Record. 2672(40):395–407.

Kouchaki, S., Roshani, H., Prozzi, J. A., Garcia, N. Z., and Hernandez, J. B. (2018). Transpor. Res. Record. 2672(40):395–407.

Kowalski, K. J., McDaniel, R. S., and Olek. J. 2010. Joint Transportation Research Program. Indiana: Purdue University.

Kummer, H. (1966). Unified theory of rubber and tire friction. engineering research bulletin B -94, pp. 100–101. Pennsylvania (U S): The Pennsylvania State University.

Kummer, H. and Meyer, W. (1963) Penn State Road Surface Friction Tester as Adapted to routine measurement of Pavement Skid Resistance. Road Surface Properties, 42nd Annual Meeting. 1963:1–31.

Leu, M. C. and Henry, J. J. 1978. Prediction of skid resistance as a function of speed from pavement texture measurements. Transport. Res. Record. 666:7–13.

Lindner, M. (2004). Experimental and analytical investigation of rubber friction. Safety. 200:300.

McCarthy, R., de León Izeppi, E., Flintsch, G. W., and McGhee, K. K. 2018. Comparison of locked wheel and continuous friction measurement equipment. 97th Annual Meeting Transportation Research Board, Washington, DC.

Masad, E., Rezaei, A., Chowdhary, A., and Harris, P. 2008. Taxas (U S): Texas Department of Transportation.

Masad, E., Rezaei, A., Chowdhury, A., and Harris, P. (2009). Predicting Asphalt Mixture Skid Resistance Based on Aggregate Characteristics. Transport. Res. Record J. Transport. Res. Board. 24–33.

Mataei, B., Zakeri, H., Zahedi, M., and Nejad, F. M. (2016). Pavement friction and skid resistance measurement methods: A literature review. Open J. Civ. Eng. 6(04):537.

Measurement Tests for Pavement Preservation Treatments. Transport. Res. Record. 2209(1):34–40.

Moore, D. F. (1972). The Friction and lubrication of elastomers. Oxford: Pergamon Press.

Noyce, D. A., Bahia, H. U., Yambo, J. M., and Kim, G. 2005. Draft literature review and state surveys. Madison, Wisconsin: Midwest Regional University Transportation Center(UMTRI).

Ong, G. P. and Fwa, T. F. (2010). Mechanistic interpretation of braking distance specifications and pavement friction requirements. Transp. Res. Record. 2155(1):145–157.

Ong, G. P., Fwa, T. F., and Guo, J. (2005). Modeling hydroplaning and effects of pavement microtexture. Transport. Res. Record. 1905(1):166–176.

Ongel, A., Lu, Q., and Harvey, H. 2009. Frictional properties of asphalt concrete mixes, Proc. Inst. Civ. Eng. Transp. 162 (1):19–26.

Persson, B. N. 2000. Sliding Friction: Physical Principles and Applications. Berlin: Springer Science & Business Media.

Saito, K., Horiguchi, T., Kasahara, A., Abe, H., and Henry, J. J. (1996) Development of portable tester for measuring skid resistance and its speed dependency on pavement surfaces. J. Transpor. Res. Board. 1536:45–51.

Smith, R. H. 2008. Analyzing friction in the design of rubber products and their paired surfaces. Boca Raton: CRC Press.

Wallman, C.-G. and Åström, H. (2001) Friction measurement methods and the correlation between road friction and traffic safety. Linköping, Sweden: Swedish National Road and Transport Research Institute.

Wilson, D. J. 2013. The effect of rainfall and contaminants on road pavement skid resistance. New Zealand Trans. Agency Research Report.515.

Yan, Y., Ran, M., Sandberg, U., and Zhou, X., and Shenqing, X. (2020). Spectral techniques applied to evaluate pavement friction and surface texture. Coatings. 10(4):424.

Yu, M., Xiao, B., You, Z., Wu, G., Li, X., and Ding, Y. (2010). Construct. Build. Mater. 258:119492.

161 Modal analysis of SiC reinforced Al 7075 composites

Srinidhi Acharya S. R.[a], and Suresh P. M.[b]

Department of Mechanical Engineering, ACS College of Engineering, Bangalore, VTU Research Centre, Belagavi, India

Abstract

Aluminium (Al) reinforced with silicon carbide (SiC) plays a key role in various engineering sector because of its advanced applications in automobile, aerospace, mechanical and civil engineering structures. The desirable properties like specific strength, weight to strength ratio, Stiffness, resistance to corrosion, larger fatigue life have become the widely accepted materials amongst researchers. The addition of SiC reinforcements in terms of various weight percentages (210%) enhances properties in terms of elastic, hardness thermal and vibration characteristics. The Al-SiC metal matrix composites (MMC's) is manufactured by stir casting technique which is one of the commonly used by researchers because of lower cost production.

Modal analysis is an excellent technique which is used to determine the dynamic responses of any structures. The structure vibrates at greater amplitude leading to resonance and this is essential to understand the frequency at resonance, damping characteristics and various mode shapes of the MMC's.

Majority of the work carried out is based on the fabrication of Al with various reinforcements and analysing the mechanical properties and comparing with the base metal. The dynamics behaviour of the MMC's is sparingly done that too on Al 7075 with SiC. This paper elucidates at the fabrication of Aluminium 7075 with SiC and to determine the dynamic behaviour using signal analyser, accelerometer sensor and an impact hammer. Fast Fourier transform (FFT) analyser and modal analysis software are have been used for the experimental analysis. Further to validate the results the finite element method (FEM) is employed using ANSYS. Test results reveal better correlation. The results indicated that the Al with 8% of SiC reinforcements leads to better damping characteristics.

Keywords: Fabrication, fast Fourier transform, finite element modal Analysis, metal matrix composites, modal testing.

Introduction

Metal matrix composite (MMC) is a combination of matrix and reinforcement which are fused at macroscopic or microscopic level. MMC's are obtained by scattering reinforcing material into a matrix metal. This dispersion at various levels gives rise to newer product with enhanced strength, stiffness, surface hardness and resistance to corrosion. Al7075 and silicon carbide were fabricated through stir casting process. The schematic representation of stir casting process is illustrated in the figure 161.1. X-ray diffraction (XRD) and micro structural scanning electron microscopy (SEM) examination was conducted to reveal the presence of SiC (2–10% wt) in Aluminium 7075 indicating the successful fabrication of MMC using stir casting process (Srinidhi and Sresh, 2021). Aluminum-Beryl MMC's was fabricated by varying beryl from 0–14 weight percentage. Stir casting technique was adopted to fabricate MMC (Sagar et al., 2018).

Al 2024 with beryl as the reinforcement was concocted using stir casting method. Various mechanical properties hardness, Toughness and strength in tensile were examined. Also, micro structural studies and XRD tests were done and the results showed the uniform and homogeneous dispersion of beryl particles (Sagar et al., 2018). Vibration attributes on Al5083 with various weight percentages of fly ash and SiC was studied. The FFT analyser were used to examine the vibration characterisation of MMC's fabricated by stir casting method. The dynamic response of MMC's showed that 2% of fly ash and 9% of SiC showed good damping behaviour. The experimental frequency and finite elemental Analysis results were in close relation (Ramanathan and Santhosh, 2019). Aluminium 5083 was reinforced with multi wall carbon nano tube of different diameters and pointed out that the mechanical properties were improved. Free vibration analysis using FEM was done to examine the natural frequency of the MMC's. Modal analysis results showed that the variations of frequency for MMC's when compared to base alloy. This variation was due to the addition of nano tubes (Samuel Ratna Kumar et al., 2017). Free vibration of A357 and SiC of dual Particle Size (3 wt. % coarse and fine powder, 4 wt. % coarse and 2 wt. % fine) are done using FEM. The effect on natural frequencies of A357/DPS-SiC was carried out considering weight fraction, boundary conditions, aspect ratio. The A357/DPS-SiC with 4wt % coarse and 2 wt% fine SiC revealed that the maximum natural frequency which leads to better rigidity and elastic modulus (Lakshmikanthan et al., 2021). Dynamic

---

[a]sri1660@gmail.com; [b]spm_dvg@yahoo.co.in

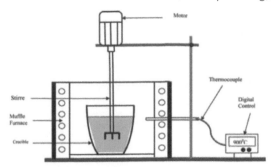

Figure 161.1 Schematic view of stir casting process (Srinidhi and Sresh, 2021; Sagar et al., 2018; 2020)

characteristics of two storied metallic component was carried using signal analyser, acceleration sensor and impact hammer. Frequency response functions (FRF's) were analysed using modal analysis experimental set up (Chandravanshi and Mukhopadhyay, 2013). Modal analysis experiment was performed on car roof with and without dampers for free-free boundary conditions. The test was performed using impact hammer, accelerometer and FFT analyser (Chandru and Suresh, 2017). Allien et al., (2019) aimed at analysing the effect of SiC reinforcement in Al 6082 and Al 7075 matrix. MMC's of Al6082 and Al7075 with SiC (0, 1, 2, 3, 4, 5, 7.5, 10, 15 and 20) different weight percentages were fabricated. The percentage damping ratio and natural frequencies of the MMC's were examined. The results showed that strength and stiffness were enhanced with increasing percentage of SiC.The MMC's with 15% SiC/Al7075 composite had better damping characteristics. Taj et. al., (2017) conducted experimental and analytical modal analysis Aluminium Graphite MMC's using FFT analyser. ANSYS was used to validate the experimental results. Analytical and experimental results were observed and final conclusion was obtained. The test results showed a minimum deviation of 1.210.2%. Damping ratio factor ranged from 0.0982 to 0.0313 which indicates the decreasing of density as the reinforcement is increased. Ramu and Reddy (2017) characterised dynamic responses like mode shape and natural frequencies of Al6061 flat plates. Experimental and simulation method of Modal analysis were conducted. The Aluminium Plate attached with thin tire tube of different dimensions which acts as damper was selected as the specimen for analysis. The initial five natural frequencies along with mode shapes were arrived. The test results clearly revealed that frequencies of Al flat plate and the plate covered with tire tube sheets have variations. This variation is because of difference in mass and property of the components. The plate covered with 3/4th tire tube sheets leads to better damping percentage ratio.

Kumbhar et. al. (2018) selected Al 6061 as the base materials with silicon carbide powders as the reinforcement. The influence of reinforcement on vibration characterization was examined. MMC's were fabricated by changing SiC (0, 3, 6, and 9 percentage by weight) adopting stir casting method. The results clearly showed that the addition of SiC in aluminium matrix increases natural frequency. Al6061 with SiC 9 wt.% showed maximum natural frequency leading to better damping characteristics. Ericson and Parker (2013) conducted a modal analysis for various shape accompanied by the specific natural frequency and damping factors. Kumar et. al. (2013) concluded that the FEM method is an advanced method for analysing vibration characteristics. The CFRP raft frame structure was selected and natural frequencies, mode shapes and damping factors for the first three modes were analysed.

Majority of the work done on the vibration analysis of composite plates referred in the literature review is either experimental or analytical methods. The present study aims at deriving the modal analysis and mode shapes of MMC's by experimental work for Al7075 reinforced with various proportions of SiC for fixed free boundary condition. Also to understand the dynamic responses of the MMC's using vibration dynamic experimental set up which includes Impact hammer, accelerometer and data signal analyser. fast Fourier transforms (FFT) analysers and frequency response functions (FRFs) are setup is utilised for experimental analysis and modal analysis respectively. Further, the results are validated using FEM. The experimental and theoretical results were compared to obtain the final results. Modal analysis is a newer and advanced strong tool adopted to analyse the dynamic responses of MMC's. The fabricated MMC's vibrates at its peak amplitude. Hence, it is important to understand the modal vales, mode shapes and damping factors of various MMC's in order to improve its strength for industry driven designs. The modal analysis tool with high configuration computer and digital analysis system is used to analyse the behaviour of dynamic responses. This analysis of dynamic responses of fabricated MMC's plays a magnificent role in the future applications of structural composites.

Objective of the Work

The present study involves the fabrication of Aluminium 7075 with several percentage (by weight) of Silicon Carbide (SiC) by stir casting method and the influence of Silicon Carbide particles on the vibration behaviour as less work is carried out, which thereby has opened up new avenues for study of vibration characteristics of this MMC's. The Al-SiC plates of dimensions 150 × 100 × 8 mm plates were fabricated and the following major research aspects were conducted.

- Modal analysis of Hybrid Metal Matrix Composites (MMC's)
- Influence of reinforcements on the dynamic behaviour.

The critical measures of frequency response (f) and damping factor (ξ) are obtained. Considering the scope of the work, the research work has been accomplished, with an important objective of exploring the use of Silicon carbide as the reinforcement in Al7075 matrix for enhanced vibration characteristics.

Material Selection

As per the modal test and analysis requirements of the components, Aluminium 7075 is used as the matrix and Silicon carbide (SiC) is used as the reinforcement. Aluminium 7000 series with zinc as the major element as shown in Table 161.1 is widely accepted Metal Matrix at various levels. The properties like increased ductility, high strength, greater toughness and high young's modulus etc makes as one of the extensively used MMC's. Table 161.1 shows the compositions of Al 7075.

The reinforcements are generally dispersed into the matrix material in order to strengthen the mechanical properties. Silicon carbide (SiC), also known as carborundum is a semiconductor in the form of particulates consists greater amount of silicon and carbon. It's a rare mineral which exists naturally. The various proportions of Silicon carbide are mixed with Aluminium 7075 using stir casting technique. The cast MMC's are obtained and the details are tabulated in Table 161.2.

Modal Analysis

Alnefaie (2009) clearly concluded that the modal analysis determines the natural frequency of any given sample or structure. These vibration responses, Mode shapes and natural frequency are of important parameters to withstand the dynamic load. During modal analysis the samples are dispersed as a model which is composed of finite elements (De Silva, 2007).

The unit mass matrix [M] and the unit stiffness matrix [K] are applied. The impact is f(t) and the unit damping matrix is [C]. The vibration differential equation of is given by

$$[M]\{\ddot{u}\} + [c]\{\dot{u}\} + [K]\{u\} = \{f(t)\} \tag{1}$$

Table 161.1 Aluminium 7075 compositions

Compositions	Percentage
Zinc	5.62–6.18
Magnesium	2.13–2.49
Copper	1.19–1.62
Silicon, manganese, titanium, chromium.	< 1

Table 161.2 Details of stir cast specimens

Specimen Designation	Al-7075 (Weight %)	SiC (Weight %)
Specimen 1	100	0
Specimen 2	98	2
Specimen 3	96	4
Specimen 4	94	6
Specimen 5	92	8
Specimen 6	90	10

frequency domain for equation of motion using Fourier transform is written as

$$\left[1 - \left(\frac{\omega}{\omega_n}\right)^2 + 2\xi\frac{\omega}{\omega_n}\right] X = F \tag{2}$$

Dynamic responses at any point due to external load are expressed as the linear equation as,

$$x_i(\omega) = \varphi_{i1} q_1(\omega) + \varphi_{i2} q_2(\omega) + \ldots + \varphi_{iN} q_N(\omega) = \sum_{r=1}^{N} \varnothing_{ir} q_r(\omega) \tag{3}$$

where φ_{IN} *represents* r^{th} *order modal coefficient.*

$$q_r = \frac{F_r}{K_r - \omega^2 M_r + j\omega C_r} \tag{4}$$

$$F_r = \varnothing_r^i F(\omega) = \sum_{j=1}^{N} \varphi_{jr} f_j(\omega), \quad (j = 1,2\ldots N) \tag{5}$$

The dynamic frequency response measured between the measured point m and the last excitation point l is given as

$$H_{ip}(\omega) = \frac{x_m(\omega)}{f_l(\omega)} = \sum_{r=1}^{N} \frac{\varnothing_{mr}\varnothing_{lr}}{K_r - \omega^2 M_r + j\omega C} \tag{6}$$

Upon rearranging, final transfer function is obtained as given as

$$H_{ip}(\omega) = \sum_{r=1}^{N} \frac{\varnothing_{mr}\varnothing_{lr}}{M_r\left[(\omega_r^2 - \omega^2) + j2\zeta_r\omega_r\omega\right]} = \sum_{r=1}^{N} \frac{1}{M_{er}\left[(\omega_r^2 - \omega^2) + j2\zeta_r\omega_r\omega\right]} \tag{7}$$

$$\zeta_r = \frac{C_r}{2M_r\omega_r} \tag{8}$$

$$M_{er} = \frac{M_r}{\varnothing_{mr}\varnothing_{lr}} \tag{9}$$

Young's Modulus and Density of MMC's

Thus, Hybrid Metal Matrix Composites (HMMC's) is analyzed using fast Fourier transformation (FFT) analysis and validated with the simulation using ANSYS software. The results of vibration tests are intricately analysed and tabulated. The mechanical dimensions of the materials are shown in Table 161.3. Using Hashin Shtrikman equation, the young's modulus of the MMC's is calculated using

$$E_{CM} = \frac{E_{ma}\left[E_{ma}(1-\epsilon_{rf}) + E_{rf}(\epsilon_{rf}+1)\right]}{E_r(1-\epsilon_{rf}) + E_{ma}(\epsilon_{rf}+1)} \tag{10}$$

E_{CM} = Young's modulus for Composite Material, E_{ma} = Young's modulus for the matrix element, E_{rf} = Young's modulus for the reinforcement, ϵ_{rf} = percentage of reinforcement.

For Example: E_{ma}(Aluminium) = 70 GPa; E_{rf} (Silicon Carbide) = 400 GPa ; ϵ_{rf} = 0.1 (for 10% of SiC reinforcement).

$$E_{CM} = \frac{70\left[70(1-0.1) + 400(0.1+1)\right]}{400(1-0.1) + 70(0.1+1)}$$

Therefore, analytical young's modulus value of MMC a 10% reinforcement is

E_{CM} = 80.57 GPa

Similarly, the young's modulus for other MMC's is tabulated in Table 161.3.
The density of the MMC's are calculated using the equation

$$\rho_{CM} = \rho_{Al}V_{Al} + \rho_{SiC}V_{SiC} \tag{11}$$

ρ_{CM} = Density of composite material; ρ_{Al} = Density of Aluminium = 2.7g/cm³;
ρ_{SiC} = Density of silicon carbide = 3.21 g/cm³.
V_{Al} and V_{SiC} are the volume fraction of aluminium and silicon carbide respectively.
    For Example: Volume fraction of reinforcement (SiC) is taken as 10%.

$$\rho_{CM} = (2.7 \times 0.90) + (3.21 \times 0.1) = 2.751 \text{ g/cm}^3$$

Similarly, the density of other MMC's is tabulated in Table 161.3.

Experimental Modal Analysis

During the analysis, the MMC's are divided into equal grids by drawing horizontal and vertical lines. The intersection points are called as nodes. Each node is designated with a number (node 1, node 2, node 3...... last node 35) as shown in Figure 161.2.a. The MMC Plates ate fixed at one end (which acts as cantilever) using fixtures and given a constant torque with the help of a Torque wrench. The modal analysis is carried out using data acquisition (DAQ) system with eight-channel USB-based voltage unit designed by Dewesoft as shown in figure 161.2.b. DAQ system collects the signals, analyses the signals and gives the output response on the display unit. For the excitation unidirectional impact test hammer with a scaling factor of 22.7 mV/N attached with an acceleration sensor is used. During analysis the acceleration sensor is fixed on a node 35 (Last Node on the MMC) using a beeswax. A small excitation is given to MMC's impact hammer on all nodes to measure the responses. The results obtained are then imported into modal analysis software and are processed for frequency response curve. The experimental modal analysis obtained on the display unit for various modes for specimen 1 is tabulated in the Table 161.4. Also, the five mode shapes of MMC's are shown in Figure 161.3.
    Similarly, the experimental modal analysis obtained on the display unit for various modes for Specimen 2 is tabulated in the Table 161.5.

Table 161.3 Plate dimensions and details

Specimen	Specimen dimensions (mm)			M (kg)	E (N/mm²)	ρ (kg/mm³)
	l	b	t			
Specimen1	150.3	100.2	8.0	0.327	70	2.7
Specimen2	150.2	100.2	8.1	0.332	72	2.7102
Specimen3	150.6	100.7	8.2	0.306	74.04	2.7204
Specimen4	150.8	100.3	8.3	0.324	76.15	2.7306
Specimen5	150.2	100.9	8.2	0.317	78.33	2.7408
Specimen6	150.9	100.2	8.1	0.325	80.57	2.751

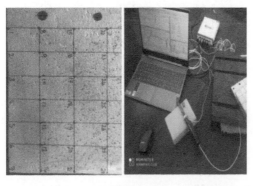

Figure 161.2 (a) MMC diving in to equal grids; (b) MMC with DAQ system

Table 161.4 Experimental modal analysis values for specimen 1 (100% aluminium)

Mode shape	1	2	3	4	5
frequency (Hz)	5.021	20.03	29.92	32.06	65.08

Similarly, the experimental modal analysis obtained on the display unit for various modes for Specimen3, 4, 5 and 6 are tabulated in the Table 161.6. The results obtained were validated using finite elemental analysis (FEA) Software. ANSYS was used for validation process. The model of the geometry was obtained and made as fixed free conditions by constraining all degree of freedom (DOF). The model is meshed with solid 124 elements with a total of 1128 nodes. The ANSYS modal value and modal shapes for all MMCs' are shown in Table 161.6 and Figure 161.4.

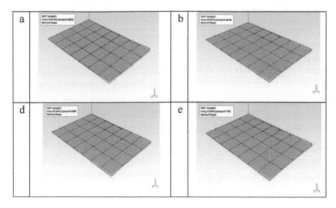

Figure 161.3 Experimental modal shapes of sample 1 (a) Mode 1; (b) Mode 2; (c) Mode 3; (d) Mode 4

Table 161.5 Experimental modal analysis values for specimen 2 (98% Al 7075 + 2% SiC)

Mode shape	1	2	3	4	5
frequency (Hz)	5.745	21.157	29.3585	33.1553	66.640

Table 161.6 Experimental and finite elemental frequency of Al-SiC specimens

Specimen	Mode shapes	F_1	F_2	$|(F_1-F_2)|$	% error	ξ
	1	5.02	5.74	0.72	14.34	0.486
	2	20.0	21.1	1.12	5.59	0.503
Specimen 1	3	29.9	29.3	0.56	1.87	0.568
	4	32.0	33.1	1.09	3.39	0.73
	5	65.0	66.6	1.56	2.39	0.81
Specimen 2	1	5.68	5.81	0.12	2.16	0.773
	2	20.8	21.4	0.57	2.75	1.2
	3	30.8	29.7	1.11	3.60	1.68
	4	34.9	33.5	1.37	3.92	1.91
	5	66.1	67.4	1.28	1.93	2.18
Specimen 3	1	5.71	5.88	0.17	2.97	0.405
	2	21.2	21.6	0.43	2.03	0.98
	3	29.8	30.0	0.23	0.77	1.25
	4	33.0	33.9	0.96	2.90	1.79
	5	67.0	68.2	1.18	1.75	2.36
Specimen 4	1	5.89	5.95	0.06	1.15	0.531
	2	21.7	21.9	0.22	1.02	0.524
	3	28.6	30.3	1.7	5.93	0.92
	4	35.9	34.3	1.57	4.37	1.94
	5	70.6	69.11	1.54	2.17	2.43
Specimen 5	1	5.99	6.028	0.03	0.50	1.1
	2	22.0	22.20	0.11	0.50	1.06
	3	29.8	30.8	0.91	3.04	1.62
	4	34.8	34.79	0.05	0.16	1.97
	5	69.8	69.98	0.09	0.12	2.61

Table 161.6 Experimental and finite elemental frequency of Al-SiC specimens

Specimen	Mode shapes	F_1	F_2	$|(F_1-F_2)|$	% error	ξ
Specimen 6	1	6.09	6.1	0.01	0.16	1.57
	2	22.1	22.47	0.295	1.33	0.68
	3	31.8	31.88	0.01	0.03	1.09
	4	35.1	35.22	0.04	0.11	1.94
	5	70.69	70.79	0.77	0.01	2.81

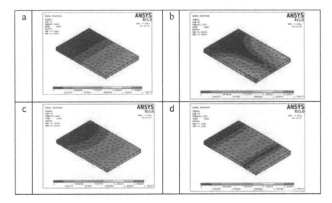

Figure 161.4 Finite elemental modal shapes of sample 1 (a) Mode 1; (b) Mode 2; (c) Mode 3; (d) Mode 4

Results and Discussion

The experimental natural frequency obtained from the FFT analyser and finite elemental frequency obtained from ANSYS for each mode shape and for all Specimens is tabulated in detail in Table 161.6 and the modal shapes are shown in Figure 161.3, 161.4, 161.5 and 161.6.

For Specimen 1 (100% Al), the experimental natural frequency for mode 1 is 5.02 Hz and raises up to 65.08 for mode 5. While the finite elemental frequency obtained from ANSYS for mode 1 is 5.74 Hz and for mode 5 is 66.64 Hz which clearly indicates that the natural frequency for all the mode shapes are in close approximation. The damping factor is 0.486 for mode 1 and 0.81 for mode 5.

For Specimen 2 (98% Al+2%SiC), the experimental natural frequency for mode 1 is 5.687 Hz and raises up to 66.18 Hz for mode 5. While the finite elemental frequency obtained from ANSYS for mode 1 is 5.815Hz and for mode 5 is 67.46 Hz which clearly indicates that the natural frequency for all the mode shapes are in close approximation. Further, the damping factor is 0.77 for mode 1 and 2.81 for mode 5.

For Specimen 3 (96% Al+4%SiC), the experimental natural frequency for mode 1 is 5.713Hz and raises up to 67.06 Hz for mode 5. While the finite elemental frequency obtained from ANSYS for mode 1 is 5.883Hz and for mode 5 is 68.24 Hz which clearly indicates that the natural frequency for all the mode shapes are in close approximation. Further, the damping factor is 0.405 for mode 1 and 2.36 for mode 5.

For Specimen 4 (94% Al+6%SiC), the experimental natural frequency for mode 1 is 5.89 Hz and raises up to 70.65Hz for mode 5. While the finite elemental frequency obtained from ANSYS for mode 1 is 5.958 Hz and for mode 5 is 69.11Hz which clearly indicates that the natural frequency for all the mode shapes are in close approximation. Further, the damping factor is 0.531 for mode 1 and 2.43 for mode 5.

For Specimen 5 (92% Al+8%SiC), the experimental natural frequency for mode 1 is 5.998 Hz and raises up to 69.89Hz for mode 5. While the finite elemental frequency obtained from ANSYS for mode 1 is 6.08Hz and for mode 5 is 69.98 Hz which clearly indicates that the natural frequency for all the mode shapes are in close approximation. Further, the damping factor is 1.1 for mode 1 and 2.61 for mode 5.

For Specimen 6 (90% Al+10%SiC), the experimental natural frequency for mode 1 is 6.09 Hz and raises up to 70.02 Hz for mode 5. While the finite elemental frequency obtained from ANSYS for mode 1 is 6.1Hz and for mode 5 is 70.02Hz which clearly indicates that the natural frequency for all the mode shapes are in close approximation. Further, the damping factor is 1.57 for mode 1 and 2.81 for mode 5.

From Table 161.6, it is clear that the percentage error is reducing for the increasing weight percentage of silicon carbide.

Conclusions

1. The young's modulus and density increased as weight percentage of reinforcement is increased which makes the MMC stronger and stiffener.

2. FFT Analyzer elucidated on natural frequency and damping values for various MMC's.
3. Using finite element analysis and modal analysis, the vibration characteristics of Al-SiC MMC's was analysed and was verified.
4. Al 7075 MMC with 8% SIC yielded better damping characteristics.
5. The decreasing percentage error and increasing damping values concludes that MMC's are fabricated successfully using stir casting technique.
6. From the analysis it is clear that the addition of silicon carbide strengthen the damping ratio for better dynamic characteristics since SiC are hard ceramics particulates doesn't permit the waves to transmit from one atomic other.
7. The increased damping (%) of structure when loaded with increasing percentage of reinforcement (SiC) is an indication of high vibration amplitude of the MMC's.

References

Allien, V. J., Kumar, H. and Desai, V. (2019). Dynamic analysis and optimization of SiC reinforced Al6082 and Al7075 MMCs. Mater. Res. Express. 6.

Alnefaie, K. (2009). Finite element modeling of composite plates with internal delamination. Compos Struct. 90:21–27.

Chandravanshi, M. L. and Mukhopadhyay, A. K. (2013). Modal analysis of structural vibration. Proceedings of the International Mechanical Engineering Congress & Exposition., IMECE 2013.

Chandru B. T., Suresh, P. M. (2017). Finite Element and experimental modal analysis of car roof with and without damper. Mater. Today: Proc. 4(10):11237–11244.

De Silva, C. W. 2007. Vibration damping, control, and design. Boca Raton:London.

Ericson, T. M. and Parker, R. G. (2013). Planetary gear modal vibration experiments and correlation against lumped-parameter and finite element models. J. Sound Vib. 332:2350–2375.

Kumar, A., Dwivedi, A., and Paliwal. V. (2014). Free vibration analysis of Al 2024 wind turbine blade designed for Uttarakhand region based on FEA. Proc Technol. 14:336–347.

Kumbhar., A. P., Vyavahare, R. T., and Kulkarni, S. G. (2018). Vibration response and mechanical properties characterization of aluminum alloy 6061/Sic composite. AIP Conference Proceedings 1966. https://doi.org/10.1063/1.5038715

Lakshmikanthan, A., Mahesh, V., Prabhu, R. T., Patel, M. G. C., and Bontha, S. (2021). Free vibration analysis of a357 alloy reinforced with dual particle size silicon carbide metal matrix composite plates using finite element method. Archives Foundry Eng. 21(1):101–112.

Ramanathan, C. and Santhosh, N. (2019). Vibration characterization of reinforced aluminium composite plates. J. Eng. Sci. Technol. 2(1):71–88.

Ramu, S. B. and Reddy, Y. V. M. (2017). Vibrational Analysis of Aluminum 6061 Plate. Int. J. Innov. Res. Sci. Eng. Technol. 15038–15045. doi: 10.15680/IJIRSET.2017.0607327.

Sagar, K. G., Suresh, P. M., and Suresh, R. (2018). Wear behavior of aluminum alloy reinforced with beryl metal matrix composites using Taguchi method. Mater. Today Proc. 5(11):24497–24504. doi: 10.1016/j.matpr.2018.10.246.

Sagar, K. G., Suresh, P. M., and Nataraj, J. R. (2020). Effect of beryl reinforcement in Aluminum 2024 on mechanical properties. J. Inst. Eng.: C. 101(3). 507–516. doi: 10.1007/s40032-020-00554-x.

Samuel Ratna Kumar, P. S., Robinson Smart, D. S. John Alexis, S., Sangeetha, N., and Ramanathan, S. (2017). Modal analysis of MWCNT reinforced Aa5083 composite material. Int. J. Civil Eng. Technol. 8(9):167–177.

Srinidhi, A. S. R. and Sresh, P. M. (2021). Suresh. Fabrication and micro structural characterization of Al 7075 reinforced with various proportions of SiC. Mater. Today: Proc. 49(3):638643. https://doi.org/10.1016/j.matpr.2021.05.162

Taj, A., Doddamani, S, and Vijaykumar, T. N. (2017). Vibrational analysis of aluminium graphite metal matrix composite. (2017). Int. J. Eng. Res. 6(4). doi: 10.17577/ijertv6is040720.

162 Early detection of diabetic retinopathy using vessel segmentation based on deep neural network

S. V. Deshmukh[1,a] and Apash Roy[2,b]

[1]Department of Computer Science and Engineering, Lovely Professional University (LPU), Jalandhar, Punjab, India

[2]Department of Computer Science and Application, Lovely Professional University (LPU), Jalandhar, Punjab, India

Abstract

Diabetic retinopathy (DR) is the widespread cause of preventable vision destruction, most distressing the working people in the entire humanity. Blood vessel segmentation is useful in the early detection of retinal fundus images which is more important to identifying the various eye-related diseases. A lot of previous research is done in the field of vessel segmentation with different consequences. The existing methods require more sufficient outcomes which can be useful in pathological research. In this paper, U-net-based various models are proposed to identify vessel segmented images from the original retinal images. The process of the proposed model is divided into four steps: pre-processing; u-net-base model building; model testing; model evaluation. Original retinal images are fed into pre-processing, where image augmentation is performed to enhance the size of training data. Augmented images are fed into the u-net-based model to train the model, which contains encoder-convolution-decoder layers which form U shape. Three models have been proposed where the number of channels in each layer is consistently changed. The model applies to test images and got the segmented images. Model evaluation is performed using sensitivity, specificity, and accuracy on datasets of DRIVE, STARE, and HRF. Proposed models are assessed with state-of-the-art methods. Model 2 and model 3 are performed best in terms of sensitivity. Model3 achieved 82.85% sensitivity on HRF, model 2 achieved 94.3% on DRIVE and 90.6% on the STARE dataset which is higher than existing methods.

Keywords: Deep Learning, diabetic retinopathy, retinal images, U-net architecture, vessel segmentation.

Introduction

Diabetic retinopathy (DR) is the most usual reason for vision loss in people between the ages of 25 and 74 (Fraser et al., 2017). At least 90% of verified DR cases can be avoided with proper eye care and screening (Griffin et al., 2019). A person who has had diabetes for many years is more likely to get DR. It changes the lens curvature, causing visual issues, and is one of the most serious eye disorders (Shakeel et al., 2019; Smith-Morris et al., 2020). Furthermore, DR is one of the primary sources of blindness, and its slow growth makes it treatable if detected early; nevertheless, if detected late, it can damage the human eye's retina, resulting in permanent blindness (Acharya et al., 2009). DR is caused by changes in the blood vessels in the retina. The retina is a light-sensitive thin inner lining at the back of the eye. When the glucose level in the blood is increased then it can harm blood vessels. These blood vessels thicken and leaks might occur, which can lead to vision loss. Generally, diabetic retinopathy is divided into four stages: mild, average, severe non-proliferative, and proliferative. When balloon-like swelling occurs in small sections of the blood vessels in the retina, then it considers in the first stage, which is mild non-proliferative (Shanthi and Sabeenian, 2019). When some of the blood vessels in the retina get blocked then it considers in the second stage, which is termed as average non-proliferative retinopathy. Severe non-proliferative retinopathy, the third stage, causes more blocked blood vessels, resulting in parts of the retina receiving insufficient blood flow. Proliferative retinopathy is the fourth and final stage. The sickness has progressed to this level. In the retina, new blood vessels will start to form, but they will be unstable and distorted. As a result, they may leak blood, causing vision loss and possibly blindness 9Hoover et al., 2000).

Diabetic patients' normal retinal examinations are helpful in the early detection of DR, which reduces the risk of blindness. Retinal screening, on the other hand, takes time and necessitates the use of highly skilled ophthalmologists to examine the fundus images probing the retinal vessels. The optic disc and other probable abnormalities that are present in an abnormal image can be distinguished by segmenting the

[a]jaykantdeshmukh@gmail.com; [b]apash.23550@lpu.co.in

retinal blood vessels. The examination of blood vessel diameter, vessel density, tortuosity, and neovascularisation are aided by automated segmentation of blood vessels. Neovascularisation is the creation of new, weak vessels.

During the early stages, generally, people don't notice any changes in their vision, but indications such as blurred vision and color vision begin to occur with time. Because DR goes undetected until vision distortion emerges as a result of this disease, effective treatment must be provided at an early stage. Retinal blood vessel segmentation is an essential task because of the clinical importance of early identification of DR. It is critical to separate regions corresponding to distinct eye parts from fundal images for automatic screening of eye disorders. In this scenario, segmenting the blood vessel network is a tough task. The blood vessel network, which varies in density and fineness of structure, runs the length of the fundal image. Changes in illumination, color, and disease make segmenting blood vessels even more challenging.

Several researchers have studied the problem of classification of blood vessels in basic images over the past few decades. Due to differences in editing techniques, image textures, colors, pathological images, and their quality, most current solutions rely on a number of tools for specific image processing that are difficult to generalise. In recent years, machine learning and deep learning-based technologies have become increasingly used for a wide range of automation-based applications, including object-finding and image-splitting on natural images. In medical applications, however, networks must be updated to adjust for challenges such as structural features and spatial resolution that are specific to each diagnostic instrument. The differences in picture resolution and pathology were considered by several researchers. They recommended that the blood vessel network be segmented among databases using a customised CNN architecture that was fine-tuned.

This research is focused on segmenting the blood vessels for the early detection of diabetic retinopathy. The outline of our contributions are as follows:

1. Apply the pre-script-based on the original dataset using augmentation.
2. Proposed model variations in U-NET architecture.
3. Proposed a script-based algorithm for vessel segmentation.
4. Designed models were evaluated using the DRIVE dataset, the STARE dataset, and the HRF dataset. The results of this work are compared with the state of art.

The organisation of the paper is as follows; In section 2 presents a brief description of previous related work and various machine and deep learning techniques utilised for segmentation. Section 3 presents a complete description of the proposed model. Section 4, presents the measure of the performance analysis of proposed models. Section 4 presents the comparative results of models and discussion. and finally, conclude the proposed work with future scope.

Related Work

In previous years, so many of the algorithms incorporating several strategies for autonomous retinal vessel segmentation have been presented. Researchers have conducted various related research, which is discussed in this section. Various methods for blood vessel segmentation in fundal imaging have been reported (Niemeijer et al., 2004; Fraz et al., 2012; Zhang et al., 2014; Salamat et al., 2019; Soares et al., 2006).

To take out the blood vessels, Soares et al. (2006) represented each pixel with feature vectors and recorded it at several scales. Although this method has produced AUC scores of 96% on the DRIVE database, the pre-processing entails changing pixel intensities defined by the camera aperture and ROI selection. As a result of these preprocessing concerns, the algorithm's generalisability may suffer (Budai et al., 2013). Hoover et al. (2000) proposed segmenting the vessel network by combining local and global vessel characteristics. However, various assumptions are made in this work. The Gaussian function is thought to represent a blood vessel profile template. Fixed width and orientation assumptions are assumed (Hoover et al., 2000). In comparison to the original Frangi approach, Budai et al. (2013) presented an approach that reduced calculation time achieved excellent accuracy, and increased sensitivity. This method includes techniques for avoiding issues such as specular reflexes in thick arteries. The presented approach is compared to baseline methods utilising the STARE and DRIVE datasets. The results reveal a high level of accuracy above 94% and cheap computational requirements. Baseline approaches are outperformed by this (Budai et al., 2013). The fundus image segmentation method is described by Roychowdhury et al. (2014), is based on feature retrieval of the major vessel and sub-image categorisation. A high-pass filter is used to obtain a binary picture, which is then used to reconstruct the improved image of the vascular region. The primary blood vessel is identified as the common region in binary image extraction, and the pixels of the remaining images are categorised using a Gaussian mixture model classifier (Roychowdhury et al., 2014). In general, the supervised technique outperforms the unsupervised method in terms of performance. To achieve the goal of segmentation, the supervised technique classifies pixels (You et al., 2011). This method frequently uses the retrieved feature vector to train the classifier for vessel and non-vessel pixel detection. Franklin and Rajan (2014) proposed splitting picture pixels into the vessel and non-vessel points based on the size of the retinal blood vessels and then identifying and segmenting the retinal vessels using a multi-layered neural network.

For finding new retinal vessels, Welikala et al. (2015) proposed a two-vessel segmentation approach. The approach was used to create two binary vessel maps, each with important features. Each binary vessel diagram's local morphs were used to generate two more 4-D feature vectors. Instead, the software used these findings to conclude. Ronneberger et al. (2015), created the U-Net model based on fully convolutional networks (FCNs) (Long et al., 2015), which has had considerable success in the field of cell segmentation in microscopic tissue sections since the U-Net delivers acceptable results. The approach has been employed in a variety of semantic segmentation applications, including satellite image segmentation and industrial fault identification. Data augmentation can be used to train tiny samples of data, particularly medical data. The introduction of U-Net to deep learning for medical imaging with fewer samples has proven to be extremely beneficial. An encoding and a decoding path are both parts of the U-Net system. The pooling layer pulls abstract features from the encoding path, which gradually reduces position information. During the up-sampling phase, the local pixel features recovered from the decoding path will be integrated with the new feature map to maintain as much important feature information from the previous down-sampling procedure as possible for accurate positioning Ronneberger et al. (2015). Various CNN-based techniques suggested for retinal blood vessel segmentation Maji et al. (2016), Liskowski and Krawiec (2016), Maninis et al. (2016). Zhang et al. (2014) and Mo and Zhang (2017) employed an FCN with some additional classifiers to segment the images into the vessel and non-vessel pixels. They employed the transfer learning method to train the FCN model to overcome the limited number of accessible examples. They assessed the system using the DRIVE, STARE, and CHASE criteria (Zhang et al., 2014); Mo and Zhang, 2017; Dasgupta and Singh, 2017). The blood vessel segmentation issue was reformulated as a boundary detection problem by Fu et al. (2016), who developed the deep vessel approach by a combination of CRF and CNN as RNN (Fu et al., 2016). To extract vessels from fundus pictures, Lu et al. (2018), presented a unique coarse-to-fine fully convolutional neural network (CF-FCN). By exploiting the spatial relationship between pixels in fundus images, the CF-FCN aims to make full use of the original data information and compensate for the neural network's coarse output. Our trials on the datasets have confirmed the efficacy and efficiency of our CF-FCN. On DRIVE datasets, it obtains 0.7941 sensitivity, 0.9870 specificities, 0.9634 accuracies, and 0.9787 Area Under Receiver Operating Characteristic Curve (AUC) Lu et al. (2018). Jebaseeli et al. (2019), proposed methodology for improving the quality of segmentation results in morbid retinal pictures. For removing the background from the source image and enhancing the foreground blood vessel pixels, this system employs contrast limited adaptive histogram equalization (CLAHE). They offered a deep learning-based support vector machine (DLBSVM) for blood vessel classification and extraction. The recommended procedures are evaluated using the fundus imaging datasets (Jebaseeli et al., 2019).

Proposed Methodology

In the proposed work, U-Net architecture is used for vessel segmentation. The novelty of this research is to propose the three deep learning models. The model is based on basic U-Net architecture which contains only three layers. After that modify this model with one layer and again modify the model with one layer so we get three models. After comparing all three models we observed that the second model gives the best result as compared to the first and third. Figure 162.1 shows the architecture of the proposed model in which the vessel segmentation process is stated. The process is divided into four steps: pre-processing; U-Net-b model building; model testing; model evaluation. Original retinal images are fed into pre-processing, where image augmentation is performed to enhance the size of training data. Augmented images are fed into the u-net-model to train the model, which contains encoder-convolution-decoder layers in the U-shape. The model applies to test images and got the segmented images. Model evaluation is performed using specificity, sensitivity, and accuracy. The complete process is mentioned in detail in future sub-sections.

Data preprocessing using image augmentation

Augmentation is a technique for increasing the number of images in a collection by transforming and cropping present images or forming new artificial images. This approach may aid in reducing model over-fitting as well as the issue of class imbalance. It aids in the development of a generously proportioned training dataset and additional robust models. Five different formations of origin-based mage are generated through different techniques such as horizontal flip, vertical flip, elastic transform, grid distortion, and optical distortion. Figure 162.2 shows sample images of augmentation for every step which are identified

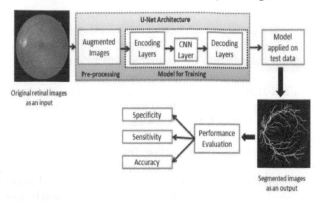

Original retinal images
as an input

Figure 162.1 Architecture of proposed model

Figure 162.2 (a) Horizontal flip; (b) vertical flip; (c) grid distortion; (d) elastic transform; (e) optical distortion

through experimentation. Horizontal flip augmentation is the process of horizontally flipping the complete rows and columns of images pixels and vertical flip augmentation is the process of vertically reversing the complete rows and columns of images pixels.

Proposed model training and testing

The augmentation process gives the five variants of original images which are fed into a deep neural-based model for training. Variations in models of U-net architecture are proposed and performed in this paper. Three models have been proposed where the number of channels in each layer is consistently changed. The model contains an encoder block, decoder block, and convolution block. The encoding block progressively reduces position values and information and identifies abstract features using the pooling layer. Precise positioning is performed through the pixel features which are local and identified from the decoding layer. These local features collectively participate with the novel feature map when performing the up-sampling method to maintain some significant feature information from the preceding down-sampling method. After this input images are fed to a convolution block with a dimension of 512 and performed the batch normalisation and max-pooling operation. Max pooling operation can reduce the error and retain the texture information of the images. The decoder block performs the up-sampling and reduces the image dimensions to identify the dimension of the original image which is 64. Decoder layers perform the convolution transpose and concatenate the skip connection feature vectors. Model 2 contains four encoder layers, one convolution layer, and four decoder layers. Model 3 contains five encoder layers, one convolution layer, and five decoder layers. These three models are trained on the training dataset and applied to test data for predicting segmented images. Figure 162.3 shows the complete working flow of the proposed model.

Experimental Setup and Evaluation

Proposed models are experimented on the TensorFlow framework using Keras API for deep learning methods. Models are tested using different hyperparameters such as Adam optimiser, dice loss function, 50 number of epochs, and batch size 32. The learning rate is 1e-4 and 1e-5. For a wide range of issues, the Adam optimiser is the best option. As a result, the Adam optimiser is used. The categorical cross-entropy loss function L is reduced for 10 epochs in the proposed methodology. The loss function for categorical cross-entropy is defined in Equation 1 (Dasgupta and Singh, 2017). The learning rate is 1e-4. Experiments were performed for epoch 10 with batch size 4. In equation 1, y is the original image and ŷ is the predicted

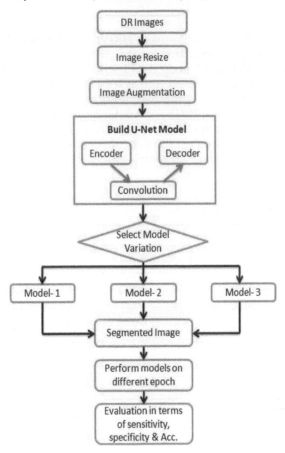

Figure 162.3 Flowchart for vessel segmentation

image and Q is the number r of training samples and R is the number of categories. Hyper-parameter setup is shown in Table 162.1.

$$L(y,\hat{y}) = -\sum_{r=0}^{R}\sum_{q=0}^{Q}\left(y_{q,r},\log\left(y_{q,r}\right)\right) \qquad (1)$$

Data set used

DRIVE: It contains 40 retinal images, of which 20 images are for training data and 20 images for testing data. This dataset can be used for the classification and vessel segmentation of DR images. After applying augmentation, the dataset is converted into 120 images (Niemeijer et al., 2017).

STARE: It contains 40 retinal images, of which 20 images are for training data and 20 images for testing data. This dataset can be used for the classification and vessel segmentation of DR images. After applying augmentation, the dataset is converted into 120 images.

HRF: It consists of 45 fundus images, where 15 images belong to healthy patients, 15 images belong to diabetic retinopathy patients and the remaining 15 images belong to glaucomatous patients (Köhler et al., 2013). In this research, only 15 images of DR patients are used for vessel segmentation which is converted into 90 images after augmentation.

Evaluation

There are a variety of performance parameters that can be used to measure the effectiveness of the deep learning models. Following are some parameters used in the proposed model to evaluate the performance.

Sensitivity (Spec) = TP/ TP + FN

Specificity (Sens) = TN/ TN + FP

Accuracy (Acc) = TP + TN/ (TP + FN + TN + FP)

Results and Discussion

U-net-based proposed models are extracting vessel segmented images from DRIVE, STARE, and HRF retinal datasets. It can be observed that more thin vessels are extracted for the HRF dataset using proposed models. Performance comparisons of model 1, model 2, and model 3 on DRIVE, STARE, and HRF datasets are presented in Table 162.1 of the three-evaluation parameters such as sensitivity, specificity, and accuracy. It can be observed that the sensitivity of the HRF dataset is improved using model3 as compared to model 1 and model 2, which is increased up to 14%. For the DRIVE dataset, sensitivity is increased by 28% using model 2 as compared to model1 and 3% using model 3 as compared to model1 approximately. Similarly, for the STARE dataset, sensitivity is increased by 19% using model 2 as compared to model1 and 16% as compared to model3. Specificity and accuracy for three datasets are better using all

Table 162.1 Performance evaluation of three models based on three datasets

Dataset	Model-1			Model-2			Model-3		
	Spec	*Sens*	*Acc*	*Spec*	*Sens*	*Acc*	*Spec*	*Sens*	*Acc*
HRF	93.91	68.45	94.73	92.74	68.72	94.46	92.52	82.85	94.12
DRIVE	92.05	66.92	94.53	95.80	94.32	94.88	94.07	91.09	94.49
STARE	95.48	71.41	93.74	94.61	90.06	94.09	94.79	74.99	93.62

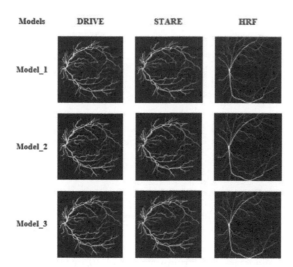

Figure 162.4 Sample images of segmented results of proposed models of DRIVE, STARE, and HRF dataset

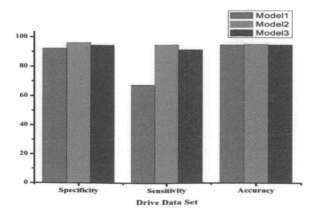

Figure 162.5 Assessment of proposed models using sensitivity, accuracy, and specificity on DRIVE dataset

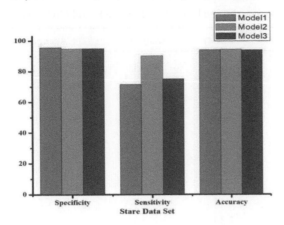

Figure 162.6 Assessment of proposed models using sensitivity, accuracy, and specificity on STARE dataset

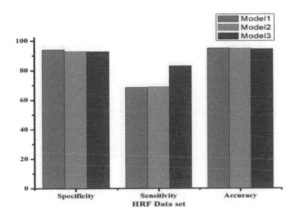

Figure 162.7 Assessment of proposed models using sensitivity, accuracy, and specificity on HRF dataset

three models. The specificity of the STARE dataset is best using model 1. Figures 162.5–162.7 show the comparative and performance analysis of three models on three datasets DRIVE, STARE, and HRF. It can be stated that model 3 performed best on the HRF dataset and model 2 performed best on DRIVE and STARE datasets. As encoder and decoder layers are increased in models which give more significant results.

The performance of the proposed models is also compared with the state-of-the-art methods of vessel segmentation of retinal images. The comparison of methods is shown in Table 162.2 for DRIVE, STARE, and HRF datasets. In the DRIVE dataset, Hua et al.'s (2019) method achieved better specificity as compared to other methods. However, proposed model 2 achieved 1.61% lower specificity than Hua et al. (2019) method. In Maninis et al. (2016), the method has lower specificity as compared to other methods because it generated noise in segmented images, and the segmented vessels are additional thicker than the original. Tian's method is also performed better in terms of all three measurements, specifically sensitivity and accuracy. The sensitivity of Tian's method is best among all previous methods mentioned in Table 162.4, however proposed model 2 achieved 7.93% higher sensitivity than Tian. Few surrounding pixels are also identified as vascular pixels so that the sensitivity is elevated and the specificity is near to the lower side. For the STARE dataset, Zhang et al. (2014) and Mo and Zhang (2017) methods achieved the best results for specificity and accuracy, but the sensitivity of proposed model 2 is 9% higher and performed best among mentioned methods. For the HRF dataset, park's method is performed best in terms of specificity and accuracy but the sensitivity of model 3 is 13.37% higher than park's method. The accuracy of the proposed models is not proper because of blurred vessel segmentation images in results. Image augmentation and cost-sensitive loss functions are also supplied appreciably to the final segmented results.

Table 162.2 Assessment of proposed models with state of art methods on DRIVE, STARE, and HRF dataset

Dataset	Ref.	Specificity	Sensitivity	Accuracy
DRIVE	Maninis et al. (2016)	91.15	82.61	95.41
	Vlachos and Dermatas (2010)	95.51	74.68	92.85
	Azzopardi et al. (2015)	97.04	76.55	94.42
	Tian et al. (2020)	96.65	86.39	95.80
	Hua et al. (2019)	97.41	79.32	95.11
	Proposed_M1	92.05	66.92	94.53
	Proposed_M2	95.80	94.32	94.88
	Proposed_M3	94.07	91.09	94.49
STARE	Maninis et al. (2016)	98.44	81.47	96.76
	Vlachos and Dermatas (2010)	97.30	69.96	94.40
	Azzopardi et al. (2015)	97.01	77.16	95.63
	Proposed_M1	95.48	71.41	93.74
	Proposed_M2	94.61	90.06	94.09
	Proposed_M3	94.79	74.99	93.62
HRF	Park et al. (2020)	99.31	69.48	97.00
	Li et al. (2020)	97.43	75.24	96.23
	Proposed_M1	93.91	68.45	94.73
	Proposed_M2	92.74	68.72	94.46
	Proposed_M3	92.52	82.85	94.12

Conclusion

In this paper, blood vessel segmentation is performed using u-net architecture. U-net model variations are proposed and experimented on DRIVE, STARE, and HRF datasets. A script-based algorithm is also proposed for vessel segmentation. The performance comparison of model1, model 2, and model 3 on DRIVE, STARE, and HRF datasets are presented using three evaluation metrics sensitivity, specificity, and accuracy. It is identified that model 3 performed best on the HRF dataset and model 2 performed best for DRIVE and STARE datasets. As encoder and decoder layers are increased in models which give more significant results. The performance of the proposed models is also compared with the state-of-the-art methods. The sensitivity of proposed model 2 and model 3 is higher than the existing methods on all datasets. The accuracy of the proposed models is not proper because of blurred vessel segmentation images in results. In further work, the convolution layer can also increase for more refined results. More experiments on hyper-parameters can also be performed in future work for more variations in results. The resulted segmented images can be used for diabetes retinopathy classification and feature extraction.

References

Acharya, U. R., Lim, C. M., Ng, E. Y. K., Chee, C., and Tamura, T. (2009). Computer-based detection of diabetic retinopathy stages using digital fundus images. Proc. Inst. Mech. Eng. Part H: J. Eng. Med. 223:545–553.

Azzopardi, G., Strisciuglio, N., Vento, M., and Petkov, N. (2015). Trainable COSFIRE filters for vessel delineation with application to retinal images. Med. Image Anal. 19:46–57.

Budai, A., Bock, R., Maier, A., Hornegger, J., and Michelson, G. (2013). Robust vessel segmentation in fundus images. Int. J. Biomed. Imag.

Dasgupta, A. and Singh, S. (2017). A fully convolutional neural network-based structured prediction approach towards the retinal vessel segmentation. 2017 IEEE 14th International Symposium on Biomedical Imaging (ISBI 2017), (pp. 248–251). IEEE.

Franklin, S. W. and Rajan, S. E. (2014). Computerised screening of diabetic retinopathy employing blood vessel segmentation in retinal images. Biocybern. Biomed. Eng. 34:117–124.

Fraser, C. E., D'amico, D. J., Nathan, D., Trobe, J., and Mulder, J. (2017). Diabetic retinopathy: Classification and clinical features. UpToDate. UpToDate, Waltham, MA. (Accessed on February 24, 2016.)

Fraz, M. M., Remagnino, P., Hoppe, A., Uyyanonvara, B., Rudnicka, A. R., Owen, C. G., and Barman, S. A. (2012). Blood vessel segmentation methodologies in retinal images–a survey. Comput. Methods Programs Biomed. 108:407–433.

Fu, H., Xu, Y., Lin, S., Wong, D. W. K., and Liu, J. (2016). Deepvessel: Retinal vessel segmentation via deep learning and conditional random field. International Conference on Medical Image Computing and Computer-Assisted Intervention. (pp. 132–139) Springer.

Griffin, S. J., Rutten, G. E., Khunti, K., Witte, D. R., Lauritzen, T., Sharp, S. J., Dalsgaard, E.-M., Davies, M. J., Irving, G. J., and Vos, R. C. (2019). Long-term effects of intensive multifactorial therapy in individuals with screen-detected type 2 diabetes in primary care: 10-year follow-up of the ADDITION-Europe cluster-randomised trial. Lancet Diabetes End. 7:925–937.

Hoover, A., Kouznetsova, V., and Goldbaum, M. (2000). Locating blood vessels in retinal images by piecewise threshold probing of a matched filter response. IEEE Trans. Med. Imaging 19:203–210.

Hua, C.-H., Huynh-The, T., and Lee, S. (2019). Retinal vessel segmentation using round-wise features aggregation on bracket-shaped convolutional neural networks. 2019 41st Annual International Conference of the IEEE Engineering in Medicine and Biology Society (EMBC), (pp. 36–39). IEEE.

Jebaseeli, T. J., Durai, C. A. D., and Peter, J. D. (2019). Retinal blood vessel segmentation from diabetic retinopathy images using tandem PCNN model and deep learning based SVM. Optik 199:163328.

Köhler, T., Budai, A., Kraus, M. F., Odstrčilik, J., Michelson, G., and Hornegger, J. (2013). Automatic no-reference quality assessment for retinal fundus images using vessel segmentation. Proceedings of the 26th IEEE International Symposium on Computer-Based Medical Systems, 2013. (pp. 95–100). IEEE.

Li, L., Verma, M., Nakashima, Y., Nagahara, H., and Kawasaki, R. (2020). Iternet: Retinal image segmentation utilizing structural redundancy in vessel networks. Proceedings of the IEEE/CVF Winter Conference on Applications of Computer Vision, 2020. (pp. 3656–3665).

Liskowski, P. and Krawiec, K. (2016). Segmenting retinal blood vessels with deep neural networks. IEEE Trans. Med. Imag. 35:2369–2380.

Long, J., Shelhamer, E., and Darrell, T. (2015). Fully convolutional networks for semantic segmentation. Proceedings of the IEEE Conference on Computer Vision and Pattern Recognition, (pp. 3431–3440).

Lu, J., Xu, Y., Chen, M. and Luo, Y. (2018). A coarse-to-fine fully convolutional neural network for fundus vessel segmentation. Symmetry 10:607.

Maji, D., Santara, A., Mitra, P., and Sheet, D. (2016). Ensemble of deep convolutional neural networks for learning to detect retinal vessels in fundus images. arXiv preprint arXiv:1603.04833.

Maninis, K.-K., Pont-Tuset, J., Arbeláez, P., and Van Gool, L. (2016). Deep retinal image understanding. International Conference on Medical Image Computing and Computer-Assisted Intervention, 2016. (pp. 140–148) Springer.

Mo, J. and Zhang, L. (2017). Multi-level deep supervised networks for retinal vessel segmentation. Int. J. Comput. Assist. Radiol. Surgery 12:2181–2193.

Niemeijer, M., Staal, J., Ginneken, B., Loog, M., and Abramoff, M. (2017). DRIVE: digital retinal images for vessel extraction; 2004. WebLink: http://www. isi. uu. nl/Research/Databases/DRIVE.

Niemeijer, M., Staal, J., Van Ginneken, B., Loog, M., and Abramoff, M. D. (2004). Comparative study of retinal vessel segmentation methods on a new publicly available database. Med. Imag. 648–656:

Park, K.-B., Choi, S. H., and Lee, J. Y. (2020). M-gan: Retinal blood vessel segmentation by balancing losses through stacked deep fully convolutional networks. IEEE Access 8:146308–146322.

Ronneberger, O., Fischer, P., and Brox, T. (2015) U-net: Convolutional networks for biomedical image segmentation. International Conference on Medical image computing and computer-assisted Intervention, 2015. (pp. 234–241), Springer.

Roychowdhury, S., Koozekanani, D. D., and Parhi, K. K. (2014). Blood vessel segmentation of fundus images by major vessel extraction and sub-image classification. IEEE J. Biomed. Health Inform. 19:1118–1128.

Salamat, N., Missen, M. M. S., and Rashid, A. (2019). Diabetic retinopathy techniques in retinal images: A review. Artif. Intell. Med. 97:168–188.

Shakeel, P. M., Baskar, S., Sampath, R., and Jaber, M. M. (2019). Echocardiography image segmentation using feed-forward artificial neural network (FFANN) with fuzzy multi-scale edge detection (FMED). Int. J. Signal Imaging Syst. Eng. 11:270–278.

Shanthi, T. and Sabeenian, R. (2019). Modified alexnet architecture for classification of diabetic retinopathy images. Comput. Electrical Eng. 76:56–64.

Smith-Morris, C., Bresnick, G. H., Cuadros, J., Bouskill, K. E., and Pedersen, E. R. (2020). Diabetic retinopathy and the cascade into vision loss. Med. Anthropol. 39:109–122.

Soares, J. V., Leandro, J. J., Cesar, R. M., Jelinek, H. F., and Cree, M. J. (2006). Retinal vessel segmentation using the 2-D Gabor wavelet and supervised classification. IEEE Trans. Med. Imag. 25:1214–1222.

Tian, C., Fang, T., Fan, Y., and Wu, W. (2020). Multi-path convolutional neural network in fundus segmentation of blood vessels. Biocybern. Biomed. Eng. 40:583–595.

Vlachos, M. and Dermatas, E. (2010). Multi-scale retinal vessel segmentation using line tracking. Comput. Med. Imag. Graphics 34:213–227.

Welikala, R. A., Fraz, M. M., Dehmeshki, J., Hoppe, A., Tah, V., Mann, S., Williamson, T. H., and Barman, S. A. (2015). Genetic algorithm-based feature selection combined with dual classification for the automated detection of proliferative diabetic retinopathy. Comput. Med. Imag. Grap. 43:64–77.

You, X., Peng, Q., Yuan, Y., Cheung, Y.-M., and Lei, J. (2011). Segmentation of retinal blood vessels using the radial projection and semi-supervised approach. Pattern Recognit. 44:2314–2324.

Zhang, J., Li, H., Nie, Q., and Cheng, L. (2014). A retinal vessel boundary tracking method based on Bayesian theory and multi-scale line detection. Comput. Med. Imaging Graphics 38:517–525.

163 A systematic study on revelation and examination of COVID-19 using radiology

Mirza Qadir Baig[a], Reena Thakur[b], and Mona Mulchandani[c]

Jhulelal Intitute of Technology, Nagpur, India

Abstract

The key issues in the present COVID-19 pandemic are early revelation and examination of COVID-19, despite their widespread use in diagnostic centres, diagnostic procedures based on radiological scans have flaws when it comes to the disease's novelty. As a result, machine-learning models are commonly used to evaluate radiological pictures by medical and computer researchers. Methods and Materials from November 1, 2019, to November 1, 2019, an investigation review was undertaken by examining the three databases of Scopus, PubMed, and Web Science. Depending on a search technique, through July 20, 2020, a total of 170 articles were mined, and the 38 pieces were brought as the research populace and after relating the presence and rejection criteria. This paper gives an indication of the present state of all prototypes for COVID-19 identification and revelation using radiological modalities and deep learning processing. Deep learning-based models, according to the findings, have a remarkable capacity for providing a precise and economical system for the revelation and examination of COVID-19, and their usage in the managing of sense modality would result in considerable rise in thoughtfulness and specific values. The revelation and examination of this condition and provides an once-in-a-lifetime opportunity for individuals to obtain quick, affordable, and secure investigative services.

Keywords: COVID-19 pandemic, diagnostic procedures, machine learning, diagnostic services and deep learning.

Introduction

Following the eruption of an unidentified virus in China until recently 2019, not many people feel affected in a regional marketplace. The virus was first unidentified, but experts identified its indications as being comparable to flu and corona virus contamination. The viral infection was identified, and it is termed 'COVID-19' by the World Health Organizations (WHO) (Wang et al., 2020). The COVID-19 virus criss-crossed geographical confines in a short amount of time, wreaking havoc on the global population's health, economics, and welfare. Until January 5, 2021, according to World meters statistics COVID-19 infected about 86 million individuals globally, with over 1,870,000 people dying because of the disease. Among one of the method's downsides is the necessity for a research test centre kit, which many countries find impractical to get during emergencies and pandemics. This method, like numerous investigative and laboratory procedures used in health care, is not miscalculation-free and subjective Nasal and throat mucosa sampling needs a trained laboratory technician and is an unpleasant procedure, which is why many patients decline to have their nasal and throat mucosa sampled. More importantly, several investigations have confirmed the RTPCR litmus test has a limited compassion. The sensitivity of this diagnostic approach has been indicated in between 31–61% in numerous investigations, implying a decline in the correctness of COVID-19 identification in many cases. Its false-negative rate and inconsistent outcomes have also been mentioned in several research.

CT-scan images have a high understanding in identifying and detecting COVID-19 patients when compared to RT- PCR, but a low specificity. This suggests that CT scans are more precise in COVID-19 instances but less precise in nonviral pneumonia instances. Association and ground glass opacities were not seen in 15% of CT-scan images in study performed on the finding of patients in Wuhan, China. Based on their CT scan findings, 15% of convincing cases of COVID-19 were missed in diagnosis as impeccably nutritious. Only 11 of the 19 persons with COVID-19 who had GGO with amalgamation had GGO, demonstrating that there was no consolidation or illness. COVID-19 was difficult, if not impossible, to identify in many cases despite the occurrence of amalgamation prior to initiating GGO. All these instances revealed a flaw in the use of CT scans to diagnose COVID-19.

Even though chest CT scans are effective at identifying COVID-19 associated lung injury, there are some drawbacks to using this problem-solving technique. Despite the WHO's advice, some individuals' upper

[a]bmqadir.8@gmail.com; [b]r.thakur@jit.org.in; [c]mona.mulchandani@jit.org.in

body CT scans are usual at the start of the virus. For the duration of the illness, cans are the most consumed. CT examinations shouldn't be utilized as in the beginning line of identification, according to the American College of Radiology.

Challenges such as the threat of virus transmission when using a CT-scan gadget and its great expense can cause major obstacles for patients and health care organizations, hence it is recommended that CT-scans be switched with CXR skiagraphy if medicinal imaging is required. Several investigations have shown that CXR imaging is ineffective in identifying COVID-19 and distinguishing it on another forms of pneumonia. Regardless of the low exactness of COVID-19 X ray identification, it does have a few advantages. Various studies along the application of DL in the interpretation of radiology photographs have been done to conquer the shortcomings of COVID-19 analytic tests utilising radiological pictures.

Aim and Objective

The coronavirus 2019 (COVID-19) outbreak that had a significant impact on the global health and the global economy. The ability to access COVID-19 from any technological device, such as a cell phone, can be quite beneficial. The goal of this all research was to find the COVID-19 in the X-ray images. The most frequent medical imaging techniques was utilised in the diagnosis of lung disease are chest radiography (CXR) and the computed tomography (CT) images. Despite the fact that CT imaging are commonly utilised in the diagnosis of COVID-19, expenditures of 5–78 and radiation exposure are key concerns. CXR images are favoured over CT scans because they expose patients to less radiation and are more readily available. As a result, CXR pictures were employed in this investigation to diagnose COVID-19 automatically. The model was built utilising end-to-end architecture and no feature reduction techniques were used. COVID-Net, a convolutional neural network (DCNN) created using a machine-tested test technique, was proposed by Wang and Wong. COVID-19 was detected in common and non- COVID-19 disease utilising a multi-stage classification method employing CXR pictures. The model's success was determined using both the qualitative and quantitative analyses. From obtaining COVID-19, the improved model was able to achieve 91% sensitivity. Panwar et al. developed nCOVnet, a transfer learning- based method for swiftly detecting COVID-19 in CXR images. PROPOSED WORK PLAN Despite their high speculative performance and self-study capacity, SOM and LWL versions attain human like precision in picture definition and extrapolation concerns. The most important goal of our framework is to provide segregating material and a rapid symptomatic approach that can be utilised to distinguish fresh COVID-19 X rays. Physicians may find this process useful as a therapeutic option that can be used depending on the type of illness and can deliver quick results. The proposed framework, the SOM-LWL system architecture, and the unequal X-ray database solution are described in the following sections. Python and the PyTorch will be used to build the framework. Creating a public open data collection for chest X-ray and CT pictures of COVID-19 and other viral pneumonia patients (MERS, SARS, and ARDS.). Data will be gathered from both public and private sources, including hospitals and doctors. Chest X-ray 14 is a medical imaging database which contains 112,120 forward-looking X-ray pictures of the 30,805 patients with the common disease labels in the fourteen mines, it adds six more asthma disorders to Chest X-ray 8: oedema, emphysema, fibrosis, pleural thickening, and hernia. TO BE APPLIED RESEARCH METHODOLOGY We can show that nice textures and statistical groups may be a big visual factor by looking at X-rays. Over the last decade, some researchers have begun to employ textural and mathematical traits to discover models of differentiation difficulties. A third-party specialist re-evaluated the exam set to account for any grading issues. It is not necessary to have knowledge of issue classes or procedures that support the concepts of handicrafts. This function isn't necessary. Although unpublished descriptions have certain clear characteristics, we should be aware that handmade features have some characteristics that can make them highly beneficial in dealing with a variety of functions. Some of these advantages is that the hand-crafted characteristics are exceptionally durable, as some practices are frequently used to record problem-related trends in a conclusive manner. A more accurate description of the patterns formed by hand-made elements of photographs is achievable rather than employing unverified elements. Both individuals with no symptoms and those with indicators of COVID-19 testing benefit from AI- based X-ray testing. The database is based on photographs of pneumonia sufferers' chests. Using photos from a variety of open access sources, Cohen created a COVID-19 9 X-ray imaging website. This database is updated on a regular basis with photographs given by scientists from various places. There are presently 127 X-ray images that are obtained with the COVID-19 on the website. Figure 163.1 shows various COVID-19 samples retrieved from X-ray databases as an example. We use RISE to create excellent speculation maps for our model to illustrate the results appropriately. The goal of this observation was to conduct more tests to rule out model over-exertion and check that the regions of focus were compatible with the pertinent aspects from the radiologist's perspective.

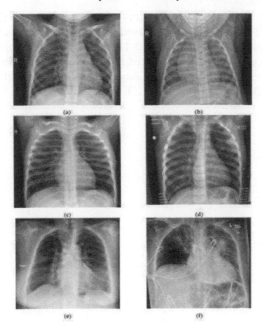

Figure 163.1 Chest x-ray sample

Methods and Materials

Deep learning

Salakhutdinov and Hinton who circulated a scientific article in the journal knowledge in 2006 which was ushered in the era of DL. They explained that if a nervous system with the concealed layers was essential in enhancing the include knowledge power. So, these procedures can enhance the categorization precision of the numerous sorts of information. The detection of muscle emaciated abnormalities and the classification of infections were two of the most common uses of DL in radiology practices. A structured review strategy was used to find studies related to the detection and identification of COVID-19 for this study. Using past studies and the authors' judgments, a systematic search approach was established.

Search criteria

- What modalities can be utilised in conjunction with DL to help detect and analyse COVID-19?
- How effective are several forms of DL and their structural design in fostering COVID-19 identification in relationship to one another?

Data extraction

In information mining forms, appropriate research, explanations of their techniques, and their consequences were noted. So, the Figure 163.2 was utilised to manage data collection and the mining. The initial details of the approaches were documented in information mining layers to recognize procedures and DL procedures.

At first, 140 extracts and the full text edition papers remained evaluated, and 32 research which met the inclusion criteria where they chose. Because of the virus's novelty, all the select publications were printed in late 2019. In pandemics, picture based analytical approaches are important for diagnosis touched instances. The CT scan and the CXR are the two of the extremely popular radiological modalities which is used to identify and detect the COVID-19.

Radiological imaging was evaluated in all the studies reviewed here to analyse COVID-19 with DL. Examining existing books and dictionaries, as well as consulting radiologists and epidemiologists, Part of a genuine object that can be examined or whose presence can be demonstrated or denied is defined as detection. The Discovery is regarded as a prerequisite to the identification in the medical texts. Similarly, the numerous studies have used these two terms interchangeably in the case of the COVID-19, yet they were clinically distinct. In this investigation, detection was defined as distinguishing COVID-19 infected cases from non COVID-19 infected cases by separating these two words. In this regard, it was found that 13 paragraphs employed DL to identify COVID-19 after assessing the removed articles. Several articles, on the other hand, have used DL algorithms to diagnose COVID-19. COVID-19 was correctly identified among the several kinds of pneumonia in these instances. Some research looked at radiology modalities to see whether they could detect and diagnose it at the same time. The experiments on COVID-19 detection and diagnosis are shown in Figures 163.3 and 163.4. As previously stated, CT scan pictures have a low

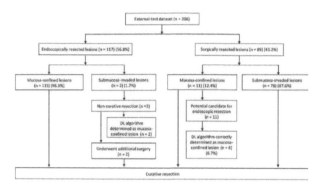

Figure 163.2 Data mining flow diagram

A)

	COVID-19 (n)	Flu-like (n)	Total
Positive algorithm	22 Sensitivity: 57.9% PPV: 91.7%	2	21
Negative algorithm	16	14 Specificity: 87.5% NPV: 46.7%	16
Total	38	16	54

B)

	Severe (n)	Non-severe (n)	Total
Positive algorithm	10 Sensitivity: 83.3% PPV: 45.5%	12	21
Negative algorithm	2	14 Specificity: 53.8% NPV: 87.5%	16
Total	12	27	38

Figure 163.3 The experimental tabular data

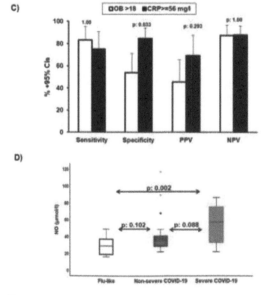

Figure 163.4 The experimental bar chart data

specificity for detecting COVID-19 patients, which is a diagnostic disadvantage. Much research has pursued to improve these procedures in the analysis of CT-scan pictures using DL procedures, corresponding to this study. The removal and choice of elements submerged in the images exists to be the key to these approaches' efficacy in detecting COVID19-induced lung lesions. Even though DL algorithms improved COVID-19 detection and diagnosis, one of the main difficulties of this modality in COVID-19 diagnosis was the dearth of this gear in all therapeutic and analytical institutes. Additionally, many COVID-19 affected role needed multiple CT-scans of their upper body. Emission exposure during CT-scans affects major health difficulties for people. Furthermore, because the CT-scan tunnel contamination, there is the risk of virus transfer from the first patient to another.

As a result, many scientists and doctors have relied on simple radiographic pictures or X-rays to detect COVID-19. Nevertheless, these pictures lack the necessary decision and precision to detect COVID-19 from the beginning, as well as they have a lot of disadvantages in this regard. As a result, artificial intelligence researchers went to the support of experimental experts and appointed deep learning (DL) as a persuasive skill to improve the accuracy of COVID-19 identification using X ray images. This method can identify affected role with COVID-19 and harvesting contagious lung nerves owing to the kind of DL in the mining of picture characteristics. To evaluate these photos, various studies used a variety of DL methods. CT- scans were more widespread in the early period of the COVID-19 plague, but X rays became more normal as time turned on. As a result, the emphasis of research paper shifted from CT-scan to radiographic image analysis.

The architecture used by deep neural networks is one of the most important elements in terms of their efficacy. Deep neural network architectures show a remarkable capacity to operate a wide array of purposes for various information types (Oliveira et al., 2020). COVID-19 has been used in a variety of research with various DL designs. Several of these investigations examined their diagnosis rates in the identification of COVID-19 utilising various types of designs. Figure 163.4 shows that the frequency of CNN designs to be utilised in the examined studies. The architectures shown in this diagram are either distinct editions of the same architecture or a family of the same architecture. However, some studies found that the ResNet-50 design was the most effective in detection and diagnosis COVID-19, while others found that other Res Net copies were more useful in evaluating radiological pictures for COVID-19 analysis.

The planned method has been employed in a variety of research using highly known or state of art models. Specific research, on the other hand, have presented their own customised algorithm and architecture, which are not based on well-known architectures.

Discussion

This organised study looked at findings to help investigators explore and create learning-based activities established on artificial intelligence (AI) for COVID-19 revelation and examination. To our knowledge,

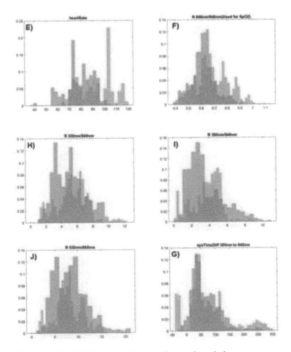

Figure 163.5 The experimental graphical data

some of the highly widespread findings on the revelation and examination of this condition is the present review, which looked at several DL approaches for analysing radiological images. The current study offered current data on DL set of rules and their use in COVID-19 radiographic image analysis. Much research has demonstrated that using DL set of rules can expand the rate of metrical individualities in CT-scanned imageries, as a result, using this low-cost and accessible technique to diagnose COVID-19 should be regarded a reliable strategy We can uncover the cheapest and safest imaging approaches to avoid COVID-19 spread by ornamental imaging tactics with artificial intelligence (AI) technologies. Permitting to a survey of published journals, using DL set of rules under the guidance of a radiologist to identify this virus boosted efficacy and diminished symptomatic blunders in numerous instances of pneumonia, chiefly COVID-19. All inquiries using the X-ray sense modality demonstrated a warmth average of greater than 95%, a specificity mean of greater than 91%, and a higher analytic rate than that acknowledged in old-style manuals and approaches.

The problem can likewise be extrapolated that in instance of COVID-19, the quality of belonging in CT scan images developed by the deep learning technique was on average higher than 92%, indicating that the DL approach is more efficient in terms of specificity than prior texts in many cases. In many situations, the quality of belonging of DL techniques in COVID-19 CT scan pictures was higher than or equivalent to that of traditional diagnostic approaches. In certain instances, different algorithms were utilized in addition to the CNN algorithm. Without altering the parameters of the CNN architectures used in these experiments, it is impossible to access their capability to reveal and examine COVID-19. This study backs up the premise that DL algorithms are an encouraging practice to improve healthcare and diagnostic and treatment outcomes

References

Mahmud, T., Rahman, M. A., and Fattah, S. A. (2020). Computers in Biology and Medicine. 122.

Hasan, A. M, Al-Jawad M. M., Jalab, H. A., Shaiba, H., Ibrahim, R. W., and Al- Shamasneh, A. R. (2020). Classification of COVID19 coronavirus, pneumonia Francis. 22(5):517.

Oliveira, B. A., Oliveira, L. C., Sabin, E. C., and Okay, T. S. (2020). SARS-CoV-2 and the COVID-19 disease. SARS-CoV-2 and the COVID.19 disease: a mini review on diagnostic methods. Rev. Inst. Med. Trop. Sao. Paulo. 29(62):e44. doi: 10.1590/S1678-99462020620

164 The importance and limitations of big data technologies in education

Ruth Chweya[1,a], Samuel-Soma M. Ajibade[2,b], and Ayodele John Melbury[3,c]

[1]Department of Computing, School of Information Science and Technology, Kisii University, Kisii, Kenya

[2]Computer Engineering Department, Istanbul Ticaret Universitesi, Istanbul, Turkey

[3]Dept of Mission Planning and Satellite Data Management, NSRDA, Abuja, Nigeria

Abstract

Currently, there is a great rise in the utilisation of digital technologies for enhancing teaching and learning procedures in education. In the age of big data, universities are on the rise to utilise big data technologies to improve the learning experience for both learners and instructors. The key objective of this paper is to review the benefits and constraints of big data technologies for education. Big data being a rising innovation, its challenge, and constraints in education are yet to be fully exploited. This article aims to provide knowledge about the utilisation of big data in learning institutions by examining its significance. Besides, significant utilisation of big data is only possible through understanding its benefit. Big data can improve learning activity in educational institutions, enhance decision making, offer a competitive advantage, and provide more knowledge to learners. It provisions key benefits to change the educational environments from traditional to innovative areas.

Keywords: big data, education institutions, instructions, learning.

Introduction

Currently, there has been a surge in the growth of big data (Chaurasia et al., 2016). For instance, 2015 experienced a market upscale of 31.4% leading to a $4.50 billion growth globally. According to research, the data amount grows nine times within five years (Klašnja-Milićević et al., 2017). This is because, big data results in competitive advantage and market sustainability of companies and organisations (Chaurasia et al., 2016). Greater work has been done in big data in the private areas but public areas like education lack profound studies. Hence, the utilisation of big data as an influence in learning has attracted research and several discussions (Sedkaoui and Khelfaoui, 2019). There are a lot of alterations in the educational environments due to greater enhancements in technologies (Chweya et al., 2019). And big data is among the most examined facts in the present information technologies with greater contribution to learning platforms in education (Nazarenko and Khronusova, 2017). Big Data entails a summary including volume, variety, velocity, and value (Klašnja-Milićević et al., 2017). But the data obtained in the learning environments, which possesses a magnificent substance is greatly unutilised (Sedkaoui and Khelfaoui, 2019). The reason could be because its significance is yet to be fully comprehended.

The growth in new technologies like IoT, big data, sensory gadgets, learner cards, among others, are entirely altering the notion of learning and instruction (Klašnja-Milićević et al., 2017). Worldwide, there is enormous pressure to alter activities and the structure of learning environments to accommodate the new agenda regarding international demands (Huda et al., 2016). Learning environments are under accreditation scrutiny to establish new methods to enhance and trace learner achievement among other educational procedures (Tulasi, 2013). This has pushed learning institutions to re-evaluate pragmatic insights from data and to initiate tactics to meet the latest demands. Hence, several universities are now taking advantage of the current innovations in an online study (Huda et al., 2016). The present eLearning innovations focus on enhancing the learning experience for students and instructors (Reidenberg and Schaub, 2018). The outcome is a supported learner experience as per personal requirements, progress monitoring and tracking, and trace of any warning signals linked to learners (Marquez et al., 2016; Reidenberg and Schaub, 2018; Tan et al., 2018; Chweya et al., 2020). As education is being digitalised, the online pursuits result in a digital collection of details (Wang, 2016). Big data is among the most examined facts in the present information technologies and has a greater contribution to learning platforms in education (Nazarenko and Khronusova, 2017). This is because, utilisation of big data informs educational institutions about the

---

[a]ruthchweya@gmail.com; [b]asamuel@ticaret.edu.tr; [c]ayodelemelbury@gmail.com

upcoming innovation (Daniel, 2015). The usage of the data also allows universities to understand students and improve learning (Alyoussef and Al-Rahmi, 2022). Key significant to education are the learners and instructors, the technologies and pursuits, accomplishment, and execution (Macfadyen et al., 2014). Hence, there is a need to maximise the potential of data gathering to furnish its significance in online studies (Huda et al., 2016).

According to studies, big data provisions for collection, assembly, and evaluation of huge details (Huda et al., 2017). As learning environments are easily accessible via the internet, learners have the possibility of reaching their courses anywhere to learn (Sin and Muthu, 2015). Through the learning management structures, there is creation of huge data for utilisation in growing the learner environment. According to Sin and Muthu (2015), the need to use big data technologies in educational institutions results from the enormous details from student activities and the institutions themselves. Specifically, learning institutions create huge data daily that provision for extraction for significant tasks for stakeholders (Huda et al., 2016). For instance, higher learning institutions gather huge amounts of details regarding their learners, courses, and facilities (Williamson, 2018; Ajibade and Adediran, 2016). However, to date, these details have failed to be linked intelligently to enhance learning and to inform instruction practice (Macfadyen et al., 2014). This happens amidst attestation from various environments for instance retail, health care, and marketing that efficient utilisation of big data has prospects to impact the structures and outcomes. Further, universities are endlessly being entangled in digital innovation networks and expert technical procedures leading to smarter institutions (Sin and Muthu, 2015). Therefore, faster incorporation and access to data will create value and bring greater significance (Macfadyen et al., 2014). Besides, this will assist and enhance the learner's general study experience (Daniel, 2015; Sin and Muthu, 2015). Data infrastructures will also make the institutions smarter, enhance learner experience through personalised assistance and provide support for every learner (Huda et al., 2016; Ajibade and Adediran, 2016).

Several environments have incorporated big data with limited work done in higher education (Chaurasia et al., 2016). Research has also been conducted on big data in education and learning. For instance, research has explored utilising the recommender systems for big data in education (Dwivedi and Roshni, 2017). From their study, the structure helps filter out for recommendations on courses to learners. Another study examined the structural effect of big data-driven education (Zeide, 2017). The outcome reveals that with technological mediation, there is an impact in learning environments. Besides, big data tools assist to map content, establishing required metrics, and lead to required learning outcomes of instruction. From research by Pardos, (2017) on big data in education and models, it examines the models for required testing, computer tutor, and online study. It argues the need for connectionist paradigms to model and reaffirms ethical responsibilities for big data in education. Accordingly Williamson (2018) evaluates the establishment of big data infrastructure for smart universities. The study examines programs, measures, and visual analytics innovations that detail the infrastructure. The outcome is market reformations for better universities.

This study puts into consideration ways in which big data technologies can enhance and support the improvement of learning and teaching in educational institutions. The authors emphasise the significance resulting from the utilisation of Big data technologies in institutions. Further, the constraints are mentioned for improvement. From the outcome, it is acknowledged that educational institutions are complex structures hence the need to understand the benefits resulting from Big data before consideration. The goal is to provide prior knowledge and an understanding of big data impact for learning environments.

Overview of Big Data

There is huge growth of data in the world with expectations of up to 16 zettabytes by 2020 (Cavanillas et al., 2016). The utilisation of sensors is also playing a role in the growth of large streams of data. Besides, big data is furnishing new procedures of extracting impactful details from the new information. Big data has currently been appreciated by several firms and governments due to many added values and competitive benefit (Benjelloun et al., 2015). Hence, several firms are growing their key businesses on the possibility to gather and evaluate information for more insight and knowledge (Cavanillas et al., 2016). This shows that big data is now linked to practically every human activity facet (Demchenko et al., 2014). Big data handles different kinds of data, having a variety of magnitude, with different variability and complexity in various ways (Benjelloun et al., 2015; Demchenko et al., 2014; Ajibade et al., 2018). The emergence of big data was to handle the billions of data from wearables, sensory gadgets among others (Lee, 2017). It creates a way of gathering various details, and evaluating them for provision of meaningful data to clients through a reliable data execution technology (Liu et al., 2016). Big data also encapsulates tools plus operations for evaluating, executing, and directing the immense, hard, and progressing data sets (Alharthi et al., 2017). In the end, the data requires quick analyses, decisions and actions for keeping track with emerging trends that change continuously (Shah, 2022).

Need for Big Data and its Shortcomings in Education

The education environment has had great transformation with the rise of the internet (Wang, 2016). It is anticipated that digital gadgets will be prevalent in the learning environment. There have been great enhancements in learning and pedagogy leading to increase in new study surroundings (Sin and Muthu, 2015). The current learners are involved in utilisation of online tools like chats, instant message, discussion groups and other learning management structures like flipped classes, Moodle, among others. Further, there is also utilisation of mobile gadgets to reach learning anywhere through the internet. As educations is being digitised, the online pursuits are establishing huge digital chunks of data (Wang, 2016). The outcome is huge details that are speculated to be significant in improving the study environment, assist learners in studying, and enhance the entire study encounter (Sin and Muthu, 2015). Besides, there is also huge data emanating from learning institutions through the utilisation of course, class and learner management applications. This shows that big data has established an area in education with further prediction for greater implementation in learning institutions (Reyes, 2015). However, there is still limited effort to utilise big data in learning environments even amidst great benefits (Wang, 2016). The presence of concerns like security, storage and privacy are some issues that need further examination (Daniel, 2015; Nazarenko and Khronusova, 2017).

Big data has had utilisation in various sectors like transport, tourism, research, health, and politics (Benjelloun et al., 2015). Further, it is expected to show significance in all areas also like media, finance and insurance, telecom and entertainment, retail, manufacturing, among the rest (Cavanillas et al., 2016). Its utilisation in education allows usage of new notions and current study concepts to solve education difficulties (Birjali et al., 2016). This section examines the reasons for utilizing big data in education and further the shortcomings.

Reasons for Big Data in Education

One interesting aspect of the current teaching and study environments is its faster and ever-changing utilisation of innovations. Big data is among the notions that have attracted much study in learning areas. This section examines some benefits of utilising big data in educational surroundings.

Informed decisions: With the rise in the number of learners, there is also an increase in the generated data volume (Dahdouh et al., 2018). This necessitates the utilisation of big data technologies to deal with the greater data amounts for better-informed decisions. Big data is speculated to provide benefits for complex skills assessment, personalising study, enhancing, and improving education quality (Nazarenko and Khronusova, 2017). For instance, as educational information is broadly available, there is a possibility of utilizing data-driven techniques that support learners' activities to provide recommendations, extricate knowledge frameworks, or also to help instructors in extracting educational material (Lynch, 2017). According to study, big data can be used for managing education systems and further make intelligent decisions (Yang, 2022).

Better analytics before market entry: From research, much of the benefit comes due to strained budgets for institutions with a need for awareness in spending (Chaurasia et al., 2016). For instance, market exploration approaches are utilised in trade to examine merchandise' entry to various trade sectors (Kalota, 2015). Similarly, learning institutions can gather details and employ analytics before market entry. This assists in reliable future predictions. It will further enhance decision making in institutions (Klašnja-Milićević et al., 2017; Tulasi, 2013). Through learning analytics, big data contains potentiality to informal educational policy making and implementation (Wang, 2016). The online comments and messages help to understand the learner's needs and address them to improve institutions at large. The endpoint is a growing institution with challenges handled efficiently (Tulasi, 2013). For instance, study points that learning analytics helps understand a learner and their progress and provide feedback for improvement at the end (Beerkens, 2022).

Improved study and assessment process: Big data offers instructors with more materials and helps in the improvement of study and assessment procedures. Accordingly, big data supports instruction procedures through provisioning for dependable data sources (Huda et al., 2016). This will provide for learner interaction, engagement, and knowledge transmission to all. Further, this will help in the prediction of student dropout through learner achievement (Kalota, 2015; Klašnja-Milićević et al., 2017). The achievement rate of learners can be earlier enhanced (Tulasi, 2013). For instance, study shows that higher education institutions can help improve effectiveness through the collected information and further testing the progress of each student (Tasmin et al., 2022).

Enhancement of learner experience: Research conducted by (Klašnja-Milićević et al., 2017) points out that Big Data can help enhance a student's study experience. This is possible through the collection of details, evaluation, and reporting of data regarding the learner context (Wang, 2016). With the utilisation of massive open online courses, this creates a possibility for data gathering and analysis for prediction of learner achievement (Kalota, 2015). This will enable the prediction of learner dropout. For instance, the traditional way of measuring learner behaviour was through home assignments and examination but big data can help this evaluation for timely and accurate evaluation (Khan and Khojah, 2022).

Shortcomings on big data in education

Storage issues: As the educational environment is growing to meet the needs of the current learners, big data also furnishes unique issues for the education shareholders. Some studies reveal that big data suffers from storage issues (Sin and Muthu, 2015). Therefore, with limits in the hard disks utilised presently, big data details are only able to be stored after compression. Besides these data possess various structures and different sizes while necessitating more processing time and resources. The details must then be split and executed in different computers through the network. This brings the data interpretation issue because of the huge data necessitating better techniques than the traditional reports.

Privacy issues: Big data suffers from privacy issues (Daniel, 2015; Ajibade et al., 2019; Wang, 2016; Nazarenko and Khronusova, 2017; Kalota, 2015). Improving education while protecting the privacy of learners is an important societal value (Reidenberg and Schaub, 2018). For instance, most details gathered in learning institutions entail personal data that necessitate individual approval (Wang, 2016). Also, to track the activity of a leaner demand for individual details like temperament. But learners are not ready to provide these details to university management. As a result, this gathering of educational details heightens concerns on the ethics linked with details ownership, privacy, security, and utilisation ethics (Daniel, 2015). Again, there exist incidences of accountability linked with the utilisation of learner details for prediction. As there are greater efforts to digitise all data in learning environments, there is a lack of measures on how to prevent abuse and misuse of the details gathered (Nazarenko and Khronusova, 2017). Continual learning means continuous tracking of learner details posing greater consequences for students. Hence, there is a need to examine policies to protect and hinder learner details from abuse and misuse. This is because, as much as data helps to bring insights for improvement, privacy can affect student study and social growth (Reidenberg and Schaub, 2018).

Veracity: Another issue is the veracity of Big data. This covers many different elements with various contrast regarding exploration, precision, and swiftness of the details (Wang, 2016). The mentioned details fail to undergo analysis through just any gadget. Therefore, to generally manage and examine massive details, there is a need for extra tools to settle this difficulty on the complexity of big data (Huda et al., 2016). Therefore, organisations are required to transfer details to be interpreted as per the necessities and demands of instruction, learning materials, and even healthcare, among others.

Biasness in details gathering: research points out that employing big data results in hidden biases implicit to the gathering, evaluation, and description of the details (Daniel, 2015). This means there cannot be ultimate dependence on big data solely minus different sources of evidence, for instance, experience. The outcome will be deceptive and impede some people in an institution.

Conclusions

This paper highlights that big data technology has significant benefits for educational institutions. Universities are key in the provision of knowledge and skills to learners in various ways. Hence, there is speculation about big data altering the operations of higher learning through the utilisation of various tools to gather, process, and exhibit details for decision making. This study agrees that there is a greater amount of data resulting from learners, instructors, and other staff in educational institutions. This data can be utilised for meaningful insights and learner achievement. Unfortunately, fewer institutions are incorporating newer innovations in their environments even amidst discussed opportunities. The growth of big data warrants an understanding of the benefits and challenges to fully change the education environment. The unforeseen benefits and difficulties have not been fully exhausted in this study and big data is continually evolving. Hence, there is a need for endless examinations on big data to grow learning institutions. This will help the capitalisation of this technology by stakeholders in learning environments. It is suggested that learning institutions should incorporate the mentioned technology in their environments to achieve quality learning. This will in turn reform and alter the educational environment in the way instruction is delivered, learner-instructor communication, and development of study materials. There is also needed to find ways of handling issues like privacy, ethics, and security concerning big data.

References

Ajibade, S. S. and Adediran, A. (2016). An overview of big data visualisation techniques in data mining. Int. J. Comput. Sci. Inf. Technol. Res. 4(3):105–113.

Ajibade, S. S. M., Ahmad, N. B., and Shamsuddin, S. M. (2018). A data mining approach to predict academic performance of students using ensemble techniques. In International conference on intelligent systems design and applications (pp. 749–760). Springer, Cham.

Ajibade, S. S. M., Ahmad, N. B., and Shamsuddin, S. M. (2019). An heuristic feature selection algorithm to evaluate academic performance of students. In 2019 IEEE 10th control and system graduate research colloquium (ICSGRC), (pp. 110–114). IEEE.

Alharthi, A., Krotov, V., and Bowman, M. (2017). Addressing barriers to big data. Bus. Horiz. 60(3):285–292. doi:https://doi.org/10.1016/j.bushor.2017.01.002.

Alyoussef, I. Y. and Al-Rahmi, W. M. (2022). Big data analytics adoption via lenses of technology acceptance model: Empirical study of higher education. Entrepreneurship and Sustain. Issues 9(3):399–413.

Beerkens, M. (2022). An evolution of performance data in higher education governance: a path towards a 'big data'era? Qual. High. Educ. 28(1):29–49.

Benjelloun, F. Z., Lahcen, A. A., and Belfkih, S. (2015). An overview of big data opportunities, applications and tools. Paper presented at the 2015 Intelligent Systems and Computer Vision (ISCV).

Birjali, M., Beni-Hssane, A., and Erritali, M. (2016). Learning with big data technology: The future of education. In Paper presented at the international afro-european conference for industrial advancement.

Cavanillas, J. M., Curry, E., and Wahlster, W. (2016). The big data value opportunity New horizons for a data-driven economy, (pp. 3–11): Springer, Cham.

Chaurasia, P., McClean, S. I., Nugent, C. D., Cleland, I., Zhang, S., Donnelly, M. P., and Tschanz, J. (2016). Modelling assistive technology adoption for people with dementia. J. Biomed. Inform. 63(Suppl. C):235–248. doi:https://doi.org/10.1016/j.jbi.2016.08.021

Chweya, R., Ajibade, S. S. M., Buba, A. K., and Samuel, M. (2020). IoT and big data technologies: opportunities and challenges for higher learning. Int. J. Recent Technol. Eng. (IJRTE) 9(2):909–913.

Chweya, R., Ibrahim, O., and Nilashi, M. (2019). IoT in higher learning institutions: opportunities and challenges. J. Soft Comput. Decis. Support Syst. 6(6):1–8.

Dahdouh, K., Dakkak, A., Oughdir, L., and Messaoudi, F. (2018). Big data for online learning systems. Educ. Inf. Technol. 23(6):783–2800.

Daniel, B. (2015). Big Data and analytics in higher education: Opportunities and challenges. Br. J. Educ. Technol. 46(5):904–920.

Demchenko, Y., Gruengard, E., and Klous, S. (2014). Instructional model for building effective big data curricula for online and campus education. In Paper presented at the 2014 IEEE 6th international conference on cloud computing technology and science.

Dwivedi, S. and Roshni, V. K. (2017). Recommender system for big data in education. In Paper presented at the 2017 5th National Conference on E-Learning & E-Learning Technologies (ELELTECH).

Huda, M., Anshari, M., Almunawar, M. N., Shahrill, M., Tan, A., Jaidin, J. H., and Masri, M. (2016). Innovative teaching in higher education: the big data approach. TOJET.

Huda, M., Maseleno, A., Shahrill, M., Jasmi, K. A., Mustari, I., and Basiron, B. (2017). Exploring adaptive teaching competencies in big data era. Int. J. Emerg. Technol. Learn. (iJET) 12(03):68–83.

Kalota, F. (2015). Applications of big data in education. Int. J. Educ. Pedagog. Sci. 9(5):1607–1612.

Khan, M. A. and Khojah, M. (2022). Artificial intelligence and big data: The advent of new pedagogy in the adaptive e-learning system in the higher educational institutions of Saudi Arabia. Education Research International.

Klašnja-Milićević, A., Ivanović, M., and Budimac, Z. (2017). Data science in education: Big data and learning analytics. Comput. Appl. Eng. Educ. 25(6):1066–1078.

Lee, I. (2017). Big data: Dimensions, evolution, impacts, and challenges. Bus. Horiz. 60(3):293–303. doi:https://doi.org/10.1016/j.bushor.2017.01.004

Liu, R. H., Kuo, C. F., Yang, C. T., Chen, S. T., and Liu, J. C. (2016). On construction of an energy monitoring service using big data technology for smart campus. In Paper presented at the 2016 7th international conference on cloud computing and big data (CCBD).

Lynch, C. F. (2017). Who prophets from big data in education? New insights and new challenges. Theory Res. Educ. 15(3):249–271.

Macfadyen, L. P., Dawson, S., Pardo, A., and Gaševic, D. (2014). Embracing big data in complex educational systems: The learning analytics imperative and the policy challenge. Res. Pract. Assessment 9:17–28.

Marquez, J., Villanueva, J., Solarte, Z., and Garcia, A. (2016). IoT in education: integration of objects with virtual academic communities. Paper presented at the World CIST (1).

Nazarenko, M. A. and Khronusova, T. V. (2017). Big data in modern higher education. Benefits and criticism. In Paper presented at the 2017 International conference quality management, transport and information security, information technologies (IT & QM & IS).

Pardos, Z. A. (2017). Big data in education and the models that love them. Curr. Opin. Behav. Sci. 18:107–113. doi:https://doi.org/10.1016/j.cobeha.2017.11.006.

Reidenberg, J. R. and Schaub, F. (2018). Achieving big data privacy in education. Theory Res. Educ. 16(3):263–279.

Reyes, J. A. (2015). The skinny on big data in education: Learning analytics simplified. TechTrends 59(2):75–80.

Sedkaoui, S. and Khelfaoui, M. (2019). Understand, develop and enhance the learning process with big data. Inf. Discov. Deliv. 47(2).

Shah, T. H. (2022). Big data analytics in higher education. In Research Anthology on Big Data Analytics, Architectures, and Applications, (pp. 1275–1293).

Sin, K. and Muthu, L. (2015). Application of big data in education data mining and learning analytics--a literature review. ICTACT J. Soft Comput. 5(4).

Tan, P., Wu, H., Li, P., and Xu, H. (2018). Teaching management system with applications of RFID and IoT technology. Educ. Sci. 8(1):26.

Tasmin, R., Huey, T. L., Nda, R. M., and Jaafar, I. (2022). What does it take to adopt big data management approach at malaysian higher education institutions. doi: 10.5171/2022.924024

Tulasi, B. (2013). Significance of big data and analytics in higher education. Int. J. Comput. Appl. 68(14).

Wang, Y. (2016). Big opportunities and big concerns of big data in education. TechTrends 60(4):381–384.

Williamson, B. (2018). The hidden architecture of higher education: building a big data infrastructure for the 'smarter university'. Int. J. Educ. Technol. High. Educ. 15(1):1–26.

Yang, Q. (2022). Modeling and analysis of the impact of big data on the development of education network. Discrete Dynamics in Nature and Society.

Zeide, E. (2017). The structural consequences of big data-driven education. Big Data 5(2):164–172.

165 Optimisation of process parameters of CNC milling: An approach

Ajay Bonde[a], Dikshant Kamble[b], Shashaank Laad[c], Vedant Barhate[d], Yash Kawalkar[e], and Chandrakant Kshirsagar[f]

Department of Mechanical Engineering, Yeshwantrao Chavan College of Engineering, Nagpur, India

Abstract

This paper deals with the literature reviews carried out for parametric optimisation in CNC milling. The optimisation of milling process is targeted with the view point of better surface finish. Various optimisation methods involved are factorial design method (FDM), Taguchi design of experiments (DOE), response surface methodology (RSM), artificial neural network (ANN), genetic algorithm (GA), particle swarm optimisation (PSO), fuzzy sets, etc. All these techniques are in current use with present day research. The overall work aims at finding a set of optimal process parameters to enhance the surface finish, conversely reducing surface roughness. Some influencing factors on the basis of literature review are surface roughness, feed, material removal rate (MRR), tool wear. These studies are helpful to further plan and devise the optimisation techniques.

Keywords: Artificial neural network, genetic algorithm, machining parameters, material removal rate, particle swarm optimisation, tool wear rate.

Introduction

The process of metal cutting is manufacturing industrial operations that remove undesirable elements from metal objects to produce the desired forms. Traditional chip formation procedures such as turning, milling, boring, and drilling were used to remove the materials. Metal is abolished as a plastically deformed chip of substantial size in these techniques (Sahare et al., 2018). Numerical, artificial neural networks (ANN) and other soft computing methods are now frequently employed for modelling and optimising industrial technology performance. In production contexts, the economy of machining operations directly influences the product cost. The goal of this study is to determine the best process parameters for various milling operations because of its adaptability and efficiency. Milling is widely used in aerospace, variety of other industries and from large manufacturers to small tool. Aside from basic cutting process parameters such as cutting environment, tool type, feed rate, tool geometry, cutting speed, and, depth of cut play a significant role in determining the performance of quality specification in metal cutting processes, particularly turning and milling processes. Cutting tool geometries (nose radius, relief angle, rake angle and helix angle) and machining circumstances (spindle speed, feed rate, and depth of cut) have a significant impact on machining performance. Incorrectly chosen parameters for machining and an appallingly designed cutting tool degrade surface smoothness as well decreases stability and hence creating thermal distortion due to high temperature variations at the cutting area and quick tool wear. Productivity, the accuracy of machined parts, product and cutting tool integrity are the key features of elegant manufacturing processes. Lower production, labour and maintenance costs are all benefits of a well-designed manufacturing process.

Literature Review on Optimisation of CNC Milling Machine

Surface roughness

This work analyses surface quality, resource cutting conditions, and material removal rate during upscale end milling of Al2024-T4 aluminium in order to create an adequate roughness prediction model with the improvement in the cutting parameters using Taguchi and ANN (Sahare et al., 2018). The predicted results of the ANN are found to be in consistent with the experimental data. As a result, it was able to demonstrate

[a]ajaysbonde04@gmail.com; [b]dikshantkamble98@gmail.com; [c]shashanklaad47@gmail.com; [d]vedantbarhate@gmail.com; [e]yashrkawalkar@gmail.com; [f]cuk532000@gmail.com

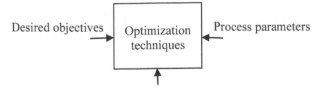

Figure 165.1 Optimisation-based approach

its effectiveness in optimizing the final milling process. Ghani et al. (2004) hint that cutting speed, feed rate, and depth of cut are the milling parameters that are assessed. The influence of these milling parameters is investigated using a Pareto analysis of variance (ANOVA), signal-to-noise (S/N) ratio, and orthogonal array. In general using a high cutting speed, moderate feed rate, and shallow depth of cut results in reduced cutting force and superior surface roughness. Pare et al. (2011) also suggest that optimum set of input variable values is determined using the particle swarm optimisation technique. Conclusions are equated to those obtained using genetic algorithm optimisation in the literature. The results achieved by PSO using rake angle in degrees and radians are the same in both situations, indicating that the application of PSO is independent of the rake angle unit utilised. The results achieved by Particle Swarm Optimisation are superior to those obtained by Zain et al. using GA, indicating that PSO is better suited to this type of problem. Lakshmi (2012) Response surface approach was used to model and optimise the average surface roughness values gained when milling EN24 grade steel with a hardness of 260 BHN using solid coated carbide tools. Cut depth (D), cutting speed (V) and feed rate (f) are the input variables (d). Material removal rates and surface roughness are the output variables. RSM is an effective tool for predicting the finish of a surface during operation. Surface roughness values predicted using quadratic second-order models. They were relatively close to the true results found in the studies. The cutting speed (v), depth of cut (d), and cutting feed (f) are the machining specification investigated in this research, with the response factor calculated as value of the matching surface quality (Daniyan et al., 2019). The model shows that all of the model elements, such as spindle speed, depth of cut, and cutting speed and feed, have a considerable impact on the values of surface roughness. It suggests the importance of choosing the proper values for the cutting specifications. Generated module can be used to forecast the span of surface quality that can be achieved by combining the process parameters. The parameters impacting surface roughness in the given machining process include depth of cut, speed, and feed rate (Ghalme et al., 2016). The design of experiment (DOE)-Taguchi technique tool is used to propose experiments and examine results. Seçgin and Sogut (2021) this work intends to define optimisation parameters for the milling process of AL 6061-T6 aluminium alloy used in the aviation industry based on machining processes. The Taguchi method was used to figure out what the best specifications were. Additionally, variance analysis was used to assess the impact of machining factors on surface roughness.

Nadafa and Shinde (2020) Used 27 experimental runs based upon L27 orthogonal array of Taguchi method were performed to optimise the process parameters on CNC milling for mild steel IS 2062:2011 E250 Gr.A. The machining performed on VMC was with a 63 mm cutter diameter to optimise the surface roughness. Panshetty et al. (2016) also proposed his research that has used Taguchi's method that reduces the number of experiments and is used for enhancing machining parameters. Confirmed on the examination done process parameters were optimised and ANOVA was performed to observe the relative magnitude of every factor on the objective function. Finally, it concluded the surface roughness and other output factors were optimised. Pang et al. (2013) this research article also introduces the execution of the Taguchi optimisation method to improve the cutting parameters for the end milling process of hallo site nanotubes. The results from this research paper tend to show the implications of the Taguchi method that can find the foremost combination of machining parameters that anticipate the good machining response conditions (Ribeiro et al., 2017). Also in his research, the conclusion showed that the radial cutting depth and the interaction of radial and axial depth of cut are the most relevant parameters used for minimizing the surface roughness for optimisation of machining parameters. Singh and Sultan (2018) have also proposed in their paper that has focused its importance on the sustainable modelling and optimisation of the CNC milling process. The factors which are considered to make a specific job were material utilised, time taken to complete that job, and energy required. Different consequences of material selection on sustainable components have been studied for various processes.

Vardhan et al. (2019) to ensure the analysis of experiment having input parameters during Process of milling on CNC Milling Machine of P20 steel. It enhances the surface roughness by lower feeding of wear rate at the cryogenic condition, which magnifies tool life and work Rate. Kriswanto et al. (2020) find that the variety of milling input parameters based on the L9 orthogonal array of Taguchi Method and Master CAM software. Based on the NC program, the S/N ratio analysis feed providing is less by which they are getting better surface Quality and greater in machining time value of the Bohler m303 extra material by Taguchi Method. In this experiment study, the consequences of the methodology of Taguchi L18 Orthogonal array to determine surface roughness and cutting force is discussed (Karabulut, 2015). On the basis of this study, analytical model was given by ANN and Regression Analysis. Here Author also sums up values of surface roughness which are enriched by escalating hardness in milling Al2o3MMC. Naidu et al. (2014) With the help of Taguchi methodology enhanced surface roughness by examining control factor carries out with help of L8 Orthogonal array for EN-31 alloy steel material . They expected S/N ratio by check-in predicated limit value. Lestari et al. (2019) in milling, UHMWPE ace tabular cup adopted Taguchi , RSM and Optimise specifications for yielding the minimum surface roughness. For smooth surface roughness of 0.143μm also found different machining components which may decrease machining cost.

Sridhar and Sellamani (2020) this research paper represents that End milling is performed on Al 356 / SiC metal matrix composites using a high-speed steel end mill cutter. From ANOVA analysis, it has been concluded that the most important variable process that influences are , an increase in the temperature of a rake. It is followed by a cutting speed, a depth of cut nose angle and a helix angle. Jain and Parashar (2020) For face milling of En-8 steel , the author has used a particle swarm optimisation algorithm to investigate various process parameters like depth of cut, spindle speed, feed rate, etc. Shelar and Shaikh (2017) three aspects and three characteristics are studied in this study, as well as the L27 orthogonal array should be made using two different inlet holes. In this refinement of the milling process, the Taguchi methodology is being used to make Ai-Si 316 stainless steel. This research paper represented by Ramya et al. (2017) presents the solution of real-life issues related to raw material extraction and metal from a piece of work. By using Taguchi methodology this experimentation is carried out. Shaik and Srinivas (2017) proposes a multi-purpose approach based on genetic algorithms using test data to simultaneously reduce tool vibration amplitudes and the complexity of the local function. The perfect combination of process variability is also confirmed by the radial neural network model. The interaction results presented that the lesser the feed and the more the spinning speed, reduce the stiffness and vibration levels of the amplitude tool.

Karabulut et al. (2018) carried out detailed information of studies conducted on aluminium 7039 based composites to optimise parameters for optimum surface finish. The experimentation was conducted using Taguchi L18 array, ANOVA, and ANN. Feed rate and cutting speed was shown that they were the key contributors to an increase in surface quality and optimizing the milling process. Wu and Lei (2019) used the same set of input parameters in the prediction of surface roughness of S45C steel. This work was conducted using the methods of vibration signal analysis and ANN. The features of signals were obtained through statistical computation which was later used as inputs for the layers of ANN inner layers. It was concluded that the surface roughness was affected by not only cutting parameters but also through vibration signals. Mia (2018) addressed the sustainability challenges in machining (MQL) by examining cutting energy, surface finish, and Minimum Quantity Lubrication. The input parameters were oriented by full factorial DOE and further optimised using the Taguchi method and RSM to optimise surface roughness. This study also showed that surface roughness is mostly influenced by MQL flow rate, followed by cutting speed, according to an ANOVA. Sathish et al. (2020) organised the milling operation of aluminium alloy AA6063 with SiC composites to reduce surface roughness. It was observed that at low cutting speeds, the surface roughness value is large but as the cutting speed increases, the surface roughness value drops when the analysis of variance (ANOVA) method was used to plot the relation between input and output parameters. It was also implemented during the helical milling of carbon fibre reinforced polymer using tungsten carbide tool. Amini et al. (2019) hints at the surface response models from the Regression method revealed a more optimised surface roughness quality when there was an increase as well reduction in cutting velocity and feed rate respectively. Sukumar et al. (2014) optimised the process of face milling for the material AL 6061. In this experimentation, the Taguchi Orthogonal Array was designed to discover sets of inputs and evaluate them against surface roughness. The S/N ratios were used to locate the optimum combination of inputs. The experimental values of parameters were unearthed with the full factorial method and analysed using ANN. Optimal conditions were established by comparing these values with S/N ratios. Hazir et al. (2018) minimised the surface roughness in a similar sense using the carbide tool. The factors were determined using the Pareto, ANOVA analysis and evaluated using multiple regression analysis. Surface roughness can be minimised by employing a faster spindle, feed rate, a smaller tool radius, a lower depth of cut and tangential cutting orientations, according to the research. Mohammad et al. (2016) concluded the performance by Surface roughness, MRR per minutes and utilised the rate for orthogonal array by Taguchi Method to reduce the process limit on CNC milling for EN19 andEN24 Alloy.

Material removal rate

Vardhana et al. (2017) have worked on optimizing input parameters of CNC milling of P20 steel by using the Response surface method and Taguchi method such as Feed, Axial depth, Cutting speed, Nose radius etc. The optimised results were feed = 0.18 mm/minute, cutting speed = 95 m/minute, nose radius = 0.8 mm axial, radial depth of cut = 0.69 mm, depth of cut = 1.5 mm. It gives the output to be surface quality Ra = 0.615 μm and MRR = 298.5cm³/minute. Gupta et al. (2017) also pointed that manufacturing parameters responsible for exactness in surface irregularities, MRR and TWR. It was observed that, increase in depth of cut and feed increases MRR also. It was observed that, decrease in TWR as step over ratio increase. Daniyan et al. (2019) hints that the milling method was designed in this study is to enhance the impacts of machine specifications, such as the depth of cut, feed rate, width of cut and cutting pressures during milling of effective AISI P20. The Response Surface Methodology and the Complete Abacus Environment (CAE) were used to carry out the numerical design (RSM). The results indicate that the tool can conduct the operation of cutting to its maximum potential and that the cutting speed, feed rate are important specifications which determine the rate of surface quality and MRR. This study has applications in the manufacturing industry, proposes a mathematical simulation equation for milling process enhancement for the purpose of lower total production costs while maintaining product quality and increasing productivity. Pandey et al. (2013) described the process development when it comes to MRR and surface termination, using Taguchi's L18 orthogonal array. It had been desirable that sound to noise measurement and Taguchi parameter design is a reliable, systematic, simple and highly efficient tool to improve the many functional features of the CNC milling process parameters. Analysis by Mehta and Kumar (2018), it was described the parameters that are most effective in determining the hardness of space in a composite compound used. Surface Roughness is compared to the latest development method. In the case of linear regression, the improvement in outcomes was 20.5% and in indirect retreat analysis it was 19%. Preparations had been made to reduce local shock and side effects by increasing MRR.

Daniyan et al. (2021) conducted a research study to improve the whole machining process in terms of machining cost and economics, as well as the manufacturing process' overall sustainability for end milling of AA6063. The Response Surface Methodology (RSM) was employed to process the machining parameters for optimum value whose results were compared with physical experiments performed on DMU80 mono BLOCK Deckle Maho 5-axis milling machine to create a mathematical model predicting the finest values of Material Removal Rate during the process. The optimisation of end milling of AL7075 matrix components was achieved through the maximisation of the material removal rate. In this, Kumar et al. (2020) examined the machining characteristics by designing the constraints using Taguchi Orthogonal Array and further analysing them using ANOVA. The Grey Relational Analysis was also performed to reduce the multi-response problem. The study specified that, higher inputs led to higher Material Removal Rate.

Tool wear

In this article, the researchers have proposed a substantial experimental and micro structural analysis that studies the tungsten carbide end mill cutter in P20 mould steel with extensive cryogenic treatment soaking period (Mukkoti et al., 2020). A combined RSM and NSGA – II method for optimising process parameters in CNC milling on P20 steel has been proposed. Cryogenic treatment has a significant impact on performance indicators such as depth of cut, feed, and cutting speed. This also helped in optimising the tool wear rate. Premnath et al. (2012) proposed about the tool wear in his research paper that represents the enhancement of the machining parameters on milling of a hybrid metal matrix composite using tungsten carbide insert. Various mathematical models were developed for testing the adequacy and optimsing tool wear. Tran et al. (2021) studied milling process-based milling parameters enhancing surface roughness and cutting parameters. After investigation they found at Sof 4989.739, Fof1182.38 mm/minute and to f0.0674 mm get approx. 0.260586713 μm and 1012.76775 seconds respectively by sensitivity and ANOV analysis. Research paper gives knowledge of computer-based axis milling machine designing processes (Abd Rahman et al., 2021). The goal of the study is to find the best combination of geometrical specifications like nose radius, rake angle, and helix angle as well as machining parameters, like depth of cut, feed rate, and cutting speed to achieve the lowest tool wear and surface roughness in end milling of Al 356/SiC metal matrix composites (MMCs) with a high-speed steel end mill cutter (Rajeswari and Sivasakthivel, 2018). Research by Daniyan et al. (2019) states, the process of milling was proposed to enhance the machining effects on its specifications. Using response surface methodology (RSM) and complete abacus environment (CAE), the numerical design was performed, therefore stating the importance of proper milling process design and control.

Process optimisation

Campatelli et al. (2014) focus on machining centre efficiency and present an experimental approach for evaluating and optimizing process parameters to reduce power consumption in a milling operation on a current CNC machine. Cutting speed, axial and radial depth of cut, and feed rate are among the factors assessed. Several distinct power consumption factors throughout the machining process are highlighted in the model for the NMV1500DCG milling machine. The first and most crucial conclusion is that the machines idle or basic state consumes the majority of its power. The developed model emphasises the importance of increasing the material removal rate as much as possible to reduce the machine's environmental footprint by choosing a cutting speed, chip section and feed rate. It should be as high as possible while trying to remain compatible with the tool's feasible working parameters. Nguyen et al. (2020) intends to improve the milling process energy power factor (PO), surface roughness (Ra), and efficiency (EF) by optimizing machining inputs. The milling speed (V), tool radius(r), feed (f), and depth of cut (a) are all taken into account. For the stainless steel 304, the machining operations were carried out on a vertical milling machine in a dry environment. The findings revealed that a, f, V, and r have the most influence on milling reactions. When compared to the initial parameter settings, the lowering in surface roughness is around 39.18%, while the modification in PO and EF are 26.47% and 22.61%, respectively. Kant and Sangwan (2014) paper offers a multi-objective prediction model for machining power usage and surface roughness minimisation of AISI 1045 steel using principal component analysis, response surface technique and grey relational analysis to find the best machining parameters. This statistical model suggests depth of cut, cutting speed and feed rate to be the main factors for reducing power consumption. Malghan et al. (2017) carried out several tests during face milling of aluminium matrix composite AA6061. The trials, when performed using the desirability function and ANOVA analysis indicated that, low feed rate, depth of cut and high spindle speed resulted in better surface quality. Also, it was noticed that the Particle Swarm Optimisation method was more accurate in providing results than the desirability function. Subagio et al. (2019) hints the pattern on 3-axis CNC milling machine driven by servo motors with spindle rotation around 12000 rpm. They have to improve the design of the machine and get result for enhancing accuracy. Deviation on each axis of 0.033 for X, 0.102 for Y, 0.063 for Z and 0.096 mm of flatness for 3-axis CNC milling machine. Kumar ct al. (2021) this paper focuses on an improved multi-purpose solution for site stability, object removal rate, power consumption. Tool life are determined and guaranteed by validation performance.

Aggarwal and Xirouchakis (2013) studied the conditions required to optimise machining parameters during the process of pocket milling of Certal (AW 7072). This study was conducted using GA-Opt Mill which is a type of genetic algorithm-based technique through which they minimised pocket milling time. This system assisted them to carry out a flexible analysis while maintaining one of the parameters constant as per their constraints. The system also helped to minimise the static deflection which is otherwise caused due to high cutting forces. It could cause damage to tool life and will be resulted in to reduction of the overall productivity.

Table 165.1

Sr no.	Details of Latest Literature Review Paper				
	Year	Author	Tool and Material	Input parameter	Output parameter
1	2020	Vishnu Vardhan Mukkoti, Chinmaya, Dhanraj B Sankaraiah Gandla , Pallab Sarkar , Srinivasa Rao P and Prasad Mohanty Mukkoti et al. (2020)	P20 mold steel, Tungsten carbide end mill cutter	Cutting speed, feed rate, depth of cut	Material removal rate (MRR), Surface roughness (Ra), and Tool wear rate (TWR)
2	2020	T. Sathish, N. Sabarirajan, S. Karthick, Sathish et al. (2020)	Aluminium Alloy AA6063 + SiC, Coated Carbide	Depth of cut, feed rate, and Cutting Speed,	Surface Roughness
3	2021	Kriswanto1, WSumbodo1, W Aryadi1, and J Jamari Kriswanto et al. (2020)	The Bohler M303 Extra, Carbide Coating End Mill	Depth of cut Feed Rate and Spindle Speed	Surface Quality and Machining Time
4	2021	Chi Thien Tran,1 Minh Phung Dang,1 HieuGiang Le,1 Ngoc Le Chau,2 and Thanh-Phong Dao Tran et al. (2021)	SIMOLD 2083 Alloy Mold Core, Coated Carbide	Depth of cut, Feed rate, Spindle speed, Tool radius, and cutting directions	Surface Roughness, Milling Time
5	2021	AbdRahman Z.1, Mohamed S.B2, Zulkifli A.R.3, Kasim M.S4, W.N.F.Mohamad5 Abd Rahman et al. (2021)	UHMW-PE Block Material, End Milling Tool	Cutting Depth, Cutting Speed, Safety Features, and Feed Rate	Surface Roughness and Materials Removal Rate

Objective

The major goal of our research is to find the factor or levels that will meet the necessary standards at the same time. The optimal combination of variables that generates the desired response will assist in describing a response that is close to the optimum level. Another goal of this study is to see how a specific reaction changes when the level of the elements changes over the stated points of interest. This study is also being carried out to gain a quantitative understanding of the region, based on the assumption that several attributes are not involved in the process. This will not only contribute to the estimation of product qualities throughout the region but will also assist in the identification of the combinations essential for process stability.

Research gap

After reviewing the research papers, many important insights were derived on the issues related to our study. It had been observed that very less work was done on modern tools as well technology and the use of traditional techniques is still more prevalent in the manufacturing industry. It was also discovered that the hand field tools were given more focus most of the time as compared to the automated machines. The majority of studies in this field were focused on prediction rather than determining the actual roughness of the surface. It is the tool geometry, speed, feed and depth of cut that largely contribute to milling process. Obviously they influence the surface roughness of work piece under machining. Increment in tool wear and abrasive forces will be resulted in to deprived surface roughness. This fact is grossly ignored because it is time transitive. Such values must be recomputed as in process parameters.

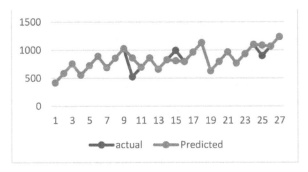

Figure 165.2 Rate of material removal: Actual and predicted values
Source: Sahare et al. (2018)

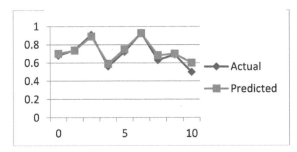

Figure 165.3 Experimental and predicted surface roughness values
Source: Daniyan et al. (2019)

Graphs representing data from literature review-

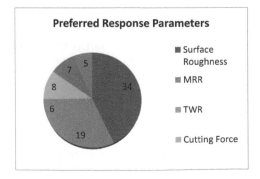

Figure 165.4 Output parameters

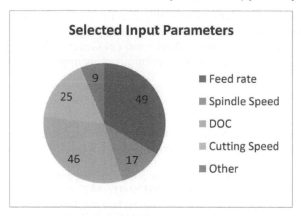

Figure 165.5 Input parameters

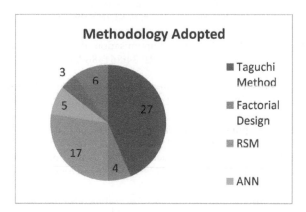

Figure 165.6 Tools and techniques

Discussion and Conclusion

An investigation of milling process is done in current research work according to structure research of milling centre components and response surface methodology. This can be used to calculate the value of surface roughness for a given machining parameter. It will assist in the selection of operating parameters when a specific surface finish is required. It has been shown that the predicted results are better for different materials used here.

Future research work can be carried out through comparing the results of finite element analysis by means of other software performed under different loading conditions. The future work can also take a different approach by including additional variables such as approach angle, rake angle, coating material, the effect of coolant as well as lubrication conditions. The conducted studies can be compared with the help of conventional and wiper inserts for improved surface finish for different types of materials. The effectiveness of this study can be further increased by employing a non-contact type roughness measurement system using machine vision. The techniques of artificial neural network (ANN) and genetic algorithm can be used to construct and refine empirical models as well as models created by the full factorial design as well. They can be compared by using a different set of methods for optimised results. It was concluded that the consumption of energy and the wastage of resources were affected by multiple factors. From analysing various research papers it had been found that the researchers have taken depth of cut, spindle speed and feed rate as input specifications. Depth of cut, feed rate, and spindle speeds are the important specifications while learning the impact of process variables on the desired response characteristics.

Quality is the most important characteristic in manufacturing sectors. Most of the researchers have taken surface roughness as the quality factor. Researchers tried to enhance the performance attribute according to the query identified. Taguchi combined with ANOVA are good methods to analyse the conclusion data. Taguchi helps to find the optimised sequence and the ANOVA technique helps to find which parameters are most genuine as well their percent contribution. Soft computing tools had been utilised for a rough calculation of the tool wear and tool life. The results aren't as impressive as they were in the case of the surface roughness supposition. The very stochastic nature of tool wear and tool life as well as the difficulty in identifying quantitative specification with which tool wear can be well-correlated, are the reasons for this. Soft computing enhancing ways like Genetic Algorithm has been used in machining area for determination of enhancement of the internal parameters of neural networks. Application of Genetic Algorithm for machining has attained maturity. Particle swarm optimisation, a newer technique, may appear as a better substitute for the genetic algorithm.

References

Abd Rahman, Z., Mohamed, S. B., Zulkifli, A. R., Kasim, M. S., and Mohamad, W. N. F. (2021). Design and fabrication of a PC-Based 3 axis CNC milling machine. Int. J. Eng. Trends Technol. 69(9):1–13. ISSN: 2231-5381, doi:https://doi.org/10.14445/22315381/IJETT-V69I9P201.

Aggarwal, S. and Xirouchakis, P. (2013). Selection of optimal cutting conditions for pocket milling using genetic algorithm. Int. J. Adv. Manuf. Technol. 66:1943–1958. https://doi.org/10.1007/s00170-012-4472-x.

Amini, S., Baraheni, M., and Hakimi, E. (2019). Enhancing dimensional accuracy and surface integrity by helical milling of carbon fiber reinforced polymers. Int. J. Lightweight Mater. Manuf. 2(4):362–372. ISSN 2588-8404.https://doi.org/10.1016/j.ijlmm.2019.03.001.

Campatelli, G., Lorenzini, L., and Scippa, A. (2014). Optimisation of process parameters using a response surface method for minimizing power consumption in the milling of carbon steel. J. Clean. Prod. 66:309–316. ISSN 0959-6526, https://doi.org/10.1016/j.jclepro.2013.10.025.

Daniyan, I. A., Tlhabadira, I., Daramola, O. O., and Mpofu, K. (2019). Design and optimisation of machining parameters for effective AISI P20 removal rate during milling operation. Procedia CIRP 84:861–867. ISSN 2212-8271, https://doi.org/10.1016/j.procir.2019.04.31

Daniyan, I. A., Tlhabadira, I., Mpofu, K., and Adeodu, A. O. (2021). Process design and optimisation for the milling operation of aluminium alloy (AA6063 T6). Mater. Today: Proc. 38(Part 2):536–543. ISSN 2214-7853.https://doi.org/10.1016/j.matpr.2020.02.396.

Ghalme, S., Mankar, A., and Bhalerao, Y. J. (2016). Parameter optimisation in milling of glass fiber reinforced plastic (GFRP) using DOE-Taguchi method. Springer Plus 5:376. https://doi.org/10.1186/s40064-016-3055-y.

Ghani, J. A., Choudhury, I. A., and Hassan, H. H. (2004). Application of taguchi method in the optimisation of end milling parameters. J. Mater. Process. Technol. 145(1):84–92. ISSN 0924-0136, https://doi.org/10.1016/S0924-0136(03)00865-3.

Gupta, A., Soni, P. K., and Krishna, C. M. (2017). modelling and analysis of CNC milling process parameters on Al3030 based composite. Int. Conf. Recent Adv. Mater. Manuf. Technol. 346:28–29. doi:https://doi.org/10.1088/1757-899X/346/1/012073.

Hazir, E., Erdinler, E. S., and Koc, K. H. (2018). Optimisation of CNC cutting parameters using design of experiment (DOE) and desirability function. J. For. Res. 29:423–1434. https://doi.org/10.1007/s11676-017-0555-8.

Jain, S. and Parashar, V. (2020). Optimisation of CNC face milling process parameters using response surface methodology and particle swarm optimisation algorithm. Int. J. Mech. Prod. Eng. Res. Dev. (IJMPERD) 10(3):645–654. https://doi.org/10.1016/j.aej.2017.02.006.

Kant, G. and Sangwan, K. S. (2014). Prediction and optimisation of machining parameters for minimizing power consumption and surface roughness in machining. J. Clean. Prod. 83:151–164. ISSN 0959-6526. https://doi.org/10.1016/j.jclepro.2014.07.073.

Karabulut, S. (2015). Optimisation of surface roughness and cutting force during AA7039/Al2O3 metal matrix composites milling using neural networks and Taguchi method. Department of Mechanical Program, Hacettepe University, 06935 Ankara, Turkey, (Vol. 66, pp. 139–149). DOI: http://dx.doi.org/10.1016/j.measurement.2015.01.027.

Karabulut, S., Gökmen, U., and Çinici, H. (2018). Optimisation of machining conditions for surface quality in milling AA7039-based metal matrix composites. Arab. J. Sci. Eng. 43:1071–1082. ISSN 10.1007.https://doi.org/10.1007/s13369-017-2691-z.

Kriswanto, M., Sumbodo, W., Aryadi, W., and Jamari, J. (2020). Optimisation of milling parameters to increase surface quality and machining time of the bohler m303 extra. IOP Conf. Series: Earth Environ. Sci. 700:24. doi:https://iopscience.iop.org/article/10.1088/1755-1315/700/1/0120094.

Kumar, M. B., Sathiya, P., and Parameshwaran, R. (2020). Parameters optimisation for end milling of Al7075–ZrO₂–C metal matrix composites using GRA and ANOVA. Trans. Indian Inst. Met. 73:2931–2946. https://doi.org/10.1007/s12666-020-02089-2.

Kumar, R. S. Kumar, S. S., Murugan, K., Guruprasad, B., Manavalla, S., Madhu, S., Hariprabhu, M., Balamuralitharan, S., and Venkatesa Prabhu, S. (2021). Optimisation of CNC end milling process parameters of low-carbon mold steel using response surface methodology and grey relational analysis. Adv. Mater. Sci. Eng. 2021:1–11. https://doi.org/10.1155/2021/4005728.

Lakshmi, V. V. K. (2012). Modeling and optimisation of process parameters during end milling of hardened steel. Int. J. Eng. Res. 2:674–679.

Lestari, W. D., Ismail, R., Jamari, J., and Bayuseno, A. P. (2019) Optimisation of CNC milling parameters through the Taguchi and RSM methods for surface roughness of UHMWPE acetabular cup. Int. J. Mech. Eng. Technol. (IJMET) 10. ISSN Online: 09766359, https://iaeme.com/MasterAdmin/Journal_uploads/IJMET/VOLUME_10_ISSUE_2/IJMET_10_02_182.pdf.

Malghan, R. L., Karthik, M. C., Rao, A. K. S., Rao, S. S., and D'Souza, R. J. (2017). Application of particle swarm optimisation and response surface methodology for machining parameters optimisation of aluminium matrix composites in milling operation. J. Braz. Soc. Mech. Sci. Eng. 39(9). ISSN 10.1007.https://doi.org/10.1007/s40430-016-0675-7.

Mehta, A. and Kumar, M. (2018). Milling machine parameter optimisation by grey wolf optimisation (GWO). Int. J. Res. Electr. Comput. Eng. 6(2):1563–1567. ISSN: 2348-2281. http://nebula.wsimg.com/252d6c299d9b73f9bc-142fa9c3045dc3?AcccssKeyId=DFB1BA3CED7E7997D5B1&disposition=0&alloworigin=1.

Mia, M. (2018). Mathematical modeling and optimisation of MQL assisted end milling characteristics based on RSM and Taguchi method. Measurement 121:249–260. ISSN 0263-2241. https://doi.org/10.1016/j.measurement.2018.02.017.

Mohammad, M. F., Praveen Kumar, B., and Madhavan, P. L. (2016). Multi-objective optimisation of CNC milling parameters using taguchi method for EN19 & EN24 Alloy. Int. J. TechnoChem Res. 02(01):11–18. ISSN:2395-4248, https://www.researchgate.net/profile/Fakkir-Mohamed-M.

Mukkoti, V. V., Mohanty, C. P., Gandla, S., Sarkar, P., Rao, P. S., and Dhanraj, B. (2020). Optimisation of process parameters in CNC milling of P20 steel by cryo-treated tungsten carbide tools using NSGA-II, (Vol. 8(1-22):1790436). Taylor & Francis. ISSN 10.1080/21693277, https://doi.org/10.1080/21693277.2020.1790436.

Nadafa, S. S. and Shinde, M. Y. (2020). Optimisation of process parameters on CNC milling machine for mild steel IS 2062:2011 E250 Gr. A with AlTiN coated tool insert in wet condition. IOP Conf. Series: Mater. Sci. Eng. 748:012030. ISSN 10.1088/1757-899X. https://doi.org/10.1088/1757-899X/748/1/012030.

Naidu, G. G., Vishnu, A. V., and Raju, G. J. (2014). Optimisation of process parameters for surface roughness in milling of en-31 steel material using Taguchi robust design methodology. Int. J. Mech. Prod. Eng. 2(9). ISSN: 2320-2092, https://www.researchgate.net/publication/319127988.

Nguyen, T. T., Nguyen, T. A., and Trinh, Q. H. (2020). Optimisation of milling parameters for energy savings and surface quality. Arab. J. Sci. Eng. 45:9111–9125. https://doi.org/10.1007/s13369-020-04679-0.

Subagio, D. G., Subekti, R. A., Saputra, H. M., Rajani, A., and Sanjaya, K. H. (2019). Three-axis deviation analysis of CNC milling machine. Res. Centre Electr. Power Mechatron. 10:93–101. e-ISSN: 2088-6985, DOI: https://dx.doi.org/10.14203/j.mev.2019.v10.93-101.

Pandey, P., Sinha, P. K., Kumar, V., and Tiwari, M. (2013). Process parametric optimisation of CNC vertical milling machine using taguchi technique in varying condition. IOSR J. Mech. Civil Eng. (IOSR-JMCE), 6(5):34–42. ISSN: 2320-334X, https://doi.org/10.9790/1684-0653442.

Pang, J. S., Ansari, M. N. M., Zaroog, O. S. Ali, M. H., and Sapuan, S. M. (2013). Taguchi design optimisation of machining parameters on the CNC end milling process of halloysite nanotube with aluminium reinforced epoxy matrix (HNT/Al/Ep) hybrid composite, (Vol. 10(2):138–144), Taylor & Francis Group. ISSN 10.1016. https://doi.org/10.1016/j.hbrcj.2013.09.007.

Panshetty, S. S., Bute, P. V., Patil, R. R., and Satpute, J. B. (2016). Optimisation of process parameters in milling operation by taguchi's technique using regression analysis. IJSTE - Int. J. Sci. Technol. Eng. 2(11). ISSN (online): 2349-784X.https://www.researchgate.net/publication/302585051_.

Pare, V., Agnihotri, G., and Krishna, C. M. (2011). Optimisation of cutting conditions in end milling process with the approach of particle swarm optimisation. Int. J. Mech. Ind. Eng. 1(2): Article 9. doi: http://doi.org/10.47893/ijmic.2011.1025.

Premnath, A., Alwarsamy, T. and Rajmohan, T. (2012). Experimental investigation and optimisation of process parameters in milling of hybrid metal matrix composites, (Vol. 27 pp. 1035–1044), Taylor & Francis Group, ISSN 10426914. https://doi.org/10.1080/10426914.2012.677911.

Rajeswari, S. and Sivasakthivel, P. S. (2018). Optimisation of milling parameters with multi-performance characteristic on Al/SiC metal matrix composite using the grey-fuzzy logic algorithm. Multidiscip. Model. Mater. Struct. 14(2):284–305. https://doi.org/10.1108/MMMS-04-2017-0027.

Ramya, G., Kumar, B. S., and Gopikrishna, N. (2017). Optimisation of milling process parameters using taguchi parameter design approach. Int. J. Nanotechnol. Appl. Res. 11(4):293–304. (India Publications), ISSN 0973-631X, https://www.researchgate.net/publication/323079494_Optimisation_of_Milling_Process_Parameters_using_Taguchi_Parameter_Design_Approach.

Ribeiroa, J. E., Césara, M. B., and Lopes, H. (2017). Optimisation of machining parameters to improve the surface quality, (Vol. 5, pp. 355–362). https://doi.org/10.1016/j.prostr.2017.07.182.

Sahare, S. B., Untawale, S. P., Chaudhari, S. S., Shrivastava, R. L., and Kamble, P. D. (2018). Optimisation of end milling process for Al2024-T4 aluminium by combined Taguchi and artificial neural network process. Adv. Intell. Syst. Comput. 584:525–535. doi:10.1007/978-981-10-5699-4_49.

Sathish, T., Sabarirajan, N. and Karthick, S. (2020). Machining parameters optimisation of aluminium alloy 6063 with reinforcement of SiC composites. Mater. Today: Proc. 33:2559–2563.7853.https://doi.org/10.1016/j.matpr.2019.12.085.

Seçgin, Ö. and Sogut, M. Z. (2021). Surface roughness optimisation in milling operation for aluminium alloy (Al 6061-T6) in aviation manufacturing elements. Aircr. Eng. Aerosp. Technol. 93(8):1367–1374. https://doi.org/10.1108/AEAT-05-2021-0146.

Shaik, J. H. and Srinivas, J. (2017). Optimal selection of operating parameters in end milling of Al-6061 work materials using multi-objective approach. Mech. Adv. Mater. Mod. Process. 3:1–11. (Springer). https://doi.org/10.1186/s40759-017-0020-6.

Shelar, A. B. and Shaikh, A. M. (2017). Optimisation of CNC milling process by using different coatings - a review. Int. Adv. Res. J. Sci. Eng. Technol. 4(1):143–147. ISSN 2394-1588, https://doi.org/10.17148/IARJSET/NCDMETE.2017.33.

Singh, K. and Sultan, I. A. (2018). A computer-aided sustainable modelling and optimisation analysis of CNC milling and turning processes. J. Manuf. Mater. Process. 2. ISSN 10.3390. http://doi.org/10.3390/jmmp2040065.

Sridhar, S. and Sellamani, R. (2020). Investigation of input variables on temperature rise while end milling Al/SiC metal matrix composite. World J. Eng. 17(4):599–607. (India, Emerald Publishing Limited), ISSN 1708-5284 https://doi.org/10.1108/WJE-01-2020-0031.

Sukumar, M. S., Venkata Ramaiah, P. and Nagarjuna, A. (2014). Optimisation and prediction of parameters in face milling of Al-6061 using taguchi and ANN approach. Procedia Eng. 97:365–371. ISSN 1877-7058.https://doi.org/10.1016/j.proeng.2014.12.260.

Tran, C. T., Dang, M. P., Le, H. G., Le Chau, N. and Dao, T. P. (2021). Optimisation of CNC milling parameters for complex 3D surfaces of SIMOLD2083 alloy mold core utilizing multi objective water cycle algorithm. Hindawi, Math. Probl. Eng. 2021:1–11. Article ID:9946404, DOI: https://doi.org/10.1155/2021/9946404.

Vardhan, M. V., Mohanty, C. P. and Dhanraj, B. (2019). Experimental studyon parameters of P-20 steel in CNC milling machine. In International Conference on Multifunctional Materials (ICMM-2019), (Vol. 01), https://doi.org/10.1088/1742-6596/1495/1/012027.

Vardhana, M. V., Sankaraiahb, G., Yohanc, M. and Rao, H. J. (2017). Optimisation of parameters in CNC milling of P20 steel using response surface methodology and Taguchi method. Mater. Today: Proc. 4:9163–9169. (Elsevier Ltd.). ISSN 2214.7853. https://doi.org/10.1016/j.matpr.2017.07.273.

Wu, T. Y. and Lei, K. W. (2019). Prediction of surface roughness in milling process using vibration signal analysis and artificial neural network. Int. J. Adv. Manuf. Technol. 102:305–314. ISSN 10.1007 https://doi.org/10.1007/s00170-018-3176-2.

166 Comparative study on destructive and non destructive test on concrete using fly ash as one of the ingredient

V. G. Meshram[1,a], S. B. Borghate[2,b], and Manoj Patil[3,c]

[1]Department of Civil Engineering Yeshwantrao Chavan Collage of Engineering, Nagpur, India

[2]Department of Applied Mechanics, Visvesarayya National Institute of Technology, Nagpur, India

[3]Jawaharlal Nehru Medical College, Datta Meghe Institute of Medical Sciences, Sawangi (M), Wardha, India

Abstract

Periodic structure testing is necessary for proper civil infrastructure upkeep. Non-destructive testing procedures such as the Rebound hammer test and ultrasonic pulse velocity test are now commonly employed for this purpose. This paper aims at finding the correlation between the destructive and non-destructive testing on concrete using fly ash. Concrete cubes of grades M20, M25, M30 and M35 were casted for this purpose. As a cement substitute, 10 percent and 15 percent fly ash were employed. Compressive strength testing was used in the destructive tests, while the Rebound hammer test and Ultrasonic Pulse Velocity Test were used in the non-destructive tests. There was a correlation discovered between the destructive and non-destructive test results. The findings show that the non destructive tests can fairly be used for the assessment of strength of structure.

Keywords: Compressive strength, fly ash concrete, pulse velocity, rebound hammer, ultrasonic.

Introduction

Civil Engineering Structures are designed to last up to 100 years. The structures, on the other hand, are being harmed by the environment. Some waste materials and recycling materials, such as fly ash, sugar cane bagasse ash, rice husk ash, silico manganese slag, and others, are now frequently employed in concrete. As a result, frequent structural testing is essential for proper civil engineering infrastructure maintenance. The most often used criterion for assessing the quality of a structure is its compressive strength. At the initial stage, the traditional compressive strength test which is the destructive test is best suited. The destructive compressive strength test (DT), on the other hand, is not used to evaluate existing concrete structures. As a result, non-destructive testing procedures (NDT) must be used. Now s days, various Non-destructive testing procedures have been developed and proved to be successful for assessing structure strength at any stage. The most widely used NDT methods are the rebound hammer test and ultrasonic pulse velocity test.

An experimental investigation on combined NDT and DT methodologies was conducted by a group of researchers. According to Aydin and Saribiyik (2010), Shang et al. (2012), and Samson et al.(2011), the rebound hammer test can only be used to assess the compressive strength of concrete. However, Shariati et al. (2011), Nacer and Hannachi (2012), and Patil et al. (2015) conducted studies using a combination of methodologies and concluded that the combined methods were more accurate for estimating concrete compressive strength.

Mulik et al. (2015) and Kumavat et al. (2017) have shown graphically the relationship between the compressive strength obtained by the destructive and nondestructive test. For rebound value they have obtained the correlation coefficient in the range of 0.3 to 0.789 and for UPV value, the correlation coefficient was found in the range of 0.3550.672.

When compared to the core test findings, Benyahia et al. (2017) discovered that estimating strength based on correlation curves provided by equipment manufacturers or from the literature is sometimes confusing and yields inconsistent results. As a result, correct correlations for concrete manufactured with local materials and under local environmental conditions are required. It is also necessary to identify and quantify the impacts of site, curing, and age on the response of NDT measurements and concrete strengths.

Rohit et al. (2012), Jain et al.(2013), Reddy (2014), Supe and Gupta (2014), Konapure and Richardrobin (2015) Lopez et al. (2016), and Kumavat et al.(2017) studied the influence of concrete age on the compressive

---

[a]meshramvg@gmail.com; [b]sbborghate@apm.vnit.ac.in line; [c]patil98dent@gmail.com

strength obtained by DT and NDT. The authors discovered a direct link between concrete age and compressive strength as measured by ultrasonic pulse velocity and rebound number. As the age of concrete increases, the results of ultrasonic pulse velocity and rebound number also increases proportionately. It was also observed that, the results can be shown in the form of equation relating these two parameters. It was also observed that, the results given by rebound hammer test are more realistic as compared to ultrasonic pulse velocity test.

Shakir Mishhadani et al. (2012) looked at the impact of concrete age on the measurement of nondestructive tests performed over a 90- to 365-day period and discovered the average rate of growth in the rebound number is 14.75%, whereas the average rate of increase in ultrasonic pulse velocity is 2.4%.

From the literatures cited above, it can be seen that, reliability of non-destructive test is the basic problems. This can be overcome, if good correlation is found between the destructive and non-destructive test. This paper aims at finding this relationship for the concrete using 10% and 15% cement replacement by fly ash for M20, M25, M30 and M35 grade of concrete. An attempt is also made to corelate two most widely used methods of NDTs.

Methodology

Materials and mix design

The materials utilised in this experiment include 43 grade ordinary Portland cement as per IS 8112-1989, coarse aggregates (CA) and fine aggregates (FA) as per IS 383-1970 and fly ash as per IS 3812 (Part 1):2003.

The properties of the ingredients are as follows:-

1) Maximum size of aggregate = 20 mm.
2) Cement type = OPC 53 grade cement.
3) Specific gravity of cement = 3.15
4) Specific gravity of fine aggregate = 2.77
5) Specific gravity of coarse aggregate = 2.8
6) Specific gravity of fly ash = 2.2
6) Water absorption of coarse aggregate = 0.5%
7) Water absorption of fine aggregate = 1.1%
8) Free moisture content:
   i) Coarse aggregate = Nil.
   ii) Fine aggregate = 1%

For four grades, M20, M25, M30 and M35, the mix design for concrete cubes was carried out according to IS 10262-2009. Fly ash was used to replace cement in all concrete grades to the level of 10% and 15%. The material quantity required for various mix proportions is given in Table 166.1.

Tests performed

i.   Compressive Strength Test: After 3, 28, 56, and 90 days of curing, concrete cubes of size 150 mm were tested for compressive strength according to IS 516-1959.
ii. Ultrasonic Pulse Velocity Test: The Ultrasonic Pulse Velocity test was carried out on concrete cubes of size 150 mm in accordance with IS: 13311 (Part 1) – 1992.
iii. Rebound Hammer Test: The Rebound Hammer Test was carried out on concrete cubes of size 150 mm in accordance with IS: 13311 (Part 2) – 1992.

Table 166.1 Material quantity for various mix proportions

Grade of Concrete	% FA	Cement (kg/m³)	Water (kg/m³)	C.A. (kg/m³)	F.A. (kg/m³)	Fly ash (kg/m³)
M20	10	354.6	214.15	1220.48	601.60	39.4
	15	334.9	214.15	1220.48	601.60	59.1
M25	10	393.3	213.79	1176.14	605.31	43.7
	15	371.4	213.79	1176.14	605.31	65.5
M30	10	442.8	213.34	1126.83	605.21	49.2
	15	418.2	213.34	1126.83	605.21	73.8
M35	10	505.8	212.76	1069.22	599.03	56.2
	15	477.7	212.76	1069.22	599.03	84.3

Result and Discussion

The test results of all tests are shown graphically from Figure 166.1 to 166.16.

Correlation between age of concrete and various parameters

The variation of compressive strength of concrete with respect to age of concrete is shown from Figures 166.1 and 166.2. The variation of ultrasonic pulse velocity with respect to age of concrete is shown from Figures 166.3 and 166.4. The variation of Rebound Hammer Strength with respect to age of concrete is shown from Figures 166.5 and 166.6.

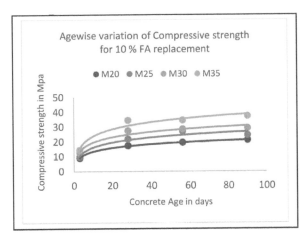

Figure 166.1 Age-wise variation of compressive strength for 10% FA replacement

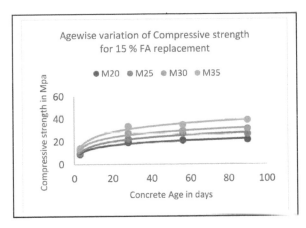

Figure 166.2 Age-wise variation of compressive strength for 15% FA replacement

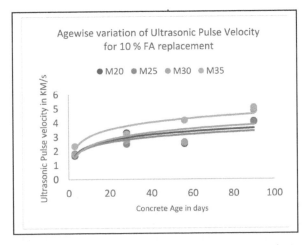

Figure 166.3 Age-wise variation of ultrasonic pulse velocity for 10% FA replacement

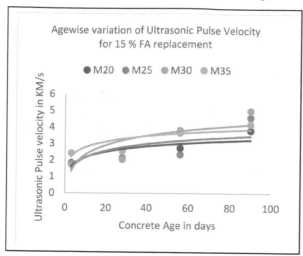

Figure 166.4 Age-wise variation of ultrasonic pulse velocity for 15% FA replacement

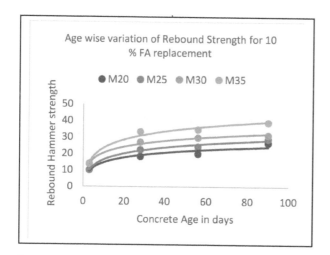

Figure 166.5 Age-wise variation of rebound hammer strength for 10% FA replacement

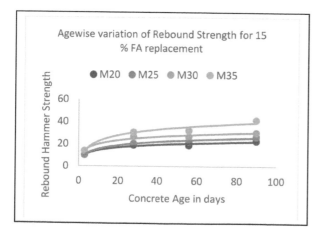

Figure 166.6 Age-wise variation of rebound hammer strength for 15% FA replacement

The equation of correlation and the corresponding **R²** value between age of concrete and Compressive strength, rebound hammer strength and ultrasonic pulse velocity is shown in Table 166.1.

Figures 166.1–166.6, show that the compressive strength, rebound hammer strength and ultrasonic pulse velocity of fly ash concrete are all related to the concrete's age. These parameters are shown to increase correspondingly as the age of concrete increases from 3 to 90 days. The correlation's R² value indicates that the compressive strength of concrete at any age can be found using the equations given in Table 166.1.

From the Table 166.1, the average **R²** value is found to be 0.953 for 10 % Fly ash while it is 0.978 for 15% Fly ash for compressive strength. For Ultrasonic Pulse Velocity, the average **R²** value is found to be 0.7121 for 10% Fly ash while it is 0.637 for 15% Fly ash. For Rebound Hammer Strength, the average **R²** value is found to be 0.968 for 10 % Fly ash while it is 0.966 for 15% Fly ash. This shows that though the NDT results for higher percentage of fly ash are on lower side, the difference is negligible. This show that using fly ash in concrete has no major effect on the results of non-destructive test.

Correlation between compressive strength and ultrasonic pulse velocity

The ultrasonic pulse velocity variation with respect to compressive strength of concrete for various curing period of concrete is shown from Figures 166.7–166.11.

From Figures 166.7–166.10, it is found that the R^2 value for 3, 28, 56 and 90 days is 0.6688, 0.4389, 0.5821 and 0.2966 respectively. From Figure 166.11, it is found that the correlation is found to be weak ($R^2 = 0.4918$) for all curing period days.

The concrete is a heterogeneous material and its strength is largely affected by w/c, type and size of aggregates and other variables. The results of Ultrasonic Pulse Velocity test also depends on various factors such as age of structure, plastering, access for the test etc.(Trtnik et al., 2009).

As a result, it is not possible to deduce the direct relationship between compressive strength and ultrasonic pulse velocity within the limited data size.

Table 166.2 Correlation between age of concrete and various parameters

S. N.	FA %	Concrete Grade	Compressive strength		Ultrasonic pulse velocity		Rebound hammer strength	
			Equation	R^2	Equation	R^2	Equation	R^2
1	10	M20	y = 3.5436 ln(x) + 5.2297	0.9986	y = 0.5712 ln(x) + 1.0448	0.6827	y = 4.3628 ln(x) + 4.619	0.9175
2	10	M25	y = 4.8364 ln(x) + 5.05	0.9265	y = 0.5199 ln(x) + 1.0648	0.6876	y = 5.3506 ln(x) + 4.0843	0.9871
3	10	M30	y = 5.1141 ln(x) + 7.7329	0.9421	y = 0.6837 ln(x) + 0.773	0.6126	y = 5.309 ln(x) + 8.2217	0.9899
4	10	M35	y = 6.7963 ln(x) + 8.0769	0.9463	y = 0.7269 ln(x) + 1.338	0.8655	y = 7.2928 ln(x) + 6.6286	0.9781
5	15	M20	y = 3.9276 ln(x) + 4.986	0.9768	y = 0.4769 ln(x) + 1.0879	0.6796	y = 3.5756 ln(x) + 6.1687	0.9577
6	15	M25	y = 5.344 ln(x) + 4.0409	0.9785	y = 0.5902 ln(x) + 0.8109	0.4916	y = 4.6579 ln(x) + 4.7432	0.9847
7	15	M30	y = 5.7013 ln(x) + 6.6959	0.9817	y = 0.834 ln(x) + 0.4162	0.6829	y = 4.772 ln(x) + 8.8786	0.964
8	15	M35	y = 7.2928 ln(x) + 6.6286	0.9781	y = 0.4915 ln(x) + 1.6557	0.6939	y = 7.6696 ln(x) + 4.673	0.9597

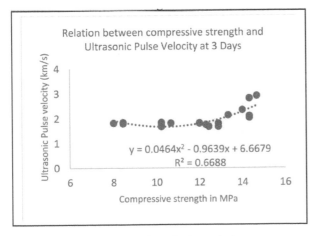

Figure 166.7 Ultrasonic pulse velocity variation wrt. compressive strength at 3 days

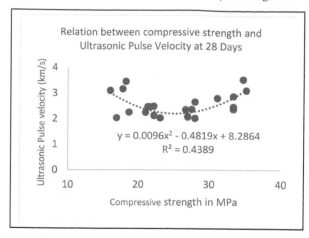

Figure 166.8 Ultrasonic pulse velocity variation wrt. compressive strength at 28 days

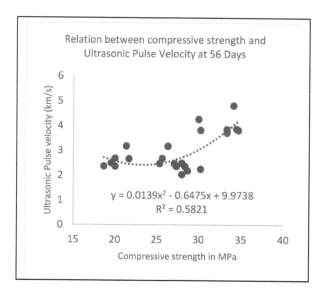

Figure 166.9 Ultrasonic pulse velocity variation wrt. compressive strength at 56 days

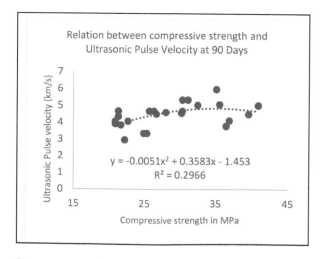

Figure 166.10 Ultrasonic pulse velocity variation wrt. compressive strength at 90 days

Correlation between compressive strength and rebound hammer strength

The variation of rebound hammer strength and compressive strength of concrete for various curing period of concrete from Figures 166.12–166.16.

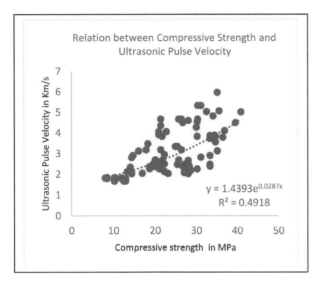

Figure 166.11 Ultrasonic pulse velocity variation wrt. compressive strength for all curing days

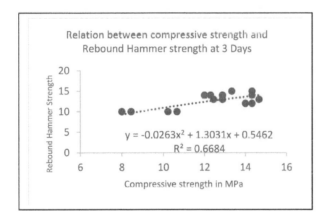

Figure 166.12 Rebound Hammer strength variation wrt. compressive strength at 3 days

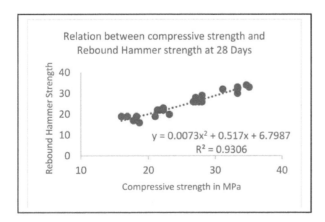

Figure 166.13 Rebound Hammer strength variation wrt. compressive strength at 28 days

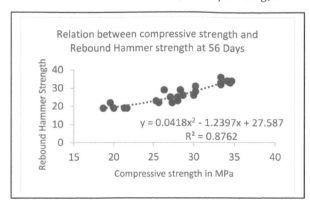

Figure 166.14 Rebound Hammer strength variation wrt. compressive strength at 56 days

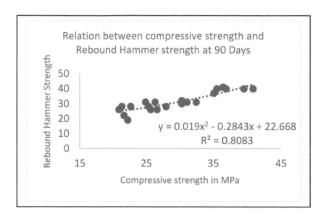

Figure 166.15 Rebound Hammer strength variation wrt. compressive strength at 90 days

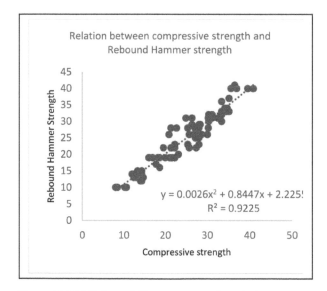

Figure 166.16 Rebound Hammer strength variation wrt. compressive strength for all curing days

From Figures 166.12–166.15, it is found that the R^2 value for 3, 28, 56 and 90 days is 0.6684, 0.9306, 0.8762 and 0.8083 respectively. The correlation is found to be on lower side for 3 days. The reason behind may be that at early age, the effects of improper compaction and insufficient curing affect the Rebound hammer test results. [18]

However, for 28, 56 and 90 days, the R^2 value is in the range of 0.9. From Figure 166.16, it is observed that the direct relationship can be established (R^2 value > 0.92) between the compressive strength and rebound hammer strength of fly ash concrete.

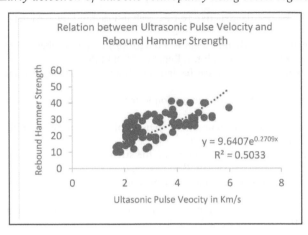

Figure 166.17 Rebound Hammer strength variation wrt. ultrasonic pulse velocity

Correlation between ultrasonic pulse velocity and rebound hammer strength

The variation of rebound hammer strength and ultrasonic pulse velocity of concrete is shown in Figure 166.17.

From Figure 166.7, it is found that the relationship between the most popular types of NDT can be established, though the coefficient of determination (R^2 value is 0.5033) is on lower side.

Conclusions

Based on the experimental findings and graphical depiction, it can be inferred that:

1. There is a direct relationship between concrete age and compressive strength as measured by ultrasonic pulse velocity and rebound number.
2. The ultrasonic pulse velocity and rebound number rise as the age of concrete increases.
3. The addition of fly ash has no major effect on the rebound hammer test or the ultrasonic pulse velocity test results.
4. The age of concrete is closely connected to its compressive strength as determined by the Rebound Hammer test. The correlation's R^2 value is approximately 0.92. It has been found that the compressive strength of concrete can be determined using these correlation equations at any time.
5. When predicting concrete compressive strength, the results of the rebound hammer test are more promising than those of the ultrasonic pulse velocity

References

Aydin, F. and Saribiyik, M. (2010). Correlation between Schmidt Hammer and destructive Compressive testing for Concretes in existing buildings. Scient. Res. Essays. 5:16441648.

Benyahia, K. A., Ghrici, M., Kenai, S., Breysse, D., and Sbartai, Z. M. (2017). Analysis of the relationship between Nondestructive and Destructive Testing of low concrete strength in new structures. Asian J. Civil Eng. (Build. Housing). 18(2):191205.

Jain, A., Kathuria, A., Kumar, A., Verma, Y., and Krishna, M., (2013). Combined use of non-destructive tests for assessments of strength of concrete in structure. 54:241251.

Konapure, G. and Richardrobin, J. (2015). Relationship between non-destructive testing of rebound hammer and destructive testing. Int. J. Current Eng. Technol. 5.

Kumavat, R., Patel, V., Tapkire, G., and Patil R., (2017). Utilization of Combined NDT in the Concrete Strength Evaluation of Concrete Specimen from existing building. Int. J. Innov. Res. Sci. Eng. Technol.6(1).

Kumavat, H. R., Chandak, N. R., and Patil, I. T. (2021). Factors influencing the performance of rebound hammer used for non-destructive testing of concrete members: A review. Construct. Mater. 14:112

Lopez, Y., Vannali, L., and Jose, V. (2016). Concrete compressive strength estimation by means of non-destructive testing: A case study. Open J. Civil Eng. 6.

Mulik, V., Minal, R., Deep, S., Vijay, D., Vishal, S., and Shweta, P. (2015). The use of Combined Non- destructive Testing in the Concrete Strength Assessment from Laboratory Specimens and Existing Buildings. Int. J. Current Eng. Scient. Res. 2(5).

Nacer M. and Hannachi S., (2012). Application of the Combined Method for evaluating the Compressive Strength of Concrete in situ. Open J. Civil Eng. 2:1621.

Patil, H., Khairnar, D., and Thube, R., (2015). Comparative study of effect of Curing on Compressive strength of Concrete by using NDT &DT. Int. J. Sci. Adv. Res. Technol. 1(6).

Reddy, K., (2014). Assessment of strength of Concrete by Non-Destructive Testing Techniques. Int. J. Eng. Manag. Res. 4(3):248256.

Rohit, K., Patel, I., and Modhera, C. (2012). Comparative Study on Flexural Strength of Plain and Fiber reinforced HVFA Concrete by Destructive and Non-destructive Techniques. Int. J. Eng. Sci. 1(2):4248.

Samson D., Omoniyi, M. T., (2014). Correlation between Non-destructive Testing and Destructive Testing of Concrete. Int. J. Eng. Sci. Inven. 3(9):1217.

Shakir, A., Al-Mishhadani, H. J., and Mushtaq, S. R. (2012). Effect of age on nondestructive tests results for existing concrete. Iraqi J. Mech. Mater. Eng. 12(4):631646.

Shang, H., Ting Y., and Yang L. (2012). Experimental study on the compressive strength of big mobility concrete with non-destructive testing method. Adv. Mater. Sci. Eng. 2.

Shariati, M., Hafizah, N., Mehdi, H., Shafigh, P., and Sinaei, H. (2011). Assessing the strength of reinforced concrete structures through UPV and Schmidt Rebound Hammer tests. Scient. Res. Essays. 6:213220.

Supe, J. and Gupta, M., (2014). Predictive Model of Compressive Strength for Concrete situ. Int. J. Struct. Civil Eng. Res. 3.

Trtnik, G., Kavcic, F., and Turk, G. (2009). Prediction of concrete strength using ultrasonic pulse velocity and artificial neural networks. Ultrasonics. 49(1):5360. doi: 10.1016/j.ultras.2008.05.001

167 Effect of varying sizes of rubber particles in concrete

Dhiraj Agrawal[1,a], U. P. Waghe[1,b], A. V. Patil[1,c], and N. U. Thakare[2,d]

[1]Department of Civil Engineering, Yeshwantrao Chavan College of Engineering (YCCE), Nagpur, India

[2]Department of Civil Engineering, G. H. Raisoni Institute of Engineering & Technology, Nagpur, India

Abstract:

Effect of adding rubber particles to concrete by means of fractional substitution for fine aggregate and testing their mechanical properties is presented in this work. Crumb rubber (CR) and rubber powder (RP) were the two diverse rubber particles that were studied. In this study, rubber particles with sizes ranging from 2.361.18 mm and rubber powder with a size of 30 (micron) were used. Mechanical properties for samples of concrete combinations with variable proportion of crumb rubber were determined. It was decided to swap fine aggregates with unique rubber particles with an incremental exchange rate of 5% and up to 20% of fine aggregates. According to the findings of this investigation, rubber in powdered form has a negative impact on concrete. The inclusion of rubber in concrete boosted the flexural and splitting tensile strengths. Abrasion resistance was optimum at 15% and then reduced. Impact resistance is increased up to 10% replacement.

Keywords: Compressive strengths, crumb rubber, flexural strengths, particle size, rubberized concrete.

Introduction

Concrete is broadly used material after the water on this globe. As it witnessed the infrastructural development of a nation. Reviewing the utilization of concrete, it can be concluded that the demand for its constituents is at its peak. As almost all the ingredients of concrete are based on the natural resource materials and almost every part of the world is facing acute shortage of these raw materials used for preparation of concrete. At the same time the automobile industry has grown rapidly in last three decades.

Infrastructural development of any nation plays very vital role in building the nation as global power. Concrete structures are widely used for infrastructural growth, as concrete is used for constructing roads, bridges, buildings, dams, canals, etc. Inspecting at the consumption of concrete at present, it can be observed that the demand of constituents of concrete is very high. Cement is used as the key binder material of concrete, a wide-ranging experiment have been carried out worldwide for finding the partial/complete replacement of cement in concrete by using distinct industrial by-products to overcome the problems like carbon dioxide emission during manufacturing of cement, to keep control on prices of cement and other environmental hazards. The two other important constituents of concrete i.e. coarse aggregates and fine aggregates are natural resources and its huge demand in preparing concrete caused acute shortage of them. The research conducted on finding the alternative for coarse and fine aggregates was limited in comparison to the research conducted for investigating the alternative of cement. At present it is essential to discover the alternative of coarse and fine aggregates from industrial wastes.

The use of vehicles has expanded dramatically in the previous two decades, posing a challenge with tyre disposal. As a result of the build-up of these tyres, there is a landfill problem as well as environmental and health risks. Pyrolysis is usually done on discarded tyres, but it also generates carbon dioxide, which harms the environment. Fine or coarse aggregates can be partially replaced with shredded rubber, crumb rubber, or rubber powder. Rubber particles were operated as a full or part replacement for aggregates to concrete by a few researchers, with promising results. For example, Al-Tayeb et al. (2013)., developed amalgam rubberized concrete with recycled fine crumb rubber replacing some quantity of the sand. Test results showed that hybrid rubberised concrete had improved flexural impact performance under dynamic loading, as well as increased toughness and deformation ability. The experimental investigation on the rubber waste as a fractional substitution for coarse particles in concrete was given by Chen Bing and Ning and Liu (2014). They also came to conclude that rubberised concrete had acceptable workability when its strengths were reduced. Concrete's specific gravity was also lowered, perhaps resulting in lightweight concrete. The study was directed by Youssf et al. (2014) in order to illustrate potential usage of crumb rubber concrete in structural columns for seismic zones. The fine aggregate is gradually replaced from 0% to 20%, with a 5% increment in between. They suggested that, as compared to the control mix, crumb rubber concrete has increased ductility, promising energy

[a]dgagrawal@ycce.edu; [b]udaywaghe@yahoo.com; [c]avpatil@ycce.edu; [d]nitinthakare2455@gmail.com

dissipation, and improved damping ratio, making it a potential candidate for seismic resistant constructions. Moustafa and El Gawady (2015) conducted an experiment in which leftover tyre rubber was used to partially substitute sand up to 30%. It was discovered that increasing the amount of rubber added to concrete improved its dynamic qualities. Thomas et al. (2015) tested rubberized concrete in a hostile environment to see how it behaved. High-strength rubberized concrete, according to the authors, can be used in places where acid attack is widespread. They also came to the conclusion that rubberised concrete was extremely resistant to harsh environments. Experiments by Xue and Shinozuka (2013); Pham et al. (2018) aimed at improving the capacity of energy absorption for concrete built with recycled rubber as a part switch for fine aggregate. According to the findings, the coefficient of damping for rubberized concrete was raised by 62% when compared to regular concrete, and thus the seismic reaction quickening of rubberized concrete was reduced by 27% when linked to traditional concrete structures. The experiment was carried out by Sukontasukkul et al. (2013), who came to the conclusion that rubberized concrete can be employed as a cushion layer of concrete in bulletproof walls. The maximum energy was absorbed by the rubberised concrete layer on the front, and so the damage was much reduced. The investigation on generating lightweight aggregate concrete was carried out by Hunag et al. (2015). In comparison to regular concrete, the workability of various lightweight aggregate concretes was determined to be satisfactory. The addition of a sufficient amount of recycled resources improves the durability of lightweight aggregate concrete. Jing et al. (2015) tested the mechanical assets of rubber particles to see if they could be used in lightweight aggregate concrete. They came to various conclusions, such as the fact that the elastic modulus of concrete was lessened because of inclusion of rubber waste, which give the concrete flexibility. The rubberized concrete's compressive, flexural, and compressive strengths were all significantly reduced. Angelin et al. (2015) presented laboratory work on consumption of discarded tyre rubber in fibre form, in cement mortars, demonstrating that strength and are interrelated. With the addition of rubber to cement paste, the porosity and water absorption were dramatically increased. Application of rubberized concrete in structural columns subjected to seismic loads was described by Youssf et al. (2015). Despite a 28% loss in compressive strength, energy dissipation along with hysteretic damping were boosted by 150% and 13%, respectively, as a result of this experiment. Flexural performance of steel fibre strengthened concrete was studied by Osama et al.(2015). The findings of this study showed that incorporating steel fibre and crumb rubber in concrete boosts the elastic property of concrete. Gupta et al. (2015) claimed that using scrap tyre rubber fibre and silica fume, they were able to raise the impact resistance and ductileness of concrete casted. From 025%, with a 5% increase each time. Silica fume was united as partial switch to cement in three distinct sets, ranging from 015% with a 5% increase. The study was conducted for three different water- cement ratios.

As a result of these studies, this study compares properties of rubberised concrete to regular concrete through different sizes of rubber particles. The mechanical qualities of nine distinct rubberised concrete (RC) mixtures ranging from 020% volumetric partial shift of fine aggregate, with a 5% augmentation of crumb rubber in each mix, were compared. The impact of sizes of rubber particle was tested using two distinct sizes. Along with the cement, Alco fine was utilised as a binder powder. Compressive, flexural, and splitting tensile strength, and acid action tests are performed on samples. Before casting specimens, constituents of concrete were tested according to Indian standards.

Investigational Process

Material

Cement having specific gravity 3.15, with cement fineness employed in this investigation was determined to be 2% in accordance to BIS 4031-1(1996). The cement is found to have a 31% consistency.

Apart from these physical qualities, the chemical structure of cement was determined according to BIS 4031-4 (1988). Table 167.1 displays the chemical conformation of cement. crushed aggregates having

Table 167.1 Chemical configuration of cement

Contents of cement	Chemical composition (%)
SiO_2	20.46
CaO	62.88
Fe_2O_3	2.87
Al_2O_3	4.96
MgO	2.36
K_2O	0.76
Loss on ignition	1.95

Specific gravity 2.74, while specific gravity of sand is 2.65. Crumb rubber has a specific gravity of 1.1 and is obtained in crushed form without steel wires from the rubber industry. For one set of specimens, the size of the CR was smaller than 3 mm, and RP was utilized to assess variation in properties of hardened rubberised concrete by varying particle sizes in comparison to typical conventional concrete.

Combinations of concrete

Concrete combinations were casted together with CR of sizes 2 to 4.75 mm and RP size ranging between 0.7 mm to 2 mm. Mix design for M-60 grade concrete is carried out in accordance to BIS 10262 (2019). Sand was partially replaced from 0% to 20% with an interval of 5%. Fine aggregates are replaced by volume. Water- cement ratio was taken as 0.33. Alccofine 1203 is used as cementitious powder along with cement as per the codal provision. Coarse aggregate has sizes in between 1020 mm, in reference to the clause given in code. Viscoflux -5507 Super plasticiser is used to achieve the desired workability. The constituents of concrete were dry mixed before adding the water to it. The slump value is kept constant at 120 mm for all specimens.

Mechanical Properties of Concrete

For each concrete test, three specimens were cast and tested. Rubberised concrete and regular concrete were both tested in accordance with Indian standards. BIS 516 (1959) was recommended for completing concrete tests. Six specimens of each mix percentage were constructed and tested for 28 and 56 days to determine compressive strength along with this, flexural strength tests was performed for each incremental mixing proportion of crumb rubber and rubber powder in standard concrete. For the flexural strength test, the sample size was 100*100*500 mm. loading is given to a rectangular specimen, and fracture distance is measured. Calculations are then made according to the equations based on the distance of fracture mentioned in BIS 5816 (1999). Splitting tensile strength of cylinders with diameter of 150 mm and 300 mm length are manufactured and tested on compressive testing machine. As per ASTM 267 (2020), acidic action tests were performed on all rubberized concrete and conventional concrete specimens.

Compressive strength test:

Compressive strength for concrete with CR and RP is demonstrated in Figures 167.1 and 167.2 correspondingly. Compressive strengths of modified concretes were judged with compressive strengths of normal concrete. Figure 167.1 shows that after using rubber as a fine aggregate substitute for more than 10% of time, compressive strength for concrete with CR reduced for 28 and 56 days. Compressive strength for concrete with RP is compared in respect to conventional concrete in Figure 167.2. The strength of the concrete made using RP was found to be lower than that of the concrete made with CR. Figure 167.3 was used to distinguish the comparison. Figure 167.3 shows design strength was attained by using CR to partially replace sand up to 5%.

Flexural strength test:

Flexural strength for concrete with CR and RP were compared to each other and to regular concrete. Figure 167.4 shows that using crumb rubber instead of fine aggregate increases flexural strength. The flexural strength was reduced as more rubber particles were added.

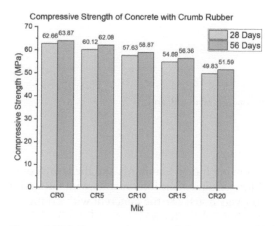

Figure 167.1 Compressive strength for concrete with CR

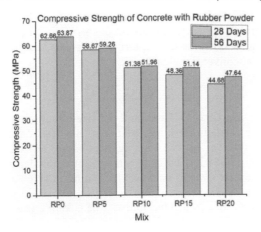

Figure 167.2 Compressive strength for concrete with RP

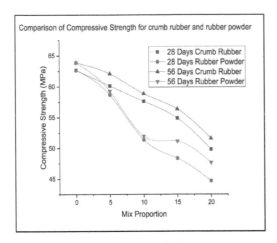

Figure 167.3 Comparative study for compressive strength for concrete with CR and RP

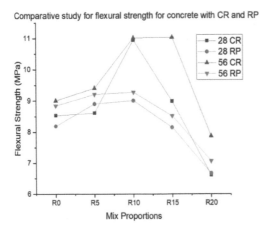

Figure 167.4 Comparative study for flexural strength for concrete with CR and RP

Splitting tensile strength:

Splitting tensile strength was executed on cylinders by following Indian Standards and the results obtained have shown that the split tensile strength was improved for certain extent for concrete with CR and RP as compared to the orthodox concrete. Maximum variation was noted for the concrete with crumb rubber at 56th day. Splitting tensile strength for concrete with rubber elements was lessened gradually as more amount rubber is added.

Acidic action test

In the acidic action test, a concrete specimen is cured for 28 days, and dried for 24 hours subsequently. After that, a plastic tank that can withstand the effects of acid is utilized, and 3% sulphuric acid is poured into a water-filled tank. Allow 60 days for the tank to settle. Following that, the specimen is removed, and several tests are carried out. Because of the acidic activity, there was a linear decline in 60-day compressive strength tests for concrete including CR and RP. Figure 167.5 shows the acidic action effect on rubberized concrete.

Conclusion

- The addition of rubber weakens bonding between cement and rubber particles, and subsequently into loss in compressive strength.
- 28-day compressive strength for concrete with CR was reduced by 20.5% for a 20% sand replacement in comparison to regular concrete. For 20% sand replacement, concrete with RP was reduced by 28.70%. Over 20% sand replacement, the compressive strength for concrete with CR and RP was lowered by 20.3% and 25.5% for 56 days, respectively.
- Rubber added to the mix improved flexural strength slightly, but only to the extent that it replaced 15% of the fine aggregates. The decrement ratio for concrete with RP is higher than concrete with CR.
- Split tensile strength was also boosted to a certain level before being reduced to allow for greater rubber incorporation in concrete.
- Compressive strengths were reduced for both type of rubberised concrete but the decrement ratio for concrete with RP is more as related to the concrete with CR.
- Finer rubber particles have affected the strength aggressively as compared to the crumb rubber, so use crumb rubber is recommended over rubber powder.

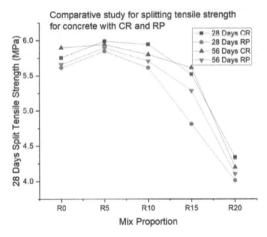

Figure 167.5 Comparative study for splitting tensile strength for concrete with CR and RP

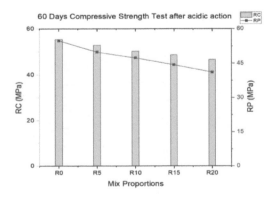

Figure 167.6 Comparison of acidic action test on rubberised concrete

References

Al-Tayeb, M. M., Bakar, B. H. A., Ismail, H., and Akil, H. M. (2013). Effect of partial replacement of sand by recycled fine crumb rubber on the performance of hybrid rubberized-normal concrete under impact load: experiment and simulation. J. Cleaner Produc. 284–289.

Angelin, A. F., Andrade, M. F. F., Bonatti, R., Lintz, R. C. C., Gachet-Barbosa, L. A., and Osório, W. R.(2015). Effects of spheroid and fiber-like waste-tire rubbers on interrelation of strength-to-porosity in rubberized cement and mortars. Construct. Build. Mater. 95:525–536.

ASTM C 267, (2020). Standard Test Methods for Chemical Resistance of Mortars, Grouts, and Monolithic Surfacing and Polymer Concretes.

BIS 4031 (Part-1), (1996). Method of Physical Tests for Hydraulic cement. Bureau of Indian Standards, New Delhi.

BIS 4031 (Part 4), (1988). Determination of Consistancy of Standard Cement Paste. New Delhi.

BIS 10262, 2019. Concrete Mix Proportioning Guidelines (Second Revision). New Delhi.

BIS 516, 1959. Methods of Tests for Stengths of Concrete. New Delhi

BIS 5816, 1999. Splitting Tensile strength of Concrete - Method of Test. New Delhi.

Gupta, T., Sharma, R. K., and Chaudhary, S. (2015). Impact resistance of concrete containing waste rubber fiber and silica fume. Int. J. Impact Eng. 83:76–87.

Hunag, L. J., Wang, H. Y., and Wang, S. Y. (2015). A study of the durability of recycled green building materials in lightweight aggregate concrete. Construct. Build. Mater. 96:353–359.

Jing, L., Zhou, T., Du, Q., and Wu, H. (2015). Effects of rubber particles on mechanical properties of lightweight aggregate concrete. Construct. Build. Mater. 91:145–149.

Moustafa, A. and ElGawady, M. A. (2015). Mechanical properties of high strength concrete with scrap tire rubber. Construct. Build. Mater. 93:249–256.

Ning, C. and Bing, L. (2014). Effect of partial replacement of sand by recycled fine crumb rubber on the performance of hybrid rubberized-normal concrete under impact load: experiment and simulation. J. Mater. Civ. Eng. 17.

Osama, A. A., ASCE1, O. A. M., and Hussein, Z. S. (2015). Flexural Behavior of Steel Fiber-Reinforced Rubberized Concrete. J. Mater. Civil Eng. 28(1).

Pham, T. M., Zhang, X., Elchalakani, M., Karrech, A., Hao, H., and Ryan, A. (2018). Dynamic response of rubberized concrete columns with and without FRP confinement subjected to lateral impact. Construct. Build. Mater. 186:207–218.

Sukontasukkul, P., Jamnam, S., Rodsin, K., and Banthia, N. (2013). Use of rubberized concrete as a cushion layer in bulletproof fiber reinforced concrete panels. Construct. Build. Mater. 41:801–811.

Thomas, B. K., Gupta, R. C., Mehra, P., and Kumar, S. (2015). Performance of high strength rubberized concrete in aggressive environment. Construct. Build. Mater. 83:320–326.

Xue, J. and Shinozuka, M. (2013). Rubberized concrete: A green structural material with enhanced energy-dissipation capability. Construct. Build. Mater. 42:196–204.

Youssf, O., ElGawady, M. A., Mills a, J. E., and Maa, X. (2014). An experimental investigation of crumb rubber concrete confined by fibre reinforced polymer tubes. *Construct. Build. Mater.* 53:522–532.

Youssf, O., ElGawady, M. A., and Mills, J. E. (2015). Experimental Investigation of Crumb Rubber Concrete Columns under Seismic Loading. Structures. 3:13–27.

168 Detecting home violence related tweets using machine learning techniques during the Covid-19

Saleem Adeeba[a], Kuhaneswaran Banujan[b], and B.T.G.S. Kumara[c]

Department of Computing and Information Systems, Sabaragamuwa University of Sri Lanka, Belihuloya, Sri Lanka

Abstract

Home violence (HV) is one of the most common forms of violence, and it has become a major global problem around everyone. People are becoming more reliant on social media platforms like Twitter, Instagram, Facebook, YouTube, etc. In a naturalistic atmosphere, they offer their ideas and opinions about daily events and happenings on these sites. Twitter is a real-time social media platform that allows users worldwide to connect via public and private messages chronologically structured on each account. In our research, we proposed a method to detect HV related tweets. Thus, if posts are associated with HV, that detect the post as HV related. More than 10,000 tweets were collected from 2019 May to 2021 April using Twitter API. After, the data pre-processing step is applied to the dataset to clean the data. Once the data set were cleaned, the word-embedding technique was used for pre-processed data set in data preparation. Then, to construct the model, the data set was split into training and testing data set, and different ML models were applied. This research uses four different ML models to detect the HV related tweets. Also, this research provides an approach to compare different 4 ML models to find an effective algorithm for detecting HV-related tweets. The artificial neural network (ANN) model has higher accuracy than other models. Accuracy, recall, precision, F1-score, and AUC for ANN are 88.12 %, 93.03%, 92.49%, 92.76%, and 90% respectively.

Keywords: Artificial Neural Network, home violence, deep learning, Covid 19, Twitter.

Introduction

A demonstration of actual power that causes or is planned to cause injury is alluded to as viciousness or violence. Viciousness can cause physically, psychologically, or both types of harm. Viciousness is a somewhat predominant human conduct that might be seen worldwide. World Health Organization (WHO) reported that about 1.6 million people die due to violence worldwide every year. Violence is one of the top causes of death among people aged 15–44 years old worldwide, accounting for 14% of male deaths and 7% of female deaths (Daher, 2003). A lot more individuals are hurt and experience the ill effects of different physical, sexual, conceptive, and psychological well-being issues for each individual who dies because of brutality. Furthermore, violence has a significant economic impact on countries, costing billions of dollars in health care, law enforcement, and lost productivity each year.

Abusive conduct can happen at any stage in life, albeit more seasoned youngsters and youthful grown-ups will probably do so. Viciousness has an assortment of hurtful results for individuals who see or experience it, and children are particularly vulnerable. Fortunately, some strategies for avoiding and reducing violence have been successful (Perkins, 1997).

Homicide (the killing of another human being, sometimes for legal reasons), assault (physically attacking another person with the purpose to cause harm), robbery (forcibly taking something from another person), and rape are the four basic kinds of violent crimes (forcible sexual intercourse with another person). Other types of violence, such as home violence (HV), overlap with these primary types (violent behaviour between relatives, usually spouses) (Widom, 1989).

As a result of the aforementioned paragraph, HV is one influencer that influences all other types. Therefore, focusing on HV is a must. The HV encompasses a wide range of acts of abuse, including physical, sexual, emotional, and controlling behaviour in a close relationship. HV is one of the most common forms of violence, and it has become a major global problem around everyone. The WHO has been reported that 35% of women globally have suffered intimate partner violence (IPV). The IPV encompasses physical violence and sexual, verbal, psychological, and financial violence. Further, HV is restricted to women and extended to children, elders, and any gender (Subramani et al., 2017).

According to the WHO, HV has severe consequences for its victims' mental and physical health. To tackle the issues on HV, WHO has been provided with several techniques for prevention and mitigation.

[a]sadeeba@std.appsc.sab.ac.lk; [b]bhakuha@appsc.sab.ac.lk; [c]btgsk2000@gmail.com

Such as women's social and economic empowerment is promoted through media and advocacy efforts that create awareness and knowledge, and; domestic violence crisis service (DVCS) provides early intervention services for at-risk families as well as increased access to complete service response for survivors are two of them (Subramani et al., 2019; Subramani et al., 2018).

Further, along with COVID-19, HV has a hidden pandemic, consists of IPV or domestic violence (DV), youngster misuse, and senior maltreatment. HV is on the ascent, with ladies and kids being excessively impacted and powerless at this period. Thus, the COVID-19 pandemic asked people to stay at home to decrease the spread of the coronavirus. Even though these measures effectively prevent infection, they have several detrimental societal repercussions, including mental pressure, joblessness, ageism and expanded paces of violence against ladies, children and the elder (Xue et al., 2020).

Most of the people on social media post HV from time to time. Thus, posts can be awareness, health-issue, real HV scenarios, etc. It is indicated that social media is an essential source of information regarding any type of news. People are becoming more reliant on social media platforms like Twitter, Instagram, Facebook, YouTube, etc. In a naturalistic atmosphere, they offer their ideas and opinions about daily events and happenings on these sites. Millions of people worldwide have used it to share their everyday thoughts and events on various themes, including politics, crimes, sports, dramas, movies, violence, health, etc. As a result, many user data is generated on various issues, including socially stigmatised contexts and taboo themes like HV (Subramani et al., 2017; Shaikh et al., 2015)

Moreover, before and after the COVID-19 pandemic, social media have become one of the most important platforms for reporting real-time happenings. Social media should promote public awareness, exchange best practices, and support the COVID-19 pandemic.

Twitter is an ongoing online media stage that permits clients worldwide to associate using public and private messages sequentially organised for every client (Xue et al., 2020). Tweets were used in many research to assess an HV (Shaikh et al., 2015). Also, nowadays, Twitter has become a primary and valuable source to use in research on machine learning (ML) (Xue et al., 2020).

The ML is the study of computer algorithms that can learn and develop based on their experience and data. It's often seen as part of artificial intelligence. ML models apply sample data to create models that can make forecasts without being definitively programmed. Globally, it's used in vast fields. Such as health, education, agriculture, Information technology, etc. Modrek and Chakalov looked at #MeToo tweets in the US and found that ML approaches can help interpret the massive sexual assault self-revelations on the Twitter platform.

The main objective of this paper is to utilise ML technologies to detect HV related posts on social media. Thus, if posts are associated with HV, that detect the post as HV related. This research paper proposes an HV Twitter posts detection model based on ML algorithms such as Artificial Neural Network ANN (ANN), Naive Bayes (NB), Decision Tree (DT), and Logistic Regression (LR).

Literature Review and Related Works

Twitter is a social media tool for exchanging ideas, a source of inspiration, a location to gather knowledge or see what your pals are up to. Twitter permits clients to compose and understand tweets, which are instant messages with a limit of 280 characters. Thus, Twitter can be a wonderful spot to share limitless thoughts and data, like current status, photographs, and recordings with a portrayal, permitting many clients' characters to be recognised via their profiles (Virra et al., 2019).

Twitter users can not only express themselves by writing a tweet, but they can also connect with others by retweeting or replying (Khatua et al., 2018)Therefore, the Twitter platform is the best source for conducting research in various fields (Suvarna et al., 2020). Such as health, agriculture, politics, sports, environment, etc.

In (Zhao et al., 2020), authors who had investigated to evaluate a psychological well-being signal among sexual and orientation minorities using the information of Twitter. In (Hsieh et al., 2012), this paper discusses Twitter as the experiment platform. It collects a large-scale dataset of Twitter tweets for 18 popular sports games from four different sports in 2011 to detect live semantic sports highlights.

In (Mousavi and Gu, 2015), the authors use the variation in Congress's decision to join Twitter to determine the influence of joining Twitter on voting behaviour. Further, they gathered the members' voting records, the date they opened their Twitter accounts, and the political orientation of their constituents.

Authors Pimolrat Ounsrimuang, Supakit Nootyaskool discusses establishing a system that classifies text messages from social media/Twitter for the traffic services organisation to consider and take action. (Ounsrimuang and Nootyaskool, 2019).

Above mentioned literature indicates the importance of the Twitter platform in different fields to accomplish research works for different contexts. HV, also known as IPV, is a type of violence that occurs between intimate partners, and it's a significant problem around the world. DVCS was established to provide HV victims with various services, including a crisis hotline, counselling, advocacy, and emergency shelter. It has improved the mental health condition of the victims (Subramani et al., 2018) . On the other hand, victims may not always make efficient support services because they must actively seek them out. Because of social, monetary, and strict limitations, numerous casualties have picked not to contact DVCS gatherings and admit their quandary.

The widespread use of Twitter has cast doubt on the notion of violence as a private matter. By bringing issues to light through sharing information and introducing stories to general society, Twitter has been utilised to help avoid violence. The benefits of Twitter in information distribution have been used for various applications, including crisis planning. Nonetheless, the expected advantages of Twitter in recognising and conveying quick help to HV victims in desperate need have yet to be realised (Subramani et al., 2018). The following related works discuss that Twitter is essential in conducting HV-related research using ML algorithms.

In (McCauley et al., 2018), research authors conclude that social media platforms can effectively engage public discourse about the realities of IPV and space for looking for and offering social help concerning this fundamental ladies' medical problem. A similar topic addressed in (Storer et al., 2021), this research looks into Twitter users' reasons to stay in violent relationships. In this research, the review test (n = 3,086) comprises an arbitrary example of 61,725 English-language tweets from around the world that utilised the hashtags #WhyIStayed and #WhyILeft. The authors used theme content analysis algorithms to examine all of the tweets.

In (Khatua et al., 2018), the author's objective is to determine what factors are linked to these sexual attacks. To accomplish this reason, the creators utilised the Twitter API to separate 0.7 million tweets with the hashtag #MeToo. They utilised profound learning-based text categorisation methods to concentrate on the extent of sexual viciousness in various geo-areas. As per their discoveries, rapes by relatives at home are a more prominent concern than badgering by outsiders in open regions.

Moreover, in (Xue et al., 2020), the authors direct an enormous scope assessment of public discourse on Twitter about family viciousness and the COVID-19 plague. Authors identified important themes, subjects, and representative tweets using the ML technique Latent Dirichlet Allocation (LDA).

As a result of the literature mentioned above, none of the research fails to detect the HV related posts using ML models. But this research paper identified a method to detect HV related posts using ML models with the help of Twitter media.

Table 168.1 shows the comparison of existing four different works with this research study in context of (i) source of the media where the data is collected through (ii) the keywords used for the particular research in order to extract the dataset (iii) the research focus some particular country relevant to the subject (iv) Main focus area of the research study and (v) Technologies that utilised to achieve the objective of the researches. As results most of the existing researches were used the Twitter as a source to conduct their studies and some of the research had used Reddit. According to the focus of the research area (area focused) keywords are vary in each (keywords used). Therefore, this research was used HV related keywords to extract the dataset from the Twitter. Also, in this research, research area includes all the areas of the existing studies. Further, it is utilized different technologies which not utilized by existing related studies.

Proposed approach

Figure 168.1 depicts the proposed method for the detection of HV related posts. Twitter is retrieved using Twitter API from the postman workspace in the data collection step. After, the data pre-processing step is applied to the data set to clean the data. Once the data set were cleaned, the word-embedding technique was used for pre-processed data set in data preparation. Then, in the constructed model, the data set was split into training and testing data set, and different ML models and classifiers were used to detect HV related posts.

Data collection

Recently Twitter posts collection is not an easy activity. It's associated with several processes or steps. Initially, Consumer Key (API Key), Consumer Secret (API Secret), Access Token, Access Token Secret, Bearer token were needed to authorise to retrieve Twitter posts. It had been done through the Twitter developer portal.

To collect tweets, a Twitter dev environment was used. That allows users to collect archive tweets with a limited number of getting requests to free-trial versions. Therefore 1.1 search/full archive API endpoint was used to collect the HV tweets. This process was done postman by providing HV related keywords as a query parameter.

The following keywords were used to collect HV related Twitter posts such as alcohol violence, child abuse, Child maltreatment, domestic violence, elder abuse, sex violence, family violence, gender-based

Table 168.1 Comparison of existing studies

Paper	Source	Keywords used	Geo-location	Area Focused	Technologies utilised
(McCauley et al., 2018)	Twitter	#MaybeHeDoesntHitYou	World-wide	IPV	NCapture
(Homan et al., 2020)	Twitter	#WhyIStayed and #WhyILeft	World-wide	IPV	Topsy API, ML predictive models including SVM
(Soldevilla and Flores, 2021)	Twitter, Reddit	Collected through previous researches	World-wide	GBV	BERT
(Najadat et al., 2021)	Twitter	#metoo	Arabic countries	VAW	SVM, K-NN, DT, NB
This research	Twitter	Child abuse, domestic violence, elder abuse, gender-based violence, IPV etc	Worldwide	HV includes IPV, VAW, GBV etc.	Postman, ANN, NB, DT and LR

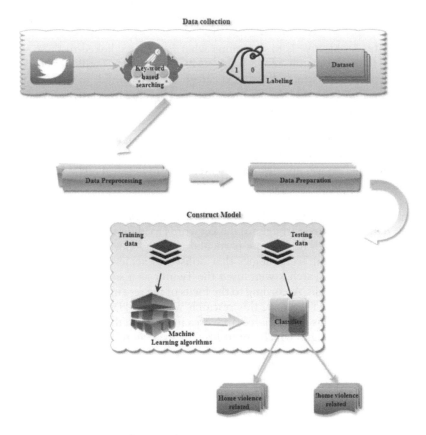

Figure 168.1 Proposed approach

violence, IPV, IPV and violence against women. More than 10,000 tweets were collected from 2019 May to 2021 April. Then, tweets texts were labelled based on related to HV used to create training and testing set of data.

Data pre-processing

Data pre-processing is cleaning the data that data will use to apply for further analysis. In case retrieval of tweets contents at first, the place contains dirty data. Thus, the text included special characteristics such as @, website links, hashtags, and so on that will overload the processing activities on the next steps. Therefore, cleaning a text is necessary for the pre-processing data module and the next moves.

Data pre-processing module was carried out in the following steps for collected Twitter posts. Further, the Neat text library, a natural language processing (NLP) package for cleaning contents/texts, and natural language toolkit is used to complete the pre-processing module.

Removal of 'RT' Twitter post

Old tweet contents had 'RT' (Re-tweets) words at the beginning of each Twitter content. An RT is a tweet that has been re-posted. The RT function on Twitter allows you and others to share a tweet with all of your followers rapidly. Therefore, the Twitter posts associated with 'RT' texts were removed to reduce the overloading process for the next steps.

Removal of stops words

Stop words can be simple words that won't provide any meaning full/help in further analysis. Removing a stops word over the loading process will be reduced. 'him', 'of', 'own' are examples of stop words that help NLP tasks. Therefore, stops were removed on tweet contents.

Removal of user handles

Tweets text were included user handles in their contents. Such as '@_your_welcome' are removed by pre-processing as the third step.

Removal of unwanted contents

Unwanted contents such as URLs, HTML, hashtags, emojis, extra white space, special characterics, quotes, phone numbers and expressions were removed to avoid the overloading process as the fourth step.

Normalisation and lemmatisation

Normalisation and lemmatisation were carried out in the fifth pre-processing step. For example, normalisation changed the tweets into simple cases, and lemmatisation changed like 'announced funding' tweet content to 'announce fund'.

Data preparation

Once tweets are pre-processed data preparation module was carried out to apply ML algorithms. This means data are split into training and testing data set using python ski learn library to apply ML algorithms. A total of 30% of the data set are split into testing, and 70% are split into training.

Then Term Frequency, Inverse Document Frequency (TF-IDF) is used to transform tweets text into numerical ways. Because ML algorithms and classifiers expect numerical factors rather than text documents, also, it can be done through a counter vectoriser, a bag of words and one-hot encoding.

Construct model

In this study, we use four different types of ML algorithms to predict HV related posts. ANN make use of learning algorithms that can make adjustments - or learn - on their own as fresh information is received. As a result, they're a great tool for non-linear statistical data modelling. Learning at a deeper level ML and the broader field of artificial intelligence (AI) technology rely heavily on ANNs. An ANN is made up of three or more interconnected layers. Input neurons make up the first layer. These neurons send input to deeper layers, which then deliver the final output data to the final output layer.

In this research ANN approach design consists of three layers. Thus, the input layer, first hidden layer and output layer. The following Table 168.2 shows the configuration for the ANN.

Further, LR, NB, and DT are ML models applied to the data training and testing set after data preparation. LR, NB, DT are the traditional ML models that help to analysis the classification problems. In this research, these traditional algorithms were utilized to compare the results or performance with an ANN.

Results and Discussion

The research was performed on Windows operating system with Intel(R) Core (TM) i7-8550U CPU @ 1.80GHz 1.99 GHz and 8 GB RAM computer. Python programming language is used to implement pre-processing and implement the four different ML models. HV-related tweets were collected through Tweet API using postman. Each tweet was labelled to train and test the data.

Data pre-processing

Data pre-processing is a process to get the useable information to train the data set for the ML model, which was explained in detail in the previous section that helped clean the tweets. The following example explains the tweets before and after it's clean.

Before pre-processing

> *"RT @SenatorTomUdall: There are 5,000+ known cases of missing Native women. 55% of Native women have experienced domestic violence.56%â€¡"*

After pre-processing

> *"know case miss Native women Native women experience domestic violence."*

Performance of proposed approach

More than 10,000 original tweets were retrieved and 30% of the data set contributed to check the performance or evaluation of each trained ML model. On other hand 70 % of data set was utilised to trained each ML model.

The following Table 168.3 compares four different ML models to find the most effective algorithm for detecting the HV-related posts. The comparison is based on accuracy, recall, precision, F1 score, Mean Square Error (MSE), Mean Absolute error (MAE) and Area under the curve (AUC).

The most important criterion for model evaluation is accuracy. The number of tuples correctly classified is accuracy. As a result of comparing accuracy, ANN has given higher accuracy as 88.12% than other models. Figure 168.2 shows the accuracy of the training and testing set of data for the ANN model. The final ANN construct model achieves 88% accuracy for the testing data set and 94% accuracy for the training data set.

Further, we used to recall, F1 score, precision, MSE, MAE, and AUC to evaluate each model using the aforementioned algorithms. Recall is the method of correctly identifying the true positives. As a result,

Table 168.2 Configuration parameters for ANN

Parameter	value
Epochs	250
Batch size	20
Optimizer	sgd
loss	binary_crossentropy
Activation (first hidden layer)	ReLu
Activation (output layer)	Sigmoid

Table 168.3 Comparison of different four ML models

	Accuracy %	Recall %	Precision %	F1 score %	MSE	MAE	AUC
NB	64.92	85.60	68.65	76.20	0.35	0.35	0.55
LR	86.23	86.08	99.21	92.18	0.13	0.13	0.87
DT	87.04	90.93	93.48	92.19	0.12	0.12	0.78
ANN	88.12	93.03	92.49	92.76	0.08	0.16	0.90

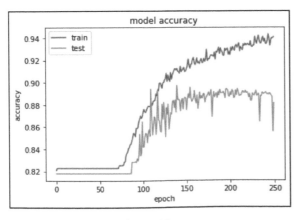

Figure 168.2 Accuracy for ANN

ANN has given a high recall value. F1 score is a component of recall and precision. In conclusion, the F1 score is high for ANN as 92.49% between each ML algorithm.

The normal distinction between the Original Values and the Predicted Values is the MAE. It tells us how far the forecasts differed from the actual result. The sole difference between MSE and MAE is that MSE takes the average of the square of the difference between the actual and projected values. Figure 168.3 shows the MAE for the ANN dataset.

The area under the curve (AUC) sums up the ROC bend that actions a classifier's capacity to recognise classes. The AUC demonstrates how well the model recognises positive and negative classes. The more noteworthy the AUC, the better the ROC As a result, the AUC value is higher to ANN as 0.90 than the other three models.

From the evaluation among ML models, ANN has performed well to detect the HV posts on social media.

Model error is calculated by loss function in ANN. Figure 168.4 shows the loss function for the ANN, and this research was utilised the binary_crossentropy loss function. The final ANN construct model achieved a 0.32 loss function for the testing data set and a 0.2 loss function for the training data set.

Moreover, Figures 168.2–168.4 indicate that the data training and testing set are aligned. That has given results as ANN models are trained well

Further, the signal processing community developed the Receiver Operating Characteristics (ROC) curve to assess a human operator's ability to differentiate informative radar signals from background noise. The compromise between the actual positive rate and the misleading positive rate for a proactive model using different likelihood edges is summed up by ROC Curves. The following equations explain the true positive and false positive rates in more detail. Moreover, Figure 168.5 shows the ROC graph of ANN. Curves that bow up to the top left of the plot typically indicate skilled models.

$$True\,positive\,rate = \frac{TP}{TP + FN} \tag{1}$$

$$False\,positive\,rate = \frac{FP}{FP + TN} \tag{2}$$

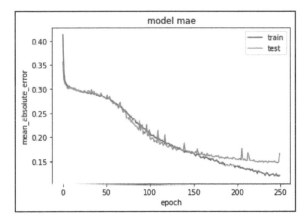

Figure 168.3 MAE for ANN

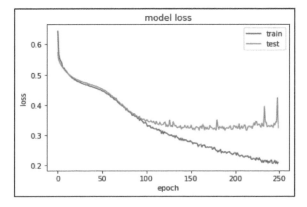

Figure 168.4 The loss function for ANN

The confusion network is perhaps the most basic and clear measurement for deciding the model's rightness and exactness. It's utilised to solve grouping issues where the result can be isolated into two classes. Although it is not a performance measure in and of itself, the confusion matrix is used in practically all performance measurements. Following Figure 168.6 shows the confusion matrix for the ANN.

The following examples show HV-related Twitter posts correctly classified by ANN trained model.

'Asexual, Agender, and Aromantic people are LGBTQIA+, and experience hate violence, asexual violence, and IPV at the same rates as out community members. We must uplift their visibility and narratives! Your matter. #ripbianca https://t.co/VJvkRdErFv' =[true]

"RT @TonyShaw22: Susie OBriens article today on the Bartel split up was a disgrace. The article implicated you canâ€™t have a stance against viâ€¦" =[false]

"@ddirkenheimer @LewisUngit Most people in survivor networks use the word traditional or fundamentalist because itâ€™s the strictness or rigidity that affects the dynamic, not theology. For example, high rates of child abuse exist in Traditional Mass Catholic, Nazi & Muslim cults." =[true]

Conclusion and Future Work

This research paper discusses the effectiveness of four types of ML algorithms to detect HV related posts with the help of the Twitter platform. To assess the performance of our method, we are used to collecting information from 2019 May to 2021 April. It's indicated that HV spreads whole over the world. As a result, the artificial neural network (ANN) model has given higher accuracy than other models. Accuracy,

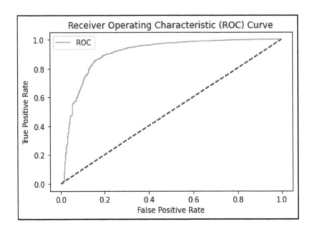

Figure 168.5 ROC graph for ANN

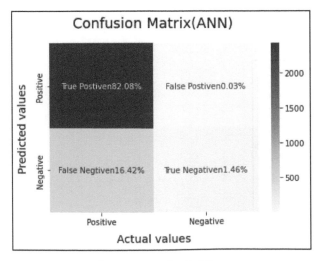

Figure 168.6 Confusion matrix for ANN

recall, precision, F1-score, MSE and AUC for ANN are 88.12%, 93.03%, 92.49%, 92.76%, 8% and 90% respectively.

In future work, we plan to evaluate the HV-related posts using ANN with different embedding techniques. In addition, we are planning to analyse the HV before and during the Covid19

References

Daher, M. (2003). World report on violence and health Le Journal Médical Libanais. Lebanese Medical J. 51:59–63.

Homan, C. M. Schrading, J. N. Ptucha, R. W. Cerulli, C., and Alm, C. O. (2020). Quantitative methods for analysing intimate partner violence in microblogs: observational study. J. Med. Inter. Res. 22:e15347.

Hsieh, L.-C., Lee, C.-W., Chiu, T.-H., and Hsu, W. (2012). Live semantic sport highlight detection based on analysing tweets of twitter in 2012 IEEE International Conference on Multimedia and Expo. 949–954.

Khatua, A., Cambria, E., and Khatua, A. (2018). Sounds of silence breakers: Exploring sexual violence on twitter. In 2018 IEEE/ACM International Conference on Advances in Social Networks Analysis and Mining (ASONAM). 397–400.

McCauley, H. L., Bonomi, A. E., Maas, M. K., Bogen, K. W., and O'Malley, T. L. (2018). MaybeHeDoesntHitYou: Social media underscore the realities of intimate partner violence. J. Wom. Hea. 27:885–891.

Mousavi, R. and Gu, B. (2015). The impact of Twitter adoption on decision making in politics. In 2015 48th Hawaii International Conference on System Sciences. 4854–4863.

Najadat, H., Zyout, M. M., and Bashabsheh, E. A. (2021). Sentiment analysis of arabic tweets about violence against women using machine learning. Indo. J. Comp. Sci. 10.

Ounsrimuang, P. and Nootyaskool, S. (2019). Classifying vehicle traffic messages from twitter to organise traffic services. In 2019 IEEE 6th International Conference on Industrial Engineering and Applications (ICIEA). 705–708.

Perkins, C. A. (1997). Age patterns of victims of serious violent crime. US Department of Justice, Office of Justice Programs, Bureau of Justice.

Shaikh, M., Salleh, N., and Marziana, L. (2015). Social networks content analysis for peacebuilding application. In Pattern Analysis, Intelligent Security and the Internet of Things, ed. A., Abraham, Muda, A. K., and Choo, Y. H., (pp. 193200). Switzerland: Springer International Publishing.

Soldevilla, I and Flores, N. (2021). Natural language processing through BERT for identifying gender-based violence messages on social media. In 2021 IEEE International Conference on Information Communication and Software Engineering (ICICSE). 204–208.

Subramani, S. Michalska, S., Wang, H., Du, J. Zhang, Y., and Shakeel, H. (2019). Deep learning for multi-class identification from domestic violence online posts. IEEE Access. 7:46210–46224.

Storer, H. L., Rodriguez, M., and Franklin, R. (2021). Leaving was a process, not an event: the lived experience of dating and domestic violence in 140 characters. J. Int. Viol. 36:NP6553–NP6580.

Subramani, S., Vu, H. Q., and Wang, H. (2017). Intent classification using feature sets for domestic violence discourse on social media. In 2017 4th Asia-Pacific World Congress on Computer Science and Engineering (APWC on CSE). pp. 129–136.

Subramani, S., Wang, H. Vu, H. Q., and Li, G. (2018). Domestic violence crisis identification from facebook posts based on deep learning. IEEE Access. 6:54075–54085.

Suvarna, A., Bhalla, G. Kumar, S., and Bhardwaj, A. (2020). Identifying Victim Blaming Language in Discussions about Sexual Assaults on Twitter in International Conference on Social Media and Society. 156–163.

Virra, K. Andreswari, R., and Hasibuan, M. A. (2019). Sentiment Analysis of Social Media Users Using Naïve Bayes, Decision Tree, Random Forest Algorithm: A Case Study of Draft Law on the Elimination of Sexual Violence (RUU PKS). In 2019 International Conference on Sustainable Engineering and Creative Computing (ICSECC). 239–244.

Widom, C. S. (1989). Does violence beget violence? A critical examination of the literature. Psychol. Bulletin. 106:3.

Xue, J. Chen, J. Chen, C., Hu, R., and Zhu, T. (2020). The hidden pandemic of family violence during COVID-19: unsupervised learning of tweets. J. Medical Internet Res. 22:e24361.

Zhao, Y., Guo, Y., He, X.,Wu, Y., Yang, X., and Prosperi, M. (2020) Assessing mental health signals among sexual and gender minorities using Twitter data. Health Infor. J. 26:765–786.

169 Orthopedic implants: Additive manufacturing, failure cause, effect, remedy and challenges

Rakesh Kumar[1,a] and Santosh Kumar[2,b]

[1]Department of Regulatory Affair, Auxein Medical Pvt. Ltd., Sonipat, Haryana, Punjab, India

[2]Department of Mechanical Engineering, Chandigarh Group of Colleges, Landran, Mohali, India

Abstract

Additive manufacturing (AM) is also known as 3-D printing, layered manufacturing, freeform fabrication is a commercially growing technology in the recent era. This technology offered lots of benefits such as capable of fabricating complex as well as customisation shaped objects using wide variety of materials with very less wastage. In addition, additive manufacturing avoid the use of coolant, fixture, tooling, less energy consumption and reduced environmental impact. Because of these advantages, AM is an excellent choice for next-generation orthopaedic implant design and fabrication. Hence, the aim of this manuscript is to provide an overview of distinct additive manufacturing techniques of orthopaedic implants and biomaterials. Further, the distinct types of orthopaedic implant, their cause and effect of failure along with remedial action are discussed. Finally, the various challenges meet by additive manufacturing in orthopaedics are discussed.

Keywords: Additive manufacturing, applications, challenges, failure, materials, orthopaedic implants, preventive measure.

Introduction

The manufacturing processes are generally categorised into three groups called subtractive manufacturing (machining, milling, and boring), additive manufacturing (FDM, SLS, SLM, SLA, IJP, EBM) and fabricating (welding, riveting, soldering and brazing) (Uddin et al., 2018). Additive or Generative manufacturing is a part fabricating technology, which utilised 3-D CAD data by addition of material in layer fashion having same thickness put over one another. But, even the layer thickness also produce a stair stepping effect on the components. In simple word AM constructing the physical model tooling and finished parts in layer by layer fashion from CAD data. Layer by layer systems, generally based on horizontal as well as thin cross section that is taken from a three dimensional computer model, fabricates together powder, sheet/plate and liquid materials to fabricate a component of plastic, metal, composite and ceramic) (Wong and Hernandez, 2012). Over the period of last twenty years, an additive manufacturing technology developed more rapidly and recently it is available for commercial application. Additive manufacturing process technology fabricate complex designed components at a low cost, easy to change design, create any geometrical shape without restriction, suitable for low melting point materials (polymer) etc. (Chhabra and Singh, 2012; Ngo et al., 2018; Bandyopadhyay and Bose, 2020). The demand of 3-D printing is recently increasing more rapidly. The annual growth rate for the industries is shown in Figure 169.1.

The some examples of expanded and novel aspect of the 2021 edition consist of food 3-D printing, techniques of AM component analysis/inspection, pandemic impact on the 3-D printing industries, medicine etc. The global market for AM part and services was approx. 12.6 billion as per United States dollars in 2020. Further, the market size of AM from 2020–2026 is shown in Figure 169.2.

AM is firstly introduced in 1987 to the market. Then, it was developed, with the use of stereo lithography (SLA) process. RP quickly make a prototype for testing purposes. The simple word RP is a technique utilises to quickly fabricate/manufacture parts based on three dimensional CAD data (Bahnini et al., 2018). In general RP is most wisely utilised in the areas of medical, manufacturing and consumer products. RP provides time and cost reductions during model manufacturing as well as part testing stages of innovation. Although, in case of conventional manufacturing when manufacturing one kind of product, it is very costly, to use traditional manufacturing. Recently lots of RP process methods are applicable in the market but all are utilise same steps to fabricate the parts as mentioned in Figure 169.3.

Earlier, conventional orthopaedic implants developed through forging or castings do not have the structural properties of human bone. However, in some cases, they might be similar to the overall shape of bone. But, at the same time, they lack the complex micro-architecture of human bone. To overcome this

[a]rk05691@gmail.com; [b]santoshdgc@gmail.com

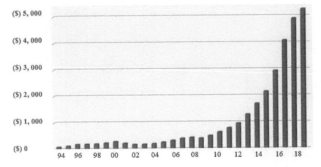

Figure 169.1 Wohlers report 2021 on 3-D printing and AM (Global State of the Industry) Wohlers 2021)

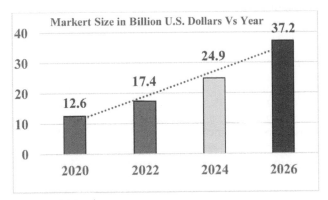

Figure 169.2 Worldwide AM parts and services market size (2020–2026)

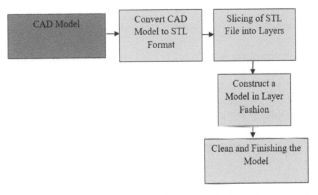

Figure 169.3 General steps used in 3-D printing (Kumar et al., 2021b)

difficulty, additive manufacturing is employed to develop distinct types of patient-specific orthopaedic implants. Additive manufacturing techniques exhibits unique benefits to economically fabricate low volume batches of product having complex geometry. Because, AM techniques do not require jig and fixture. In addition, this technology enables the introduction of porous structures for biomedical implant fixation and bone ingrowth. Overall, additive manufacturing is an ideal manufacturing process for orthopaedic implants (Javaid and Haleem, 2018). Hence, this paper provides an overview of distinct AM techniques and material used for medical applications. Further, various additively manufactured biomaterials are discussed. Thereafter, the distinct types of orthopaedic implant and their cause and effects of failure including remedial action are discussed. Finally, various challenges meet by AM techniques in orthopaedics are discussed, so that this manuscript could become useful for the futuristic researchers working in the field of orthopaedics.

Orthopaedic Implant

Those materials that are utilised for hard tissue applications to replace joints and bone are known as orthopaedic implant (Auxien Medical). These implants are divided into two groups as shown in Figure 169.4.

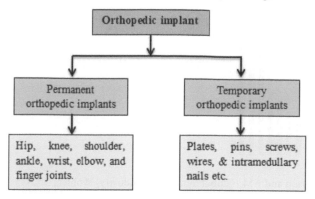

Figure 169.4 Orthopaedic implants: (a) Permanent; (b) temporary

Need and Global market of orthopaedic implant

"According to published reports by the Bureau of Indian standard (BIS) research, the global market for orthopaedic implant a product in 2016 as per United States dollar was 40.20 billion. The industry sees the annual growth of 6.1% compared with last year, which means the market is expected to reach 61.02 billion dollars by 2023. Hence, the need of implant products is significant and the quality of implant products should be rigorously controlled and supervised" (Global Orthopedics Devices Market).

Current, the demands of orthopaedic implant are increasing day by day due to following 4 major reasons:

a.   Increasing number of road accidents
b.   As people begin to age, they may start experiencing pains in their bones and joints. In case, the pain is not managed with medicine, an orthopaedic implant may be deemed the best solution.
c.   Back pain treatment.
d.   Orthopaedic injuries such as shoulder dislocation and wrist fracture etc. (Ribeiro et al., 2012).

The main aim of orthopaedic implants is to give back the functionality and structural integrity of damaged bones and joints. "Thus, to develop a safe implant without inducing rejection for long lifetime, the biomaterials should exhibit desirable mechanical properties, high resistance against wear, corrosion, biocompatibility, and Osseo integration" (Davis, 2003; Chen and Thouas, 2014). However, the selection of the proper materials for orthopaedic implant depends on the definite applications. Various metallic alloys, polymers and ceramics are commonly utilised in orthopaedic implants. These materials exhibit distinct chemical, physical and biological characteristics that cater to particular applications.

Additive Manufacturing Techniques

Presently, a variety of material (metals, ceramics, polymers) is processed by using distinct additive manufacturing. The summary of distinct AM techniques and materials used by various researchers for distinct medical applications is given in Table 169.1.

Additively manufactured biomaterials

"The biomaterials are mainly divided into three categories: metallic, polymers and bio-ceramics. All these biomaterials must fulfil all the manufacturing, clinical, and economic necessities in order to be utilised for distinct orthopaedic implants. In addition, to satisfy the clinical needs, the developed implant should be biocompatible and have better or comparable mechanical characteristics with bone. Further, the implant should have good resistance against corrosion in the biological environment of the human body ((Kumar et al., 2021a; Tilton et al., 2018).

Metallic biomaterials

Metallic materials or conventional biomaterials are mainly designed to exhibit internal support to biological tissue and they are being utilised widely in orthopaedic fixation, joint replacement, dental implant etc. It generally utilised in load-bearing implants owing to superior strength and corrosion resistance (Williams, 2009). The actual images of orthopaedic implants are depicted in Figure 169.5 (a–d) respectively.

Table 169.1 Various AM techniques principle, materials and their medical applications

AM Techniques	Principle	Materials	Medical Application	Ref.
EBM	It utilises electron beam for melting of powder material and is fused then together on the building bed.	Metal powder Titanium	Implant fabrication	Wysocki et al. (2017); Gibbons and Hansell (2005)
SLM	The high potential-density laser is utilised to liquefy & hence, solidify the printing materials.	Steel, Fe, Ti Ni, Al based alloy.	Implant and dental fabrication	Gokuldoss et al. (2017)
SLS	This process utilised CO_2 laser that is highly costly, for sintering the powder material, which sprayed on the bad and thereafter laser is focused to develop a specific shape.	Metal, Ceramic, Thermoplastics, Powder Plastic	Human anatomy Customised implants for training	Sing et al. (2017); Deckers et al. (2016)
FDM	This method uses a plastic material (ABS/PLA/PC/Nylon) that are fed into the system after heating into the meltor, and then this molten state material is extruded through a nozzle with greater pressure. Finally fabricate the object in layer fashion. FDM components are a less costly than other AM printing methods such as SLA or SLS	Polycarbonate, Acrylonitrile butadiene styrene (ABS), polyesters, polypropylene, Wax etc.	Maxillofacial surgery	Armillotta et al. (2017); Sing (2013);] Mohamed (2015); Gibson et al. (2010)
SLA	This process is based on the selective curing of a photopolymer (called resin) with a UV laser. In the machine's constructing space, a thin coating of liquid resin (50-100 microns thick) is created. The Laser creates a pattern on that layer, curing just the desired form in the model's initial layer.	Acrylate, photopolymer, Plastic Glass, Ceramic Epoxy, resin	Anatomical model Prosthetics	Zhou et al (2013); Bens et al. (2007)
IJP	This printing method utilised ink of tiny droplets for print and is similar to other AM techniques as it is adaptive to distinct ranges solid or liquid materials, which high grade resolution. As the printing process occur the liquid droplets of materials quickly solidifies to makes a layer of part as needed	Power, Liquid binder	Medical education and Surgical planning	Guo et al. (2017); Yang et al. (2018); Mancanares et al. (2015)

Here, EBM = Electron beam melting; SLM = Selective laser melting; LS = Selective laser sintering; FDM = Fusion deposition modelling; SLA = Stereolithography; IJP = Ink jet printing

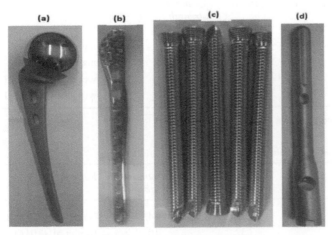

Figure 169.5 (a) Austin Moore hip prosthesis Ti; (b) Lateral distal fibula plate 4 holes right SS; (c) Wise lock screw self-tapping (Hex head) SS (d) Ajax nail (titanium)

The mechanical characteristics of commonly utilised metallic biomaterials are compared with cortical bone are summarised in Table 169.2.

Stainless steel

Stainless steel is an orthopedic implant material of elastic modulus is 190 GPa, Yield strength 221–1213 MPa and maximum ultimate tensile strength is 586–1351 MPa. It is a less costly than other implant materials such as Ti alloy, but the higher percentage of iron content as well as too high weight is the major limitation of this implant material (Ratner et al., 2004). In addition, the mechanical characteristics of the metallic AM parts is greatly depends upon the manufacturing technique utilised as given in Table 169.3.

Titanium and titanium alloys (Ti6Al4V):

It is a light weight material, whose elastic modulus (110 for Ti and 116 for Ti alloy) is very near to cortical bone (15–30) as compared to SS material. The Ti alloy is an alpha + beta alloy that comprises 6% Al and 4% vanadium. Ti alloy have lower modulus of elasticity, superior biocompatibility and higher corrosion resistant then SS and Co alloy. The yield strength for Ti and Ti alloy is 485 and 896-1034Mpa, whereas max. Tensile strength is 760 and 965–1103 MPa respectively. As per ISO 5832-3 the alloying element in Ti6Al4V, the Al is 6%, 4%V, 0.08%C, 0.2% oxygen and balance is titanium (Park and Lakes, 2007).

Co-Cr alloys:

It is also called implant material. The elastic modulus, yield strength, and max. tensile strength is lies b/w 210–253 GPa, 448–1606 MPa and 655–1896 GPa respectively. These materials have high resistance against wear, high strength etc. hence it is used in dental and orthopaedic implant application. Co Cr is a workable material but, owing to the low value of ductility and high value of elastic modulus it can exhibit material removal threat. The knee replacement is the main application of Co Cr alloy. (Gaytan et al., 2010).

The mechanical characteristics of CoCr alloy fabricated by SLM, EBM and casting are given in Table 169.4.

Table 169.2 Mechanical properties of metallic biomaterials

Materials	Young modulus (GPa)	Maximum tensile strength (MPa)	Yield Strength (MPa)	Ref.
Stainless steel	190	566 to 1350	221 to 1213	Ratner et al. (2004)
Cobalt-chromium alloys	210 to 253	655 to 1896	448 to 1606	
Titanium	110	760	485	
Ti-6Al-4 V	116	965 to 1103	896 to 1034	
Cortical bone	15 to 30	70 to 150	30 to 70	

Table 169.3 Mechanical properties of metallic implant fabricated by AM techniques

Material	Process	Yield strength (MPa)	% Elongation	Maximum tensile strength (MPa)	Ref.
SS316 L	L-PBF	487	49	594	Zhong et al., 2016
SS316 L	EBM	253	59	509	Zhong et al., 2017
SS316 L	Forged	1241	1262	1344	Mower and Long (2016)
SS316 L	Wrought	245	30	563	Mower and Long (2016)

*L-PBF: Laser powder bed fusion, **EBM**: Electron beam melting

Table 169.4 Mechanical characteristics of Co-Cr alloy fabricated by SLM, EBM and casting

Material	Process	Yield strength (MPa)	% Elongation	Maximum tensile strength (MPa)	Ref.
Co-Cr-Mo	EBM	510	36	1450	Gaytan et al. (2010)
Co-Cr alloy	L-PBF	580	32	1050	Kim et al. (2016)
Co-Cr alloy	Cast	540	10	800	Kim et al. (2016)

Tantalum:

Porous Tantalum has superior chemical resistance, nontoxic, good biocompatible, and have a good mechanical properties. But when we compares with Ti alloy and stainless steel implant the implant strength is less, greater density and not a cost effective, difficult to process, for implants & grafts. Thus it is less used in orthopaedic implant application Vasilescu et al., 2012; Javaid and Haleem, 2021.

Other biomaterials

Polyether ketone (PEEK):

It is a polymer material mainly used for spinal implant (cages), because it is a nontoxic, extremely light weight, high temperature resistant, high toughness, high strength, and excellent sterilisation performance. However, very high cost. "In addition, the PEEK has a bulk young modulus of elasticity 3.5 GPa which can be tailored to match or near to the bone structure by changing its chemical composition. Bulk implants are generally made of this material that revealed to be bio-inert in both soft and hard tissue. But, not allow bone in-growth. Owing to this reason the coating microarchitecture enhancement and surface modification of PEEK implants are utilised to enhance their Osseo integration as well as bioactive and characteristics" (Kurtz, 2011; Adhikari et al., 2018).

PLA:

PLA material is a biodegradable, use low temperature range (190°C to 230°C), less shrinkage etc. However, it is highly complex material to manipulate because of high solidification rate. (Mittal, 2012).

Ceramics:

"The ceramic materials (zirconia and alumina) have greater fatigue stress, bioactive as well as superior tribological characteristics. Owing to this it is used for total joint replacement mainly femoral heads. The poor interfacial bonding in the implant & tissue may further aseptic loosening of the implant at the site and ultimately failure of the device" (Wauthle et al., 2015; Campbell, 2003).

Problems and Defects in Traditional Manufacturing

Traditional manufacturing technologies are not flexible to fabricate implants or medical instruments/tools having complex shape in less time (Vasilescu et al., 2012; Javaid and Haleem, 2021). In addition, it is observed that most of the cases the yield strength, maximum tensile strength and percentage elongation of parts produced by traditional method is less than part fabricated by AM. In addition, the life of orthopaedic implants life is severely hampered by distinct factors that include corrosion, infections, excessive inflammation, wearing, severe toxicity, poor osseointegration as well as foreign body effects. The possible causes of implant failure, their effects and remedial action are described in the Table 169.5.

Challenges Meet by am in Orthopaedics

Additive manufacturing technology has capability to fabricate multiple types of material and distinct colours which are useful for surgeon and doctors towards analysing the defects. This technology decreases cost, product development time and enhance the communication b/w patient and surgeon. This technology meets distinct challenges in orthopaedics as given below:

a.   AM techniques is used for facilitate complex surgery design and medical education (Cai, 2015; Verma et al., 2016)

b.   MRI scan and CT scan are utilised to develop 3 dimensional image of bone AND fabricate bones by AM techniques (Lal et al., 2017).

c.   AM is easy method to develop model/artificial device which easily replace damaged body part lost by diseases or accidents (Chung et al., 2014).

d.   AM is enable to develop prosthesis and orthotic devices of patient specific anatomy (Kim et al., 2016a; Bagaria et al., 2011).

e.   "This technology is cost effective for the production of low volume customised parts having complex geometry".

f.   Low wastage and expensive tooling is not needed to develop an implant (Kumar et al., 2016).

g.   "Additive manufacturing have capability to print complex shape product, medical tools, implants and instruments with high accuracy, which is not previously possible with traditional manufacturing (Schmauss et al., 2013).

Table 169.5 Possible cause of implant failure

S. No.	Defect	Causes and Effects	Remedies	Ref.
1	Corrosion	When there is change in oxygen tension or pH in tissue, then damage of oxide layer take place. This results in corrosion. "Corrosion weaken the implanted metal, alters the surface of the metal and movement of metal ions into the human body	Surface modification of orthopaedic implants, which is considered to be the best solution to combat corrosion & to improve the life of the implants.	Kurtz (2011), Adhikari et al. (2018), Mittal (2012)
2	Infection	Bacterial adhesion is the first and most important step in implant infection. It is a complex process influenced by environmental factors, bacterial properties, material surface properties and by the presence of serum or tissue proteins. "Infection in orthopaedic surgery is disaster for the surgeon and patient. Because, it increase antibiotic use, prolonged hospital stay, repeated debridement, mortality and morbidity"	The incidence of infection can be minimised by using modern theatre facility and aseptic measure	Mittal (2012), Ribeiro et al. (2012)
3	Surgical failure	Mixing of implant (means mix and match implants from distinct manufacturers in fracture fixation. This leads to high risk of jamming, corrosion, and loosening.	Surgical failure can be reduced by using implant or instruments from one manufactures.	Mittal (2012)
4	Fatigue failure	This failure occurs in implant owing to repetitive loading on device.	"When surgeon inserts an implant he must realise that he is entering a race b/w healing of fracture and fatigue of implant".	Mittal (2012)
5	Wear	"Wear is the main cause of osteolysis in total hip arthroplasty. Submicron particles migrate into the effective joint space and stimulate a foreign-body response resulting in bone loss". Wear reduce the quality of life of patients and longevity of implants."	Surface modification by ion implantation techniques can be used	Zhu et al. (2001), Sreekanth and Kanagaraj (2013)

Am Design Consideration

It include patient specific design procedures, porosity, clinical use, patient variability, shoulder and nee joint replacements, fracture fixation, and surgical defects (Kumar et al., 2020; Sidhu et al., 2020; Bedi et al., 2019; Kumar and Kumar, 2021; Kumar and Kumar, 2022; Kumar et al., 2019; Bedi et al., 2019; Kumar et al., 2018; Jardini et al., 2014).

Conclusions

Presently AM techniques are utilised efficiently in orthopaedics for a replica of the bone. In future AM will experience a rapid translation in orthopaedic owing to its capability to manufacture complex part, customisation and geometric freedom. In addition, the success of orthopaedic implant is mainly depends on the selection of the materials. However, corrosion, infection, wear and fatigue are still some common cause of premature failure of implant. Hence, to assure long term performance of orthopaedic implants, desirable chemical, physical and biological properties are required especially for permanent implants. Thus, to fulfil the ever increasing demand, better & new biomaterials are being developed continuously.

Acknowledgment

The authors are highly thankful to Auxein Medical Pvt. Limited, & CGC Landran Mohali, Punjab.

References

Adhikari, J., Saha, P., and Sinha, A. (2018). 14 - Surface modification of metallic bone implants—Polymer and polymer-assisted coating for bone in-growth. In Woodhead publishing series in biomaterials, fundamental biomaterials:

Metals, eds. P. Balakrishnan, M. S. Sreekala, and S. Thomas, (pp. 299–321), Woodhead Publishing, https://doi.org/10.1016/B978-0-08-102205-4.00014-3.

Armillotta, A., Bianchi, S., Cavallaro, M., and Minnella, S. (2017). Edge quality in fused deposition modeling: II. experimental verification. Rapid Prototyp. J. 19:58–63.

Auxien Medical Products of medical auxien Sonipat. https://www.auxein.com/products/.

Bagaria, V., Deshpande, S., and Rasalkar, D. D. (2011). Use of rapid prototyping and three-dimensional reconstruction modeling in the management of complex fractures. Eur. J. Radiol. 80:814–820.

Bahnini, I., Rivette, M., Rechia, A., Siadat, A., and Elmesbahi, A. (2018). Additive manufacturing technology: the status, applications, and prospects. Int. J. Adv. Manuf. Technol. 97(1):147–161.

Bandyopadhyay, A. and Bose, S. (2020). Additive manufacturing. (2nd ed.), (pp. 1–463). New York (N Y): CRC Press.

Bedi, T. S., Kumar, S., and Kumar, R. (2019). Corrosion performance of hydroxyapaite and hydroxyapaite/titania bond coating for biomedical applications. Mater. Res. Express 7:015402. https://doi.org/10.1088/2053-1591/ab5cc5.

Bens, A., Seitz, H., Bermes, G., Emons, M., Pansky, A., Roitzheim, B., Tobiasch, E., and Tille, C. (2007). Nontoxic flexible photopolymers for medical stereolithography technology. Rapid Prototyp. J. 13:38–47.

Cai, H. (2015). Application of 3D printing in orthopedics: status quo and opportunities in China. Ann. Transl. Med. 3(Suppl. 1):S12.

Campbell, A. A. (2003). Bioceramics for implant coatings. Mater. Today 6:26–30.

Chen, Q. and Thouas, G. (2014). Biomaterials: A basic introduction. Boca Raton, FL: CRC Press.

Chhabra, M. and Singh, R. (2012). Obtained desired surface roughness of castings produced using Zcast direct metal casting process through Taguchi's experimental approach. Rapid Prototyp. J. 18(6):458–471.

Chung, K. J., Hong-do, Y., Kim, Y. T., Yang, I., Park, Y. W. and Kim, H. N. (2014). Preshaping plates for minimally invasive fixation of calcaneal fractures using a real-size 3D-printed model as a preoperative and intraoperative tool. Foot Ankle. Int. 35:1231–1236.

Davis, J. R. (2003). Handbook of materials for medical devices. Materials Park, OH: ASM International. https://www.asminternational.org/documents/10192/1849770/06974G_Frontmatter.pdf.

Gaytan, S. M., Murr, L. E., and Martinez, E. (2010). Comparison of microstructures and mechanical properties for solid and mesh cobalt-base alloy prototypes fabricated by electron beam melting. Metall. Mater. Trans. A Phys. Metall. Mater. Sci. 41:3216–27.

Gibbons, G. J. and Hansell, R. G. (2005). Direct tool steel injection mould inserts through the arcam EBM free form fabrication process. Assem. Autom. 25(4):300–305.

Gibson, I., Rosen, D., and Stucker, B. (2010). Additive manufacturing technologies. Rapid prototyping to direct digital manufacturing, https://link.springer.com/book/10.1007/978-1-4419-1120-9.

Global Orthopedics Devices Market. Global Orthopedics devices market to reach $61.02 Billion by 2023, Reports BIS Research. https://markets.businessinsider.com/news/stocks/global-orthopedics-devices-market-to-reach-61-02-billion-by-2023-reports-bis-research-1002206210.

Gokuldoss, P. K., Kolla, S. and Eckert, J. (2017). Additive manufacturing processes: Selective laser melting, electron beam melting and binder jetting-selection guidelines. MDPI 10(6):1–12. https://doi.org/10.3390/ma10060672.

Guo, Y., Patanwala, H. S., Bognet, B., Ma, A. W. K. (2017). Inkjet and inkjet-based 3D printing: connecting fluid properties and printing performance. Rapid Prototyp. J. 23(3):562–576.

Jardini, A. L., Larosa, M. A., AND Filho, R. M. (2014). Cranial reconstruction: 3D biomodel and custom-built implant created using additive manufacturing. J. CranioMaxillofac Surg. 42:1877.

Javaid, M. and Haleem, A. (2018). Additive manufacturing applications in orthopaedics: A review. J. Clin. Orthop. Trauma. 9(3):202–206.

Javaid, M. and Haleem, A. (2021). 3D bioprinting applications for the printing of skin: A brief study. Sens. Int. 2:100123.

Kim, G. B., Lee, S., and Kim, H. (2016a). Three-dimensional printing: Basic principles and applications in medicine and radiology. Korean J. Radiol. 17:182–197.

Kim, H. R., Jang, S. H., and Kim, Y. K. (2016b). Microstructures and mechanical properties of Co-Cr dental alloys fabricated by three CAD/CAM-based processing techniques. Materials 9:596.

Kumar, L., Tanveer, Q., Kumar, V., Javaid, M. and Haleem, A. (2016). Developing low cost 3d printer. Int. J. Appl. Sci. Eng. Res. 5:433–447.

Kumar, R. and Kumar, S. (2022). Implant material specific properties and corrosion testing procedure: A study. I-Manager's J. Future Eng. Technol. 17(1):1–10.

Kumar, R., Kumar, M. and Chohan, J. S. (2021a). Material-specific properties and applications of additive manufacturing techniques: a comprehensive review, Bull. Mater. Sci. 44:181. https://doi.org/10.1007/s12034-021-02364-y.

Kumar, R., Kumar, M., and Chohan, J. S. (2021b). The role of additive manufacturing for biomedical applications: A critical review, J. Manuf. Process. 64:828–850. https://doi.org/10.1016/j.jmapro.2021.02.022.

Kumar, R., Kumar, R., Kumar, S. and Goyal, N. (2020). Trending applications and mechanical properties of 3D printing: A review. I Manager J. Mech. Eng.11(1):1–18.

Kumar, S. and Kumar, R. (2021). Influence of processing conditions on the properties of thermal sprayed coating: a review. Surf. Eng.37(11):1–35. DOI: 10.1080/02670844.2021.1967024.

Kumar, S., Kumar, M. and Handa, A. (2018). Combating hot corrosion of boiler tubes- A study. J. Eng. Fail. Anal. 94:379–395. https://doi.org/10.1016/j.engfailanal.2018.08.004.

Kumar, S., Kumar, M. and Handa, A. (2019). Comparative study of high temperature oxidation behavior of wire Arc sprayed Ni-Cr and Ni-Al coatings. Eng. Fail. Anal. 106:104173–104189.

Kurtz, S. M. (2011). PEEK biomaterials handbook. Oxford: William Andrew.

Lal, H., Kumar, L., Kumar, R,, Boruah, T., Jindal. P. K. and Sabharwal, V. K. (2017). Inserting pedicle screws in lumbar spondylolisthesis – the easy bone conserving way. J. Clin. Orthop. Trauma. 8(2):156–164. DOI: 10.1016/j.jcot.2016.11.010.

Mancanares, C. G., Zancul, E. D. S., Silva, J. C. D. and Miguel, P. A. C. (2015). Additive manufacturing process selection based on parts' selection criteria. Int. J. Adv. Manuf. Technol. 80(5):1007–1014.

Mittal, A. (2012). Implant failure. https://www.slideshare.net/mittal87/implant-failure-12655431.

Mohamed, O. A., Masood, S. H. and Bhowmik, J. L. (2015). Optimisation of fused deposition modeling process parameters: A review of current research and future prospects. Adv. Manuf. 3:42–52. https://doi.org/10.1007/s40436-014-0097-7.

Mower, T. M., and Long, M. J. (2016). Mechanical behavior of additive manufactured, powder-bed laser-fused materials. Mater. Sci. Eng. A. 651(0921–5093):198–213.

Ngo, T. D., Kashani, A., and Imbalzano, G. (2018). Additive manufacturing (3D printing): A review of materials, methods, applications and challenges, Compos. Part B: Eng. 143:172–196. https://doi.org/10.1016/j.compositesb.2018.02.012.

Park, J. and Lakes, R. D. (2007). Biomaterials: An introduction. New York: Springer Science & Business Media.

Ratner, B. D., Hoffman, A. S., Schoen, F. J., and Lemons, J. E. (2004). Biomaterials science: An introduction to materials in medicine. New York,(N Y): Academic Press.

Ribeiro, M., Monteiro, F. J., and Ferraz, M. P. (2012). Infection of orthopedic implants with emphasis on bacterial adhesion process and techniques used in studying bacterial-material interactions. Biomatter 2(4):176–194. https://doi.org/10.4161/biom.22905.

Schmauss, D., Gerber, N., and Sodian, R. (2013).Three-dimensional printing of models for surgical planning in patients with primary cardiac tumors. J. Thorac. Cardiovasc. Surg. 145:1407–1408.

Sidhu, H. S., Kumar, S., Kumar, R., and Singh, S. (2020). Experimental investigation on design and analysis of prosthetic leg. J. Xidian Univ. 14(5):4486–4501. https://doi.org/10.37896/jxu14.5/491.

Sing, S. L., Yeong, W. Y., Wiria, F. E., Tay, B. Y., Zhao, Z., Zhao, L., Tian, Z. and Yang, S. (2017). Direct selective laser sintering and melting of ceramics: a review. Rapid Prototyp J. 23:611–623.

Singh, R. (2013). Some investigations for small sized product fabrication with FDM for plastic components, Rapid Prototyp. J. 19(1):58–63.

Sreekanth, P. S. R. and Kanagaraj, S. (2013). Wear of biomedical implants. In: Menezes, P., Nosonovsky, M., Ingole, S., Kailas, S., Lovell, M. (eds) Tribology for Scientists and Engineers. Springer, New York, NY. pp. 657–674. http://dx.doi.org/10.1007/978-1-4614-1945-7_20.

Tilton, M., Lewis, G. S., and Manogharan, G. P. (2018). Additive manufacturing of orthopedic implants, In Orthopedic biomaterials, eds. B. Li and T. Webster, Springer International Publishing AG, Part of Springer Nature. https://doi.org/10.1007/978-3-319-89542-0_2.

Uddin, S. Z., Murr, L. E., and Terrazas, C. A. (2018). Processing and characterisation of crack-free aluminum 6061 using high-temperature heating in laser powder bed fusion additive manufacturing. Addit. Manuf. 22:405–415. https://doi.org/10.1016/j.addma.2018.05.047.

Vasilescu, C., Drob, S., Neacsu, E., and Rosca, J. M. (2012). Surface analysis and corrosion resistance of a new titanium base alloy in simulated body fluids. Corros Sci. 65:431–40.

Verma, T., Sharma, A., Sharma, A., and Maini, L. (2016). Customised iliac prosthesis for reconstruction in giant cell tumour: A unique treatment approach. J. Clin. Orthop. Trauma. 10.1016/j.jcot.2016.10.001.

Wauthle, R., Van Der Stok, J., and Yavari, S. A. (2015). Additively manufactured porous tantalum implants. Acta Biomater. 14:217.

Williams, D. F. (2009). On the nature of biomaterials. Biomaterials 30:5897.

Wohlers, T. (2021). Wohlers report 2021, 3D printing and additive manufacturing global state of the industry, https://www.fastenernewsdesk.com/28315/wohlers-report-2021-3d-printing-and-additive-manufacturing-global-state-of-the-industry/.

Wong, K. V. and Hernandez, A. (2012). A review of additive manufacturing. Int. Sch. Res. Notices Article ID 208760 https://doi.org/10.5402/2012/208760.

Wysocki, B., Maj, P., Sitek, R., Buhagiar, P., Kurzydłowski, K. J., and Swieszkowski, W. (2017). Laser and electron beam additive manufacturing methods of fabricating titanium bone implants. Appl. Sci. 7:1–20.

Yang, M., Lv, X., Liu, X., and Zhang, J. (2018). Research on color 3D printing based on color adherence. Rapid Prototyp. J. 24(1):37–45.

Zhong, Y., Liu, L., Wikman, S., Cui, D., and Shen, Z. (2016). Intragranular cellular segregation network structure strengthening 316L stainless steel prepared by selective laser melting. J Nucl. Mater. 470:170.

Zhong, Y., Rännar, L.-E., and Liu, L. (2017). Additive manufacturing of 316L stainless steel by electron beam melting for nuclear fusion applications. J. Nucl. Mater. 486:234.

Zhou, C., Chen, Y., Yang, Z., and Khoshnevis, B. (2013). Digital material fabrication using mask image projection based stereolithography. Rapid Prototyp. J. 19:153–165.

Zhu, Y. H., Chiu, K. Y., and Tang, W. M. (2001). Review article: polyethylene wear and osteolysis in total 408 hip arthroplasty. J. Orthop. Surg. 9:91–99.

170 Information security enhancement by increased randomness of stream ciphers in GSM

Ram Prakash Prajapat[1,a], Rajesh Bhadada[1,b], and Giriraj Sharma[2,c]

[1]MBM Engineering College, Jodhpur, India

[2]SDE, BSNL Kota-324002, India

Abstract

Information security is a crucial issue and needs to be addressed efficiently. Lots of researches are being carried out in this field now a day. Encryption of the original information is used to ensure privacy during exchange of information. In Global System for Mobile (GSM) standard, when any new or existing user requests to get access to the resources of the network, an authentication process takes place, for each location update. Once the voice traffic initiates after signaling, encryption comes into the picture to ensure privacy during the call. Here in this process, the plaintext is encrypted in to cipher-text using stream ciphers. Ciphers are basically pseudo-random number (bits) generators. For stronger security, strong ciphers with strong randomness are required.

Linear Feedback Shift Registers (LFSRs) based A5 algorithm family is used for encryption in GSM. The A5/1 cipher is stronger and most widely used. While A3 is used for authentication and A8 supports both processes by generating K_c cipher key. A triplet is generated under challenge-response mechanism for authentication. The configuration of A5 was never shared by the developer of GSM, ETSI. However, it got reverse engineered & became public. Many crypto attacks took place like plain-text attacks, Rainbow Table attacks, Bias Birthday attacks and Randomsubgraph attacks.

Some ways are proposed in this paper to ensure better security by enhancing the randomness of the generated bit stream being used for encryption. These are incorporation of user's current location i.e. the CGI, reuse of already generated 32 bits of Signed Response during authentication process and conversion of linear FSRs into nonlinear FSRs. Statistical Test Suite is used to test the various properties of random bit stream and an attempt has been made to achieve better randomness, hence more security.

Keywords: A5/1 stream cipher, encryption, NIST test suite, randomness, security.

Introduction

Although with the advancement of technology, many vulnerabilities of security threats of GSM have been addressed in EDGE, 3G (HPA/HSDPA) and 4G (LTE) but this is still relevant as large number of people use GSM specially in rural areas. In addition to this, stream ciphers are used in many other wireless applications like modems/routers, smart appliances and security devices.

Few algorithms like A3, A8 and A5 are used in GSM for authentication and encryption process over A_{bis} air interface between user mobile (MS) and base station (BS). The details of inputs, outputs and use of these algorithms are described in Table 170.1. Here the SRES is Signed Response of 32 bits, K_c is a 64 bits Cipher Key and RAND is a 128 bits random number. The combination of K_c, RAND and SRES is called "triplet" (Figure 170.1).

A5 is mainly responsible for encryption as shown in Figure 170.2 below. Here by using Cipher Key K_c of 64 bit along with TDMA Frame number F_n, a 228 bits pseudo random number PRAN is generated which is XORed with 228 bits plain-text in bit-by-bit manner to get cipher-text. This cipher form of information after encryption is finally transmitted over the air interface between the user mobile station and base station.

Two stage security, i.e. "Authentication" and "Encryption" is implemented in GSM. Initially, the access of the network resources is granted to any new or existing subscriber on its request after authentication process on every location update.

During this "Authentication Process", the core network challenges MS and in response to this, MS sends SRES. This is matched with the SRES available with itself and grants the access on matching only as described in Figure 170.3. After getting the access of the network, the encryption process takes place to ensure the privacy during the call. Here in this process, the plaintext is encrypted in to cipher-text using stream ciphers of A5/1 algorithm. In the same way, decryption occurs at the other end to reconstruct the original information.

---

[a]prajapat@rediffmail.com; [b]rajesh_bhadada@rediffmail.com; [c]sharma4176@gmail.com

Table 170.1 Algorithms used for information security in GSM

Algorithm	Inputs	Output	Purpose
A3	K_i (128 Bits) RAND (128 Bits)	SRES (32 Bits)	Authentication Process
A8	K_i (128 Bits) RAND (128 Bits)	K_c (64 Bits)	Cipher Key Generation (K_c)
A5	K_c (64 Bits) TDMA Frame No. (22 Bits) Plain Text (228 Bits)	Cipher Text (228 Bits)	Encryption Process (Voice & Data)

Figure 170.1 Triplet details

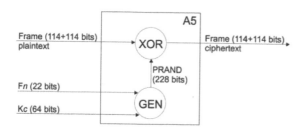

Figure 170.2 Encryption by A5/1 in GSM

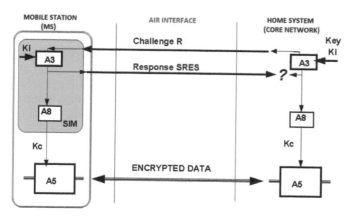

Figure 170.3 Authentication and encryption process

As the transmission of information is bursty in nature in GSM, the 114 bit frame sequence in downlink (BS to MS) & the same way a 114 bit frame sequence in uplink (MS to BS) is transmitted every 4.6 milliseconds. K_c is produced and mixed with a publically known TDMA frame number Fn for each frame for every new voice call.

The configuration of A5 was never shared by the ETSI (GSM Recommendations 02.09, 1993). Although, it got reverse engineered and became available to all. Many crypto attacks occured by cryptanalysts to reveal the internal structure of this algorithm and hackers and intruders managed to decrypt the transmitted information by users. Golic (1994) and Golić (1997) presented the functional design in the year 1994 and later Briceno (1999) revealed the entire design by reverse engineering in the year 1999.

Biryukov et al. (2001), Ekdahl and Johansson (2003), Biham and Dunkelman (2000), Maximov et al. (2005) and Barkan and Biham (2006) contributed lot. Very recently, Upadhyay et al. (2014), Sadkhan et al. (2018), Sadkhan and Hamza (2019), Sadkhan (2020) and Yerukala et al. (2020) made some modifications by changing of feedback taps in A5/1 stream cipher to improve the randomisation property to make it robust to attacks.

Many cryptanalysts proved that due to weaknesses of this stream cipher, information security can be compromised in GSM (Prajapat et al., 2021). These weaknesses are:

(i)   Weak Linear Complexity (LC)
(ii) Poor clocking system (Majority Function Rule)
(iii) Clocking period is too short
(iv) Poor clocking taps selection
(v)   Collision issue

Using an improved clocking system with a combinational function of high correlation immunity and high algebraic degree, the security can be increased (Biryukov et al., 2001; Maximov et al., 2005).

Logisim simulator (primarily developed by Dr. Carl Burch) is used to realise the structure of the proposed A5/1 algorithm and its randomness parameters are analysed by NIST Suite (Burch, 2002). The brief information of this research paper is as below:

Section II: Internal structure of A5/1
Section III: Proposed modifications
Section IV: Observations and randomness analysis
Section IV: Conclusions

Internal structure of A5/1

It has three Linear Feedback Shift Registers (LFSRs) of different bit lengths to generate a pseudo random binary stream. The total bit length of this cipher is 64 bits in which LFSR-1 (R1) has 19 bits, LFSR-2 (R2) has 22 bits, and LFSR-3 (R3) has 23 bits as depicted in Figure 170.4.

Following its own feedback polynomial, periodic bit sequence is generated by these registers. The feedback bit positions are predefined at 13, 16, 17, 18 and 20, 21 and 7, 20, 21, 22 for R1, R2 and R3 respectively.

Similarly, the single clocking taps of these registers are also predefined at tap 8, 10, and 10 for R1, R2 and R3, respectively. Bit length, clocking bit, tap bits and primitive polynomial are shown in Table 170.2.

The clocking mechanism of each register is decided by Majority Rule as shown in the truth table (Table 170.3) below. For each cycle, only those registers will be clocked and updated whose clocking bit values have majority. The majority value m is decided by m = *maj* (C1,C2,C3), here C1, C2 and C3 are the clocking bits of all three registers.

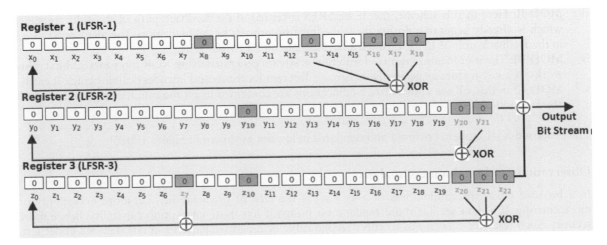

Figure 170.4 Internal design of A5/1

Table 170.2 Information table for LFSRs

LFSR	Length (in bits)	Clocking Bit	Tap Bits	Primitive Feedback Polynomials
R1	19	8	13, 16, 17, 18	$x^{19}+x^{18}+x^{17}+x^{14}+1$
R2	22	10	20, 21	$x^{22}+x^{21}+1$
R3	23	10	7, 20, 21, 22	$x^{23}+x^{22}+x^{21}+x^{8}+1$

Table 170.3 Majority rule truth table

Clocking Bit			Majority Function	Clocked LFSR		
R1	R2	R3		R1	R2	R3
0	0	0	0	Yes	Yes	Yes
0	0	1	0	Yes	Yes	No
0	1	0	0	Yes	No	Yes
0	1	1	1	No	Yes	Yes
1	0	0	0	No	Yes	Yes
1	0	1	1	Yes	No	Yes
1	1	0	1	Yes	Yes	No
1	1	1	1	Yes	Yes	Yes

In simple words, "at least two out of three" is the majority rule i.e. the majority among these bits. If two or more clocking bits are 1, the majority value m will be 1, and similarly if two or more are 0, the majority value m will be 0.

Thus, in this mechanism, two or more, whose clocking bit is the equal to m, will be clocked at each clock cycle. Every register has the clocking probability of 0.75 and non-clocking probability of 0.25. The majority rule function can be realised using logic gates as shown below in Figure 170.5 (Logisim simulator).

All three registers are reset by setting a zero value. In next 86 cycles, K_c and Fn are loaded bit by bit with regular clocking. The output ignored during initial stage of the first 100 clock cycles and during this period the irregular clocking continues for all three LFSRs as per the majority rule. Now the required random bit stream of 228 bits is obtained for 228 clock cycles (Prajapat et al., 2021). The same steps are repeated for the next frame.

Proposed modified cipher

To overcome some of the problems mentioned above in previous section, the following schemes/modifications are proposed to increase the randomness:

i.   **MOD-I:** Here in this scheme, the nonlinearity is introduced in the feedback path of the shift registers by adding universal gates. Thus, the LFSRs have been converted into NLFSRs.
ii.   Polynomial equations of this MOD-I scheme will have the impact of NAND and NOR logic gates in it.
iii. **MOD-II:** Here in this scheme, the 32 bit SRES is reused in the feedback path of the shift registers, which is already generated during the authentication process by A3 algorithm. The SRES is XOR'ed in the feedback unit of LFSR through a NAND gate on bit by bit basis.
iv.   **MOD-III:** The user location is mixed with the bit stream generated using XOR (only last 32 bits). This works as a key feature as generally users have different locations and intruders cannot crack it easily.
v.   **MOD-C:** In this, all the above three modifications are combined to get the simultaneous impact of all above modifications.

The proposed A5/1 cipher is realised and simulated in logisim as shown in Figure 170.10.

Observations and randomness analysis

In a broader sense, the randomness of a data set is the lack of predictability, i.e. more uncertainty or more entropy. If a data set has more randomness means it has more encryption capability, hence more security. Such tests are carried out to check recognizable or repetitive patterns in any data set under test.

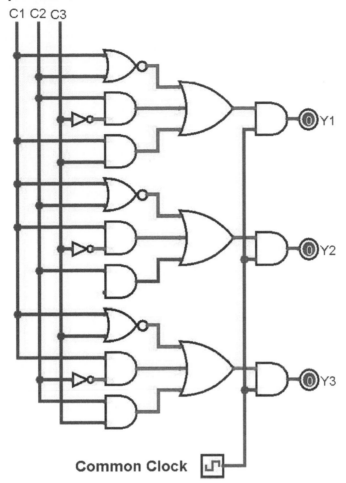

Figure 170.5 Realization majority function in Logisim

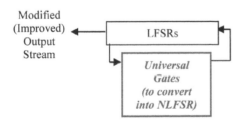

Figure 170.6 Proposed MOD-I scheme

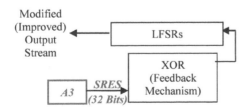

Figure 170.7 Proposed MOD-II scheme

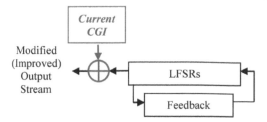

Figure 170.8 Proposed MOD-III scheme

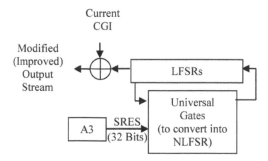

Figure 170.9 Proposed scheme of MOD-C

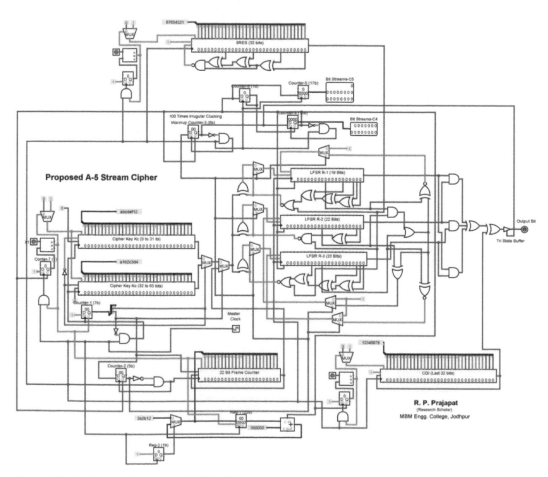

Figure 170.10 Proposed scheme of MOD-C in Logisim

Randomness is related to the theory of information entropy, probability, and chance. Entropy is a measuring tool for randomness.

NIST Suite is a statistical test suite based on Linux operating system and is used for statistical parameters testing of the output bit stream of both the actual and proposed scheme of A5/1 cipher. This test suite is also used in various cryptographic applications (Burch, 2002). Performance comparison has also been done between these two with respect to various parameters. Following are the main tests which were carried out for the randomness test:

(i) **The frequency test (Monobit and within a block)**: It tests the balance of 0's and 1's in the bit stream. The equilibrium between 0's and 1's should be maintained for a perfectly random data set. Therefore, the probability of availability of 0's and 1's should be close to 0.5 (Burch, 2002).

(ii) **The cumulative sums test (Cusums)**: It tests the randomness of a sequence of 0's and 1's which are called "random walks" or "partial sequences". It tells that the sum of the partial sequences is too large or too small (Burch, 2002).

(iii) **The runs test**: It analyses the occurrence of similar patterns that are separated by different patterns (Burch, 2002).

(iv) **The DFT (Spectral) test**: It finds out the patterns which are periodic in nature in a random bit sequence. The repetitive or periodic patterns close to each other are detected in this test (Burch, 2002).

(v) **The serial test**: It detects the pair or patterns like 00, 01, 10, 11, 100 and 101 etc and checks the balances with its complimentary pair/pattern (Burch, 2002).

(vi) **The linear complexity test**: It is directly related to the bit length of the LFSRs used to generate the random bit stream (Burch, 2002).

The *P-value* parameter is defined for all these tests in NIST suite. It shows the probability of a bit stream of being random in nature. For an ideal random bit set, this *P-value* is 1 and if it is 0, then the bit stream is completely nonrandom. Thus, a higher value or close to 1 is desirable.

Different sizes of data (up to 10,000 blocks of 114 bit i.e. 10,000 x 114 =10, 00,000 consecutive bits) are used during the statistical tests both for the actual and proposed cipher (Prajapat et al., 2021). The observations of different tests conducted upon the generated bit stream are as follows:

As the LFSRs of the actual cipher have been converted into NLFSRs in MOD-I scheme, we see a slight increase in the *P-values* of various tests. The 32 bit SRES is reused in feedback path of the shift registers in MOD-II and again slight increment in the *P-values* of various tests. As the last 32 bits of current location (CGI) of the user incorporated in output bit stream the slight increment in the *P-values* of various tests can be observed I MOD-III. Because all proposed modifications are implemented here simultaneously in MOD-C, a huge increase in the *P-values* of various tests can be observed.

The details of observations of different sizes of the data set are also provided in Table 170.5 for this MOD-C scheme.

Conclusions

The improvements in *P-values* of various tests of all four modifications (MOD-I, MOD-II, MOD-III and MOD-C) have been described in different tables of previous section. Based on these test results, it is stated that there is slight improvement in the randomness of generated bit stream of all three MODs but when all MODs are combined simultaneously in MOD-C a huge improvement can be observed in the *P-values* of cumulative sum test, frequency test and most importantly in Linear complexity test. There is a good balance between 0's and 1's (results of frequency test), better random walks (results of cumulative test), less occurrence of similar patterns and that too are well separated by different patterns (results of run tests), low

Table 170.4 Observations of the proposed schemes

Test Parameter	Actual A5/1 (P-Value)	Proposed MOD-I Scheme (P-Value)	Proposed MOD-II Scheme (P-Value)	Proposed MOD-III Scheme (P-Value)	Proposed MOD-C Scheme (P-Value)
Frequency	0.46	0.55	0.58	0.57	0.77
Block Frequency	0.82	0.87	0.87	0.86	0.92
Cumulative Sum	0.44	0.61	0.62	0.64	0.80
Runs	0.90	0.91	0.92	0.91	0.94
Spectral DFT	0.95	0.96	0.95	0.96	0.98
Serial	0.96	0.97	0.96	0.97	0.98
Linear Complexity	0.79	0.82	0.81	0.82	0.88

Table 170.5 Statistical test results by NIST test suite (A: Actual A5/1 & P: Proposed A5/1)

Bit Stream	Bits	MOD-C Scheme													
		Frequency		Block Frequency		Cumulative Sum		Runs		DFT		Serial		Linear Complexity	
		A	P	A	P	A	P	A	P	A	P	A	P	A	P
100	114	0.43	0.76	0.80	0.89	0.42	0.77	0.83	0.89	0.92	0.96	0.91	0.95	0.73	0.84
500	114	0.47	0.79	0.79	0.86	0.41	0.76	0.82	0.88	0.93	0.96	0.92	0.96	0.74	0.85
1000	114	0.49	0.81	0.82	0.91	0.43	0.78	0.83	0.89	0.92	0.97	0.95	0.98	0.75	0.86
5000	114	0.48	0.80	0.80	0.90	0.43	0.79	0.85	0.91	0.93	0.98	0.93	0.96	0.76	0.87
8000	114	0.43	0.78	0.81	0.90	0.41	0.77	0.89	0.93	0.94	0.98	0.96	0.98	0.76	0.88
10000	114	0.46	0.77	0.82	0.92	0.44	0.80	0.90	0.94	0.95	0.98	0.96	0.98	0.79	0.88
Average		0.46	0.79	0.81	0.90	0.42	0.78	0.85	0.91	0.93	0.97	0.94	0.97	0.76	0.86

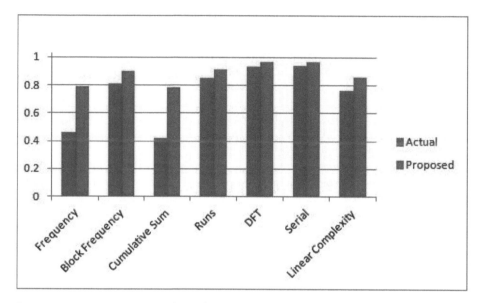

Figure 170.11 Comparative analysis of test results of MOD-C scheme

periodic patterns (results of DFT test) and enhanced entropy (results of linear complexity test) in random bit stream generated by proposed cipher.

An effort is made in this paper to improve the randomness by making three modifications MOD-I, II, and III and then comparing the NIST test results. These modifications are the incorporation of nonlinearity I feedback path of LFSRs, reusing the SRES of A3, and inclusion of current CGI of the user, respectively to improve the entropy. After that, all three modifications are implemented simultaneously in a combined manner to achieve better results. These are the major improvements and contributions in the proposed cipher scheme.

The weakness issues mentioned in early part of this paper are addressed significantly by these proposed schemes. Simulation and testing results confirmed it.

The test results of MOD-C scheme for different data sizes as shown in Table 170.5 and depicted in graphical form in Figure 170.11 above. The *P-values* of all the tests of proposed A5/1 have been increased by a significant value in comparison with the original A5/1 cipher. For better randomness, higher *P-values* are desired and better randomness means stronger encryption and enhanced security. Therefore, as per the observations and test results, it is concluded that the proposed scheme of cipher is having better security against the cryptographic attacks with respect to the actual A5/1 cipher due to increased randomness (at the cost of slight incremet in hardware).

References

Barkan, E. and Biham, E. (2006). Conditional estimators: an effective attack on A5/1. In Proceedings of SAC 2005, LNCS. 3897, 1–19. Berlin, Heidelberg: Springer.

Barkan, E., Biham, E., and Keller, N. (2006). Instant Ciphertext-Only Cryptanalysis of GSM Encrypted Communication. J Cryptol. 21:392–429.

Biham, E. and Dunkelman, O. (2000). Cryptanalysis of the A5/1 GSM stream cipher, In Progress in cryptology, proceedings of INDOCRYPT'00, LNCS, pp. 43–51. Berlin, Heidelberg: Springer.

Biryukov, A., Shamir, A. and Wagner, D. (2001). Real time cryptanalysis of A5/1 on a PC, In Advances in cryptology, proceedings of fast softwareencryption'00, LNCS, (pp. 1–18), Berlin, Heidelberg: Springer.

Briceno, M. Goldberg, I., and Wagner, D. (1999). A pedagogical implementation of the GSM A5/1 and A5/2 voice privacy encryption algorithms. GSM A5 Files Published on Cryptome. http://cryptome.org/gsm-a512.htm.

Burch, C. (2002). Logisim: a graphical system for logic circuit design and simulation. J. Educ. Res. Comput. 2(1):5–16.

Ekdahl, P. and Johansson, T. (2003). Another attack on A5/1. IEEE Trans. Inf. Theory. 49:284–289.

Erguler, I. and Anarim, E. (2005). A modified stream generator for the GSM encryption algorithms A5/1 and A5/2. In 13th European Signal Processing Conference (EUSIPCO'05), September, 2005.

Firoozjaei, M. D. and Vahidi, J. (2012). Implementing Geo-encryption in GSM cellular network. In IEEE, 2012, (pp. 299–302).

Golić, J. D. (1994). On the security of shift register based key stream generators. In Fast Software Encryption-Cambridge' 93, *Lecture Notes in Computer Science*, e. d. R. J. Anderson ed. 90–100. Berlin, Heidelberg: Springer.

Golić, J. D. (1997). Cryptanalysis of alleged A5 stream cipher. In Advances in cryptology, proceedings of EUROCRYPT'97, LNCS. 1233:239–255. Berlin, Heidelberg: Springer.

Goswami, S., Laha, S., Chakraborty, S., and Dhar, A. (2012). Enhancement of GSM security using elliptic curve cryptography algorithm In 3rd Int. Conference on Intelligent Systems Modelling and Simulation, IEEE, 2012 (pp. 639–644).

GSM Recommendations 02.09 (1993). Security aspects by european telecommunications standards institute (ETSI).

Jawad, N. H. (2017). Simulation and Developed A5/3. In International Conference on Current Research in Computer Science and Information Technology (ICCIT), Slemani - Iraq, 2017.

Jurecek, M., Bucek, J., and Lórencz, R. (2019). Side-channel attack on the A5/1 stream cipher. In 22nd Euromicro Conference on Digital System Design (DSD), 2019.

Kasim, B. and Ertaul, L. (2005). GSM security. In International Conference on Wireless Networks (ICWN). 555561.

Komninos, N., Honary, B., and Darnel1, M. (2002). Security enhancements for A5/1 without loosing hardware efficiency in future. In 3G Mobile Communication Technologies Conference. pp. 324–328.

Maximov, A., Johansson, T., and Babbage, S. (2005). An improved correlation attack on A5/1. In Proceedings of SAC 2004, LNCS. 3357:1–18.

Pankaj, Singh, A. S., and Bora, B. S. (2016). Design of enhanced pseudo-random sequence generator usable in GSM communication. In IEEE WiSPNET 2016 conference. 1–812.

Prajapat, R. P., Bhadada, R., and Sharma, G. (2021). Security enhancement of A5/1 stream cipher in GSM communication & its randomness analysis. In IEEE 6th International Forum on Research and Technology for Society and Industry (RTSI). 6–9.

Qian, Y., Ye, F., and Chen, H. H. (2022). Security in wireless communication networks (SWCN). USA: Wiley-IEEE Press.

Rukhin, A., Soto, J., Nechvatal, J., Smid, M., Barker, E., Leigh, S., Levenson, M., Vangel, M., Banks, D., Heckert, A., Dray, J., and Vo, S. (2010). A statistical test suit for random and pseudorandom number generators for cryptographic applications.

Sadkhan, S. B. (2020). A proposed development of clock control stream cipher based on suitable attack. In 1st. Information Technology to Enhance e-learning and Other Application (IT-ELA).

Sadkhan, S. B. and Hamza, Z. (2019). Proposed enhancement of A5/1 stream cipher. In 2nd International Conference on Engineering Technology and its Applications (IICETA).

Sadkhan, S. B. and Yaseen, B. S. (2018). A DNA-sticker algorithm for cryptanalysis LFSRs and NLFSRs based stream cipher. In International Conference on Advanced Science and Engineering (ICOASE), Iraq. doi: 10.1109/ICOASE.2018.8548888

Sadkhan, S. B. and Hamza, J. (2019). Proposed enhancement of A5/1 stream cipher. In second International Conference on Engineering Technology and their Applications 2019-IICET2019-IRAQ,. doi: 10.1109/IICETA47481.2019.9013008

Upadhyay, D., Sharma, P., and Valiveti, S. (2014). Randomness analysis of A5/1 stream cipher for secure mobile communication. Int. J. Comp. Sci. Commun. 5:95–100.

Yerukala, N., Prasad, V. K., and Apparao, A. (2020). Performance and Statistical analysis of stream ciphers in GSM communications. J. Commun. Softw. Syst. 16(1).

171 Optimisation of machining parameters in turning of Inconel 718 using Al_2O_3-ZrO_2 ceramic cutting tool

Hariketan Patel[a], Hiralal Patil[b], and Dhaval Chaudhari[c]

Mechancal Engineering Deprartmentept, GIDC Degree Engineerng College, Abrama, Navsari, India

Abstract

The present work is carried out to investigate the significance of process parameters like cutting speed, feed rate, and depth of cut in turning of Inconel 718. The experiments are carried out using Al_2O_3-ZrO_2 (Alumina-Zirconia) ceramic cutting tool insert under dry machining environment according to Taguchi's L_{25} orthogonal array. The weights to each criterion are defined through modified CRITIC (CRITIC-M) method. The multi objective optimisation technique TOPSIS is applied to optimise the process parameter. The optimum machining condition has cutting speed 3875 rpm, feed rate 0.25 mm/rev., and depth of cut 0.4 mm. A quadratic mathematical model for each criterion is developed using response surface methodology (RSM). A confirmation test is performed that validates the result obtained through the multi-objective optimisation.

Keywords: Al_2O_3-ZrO_2 (Alumina-Zirconia) ceramic cutting tool insert, modified CRITIC (CRITIC-M), response surface methodology (RSM), TOPSIS.

Introduction

Machining processes are the most popular in manufacturing industries (Pervaiz et al., 2014). Numerous studies have been conducted in recent years to enhance the overall performance and effectiveness of machining operations. The investigations aim to attempt optimal parameters to meet the desire surface quality, minimisation of cutting forces, low tool wear and high material removal rate (Trung et al., 2021).

Inconel 718 poses excellent characteristics and widely adopted to manufacture aerospace parts and rotary components of gas turbines (Kumar, 2019). Poor machinability of Inconel 718 is due to its excellent characteristics such as high strength, high melting temperature, and high thermal sustainability (Hao et al., 2011). A very high tool wear is noticed during machining of Inconel 718, as the cutting tool approaches to critical value, it has direct impact on a machined surface quality (Jafarian et al., 2016). As Inconel 718 is termed as difficult to cut, it is critical to employ an adequate cutting tool for machining. According to Pervaiz et al. (2014), the cutting tool material and geometry are the critical aspect in order to have excellent machining outcome. Khidhir and Mohamed (2010) studied the effect of different cutting variables, as well as insert geometry, on surface quality and tool life. Tebassi et al. (2017) optimised the machining of Inconel 718 and found that the feed rate and the cutting speed were the two most important factors influencing surface quality. During machining of Inconel 718, the focus kept on optimizing machining parameters to attain an optimal surface quality, minimum cutting force and power with high productivity. The cutting force and product of feed rate and depth of cut was seen to be the most influencing parameters.

Multi-objective optimisation process are important tools for solving decision making challenges, such as distinct and multifunctional goals in various situations (Kumar and Parameshwaran, 2020). The availability of more than one measure to determine possibilities adds to the complexity of a choice-based challenges (Pohekar and Ramachandran, 2004). The strategy is greatly influenced by the preferences of the parameter selections (Ayrim et al., 2018). The Technique for Order Preferences by Similarity to Ideal Solution (TOPSIS) methodology is applied to determine the favourable and unfavourable ideal alternatives from the evaluation criteria. During drilling of hybrid polymer composite, the TOPSIS was effectively employed to choose the parameters. TOPSIS was used efficiently to determine the optimum cutting fluid which is advantageous for machining process and promotes surface quality, tool life and productivity (Prasad and Chakraborty, 2018). The TOPSIS method is extensively used to optimise the process parameters of several machining applications (Shukla et al., 2017).

[a]hari.mech.patel@gmail.com; [b]hspatil28@gmail.com; [c]er.dhaval007@gmail.com

In the present study, machining experiment trials are performed on Inconel 718 using Al_2O_3-ZrO_2 (Alumina-Zirconia) ceramic cutting tool insert by setting the process parameters with different combination sets. The turning process parameters are cutting speed, feed rate and department of cut, all are set at five different levels. The measure responses are surface roughness, material removal rate, tool flank wear, and tool wear loss. The experiments are carried out as per Taguchi's L_{25} orthogonal array. The weights to each machining parameters are assigned by modified CRITIC (CRITIC-M) approach and for multi-optimisation of machining parameters, TOPSIS methodology is employed. The quadratic mathematical models are developed by using response surface methodology (RSM) for each response measure. Finally, the result obtained by the multi-objective optimisation is to be validated by performing the confirmation test.

Materials and methods

A. Experimental setup:

Inconel 718 (Plus Metals, Mumbai, India) having diameter of 40 mm and length 200 mm was used as a work piece material. Its chemical and mechanical properties are shown in Tables 171.1 and 171.2 respectively.

The cutting experiments were performed on a CNC lathe (make: GEDEE WEILWE, Model: Uniturn 500) having maximum spindle power of 7.5 kW and spindle speed of maximum 4000 rpm. The Al_2O_3-ZrO_2 (Alumina-Zirconia) ceramic cutting tool insert (make: Union materials, South Korea) rhombic shape with

Table 171.1 Chemical compositions of Inconel 718

Elements	%wt	Elements	%wt
Nickel (Ni)	53.14	Titanium (Ti)	0.86
Iron (Fe)	18.61	Aluminum (Al)	0.60
Chromium (Cr)	17.63	Carbon (C)	0.04
Niobium (Nb)	4.82	Sulphur (S)	0.03
Molybdenum (Mo)	3.09	Copper (Cu)	0.02

Table 171.2 Mechanical properties of Inconel 718

Properties	
Modulus of Elasticity (GPa)	210
Ultimate Tensile Strength (MPa)	1375
Poison's Ratio	0.25–0.35
Hardness (HRc)	54

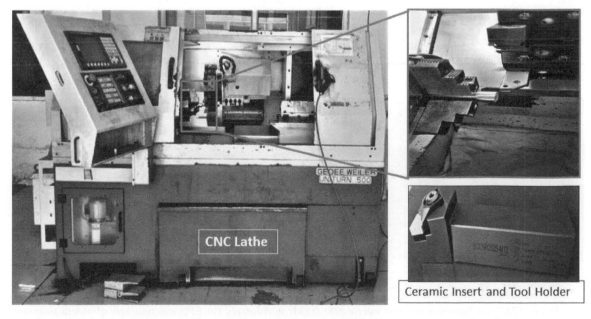

Figure 171.1 Experiment setup for machining operation using Al_2O_3-ZrO_2 ceramic cutting tool insert

Table 171.3 Properties of Al_2O_3-ZrO_2 ceramic cutting tool

Colour	White
Density (g/cm³)	4
Vickers Hardness	1800
Fracture Toughness (MPa-m$^{0.5}$)	4.50

ISO designation CNGN 120408 E40, 0° rake angle and without chip breaker was used as a cutting. The experiment setup is shown in Figure 171.1. The cutting operation was performed under dry environment conditions. For each cutting operation fresh cutting tip was used. Its properties are shown in Table 171.3.

The SEM micrograph and EDS spectrograph of Al_2O_3-ZrO_2 ceramic cutting tool is shown in Figure 171.2

The surface roughness was measured using a TIME TR-200 roughness tester. To eliminate uncertainty, repetitions of surface roughness measurement performed three times and the average roughness was accounted. The tool flank wear was measured using Tool maker's microscope (make: Radical RT- 500). The tool wear loss was measured using laboratory weight scale (make: Scale-Tec CWS Series, Precision: ± 0.001g). The material removal rate (MRR) was calculated trough conventional method.

Design of experiments:

The design of experiment (DOE) is a typical method for conducting experiments in the most efficient manner possible in order to examine the impact of input parameters. In this experimental work cutting speed (rpm), feed rate (mm/rev.) and depth of cut (mm) are opted as input parameter with each parameter having different five levels as presented in Table 171.4.

All input settings are chosen from a range specified by the manufacturer and from previous literature. The experiments were planned according to Taguchi's L_{25} orthogonal array as shown in Table 171.5. All experiments were performed randomly to eliminate any ambiguity in the experiments.

The focus of this research is to determine optimum cutting speed, depth of cut and feed rate simultaneously to attain minimum surface roughness, tool flank wear and tool wear loss while maximum material removal rate. As a result, TOPSIS methodology is opted to reach a judgement on this multi-response problem.

Table 171.4 Machining parameters with their levels

Machining parameters	Cutting Speed (rpm)	Feed Rate (mm/rev.)	Depth of Cut (mm)
Symbol	υ	f	d
Level 1	3500	0.05	0.1
Level 2	3625	0.1	0.2
Level 3	3750	0.15	0.3
Level 4	3875	0.2	0.4
Level 5	4000	0.25	0.5

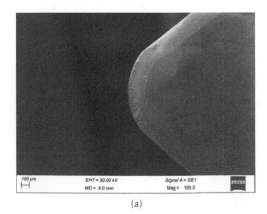
(a)

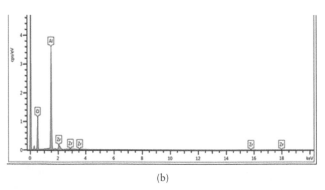
(b)

Figure 171.2 (a) Scanning electron microscope (SEM) micrograph; (b) EDS spectrograph of alumina-zirconia (Al_2O_3-ZrO_2) ceramic insert tool

Table 171.5 Experiment runs as per L$_{25}$ orthogonal array and the measured responses

Exp. No.	Cutting Speed (rpm)	Feed Rate (mm/rev.)	Depth of cut (mm)
1	3500	0.05	0.1
2	3500	0.1	0.2
3	3500	0.15	0.3
4	3500	0.2	0.4
5	3500	0.25	0.5
6	3625	0.05	0.4
7	3625	0.1	0.5
8	3625	0.15	0.1
9	3625	0.2	0.2
10	3625	0.25	0.3
11	3750	0.05	0.2
12	3750	0.1	0.3
13	3750	0.15	0.4
14	3750	0.2	0.5
15	3750	0.25	0.1
16	3875	0.05	0.5
17	3875	0.1	0.1
18	3875	0.15	0.2
19	3875	0.2	0.3
20	3875	0.25	0.4
21	4000	0.05	0.3
22	4000	0.1	0.4
23	4000	0.15	0.5
24	4000	0.2	0.1
25	4000	0.25	0.2

C. Modified CRITIC Method (CRITIC-M) (Žižovic et al., 2020):

Defining weights of the responses is the major critical aspect of the multi criteria analysis. The Criteria Importance Through Correlation (CRITIC) is extensively utilised and well-known method (Xu et al., 2020). It is a statistical approach that determines criteria variations by using the standard deviations of ranking attributes of alternative per column including the correlational coefficient of all associated columns (Slebi-Acevedo et al., 2020). Žižovic et al. (2020) proposed the modified CRITIC (CRITIC-M) approach by ascertained shortcoming of the conventional CRITIC approach. The CRITIC-M are intended to get more definitive weight coefficient.

Step 1. Create a decision matrix $X = [\zeta_{ij}]$, which include the output responses.

$$X = \begin{bmatrix} A_{11} & A_{12} & \cdots & A_{1n} \\ A_{21} & A_{22} & \cdots & A_{2n} \\ \vdots & \vdots & & \vdots \\ A_{m1} & A_{m2} & \cdots & A_{mn} \end{bmatrix} \tag{1}$$

Step 2. Normalise the decision matrix: In first phase the elements of the matrix X are normalised using (2).

$$\zeta_{ij}^{*} = \frac{\zeta_{ij}}{\zeta_j^{\max.}} \tag{2}$$

Where, $\zeta_j^{\max} = \max_j\{\zeta_{1j}, \zeta_{2j}, \ldots, \zeta_{mj}\}$; i= 1, 2,...., n: j = 1, 2, ..., m;
In second phase normalise the value using (3)

$$\zeta_{ij} = -\zeta_{ij}^{*} + \zeta_j^{*\max.} + \zeta_j^{*\min.} \tag{3}$$

Where, i= 1, 2..., n: j= 1, 2, ..., m;

$$\zeta_j^{*\max.} = \max_j\{\zeta_{1j}^{*}, \zeta_{2j}^{*}, \ldots, \zeta_{mj}^{*}\}; \quad \zeta_j^{\min.} = \min_j\{\zeta_{1j}^{*}, \zeta_{2j}^{*}, \ldots, \zeta_{mj}^{*}\}$$

This approach reduces the root-mean-square (RMS) variance, and the final weight coefficient effectively address the correlation between the variables in the original decision matrix.

Step 3. From the elements of normalised matrix, determine the standard deviations (σ). Define standard deviations (σ_j), for each criterion $C_j (j = 1,2...n)$.

Step 4. Create the matrix of linear correlations $R = [r_{jk}]_{n \times n}$. The vector $\zeta_j = (\zeta_{1j}, \zeta_{2j}... \zeta_{mj})$ defined and calculate the linear vector correlation ζ_j and ζ_k for criteria C_j from the normalised matrix $X = [\zeta_{ij}]$. The sum of linear correlations for each criterion yields the following measure of criterion conflicts using (4):

$$\theta_j = \sum_{k=1}^{n} (1 - r_{jk})$$
(4)

Calculate the measures of data in the jth response using the formula (5):

$$C_j = \sigma_j * \sum_{k=1}^{n} (1 - r_{jk})$$
(5)

Step 5: Compute criterion weight coefficients based on standard vector deviation values as expression (6).

$$w_j = \frac{\dfrac{\overline{\zeta_j}}{1 - \overline{\zeta_j}} * \sigma_j}{\sum_{j=1}^{n} \dfrac{\overline{\zeta_j}}{1 - \overline{\zeta_j}} * \sigma_j}$$
(6)

Where ($\overline{\zeta_j}$) is average of ζ_j^* of each criterion; σ_j is standard deviations.

Technique for order prefernce by similarity of ideal solution (TOPSIS) (Varatharajulu et al., 2021)

Step 1. Normalise the decision matrix is calculated as (7)

$$N_{ij} = \frac{x_{ij}}{\sqrt{\sum_{i=1}^{n} x_{ij}^2}}$$
(7)

Step 2. Using the weights calculated by the CRITIC technique, the weighted normalisation matrix is computed as shown in (7)

$$V_j = w_j * \frac{x_{ij}}{\sqrt{\sum_{i=1}^{n} x_{ij}^2}}$$
(8)

Step 3. Determine positive $V^+ = (V_1^+, V_2^+, ... V_n^+)$ and negative $V^- = (V_1^-, V_2^-, ... V_n^-)$ ideal solution.

$$V^+ \begin{cases} \max_j V_{ij} \; for \; j = 1,...,k \\ \min_j V_{ij} \; for \; j = k+1,...,n \end{cases}$$

$$V^- \begin{cases} \min_j V_{ij} \; for \; j = 1,...,k \\ \max_j V_{ij} \; for \; j = k+1,...,n \end{cases}$$

Step 4. The distance of every criterion value from V^+ and V^- is calculated as (9) and (10):

$$S_i^+ = \left[\sum_{j=1}^{n} \left(V_{ij} - V^+ \right)^2 \right]^{0.5} \quad i = 1,2,...,m$$
(9)

$$S_i^- = \left[\sum_{j=1}^{n} \left(V_{ij} - V^- \right)^2 \right]^{0.5} \quad i = 1,2,...,m$$
(10)

Step 5. The relative closeness to the ideal solution is calculated as (11):

$$K_i = \frac{S_i^-}{S_i^+ + S_i^-} \qquad (11)$$

i= 1, 2,...m

Step 6. Define ranking as the value of relative closeness.

Response surface methodology (RSM):

The strong correlation amongst the significant variables and the performance indicators which constitute the domain is nonlinear. To integrate the correlation quadratic response surface mathematical model is used that determine more precise approximations. The linear, square and interactive correlations are considered for statistical results.

$$Y = \alpha_1 + \alpha_2 \upsilon + \alpha_3 f + \alpha_4 d + \alpha_5 \upsilon^2 + \alpha_6 f^2 + \alpha 7 d^2 + \alpha_8 \upsilon f + \alpha_9 \upsilon d + \alpha_{10} f d \qquad (11)$$

Where $\alpha_1 + \alpha_2$, ... α_{10} are coefficient of each variables.

Result and discussion

Experiment results

The experiments were planned according to Taguchi's L_{25} orthogonal array as shown in Table 171.5 and the output results are as shown in Table 171.6.

The objective desirability is to minimise the surface roughness, the minimum surface roughness is 0.965 μm, so experiment no 24 is desirable for minimum surface roughness as shown in Table 171.7. With this experiment desirability of tool flank wear is also matches, but here the tool flank wear is not minimum and the material removal rate is not maximum to attain its desirability. Similarly, for experiment no 1 tool flank wear is desirable but other variables are not attaining their desire output.

According to the above observations, it is critical to adopt any one experiment trials out of tis 25 runs to ensure the required desirability of all at the same time. As a result, in order to achieve the low surface

Table 171.6 Experiment results

Exp No	Surface Roughness (μm)	Material removal rate (mm³/sec)	Tool flank wear (mm)	Tool wear loss (gms)
1	1.153	0.836	1.453	0.350
2	1.162	3.188	1.462	0.351
3	1.179	6.728	1.479	0.353
4	1.196	11.433	1.496	0.356
5	1.212	17.452	1.512	0.359
6	1.132	3.780	1.382	0.364
7	1.159	9.906	1.409	0.367
8	1.118	2.716	1.368	0.363
9	1.161	7.788	1.411	0.364
10	1.175	15.371	1.425	0.366
11	1.077	2.191	1.277	0.376
12	1.085	7.103	1.285	0.377
13	1.109	14.700	1.309	0.380
14	1.136	25.089	1.336	0.385
15	1.114	5.654	1.314	0.376
16	1.075	6.786	1.225	0.390
17	1.053	2.653	1.203	0.388
18	1.061	8.252	1.211	0.390
19	1.086	16.943	1.236	0.392
20	1.103	29.333	1.253	0.397
21	1.044	4.712	1.144	0.401
22	1.062	13.368	1.162	0.404
23	1.082	25.819	1.182	0.408
24	0.965	6.508	1.065	0.401
25	1.015	17.024	1.115	0.403

Table 171.7 Desirability of each criterion

Output Responses	Minimum	Maximum	Desirability	Exp. No.
Surface Roughness	0.965	1.212	Minimize	24
Material removal rate	0.836	29.333	Maximize	20
Tool flank wear	1.065	1.512	Minimize	24
Tool wear loss	0.350	0.408	Minimize	1

Table 171.8 The normalized decision matrix

Exp No	Surface Roughness (μm)	Material removal rate (mm³/sec)	Tool flank wear (mm)	Tool wear loss (gms)
1	0.845	1.000	0.743	1.000
2	0.837	0.920	0.737	0.998
3	0.823	0.799	0.726	0.993
4	0.809	0.639	0.715	0.986
5	0.796	0.434	0.704	0.979
6	0.862	0.900	0.790	0.966
7	0.840	0.691	0.772	0.958
8	0.874	0.936	0.800	0.968
9	0.838	0.763	0.771	0.966
10	0.827	0.504	0.762	0.960
11	0.908	0.954	0.860	0.938
12	0.901	0.786	0.854	0.933
13	0.881	0.527	0.839	0.926
14	0.859	0.173	0.821	0.915
15	0.877	0.836	0.835	0.937
16	0.909	0.797	0.894	0.902
17	0.927	0.938	0.909	0.907
18	0.921	0.747	0.903	0.903
19	0.900	0.451	0.887	0.896
20	0.886	0.028	0.876	0.884
21	0.935	0.868	0.948	0.875
22	0.920	0.573	0.936	0.868
23	0.903	0.148	0.923	0.857
24	1.000	0.807	1.000	0.875
25	0.959	0.448	0.967	0.870

Table 171.9 The standard deviations

	Surface Roughness (μm)	Material removal rate (mm³/sec)	Tool flank wear (mm)	Tool wear loss (gms)
σ_j	0.049	0.269	0.085	0.045

Table 171.10 Weight coefficient for each criterion

	Surface Roughness (μm)	Material removal rate (mm³/sec)	Tool flank wear (mm)	Tool wear loss (gms)
w_j	0.1867	0.2760	0.2261	0.3113

roughness, low tool flank wear, low tool wear loss and high material removal rate, it must be performed by multi-objective optimisation. This aspect will be discussed in the following sections.

Modified CRITIC (CRITIC-M)

Step 1. A decision matrix X= [ζ_{ij}], which include the output responses as shown in Table 171.6.
Step 2. The normalised decision matrix from (2) and (3) is presented in Table 171.8.
Step 3. The standard deviations (σ), from the elements of normalised matrix is presented in Table 171.9.
Step 4. Create the matrix of linear correlations $R = [r_{jk}]_{nxn}$.

Table 171.11 Normalized decision matrix

Exp No	Surface Roughness (μm)	Material removal rate (mm³/sec)	Tool flank wear (mm)	Tool wear loss (gms)
1	0.208	0.013	0.221	0.185
2	0.209	0.049	0.222	0.185
3	0.212	0.102	0.225	0.186
4	0.215	0.174	0.228	0.188
5	0.218	0.266	0.230	0.189
6	0.204	0.058	0.210	0.192
7	0.209	0.151	0.214	0.194
8	0.201	0.041	0.208	0.192
9	0.209	0.119	0.215	0.192
10	0.212	0.234	0.217	0.193
11	0.194	0.033	0.194	0.198
12	0.195	0.108	0.196	0.199
13	0.200	0.224	0.199	0.201
14	0.205	0.382	0.203	0.203
15	0.201	0.086	0.200	0.198
16	0.194	0.103	0.186	0.206
17	0.190	0.040	0.183	0.205
18	0.191	0.126	0.184	0.206
19	0.196	0.258	0.188	0.207
20	0.199	0.447	0.191	0.210
21	0.188	0.072	0.174	0.212
22	0.191	0.204	0.177	0.213
23	0.195	0.393	0.180	0.216
24	0.174	0.099	0.162	0.212
25	0.183	0.259	0.170	0.213

Step 5: Based on standard vector deviation values, using (6), the final calculated weight coefficient for each criterion is as shown in Table 171.10.

Technique for order prefernce by similarity of ideal solution (TOPSIS):

Step 1. The normalise decision matrix using (7) is presented in Table 171.11.
Step 2. The weighted normalisation matrix is computed as shown in (8).
Step 3. Determine the positive and negative ideal solutions.
Step 4. The distance of every criterion value from and is calculated as (9) and (10) and presented in Table 171.12.
Step 5. The relative closeness to the ideal solution is calculated as (11) and shown in Table 171.12.

The grater magnitude of serves the improved response and based on that ranking to each experiment run is given as shown in Table 171.12. The obtained statistics revealed that the overall experiment ranking alternative is 25- 22- 17- 10- 4- 20- 11- 23- 13- 7- 24- 15- 8- 3- 18- 16- 21- 12- 6- 1- 19- 9- 2- 14- 5. The optimal sequence of the experiment run as 20 > 23 > 14 > 5 > 25 > 19 > 10 > 13 > 22 > 4 > 7 > 18 > 9 > 24 > 12 > 16 > 3 > 15 > 21 > 6 > 17 > 2 > 8 > 11 >1. It implies that experiment run 20 has the finest combinations of input parameters while it is worst in experiment run 1. The best experiment trial 20 and experiment trial 1 have relative closeness of 100% and 8.34% respectively. For optimum responses the best combination experiment trial is 20 followed by trial run 23 and 14. The cutting speed of 3875 rpm, feed rate of 0.25 mm/rev., and depth of cut of 0.4 mm are found to be the optimum set. As a consequence, it is recommended that the above sequence be used to minimise the surface roughness, tool flank wear and tool wear loss at the enhanced material removal rate.

Modeling through response surface metodology (RSM)

In this study, to assess the non-linearity effect of different factors, each factor was explored at different level. As a result, a second order mathematical polynomial predictive response surface model for surface

Table 171.12 The distance of every criterion value and relative closeness

Exp no	S_i^+	S_i^-	K_i	Rank
1	0.121	0.010	0.076	25
2	0.111	0.014	0.111	22
3	0.096	0.026	0.215	17
4	0.077	0.045	0.371	10
5	0.053	0.070	0.571	4
6	0.108	0.015	0.124	20
7	0.083	0.039	0.320	11
8	0.113	0.012	0.099	23
9	0.092	0.030	0.249	13
10	0.060	0.062	0.505	7
11	0.114	0.012	0.096	24
12	0.094	0.028	0.231	15
13	0.062	0.059	0.486	8
14	0.022	0.102	0.825	3
15	0.100	0.022	0.182	18
16	0.095	0.027	0.224	16
17	0.112	0.015	0.114	21
18	0.089	0.033	0.273	12
19	0.053	0.069	0.564	6
20	0.011	0.120	0.915	1
21	0.104	0.021	0.171	19
22	0.068	0.054	0.444	9
23	0.018	0.106	0.851	2
24	0.096	0.030	0.235	14
25	0.053	0.070	0.570	5

roughness, material removal rate, tool flank wear and tool wear loss with cutting speed (υ), feed rate (f), and depth of cut (d) was developed as (12), (13), (14), and (15):

$$\begin{aligned} \text{Surface Roughness} = {}& 1.10951 - 0.07253*\upsilon + 0.01772* f + 0.02173*d - 0.00217*\upsilon*\upsilon \\ & + 0.00349*f*f - 0.00376*d*d - 0.001676* \upsilon*f + 0.02303*\upsilon*d \\ & + 0.00347*f*d \end{aligned} \tag{12}$$

$$\begin{aligned} \text{Material Removal Rate} = {}& 11.666 - 4.394*\upsilon + 7.38* f + 7.495*d - 0.959*\upsilon*\upsilon - 0.58*f*f \\ & - 0.566*d*d + 3.173* \upsilon*f + 3.184*\upsilon*d + 4.779*f*d \end{aligned} \tag{13}$$

$$\begin{aligned} \text{Tool Flank Wear} = {}& 1.30951 - 0.17253*\upsilon + 0.01772* f + 0.02173*d - 0.00217*\upsilon*\upsilon \\ & - 0.00376*f*f - 0.00349*d*d - 0.01676*\upsilon*f + 0.02303*\upsilon*d + 0.00347*f*d \end{aligned} \tag{14}$$

$$\begin{aligned} \text{Tool Wear Loss} = {}& 0.378597 + 0.025842*\upsilon + 0.002213* f + 0.003387*d - 0.00031*\upsilon*\upsilon \\ & - 0.00046*f*f + 0.000548*d*d + 0.000651*\upsilon*f + 0.000509*\upsilon*d \\ & + 0.002203*f*d \end{aligned} \tag{15}$$

Figure 171.3 (A–D) displays the comparison between experiment values and predicted values for each response. It can be observed from these graphs, there is a close agreement between the experimental data and predicted RSM model and no unexpected deviations. As a result, the quadratic mathematical RSM model can be utilised to achieve the desirability of the output response measures.

Confirmation test:

The optimum parameter setting obtained from the multi-objective optimisation is: cutting speed 3875 rpm, feed rate 0.25 mm/rev., and depth of cut 0.4 mm. These parameter combinations are treated as an input parameter and output responses are measured, as presented in Table 171.13.

The validation of the proposed inputs is confirmed by the test results, which are contemporaneous with the optimum solution.

Optimization of machining parameters in turning of Inconel 718   1517

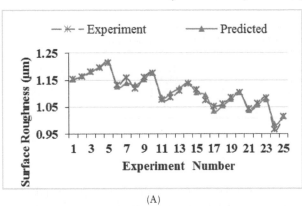

(A)

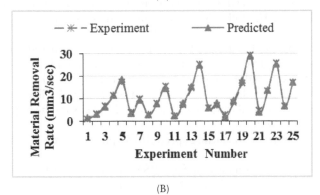

(B)

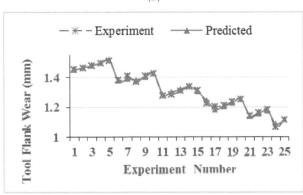

(C)

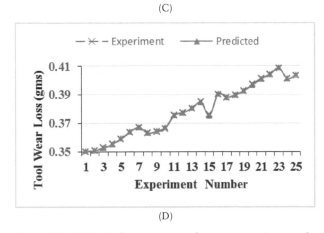

(D)

Figure 171.3 (A–D) the comparison between experiment values and predicted values for each response

Table 171.13 Confirmation test results

Response measures	Optimum value	Confirmation test value
Surface Roughness (µm)	1.103	1.065
Material removal rate (mm^3/sec)	29.333	29.297
Tool flank wear (mm)	1.253	1.212
Tool wear loss (gms)	0.397	0.403

Conclusions

In this current study, modified CRITIC (CRITIC-M) for weight criteria for each parameter and TOPSIS is applied for multi-objective optimisation to carry out the machining operation of Inconel 718 using Al_2O_3-ZrO_2 (Alumina-Zirconia) ceramic cutting tool insert. The following conclusions can be drawn:

- The TOPSIS approach opted the best possible combinations of machining parameters based on the values of the relative closeness; the ranking alternative is 25- 22- 17- 10- 4- 20- 11- 23- 13- 7- 24- 15- 8- 3- 18- 16- 21- 12- 6- 1- 19- 9- 2- 14- 5. The optimal sequence of the experiment run as 20 > 23 > 14 > 5 > 25 > 19 > 10 > 13 > 22 > 4 > 7 > 18 > 9 > 24 > 12 > 16 > 3 > 15 > 21 > 6 > 17 > 2 > 8 > 11 >1.
- The best combinations of input parameters are obtained in an experiment run 20 to ensure the required desirability of all response measures at the same time.
- The optimum identified input machining parameters are cutting speed 3875 rpm, feed rate 0.25 mm/rev., and depth of cut 0.4 mm.
- A second order mathematical polynomial predictive response surface (RSM) model for surface roughness, material removal rate, tool flank wear and tool wear loss with cutting speed (v), feed rate (f), and depth of cut (d) are developed. All models give a close agreement between the experimental data and predicted.
- The optimum combination set obtained through TOPSIS is validated through confirmation test that satisfy the desirability of each responses.

This work can be further extended by performing the experiments in wet and MQL conditions. Other machining parameters like rake angle, nose radius, tool shape, cutting tool material can be included. Consideration of cutting forces, torque, chip morphology and many other response parameters would be a recommended implementation.

References

Ayrim, Y., Atalay, K. D., and Can, G. F. (2018). International J. Inf. Technol. Decis. Mak. 17(3):857–882.
Hao, Z., Gao, D., Fan, Y., and Han, R. (2011). Int. J. Mach. Tools Manuf. 51(12):973–979.
Jafarian, F., Umbrello, D., Golpayegani, S., and Darake, Z. (2016). Mater. Manuf. Process. 31(13):1683–1691.
Khidhir, B. A. and Mohamed, B. (2010). J. Mech. Sci. Technol. 24(5):1053–1059.
Kumar, M. B. and Parameshwaran, R. (2020). Int. J. Ser. Oper. Manag. 37(2):170–196.
Kumar, S. (2019). Int. J. Syst. Assur. Eng. Manag..
Pervaiz, S., Rashid, A., Deiab, I. and Nicolescu, M. (2014). Mater. Manuf. Process. 29(3):219–252.
Pohekar, S. D. and Ramachandran, M. (2004). Renew. Sustain. Energy Rev. 8(4):365–381.
Prasad, K. and Chakraborty, S. (2018). Decis. Sci. Lett. 7(3):273–286.
Shukla, A., Agarwal, P., Rana, R. S. and Purohit, R. (2017). Mater. Today: Proc. 4(4):5320–5329.
Slebi-Acevedo, C. J., Silva-Rojas, I. M., Lastra-González, P., Pascual-Muñoz, P. and Castro-Fresno, D. (2020). Constr. Build. Mater. 233.
Tebassi, H., Yallese, M. A., Meddour, I., Girardin, F. and Mabrouki, T. (2017). Period. Polytech. Mech. Eng. 61(1):1–11.
Trung, D. D., Nguyen, N. T. and Van Duc, D. (2021). EUREKA, Phys. Eng. 2021(2):52–65.
Varatharajulu, M., Kumar, M. B., Duraiselvam, M., Jayaprakash, G. and Baskar, N. (2021). J. Magnes. Alloys.
Xu, C., Ke, Y., Li, Y., Chu, H. and Wu, Y. (2020). Energy Conv. Manag. 215(January).
Žižovic, M., Miljkovic, B. and Marinkovic, D. (2020). Decis. Mak: Appl. Manag. Eng. 3(2):149–161.

172 Investigations into metallurgical characterisation and machining of cryotreated tungsten carbide end mill

Narendra Bhople[1,a], Sachin Mastud[2,b], Sandeep Jagtap[3,c], and Niloy Nath[3,d]

[1]Production Engineering, Veermata Jijabai Technological Institute, Mumbai, India

[2]Mechanical Engineering, Verrmata Jijabai Technological Institute, Mumbai, India

[3]Mechanical Engineering, JSPM's Rajarshi Shahu college of Engineering, Pune, India

Abstract

The trend of micro-manufacturing has been increasing from last decade. Researcher realised the need of non-MEMS (Micro electro-mechanical system) techniques due to the limitation of MEMS such as, types of material processing and manufacturing of complex three dimensional components. Now day's mechanical micro cutting is the preferred manufacturing process for the production of various micro-components. Micro-cutting has potential to overcome the limitations of non-MEMS techniques, however the short tool life of micro tool is the big challenge to improve micro-cutting process and manufacturing cost. The conventional tool life improvement techniques have some limitation in micro-cutting. Therefore in the present work, cryogenic treatment (CT) has been used to treat tungsten carbide-cobalt (WC-Co) end mills. To study the effect of cryotreatment metallurgical characterisation carried out by various tests. Moreover the performance of cryotreated tool evaluated by measuring cutting force, surface roughness and tool wear while machining titanium material. It is found the DCT (deep cryotreated tool) for 20 hr performed good in terms of cutting force and surface roughness.

Keywords: cryogenic treatment, cutting force, micro-milling, surface roughness, Ti-6Al-4V, tungsten carbide.

Introduction

The trend of miniaturization started in the beginning of third millennium in biomedical, automobile, electronics, aviation and others sectors of the industry (Camara et al., 2012). Initially the MEMS or lithography based processes had wide use to produce semiconductors and micro electric components like sensors, actuators etc. from silicon and limited range of material. In last two decade researchers have been invented non-MEMS techniques which can be use to process almost all type of material and to produce complex three dimensional components. Mechanical Micro-cutting is one of the important type of non- MEMS techniques (Takacs et al., 2012; Piljek et al., 2014; Chae et al., 2005).

Many author defined micromachining on the basis of tool diameter or burr size. 'Tool diameter falls in the range of 1 to 999 μm or if undeformed chip thickness is comparable to tool edge radius or material grain size'. Aramcharoen et al. (2008), Aramcharoen and Mativenga (2008) Reduced flexural stiffness of cutting tool is the main constrain for miniaturization of the cutting process. The whole process of machining significantly affected by size effect, material inhomogeneity and minimum chip thickness. Generally the failure of the tool occurred due to low wear resistance, generation of internal stresses while sintering, chip-off of edge radius and increase in specific cutting energy (Gandarias, 2009; Aramcharoen and Mativenga, 2008). The value of cutting forces significantly increases due to increase in radius and this promotes the ploughing mechanism (Cardoso and Davim, 2012; Ajish and Govindan, 2014). The cost of micro-tools is comparatively higher than that of conventional tools (Carou et al., 2017). Also to machine material like Ti-6Al- 4V and hardened steel is challenging in micromachining.

Tool life improvement techniques such as application of hard coating, use of cutting fluid and hybrid machining has regular use in conventional machining. These techniques have some limitations in micromachining. Transportation of cutting fluid in such narrow zone is difficult and also cutting fluid may affect the provided feed and depth of (Bissacco et al., 2005). Application of coating increases the cutting edge radius and it will be useless once it get removed from the surface. Cryogenic treatment (CT) of cutting tool can be a good alternative to improve the life of the micro tools. The metallurgical changes improve the mechanical and physical properties of tool without affecting its geometry.

CT also called as subzero treatment. This process can be used to treat various metals, composites at low processing cost. CT can improve the properties of tungsten carbide (WC) tools by producing additional

[a]nrbhople11@gmail.com; [b]samastud@me.vjti.ac.in; [c]sjjagatap_mech@jspmrscoe.edu.in; [d]nknath_mech@jspmrscoe.edu.in

eta (η) carbide in matrix along with some metallurgical changes. The desired properties can be achieved by selecting proper cryo-cycle. CT gives refined, uniform and dense microstructure this leads to improvement in physical and mechanical properties of carbide tools. The temperature range for shallow cryogenic treatment (termed as SCT) is −80°C to −140°C, while for deep cryogenic treatment (termed as DCT) is −140°C to −196°C (Akincioglu et al., 2015). In Figure 172.1B and G indicates the soaking temperature for SCT and DCT, whereas B-C and G-H is soaking period, D-E-F and I-J-K represents the tempering process.

In conventional machining cryotreated tools performed better than untreated and some coated tools. The application of cryotreated tools while turning improve wear resistance of WC tools up to 29% in flank wear, 67% in crater wear and 81% in notch wear compared to untreated tools (Ozbek et al., 2014). The cryotreated micromilling cutters performed better in wet as well as in dry condition. Coated tools after cryotreatment performed better than that of only coated tools (Thamizhmanii et al., 2011). The desired properties can be imparted to cutting tools by altering the cryotreatment parameters for the given cutting tool material (Gill et al., 2010). Limited literature has been found, related to application of cryotreated tool in micromachining. The main objective this study is to study the effects of cryogenic treatment on WC micro end mill. Therefore in the present work cryogenic treatment conducted at two different soaking temperature (SCT and DCT) and period. The metallurgical characterization of tool conducted by Vickers micro-hardness test, scanning electron microscopy (SEM) and X-ray diffraction (XRD) test. Further the performance of cryotreated tool evaluated by measuring cutting force, surface roughness and type of tool wear while milling Ti-6Al-4V material. DCT 20 hr tool produced lower cutting forces and roughness, this mainly attributed to the formation additional η carbide and its precipitation.

Experimental details

Tungsten carbide is preferable cutting tool material to machine titanium and ferrous based material. Tungsten Carbide has higher hot hardness, low cost and suitable to machine Titanium based material. In the present work, Axis made two flute micro tungsten carbide end mills (E9661), with 10% of cobalt binder have been used. Figure 172.2(a) shows the dimensions of micro end mill and Figure 172.2(b) is actual tool.

In the literature it has been found that CT improves the properties of WC cutting tools. The effect of cryogenic treatment is strongly associated with the selected cryo parameters. To cover the overall study both SCT and DCT temperatures are taken into consideration. Slow cooling rate (−1°C per minute) has been selected to avoid micro cracking of tool surface. Tungsten carbide end mills cryotreated at −80°C (SCT) and − 180°C (DCT) respectively for 20 Hr. Table 172.1 shows the cryogenic treatment parameters used in present study, untreated tool kept for comparison purpose. Similarly Table 172.2 shows the chemical composition of Ti-6Al-4V workpiece. Liquid nitrogen has been used for cryotreatment of cutting tools followed by tempering treatment at 150°C for 2 hour. The purpose of tempering is to relieve the internal stresses of tools and slightly soften the cobalt phase. The entire cryogenic set up shown in Figure

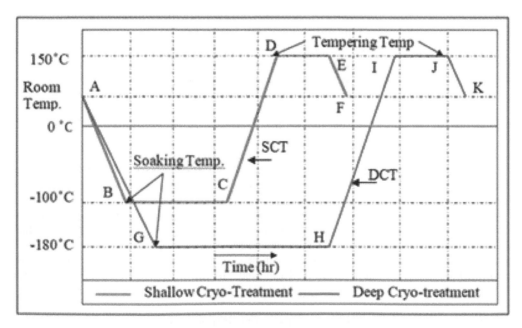

Figure 172.1 Typical cryogenic treatment process

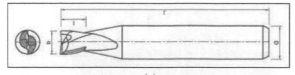

(a)
d=0.8, l=2.4, D=3, L= 38 (All Dimensions are in mm.)

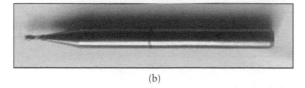

(b)

Figure 172.2 (a) Geometry of micro end mill cutter; (b) actual end milling tool

Table 172.1 Applied cryogenic cycles

Sr. No.	Soaking Temperature (⁰C)	Soaking Period (Hrs.)	Tempering Temperature and Time
1	untreated	–	–
2	–80°C	20 hr	150°C for 2 hr
3	–180°C	20 hr	150°C for 2 hr

Table 172.2 Chemical composition of Ti-6Al-4V

Material	Ti	Al	V	Fe	O	C	N
Ti-6Al-4V (%)	89.75	6.08	3.90	0.16	0.13	0.008	0.007

172.3. High speed micro-milling set up available at machine tool lab IIT Bombay has been used to conduct machining test shown in Figure 172.3(b).

To study the effects of CT on tungsten carbide micro end mill, different tests have been conducted. After completion of cryo process samples were prepared according to ASTM standard B 665-03. To avoid the phase transformation wire cut EDM process has been used to cut the tool shank. Hardness of the tool measured by Vickers micro hardness tester with load of 0.5 kg according to ASTM E384-09 standard. Three readings were taken from core to surface with equal distance of 500 μm and average of three values taken as final hardness. Hardness of tool measured at Phoenix Testing and Consultancy Services Bhosari .FEI Nova Nano SEM 450 field emission scanning electron microscope was used to observe the structure of micro end mills and to study the wear mechanism by using highly magnified images. SEM facility available at Department of Chemistry, SPPU, Pune has been used for detailed characterization. Set up of SEM shown in Figure 172.4(a). In the present work XRD tests are conducted to identify the amount and type of structural changes occurred after CT. XRD test conducted to identify the phase changes. XRD test facility available at Department of Metallurgy, IIT Bombay has been used. High score software used to identify the phases. Figure 172.4(b) shows XRD set up.

Metallurgical characterisation of micro end mill

To study the effects of cryogenic treatment Vickers micro-hardness, SEM and XRD test has been conducted, below section present the result obtained in respective test.

Vicker micro hardness test

The effect of shallow cryogenic treatment for 20 hr soaking period (termed as SCT 20) and deep cryogenic treatment for 20 hr (termed as DCT 20) on hardness is shown in Figure 172.5. It is found the value of hardness for both SCT and DCT tool is significantly higher than untreated tool. The hardness value increased from core to surface. The hardness of DCT 20 hr tool is highest, it is increased by 8% compare to untreated tool. Formation of additional η carbide and its precipitation might be attributed for this. Newly formed

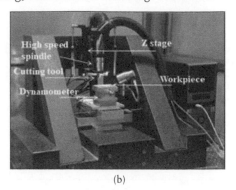

(a) (b)

Figure 172.3 (a) Chamber of cryogenic treatment with liquid nitrogen cylinder; (b) High speed micro milling setup

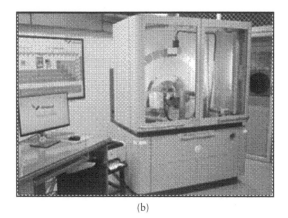

(a) (b)

Figure 172.4 (a) Setup of scanning electron microscope test; (b) Setup of x-ray diffraction test

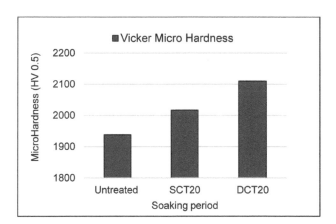

Figure 172.5 Effect of SCT and DCT on hardness of WC-Co

carbide are fine and it fill the voids in matrix [gill2012]. This denser matrix after cryotreatment produced resistance to form indentation on SCT and DCT tool compare to untreated tool. The Vicker test indicates soaking temperature play a significant role to decide the hardness of tool. Similar results of higher hardness at deep soaking temperature are observed by (Ozbek et al., 2014).

Scanning electron microscopy test

The grains of WC and binder cobalt shown in Figure 172.6. Figures 172.6(a-c) are the microstructures of untreated, SCT 20 and DCT 20 respectively at 30000 X. The proportion of α phase is higher in WC tool and spread into entire bulk of the material. Figure 172.6(a) shows in untreated tool the size of WC grain is long rectangular structure and it is non-uniform. Average grain size of WC untreated tool is 1008 nm. WC

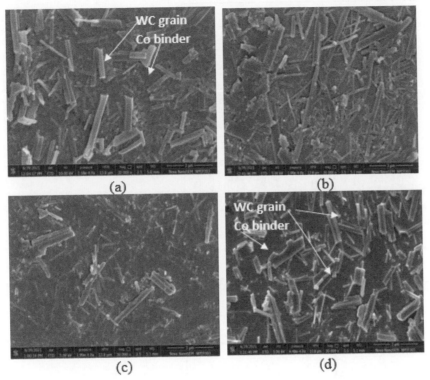

Figure 172.6 Microstructure of WC-Co tool (a) untreated; (b) SCT 20 (g) DCT 20

grain in untreated tool look like loosely bonded with cobalt matrix. Figure 172.6(b) shows the size of long rectangular WC grain reduced and it is approximately uniform in SCT 20. It also shows the densification of matrix phase i.e. cobalt occurred. The average grain size of WC grain at SCT 20 is 878 nm. Figure 172.6(c) is the structure of DCT 20 WC tool, it is found the large WC gains are almost eliminated in DCT 20. DCT brought WC grain closer to each other and formed compact structure compared to untreated and shallow cryotreatment. The average grain size of DCT 20 WC grain is 575.3 nm. The densification of cobalt has been occurred and it holding WC grain strongly, it causes the locking of WC in the cobalt matrix. The densification of cobalt also might be attributed for increased hardness in all cryotreated tools.

X-ray diffraction test:

X-ray diffraction method has been used to examine the structural changes occurred after cryotreatment in WC micro end mills. Figures 172.7(a–c) shows the XRD spectrum of untreated, SCT 20 and DCT 20 tool respectively.

Figure 172.7(a) indicate presence of η carbide at small extent in untreated tool (green colour peak around 43°). Spectrum of SCT and DCT tools indicates formation of extra eta carbide along with previously present carbides. The proportion of η carbide increased in both SCT and DCT tools, the notable rise is in case of DCT (green colour peak around 43°) shown in Figure 172.7(c). Therefore it can be conclude that cryotreatment promotes the formation of additional η carbide. In the present study DCT is effective to produce additional η carbide than SCT. Two types of η carbide phases may form after cryotreatment, binary phase W_2C and ternary phase which may obtain in Co_3W_3C or Co_6W_6C form. In the XRD test we found the presence of ternary η phase i.e. Co_6W_6C which is hard in nature. According to Gill et al. (2012) Co_6W_6C present at the outer surface or on the substrate of tool and it has does embrittle the structure. Due to size limitation it is difficult to conduct pin-on-disc test on micro tools. However, the pin-on-disc test performed on cryotreated conventional size WC tool shows increased wear resistance (Gill et al., 2012). Therefore it is expected, the formation of Co_6W_6C ternary η phase can contribute to increase the tool life by reducing wear while machining.

Applications of cryotreated tools in micromilling

Only metallurgical characterization is not sufficient to predict the suitability of cutting tool in machining. Therefore to test its suitability, all tool subjected to micromilling test. This section presents the results obtained in machining test, in terms of cutting force, surface roughness and tool wear mechanism.

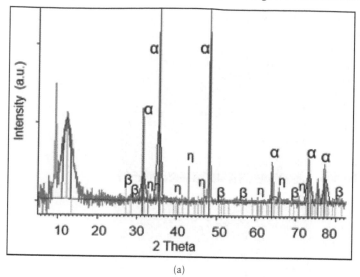

(a)

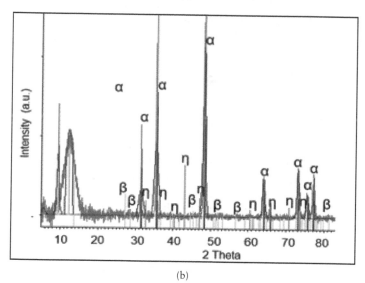

(b)

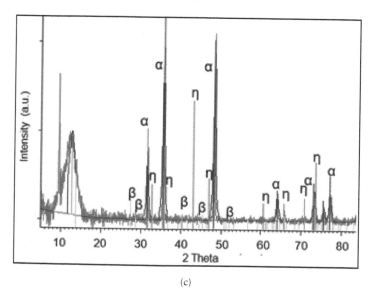

(c)

Figure 172.7 XRD spectrum of WC-Co tool (a) untreated; (b) SCT 20 (g) DCT 20

Measurement of cutting forces:

All tools (untreated and cryotreated) used to machine Ti-6Al-4V, selected cutting parameters were spindle speed 100000 rpm, feed rate 4 μm/flute and depth of cut 150 μm in dry cutting condition. In Figure 172.8 Fy indicates the cutting force normal to feed and and Fx is cutting force along the feed direction. The values of Fy found higher than that of Fx in all cases. In milling higher swept angle is required to shear the material basically this attributed to higher value of Fy than Fx.

Figure 172.8 shows the value of cutting forces (Fy and Fx) is significantly higher for SCT 20 tool and DCT 20 tool has generated lowest cutting forces. Untreated tool indicates the moderate value of cutting forces. In pervious section, XRD results suggest the cryogenic improve the hardness and wear resistance. Therefore the increased cutting forces in case of SCT 20 must be due generation of built-up-edge (BUE). Low thermal conductivity of Titanium generates high temperature in cutting zone, this leads to formation of (BUE). Untreated tool has low hardness and Pin-on-disc wear test conducted by Gill et al. (2012) shows untreated tool has low wear resistance compare to cryotreated tool (Gill et al., 2012). Therefore in this case untreated tool subjected to high wear, this attributed high value of cutting force. Deep cryogenic treatment re-arranged tungsten particles this might be leads to improve thermal conductivity and reduced probability of BUE. The formation of BUE for DCT 20 cannot deny, but sharp cutting edge than untreated and SCT 20 leads to low cutting forces.

Measurement surface roughness (Ra)

Surface roughness not only limited to aesthetic appearance but properties like corrosion, fatigue life, thermal conductivity and electrical conductivity strongly depends on surface roughness. Ra value measured at two location in slot, initially at the starting and second at last of the slot and average of that taken as final Ra value. Figure 172.9 shows the Ra value obtained for respective tool. The roughness produced by DCT 20 tool is quite low than untreated and SCT 20 tool this attributed to the increased wear resistance due to formation additional η carbide. The Ra value of SCT 20 tool might be affected due to the high heat generation in cutting zone. The cutting edge radius increased due to lower wear resistance increased Ra value for untreated tool.

Tool wear study of mircro end mills

To study the wear mechanism SEM and Energy Dispersive Spectroscopy (EDS) test have been conducted. SEM used to produce highly magnified images of cutting edge and EDS to confirm the workpiece deposition.

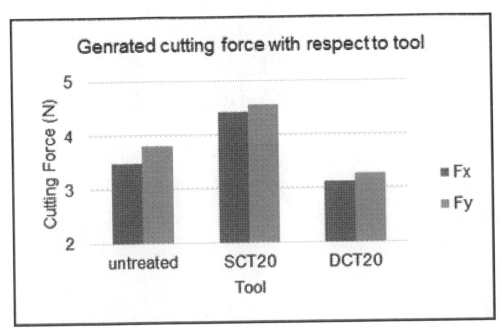

Figure 172.8 Cutting forces of untreated and cryotreated micro end mill

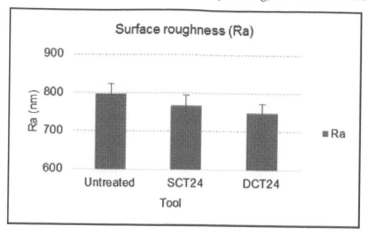

Figure 172.9 Effect of cryotreatment on surface roughness (nm Ra)

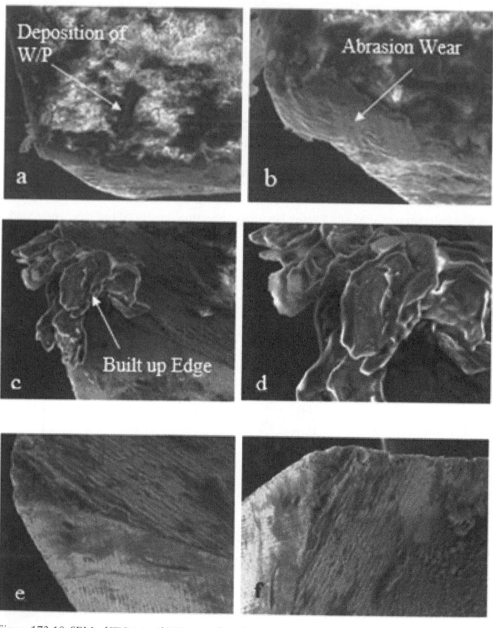

Figure 172.10 SEM of WC (a) – (b) Untreated tool (c) – (d) SCT 20 and (e) – (f) DCT20

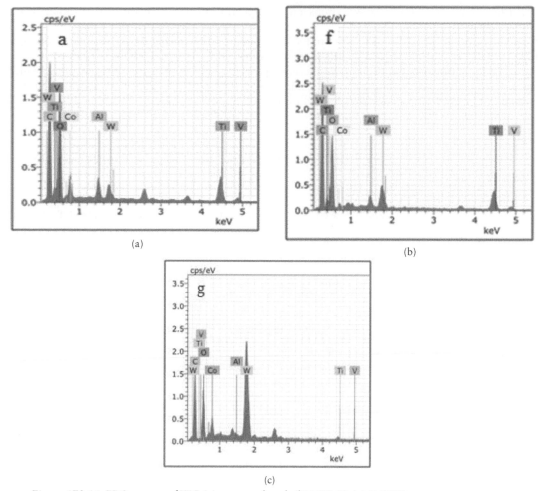

Figure 172.11 EDS spectra of WC (a) untreated tool; (b) SCT 20; (c) DCT20

Figure 172.10(a) shows in case of untreated tool the workpiece deposited beside the cutting edge, similarly Figure 172.10(b) shows the increased edge radius due to abrasion wear. It is found untreated tool experienced high wear while machining Titanium. Figure 172.10(c) clearly shows the formation of BUE for SCT 20 tool, this attributed to the thermal softening of the workpiece. Figures 10(e–f) shows the cutting edges of DCT 20 tool. The wear of DCT 20 tools is significantly less and almost eliminated BUE. XRD confirmed the additional η carbide formation in DCT 20 tool and SEM shows the elimination of large WC grains. Formation of Co_6W_6C ternary η carbide in metal matrix retained cutting edge of DCT 20 Tool. According to Gill et al. (2012) cryotreatment increases the thermal conductivity. The structural changes occurred at DCT 20 tool leads to improve thermal conductivity and to reduce BUE. Here it can be concluded cryotreatment of micro tool is good option to retain sharp cutting edge.

In EDS test mainly the area of cutting edge taken into consideration to check the workpiece deposition. Low thermal conductivity of Ti-6Al-4V affecting machining process. Figure 172.11(a) shows EDS spectra of untreated tool, the high peaks of Ti and Al indicates notable deposition of workpiece. Similarly, Figure 172.11(b) suggest the deposition for SCT 20 tool. In Figure 11(c), spectra of DCT 24 tool shows very less deposition of workpiece material of cutting too, it is almost negligible. This attributed to the improved thermal conductivity of DCT 24 tool.

Conclusion

In the present work, the effect of shallow and deep cryogenic treatment on tungsten carbide end mill has been studied. Vicker micro-hardness, SEM and XRD test used to conduct metallurgical tests. Moreover, all the tools used in micro-milling process. Based on the conducted study following conclusion can be drawn.

- Cryogenic treatment increases the hardness of WC tools, this credited to formation of additional Co_6W_6C ternary η carbide in metal matrix. Deep cryogenic treatment increased hardness of WC by 8% compare to untreated tool.

- Cryotreatment eliminated the large WC grains and formed uniform grain structure, interlocking of WC grains also occurred.
- Untreated tool subjected to abrasion wear and SCT 20 tool experienced significant BUE formation this lead higher value of cutting force and surface roughness for both than DCT 20 tool.
- In case of DCT 20 tool, the significant rise in ternary η carbide leads to increase wear resistance, which retained sharp cutting edge compare to untreated and SCT 20 tool. The densification of cobalt also contributed to holds WC grain more tightly which leads to reduce tool wear.

References

Ajish, T. N. and Govindan, P. (2014). Effects and challenges of tool wears in micro milling by using different tool coated materials. IJRMET. 4:2249–5762.

Akincioglu, S., Gokkaya, H., and Uygur, I. (2015). A review of cryogenic treatment on cutting tools. Int. J. Manuf. Technol. 78:1609–1627.

Aramcharoen, A. and Mativenga, P. T. (2008). Size effect and tool geometry in micromilling of tool steel. Precis. Eng. 33:402–407.

Aramcharoen, A., Mativenga, P. T., Yang, S., Cooke, K. E., and Teer, D. G. (2008). Evaluation and selection of hard coatings for micro milling of hardened tool steel. Int. J. Mach. Tools Manuf. 48:1578–1584.

Bissacco, G., Hansen, H. N., and De Chiffre, L. (2005). Micromilling of hardened tool steel for mould making applications. J. Mater. Process. Technol. 167:201–207.

Camara, M. A., Campos Rubio, J. C., Abrao, A. M., and Davim, J. P. (2012). State of the art of micromilling of material, A review. J. Mater. Sci. Techol. 28:673–685.

Cardoso, P. and Davim, J. P. (2012). A brief review on micromachining of material. Rev. Adv. Mater. Sci. 30:98–102.

Carou, D., Rubio, E. M., Herrera, J., Lauro, C. H., and Davim, J. M. (2017). Latest advances in the micro-milling of titanium alloys: a review. Procedia Manuf. 13:275–282.

Chae, J., Park, S., and Freiheit, T. (2005). Investigation of micro-cutting operation, Int. J. Mach. Tools Manuf. 46:313–332.

Gandarias, E. (2009). Micromilling technology: Global review. Courtesy of global thesis.

Gill, S. S., Singh, H., Singh, R., and Singh. J. (2010). Cryo-processing of cutting tool material-a review. Int. J. Adv. Manuf. Technol. 48:175–192.

Gill, S., Singh, J., Singh, H., and Singh, R. (2012). Metallurgical and mechanical characteristics of cryogenically treated tungsten carbide (WC-Co). Int. J. Adv. Manuf. Technol. 58:119–131.

Ozbek, N. A., Cicek, A., Gulesin, M., and Ozbek, O. (2014). Investigation of the effect of cryogenic treatment applied at different holding times to cemented carbide inserts on tool wear. Int. J. Mach. Tools Manuf. 86:34–43.

Piljek, P., Keran, Z., and Math, M. (2014). Micromachining – review of literature from 1980 to 2010. Interdiscip. Description of Complex Sys. 1:1–27.

Takacs, M., Vero, B. and Meszaros, I. (2012). Micromilling of metallic material. J. Mater. Proc. Technol. 138:152-155.

Thamizhmanii, S., Mohd, Nagib. and Sulaiman, H. (2011). Performance of deep cryogenically treated and non-treated PVD inserts in milling. J. Achiev. Mater. Manuf. Eng. 49:460-466.

173 Numerical and experimental topology optimisation of crane hook

Avilasha B. G.[1,a] and Ramakrishna, D. S.[2,b]

[1]Assistant Professor, Mechanical Engineering Department, Dayanandasagar College of Engineering, Bengaluru, India

[2]Shimoga, India

Abstract

Resource constraints have necessitated design and development of machines and structures that are more efficient and have minimum weight. Optimisation procedures are widely used for this purpose. Topology optimisation technique has wider area of applications, like design and compliant mechanisms. Extensive studies have been carried out employing computational method with finite element analysis to carry out topology optimisation. The computational method essentially consists of density approach which assigns 0 (no material) and 1 (with material) at low stress and high stress region respectively. Final topology of the machine or structural member is arrived at after a number of iterations. However, the scope for using an experimental methods like photoelasticity for topology optimisation has not received due attention. During this research, experimental photo elasticity approach has been used to carry out topology optimisation. Photo elastic method, being a whole field technique has been widely used for stress analysis purpose and can be used with advantage to carry out topology optimisation. When the photo elastic model of a machine or structural member under load is viewed in the polariscope, the isochromatic fringe pattern observed from the model can be used to estimate the stress levels at a given location. The elements which of high and low stress can be forecast with reasonable accuracy. The sequential removal of material from the low stress region results in a topologically optimised model.

For the purpose of this study, both numerical and experimental analysis is carried out on a crane hook under tensile loading. Optimisation analysis of crane hook is carried out using Hyperwork, Optitruct software. Final geometry is arrived at after 21 iterations. The weight reduction obtained from numerical analysis is 23.52%. The photoelasticity approach required only 7 iterations to arrive at the final geometry. Weight reduction obtained by experimental approach is 19.15%. The stress level in the experimentally obtained model is verified by carrying out finite element analysis. The weight reduction and optimisation is achieved without significant reduction in strength of the crane hook.

Keywords: Numerical analysis, photo elasticity, stress analysis, topology optimisation.

Introduction

The paucity of natural resources, increasing cost of raw materials leads to scope for the machines and structures to be of insubstantial, having fewer numbers of parts and better performance. Design optimisation plays an important role in fulfilling these objectives. The process of topology optimisation is carried out using a numerical analysis and algorithm analysis. Based on the type of design variables used Bendsoe and Kikuchi (1988), optimisation methods can be grouping as parameter, shape and topology optimisation. While the concept of parameter Bendsøe (1989) optimisation is to change the model parameters, the concept of shape optimisation involves moving part boundaries and constraints. However, the concept of topology optimisation is to change the density of material regions to form shape and topology that gives the optimal distribution of material in a given design domain fulfilling a predefined criterion. It is an iterative finite element analysis involving material removal in each successive iteration (Bendsøe, 1995). Besides allowing for size and shape changes, topology optimisation allows voids to appear and disappear to obtain an optimal design. The topology optimisation is used for, obtaining optimal design of structures and machine components and Design of compliant mechanisms. When used for optimal design of structures, the neighborhood of structures that have least contribution to the overall stress level or stiffness are identified and removed. The topology optimisation can be implemented for keep down the weight and making the component more compliance. Flipside this can be used for compliant mechanism.

The topology optimisation became an important tool for every area where there is scope for material reduction. The conceptual design of mechanical structures and optimizing the material distribution without compromising the strength of the component can be achieved by using the topology optimisation method.

---

[a]avilashabgr9@gmail.com; [b]ramak_ds@yahoo.com

The photoelasticity experimental method is one of the recommended methods to achieve the optimisation. It is prominent for the design and development of new products to find the best possible topology or layout for given design objectives (Bendsoe and Kikuchi, 1988).

The review of topology optimisation proved that the work achieved are only numerical methods and algorithms, so this research work explores the contribution of experimental methods of topology optimisation, on evolving the multiple light-in-weight topologies of compliant mechanism, tracing user-defined path the evolutionary algorithm (NSGA-II) is customized to efficiently deal with the constraint bi-objective, non-linear and discrete problem of compliant mechanisms (París et al., 2009).

The topology optimisation can be done using different methods like level set method, density method, SIMP method and stress limitation methods, with stress constraints was initially considered in references. Behind that stress limitation formulation was further extended to consider differing stress constraints limits and to improve the solution efficiency using different strategies Holmberg et al. (2013), Cheng and Olhoff (1982) as present work concentrates on topology optimisation for weight reduction, the density method is used. The literature survey indicates that the topology optimisation is an effective tool for mechanical structures to get maximum output from the minimum weight ratio. However, experimental methods for topology optimisation have not received much awareness. Photo elasticity method, being a whole field method of stress analysis can be used with advantage to carry out topology optimisation studies on machine or structural members.

Theory of topology optimisation

There are many optimisation techniques has been used from last decade, the optimisation techniques can be adopted to load carrying members in aerospace application, surgical tools and other applications. The research interest is to remove the material from low stress region and make into compliance so that the stress study is takes vital role in the optimisation. So far, the FEM, mathematical relations and software's are used for stress analysis and the optimisation. The topology optimisation Duysinx (1995) as shown in Figure 173.1.

The topology optimisation can be achieved by assigning the design space Ω. The low stress area, the material value given is 0 and high stress area 1, from low stress region the material will be removed by density method. Topology optimisation is using the numerical FEM method. The model has been created and the optimisation done by assigning design space and constraints. The final optimised structure can be obtained.

The basic methods of optimisation are achieved by boundary and material constraints. The optimisation can be accomplished by stress constraint method, density, computational methods, strain constraint and other few methods.

Topology has only been considered as a mathematical definition. The depiction of structural, shall be adapted to topology optimisation (París et al., 2009) E = ½rE0

The E is the composite's young's modulus,
½ is the density of material,
E0 is the young's modulus, and
r is the exponent of the power law.

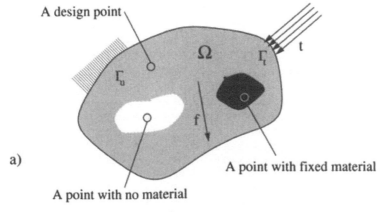

Figure 173.1 Topology optimisation concept

which behaves like a penalty factor. Thereby, the optimisation is accomplished by finding the densities of the material (París et al., 2009).

As utmost structural analyses are performed with the finite element method, the density distribution is also defined with the finite element interpolation. The lower density for each finite element. Thereby, the discretised finite element optimisation problem has as many design variables as the number of finite elements in the mesh, in each element the stresses are studied and for lower stress the 0-density assigned.

In this dissertation work the density method adopted for topology optimisation. The low stress region assigned by density value 0 and the maximum stress regions assigned by density value 1. The objective function is the total mass of the design (Duysinx and Bendsøe, 1998). The topology optimisation is tool which is open to get the optimised shape using existing data and new design and non-design space. The different methods are adopted for topology optimisation for optimal shape. The density method uses the element density of the component by considering the stress level in the component. The low stress regions are the focus for optimisation, these elements assigned by the 0-density value and the high stress regions are retained with the density value 1.by this approach where the density value 0 these locations the material removed and the compliance has achieved. The density approach applied only to design space. The non-design space retained same.

$$F(\rho)\int \rho \, d\Omega$$

The discussed earlier, the density method of optimisation problem formulation explained, let us assume that the objective shall be to minimise the material volume of the body (Tvergaard, 1975).

$$F = \int_{V}^{V} \rho.dV \quad dv = \text{minimum}$$

Where ρ is a volume density of material which is variable on the structural domain ($\rho = 1$ for solid material). The finite element material models that cover the complete range of values from 0 over intermediate values to 1 and this provides the topology optimisation concept.

Topology optimisation using numerical analysis

During this research work hyper work 'Optitruct', has been used for the topology optimisation. The crane hook application is selected for the work. It is subjected to tensile loading. The design space, non-design space and the boundary conditions are specified in the optitruct for solving the problem as shown in Figure 173.2. The design and non-design space has been decided by considering the loading and low stress regions in crane hook. The upper circular part and tip portion of the hook constrained by x, y, and z direction displacements, the load 1000N as applied at inner circumference of the hook. The properties have assigned for design and non-design space, the element used for meshing is tetra element. The Optitruct tool has run for these inputs, the 25 iterations have processed and analysed for the optimal topology of crane hook.

The optimisation consists of determining whether each element in the crane hook should contain material or not .to achieve this, the density of material within each finite element is used as a design variable, the limits 1 and 0. The positioning of material in the structure is crucial for its optimality so, the selected iterations are shown which optimal convergence reached. The material removed from the design space of

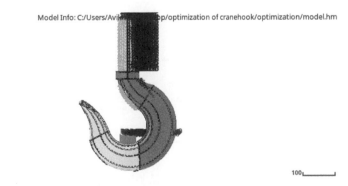

Model Info: C:/Users/Avi......p/optimization of cranehook/optimization/model.hm

Figure 173.2 Design, non-design space and boundary conditions

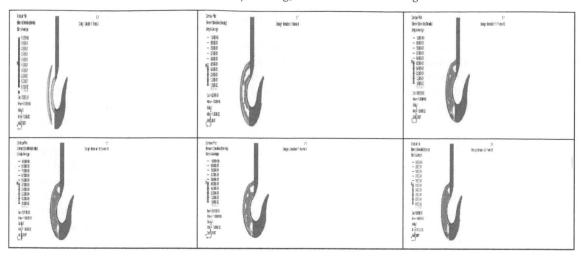

Figure 173.3 OptiStruct optimisation results

the crane hook where the stress is low. The iteration 3 has considered for the study. Each iteration shows the densities of the crane hook, the densities value varies between 1 to 0, the value 1 indicates retained the material, less than 1 and 0 indicates material removed. The results from Optisruct as shown in Figure 173.3. The stress and displacements values are tabulated.

The convergence criteria achieved for optimal solution. The convergence of the component is achieved after optimisation has achieved. so, the convergence history for all the iteration studies to get optimal solution. The graph shows that the convergence achieved at 10th to 20th iteration. The convergence history shown in Fgure 173.4.

The following data are observed from the topology optimisation using Optisruct

The initial mass of the cranehook – 17 kg
The mass of the crane hook after optimisation – 13 kg
The displacement before optimisation – 0.26 mm
The displacement after optimisation – 0.320 mm
The initial von-mises stress of crane hook – 15.65 Mpa
The stress after optimisation – 16 Mpa

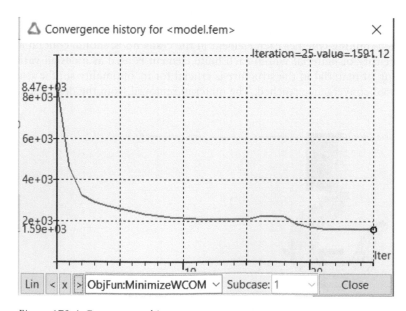

Figure 173.4 Convergence history

Photoelasticity approach for topology optimisation

The experimental method for topology optimisation is photo elastic method. It is a non-destructive and stress-analysis technique. The method based on an opto-mechanical called birefringence property. Many transparent polymer materials like epoxy resin possess this property of birefringence. The method is used for stress analysis of two-dimensional plane problems. When a photo elastic model of a machine or structural member is loaded and viewed in a circular polariscope arrangement, isochromatic fringes are observed. The isochromatic fringes are the contours having constant principal stress difference. Every isochromatic fringe is identified by its fringe order N.

The photo elasticity experimental approach for the topology optimisation consists of following steps, as shown in flowchart in Figure 173.5.

Initially the material selection has to be done because there are many types of birefringence property polymers available. The epoxy resin has been used for the model. The material used for calibration for determining its material fringe value f_σ. These values can be found by calibration method. With P as applied load, D the diameter and N as isochromatic fringe order at the center of the circular disc calibration model, the material fringe value for the material is obtained from mathematical relation.

$$f_\sigma = \frac{8P}{\pi DN}$$

With h as the model thickness, the principal stress difference at any point is given by

$$\sigma_1 - \sigma_2 = \frac{Nf_\sigma}{h}$$

The knowledge of isochromatic fringe order N at a point will help in estimating the stress level at the location. At any location, if the isochromatic fringe order is zero, both principal stresses at the location are zero or they are of equal magnitude. This information together with the preliminary knowledge of stress distribution due to the applied load on the member, region of low stress can be predicted. Topology optimisation is begun with removal of material from this low stress region. The model with material removed from low stress region is viewed in the polariscope and the iterations are continued till the optimisation criteria are fulfilled.

Experimental optimisation

The photo-elasticity approach as explained in the previous section has been employed for topology optimisation. The experiments are carried out on an epoxy resin circular disc model to calibrate the material and to find out the material fringe constant. The diameter of the specimen is 53 mm, thickness 5 mm, it is

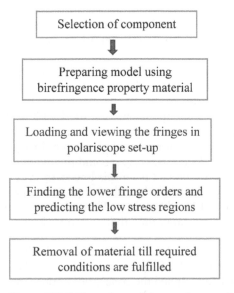

Figure 173.5 Flow chart of the topology optimisation using photoelasticity

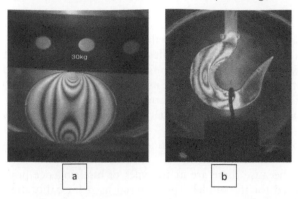

Figure 173.6 (a) Calibration of photo elastic material; (b) Isochromatic fringes

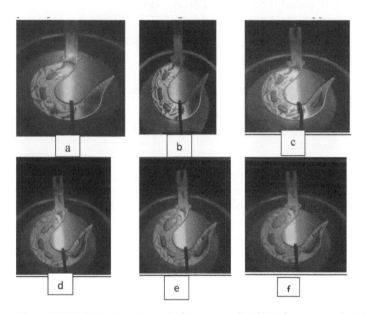

Figure 173.7 (a) 1st iteration - 2 slots are made; (b) 3 slots are made; (c) the material has been removed from the tip of the crane hook; (d) the upper slot dimension enlarged towards the fixed end; (e) the upper slot dimension enlarged towards the diametral surfaces; (f) the upper and middle slots dimension enlarged

subjected to a diametric compression loading and is viewed in a photo-elastic, circular polariscope arrangement. The experimental set up is shown in Figure 173.6(a). Fringe constant of the photo-elastic material is determined as f_σ = 9.6 N/mm/fringe.

A crane hook model made of the photo-elastic material and having same dimensions as that of the model used for numerical analysis is used for experimentation. The hook fixed at the upper end in the loading frame and is subjected to tensile loading as shown in Figure 173.6(b).

A circular polariscope with dark field arrangement is used. The experiment is started with no-load condition and isochromatic fringe pattern is observed as the applied load is gradually increased. Zero order fringe is seen at the point of intersection of neutral axis and the horizontal line passing through the center of curvature of the hook. This is the location where the induced stress is zero and the region in the neighborhood will be low stress region. To start with, material is removed from this low stress region by introducing slots as shown in Figure 173.7(a). After viewing the model in the polariscope, more material is removed by introducing additional slots as shown in Figures 173.7(b) and (c). These slots are further enlarged as shown in Figures 173.7(d), (e) and (f), ensuring at the same time that the stress level at the critical point is within the permissible value.

The fringe patterns are observed from inside circumference to outer layer of the crane hook, the order of the fringes N is identified to achieve the optimisation, the lower order fringes are the location where the difference of principal stress is low, these regions are identified and applied the concept of optimisation.

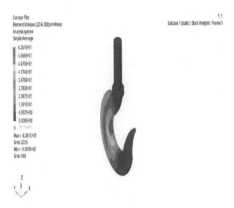

Figure 173.8 FEM analysis before optimisation

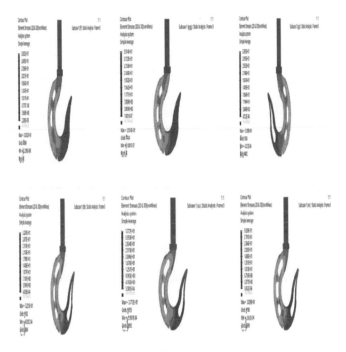

Figure 173.9 FEM analysis of iterations

For validating experimentally obtained topology, finite element analysis is carried out on model having same geometry, dimensions and loaded in the same way as applied while conducting experiment. The finite element method stress distribution is shown in Figures 173.8 and 173.9. The displacement and stress values are noted down for crane hook before optimisation. The displacement value is 0.21 mm and the stress value is 15.65 Mpa. Finite element Results validated with experimental results and results shows with minimum error.

The stress induced in the crane hook before optimisation and after optimisation should be same as the present work concentrate on the optimisation without comprising the strength of the component. The stress results are compared in each iteration and the results are appealing. The low stress regions are identified by FEM stress analysis and these results are compared with the Optisruct and photoelasticity results.

Results and discussions

The photo-elastic experiment has been carried out for topology optimisation of crane hook. The Optisruct software, optimal shape results have taken as reference for the experiment. The number of iterations is carried out to get the optimized crane hook. The experimental photo elastic method allows the user to analyse the lower order fringes and predicting the stress level and assigning the density value. The FEM analysis carried out for identifying the low stress region in crane hook followed with photoelasticity experiment.

The 1st iteration carried out without removal of material from the crane hook, the specimen loaded with tensile force, one end of the specimen fixed and another end subjected to tensile load till 98.1 N.

The 0-order fringe which the $\sigma_1 - \sigma_2 = 0$ so, the stress at 0 order fringe is minimum or 0, the material can be removed from this location of crane hook, The deformation and stress values are tabulated. Deformation was noted down 0.1 mm and the stress 15.4 Mpa. by observing the isochromatic fringes with 0th -order fringe, the number of iterations is carried out by removing material from specimen in each iteration. The experiments conducted for 7 iterations by removing the material from different location of the crane hook in each iteration. The iterations are as shown in the Figure 173.7.

The Figure 173.9 shows that, the isochromatic fringes are increased with each iteration, the lower order fringes are found at the neutral axis so the material removed by making the slots initially, later by conducting number of iteration and predicting the stress levels, the slots are widen and also by experiment also observed that at the tip of the crane hook doesn't have any high order fringes and also stress level is very low, so the material removed from tip, the experiment conducted and no change in the fringe pattern. As research focus on this fringe order which is key parameter for topology optimisation. From each iteration the stress distribution has observed carefully and by making the prediction of lower fringe orders the optimisation has done. The Finite element analysis is carried out for each iteration of experimental model and compared the stress and displacements.

The iterations show that change in stress distribution as the material is removed from the crane hook. The specimen studied for stress distribution $\sigma_1 - \sigma_2$, the difference of principle stress will increase with the fringe order so, lower the fringe order, lower the stress value. The material removed from 0- order fringe locations and the optimisation are achieved. By observing the fringe order and fringe pattern can be decide whether the material can be removed or retained from the specimen. The topology optimisation is achieved without compromising the strength of crane hook. The crane hook became more compliance. The results from the 7th iteration are tabulated and compared to achieve the optimisation.

The finite element analysis of all the iterations is as shown in the Figure 173.9. The analysis shows that even after removal of material there is only small change in the stress level and also displacement value so, the FEM analysis supports the results from experimental results. The blue area indicates low stress regions, by comparing the stress distribution from FEM and photoelasticity the topology optimisation achieved. The comparison between experimental and FEM results are as shown in Table 173.1.

The table clearly shows that the comparison of stress distribution in experiment and FEM analysis of each iteration. The deformation and stress values are converged. The deformation and stress do not change even after removal of material from low stress regions. The results are compared with the Optisruct results, Hyperwork Optitruct software predicted the topology optimized shape for crane hook. The weight reduction of crane hook was analysed, the photoelasticity results are compared with the Optitruct results. The comparison results are shown in Table 173.2.

Table 173.1 The comparison results for FEM and photoelasticity experiment

Iteration	Photoelasticity Experiment				FEM	
	Deformation mm	Stress Mpa	Load Kg	Weight Gram	Deformation mm	Stress Mpa
1	1	17.17	10a	20.84	2.3	13.2
2	1	17.17		20.02	2.4	15.1
3	2	17.17		19.71	2.5	15.22
4	2	17.22		19.28	2.6	15.33
5	2.2	17.36		17.69	2.8	16.1
6	2.2	17.86		17.26	3	16.12
7	2.2	17.86		16.85	3.1	16.12

Table 173.2 Comparison results for before and after optimisation

Parameter	Photoelasticity experiment		Optisruct Results	
	Before Optimization	After Optimization	Before Optimization	After Optimization
Weight	20.848 gram	16.855 gram	17 kg	13 kg
Displacement mm	1	2.2	0.26	0.32
Stress MPa	17.176	17.95	15.65	16.1
% Reduction weight	19.15		23.82	

The above results clearly show that the 19.15% weight reduction in photoelasticity method and by using Optisruct 23.82%, so results are convergence. The photoelasticity technique can be used for topology optimisation.

Conclusion

The present scenario the material is important factor with respect to economy and efficiency of components. The tool which is useful for the Topology optimisation for redesigns the component by considering the design space. The topology optimisation carried out by using the software's and algorithms. These methods do not allow the user to remove the material from the non-design space regions. The non-design space is which having fixtures, load and also few shapes which does not want to change in the component will be the non-design space. The optimisation carried out only in design space so, there is a scope for optimisation even after solution done. To rectify this the photoelasticity approach has implemented for topology optimisation, the approach allows the component to optimise in both design and non-design space, this allows freedom to optimise the whole component, this makes the component more compliance. To manifest this the experimental method developed and evaluated for Topology optimistion. The photoelasticity results are verified by numerical analysis using Optisruct and the results are appealing. The topology optimisation from 'Optisruct' by density method has done as conceptual topology optimisation, present research work results concluding that, the photoelasticity method can be used as topology optimisation tool for predicting the optimal solution for the components. This achieved by finding the order of fringes in the photo elastic model, the lower order fringes predicts that the $\sigma_1 - \sigma_2$ will be minimal so, the material removed from these places does not affects the performance and stress induced. The number of iterations conducted to check the feasibility of using the photoelasticity experiment.

As discussed in the last section about the comparison of results clearly justifies the outcome from photo elastic experiment and Optisruct results. The weight reduction is about 19.15% and the stress induced is same as before optimisation, from all these considerable results can conclude that the photo elastic experiment can be used as one of the tools for the topology optimisation.

References

Bendsøe, M. P. (1989). Optimal shape design as a material distribution problem. Struct. Optima. 1:193–202.

Bendsøe, M. P. (1995). Optimization of structural topology, shape, and material. Berlin, Heidelberg: Springer.

Bendsøe, M. P. (1999). Variable-topology optimization: Status and challenges. In Proceedings European Conference on Computational Mechanics—ECCM '99, ed. W. Wunderlich, pp. 21. Munich, Germany: CD-Rom, Tech University of Munich, Germany.

Bendsøe, M. P. and Díaz, A. R.(1999). A method for treating damage related criteria in optimal topology design of continuum structures. Struct. Optim. 16(23):108–115.

Bendsøe, M. P. and Kikuchi, N. (1988). Generating optimal topologies in structural design using a homogenization method. Comput. Method. Appl. Mech. Eng. 71:197–224.

Bhima Raju, H. S. (2015). Design & analysis of static stresses for leaf springs using photoelasticity & numerical methods. Int. J. Eng. Technol. Comput. Res. 3(4):225–236.

Cheng, K. T. and Olhoff, N. (1982). Regularized formulation for optimal design of axisymmetric plates. Int. J. Solid. Struc. 18(2):153–170.

Duysinx, P. (1995). Optimization topologies: Du milieu a la structure elastique. PhD Thesis, Faculty of Applied Sciences. Belgium: University of Liege.

Duysinx, P. and Bendsøe, M. P. (1998). Topology optimization of continuum structures with local stress constraints. Int. J. Numer. Methods Eng. 43:1453–1478.

Flores-Johnson, E. A., Vázquez-Rodríguez, J. M., Herrera-Franco, P. J., and González-Chi, P. I. (2011). Photo elastic evaluation of fiber surface-treatments on the interfacial performance of a polyester fiber/epoxy model composite. Compos. Part A: Appl. Sci. Manuf. 42(8):1017–1024.

Holmberg, E., Torstenfelt, B., and Klarbring, A. (2013). Stress constrained topology optimization. Struc. Multidisc. Optim. 48(1):3347.

Hughes, T. J. R. (1987). The finite element method: linear static and dynamic finite element analysis. Englewood Cliffs: Prentice-Hall.

Jadhav Mahesh, V., Zoman Digambar, B. and Kharde, Y. R. (2012) Performance analysis of two mono leaf spring used for maruti 800, in Vehicle. Int. J. Innov. Technol. Exploring Eng.

Liang, Q. Q., Xie, Y. M., and Steven, G. P. (1999). Optimal selection of topologies for the minimum-weight design of continuum structures with stress constraints. Proc. Inst. Mech. Eng. C, J. Mech. Eng. Sci. 213(8):755–762.

Ma, Z. D. and Kikuchi, N. (2006). Multidomain topology optimization for structural and material designs. J. Appl. Mech. 73:565.

Navarrinal., F., Muiños, I., Colominas, I., and Casteleiro, M. (2004). Topology optimization of structures: a minimum weight approach with stress constraints. Adv. Eng. Softw. 36:599–606.

París, J., Navarrina, F., Colominas, I., and Casteleiro, M. (2009). Topology optimization of continuum structures with local and global stress constraints. Struc. Multidisc. Optim. 39:419–437.

Ramesh, K. and Ramakrishnan, V. (2016). Digital photoelasticity of glass: a comprehensive review. Opt. Lasers Eng. 87:59–74.

Ramesh, K., Vivek, R., Dora, P. K., and Sanyal, D. (2013). A simple approach to photo elastic calibration of glass using digital photoelasticity. J. Non-Cryst. Solids. 378:7–14.

Sampson, R. C. (1970). A stress-optic law for photo elastic analysis of orthotropic composites. Exp. Mech. 10.5:210–215.

Tvergaard, V. (1975). On the optimum shape of a fillet in a flat bar with restrictions. In Optimization in Structural Design, eds, A. Sawczuk and Z. Mroz, New York: Springer, (pp. 181–195); also, in: Proceedings IUTAM Symposium on Optimization in Structural Design. Berlin: Springer.